STATISTICAL PARAMETRIC MAPPING

STATISTICAL PARAMETRIC MAPPING

The Analysis of Functional Brain Images

Edited by

Karl Friston
John Ashburner
Stefan Kiebel
Thomas Nichols
William Penny

ELSEVIER

AMSTERDAM • BOSTON • HEIDELBERG • LONDON • NEW YORK • OXFORD
PARIS • SAN DIEGO • SAN FRANCISCO • SINGAPORE • SYDNEY • TOKYO

Academic Press is an imprint of Elsevier

Academic Press is an imprint of Elsevier
84 Theobald's Road, London WC1X 8RR, UK
Radarweg 29, PO Box 211, 1000 AE Amsterdam, The Netherlands
30 Corporate Drive, Suite 400, Burlington, MA 01803, USA
525 B Street, Suite 1900, San Diego, CA 92101-4495, USA

First edition 2007
Reprinted 2007, 2008 (twice)

British Library Cataloguing in Publication Data
A catalogue record for this book is available from the British Library

Library of Congress Catalog Number: 2006933665

ISBN: 978-0-12-372560-8

For information on all Academic Press publications visit our website at www.elsevierdirect.com

Printed and bound in the United Kingdom

Transferred to Digital Print 2010

Contents

Part 6 Biophysical models

Part 7 Connectivity

Appendices

Acknowledgements

This book was written on behalf of the SPM co-authors, who at the time of writing, include:

Jesper Andersson
John Ashburner
Nelson Trujillo-Barreto
Matthew Brett
Christian Büchel
Olivier David
Guillaume Flandin
Karl Friston
Darren Gitelman
Daniel Glaser
Volkmar Glauche
Lee Harrison
Rik Henson
Andrew Holmes
Stefan Kiebel
James Kilner
Járâmie Mattout
Tom Nichols
Will Penny
Christophe Phillips
Jean-Baptise Poline
Klaas Stephan

We are also deeply indebted to many colleagues who have developed imaging methodology with us over the years. Though there are too many to name individually we would especially like to thank *Keith Worsley* and colleagues at McGill. We are also very grateful to the many imaging neuroscientists who have alpha-tested implementations of our methodology and the many researchers who regularly make expert contributions to the SPM email list. We would especially like to thank *Jenny Crinion, Alex Hammers, Bas Neggers, Uta Noppeney, Helmut Laufs, Torben Lund, Marta Garrido, Christian Gaser, Marcus Gray, Andrea Mechelli, James Rowe, Karsten Specht, Marco Wilke, Alle Meije Wink and Eric Zarahn.*

Finally, any endeavour of this kind is not possible without the patience and support of our families and loved ones, to whom we dedicate this book.

Part 1

Introduction

CHAPTER

1

A short history of SPM

K. Friston

INTRODUCTION

For a young person entering imaging neuroscience it must seem that the field is very large and complicated, with numerous approaches to experimental design and analysis. This impression is probably compounded by the abundance of TLAs (three-letter-acronyms) and obscure terminology. In fact, most of the principles behind design and analysis are quite simple and had to be established in a relatively short period of time at the inception of brain mapping. This chapter presents an anecdotal perspective on this period. It serves to explain why some ideas, like *t*-maps or, more technically, statistical parametric maps, were introduced and why other issues, like global normalization, were crucial, even if they are not so important nowadays.

The history of human brain mapping is probably shorter than many people might think. Activation studies depend on imaging changes in brain state within the same scanning session. This was made possible using short-half-life radiotracers and positron emission tomography (PET). These techniques became available in the eighties (e.g. Herscovitch *et al.*, 1983) and the first activation maps appeared soon after (e.g. Lauter *et al.*, 1985; Fox *et al.*, 1986). Up until this time, regional differences among brain scans had been characterized using hand-drawn regions of interest (ROI), reducing hundreds of thousands of voxels to a handful of ROI measurements, with a somewhat imprecise anatomical validity. The idea of making voxel-specific statistical inferences, through the use of statistical parametric maps, emerged in response to the clear need to make inferences about brain responses without knowing where those responses were going to be expressed. The first *t*-map was used to establish functional specialization for colour processing in 1989 (Lueck *et al.*, 1989). The underlying methodology was described in a paper entitled: 'The relationship between global and local changes in PET scans' (Friston *et al.*, 1990). This may seem an odd title to introduce statistical parametric mapping (SPM) but it belies a key motivation behind the approach.

Statistical maps versus regions of interest

Until that time, images were usually analysed with analysis of variance (ANOVA) using ROI averages. This approach had become established in the analysis of autoradiographic data in basic neuroscience and metabolic scans in human subjects. Critically, each region was treated as a level of a factor. This meant that the regional specificity of a particular treatment was encoded in the region by treatment interaction. In other words, a main effect of treatment *per se* was not sufficient to infer a regionally specific response. This is because some treatments induced a global effect that was expressed in all the ROIs. Global effects were, therefore, one of the first major conceptual issues in the development of SPM. The approach taken was to treat global activity as a confound in a separate analysis of covariance (ANCOVA) *at each voxel*, thereby endowing inference with a regional specificity that could not be explained by global changes. The resulting SPMs were like X-rays of region-specific changes and, like X-rays, are still reported in maximum-intensity projection format (known colloquially as glass-brains). The issue of regional versus global changes and the validity of global estimators were debated for several years, with many publications in the specialist literature. Interestingly, it is a theme that enjoyed a reprise with the advent of functional magnetic resonance imaging (fMRI) (e.g. Aguirre *et al.*, 1998) and still attracts some research interest today.

Adopting a voxel-wise ANCOVA model paved the way for a divergence between the mass-univariate approach used by SPM (i.e. a statistic for each voxel) and multivariate models used previously. A subtle but

Statistical Parametric Mapping, by Karl Friston et al.
ISBN–13: 978-0-12-372560-8 ISBN–10: 0-12-372560-7

important motivation for mass-univariate approaches was the fact that a measured haemodynamic response in one part of the brain may differ from the response in another, *even if the underlying neuronal activation was exactly the same*. This meant that the convention of using region-by-condition interactions as a test for regionally specific effects was not tenable. In other words, even if one showed that two regions activated differently in terms of measured haemodynamics, this did not mean there was a regionally specific difference at the neuronal or computational level. This issue seems to have escaped the electroencephalography (EEG) community, who still use ANOVA with region as a factor, despite the fact that the link between neuronal responses and channel measurements is even more indeterminate than for metabolic imaging. However, the move to voxel-wise, whole-brain analysis entailed two special problems: the problem of registering images from different subjects so that they could be compared on a voxel-by-voxel basis and the multiple-comparisons problem that ensued.

Spatial normalization

The pioneering work of the St Louis group had already established the notion of a common anatomical or stereotactic space (Fox *et al.*, 1988) in which to place subtraction or difference maps, using skull X-rays as a reference. The issue was how to get images into that space efficiently. Initially, we tried identifying landmarks in the functional data themselves to drive the registration (Friston *et al.*, 1989). This approach was dropped almost immediately because it relied on landmark identification and was not a hundred per cent reproducible. Within a year, a more reliable, if less accurate, solution was devised that matched images to a template without the need for landmarks (Friston *et al.*, 1991a). The techniques for spatial normalization using template- or model-based approaches have developed consistently since that time and current treatments regard normalization as the inversion of generative models for anatomical variation that involve warping templates to produce subject-specific images (e.g. Ashburner and Friston, 2005).

Topological inference

Clearly, performing a statistical test at each voxel engendered an enormous false positive rate when using unadjusted thresholds to declare activations significant. The problem was further compounded by the fact that the data were not spatially independent and a simple Bonferroni correction was inappropriate (PET and SPECT (single photon emission computerized tomography) data are inherently very smooth and fMRI had not been invented at this stage). This was the second major theme that occupied people trying to characterize functional neuroimaging data. What was needed was a way of predicting the probabilistic behaviour of SPMs, under the null hypothesis of no activation, which accounted for the smoothness or spatial correlations among voxels. From practical experience, it was obvious that controlling the false positive rate of voxels was not the answer. One could increase the number of positive voxels by simply making the voxels smaller but without changing the topology of the SPM. It became evident that conventional control procedures developed for controlling family-wise error (e.g. the Bonferroni correction) had no role in making inferences on continuous images. What was needed was a new framework in which one could control the false positive rate of the regional effects themselves, noting a regional effect is a topological feature, not a voxel.

The search for a framework for topological inference in neuroimaging started in the theory of stochastic processes and level-crossings (Friston *et al.*, 1991b). It quickly transpired that the resulting heuristics were the same as established results from the theory of random fields. Random fields are stochastic processes that conform very nicely to realizations of brain scans under normal situations. Within months, the technology to correct p-values was defined within random field theory (Worsley *et al.*, 1992). Although the basic principles of topological inference were established at this time, there were further exciting mathematical developments with extensions to different sorts of SPMs and the ability to adjust the p-values for small bounded volumes of interest (see Worsley *et al.*, 1996). Robert Adler, one of the world's contemporary experts in random field theory, who had abandoned it years before, was understandably very pleased and is currently writing a book with a protégé of Keith Worsley (Adler and Taylor, in preparation).

Statistical parametric mapping

The name 'statistical parametric mapping' was chosen carefully for a number of reasons. First, it acknowledged the TLA of 'significance probability mapping', developed for EEG. Significance probability mapping involved creating interpolated pseudo-maps of p-values to disclose the spatiotemporal organization of evoked electrical responses (Duffy *et al.*, 1981). The second reason was more colloquial. In PET, many images are derived from the raw data reflecting a number of different physiological parameters (e.g. oxygen metabolism, oxygen extraction fraction, regional cerebral blood flow etc.). These were referred to as parametric maps. All parametric maps are non-linear functions of the original data. The distinctive thing about *statistical* parametric maps is that they have a known distribution under the null hypothesis. This is because they are predicated on a statistical model of the data (as opposed to a physiological parametric model).

One important controversy, about the statistical models employed, was whether the random fluctuations or error variance was the same from brain region to brain region. We maintained that it was not (on common sense grounds that the frontal operculum and ventricles were not going to show the same fluctuations in blood flow) and adhered to voxel-specific estimates of error. For PET, the Montreal group considered that the differences in variability could be discounted. This allowed them to pool their error variance estimator over voxels to give very sensitive SPMs (under the assumption of stationary error variance). Because the error variance was assumed to be the same everywhere, the resulting *t*-maps were simply scaled subtraction or difference maps (see Fox *et al.*, 1988). This issue has not dogged fMRI, where it is generally accepted that error variance is voxel-specific.

The third motivation for the 'statistical parametric mapping' was that it reminded people they were using parametric statistics that assume the errors are additive and Gaussian. This is in contradistinction to non-parametric approaches that are generally less sensitive, more computationally intensive, but do not make any assumptions about the distribution of error terms. Although there are some important applications of non-parametric approaches, they are generally a specialist application in the imaging community. This is largely because brain imaging data conform almost exactly to parametric assumptions by the nature of image reconstruction, post-processing and experimental design.

THE PET YEARS

In the first few years of the nineties, many landmark papers were published using PET and the agenda for a functional neuroimaging programme was established. SPM proved to be the most popular way of characterizing brain activation data. It was encoded in Matlab and used extensively by the MRC Cyclotron Unit at the Hammersmith Hospital in the UK and was then distributed to collaborators and other interested units around the world. The first people outside the Hammersmith group to use SPM were researchers at NIH (National Institutes of Health, UDA) (e.g. Grady *et al.*, 1994). Within a couple of years, SPM had become the community standard for analysing PET activation studies and the assumptions behind SPM were largely taken for granted. By this stage, SPM was synonymous with the general linear model and random field theory. Although originally framed in terms of ANCOVA, it was quickly realized that any general linear model could be used to produce an SPM. This spawned a simple taxonomy of experimental designs and their associated statistical models. These were summarized in terms of subtraction or categorical designs, parametric designs and factorial designs (Friston *et al.*, 1995a). The adoption of factorial designs was one of the most important advances at this point. The first factorial designs focused on adaptation during motor learning and studies looking at the interaction between a psychological and pharmacological challenge in psychopharmacological studies (e.g. Friston *et al.*, 1992). The ability to look at the effect of changes in the level of one factor on activations induced by another led to a rethink of cognitive subtraction and pure insertion and the appreciation of context-sensitive activations in the brain. The latitude afforded by factorial designs is reflected in the fact that most studies are now multifactorial in nature.

THE fMRI YEARS

In 1992, at the annual meeting of the Society of Cerebral Blood Flow and Metabolism in Miami, Florida, Jack Belliveau presented, in the first presentation of the opening session, provisional results using photic stimulation with fMRI. This was quite a shock to the imaging community that was just starting to relax: most of the problems had been resolved, community standards had been established and the way forward seemed clear. It was immediately apparent that this new technology was going to reshape brain mapping radically, the community was going to enlarge and established researchers were going to have to re-skill. The benefits of fMRI were clear, in terms of the ability to take many hundreds of scans within one scanning session and to repeat these sessions indefinitely in the same subject. Some people say that the main advances in a field, following a technological breakthrough, are made within the first few years. Imaging neuroscience must be fairly unique in the biological sciences, in that exactly five years after the inception of PET activation studies, fMRI arrived. The advent of fMRI brought with it a new wave of innovation and enthusiasm.

From the point of view of SPM, there were two problems, one easy and one hard. The first problem was how to model evoked haemodynamic responses in fMRI time-series. This was an easy problem to resolve because SPM could use any general linear model, including convolution models of the way haemodynamic responses were caused (Friston *et al.*, 1994). Stimulus functions encoding the occurrence of a particular event or experimental state (e.g. boxcar-functions) were simply convolved with a haemodynamic response function (HRF) to form regressors in a general linear model (*cf* multiple linear regression).

Serial correlations

The second problem that SPM had to contend with was the fact that successive scans in fMRI time-series were not independent. In PET, each observation was statistically independent of its precedent but, in fMRI coloured time-series, noise rendered this assumption invalid. The existence of temporal correlations originally met with some scepticism, but is now established as an important aspect of fMRI time-series. The SPM community tried a series of heuristic solutions until it arrived at the solution presented in Worsley and Friston (1995). This procedure, also known as 'pre-colouring', replaced the unknown endogenous autocorrelation by imposing a known autocorrelation structure. Inference was based on the Satterthwaite conjecture and is formally identical to the non-specificity correction developed by Geisser and Greenhouse in conventional parametric statistics. An alternative approach was 'pre-whitening' which tried to estimate a filter matrix from the data to de-correlate the errors (Bullmore *et al.*, 2001). The issue of serial correlations, and more generally non-sphericity, is still important and attracts much research interest, particularly in the context of maximum likelihood techniques and empirical Bayes (Friston *et al.*, 2002).

New problems and old problems

The fMRI community adopted many of the developments from the early days of PET. Among these were the use of the standard anatomical space provided by the atlas of Talairach and Tournoux (1988) and conceptual issues relating to experimental design and interpretation. Many debates that had dogged early PET research were resolved rapidly in fMRI; for example, 'What constitutes a baseline?' This question, which had preoccupied the whole community at the start of PET, appeared to be a non-issue in fMRI with the use of well-controlled experimental paradigms. Other issues, such as global normalization were briefly revisited, given the different nature of global effects in fMRI (multiplicative) relative to PET (additive). However, one issue remained largely ignored by the fMRI community. This was the issue of adjusting p-values for the multiplicity of tests performed. While people using SPM quite happily adjusted their p-values using random field theory, others seemed unaware of the need to control false positive rates. The literature now entertained reports based on uncorrected p-values, an issue which still confounds editorial decisions today. It is interesting to contrast this, historically, with the appearance of the first PET studies.

When people first started reporting PET experiments there was an enormous concern about the rigor and validity of the inferences that were being made. Much of this concern came from outside the imaging community who, understandably, wanted to be convinced that the 'blobs' that they saw in papers (usually *Nature* or *Science*) reflected true activations as opposed to noise. The culture at that time was hostile to capricious reporting and there was a clear message from the broader scientific community that the issue of false positives had to be resolved. This was a primary motivation for developing the machinery to adjust p-values to protect against family-wise false positives. In a sense, SPM was a reaction to the clear mandate set by the larger community, to develop a valid and rigorous framework for activation studies. In short, SPM was developed in a culture of scepticism about brain mapping that was most easily articulated by critiquing its validity. This meant that the emphasis was on specificity and reproducibility, as opposed to sensitivity and flexibility. Current standards for reporting brain mapping studies are much more forgiving than they were at its beginning, which may explain why recent developments have focused on sensitivity (e.g. Genovese *et al.*, 2002).

The convolution model

In the mid-nineties, there was lots of fMRI research; some of it was novel, some recapitulating earlier findings with PET. From a methodological point of view, notable advances included the development of event-related paradigms that furnished an escape from the constraints imposed by block designs and the use of retinotopic mapping to establish the organization of cortical areas in human visual cortex. This inspired a whole sub-field of cortical surface mapping that is an important endeavour in early sensory neuroimaging. For SPM there were three challenges that needed to be addressed:

Temporal basis functions

The first involved a refinement of the models of evoked responses. The convolution model had become a cornerstone for fMRI with SPM. The only remaining issue was the form of the convolution kernel or haemodynamic response function that should be adopted and whether the form changed from region to region. This was resolved simply by convolving the stimulus function with not one response function but several [basis functions]. This meant that one could model condition, voxel and subject-specific haemodynamic responses using established approaches. Temporal basis functions (Friston *et al.*, 1995b) were important because they allowed one to define a family of HRFs that could change their form from voxel to voxel. Temporal basis functions found an important application in the analysis of event-related fMRI. The general acceptance of the convolution model was consolidated by the influential paper of Boynton a year later (Boynton *et al.*, 1996). However, at this time, people were starting to notice some non-linearities in fMRI responses (Vazquez and Noll, 1998) that were formulated, in the context of SPM, as a Volterra series expansion of the stimulus function (Friston *et al.*, 1998). This was simple because the Volterra series can be formulated as another linear model (compare with a Taylor expansion). These Volterra characterizations would later be used to link empirical data and balloon models of haemodynamic responses.

Efficiency and experimental design

The second issue that concerned the developers of SPM arose from the growing number and design of event-related fMRI studies. This was the efficiency with which responses could be detected and estimated. Using an analytical formulation, it was simple to show that the boxcar paradigms were much more efficient that event-related paradigms, but event-related paradigms could be made efficient by randomizing the occurrence of particular events such that they 'bunched' together to increase experimental variance. This was an interesting time in the development of data analysis techniques because it enforced a signal processing perspective on the general linear models employed.

Hierarchical models

The third area motivating the development of SPM was especially important in fMRI and reflects the fact that many scans can be obtained in many individuals. Unlike in PET, the within-subject scan-to-scan variability can be very different from the between-subject variability. This difference in variability has meant that inferences about responses in a single subject (using within-subject variability) are distinct from inferences about the population from which that subject was drawn (using between-subject variability). More formally, this distinction is between fixed- and random-effects analyses. This distinction speaks to hierarchical observation models for fMRI data. Because SPM only had the machinery to do single-level (fixed-effects) analyses, a device was required to implement random-effects analyses. This turned out to be relatively easy and intuitive: subject-specific effects were estimated in a first-level analysis and the contrasts of parameter estimates (e.g. activations) were then re-entered into a second-level SPM analysis (Holmes and Friston, 1998). This recursive use of a single-level statistical model is fortuitously equivalent to multilevel hierarchical analyses (compare with the summary statistic approach in conventional statistics).

Bayesian developments

Understanding hierarchical models of fMRI data was important for another reason: these models support empirical Bayesian treatments. Empirical Bayes was one important component of a paradigm shift in SPM from classical inference to a Bayesian perspective. From the late nineties, Bayesian inversion of anatomical models had been a central part of spatial normalization. However, despite early attempts (Holmes and Ford, 1993), the appropriate priors for functional data remained elusive. Hierarchical models provided the answer, in the form of empirical priors that could be evaluated from the data themselves. This evaluation depends on the conditional dependence implicit in hierarchical models and brought previous maximum likelihood schemes into the more general Bayesian framework. In short, the classical schemes SPM had been using were all special cases of hierarchical Bayes (in the same way that the original ANCOVA models for PET were special cases of the general linear models for fMRI). In some instances, this connection was very revealing, for example, the equivalence between classical covariance component estimation using restricted maximum likelihood (i.e. ReML) and the inversion of two-level models with expectation maximization (EM) meant we could use the same techniques used to estimate serial correlations to estimate empirical priors on activations (Friston *et al.*, 2002).

The shift to a Bayesian perspective had a number of motivations. The most principled was an appreciation that estimation and inference corresponded to Bayesian inversion of generative models of imaging data. This placed an emphasis on generative or forward models for fMRI that underpinned work on biophysical modelling of haemodynamic responses and, indeed, the framework entailed by dynamic causal modelling (e.g. Friston *et al.*, 2003; Penny *et al.*, 2004). This reformulation led to more informed spatiotemporal models for fMRI (e.g. Penny *et al.*, 2005) that effectively estimate the optimum smoothing by embedding spatial dependencies in a hierarchical model. It is probably no coincidence that these developments coincided with the arrival of the Gatsby Computational Neuroscience Unit next to the Wellcome Department of Imaging Neuroscience. The Gatsby housed several experts in Bayesian inversion and machine learning and the Wellcome was home to many of the SPM co-authors.

The second motivation for Bayesian treatments of imaging data was to bring the analysis of EEG and fMRI data into the same forum. Source reconstruction in EEG and MEG (magnetoencephalography) is an ill-posed problem that depends explicitly on regularization or priors on the solution. The notion of forward models in EEG-MEG, and their Bayesian inversion had been well established for decades and SPM needed to place fMRI on the same footing.

THE MEG-EEG YEARS

At the turn of the century people had started applying SPM to source reconstructed EEG data (e.g. Bosch-Bayard *et al.*, 2001). Although SPM is not used widely for the analysis of EEG-MEG data, over the past five years most of the development work in SPM has focused on this modality. The motivation was to accommodate different modalities (e.g. fMRI-EEG) within the same analytic and anatomical framework. This reflected the growing appreciation that fMRI and EEG could offer complementary constraints on the inversion of generative models. At a deeper level, the focus had shifted from generative models of a particular modality (e.g. convolution models for fMRI) and towards models of neuronal dynamics that could explain any modality. The inversion of these models corresponds to true multimodal fusion and is the aim of recent and current developments within SPM.

In concrete terms, this period saw the application of random field theory to SPMs of evoked and induced responses, highlighting the fact that SPMs can be applied to non-anatomical spaces, such as space-peristimulus-time or time-frequency (e.g. Kilner *et al.*, 2005). It has seen the application of hierarchical Bayes to the source reconstruction problem, rendering previous heuristics, like L-curve analysis, redundant (e.g. Phillips *et al.*, 2002) and it has seen the extension of dynamic causal modelling to cover evoked responses in EEG-MEG (David *et al.*, 2006).

This section is necessarily short because the history of SPM stops here. Despite this, a lot of the material in this book is devoted to biophysical models of neuronal responses that can, in principle, explain any modality. Much of SPM is about the inversion of these models. In what follows, we try to explain the meaning of the more important TLAs entailed by SPM (Table 1-1).

TABLE 1-1 Some common TLAs

TLA Three letter acronym	**ERP** Event-related potential
SPM Statistical parametric map(ping)	**ERF** Event-related field
GLM General linear model	**MMN** Mis-match negativity
RFT Random field theory	**PPI** Psychophysiological interaction
VBM Voxel-based morphometry	**DCM** Dynamic causal model
FWE Family-wise error	**SEM** Structural equation model
FDR False discovery rate	**SSM** State-space model
IID Independent and identically distributed	**MAR** Multivariate autoregression
MRI Magnetic resonance imaging	**LTI** Linear time invariant
PET Positron emission tomography	**PEB** Parametric empirical Bayes
EEG Electroencephalography	**DEM** Dynamic expectation maximization
MEG Magnetoencephalography	**GEM** Generalized expectation maximization
HRF Haemodynamic response function	**BEM** Boundary-element method
IRF Impulse response function	**FEM** Finite-element method
FIR Finite impulse response	

REFERENCES

Adler RJ, Taylor JE Random fields and geometry. In preparation - To be published by Birkhauser.

Aguirre GK, Zarahn E, D'Esposito M (1998) The inferential impact of global signal covariates in functional neuroimaging analyses. *NeuroImage* **8**: 302–06

Ashburner J, Friston KJ (2005) Unified segmentation. *NeuroImage* **26**: 839–51

Bosch-Bayard J, Valdes-Sosa P, Virues-Alba T *et al.* (2001) 3D statistical parametric mapping of EEG source spectra by means of variable resolution electromagnetic tomography (VARETA). *Clin Electroencephalogr* **32**: 47–61

Boynton GM, Engel SA, Glover GH *et al.* (1996) Linear systems analysis of functional magnetic resonance imaging in human V1. *J Neurosci* **16**: 4207–21

Bullmore ET, Long C, Suckling J *et al.* (2001) Colored noise and computational inference in neurophysiological (fMRI) time series analysis: resampling methods in time and wavelet domains. *Hum Brain Mapp* **12**: 61–78

David O, Kiebel SJ, Harrison LM *et al.* (2006). Dynamic causal modeling of evoked responses in EEG and MEG. *NeuroImage*, **30**(4): 1255–7

Duffy FH, Bartels PH, Burchfiel JL (1981) Significance probability mapping: an aid in the topographic analysis of brain electrical activity. *Electroencephalogr Clin Neurophysiol* **51**: 455–62

Fox PT, Mintun MA, Raichle ME *et al.* (1986) Mapping human visual cortex with positron emission tomography. *Nature* **323**: 806–9

Fox PT, Mintun MA, Reiman EM *et al.* (1988) Enhanced detection of focal brain responses using intersubject averaging and change-distribution analysis of subtracted PET images. *J Cereb Blood Flow Metab* **8**: 642–53

Friston KJ, Passingham RE, Nutt JG *et al.* (1989) Localisation in PET images: direct fitting of the intercommissural (AC-PC) line. *J Cereb Blood Flow Metab* **9**: 690–705

Friston KJ, Frith CD, Liddle PF *et al.* (1990) The relationship between global and local changes in PET scans. *J Cereb Blood Flow Metab* **10**: 458–66

Friston KJ, Frith CD, Liddle PF *et al.* (1991a) Plastic transformation of PET images. *J Comput Assist Tomogr* **15**: 634–39

Friston KJ, Frith CD, Liddle PF *et al.* (1991b) Comparing functional (PET) images: the assessment of significant change. *J Cereb Blood Flow Metab* **11**: 690–09

Friston KJ, Frith C, Passingham RE *et al.* (1992) Motor practice and neurophysiological adaptation in the cerebellum: a positron tomography study. *Proc R Soc Lond Series B* **248**: 223–28

Friston KJ, Jezzard PJ, Turner R (1994) Analysis of functional MRI time-series. *Hum Brain Mapp* **1**: 153–71

Friston KJ, Holmes AP, Worsley KJ *et al.* (1995a) Statistical parametric maps in functional imaging: a general linear approach. *Hum Brain Mapp* **2**: 189–210

Friston KJ, Frith CD, Turner R *et al.* (1995b) Characterizing evoked hemodynamics with fMRI. *NeuroImage* **2**: 157–65

Friston KJ, Josephs O, Rees G *et al.* (1998) Nonlinear event-related responses in fMRI. *Magnet Reson Med* **39**: 41–52

Friston KJ, Glaser DE, Henson RN *et al.* (2002) Classical and Bayesian inference in neuroimaging: applications. *NeuroImage* **16**: 484–512

Friston KJ, Harrison L, Penny W (2003) Dynamic causal modelling. *NeuroImage* **19**: 1273–302

Genovese CR, Lazar NA, Nichols T (2002) Thresholding of statistical maps in functional neuroimaging using the false discovery rate. *NeuroImage* **15**: 870–78

Grady CL, Maisog JM, Horwitz B *et al.* (1994) Age-related changes in cortical blood flow activation during visual processing of faces and location. *J Neurosci* **14**:1450–62

Herscovitch P, Markham J, Raichle ME (1983) Brain blood flow measured with intravenous H2(15)O. I. Theory and error analysis. *J Nucl Med* **24**: 782–89

Holmes A, Ford I (1993) A Bayesian approach to significance testing for statistic images from PET. In *Quantification of brain function, tracer kinetics and image analysis in brain PET*, Uemura K, Lassen NA, Jones T *et al.* (eds). *Excerpta Medica, Int. Cong.* Series no. **1030**: 521–34

Holmes AP, Friston KJ (1998) Generalisability, random effects and population inference. *NeuroImage* **7**: S754

Kilner JM, Kiebel SJ, Friston KJ (2005) Applications of random field theory to electrophysiology. *Neurosci Lett* **374**: 174–78

Lauter JL, Herscovitch P, Formby C *et al.* (1985) Tonotopic organization in human auditory cortex revealed by positron emission tomography. *Hear Res* **20**: 199–205

Lueck CJ, Zeki S, Friston KJ *et al.* (1989) The colour centre in the cerebral cortex of man. *Nature* **340**: 386–89

Penny WD, Stephan KE, Mechelli A *et al.* (2004) Comparing dynamic causal models. *NeuroImage* **22**: 1157–72

Penny WD, Trujillo-Barreto NJ, Friston KJ (2005) Bayesian fMRI time series analysis with spatial priors. *NeuroImage* **24**: 350–62

Phillips C, Rugg MD, Friston KJ (2002) Systematic regularization of linear inverse solutions of the EEG source localization problem. *NeuroImage* **17**: 287–301

Talairach P, Tournoux J (1988) *A stereotactic coplanar atlas of the human brain.* Thieme, Stuttgart

Vazquez AL, Noll DC (1998) Nonlinear aspects of the BOLD response in functional MRI. *NeuroImage* **7**: 108–18

Worsley KJ, Evans AC, Marrett S *et al.* (1992) A three-dimensional statistical analysis for CBF activation studies in human brain. *J Cereb Blood Flow Metab* **12**: 900–18

Worsley KJ, Friston KJ (1995) Analysis of fMRI time-series revisited – again. *NeuroImage* **2**: 173–81

Worsley KJ, Marrett S, Neelin P *et al.* (1996) A unified statistical approach of determining significant signals in images of cerebral activation. *Hum Brain Mapp* **4**: 58–73

CHAPTER

2

Statistical parametric mapping

K. Friston

INTRODUCTION

This chapter summarizes the ideas and procedures used in the analysis of brain imaging data. It provides sufficient background to understand the principles of experimental design and data analysis and serves to introduce the main themes covered by subsequent chapters. These chapters have been organized into six parts. The first three parts follow the key stages of analysis: image transformations, modelling, and inference. These parts focus on identifying, and making inferences about, regionally specific effects in the brain. The final three parts address biophysical models of distributed neuronal responses, closing with analyses of functional and effective connectivity.

Characterizing a regionally specific effect rests on estimation and inference. Inferences in neuroimaging may be about differences expressed when comparing one group of subjects to another or, within subjects, changes over a sequence of observations. They may pertain to structural differences (e.g. in voxel-based morphometry) (Ashburner and Friston, 2000) or neurophysiological measures of brain functions (e.g. fMRI or functional magnetic resonance imaging). The principles of data analysis are very similar for all of these applications and constitute the subject of this and subsequent chapters. We will focus on the analysis of fMRI time-series because this covers many of the issues that are encountered in other modalities. Generally, the analyses of structural and PET (positron emission tomography) data are simpler because they do not have to deal with correlated errors from one scan to the next. Conversely, EEG and MEG (electro- and magnetoencephalography) present special problems for model inversion, however, many of the basic principles are shared by fMRI and EEG, because they are both caused by distributed neuronal dynamics. This chapter focuses on the design and analysis of neuroimaging studies. In the next chapter, we will look at conceptual and mathematical models that underpin the operational issues covered here.

Background

Statistical parametric mapping is used to identify regionally specific effects in neuroimaging data and is a prevalent approach to characterizing functional anatomy, specialization and disease-related changes. The complementary perspective, namely functional integration, requires a different set of approaches that examine the relationship among changes in one brain region relative to changes in others. Statistical parametric mapping is a voxel-based approach, employing topological inference, to make some comment about regionally specific responses to experimental factors. In order to assign an observed response to a particular brain structure, or cortical area, the data are usually mapped into an anatomical space. Before considering statistical modelling, we deal briefly with how images are realigned and normalized into some standard anatomical space. The general ideas behind statistical parametric mapping are then described and illustrated with attention to the different sorts of inferences that can be made with different experimental designs.

EEG, MEG and fMRI data lend themselves to a signal processing perspective. This can be exploited to ensure that both the design and analysis are as efficient as possible. Linear time invariant models provide the bridge between inferential models employed by statistical mapping and conventional signal processing approaches. We will touch on these and develop them further in the next chapter. Temporal autocorrelations in noise processes represent another important issue, especially in fMRI, and approaches to maximizing efficiency in the context of serially correlated errors will be discussed. We will also consider event and epoch-related designs in terms of efficiency. The chapter closes by looking at the distinction

Statistical Parametric Mapping, by Karl Friston et al.
ISBN–13: 978-0-12-372560-8 ISBN–10: 0-12-372560-7

between fixed and random-effect analyses and how this relates to inferences about the subjects studied or the population from which these subjects came.

In summary, this chapter reviews the three main stages of data analysis: spatial or image transforms, modelling and inference; these are the areas covered in the first three parts of this book and are summarized schematically in Plate 1 (see colour plate section). We then look at experimental design in light of the models covered in earlier parts. The next chapter deals with different models of distributed responses and previews the material covered in the final three parts of this book.

SPATIAL TRANSFORMS AND COMPUTATIONAL ANATOMY

A central theme in this book is the inversion of forward or generative models of how data are caused. We will see this in many different contexts, from the inversion of linear models of fMRI time-series to the inversion of dynamic causal models of distributed EEG responses. Image reconstruction, in imaging modalities like PET and fMRI, can be regarded as inverting a forward model of how signals, deployed in anatomical space, conspire to produce measured signals. In other modalities, like EEG and MEG, this inversion, or source reconstruction, can be a substantial problem in its own right. In most instances, it is expedient to decompose the inversion of forward spatiotemporal models into spatial and temporal parts. Operationally, this corresponds to reconstructing the spatial signal at each time point and then inverting a temporal model of the time-series at each spatial source (although we will consider full spatiotemporal models in Chapters 25 and 26). This view of source or image reconstruction as model inversion can be extended to cover the inversion of anatomical models describing anatomical variation within and between subjects. The inversion of these models corresponds to registration and normalization respectively. The aim of these anatomical inversions or transformations is to remove or characterize anatomical differences. Chapters 4 to 6 deal with the inversion of anatomical models for imaging modalities. Figure 2.1 shows an example of a generative model for structural images that is presented in Chapter 6. Chapters 28 and 29 deal with the corresponding inversion for EEG and MEG data.

This inversion corresponds to a series of spatial transformations that try to reduce unwanted variance components in the voxel time-series. These components are induced by movement or shape differences among a series of scans. Voxel-based analyses assume that data from a particular voxel derive from the same part of the brain. Violations of this assumption will introduce artefactual changes in the time-series that may obscure changes, or differences, of interest. Even single-subject analyses usually proceed in a standard anatomical space, simply to enable reporting of regionally-specific effects in a frame of reference that can be related to other studies. The first step is to realign the data to undo the effects of subject movement during the scanning session (see Chapter 4). After realignment, the data are then transformed using linear or non-linear warps into a standard anatomical space (see Chapters 5 and 6). Finally, the data are usually spatially smoothed before inverting the temporal part of the model.

Realignment

Changes in signal intensity over time, from any one voxel, can arise from head motion and this represents a serious confound, particularly in fMRI studies. Despite restraints on head movement, cooperative subjects still show displacements of up several millimetres. Realignment involves estimating the six parameters of an affine 'rigid-body' transformation that minimizes the differences between each successive scan and a reference scan (usually the first or the average of all scans in the time series). The transformation is then applied by re-sampling the data using an interpolation scheme. Estimation of the affine transformation is usually effected with a first-order approximation of the Taylor expansion of the effect of movement on signal intensity using the spatial derivatives of the images (see below). This allows for a simple iterative least square solution that corresponds to a Gauss-Newton search (Friston *et al.*, 1995a). For most imaging modalities this procedure is sufficient to realign scans to, in some instances, a hundred microns or so (Friston *et al.*, 1996a). However, in fMRI, even after perfect realignment, movement-related signals can still persist. This calls for a further step in which the data are *adjusted* for residual movement-related effects.

Adjusting for movement-related effects

In extreme cases, as much as 90 per cent of the variance in fMRI time-series can be accounted for by the effects of movement *after* realignment (Friston *et al.*, 1996a). Causes of these movement-related components are due to movement effects that cannot be modelled using a linear model. These non-linear effects include: subject movement between slice acquisition, interpolation artefacts (Grootoonk *et al.*, 2000), non-linear distortion due to magnetic field inhomogeneities (Andersson *et al.*, 2001) and spin-excitation history effects (Friston *et al.*, 1996a). The

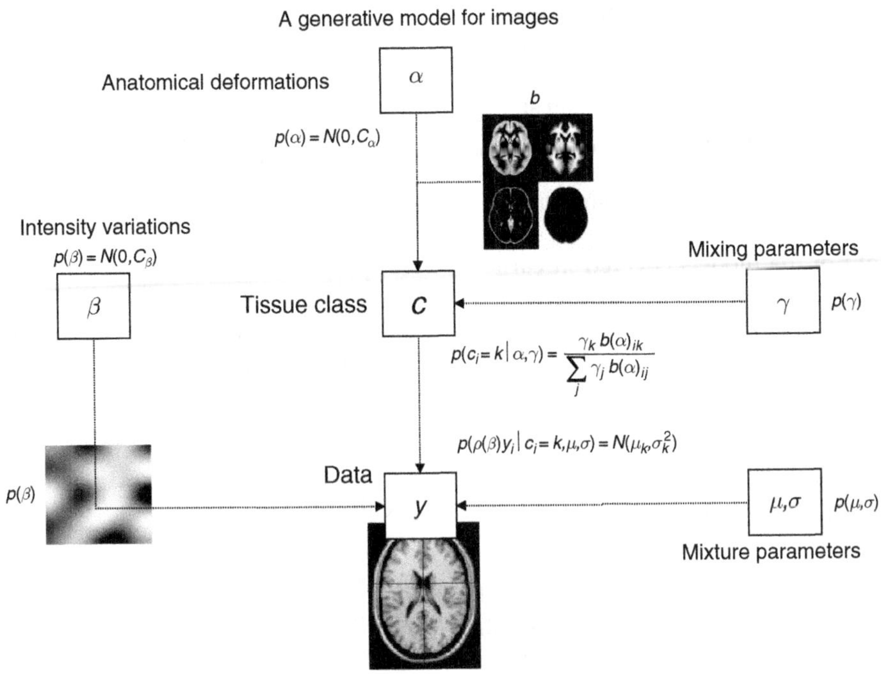

FIGURE 2.1 A graphical model describing the generation of an image. The boxes or 'nodes' represent quantities required to generate an image and the lines or 'edges' encode conditional dependencies among these quantities. This graphical description is a useful way to describe a generative model and makes all the conditional dependencies explicit. In this example, one starts by sampling some warping parameters α from their prior density $p(\alpha)$. These are used to resample (i.e. warp) a series of tissue-class maps to give $b(\alpha)_{ik}$ for each voxel and tissue class. The warping parameters model subject-specific anatomical deformations. Mixing parameters γ are then selected from their prior density $p(\gamma)$; these control the relative proportions of different tissue-classes over the brain. The mixing parameters scale the tissue-class maps to provide a density from which a voxel-specific tissue-class c_i is sampled. This specifies a mixture of Gaussians from which the voxel intensity is sampled. This mixture is specified in terms of the expectations μ and variances σ of their constituent Gaussians that are sampled from the prior density $p(\mu, \sigma)$. The final stage of image construction is to scale the voxel values with some slowly varying intensity field whose parameters β are sampled from their prior $p(\beta)$. The resulting image embodies random effects expressed at the level of anatomical deformation, amount of different tissue types, the expression of those tissues in the measurement, and image-specific inhomogeneities. Inversion of this generative model implicitly corrects for intensity variations, classifies each voxel probabilistically (i.e. segments) and spatially normalizes the image. Critically, this inversion accounts properly for all the conditional dependencies among the model's parameters and provides the most likely estimates given the data (see Chapter 6 for details of this model and its inversion).

latter can be pronounced if the repetition time approaches T1 making the current signal a function of movement history. These effects can render the movement-related signal a non-linear function of displacement in the n-th and previous scans:

$$y_n = f(x_n, x_{n-1}, \ldots)$$

By assuming a sensible form for this function, one can include these effects in the temporal model, so that they are explained away when making inferences about activations. This relies on accurate displacement estimates from the realignment and assumes activations are not correlated with the movements (any component that is correlated will be explained away).

The form for $f(x_n, x_{n-1}, \ldots)$, proposed in Friston *et al.* (1996a), was a non-linear autoregression model that used polynomial expansions to second order. This model was motivated by spin-excitation history effects and allowed displacement in previous scans to explain movement-related signal in the current scan. However, it is also a reasonable model for other sources of movement-related confounds. Generally, for repetition times (TR) of several seconds, interpolation artefacts supersede (Grootoonk *et al.*, 2000) and first-order terms, comprising an expansion of the current displacement in terms of periodic basis functions, are sufficient.

This section has considered *spatial* realignment. In multislice acquisition, different slices are acquired at different times. This raises the possibility of *temporal* realignment to ensure that the data from any given volume were sampled at the same time. This is usually performed using interpolation over time and only when the TR is sufficiently small to permit interpolation. Generally, timing effects of this sort are not considered problematic because they manifest as artefactual latency differences in evoked responses from region to region. Given that biological latency differences are in the order of a few seconds, inferences about these differences are only made when comparing different trial types at the *same* voxel. Provided the effects of latency differences are modelled (see Chapter 14) temporal realignment is unnecessary in most applications.

Spatial normalization

In realignment, the generative model for within-subject movements is a rigid-body displacement of the first image. The generative model for spatial normalization is a canonical image or template that is distorted to produce a subject-specific image. Spatial normalization inverts this model by undoing the warp using a template-matching procedure. We focus on this simple model here, but note that more comprehensive models can be adopted (see Figure 2.1 and Chapter 6).

After realigning the data, a mean image of the series, or some other co-registered (e.g. a T1-weighted) image, is used to estimate some warping parameters that map it onto a template that already conforms to some standard anatomical space (e.g. Talairach and Tournoux, 1988). This estimation can use a variety of models for the mapping, including: a twelve-parameter affine transformation, where the parameters constitute a spatial transformation matrix; low-frequency basis functions, usually a discrete cosine set or polynomials, where the parameters are the coefficients of the basis functions employed; or a vector field specifying the mapping for each control point (e.g. voxel). In the latter case, the parameters are vast in number and constitute a vector field that is bigger than the image itself. Estimation of the parameters of all these models can be accommodated in a Bayesian framework, in which one is trying to find the warping parameters θ that have the maximum posterior probability $p(\theta|y)$ given the data y, where $p(\theta|y)p(y) = p(y|\theta)p(\theta)$. Put simply, one wants to find the deformation that is most likely given the data. This deformation can be found by maximizing the probability of getting the data, given the current parameters, times the probability of those parameters. In practice, the deformation is updated iteratively using a Gauss-Newton scheme to maximize $p(\theta|y)$. This involves jointly minimizing the likelihood and prior potentials $H(y|\theta) = \ln p(y|\theta)$ and $H(\theta) = \ln p(\theta)$. The likelihood potential is generally taken to be the sum of squared differences between the template and deformed image and reflects the probability of actually getting that image if the transformation was correct. The prior potential can be used to incorporate prior information or constraints on the warp. Priors can be determined empirically or motivated by constraints on the mappings. Priors play a more essential role as the number of parameters specifying the mapping increases and are central to high-dimensional warping schemes (Ashburner *et al.*, 1997 and see Chapter 5).

In practice, most people use an affine or spatial basis function warps and iterative least squares to minimize the posterior potential. A nice extension of this approach is that the likelihood potential can be refined and taken as the difference between the index image and a mixture of templates (e.g. depicting grey, white and skull tissue partitions). This models intensity differences that are unrelated to registration differences and allows different modalities to be co-registered (see Friston *et al.*, 1995a; Figure 2.2).

A special consideration is the spatial normalization of brains that have gross anatomical pathology. This pathology can be of two sorts: quantitative changes in the amount of a particular tissue compartment (e.g. cortical atrophy), or qualitative changes in anatomy involving the insertion or deletion of normal tissue compartments (e.g. ischaemic tissue in stroke or cortical dysplasia). The former case is, generally, not problematic in the sense that changes in the amount of cortical tissue will not affect its optimum spatial location in reference to some template (and, even if it does, a disease-specific template is easily constructed). The second sort of pathology can introduce bias in the normalization (because the generative model does not have a lesion) unless special precautions are taken. These usually involve imposing constraints on the warping to ensure that the pathology does not bias the deformation of undamaged tissue. This involves hard constraints implicit in using a small number of basis functions or soft constraints implemented by increasing the role of priors in Bayesian estimation. This can involve decreasing the precision of the data in the region of pathology so that more importance is afforded to the priors (cf. lesion masking). An alternative strategy is to use another modality that is less sensitive to the pathology as the basis of the spatial normalization procedure.

Registration of functional and anatomical data

It is sometimes useful to co-register functional and anatomical images. However, with echo-planar imaging, geometric distortions of T2* images, relative to anatomical T1-weighted data, can be a serious problem because of the very low frequency per point in the phase encoding direction. Typically, for echo-planar fMRI, magnetic field inhomogeneity, sufficient to cause de-phasing of 2π through the slice, corresponds to an in-plane distortion of a voxel. Un-warping schemes have been proposed to correct for the distortion effects (Jezzard and Balaban, 1995). However, this distortion is not an issue if one spatially normalizes the functional data.

Spatial smoothing

The motivations for smoothing the data are fourfold. By the matched filter theorem, the optimum

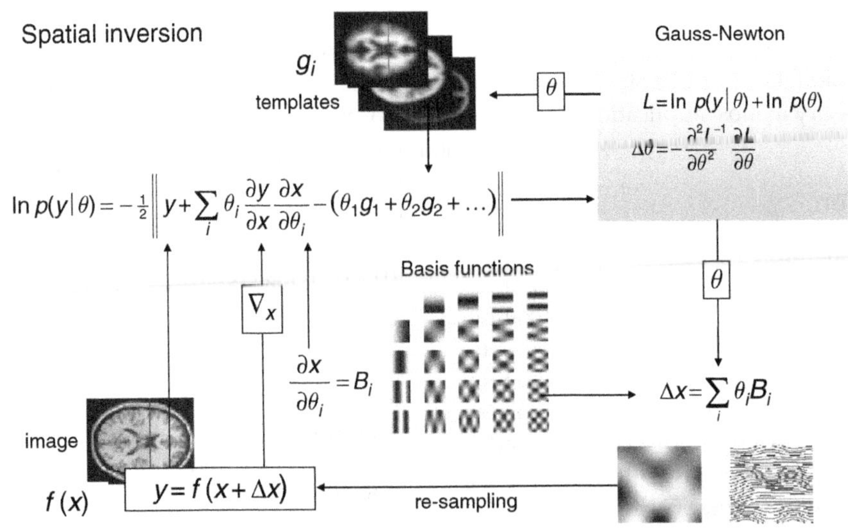

FIGURE 2.2 Schematic illustrating a Gauss-Newton scheme for maximizing the posterior probability $p(\theta|y)$ of the parameters required to spatially normalize an image. This scheme is iterative. At each step, the conditional estimate of the parameters obtains by jointly minimizing the likelihood and the prior potentials. The former is the difference between a resampled (i.e. warped) version $y = f(x + \Delta x)$ of the image $f(x)$ and the best linear combination of some templates $g(x)_1, \ldots$. These parameters are used to mix the templates and resample the image to reduce progressively both the spatial and intensity differences. After convergence, the resampled image can be considered normalized.

smoothing kernel corresponds to the size of the effect that one anticipates. The spatial scale of haemodynamic responses is, according to high-resolution optical imaging experiments, about 2–5 mm. Despite the potentially high resolution afforded by fMRI, an equivalent smoothing is suggested for most applications. By the central limit theorem, smoothing the data will render the errors more normal in their distribution and ensure the validity of inferences based on parametric tests. When making inferences about regional effects using random field theory (see below) the assumption is that the error terms are a reasonable lattice representation of an underlying continuous scalar field. This necessitates smoothness to be substantially greater than voxel size. If the voxels are large, then they can be reduced by sub-sampling the data and smoothing (with the original point spread function) with little loss of intrinsic resolution. In the context of inter-subject averaging it is often necessary to smooth more (e.g. 8 mm in fMRI or 16 mm in PET) to project the data onto a spatial scale where homologies in functional anatomy are expressed among subjects.

Summary

Spatial registration and normalization can proceed at a number of spatial scales depending on how one parameterizes variations in anatomy. We have focused on the role of normalization to remove unwanted differences to enable subsequent analysis of the data. However, it is important to note that the products of spatial normalization are twofold: a spatially normalized image and a deformation field (Plate 2). This deformation field contains important information about anatomy, in relation to the template used in the normalization procedure. The analysis of this information forms a key part of computational neuroanatomy. The tensor fields can be analysed directly (deformation-based morphometry – Ashburner *et al.*, 1998; Chung *et al.*, 2001) or used to create maps of specific anatomical attributes (e.g. compression, shears etc.). These maps can then be analysed on a voxel by voxel basis (tensor-based morphometry). Finally, the normalized structural images can themselves be subject to statistical analysis after some suitable segmentation procedure. This is known as *voxel-based morphometry*. Voxel-based morphometry is the most commonly used voxel-based neuroanatomical procedure and can easily be extended to incorporate tensor-based approaches (see Chapters 6 and 7).

STATISTICAL PARAMETRIC MAPPING AND THE GENERAL LINEAR MODEL

Functional mapping studies are usually analysed with some form of statistical parametric mapping. Statistical parametric mapping entails the construction of continuous statistical processes to test hypotheses about regionally specific effects (Friston *et al.*, 1991). Statistical parametric maps (SPMs) are images or fields with values that are, under the null hypothesis, distributed according to a known probability density function, usually the Student's t or F-distributions. These are known colloquially as t- or F-maps. The success of statistical parametric mapping is due largely to the simplicity of the idea. Namely, one analyses each and every voxel using any standard (univariate) statistical test. The resulting statistical parameters are assembled into an image – the SPM. SPMs are interpreted as continuous statistical processes by referring to the probabilistic behaviour of random fields (Adler, 1981; Worsley *et al.*, 1992, 1996; Friston *et al.*, 1994). Random fields model both the univariate probabilistic characteristics of an SPM and any non-stationary spatial covariance structure. 'Unlikely' topological features of the SPM, like peaks, are interpreted as regionally specific effects, attributable to the experimental manipulation.

Over the years, statistical parametric mapping has come to refer to the conjoint use of *the general linear model* (GLM) and *random field theory* (RFT) theory to analyse and make classical inferences about topological features of the statistical parametric maps (SPM). The GLM is used to estimate some parameters that explain continuous data in exactly the same way as in conventional analyses of discrete data (see Part 2). RFT is used to resolve the multiple comparison problem that ensues when making inferences over the volume analysed (see Part 3). RFT provides a method for adjusting p-values for the search volume and plays the same role for continuous data (i.e. images) as the Bonferroni correction for a number of discontinuous or discrete statistical tests.

The approach was called SPM for three reasons:

1 To acknowledge significance *p*robability *m*apping, where interpolated pseudo-maps of p-values are used to summarize the analysis of multichannel event-related potential (ERP) studies.
2 For consistency with the nomenclature of parametric maps of physiological or physical parameters (e.g. parametric maps of regional cerebral blood flow (rCBF) or volume).
3 In reference to the *parametric* statistics that populate the maps.

Despite its simplicity, there are some fairly subtle motivations for the approach that deserve mention. Usually, given a response or dependent variable, comprising many thousands of voxels, one would use *multivariate* analyses as opposed to the *mass-univariate* approach that SPM represents. The problems with multivariate approaches are that:

1 they do not support inferences about regionally specific effects (i.e. topological features with a unique localizing attribute)
2 they require more observations than the dimension of the response variable (i.e. need more scans than voxels)
3 even in the context of dimension reduction, they are less sensitive to focal effects than mass-univariate approaches.

A heuristic, for their relative lack of power, is that multivariate approaches estimate the model's error covariances using lots of parameters (e.g. the covariance between the errors at all pairs of voxels). Conversely, SPM characterizes spatial covariance with a smoothness parameter, for each voxel. In general, the more parameters (and hyperparameters) an estimation procedure has to deal with, the more variable the estimate of any one parameter becomes. This renders inferences about any single estimate less efficient.

Multivariate approaches consider voxels as different levels of an experimental or treatment factor and use classical analysis of variance, not at each voxel but by considering the data sequences from all voxels together, as replications over voxels. The problem here is that regional changes in error variance, and spatial correlations in the data, induce profound non-sphericity[1] in the error terms. This non-sphericity would again require large numbers of parameters to be estimated for each voxel using conventional techniques. In SPM, the non-sphericity is parameterized in a parsimonious way with just two parameters for each voxel. These are the error variance and smoothness estimators (see Part 3). This minimal parameterization lends SPM a sensitivity that surpasses multivariate approaches. SPM can do this because RFT implicitly harnesses constraints on the non-sphericity implied by the continuous (i.e. analytic) nature of the data. This is something that conventional multivariate and equivalent univariate approaches cannot accommodate, to their cost.

Some analyses use statistical maps based on non-parametric tests that eschew distributional assumptions about the data (see Chapter 21). These approaches are generally less powerful (i.e. less sensitive) than parametric approaches (see Aguirre *et al.*, 1998). However, they have an important role in evaluating the assumptions

[1] Sphericity refers to the assumption of identically and independently distributed error terms (IID). Under IID assumptions the probability density function of the errors, from all observations, has spherical iso-contours, hence *sphericity*. Deviations from either of the IID criteria constitute non-sphericity. If the error terms are not identically distributed then different observations have different error variances. Correlations among errors reflect dependencies among the error terms (e.g. serial correlation in fMRI time series) and constitute the second component of non-sphericity. In neuroimaging both spatial and temporal non-sphericity can be quite profound.

behind parametric approaches and may supervene in terms of sensitivity when these assumptions are violated (e.g. when degrees of freedom are very small and voxel sizes are large in relation to smoothness).

In Part 4 we consider Bayesian alternatives to classical inference with SPMs. This rests on conditional inferences about an effect, given the data, as opposed to classical inferences about the data, given the effect is zero. Bayesian inferences on continuous fields or images use posterior probability maps (PPMs). Although less commonly used than SPMs, PPMs are potentially useful, not least because they do not have to contend with the multiple comparisons problem induced by classical inference. In contradistinction to SPM, this means that inferences about a given regional response do not depend on inferences about responses elsewhere. Next we consider parameter estimation in the context of the GLM. This is followed by an introduction to the role of RFT when making classical inferences about continuous data.

The general linear model

Statistical analysis of imaging data corresponds to inverting generative models of the data to partition observed responses into components of interest, confounds and error. Inferences are then pursued using statistics that compare interesting effects and the error. This classical inference can be regarded as a direct comparison of the variance due to an interesting experimental manipulation with the error variance (compare with the F-statistic and other likelihood ratios). Alternatively, one can view the statistic as an estimate of the response, or difference of interest, divided by an estimate of its standard deviation. This is a useful way to think about the t-statistic.

A brief review of the literature may give the impression that there are numerous ways to analyse PET and fMRI time-series with a diversity of statistical and conceptual approaches. This is not the case. With very few exceptions, every analysis is a variant of the general linear model. This includes simple t-tests on scans assigned to one condition or another, correlation coefficients between observed responses and boxcar stimulus functions in fMRI, inferences made using multiple linear regression, evoked responses estimated using linear time invariant models, and selective averaging to estimate event-related responses in fMRI. Mathematically, they are all formally identical and can be implemented with the same equations and algorithms. The only thing that distinguishes among them is the design matrix encoding the temporal model or experimental design. The use of the correlation coefficient deserves special mention because of its popularity in fMRI (Bandettini *et al.*, 1993). The significance of a correlation is identical to the significance of the equivalent t-statistic testing for a regression of the data on a stimulus function. The correlation coefficient approach is useful but the inference is effectively based on a limiting case of multiple linear regression that obtains when there is only one regressor. In fMRI, many regressors usually enter a statistical model. Therefore, the t-statistic provides a more versatile and generic way of assessing the significance of regional effects and is usually preferred over the correlation coefficient.

The general linear model is an equation $Y = X\beta + \varepsilon$ that expresses the observed response variable in terms of a linear combination of explanatory variables X plus a well behaved error term (Figure 2.3 and Friston *et al.*, 1995b). The general linear model is variously known as 'analysis of covariance' or 'multiple regression analysis' and subsumes simpler variants, like the 't-test' for a difference in means, to more elaborate linear convolution models such as finite impulse response (FIR) models. The matrix that contains the explanatory variables (e.g. designed effects or confounds) is called the *design matrix*. Each column of the design matrix corresponds to an effect one has built into the experiment or that may confound the results. These are referred to as explanatory variables, covariates or regressors. The example in Plate 1 relates to an fMRI study of visual stimulation under four conditions. The effects on the response variable are modelled in terms of functions of the presence of these conditions (i.e. boxcars smoothed with a haemodynamic response function) and constitute the first four columns of the design matrix. There then follows a series of terms that are designed to remove or model low-frequency variations in signal due to artefacts such as aliased biorhythms and other drift terms. The final column is whole brain activity. The relative contribution of each of these columns is assessed using standard maximum likelihood and inferences about these contributions are made using t or F-statistics, depending upon whether one is looking at a particular linear combination (e.g. a subtraction), or all of them together. The operational equations are depicted schematically in Figure 2.3. In this scheme, the general linear model has been extended (Worsley and Friston, 1995) to incorporate intrinsic non-sphericity, or correlations among the error terms, and to allow for some temporal filtering of the data with the matrix S. This generalization brings with it the notion of *effective degrees of freedom*, which are less than the conventional degrees of freedom under IID assumptions (see footnote). They are smaller because the temporal correlations reduce the effective number of independent observations. The statistics are constructed using the approximation of Satterthwaite. This is the same approximation used in classical non-sphericity corrections such as the Geisser-Greenhouse correction. However, in the

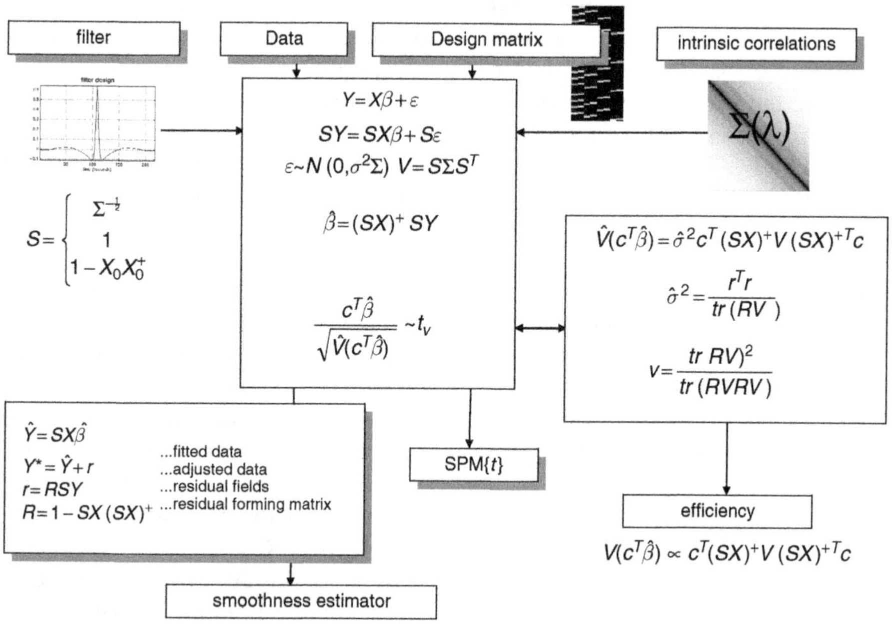

FIGURE 2.3 The general linear model. The general linear model is an equation expressing the response variable Y in terms of a linear combination of explanatory variables in a design matrix X and an error term with assumed or known autocorrelation Σ. The data can be filtered with a convolution or residual forming matrix (or a combination) S, leading to a generalized linear model that includes [intrinsic] serial correlations and applied [extrinsic] filtering. Different choices of S correspond to different estimation schemes. The parameter estimates obtain in a least squares sense using the pseudo-inverse (denoted by +) of the filtered design matrix. Generally, an effect of interest is specified by a vector of contrast weights c that give a weighted sum or compound of parameter estimates referred to as a *contrast*. The t-statistic is simply this contrast divided by its standard error (i.e. square root of its estimated variance). The ensuing t-statistic is distributed with v degrees of freedom. The equations for estimating the variance of the contrast and the degrees of freedom are provided in the right-hand panel. Efficiency is simply the inverse of the variance of the contrast. These expressions are useful when assessing the relative efficiency of different designs encoded in X. The parameter estimates can be examined directly or used to compute the fitted responses (see lower left panel). Adjusted data refer to data from which fitted components (e.g. confounds) have been removed. The residuals r, obtain from applying the residual-forming matrix R to the data. These residual fields are used to estimate the smoothness of the component fields of the SPM and are needed by random field theory (see Figure 2.4).

Worsley and Friston (1995) scheme, this approximation is used to construct the statistics and appropriate degrees of freedom, not simply to provide a *post hoc* correction to the degrees of freedom.

There is a special and important case of temporal filtering. This is when the filtering de-correlates (i.e. whitens) the error terms by using $S = \Sigma^{-1/2}$. This is the filtering scheme used in current implementations of the SPM software and renders the ordinary least squares (OLS) parameter estimates maximum likelihood (ML) estimators. These are optimal in the sense that they are the minimum variance estimators of all unbiased estimators. The estimation of $S = \Sigma^{-1/2}$ uses expectation maximization (EM) to provide restricted maximum likelihood (ReML) estimates of $\Sigma = \Sigma(\lambda)$ in terms of hyperparameters λ corresponding to variance components (see Chapter 11 and Chapter 24 for an explanation of EM). In this case, the effective degrees of freedom revert to the maximum that would be attained in the absence of temporal correlations or non-sphericity.

Contrasts

The equations summarized in Figure 2.3 can be used to implement a vast range of statistical analyses. The issue is therefore not the mathematics but the formulation of a design matrix appropriate to the study design and inferences that are sought. The design matrix can contain both covariates and indicator variables. Each column has an associated unknown or free parameter β. Some of these parameters will be of interest (e.g. the effect of a particular sensorimotor or cognitive condition or the regression coefficient of haemodynamic responses on reaction time). The remaining parameters will be of no interest and pertain to confounding effects (e.g. the effect of being a particular subject or the regression slope of voxel activity on global activity). Inferences about the parameter estimates are made using their estimated variance. This allows one to test the null hypothesis, that all the estimates are zero, using the F-statistic to give an SPM{F} or that some particular linear combination (e.g. a subtraction) of the estimates is zero using an SPM{t}. The

t-statistic obtains by dividing a contrast or compound (specified by contrast weights) of the ensuing parameter estimates by the standard error of that compound. The latter is estimated using the variance of the residuals about the least-squares fit. An example of a contrast weight *vector* would be $[-1\ 1\ 0 \ldots\ldots]$ to compare the difference in responses evoked by two conditions, modelled by the first two condition-specific regressors in the design matrix. Sometimes several contrasts of parameter estimates are jointly interesting. For example, when using polynomial (Büchel *et al.*, 1996) or basis function expansions of some experimental factor. In these instances, the SPM$\{F\}$ is used and is specified with a *matrix* of contrast weights that can be thought of as a collection of 't-contrasts' (see Chapter 9 for a fuller explanation). An F-contrast may look like:

$$\begin{bmatrix} -1 & 0 & 0 & 0 & \ldots \\ 0 & 1 & 0 & 0 & \ldots \end{bmatrix}$$

This would test for the significance of the first *or* second parameter estimates. The fact that the first weight is negative has no effect on the test because the F-statistic is based on sums of squares.

In most analyses, the design matrix contains indicator variables or parametric variables encoding the experimental manipulations. These are formally identical to classical analysis of covariance (i.e. ANCOVA) models. An important instance of the GLM, from the perspective of fMRI, is the linear time-invariant (LTI) model. Mathematically, this is no different from any other GLM. However, it explicitly treats the data-sequence as an ordered time-series and enables a signal processing perspective that can be very useful (see next chapter and Chapter 14).

TOPOLOGICAL INFERENCE AND THE THEORY OF RANDOM FIELDS

Classical inferences using SPMs can be of two sorts, depending on whether one knows where to look in advance. With an anatomically constrained hypothesis, about effects in a particular brain region, the uncorrected p-value associated with the height or extent of that region in the SPM can be used to test the hypothesis. With an anatomically open hypothesis (i.e. a null hypothesis that there is no effect anywhere in a specified volume of the brain), a correction for multiple dependent comparisons is necessary. The theory of random fields provides a way of adjusting the p-value that takes into account the fact that neighbouring voxels are not independent by virtue of continuity in the original data. Provided the data are smooth the RFT adjustment is less severe (i.e. is more sensitive) than a Bonferroni correction for the number of voxels. As noted above, RFT deals with the multiple comparisons problem in the context of continuous, statistical fields, in a way that is analogous to the Bonferroni procedure for families of discrete statistical tests. There are many ways to appreciate the difference between RFT and Bonferroni corrections. Perhaps the most intuitive is to consider the fundamental difference between an SPM and a collection of discrete t-values. When declaring a peak or cluster of the SPM to be significant, we refer collectively to all the voxels associated with that feature. The false positive rate is expressed in terms of peaks or clusters, under the null hypothesis of no activation. This is not the expected false positive rate of voxels. One false positive peak may be associated with hundreds of voxels, if the SPM is very smooth. Bonferroni correction controls the expected number of false positive *voxels*, whereas RFT controls the expected number of false positive *peaks*. Because the number of peaks is always less than the number of voxels, RFT can use a lower threshold, rendering it much more sensitive. In fact, the number of false positive voxels is somewhat irrelevant because it is a function of smoothness. The RFT correction discounts voxel size by expressing the search volume in terms of smoothness or resolution elements (*Resels*) (Figure 2.4). This intuitive perspective is expressed formally in terms of differential topology using the *Euler characteristic* (Worsley *et al.*, 1992). At high thresholds the Euler characteristic corresponds to the number of peaks above threshold.

There are only two assumptions underlying the use of the RFT correction:

1 the error fields (but not necessarily the data) are a reasonable lattice approximation to an underlying random field with a multivariate Gaussian distribution
2 these fields are continuous, with a differentiable and invertible autocorrelation function.

A common misconception is that the autocorrelation function has to be Gaussian. It does not. The only way in which these assumptions can be violated is if:

1 the data are not smooth, violating the reasonable lattice assumption or
2 the statistical model is mis-specified so that the errors are not normally distributed.

Early formulations of the RFT correction were based on the assumption that the spatial correlation structure was wide-sense stationary. This assumption can now be relaxed due to a revision of the way in which the smoothness estimator enters the correction procedure (Kiebel *et al.*, 1999). In other words, the corrections retain their validity, even if the smoothness varies from voxel to voxel.

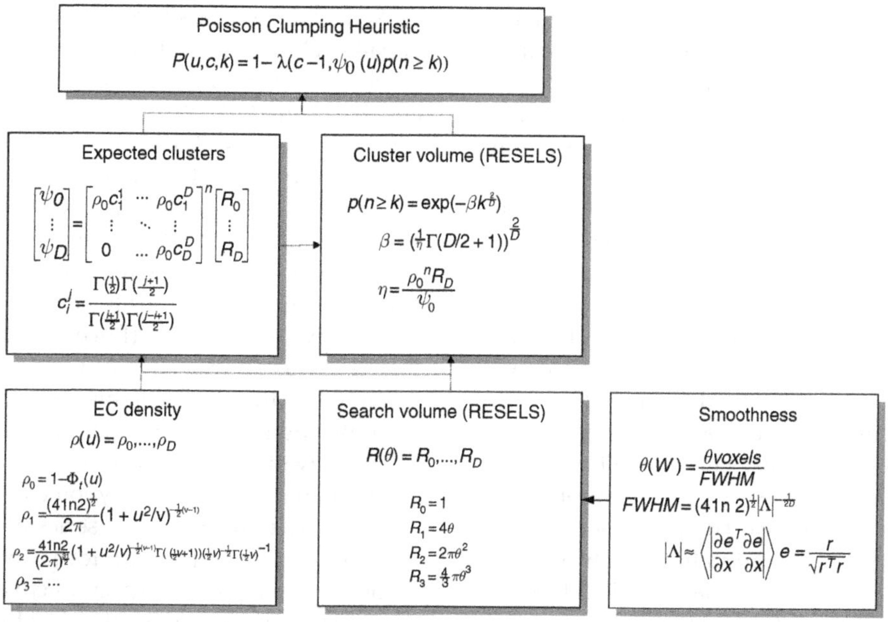

FIGURE 2.4 Schematic illustrating the use of random field theory (RFT) in making inferences about SPMs. If one knew precisely where to look, then inference can be based on the value of the statistic at the specified location in the SPM. However, generally, one does not have a precise anatomical *prior*, and an adjustment for multiple dependent comparisons has to be made to the *p*-values. These corrections use distributional approximations from RFT. This schematic deals with a general case of n SPM$\{t\}$ whose voxels all survive a common threshold u (i.e. a conjunction of n component SPMs). The central probability, upon which all peak, cluster or set-level inferences are made, is the probability $P(u, c, k)$ of getting c or more clusters with k or more resels (resolution elements) above this threshold. By assuming that clusters behave like a multidimensional Poisson point-process (i.e. the Poisson clumping heuristic) $P(u, c, k)$ is determined simply; the distribution of c is Poisson with an expectation that corresponds to the product of the expected number of clusters, of any size, and the probability that any cluster will be bigger than k resels. The latter probability depends on the expected number of resels per cluster η. This is simply the expected suprathreshold volume, divided by the expected number of clusters. The expected number of clusters ψ_0 is estimated with the Euler characteristic (EC) (effectively the number of blobs minus the number of holes). This depends on the EC density for the statistic in question (with degrees of freedom v) and the resel counts. The EC density is the expected EC per unit of D-dimensional volume of the SPM where the volume of the search is given by the resel counts. Resel counts are a volume measure that has been normalized by the smoothness of the SPMs component fields, expressed in terms of the full width at half maximum (FWHM). This is estimated from the determinant of the variance-covariance matrix of the first spatial derivatives of e, the normalized residual fields r (from Figure 2.3). In this example equations for a sphere of radius θ are given. ϕ denotes the cumulative density function for the statistic in question. (See Appendix 6 for technical details.)

Anatomically closed hypotheses

When making inferences about regional effects (e.g. activations) in SPMs, one often has some idea about where the activation should be. In this instance, a correction for the entire search volume is inappropriate. However, a problem remains in the sense that one would like to consider activations that are 'near' the predicted location, even if they are not exactly coincident. There are two approaches one can adopt: pre-specify a small search volume and make the appropriate RFT correction (Worsley *et al.*, 1996); or use the uncorrected p-value based on spatial extent of the nearest cluster (Friston, 1997). This probability is based on getting the observed number of voxels, or more, in a given cluster (conditional on that cluster existing). Both these procedures are based on distributional approximations from RFT.

Anatomically open hypotheses and levels of inference

To make inferences about regionally specific effects, the SPM is thresholded using some height and spatial extent thresholds that are specified by the user. Corrected p-values can then be derived that pertain to:

1 The number of activated regions (i.e. number of clusters above the height and volume threshold). These are *set-level inferences*.
2 The number of activated voxels (i.e. volume) comprising a particular region. These are *cluster-level inferences*.
3 The p-value for each peak within that cluster, i.e. *peak-level inferences*.

These p-values are corrected for the multiple dependent comparisons and are based on the probability of obtaining c, or more, clusters with k, or more, voxels, above a threshold u in an SPM of known or estimated smoothness. This probability has a reasonably simple form (see Figure 2.4 for details).

Set-level refers to the inference that the number of clusters comprising an observed activation profile is highly unlikely to have occurred by chance and is a statement about the activation profile, as characterized by its constituent regions. Cluster-level inferences are a special case of set-level inferences that obtain when the number of clusters $c = 1$. Similarly, peak-level inferences are special cases of cluster-level inferences that result when the cluster can be small (i.e. $k = 0$). Using a theoretical power analysis (see Friston *et al.*, 1996b and Chapter 19) of distributed activations, one observes that set-level inferences are generally more powerful than cluster-level inferences and that cluster-level inferences are generally more powerful than peak-level inferences. The price paid for this increased sensitivity is reduced localizing power. Peak-level tests permit individual maxima to be identified as significant, whereas cluster and set-level inferences only allow clusters or sets of clusters to be declared significant. It should be remembered that these conclusions, about the relative power of different inference levels, are based on distributed activations. Focal activation may well be detected with greater sensitivity using tests based on peak height. Typically, people use peak-level inferences and a spatial extent threshold of zero. This reflects the fact that characterizations of functional anatomy are generally more useful when specified with a high degree of anatomical precision.

EXPERIMENTAL AND MODEL DESIGN

This section considers the different sorts of designs that can be employed in neuroimaging studies. Experimental designs can be classified as *single factor* or *multifactor* designs; within this classification the levels of each factor can be *categorical* or *parametric*. We will start by discussing categorical and parametric designs and then deal with multifactor designs. We then move on to some more technical issues that attend the analysis of fMRI experiments. These are considered in terms of model design, using a signal processing perspective.

Categorical designs, cognitive subtraction and conjunctions

The tenet of cognitive subtraction is that the difference between two tasks can be formulated as a separable cognitive or sensorimotor component and that regionally specific differences in haemodynamic responses, evoked by the two tasks, identify the corresponding functionally selective area. Early applications of subtraction range from the functional anatomy of word processing (Petersen *et al.*, 1989) to functional specialization in extrastriate cortex (Lueck *et al.*, 1989). The latter studies involved presenting visual stimuli with and without some sensory attribute (e.g. colour, motion etc.). The areas highlighted by subtraction were identified with homologous areas in monkeys that showed selective electrophysiological responses to equivalent visual stimuli.

Cognitive conjunctions (Price and Friston, 1997) can be thought of as an extension of the subtraction technique, in the sense that they combine a series of subtractions. In subtraction, one tests a single hypothesis pertaining to the activation in one task relative to another. In conjunction analyses, several hypotheses are tested, asking whether the activations, in a series of task pairs, are collectively significant (cf. an F-test). Consider the problem of identifying regionally specific activations due to a particular cognitive component (e.g. object recognition). If one can identify a series of task pairs whose differences have only that component in common, then the region which activates, in all the corresponding subtractions, can be associated with the common component. Conjunction analyses allow one to demonstrate the context-invariant nature of regional responses. One important application of conjunction analyses is in multisubject fMRI studies, where generic effects are identified as those that are jointly significant in all the subjects studied (see below).

Parametric designs

The premise behind parametric designs is that regional physiology will vary systematically with the degree of cognitive or sensorimotor processing or deficits thereof. Examples of this approach include the PET experiments of Grafton *et al.* (1992) that demonstrated significant correlations between haemodynamic responses and the performance of a visually guided motor tracking task. On the sensory side, Price *et al.* (1992) demonstrated a remarkable linear relationship between perfusion in periauditory regions and frequency of aural word presentation. This correlation was not observed in Wernicke's area, where perfusion appeared to correlate, not with the discriminative attributes of the stimulus, but with the presence or absence of semantic content. These relationships or *neurometric functions* may be linear or non-linear. Using polynomial regression, in the context of the GLM, one can identify non-linear relationships between stimulus parameters (e.g. stimulus duration or presentation rate) and evoked responses. To do this one usually uses an SPM{F} (see Büchel *et al.*, 1996).

The example provided in Figure 2.5 illustrates both categorical and parametric aspects of design and analysis. These data were obtained from an fMRI study of visual motion processing using radially moving dots. The stimuli were presented over a range of speeds using *isoluminant* and *isochromatic* stimuli. To identify areas involved in visual motion, a stationary dots condition was subtracted from the moving dots conditions (see the contrast weights in the upper right). To ensure significant motion-sensitive responses, using colour and luminance cues, a conjunction of the equivalent subtractions was assessed under both viewing contexts. Areas V5 and V3a are seen in the ensuing SPM$\{t\}$. The t-values in this SPM are simply the minimum of the t-values for each subtraction. Thresholding this SPM ensures that all voxels survive the threshold u in each subtraction separately. This *conjunction* SPM has an equivalent interpretation; it represents the intersection of the excursion sets, defined by the threshold u, of each *component* SPM. This intersection is the essence of a conjunction. The expressions in Figure 2.4 pertain to the general case of the minimum of n t-values. The special case where $n = 1$ corresponds to a conventional SPM$\{t\}$.

The responses in left V5 are shown in the lower panel of Figure 2.5 and speak to a compelling inverted 'U' relationship between speed and evoked response that peaks

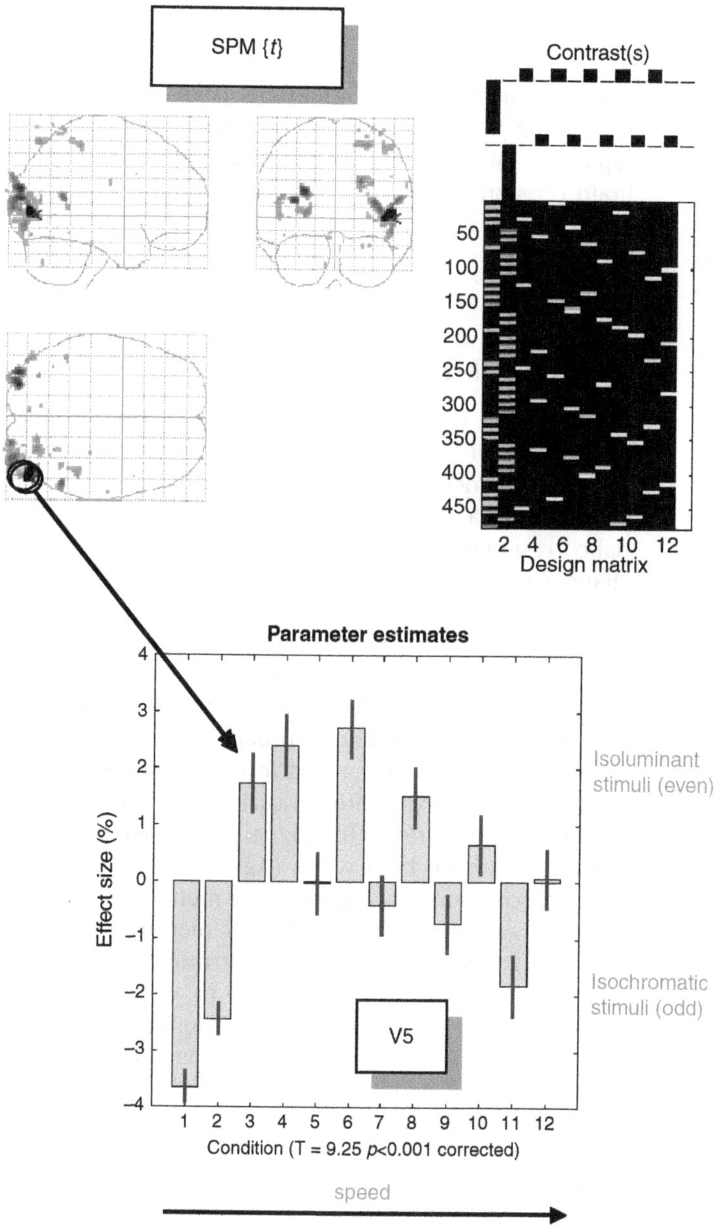

FIGURE 2.5 Top right: design matrix: this is an image representation of the design matrix. Contrasts: these are the vectors of contrast weights defining the linear compounds of parameters tested. The contrast weights are displayed over the column of the design matrix that corresponds to the effects in question. The design matrix here includes condition-specific effects (boxcar-functions convolved with a haemodynamic response function). Odd columns correspond to stimuli shown under isochromatic conditions and even columns model responses to isoluminant stimuli. The first two columns are for stationary stimuli and the remaining columns are for conditions of increasing speed. The final column is a constant term. Top left: SPM$\{t\}$: this is a maximum intensity projection conforming to the standard anatomical space of Talairach and Tournoux (1988). The values here are the minimum t-values from both contrasts, thresholded at $p = 0.001$ uncorrected. The most significant conjunction is seen in left V5. Lower panel: plot of the condition-specific parameter estimates for this voxel. The t-value was 9.25 (p <0.001 corrected – see Figure 2.4).

at around eight degrees per second. It is this sort of relationship that parametric designs try to characterize. Interestingly, the form of these speed-dependent responses was similar using both stimulus types, although luminance cues are seen to elicit a greater response. From the point of view of a factorial design there is a *main effect* of cue (isoluminant versus isochromatic), a main effect of speed, but no speed by cue *interaction*.

Clinical neuroscience studies can use parametric designs by looking for the neuronal correlates of clinical (e.g. symptom) ratings over subjects. In many cases, multiple clinical scores are available for each subject and the statistical design can usually be seen as a multilinear regression. In situations where the clinical scores are correlated, principal component analysis or factor analysis is sometimes applied to generate a new, and smaller, set of explanatory variables that are orthogonal to each other. This has proved particularly useful in psychiatric studies where syndromes can be expressed over a number of different dimensions (e.g. the degree of psychomotor poverty, disorganization and reality distortion in schizophrenia; see Liddle *et al.*, 1992). In this way, regionally specific correlates of various symptoms may point to their distinct pathogenesis in a way that transcends the syndrome itself. For example, psychomotor poverty may be associated with left dorso-lateral prefrontal dysfunction, irrespective of whether the patient is suffering from schizophrenia or depression.

Factorial designs

Factorial designs are more prevalent than single-factor designs because they enable inferences about interactions. At its simplest, an interaction represents a change in a change. Interactions are associated with factorial designs where two or more factors are combined in the same experiment. The effect of one factor, on the effect of the other, is assessed by the interaction. Factorial designs have a wide range of applications. An early application, in neuroimaging, examined adaptation and plasticity during motor performance by assessing time by condition interactions (Friston *et al.*, 1992a). Psychopharmacological activation studies are further examples of factorial designs (Friston *et al.*, 1992b). In these studies, cognitively evoked responses are assessed before and after being given a drug. The interaction term reflects the pharmacological modulation of task-dependent activations. Factorial designs have an important role in the context of cognitive subtraction and additive factors logic by virtue of being able to test for interactions, or context-sensitive activations, i.e. to demonstrate the fallacy of pure-insertion (see Friston *et al.*, 1996c). These interaction effects can sometimes be interpreted as the integration of the two or more [cognitive] processes or the modulation of one [perceptual] process by another. Figure 2.6 shows an example which takes an unusual perspective on the modulation of event-related responses as the interaction between stimulus presentation and experiential context.

From the point of view of clinical studies, interactions are central. The effect of a disease process on sensorimotor or cognitive activation is simply an interaction and involves replicating a subtraction experiment in subjects with and without the pathology. Factorial designs can also embody parametric factors. If one of the factors has a number of parametric levels, the interaction can be expressed as a difference in regression slope of regional activity on the parameter, under both levels of the other [categorical] factor. An important example of factorial designs, that mix categorical and parameter factors, are those looking for *psychophysiological interactions*. Here the parametric factor is brain activity measured in a particular brain region. These designs have proven useful in looking at the interaction between bottom-up and top-down influences within processing hierarchies in the brain (Friston *et al.*, 1997). This issue will be addressed below and in Part 6, from the point of view of effective connectivity.

Designing fMRI studies

In this section, we consider fMRI time-series from a signal processing perspective with particular focus on optimal experimental design and efficiency. fMRI time-series can be viewed as a linear admixture of signal and noise. Signal corresponds to neuronally mediated haemodynamic changes that can be modelled as a convolution of some underlying neuronal process, responding to changes in experimental factors, by a haemodynamic response function. Noise has many contributions that render it rather complicated in relation to some neurophysiological measurements. These include neuronal and non-neuronal sources. Neuronal noise refers to neurogenic signal not modelled by the explanatory variables and has the same frequency structure as the signal itself. Non-neuronal components have both white (e.g. Johnson noise) and coloured components (e.g. pulsatile motion of the brain caused by cardiac cycles and local modulation of the static magnetic field by respiratory movement). These effects are typically low-frequency or wide-band (e.g. aliased cardiac-locked pulsatile motion). The superposition of all these components induces temporal correlations among the error terms (denoted by Σ in Figure 2.3) that can affect sensitivity to experimental effects. Sensitivity depends upon the relative amounts of signal and noise and the efficiency of the experimental design. Efficiency is simply a measure of how reliable the parameter

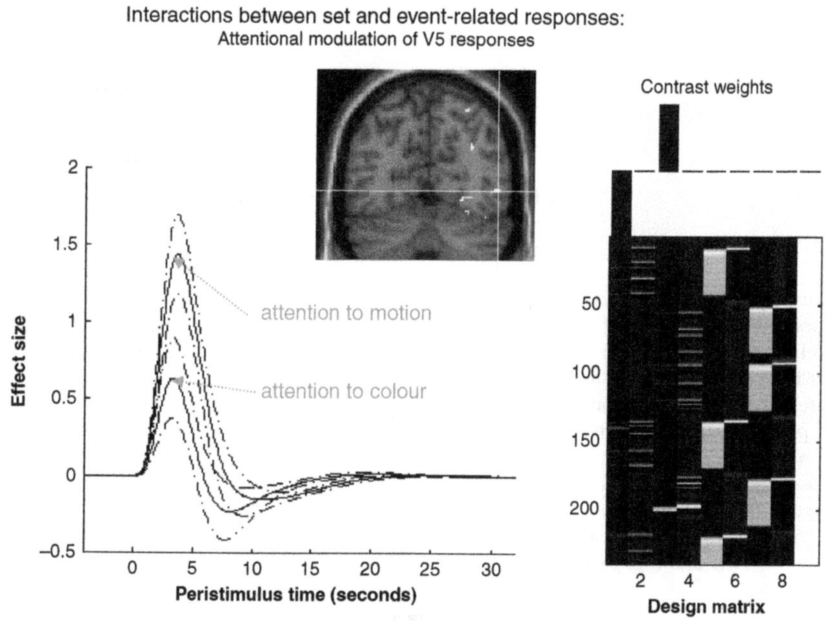

FIGURE 2.6 Results showing attentional modulation of visually-evoked responses. Subjects viewed stationary monochromatic stimuli that occasionally changed colour and moved at the same time. These compound events were presented under two levels of attentional set (attention to colour and attention to motion). The event-related responses are modelled, in an attention-specific fashion by the first four regressors (stick-functions convolved with a haemodynamic response function and its derivative) in the design matrix on the right. The main effects of attention are modelled as similarly convolved boxcars. The interaction between attentional set and visually evoked responses is simply the difference in evoked responses under both levels of attention and is tested for with the appropriate contrast weights (upper right). Only the first 256 rows of the design matrix are shown. The most significant modulation of evoked responses, under attention to motion, was seen in left V5 (insert). The fitted responses and their standard errors are shown on the left as functions of peristimulus time.

estimates are and can be defined as the inverse of the variance of a contrast of parameter estimates (see Figure 2.3). There are two important considerations that arise from this perspective on fMRI time-series: the first pertains to optimal experimental design and the second to optimum de-convolution of the time-series to obtain the most efficient parameter estimates.

The haemodynamic response function and optimum design

As noted above, an LTI model of neuronally mediated signals in fMRI suggests that only those experimentally induced signals that survive convolution with the haemodynamic response function (HRF) can be estimated with any efficiency. By convolution theorem, the frequency structure of experimental variance should therefore be designed to match the transfer function of the HRF. The corresponding frequency profile of this transfer function is shown in Figure 2.7 (solid line). It can be seen that frequencies around 0.03 Hz are optimal, corresponding to periodic designs with 32-second periods (i.e. 16-second epochs). Generally, the first objective of experimental design is to comply with the natural constraints imposed by the HRF and ensure that experimental variance occupies these intermediate frequencies.

Serial correlations and filtering

This is quite a complicated but important area. Conventional signal processing approaches dictate that whitening the data engenders the most efficient parameter estimation. This corresponds to filtering with a convolution matrix $S = K^{-1}$ that is the inverse of the intrinsic convolution matrix, i.e. $KK^T = \Sigma$ (see Figure 2.3). This whitening strategy renders the least-square estimator in Figure 2.3 equivalent to the ML or Gauss-Markov estimator. However, one generally does not know the form of the intrinsic correlations, which means they have to be estimated. This estimation usually proceeds using a restricted maximum likelihood (ReML) estimate of the serial correlations, among the residuals, that properly accommodates the effects of the residual-forming matrix and associated loss of degrees of freedom. However, using this estimate of the intrinsic non-sphericity to form a Gauss-Markov estimator at each voxel is not easy. First, the estimate of non-sphericity can itself be imprecise leading to bias in the standard error (Friston *et al.*, 2000). Second, ReML estimation requires a computationally

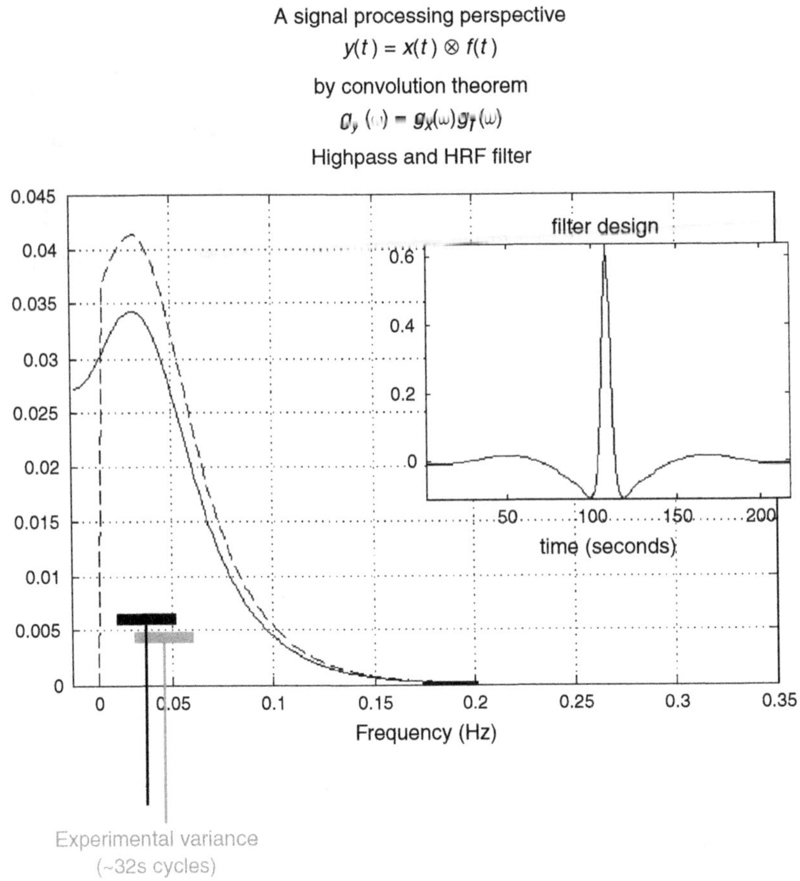

FIGURE 2.7 Transfer function of a canonical haemodynamic response function (HRF), with (broken line) and without (solid line) the application of a highpass filter. This transfer function corresponds to the spectral density of a white-noise process after convolution with the HRF and places constraints on the frequencies that survive haemodynamic convolution. This follows from convolution theorem (summarized in the equations). The insert is the filter expressed in time, corresponding to the spectral density that obtains after convolution with the HRF and highpass filtering.

prohibitive iterative procedure at every voxel. There are a number of approaches to these problems that aim to increase the efficiency of the estimation and reduce the computational burden. The approach adopted in current versions of our software is to use ReML estimates based on all voxels that respond to experimental manipulation. This affords very efficient hyperparameter estimates[2] and, furthermore, allows one to use the same matrices at each voxel when computing the parameter estimates.

Although we usually make $S = \Sigma^{-1/2} = K^{-1}$, using a first-pass ReML estimate of the serial correlations, we will deal with the simpler and more general case where S can take any form. In this case, the parameter estimates are *generalized* least square (GLS) estimators. The GLS estimator is unbiased and, luckily, is identical to the Gauss-Markov estimator if the regressors in the design matrix are periodic.[3] After GLS estimation, the ReML estimate of $V = S\Sigma S^T$ enters into the expressions for the standard error and degrees of freedom provided in Figure 2.3.

fMRI noise has been variously characterized as a $1/f$ process (Zarahn *et al.*, 1997) or an autoregressive process (Bullmore *et al.*, 1996) with white noise (Purdon and Weisskoff, 1998). Irrespective of the exact form these serial correlations take, treating low-frequency drifts as fixed effects can finesse the hyperparameterization of serial correlations. Removing low frequencies from the time-series allows the model to fit serial correlations over a more restricted frequency range or shorter time spans. Drift removal can be implemented by including drift terms in the design matrix or by including the implicit residual forming matrix in S to make it a highpass filter. An example of a highpass filter with a highpass cut-off of 1/64 Hz is shown in the inset of Figure 2.7. This filter's transfer function (the broken line in the main panel) illustrates the frequency structure of neurogenic signals after highpass filtering.

Spatially coherent confounds and global normalization

Implicit in the use of highpass filtering is the removal of low-frequency components that can be regarded as confounds. Other important confounds are signal

[2] The efficiency scales with the number of voxels.

[3] More exactly, the GLS and ML estimators are the same if the design matrix is spanned by the eigenvectors of the Toeplitz autocorrelation matrix Σ.

components that are artefactual or have no regional specificity. These are referred to as global confounds and have a number of causes. These can be divided into physiological (e.g. global perfusion changes in PET) and non-physiological (e.g. transmitter power calibration or receiver gain in fMRI). The latter generally scale the signal before the MRI sampling process. Other non-physiological effects may have a non-scaling effect (e.g. Nyquist ghosting, movement-related effects etc.). In PET, it is generally accepted that regional changes in rCBF, evoked neuronally, mix additively with global changes to give the measured signal. This calls for a global normalization procedure where the global estimator enters into the statistical model as a confound. In fMRI, instrumentation effects that scale the data motivate a global normalization by proportional scaling, using the whole brain mean, before the data enter into the statistical model.

It is important to differentiate between global confounds and their estimators. By definition, the global mean over intracranial voxels will subsume all regionally specific effects. This means that the global estimator may be partially collinear with effects of interest, especially if evoked responses are substantial and widespread. In these situations, global normalization may induce apparent deactivations in regions not expressing a physiological response. These are not artefacts in the sense that they are real, relative to global changes, but they have less face validity in terms of the underlying neurophysiology. In instances where regionally specific effects bias the global estimator, some investigators prefer to omit global normalization. Provided drift terms are removed from the time-series, this is generally acceptable because most global effects have slow time constants. However, the issue of normalization-induced deactivations is better circumnavigated with experimental designs that use well-controlled conditions, which elicit differential responses in restricted brain systems.

Non-linear system identification approaches

So far, we have only considered linear models and first-order HRFs. Another signal processing perspective is provided by non-linear system identification (Vazquez and Noll, 1998). This section considers non-linear models as a prelude to the next subsection on event-related fMRI, where non-linear interactions among evoked responses provide constraints for experimental design and analysis. We have described an approach to characterizing evoked haemodynamic responses in fMRI based on non-linear system identification, in particular the use of Volterra series (Friston *et al.*, 1998). This approach enables one to estimate Volterra kernels that describe the relationship between stimulus presentation and the haemodynamic responses that ensue. Volterra series are essentially high-order extensions of linear convolution models. These kernels therefore represent a non-linear characterization of the HRF that can model the responses to stimuli in different contexts and interactions among stimuli. In fMRI, the kernel coefficients can be estimated by using a second order approximation to the Volterra series to formulate the problem in terms of a general linear model and expanding the kernels in terms of temporal basis functions (see Chapter 27). This allows the use of the standard techniques described above to estimate the kernels and to make inferences about their significance on a voxel-specific basis using SPMs.

One important manifestation of non-linear effects, captured by second order kernels, is a modulation of stimulus-specific responses by preceding stimuli that are proximate in time. This means that responses at high stimulus presentation rates saturate and, in some instances, show an inverted U behaviour. This behaviour appears to be specific to blood oxygenation-level-dependent (BOLD) effects (as distinct from evoked changes in cerebral blood flow) and may represent a *haemodynamic refractoriness*. This effect has important implications for event-related fMRI, where one may want to present trials in quick succession.

The results of a typical non-linear analysis are given in Figure 2.8. The results in the right panel represent the average response, integrated over a 32-second train of stimuli as a function of stimulus onset asynchrony (SOA) within that train. These responses are based on the kernel estimates (left hand panels) using data from a voxel in the left posterior temporal region of a subject obtained during the presentation of single words at different rates. The solid line represents the estimated response and shows a clear maximum at just less than one second. The dots are responses based on empirical data from the same experiment. The broken line shows the expected response in the absence of non-linear effects (i.e. that predicted by setting the second order kernel to zero). It is clear that non-linearities become important at around two seconds leading to an actual diminution of the integrated response at sub-second SOAs. The implication of this sort of result is that SOAs should not really fall much below one second and at short SOAs the assumptions of linearity are violated. It should be noted that these data pertain to single word processing in auditory association cortex. More linear behaviours may be expressed in primary sensory cortex where the feasibility of using minimum SOAs, as low as 500 ms, has been demonstrated (Burock *et al.*, 1998). This lower bound on SOA is important because some effects are detected more efficiently with high presentation rates. We now consider this from the point of view of event-related designs.

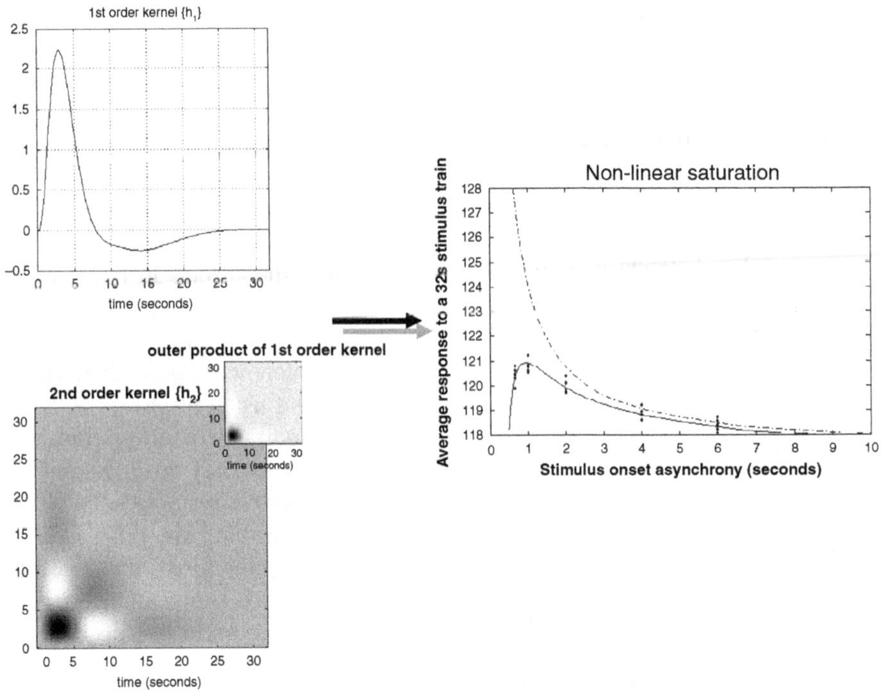

FIGURE 2.8 Left panels: Volterra kernels from a voxel in the left superior temporal gyrus at −56, −28, 12 mm. These kernel estimates were based on a single-subject study of aural word presentation at different rates (from zero to ninety words per minute) using a second order approximation to a Volterra series expansion modelling the observed haemodynamic response to stimulus input (a delta function for each word). These kernels can be thought of as a characterization of the second order haemodynamic response function. The first order kernel (h_1 – upper panel) represents the (first order) component usually presented in linear analyses. The second order kernel (h_2 – lower panel) is presented in image format. The colour scale is arbitrary; white is positive and black is negative. The insert on the right represents $h_1 h_1{}^T$, the second order kernel that would be predicted by a simple model that involved linear convolution with h_1 followed by some static non-linearity. Right panel: integrated responses over a 32-second stimulus train as a function of SOA. Solid line: estimates based on the non-linear convolution model parameterized by the kernels on the left. Broken line: the responses expected in the absence of second order effects (i.e. in a truly linear system). Dots: empirical averages based on the presentation of actual stimulus trains.

Event and epoch-related designs

A crucial distinction in experimental design for fMRI is that between epoch and event-related designs. In single photon emission computerized tomography (SPECT) and positron emission tomography (PET) only epoch-related responses can be assessed because of the relatively long half-life of the radiotracers used. However, in fMRI there is an opportunity to measure event-related responses, not unlike the paradigm used in electroencephalography (EEG) and magnetoencephalography (MEG). An important issue, in event-related fMRI, is the choice of inter-stimulus interval or more precisely SOA. The SOA, or the distribution of SOAs, is a critical factor and is chosen, subject to psychological or psychophysical constraints, to maximize the efficiency of response estimation. The constraints on the SOA clearly depend upon the nature of the experiment but are generally satisfied when the SOA is small and derives from a random distribution. Rapid presentation rates allow for the maintenance of a particular cognitive or attentional set, decrease the latitude that the subject has for engaging alternative strategies, or incidental processing, and allows the integration of event-related paradigms using fMRI and electrophysiology. Random SOAs ensure that preparatory or anticipatory factors do not confound event-related responses and ensure a uniform context in which events are presented. These constraints speak of the well-documented advantages of event-related fMRI over conventional blocked designs (Buckner *et al.*, 1996; Clark *et al.*, 1998).

In order to compare the efficiency of different designs, it is useful to have a common framework that encompasses all of them. The efficiency can then be examined in relation to the parameters of the designs. Designs can be *stochastic* or *deterministic* depending on whether there is a random element to their specification. In stochastic designs (Heid *et al.*, 1997) one needs to specify the probabilities of an event occurring at all times those events could occur. In deterministic designs, the occurrence probability is unity and the design is completely specified by the times of stimulus presentation or trials. The distinction between stochastic and deterministic designs pertains to how a particular realization or stimulus sequence is created. The efficiency afforded by a particular event sequence is a function of the event sequence itself, and not of the process generating the sequence (i.e. deterministic or stochastic). However, within stochastic designs, the design matrix X, and associated efficiency, are random variables and the *expected* or average efficiency, over realizations of X is easily computed.

In the framework considered here (Friston *et al.*, 1999a), the occurrence probability p of any event occurring is specified at each time that it could occur (i.e. every SOA or stimulus onset asynchrony). Here p is a vector with an element for every SOA. This formulation engenders the distinction between *stationary* stochastic designs, where the occurrence probabilities are constant and *non-stationary* stochastic designs, where they change over time. For deterministic designs, the elements of p are 0 or 1, the presence of a 1 denoting the occurrence of an event. An example of p might be the boxcars used in conventional block designs. Stochastic designs correspond to a vector of identical values and are therefore stationary in nature. Stochastic designs with temporal modulation of occurrence probability have time-dependent probabilities varying between 0 and 1. With these probabilities the expected design matrices and expected efficiencies can be computed. A useful thing about this formulation is that by setting the mean of the probabilities p to a constant, one can compare different deterministic and stochastic designs given the same number of events. Some common examples are given in Figure 2.9 (right panel) for an SOA of one second and 32 expected events or trials over a 64 second period (except for the first deterministic example with four events and an SOA of 16 seconds). It can be seen that the least efficient is the sparse deterministic design (despite the fact that the SOA is roughly optimal for this class), whereas the most efficient is a block design. A slow modulation of occurrence probabilities gives high efficiency while retaining the advantages of stochastic designs and may represent a useful compromise between the high efficiency of block designs and the psychological benefits and latitude afforded by stochastic designs. However, it is important not to generalize these conclusions too far. An efficient design for one effect may

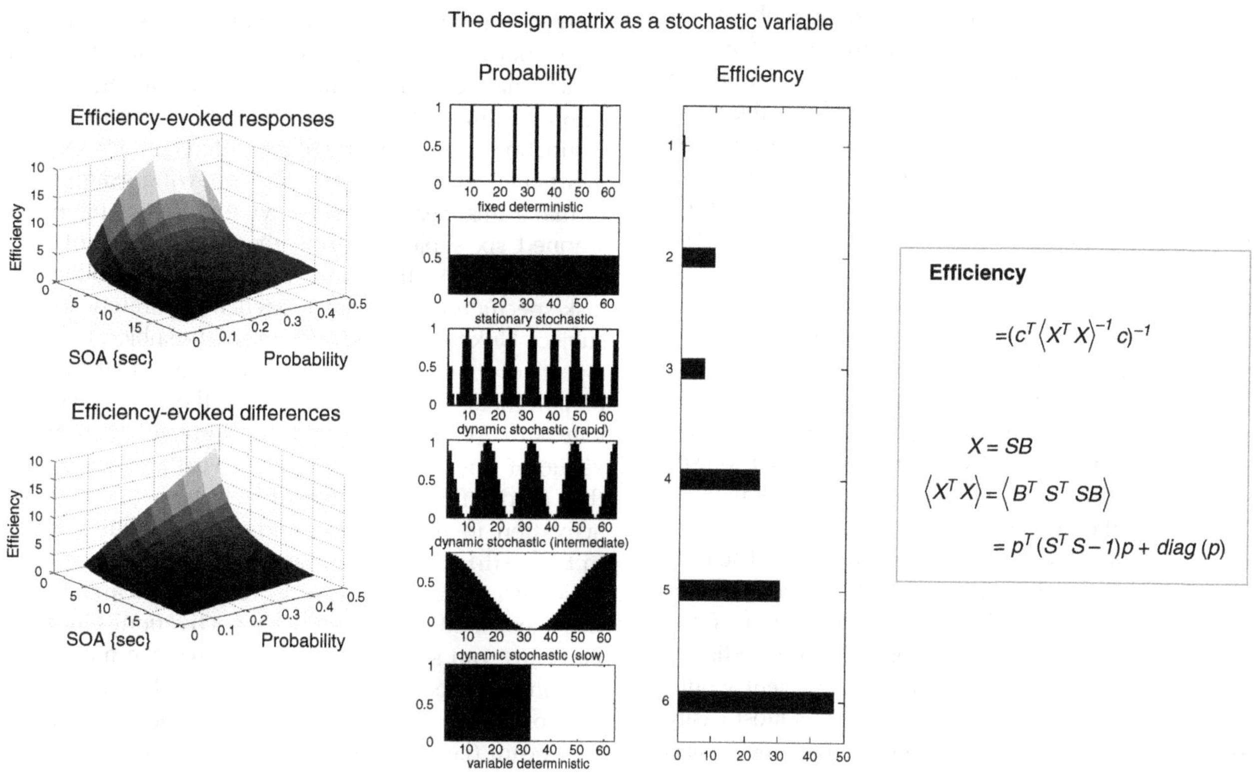

FIGURE 2.9 Efficiency as a function of occurrence probability p for a model X formed by post-multiplying S (a matrix containing n columns, modelling n possible event-related responses every SOA) by B. B is a random binary vector that determines whether the n-th response contributes to the design matrix $X = SB$ where $\langle B\rangle = p$. Right panels: a comparison of some common designs. A graphical representation of the occurrence probability as a function of time (seconds) is shown on the left and the corresponding efficiency is shown on the right. These results assume a minimum SOA of one second, a time-series of 64 seconds and a single trial-type. The expected number of events was 32 in all cases (apart from the first). Left panels: efficiency in a stationary stochastic design with two event types both presented with increasing probability every SOA. The upper graph is for a contrast testing for the response evoked by one trial type and the lower graph is for a contrast testing for differential responses.

not be the optimum for another, even within the same experiment. This can be illustrated by comparing the efficiency with which evoked responses are detected and the efficiency of detecting the difference in evoked responses elicited by two sorts of trials.

Consider a stationary stochastic design with two trial types. Because the design is stationary, the vector of occurrence probabilities, for each trial type, is specified by a single probability. Let us assume that the two trial types occur with the same probability p. By varying p and SOA one can find the most efficient design depending upon whether one is looking for evoked responses *per se* or differences among evoked responses. These two situations are depicted in the left panels of Figure 2.9. It is immediately apparent that, for both sorts of effects, very small SOAs are optimal. However, the optimal occurrence probabilities are not the same. More infrequent events (corresponding to a smaller $p = 1/3$) are required to estimate the responses themselves efficiently. This is equivalent to treating the baseline or control condition as any other condition (i.e. by including null events, with equal probability, as further event types). Conversely, if we are only interested in making inferences about the differences, one of the events plays the role of a null event and the most efficient design ensues when one or the other event occurs (i.e. $p = 1/2$). In short, the most efficient designs obtain when the events subtending the differences of interest occur with equal probability.

Another example of how the efficiency is sensitive to the effect of interest is apparent when we consider different parameterizations of the HRF. This issue is sometimes addressed through distinguishing between the efficiency of response *detection* and response *estimation*. However, the principles are identical and the distinction reduces to how many parameters one uses to model the HRF for each trial type (one basis function is used for detection and a number are required to estimate the shape of the HRF). Here the contrasts may be the same but the shape of the regressors will change depending on the temporal basis set employed. The conclusions above were based on a single canonical HRF. Had we used a more refined parameterization of the HRF, say using three-basis functions, the most efficient design to estimate one basis function coefficient would not be the most efficient for another. This is most easily seen from the signal processing perspective where basis functions with high-frequency structure (e.g. temporal derivatives) require the experimental variance to contain high-frequency components. For these basis functions a randomized stochastic design may be more efficient than a deterministic block design, simply because the former embodies higher frequencies. In the limiting case of finite impulse response (FIR) estimation, the regressors become a series of stick functions all of which have high frequencies. This parameterization of the HRF calls for high frequencies in the experimental variance. However, the use of FIR models is contraindicated by model selection procedures (see Chapter 14) that suggest only two or three HRF parameters can be estimated with any efficiency. Results that are reported in terms of FIRs should be treated with caution because the inferences about evoked responses are seldom based on the FIR parameter estimates. This is precisely because they are estimated inefficiently and contain little useful information.

INFERENCE IN HIERARCHICAL MODELS

In this section, we consider some issues that are generic to brain mapping studies that have repeated measures or replications over subjects. The critical issue is whether we want to make an inference about the effect in relation to the *within-subject variability* or with respect to the *between-subject variability*. For a given group of subjects, there is a fundamental distinction between saying that the response is significant relative to the precision[4] with which that response is measured and saying that it is significant in relation to the inter-subject variability. This distinction relates directly to the difference between *fixed-* and *random*-effect analyses. The following example tries to make this clear. Consider what would happen if we scanned six subjects during the performance of a task and baseline. We then construct a statistical model where task-specific effects were modelled separately for each subject. Unknown to us, only one of the subjects activated a particular brain region. When we examine the contrast of parameter estimates, assessing the mean activation over all subjects, we see that it is greater than zero by virtue of this subject's activation. Furthermore, because that model fits the data extremely well (modelling no activation in five subjects and a substantial activation in the sixth), the error variance, on a scan-to-scan basis, is small and the t-statistic is very significant. Can we then say that the group shows an activation? On the one hand, we can say, quite properly, that the mean group response embodies an activation but, clearly, this does not constitute an inference that the group's response is significant (i.e. that this sample of subjects shows a consistent activation). The problem here is that we are using the *scan-to-scan* error variance and this is not necessarily appropriate for an inference about group responses. To make the inference that the group showed a significant activation, one would have to assess the variability in

[4] Precision is the inverse of the variance.

activation effects from *subject to subject* (using the contrast of parameter estimates for each subject). This variability now constitutes the proper error variance. In this example, the variance of these six measurements would be large relative to their mean and the corresponding t-statistic would not be significant.

The distinction between the two approaches above relates to how one computes the appropriate error variance. The first represents a fixed-effects analysis and the second a random-effects analysis (or more exactly a mixed-effects analysis). In the former, the error variance is estimated on a scan-to-scan basis, assuming that each scan represents an independent observation (ignoring serial correlations). Here the degrees of freedom are essentially the number of scans (minus the rank of the design matrix). Conversely, in random-effects analyses, the appropriate error variance is based on the activation from subject to subject where the effect *per se* constitutes an independent observation and the degrees of freedom fall dramatically to the number of subjects. The term 'random effect' indicates that we have accommodated the randomness of different responses from subject to subject. Both analyses are perfectly valid but only in relation to the inferences that are being made: inferences based on fixed-effects analyses are about the particular subject(s) studied. Random-effects analyses are usually more conservative but allow the inference to be generalized to the population from which the subjects were selected.

Random-effects analyses

The implementation of random-effects analyses in SPM is fairly straightforward and involves taking the contrasts of parameters estimated from a *first-level* (within-subject) analysis and entering them into a *second-level* (between-subject) analysis. This ensures that there is only one observation (i.e. contrast) per subject in the second-level analysis and that the error variance is computed using the subject-to-subject variability of estimates from the first level. This is also known as a summary statistic approach and, in the context of fully balanced designs is formally identical to mixed-effects analysis. The nature of the inference made is determined by the contrasts that enter the second level (Figure 2.10). The second-level design matrix simply tests the null hypothesis that the contrasts are zero (and is usually a column of ones, implementing a single-sample t-test).

The reason this multistage procedure emulates a full mixed-effects analysis, using a hierarchical observation model (see Chapters 11 and 12), rests upon the fact that the design matrices for each subject are the same (or sufficiently similar). In this special case, the estimator of the

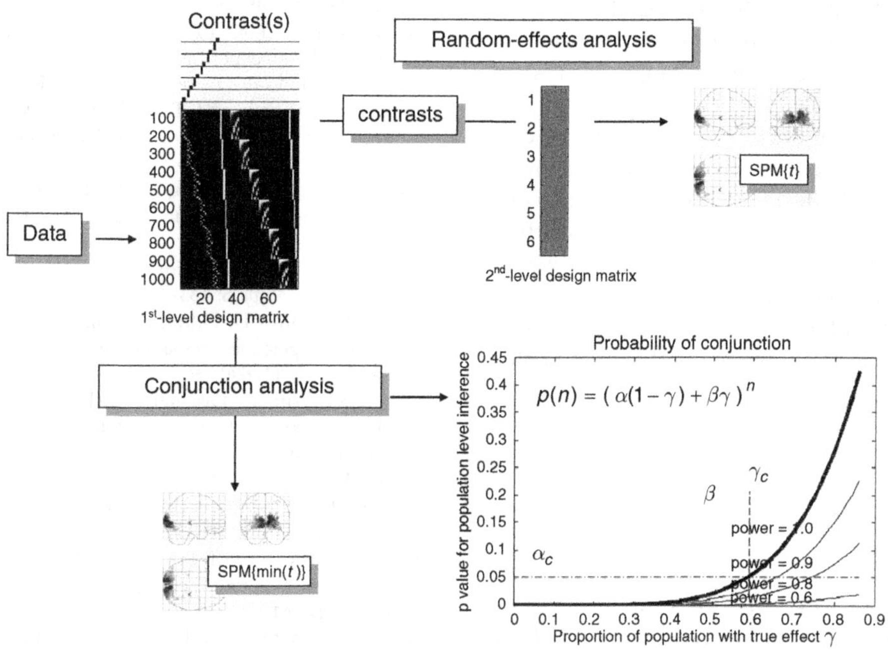

FIGURE 2.10 Schematic illustrating the implementation of random-effect and conjunction analyses for population inference. The lower right graph shows the probability $p(n) = (\alpha(1-\gamma)+\beta\gamma)^n$ of obtaining a conjunction over n subjects, conditional on a certain proportion γ of the population expressing the effect, for a test with specificity of $\alpha = 0.05$, at several sensitivities ($\beta = 1$, 0.9, 0.8 and 0.6). The broken lines denote the critical specificity for population inference α_c and the associated proportion of the population γ_c (see Friston *et al.*, 1999b for details).

variance at the second level contains the right mixture of within- and between-subject error. It is important to appreciate this because the efficiency of the design at the first level percolates to higher levels. It is therefore important to use efficient strategies at all levels in a hierarchical design.

Conjunction analyses and population inferences

In some instances, a fixed-effects analysis is more appropriate, particularly when reporting single-case studies. With a series of single cases, it is natural to ask what are common features of functional anatomy (e.g. the location of V5) and what aspects are subject specific (e.g. the location of ocular dominance columns)? One way to address commonalties is to use a conjunction analysis over subjects. It is important to understand the nature of the inference provided by conjunction analyses, because there has been some confusion (see Nichols *et al.*, 2005; Friston *et al.*, 2005). Imagine that, in sixteen subjects the activation in V5, elicited by a motion stimulus, was greater than zero. The probability of this occurring by chance, in the same area, is extremely small and is the p-value returned by a conjunction analysis using a threshold of $p = 0.5$ (i.e. $t = 0$) for each subject. This constitutes evidence that V5 is engaged by motion. However, it is not an assertion that each subject activated significantly (we only require the t-value to be greater than zero for each subject). In other words, a significant conjunction is not a conjunction of significance.

The motivations for conjunction analyses, in the context of multisubject studies, are twofold. They provide an inference, in a fixed-effects context, testing the null hypotheses of no activation in any of the subjects, which can be much more sensitive than testing for the average activation. Second, they can be used to make inferences about the population in terms of confidence intervals on the proportion of subjects showing an effect (see Friston *et al.*, 1999b).

CONCLUSION

In this chapter, we have reviewed the main components of image analysis and have introduced the tenets of statistical parametric mapping. We have also considered the design of experiments and their statistical models, with a special focus on fMRI. This chapter has covered the key operational issues in identifying regionally specific effects in neuroimaging. In the next chapter, we look at models for neuroimaging from a broader perspective and address the functional integration of distributed responses in the brain.

REFERENCES

Adler RJ (1981) *The geometry of random fields*. Wiley, New York

Aguirre GK, Zarahn E, D'Esposito M (1998) A critique of the use of the Kolmogorov-Smirnov (KS) statistic for the analysis of BOLD fMRI data. *Mag Res Med* **39**: 500–05

Andersson JL, Hutton C, Ashburner J *et al.* (2001) Modeling geometric deformations in EPI time series. *NeuroImage* **13**: 903–19

Ashburner J, Friston KJ. (2000) Voxel-based morphometry – the methods. *NeuroImage* **11**: 805–21

Ashburner J, Neelin P, Collins DL *et al.* (1997) Incorporating prior knowledge into image registration. *NeuroImage* **6**: 344–52

Ashburner J, Hutton C, Frackowiak R *et al.* (1998) Identifying global anatomical differences: deformation-based morphometry. *Hum Brain Mapp* **6**: 348–57

Bandettini PA, Jesmanowicz A, Wong EC *et al.* (1993) Processing strategies for time course data sets in functional MRI of the human brain. *Mag Res Med* **30**: 161–73

Büchel C, Wise RJS, Mummery CJ *et al.* (1996) Non-linear regression in parametric activation studies. *NeuroImage* **4**: 60–66

Buckner R, Bandettini P, O'Craven K *et al.* (1996) Detection of cortical activation during averaged single trials of a cognitive task using functional magnetic resonance imaging. *Proc Natl Acad Sci USA* **93**: 14878–83

Bullmore ET, Brammer MJ, Williams SCR *et al.* (1996) Statistical methods of estimation and inference for functional MR images. *Mag Res Med* **35**: 261–77

Burock MA, Buckner RL, Woldorff MG *et al.* (1998) Randomized event-related experimental designs allow for extremely rapid presentation rates using functional MRI. *NeuroReport* **9**: 3735–39

Chung MK, Worsley KJ, Paus T *et al.* (2001) A unified statistical approach to deformation-based morphometry. *NeuroImage* **14**: 595–606

Clark VP, Maisog JM, Haxby JV (1998) fMRI study of face perception and memory using random stimulus sequences. *J Neurophysiol* **76**: 3257–65

Friston KJ (1997) Testing for anatomical specified regional effects. *Hum Brain Mapp* **5**: 133–36

Friston KJ, Frith CD, Liddle PF *et al.* (1991) Comparing functional (PET) images: the assessment of significant change. *J Cereb Blood Flow Metab* **11**: 690–99

Friston KJ, Frith C, Passingham RE *et al.* (1992a) Motor practice and neurophysiological adaptation in the cerebellum: a positron tomography study. *Proc Roy Soc Lon Series B***248**: 223–28

Friston KJ, Grasby P, Bench C *et al.* (1992b) Measuring the neuromodulatory effects of drugs in man with positron tomography. *Neurosci Lett* **141**: 106–10

Friston KJ, Worsley KJ, Frackowiak RSJ *et al.* (1994) Assessing the significance of focal activations using their spatial extent. *Hum Brain Mapp* **1**: 214–20

Friston KJ, Ashburner J, Frith CD *et al.* (1995a) Spatial registration and normalization of images. *Hum Brain Mapp* **2**: 165–89

Friston KJ, Holmes AP, Worsley KJ *et al.* (1995b) Statistical parametric maps in functional imaging: a general linear approach. *Hum Brain Mapp* **2**: 189–210

Friston KJ, Williams S, Howard R *et al.* (1996a) Movement related effects in fMRI time series. *Mag Res Med* **35**: 346–55

Friston KJ, Holmes A, Poline J-B *et al.* (1996b) Detecting activations in PET and fMRI: levels of inference and power. *NeuroImage* **4**: 223–35

Friston KJ, Price CJ, Fletcher P *et al.* (1996c) The trouble with cognitive subtraction. *NeuroImage* **4**: 97–104

Friston KJ, Büchel C, Fink GR *et al.* (1997) Psychophysiological and modulatory interactions in neuroimaging. *NeuroImage* **6**: 218–29

Friston KJ, Josephs O, Rees G *et al.* (1998) Non-linear event-related responses in fMRI. *Mag Res Med* **39**: 41–52

Friston KJ, Zarahn E, Josephs O *et al.* (1999a) Stochastic designs in event-related fMRI. *NeuroImage* **10**: 607–19

Friston KJ, Holmes AP, Price CJ *et al.* (1999b) Multisubject fMRI studies and conjunction analyses. *NeuroImage* **10**: 385–96

Friston KJ, Josephs O, Zarahn E *et al.* (2000) To smooth or not to smooth? Bias and efficiency in fMRI time-series analysis. *NeuroImage* **12**: 196–208

Friston KJ, Penny WD and Glaser DE (2005) Conjunction revisited. *NeuroImage* **25**(3): 661–7

Grafton S, Mazziotta J, Presty S *et al.* (1992) Functional anatomy of human procedural learning determined with regional cerebral blood flow and PET. *J Neurosci* **12**: 2542–48

Grootoonk S, Hutton C, Ashburner J *et al.* (2000) Characterization and correction of interpolation effects in the realignment of fMRI time series. *NeuroImage* **11**: 49–57

Heid O, Gönner F, Schroth G (1997) Stochastic functional MRI. *NeuroImage* **5**: S476

Jezzard P, Balaban RS (1995) Correction for geometric distortion in echo-planar images from B0 field variations. *Mag Res Med* **34**: 65–73

Kiebel SJ, Poline JB, Friston KJ *et al.* (1999) Robust smoothness estimation in statistical parametric maps using standardized residuals from the general linear model. *NeuroImage* **10**: 756–66

Liddle PF, Friston KJ, Frith CD *et al.* (1992) Cerebral blood-flow and mental processes in schizophrenia. *Roy Soc Med* **85**: 224–27

Lueck CJ, Zeki S, Friston KJ *et al.* (1989) The color centre in the cerebral cortex of man. *Nature* **340**: 386–89

Nichols T, Brett M, Andersson J, Wager T, Poline JB (2004) Valid conjunction interference with the minimum statistic. *NeuroImage* **25**(3): 653–60

Petersen SE, Fox PT, Posner MI *et al.* (1989) Positron emission tomographic studies of the processing of single words. *J Cog Neurosci* **1**: 153–70

Price CJ, Friston KJ (1997) Cognitive conjunction: a new approach to brain activation experiments. *NeuroImage* **5**: 261–70

Price CJ, Wise RJS, Ramsay S *et al.* (1992) Regional response differences within the human auditory cortex when listening to words. *Neurosci Lett* **146**: 179–82

Purdon PL, Weisskoff RM (1998) Effect of temporal autocorrelation due to physiological noise and stimulus paradigm on voxel-level false-positive rates in fMRI. *Hum Brain Mapp* **6**: 239–495

Talairach P, Tournoux J (1988) A stereotactic coplanar atlas of the human brain. Thieme, Stuttgart

Vazquez AL, Noll CD (1998) Non-linear aspects of the BOLD response in functional MRI. *NeuroImage* **7**: 108–18

Worsley KJ, Friston KJ (1995) Analysis of fMRI time-series revisited – again. *NeuroImage* **2**: 173–81

Worsley KJ, Evans AC, Marrett S *et al.* (1992) A three-dimensional statistical analysis for rCBF activation studies in human brain. *J Cereb Blood Flow Metab* **12**: 900–18

Worsley KJ, Marrett S, Neelin P *et al.* (1996) A unified statistical approach or determining significant signals in images of cerebral activation. *Hum Brain Mapp* **4**: 58–73

Zarahn E, Aguirre GK, and D'Esposito M (1997) Empirical analyses of BOLD fMRI statistics: I Spatially unsmoothed data collected under null-hypothesis conditions. *NeuroImage* **5**: 179–97

CHAPTER

3

Modelling brain responses

K. Friston and K. Stephan

INTRODUCTION

In the previous chapter, we focused on the practical issues encountered in the analysis of neuroimaging data. In this chapter, we look at modelling in a more principled way; placing statistical parametric mapping in the larger context of modelling distributed brain responses. Inferences about the functional organization of the brain rest on models of how measurements of evoked responses are caused. These models can be quite diverse, ranging from conceptual models of functional anatomy to mathematical models of neuronal and haemodynamics. The aim of this chapter is to introduce the key models used in imaging neuroscience and how they relate to each other. We start with anatomical models of functional brain architectures, which motivate some of the fundaments of neuroimaging. We then reprise basic statistical models (e.g. the general linear model) used for making classical and Bayesian inferences about *where* neuronal responses are expressed. By incorporating biophysical constraints, these basic models can be finessed and, in a dynamic setting, rendered causal. This allows us to infer *how* interactions among brain regions are mediated. The chapter serves to introduce the themes covered in the final three parts of this book.

We will review models of brain responses starting with the general linear model of functional magnetic resonance imaging (fMRI) data discussed in the previous chapter. This model is successively refined until we arrive at neuronal mass models of electroencephalographic (EEG) responses. The latter models afford mechanistic inferences about how evoked responses are caused, at the level of neuronal subpopulations and the coupling among them.

Overview

Neuroscience depends on conceptual, anatomical, statistical and causal models that link ideas about how the brain works to observed neuronal responses. Here we highlight the relationships among the sorts of models that are employed in imaging. We will show how simple statistical models, used to identify *where* evoked brain responses are expressed (cf. neo-phrenology) can be elaborated to provide models of *how* neuronal responses are caused (e.g. dynamic causal modelling). We will review a series of models that cover conceptual models, motivating experimental design, to detailed biophysical models of coupled neuronal ensembles that enable questions to be asked at a physiological and computational level.

Anatomical models of functional brain architectures motivate the fundaments of neuroimaging. In the first section, we review the distinction between functional *specialization* and *integration* and how these principles serve as the basis for most models of neuroimaging data. The next section turns to simple statistical models (e.g. the general linear model) used for making classical and Bayesian inferences about functional specialization in terms of where neuronal responses are expressed. By incorporating biological constraints, simple observation models can be made more realistic and, in a dynamic framework, causal. This section concludes by considering the biophysical modelling of haemodynamic responses. All the models considered in this section pertain to regional responses. In the final section, we focus on models of distributed responses, where the interactions among cortical areas or neuronal subpopulations are modelled explicitly. This section covers the distinction between *functional and effective connectivity* and reviews dynamic causal modelling of functional

Statistical Parametric Mapping, by Karl Friston et al.
ISBN–13: 978-0-12-372560-8 ISBN–10: 0-12-372560-7

integration, using fMRI and EEG. We conclude with an example from ERP (event-related potential) research and show how the mismatch negativity (MMN) can be explained by changes in coupling among neuronal sources that may underlie perceptual learning.

ANATOMICAL MODELS

Functional specialization and integration

From a historical perspective, the distinction between functional specialization and functional integration relates to the dialectic between *localizationism* and *connectionism* that dominated thinking about brain function in the nineteenth century. Since the formulation of phrenology by Gall, who postulated fixed one-to-one relations between particular parts of the brain and specific mental attributes, the identification of a particular brain region with a specific function has become a central theme in neuroscience. Somewhat ironically, the notion that distinct brain functions could, at least to some degree, be localized in the brain was strengthened by early scientific attempts to refute the phrenologists' claims. In 1808, a scientific committee of the Athénée at Paris, chaired by Cuvier, declared that phrenology was an unscientific and invalid theory (Staum, 1995). This conclusion, which was not based on experimental results, may have been enforced by Napoleon Bonaparte (who, allegedly, was not amused after Gall's phrenological examination of his own skull did not give the flattering results he expected). During the following decades, lesion and electrical stimulation paradigms were developed to test whether functions could indeed be localized in animal models. Initial lesion experiments on pigeons by Flourens gave results that were incompatible with phrenologist predictions, but later experiments, including stimulation experiments in dogs and monkeys by Fritsch, Hitzig and Ferrier, supported the idea that there was a relation between distinct brain regions and certain cognitive or motor functions. Additionally, clinicians like Broca and Wernicke showed that patients with focal brain lesions in particular locations showed specific impairments. However, it was realized early on that, in spite of these experimental findings, it was generally difficult to attribute a specific function to a cortical area, given the dependence of cerebral activity on the anatomical connections between distant brain regions; for example, a meeting that took place on August 4th 1881 addressed the difficulties of attributing function to a cortical area, given the dependence of cerebral activity on underlying connections (Phillips *et al.*, 1984). This meeting was entitled 'Localisation of function in the cortex cerebri'. Goltz (1881), although accepting the results of electrical stimulation in dog and monkey cortex, considered that the excitation method was inconclusive, in that the movements elicited might have originated in related pathways, or current could have spread to distant centres. In short, the excitation method could not be used to infer functional localization because localizationism discounted interactions, or functional integration among different brain areas. It was proposed that lesion studies could supplement excitation experiments. Ironically, it was observations on patients with brain lesions some years later (see Absher and Benson, 1993) that led to the concept of *disconnection syndromes* and the refutation of localizationism as a complete or sufficient explanation of cortical organization. Functional localization implies that a function can be localized in a cortical area, whereas specialization suggests that a cortical area is specialized for some aspects of perceptual or motor processing, and that this specialization is anatomically *segregated* within the cortex. The cortical infrastructure supporting a single function may then involve many specialized areas whose union is mediated by the functional integration among them. In this view, functional specialization is only meaningful in the context of functional integration and vice versa.

Functional specialization and segregation

The functional role of any component (e.g. cortical area, sub-area or neuronal population) of the brain is defined largely by its connections. Certain patterns of cortical projections are so common that they could amount to rules of cortical connectivity. 'These rules revolve around one, apparently, overriding strategy that the cerebral cortex uses – that of functional segregation' (Zeki, 1990). Functional segregation demands that cells with common functional properties be grouped together. This architectural constraint necessitates both convergence and divergence of cortical connections. Extrinsic connections among cortical regions are not continuous but occur in patches or clusters. This patchiness has, in some instances, a clear relationship to functional segregation. For example, V2 has a distinctive cytochrome oxidase architecture, consisting of thick stripes, thin stripes and inter-stripes. When recordings are made in V2, directionally selective (but not wavelength or colour selective) cells are found exclusively in the thick stripes. Retrograde (i.e. backward) labelling of cells in V5 is limited to these thick stripes. All the available physiological evidence suggests that V5 is a functionally homogeneous area that is specialized for visual motion. Evidence of this nature supports the notion that patchy connectivity is the anatomical infrastructure that mediates functional segregation and specialization. If it is the case that neurons in a given cortical area share a common responsiveness, by virtue of their extrinsic connectivity, to some sensorimotor or cognitive attribute, then this functional segregation is also an anatomical one.

In summary, functional specialization suggests that challenging a subject with the appropriate sensorimotor attribute or cognitive process should lead to activity changes in, and only in, the specialized areas. This is the anatomical and physiological model upon which the search for regionally specific effects is based. We will deal first with models of regionally specific responses and return to models of functional integration later.

STATISTICAL MODELS

Statistical parametric mapping

Functional mapping studies are usually analysed with some form of statistical parametric mapping. As described in the previous chapter, statistical parametric mapping entails the construction of continuous statistical processes to test hypotheses about regionally specific effects (Friston *et al.*, 1991). Statistical parametric mapping uses the general linear model (GLM) and random field theory (RFT) to analyse and make classical inferences. Parameters of the GLM are estimated in exactly the same way as in conventional analysis of discrete data. RFT is used to resolve the multiple-comparisons problem induced by making inferences over a volume of the brain. RFT provides a method for adjusting p-values for the search volume of a statistical parametric map (SPM) to control false positive rates. It plays the same role for continuous data (i.e. images or time-series) as the Bonferroni correction for a family of discontinuous or discrete statistical tests.

We now consider the Bayesian alternative to classical inference with SPMs. This rests on conditional inferences about an effect, given the data, as opposed to classical inferences about the data, given the effect is zero. Bayesian inferences about effects that are continuous in space use posterior probability maps (PPMs). Although less established than SPMs, PPMs are potentially very useful, not least because they do not have to contend with the multiple-comparisons problem induced by classical inference (see Berry and Hochberg, 1999). In contradistinction to SPM, this means that inferences about a given regional response do not depend on inferences about responses elsewhere. Before looking at the models underlying Bayesian inference, we briefly review estimation and classical inference in the context of the GLM.

The general linear model

Recall from Chapter 2 that the general linear model:

$$y = X\beta + \varepsilon \qquad 3.1$$

expresses an observed response y in terms of a linear combination of explanatory variables in the design matrix X plus a well-behaved error term. The general linear model is variously known as analysis of variance or multiple regression and subsumes simpler variants, like the t-test for a difference in means, to more elaborate linear convolution models. Each column of the design matrix models a cause of the data. These are referred to as explanatory variables, covariates or regressors. Sometimes the design matrix contains covariates or indicator variables that take values of zero or one to indicate the presence of a particular level of an experimental factor (cf. analysis of variance – ANOVA). The relative contribution of each of these columns to the response is controlled by the parameters β. Inferences about the parameter estimates are made using t or F-statistics, as described in the previous chapter.

Having computed the statistic, RFT is used to assign adjusted p-values to topological features of the SPM, such as the height of peaks or the spatial extent of blobs. This p-value is a function of the search volume and smoothness. The intuition behind RFT is that it controls the false positive rate of peaks corresponding to regional effects. A Bonferroni correction would control the false positive rate of voxels, which is inexact and unnecessarily severe. The p-value is the probability of getting a peak in the SPM, or higher, by chance over the search volume. If sufficiently small (usually less than 0.05) the regional effect is declared significant.

Classical and Bayesian inference

Inference in neuroimaging is restricted largely to classical inferences based upon statistical parametric maps. The statistics that comprise these SPMs are essentially functions of the data. The probability distribution of the chosen statistic, under the null hypothesis (i.e. the null distribution) is used to compute a p-value. This p-value is the probability of obtaining the statistic, or the data, given that the null hypothesis is true. If sufficiently small, the null hypothesis is rejected and an inference is made. The alternative approach is to use Bayesian or conditional inference based upon the posterior distribution of the activation given the data. This necessitates the specification of priors (i.e. the probability distribution of the activation). Bayesian inference requires the posterior distribution and therefore rests upon a posterior density analysis. A useful way to summarize this posterior density is to compute the probability that the activation exceeds some threshold. This represents a Bayesian inference about the effect, in relation to the specified threshold. By computing posterior probability for each voxel, we can construct PPMs that are a useful complement to classical SPMs.

The motivation for using conditional or Bayesian inference is that it has high face validity. This is because the inference is about an effect, or activation, being greater than some specified size that has some meaning in relation to underlying neurophysiology. This contrasts with classical inference, in which the inference is about the effect being significantly different from zero. The problem for classical inference is that trivial departures from the null hypothesis can be declared significant, with sufficient data or sensitivity. From the point of view of neuroimaging, posterior inference is especially useful because it eschews the multiple-comparisons problem. In classical inference, one tries to ensure that the probability of rejecting the null hypothesis incorrectly is maintained at a small rate, despite making inferences over large volumes of the brain. This induces a multiple-comparisons problem that, for spatially continuous data, requires an adjustment or correction to the p-value using RFT as mentioned above. This correction means that classical inference becomes less sensitive or powerful with large search volumes. In contradistinction, posterior inference does not have to contend with the multiple-comparisons problem because there are no false positives. The probability that activation has occurred, given the data, at any particular voxel is the same, irrespective of whether one has analysed that voxel or the entire brain. For this reason, posterior inference using PPMs represents a relatively more powerful approach than classical inference in neuroimaging.

Hierarchical models and empirical Bayes

PPMs require the posterior distribution or conditional distribution of the activation (a contrast of conditional parameter estimates) given the data. This posterior density can be computed, under Gaussian assumptions, using Bayes' rule. Bayes' rule requires the specification of a likelihood function and the prior density of the model parameters. The models used to form PPMs and the likelihood functions are exactly the same as in classical SPM analyses, namely the GLM. The only extra information that is required is the prior probability distribution of the parameters. Although it would be possible to specify those using independent data or some plausible physiological constraints, there is an alternative to this fully Bayesian approach. The alternative is *empirical Bayes* in which the prior distributions are estimated from the data. Empirical Bayes requires a *hierarchical observation model* where the parameters and hyperparameters at any particular level can be treated as priors on the level below. There are numerous examples of hierarchical observation models in neuroimaging. For example, the distinction between fixed- and mixed-effects analyses of multisubject studies relies upon a two-level hierarchical model. However, in neuroimaging, there is a natural hierarchical observation model that is common to all brain mapping experiments. This is the hierarchy induced by looking for the same effects at every voxel within the brain (or grey matter). The first level of the hierarchy corresponds to the experimental effects at any particular voxel and the second level comprises the effects over voxels. Put simply, the variation in a contrast, over voxels, can be used as the prior variance of that contrast at any particular voxel. Hierarchical linear models have the following form:

$$
\begin{aligned}
y &= X^{(1)}\beta^{(1)} + \varepsilon^{(1)} \\
\beta^{(1)} &= X^{(2)}\beta^{(2)} + \varepsilon^{(2)} \\
\beta^{(2)} &= \ldots
\end{aligned}
\qquad 3.2
$$

This is exactly the same as Eqn. **3.1**, but now the parameters of the first level are generated by a supraordinate linear model and so on to any hierarchical depth required. These hierarchical observation models are an important extension of the GLM and are usually estimated using expectation maximization (EM) (Dempster *et al.*, 1977). In the present context, the response variables comprise the responses at all voxels and $\beta^{(1)}$ are the treatment effects we want to make an inference about. Because we have invoked a second level, the first-level parameters embody random effects and are generated by a second-level linear model. At the second level, $\beta^{(2)}$ is the average effect over voxels and $\varepsilon^{(2)}$ its voxel-to-voxel variation. By estimating the variance of $\varepsilon^{(2)}$ one is implicitly estimating an empirical prior on the first-level parameters at each voxel. This prior can then be used to estimate the posterior probability of $\beta^{(1)}$ being greater than some threshold at each voxel. An example of the ensuing PPM is provided in Figure 3.1 along with the classical SPM.

In summary, we have seen how the GLM can be used to test hypotheses about brain responses and how, in a hierarchical form, it enables empirical Bayesian or conditional inference. Next, we deal with dynamic systems and how they can be formulated as GLMs. These dynamic models take us closer to how brain responses are actually caused by experimental manipulations and represent the next step towards causal models of brain responses.

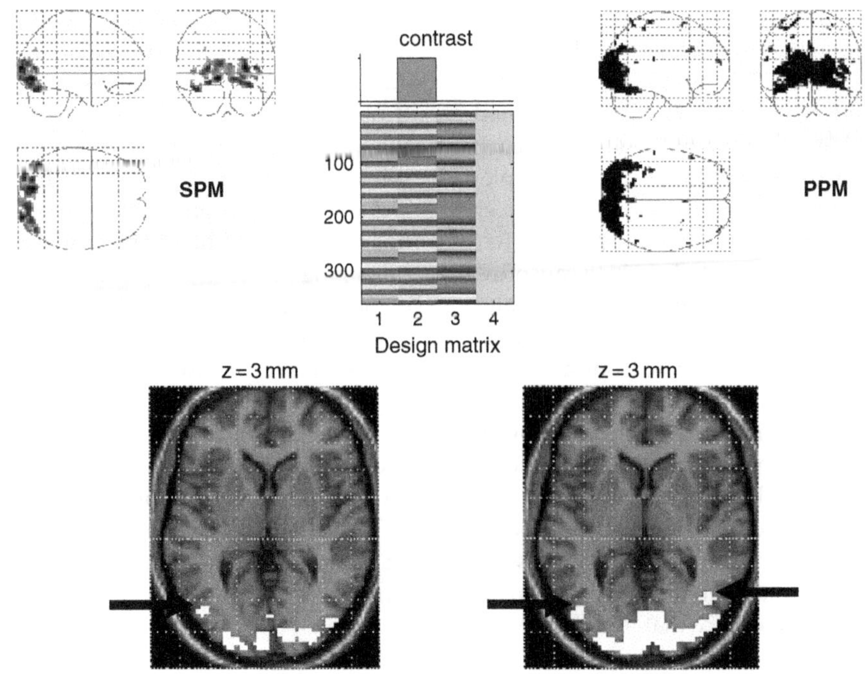

FIGURE 3.1 SPM and PPM for an fMRI study of attention to visual motion. The display format in the lower panel uses an axial slice through extrastriate regions but the thresholds are the same as employed in the maximum intensity projections (upper panels). Upper right: the activation threshold for the PPM was 0.7 a.u., meaning that all voxels shown had a 90 per cent chance of an activation of 0.7 per cent or more. Upper left: the corresponding SPM using an adjusted threshold at $p = 0.05$. Note the bilateral foci of motion-related responses in the PPM that are not seen in the SPM (grey arrows). As can be imputed from the design matrix (upper middle panel), the statistical model of evoked responses comprised boxcar regressors convolved with a canonical haemodynamic response function. The middle column corresponds to the presentation of moving dots and was the stimulus attribute tested by the contrast.

Dynamic models

Convolution models and temporal basis functions

In Friston *et al.* (1994) the form of the haemodynamic impulse response function (HRF) was estimated using a least squares de-convolution and a linear time invariant model, where evoked neuronal responses are *convolved* or smoothed with an HRF to give the measured haemodynamic response (see also Boynton *et al.*, 1996). This simple linear convolution model is the cornerstone for making statistical inferences about activations in fMRI with the GLM. An impulse response function is the response to a single impulse, measured at a series of times after the input. It characterizes the input-output behaviour of the system (i.e. voxel) and places important constraints on the sorts of inputs that will excite a response.

Knowing the form of the HRF is important for several reasons, not least because it furnishes better statistical models of the data. The HRF may vary from voxel to voxel and this has to be accommodated in the GLM. To allow for different HRFs in different brain regions, temporal basis functions were introduced (Friston *et al.*, 1995) to model evoked responses in fMRI and applied to event-related responses in Josephs *et al.* (1997) (see also Lange and Zeger, 1997). The basic idea behind temporal basis functions is that the haemodynamic response, induced by any given trial type, can be expressed as the linear combination of (basis) functions of peristimulus time. The convolution model for fMRI responses takes a stimulus function encoding the neuronal responses and convolves it with an HRF to give a regressor that enters the design matrix. When using basis functions, the stimulus function is convolved with each basis function to give a series of regressors. Mathematically, we can express this model as:

$$\begin{array}{lcl} y(t) = X\beta + \varepsilon & & y(t) = u(t) \otimes h(t) \\ & \Leftrightarrow & \\ X_i = T_i(t) \otimes u(t) & & h(t) = \beta_1 T_1(t) + \beta_2 T_2(t) + \dots \end{array} \quad 3.3$$

where $\otimes$ means convolution. This equivalence shows how any convolution model (right) can be converted into a GLM (left), using temporal basis functions. The parameter estimates β_i are the coefficients or weights that determine the mixture of basis functions of time $T_i(t)$ that models $h(t)$, the HRF for the trial type and voxel in question. We find the most useful basis set to be a canonical HRF and its derivatives with respect to the key parameters that determine its form (see below). Temporal basis functions are important because they provide a graceful transition between conventional multilinear regression models with one stimulus function per condition and finite impulse response (FIR) models with a parameter for each time point following the onset of a condition or trial type. Plate 3 (see colour plate section) illustrates this graphically (see plate caption). In short, temporal basis functions offer useful constraints on the form of the estimated response that retain the flexibility of FIR models and the efficiency of single regressor models.

Biophysical models

Input-state-output systems

By adopting a convolution model for brain responses in fMRI, we are implicitly positing a dynamic system that converts neuronal responses into observed haemodynamic responses. Our understanding of the biophysical and physiological mechanisms that underpin the HRF has grown considerably in the past few years (e.g. Buxton and Frank 1997; Mandeville *et al.*, 1999). Figure 3.2 shows some simulations based on the haemodynamic model described in Friston *et al.* (2000). Here, neuronal activity induces some autoregulated vasoactive signal that causes transient increases in regional cerebral blood flow (rCBF). The resulting flow increases dilate a venous balloon, increasing its volume and diluting venous blood to decrease deoxyhaemoglobin content. The blood oxygenation-level-dependent (BOLD) signal is roughly proportional to the concentration of deoxyhaemoglobin and follows the rCBF response with about a one second delay. The model is framed in terms of differential equations, examples of which are provided in the left panel.

Notice that we have introduced variables, like volume and deoxyhaemoglobin concentrations, that are not actually observed. These are referred to as the *hidden states* of input-state-output models. The state and output equations of any analytic dynamical system are:

$$\dot{x}(t) = f(x, u, \theta)$$
$$y(t) = g(x, u, \theta) + \varepsilon \qquad 3.4$$

The first line is an ordinary differential equation and expresses the rate of change of the states as a parameterized function of the states and inputs. Typically, the inputs u(t) correspond to designed experimental effects (e.g. the stimulus function in fMRI). There is a fundamental and causal relationship (Fliess *et al.*, 1983) between the outputs and the history of the inputs in Eqn. **3.4**. This relationship conforms to a Volterra series, which expresses the output as a generalized convolution of the input, critically without reference to the hidden states x(t). This series is simply a functional Taylor expansion of the outputs with respect to the inputs (Bendat, 1990). The reason it is a functional expansion is that the inputs

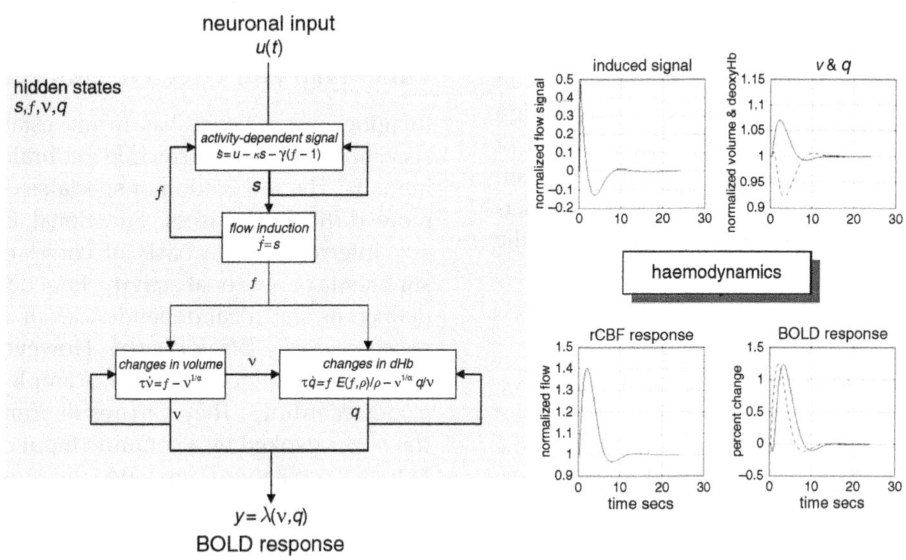

FIGURE 3.2 Right: haemodynamics elicited by an impulse of neuronal activity as predicted by a dynamical biophysical model (left). A burst of neuronal activity causes an increase in flow-inducing signal that decays with first order kinetics and is downregulated by local flow. This signal increases rCBF, which dilates the venous capillaries, increasing volume v. Concurrently, venous blood is expelled from the venous pool decreasing deoxyhaemoglobin content q. The resulting fall in deoxyhaemoglobin concentration leads to a transient increase in BOLD (blood oxygenation-level-dependent) signal and a subsequent undershoot. Left: haemodynamic model on which these simulations were based (see Friston *et al.*, 2000 and Chapter 27 for details).

are a function of time. (For simplicity, here and in and Eqn. **3.7**, we deal with only one experimental input.)

$$y(t) = \sum_i \int_0^t \dots \int_0^t \kappa_i(\sigma_1, \dots, \sigma_i)$$
$$\times u(t-\sigma_1), \dots, u(t-\sigma_i) d\sigma_1, \dots, d\sigma_i$$
$$\kappa_i(\sigma_1, \dots, \sigma_i) = \frac{\partial^i y(t)}{\partial u(t-\sigma_1), \dots, \partial u(t-\sigma_i)} \qquad 3.5$$

where $\kappa_i(\sigma_1, K, \sigma_i)$ is the i-th order kernel. In Eqn. **3.5** the integrals are restricted to the past. This renders the system causal. The key thing here is that Eqn. **3.5** is simply a convolution and can be expressed as a GLM, as in Eqn. **3.3**. This means that we can take a neurophysiologically realistic model of haemodynamic responses and use it as an observation model to estimate parameters using observed data. Here the model is parameterized in terms of kernels that have a direct analytic relation to the original parameters θ of the biophysical system. The first-order kernel is simply the conventional HRF. High-order kernels correspond to high-order HRFs and can be estimated using basis functions as described above. In fact, by choosing basis functions according to:

$$A(\sigma)_i = \frac{\partial \kappa(\sigma)_1}{\partial \theta_i} \qquad 3.6$$

one can estimate the biophysical parameters because, to a first-order approximation, $\beta_i = \theta_i$. The critical step we have taken here is to start with a dynamic causal model of how responses are generated and construct a general linear observation model that allows us to estimate and infer things about the parameters of that model. This is in contrast to the conventional use of the GLM with design matrices that are not informed by a forward model of how data are caused. This approach to modelling brain responses has a much more direct connection with underlying physiology and rests upon an understanding of the underlying system.

Non-linear system identification

Once a suitable causal model has been established (e.g. Figure 3.2), we can estimate second-order kernels. These kernels represent a non-linear characterization of the HRF that can model interactions among stimuli in causing responses. One important manifestation of the non-linear effects, captured by the second-order kernels, is a modulation of stimulus-specific responses by preceding stimuli that are proximate in time. This means that responses at high stimulus presentation rates saturate and, in some instances, show an inverted U behaviour. This behaviour appears to be specific to BOLD effects (as distinct from evoked changes in cerebral blood flow) and may represent a *haemodynamic refractoriness*. This effect has important implications for event-related fMRI, where one may want to present trials in quick succession. (See Figure 2.8 in the previous chapter for an example of second-order kernels and the implications for haemodynamic responses.)

In summary, we started with models of regionally specific responses, framed in terms of the general linear model, in which responses were modelled as linear mixtures of designed changes in explanatory variables. Hierarchical extensions to linear observation models enable random-effects analyses and, in particular, empirical Bayes. The mechanistic utility of these models is realized though the use of forward models that embody causal dynamics. Simple variants of these are the linear convolution models used to construct explanatory variables in conventional analyses of fMRI data. These are a special case of generalized convolution models that are mathematically equivalent to input-state-output systems comprising hidden states. Estimation and inference with these dynamic models tells us something about *how* the response was caused, but only at the level of a single voxel. The next section retains the same perspective on models, but in the context of distributed responses and functional integration.

MODELS OF FUNCTIONAL INTEGRATION

Functional and effective connectivity

Imaging neuroscience has firmly established functional specialization as a principle of brain organization in humans. The integration of specialized areas has proven more difficult to assess. Functional integration is usually inferred on the basis of correlations among measurements of neuronal activity. Functional connectivity is defined as statistical dependencies or correlations *among remote neurophysiological events*. However, correlations can arise in a variety of ways: for example in multiunit electrode recordings, they can result from stimulus-locked transients evoked by a common input or reflect stimulus-induced oscillations mediated by synaptic connections (Gerstein and Perkel, 1969). Integration within a distributed system is usually better understood in terms of effective connectivity; effective connectivity refers explicitly to *the influence that one neural system exerts over another*, either at a synaptic (i.e. synaptic efficacy) or population level. It has been proposed that 'the [electrophysiological] notion of effective connectivity should be understood as the experiment- and time-dependent, simplest possible circuit diagram that would replicate the observed timing relationships between the recorded neurons' (Aertsen and Preißl, 1991). This speaks of two important points: effective connectivity is dynamic, i.e. activity-dependent and it depends upon a model of the interactions. The estimation procedures employed in functional neuroimaging can be divided into linear non-dynamic models (e.g. McIntosh and Gonzalez-Lima, 1994) or non-linear dynamic models.

There is a necessary link between functional integration and multivariate analyses because the latter are necessary to model interactions among brain regions. Multivariate approaches can be divided into those that are inferential in nature and those that are data-led or exploratory. We will first consider multivariate approaches that look at functional connectivity or covariance patterns (and are generally exploratory) and then turn to models of effective connectivity (that allow for inference about their parameters).

Eigenimage analysis and related approaches

In Friston *et al.* (1993), we introduced voxel-based principal component analysis (PCA) of neuroimaging time-series to characterize distributed brain systems implicated in sensorimotor, perceptual or cognitive processes. These distributed systems are identified with principal components or *eigenimages* that correspond to spatial modes of coherent brain activity. This approach represents one of the simplest multivariate characterizations of functional neuroimaging time-series and falls into the class of exploratory analyses. Principal component or eigenimage analysis generally uses singular value decomposition (SVD) to identify a set of orthogonal spatial modes that capture the greatest amount of variance expressed over time. As such, the ensuing modes embody the most prominent aspects of the variance-covariance structure of a given time-series. Noting that covariance among brain regions is equivalent to functional connectivity renders eigenimage analysis particularly interesting because it was among the first ways of addressing functional integration (i.e. connectivity) with neuroimaging data. Subsequently, eigenimage analysis has been elaborated in a number of ways. Notable among these is canonical variate analysis (CVA) and multidimensional scaling (Friston *et al.*, 1996a, b). Canonical variate analysis was introduced in the context of MANCOVA (multiple analysis of covariance) and uses the generalized eigenvector solution to maximize the variance that can be explained by some explanatory variables relative to error. CVA can be thought of as an extension of eigenimage analysis that refers explicitly to some explanatory variables and allows for statistical inference.

In fMRI, eigenimage analysis (e.g. Sychra *et al.*, 1994) is generally used as an exploratory device to characterize coherent brain activity. These variance components may, or may not, be related to experimental design. For example, endogenous coherent dynamics have been observed in the motor system at very low frequencies (Biswal *et al.*, 1995). Despite its exploratory power, eigenimage analysis is limited for two reasons. First, it offers only a linear decomposition of any set of neurophysiological measurements and second, the particular set of eigenimages or spatial modes obtained is determined by constraints that are biologically implausible. These aspects of PCA confer inherent limitations on the interpretability and usefulness of eigenimage analysis of biological time-series and have motivated the exploration of non-linear PCA and neural network approaches.

Two other important approaches deserve mention here. The first is independent component analysis (ICA). ICA uses entropy maximization to find, using iterative schemes, spatial modes or their dynamics that are approximately *independent*. This is a stronger requirement than *orthogonality* in PCA and involves removing high-order correlations among the modes (or dynamics). It was initially introduced as *spatial* ICA (McKeown *et al.*, 1998) in which the independence constraint was applied to the modes (with no constraints on their temporal expression). More recent approaches use, by analogy with magneto- and electrophysiological time-series analysis, *temporal* ICA where the dynamics are enforced to be independent. This requires an initial dimension reduction (usually using conventional eigenimage analysis). Finally, there has been an interest in cluster analysis (Baumgartner *et al.*, 1997). Conceptually, this can be related to eigenimage analysis through multidimensional scaling and principal coordinate analysis.

All these approaches are interesting, but they are not used very much. This is largely because they tell you nothing about how the brain works nor allow one to ask specific questions. Simply demonstrating statistical dependencies among regional brain responses or endogenous activity (i.e. demonstrating functional connectivity) does not address how these responses were caused. To address this one needs explicit models of integration or more precisely, effective connectivity.

Dynamic causal modelling with bilinear models

This section is about modelling interactions among neuronal populations, at a cortical level, using neuroimaging time-series and dynamic causal models that are informed by the biophysics of the system studied. The aim of dynamic causal modelling is to estimate, and make inferences about, the coupling among brain areas and how that coupling is influenced by experimental changes (e.g. time or cognitive set). The basic idea is to construct a reasonably realistic neuronal model of interacting cortical regions or nodes. This model is then supplemented with a forward model of how neuronal or synaptic activity translates into a measured response (see previous section). This enables the parameters of the neuronal model (i.e. effective connectivity) to be estimated from observed data.

Intuitively, this approach regards an experiment as a designed perturbation of neuronal dynamics that are promulgated and distributed throughout a system of coupled anatomical nodes to change region-specific neuronal activity. These changes engender, through a measurement-specific forward model, responses that are used to identify the architecture and time constants of the system at a neuronal level. This represents a departure from conventional approaches (e.g. structural equation modelling and autoregression models; McIntosh and Gonzalez-Lima, 1994; Büchel and Friston, 1997), in which one assumes the observed responses are driven by endogenous or intrinsic noise (i.e. innovations). In contrast, dynamic causal models assume the responses are driven by designed changes in inputs. An important conceptual aspect of dynamic causal models pertains to how the experimental inputs enter the model and cause neuronal responses. Experimental variables can elicit responses in one of two ways. First, they can elicit responses through direct influences on specific anatomical nodes. This would be appropriate, for example, in modelling sensory evoked responses in early visual cortices. The second class of input exerts its effect vicariously, through a modulation of the coupling among nodes. These sorts of experimental variables would normally be more enduring; for example attention to a particular attribute or the maintenance of some perceptual set. These distinctions are seen most clearly in relation to particular forms of causal models used for estimation, e.g. the bilinear approximation:

$$\begin{aligned} \dot{x} &= f(x, u) \\ &= Ax + uBx + Cu \\ y &= g(x) + \varepsilon \\ A &= \frac{\partial f}{\partial x} \quad B = \frac{\partial^2 f}{\partial x \partial u} \quad C = \frac{\partial f}{\partial u} \end{aligned} \qquad 3.7$$

where $\dot{x} = \partial x / \partial t$. This is an approximation to any model of how changes in neuronal activity in one region x_i are caused by activity in the other regions. Here the output function $g(x)$ embodies a haemodynamic convolution, linking neuronal activity to BOLD, for each region (e.g. that in Figure 3.2). The matrix A represents the coupling among the regions in the absence of input $u(t)$. This can be thought of as the latent coupling in the absence of experimental perturbations. The matrix B is effectively the change in latent coupling induced by the input. It encodes the input-sensitive changes in A or, equivalently, the modulation of coupling by experimental manipulations. Because B is a second-order derivative it is referred to as *bilinear*. Finally, the matrix C embodies the extrinsic influences of inputs on neuronal activity. The parameters $\theta = A, B, C$ are the connectivity or coupling matrices that we wish to identify and define the functional architecture and interactions among brain regions at a neuronal level.

Because Eqn. **3.7** has exactly the same form as Eqn. **3.4**, we can express it as a GLM and estimate the parameters using EM in the usual way (see Friston *et al.*, 2003). Generally, estimation in the context of highly parameterized models like DCMs requires constraints in the form of priors. These priors enable conditional inference about the connectivity estimates. The sorts of questions that can be addressed with DCMs are now illustrated by looking at how attentional modulation is mediated in sensory processing hierarchies in the brain.

DCM and attentional modulation

It has been established that the superior posterior parietal cortex (SPC) exerts a modulatory role on V5 responses using Volterra-based regression models (Friston and Büchel, 2000) and that the inferior frontal gyrus (IFG) exerts a similar influence on SPC using structural equation modelling (Büchel and Friston, 1997). The example here shows that DCM leads to the same conclusions but starting from a completely different construct. The experimental paradigm and data acquisition are described in Figure 3.3. This figure also shows the location of the regions that entered the DCM. These regions were based on maxima from conventional SPMs testing for the effects of photic stimulation, motion and attention. Regional time courses were taken as the first eigenvariate of 8 mm spherical volumes of interest centred on the maxima shown in the figure. The inputs, in this example, comprise one sensory perturbation and two contextual inputs. The sensory input was simply the presence of photic stimulation and the first contextual one was presence of motion in the visual field. The second contextual input, encoding attentional set, was one during attention to speed changes and zero otherwise. The outputs corresponded to the four regional eigenvariates in (Figure 3.3, left panel). The intrinsic connections were constrained to conform to a hierarchical pattern in which each area was reciprocally connected to its supraordinate area. Photic stimulation entered at, and only at, V1. The effect of motion in the visual field was modelled as a bilinear modulation of the V1 to V5 connectivity and attention was allowed to modulate the backward connections from IFG and SPC.

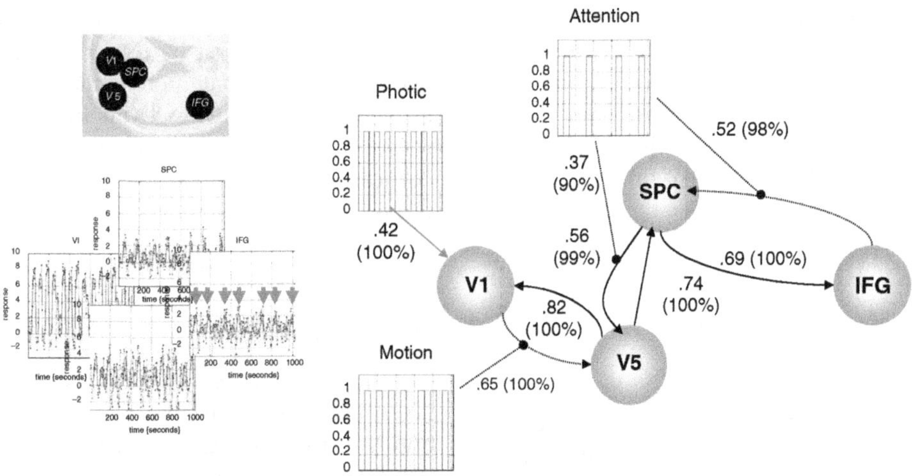

FIGURE 3.3 Results of a DCM analysis of attention to visual motion with fMRI. Right panel: functional architecture based upon the conditional estimates shown alongside their connections, with the per cent confidence that they exceeded threshold in brackets. The most interesting aspects of this architecture involve the role of motion and attention in exerting bilinear effects. Critically, the influence of motion is to enable connections from V1 to the motion-sensitive area V5. The influence of attention is to enable backward connections from the inferior frontal gyrus (IFG) to the superior parietal cortex (SPC). Furthermore, attention increases the influence of SPC on V5. Dotted arrows connecting regions represent significant bilinear effects in the absence of a significant intrinsic coupling. Left panel: fitted responses based upon the conditional estimates and the adjusted data are shown for each region in the DCM. The insert (upper left) shows the location of the regions.

Subjects were studied with fMRI under identical stimulus conditions (visual motion subtended by radially moving dots) while manipulating the attentional component of the task (detection of velocity changes). The data were acquired from a normal subject at 2 Tesla. Each subject had four consecutive 100-scan sessions comprising a series of 10-scan blocks under five different conditions, D F A F N F A F N S. The first condition (D) was a dummy condition to allow for magnetic saturation effects. F (Fixation) corresponds to a low-level baseline where the subjects viewed a fixation point at the centre of a screen. In condition A (Attention), subjects viewed 250 dots moving radially from the centre at 4.7 degrees per second and were asked to detect changes in radial velocity. In condition N (No attention), the subjects were asked simply to view the moving dots. In condition S (Stationary), subjects viewed stationary dots. The order of A and N was swapped for the last two sessions. In all conditions subjects fixated the centre of the screen. During scanning there were no speed changes. No overt response was required in any condition.

The results of the DCM are shown in Figure 3.3 (right panel). Of primary interest here is the modulatory effect of attention that is expressed in terms of the bilinear coupling parameters for this input. As expected, we can be highly confident that attention modulates the backward connections from IFG to SPC and from SPC to V5. Indeed, the influences of IFG on SPC are negligible in the absence of attention (dotted connection). It is important to note that the only way that attentional manipulation can affect brain responses is through this bilinear effect. Attention-related responses are seen throughout the system (attention epochs are marked with arrows in the plot of IFG responses in the left panel). This attentional modulation is accounted for, sufficiently, by changing just two connections. This change is, presumably, instantiated by instructional set at the beginning of each epoch.

The second thing this analysis illustrates is how functional segregation is modelled in DCM. Here one can regard V1 as 'segregating' motion from other visual information and distributing it to the motion-sensitive area, V5. This segregation is modelled as a bilinear 'enabling' of V1 to V5 connections when, and only when, motion is present. Note that, in the absence of motion, the latent V1 to V5 connection was trivially small (in fact the estimate was −0.04). The key advantage of entering motion through a bilinear effect, as opposed to a direct effect on V5, is that we can finesse the inference that V5 shows motion-selective responses with the assertion that these responses are mediated by afferents from V1. The two bilinear effects above represent two important aspects of functional integration that DCM is able to characterize.

Structural equation modelling as a special case of DCM

The central idea behind dynamic causal modelling is to treat the brain as a deterministic non-linear dynamic system that is subject to inputs and produces outputs. Effective connectivity is parameterized in terms of coupling among unobserved brain states (e.g. neuronal activity in different regions). The objective is to estimate these parameters by perturbing the system and measuring the response. This is in contradistinction to established methods for estimating effective connectivity from neurophysiological time-series, which include structural equation modelling and models based on multivariate autoregressive processes. In these models, there is no designed perturbation and the inputs are treated as unknown and stochastic. Furthermore, the inputs are often assumed to express themselves instantaneously such that, at the point of observation the change in states is zero. From Eqn. **3.7**, in the absence of bilinear effects we have:

$$\begin{aligned} \dot{x} &= 0 \\ &= Ax + Cu \\ x &= -A^{-1}Cu \end{aligned} \quad \mathbf{3.8}$$

This is the regression equation used in structural equation modelling (SEM) where $A = D - I$ and D contains the off-diagonal connections among regions. The key point here is that A is estimated by assuming $u(t)$ is some random innovation with known covariance. This is not really tenable for designed experiments when $u(t)$ represent carefully structured experimental inputs. Although SEM and related autoregressive techniques are useful for establishing dependence among regional responses, they are not surrogates for informed causal models based on the underlying dynamics of these responses.

In this section, we have covered multivariate techniques ranging from eigenimage analysis that does not have an explicit forward or causal model to DCM that does. The bilinear approximation to any DCM has been illustrated through its use with fMRI to study attentional modulation. The parameters of the bilinear approximation include first-order effective connectivity A and its experimentally-induced changes B. Although the bilinear approximation is useful, it is possible to model coupling among neuronal subpopulations explicitly. We conclude with a DCM that embraces a number of neurobiological facts and takes us much closer to a mechanistic understanding of how brain responses are generated. This example uses responses measured with EEG.

Dynamic causal modelling with neural-mass models

Event-related potentials (ERPs) have been used for decades as electrophysiological correlates of perceptual and cognitive operations. However, the exact neurobiological mechanisms underlying their generation are largely unknown. In this section, we use neuronally plausible models to understand event-related responses. Our example shows that changes in connectivity are sufficient to explain certain ERP components. Specifically, we will look at the MMN, a component associated with rare or unexpected events. If the unexpected nature of rare stimuli depends on learning which stimuli are frequent, then the MMN must be due to plastic changes in connectivity that mediate perceptual learning. We conclude by showing that advances in the modelling of evoked responses now afford measures of connectivity among cortical sources that can be used to quantify the effects of perceptual learning.

Neural-mass models

The minimal model we have developed (David *et al.*, 2006) uses the connectivity rules described in Felleman and Van Essen (1992) to assemble a network of coupled sources. These rules are based on a partitioning of the cortical sheet into supra-, infra-granular layers and granular layer (layer 4). Bottom-up or forward connections originate in agranular layers and terminate in layer 4. Top-down or backward connections target agranular layers. Lateral connections originate in agranular layers and target all layers. These long-range or extrinsic cortico-cortical connections are excitatory and arise from pyramidal cells.

Each region or source is modelled using a neural mass model described in David and Friston (2003), based on the model of Jansen and Rit (1995). This model emulates the activity of a cortical area using three neuronal subpopulations, assigned to granular and agranular layers. A population of excitatory pyramidal (output) cells receives inputs from inhibitory and excitatory populations of interneurons, via intrinsic connections (intrinsic connections are confined to the cortical sheet). Within this model, excitatory interneurons can be regarded as spiny stellate cells found predominantly in layer 4 and in receipt of forward connections. Excitatory pyramidal cells and inhibitory interneurons are considered to occupy agranular layers and receive backward and lateral inputs (Figure 3.4).

To model event-related responses, the network receives inputs via input connections. These connections are exactly the same as forward connections and deliver inputs to the spiny stellate cells in layer 4. In the present context, inputs $u(t)$ model sub-cortical auditory inputs. The vector C controls the influence of the input on each source. The lower, upper and leading diagonal matrices A^F, A^B, A^L encode forward, backward and lateral connections respectively. The DCM here is specified in terms of the state equations shown in Figure 3.4 and a linear output equation:

$$\begin{aligned} \dot{x} &= f(x, u) \\ y &= Lx_0 + \varepsilon \end{aligned} \quad \mathbf{3.9}$$

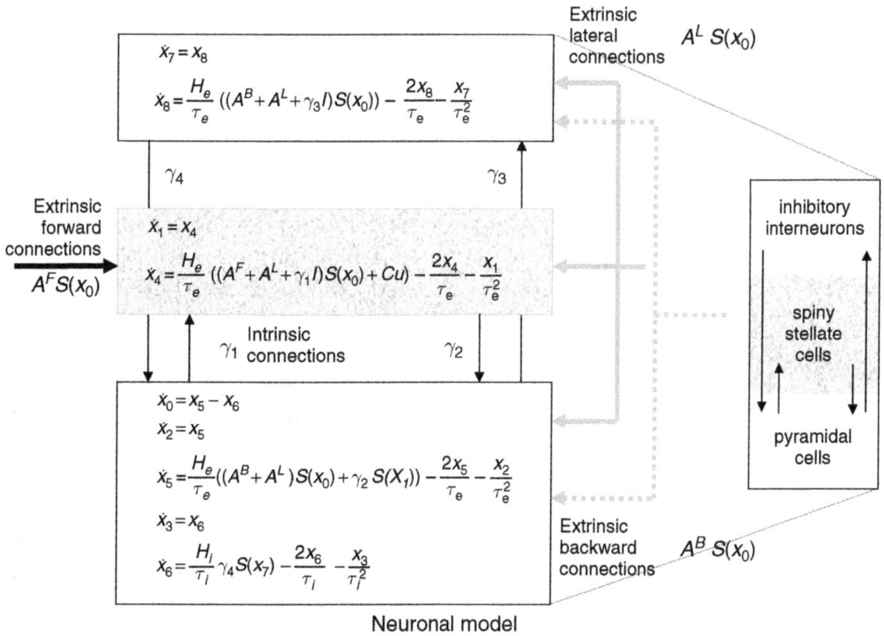

FIGURE 3.4 Schematic of the DCM used to model electrical responses. This schematic shows the state equations describing the dynamics of sources or regions. Each source is modelled with three subpopulations (pyramidal, spiny stellate and inhibitory interneurons) as described in Jansen and Rit (1995) and David and Friston (2003). These have been assigned to granular and agranular cortical layers which receive forward and backward connections respectively.

where x^0 represents the transmembrane potential of pyramidal cells and L is a lead field matrix coupling electrical sources to the EEG channels. This should be compared to the DCM above for haemodynamics; here the equations governing the evolution of neuronal states are much more complicated and realistic, as opposed to the bilinear approximation in Eqn. **3.7**. Conversely, the output equation is a simple linearity, as opposed to the nonlinear observer used for fMRI. As an example, the state equation for the inhibitory subpopulation is:

$$\dot{x}_7 = x_8$$
$$\dot{x}_8 = \frac{H_e}{\tau_e}((A^B + A^L + \gamma_3 I)S(x_0)) - \frac{2x_8}{\tau_e} - \frac{x_7}{\tau_e^2} \qquad \textbf{3.10}$$

Propagation delays on the extrinsic connections have been omitted for clarity here and in Figure 3.4.

Within each subpopulation, the evolution of neuronal states rests on two operators. The first transforms the average density of presynaptic inputs into the average postsynaptic membrane potential. This is modelled by a linear transformation with excitatory and inhibitory kernels parameterized by $H_{e,i}$ and $\tau_{e,i}$. $H_{e,i}$ control the maximum postsynaptic potential and $\tau_{e,i}$ represent a lumped rate-constant. The second operator S transforms the average potential of each subpopulation into an average firing rate. This is assumed to be instantaneous and is a sigmoid function. Interactions among the subpopulations depend on constants $\gamma_{1,2,3,4}$, which control the strength of intrinsic connections and reflect the total number of synapses expressed by each subpopulation. In Eqn. **3.10**, the top line expresses the rate of change of voltage as a function of current. The second line specifies how current changes as a function of voltage, current and presynaptic input from extrinsic and intrinsic sources. Having specified the DCM in terms of these equations, one can estimate the coupling parameters from empirical data using EM as described above.

Perceptual learning and the MMN

The example shown in Figure 3.5 is an attempt to model the MMN in terms of changes in backward and lateral connections among cortical sources. In this example, two [averaged] channels of EEG data were modelled with three cortical sources. Using this generative or forward model, we estimated differences in the strength of these connections for rare and frequent stimuli. As expected, we could account for detailed differences in the ERPs (the MMN) by changes in connectivity (see Figure 3.5 for details). Interestingly, these differences were expressed selectively in the lateral connections. If this model is a sufficient approximation to the real sources, these changes are a non-invasive measure of plasticity, mediating perceptual learning, in the human brain.

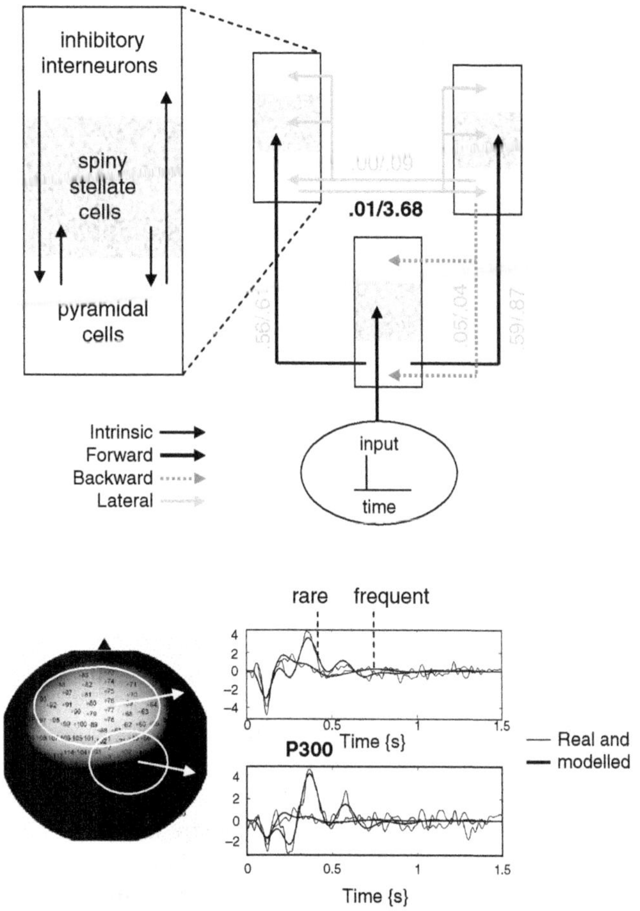

FIGURE 3.5 Summary of a DCM analysis of event-related potentials (ERPs) elicited during an auditory oddball paradigm, employing rare and frequent pure tones. Upper panel: schematic showing the architecture of the neuronal model used to explain the empirical data. Sources were coupled with extrinsic cortico-cortical connections following the rules of Felleman and van Essen (1992). The free parameters of this model included intrinsic and extrinsic connection strengths that were adjusted best to explain the data. In this example, the lead field was also estimated, with no spatial constraints. The parameters were estimated for ERPs recorded during the presentation of rare and frequent tones and are reported beside their corresponding connection (frequent/rare). The most notable finding was that the mismatch response could be explained by a selective increase in lateral connection strength from 0.1 to 3.68 Hz (highlighted in bold). Lower panel: the channel positions (left) and ERPs (right) averaged over two subsets of channels (circled on the left). Note the correspondence between the measured ERPs and those generated by the model (see David *et al.*, 2006 for details).

Auditory stimuli, 1000 or 2000 Hz tones with 5 ms rise and fall times and 80 ms duration, were presented binaurally. The tones were presented for 15 minutes, every 2 s in a pseudo-random sequence with 2000 Hz tones occurring 20 per cent of the time and 1000 Hz tones occurring 80 per cent of the time. The subject was instructed to keep a mental record of the number of 2000 Hz tones (non-frequent target tones). Data were acquired using 128 EEG electrodes with 1000 Hz sample frequency. Before averaging, data were referenced to mean earlobe activity and band-pass filtered between 1 and 30 Hz. Trials showing ocular artefacts and bad channels were removed from further analysis.

CONCLUSION

In this chapter, we have reviewed some key models that underpin image analysis and have touched briefly on ways of assessing specialization and integration in the brain. These models can be regarded as a succession of modelling endeavours that draw more and more on our understanding of how brain-imaging signals are generated, both in terms of biophysics and the underlying neuronal interactions. We have seen how hierarchical linear observation models encode the treatment effects elicited by experimental design. General linear models based on convolution models imply an underlying dynamic input-state-output system. The form of these systems can be used to constrain convolution models and explore some of their simpler non-linear properties. By creating observation models based on explicit forward models of neuronal interactions, one can model and assess interactions among distributed cortical areas and make inferences about coupling at the neuronal level. The next years will probably see an increasing realism in the dynamic causal models introduced above. These endeavours are likely to encompass fMRI signals enabling the conjoint modelling, or fusion, of different modalities and the marriage of computational neuroscience with the modelling of brain responses.

REFERENCES

Absher JR, Benson DF (1993) Disconnection syndromes: an overview of Geschwind's contributions. *Neurology* **43**: 862–67

Aertsen A, Preißl H (1991) Dynamics of activity and connectivity in physiological neuronal networks. In *Non linear dynamics and neuronal networks*, Schuster HG (ed.). VCH Publishers Inc., New York, pp 281–302

Baumgartner R, Scarth G, Teichtmeister C *et al.* (1997) Fuzzy clustering of gradient-echo functional MRI in the human visual cortex. Part 1: reproducibility. *J Mag Res Imaging* **7**: 1094–101

Bendat JS (1990) *Nonlinear system analysis and identification from random data*. John Wiley and Sons, New York

Berry DA, Hochberg Y (1999) Bayesian perspectives on multiple comparisons. *J Stat Plann Inference* **82**: 215–227

Biswal B, Yetkin FZ, Haughton VM *et al.* (1995) Functional connectivity in the motor cortex of resting human brain using echo-planar MRI. *Mag Res Med* **34**: 537–41

Boynton GM, Engel SA, Glover GH *et al.* (1996) Linear systems analysis of functional magnetic resonance imaging in human V1. *J Neurosci* **16**: 4207–21

Büchel C, Friston KJ (1997) Modulation of connectivity in visual pathways by attention: cortical interactions evaluated with structural equation modelling and fMRI. *Cereb Cortex* **7**: 768–78

Buckner RL, Koutstaal W, Schacter DL, *et al.* (1998) Functional-anatomic study of episodic retrieval. II. Selective averaging of event-related fMRI trials to test the retrieval success hypothesis. *NeuroImage* **7**: 163–75

Buxton RB, Frank LR (1997) A model for the coupling between cerebral blood flow and oxygen metabolism during neural stimulation. *J Cereb Blood Flow Metab* **17**: 64–72

David O, Friston KJ (2003) A neural mass model for MEG/EEG: coupling and neuronal dynamics. *NeuroImage* **20**: 1743–55

David O, Kiebel SJ, Harrison LM *et al.* (2006) Dynamic causal modelling of evoked responses in EEG and MEG. Neuroimage. Feb 8; (Epub ahead of print)

Dempster AP, Laird NM, Rubin DB (1977) Maximum likelihood from incomplete data via the EM algorithm. *J Roy Stat Soc Series* B **39**: 1–38

Felleman DJ, Van Essen DC (1992) Distributed hierarchical processing in the primate cerebral cortex. *Cerebr Cortex* **1**: 1–47

Fliess M, Lamnabhi M, Lamnabhi-Lagarrigue F (1983) An algebraic approach to nonlinear functional expansions. *IEEE Trans Circuits Syst* **30**: 554–70

Friston KJ, Büchel C (2000) Attentional modulation of effective connectivity from V2 to V5/MT in humans. *Proc Natl Acad Sci USA* **97**: 7591–96

Friston KJ, Frith CD, Liddle PF *et al.* (1991) Comparing functional (PET) images: the assessment of significant change. *J Cereb Blood Flow Metab* **11**: 690–99

Friston KJ, Frith CD, Liddle PF *et al.* (1993). Functional connectivity: the principal component analysis of large data sets. *J Cereb Blood Flow Metab* **13**: 5–14

Friston KJ, Jezzard P, Turner M (1994) Analysis of functional MRI time-series. *Human Brain Mapping* **1**(2): 153–71

Friston KJ, Frith CD, Turner R *et al.* (1995) Characterising evoked hemodynamics with fMRI. *NeuroImage* **2**: 157–65

Friston KJ, Poline J-B, Holmes AP *et al.* (1996a) A multivariate analysis of PET activation studies. *Hum Brain Mapp* **4**: 140–51

Friston KJ, Frith CD, Fletcher P *et al.* (1996b) Functional topography: multidimensional scaling and functional connectivity in the brain. *Cereb Cortex* **6**: 156–64

Friston KJ, Mechelli A, Turner R *et al.* (2000) Nonlinear responses in fMRI: the Balloon model, Volterra kernels, and other hemodynamics. *NeuroImage* **12**: 466–77

Friston KJ, Harrison L, Penny W. (2003) Dynamic causal modelling. *NeuroImage* **19**:1273–302

Gerstein GL, Perkel DH, Taylor JG (2001) Neural modelling and functional brain imaging: an overview. *Neural Netw* **13**: 829–46

Gerstein GL, Perkel DH (1969) Simultaneously recorded trains of action potentials: analysis and functional interpretation. *Science* **16**: 828–30

Jansen BH, Rit VG (1995). Electroencephalogram and visual evoked potential generation in a mathematical model of coupled cortical columns. *Biol Cybern* **73**: 357–66

Josephs O, Turner R, Friston K (1997) Event-related fMRI. *Human Brain Mapping* **5**(4): 243–8

Lange N, Zeger SL (1997) Non-linear Fourier time series analysis for human brain mapping by functional magnetic resonance imaging (with discussion) J *Roy Stat Soc Ser C* **46**: 1–29

Mandeville JB, Marota JJ, Ayata C *et al.* (1999) Evidence of a cerebrovascular postarteriole Windkessel with delayed compliance. *J Cereb Blood Flow Metab* **19**: 79–89

McIntosh AR, Gonzalez-Lima F (1994) Structural equation modelling and its application to network analysis in functional brain imaging. *Hum Brain Mapp* **2**: 2–22

McKeown M, Jung T-P, Makeig S *et al.* (1998) Spatially independent activity patterns in functional MRI data during the Stroop colour naming task. *Proc Natl Acad Sci* **95**: 803–10

Phillips CG, Zeki S, Barlow HB (1984) Localisation of function in the cerebral cortex. Past present and future. *Brain* **107**: 327–61

Staum M (1995) Physiognomy and phrenology at the Paris Athénée. *J Hist Ideas* **56**: 443–62

Sychra JJ, Bandettini PA, Bhattacharya N *et al.* (1994) Synthetic images by subspace transforms. I Principal component images and related filters. *Med Physics* **21**: 193–201

Zeki S (1990) The motion pathways of the visual cortex. In *Vision: coding and efficiency*, Blakemore C (ed.). Cambridge University Press, Cambridge, pp 321–45

Part 2

Computational anatomy

CHAPTER

4

Rigid Body Registration

J. Ashburner and K. Friston

INTRODUCTION

Rigid body registration is one of the simplest forms of image registration, so this chapter provides an ideal framework for introducing some of the concepts that will be used by the more complex registration methods described later. The shape of a human brain changes very little with head movement, so rigid body transformations can be used to model different head positions of the same subject. Registration methods described in this chapter include within modality, or between different modalities such as positron emission tomography (PET). and magnetic resonance imaging (MRI). Matching of two images is performed by finding the rotations and translations that optimize some mutual function of the images. Within-modality registration generally involves matching the images by minimizing the mean squared difference between them. For between-modality registration, the matching criterion needs to be more complex.

Image registration is important in many aspects of functional image analysis. In imaging neuroscience, particularly for functional MRI (fMRI), the signal changes due to any haemodynamic response can be small compared to apparent signal differences that can result from subject movement. Subject head movement in the scanner cannot be completely eliminated, so retrospective motion correction is performed as a preprocessing step. This is especially important for experiments where subjects may move in the scanner in a way that is correlated with the different conditions (Hajnal *et al.*, 1994). Even tiny systematic differences can result in a significant signal accumulating over numerous scans. Without suitable corrections, artefacts arising from subject movement correlated with the experimental paradigm may appear as activations. A second reason why motion correction is important is that it increases sensitivity. The *t*-test is based on the signal change relative to the residual variance. The residual variance is computed form the sum of squared differences between the data and the linear model to which it is fitted. Movement artefacts add to this residual variance, and so reduce the sensitivity of the test to true activations.

For studies of a single subject, sites of activation can be accurately localized by superimposing them on a high resolution structural image of the subject (typically a T1-weighted MRI). This requires registration of the functional images with the structural image. As in the case of movement correction, this is normally performed by optimizing a set of parameters describing a rigid body transformation, but the matching criterion needs to be more complex because the structural and functional images normally look very different. A further use for this registration is that a more precise spatial normalization can be achieved by computing it from a more detailed structural image. If the functional and structural images are in register, then a warp computed from the structural image can be applied to the functional images.

Another application of rigid registration is within the field of morphometry, and involves identifying shape changes within single subjects by subtracting coregistered images acquired at different times. The changes could arise for a number of different reasons, but most are related to pathology. Because the scans are of the same subject, the first step for this kind of analysis involves registering the images together by a rigid body transformation.

At its simplest, image registration involves estimating a mapping between a pair of images. One image is assumed to remain stationary (the reference image), whereas the other (the source image) is spatially transformed to match it. In order to transform the source to match the reference, it is necessary to determine a mapping from each voxel position in the reference to a corresponding position in the source. The source is then resampled at the new positions. The mapping can be thought of as a function of a set of estimated

Statistical Parametric Mapping, by Karl Friston et al.
ISBN–13: 978-0-12-372560-8 ISBN–10: 0-12-372560-7

transformation parameters. A rigid body transformation in three dimensions is defined by six parameters: three translations and three rotations.

There are two steps involved in registering a pair of images together. There is the *registration* itself, whereby the set of parameters describing a transformation is estimated. Then there is the *transformation*, where one of the images is transformed according to the estimated parameters. Performing the registration normally involves iteratively transforming the source image many times, using different parameters, until some matching criterion is optimized.

First of all, this chapter will explain how images are transformed via the process of resampling. This chapter is about rigid registration of images, so the next section describes the parameterization of rigid body transformations as a subset of the more general affine transformations. The final two sections describe methods of rigid body registration, in both intra- and inter-modality contexts. Intra-modality registration implies registration of images acquired using the same modality and scanning sequence or contrast agent, whereas inter-modality registration allows the registration of different modalities (e.g. T1- to T2-weighted MRI, or MRI to PET).

RE-SAMPLING IMAGES

An image transformation is usually implemented as a 'pulling' operation (where pixel values are pulled from the original image into their new location) rather than a 'pushing' one (where the pixels in the original image are pushed into their new location). This involves determining for each voxel in the transformed image, the corresponding intensity in the original image. Usually, this requires sampling between the centres of voxels, so some form of interpolation is needed.

Simple interpolation

The simplest approach is to take the value of the closest voxel to the desired sample point. This is referred to as *nearest neighbour* or *zero-order hold* resampling. This has the advantage that the original voxel intensities are preserved, but the resulting image is degraded quite considerably, resulting in the resampled image having a 'blocky' appearance.

Another approach is to use *trilinear interpolation* (*first-order hold*) to resample the data. This is slower than nearest neighbour, but the resulting images are less 'blocky'. However, trilinear interpolation has the effect of losing some high frequency information from the image.

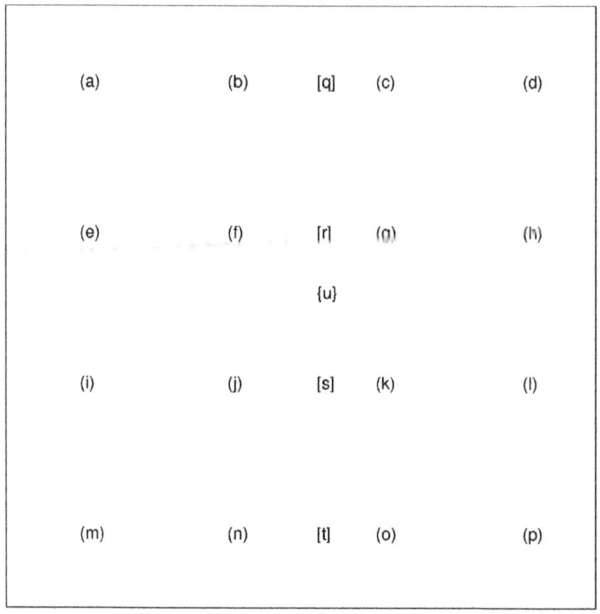

FIGURE 4.1 Illustration of image interpolation in two dimensions. Points a through to p represent the original regular grid of pixels. Point u is the point whose value is to be determined. Points q to t are used as intermediates in the computation.

Figure 4.1 will now be used to illustrate bilinear interpolation (the two dimensional version of trilinear interpolation). Assume that there is a regular grid of pixels at coordinates x_a, y_a to x_p, y_p, having intensities v_a to v_p, and that the point to re-sample is at u. The value at points r and s are first determined (using linear interpolation) as follows:

$$v_r = \frac{(x_g - x_r)v_f + (x_r - x_f)v_g}{x_g - x_f}$$
$$v_s = \frac{(x_k - x_s)v_j + (x_s - x_j)v_k}{x_k - x_j} \qquad 4.1$$

Then v_u is determined by interpolating between v_r and v_s:

$$v_u = \frac{(y_u - y_s)v_r + (y_r - y_u)v_s}{y_r - y_s} \qquad 4.2$$

The extension of the approach to three dimensions is trivial.

Windowed sinc interpolation

The optimum method of applying rigid body transformations to images with minimal interpolation artefact is to do it in Fourier space. In real space, the interpolation method that gives results closest to a Fourier interpolation is *sinc* interpolation. This involves convolving the

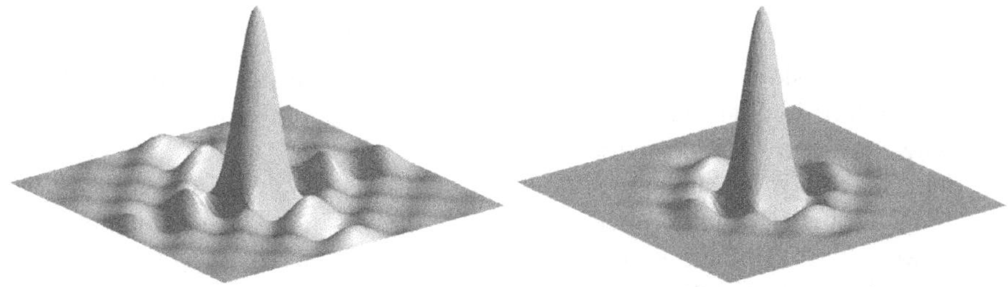

FIGURE 4.2 Sinc function in two dimensions, both with (right) and without (left) a Hanning window.

image with a sinc function centred on the point to be re-sampled. To perform a pure sinc interpolation, every voxel in the image should be used to sample a single point. This is not feasible due to speed considerations, so an approximation using a limited number of nearest neighbours is used. Because the sinc function extends to infinity, it is often truncated by modulation with a Hanning window (Figure 4.2). Because the function is separable, the interpolation is performed sequentially in the three dimensions of the volume. For one dimension the windowed sinc function using the I nearest neighbours would be:

$$\sum_{i=1}^{I} v_i \frac{\frac{sin(\pi d_i)}{\pi d_i}\frac{1}{2}(1+cos(2\pi d_i/I))}{\sum_{j=1}^{I}\frac{sin(\pi d_j)}{\pi d_j}\frac{1}{2}(1+cos(2\pi d_j/I))} \quad 4.3$$

where d_i is the distance from the centre of the ith voxel to the point to be sampled, and v_i is the value of the ith voxel.

Generalized interpolation

The methods described so far are all classical interpolation methods that locally convolve the image with some form of interpolant. Much more efficient re-sampling can be performed using *generalized interpolation* (Thévenaz *et al.*, 2000). Generalized interpolation methods model an image as a linear combination of basis functions with local support, typically *B-splines* or *o-Moms* (maximal-order interpolation of minimal support) basis functions (Figure 4.3). Before resampling begins, an image of basis function coefficients is produced, which involves a very fast deconvolution (Unser *et al.*, 1993a,b). The separability of the basis functions allow this to be done sequentially along each of the dimensions. Resampling at each new point then involves computing the appropriate linear combination of basis functions, which can be thought of as a local convolution of the basis function coefficients. Again, this is done sequentially because of the separability of the bases.

B-splines are a family of functions of varying degree. Interpolation using B-splines of degree 0 or 1 (first and second order) is identical to nearest neighbour[1] or linear interpolation respectively. B-splines of degree n are given by:

$$\beta^n(x) = \sum_{j=0}^{n} \frac{(-1)^j(n+1)}{(n+1-j)!j!} \max\left(\frac{n+1}{2}+x-j, 0\right)^n \quad 4.4$$

An nth degree B-spline has a local support of $n+1$, which means that during the final resampling step, a linear combination of $n+1$ basis functions are needed to compute an interpolated value. o-Moms are derived from B-splines, and consist of a linear combination of the B-spline and its derivatives. They produce the most accurate interpolation for the least local support, but lack some of the B-splines' advantages. Unlike the o-Moms' functions, a B-spline of order n is $n-1$ times continuously differentiable.

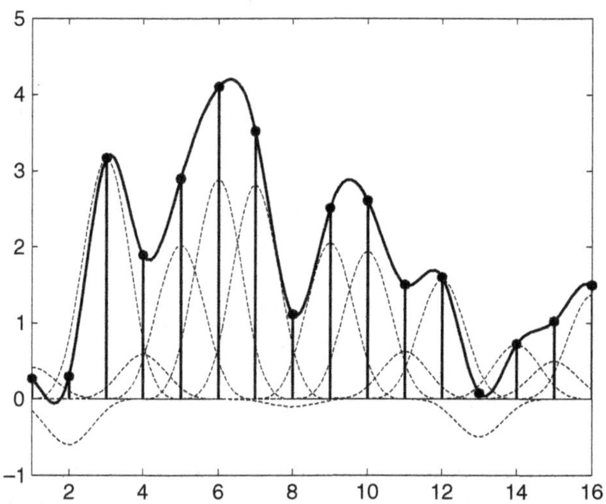

FIGURE 4.3 This figure illustrates a one dimensional B-spline representation of an image, where the image is assumed to be composed of a linear combination of B-spline basis functions. The dotted lines are the individual basis functions, which sum to produce the interpolated function (solid line).

[1] Except with a slightly different treatment exactly in the centre of two voxels.

Fourier methods

Higher order interpolation is slow when many neighbouring voxels are used, but there are faster ways of interpolating when doing rigid body transformations. Translations parallel to the axes are trivial, as these simply involve convolving with a translated delta function. For translations that are not whole numbers of pixels, the delta function is replaced by a sinc function centred at the translation distance. The use of fast Fourier transforms means that the convolution can be performed most rapidly as a multiplication in Fourier space. It is clear how translations can be performed in this way, but rotations are less obvious. One way that rotations can be effected involves replacing them by a series of shears (Eddy *et al.*, 1996; Cox and Jesmanowicz, 1999) (see later). A shear simply involves translating different rows or columns of an image by different amounts, so each shear can be performed as a series of one dimensional convolutions.

RIGID BODY TRANSFORMATIONS

Rigid body transformations consist of only rotations and translations, and leave given arrangements unchanged. They are a subset of the more general affine[2] transformations. For each point (x_1, x_2, x_3) in an image, an affine mapping can be defined into the coordinates of another space (y_1, y_2, y_3). This is expressed as:

$$\begin{aligned} y_1 &= m_{11}x_1 + m_{12}x_2 + m_{13}x_3 + m_{14} \\ y_2 &= m_{21}x_1 + m_{22}x_2 + m_{23}x_3 + m_{24} \\ y_3 &= m_{31}x_1 + m_{32}x_2 + m_{33}x_3 + m_{34} \end{aligned} \qquad 4.5$$

which is often represented by a simple matrix multiplication ($\mathbf{y} = \mathbf{Mx}$):

$$\begin{bmatrix} y_1 \\ y_2 \\ y_3 \\ 1 \end{bmatrix} = \begin{bmatrix} m_{11} & m_{12} & m_{13} & m_{14} \\ m_{21} & m_{22} & m_{23} & m_{24} \\ m_{31} & m_{32} & m_{33} & m_{34} \\ 0 & 0 & 0 & 1 \end{bmatrix} \begin{bmatrix} x_1 \\ x_2 \\ x_3 \\ 1 \end{bmatrix} \qquad 4.6$$

The elegance of formulating these transformations in terms of matrices is that several of them can be combined simply by multiplying the matrices together to form a single matrix. This means that repeated re-sampling of data can be avoided when reorienting an image. Inverse affine transformations are obtained by inverting the transformation matrix.

[2] Affine means that parallel lines remain parallel after the transformation.

Translations

If a point $\mathbf{x}$ is to be translated by $\mathbf{q}$ units, then the transformation is simply:

$$\mathbf{y} = \mathbf{x} + \mathbf{q} \qquad 4.7$$

In matrix terms, this can be considered as:

$$\begin{bmatrix} y_1 \\ y_2 \\ y_3 \\ 1 \end{bmatrix} = \begin{bmatrix} 1 & 0 & 0 & q_1 \\ 0 & 1 & 0 & q_2 \\ 0 & 0 & 1 & q_3 \\ 0 & 0 & 0 & 1 \end{bmatrix} \begin{bmatrix} x_1 \\ x_2 \\ x_3 \\ 1 \end{bmatrix} \qquad 4.8$$

Rotations

In two dimensions, a rotation is described by a single angle. Consider a point at coordinate (x_1, x_2) on a two-dimensional plane. A rotation of this point to new coordinates (y_1, y_2), by θ radians around the origin, can be generated by the transformation:

$$\begin{aligned} y_1 &= cos(\theta)x_1 + sin(\theta)x_2 \\ y_2 &= -sin(\theta)x_1 + cos(\theta)x_2 \end{aligned} \qquad 4.9$$

This is another example of an affine transformation. For the three-dimensional case, there are three orthogonal planes that an object can be rotated in. These planes of rotation are normally expressed as being around the axes. A rotation of q_1 radians about the first (x) axis is performed by:

$$\begin{bmatrix} y_1 \\ y_2 \\ y_3 \\ 1 \end{bmatrix} = \begin{bmatrix} 1 & 0 & 0 & 0 \\ 0 & cos(q_1) & sin(q_1) & 0 \\ 0 & -sin(q_1) & cos(q_1) & 0 \\ 0 & 0 & 0 & 1 \end{bmatrix} \begin{bmatrix} x_1 \\ x_2 \\ x_3 \\ 1 \end{bmatrix} \qquad 4.10$$

Similarly, rotations about the second (y) and third (z) axes are carried out by the following matrices:

$$\begin{bmatrix} cos(q_2) & 0 & sin(q_2) & 0 \\ 0 & 1 & 0 & 0 \\ -sin(q_2) & 0 & cos(q_2) & 0 \\ 0 & 0 & 0 & 1 \end{bmatrix} \text{ and } \begin{bmatrix} cos(q_3) & sin(q_3) & 0 & 0 \\ -sin(q_3) & cos(q_3) & 0 & 0 \\ 0 & 0 & 1 & 0 \\ 0 & 0 & 0 & 1 \end{bmatrix}.$$

Rotations are combined by multiplying these matrices together in the appropriate order. The order of the operations is important. For example, a rotation about the first axis of $\pi/2$ radians followed by an equivalent rotation about the second would produce a very different result to that obtained if the order of the operations was reversed.

Zooms

The affine transformations described so far will generate purely rigid body mappings. Zooms are needed to change the size of an image, or to work with images whose voxel sizes are not isotropic, or differ between images. These represent scalings along the orthogonal axes, and can be represented via:

$$\begin{bmatrix} y_1 \\ y_2 \\ y_3 \\ 1 \end{bmatrix} = \begin{bmatrix} q_1 & 0 & 0 & 0 \\ 0 & q_2 & 0 & 0 \\ 0 & 0 & q_3 & 0 \\ 0 & 0 & 0 & 1 \end{bmatrix} \begin{bmatrix} x_1 \\ x_2 \\ x_3 \\ 1 \end{bmatrix} \quad \textbf{4.11}$$

A single zoom by a factor of -1 will flip an image. Two flips in different directions will merely rotate it by π radians (a rigid body transformation). In fact, any affine transformation with a negative determinant will render the image flipped.

Shears

Shearing by parameters q_1, q_2 and q_3 can be performed by the following matrix:

$$\begin{bmatrix} 1 & q_1 & q_2 & 0 \\ 0 & 1 & q_3 & 0 \\ 0 & 0 & 1 & 0 \\ 0 & 0 & 0 & 1 \end{bmatrix} \quad \textbf{4.12}$$

A shear by itself is not a rigid body transformation, but it is possible to combine shears in order to generate a rotation. In two dimensions, a matrix encoding a rotation of θ radians about the origin can be constructed by multiplying together three matrices that effect shears (Eddy *et al.*, 1996):

$$\begin{bmatrix} cos(\theta) & sin(\theta) & 0 \\ -sin(\theta) & cos(\theta) & 0 \\ 0 & 0 & 1 \end{bmatrix} \equiv$$

$$\begin{bmatrix} 1 & tan(\theta/2) & 0 \\ 0 & 1 & 0 \\ 0 & 0 & 1 \end{bmatrix} \begin{bmatrix} 1 & 0 & 0 \\ sin(\theta) & 1 & 0 \\ 0 & 0 & 1 \end{bmatrix} \begin{bmatrix} 1 & tan(\theta/2) & 0 \\ 0 & 1 & 0 \\ 0 & 0 & 1 \end{bmatrix} \quad \textbf{4.13}$$

Rotations in three dimensions can be decomposed into four shears (Cox and Jesmanowicz, 1999). As shears can be performed quickly as one dimensional convolutions, then these decompositions are very useful for doing accurate and rapid rigid body transformations of images.

Parameterizing a rigid body transformation

When doing rigid registration of a pair of images, it is necessary to estimate six parameters that describe the rigid body transformation matrix. There are many ways of parameterizing this transformation in terms of six parameters ($\mathbf{q}$). One possible form is:

$$\mathbf{M} = \mathbf{TR} \quad \textbf{4.14}$$

where

$$\mathbf{T} = \begin{bmatrix} 1 & 0 & 0 & q_1 \\ 0 & 1 & 0 & q_2 \\ 0 & 0 & 1 & q_3 \\ 0 & 0 & 0 & 1 \end{bmatrix} \quad \textbf{4.15}$$

and

$$\mathbf{R} = \begin{bmatrix} 1 & 0 & 0 & 0 \\ 0 & cos(q_4) & sin(q_4) & 0 \\ 0 & -sin(q_4) & cos(q_4) & 0 \\ 0 & 0 & 0 & 1 \end{bmatrix} \begin{bmatrix} cos(q_5) & 0 & sin(q_5) & 0 \\ 0 & 1 & 0 & 0 \\ -sin(q_5) & 0 & cos(q_5) & 0 \\ 0 & 0 & 0 & 1 \end{bmatrix}$$

$$\times \begin{bmatrix} cos(q_6) & sin(q_6) & 0 & 0 \\ -sin(q_6) & cos(q_6) & 0 & 0 \\ 0 & 0 & 1 & 0 \\ 0 & 0 & 0 & 1 \end{bmatrix} \quad \textbf{4.16}$$

Sometimes it is desirable to extract transformation parameters from a matrix. Extracting these parameters $\mathbf{q}$ from $\mathbf{M}$ is relatively straightforward. Determining the translations is trivial, as they are simply contained in the fourth column of $\mathbf{M}$. This just leaves the rotations:

$$\mathbf{R} = \begin{bmatrix} c_5c_6 & c_5s_6 & s_5 & 0 \\ -s_4s_5c_6 - c_4s_6 & -s_4s_5s_6 + c_4c_6 & s_4c_5 & 0 \\ -c_4s_5c_6 + s_4s_6 & -c_4s_5s_6 - s_4c_6 & c_4c_5 & 0 \\ 0 & 0 & 0 & 1 \end{bmatrix} \quad \textbf{4.17}$$

where s_4, s_5 and s_6 are the sines, and c_4, c_5 and c_6 are the cosines of parameters q_4, q_5 and q_6 respectively. Therefore, provided that c_5 is not zero:

$$\begin{aligned} q_5 &= sin^{-1}(r_{13}) \\ q_4 &= atan2(r_{23}/cos(q_5), r_{33}/cos(q_5)) \\ q_6 &= atan2(r_{12}/cos(q_5), r_{11}/cos(q_5)) \end{aligned} \quad \textbf{4.18}$$

where *atan2* is the four quadrant inverse tangent.

Working with volumes of differing or anisotropic voxel sizes

Voxel sizes need to be considered during image registration. Often, the images (say **f** and **g**) will have voxels that are anisotropic. The dimensions of the voxels are also likely to differ between images of different modalities. For simplicity, a Euclidean space is used, where measures of distance are expressed in millimetres. Rather than transforming the images into volumes with cubic voxels that are the same size in all images, one can simply define affine transformation matrices that map from voxel coordinates into this Euclidean space. For example, if image **f** is of size 128 × 128 × 43 and has voxels that are 2.1 *mm* × 2.1 *mm* × 2.45 *mm*, the following matrix can be defined:

$$\mathbf{M}_f = \begin{bmatrix} 2.1 & 0 & 0 & -135.45 \\ 0 & 2.1 & 0 & -135.45 \\ 0 & 0 & 2.45 & -53.9 \\ 0 & 0 & 0 & 1 \end{bmatrix} \quad (4.19)$$

This transformation matrix maps voxel coordinates to a Euclidean space whose axes are parallel to those of the image and distances are measured in millimetres, with the origin at the centre of the image volume (i.e. $\mathbf{M}_f[64.5\ 64.5\ 22\ 1]^T = [0\ 0\ 0\ 1]^T$. A similar matrix can be defined for **g** ($\mathbf{M}_g$). Because modern MR image formats, such as DICOM, generally contain information about image orientations in their headers, it is possible to extract this information to compute automatically values for $\mathbf{M}_f$ or $\mathbf{M}_g$. This makes it possible to register easily images together that were originally acquired in completely different orientations.

The objective of a rigid body registration is to determine the affine transformation that maps the coordinates of image **g** to that of **f**. To accomplish this, a rigid body transformation matrix $\mathbf{M}_r$ is determined, such that $\mathbf{M}_f^{-1}\mathbf{M}_r^{-1}\mathbf{M}_g$ will map from voxels in **g** to those in **f**. The inverse of this matrix maps from **f** to **g**. Once $\mathbf{M}_r$ has been determined, $\mathbf{M}_f$ can be set to $\mathbf{M}_r\mathbf{M}_f$. From there onwards the mapping between the voxels of the two images can be achieved by $\mathbf{M}_f^{-1}\mathbf{M}_g$. Similarly, if another image (**h**) is also registered with **g** in the same manner, then not only is there a mapping from **h** to **g** (via $\mathbf{M}_g^{-1}\mathbf{M}_h$), but there is also one from **h** to **f**, which is simply $\mathbf{M}_f^{-1}\mathbf{M}_h$ (derived from $\mathbf{M}_f^{-1}\mathbf{M}_g\mathbf{M}_g^{-1}\mathbf{M}_h$).

Left- and right-handed coordinate systems

Positions in space can be represented in either a left- or right-handed coordinate system (Figure 4.4), where one system is a mirror image of the other. For example,

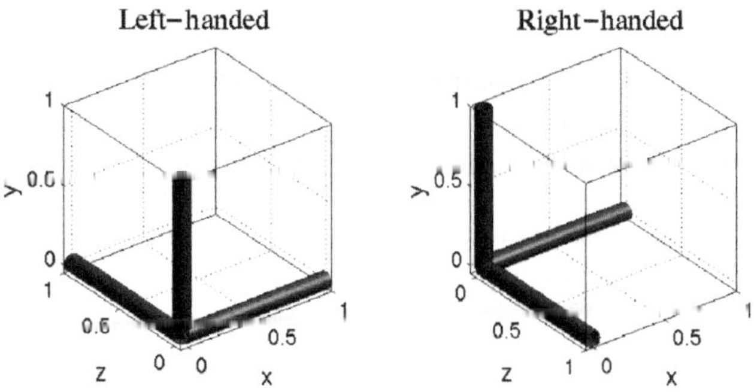

FIGURE 4.4 Left- and right-handed coordinate systems. The thumb corresponds to the x-axis, the index finger to the y-axis and the second finger to the z-axis.

the system used by the atlas of Talairach and Tournoux (1988) is right-handed, because the first dimension (often referred to as the x direction) increases from left to right, the second dimension goes from posterior to anterior (back to front) and the third dimension increases from inferior to superior (bottom to top). The axes can be rotated by any angle, and they still retain their handedness. An affine transformation mapping between left and right-handed coordinate systems has a negative determinant, whereas one that maps between coordinate systems of the same kind will have a positive determinant. Because the left and right sides of a brain have similar appearances, care must be taken when reorienting image volumes. Consistency of the coordinate systems can be achieved by performing any reorientations using affine transformations, and checking the determinants of the matrices.

Rotating tensors

Diffusion tensor imaging (DTI) is becoming increasingly useful. These datasets are usually stored as six images containing a scalar field for each unique tensor element. It is worth noting that a rigid body transformation of a DTI dataset is not a simple matter of rigidly rotating the individual scalar fields.[3] Once these fields have been resampled, the tensor represented at every voxel position needs to be rotated. A 3 × 3 tensor **T** can be rotated by a 3 × 3 matrix **R** by $\mathbf{T}' = \mathbf{RTR}^T$.

If DTI volumes are to be transformed using more complex warping models, then the local derivatives of the deformations (Jacobian matrices) need to be computed at each voxel. Suitable transformations can then

[3] It is worth noting that some interpolation methods are unsuitable for resampling the raw scalar fields, as the introduction of sampling errors can cause the positive definite nature of the tensors to be lost.

be extracted from these derivatives, and applied to each element of the re-sampled tensor field (Alexander *et al.*, 1999, 2001).

WITHIN-MODALITY RIGID REGISTRATION

Whenever several images of the same subject have been acquired, it is extremely useful to have them all in register. Some of the simple benefits of this include allowing images to be averaged in order to increase signal to noise, or to subtract one image from another to emphasize differences between the images. Rigid[4] registration is normally used for retrospectively registering images of the same subject that have been collected at different times. Even if images were acquired during the same scanning session, the subject may have moved slightly between acquisitions.

The most common application of within-modality registration in functional imaging is to reduce motion artefacts by realigning the volumes in image time-series. The objective of realignment is to determine the rigid body transformations that best map the series of functional images to the same space. This can be achieved by minimizing the mean squared difference between each of the images and a reference image, where the reference image could be one of the images in the series. For slightly better results, this procedure could be repeated, but instead of matching to one of the images from the series, the images would be registered to the mean of all the realigned images. Because of the non-stationary variance in the images, a variance image could be computed at the same time as the mean, in order to provide better weighting for the registration. Voxels with a lot of variance should be given lower weighting, whereas those with less variance should be weighted more highly.

Within-modality image registration is also useful for looking at shape differences of brains. Morphometric studies sometimes involve looking at changes in brain shape over time, often to study the progression of a disease such as Alzheimer's, or to monitor tumour growth or shrinkage. Differences between structural MR scans acquired at different times are identified, by first coregistering the images and then looking at the difference between the registered images. Rigid registration can also be used as a preprocessing step before using non-linear registration methods for identifying shape changes (Freeborough and Fox, 1998).

[4] Or affine registration if voxel sizes are not accurately known.

Image registration involves estimating a set of parameters describing a spatial transformation that 'best' match the images together. The goodness of the match is based on an *objective function*, which is maximized or minimized using some *optimization algorithm*. This section deals with registering images that have been collected using the same (or similar) modalities, allowing a relatively simple objective function to be used. In this case, the objective function is the mean squared difference between the images. The more complex task of registering images with different contrasts will be dealt with later.

Optimization

The objective of optimization is to determine the values for a set of parameters for which some function of the parameters is minimized (or maximized). One of the simplest cases involves determining the optimum parameters for a model in order to minimize the mean squared difference between a model and a set of real world data. Normally there are many parameters and it is not possible to search exhaustively through the whole parameter space. The usual approach is to make an initial parameter estimate, and begin iteratively searching from there. At each iteration, the model is evaluated using the current parameter estimates, and the objective function computed. A judgement is then made about how the parameter estimates should be modified, before continuing on to the next iteration. The optimization is terminated when some convergence criterion is achieved (usually when the objective function stops decreasing, or its derivatives with respect to the parameters become sufficiently small).

The registration approach described here is essentially an optimization. One image (the source image) is spatially transformed so that it matches another (the reference image) by minimizing the mean squared difference. The parameters that are optimized are those that describe the spatial transformation (although there are often other nuisance parameters required by the model, such as intensity scaling parameters). A good algorithm to use for rigid registration (Friston *et al.*, 1995; Woods *et al.*, 1998) is *Gauss-Newton* optimization, and it is illustrated here.

Suppose that $b_i(\mathbf{q})$ is the function describing the difference between the source and reference images at voxel i, when the vector of model parameters have values $\mathbf{q}$. For each voxel, a first approximation of Taylor's theorem can be used to estimate the value that this difference will take if the parameters $\mathbf{q}$ are decreased by $\mathbf{t}$:

$$b_i(\mathbf{q}-\mathbf{t}) \simeq b_i(\mathbf{q}) - t_1\frac{\partial b_i(\mathbf{q})}{\partial q_1} - t_2\frac{\partial b_i(\mathbf{q})}{\partial q_2}\ldots \qquad 4.20$$

This allows the construction of a set of simultaneous equations (of the form $\mathbf{At} \simeq \mathbf{b}$) for estimating the values that $\mathbf{t}$ should assume to in order to minimize $\sum_i b_i(\mathbf{q}-\mathbf{t})^2$:

$$\begin{bmatrix} \frac{\partial b_1(\mathbf{q})}{\partial q_1} & \frac{\partial b_1(\mathbf{q})}{\partial q_2} & \cdots \\ \frac{\partial b_2(\mathbf{q})}{\partial q_1} & \frac{\partial b_2(\mathbf{q})}{\partial q_2} & \cdots \\ \vdots & \vdots & \ddots \end{bmatrix} \begin{bmatrix} t_1 \\ t_2 \\ \vdots \end{bmatrix} \simeq \begin{bmatrix} b_1(\mathbf{q}) \\ b_2(\mathbf{q}) \\ \vdots \end{bmatrix} \quad 4.21$$

From this, an iterative scheme can be derived for improving the parameter estimates. For iteration n, the parameters $\mathbf{q}$ are updated as:

$$\mathbf{q}^{(n+1)} = \mathbf{q}^n - (\mathbf{A}^T\mathbf{A})^{-1}\mathbf{A}^T\mathbf{b} \quad 4.22$$

where $\mathrm{A} = \begin{bmatrix} \frac{\partial b_1(\mathbf{q})}{\partial q_1} & \frac{\partial b_1(\mathbf{q})}{\partial q_2} & \cdots \\ \frac{\partial b_2(\mathbf{q})}{\partial q_1} & \frac{\partial b_2(\mathbf{q})}{\partial q_2} & \cdots \\ \vdots & \vdots & \ddots \end{bmatrix}$ and $\mathbf{b} = \begin{bmatrix} b_1(\mathbf{q}) \\ b_2(\mathbf{q}) \\ \vdots \end{bmatrix}$.

This process is repeated until convergence. There is no guarantee that the best global solution will be reached, because the algorithm can get caught in a local minimum. To reduce this problem, the starting estimates for $\mathbf{q}$ should be set as close as possible to the optimum solution. The number of potential local minima can also be decreased by working with smooth images. This also has the effect of making the first order Taylor approximation more accurate for larger displacements. Once the registration is close to the final solution, it can continue with less smooth images.

In practice, $\mathbf{A}^T\mathbf{A}$ and $\mathbf{A}^T\mathbf{b}$ from Eqn. **4.22** are often computed 'on the fly' for each iteration. By computing these matrices using only a few rows of $\mathbf{A}$ and $\mathbf{b}$ at a time, much less computer memory is required than would be needed for the whole of matrix $\mathbf{A}$. Also, the partial derivatives $\partial b_i(\mathbf{q})/\partial q_j$ can be rapidly computed from the gradients of the images using the chain rule (see Woods, 1999 for detailed information).

It should be noted that $\mathbf{A}^T\mathbf{b}$ corresponds to the first derivatives of the objective function with respect to the parameters, and $\mathbf{A}^T\mathbf{A}$ approximately corresponds to the second derivatives (one half of the Hessian matrix, often referred to as the curvature matrix – see Press *et al.*, 1992, Section 15.5 for a general description, or Woods, 1999, 2000 for more information related to image registration). Another way of thinking about the optimization is that it fits a quadratic function to the error surface at each iteration. Successive parameter estimates are chosen such that they are at the minimum point of this quadratic (illustrated for a single parameter in Figure 4.5).

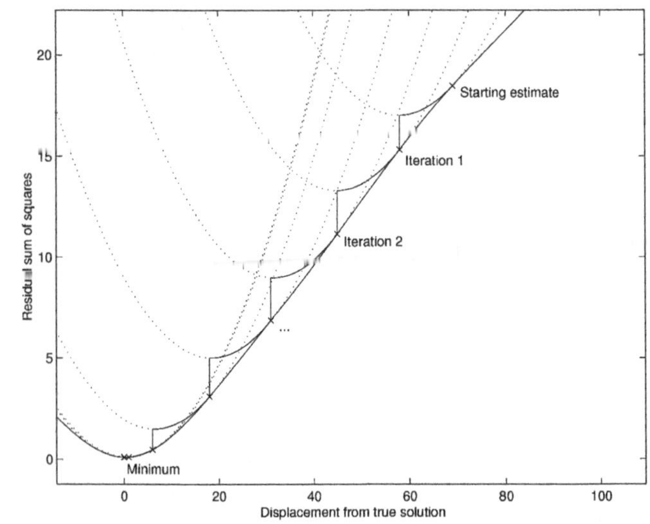

FIGURE 4.5 The optimization can be thought of as fitting a series of quadratics to the error surface. Each parameter update is such that it falls at the minimum of the quadratic.

Implementation

This section is about estimating parameters that describe a rigid body transformation, but the principles can be extended to models that describe non-linear warps. To register a source image $\mathbf{f}$ to a reference image $\mathbf{g}$, a six parameter rigid body transformation (parameterized by q_1 to q_6) would be used. To perform the registration, a number of points in the reference image (each denoted by $\mathbf{x}_i$) are compared with points in the source image (denoted by $\mathbf{Mx}_i$, where $\mathbf{M}$ is the rigid body transformation matrix constructed from the six parameters). The images may be scaled differently, so an additional intensity scaling parameter (q_7) may be included in the model. The parameters ($\mathbf{q}$) are optimized by minimizing the mean squared difference[5] between the images according to the algorithm described in the previous section. The function that is minimized is:

$$\sum_i (f(\mathbf{Mx}_i) - q_7 g(\mathbf{x}_i))^2$$

where $\mathbf{M} = \mathbf{M_f}^{-1}\mathbf{M_r}^{-1}\mathbf{M_g}$, and $\mathbf{M_r}$ is constructed from parameters $\mathbf{q}$. Vector $\mathbf{b}$ is generated for each iteration as:

$$\mathbf{b} = \begin{bmatrix} f(\mathbf{Mx}_1) - q_7 g(\mathbf{x}_1) \\ f(\mathbf{Mx}_2) - q_7 g(\mathbf{x}_2) \\ \vdots \end{bmatrix} \quad 4.23$$

[5] Inevitably, some values of $\mathbf{Mx}_i$ will lie outside the domain of $\mathbf{f}$, so nothing is known about what the image intensity should be at these points. The computations are only performed for points where both $\mathbf{x}_i$ and $\mathbf{Mx}_i$ lie within the field of view of the images.

Each column of matrix **A** is constructed by differentiating **b** with respect to parameters q_1 to q_7:

$$\mathbf{A} = \begin{bmatrix} \frac{\partial f(\mathbf{Mx}_1)}{\partial q_1} & \frac{\partial f(\mathbf{Mx}_1)}{\partial q_2} & \cdots & \frac{\partial f(\mathbf{Mx}_1)}{\partial q_6} & -g(\mathbf{x}_1) \\ \frac{\partial f(\mathbf{Mx}_2)}{\partial q_1} & \frac{\partial f(\mathbf{Mx}_2)}{\partial q_2} & \cdots & \frac{\partial f(\mathbf{Mx}_2)}{\partial q_6} & -g(\mathbf{x}_2) \\ \vdots & \vdots & \ddots & \vdots & \vdots \end{bmatrix} \quad \mathbf{4.24}$$

Because non-singular affine transformations are easily invertible, it is possible to make the registration more robust by also considering what happens with the inverse transformation. By swapping around the source and reference image, the registration problem also becomes one of minimizing:

$$\sum_j (g(\mathbf{M}^{-1}\mathbf{y}_j) - q_7^{-1} f(\mathbf{y}_j))^2$$

In theory, a more robust solution could be achieved by simultaneously including the inverse transformation to make the registration problem symmetric (Woods *et al.*, 1998). The objective function would then be:

$$\lambda_1 \sum_i (f(\mathbf{Mx}_i) - q_7 g(\mathbf{x}_i))^2 + \lambda_2 \sum_j (g(\mathbf{M}^{-1}\mathbf{y}_j) - q_7^{-1} f(\mathbf{y}_j))^2 \quad \mathbf{4.25}$$

Normally, the intensity scaling of the image pair will be similar, so equal values for the weighting factors (λ_1 and λ_2) can be used. Matrix **A** and vector **b** would then be formulated as:

$$\mathbf{b} = \begin{bmatrix} \lambda_1^{\frac{1}{2}} (f(\mathbf{Mx}_1) - q_7 g(\mathbf{x}_1)) \\ \lambda_1^{\frac{1}{2}} (f(\mathbf{Mx}_2) - q_7 g(\mathbf{x}_2)) \\ \vdots \\ \lambda_2^{\frac{1}{2}} (g(\mathbf{M}^{-1} - \mathbf{y}_1) - q_7^{-1} f(\mathbf{y}_1)) \\ \lambda_2^{\frac{1}{2}} (g(\mathbf{M}^{-1} - \mathbf{y}_2) - q_7^{-1} f(\mathbf{y}_2)) \\ \vdots \end{bmatrix} \quad \mathbf{4.26}$$

and

$$\mathbf{A} = \begin{bmatrix} \lambda_1^{\frac{1}{2}} \frac{\partial f(\mathbf{Mx}_1)}{\partial q_1} & \cdots & \lambda_1^{\frac{1}{2}} \frac{\partial f(\mathbf{Mx}_1)}{\partial q_6} & -\lambda_1^{\frac{1}{2}} g(\mathbf{x}_1) \\ \lambda_1^{\frac{1}{2}} \frac{\partial f(\mathbf{Mx}_2)}{\partial q_1} & \cdots & \lambda_1^{\frac{1}{2}} \frac{\partial f(\mathbf{Mx}_2)}{\partial q_6} & -\lambda_1^{\frac{1}{2}} g(\mathbf{x}_2) \\ \vdots & \ddots & \vdots & \vdots \\ \lambda_2^{\frac{1}{2}} \frac{\partial f(\mathbf{M}^{-1}\mathbf{y}_1)}{\partial q_1} & \cdots & \lambda_2^{\frac{1}{2}} \frac{\partial g(\mathbf{M}^{-1}\mathbf{y}_1)}{\partial q_6} & \lambda_1^{\frac{1}{2}} q_7^{-2} f(\mathbf{y}_1) \\ \lambda_2^{\frac{1}{2}} \frac{\partial g(\mathbf{M}^{-1}\mathbf{y}_2)}{\partial q_1} & \cdots & \lambda_2^{\frac{1}{2}} \frac{\partial g(\mathbf{M}^{-1}\mathbf{y}_2)}{\partial q_6} & \lambda_2^{\frac{1}{2}} q_7^{-2} f(\mathbf{y}_2) \\ \vdots & \ddots & \vdots & \vdots \end{bmatrix} \quad \mathbf{4.27}$$

Residual artefacts from PET and fMRI

Even after realignment, there may still be some motion related artefacts remaining in functional data. After retrospective realignment of PET images with large movements, the primary source of error is due to incorrect attenuation correction. In emission tomography methods, many photons are not detected because they are attenuated by the subject's head. Normally, a transmission scan (using a moving radioactive source external to the subject) is acquired before collecting the emission scans. The ratio of the number of detected photon pairs from the source, with and without a head in the field of view, produces a map of the proportion of photons that are absorbed along any line-of-response. If a subject moves between the transmission and emission scans, then the applied attenuation correction is incorrect because the emission scan is no longer aligned with the transmission scan. There are methods for correcting these errors (Andersson *et al.*, 1995), but they are beyond the scope of this book.

In fMRI, there are many sources of motion related artefacts. The most obvious ones are:

- Interpolation error from the re-sampling algorithm used to transform the images can be one of the main sources of motion related artefacts.
- When MR images are reconstructed, the final images are usually the modulus of the initially complex data. This results in voxels that should be negative being rendered positive. This has implications when the images are re-sampled, because it leads to errors at the edge of the brain that cannot be corrected however good the interpolation method is. Possible ways to circumvent this problem are to work with complex data, or apply a low pass filter to the complex data before taking the modulus.
- The sensitivity (slice selection) profile of each slice also plays a role in introducing artefacts (Noll *et al.*, 1997).
- fMRI images are spatially distorted, and the amount of distortion depends partly upon the position of the subject's head within the magnetic field. Relatively large subject movements result in the brain images changing shape, and these shape changes cannot be corrected by a rigid body transformation (Jezzard and Clare, 1999; Andersson *et al.*, 2001).
- Each fMRI volume of a series is currently acquired a plane at a time over a period of a few seconds. Subject movement between acquiring the first and last plane of any volume is another reason why the images may not strictly obey the rules of rigid body motion.
- After a slice is magnetized, the excited tissue takes time to recover to its original state, and the amount of recovery that has taken place will influence the intensity of the tissue in the image. Out of plane movement

will result in a slightly different part of the brain being excited during each repeat. This means that the spin excitation will vary in a way that is related to head motion, and so leads to more movement related artefacts (Friston *et al.*, 1996).

- Nyquist ghost artefacts in MR images do not obey the same rigid body rules as the head, so a rigid rotation to align the head will not mean that the ghosts are aligned. The same also applies to other image artefacts, such as those arising due to chemical shifts.
- The accuracy of the estimated registration parameters is normally in the region of tens of microns. This is dependent upon many factors, including the effects just mentioned. Even the signal changes elicited by the experiment can have a slight effect (a few microns) on the estimated parameters (Freire and Mangin, 2001).

These problems cannot be corrected by simple image realignment and so may be sources of possible stimulus correlated motion artefacts. Systematic movement artefacts resulting in a signal change of only one or two per cent can lead to highly significant false positives over an experiment with many scans. This is especially important for experiments where some conditions may cause slight head movements (such as motor tasks, or speech), because these movements are likely to be highly correlated with the experimental design. In cases like this, it is difficult to separate true activations from stimulus correlated motion artefacts. Providing there are enough images in the series and the movements are small, some of these artefacts can be removed by using an analysis of covariation (ANCOVA) model to remove any signal that is correlated with functions of the movement parameters (Friston *et al.*, 1996). However, when the estimates of the movement parameters are related to the experimental design, it is likely that much of the true fMRI signal will also be lost. These are still unresolved problems.

BETWEEN-MODALITY RIGID REGISTRATION

The combination of multiple imaging modalities can provide enhanced information that is not readily apparent on inspection of individual image modalities. For studies of a single subject, sites of activation can be accurately localized by superimposing them on a high resolution structural image of the subject (typically a T1-weighted MRI). This requires registration of the functional images with the structural image. A further possible use for this registration is that a more precise spatial normalization can be achieved by computing it from a more detailed structural image. If the functional and structural images are in register, then a warp computed from the structural image can be applied to the functional images. Normally a rigid body model is used for registering images of the same subject, but because fMRI images are usually severely distorted – particularly in the phase encode direction (Jezzard and Clare, 1999; Jezzard, 2000) – it is often preferable to do non-linear registration (Kybic *et al.*, 2000; Studholme *et al.*, 2000). Rigid registration models require voxel sizes to be accurately known. This is a problem that is particularly apparent when registering images from different scanners.

Two images from the same subject acquired using the same modality or scanning sequences generally look similar, so it suffices to find the rigid body transformation parameters that minimize the sum of squared differences between them. However, for coregistration between modalities there is nothing quite so obvious to minimize, as there is no linear relationship between the image intensities (Figure 4.6).

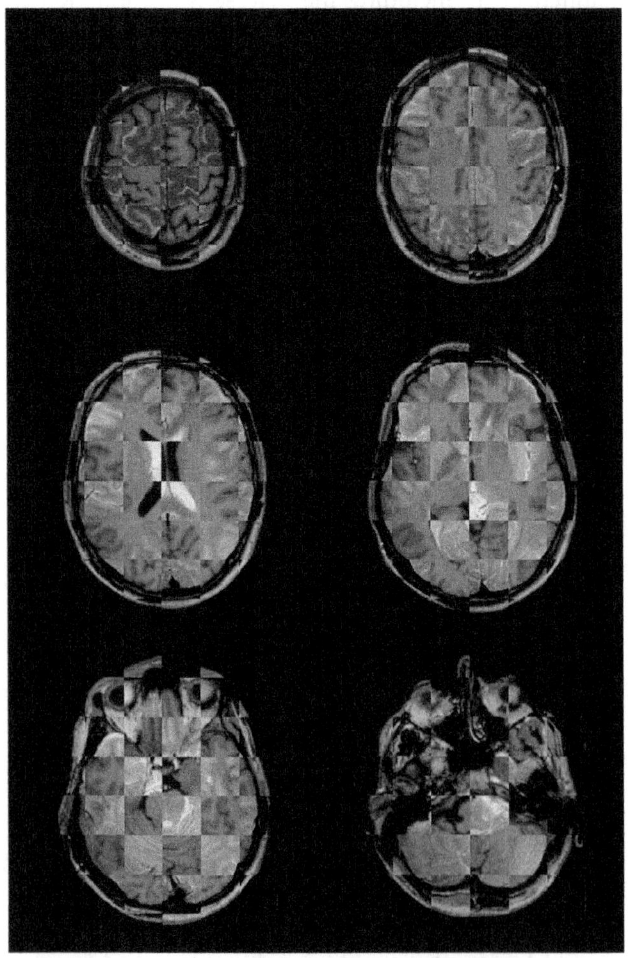

FIGURE 4.6 An example of T1- and T2-weighted MR images registered using mutual information. The two registered images are shown interleaved in a chequered pattern.

Older methods of registration involved the manual identification of homologous landmarks in the images. These landmarks are aligned together, thus bringing the images into registration. This is time-consuming, requires a degree of experience, and can be rather subjective. One of the first widely used semiautomatic coregistration methods was that known as the 'head-hat' approach (Pelizzari *et al.*, 1988). This method involved extracting brain surfaces of the two images, and then matching the surfaces together. There are also a number of other between-modality registration methods that involve partitioning the images, or finding common features, and then registering them together, but they are beyond the scope of this chapter.

The first intensity based inter-modal registration method was *AIR* (Woods *et al.*, 1993), which has been widely used for a number of years for registering PET and MR images. This method uses a variance of intensity ratios (VIR) objective function, and involves dividing the MR images into a number of partitions based on intensity. The registration is approximately based on minimizing the variance of the corresponding PET voxel intensities for each partition. It makes a number of assumptions about how the PET intensity varies with the MRI intensity, which are generally valid within the brain, but do not work when non-brain tissue is included. Because of this, the method has the disadvantage of requiring the MR images to be edited to remove non-brain tissue. For reviews of a number of inter-modality registration approaches, see Zuk and Atkins, 1996 and Hill *et al.*, 2001.

Information theoretic approaches

The most recent voxel-similarity measures to be used for inter-modal (as well as intra-modal (Holden *et al.*, 2000)) registration have been based on *information theory*. These measures are based on joint probability distributions of intensities in the images, usually discretely represented in the form of 2D joint histograms, which are normalized to sum to one.

The first information theoretic measure to be proposed was the entropy of the joint probability distribution (Studholme *et al.*, 1995), which should be minimized when the images are in register:

$$H(\mathbf{f},\mathbf{g}) = -\int_{-\infty}^{\infty}\int_{-\infty}^{\infty} P(\mathbf{f},\mathbf{g}) \log P(\mathbf{f},\mathbf{g}) d\mathbf{f} d\mathbf{g} \qquad 4.28$$

The discrete representation of the probability distributions is from a joint histogram, which can be considered as an I by J matrix $\mathbf{P}$. The entropy is then computed from the histogram according to:

$$H(\mathbf{f},\mathbf{g}) = \sum_{j=1}^{J}\sum_{i=1}^{I} p_{ij} \log p_{ij} \qquad 4.29$$

In practice, the entropy measure was found to produce poor registration results, but shortly afterwards, a more robust measure of registration quality was introduced. This was based on *mutual information* (MI) (Collignon *et al.*, 1995; Wells *et al.*, 1996) (also known as *Shannon information*), which is given by:

$$I(\mathbf{f},\mathbf{g}) = H(\mathbf{f}) + H(\mathbf{g}) - H(\mathbf{f},\mathbf{g}) \qquad 4.30$$

where $H(\mathbf{f},\mathbf{g})$ is the joint entropy of the images, and $H(\mathbf{f})$ and $H(\mathbf{g})$ are their marginalized entropies given by:

$$H(\mathbf{f}) = -\int_{-\infty}^{\infty} P(\mathbf{f}) \log P(\mathbf{f}) d\mathbf{f} \qquad 4.31$$

$$H(\mathbf{g}) = -\int_{-\infty}^{\infty} P(\mathbf{g}) \log P(\mathbf{g}) d\mathbf{g} \qquad 4.32$$

MI is a measure of dependence of one image on the other, and can be considered as the distance (Kullback-Leibler divergence) between the joint distribution ($P(\mathbf{f},\mathbf{g})$) and the distribution assuming complete independence ($P(\mathbf{f})P(\mathbf{g})$). When the two distributions are identical, this distance (and the mutual information) is zero. After rearranging, the expression for MI becomes:

$$\begin{aligned} I(\mathbf{f},\mathbf{g}) &= KL(P(\mathbf{f},\mathbf{g}) \| P(\mathbf{f})P(\mathbf{g})) \\ &= \int_{-\infty}^{\infty}\int_{-\infty}^{\infty} P(\mathbf{f},\mathbf{g}) \log\left(\frac{P(\mathbf{f},\mathbf{g})}{P(\mathbf{f})P(\mathbf{g})}\right) d\mathbf{f} d\mathbf{g} \end{aligned} \qquad 4.33$$

It is assumed that the MI between the images is maximized when they are in register (Figure 4.7). Another information theoretic measure (Studholme *et al.*, 1999) that can be used for registration is:

$$\tilde{I}(\mathbf{f},\mathbf{g}) = \frac{H(\mathbf{f}) + H(\mathbf{g})}{H(\mathbf{f},\mathbf{g})} \qquad 4.34$$

Another useful measure (Maes *et al.*, 1997) is:

$$\tilde{I}(\mathbf{f},\mathbf{g}) = 2H(\mathbf{f},\mathbf{g}) - H(\mathbf{f}) - H(\mathbf{g}) \qquad 4.35$$

and also the *entropy correlation coefficient* (Maes *et al.*, 1997) (see Press *et al.*, 1992, for more information):

$$U(\mathbf{f},\mathbf{g}) = 2\frac{H(\mathbf{f}) + H(\mathbf{g}) - H(\mathbf{f},\mathbf{g})}{H(\mathbf{f}) + H(\mathbf{g})} \qquad 4.36$$

Implementation details

Generating a joint histogram involves scanning through the voxels of the reference image and finding the corresponding points of the source. The appropriate bin in the histogram is incremented by one for each of these

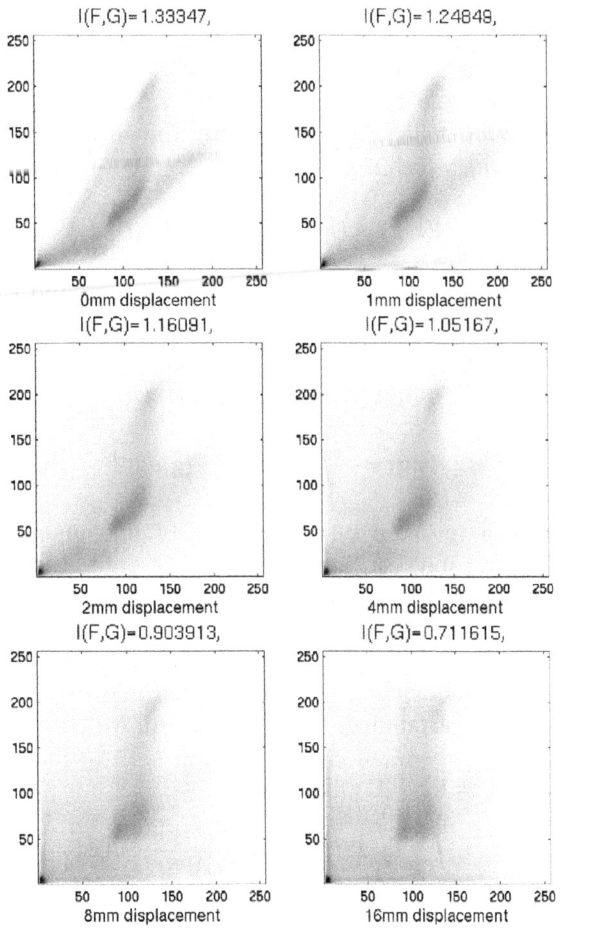

FIGURE 4.7 An illustration of how the joint histogram of an image pair changes as they are displaced relative to each other (note that the pictures show $\log(1+N)$, where N is the count in each histogram bin). The MI of the images is also shown.

point pairs. Pairs are ignored if the corresponding voxel is unavailable because it lies outside the image volume. The coordinate of the corresponding point rarely lies at an actual voxel centre, meaning that interpolation is required.

Many developers use *partial volume interpolation* (Collignon *et al.*, 1995), rather than interpolating the images themselves, but this can make the MI objective function particularly susceptible to interpolation artefact (Figure 4.8). The MI tends to be higher when voxel centres are sampled, where one is added to a single histogram bin. MI is lower when sampling in the centre of the eight neighbours, as an eighth is added to eight bins. These artefacts are especially prominent when fewer point pairs are used to generate the histograms.

A simpler alternative is to interpolate the images themselves, but this can lead to new intensity values in the histograms, which also cause interpolation artefact. This artefact largely occurs because of aliasing after integer represented images are rescaled so that they have values between zero and $I-1$, where I is the number of bins in the histogram (Figure 4.9). If care is taken at this stage, then interpolation of the image intensities becomes less of a problem. Another method of reducing these artefacts is to not sample the reference image on a regular grid, by (for example) introducing a random jitter to the sampled points (Likar and Pernuš, 2001).

Histograms contain noise, especially if a relatively small number of points are sampled in order to generate them. The optimum binning to use is still not fully resolved, and is likely to vary from application to application, but most researchers use histograms ranging between about 16×16 and 256×256. Smoothing a histogram has a similar effect to using fewer bins. Another

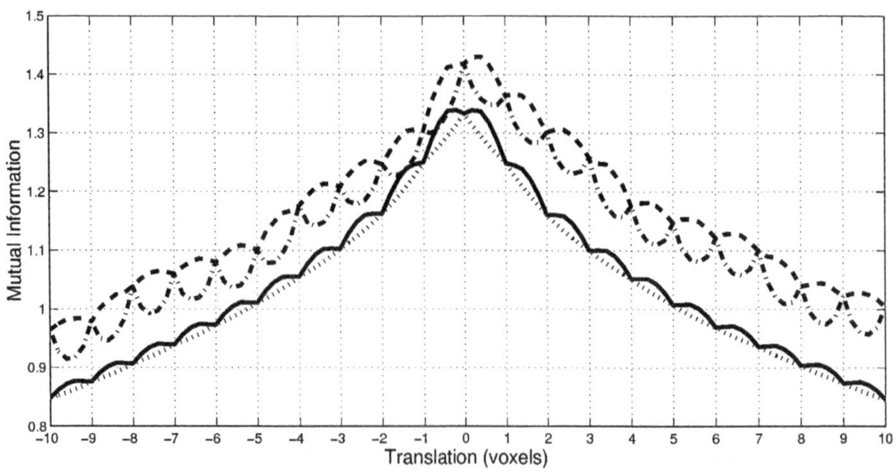

FIGURE 4.8 The mutual information objective function can be particularly susceptible to interpolation artefacts. This figure shows a plot of the MI between two images when they are translated with respect to each other. The dotted and dot-dashed lines show it computed using partial volume interpolation at high and lower sampling densities respectively. The solid and dashed lines show MI computed by interpolating the images themselves (solid indicates high sampling density, dashed indicates lower density).

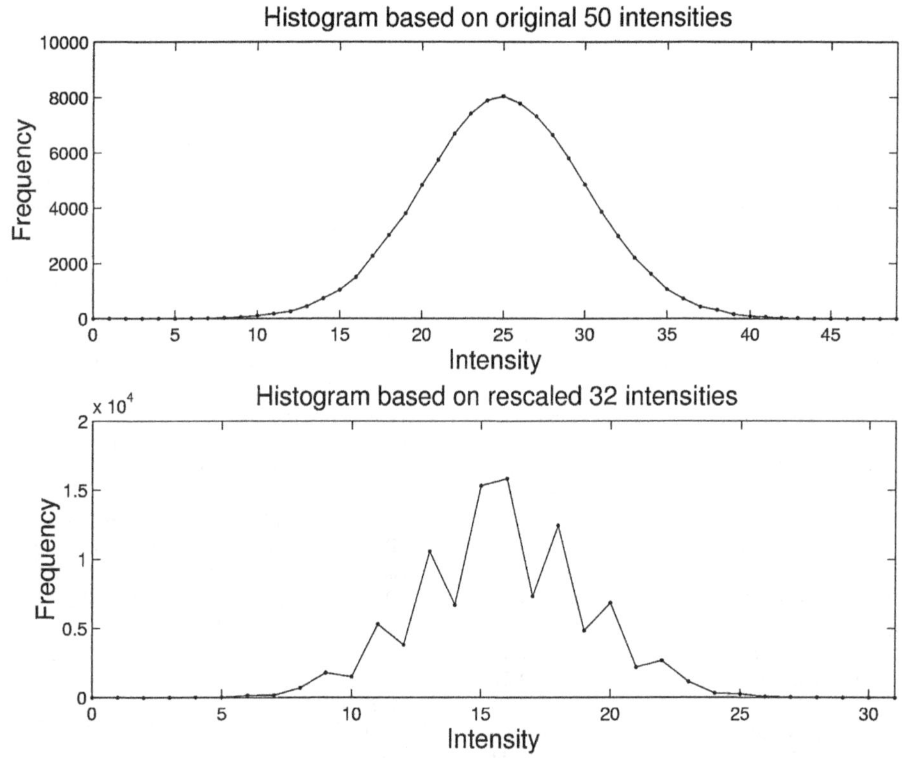

FIGURE 4.9 Rescaling an image can lead to aliasing artefacts in its histogram. Above: histogram based on original integer intensity values, simulated to have a Gaussian distribution. Below: the histogram after the intensities are rescaled shows aliasing artefacts.

alternative is to use a continuous representation of the joint probability distribution, such as a Parzen window density estimate (Wells *et al.*, 1996), or possibly even a Gaussian mixture model representation.

A method of optimization based on the first and second derivatives of the objective function was introduced earlier. Similar principles have been applied to minimizing the VIR objective function (Woods *et al.*, 1993), and also to maximizing MI (Thévenaz and Unser, 2000).[6] However, the most widely adopted scheme for maximizing MI is Powell's method (see Press *et al.*, 1992), which involves a series of successive line searches. Failures occasionally arise if the voxel similarity measure does not vary smoothly with changes to the parameter estimates. This can happen because of interpolation artefact, or if insufficient data contribute to the joint histogram. Alternatively, the algorithm can get caught within a local optimum, so it is important to assign starting estimates that approximately register the images. The required accuracy of the starting estimates depends on the particular images, but an approximate figure for many brain images with a good field of view would be in the region of about 5 cm for translations and 15° for rotations.

[6] This paper uses Levenberg-Marquardt optimization (Press *et al.*, 1992), which is a stabilized version of the Gauss-Newton method.

REFERENCES

Alexander DC, Gee JC, Bajcsy R (1999) Strategies for data reorientation during non-rigid transformations of diffusion tensor images. In *Proc Medical Image Computing and Computer-Assisted Intervention (MICCAI)*, Taylor C, Colchester A (eds), vol. 1679 of *Lecture Notes in Computer Science*. Springer-Verlag, Berlin and Heidelberg, pp. 463–72

Alexander DC, Pierpaoli C, Basser PJ *et al.* (2001) Spatial transformations of diffusion tensor magnetic resonance images. *IEEE Trans Med Imag* **20**: 1131–39

Andersson JLR, Hutton C, Ashburner J *et al.* (2001) Modeling geometric deformations in EPI time series. *NeuroImage* **13**: 903–19

Andersson JLR, Vagnhammar BE, Schneider H (1995) Accurate attenuation correction despite movement during PET imaging. *J Nucl Med* **36**: 670–78

Collignon A, Maes F, Delaere D *et al.* (1995) Automated multimodality image registration based on information theory. In *Proc Information Processing in Medical Imaging (IPMI)*, Bizais Y, Barillot C, Di Paola R (eds). Kluwer Academic Publishers, Dordrecht

Cox RW, Jesmanowicz A (1999) Real-time 3D image registration for functional MRI. *Magn Res Med* **42**: 1014–18

Eddy WF, Fitzgerald M, Noll DC (1996) Improved image registration by using Fourier interpolation. *Mag Res Med* **36**: 923–31

Freeborough PA, Fox NC (1998) Modelling brain deformations in alzheimer disease by fluid registration of serial MR images. *J Comput Assist Tomogr* **22**: 838–43

Freire L, Mangin J-F (2001) Motion correction algorithms of the brain mapping community create spurious functional activations. In *Proc Information Processing in Medical Imaging (IPMI)*, Insana MF, Leahy RM (eds), vol. 2082 of *Lecture Notes in Computer Science*. Springer-Verlag, Berlin and Heidelberg, pp 246–58

Friston KJ, Ashburner J, Frith CD *et al.* (1995) Spatial registration and normalization of images. *Hum Brain Mapp* **2**: 165–89

Friston KJ, Williams S, Howard R *et al.* (1996) Movement-related effects in fMRI time-series. *Mag Res Med* **35**: 346–55

Hajnal JV, Mayers R, Oatridge A *et al.* (1994) Artifacts due to stimulus correlated motion in functional imaging of the brain. *Mag Res Med* **31**: 289–91

Hill DLG, Batchelor PG, Holden M *et al.* (2001) Medical image registration. *Phys Med Biol* **46**: R1–R45

Holden M, Hill DLG, Denton ERE *et al.* (2000) Voxel similarity measures for 3-D serial MR brain image registration. *IEEE Trans Med Imag* **19**(2): 94–102

Jezzard P (2000) *Handbook of medical imaging*, Chap. 26. Academic Press, San Diego, pp 425–38

Jezzard P, Clare S (1999) Sources of distortion in functional MRI data. *Hum Brain Mapp* **8**(2): 80–85

Kybic J, Theévenaz P, Nirkko A *et al.* (2000) Unwarping of unidirectionally distorted EPI images. *IEEE Trans Med Imag* **19**(2): 80–93

Likar B, Pernuš F (2001) A heirarchical approach to elastic registration based on mutual information. *Image Vision Comput* **19**: 33–44

Maes F, Collignon A, Vandermeulen D *et al.* (1997) Multimodality image registration by maximisation of mutual information. *IEEE Trans Med Imag* **16**: 187–97

Noll DC, Boada FE, Eddy WF (1997) A spectral approach to analyzing slice selection in planar imaging: optimization for through-plane interpolation. *Mag Res Med* **38**: 151–60

Pelizzari CA, Chen GTY, Spelbring DR *et al.* (1988) Accurate three-dimensional registration of CT, PET and MR images of the brain. *Comput Assist Tomogr* **13**: 20–26

Press WH, Teukolsky SA, Vetterling WT *et al.* (1992) *Numerical Recipes in C*, 2nd edn. Cambridge University Press, Cambridge

Studholme C, Constable RT, Duncan JS (2000) Accurate alignment of functional EPI data to anatomical MRI using a physics-based distortion model. *IEEE Trans Med Imag* **19**(11): 1115–27

Studholme C, Hill DLG, Hawkes DJ (1995) Multiresolution voxel similarity measures for MR-PET coregistration. In *Proc Information Processing in Medical Imaging (IPMI)*, Bizais Y, Barillot C, Di Paola R (eds). Kluwer Academic Publishers, Dordrecht.

Studholme C, Hill DLG, Hawkes DJ (1999) An overlap invariant entropy measure of 3D medical image alignment. *Pattern Recognit* **32**: 71–86

Talairach J, Tournoux P (1988) *Coplanar stereotaxic atlas of the human brain*. Thieme Medical, New York

Thévenaz P, Blu T, Unser M (2000) Interpolation revisited. *IEEE Trans Med Imag* **19**(7): 739–58

Thévenaz P, Unser M (2000) Optimization of mutual information for multiresolution image registration. *IEEE Trans Image Process* **9**(12): 2083–99

Unser M, Aldroubi A, Eden M (1993a) B-spline signal processing: Part I – theory. *IEEE Trans Signal Process* **41**(2): 821–33

Unser M, Aldroubi A, Eden M (1993b) B-spline signal processing: Part II – efficient design and applications. *IEEE Transact Signal Process* **41**(2): 834–48

Wells WM III, Viola P, Atsumi H *et al.* (1996) Multi-modal volume registration by maximisation of mutual information. *Med Image Anal* **1**(1): 35–51

Woods RP (1999) *Brain Warping*, chap. 20. Academic Press, San Diego, pp 365–76

Woods RP (2000) *Handbook of Medical Imaging*, chap. 33. Academic Press, San Diego, pp 529–36

Woods RP, Grafton ST, Holmes CJ *et al.* (1998) Automated image registration: I. General methods and intrasubject, intramodality validation. *Comput Assist Tomogr* **22**: 139–52

Woods RP, Mazziotta JC, Cherry SR (1993) MRI-PET registration with automated algorithm. *Comput Assist Tomogr* **17**: 536–46

Zuk TD, Atkins MS (1996) A comparison of manual and automatic methods for registering scans of the head. *IEEE Trans Med Imag* **15**(5): 732–44

CHAPTER

5

Non-linear Registration

J. Ashburner and K. Friston

INTRODUCTION

This chapter provides an overview of the ideas underlying non-linear image registration and explains the principles behind the spatial normalization in the statistical parametric mapping (SPM) software. The previous chapter described rigid body approaches for registering brain images of the same subject, which assume that there are no differences among the shapes of the brains. This is often a reasonable assumption to make for intra-subject registration, but this model is not appropriate for matching brain images from different subjects. In addition to estimating an unknown pose and position, inter-subject registration approaches also need to model the different shapes of the subjects' heads or brains.

There are also cases when within subject registration of brain images may need to account for different shapes. This can be because there have been real changes in a brain's anatomy over time. Such changes can arise as a result of growth, ageing, disease and surgical intervention. There can also be shape differences that are purely due to artefacts in the images. For example, functional magnetic resonance imaging (fMRI) data usually contain quite large distortions (Jezzard and Clare, 1999), which means that accurate intra-subject alignment with an anatomical image can be problematic. In an fMRI study, if a match between anatomical and functional images is poor, then this will lead to mis-localization of brain activity. Interactions between image distortion and the orientation of a subject's head in the scanner can also cause other problems because purely rigid body alignment does not take this into account (Andersson *et al.*, 2001).

The main application for non-linear image registration within the SPM software is *spatial normalization*. This involves warping images from a number of individuals into roughly the same standard space to allow signal averaging across subjects. In functional imaging studies, spatial normalization is useful for determining what happens generically over individuals. A further advantage of using spatially normalized images is that activation sites can be reported according to their Euclidean coordinates within a standard space (Fox, 1995). The most commonly adopted coordinate system within the brain imaging community is that described by Talairach and Tournoux (1988), although new standards are now emerging that are based on digital atlases (Evans *et al.*, 1993, 1994; Mazziotta *et al.*, 1995).

Another application for non-linear registration is for assisting in image segmentation and for generating individualized brain atlases (Collins *et al.*, 1995; Tzourio-Mazoyer *et al.*, 2002). If a template image is warped to match an individual's brain image, then other data that are in alignment with the template can be overlayed on to that image. These additional data can be predefined labels or probability maps. The next chapter says more on this subject.

Sometimes, the actual shape differences among brains are of interest in their own right. There is now an enormous literature on comparing anatomy (Dryden and Mardia, 1998; Kendall *et al.*, 1999; Miller, 2004). Most of these approaches require some sort of representation of the relative shapes of the brains, which can be derived using non-linear registration.

Methods of registering images can be broadly divided into *label based* and *intensity based*. Label based techniques identify homologous features (labels) in the source and reference images and find the transformations that best superpose them. The labels can be points, lines or surfaces. Homologous features are often identified manually, but this process is time-consuming and subjective. Another disadvantage of using points as landmarks is that there are very few readily identifiable discrete points in the brain. Lines and surfaces are more readily identified, and in many instances they can be extracted automatically (or at least semiautomatically). Once they are

Statistical Parametric Mapping, by Karl Friston et al.
ISBN–13: 978-0-12-372560-8 ISBN–10: 0-12-372560-7

identified, the spatial transformation is effected by bringing the homologies together. If the labels are points, then the required transformations at each of those points is known. Between the points, the deforming behaviour is not known, so it is forced to be as 'smooth' as possible. There are a number of methods for modelling this smoothness. The simplest models include fitting splines through the points in order to minimize *bending energy* (Bookstein, 1989, 1997). More complex forms of interpolation, such as viscous fluid models, are often used when the labels are surfaces (Davatzikos, 1996; Thompson and Toga, 1996).

Intensity (non-label) based approaches identify a spatial transformation that optimizes some voxel-similarity measure between a source and reference image, where both are treated as unlabelled continuous processes. The matching criterion is often based upon minimizing the sum of squared differences or maximizing the correlation between the images. For this criterion to be successful, it requires the reference to appear like a warped version of the source image. In other words, there must be correspondence in the grey levels of the different tissue types between the source and reference images. In order to warp together images of different modalities, a few intensity based methods have been devised that involve optimizing an information theoretic measure (Studholme *et al.*, 2000; Thévenaz and Unser, 2000).

Intensity matching methods are usually very susceptible to poor starting estimates, so more recently a number of hybrid approaches have emerged that combine intensity based methods with matching user defined features (typically sulci). Registration methods usually search for the single most probable realization of all possible transformations. Robust optimization methods that almost always find the global optimum would take an extremely long time to run with a model that uses millions of parameters. These methods are simply not feasible for problems of this scale. However, if sulci and gyri can be easily labelled from the brain images, then robust methods can be applied in order to match the labelled features. Robust methods become more practical when the amount of information is reduced to a few key features. The robust match can then be used to bias the registration (Joshi *et al.*, 1995; Davatzikos, 1996; Thompson and Toga, 1996), therefore increasing the likelihood of obtaining the global optimum.

The next section of this chapter will discuss *objective functions*, which are a measure of how well images are registered. Registering images involves estimating some mapping from the domain of one image to the range of another, where the measure of 'goodness' is the objective function. The mapping is parameterized in some way, and this is the subject of the *deformation models* section. A procedure for estimating the optimal parameters is introduced in *estimating the mappings*, along with some strategies for improving the internal consistency of the registration models. Then the *spatial normalization in the SPM software* is described briefly, before finishing with a discussion about *evaluation strategies* for non-linear registration approaches.

OBJECTIVE FUNCTIONS

Image registration procedures use a mathematical model to explain the data. Such a model will contain a number of unknown parameters that describe how an image is deformed. The objective is usually to determine the single 'best' set of values for these parameters. The measure of 'goodness' is known as the *objective function*. The aim of the registration is usually to find the most probable deformation, given the data. In such cases, the objective function is a measure of this probability. A key element of probability theory is Bayes' theorem, which states:

$$P(\theta, D) = P(\theta|D)P(D) = P(D|\theta)P(\theta) \qquad 5.1$$

This can be rearranged to give:

$$P(\theta|D) = \frac{P(D|\theta)P(\theta)}{P(D)} \qquad 5.2$$

This *posterior probability* of the parameters given the image data ($P(\theta|D)$) is proportional to the probability of the image data given the parameters ($P(D|\theta)$ – the *likelihood*), times the *prior probability* of the parameters ($P(\theta)$). The probability of the data ($P(D)$) is a constant. The objective is to find the most probable parameter values, and not the actual probability density, so this factor is ignored. Bayes' theorem is illustrated in Plate 4 (see colour plate section).

The single most probable estimate of the parameters is known as the *maximum a posteriori* (MAP) estimate.[1] There is a monotonic relationship between a value and its logarithm. In practice, the objective function is normally the logarithm of the posterior probability (in which case it is maximized) or the negative logarithm (which is minimized). It can therefore be considered as the sum of two terms: a likelihood term, and a prior term.

$$-\log P(\theta, D) = -\log P(D|\theta) - \log P(\theta) \qquad 5.3$$

[1] True Bayesians realize that such a point estimate is slightly arbitrary as the mode of the probability density may change if the parameterization is changed, but it is a commonly accepted approach within the image registration field.

Likelihood term

The likelihood term is a measure of the probability of observing an image given some set of model parameters. A simple example would be where an image is modelled as a warped version of a template image, but with Gaussian random noise added. We denote the intensity of the ith voxel of the image by g_i, the parameters by $\boldsymbol{\alpha}$ (a column vector of M elements), and the intensity of the ith voxel of the deformed template image by $f_i(\boldsymbol{\alpha})$. If the variance of the Gaussian noise is σ^2, then the probability of observing g_i given the parameters is:

$$P(g_i|\boldsymbol{\alpha}) = (2\pi\sigma^2)^{-\frac{1}{2}} \exp\left(-\frac{(g_i - f_i(\boldsymbol{\alpha}))^2}{2\sigma^2}\right) \qquad 5.4$$

If the Gaussian noise at each voxel is identically and independently distributed (IID), then the probability of the whole image $\mathbf{g}$ is obtained from the product of the probabilities at each of the I voxels.

$$P(\mathbf{g}|\boldsymbol{\alpha}) = \prod_{i=1}^{I} (2\pi\sigma^2)^{-\frac{1}{2}} \exp\left(-\frac{(g_i - f_i(\boldsymbol{\alpha}))^2}{2\sigma^2}\right) \qquad 5.5$$

Therefore, the negative log-likelihood is:

$$-\log P(\mathbf{g}|\boldsymbol{\alpha}) = \frac{I}{2}\log(2\pi\sigma^2) + \sum_{i=1}^{I} \frac{(g_i - f_i(\boldsymbol{\alpha}))^2}{2\sigma^2} \qquad 5.6$$

If I and σ^2 are constant, then maximizing $P(\mathbf{g}|\boldsymbol{\alpha})$ can be achieved by minimizing the following sum of squared difference:

$$\mathcal{E}(\boldsymbol{\alpha}) = \frac{1}{2\sigma^2}\sum_{i=1}^{I} (g_i - f_i(\boldsymbol{\alpha}))^2 \qquad 5.7$$

There is often a slight complication though, and that is that I may not be constant. It is possible that parts of $\mathbf{g}$ correspond to a region that falls outside the field of view of $\mathbf{f}$ (i.e. $f_i(\boldsymbol{\alpha})$ is not defined at certain voxels). Different parameter estimates result in different deformations and, hence, different numbers of voxels without corresponding regions in the template. For this reason, it is common simply to minimize the mean squared difference instead. The motivation for this approach is that it assumes the residuals that should arise from the missing data are drawn from the same distribution as those that can be computed. Another issue is that the residual differences are rarely IID in practice, as there are usually correlations among neighbouring voxels. If this non-sphericity is not taken into consideration, then the model will overestimate the likelihood.

The mean-squared difference objective function makes a number of assumptions about the data. If the data do not meet these assumptions then the probabilities may not accurately reflect the goodness of fit, and the estimated deformations will be poor. In particular, it assumes that the image resembles a warped version of the template. Under some circumstances, it may be better to model spatially varying variances, which would effectively weight different regions to a greater or lesser extent. For example, if matching a template to a brain image containing a lesion, then the mean squared difference around the lesion should contribute little or nothing to the objective function (Brett *et al.*, 2001). This is achieved by assigned lower weights (higher σ^2) in these regions, so that they have much less influence on the final solution.

In addition to modelling non-linear deformations of the template, there may also be additional parameters within the model that describe intensity variability.[2] A very simple example would be the inclusion of an additional intensity scaling parameter, but the models can be much more complicated. There are many possible objective functions, each making a different assumption about the data and requiring different parameterizations of the template intensity distribution. These include the information theoretic objective functions mentioned in Chapter 4, as well as the objective function described in Chapter 6. There is no single universally best criterion to use for all data.

Prior term

This term reflects the prior probability of a deformation occurring – effectively biasing the deformations to be realistic. If one considers a model whereby each voxel can move independently in three dimensions, then there would be three times as many parameters to estimate as there are observations. This would simply not be achievable without *regularizing* the parameter estimation by modelling a prior probability.

The prior term is generally based on some measure of deformation smoothness. Smoother deformations are deemed to be more probable – *a priori* – than deformations containing a great deal of detailed information. Usually, the model parameters are such that they can be assumed to be drawn from a multivariate Gaussian distribution. If the mean of the distribution of the parameters $\boldsymbol{\alpha}$ (a column vector of length M) is $\boldsymbol{\alpha}_0$ and the covariance matrix describing the distribution is $\mathbf{C}_\alpha$, then:

$$P(\boldsymbol{\alpha}) = (2\pi)^{-\frac{M}{2}} |\mathbf{C}_\alpha|^{-\frac{1}{2}} \exp\left(-\frac{1}{2}(\boldsymbol{\alpha}-\boldsymbol{\alpha}_0)^T \mathbf{C}_\alpha^{-1}(\boldsymbol{\alpha}-\boldsymbol{\alpha}_0)\right) \qquad 5.8$$

[2] Ideally, the likelihood would be marginalized with respect to these additional parameters, but this is usually intractable in practice.

By taking the negative logarithm of this probability, we obtain an expression that can be compared with that of Eqn. **5.6**.

$$-\log P(\boldsymbol{\alpha}) = \frac{M}{2}\log(2\pi) + \frac{1}{2}\log|\mathbf{C}_\alpha| + \frac{1}{2}(\boldsymbol{\alpha}-\boldsymbol{\alpha}_0)^T \mathbf{C}_\alpha^{-1}(\boldsymbol{\alpha}-\boldsymbol{\alpha}_0) \qquad \textbf{5.9}$$

In most implementations, the mean of the probability distribution is zero, and the inverse of the covariance matrix has a simple numerical form. The expression $\frac{1}{2}\boldsymbol{\alpha}^T\mathbf{C}_\alpha^{-1}\boldsymbol{\alpha}$ is often thought of as an 'energy density' ($\mathcal{H}(\boldsymbol{\alpha})$). Commonly used forms for this are the *membrane energy* (Amit *et al.*, 1991; Gee *et al.*, 1997b), *bending energy* (Bookstein, 1997) or *linear-elastic energy* (Miller *et al.*, 1993; Christensen *et al.*, 1996a; Davatzikos, 1996). The form of the prior used by the registration will influence the estimated deformations. This is illustrated by Figure 5.1.

The simplest model used for linear regularization is based upon minimizing the membrane energy of a vector function $\mathbf{u}(\mathbf{x}, \boldsymbol{\alpha})$ over the domain of the image (see next section). If $\mathbf{u}$ is a linear function of $\boldsymbol{\alpha}$, then this can be represented by $\frac{1}{2}\boldsymbol{\alpha}^T\mathbf{C}_\alpha^{-1}\boldsymbol{\alpha} = \mathbf{u}^T\mathbf{L}^T\mathbf{L}\mathbf{u}$, where $\mathbf{L}$ is a matrix of differential operators. The membrane energy model is also known as the Laplacian model.

$$\mathcal{H}(\boldsymbol{\alpha}) = \frac{\lambda}{2}\int_{x\in\Omega}\sum_{i=1}^{3}\sum_{j=1}^{3}\left(\frac{\partial u_i(\mathbf{x},\boldsymbol{\alpha})}{\partial x_j}\right)^2 d\mathbf{x} \qquad \textbf{5.10}$$

The bending energy (biharmonic or thin plate model) would be given by:

$$\mathcal{H}(\boldsymbol{\alpha}) = \frac{\lambda}{2}\int_{x\in\Omega}\sum_{i=1}^{3}\sum_{j=1}^{3}\sum_{k=1}^{3}\left(\frac{\partial^2 u_i(\mathbf{x},\boldsymbol{\alpha})}{\partial x_j \partial x_k}\right)^2 d\mathbf{x} \qquad \textbf{5.11}$$

The linear elastic energy is given by:

$$\mathcal{H}(\boldsymbol{\alpha}) = \frac{1}{2}\int_{x\in\Omega}\sum_{j=1}^{3}\sum_{k=1}^{3}\left(\lambda\left(\frac{\partial u_j(\mathbf{x},\boldsymbol{\alpha})}{\partial x_j}\right)\left(\frac{\partial u_k(\mathbf{sx},\boldsymbol{\alpha})}{\partial x_k}\right) + \frac{\mu}{2}\left(\frac{\partial u_j(\mathbf{x},\boldsymbol{\alpha})}{\partial x_k} + \frac{\partial u_k(\mathbf{x},\boldsymbol{\alpha})}{\partial x_j}\right)^2\right) d\mathbf{x} \qquad \textbf{5.12}$$

In reality, the true amount of anatomical variability is very likely to differ from region to region (Lester *et al.*, 1999), so a non-stationary prior probability model could, in theory, produce more accurate estimates. If the true prior probability distribution of the parameters is known (somehow derived from a large number of subjects), then $\mathbf{C}_\alpha$ could be an empirically determined covariance matrix describing this distribution. This approach would have the advantage that the resulting deformations are more typically 'brain like', and so increase the face validity of the approach.

In principle, the hyperparameters (e.g. λ and σ^2) of the registration model could be estimated empirically using *restricted maximum likelihood* (ReML – also known as *type II maximum likelihood, ML-II* or *parametric empirical*

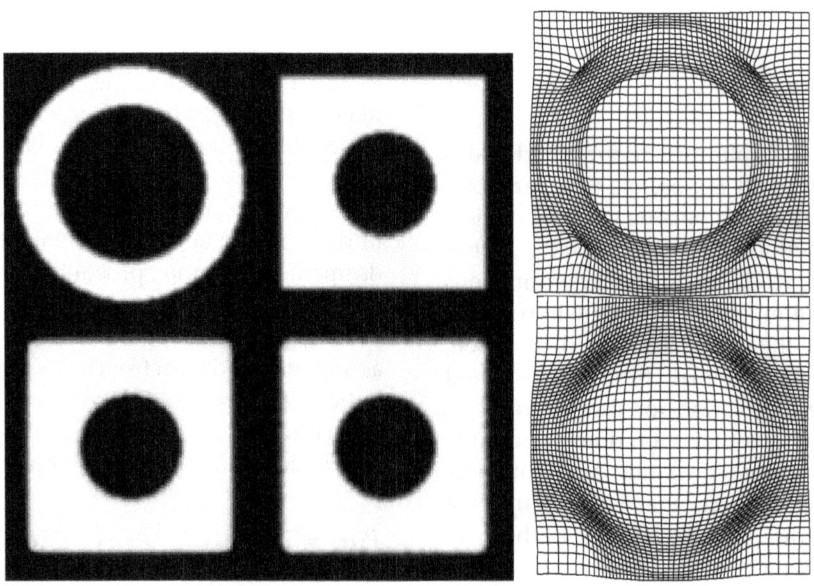

FIGURE 5.1 This figure illustrates the effect of different types of regularization. The top row on the left shows simulated 2D images of a circle and a square. Below these is the circle after it has been warped to match the square, using both membrane and bending energy priors. These warped images are almost visually indistinguishable, but the resulting deformation fields using these different priors are quite different. These are shown on the right, with the deformation generated with the membrane energy prior shown above the deformation that used the bending energy prior.

Bayes) with *Laplace approximation* schemes similar to those mentioned in the later chapters. In practice, however, values for these hyperparameters are often imputed in a way that depends upon a subjective belief in what an optimal trade-off between the likelihood and prior terms should be.

Deformation models

At its simplest, image registration involves estimating a smooth, continuous mapping between the points in one image, and those in another. This mapping allows one image to be re-sampled so that it is warped (deformed) to match another (Figures 5.2 and 5.3). There are many ways of modelling such mappings, but these fit into two broad categories of parameterization (Miller *et al.*, 1997).

- The *small deformation* framework does not necessarily preserve topology[3] – although if the deformations are relatively small, then it may still be preserved.
- The *large deformation* framework generates deformations (*diffeomorphisms*) that have a number of elegant mathematical properties, such as enforcing the preservation of topology.

Both of these approaches require some model of a smooth vector field. Such models will be illustrated with the simpler small deformation framework, before briefly introducing the principles that underlie the large deformation framework.

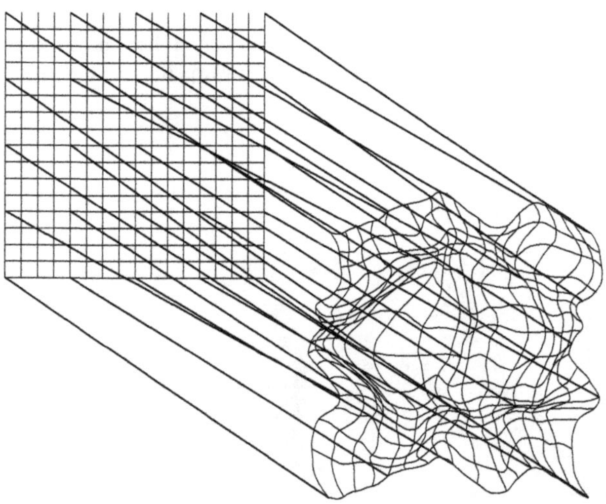

FIGURE 5.2 This figure illustrates a hypothetical deformation field that maps between points in one image and those in another. This is a continuous function over the domain of the image.

[3] The word 'topology' is used in the same sense as in 'Topological properties of smooth anatomical maps' (Christensen *et al.*, 1995). If spatial transformations are not one-to-one and continuous, then the topological properties of different structures can change.

Small deformation approaches

First of all, some simple notation is introduced. Each coordinate within an image is a vector of three elements. For example, the coordinate of the ith voxel could be denoted by $\mathbf{x}_i = [x_{i1}\ x_{i2}\ x_{i3}]$. For an image $\mathbf{f}$, the ith voxel may be indicated by f_i or by $f(\mathbf{x}_i)$. Similarly, a point in a deformation field can also be denoted as a vector. The ith voxel of an image deformed this way could be denoted by $g_i(\boldsymbol{\alpha})$ or $g(\mathbf{y}(\mathbf{x}_i, \boldsymbol{\alpha}))$. The images are treated as continuous functions of space. Reading off the value at some arbitrary point involves interpolating between the original voxels. For many interpolation methods, the functions are parameterized by linear combinations of basis functions, such as B-spline bases, centred at each original voxel. Similarly, the deformations themselves can be parameterized by a linear combination of smooth, continuous basis functions.

$$\mathbf{y}(\mathbf{x}_i, \boldsymbol{\alpha}) = \sum_{m=1}^{M} \alpha_m \boldsymbol{\phi}_m(\mathbf{x}_i) \qquad 5.13$$

Most models treat the basis functions as scalar fields representing deformations in the three orthogonal directions[4] (Figure 5.4) such that:

$$\begin{aligned} y_1(\mathbf{x}_i, \boldsymbol{\alpha}) &= \sum_{m=1}^{M} \alpha_{m1} \phi_{m1}(\mathbf{x}_i) \\ y_2(\mathbf{x}_i, \boldsymbol{\alpha}) &= \sum_{m=1}^{M} \alpha_{m2} \phi_{m2}(\mathbf{x}_i) \\ y_3(\mathbf{x}_i, \boldsymbol{\alpha}) &= \sum_{m=1}^{M} \alpha_{m3} \phi_{m3}(\mathbf{x}_i) \end{aligned} \qquad 5.14$$

A potentially enormous number of parameters are required to describe the non-linear transformations that warp two images together (i.e. the problem can be very high-dimensional). However, much of the spatial variability can be captured using just a few parameters. Sometimes only an affine transformation is used to register approximately images of different subjects. This accounts for differences in position, orientation and overall brain dimensions (Figure 5.5), and often provides a good starting point for higher-dimensional registration models. The basis functions for such a transform are

[4] Exceptions are some models that use linear elastic regularization (Christensen *et al.*, 1996a).

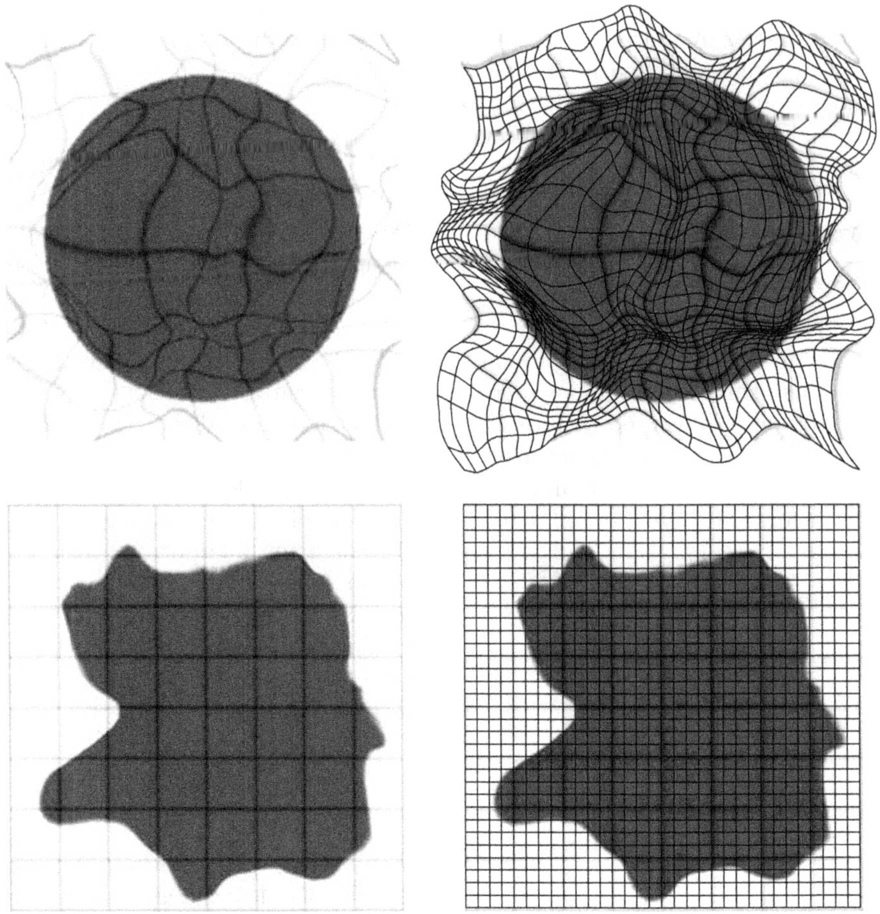

FIGURE 5.3 This figure illustrates a deformation field that brings the top left image into alignment with the bottom left image. At the top right is the image with the deformation field overlayed, and at the bottom right is this image after it has been warped. Note that in this example, the deformation wraps around at the boundaries.

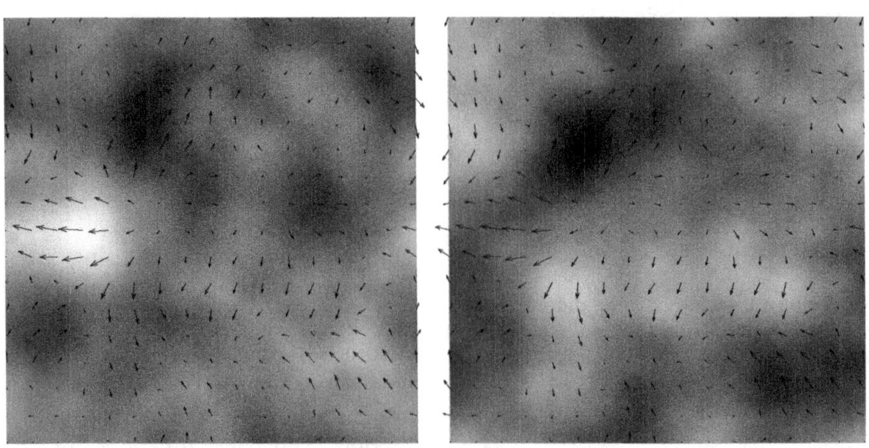

FIGURE 5.4 This figure illustrates 2D displacements generated from two scalar fields. On the left are the horizontal displacements, where dark areas indicate displacements to the right, and light areas indicate displacements to the left. Vertical displacements are shown on the right, with dark indicating upward displacements and light indicating downward. These displacement fields are modelled by linear combinations of basis functions. The superimposed arrows show the combined directions of displacement.

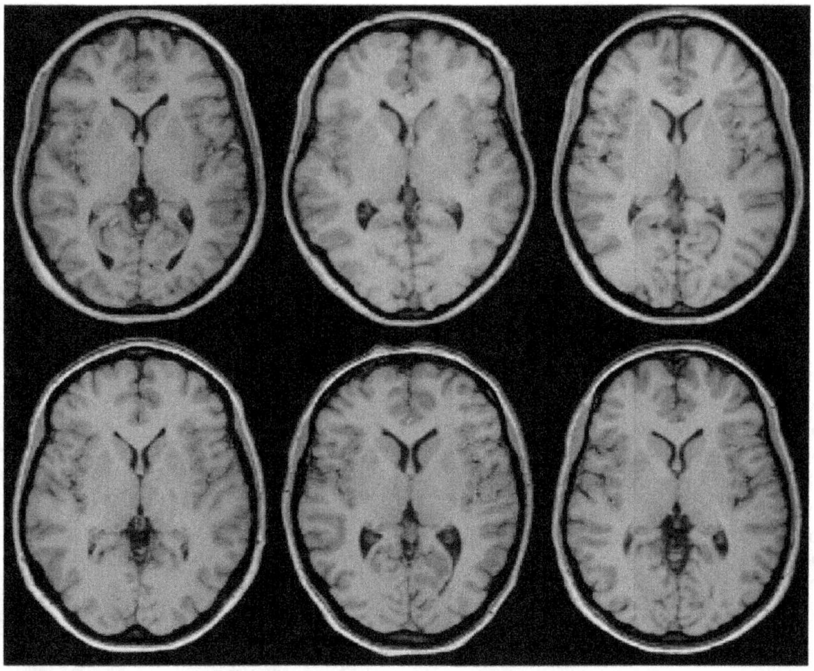

FIGURE 5.5 Images of six subjects registered using a 12-parameter affine registration (see also Figure 5.9). The affine registration matches the positions and sizes of the images.

coordinates themselves, leading to the following parameterization in 3D:

$$\begin{aligned} y_1(\mathbf{x}, \boldsymbol{\alpha}) &= \alpha_1 x_1 + \alpha_2 x_2 + \alpha_3 x_3 + \alpha_4 \\ y_2(\mathbf{x}, \boldsymbol{\alpha}) &= \alpha_5 x_1 + \alpha_6 x_2 + \alpha_7 x_3 + \alpha_8 \\ y_3(\mathbf{x}, \boldsymbol{\alpha}) &= \alpha_9 x_1 + \alpha_{10} x_2 + \alpha_{11} x_3 + \alpha_{12} \end{aligned} \quad 5.15$$

Low spatial frequency global variability of head shape can be accommodated by describing deformations by a linear combination of a few low frequency basis functions. One widely used basis function registration method is part of the AIR package (Woods *et al.*, 1998a,b), which uses polynomial basis functions (Figure 5.6) to model shape variability. These basis functions are a simple extension to those used for parameterizing affine transformations. For example, a two-dimensional third order polynomial mapping is:

$$\begin{aligned} y_1(\mathbf{x}, \boldsymbol{\alpha}) = \alpha_1 &+ \alpha_2 x_1 + \alpha_3 x_1^2 + \alpha_4 x_1^3 \\ \alpha_5 x_2 &+ \alpha_6 x_1 x_2 + \alpha_7 x_1^2 x_2 \\ \alpha_8 x_2^2 &+ \alpha_9 x_1 x_2^2 \\ \alpha_{10} x_2^3 & \\ y_2(\mathbf{x}, \boldsymbol{\alpha}) = \alpha_{11} &+ \alpha_{12} x_1 + \alpha_{13} x_1^2 + \alpha_{14} x_1^3 \\ \alpha_{15} x_2 &+ \alpha_{16} x_1 x_2 + \alpha_{17} x_1^2 x_2 \\ \alpha_{18} x_2^2 &+ \alpha_{19} x_1 x_2^2 \\ \alpha_{20} x_2^3 & \end{aligned} \quad 5.16$$

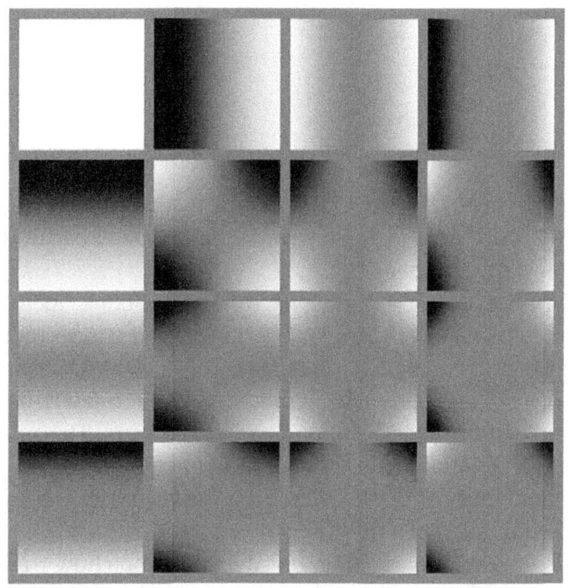

FIGURE 5.6 Polynomial basis functions. Note that additional basis functions not shown in Eqn. **5.16** are included here.

Other models parameterize a displacement field, which is added to an identity transform:

$$\mathbf{y}(\mathbf{x}_i, \boldsymbol{\alpha}) = \mathbf{x}_i + \sum_{m=1}^{M} \alpha_m \boldsymbol{\phi}_m(\mathbf{x}_i) \quad 5.17$$

In such parameterizations, the inverse transformation is sometimes approximated by subtracting the

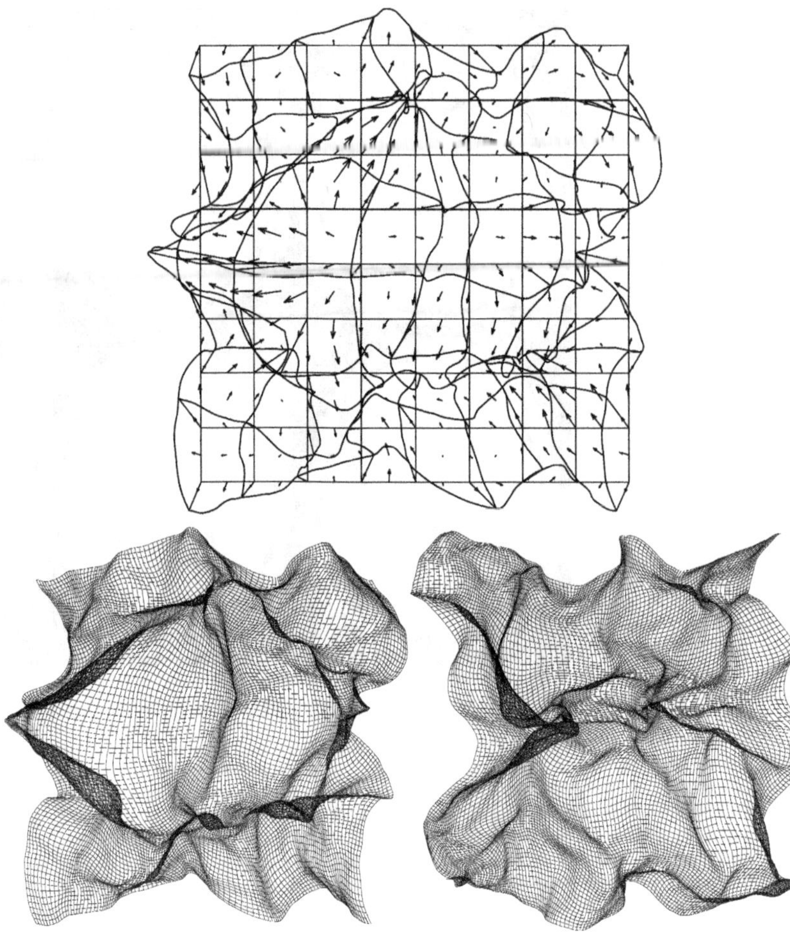

FIGURE 5.7 This figure illustrates the small deformation setting. At the top is an illustration of a displacement added to an identity transform. Below this is a forward and inverse deformation generated within the small deformation setting. Note that the one-to-one mapping is lost because the displacements are too large, and the mappings are not accurate inverses of each other.

displacement (Figure 5.7). It is worth noting that this is only a very approximate inverse, which fails badly for larger deformations.

Families of basis functions for such models include Fourier bases (Christensen, 1999), sine and cosine transform basis functions (Christensen, 1994; Ashburner and Friston, 1999) (Figures 5.8 and 5.9). These models usually use in the order of about 1000 parameters. The small number of parameters will not allow every feature to be matched exactly, but it will permit the global head shape to be modelled rapidly.

The choice of basis functions depends upon how translations at borders should behave (i.e. the *boundary conditions*). If points at the borders over which the transform is computed are not required to move in any direction, then the basis functions should consist of the lowest frequencies of the three-dimensional sine transform. If the borders are allowed to move freely, then a three-dimensional cosine transform is more appropriate. Fourier transform basis functions could be used if the displacements are to wrap around (*circulant* boundary conditions). All these transforms use the same set of basis functions to represent warps in each of the directions. Alternatively, a

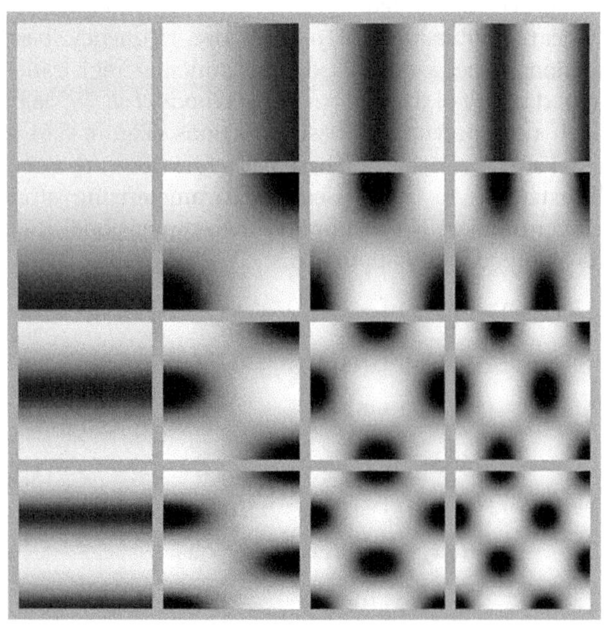

FIGURE 5.8 Cosine transform basis functions.

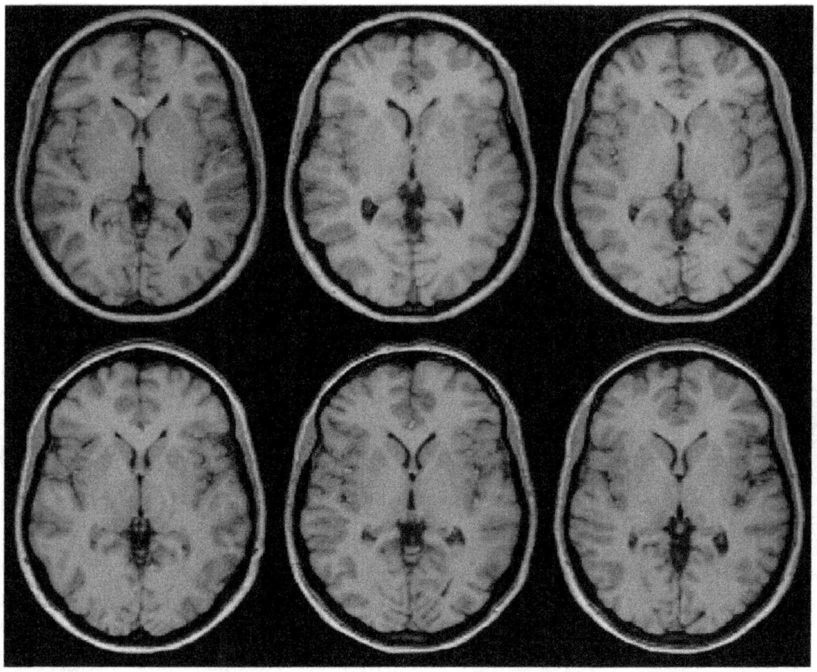

FIGURE 5.9 Six subjects' brains registered with both affine and cosine transform basis function registration (see also Figure 5.8), as implemented by SPM. The basis function registration estimates the global shapes of the brains, but is not able to account for high spatial frequency warps.

mixture of cosine and sine transform basis functions can be used to constrain translations at the surfaces of the volume to be parallel to the surface only (*sliding* boundary conditions). By using a different combination of basis function sets, the corners of the volume can be fixed and the remaining points on the surface can be free to move in all directions (*bending* boundary conditions) (Christensen, 1994). These various boundary conditions are illustrated in Figure 5.10.

Radial basis functions are another family of parameterizations which are often used in conjunction with an affine transformation. Each radial basis function is centred at some point, and the amplitude is then a function of the distance from that point. Thin-plate splines are one of the most widely used radial basis functions for image warping, and are especially suited to manual landmark matching (Bookstein, 1989, 1997; Glasbey and Mardia, 1998). The landmarks may be known, but interpolation is needed in order to define the mapping between these known points. By modelling it with thin-plate splines, the mapping function has the smallest bending energy. Other choices of basis function reduce other energy measures, and these functions relate to the convolution filters that are sometimes used for fast image matching (Bro-Nielsen and Gramkow, 1996; Beg *et al.*, 2005).

B-spline bases are also used for parameterizing displacements (Studholme *et al.*, 2000; Thévenaz and Unser, 2000) (Figure 5.11). They are related to the radial basis functions in that they are centred at discrete points, but the amplitude is the product of functions of distance in the three orthogonal directions (i.e. they are separable). The separability and local support of these basis functions

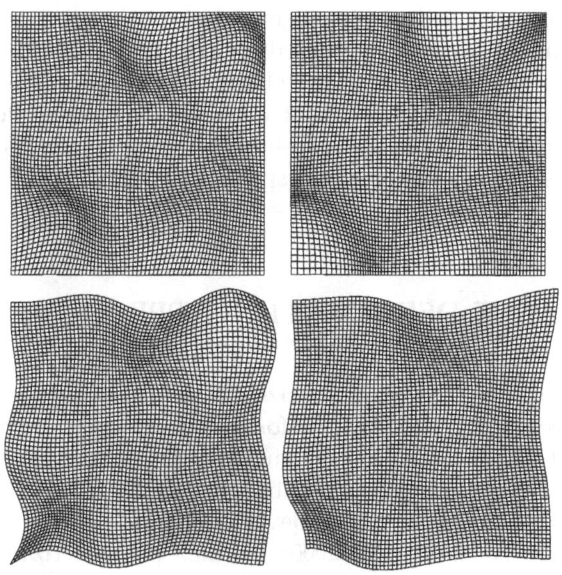

FIGURE 5.10 Different boundary conditions. Above left: fixed boundaries (generated purely from sine transform basis functions). Above right: sliding boundaries (from a mixture of cosine and sine basis functions). Below left: bending boundaries (from a different mixture of cosine and sine basis functions). Below right: free boundary conditions (purely from cosine basis functions).

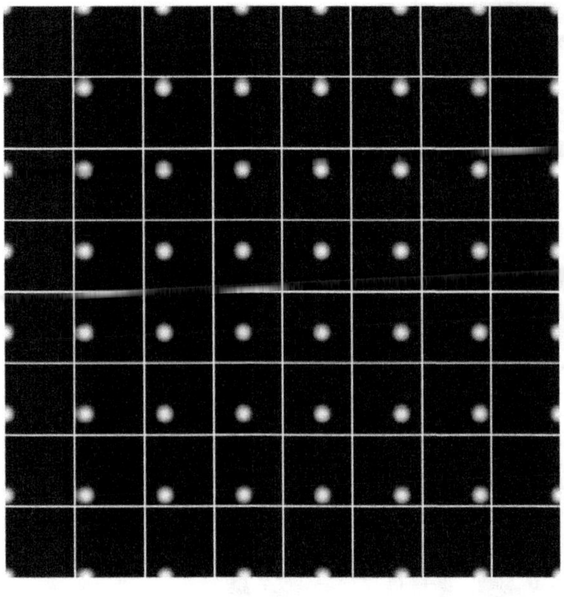

FIGURE 5.11 B-spline basis functions.

confers certain advantages in terms of being able rapidly to generate displacement fields through a convolution-like procedure.

Very detailed displacement fields can be generated by modelling an individual displacement at each voxel. These may not appear to be basis function approaches, but the assumptions within the registration models often assume that the fields are tri-linearly interpolated. This is the same as a first degree B-spline basis function model.

LARGE DEFORMATION APPROACHES

Approaches that parameterize the deformations themselves do not necessarily enforce a one-to-one mapping, particularly if the prior is a multivariate Gaussian probability density. Small deformation models that incorporate this constraint have been devised (Edwards *et al.*, 1998; Ashburner *et al.*, 1999, 2000), but their parameterization is still essentially within a small deformation setting.

The key element of the large-deformation or *diffeomorphic*[5] setting is that the deformations are generated by the composition of a number of small deformations (i.e. warped warps). A composition of two functions is essentially taking one function of the other in order to produce a new function. For two functions, $\mathbf{y}_2$ and $\mathbf{y}_1$ this would be denoted by:

$$(\mathbf{y}_2 \circ \mathbf{y}_1)(\mathbf{x}) = \mathbf{y}_2(\mathbf{y}_1(\mathbf{x})) \quad \textbf{5.18}$$

For deformations, the composition operation is achieved by re-sampling one deformation field by another. Providing the original deformations are small enough, then they are likely to be one-to-one. A composition of a pair of one-to-one mappings will produce a new mapping that is also one-to-one. Multiple nesting can also be achieved, so that large one-to-one deformations can be obtained from the composition of many very small deformations.

$$\begin{aligned}(\mathbf{y}_3 \circ \mathbf{y}_2 \circ \mathbf{y}_1)(\mathbf{x}) &= ((\mathbf{y}_3 \circ \mathbf{y}_2) \circ \mathbf{y}_1)(\mathbf{x}) \\ &= (\mathbf{y}_3 \circ (\mathbf{y}_2 \circ \mathbf{y}_1))(\mathbf{x}) = \mathbf{y}_3(\mathbf{y}_2(\mathbf{y}_1(\mathbf{x}))) \quad \textbf{5.19}\end{aligned}$$

The early diffeomorphic registration approaches were based on the *greedy* 'viscous fluid' registration method of Christensen and Miller (Christensen *et al.*, 1994, 1996b). In these models, finite difference methods are used to solve the partial differential equations that model one image as it 'flows' to match the shape of the other. At the time, the advantage of these methods was that they were able to account for large displacements while ensuring that the topology of the warped image is preserved. They also provided a useful foundation from which the later methods arose. Viscous fluid methods require the solutions to large sets of partial differential equations (see Chapter 19 of Press *et al.*, 1992). The earliest implementations were computationally expensive because solving the equations used successive over-relaxation. Such relaxation methods are inefficient when there are large low frequency components to estimate. Since then, a number of faster ways of solving the differential equations have been devised (Modersitzki, 2003). These include the use of Fourier transforms to convolve with the impulse response of the linear regularization operator (Bro-Nielsen and Gramkow, 1996), or by convolving with a separable approximation (Thirion, 1995). Another fast way of solving such equations would be to use a multigrid method (Haber and Modersitzki, 2006), which efficiently makes use of relaxation methods over various spatial scales.

The greedy algorithm works in several stages. In the first stage, a heavily regularized small deformation ($\mathbf{y}_1$) would be estimated that brings one image f into slightly better correspondence with the other g. A deformed version of this image would then be created by $f_1 = f(\mathbf{y}_1)$. Then another small deformation is estimated ($\mathbf{y}_2$) by matching f_1 with g, and a second deformed version of f created by $f_2 = f(\mathbf{y}_2 \circ \mathbf{y}_1)$. This procedure would be repeated, each time generating a warped version of f that is closer to g. Although the likelihood will be increased

[5] A diffeomorphism is a globally one-to-one (bijective) smooth and continuous mapping with derivatives that are invertible (i.e. non-zero Jacobian determinant).

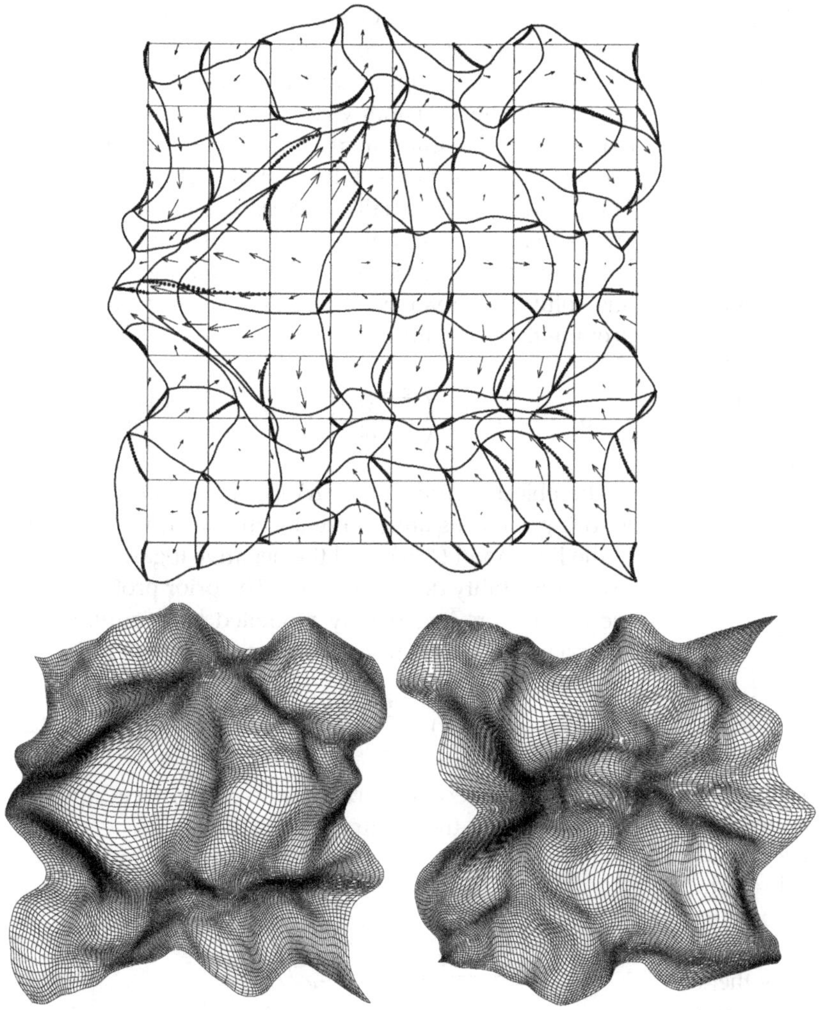

FIGURE 5.12 This figure illustrates the large deformation setting. In this setting, the deformations follow a curved trajectory (just as a rigid-body rotation follows a curved trajectory). Below is an example of a forward and inverse deformation. Unlike those shown in Figure 5.6, these are one-to-one mappings, and the transforms are actually inverses of each other.

by such a greedy algorithm, it will not produce the smoothest possible deformation. Such a warping strategy does not maximize any clearly defined prior probability of the parameters.

More recent algorithms for large deformation registration do aim to maximize both the log-likelihood and log-prior. For example, the LDDMM (large deformation diffeomorphic metric mapping) algorithm (Beg *et al.*, 2005) does not fix the deformation parameters once they have been estimated. It continues to update them using a gradient descent algorithm such that the log-prior term is properly optimized. Such approaches essentially parameterize the model by velocities, and compute the deformation as the medium warps over unit time (Joshi and Miller, 2000; Miller *et al.*, 2006).

Diffeomorphic warps can also be inverted (Figure 5.12). If each small deformation is generated by adding a small displacement (velocity) to an identity transform, then their approximate inverses can be derived by subtracting the same displacement. A composition of these small deformation inverses will then produce the inverse of the large deformation.

In principle, a single velocity field could be used to parameterize the model,[6] which would confer certain computational advantages. In Group theory, the velocities are a *Lie algebra*, and these are *exponentiated* to produce a deformation, which is a *Lie group* (see e.g. Miller and Younes, 2001; Woods, 2003; Miller *et al.* 2006; Vaillant *et al.*, 2004). If the velocity field is assumed constant throughout, then the exponentiation can be done recursively in a way that is analogous to exponentiating a matrix (Moler and Loan, 2003) by recursive squaring. A full deformation can be computed from the square[7] of a half-way deformation, a half-way deformation can be

[6] In such a model, the motion is at constant velocity within the Lagrangian frame of reference, but variable velocity if viewed within the Eulerian frame.

[7] The use of 'square' is in the sense of a composition of a function by itself.

computed by squaring a quarter-way deformation, and so on.

Many researchers are interested in deriving metrics from deformations in order to compare the similarities of shapes (Miller *et al.*, 1997, Miller, 2004). These metrics are a measure of the difference between the shapes, and can be thought of as measures of geodesic distance. One example of the use of such distance measures would be for tackling multivariate classification problems using non-linear kernel methods (such as support vector machines). In order to be a metric, a measure must be non-negative ($dist(A, B) \geqslant 0$), symmetric ($dist(A, B) = dist(B, A)$) and satisfy the triangle inequality ($dist(A, B) + dist(B, C) \geqslant dist(A, C)$). The more recent diffeomorphic registration approaches generate suitable measures of difference between shapes, which are derived from the energy densities of the velocity fields.

ESTIMATING THE MAPPINGS

Most non-linear registration approaches search for a maximum *a posteriori* (MAP) estimate of the parameters defining the warps. This corresponds to the mode of the posterior probability density of the model parameters. There are many optimization algorithms that try to find the mode, but most of them only perform a local search. It is possible to use relatively simple strategies for fitting models with few parameters but, as the number of parameters increases, the time required to estimate them will increase dramatically. For this reason, it is common to use optimization algorithms that utilize the derivatives of the objective function with respect to the parameters, as these indicate the best direction in which to search. Some developers use schemes related to gradient descent, whereby the parameters are repeatedly changed by a tiny amount in the direction that improves the objective function. This procedure can still be quite slow – particularly when there are dependencies among the parameters. Faster algorithms have been devised, which assume that the probability densities can be approximated by a multivariate Gaussian. Their effectiveness depends on how good this approximation is.

The Levenberg-Marquardt (LM) algorithm is a very good general purpose optimization strategy (see Press *et al.*, 1992 for more information). The procedure is a local optimization, so it needs reasonable initial starting estimates. It uses an iterative scheme to update the parameter estimates in such a way that the objective function is usually improved each time. Each iteration requires the first and second derivatives of the objective function, with respect to the parameters. In the following scheme, $\mathbf{I}$ is an identity matrix and ζ is a scaling factor. The choice of ζ is a trade-off between speed of convergence, and stability. A value of zero for ζ gives the Newton-Raphson or Gauss-Newton optimization scheme, which may be unstable if the probability density is not well approximated by a Gaussian. Increasing ζ will slow down the convergence, but increase the stability of the algorithm. The value of ζ is usually decreased slightly after iterations that decrease (improve) the cost function. If the cost function increases after an iteration, then the previous solution is retained, and ζ is increased in order to provide more stability.

$$\boldsymbol{\alpha}^{(n+1)} = \boldsymbol{\alpha}^{(n)} - \left(\left.\frac{\partial^2 \mathcal{F}(\boldsymbol{\alpha})}{\partial \boldsymbol{\alpha}^2}\right|_{\boldsymbol{\alpha}^{(n)}} + \zeta \mathbf{I} \right)^{-1} \left.\frac{\partial \mathcal{F}(\boldsymbol{\alpha})}{\partial \boldsymbol{\alpha}}\right|_{\boldsymbol{\alpha}^{(n)}} \quad \mathbf{5.20}$$

The objective function $\mathcal{F}(\boldsymbol{\alpha})$ is the sum of two terms. Ignoring the constants, these are the negative logarithm of the likelihood ($\mathcal{E}(\boldsymbol{\alpha})$) and the negative logarithm of the prior probability density ($\mathcal{H}(\boldsymbol{\alpha})$). The prior probability of the parameters is generally modelled by a multivariate Gaussian density, with mean $\boldsymbol{\alpha}_0$ and covariance $\mathbf{C}_{\boldsymbol{\alpha}}$.

$$\mathcal{H}(\boldsymbol{\alpha}) = \frac{1}{2}(\boldsymbol{\alpha} - \boldsymbol{\alpha}_0)\mathbf{C}_{\boldsymbol{\alpha}}^{-1}(\boldsymbol{\alpha} - \boldsymbol{\alpha}_0) \quad \mathbf{5.21}$$

The first and second derivatives of $\mathcal{F}(\boldsymbol{\alpha})$ (see Eqn. **5.9**) with respect to the parameters are therefore:

$$\frac{\partial \mathcal{F}(\boldsymbol{\alpha})}{\partial \boldsymbol{\alpha}} = \frac{\partial \mathcal{E}(\boldsymbol{\alpha})}{\partial \boldsymbol{\alpha}} + \mathbf{C}_{\boldsymbol{\alpha}}^{-1}(\boldsymbol{\alpha} - \boldsymbol{\alpha}_0) \text{ and}$$
$$\frac{\partial^2 \mathcal{F}(\boldsymbol{\alpha})}{\partial \boldsymbol{\alpha}^2} = \frac{\partial^2 \mathcal{E}}{\partial \boldsymbol{\alpha}^2} + \mathbf{C}_{\boldsymbol{\alpha}}^{-1} \quad \mathbf{5.22}$$

If the model is very high-dimensional (more than about 4000 parameters), then storing a full matrix of second derivatives becomes difficult because of memory limitations. For this reason, it can be convenient to use sparse representations of the second derivatives, and use approaches for solving systems of sparse equations (Gilbert *et al.*, 1992). Sparse matrices of second derivatives can be obtained by parameterizing the deformations with basis functions that have local support (e.g. B-splines). A faster approach, however, would be to use matrix solvers that are especially designed for the particular class of problem. Full multigrid (FMG) approaches are especially suited to solving such equations, and a good introductory explanation of these methods can be found in Chapter 19 of Press *et al.* (1992).

Optimization problems for complex non-linear models, such as those used for image registration, can easily get caught in local optima, so there is no guarantee that the estimate determined by the algorithm is globally optimum. If the starting estimates are sufficiently close to the global optimum, then a local optimization algorithm is more likely to find the true MAP solution.

Therefore, the choice of starting parameters can influence the validity of the final registration result. One method of increasing the likelihood of achieving a good solution is gradually to reduce the value of λ relative to $1/\sigma^2$ over time.[8] This has the effect of making the registration estimate the more global deformations before estimating more detailed warps. Most shape variability is low frequency, so an algorithm can get reasonably close to a good solution using a relatively high value for λ. This also reduces the number of local minima for the early iterations. The images could also be smoother for the earlier iterations in order to reduce the amount of confounding information and the number of local minima. A review of such approaches can be found in Lester and Arridge (1999).

Internal consistency

Currently, most registration algorithms use only a single image from each subject, which is typically a T1-weighted MR image. Such images only really delineate different tissue types. Further information that may help the registration could be obtained from other data, such as diffusion weighted images (Zhang *et al.*, 2004). This should provide anatomical information more directly related to connectivity and implicitly function, possibly leading to improved registration of functionally specialized areas (Behrens *et al.*, 2006). Matching DTI images of a pair of subjects together is likely to give different deformation estimates than would be obtained through matching T1-weighted images of the same subjects. The only way to achieve an internally consistent match is through performing the registrations simultaneously, within the same model.

Another form of consistency is inverse consistency (Christensen, 1999). For example, suppose a deformation that matches brain **f** to brain **g** is estimated, and also a deformation that matches brain **g** to brain **f**. If one deformation field is not the inverse of the other, then something has to be wrong. The extreme case of an inconsistency between a forward and inverse transformation is when the one-to-one mapping between the images breaks down. The best way of ensuring such internal consistency is through using diffeomorphic matching strategies.

Many registration approaches use the gradients of only one of the images to drive the registration, rather than the gradients of both. This can be another reason why inconsistencies may arise between forward and inverse registration approaches. One way in which the registration can be made more symmetric is iteratively to match the images to their warped average. The result of this procedure would be two deformations that map 'half way'. From the appropriate compositions of these 'half way' deformations, a pair of deformations can be generated that map between the original images, and are both inverses of each other.

Sometimes, instead of simply matching a pair of images, the objective is to match the images of multiple subjects. This is sometimes done by registering all the images with a single template image. Such a procedure would produce different results depending upon the choice of template, so this is another area where internal consistency should be considered. One could consider an optimal template being some form of average (Hirani *et al.*, 2001; Avants and Gee, 2004; Davis *et al.*, 2004; Lorenzen *et al.*, 2004). Registering such a template with a brain image generally requires smaller (and therefore less error prone) deformations than would be necessary for registering to an unusually shaped template. Such averages generally lack the detail present in the individual subjects. The structures that are more difficult to match are generally slightly blurred in the average, whereas the structures that can be more reliably matched are sharper. Such an average generated from a large population of subjects would be ideal for use as a general purpose template.

SPATIAL NORMALIZATION IN THE SPM SOFTWARE

This section describes the steps involved in the algorithm that SPM uses to normalize spatially images of different subjects into roughly the same coordinate system. The coordinate system is defined by the template image (or series of images), which is usually one of the images released with the software. So far, this chapter has assumed the template is deformed to match an individual's brain image. In this section, the individual subjects' images are warped to approximate a template image. This is slightly suboptimal because it assumes the noise is in the template image rather than the individual image; the original motivation for this strategy was to avoid having to invert the estimated deformation fields.

Spatial normalization in SPM is currently implemented in a small deformation setting and warps a smoothed version of the image to a smooth template. It works by estimating the optimum coefficients for a set of bases, by minimizing the mean squared difference between the

[8] Regularize heavily to begin with and decrease the amount of regularization over time. The residuals are much further from IID in the early iterations, so the likelihood is overestimated. Increasing the regularization partially compensates for this.

template and a warped source image, while simultaneously minimizing the deviation of the transformation from its expected value. The images may be scaled differently, so an additional parameter (w) is needed to accommodate this difference. In fact, four intensity scaling parameters are used for each template image (also to model linear intensity gradients), but they will not be included in the following description. With a single scaling parameter, the minimized likelihood function is:

$$\mathcal{E} = \frac{1}{2\sigma^2} \sum_{i=1}^{I} (f(\mathbf{y}(\mathbf{x}_i, \boldsymbol{\alpha})) - wg(\mathbf{x}_i))^2 \qquad 5.23$$

A Gauss-Newton approach is used to optimize the parameters. This requires the following first derivatives:

$$\frac{\partial \mathcal{E}}{\partial \alpha_m} = \frac{1}{\sigma^2} \sum_{i=1}^{I} \frac{\partial f(\mathbf{y}(\mathbf{x}_i, \boldsymbol{\alpha}))}{\partial \alpha_m} (f(\mathbf{y}(\mathbf{x}_i, \boldsymbol{\alpha})) - wg(\mathbf{x}_i)) \qquad 5.24$$

$$\frac{\partial \mathcal{E}}{\partial w} = \frac{1}{\sigma^2} \sum_{i=1}^{I} g(\mathbf{x}_i) (wg(\mathbf{x}_i) - f(\mathbf{y}(\mathbf{x}_i, \boldsymbol{\alpha}))) \qquad 5.25$$

and second derivatives:

$$\frac{\partial^2 \mathcal{E}}{\partial \alpha_m \partial \alpha_n} = \frac{1}{\sigma^2} \sum_{i=1}^{I} \left(\frac{\partial f(\mathbf{y}(\mathbf{x}_i, \boldsymbol{\alpha}))}{\partial \alpha_m} \frac{\partial f(\mathbf{y}(\mathbf{x}_i, \boldsymbol{\alpha}))}{\partial \alpha_n} + \frac{\partial^2 f(\mathbf{y}(\mathbf{x}_i, \boldsymbol{\alpha}))}{\partial \alpha_m \partial \alpha_n} (f(\mathbf{y}(\mathbf{x}_i, \boldsymbol{\alpha})) - wg(\mathbf{x}_i)) \right) \qquad 5.26$$

$$\frac{\partial^2 \mathcal{E}}{\partial w^2} = \frac{1}{\sigma^2} \sum_{i=1}^{I} g(\mathbf{x}_i)^2 \qquad 5.27$$

$$\frac{\partial^2 \mathcal{E}}{\partial \alpha_m \partial w} = -\frac{1}{\sigma^2} \sum_{i=1}^{I} \frac{\partial f(\mathbf{y}(\mathbf{x}_i, \boldsymbol{\alpha}))}{\partial \alpha_m} g(\mathbf{x}_i) \qquad 5.28$$

In practice, the following approximate second derivatives are used, because they can be computed more easily, and they also make the algorithm more stable:

$$\frac{\partial^2 \mathcal{E}}{\partial \alpha_m \partial \alpha_n} \simeq \frac{1}{\sigma^2} \sum_{i=1}^{I} \frac{\partial f(\mathbf{y}(\mathbf{x}_i, \boldsymbol{\alpha}))}{\partial \alpha_m} \frac{\partial f(\mathbf{y}(\mathbf{x}_i, \boldsymbol{\alpha}))}{\partial \alpha_n} \qquad 5.29$$

The prior used for non-linear registration is based on the bending energy of the displacement field. The amount of spatial variability is assumed to be known *a priori* (i.e. λ in Eqn. **5.11** is assigned by the user), but the algorithm tries to determine an optimal σ^2 value for the particular image being registered. SPM is applied to a wide variety of images with different noise properties. Determining an optimal value for σ^2 may be impractical for most users, so the algorithm uses a heuristic approach to assign a suitable value. The heuristic is based on the mean squared difference between the template and warped image, but with a correction based on the number of independent voxels. This correction is computed at each iteration using the smoothness of the residuals. Fully Bayesian approaches assume that the variance associated with each voxel is already known, whereas the approach described here is a type of empirical Bayes, which attempts to estimate this variance from the residuals. Because the registration is based on smooth images, correlations between neighbouring voxels are considered when estimating the variance. This makes the same approach suitable for the spatial normalization of both high quality MR images, and low resolution noisy PET images.

In practice, it may be meaningless even to attempt an exact match between brains beyond a certain resolution. There is not a one-to-one relationship between the cortical structures of one brain and those of another, so any method that attempts to match brains exactly must be folding the brain to create sulci and gyri that do not exist. Even if an exact match is possible, because the registration problem is not convex, the solutions obtained by high-dimensional warping techniques may not be truly optimum. High-dimensional registrations methods are often very good at registering grey matter with grey matter (for example), but there is no guarantee that the registered grey matter arises from homologous cortical structures.

Also, structure and function are not always tightly linked. Even if structurally equivalent regions can be brought into exact register, it does not mean that the same is true for regions that perform the same or similar functions. For inter-subject averaging in SPM, an assumption is made that functionally equivalent regions lie in approximately the same parts of the brain. This leads to the current rationale for smoothing images from multisubject functional imaging studies prior to performing statistical analyses. Constructive interference of the smeared activation signals then has the effect of producing a signal that is roughly in an average location. In order to account for substantial fine scale warps in a spatial normalization, it is necessary for some voxels to increase their volumes considerably, and for others to shrink to an almost negligible size. The contribution of the shrunken regions to the smoothed images is tiny, and the sensitivity of the tests for detecting activations in these regions may be reduced. This is another argument in favour of spatially normalizing on a relatively global scale.

The first step in registering images from different subjects involves determining the optimum 12-parameter affine transformation. Unlike in Chapter 4 – where the images to be matched together are from the same subject – zooms and shears are needed to register heads of different shapes and sizes. Prior knowledge of the variability of head sizes and overall proportions is used to increase the robustness and accuracy of the method (Ashburner *et al.*, 1997).

The next part is a non-linear registration for correcting gross differences in head shapes that cannot be accounted for by the affine normalization alone. The non-linear warps are modelled by linear combinations of smooth cosine transform basis functions. For speed and simplicity, a relatively small number of parameters (approximately 1000) are used to describe the non-linear components of the registration (Ashburner and Friston, 1999).

Affine registration

Almost all inter-subject registration methods for brain images begin by determining the optimal affine transformation that registers the images together. This step is normally performed automatically by minimizing (or maximizing) some mutual function of the images. The objective of affine registration is to fit the source image to a template image, using a twelve-parameter affine transformation. The images may be scaled quite differently, so an additional intensity scaling parameter is included in the model.

Without constraints and with poor data, simple maximum-likelihood parameter optimization can produce some extremely unlikely transformations. For example, when there are only a few slices in the image, it is not possible for the algorithms to determine an accurate zoom in the out of plane direction. Any estimate of this value is likely to have very large errors. When a regularized approach is not used, it may be better to assign a fixed value for this difficult-to-determine parameter, and simply fit for the remaining ones.

By incorporating prior information into the optimization procedure, a smooth transition between fixed and fitted parameters can be achieved. When the error for a particular fitted parameter is known to be large, then that parameter will be based more upon the prior information. In order to adopt this approach, the prior distribution of the parameters has to be specified. This is derived from the zooms and shears determined by registering a large number of brain images to the template.

Non-linear registration

The non-linear spatial normalization approach of SPM assumes that the image has already been approximately registered with the template according to a twelve-parameter affine registration. This section illustrates how the parameters describing global shape differences (not accounted for by affine registration) between an image and template can be determined. A small-deformation framework is used, and regularization is by the bending energy of the displacement field. Further details can be found in Ashburner and Friston (1999).

The deformations are parameterized by a linear combination of about 1000 low-frequency three-dimensional cosine transform bases. The spatial transformation from $\mathbf{x}$, to $\mathbf{y}$ is:

$$\begin{aligned} y_1(\mathbf{x}, \boldsymbol{\alpha}) &= x_1 + u_1 = x_1 + \sum_{m=1}^{M} \alpha_{m1}\phi_m(\mathbf{x}) \\ y_2(\mathbf{x}, \boldsymbol{\alpha}) &= x_2 + u_2 = x_2 + \sum_{m=1}^{M} \alpha_{m2}\phi_m(\mathbf{x}) \\ y_3(\mathbf{x}, \boldsymbol{\alpha}) &= x_3 + u_3 = x_3 + \sum_{m=1}^{M} \alpha_{m3}\phi_m(\mathbf{x}) \end{aligned} \qquad 5.30$$

where α_{mk} is the mth coefficient for dimension k, and $\phi_m(\mathbf{x})$ is the mth basis function at position $\mathbf{x}$. The basis functions are separable, and each one is generated by multiplying three one-dimensional basis functions together.

$$\phi_m(\mathbf{x}) = \phi_{m_3}(x_3)\phi_{m_2}(x_2)\phi_{m_1}(x_1) \qquad 5.31$$

In one dimension, the cosine transform bases are generated by:

$$\begin{aligned} \phi_1(i) &= \frac{1}{\sqrt{I}} \qquad i = 1..I \\ \phi_m(i) &= \sqrt{\frac{2}{I}}\cos\left(\frac{\pi(2i-1)(m-1)}{2I}\right) \qquad i = 1..I, m = 2..M \end{aligned} \qquad 5.32$$

EVALUATION STRATEGIES

Validation of warping methods is a complex area. The appropriateness of an evaluation depends on the particular application that the deformations are to be used for. For example, if the application was spatial normalization of functional images of different subjects, then the most appropriate evaluation may be based on assessing the sensitivity of voxel-wise statistical tests (Gee *et al.*, 1997a; Miller *et al.*, 2005). Because the warping procedure is based only on structural information, it is blind to the locations of functional activation. If the locations of activations can be brought into close correspondence in different subjects, then it is safe to say that the spatial normalization procedure is working well. The best measure of correspondence depends on how much the images are smoothed prior to performing the statistical tests. Different registration methods will perform differently depending on the amount of smoothing used. For example, the difference in performance of high- versus low-dimensional methods will be less when lots of smoothing is used. Another application may involve identifying shape differences among populations of subjects. In this case, the usefulness of the warping algorithm would be assessed by how well the deformation fields can be used to distinguish between the populations (Lao *et al.*, 2004). These approaches can be considered as forms of cross-validation, because they assess how well the registration helps to predict additional information.

Generally, the results of an evaluation are specific only to the data used to evaluate the model. MR images vary a great deal with different subjects, field strengths, scanners, sequences etc, so a model that is good for one set of data may not be appropriate for another. For example, consider intra-subject brain registration, under the assumption that the brain behaves like a rigid body. If the scanner causes no distortion and computes the pixel sizes and slice thickness of the image volumes exactly, then the best model is a rigid body registration. If the scanner computes the voxel sizes incorrectly, then the best model may be an affine registration. If there are distortions, then the best registration approach will model distortions. Validation should therefore relate to both the data and the algorithm. The question should be about whether it is appropriate to apply a model to a dataset, given the assumptions made by the model.

An assessment of how well manually defined landmarks in real brains can be colocalized is another useful validation strategy, because it allows the models to be compared with human expertise (Hellier *et al.*, 2001, 2002, 2003). The use of simulated images with known underlying deformations is not appropriate for proper validation of non-linear registration methods. This is particularly true if the deformation model is the same for the simulations as it is for the registration, because this only illustrates whether or not the optimization strategy works. Another commonly used form of 'evaluation' involves looking at the residual difference after registration. Such a strategy would ignore the possibility of over-fitting, and tends to favour those models with less regularization.

Within a Bayesian framework, it is possible to compare models, and decide which is more appropriate for a given dataset. This involves comparing the posterior probabilities for different models, after taking the model complexities into consideration. Occam's razor[9] is implicitly incorporated by penalizing more complicated models.

[9] *Occam's razor* is the principle that one should not increase, beyond what is necessary, the number of entities required to explain anything. It is sometimes known as the the *principle of parsimony*, and has been historically linked with the philosopher, William of Ockham.

Given a choice of I alternative models, the probability of model $\mathcal{M}_i$ is given by:

$$P(\mathcal{M}_i|\mathbf{D}) = \frac{P(\mathbf{D}|\mathcal{M}_i)P(\mathcal{M}_i)}{\sum_j^I P(\mathbf{D}|\mathcal{M}_j)P(\mathcal{M}_j)} \qquad 5.33$$

where the evidence for model $\mathcal{M}_i$ is obtained by marginalizing with respect to model parameters ($\boldsymbol{\theta}$):

$$P(\mathbf{D}|\mathcal{M}_i) = \int_{\boldsymbol{\theta}} P(\mathbf{D}|\boldsymbol{\theta}, \mathcal{M}_i)P(\boldsymbol{\theta}|\mathcal{M}_i)d\boldsymbol{\theta} \qquad 5.34$$

Unfortunately, the large number of model parameters required by non-linear registration would make such an exact integration procedure computationally intractable.[10] Note also that this includes a prior probability term for each model, which many may consider as subjective. There are, however, broad criteria for assigning such priors, based on an understanding of the model assumptions, its complexity and on any inconsistencies it may contain.

In summary: to validate a warping method for a particular dataset, it must be considered in relation to other available methods. The Bayesian paradigm allows a model comparison to be made, from which the best can be selected. However, this model selection may not generalize for all data.

REFERENCES

Amit Y, Grenander U, Piccioni M (1991) Structural image restoration through deformable templates. *J Am Stat Assoc* **86**: 376–87

Andersson JLR, Hutton C, Ashburner J *et al.* (2001) Modeling geometric deformations in EPI time series. *NeuroImage* **13**: 903–19

Ashburner J, Andersson J, Friston KJ (1999) High-dimensional non-linear image registration using symmetric priors. *NeuroImage* **9**: 619–28

Ashburner J, Andersson J, Friston KJ (2000) Image registration using a symmetric prior – in three-dimensions. *Hum Brain Mapp* **9**: 212–25

Ashburner J, Friston KJ (1999) Nonlinear spatial normalization using basis functions. *Hum Brain Mapp* **7**(4): 254–66

Ashburner J, Neelin P, Collins DL *et al.* (1997) Incorporating prior knowledge into image registration. *NeuroImage* **6**: 344–52

Avants B, Gee JC (2004) Geodesic estimation for large deformation anatomical shape averaging and interpolation. *NeuroImage* **23**: S139–S150

Beg MF, Miller MI, Trouvé A *et al.* (2005) Computing large deformation metric mappings via geodesic flows of diffeomorphisms. *Int J Comput Vision* **61**: 139–57

Behrens TEJ, Jenkinson M, Robson MD *et al.* (2006) A consistent relationship between local white matter architecture and functional specialisation in medial frontal cortex. *NeuroImage* **30**: 220–7

[10] The use of Laplace approximations in conjunction with ML-II methods may be possible alternatives though.

Bookstein FL (1989) Principal warps: thin-plate splines and the decomposition of deformations. *IEEE Trans Pattern Anal Machine Intelligence* **11**: 567–85

Bookstein FL (1997) Landmark methods for forms without landmarks: morphometrics of group differences in outline shape. *Med Image Anal* **1**: 225–43

Brett M, Leff AP, Rorden C *et al.* (2001) Spatial normalization of brain images with focal lesions using cost function masking. *NeuroImage* **14**: 486–500

Bro-Nielsen M, Gramkow C (1996) Fast fluid registration of medical images. In *Proc Visualization in Biomedical Computing (VBC)*, Hhne K-H, Kikinis R, (eds), vol. 1131 of *Lecture Notes in Computer Science*. Springer-Verlag, Berlin and Heidelberg, pp 267–76

Christensen GE (1994) Deformable shape models for anatomy. Doctoral thesis, Washington University, Sever Institute of Technology

Christensen GE (1999) Consistent linear elastic transformations for image matching. In *Proc Information Processing in Medical Imaging (IPMI)*, Kuba A, Sámal M, Todd-Pokropek A (eds), vol. 1613 of *Lecture Notes in Computer Science*. Springer-Verlag, Berlin and Heidelberg, pp 224–37

Christensen GE, Rabbitt RD, Miller MI (1994) 3D brain mapping using using a deformable neuroanatomy. *Phys Med Biol* **39**: 609–18

Christensen GE, Rabbitt RD, Miller MI *et al.* (1995) Topological properties of smooth anatomic maps. In *Proc Information Processing in Medical Imaging (IPMI)*, Bizais Y, Barillot C, Di Paola R (eds). Kluwer Academic Publishers, Dordrecht.

Christensen GE, Kane AA, Marsh JL *et al.* (1996a) Synthesis of an individualized cranial atlas with dysmorphic shape. In *Proceedings of the Workshop on Mathematical Methods in Biomedical Image Analysis*.

Christensen GE, Rabbitt RD, Miller MI (1996b) Deformable templates using large deformation kinematics. *IEEE Trans Image Process* **5**: 1435–47

Collins DL, Evans AC, Holmes C *et al.* (1995) Automatic 3D segmentation of neuro-anatomical structures from MRI. In *Proc Information Processing in Medical Imaging (IPMI)*, Bizais Y, Barillot C, Di Paola R (eds). Kluwer Academic Publishers, Dordrecht

Davatzikos C (1996) Spatial normalization of 3D images using deformable models. *J Comput Assist Tomogr* **20**: 656–65

Davis B, Lorenzen P, Joshi S (2004) Large deformation minimum mean squared error template estimation for computation anatomy. In *Proc. IEEE International Symposium on Biomedical Imaging (ISBI)*, IEEE Publishing.

Dryden IL, Mardia KV (1998) *Statistical Shape Analysis*. John Wiley and Sons, Chichester.

Edwards PJ, Hill DLG, Little JA, Hawkes DJ (1998) Deformation for image-guided interventions using a three component tissue model. *Medical Image Analysis* **2**(4): 355–67

Evans AC, Collins DL, Mills SR *et al.* (1993) 3D statistical neuroanatomical models from 305 MRI volumes. In *Proc IEEE-Nuclear Science Symposium and Medical Imaging Conference*, IEEE Publishing.

Evans AC, Kamber M Collins DL *et al.* (1994) An MRI-based probabilistic atlas of neuroanatomy. In *Magnetic Resonance Scanning and Epilepsy*, Shorvon S, Fish D, Andermann F *et al.* vol. 264 of *NATO ASI Series A*, Life Sciences. Plenum Press, New York, pp 263–74

Fox PT (1995) Spatial normalization origins: objectives, applications, and alternatives. *Hum Brain Mapp* **3**: 161–64

Gee JC, Alsop DC, Aguirre GK (1997a) Effect of spatial normalization on analysis of functional data. In *Proc SPIE Medical Imaging 1997: Image Processing*, Hanson KM(ed.), California

Gee JC, Haynor DR, Le Briquer L *et al.* (1997b) Advances in elastic matching theory and its implementation. In *Proc. CVRMed-MRCAS'97*, Troccaz J, Grimson E, Mösges R (eds), vol. 1205 of *Lecture Notes in Computer Science*. Springer-Verlag, Berlin and Heidelberg

Gilbert JR, Moler C, Schreiber R (1992) Sparse matrices in MATLAB: Design and implementation. *SIAM J Matrix Anal Applic* **13**: 333–56

Glasbey CA, Mardia KV (1998) A review of image warping methods. *J Appl Stat* **25**: 155–71

Haber E, Modersitzki J (2006) A multilevel method for image registration. *SIAM J Sci Comput* **27**: 1594–607

Hellier P, Ashburner J, Corouge I *et al.* (2002) Inter subject registration of functional and anatomical data using SPM. In *Proc Medical Image Computing and Computer-Assisted Intervention (MICCAI)*, vol. 2489 of *Lecture Notes in Computer Science*. Springer-Verlag, Berlin and Heidelberg, pp 590–97

Hellier P, Barillot C, Corouge I *et al.* (2003) Retrospective evaluation of intersubject brain registration. *IEEE Transact Med Imag* **22**(9): 1120–30

Hellier P, Barillot C, Corouge I *et al.* (2001) Retrospective evaluation of inter-subject brain registration. In *Proc Medical Image Computing and Computer-Assisted Intervention (MICCAI)*, Niessen WJ, Viergever MA (eds), vol. 2208 of *Lecture Notes in Computer Science* Springer-Verlag, Berlin and Heidelberg, pp 258–65

Hirani AN, Marsden JE, Arvo J (2001) Averaged template matching equations. In *Proc Energy Minimization Methods in Computer Vision and Pattern Recognition: Third International Workshop, EMMCVPR 2001. LNCS 2134*, Figueiredo MAT, Zerubia J, Jain AK (eds). Springer-Verlag, Berlin and Heidelberg

Jezzard P, Clare S (1999) Sources of distortion in functional MRI data. *Hum Brain Mapp* **8**: 80–85

Joshi SC, Miller MI (2000) Landmark matching via large deformation diffeomorphisms. *IEEE Trans Med Imag* **9**: 1357–70

Joshi SC, Miller MI, Christensen GE *et al.* (1995) Hierarchical brain mapping via a generalized dirichlet solutions for mapping brain manifolds. *Vision Geometry* **2573**(1): 278–89

Kendall DG, Barden D, Carne TK *et al.* (1999) *Shape and shape theory*. Wiley, Chichester

Lao Z, Shen D, Xue Z *et al.* (2004) Morphological classification of brains via high-dimensional shape transformations and machine learning methods. *NeuroImage* **21**(1): 46–57

Lester H, Arridge SR (1999) A survey of hierarchical non-linear medical image registration. *Pattern Recognit* **32**: 129–49

Lester H, Arridge SR, Jansons KM *et al.* (1999) Non-linear registration with the variable viscosity fluid algorithm. In *Proc Information Processing in Medical Imaging (IPMI)*, Kuba A, Sámal M, Todd-Pokropek A (eds), vol. 1613 of *Lecture Notes in Computer Science*. Springer-Verlag, Berlin and Heidelberg, pp 238–51

Lorenzen P, Davis B, Gerig G *et al.* (2004) Multi-class posterior atlas formation via unbiased Kullback-Leibler template estimation. In *Proc Medical Image Computing and Computer-Assisted Intervention (MICCAI)*, Barillot C, Haynor, DR, Hellier P (eds), vol. 3216 of *Lecture Notes in Computer Science*. Springer-Verlag, Berlin and Heidelberg, pp 95–102

Mazziotta JC, Toga AW, Evans A *et al.* (1995) A probabilistic atlas of the human brain: theory and rationale for its development. *NeuroImage* **2**: 89–101

Miller M, Banerjee A, Christensen G *et al.* (1997) Statistical methods in computational anatomy. *Stat Meth Med Res* **6**: 267–99

Miller MI (2004) Computational anatomy: shape, growth, and atrophy comparison via diffeomorphisms. *NeuroImage* **23**: S19–S33

Miller MI, Beg MF, Ceritoglu C *et al.* (2005) Increasing the power of functional maps of the medial temporal lobe using large deformation metric mapping. *Proc Nat Acad Sci USA* **102**: 9685–90

Miller MI, Christensen GE, Amit Y *et al.* (1993) Mathematical textbook of deformable neuroanatomies. *Proc Nat Acad Sci USA* **90**: 11944–48

Miller MI, Trouvé A, Younes L (2006) Geodesic shooting for computational anatomy. *J Math Imag Vision* **24**: 209–28

Miller MI, Younes L (2001) Group actions, homeomorphisms, and matching: a general framework. *Int J Comput Vision* **41**(1): 61–84

Modersitzki J (2003) *Numerical methods for image registration.* Oxford University Press, Oxford

Moler C, Loan CV (2003) Nineteen dubious ways to compute the exponential of a matrix, twenty-five years later. *SIAM Rev* **45**(1): 3–49

Press WH, Teukolsky SA, Vetterling WT *et al.* (1992) *Numerical Recipes in C.* 2nd edn. Cambridge University Press, Cambridge

Studholme C, Constable RT, Duncan JS (2000) Accurate alignment of functional EPI data to anatomical MRI using a physics-based distortion model. *IEEE Trans Med Imag* **19**(11): 1115–27

Talairach J, Tournoux P (1988) *Coplanar stereotaxic atlas of the human brain.* Thieme Medical, New York

Thévenaz P, Unser M (2000) Optimization of mutual information for multiresolution image registration. *IEEE Trans Image Process* **9**(12): 2083–99

Thirion JP (1995) Fast non-rigid matching of 3D medical images. Tech. Rep. 2547, Institut National de Recherche en Informatique et en Automatique, available from http://www.inria.fr/rrrt/rr 2547.html

Thompson PM, Toga AW (1996) Visualization and mapping of anatomic abnormalities using a probablistic brain atlas based on random fluid transformations. In *Proc Visualization in Biomedical Computing (VBC)*, Hhne K-H, Kikinis R (eds), vol. 1131 of *Lecture Notes in Computer Science*. Springer Verlag, Berlin and Heidelberg, pp 383–92

Tzourio-Mazoyer N, Landeau B, Papathanassiou D *et al.* (2002) Automated anatomical labeling of activations in SPM using a macroscopic anatomical parcellation of the MNI MRI single-subject brain. *NeuroImage* **15**: 273–89

Vaillant M, Miller MI, Younes L *et al.* (2004) Statistics on diffeomorphisms via tangent space representations. *NeuroImage* **23**: S161–S169

Woods RP (2003) Characterizing volume and surface deformations in an atlas framework: theory, applications, and implementation. *NeuroImage* **18**: 769–88

Woods RP, Grafton ST, Holmes CJ *et al.* (1998a) Automated image registration: I. general methods and intrasubject, intramodality validation. *J Comput Assist Tomogr* **22**: 139–52

Woods RP, Grafton ST, Watson JDG *et al.* (1998b) Automated image registration: II. intersubject validation of linear and nonlinear models. *J Comput Assist Tomogr* **22**(1): 153–65

Zhang H, Yushkevich PA, Gee JC (2004) Registration of diffusion tensor images. In *Proc IEEE Computer Society Conference on Computer Vision and Pattern Recognition (CVPR'04)*, vol. 1, pp I-842–I847

C H A P T E R

6

Segmentation

J. Ashburner and K. Friston

INTRODUCTION

This chapter describes a method of segmenting magnetic resonance (MR) images into different tissue classes, using a modified Gaussian mixture model. By knowing the prior spatial probability of each voxel that is grey matter, white matter or cerebrospinal fluid, it is possible to obtain quite robust classifications. A probabilistic framework is presented that enables image registration, tissue classification, and bias correction to be combined within the same generative model. A derivation of a log-likelihood objective function for this unified model is provided. The model is based on a mixture of Gaussians, and is extended to incorporate a smooth intensity variation and non-linear registration with tissue probability maps. A strategy for optimizing the model parameters is described, along with the requisite partial derivatives of the objective function.

Segmentation of brain images usually takes one of two forms. It can proceed by adopting a tissue classification approach, or by registration with a template. The aim of this chapter is to unify these procedures into a single probabilistic framework.

- The first approach rests on tissue classification, whereby voxels are assigned to a tissue class according to their intensities. In order to make these assignments, the intensity distribution of each tissue class needs to be characterized, often from voxels chosen to represent each class. Automatic selection of representative voxels can be achieved by first registering the brain volume to some standard space, and automatically selecting voxels that have a high probability of belonging to each class. A related approach involves modelling the intensity distributions by a mixture of Gaussians, but using tissue probability maps to weight the classification according to Bayes' rule.
- The other approach involves some kind of registration, where a template brain is warped to match the brain volume to be segmented (Collins *et al.*, 1995). This need not involve matching volumes: some methods that are based on matching surfaces (MacDonald *et al.*, 2000; Pitiot *et al.*, 2004) would also fall into this category. These approaches allow regions that are predefined on the template to be overlaid, allowing different structures to be identified automatically.

A paradigm shift is evident in the field of neuroimaging methodology, away from simple sequential processing, towards a more integrated generative modelling approach. The model described in this chapter (which is implemented in SPM5) is one such example (see also Fischl *et al.*, 2004). Both approaches combine tissue classification, bias correction and non-linear warping within the same framework. Although the integrated frameworks have some disadvantages, these should be outweighed by more accurate results. The main disadvantage is that the approaches are more complex and therefore more difficult to implement. In addition, the algorithms are integrated, making it difficult to mix and match different programs within 'pipeline' procedures (Zijdenbos *et al.*, 2002; Fissell *et al.*, 2003; Rex *et al.*, 2003). A perceived disadvantage of these combined models is that execution time is longer than it would be for sequentially applied procedures. For example, optimizing two separate models with 100 parameters is likely to be faster than optimizing a combined single model with 200 parameters. However, the reason a combined model takes longer to run is because it actually completes the optimization. There are usually conditional correlations among parameters of the different models, which sequential processing discounts. The advantage of unified models is that they are more accurate, making better use of the information available in the data. Scanning time is relatively expensive, but computing time is relatively cheap. Complete

Statistical Parametric Mapping, by Karl Friston et al.
ISBN–13: 978-0-12-372560-8 ISBN–10: 0-12-372560-7

models may take longer to run, but they should add value to the raw data.

THE OBJECTIVE FUNCTION

In this section, we describe the segmentation model and how it is used to define an objective function. In the next section, we will show how this function is used to estimate the parameters of interest. The objective function, minimized by the optimum parameters, is derived from a mixture of Gaussians model. We show how this objective function can be extended to model smooth intensity non-uniformity. Tissue probability maps are used to assist the classification, and we describe how the objective function accommodates deformations of these maps, so that they best match the image to segment. The section ends by explaining how the estimated non-uniformity and deformations are constrained to be spatially smooth. The end-point of these model elaborations is a generative model whose inversion segments, spatially normalizes and intensity corrects a given image.

Mixture of Gaussians

A distribution can be modelled by a mixture of K Gaussians. This is a standard technique (see e.g. Bishop, 1995), which is used widely by many tissue classification algorithms. For univariate data, the kth Gaussian is modelled by its mean (μ_k), variance (σ_k^2) and mixing proportion (γ_k, where $\sum_{k=1}^{K}\gamma_k = 1$ and $\gamma_k \geq 0$). Fitting a mixture of Gaussians (MOG) model involves maximizing the probability of observing the I elements of data $\mathbf{y}$, given the parameterization of the Gaussians. In a simple MOG, the probability[1] of obtaining a datum with intensity y_i given that it belongs to the kth Gaussian ($c_i = k$), and that the kth Gaussian is parameterized by μ_k and σ_k^2 is:

$$P(y_i|c_i = k, \mu_k, \sigma_k) = \frac{1}{(2\pi\sigma_k^2)^{\frac{1}{2}}}\exp\left(-\frac{(y_i-\mu_k)^2}{2\sigma_k^2}\right) \qquad 6.1$$

The prior probability of any voxel, irrespective of its intensity, belonging to the kth Gaussian, given the proportion of voxels that belong to that Gaussian is simply:

$$P(c_i = k|\gamma_k) = \gamma_k \qquad 6.2$$

Using Bayes' rule, the joint probability of cluster k and intensity y_i is:

$$\begin{aligned} P(y_i, c_i = k|\mu_k, \sigma_k, \gamma_k) \\ = P(y_i|c_i = k, \mu_k, \sigma_k)P(c_i = k|\gamma_k) \end{aligned} \qquad 6.3$$

By integrating over all k Gaussians, we obtain the probability of y_i given the parameters:

$$P(y_i|\boldsymbol{\mu}, \boldsymbol{\sigma}, \boldsymbol{\gamma}) = \sum_{k=1}^{K} P(y_i, c_i = k|\mu_k, \sigma_k, \gamma_k) \qquad 6.4$$

The probability of the entire dataset $\mathbf{y}$ is derived by assuming that all elements are independent:

$$\begin{aligned} P(\mathbf{y}|\boldsymbol{\mu}, \boldsymbol{\sigma}, \boldsymbol{\gamma}) &= \prod_{i=1}^{I} P(y_i|\boldsymbol{\mu}, \boldsymbol{\sigma}, \boldsymbol{\gamma}) \\ &= \prod_{i=1}^{I}\left(\sum_{k=1}^{K}\frac{\gamma_k}{(2\pi\sigma_k^2)^{\frac{1}{2}}}\exp\left(-\frac{(y_i-\mu_k)^2}{2\sigma_k^2}\right)\right) \end{aligned} \qquad 6.5$$

This probability is maximized with respect to the unknown parameters ($\boldsymbol{\mu}$, $\boldsymbol{\sigma}$ and $\boldsymbol{\gamma}$), when the following cost function (ε) is minimized (because the two are monotonically related):

$$\begin{aligned} \varepsilon &= -\log P(\mathbf{y}|\boldsymbol{\mu}, \boldsymbol{\sigma}, \boldsymbol{\gamma}) \\ &= -\sum_{i=1}^{I}\log\left(\sum_{k=1}^{K}\frac{\gamma_k}{(2\pi\sigma_k^2)^{\frac{1}{2}}}\exp\left(-\frac{(y_i-\mu_k)^2}{2\sigma_k^2}\right)\right) \end{aligned} \qquad 6.6$$

The assumption that voxels are independent is clearly implausible, but the priors embody a certain degree of spatial dependency. This means that the conditional probability that a voxel belongs to a tissue class shows spatial dependencies, even though the likelihood in Eqn. **6.5** does not.

Intensity non-uniformity

MR images are usually corrupted by a smooth, spatially varying artefact that modulates the intensity of the image (bias). There are a number of sources of this artefact, which are reviewed by Sled *et al.* (1998). These artefacts, although not usually a problem for visual inspection, can impede automated processing of the images. Early bias correction techniques involved homomorphic filtering, but these have generally been superseded. A review of

[1] Strictly speaking, it is a probability density rather than a probability. The mathematical notation used is $P(\cdot)$ for both probabilities and probability densities. Some authors make a distinction by using $P(\cdot)$ for probabilities and $p(\cdot)$ for probability densities.

bias correction approaches is presented in Belaroussi *et al.* (2006). Most current methods can be broadly classed as those that use parametric representations of image intensity distributions (such as mixtures of Gaussians), and those that use non-parametric representations (such as histograms).

- Non-parametric models usually involve image intensity histograms. Some authors have proposed using a multiplicative model of bias, and optimizing a function that minimizes the entropy of the histogram of the bias corrected intensities. One problem with this is that the entropy is minimized when the bias field is uniformly zero, resulting in a single bin containing all the counts. This was a problem (pointed out by Arnold *et al.* (2001)) for the bias field correction in SPM99 (Ashburner and Friston, 2000), where there was a tendency for the correction to reduce the mean intensity of brain tissue in the corrected image. The constraint that the multiplicative bias should average to unity resulted in a bowl-shaped dip in the estimated bias field.

  To counter this problem, Mangin (2000) minimized the entropy of the histograms, but included an additional term in the cost function to minimize the squared difference between the original and restored image mean. A related solution was devised by Likar *et al.* (2001). In addition to modelling a multiplicative bias field, the latter method also modelled a smooth additive bias. These represent partial solutions to the problem, but are not ideal. When the width of a Gaussian (or any other distribution) is multiplied by a factor of ρ, then the entropy of the distribution is increased by $\log \rho$. Therefore, when scaling data by some value, the log of this factor needs to be considered when developing an entropy-based cost function.

  An alternative solution is to minimize the entropy of the histogram of log-transformed intensities. In addition to being generally better behaved, this also allows the bias fields to be modelled as an additive effect in log-space (Sled *et al.*, 1998). In order to work with log-transformed data, low intensity (and negative valued) voxels are excluded so that numerical problems are not introduced. This exclusion motivates a more generic model of all regional effects.
- Parametric bias correction models are often an integral part of tissue classification methods, many of which are based upon modelling the intensities of different tissues as a mixture of Gaussians. Other clustering methods can also be used, such as k-means and fuzzy c-means. Additional information is often encoded, in these approaches, using Markov random field models to embed knowledge that neighbouring voxels are likely to belong to the same tissue class. Most algorithms assume that the bias is multiplicative, but there are three commonly used models of how the bias interacts with noise.

  In the first parametric model, the observed signal (y_i) is assumed to be an uncorrupted signal (μ_i), scaled by some bias (ρ_i) with added Gaussian noise (n_i) that is independent of the bias (Pham and Prince, 1999; Shattuck *et al.*, 2001). The noise source is assumed to be from the scanner itself:

$$y_i = \mu_i/\rho_i + n_i \tag{6.7}$$

  The second model is similar to the first, except that the noise is added before the signal is scaled. In this case, the noise is assumed to be due to variations in tissue properties. This model is the one used in this chapter:

$$y_i = (\mu_i + n_i)/\rho_i \tag{6.8}$$

  A combination of the scanner and tissue noise models has been adopted by Fischl *et al.* (2004). This would probably be a better model, especially for images corrupted by a large amount of bias. The single noise source model was mainly chosen for its simplicity.

  A third approach involves log transforming the data first, allowing a multiplicative bias to be modelled as an additive effect in log-space (Wells *et al.*, 1996b; Garza-Jinich *et al.*, 1999; Van Leemput *et al.*, 1999a; Styner, 2000; Zhang *et al.*, 2001). The cost function for these approaches is related to the entropy of the distribution of log-transformed bias corrected data. As with the non-parametric model based on log-transformed data, low intensity voxels have to be excluded to avoid numerical problems. The generative model is of a form similar to:

$$\begin{aligned} \log y_i &= \log \mu_i - \log \rho_i + n_i \\ y_i &= \mu_i e^{n_i}/\rho_i \end{aligned} \tag{6.9}$$

  Sometimes these methods do not use a consistent generative model throughout, for example when alternating between the original intensities for the classification steps, and the log-transformed intensities for the bias correction (Wells *et al.*, 1996a).

In the model described here, bias correction is included in the MOG by extra parameters that account for smooth intensity variations. The field modelling the variation at element i is denoted by $\rho_i(\boldsymbol{\beta})$, where $\boldsymbol{\beta}$ is a vector of unknown parameters. Intensities from the kth cluster are assumed to be normally distributed with mean $\mu_k/\rho_i(\boldsymbol{\beta})$, and variance $(\sigma_k/\rho_i(\boldsymbol{\beta}))^2$. Therefore, the probability of obtaining intensity y_i from the kth cluster, given its parameterization is:

$$P(y_i|c_i = k, \mu_k, \sigma_k, \boldsymbol{\beta}) = \rho_i(\boldsymbol{\beta}) \frac{1}{(2\pi\sigma_k^2)^{\frac{1}{2}}} \exp\left(-\frac{(\rho_i(\boldsymbol{\beta})y_i - \mu_k)^2}{2\sigma_k^2}\right) \quad \textbf{6.10}$$

The tissue classification objective function is now:

$$\varepsilon = -\sum_{i=1}^{I} \log\left(\rho_i(\boldsymbol{\beta}) \sum_{k=1}^{K} \frac{\gamma_k}{(2\pi\sigma_k^2)^{\frac{1}{2}}} \exp\left(-\frac{(\rho_i(\boldsymbol{\beta})y_i - \mu_k)^2}{2\sigma_k^2}\right)\right) \quad \textbf{6.11}$$

The model employed here parameterizes the bias as the exponential of a linear combination of low-frequency basis functions. A small number of basis functions are used, as bias tends to be spatially smooth. Positivity is ensured by the exponential.

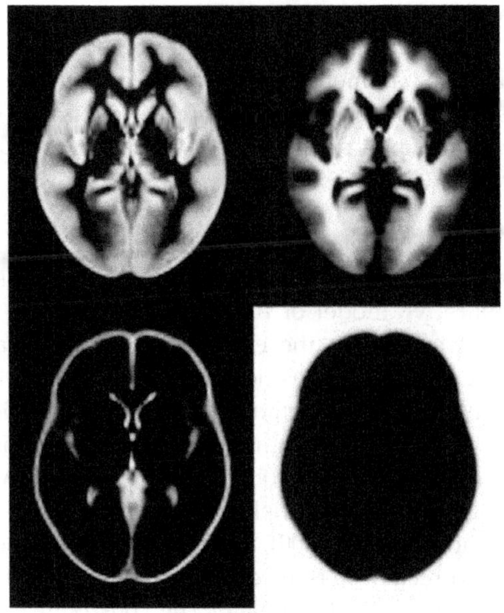

FIGURE 6.1 The tissue probability maps for grey matter, white matter, CSF and 'other'.

Spatial priors

Rather than assuming stationary prior probabilities based upon mixing proportions, additional information is used, derived from tissue probability maps from other subjects' brain images. Priors are usually generated by registering a large number of subjects together, assigning voxels to different tissue types and averaging tissue classes over subjects. The data used by SPM5 are a modified version of the ICBM Tissue Probabilistic Atlas.[2] They consist of tissue probability maps of grey and white matter, and of CSF (Figure 6.1). A fourth class is also used, which is simply one minus the sum of the first three. These maps give the prior probability of any voxel in a registered image being of any of the tissue classes – irrespective of its intensity. The implementation uses tissue probability maps for grey matter, white matter and CSF, although maps for additional tissue types (e.g. blood vessels) could also be included. The simple model of grey matter being all of approximately the same intensity could also be refined by using tissue probability maps for various internal grey matter structures (Fischl *et al.*, 2002).

The model in Eqn. **6.11** is modified to account for these spatial priors. Instead of using stationary mixing proportions ($P(c_i = k|\boldsymbol{\gamma}) = \gamma_k$), the prior probabilities are allowed to vary over voxels, such that the prior probability of voxel i being drawn from the kth Gaussian is:

$$P(c_i = k|\boldsymbol{\gamma}) = \frac{\gamma_k b_{ik}}{\sum_{j=1}^{K} \gamma_j b_{ij}} \quad \textbf{6.12}$$

[2] Avaliable from http://www.loni.ucla.edu/ICBM/ICBM_Probabilistic.html

where b_{ik} is the tissue probability for class k at voxel i. Note that $\boldsymbol{\gamma}$ is no longer a vector of true mixing proportions, but for the sake of simplicity, its elements will be referred to as such.

The number of Gaussians used to represent the intensity distribution for each tissue class can be greater than one. In other words, a tissue probability map may be shared by several Gaussians. The assumption of a single Gaussian distribution for each class does not hold for a number of reasons. In particular, a voxel may not be purely of one tissue type, and instead contain signal from a number of different tissues (partial volume effects). Some partial volume voxels could fall at the interface between different classes, or they may fall in the middle of structures, such as the thalamus, which may be considered as being either grey or white matter. Various image segmentation approaches use additional Gaussians to model such partial volume effects. These generally assume that a pure tissue class has a Gaussian intensity distribution, whereas intensity distributions for partial volume voxels are broader, falling between the intensities of the pure classes. Most of these models assume that a mixing combination of, e.g. 50/50, is just as probable as one of 80/20 (Laidlaw *et al.*, 1998; Shattuck *et al.*, 2001; Tohka *et al.*, 2004), whereas others allow a spatially varying prior probability for the mixing combination, which is dependent upon the contents of neighbouring voxels (Van Leemput *et al.*, 2001). Unlike these partial volume segmentation approaches, the model adopted here simply assumes that the intensity distribution of each class may not be Gaussian, and assigns belonging probabilities according to these non-Gaussian distributions. Selecting the optimal number of Gaussians per class is a model order selection issue, and will not be addressed here. Typical numbers of Gaussians are three for grey matter, two for white matter, two for CSF, and five for everything else.

Deformable spatial priors

The above formulation (Eqn. **6.12**) is refined further by allowing the tissue probability maps to be deformed according to parameters $\boldsymbol{\alpha}$. This allows registration to a standard space to be included within the generative model.

$$P(c_i = k|\boldsymbol{\gamma}, \boldsymbol{\alpha}) = \frac{\gamma_k b_{ik}(\boldsymbol{\alpha})}{\sum_{j=1}^{K} \gamma_j b_{ij}(\boldsymbol{\alpha})} \qquad \textbf{6.13}$$

After including the full priors, the objective function becomes:

$$\mathcal{E} = -\sum_{i=1}^{I} \log \left(\frac{\rho_i(\boldsymbol{\beta})}{\sum_{k=1}^{K} \gamma_k b_{ik}(\boldsymbol{\alpha})} \sum_{k=1}^{K} \gamma_k b_{ik}(\boldsymbol{\alpha})(2\pi\sigma_k^2)^{-\frac{1}{2}} \times \exp\left(-\frac{(\rho_i(\boldsymbol{\beta})y_i - \mu_k)^2}{2\sigma_k^2} \right) \right) \qquad \textbf{6.14}$$

There are many ways of parameterizing how the tissue probability maps could be deformed. The implementation in SPM5 uses a low-dimensional approach, which parameterizes the deformations by a linear combination of about a thousand cosine transform bases (Ashburner and Friston, 1999). This is not an especially precise way of encoding deformations, but it can model the variability of overall brain shape. Evaluations have shown that this simple model can achieve a registration accuracy comparable to other fully automated methods with many more parameters (Hellier *et al.*, 2001, 2002). This deformation means that inversion of the generative model can, implicitly, normalize images. Indeed, this is the prefered method of spatial normalization in SPM5.

Regularization

One important issue relates to the distinction between intensity variations that arise because of bias artefact due to the physics of MR scanning, and those that arise due to different tissue properties. The objective is to model the latter by different tissue classes, while modelling the former with a bias field. We know *a priori* that intensity variations due to MR physics tend to be spatially smooth, whereas those due to different tissue types tend to contain more high frequency information. A more accurate estimate of a bias field can be obtained by including prior knowledge about the distribution of the fields likely to be encountered by the correction algorithm. For example, if it is known that there is little or no intensity non-uniformity, then it would be wise to penalize large values for the intensity non-uniformity parameters. This regularization can be placed within a Bayesian context, whereby the penalty incurred is the negative logarithm of a prior probability for any particular pattern of non-uniformity. Similarly, it is possible for intensity variations to be modelled by incorrect registration. If we had some knowledge about a prior probability distribution for brain shape, then this information could be used to regularize the deformations. It is not possible to determine a complete specification of such a probability distribution empirically. Instead, the SPM5 approach (as with most other non-linear registration procedures) uses an educated guess for the form and amount of variability likely to be encountered. Without such regularization, the pernicious interactions (Evans, 1995) among the parameter estimates could be more of a problem. With the regularization terms included, fitting the model involves maximizing:

$$P(\mathbf{y}, \boldsymbol{\beta}, \boldsymbol{\alpha}|\boldsymbol{\gamma}, \boldsymbol{\mu}, \boldsymbol{\sigma}^2) = P(\mathbf{y}|\boldsymbol{\beta}, \boldsymbol{\alpha}, \boldsymbol{\gamma}, \boldsymbol{\mu}, \boldsymbol{\sigma})P(\boldsymbol{\beta})P(\boldsymbol{\alpha}) \qquad \textbf{6.15}$$

This is equivalent to minimizing:

$$\mathcal{F} = -\log P(\mathbf{y}, \boldsymbol{\beta}, \boldsymbol{\alpha}|\boldsymbol{\gamma}, \boldsymbol{\mu}, \boldsymbol{\sigma}) = \mathcal{E} - \log P(\boldsymbol{\beta}) - \log P(\boldsymbol{\alpha}) \qquad \textbf{6.16}$$

In the SPM5 implementation, the probability densities of the spatial parameters are assumed to be zero-mean multivariate Gaussians ($P(\boldsymbol{\alpha}) = N(\mathbf{0}, \mathbf{C}_\alpha)$ and $P(\boldsymbol{\beta}) = N(\mathbf{0}, \mathbf{C}_\beta)$). For the non-linear registration parameters, the covariance matrix is defined such that $\boldsymbol{\alpha}^T \mathbf{C}_\alpha^{-1} \boldsymbol{\alpha}$ gives the bending energy of the deformations (see Ashburner and Friston, 1999 for details). The prior covariance matrix for the bias is based on the assumption that a typical bias field could be generated by smoothing zero mean random Gaussian noise by a broad Gaussian smoothing kernel (about 70 mm FWHM, full width at half maximum), and then exponentiating (i.e. $\mathbf{C}_\beta$ is a Gaussian Toeplitz matrix).

OPTIMIZATION

This section describes how the objective function from Eqns **6.14** and **6.16** is minimized (i.e. how the model is inverted). There is no closed-form solution for the parameters, and optimal values for different parameters depend upon the values of others. An iterated conditional modes (ICM) approach is used. It begins by assigning starting estimates for the parameters, and then iterating until a locally optimal solution is found. Each iteration involves alternating between estimating different groups of parameters, while holding the others fixed at their current 'best' solution (i.e. conditional mode). The mixture parameters are updated using expectation maximization (EM), while holding the bias and deformations fixed at their conditional modes. The bias is estimated while holding the mixture parameters and deformation constant. Because intensity non-uniformity is very smooth, it can be described by a small number of parameters, making the Levenberg-Marquardt (LM) scheme ideal for this optimization. The deformations of the tissue probability maps are re-estimated while fixing the mixture parameters and bias field. A low-dimensional parameterization is used for the spatial deformations, so the LM strategy is also applicable here.

The model is only specified for brain, as there are no tissue probability maps for non-brain tissue (scalp etc). Because of this, there is a tendency for the approach to stretch the probability maps so that the background class contains only air, but no scalp. A workaround involves excluding extra-cranial voxels from the fitting procedure. This is done by fitting a mixture of two Gaussians to the image intensity histogram. In most cases, one Gaussian fits air, and the other fits everything else. A suitable threshold is then determined, based on a 50 per cent probability. Fitting only the intra-cranial voxels also saves time.

Mixture parameters ($\boldsymbol{\mu}$, $\boldsymbol{\sigma}$ and $\boldsymbol{\gamma}$)

It is sufficient to minimize ε with respect to the mixture parameters because they do not affect the prior or regularization terms in $\mathcal{F}$ (see Eqn. **6.16**). For simplicity, we summarize the parameters of interest by $\boldsymbol{\theta} = \{\boldsymbol{\mu}, \boldsymbol{\sigma}, \boldsymbol{\gamma}, \boldsymbol{\alpha}, \boldsymbol{\beta}\}$. These are optimized by EM (see e.g. Dempster *et al.*, 1977; Bishop, 1995 or Neal and Hinton, 1998), which can be considered as using some distribution, q_{ik}, to minimize the following upper bound on $\mathcal{E}$:

$$\varepsilon \le \varepsilon_{EM} = -\sum_{i=1}^{I} \log P(y_i|\boldsymbol{\theta}) + \sum_{i=1}^{I}\sum_{k=1}^{K} q_{ik} \log\left(\frac{q_{ik}}{P(c_i = k|y_i, \boldsymbol{\theta})}\right) \qquad 6.17$$

EM is an iterative approach, and involves alternating between an E-step (which minimizes $\mathcal{E}_{EM}$ with respect to q_{ik}), and an **M**-step (which minimizes $\mathcal{E}_{EM}$ with respect to $\boldsymbol{\theta}$). The second term of Eqn. **6.17** is a Kullback-Leibler distance, which is at a minimum of zero when $q_{ik} = P(c_i = k|y_i, \boldsymbol{\theta})$, and Eqn. **6.17** becomes an equality ($\mathcal{E} = \mathcal{E}_{EM}$).

Because q_{ik} does not enter into the first term, the E-step of iteration n consists of setting:

$$q_{ik}^{(n)} = P(c_i = k|y_i, \boldsymbol{\theta}^{(n)}) = \frac{P(y_i, c_i = k|\boldsymbol{\theta}^{(n)})}{P(y_i|\boldsymbol{\theta}^{(n)})} = \frac{p_{ik}}{\sum_{j=1}^{K} p_{ij}} \qquad 6.18$$

where

$$p_{ik} = \frac{\gamma_k b_{ik}(\boldsymbol{\alpha})}{\sum_{j=1}^{K} \gamma_j b_{ij}(\boldsymbol{\alpha})}(2\pi\sigma_k^2)^{-\frac{1}{2}} \times \exp\left(-\frac{(\rho_i(\boldsymbol{\beta})y_i - \mu_k)^2}{2\sigma_k^2}\right) \qquad 6.19$$

The M-step uses the recently updated values of $q_{ik}^{(n)}$ in order to minimize ε with respect to $\boldsymbol{\theta}$. Eqn. **6.17** can be reformulated[3] as:

$$\varepsilon = \varepsilon_{EM} = -\sum_{i=1}^{I}\sum_{k=1}^{K} q_{ik} \log P(y_i, c_i = k|\boldsymbol{\theta}) + \sum_{i=1}^{I}\sum_{k=1}^{K} q_{ik} \log q_{ik} \qquad 6.20$$

Because the second term is independent of $\boldsymbol{\theta}$, the M-step involves assigning new values to the parameters, such that the derivatives of the following are zero:

$$-\sum_{i=1}^{I}\sum_{k=1}^{K} q_{ik} \log P(y_i, c_i = k|\boldsymbol{\theta}) = \sum_{i=1}^{I}\sum_{k=1}^{K} q_{ik}\left(\log\left(\sum_{j=1}^{K}\gamma_j b_{ij}(\boldsymbol{\alpha})\right) - \log\gamma_k\right) + \sum_{i=1}^{I}\sum_{k=1}^{K} q_{ik}\left(\frac{1}{2}\log(\sigma_k^2) + \frac{1}{2\sigma^2}(\rho_i(\boldsymbol{\beta})y_i - \mu_k)^2\right) + \sum_{i=1}^{I}\sum_{k=1}^{K} q_{ik}\left(\frac{1}{2}\log(2\pi) - \log(\rho_i(\boldsymbol{\beta})b_{ik}(\boldsymbol{\alpha}))\right) \qquad 6.21$$

Differentiating Eqn. **6.21** with respect to μ_k gives:

$$\frac{\partial\mathcal{F}}{\partial\mu_k} = \frac{\partial\varepsilon}{\partial\mu_k} = \sum_{i=1}^{I}\frac{q_{ik}^{(n)}}{\sigma_k^2}(\mu_k - \rho_i(\boldsymbol{\beta})y_i) \qquad 6.22$$

This gives the update formula for μ_k by solving for $\frac{\partial\mathcal{E}}{\partial\mu_k} = 0$

$$\mu_k^{(n+1)} = \frac{\sum_{i=1}^{I} q_{ik}^{(n)}\rho_i(\boldsymbol{\beta})y_i}{\sum_{i=1}^{I} q_{ik}^{(n)}} \qquad 6.23$$

[3] Through Bayes' rule, and because $\sum_{k=1}^{K} q_{ik} = 1$, we obtain $\log P(y_i|\boldsymbol{\theta}) = \log\left(\frac{P(y_i, c_i=k|\boldsymbol{\theta})}{P(c_i=k|y_i,\boldsymbol{\theta})}\right) = \sum_{k=1}^{K} q_{ik}\log\left(\frac{P(y_i, c_i=k|\boldsymbol{\theta})}{P(c_i=k|y_i,\boldsymbol{\theta})}\right)$.

Similarly, differentiating Eqn. **6.21** with respect to σ_k^2:

$$\frac{\partial \mathcal{F}}{\partial \sigma_k^2} = \frac{\partial \mathcal{E}}{\partial \sigma_k^2} = \frac{\sum_{i=1}^{I} q_{ik}^{(n)}}{2\sigma_k^2} - \frac{\sum_{i=1}^{I} q_{ik}^{(n)}(\mu_k - \rho_i(\boldsymbol{\beta})y_i)^2}{2(\sigma_k^2)^2} \quad 6.24$$

This gives the update formula for σ_k^2:

$$(\sigma_k^2)^{(n+1)} = \frac{\sum_{i=1}^{I} q_{ik}^{(n)}(\mu_k^{(n+1)} - \rho_i(\boldsymbol{\beta})y_i)^2}{\sum_{i=1}^{I} q_{ik}^{(n)}} \quad 6.25$$

Differentiating Eqn. **6.21** with respect to γ_k:

$$\frac{\partial \mathcal{F}}{\partial \gamma_k} = \frac{\partial \varepsilon}{\partial \gamma_k} = \sum_{i=1}^{I} \frac{b_{ik}(\boldsymbol{\alpha})}{\sum_{j=1}^{K} \gamma_j b_{ij}(\boldsymbol{\alpha})} - \frac{\sum_{i=1}^{I} q_{ik}^{(n)}}{\gamma_k} \quad 6.26$$

Deriving an exact update scheme for γ_k is difficult, but the following ensures convergence:[4]

$$\gamma_k^{(n+1)} = \frac{\sum_{i=1}^{I} q_{ik}^{(n)}}{\sum_{i=1}^{I} \frac{b_{ik}(\boldsymbol{\alpha})}{\sum_{j=1}^{K} \gamma_j^{(n)} b_{ij}(\boldsymbol{\alpha})}} \quad 6.27$$

Bias ($\boldsymbol{\beta}$)

The next step is to update the estimate of the bias field. This involves holding the other parameters fixed, and improving the estimate of $\boldsymbol{\beta}$ using an LM optimization approach (see Press *et al.*, 1992 for more information). Each iteration requires the first and second derivatives of the objective function, with respect to the parameters. In the following scheme, **I** is an identity matrix and λ is a scaling factor. The choice of λ is a trade-off between speed of convergence, and stability. A value of zero for λ gives the Newton-Raphson or Gauss-Newton optimization scheme, which may be unstable. Increasing λ will slow down the convergence, but increase the stability of the algorithm. The value of λ is usually decreased slightly after iterations that decrease (improve) the cost function. If the cost function increases after an iteration, then the previous solution is retained, and λ is increased in order to provide more stability.

$$\boldsymbol{\beta}^{(n+1)} = \boldsymbol{\beta}^{(n)} - \left(\left.\frac{\partial^2 \mathcal{F}}{\partial \boldsymbol{\beta}^2}\right|_{\beta^{(n)}} + \lambda \mathbf{I} \right)^{-1} \left.\frac{\partial F}{\partial \boldsymbol{\beta}}\right|_{\beta^{(n)}} \quad 6.28$$

The prior probability of the parameters is modelled by a multivariate Gaussian density, with mean $\boldsymbol{\beta}_0$ and covariance $\mathbf{C}_{\boldsymbol{\beta}}$.

$$-\log P(\boldsymbol{\beta}) = \frac{1}{2}(\boldsymbol{\beta} - \boldsymbol{\beta}_0)\mathbf{C}_{\boldsymbol{\beta}}^{-1}(\boldsymbol{\beta} - \boldsymbol{\beta}_0) + \text{const} \quad 6.29$$

[4] The update scheme was checked empirically, and found to always reduce $\mathcal{E}$. It does not fully minimize it though, which means that this part of the algorithm is really a generalized EM.

The first and second derivatives of $\mathcal{F}$ (see Eqn. **6.16**) with respect to the parameters are therefore:

$$\frac{\partial \mathcal{F}}{\partial \boldsymbol{\beta}} = \frac{\partial \varepsilon}{\partial \boldsymbol{\beta}} + \mathbf{C}_{\boldsymbol{\beta}}^{-1}(\boldsymbol{\beta} - \boldsymbol{\beta}_0) \text{ and } \frac{\partial^2 \mathcal{F}}{\partial \boldsymbol{\beta}^2} = \frac{\partial^2 \varepsilon}{\partial \boldsymbol{\beta}^2} + \mathbf{C}_{\boldsymbol{\beta}}^{-1} \quad 6.30$$

The first and second partial derivatives of ε are:

$$\frac{\partial \varepsilon}{\partial \beta_m} = -\sum_{i=1}^{I} \frac{\partial \rho_i(\boldsymbol{\beta})}{\partial \beta_m} \times \left(\rho_i(\boldsymbol{\beta})^{-1} + y_i \sum_{k=1}^{K} \frac{q_{ik}(\mu_k - \rho_i(\boldsymbol{\beta})y_i)}{\sigma_k^2} \right) \quad 6.31$$

$$\begin{aligned}\frac{\partial^2 \varepsilon}{\partial \beta_m \beta_n} &= \sum_{i=1}^{I} \frac{\partial \rho_i(\boldsymbol{\beta})}{\partial \beta_m} \frac{\partial \rho_i(\boldsymbol{\beta})}{\partial \beta_n} \left(\rho_i(\boldsymbol{\beta})^{-2} + y_i^2 \sum_{k=1}^{K} \frac{q_{ik}}{\sigma_k^2} \right) \\ &\quad - \sum_{i=1}^{I} \frac{\partial^2 \rho_i(\boldsymbol{\beta})}{\partial \beta_m \partial \beta_n} \\ &\quad \times \left(\rho_i(\boldsymbol{\beta})^{-1} + y_i \sum_{k=1}^{K} \frac{q_{ik}(\mu_k - \rho_i(\boldsymbol{\beta})y_i)}{\sigma_k^2} \right)\end{aligned} \quad 6.32$$

The bias field is parameterized by the exponential of a linear combination of smooth basis functions:

$$\rho_i(\boldsymbol{\beta}) = \exp\left(\sum_{m=1}^{M} \beta_m \psi_{im} \right), \quad \frac{\partial \rho_i(\boldsymbol{\beta})}{\partial \beta_m} = \psi_{im} \rho_i(\boldsymbol{\beta}),$$

$$\text{and } \frac{\partial^2 \rho_i(\boldsymbol{\beta})}{\partial \beta_m \partial \beta_n} = \psi_{im} \psi_{in} \rho_i(\boldsymbol{\beta}) \quad 6.33$$

Therefore, the derivatives used by the optimization are:

$$\begin{aligned}\frac{\partial \varepsilon}{\partial \beta_m} &= -\sum_{i=1}^{I} \psi_{im} \left(1 + \rho_i(\boldsymbol{\beta}) y_i \sum_{k=1}^{K} \frac{q_{ik}(\mu_k - \rho_i(\boldsymbol{\beta})y_i)}{\sigma_k^2} \right) \\ \frac{\partial^2 \varepsilon}{\partial \beta_m \beta_n} &= \sum_{i=1}^{I} \psi_{im} \psi_{in} \left((\rho_i(\boldsymbol{\beta})y_i)^2 \sum_{k=1}^{K} \frac{q_{ik}}{\sigma_k^2} \right. \\ &\quad \left. - \rho_i(\boldsymbol{\beta}) y_i \sum_{k=1}^{K} \frac{q_{ik}(\mu_k - \rho_i(\boldsymbol{\beta})y_i)}{\sigma_k^2} \right)\end{aligned} \quad 6.34$$

Deformations ($\boldsymbol{\alpha}$)

The same LM strategy (Eqn. **6.28**) is used as for updating the bias. Schemes such as LM or Gauss-Newton are usually used only for registering images with a mean squared difference objective function, although some rare exceptions exist where LM has been applied to information-theoretic image registration (Thévenaz and Unser, 2000). The strategy requires the first and second derivatives of the cost function, with respect to the parameters that define the deformation. In order to simplify deriving the derivatives, the likelihood component of the objective function is re-expressed as:

$$\mathcal{E} = -\sum_{i=1}^{I} \log\left(\sum_{k=1}^{K} f_{ik} l_{ik}\right) - \sum_{i=1}^{I} \log \rho_i(\boldsymbol{\beta}) \qquad 6.35$$

where

$$f_{ik} = \frac{b_{ik}(\boldsymbol{\alpha})}{\sum_{j=1}^{K} \gamma_j b_{ij}(\boldsymbol{\alpha})} \qquad 6.36$$

and

$$l_{ik} = \gamma_k (2\pi\sigma_k^2)^{-\frac{1}{2}} \exp\left(-\frac{(\rho_i(\boldsymbol{\beta}) y_i - \mu_k)^2}{2\sigma_k^2}\right) \qquad 6.37$$

The first derivatives of ε with respect to α are:

$$\frac{\partial \varepsilon}{\partial \alpha_m} = -\sum_{i=1}^{I} \frac{\sum_{k=1}^{K} \frac{\partial f_{ik}}{\partial \alpha_m} l_{ik}}{\sum_{k=1}^{K} f_{ik} l_{ik}} \qquad 6.38$$

The second derivatives are:

$$\frac{\partial^2 \mathcal{E}}{\partial \alpha_m \partial \alpha_n} = \sum_{i=1}^{I} \frac{\left(\sum_{k=1}^{K} \frac{\partial f_{ik}}{\partial \alpha_m} l_{ik}\right)\left(\sum_{k=1}^{K} \frac{\partial f_{ik}}{\partial \alpha_n} l_{ik}\right)}{\left(\sum_{k=1}^{K} f_{ik} l_{ik}\right)^2} - \sum_{i=1}^{I} \frac{\sum_{k=1}^{K} \frac{\partial^2 f_{ik}}{\partial \alpha_m \partial \alpha_n} l_{ik}}{\sum_{k=1}^{K} f_{ik} l_{ik}} \qquad 6.39$$

The following is needed in order to compute derivatives of ε with respect to α:

$$\frac{\partial f_{ik}}{\partial \alpha_m} = \frac{\frac{\partial b_{ik}(\boldsymbol{\alpha})}{\partial \alpha_m}}{\sum_{j=1}^{K} \gamma_j b_{ij}(\boldsymbol{\alpha})} - \frac{b_{ik}(\boldsymbol{\alpha}) \sum_{j=1}^{K} \gamma_j \frac{\partial b_{ij}(\boldsymbol{\alpha})}{\partial \alpha_m}}{\left(\sum_{j=1}^{K} \gamma_j b_{ij}(\boldsymbol{\alpha})\right)^2} \qquad 6.40$$

The second term in Eqn. **6.39** is ignored in the optimization (Gauss-Newton approach), but it could be used (Newton-Raphson approach). These gradients and curvatures enter the update scheme as in Eqn. **6.28**.

The chain rule is used to compute derivatives of f_{ik}, based on the rate of change of the deformation fields with respect to changes of the parameters, and the tissue probability map gradients sampled at the appropriate points. Trilinear interpolation could be used as the tissue probability maps contain values between zero and one. Care is needed when attempting to sample the images with higher degree B-spline interpolation (Thévenaz *et al.*, 2000), as negative values should not occur. B-spline interpolation (and other generalized interpolation methods) require coefficients to be estimated first. This essentially involves deconvolving the B-spline bases from the image (Unser *et al.*, 1993a, b). Sampling an interpolated value in the image is then done by re-convolving the coefficients with the B-spline. Without any non-negativity constraints on the coefficients, there is a possibility of negative values occurring in the interpolated probability map.

One possible solution is to use a maximum-likelihood deconvolution strategy to estimate some suitable coefficients. This is analogous to the iterative method for maximum-likelihood reconstruction of PET (positron emission tomography) images (Shepp and Vardi, 1982), or to the way that mixing proportions are estimated within a mixture of Gaussians model. A second solution is to add a small background value to the probability maps, and take a logarithm. Standard interpolation methods could be applied to the log-transformed data, before exponentiating again. Neither of these approaches is really optimal. In practice, 3rd degree B-spline interpolation is used, but without first deconvolving. This introduces a small, but acceptable, amount of additional smoothness to the tissue probability maps.

Example

The segmentation accuracy is illustrated for data generated by the BrainWeb MR simulator (Kwan *et al.*, 1996; Cocosco *et al.*, 1997; Collins *et al.*, 1998). The simulated images were all of the same subject, had dimensions of 181 × 217 × 181 voxels of 1 × 1 × 1 mm and had 3 per cent noise (relative to the brightest tissue in the images). The contrasts of the images simulated T1-weighted, T2-weighted and proton density (PD). The T1-weighted image was simulated as a spoiled FLASH sequence, with a 30° flip angle, 18 ms repeat time, 10 ms echo time. The T2 and PD images were simulated by a dual echo spin echo, early echo technique, with 90° flip angle, 3300 ms repeat time and echo times of 35 and 120 ms. Three different levels of image non-uniformity were used: 0 per cent RF (which assumes that there is no intensity variation artefact), 40 per cent RF, and 100 per cent RF (Figure 6.2). Three components were considered: grey matter, white matter and whole brain (grey and white matter). Because the causes of the simulated images were available, it was possible to compare the segmented images with images of 'true' grey and white matter using the Dice metric, which is used widely for evaluating segmentation algorithms (e.g. Van Leemput *et al.*, 1999b; Shattuck *et al.*, 2001). The probabilities were thresholded at 0.5 in order to compute the number of misclassifications. If TP refers to the number of true positives, FP to false positives and FN to false negatives, then the Dice metric is given by:

$$\text{Dice metric} = \frac{2 \times \text{TP}}{2 \times \text{TP} + \text{FP} + \text{FN}} \qquad 6.41$$

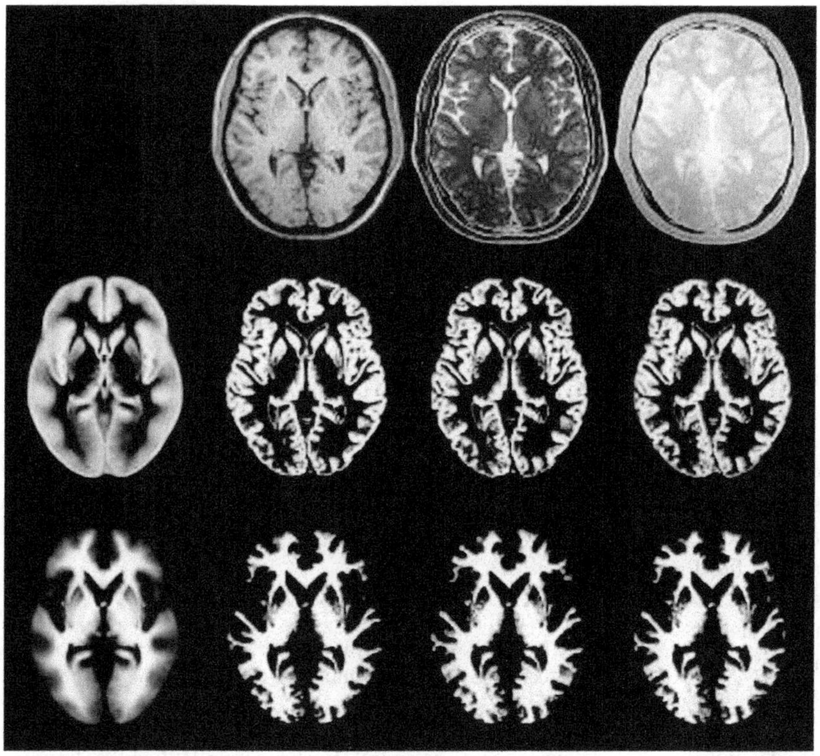

FIGURE 6.2 Results from applying the method to the BrainWeb data. The first column shows the tissue probability maps for grey and white matter. The first row of columns two, three and four show the 100 per cent RF BrainWeb T1, T2 and PD images after they are warped to match the tissue probability maps (by inverting the spatial transform). Below the warped BrainWeb images are the corresponding segmented grey and white matter.

Our results are shown in Table 6-1. Values range from zero to one, where higher values indicate better agreement.

Discussion

This chapter illustrates a framework whereby tissue classification, bias correction and image registration are integrated within the same generative model. The objective was to explain how this can be done, rather than focus on the details of a specific implementation. The same framework could be used for a more sophisticated implementation. When devising a model, it is useful to think about how that model could be used to generate data. The distribution of randomly generated data should match the distribution of any data the model has to explain. There are a number of aspects of our model that could be improved in order to achieve this goal.

The SPM5 implementation assumes that the brain consists of grey and white matter, and is surrounded by a thin layer of CSF. The addition of extra tissue probability maps should improve the model. In particular, grey matter classes for internal structures may allow them to be segmented more accurately.

It is only a single channel implementation, which can segment a single image, but is unable to make optimal use of information from two or more registered images of the same subject. Multispectral data may provide more accurate results by allowing the model to work with joint intensity probability distributions. For two registered images of the same subject, one form of objective function would use axis-aligned multivariate Gaussians (with σ_{k1}^2 and σ_{k2}^2 are diagonal elements of a 2×2 covariance matrix).

TABLE 6-1 Dice metrics computed from segmented BrainWeb images

	T1			T2			PD		
Dice metric	0%	40%	100%	0%	40%	100%	0%	40%	100%
Grey matter	0.932	0.934	0.918	0.883	0.881	0.880	0.872	0.880	0.872
White matter	0.961	0.961	0.939	0.916	0.916	0.915	0.923	0.928	0.923
Whole brain	0.977	0.978	0.978	0.967	0.966	0.965	0.957	0.959	0.955

$$
\begin{aligned}
\mathcal{E} = &-\sum_{i=1}^{I} \log \left(\frac{\rho_{i1}(\boldsymbol{\beta})\rho_{i2}(\boldsymbol{\beta})}{\sum_{k=1}^{K} \gamma_k b_{ik}(\boldsymbol{\alpha})} \right) \\
&-\sum_{i=1}^{I} \log \Bigg(\sum_{k=1}^{K} \gamma_k b_{ik}(\boldsymbol{\alpha}) \\
&\times \frac{\exp\left(-\frac{(\rho_{i1}(\boldsymbol{\beta})y_{i1}-\mu_{k1})^2}{2\sigma_{k1}^2}\right)}{(2\pi\sigma_{k1}^2)^{\frac{1}{2}}} \frac{\exp\left(-\frac{(\rho_{i2}(\boldsymbol{\beta})y_{i2}-\mu_{k2})^2}{2\sigma_{k2}^2}\right)}{(2\pi\sigma_{k2}^2)^{\frac{1}{2}}} \Bigg)
\end{aligned} \qquad 6.42
$$

Multispectral classification usually requires the images to be registered together. Another possible extension of the framework could be to include within-subject registration (Xiaohua *et al.*, 2004).

The generative model contains nothing to encode the probability that neighbouring voxels are more likely to belong to the same class. The inclusion of such priors should make the generative model more realistic. One solution could be to include a Markov random field (MRF) (Besag, 1986) in the model. Another strategy for making the model more realistic may be to have crisper tissue probability maps, and more precise warping.

Objective functions, such as the mean squared difference or cross-correlation, can only be used to register MR images generated using the same sequences, field strengths etc. An advantage that they do have over information theoretic measures (such as mutual information), is that they are also appropriate for registering to smooth averaged images. One of the benefits of the approach is that the same averaged tissue probability maps can be used to normalize spatially (and segment) images acquired with a wide range of different contrasts (e.g. T1-weighted, T2-weighted etc). This flexibility could also be considered a weakness. If the method is only to be used with images of a particular contrast, then additional constraints relating to the approximate intensities of the different tissue types could be included (Fischl *et al.*, 2002). Alternatively, the MR parameters could be estimated within the model (Fischl *et al.*, 2004), and the cluster means constrained to be more realistic. Rather than using fixed intensity distributions for the classes, a better approach would invoke some kind of hierarchical modelling, whereby prior probability distributions for the cluster parameters are used to inform their estimation.

The hierarchical modelling scheme could be extended in order to generate tissue probability maps and other priors using data from many subjects. This would involve a very large model, whereby many images of different subjects are simultaneously processed within the same hierarchical framework. Strategies for creating average (in both shape and intensity) brain atlases are currently being devised (Ashburner *et al.*, 2000; Avants and Gee, 2004; Joshi *et al.*, 2004). Such approaches could be refined in order to produce average shaped tissue probability maps and other data for use as priors.

REFERENCES

Arnold JB, Liow JS, Schaper KA *et al.* (2001) Qualitative and quantitative evaluation of six algorithms for correcting intensity nonuniformity effect. *NeuroImage* **13**: 931–43

Ashburner J, Andersson J, Friston KJ (2000) Image registration using a symmetric prior – in three-dimensions. *Hum Brain Mapp* **9**: 212–25

Ashburner J, Friston KJ (1999) Nonlinear spatial normalization using basis functions. *Hum Brain Mapp* **7**: 254–66

Ashburner J, Friston KJ (2000) Voxel-based morphometry – the methods. *NeuroImage* **11**: 805–21

Avants B, Gee JC (2004) Geodesic estimation for large deformation anatomical shape averaging and interpolation. *NeuroImage* **23**: S139–S50

Belaroussi B, Milles J, Carme S *et al.* (2006) Intensity non-uniformity correction in MRI: Existing methods and their validation. *Med Image Anal* **10**: 234–46

Besag J (1986) On the statistical analysis of dirty pictures. *J R Stat Soc Ser B* **48**: 259–302

Bishop CM (1995) *Neural networks for pattern recognition.* Oxford University Press, Oxford

Cocosco CA, Kollokian V, Kwan RK-S *et al.* (1997) Brainweb: online interface to a 3D MRI simulated brain database. *NeuroImage* **5**: S425

Collins DL, Evans AC, Holmes C *et al.* (1995) Automatic 3D segmentation of neuro-anatomical structures from MRI. In *Proc Information Processing in Medical Imaging (IPMI)*, Bizais Y, Barillot C, Di Paola R (eds). Kluwer Academic Publishers, Dordrecht

Collins DL, Zijdenbos AP, Kollokian V *et al.* (1998). Design and construction of a realistic digital brain phantom. *IEEE Trans Med Imag* **17**: 463–68

Dempster AP, Laird NM, Rubin DB (1977) Maximum likelihood from incomplete data via the EM algorithm. *J R Stat Soc Series B* **39**: 1–38

Evans AC (1995) Commentary. *Hum Brain Mapp* **2**: 165–89

Fischl B, Salat DH, Busa E *et al.* (2002) Whole brain segmentation: automated labeling of neuroanatomical structures in the human brain. *Neuron* **33**: 341–55

Fischl B, Salat DH, van der Kouwe AJW *et al.* (2004) Sequence-independent segmentation of magnetic resonance images. *NeuroImage* **23**: S69–S84

Fissell K, Tseytlin E, Cunningham D, *et al.* (2003) Fiswidgets: a graphical computing environment for neuroimaging analysis. *Neuroinformatics* **1**: 111–25

Garza-Jinich M, Yanez O, Medina V *et al.* (1999) Automatic correction of bias field in magnetic resonance images. In *Proc International Conference on Image Analysis and Processing* IEEE Computer Society, CA.

Hellier P, Ashburner J, Corouge I *et al.* (2002) Inter subject registration of functional and anatomical data using SPM. In *Proc Medical Image Computing and Computer-Assisted Intervention (MICCAI)*, vol. 2489 of *Lecture Notes in Computer Science.* Springer-Verlag, Berlin and Heidelberg

Hellier P, Barillot C, Corouge I *et al.* (2001) Retrospective evaluation of inter-subject brain registration. In *Proc Medical Image Computing and Computer-Assisted Intervention (MICCAI)*, Niessen WJ, Viergever MA (eds), vol. 2208 of *Lecture Notes in Computer Science.* Springer-Verlag, Berlin and Heidelberg, pp 258–65

Joshi S, Davis B, Jomier M *et al.* (2004) Unbiased diffeomorphic atlas construction for computational anatomy. *NeuroImage* **23**: S151–S60

Kwan RK-S, Evans AC, Pike GB (1996) An extensible MRI simulator for post-processing evaluation. In *Proc Visualization in Biomedical Computing*, Springer Verlag

Laidlaw DH, Fleischer KW, Barr AH (1998) Partial-volume bayesian classification of material mixtures in MR volume data using voxel histograms. *IEEE Trans Med Imag* **17**: 74–86

Likar B, Viergever MA, Pernuš F (2001) Retrospective correction of MR intensity inhomogeneity by information minimization. *IEEE Trans Med Imag* **20**: 1398–410

MacDonald D, Kabani N, Avis D *et al.* (2000) Automated 3-D extraction of inner and outer surfaces of cerebral cortex from MRI. *NeuroImage* **12**: 340–56

Mangin J-F (2000) Entropy minimization for automatic correction of intensity nonuniformity. In *Proc IEEE Workshop on Mathematical Methods in Biomedical Image Analysis* IEEE Computer Society, CA

Neal RM, Hinton GE (1998) A view of the EM algorithm that justifies incremental, sparse, and other variants. In *Learning in Graphical Models*, Jordan MI (ed.) Kluwer Academic Publishers, Dordrecht, pp 355–68

Pham DL, Prince JL (1999) Adaptive fuzzy segmentation of magnetic resonance images. *IEEE Trans Med Imag* **18**: 737–52

Pitiot A, Delingette H, Thompson PM *et al.* (2004) Expert knowledge-guided segmentation system for brain MRI. *NeuroImage* **23**: S85–S96

Press WH, Teukolsky SA, Vetterling WT *et al.* (1992) *Numerical Recipes in C*, 2nd edn. Cambridge University Press, Cambridge

Rex DE, Maa JQ, Toga AW (2003) The LONI pipeline processing environment. *NeuroImage* **19**: 1033–48

Shattuck DW, Sandor-Leahy SR, Schaper KA *et al.* (2001) Magnetic resonance image tissue classification using a partial volume model. *NeuroImage* **13**: 856–76

Shepp LA, Vardi Y (1982) Maximum likelihood reconstruction in positron emission tomography. *IEEE Trans Med Imag* **1**: 113–22

Sled JG, Zijdenbos AP, Evans AC (1998) A non-parametric method for automatic correction of intensity non-uniformity in MRI data. *IEEE Trans Med Imag* **17**: 87–97

Styner M (2000) Parametric estimate of intensity inhomogeneities applied to MRI. *IEEE Trans Med Imag* **19**: 153–65

Thévenaz P, Blu T, Unser M (2000) Interpolation revisited. *IEEE Trans Med Imag* **19**: 739–58

Thévenaz P, Unser M (2000) Optimization of mutual information for multiresolution image registration. *IEEE Trans Image Process* **9**: 2083–99

Tohka J, Zijdenbos A, Evans A (2004) Fast and robust parameter estimation for statistical partial volume models in brain MRI. *NeuroImage* **23**: 84–97

Unser M, Aldroubi A, Eden M (1993a) B-spline signal processing: Part I – theory. *IEEE Trans Signal Process* **41**: 821–33

Unser M, Aldroubi A, Eden M (1993b) B-spline signal processing: Part II – efficient design and applications. *IEEE Trans Signal Process* **41**: 834–48

Van Leemput K, Maes F, Vandermeulen D *et al.* (1999a) Automated model-based bias field correction of MR images of the brain. *IEEE Trans Med Imag* **18**: 885–96

Van Leemput K, Maes F, Vandermeulen D *et al.* (1999b) Automated model-based tissue classification of MR images of the brain. *IEEE Trans Med Imag* **18**: 897–908

Van Leemput K, Maes F, Vandermeulen D *et al.* (2001) A statistical framework for partial volume segmentation. In *Proc Medical Image Computing and Computer-Assisted Intervention (MICCAI)*, Niessen WJ, Viergever MA (eds) vol. 2208 of *Lecture Notes in Computer Science*. Springer-Verlag, Berlin and Heidelberg, pp 204–12

Wells WM III, Grimson WEL, Kikinis R *et al.* (1996a) Adaptive segmentation of MRI data. *IEEE Trans Med Imag* **15**: 429–42

Wells WM III, Viola P, Atsumi H *et al.* (1996b) Multi-modal volume registration by maximisation of mutual information. *Med Image Anal* **1**: 35–51

Xiaohua C, Brady M, Rueckert D (2004) Simultaneous segmentation and registration for medical image. In *Proc Medical Image Computing and Computer-Assisted Intervention (MICCAI)*, Barillot C, Haynor DR, Hellier P (eds) vol. 3216 of *Lecture Notes in Computer Science*. Springer-Verlag, Berlin and Heidelberg, pp 663–70

Zhang Y, Brady M, Smith S (2001) Segmentation of brain MR images through a hidden markov random field model and the expectation-maximization algorithm. *IEEE Trans Med Imag* **20**: 45–57

Zijdenbos AP, Forghani R, Evans AC (2002) Automatic 'pipeline' analysis of 3-D MRI data for clinical trials: application to multiple sclerosis. *IEEE Trans Med Imag* **21**: 1280–91

C H A P T E R

7

Voxel-Based Morphometry

J. Ashburner and K. Friston

INTRODUCTION

At its simplest, voxel-based morphometry (VBM) involves a voxel-wise comparison of regional grey-matter 'density' between two groups of subjects.[1] The procedure is relatively straightforward, and involves spatially normalizing and segmenting high-resolution magnetic resonance (MR) images into the same stereotaxic space. These grey-matter segments are then smoothed to a spatial scale at which differences are expressed (usually about 12 mm). Voxel-wise parametric statistical tests are performed, which compare the smoothed grey-matter images from the groups using statistical parametric mapping. Corrections for multiple comparisons are generally made using the theory of random fields.

Since the advent of magnetic resonance imaging (MRI), neuroscientists have been able to measure structures in living human brains. A large number of automated or semiautomated approaches for characterizing differences in the shape and neuroanatomical configuration of different brains have emerged due to improved resolution of anatomical scans and the development of new image processing techniques. There are many ways of identifying and characterizing structural differences among populations. Voxel-based morphometry (Wright *et al.*, 1995, 1999; Ashburner and Friston, 2000; Davatzikos *et al.*, 2001) is one such method, which has been used widely over the past decade. Its applications range from the study of progressive supranuclear palsy to neurogenetic studies of polymorphisms and age-related changes. Unlike other approaches to morphometry, VBM does not refer explicitly to anatomy; it treats images as continuous scalar measurements and tests for local differences at the appropriate spatial scale. This scale is controlled by the smoothing. Smoothing is critical for VBM because it sets a lower bound on the anatomical scale at which differences can be expressed. It does so by removing fine-scale structure that is not conserved from subject to subject. Effectively, VBM is a scale-space search for anatomical differences. There are several reasons why VBM has proved so popular: first, in contrast to conventional analyses of anatomical structures, it tests for differences anywhere in the brain. Second, VBM can be applied to any measure of anatomy, for example, grey-matter density, compression maps based on the Jacobian of deformation fields, or fractional anisotropy from diffusion weighted imaging. By choosing the data and their transformations carefully, a large range of anatomical attributes can be analysed in a simple and anatomically unbiased way.

Summary

In summary, VBM is a simple procedure that enables classical inferences about the regionally-specific effects, of experimental factors, on some structural measure. These effects are tested after discounting the large-scale anatomical differences removed by spatial normalization. Because these differences have been removed, VBM is not a surrogate for classical volume analysis of large structures or lesions (Mehta *et al.*, 2003). Furthermore, it will not replace shape analysis (Dryden and Mardia, 1998; Kendall *et al.*, 1999; Miller, 2004); VBM infers on smooth scalar fields, where constructs like 'edges' or 'shape' have no meaning. In short, VBM represents an established and effective complement to shape and volume analyses that is used widely in many basic and clinical contexts.

In what follows, we look at the various stages involved in preparing the data for VBM and then consider modelling and inference on these data.

[1] Density here refers to the relative amount of grey matter and should not be confused with cell-packing density (number of cells per unit volume of neuropil).

Statistical Parametric Mapping, by Karl Friston et al.
ISBN–13: 978-0-12-372560-8 ISBN–10: 0-12-372560-7

PREPARING THE DATA

VBM relies on good quality high-resolution MR images from different subjects, and uses these to identify correlations between disease severity and the spatial deployment of different tissue types. Many researchers assume that voxel-based morphometry can only be implemented within the statistical parametric mapping package; this is not the case. There are many software tools that could be used to perform the requisite sequence of operations. The MR images are segmented, producing images that reflect the spatial distribution of a tissue type or attribute (e.g. grey matter). To compare brains of different subjects, all the grey-matter segments are warped to the same stereotaxic space. A correction can be applied to the data that accounts for expansion and contraction during this nonlinear spatial normalization. The normalized segments are then smoothed. This makes each voxel a measure of the proportion of the brain, in a region around the voxel that is grey matter (i.e. grey-matter density). Statistical analysis using the general linear model (GLM) is used to identify regions that are significantly related to the effects under study (Friston *et al.*, 1995). The GLM is a flexible framework that allows many different tests to be applied. The output is a statistical parametric map (SPM) showing regions where tissue density differs significantly among the groups. A voxel-wise SPM comprises the result of many statistical tests, so it is necessary to correct for multiple dependent comparisons using the theory of random fields (Friston *et al.*, 1996; Worsley *et al.*,1996, 1999).

Segmentation

The images are typically partitioned into different tissue classes using one of a number of segmentation techniques. Usually, VBM involves an analysis of grey matter, but it is possible to work with other tissue classes or any other scalar measure of anatomy (e.g. fractional anisotropy). To segment the images accurately, the image needs to show clear differentiation among tissue types. Usually, high resolution T1-weighted MR images are used (Figure 7.1), although multispectral images may allow more accurate tissue classification. Different segmentation models have different requirements. For example, many approaches take smooth intensity inhomogeneities into account, so they can deal with some of the artefacts that arise in MR images. Most automated segmentation algorithms assume that the intracranial cavity contains only grey matter, white matter and cerebrospinal fluid (CSF), so they may be confounded by lesions. However, there are some that have been especially designed for dealing with such pathology (Van Leemput *et al.*, 2001), although they generally require multispectral images (e.g. aligned T1- and T2-weighted scans).

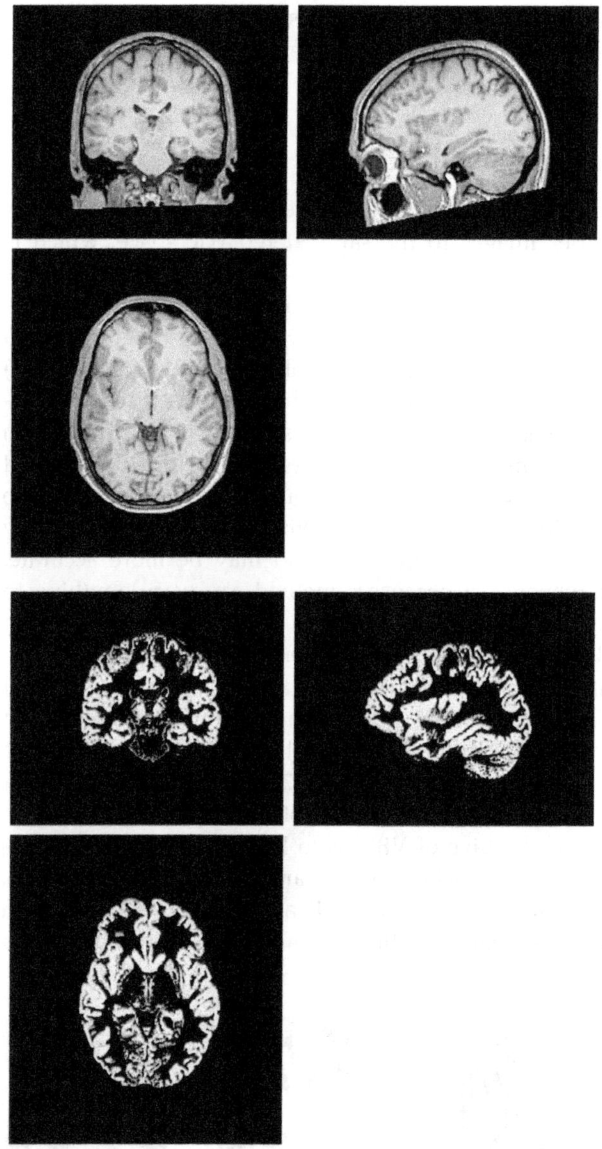

FIGURE 7.1 This figure shows the result of segmenting the grey matter from a T1-weighted MR image. Note that towards the top of the brain, the segmentation is less accurate because the grey matter is hardly visible in the original image. Similarly, in grey-matter regions such as thalamus, the intensity is closer to that of white matter, resulting in incorrect segmentation. Some partial volume voxels can be seen around the ventricles. These arise because voxels containing signal from both white matter and CSF have intensities that are close to grey matter.

Many segmentation approaches use prior spatial information in the form of tissue probability maps, which inform the segmentation algorithm about the approximate spatial distribution of the various tissues. In most cases, the tissue probability maps need to be registered with the image to segment, but the segmentation framework of SPM5 combines this registration with the tissue classification (Ashburner and Friston, 2005).

Spatial normalization

Spatial normalization involves warping all the grey-matter images to the same stereotaxic space, which is achieved by matching to a common template image (Figure 7.2). There are many non-linear registration methods that could be used for this. The unified segmentation-normalization approach (Ashburner and Friston, 2005) in SPM5 is a robust procedure because it accommodates conditional dependences among the segmentation and normalization parameters. It is a low-dimensional method, which uses about a thousand parameters to describe the deformations. Other approaches have many more degrees of freedom, and may be more accurate. However, just because a method can warp anything to anything else, does not mean that it furnishes the most likely registration (high-dimensional procedures generally overfit the data). For example, the evaluations of Hellier *et al.* (2001, 2002), which compared registration accuracy of different approaches, found that the simple spatial normalization of SPM2 was as accurate as other methods with more degrees of freedom.

The objective of VBM is to localize regions (in stereotaxic space) where there are significant differences. Accurate spatial normalization and segmentation ensures regional differences can be attributable to local effects, rather than systematic, pathology-induced mis-registration artefacts (Bookstein, 2001). However, registration can never be perfect because there is no correspondence between one subject and another at a fine spatial scale. For example, many sulci are shared between brains, but this is not the case for all. Generally, the primary sulci, which are formed earliest and tend to be quite deep, are conserved over subjects. Sulci that develop later are much more variable. Therefore, some sulci can be matched objectively, whereas others cannot. VBM accommodates the fact that the scale of anatomical correspondence, and implicitly the scale of differences in anatomy, has a lower bound by smoothing the segments (see below).

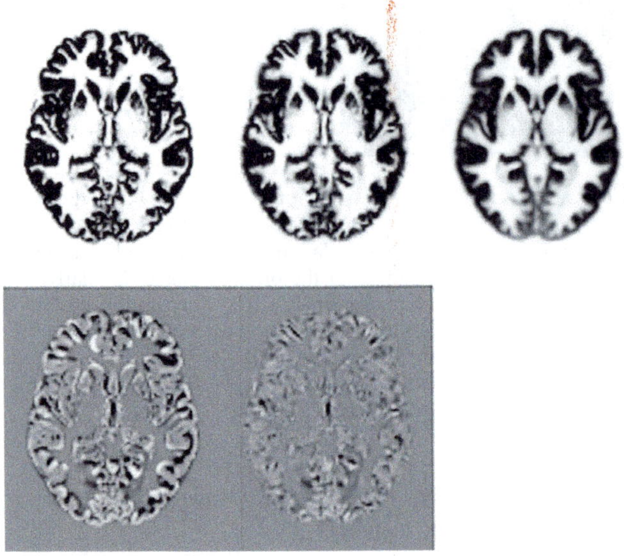

FIGURE 7.2 An image (top left) is warped to match a template (top right) to produce a spatially normalized version (top centre). For clarity, the original image was approximately aligned with the template, and the warping only done in two dimensions. The bottom row shows the difference between the template and image, both before and after the warping.

Jacobian adjustment

Non-linear spatial normalization changes the volumes of brain regions. This has implications for the interpretation of what VBM tests. The objective of VBM is to identify regional differences in the composition of brain tissue. To preserve the actual amounts of a tissue class within each voxel (Goldszal *et al.*, 1998; Davatzikos *et al.*, 2001), a further processing step can be incorporated that multiplies (modulates) the partitioned images by the relative voxel volumes before and after warping. These relative volumes are simply the Jacobian determinants of the deformation field (Figure 7.3). In the limiting case of extremely precise registration (using very high-dimensional registration), all the segments would be identical. The adjustment preserves the differences in volume of a particular tissue class (Figure 7.4).

The deformations from spatial normalization map points in a template (x_1, x_2, x_3) to equivalent points in individual source images (y_1, y_2, y_3). The derivatives of the deformation field can be thought of as matrix at each point. These are the Jacobian matrices of the deformations, and are defined by:

$$\mathrm{J} = \begin{bmatrix} \partial y_1/\partial x_1 & \partial y_1/\partial x_2 & \partial y_1/\partial x_3 \\ \partial y_2/\partial x_1 & \partial y_2/\partial x_2 & \partial y_2/\partial x_3 \\ \partial y_3/\partial x_1 & \partial y_3/\partial x_2 & \partial y_3/\partial x_3 \end{bmatrix}$$

The determinant of this matrix encodes the relative volumes of deformed and undeformed structures.

Smoothing

The warped grey-matter images are now smoothed by convolving with an isotropic Gaussian kernel. This makes the subsequent voxel-by-voxel analysis comparable to a region of interest approach, because each voxel in the

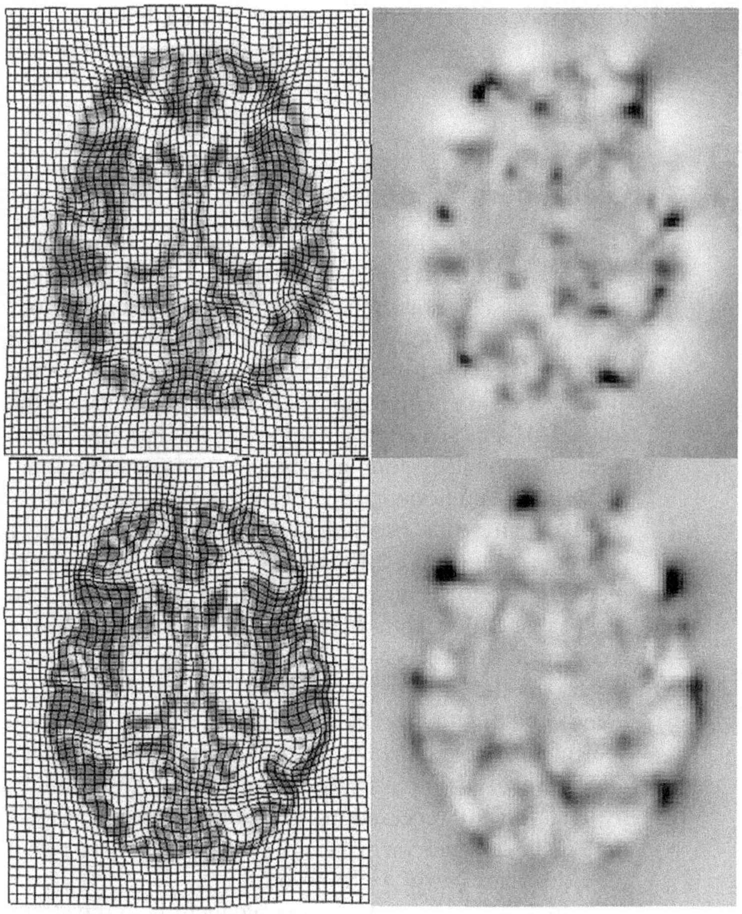

FIGURE 7.3 Warping an image results in some regions growing and others shrinking. The top-left image is a spatially normalized image with deformations overlayed (the same data as in Figure 7.2). Beside it is a map of Jacobian determinants, where darker regions indicate regions where the deformation lines are closer together. The bottom-left image is the original un-deformed image, with the deformations that would spatially normalize it overlayed. This deformation field is the inverse of the one shown above it. Beside it are the Jacobian determinants showing expansion and contraction.

smoothed images contains the average amount of grey matter from around the voxel (where the region around the voxel is defined by the form of the smoothing kernel; Figure 7.5). This is often referred to as grey-matter density, but should not be confused with cell packing density measured cytoarchitectonically. Critically, smoothing removes finescale structure from the data that is not conserved from subject to subject. This increases the sensitivity of VBM to differences that are expressed at a larger spatial scale. The smoothing conforms to the matched filter theorem, which says that the smoothing should match the scale of the difference in question. Normally, the smoothing kernel is Gaussian with a full width at half maximum (FWHM) of between 4 and 16 mm. By the central limit theorem, smoothing also has the effect of rendering the data more normally distributed, thus increasing the validity of parametric statistical tests.

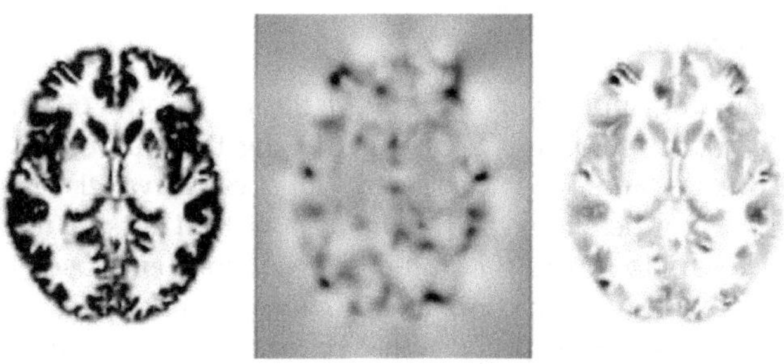

FIGURE 7.4 The image on the left is a warped grey-matter image. In the centre is a map of the Jacobian determinants from the warp (see Figure 7.3). The image on the right is the result of multiplying them together, such that the amount of grey matter is preserved.

STATISTICAL MODELLING AND INFERENCE

Voxel-wise statistical tests are performed on the preprocessed images using standard linear models (GLM). The results of these tests are an SPM (Friston *et al.*, 1995) showing significant regional effects. The GLM allows many different tests to be applied, ranging from group comparisons and identification of regional correlates of disease severity or age, to complex interactions among these effects. In the GLM, the model is expressed as

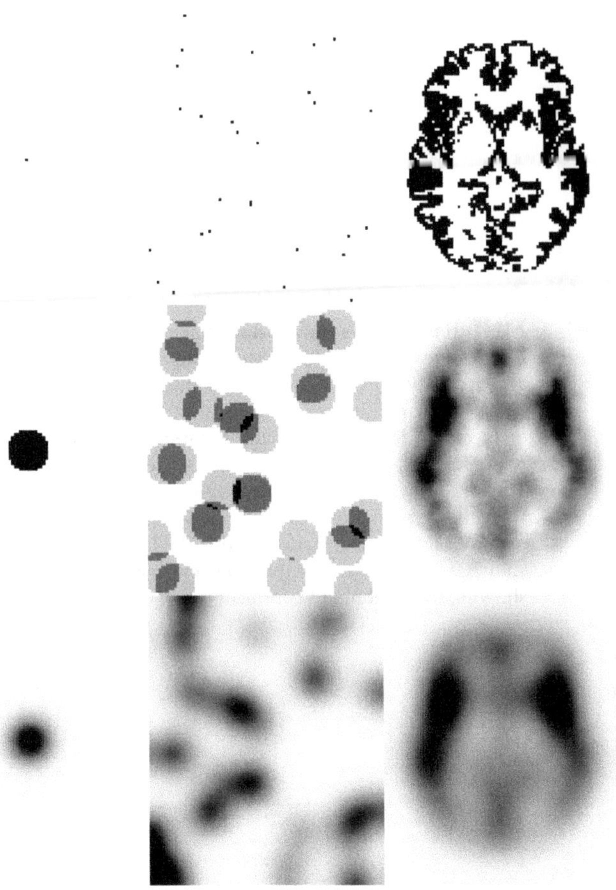

FIGURE 7.5 Smoothing effectively converts the images to maps containing a weighted count of the number of grey-matter voxels within each region. The top row shows three un-smoothed images. The middle row shows these images after they have been convolved with a circular shaped kernel. It should be clear that the result is a count of the values within each circular region. The bottom row shows the images after convolving with a Gaussian shaped kernel. The result here is a weighted count of the values around each point, where values in the centre are weighted more heavily.

a design matrix. Each row of this matrix corresponds to a scan, whereas each column is some effect that is modelled. At each voxel, the optimal linear combination of effects is computed in terms of parameter estimates. Statistics are then based on the parameter estimates at each voxel. Standard statistics (t-tests and F-tests) are used to test the hypotheses, and are valid providing the residuals, after fitting the model, are normally distributed.

Occasionally, researchers want to mix images from different scanners, or data acquired using different sequences. The effects of different types of images can generally be modelled by including confounds in the design matrix. Note that for a comparison of one group versus another, where one group is collected on one scanner, and the other group is collected on another, then modelling the scanner effect will explain away all group differences. An important confound is the global or whole measure.

Global normalization and other confounds

Global normalization is a generic issue in neuroimaging (Friston *et al.*, 1990). SPM is used to identify regionally specific (i.e. non-global) effects. This can be ensured by treating the global signal as a confound, which is explained away during inference. Because the global effect is the sum of all regional effects, this means that the sum of all regional effects, over the brain, is zero. Global normalization therefore partitions effects into regional and global components. In the context of grey matter-VBM density, this involves modelling the global grey matter as a linear confound (or by scaling each segment by its global mean). The implicit partitioning into regional and global effects finesses the characterization of anatomical differences. For example, a regional increase in grey-matter density can mean one of two things. First, it could be due to local hyperplasia, or it could be a local sparing of the region in the context of diffuse cortical atrophy everywhere else. The resolution of this ambiguity rests on inference about the global measure. In the first instance, there will be no global difference, but it would be significant in the second.

There are other confounds people include in the statistical models that remove variance in grey-matter density that could have been caused by mechanisms that are non-specific (in relation to cohort selection or experimental factors). These can be expressed in a regionally selective fashion (e.g. age or experimental factors) or globally. For example, larger brains have proportionally more white matter than grey (Zhang and Sejnowski, 2000), leading some people to treat total intracranial volume as a confound.

Inference

Following application of the GLM, the significance of any regional difference is ascertained using the parameters' estimates and their associated statistics. A voxel-wise SPM comprises the results of many statistical tests, so it is necessary to correct for multiple dependent comparisons. Without any correction, the number of false-positive results would be proportional to the number of independent tests. A Bonferroni correction would be applied if the tests were independent, but this is not the case because of the inherent spatial smoothness of the data. In practice, the effective number of independent statistical tests is determined using random field theory (RFT)

(Friston *et al.*, 1996; Worsley *et al.*, 1996, 1999). RFT provides a correction for multiple dependent comparisons that controls the rate of false-positive results. Parametric statistical tests are based on assumptions about the data. In particular, they assume that the residuals after fitting the model are normally distributed. There is a special caveat in VBM concerning the normality assumption: grey-matter segments contain values between zero and one, where most of the values are close to either of the extremes. In this case, smoothing is essential to render the residuals sufficiently normal. For group comparisons, normality can be assured with smoothing at or above 4 mm FWHM (Ashburner and Friston, 2000). For single-case comparisons, numerical studies suggest that at least 12 mm should be used (Salmond *et al.*, 2002). An alternative, for unbalanced designs like, for example single-case studies, are non-parametric methods (Holmes *et al.*, 1996; Bullmore *et al.*, 1999).

The *t* or *F*-fields produced by SPM are thresholded at some value. Regions of the field that survive the threshold (called 'clusters', 'excursion sets' or simply 'blobs') are then examined further. SPM allows inferences to be made based on the spatial extent of blobs, under the assumption that large blobs are less likely to occur by chance than small blobs. In the SPM software, implementation of RFT inferences on the extent of a cluster (but not the height of a peak) assumes that the smoothness of the residuals is roughly the same everywhere. This is assured for data that have been smoothed. However, without smoothing, large blobs that occur by chance in smooth regions may be declared significant, whereas smaller true blobs in rough regions may be missed.

Non-stationary behaviour of error terms can also misdirect the interpretation of true positive results. Spatial changes in error variance can cause blobs to move towards regions with low variance and away from regions with high residual variance (Bookstein, 2001). This explains the occasional finding of significant differences outside the brain (because the residual variance approaches zero as one moves away from the brain). Accurate localization of differences should not use the SPM. As with all applications of SPM, interpretation rests on looking at the parameter estimates that subtend the significant effect. In this instance the grey-matter difference maps provide the veridical localization.

Frequentist statistical tests cannot be used to prove a hypothesis, only to reject a null hypothesis. Any significant differences that are detected could be explained by a number of different causes, which are not disambiguated by the inference *per se*. When the null hypothesis has been rejected, it does not impute a particular explanation for the difference if there are several causes that could explain it (Figure 7.6). The preprocessing employed by VBM manipulates the data so

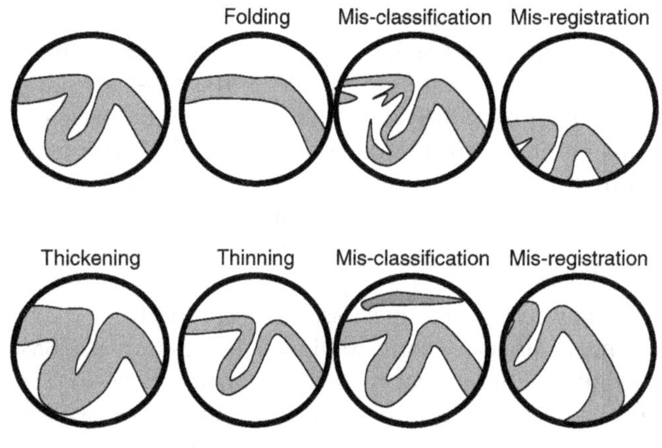

FIGURE 7.6 Significant differences determined using VBM can be explained in many ways.

that the ensuing tests are more sensitive to some causes relative to others. In particular, VBM has been devised to be sensitive to systematic differences in the density of a particular tissue class, but significant results can also arise for other reasons. A critique of VBM is that it is sensitive to systematic shape differences attributable to mis-registration from the spatial normalization step (Bookstein, 2001). This is one of a number of potential systematic differences that can arise (Ashburner and Friston, 2001). This does not mean that VBM is invalid but the explanation for the significant result may be complicated. VBM will only detect systematic group differences. This applies to mis-registration. Because the registration is driven by anatomy, any difference must be due to anatomical differences but these may not be mediated locally (e.g. misplacement of the cortex by subcortical lesions). Some explanations can be trivial (e.g. some patient groups may have to be positioned differently in the scanner) and it is important to identify confounds of this sort. In short, there may be real differences among the data, but these may not necessarily be due to reductions in regional grey matter.

VBM in clinical research

Neurodegeneration involves regional loss of tissue over time. Volumetric changes can be inferred on voxel-compression maps from high-resolution warping algorithms applied to longitudinal data (Freeborough and Fox, 1998; Thompson *et al.*, 2000). This approach has shown that the rate of tissue loss, rather than tissue distribution itself, has a better predictive validity in relation to disease progression. Comparisons of these maps between subjects requires some kind of regional averaging: either in the form of parcellation (Goldszal *et al.*, 1998) or spatial smoothing, as in VBM. Regions of compression and dilation often lie close to each other, so simple regional averaging approaches may cause differences to cancel (Scahill *et al.*, 2002). The most straightforward approach involves partitioning the compression maps into different tissue classes (cf. the RAVENS map approach (Davatzikos *et al.*, 2001)). VBM with simple intra-subject rigid registration of longitudinal data is a simpler way of achieving similar results (Ashburner and Friston, 2001). Although it is useful for research, VBM may not be the most sensitive approach for diagnosing a disease in an individual. For example, VBM may be able to say that Alzheimer patients have smaller hippocampi than controls, but it is difficult to say for certain that an individual is in the early stages of the disease, simply by examining hippocampal volume. We have contrasted generative models of regional pathology and recognition models for diagnosis and classification. Classification approaches (Davatzikos, 2004; Lao *et al.*, 2004; Davatzikos *et al.*, 2005) are generally not interested in regionally specific differences, but try to find the best non-linear function of the data, over the entire brain, which predicts diagnosis. Multivariate kernel methods, such as support-vector machines (Vapnik, 1998), relevance-vector machines (Tipping, 2001) or Gaussian process models (Williams and Barber, 1998), are established recognition models that may be useful in diagnosis and endo-phenotyping.

REFERENCES

Ashburner J, Friston KJ (2000) Voxel-based morphometry – the methods. *NeuroImage* **11**: 805–21

Ashburner J, Friston KJ (2001) Why voxel-based morphometry should be used. *NeuroImage* **14**: 1238–43

Ashburner J, Friston KJ (2005) Unified segmentation. *NeuroImage* **26**: 839–51

Bookstein FL (2001) Voxel-based morphometry should not be used with imperfectly registered images. *NeuroImage* **14**: 1454–62

Bullmore E, Suckling J, Overmeyer S *et al.* (1999) Global, voxel, and cluster tests, by theory and permutation, for a difference between two groups of structural MR images of the brain. *IEEE Trans Med Imag* **18**: 32–42

Davatzikos C (2004) Why voxel-based morphometric analysis should be used with great caution when characterizing group differences. *NeuroImage* **23**: 17–20

Davatzikos C, Genc A, Xu D *et al.* (2001) Voxel-based morphometry using the RAVENS maps: methods and validation using simulated longitudinal atrophy. *NeuroImage* **14**: 1361–69

Davatzikos C, Shen D, Gur RC, *et al.* (2005) Whole-brain morphometric study of schizophrenia revealing a spatially complex set of focal abnormalities. *Arch Gen Psychiatr* **62**: 1218–27

Dryden IL, Mardia KV (1998) *Statistical shape analysis*. John Wiley and Sons, Chichester

Freeborough PA, Fox NC (1998) Modelling brain deformations in alzheimer disease by fluid registration of serial MR images. *J Comput Assist Tomogr* **22**: 838–43

Friston KJ, Frith CD, Liddle PF *et al.* (1990) The relationship between global and local changes in PET scans. *J Cereb Blood Flow Metab* **10**: 458–66

Friston KJ, Holmes AP, Poline J-B *et al.* 1996 Detecting activations in PET and fMRI: levels of inference and power. *NeuroImage* **4**: 223–35

Friston KJ, Holmes AP, Worsley KJ *et al.* (1995) Statistical parametric maps in functional imaging: a general linear approach. *Hum Brain Mapp* **2**: 189–210

Goldszal AF, Davatzikos C, Pham DL *et al.* (1998) An image-processing system for qualitative and quantitative volumetric analysis of brain images. *J Comput Assist Tomogr* **22**: 827–37

Hellier P, Ashburner J, Corouge I *et al.* (2002) Inter subject registration of functional and anatomical data using SPM. In *Proc Medical Image Computing and Computer-Assisted Intervention (MICCAI)*, vol. 2489 of *Lecture Notes in Computer Science*. Springer-Verlag, Berlin and Heidelberg, pp 590–87

Hellier P, Barillot C, Corouge I *et al.* (2001) Retrospective evaluation of inter-subject brain registration. In *Proc Medical Image Computing and Computer-Assisted Intervention (MICCAI)*, Niessen WJ, Viergever MA (eds), vol. 2208 of *Lecture Notes in Computer Science*. Springer-Verlag, Berlin and Heidelberg, pp 258–65

Holmes AP, Blair RC, Ford I (1996) Non-parametric analysis of statistic images from functional mapping experiments. *J Cereb Blood Flow Metab* **16**: 7–22

Kendall DG, Barden D, Carne TK *et al.* (1999) *Shape and shape theory*. Wiley, Chichester

Lao Z, Shen D, Xue Z *et al.* (2004) Morphological classification of brains via high-dimensional shape transformations and machine learning methods. *NeuroImage* **21**: 46–57

Mehta S, Grabowski TJ, Trivedi Y *et al.* (2003) Evaluation of voxel-based morphometry for focal lesion detection in individuals. *NeuroImage* **20**: 1438–54

Miller MI (2004) Computational anatomy: shape, growth, and atrophy comparison via diffeomorphisms. *NeuroImage* **23**: S19–S33

Salmond CH, Ashburner J, Vargha-Khadem F *et al.* (2002) Distributional assumptions in voxel-based morphometry. *NeuroImage* **17**: 1027–30

Scahill RI, Schott JM, Stevens JM *et al.* (2002) Mapping the evolution of regional atrophy in Alzheimer's disease: unbiased analysis of fluid-registered serial MRI. *Proc Nat Acad Sci* **99**: 4703–07

Thompson PM, Giedd JN, Woods RP *et al.* (2000) Growth patterns in the developing brain detected by using continuum mechanical tensor maps. *Nature* **404**: 190–93

Tipping ME (2001) Sparse bayesian learning and the relevance vector machine. *J Machine Learning Res* **1**: 211–44

Van Leemput K, Maes F, Vandermeulen D *et al.* (2001). Automated segmentation of multiple sclerosis lesions by model outlier detection. *IEEE Trans Med Imag* **20**: 677–88

Vapnik VN (1998) *Statistical learning theory*. Wiley, New York

Williams CKI, Barber D (1998) Bayesian classification with Gaussian processes. *IEEE Trans Pattern Recogn Machine Intelligence* **20**: 1342–51

Worsley KJ, Andermann M, Koulis T *et al.* (1999) Detecting changes in non-isotropic images. *Hum Brain Mapp* **8**: 98–101

Worsley KJ, Marrett S, Neelin P *et al.* (1996) A unified statistical approach for determining significant voxels in images of cerebral activation. *Hum Brain Mapp* **4**: 58–73

Wright IC, Ellison ZR, Sharma T *et al.* (1999) Mapping of grey matter changes in schizophrenia. *Schizophren Res* **35**: 1–14

Wright IC, McGuire PK, Poline J-B *et al.* (1995) A voxel-based method for the statistical analysis of gray and white matter density applied to schizophrenia. *NeuroImage* **2**: 244–52

Zhang K, Sejnowski TJ (2000) A universal scaling law between gray matter and white matter of cerebral cortex. *Proc Natl Acad Sci* **97**.

Part 3

General linear models

CHAPTER

8

The General Linear Model

S.J. Kiebel and A.P. Holmes

INTRODUCTION

In this chapter, we introduce the general linear model. All classical analyses of functional data (electroencephalography/magnetoencephalography (EEG)/(MEG)/functional magnetic resonance imaging (fMRI)/positron emission tomography (PET)) are based on this model. Analysis comprises three parts: model specification; parameter estimation; and finally inference. Here, we focus on model specification. This chapter, and indeed this *general linear model* section, should be useful in two ways. First, working through the model equations helps one to learn about the nature of models and how they are communicated. Model specification, with an equation, or design matrix (see below), is a crucial step in the analysis of imaging data. Second, these equations entail the theory that underpins analysis. Although much of the modelling in statistical parametric mapping (SPM) is automated, one often encounters situations in which it is not obvious how to model data. We hope that this and the following chapters will help in these situations.

We assume that the data have already been preprocessed (e.g. reconstructed, normalized, and smoothed). Because SPM is a mass-univariate approach, i.e. the same model is used at each voxel, in this and the following chapters, we will focus on the model for a single voxel.

Model specification is followed by parameter estimation and finally, inference using voxel-wise statistical tests. Here, we treat inferences at a single voxel. Part 4 of this book covers inferences over many voxels and the multiple comparison problem this entails.

In the first part of this chapter, we look at the analysis of PET data. In the second part, we extend the model to accommodate fMRI data. The remaining chapters of Part 3 describe, in detail, different aspects of modelling in SPM. Chapter 9 focuses on specification of contrast weights and Chapter 10 introduces the concept of non-sphericity, which is important for modelling the error. Chapter 11 extends the general linear model by combining multiple models to form hierarchies. This approach is very useful for performing random-effects analyses (Chapter 12). Chapter 13 integrates the classical analysis of variance, which is a subdomain of the general linear model, into the SPM framework. Finally, Chapters 14, 15 and 16 treat, in detail, specific design issues for fMRI and M/EEG data.

THE GENERAL LINEAR MODEL

Before turning to the specifics of PET and fMRI, we consider the general linear model. This requires some basic matrix algebra and statistical concepts. These will be used to develop an understanding of classical hypothesis testing. Healy (1986) presents a brief summary of matrix methods relevant to statistics. Newcomers to statistical methods are directed towards Mould's excellent text *Introductory Medical Statistics* (1989), while the more mathematically experienced will find Chatfield's *Statistics for Technology* (1983) useful. Draper and Smith (1981) give a good exposition of matrix methods for the general linear model, and go on to describe regression analysis in general. The definitive tome for practical statistical experimental design is Winer *et al.* (1991). An excellent book about experimental design is Yandell (1997). A rather advanced, but very useful, text on linear models is Christensen (1996).

The general linear model – introduction

Suppose we conduct an experiment in which we measure a *response variable* (such as regional cerebral blood flow (rCBF) at a particular voxel) Y_j, where $j = 1, \ldots, J$

Statistical Parametric Mapping, by Karl Friston et al.
ISBN–13: 978-0-12-372560-8 ISBN–10: 0-12-372560-7

indexes the observation. Y_j is a random variable, conventionally denoted by a capital letter.[1] Suppose also that for each observation we have a set of $L(L < J)$ *explanatory* variables (each measured without error) denoted by x_{jl}, where $l = 1, \ldots, L$ indexes the explanatory variables. The explanatory variables may be continuous (or sometimes discrete) *covariates*, functions of covariates, or they may be *dummy* variables indicating the *levels* of an experimental *factor*.

A *general linear model* explains the response variable Y_j in terms of a linear combination of the explanatory variables plus an error term:

$$Y_j = x_{j1}\beta_1 + \cdots + x_{jl}\beta_l + \cdots + x_{jL}\beta_L + \epsilon_j \quad \textbf{8.1}$$

Here the β_l are (unknown) parameters, corresponding to each of the L explanatory variables x_{jl}. The errors ϵ_j are independent and identically distributed normal random variables with zero mean and variance σ^2, written $\epsilon_j \overset{iid}{\sim} \mathcal{N}(0, \sigma^2)$.

Examples: dummy variables

Many classical parametric statistical procedures are special cases of the general linear model. We will illustrate this by going through the equations for two well-known models.

Linear regression

A simple example is linear regression, with one continuous explanatory variable x_j for each observation $j = 1, \ldots, J$. The model is usually written as:

$$Y_j = \mu + x_j\beta + \epsilon_j \quad \textbf{8.2}$$

where the unknown parameters are μ, a *constant term* in the model, the regression slope β and $\epsilon_j \overset{iid}{\sim} \mathcal{N}(0, \sigma^2)$. This can be re-written as a general linear model by using a dummy variable, which takes the value $x_{j1} = 1$ for all j:

$$Y_j = x_{j1}\mu + x_{j2}\beta_2 + \epsilon_j \quad \textbf{8.3}$$

This has the form of Eqn. **8.1**, on replacing β_1 with μ.

Two-sample t-test

Similarly, the two-sample t-test is a special case of a general linear model: suppose Y_{j1} and Y_{j2} are two independent groups of random variables. The two-sample t-test assumes $Y_{qj} \overset{iid}{\sim} \mathcal{N}(\mu_q, \sigma^2)$, for $q = 1, 2$, and assesses the null hypothesis $\mathcal{H} : \mu_1 = \mu_2$. The index j indexes the data points in both groups. The standard statistical way of writing the model is:

$$Y_{qj} = \mu_q + \epsilon_{qj} \quad \textbf{8.4}$$

The q subscript on the μ_q indicates that there are two *levels* to the group *effect*, μ_1 and μ_2. Here, $\epsilon_{qj} \overset{iid}{\sim} \mathcal{N}(0, \sigma^2)$. This can be rewritten using two dummy variables x_{qj1} and x_{qj2} as:

$$Y_{qj} = x_{qj1}\mu_1 + x_{qj2}\mu_2 + \epsilon_{qj} \quad \textbf{8.5}$$

Which has the form of Eqn. **8.1**, after re-indexing for qj. Here the dummy variables indicate group membership, where x_{qj1} indicates whether observation Y_{qj} is from the first group, in which case it has the value 1 when $q = 1$, and 0 when $q = 2$. Similarly, $x_{qj2} = \begin{cases} 0 \text{ if } q = 1 \\ 1 \text{ if } q = 2 \end{cases}$.

Matrix formulation

In the following, we describe the matrix formulation of the general linear model and derive its least-squares parameter estimator. We then describe how one can make inferences based on a contrast of the parameters. This theoretical treatment furnishes a set of equations for the analysis of any data that can be formulated as a general linear model.

The general linear model can be expressed using matrix notation. Consider writing out Eqn. **8.1** in full, for each observation j, giving a set of simultaneous equations:

$$\begin{aligned} Y_1 &= x_{11}\beta_1 + \cdots + x_{1l}\beta_l + \cdots + x_{1L}\beta_L + \epsilon_1 \\ \vdots &= \vdots \\ Y_j &= x_{j1}\beta_1 + \cdots + x_{jl}\beta_l + \cdots + x_{jL}\beta_L + \epsilon_j \\ \vdots &= \vdots \\ Y_J &= x_{J1}\beta_1 + \cdots + x_{Jl}\beta_l + \cdots + x_{JL}\beta_L + \epsilon_J \end{aligned}$$

This has an equivalent matrix form:

$$\begin{pmatrix} Y_1 \\ \vdots \\ Y_j \\ \vdots \\ Y_J \end{pmatrix} = \begin{pmatrix} x_{11} & \cdots & x_{1l} & \cdots & x_{1L} \\ \vdots & \ddots & \vdots & \ddots & \vdots \\ x_{j1} & \cdots & x_{jl} & \cdots & x_{jL} \\ \vdots & \ddots & \vdots & \ddots & \vdots \\ x_{J1} & \cdots & x_{Jl} & \cdots & x_{JL} \end{pmatrix} \begin{pmatrix} \beta_1 \\ \vdots \\ \beta_l \\ \vdots \\ \beta_L \end{pmatrix} + \begin{pmatrix} \epsilon_1 \\ \vdots \\ \epsilon_j \\ \vdots \\ \epsilon_J \end{pmatrix}$$

[1] We talk of *random variables*, and of observations prior to their measurement, because classical (frequentist) statistics is concerned with what could have occurred in an experiment. Once the observations have been made, they are known, the residuals are known, and there is no randomness.

which can be written in matrix notation as:

$$Y = X\beta + \epsilon \qquad \textbf{8.6}$$

where Y is the column vector of observations, ϵ the column vector of error terms, and β the column vector of parameters; $\beta = [\beta_1, \ldots, \beta_l, \ldots, \beta_L]^T$. The $J \times L$ matrix X, with jl^{th} element x_{jl} is the *design matrix*. It has one row per observation, and one column (explanatory variable) per model parameter. The important point about the design matrix is that it is a near complete description of our model; the remaining model assumptions are about the distribution of errors. The design matrix encodes and quantifies our knowledge about how the expected signal was produced.

Parameter estimation

Once an experiment has been completed, we have observations of the random variables Y_j, which we denote by y_j. Usually, the simultaneous equations implied by the general linear model (with $\epsilon = 0$) cannot be solved, because the number of parameters L is typically less than the number of observations J. Therefore, some method of estimating parameters that 'best fit' the data is required. This is achieved by the method of *ordinary least squares*.

Denote a set of parameter estimates by $\tilde{\beta} = [\tilde{\beta}_1, \ldots, \tilde{\beta}_L]^T$. These parameters lead to *fitted values* $\tilde{Y} = [\tilde{Y}_1, \ldots, \tilde{Y}_J]^T = X\tilde{\beta}$, giving residual errors $e = [e_1, \ldots, e_J]^T = Y - \tilde{Y} = Y - X\tilde{\beta}$. The *residual sum-of-squares* $S = \sum_{j=1}^{J} e_j^2 = e^T e$ is the sum of the square differences between the actual and fitted values, and measures the fit of the model afforded by these parameter estimates.[2] The *least squares* estimates are the parameter estimates which minimize the residual sum-of-squares. In full:

$$S = \sum_{j=1}^{J} (Y_j - x_{j1}\tilde{\beta}_1 - \cdots - x_{jL}\tilde{\beta}_L)^2$$

This is minimized when:

$$\frac{\partial S}{\partial \tilde{\beta}_l} = 2\sum_{j=1}^{J} (-x_{jl})(Y_j - x_{j1}\tilde{\beta}_1 - \cdots - x_{jL}\tilde{\beta}_L) = 0$$

This equation is the lth row of $X^T Y = (X^T X)\tilde{\beta}$. Thus, the least squares estimates, denoted by $\hat{\beta}$, satisfy the *normal equations*:

$$X^T Y = (X^T X)\hat{\beta} \qquad \textbf{8.7}$$

[2] $e^T e$ is the L_2 norm of e – geometrically equivalent to the distance between the model and data.

For the general linear model, the least squares estimates are the *maximum likelihood estimates*, and are the *best linear unbiased estimates*.[3] That is, of all linear parameter estimates consisting of linear combinations of the data, whose expectation is the true value of the parameters, the least squares estimates have the minimum variance.

If $(X^T X)$ is invertible, which it is if, and only if, the design matrix X is of full rank, then the least squares estimates are:

$$\hat{\beta} = (X^T X)^{-1} X^T Y \qquad \textbf{8.8}$$

Overdetermined models

If X has linearly dependent columns, it is *rank deficient*, i.e. $(X^T X)$ is singular and has no inverse. In this case, the model is overparameterized: there are infinitely many parameter sets describing the same model. Correspondingly, there are infinitely many least squares estimates $\hat{\beta}$ satisfying the normal equations. We will illustrate this with an example and discuss the solution that is adopted in SPM.

One way ANOVA example

A simple example of an overdetermined model is the classic Q group one-way analysis of variance (ANOVA) model. Generally, an ANOVA determines the variability in the measured response which can be attributed to the effects of factor levels. The remaining unexplained variation is used to assess the significance of the effects (Yandell, 1997, page 4 and pages 202ff). The model for a one-way ANOVA is given by:

$$Y_{qj} = \mu + \alpha_q + \epsilon_{qj} \qquad \textbf{8.9}$$

where Y_{qj} is the jth observation in group $q = 1, \ldots, Q$. Clearly, this model does not specify the parameters uniquely: for any given μ and α_q, the parameters $\mu' = \mu + d$ and $\alpha'_q = \alpha_q - d$ give an equivalent model for any constant d. That is, the model is indeterminate up to the level of an additive constant between the constant term μ and the group effects α_q. Similarly, for any set of least squares estimates $\hat{\mu}, \hat{\alpha}_q$. Here, there is one degree of indeterminacy in the model. This means the design matrix has rank Q, which is one less than the number of parameters (the number of columns of X). If the data vector Y has

[3] Gauss-Markov theorem.

observations arranged by group, then for three groups ($Q=3$), the design matrix and parameter vectors are:

$$X = \begin{bmatrix} 1 & 1 & 0 & 0 \\ \vdots & \vdots & \vdots & \vdots \\ 1 & 1 & 0 & 0 \\ 1 & 0 & 1 & 0 \\ \vdots & \vdots & \vdots & \vdots \\ 1 & 0 & 1 & 0 \\ 1 & 0 & 0 & 1 \\ \vdots & \vdots & \vdots & \vdots \\ 1 & 0 & 0 & 1 \end{bmatrix} \quad \beta = \begin{bmatrix} \mu \\ \alpha_1 \\ \alpha_2 \\ \alpha_3 \end{bmatrix}$$

This matrix is rank deficient: the first column is the sum of the others. Therefore, in this model, one cannot test for the effect of one or more groups. However, the addition of the constant μ does not affect the differences between pairs of group effects. Therefore, *differences* in group effects are uniquely estimated, regardless of the particular set of parameter estimates used. In other words, even if the model is overparameterized, there are still linear combinations of parameters (i.e. differences between pairs of group effects) that are unique. This important concept is relevant in many designs, especially for PET and multisubject data. It will be treated more thoroughly in the *Estimable functions and contrasts* section below.

Pseudoinverse constraint

In the overdetermined case, a set of least squares estimates can be found by imposing constraints on the estimates, or by inverting (X^TX) using a pseudoinverse technique, which essentially implies a constraint. In either case, it is important to remember that the estimates obtained depend on the particular constraint or pseudoinverse method chosen. This has implications for inference: it is only meaningful to consider functions of the parameters that are not influenced by the particular constraint chosen.

Some obvious constraints are based on removing columns from the design matrix. In the one-way ANOVA example, one can remove the constant term to construct a design matrix which has linearly independent columns. However, in more complicated designs, it is not clear which columns should be removed. In particular, each experimentally induced effect could be represented by one or more regressors. This precludes the removal of columns to deal with overdetermined models.

The alternative is to use a pseudoinverse method: let $(X^TX)^-$ denote the pseudoinverse of (X^TX). Then we can use $(X^TX)^-$ in place of $(X^TX)^{-1}$ in Eqn. **8.8**. A set of least squares estimates is given by $\hat{\beta} = (X^TX)^-X^TY = X^-Y$. The pseudoinverse implemented in MATLAB is the Moore-Penrose pseudoinverse.[4] This results in least squares parameter estimates with the minimum sum-of-squares (minimum L_2 norm $||\hat{\beta}||_2$). For example, with the one-way ANOVA model, this can be shown to give parameter estimates $\hat{\mu} = \sum_{j=1}^{Q}(\overline{Y}_{q\bullet})/(1+Q)$ and $\hat{\alpha}_q = \overline{Y}_{q\bullet} - \hat{\mu}$. By $\overline{Y}_{q\bullet}$ we denote the average of Y over the observation index j, i.e. the average of the data in group q.

Using the pseudoinverse for parameter estimation in overdetermined models is the approach adopted in SPM. As mentioned above, this does not allow one to test for those linear combinations of effects for which there exist an infinite number of solutions; however, it does allow us to estimate unique mixtures without changing how X is specified.

Geometrical perspective

For some, a geometrical perspective provides a nice intuition for parameter estimation. (This section can be omitted without loss of continuity.)

The vector of observed values Y defines a single point in $\Re^J$, J-dimensional Euclidean space. $X\tilde{\beta}$ is a linear combination of the columns of the design matrix X. The columns of X are J-vectors, so $X\tilde{\beta}$ for a given $\tilde{\beta}$ defines a point in $\Re^J$. This point lies in the subspace of $\Re^J$ spanned by the columns of the design matrix, the X-space. The dimension of this subspace is rank(X). Recall that the space spanned by the columns of X is the set of points Xc for all $c \in \Re^L$. The residual sum-of-squares for parameter estimates $\tilde{\beta}$ is the distance from $X\tilde{\beta}$ to Y. Thus, the least squares estimates $\hat{\beta}$ correspond to the point in the space spanned by the columns of X that is nearest to the data Y. The perpendicular from Y to the X-space meets the X-space at $\hat{Y} = X\hat{\beta}$. It should now be clear why there are no unique least squares estimates if X is rank-deficient; in this case, any point in the X-space can be obtained by infinitely many linear combinations of the columns of X, i.e. the solution exists on a hyperplane and is not a point.

If X is of full rank, then we can define a projection matrix as $P_X = X(X^TX)^{-1}X^T$. Then $\hat{Y} = P_XY$, and geometrically P_X is a projection onto the X-space. Similarly, the residual forming matrix is $R = (I_J - P_X)$, where I_J is the identity matrix of rank J. Thus $RY = e$, and R is a projection matrix onto the space orthogonal to the X-space.

As a concrete example, consider a linear regression with only three observations. The observed data $y = [y_1, y_2, y_3]^T$ defines a point in three-dimensional Euclidean space ($\Re^3$). The model (Eqn. **8.2**) leads to a

[4] If X is of full rank, then $(X^TX)^-$ is an inefficient way of computing $(X^TX)^{-1}$.

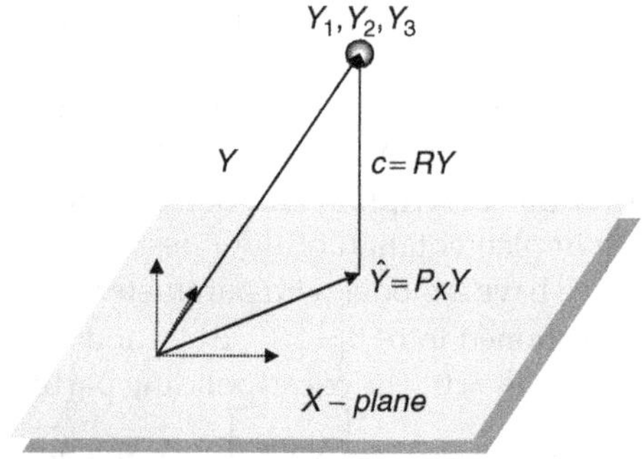

FIGURE 8.1 Geometrical perspective on linear regression. The three dimensional data Y lie in a three dimensional space. In this observation space, the (two-column) design matrix spans a subspace. Note that the axes of the design space are not aligned with the axes of the observation space. The least squares estimate is the point in the space spanned by the design matrix that has minimal distance to the data point.

design matrix $X = \begin{bmatrix} 1 & x_1 \\ 1 & x_2 \\ 1 & x_3 \end{bmatrix}$. Provided the x_js are not all the same, the columns of X span a two dimensional subspace of $\Re^3$, a plane (Figure 8.1).

INFERENCE

Here, we derive the t- and F-statistics, which are used to test for a linear combination of effects. We will also return to the issue of overdetermined models and look at which linear combinations (contrasts) we can test.

Residual sum of squares

For an independent and identical error, the residual variance σ^2 is estimated by the residual sum-of-squares divided by the appropriate degrees of freedom: $\hat{\sigma}^2 = \frac{e^T e}{J-p} \sim \sigma^2 \frac{\chi^2_{J-p}}{J-p}$ where $p = \text{rank}(X)$. (See the Appendix (8.2) for a derivation.)

Linear combinations of the parameter estimates

It is not too difficult to show that the parameter estimates are normally distributed: if X is full rank then $\hat{\beta} \sim \mathcal{N}(\beta, \sigma^2(X^TX)^{-1})$. From this it follows that for a column vector c containing L weights:

$$c^T\hat{\beta} \sim \mathcal{N}\left(c^T\beta,\ \sigma^2 c^T(X^TX)^{-1}c\right) \qquad 8.10$$

Furthermore, $\hat{\beta}$ and $\hat{\sigma}^2$ are independent (Fisher's law). Thus, prespecified hypotheses concerning linear compounds of the model parameters $c^T\beta$ can be assessed using:

$$\frac{c^T\hat{\beta} - c^T\beta}{\sqrt{\hat{\sigma}^2 c^T(X^TX)^{-1}c}} \sim t_{J-p} \qquad 8.11$$

where t_{J-p} is a Student's t-distribution with $J-p$ degrees of freedom. For example, the hypothesis $\mathcal{H} : c^T\beta = d$ can be assessed by computing

$$T = \frac{c^T\hat{\beta} - d}{\sqrt{\hat{\sigma}^2 c^T(X^TX)^{-1}c}} \qquad 8.12$$

and computing a p-value by comparing T with a t-distribution having $J-p$ degrees of freedom. In SPM, all null hypotheses are of the form $c^T\beta = 0$. Note that SPM tests based on this t-value are always one-sided.

Example – two-sample *t*-test

For example, consider the two-sample t-test. The model (Eqn. **8.4**) leads to a design matrix X with two columns of dummy variables indicating group membership and parameter vector $\beta = [\mu_1, \mu_2]^T$. Thus, the null hypothesis $\mathcal{H} : \mu_1 = \mu_2$ is equivalent to $\mathcal{H} : c^T\beta = 0$ with $c = [1, -1]^T$.

The first column of the design matrix contains n_1 1s and n_2 0s, indexing the measurements from group one, while the second column contains n_1 0s and n_2 1s for group two. Thus $(X^TX) = \begin{pmatrix} n_1 & 0 \\ 0 & n_2 \end{pmatrix}$, $(X^TX)^{-1} = \begin{pmatrix} 1/n_1 & 0 \\ 0 & 1/n_2 \end{pmatrix}$, and $c^T(X^TX)^{-1}c = 1/n_1 + 1/n_2$, giving the t-statistic (by Eqn. **8.11**):

$$T = \frac{\hat{\mu}_1 - \hat{\mu}_2}{\sqrt{\hat{\sigma}^2(1/n_1 + 1/n_2)}}$$

This is the standard formula for the two-sample t-statistic, with a Student's t-distribution of $n_1 + n_2 - 2$ degrees of freedom under the null hypothesis.

Estimable functions and contrasts

Recall that if the model is overparameterized (i.e. X is rank deficient), then there are infinitely many parameter sets describing the same model. Constraints or the use of a pseudoinverse isolate one set of parameters from infinitely many. Therefore, when examining linear compounds $c^T\beta$ of the parameters, it is imperative to consider only compounds that are invariant over the space of possible parameters. Such linear compounds are called *contrasts*. In the following, we will characterize contrasts as linear combinations having two properties, which determine whether a linear compound is a proper contrast or not.

In detail (Scheffé, 1959), a linear function $c^T\beta$ of the parameters is *estimable* if there is a linear unbiased estimate c'^TY for some constant vector of weights c'. That is $c^T\beta = E(c'^TY)$. ($E(Y)$ is the expectation of the random variable Y.) The natural estimate $c^T\hat{\beta}$ is unique for an estimable function whichever solution, $\hat{\beta}$, of the normal equations is chosen (Gauss-Markov theorem). Further: $c^T\beta = E(c'^TY) = c'^TX\beta \Rightarrow c^T = c'^TX$, so c is a linear combination of the rows of X.

A *contrast* is an estimable function with the additional property $c^T\hat{\beta} = c'^T\hat{Y} = c'^TY$. Now $c'^T\hat{Y} = c'^TY \Leftrightarrow c'^T P_XY = c'^TY \Leftrightarrow c' = P_Xc'$ (since P_X is symmetric), so c' is in the X-space. In summary, a contrast is an estimable function whose c' vector is a linear combination of the columns of X.[5]

One can test whether c is a contrast vector by combining the two properties (i) $c^T = c'^TX$ and (ii) $c' = P_Xc'$ for some vector c'. It follows that $c^T = c'^T P_XX$. Because of (i), $c^T = c^T(X^TX)^-X^TX$. In other words, c is a contrast, if it is unchanged by post-multiplication with $(X^TX)^-X^TX$. This test is used in SPM for user-specified contrasts.[6]

For a contrast it can be shown that $c^T\hat{\beta} \sim \mathcal{N}(c^T\beta, \sigma^2c'^Tc')$. Using a pseudoinverse technique, $P_X = X(X^TX)^-X^T$, so $c' = P_Xc' \Rightarrow c'^Tc' = c'^TX(X^TX)^-X^Tc' = c^T(X^TX)^-c$ since $c = c'^TX$ for an estimable function.

This shows that the distributional results given above for unique designs (Eqn. **8.10** and Eqn. **8.11**) apply to contrasts of the parameters of non-unique designs, where $(X^TX)^{-1}$ is replaced by its pseudoinverse.

In many cases, contrasts have weights that sum to zero over the levels of each factor. For example, for the one-way ANOVA with parameter vector $\beta = [\mu, \alpha_1, \ldots, \alpha_Q]^T$, the linear compound $c^T\beta$ with weights vector $c = [c_0, c_1, \ldots, c_Q]^T$ is a contrast if $c_0 = 0$ and $\sum_{q=1}^{Q} c_q = 0$.

Extra sum of squares principle; F-contrasts

The *extra sum-of-squares* principle allows one to assess general linear hypotheses, and compare models in a hierarchy, where inference is based on an F-statistic. Here, we describe the classical F-test, based on the assumption of an independent identically distributed error. In SPM, both statistics, the t- and the F-statistic, are used for making inferences.

We first present the classical F-test as found in introductory statistical texts. We will then address two critical limitations of this description and derive a more general and flexible implementation of the F-test.

Suppose we have a model with parameter vector β that can be bi-partitioned into, $\beta = [\beta_1^T, \beta_2^T]^T$, and suppose we wish to test $\mathcal{H}: \beta_1 = 0$. The corresponding partitioning of the design matrix X is $X = \begin{bmatrix} X_1 \vdots X_2 \end{bmatrix}$, and the *full model* is:

$$Y = \begin{bmatrix} X_1 \vdots X_2 \end{bmatrix} \begin{bmatrix} \beta_1 \\ \ldots \\ \beta_2 \end{bmatrix} + \epsilon$$

which, when $\mathcal{H}$ is true, reduces to the *reduced model*: $Y = X_2\beta_2 + \epsilon$. Denote the residual sum-of-squares for the full and reduced models by $S(\beta)$ and $S(\beta_2)$ respectively. The *extra sum-of-squares* due to β_1 after β_2 is then defined as $S(\beta_1|\beta_2) = S(\beta_2) - S(\beta)$. Under $\mathcal{H}$, $S(\beta_1|\beta_2) \sim \sigma^2\chi_p^2$ independent of $S(\beta)$, where the degrees of freedom are $p = \text{rank}(X) - \text{rank}(X_2)$. (If $\mathcal{H}$ is not true, then $S(\beta_1|\beta_2)$ has a non-central chi-squared distribution, still independent of $S(\beta)$.) Therefore, the following F-statistic expresses evidence against $\mathcal{H}$:

$$F = \frac{\dfrac{S(\beta_2) - S(\beta)}{p - p_2}}{\dfrac{S(\beta)}{J - p}} \sim F_{p-p_2, J-p} \tag{8.13}$$

where $p = \text{rank}(X)$ and $p_2 = \text{rank}(X_2)$. The larger F gets, the more unlikely it is that F was sampled under the null hypothesis H. Significance can then be assessed by comparing this statistic with the appropriate F-distribution. Draper and Smith (1981) give derivations.

This formulation of the F-statistic has two limitations. The first is that two (nested) models, the full and the reduced model, have to be inverted (i.e. estimated). The second limitation is that a partitioning of the design matrix into two blocks of regressors is too restrictive: one can partition X into any two sets of linear combinations of the regressors. This is particularly important when the effects of interest are encoded by a mixture of regressors (e.g. a difference). In this case, one cannot use Eqn. **8.13** to partition the design space into interesting and null subspaces. Rather, one has to re-parameterize the model such that the differential effect is modelled explicitly by a single regressor. As we will show next, this re-parameterization is unnecessary.

The key to implement F-tests that avoid these limitations lies in the notion of *contrast matrices*. A contrast matrix is a generalization of a contrast vector. Each

[5] In statistical parametric mapping, one usually refers to the vector c as the *vector of contrast weights*. Informally, we will also refer to c as the *contrast*, a slight misuse of the term.

[6] The actual implementation of this test is based on a more efficient algorithm using a singular value decomposition.

column of a contrast matrix consists of one contrast vector. Importantly, the contrast matrix specifies a partitioning of the design matrix X.

A contrast matrix c is used to specify a subspace of the design matrix, i.e. $X_c = Xc$. The orthogonal contrast to c is given by $c_0 = I_p - cc^-$. Then, let $X_0 = Xc_0$ be the design matrix of the reduced model. We wish to test the effects X_c can explain, *after* fitting the reduced model X_0. This can be achieved using a matrix that projects the data onto the subspace of X_c, which is orthogonal to X_0. We denote this subspace by X_a.

The projection matrix M due to X_a is derived from the residual forming matrix of the reduced model X_0, which is given by $R_0 = I_J - X_0X_0^-$. The projection matrix is then $M = R_0 - R$, where R is the residual forming matrix of the full model, i.e. $R = I_J - XX^-$.

The F-statistic can then be written as:

$$F = \frac{(MY)^T MY}{(RY)^T RY}\frac{J-p}{p_1} = \frac{Y^T MY}{Y^T RY}\frac{J-p}{p_1} \sim F_{p_1, J-p} \qquad 8.14$$

where p_1 is the rank of X_a. Since M projects onto a subspace within X, we can also write:

$$F = \frac{\hat{\beta}^T X^T MX\hat{\beta}}{Y^T RY}\frac{J-p}{p_1} \sim F_{p_1, J-p} \qquad 8.15$$

This equation means that we can compute an F-statistic conveniently for any user-specified contrast without a re-parameterization. In SPM, all F-statistics are based on the full model so that $Y^T RY$ is only estimated once and is stored for subsequent use. More about F-contrasts and their applications can be found in Chapter 9.

Example – one-way ANOVA

Consider a one-way ANOVA (Eqn. **8.9**), where we wish to assess the omnibus null hypothesis that all the groups are identical: $\mathcal{H} : \alpha_1 = \alpha_2 = \cdots = \alpha_Q$. Under $\mathcal{H}$ the model reduces to $Y_{qj} = \mu + \epsilon_{qj}$. Since the ANOVA model contains a constant term, μ, $\mathcal{H}$ is equivalent to $\mathcal{H} : \alpha_1 = \alpha_2 = \cdots = \alpha_Q = 0$. Thus, let $\beta_1 = (\alpha_1, \ldots, \alpha_Q)^T$, and $\beta_2 = \mu$. Eqn. **8.13** then gives an F-statistic which is precisely the standard F-statistic for a one-way ANOVA.

Alternatively, we can apply Eqn. **8.15**. The contrast matrix c is a diagonal $Q+1$-matrix with Q ones on the upper main diagonal and a zero in the $Q+1$st element on the main diagonal (Figure 8.2). This contrast matrix tests whether there was an effect due to any group, after taking into account a constant term across groups. Application of Eqn. **8.15** results in the same F-value as in Eqn. **8.13**, but without the need to invert two models.

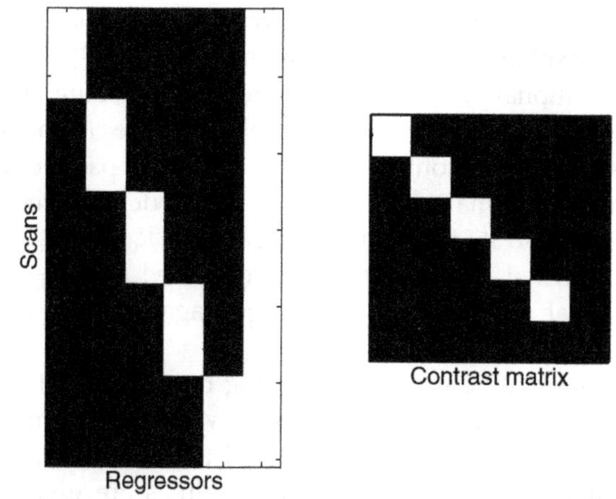

FIGURE 8.2 Example of ANOVA design and contrast matrix. Both matrices are displayed as images, where 0s are coded by black and 1s by white. Left: design matrix, where five groups are modelled by their mean and overall mean. The model is overdetermined by one degree of freedom. Right: F-contrast matrix, which tests for any group-specific deviation from the overall mean.

Adjusted and fitted data

Once an inference has been made it is usually necessary to report the nature of the effect that has been inferred. This can be in terms of the parameter estimates, or in measurement space after the estimates are projected through the design matrix. In short, one can report significant effects quantitatively using the parameter estimates or the responses that these estimates predict. Adjusted data are a useful way of summarizing effects, after uninteresting or confounding effects having been removed from the raw data Y.

How does one tell SPM which effects are of interest? The partitioning of the design matrix into interesting and uninteresting parts is based upon the same principles as the F-test developed in the preceding subsection. We can use an F-contrast for the partitioning, which is equivalent to specifying a full and reduced model. Adjusted data are the residuals of the reduced model, i.e. components that can be explained by the reduced model have been removed from the data. To compute adjusted data we need to tell SPM which part of the design matrix is of interest (to specify the reduced model). SPM then treats the part of the design matrix, which is orthogonal to the reduced model, as the effects of interest. This will be illustrated below by an example. Note that the partitioning of the design matrix follows the same logic as the F-test: first, any effect due to the reduced model is removed and only the remaining effects are taken to be of interest. An important point is that any overlap (correlation) between the reduced model and our partition of interest is *explained away* by the reduced model. In the context of adjusted data, this means that the adjusted data will not contain that component of the effects that can be explained by the reduced model.

Operationally, we compute the adjusted data using the same procedure used to calculate the F-statistic. A user-specified contrast matrix c induces a partitioning of the design matrix X. The reduced model is given by $X_0 = Xc_0$ and its residual forming matrix $R_0 = I_J - X_0X_0^-$. The adjusted data can then be computed by $\tilde{Y} = R_0Y$. Note that this projection technique again makes a re-parameterization redundant.

An alternative way of computing the adjusted data $\tilde{Y}$ is to compute the data explained by the design matrix partition orthogonal to X_0 and add the residuals of the full model, i.e. $\tilde{Y} = Y_f + e$. The residuals are given by $e = RY$, where R is the residual forming matrix of the full model, and $Y_f = MY$, where Y_f is referred to as *fitted data*. The projection matrix M is computed by $M = R_0 - R$. In other words, the fitted data are equivalent to the adjusted data minus the estimated error, i.e. $Y_f = \tilde{Y} - e$.

In SPM, both adjusted and fitted data can be plotted for any voxel. For these plots, SPM requires the specification of an F-contrast, which encodes the partitioning of the design matrix into effects of interest and no interest.

Example

Consider a one-way ANOVA with four groups (conditions). The design matrix comprises four columns, which indicate group membership. Each group has 12 measurements, so that we have 48 measurements altogether. In this example, we are interested in the average of two differences. The first difference is between groups 1 and 2 and the second difference between groups 2 and 3. If we want to test this difference with a t-statistic, the contrast vector is $c = [-1,\ 1,\ -1,\ 1]^T$. In Figure 8.3 (left), we show the data. It is easy to see that there is a difference between the average of the first two groups compared to the average of the last two groups, i.e. $c = [-1,\ -1,\ 1,\ 1]^T$. However, by visual inspection, it is hard to tell whether there is a difference between the average of groups 1 and 3 compared to the average of groups 2 and 4. This is a situation where a plot of adjusted and fitted data is helpful. First, we have to specify a reduced model. In our example, the difference is represented by the contrast vector $c = [-1,\ 1,\ -1,\ 1]^T$. The contrast matrix c_0 is given by $c_0 = I_4 - cc^-$. With c_0, we can compute X_0, R_0 and M, all of which are needed to compute the adjusted and fitted data. In Figure 8.3 (right), we show the fitted and adjusted data. In this plot it is obvious that there is a difference between groups 1 and 2 and between groups 3 and 4. This example illustrates that plots of fitted and adjusted data are helpful, when the effect of interest is obscured or confounded by other effects.

Design matrix images

SPM uses grey-scale images of the design matrix to represent linear models. An example for a single subject PET activation study with four scans, under each of three conditions, is shown in Figure 8.4. The first three columns contain indicator variables (consisting of zeroes and ones) indexing the condition. The last column contains the (mean corrected) global cerebral blood flow (gCBF) values (see below).

In the grey-scale design matrix images, −1 is black, 0 mid-grey, and +1 white. Columns containing covariates are scaled by subtracting the mean (zero for centred covariates). For display purposes regressors are divided by their absolute maximum, giving values in [−1, 1]. Design matrix blocks, containing factor by covariate interactions, are scaled such that the covariate values lie in [0,1], preserving the zeroes as mid-grey.

PET AND BASIC MODELS

Having estblished the details of the general linear model, we turn our attention to models used in functional brain mapping, discuss the practicalities of their application,

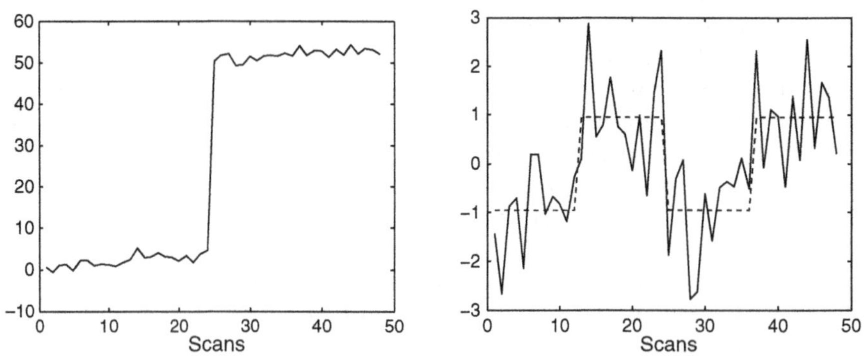

FIGURE 8.3 Adjusted and fitted data. Left: plot of raw data. Right: adjusted data (solid line); fitted data (dashed line).

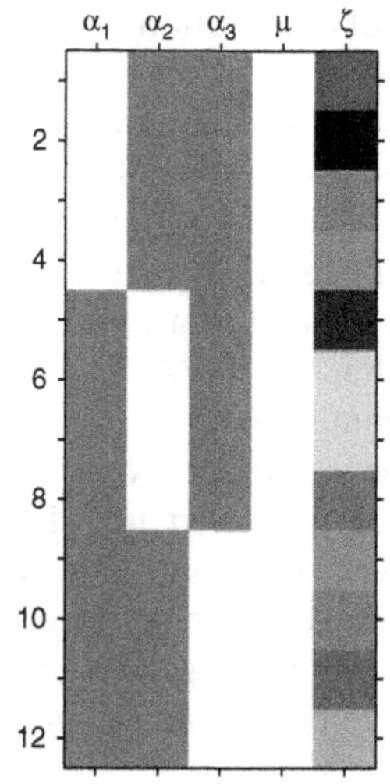

FIGURE 8.4 Single subject activation experiment, ANCOVA design. Illustrations for a three-condition experiment with four scans in each of three conditions, ANCOVA design. Design matrix image, with columns labelled by their respective parameters. The scans are ordered by condition.

and introduce some terminology used in SPM. As the approach is mass-univariate, the same model is used at every voxel simultaneously, with different parameters and error variances for each voxel. We shall concentrate on PET data, with its mature family of standard statistical experimental designs. Models of fMRI data will be presented in the next section. Although most PET experiments employ multiple subjects, many of the key concepts are readily demonstrated using a single subject design.

Global normalization

There is an important distinction between regional and global effects. Regional effects refer to the response in a single voxel or a small volume of voxels that is not mediated by global effects. Global effects have no regional specificity and can be expressed everywhere in the brain. Usually, global effects are considered as confounds (e.g. arbitrary changes in global signal due to scan-to-scan variations in radioactivity delivered). Although true global effects are generally unknown, they can be estimated using the whole-brain signal, which enters the model as a global confound. Typically, modelling global effects enhances the sensitivity and accuracy of the subsequent inference about experimentally induced regional effects.

As an example, consider a simple single subject PET experiment. The subject is scanned repeatedly under both *baseline* (control) and *activation* (experimental) conditions.

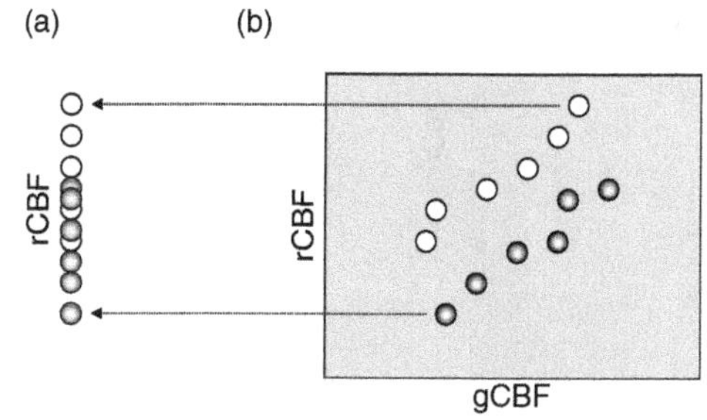

FIGURE 8.5 Single subject PET experiment, illustrative plots of rCBF at a single voxel: (a) Dot-plots of rCBF; (b) plot of rCBF versus gCBF. Both plots indexed by condition: **O** for baseline, O for active.

Inspection of regional activity (used as a measure of regional cerebral blood flow (rCBF)), at a single voxel, may not suggest an experimentally induced effect. However, after accounting for global effects (gCBF) differences between the two conditions are disclosed (Figure 8.5).

In statistical parametric mapping, the precise definition of global activity is user-dependent. The *default* definition is that global activity is the global average of image intensities of intracerebral voxels. If Y_j^k is the image intensity at voxel $k = 1, \ldots, K$ of scan j, the estimated global activity is $g_j = \overline{Y}_j^{\bullet} = \sum_{k=1}^{K} Y_j^k / K$.

Having estimated the global activity for each scan, a decision must be made about which model of global activity should be used. In SPM, there are two alternatives. The first is *proportional scaling* and the second is an analysis of covariance analysis (ANCOVA).

Proportional scaling

One way to account for global changes is scale each scan by its estimated global activity. This approach is based on the assumption that the measurement process introduces a (global) scaling of image intensities at each voxel, a gain factor. Scaling has the advantage of converting the raw data into a canonical range to give parameter estimates an interpretable scale. For PET data, the mean global value is usually chosen to be a typical gCBF of 50 ml/min/dl. The scaling factor is thus $\frac{50}{g_{\bullet}}$. We shall assume that the count rate recorded in the scanner (counts data) has been scaled into a physiologically meaningful scale. The normalized data are $Y_j'^k = \frac{50}{g_j} Y_j^k$. The model is then:

$$Y_j^k = \frac{g_j}{50}(X\beta^k)_j + \epsilon_j'^k \qquad \mathbf{8.16}$$

where $\epsilon'^k \sim \mathcal{N}(0, \sigma_k^2 \times diag((g_j/50)^2))$. The $diag()$ operator transforms a column vector to a diagonal matrix with the vector on its main diagonal and zero elsewhere. This is a weighted regression, i.e. the shape of the error covariance

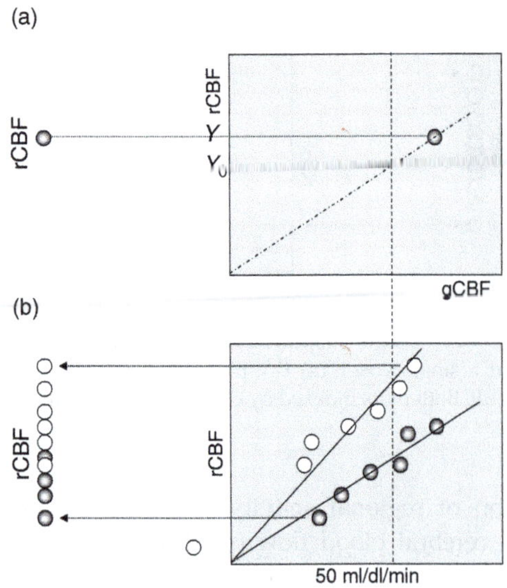

FIGURE 8.6 (a) Adjustment by proportional scaling; (b) simple single subject activation as a *t*-test on adjusted rCBF: weighted proportional regression.

matrix is no longer I_J, but a function of estimated global activity. Also note that the *jth* row of X is weighted by g_j.

The adjustment of data from Y to Y' is illustrated in Figure 8.6(a).

The ANCOVA approach

Another approach is to include the mean corrected global activity vector g as an additional regressor in the model. In this case the model (Eqn. **8.6**) becomes:

$$Y_j^k = (X\beta)_j + \zeta^k(g_j - \bar{g}_\bullet) + \epsilon_j^k \qquad 8.17$$

where $\epsilon^k \sim \mathcal{N}(0, \sigma_k^2 I_J)$ and ζ_k is the slope parameter for the global activity. In this model, the data are explained as the sum of experimentally induced regional effects and some global effects, which varies over scans. Note that the model of Eqn. **8.17** can be extended by allowing for different slopes between replications, conditions, subjects and groups.

Proportional scaling versus ANCOVA

Clearly, one cannot apply both normalization models, because proportional scaling will normalize the global mean activity such that the mean corrected g in the ANCOVA approach will be a zero vector. Proportional scaling is most appropriate for data for which there is a gain (multiplicative) factor that varies over scans. This can be a useful assumption for fMRI data (see next section). In contrast, an ANCOVA approach is appropriate if the gain does not change over scans. This is the case for PET scans using protocols which control for the administered dose. This means that a change in estimated global activity reflects a change in a subject's global activity and not a change in a global (machine specific) gain. Moreover, the ANCOVA approach assumes that regional experimentally induced effects are independent of changes in global activity. Note that the ANCOVA approach should not be used for PET data where the administered dose is not controlled for in image construction. In this case, the true underlying gCBF might be constant over scans, but the global gain factor will vary (similarly, for single photon emission computed tomography (SPECT) scans)

Implicit in allowing for changes in gCBF (either by proportional scaling or ANCOVA) when assessing condition specific changes in rCBF, is the assumption that gCBF represents the underlying background flow, about which regional differences are assessed. That is, gCBF is independent of condition. Clearly, since gCBF is calculated as the mean intracerebral rCBF, an increase of rCBF in a particular brain region must cause an increase of gCBF unless there is a corresponding decrease of rCBF elsewhere. This means that, after global normalization, regional effects are *relative* regional effects, having discounted global effects.

If gCBF varies considerably between conditions, as in pharmacological activation studies, then testing for an activation after allowing for global changes involves extrapolating the relationship between regional and global flow beyond the range of the data. This extrapolation might not be valid, as illustrated in Figure 8.7(a).

If gCBF is increased by a large activation that is not associated with a corresponding deactivation, then global normalization will make non-activated regions (whose rCBF remained constant) appear de-activated. (Figure 8.7(b) illustrates the scenario for a simple single subject activation experiment using ANCOVA.) This means it is important to qualify inferences about regional effects under global normalization, especially if the global activity shows a treatment effect. In these cases, it is useful to report an analysis of the the global signal itself, properly to establish the context in which relative regional effects are expressed.

Grand mean scaling

Grand mean scaling refers to the scaling of all scans by some factor such that the mean global activity is a (user-specified) constant over all scans. Note that this factor has no effect on inference, because it cancels in

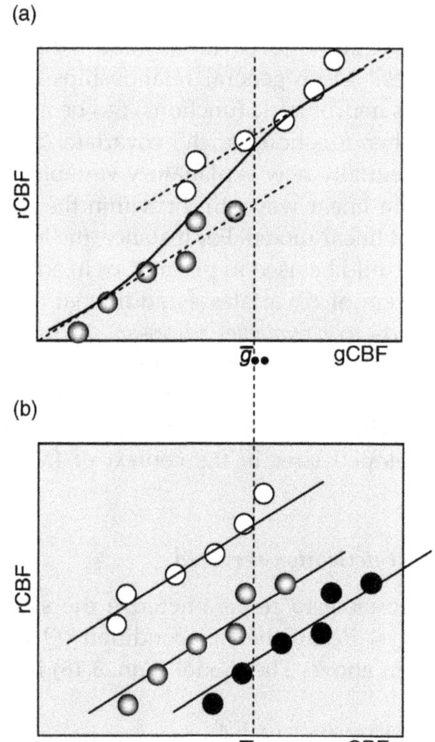

FIGURE 8.7 Single subject data, illustrative (ANCOVA) plots of rCBF versus gCBF at a single voxel showing potential problems with global changes. (a) Large change in gCBF between conditions. The apparent activation relies on linear extrapolation of the baseline and active condition regressions (assumed to have the same slope) beyond the range of the data. The actual relationship between regional and global for no activation may be given by the curve, in which case there is no activation effect. (b) Large activation inducing increase in gCBF measured as brain mean rCBF. Filled O denotes rest, O denotes active condition values if this is an activated voxel, while black O denotes active condition values where this voxel is not activated (in which case an relative deactivation is seen).

the expressions for the t- and F-statistics (Eqns. **8.12** and **8.14**). However, it does change the scaling and units of the data and parameter estimates. The units of both are rendered adimensional and refer, in a proportional sense to the grand mean after scaling. This is useful in fMRI, where the grand mean of a time-series is set to 100. This means that any activation is expressed as a per cent of whole brain mean, over scans.

PET models

In the following, we demonstrate the flexibility of the general linear model using models in the context of PET experiments. For generality, ANCOVA style models are used, with gCBF as a confounding covariate. The corresponding ANCOVA models for data adjusted by proportional scaling obtain by omitting the global term. Voxel-level models are presented in the usual statistical notation, alongside the SPM description and design matrix images. The form of contrasts for each design are indicated, and some practical issues of design specification will be discussed.

Single subject models

Single subject activation design

The simplest experimental paradigm is the single subject activation experiment. Suppose there are Q conditions, with M_q scans under condition q. Let Y_{qj}^k denote the rCBF at voxel k in scan $j = 1, \ldots, M_q$ under condition $q = 1, \ldots, Q$. The model is:

$$Y_{qj}^k = \alpha_q^k + \mu^k + \zeta^k(g_{qj} - \bar{g}_{\bullet\bullet}) + \epsilon_{qj}^k \qquad \textbf{8.18}$$

There are $Q+2$ parameters for the model at each voxel: the Q condition effects, the constant term μ^k, and the global regression effect, giving a parameter vector $\beta^k = (\alpha_1^k, \ldots, \alpha_Q^k, \mu^k, \zeta^k)^T$ at each voxel. In this model, replications of the same condition are modelled with a single effect. The model is overparameterized, having only $Q+1$ degrees of freedom, leaving $N-Q-1$ residual degrees of freedom, where $N = \sum M_q$ is the total number of scans.

Contrasts are linear compounds $c^T \beta^k$ for which the weights sum to zero over the condition effects, and give zero weight to the constant term, i.e. $\sum_{q=1}^{Q} c_q = 0$ (Figure 8.8). Therefore, linear compounds that test for a simple group effect or for an average effect over groups cannot be contrasts. However, one can test for differences between groups. For example, to test the null hypothesis

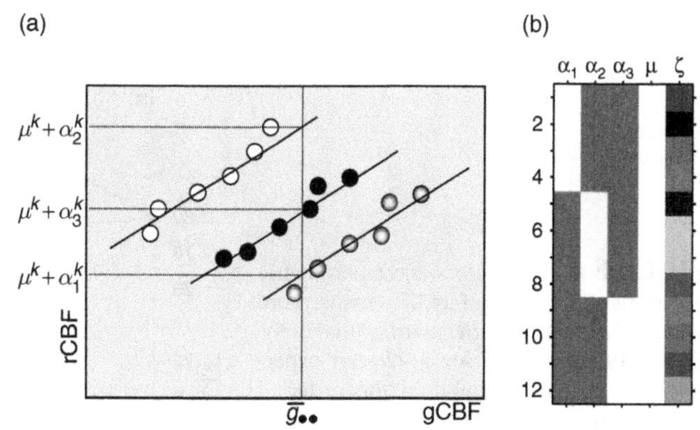

FIGURE 8.8 Single subject study, ANCOVA design. Illustration of a three-condition experiment with four scans in each of three conditions, ANCOVA design. (a) Illustrative plot of rCBF versus gCBF. (b) Design matrix image with columns labelled by their respective parameters. The scans are ordered by condition.

$\mathcal{H}^k : \alpha_1^k = (\alpha_2^k + \alpha_3^k)/2$ against the one sided alternative $\overline{\mathcal{H}}^k : \alpha_1^k > (\alpha_2^k + \alpha_3^k)/2$, the appropriate contrast weights would be $c = [1, -\frac{1}{2}, -\frac{1}{2}, 0, \ldots, 0]^T$. In other words, one tests whether the effect of group 1 is greater than the average of groups 2 and 3. Large positive values of the t-statistic express evidence against the null hypothesis, in favour of the alternative hypothesis.

Single subject parametric design

Consider the single subject parametric experiment where there is a single covariate of interest, or 'score'. For instance, the covariate may be a physiological variable, a reaction time, or a performance score. We want to find regions where rCBF is highly correlated with the covariate, taking into account global effects. Figure 8.9(a) depicts the situation. If Y_j^k is the rCBF at voxel k of scan $j = 1, \ldots, J$ and s_j is the independent covariate, then a simple ANCOVA-style model is a multiple regression with two covariates:

$$Y_j^k = \varrho^k(s_j - \bar{s}_\bullet) + \mu^k + \zeta^k(g_j - \bar{g}_\bullet) + \epsilon_j^k \qquad 8.19$$

Here, ϱ is the slope of the regression plane in the direction of increasing score, fitted separately for each voxel.

There are three model parameters, leaving $J - 3$ residual degrees of freedom. The design matrix (Figure 8.9(b)) has three columns, a column containing the (centred) score covariate, a column of dummy 1s corresponding to μ^k, and a column containing the (centred) global values.

The design is uniquely specified, so any linear combination of the three parameters is a contrast. The null hypothesis of no score effect at voxel k, $\mathcal{H}^k : \varrho^k = 0$, can be assessed against the one sided alternative hypotheses $\overline{\mathcal{H}}^k : \varrho^k > 0$ (rCBF increasing with score) with contrast weight for the effect of interest $c_1 = +1$, and against $\overline{\mathcal{H}}^k : \varrho^k < 0$ (rCBF decreasing as score increases) with contrast weight $c_1 = -1$.

This simple model assumes a linear relationship between rCBF and the covariate (and other explanatory variables). More general relationships (sometimes referred to as neurometric functions) can be modelled by including other functions of the covariate. These functions are essentially new explanatory variables which, if combined in a linear way, still fit within the framework of the general linear model. For instance, the logarithm of s_j, i.e. $\ln(s_j)$, could be used in place of, or in addition to s_j. Adding powers of covariates as additional explanatory variables leads to *polynomial regression*. More generally, a set of *basis functions* can be used to expand covariates to model, in a flexible fashion, the non-linear mapping between experimental factors and responses. This theme will be developed later in the context of fMRI, and in Chapter 14.

Single subject activation revisited

It is often possible to re-parameterize the same model in many ways. Recall the two-condition ($Q = 2$) single subject design above. The model (Eqn. **8.18**) is:

$$Y_{qj}^k = \alpha_q^k + \mu^k + \zeta^k(g_{qj} - \bar{g}_{\bullet\bullet}) + \epsilon_{qj}^k$$

This model is over-determined, so consider a sum-to-zero constraint on the condition effects. For two conditions this implies $\alpha_1^k = -\alpha_2^k$. Substituting for α_2^k, the resulting design matrix has a column containing +1s and −1s indicating the condition $q = 1$ or $q = 2$ respectively, a column of 1s for the overall mean, and a column containing the (centred) gCBF (Figure 8.10). The corresponding parameter vector is $\beta^k = [\alpha_1^k, \mu^k, \zeta^k]^T$. Clearly, this is the same design matrix as that for a parametric design with a (non-centred) 'score' covariate, indicating the condition as active or baseline (with +1 or −1 respectively). The hypothesis of no activation at voxel k, $\mathcal{H}^k : \alpha_1^k = 0$ can be tested against the one sided alternatives $\overline{\mathcal{H}}^k : \alpha_1^k > 0$

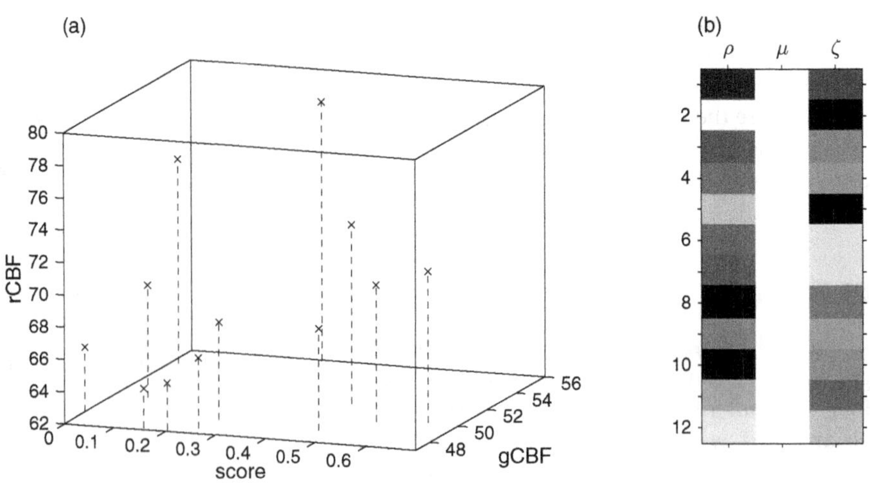

FIGURE 8.9 Single subject parametric experiment: (a) plot of rCBF versus score and gCBF. (b) Design matrix image for Eqn. **8.19**, illustrated for a 12-scan experiment. Scans are ordered in the order of acquisition.

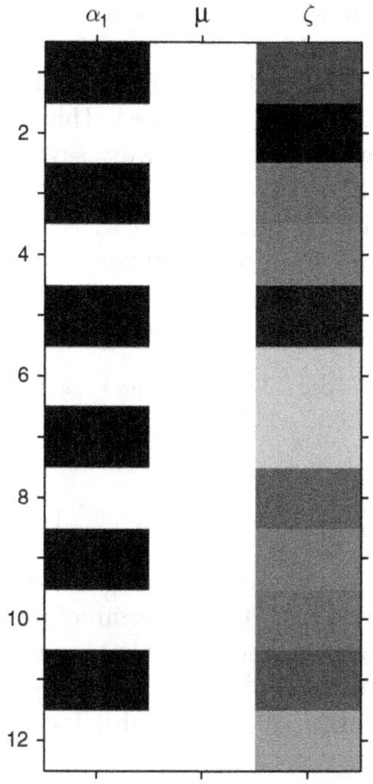

FIGURE 8.10 Example design matrix image for single subject activation study, with six scans in each of two conditions, formulated as a parametric design. The twelve scans are ordered so they alternate between baseline and activation conditions.

(activation) and $\overline{\mathcal{H}}^k : \alpha_1^k < 0$ with contrast weights for the effects of interest $c_1 = 1$ and $c_1 = -1$ respectively.

Single subject: conditions and covariates

Frequently, there are other confounding covariates in addition to gCBF that can be added to the model. For example, a linear effect of time can be modelled simply by entering the scan number as a covariate. In SPM these appear in the design matrix as additional columns adjacent to the global flow column.

Factor by covariate interactions

A more interesting experimental scenario is when a parametric design is repeated under multiple conditions in the same subject(s). A specific example would be a PET language experiment in which, during each of twelve scans, lists of words are presented. Two types of word list (the two conditions) are presented at each of six rates (the parametric component). We may be interested in locating regions where there is a difference in rCBF between conditions (accounting for changes in presentation rate), the *main* effect of condition; locating regions where rCBF increases with rate (accounting for condition), the main effect of rate; and assessing evidence for condition-specific rate effects, an interaction.[7] Let Y_{qrj}^k denote the rCBF at voxel k for the j-th measurement under rate $r = 1, \ldots, R$ and condition $q = 1, \ldots, Q$, with s_{qr} the rate covariate (some function of the rates). A suitable model is:

$$Y_{qrj}^k = \alpha_q^k + \varrho_q^k(s_{qr} - \bar{s}_{\bullet\bullet}) + \mu^k + \zeta^k(g_{qrj} - \bar{g}_{\bullet\bullet\bullet}) + \epsilon_{qrj}^k \qquad \textbf{8.20}$$

Note the q subscript on the parameter ϱ_q^k, indicating different slopes for each condition. Ignoring the global flow for the moment, the model describes two simple regressions with common error variance (Figure 8.11(a)). SPM describes such factor by covariate interactions as 'factor-specific covariate fits'. The interaction between condition and covariate effects is manifest as a different regression slope for each condition. There are $2Q+2$ parameters for the model at each voxel, $\beta^k = [\alpha_1^k, \ldots, \alpha_Q^k, \varrho_1^k, \ldots, \varrho_Q^k, \mu^k, \zeta^k]^T$, with $2Q+1$ degrees of freedom. A design matrix for the two condition example is shown in Figure 8.11(b). The factor by covariate interaction occupies the third and fourth columns, corresponding to the parameters ϱ_1^k and ϱ_2^k. Here, the covariate has been split between the columns according to condition and the remaining cells filled with zeroes.

Only the constant term and global slope are confounds, leaving $2Q$ effects of interest $\beta_1^k = [\alpha_1^k, \ldots, \alpha_Q^k, \varrho_1^k, \ldots, \varrho_Q^k]^T$. As with the activation study model, contrasts have weights that sum to zero over the condition effects. For the two-condition word presentation example, contrast weights $c_1 = [0, 0, 1, 0]^T$ test for a covariate effect in condition one, with large values indicating evidence of a positive covariate effect. Weights $c_1 = [0, 0, \frac{1}{2}, \frac{1}{2}]^T$ address the hypothesis that there is no

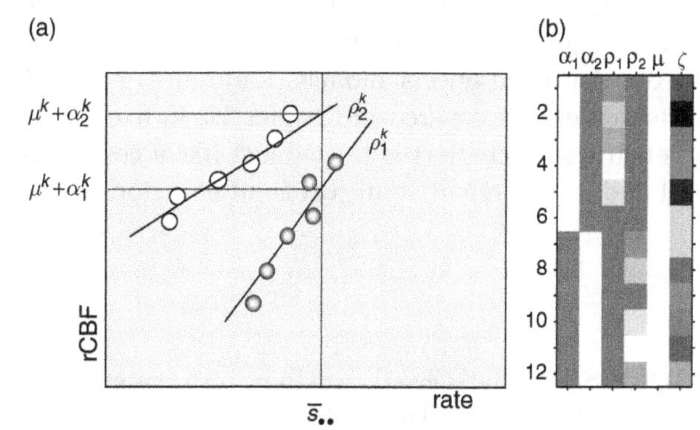

FIGURE 8.11 Single subject experiment with conditions, covariate, and condition by covariate interaction: (a) illustrative plot of rCBF versus rate. (b) Design matrix image for Eqn. **8.20**. Both are illustrated for the two-condition 12-scan experiment described in the text. The scans have been ordered by condition.

[7] Two experimental factors *interact* if the level of one affects the expression of the other.

average covariate effect across conditions, against the one-sided alternative that the average covariate effect is positive. Weights $c_1 = [0, 0, -1, +1]^T$ test the null hypothesis that there is no condition by covariate interaction, i.e. that the regression slopes are the same.

The contrast weights $c_1 = [-1, +1, 0, 0]^T$ and $c_1 = [+1, -1, 0, 0]^T$ are used to assess the hypothesis of no condition effect against appropriate one-sided alternatives. However, inference on main effects is confounded in the presence of an interaction: in the above model, both gCBF and the rate covariate were centred, so the condition effects α_q^k are the relative heights of the respective regression lines (relative to μ^k) at the mean gCBF and mean rate covariate. Clearly, if there is an interaction, then the difference in the condition effects (the separation of the two regression lines) depends on where you look at them. Were the rate covariate not centred, the comparison would be at mean gCBF and zero rate, possibly yielding a different result. More generally, the presence of an interaction means that the main effects are difficult to interpret. This difficulty is resolved by reporting the simple effects. In factorial designs, it is usual to test first for interactions. If the interactions are significant, one then proceeds to test for the appropriate simple effects. If it is not, then one reports the main effects.

Multisubject designs

Frequently, experimentally induced changes in rCBF are small, and many analyses pool data from different subjects to find significant effects. Models of data from many subjects can either treat the subject effect as a fixed quantity (giving fixed-effect models) or treat the subject effect as random (giving random-effect models). In this chapter, we consider fixed-effects models. Random- or mixed-effects models are covered in Chapter 12. In fixed-effect models the subject-effect is treated just like a condition-effect and can be regarded as just another factor.

The single-subject designs presented above can be extended to account for subject to subject differences. The simplest type of subject effect is an additive effect, otherwise referred to as a *block* effect. This implies that all subjects respond in the same way, save for an overall shift in rCBF (at each voxel). We extend our notation by adding subscript i for subjects, so Y_{iqj}^k is the rCBF at voxel k of scan j under condition q on subject $i = 1, \ldots, N$.

Multisubject activation (replications)

For instance, the single-subject activation model (Eqn. **8.18**) is extended by adding subject effects γ_i^k giving the model:

$$Y_{iqj}^k = \alpha_q^k + \gamma_i^k + \zeta^k(g_{iqj} - \bar{g}_{\bullet\bullet\bullet}) + \epsilon_{iqj}^k \qquad 8.21$$

A schematic plot of rCBF versus gCBF for this model is shown in Figure 8.12(a). In SPM terminology, this is a 'multisubject: replication of conditions' design. The parameter vector at voxel k is $\beta^k = [\alpha_1^k, \ldots, \alpha_Q^k, \gamma_1^k, \ldots, \gamma_N^k, \zeta^k]^T$. The design matrix (Figure 8.12(b)) has N columns of dummy variables corresponding to the subject effects. (Similarly, a multisubject parametric design could be derived from the single-subject case by including appropriate additive subject effects.)

Again, the model is overparameterized, although this time we have omitted the explicit constant term from the confounds, since the subject effects can model an overall level. Adding a constant to each of the condition effects and subtracting it from each of the subject effects gives the same model. Bearing this in mind, it is clear that contrasts must have weights that sum to zero over both the subject and condition effects.

Condition by replication interactions

The above model assumes that (accounting for global and subject effects) replications of the same condition give the same (expected) response. There are many reasons why this assumption may be inappropriate, such

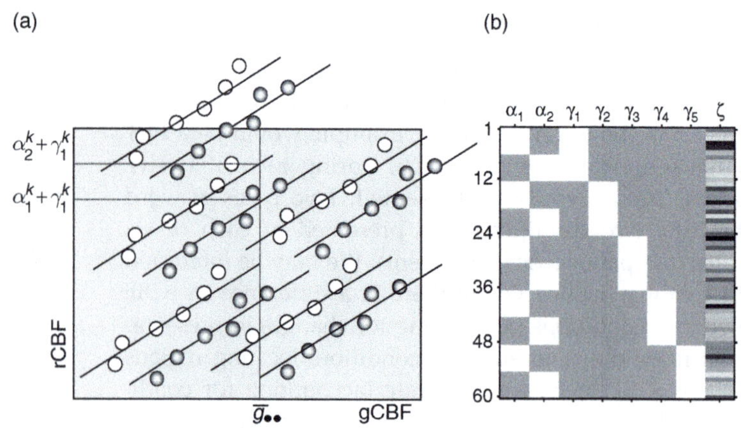

FIGURE 8.12 Multisubject activation experiment, replication of conditions (Eqn. **8.21**). Illustrations for a five-subject study, with six replications of each of two conditions per subject: (a) illustrative plot of rCBF versus gCBF. (b) Design matrix image: the first two columns correspond to the condition effects, the next five to the subject effects, the last to the gCBF regression parameter. The design matrix corresponds to scans ordered by subject, and by condition within subjects.

as learning effects or more generally effects that change as a function of time. Time effects can be modelled by including appropriate functions of the scan number as confounding covariates. With multisubject designs, we have sufficient degrees of freedom to test for replication by condition interactions. These interactions imply that the (expected) response to each condition changes with replications (having accounted for other effects in the model). Usually in statistical models, interaction terms are added to a model containing main effects (see also Chapter 13). However, this supplemented model is overparameterized such that the main effects are redundant. When they are omitted, the model is:

$$Y^k_{iqj} = \alpha\vartheta^k_{(qj)} + \gamma^k_i + \zeta^k(g_{iqj} - \bar{g}_{\bullet\bullet\bullet}) + \epsilon^k_{iqj} \qquad 8.22$$

where $\alpha\vartheta^k_{(qj)}$ is the interaction effect for replication j of condition q, the condition-by-replication effect. As with the previous model, this model is overparameterized (by one degree of freedom), and contrasts must have weights which sum to zero over the condition-by-replication effects. There are as many condition-by-replication terms as there are scans per subject. (An identical model is arrived at by considering each replication of each experimental condition as a separate condition.) If the scans are reordered such that the j-th scan corresponds to the same replication of the same condition in each subject, then the condition-by-replication corresponds to the scan number. An example design matrix for five subjects scanned twelve times is shown in Figure 8.13 where the scans have been reordered.

This is the 'classic' SPM ANCOVA implemented in the original SPM software and affords great latitude for the specification of contrasts: appropriate contrasts can be used to assess main effects, specific forms of interaction, and even parametric effects. For instance, consider the verbal fluency dataset described by Friston *et al.* (1995):[8] Five subjects were scanned twelve times; six times under two conditions, word-shadowing (condition A) and intrinsic word-generation (condition B). The scans were reordered to ABABABABABAB for all subjects. Then, a contrast with weights (for the condition-by-replication effects) of $c_1 = [-1, 1, -1, 1, -1, 1, -1, 1, -1, 1, -1, 1]^T$ assesses the hypothesis of no main effect of word-generation (against the one-sided alternative of activation). A contrast with weights of $c_1 = [5\frac{1}{2}, 4\frac{1}{2}, 3\frac{1}{2}, 2\frac{1}{2}, 1\frac{1}{2}, \frac{1}{2}, -\frac{1}{2}, -1\frac{1}{2}, -2\frac{1}{2}, -3\frac{1}{2}, -4\frac{1}{2}, -5\frac{1}{2}]^T$ is sensitive to linear decreases in rCBF over time, independent of condition, and accounting for subject effects and changes in gCBF. A contrast with weights of $c_1 = [1, -1, 1, -1, 1, -1, -1, 1, -1, 1, -1, 1]$ assesses the interaction of time and condition, subtracting the activation in the first half of the experiment from that in the latter half.

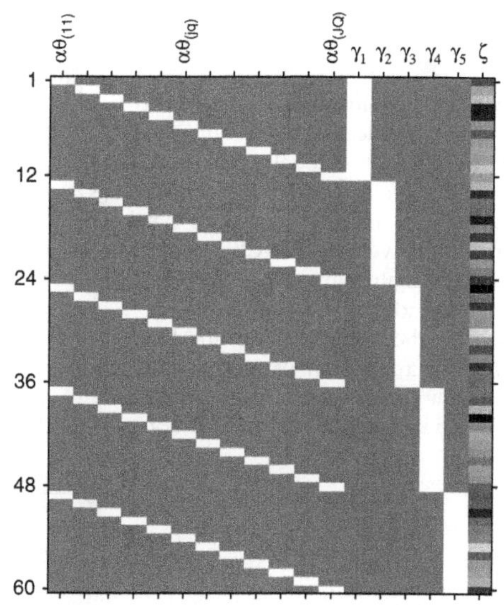

FIGURE 8.13 Multisubject activation experiment, 'classic' SPM design, where each replication of each experimental condition is considered as a separate condition (Eqn. **8.22**). Illustrative design matrix image for five subjects, each having 12 scans, the scans having been ordered by subject, and by condition and replication within subject. The columns are labelled with the corresponding parameter. The first twelve columns correspond to the 'condition' effects, the next five to the subject effects, the last to the gCBF regression parameter.

Interactions with subject

While it is (usually) reasonable to use ANCOVA style models to account for global flow, with regression parameters constant across conditions, the multisubject models considered above also assume that this regression parameter is constant across subjects. It is possible that rCBF at the same location for different subjects will respond differentially to changes in gCBF – a subject by gCBF interaction. The gCBF regression parameter can be allowed to vary from subject to subject. Extending the multisubject activation (replication) model (Eqn. **8.21**) in this way gives:

$$Y^k_{iqj} = \alpha^k_q + \gamma^k_i + \zeta^k_i(g_{iqj} - \bar{g}_{\bullet\bullet\bullet}) + \epsilon^k_{iqj} \qquad 8.23$$

Note the i subscript on the global slope term, ζ^k_i, indicating a separate parameter for each subject. A schematic plot of rCBF versus gCBF for this model and an example design matrix image are shown in Figure 8.14. In the terminology of the SPM this is an 'ANCOVA by subject'. The additional parameters are of no interest, and contrasts remain as before.

[8] This data set is available via http://www.fil.ion.ucl.ac.uk/spm/data/

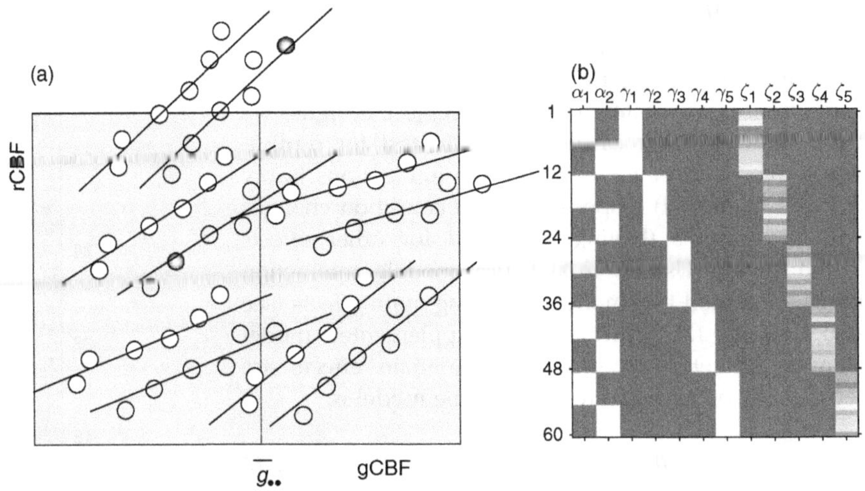

FIGURE 8.14 Multisubject activation experiment, replication of conditions, ANCOVA by subject. Model Eqn. 8.23. Illustrations for a five-subject study, with six replications of each of two conditions per subject: (a) illustrative plot of rCBF versus gCBF. (b) Design matrix image: the first two columns correspond to the condition effects, the next five to the subject effects, the last five to the gCBF regression parameters for each subject. The design matrix corresponds to scans ordered by subject, and by condition within subjects.

Multistudy designs

The last class of SPM models for PET we consider are the 'multistudy' models. In these models, subjects are grouped into two or more *studies*. The 'multistudy' designs fit separate condition effects for each study. In statistical terms, this is a *split plot* design. As an example, consider two multisubject activation studies, the first with five subjects scanned twelve times under two conditions (as described above), the second with three subjects scanned six times under three conditions. An example design matrix for a model containing study-specific condition effects, subject effects and study-specific global regression (termed 'ANCOVA by group' in SPM) is shown in Figure 8.15. The first two columns of the design matrix correspond to the condition effects for the first study, the next two to the condition effects for the second study, the next eight to the subject effects, and the last to the gCBF regression parameter. (The corresponding scans are assumed to be ordered by study, by subject within study, and by condition within subject.)

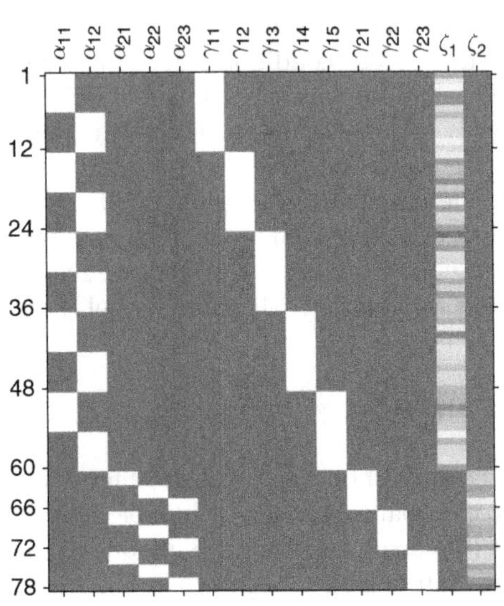

FIGURE 8.15 Design matrix image for the example multistudy activation experiment described in the text.

Contrasts for multistudy designs in SPM have weights that, when considered for each of the studies individually, would define a contrast for that study. Thus, contrasts must have weights which sum to zero over the condition effects within each study. There are three types of useful comparison. The first is a comparison of condition effects within a study; the contrast weights for this are padded with zeros for the other studies, e.g. $c_1 = [1, -1, 0, 0, 0]^T$ for the first study in our example. This has additional power, relative to an analysis of this study in isolation, since observations from the second study make the variance estimates more precise. The second is an average effect across studies; here, contrasts for a particular effect are concatenated over studies. For example, if the second study has the same conditions as the first, plus an additional condition, then the contrast could have weights $c_1 = [-1, 1, -1, 1, 0]^T$. Lastly, differences of contrasts across studies can be assessed, such as differences in activation. The contrasts weights for the appropriate main effect in each study are concatenated, with some study-contrasts negated. In our example, $c_1 = [-1, 1, 1, -1, 0]^T$ would be appropriate for locating regions where the first study activated more than the second, or where the second deactivated more than the first.

In many instances, it is appropriate to assume that the error variance between subjects or between studies is the same (i.e. homoscedasticity). For very different study populations or studies using different scanners or protocols, this assumption may not be tenable and the different variances should be modelled. We will cover the specification of this non-sphericity[9] in Chapter 10 and, briefly, in the next section on fMRI. These considerations pertain to the assumptions about the covariance components of the error but do not affect the specification of the design matrices considered in this section.

Basic models

In this section, we will discuss some of the models that are referred to in SPM as *basic models*. Typically, basic models are used for analyses at the second or between-subject level to implement mixed-effects models (see Chapter 12). For example, basic models include the one-sample t-test, the two-sample t-test, the paired t-test and a one-way ANCOVA, which are described below. For clarity, we shall drop the voxel index superscript k.

One-sample t-test

The one-sample t–test can be used to test the null hypothesis that the mean of J scans is zero. This is the simplest model available and the design matrix consists of just a constant regressor. The model is:

$$Y = x_1\beta_1 + \epsilon \tag{8.24}$$

where x_1 is a vector of ones and $\epsilon \sim N(0, \sigma^2 I_J)$. The null hypothesis is $\mathcal{H}: \beta_1 = 0$ and the alternative hypothesis is $\overline{\mathcal{H}}: \beta_1 > 0$. The t-value is computed using Eqn. **8.12** as:

$$T = \frac{\hat{\beta}_1}{\sqrt{\hat{\sigma}^2/J}} \sim t_{J-1} \tag{8.25}$$

where $\hat{\sigma}^2 = Y^T RY/(J-1)$, where R is the residual forming matrix (see above) and $Y^T RY$ are the sum of squares of the residuals. This could also be expressed as $Y^T RY = \sum_{j=1}^{J}(Y_j - \hat{Y}_j)^2$, where $\hat{Y}_j = (x_1\hat{\beta}_1)_j = \hat{\beta}_1$.

Two-sample t-test

The two-sample t-test allows one to test the null hypothesis that the means of two groups are equal. The resulting design matrix consists of three columns: the first two encode the group membership of each scan and the third models a common constant across scans of both groups. This model is overdetermined by one degree of freedom, i.e. the sum of the first two regressors equals the third regressor. Notice the difference in parameterization compared to the earlier two-sample t-test example. Nevertheless, the resulting t-value is the same for a differential contrast. Let the number of scans in the first and second groups be J_1 and J_2, where $J = J_1 + J_2$. The three regressors consists of ones and zeros, where the first regressor consist of J_1 ones, followed by J_2 zeroes. The second regressor consists of J_1 zeroes, followed by J_2 ones. The third regressor contains ones only.

Let the contrast vector be $c = [-1, 1, 0]^T$, i.e. the alternative hypothesis is $\overline{\mathcal{H}}: \beta_1 < \beta_2$. Then:

$$(X^T X) = \begin{pmatrix} J_1 & 0 & J_1 \\ 0 & J_2 & J_2 \\ J_1 & J_2 & J \end{pmatrix}.$$

This matrix is rank deficient so we use the pseudo-inverse $(X^T X)^-$ to compute the t-statistic. We sandwich $(X^T X)^-$ with the contrast and get $c^T (X^T X)^- c = 1/J_1 + 1/J_2$. The t-statistic is then given by:

$$T = \frac{\hat{\beta}_2 - \hat{\beta}_1}{\sqrt{\hat{\sigma}^2/(1/J_1 + 1/J_2)}} \sim t_{J-2} \tag{8.26}$$

and $\hat{\sigma}^2 = Y^T RY/(J-2)$. We have assumed here that we have equal variance in both groups. This assumption may not be tenable (e.g. when comparing normal subjects with patients) and we may have to take this non-sphericity into account (see Chapter 10).

Paired t-test

The model underlying the paired t-test is an extension of the two-sample t-test model. It assumes that the scans come in pairs, i.e. one scan of each pair is in the first group and the other is in the second group. The extension is that the means over pairs are not assumed to be equal, i.e. the mean of each pair is modelled separately. For instance, let the number of pairs be $N_{pairs} = 5$, i.e. the number of scans is $J = 10$. The design matrix consists of seven regressors. The first two model the deviation from the pair-wise mean within group and the last five model the pair-specific means. The model has degrees of freedom one less than the number of regressors.

Let the contrast vector be $c = [-1, 1, 0, 0, 0, 0, 0]^T$, i.e. the alternative hypothesis is $\overline{\mathcal{H}}: \beta_1 < \beta_2$. This leads to:

$$T = \frac{\hat{\beta}_2 - \hat{\beta}_1}{\sqrt{\hat{\sigma}^2/(1/J_1 + 1/J_2)}} \sim t_{J-J/2-1} \tag{8.27}$$

The difference between this and a two-sample t-test lies in the degrees of freedom $J - J/2 - 1$. The paired t-test can be a more appropriate model for a given data set, but more effects are modelled, i.e. there are fewer degrees of freedom.

[9] Non-sphericity refers to the deviation of the error covariance matrix from a diagonal shape or a shape that can be transformed into a diagonal shape. See also Chapter 10.

One-way ANCOVA

A one-way ANCOVA allows one to model group effects, i.e. the mean of each of Q groups. This model includes the one-sample and two-sample t-tests, i.e. the cases when $1 \le Q \le 2$.

Let the number of groups be $Q = 3$, where there are five scans within each group, i.e. $J_q = 5$ for $q = 1, \ldots, Q$. There are a range of different contrasts available. For instance, we could test the null hypothesis that the group means are all equal using the F-contrast as described earlier. We may want to test the null hypothesis that the mean of the first two groups is equal to the mean of the third group, i.e. $\mathcal{H} : (\beta_1 + \beta_2)/2 - \beta_3 = 0$ and our alternative hypothesis is $\overline{\mathcal{H}} : (\beta_1 + \beta_2)/2 < \beta_3$. This can be tested using a t-statistic, where $c = [-1/2, -1/2, 1, 0]^T$. The resulting t-statistic and its distribution is:

$$T = \frac{(\hat{\beta}_1 + \hat{\beta}_2)/2 - \hat{\beta}_3}{\sqrt{\hat{\sigma}^2/(1/J_1 + 1/J_2 + 1/J_3)}} \sim t_{J-Q} \qquad 8.28$$

fMRI MODELS

In this section, we describe the analysis of single-session fMRI data. Models for fMRI data are slightly more complicated than ANCOVA-like models for PET because fMRI data represent time-series. This means that the data have serial correlations, and temporal non-sphericity must be modelled. Furthermore, the data are caused by dynamical processes that call upon a convolution model for their modelling.

Historically, SPM was first developed for PET data and then generalized to handle fMRI data. In this section, we describe the extensions this entailed. This section covers linear time-series models for fMRI data, temporal or serial correlations and their estimation, temporal filtering, parameter estimation and inference.

A linear time-series model

SPM is a mass-univariate device, i.e. we use the same temporal model at each voxel. Therefore, we can describe the temporal model for fMRI data by looking at how the data from a single voxel (a time-series) are modelled. A time-series comprises sequential measures of fMRI signal intensities over the period of the experiment. Usually, fMRI data are acquired for the whole brain with a sample time of roughly 2 to 5 s, using an echo-planar imaging (EPI) sequence.

Multisubject data are acquired in sessions, with one or more sessions for each subject.[10] Here, we deal only with models for one of these sessions, e.g. a single-subject analysis. Multisubject analyses are based on hierarchial models and are described in Chapter 12.

Suppose we have a time series of N observations $Y_1, \ldots, Y_s, \ldots, Y_N$, acquired at one voxel at times t_s, where $s = 1, \ldots, N$ is the *scan number*. The approach is to model at each voxel the observed time-series as a linear combination of explanatory functions, plus an error term (where we omit voxel superscripts):

$$Y_s = \beta_1 f^1(t_s) + \cdots + \beta_l f^l(t_s) + \cdots + \beta_L f^L(t_s) + \epsilon_s \qquad 8.29$$

The L functions $f^1(.), \ldots, f^L(.)$ are a suitable set of *regressors*, designed such that linear combinations of them span the space of possible fMRI responses for this experiment, up to the level of error. Consider writing out the above Eqn. **8.29** for all time points t_s, to give a set of equations:

$$\begin{aligned}
Y_1 &= \beta_1 f^1(t_1) + \cdots + \beta_l f^l(t_1) + \cdots + \beta_L f^L(t_1) + \epsilon_1 \\
\vdots &= \vdots \\
Y_s &= \beta_1 f^1(t_s) + \cdots + \beta_l f^l(t_s) + \cdots + \beta_L f^L(t_s) + \epsilon_s \\
\vdots &= \vdots \\
Y_N &= \beta_1 f^1(t_N) + \cdots + \beta_l f^l(t_N) + \cdots + \beta_L f^L(t_N) + \epsilon_N
\end{aligned}$$

which in matrix form is:

$$\begin{pmatrix} Y_1 \\ \vdots \\ Y_s \\ \vdots \\ Y_N \end{pmatrix} = \begin{pmatrix} f^1(t_1) & \ldots & f^l(t_1) & \ldots & f^L(t_1) \\ \vdots & \ddots & \vdots & \ddots & \vdots \\ f^1(t_s) & \ldots & f^l(t_s) & \ldots & f^L(t_s) \\ \vdots & \ddots & \vdots & \ddots & \vdots \\ f^1(t_N) & \ldots & f^l(t_N) & \ldots & f^L(t_N) \end{pmatrix} \begin{pmatrix} \beta_1 \\ \vdots \\ \beta_l \\ \vdots \\ \beta_L \end{pmatrix} + \begin{pmatrix} \epsilon_1 \\ \vdots \\ \epsilon_s \\ \vdots \\ \epsilon_N \end{pmatrix} \qquad 8.30$$

or in matrix notation:

$$Y = X\beta + \epsilon \qquad 8.31$$

Here each column of the design matrix X contains the values of one of the continuous regressors evaluated at each time point t_s of fMRI time series. That is, the columns of the design matrix are the discretized regressors.

The regressors are chosen to span the space of all possible fMRI responses for the experiment in question, such that the error vector ϵ is normally distributed with zero mean. As will be discussed later, we assume a serially correlated error process ϵ.

[10] The term session will be defined later.

Proportional and grand mean scaling

Before we proceed to construction of fMRI regressors, we consider the issue of global normalization. fMRI data are known to arise from various processes that cause globally distributed confounding effects (e.g. Andersson *et al.*, 2001). A simple global confound is scanner gain. This volume-wise gain is a factor that scales the whole image and is known to vary slowly during a session. A simple way to remove the effect of gain is to estimate the gain for every image and divide all the image intensities by this estimate. This is known as *proportional scaling*.

If one does not use proportional scaling, SPM applies, by default, a session-specific scaling. This type of scaling divides each volume by a session-specific global estimator. This is known in SPM as *grand mean scaling*. Session-specific grand mean scaling is recommended, because the scaling of fMRI data can vary between sessions and could mask regional activations.

To estimate the scaling factors, SPM uses a rough estimate of the volume-wise intracerebral mean intensity. Note that both kinds of scaling render the mean global activity (either of a volume or of a session) 100. The data and a signal change can then be interpreted conveniently as a per cent of the global mean.

Constructing fMRI regressors

In the following, we describe how the regressors in Eqn. **8.30** are generated and the underlying model of blood oxygen-level-dependent (BOLD) responses that this entails. This process consists of several stages. Although most are hidden from the user, it is helpful to know about the intermediate processing steps, so that one can understand the assumptions on which the model is based. Basically, SPM needs to know two things to construct the design matrix. The first is the timing of experimental changes and the second is the expected shape of the BOLD response elicited by these changes. Given this information, SPM can then construct the design matrix.

Experimental timing

Here we describe how a design matrix for one session of functional data is generated. Let the number of scans in a session be N_{scans}, where the data be ordered according to their acquisition. In SPM, a session starts at session-time zero. This time point is when the first slice of the first scan was acquired. Session-time can be measured in scans or seconds. The duration of a session is the number of scans multiplied by the volume repetition time (RT), which is the time spent from the beginning of the acquisition of one scan to the beginning of the acquisition of the next scan. We assume that RT stays constant throughout a session. The RT and the number of scans of a given session define the start and the end of a session. Moreover, because RT stays constant throughout the experiment, one also knows the onset of each scan.

The design of the experiment is described as a series of trials or events, where each trial is associated with a trial type. Let N^m_{trials} be the number of trials of trial type m and N_{types} the number of trial types. For each trial j of trial type m, one needs to specify its onset and duration. Note that we do not need to make a distinction between event-related or blocked designs, so a trial can be either a short event or an epoch. Let the onset vector of trial type m be O^m so that O^m_j is the onset of trial j of trial type m. For example, the onset of a trial that started at the beginning of scan four is at session time three (in scans) or at session time $3 \cdot RT$ (in seconds). Let vector D^m contain the user-specified stimulus durations of each trial for trial type m.

Given all the onsets O^m and durations D^m, SPM generates an internal representation of the experiment. This representation consists of the stimulus function S^m for each trial type m. All time bins of a session are covered such that the vectors S^m represent a contiguous series of time bins. These time bins typically do not cover a time period of length RT, but a fraction of it, to provide a well-sampled discrete approximation to the stimulus functions S^m, i.e. they are over-sampled with respect to the data.

The occurrence of a stimulus in a particular time bin is represented by an indicator variable of 1 or 0 (but see later in the *Parametric modulation* section). In other words, the stimulus function is a long vector of 'switches' that encode the presence or absence of a particular trial type. In this way, stick-functions or boxcar functions of any sort can be represented.

Note that the degree of discretization of the stimulus functions is controlled by the user. Time bin size is specified as the number of time bins per RT.[11]

For example, assume the RT is 3.2 s. Then each time bin, with the default of 16 bins/RT, covers 200 ms. The length of the vector S^m is $16N_{scans}$. Note that choosing a smaller time bin size does not necessarily provide a higher temporal precision for the resulting regressors in the design matrix. This is because the BOLD response occupies a rather low-frequency band. Therefore, responses to trials a few hundred milliseconds apart are virtually indistinguishable.

[11] The effective time bin size is accessible in SPM as variable fMRI_T. Its default value is 16.

High-resolution basis functions

After the stimulus functions have been specified in terms of onsets and durations, we need to describe the shape of the expected response. This is done using temporal basis functions. The underlying assumption is that the BOLD response for a given trial type m can be generated by passing the stimulus function through a linear finite impulse response (FIR) system, whose output is the observed data Y. This is expressed by the convolution model:

$$Y = d(\sum_{m=1}^{N_{types}} h^m \otimes S^m) + \epsilon \qquad 8.32$$

where h^m is the impulse response function for trial type m. The $\otimes$ operator denotes the convolution of two vectors (Bracewell, 1986). $d(\cdot)$ denotes the down-sampling operation, which is needed to sample the convolved stimulus functions at each RT. In other words, the observed data Y are modelled by summing the output of N_{types} different linear systems. The input to the mth linear system is the stimulus function of trial type m.

The impulse response functions h^m are not known, but we assume that they can be modelled as linear combinations of some basis functions b_i:

$$Y = \sum_{m=1}^{N_{types}} \sum_{i=1}^{N_{bf}} d(b_i \beta_i^m \otimes S^m) + \epsilon \qquad 8.33$$

where β_i^m is the ith coefficient for trial type m and N_{bf} is the number of basis functions b_i. We can now move the coefficients outside the sampling operator to give:

$$Y = d([(b \otimes S^1)\beta^1 + \cdots + (b \otimes S^{N_{types}})\beta^{N_{types}}]) + \epsilon \qquad 8.34$$

where $b = \left[b_1, \ldots, b_{N_{bf}}\right]$ and $\beta^m = \left[\beta_1^{m^T}, \ldots, \beta_{N_{bf}}^{m^T}\right]^T$. Note that the convolution operates on the columns of matrix b. If we let $X = \left[(b \otimes S^1) \vdots \cdots \vdots (b \otimes S^{N_{types}})\right]$ and $\beta = \left[\beta^{1^T}, \ldots, \beta^{N_{types}^T}\right]^T$, we see that Eqn. **8.34** is a linear model, like Eqn. **8.31**. The columns of the design matrix X are given by the convolution of each of the N_{types} stimulus functions with each of the N_{bf} basis functions. Note that, although we model different impulse response functions for each trial type m, our parametrization uses the same basis functions b_i for each trial type, but different parameters $\beta_1^m, \ldots, \beta_{N_{bf}}^m$.

In summary, when we choose a specific basis set b_i, we are implicitly modelling the haemodynamic response as a linear combination of these basis functions. The question remains: which basis function set is best for fMRI data? In SPM, the default choice is a parameterized model of a canonical impulse response function. This function is a mixture of two gamma functions. To form a basis set, one usually supplements this function with its first partial derivatives with respect to its generating parameters: the onset and dispersion. This gives: b_1 the canonical response function; b_2 its partial derivative with respect to onset (time); and b_3 its partial derivative with respect to dispersion. This is referred to as the 'haemodynamic response function (HRF) with derivatives'. This set can model a BOLD response that: (i) can be slightly shifted in time with respect to the canonical form; or (ii) has a different width.

Parametric modulation

When we first introduced the stimulus functions S^m they were described as vectors consisting of ones and zeroes. However, one can assign other values to S^m by specifying a vector parametric weights that are applied to each event. This weighting allows models in which the stimulus function has an event-specific scaling. There are many applications for this parametric modulation. For instance, one can weight events by a linear function of time; this models a linear change in the responses over time. Another application is weighting of S^m with some external measure that was acquired trial-wise, e.g. reaction times. These modulated regressors allow one to test for a linear dependence between the modulating parameter and evoked responses, while taking into account the convolution with the HRF. Higher-order relationships can be modelled using polynomial expansions of the modulating parameter to give a series of stimulus functions for each trial type, each convolved with the haemodynamic basis set.

Down-sampling

In Eqn. **8.34**, a down-sampling operator d was applied to the high-resolution (continuous) regressors to the low-resolution temporal space of the data Y. Here, one has to be aware of a slight limitation for event-related data that arise due to the use of the same temporal model at each voxel.

fMRI data are acquired slice-wise, so that a small amount of time elapses from the acquisition of one slice to the next. Given standard EPI sequences, the acquisition of one slice takes roughly 60–100 ms. Therefore, an optimal sampling of the high-resolution basis functions does not exist, because any chosen sampling will only be optimal for one slice. The largest timing error is given for a slice that lies in acquisition order $\lfloor N_{slices}/2 \rfloor$ slices away from the slice for which the temporal model is exact.[12] This sampling issue is only relevant for event-related designs, which elicit BOLD responses lasting for

[12] $\lfloor x \rfloor$ denotes the nearest integer less or equal to x.

only a few seconds. For blocked designs, timing errors are small compared to epoch length so that the potential loss in sensitivity is negligible.

In SPM, there are two ways to solve this slice-timing issue. The first is to choose one time point and temporally realign all the slices to this time. This is called *slice timing correction*. However, this interpolation requires a rather short RT (<3 s), because the sampling has to be dense, in relation to the width of the BOLD response. The second option is to model latency differences with the temporal derivative of the HRF. As discussed above, the temporal derivative can model a temporal shift of the expected BOLD response. This temporal shift can not only model onset timing differences due to different slice times, but also differences due to a different vascular response onset, for example. Note that the temporal derivative can only model small shifts of up to a second (forwards or backwards in time). We generally recommend the use of the temporal derivative as part of the model to capture any potential latency differences.

Additional regressors

It is possible to add regressors to the design matrix without going through the convolution process described above. An important example is the modelling of movement-related effects. Because movement expresses itself in the data directly, and not through any haemodynamic convolution, these are added directly as explanatory variables in the usual way.

Serial correlations

fMRI data exhibit short-range serial or temporal correlations. By this we mean that the error ϵ_s at a given scan s is correlated with its temporal neighbours. This has to be modelled, because ignoring correlations leads to an inappropriate estimate of the degrees of freedom. When forming a t- or F-statistic, this would cause a biased estimate of the standard error leading to invalid tests. More specifically, when using ordinary least squares estimates, serial correlations enter through the effective degrees of freedom of the statistic's null distribution.[13] With serial correlations the effective degrees of freedom are lower than in the case of independence. Ignoring serial correlations leads to capricious and invalid tests. To derive correct tests we have to estimate the error covariance matrix by assuming some kind of non-sphericity (Chapter 10). We can then use this estimate in one of two ways.

First, we can use it to form generalized least squares estimators that correspond to the maximum likelihood estimates. This is called pre-whitening and involves decorrelating the data (and design matrix) as described in Chapters 10 and 22. In this case, the whitened errors are IID and the degrees of freedom revert to their usual value.

The alternative is to proceed with the ordinary least squares (OLS) estimates and form statistics in the usual way. One then makes a *post-hoc* adjustment to the degrees of freedom (cf. a Greenhouse-Geisser correction) to ensure the inflated estimate of the estimators' precision is accommodated when comparing the statistic obtained, with its null distribution. Current implementations of SPM use the maximum likelihood approach, which requires no *post-hoc* correction. However, to retain the connection with classical statistics in this chapter, we will describe how serial correlations are used with OLS estimates, to give the effective degrees of freedom. In both cases, one needs to estimate the serial correlations first.

Note that we are only concerned with serial correlations in the error ϵ (Eqn. **8.31**). The correlations induced by the experimental design are modelled by the design matrix X. Serial correlations in fMRI data are caused by various sources including cardiac, respiratory and vasomotor sources (Mitra *et al.*, 1997).

One model that captures the typical form of serial correlations in fMRI data is the autoregressive (order 1) plus white noise model ($AR(1) + \text{wn}$) (Purdon and Weisskoff, 1998).[14] This model accounts for short-range correlations. We only need to model short-range correlations, because we also highpass filter the data. The highpass filter removes low frequency components and long-range correlations. We refer the interested reader to Appendix 8.1 for a mathematical description of the $AR(1) + \text{wn}$ model.

Estimation of the error covariance matrix

Assuming that the $AR(1) + \text{wn}$ is an appropriate model for the fMRI error covariance matrix, we need to estimate three hyperparameters (see Appendix 8.1) at each voxel. The hyperparameterized model gives a covariance matrix at each voxel (Eqn. **8.45**). In SPM, an additional assumption is made to estimate this matrix very efficiently, which is described in the following.

The error covariance matrix can be partitioned into two components. The first component is the correlation matrix and the second is the variance. The assumption made by SPM is that the correlation matrix is the same at all voxels of interest (see Chapter 10 for further details).

[13] Effective degrees of freedom refer to the degrees of freedom of an approximation to the underlying null distribution (Worsley and Friston, 1995).

[14] The $AR(1) + \text{wn}$ is also known as the autoregressive moving-average model of order (1,1) (ARMA(1,1)).

The variance is assumed to differ between voxels. In other words, SPM assumes that the pattern of serial correlations is the same over all interesting voxels, but its amplitude is different at each voxel. This assumption seems to be quite sensible, because the serial correlations over voxels within tissue types are usually very similar. The ensuing estimate of the serial correlations is extremely precise because one can pool information from the subset of voxels involved in the estimation. This means the correlation matrix at each voxel can be treated as a known and fixed quantity in subsequent inference.

The estimation of the error covariance matrix proceeds as follows. Let us start with the linear model for voxel k:

$$Y^k = X\beta^k + \epsilon^k \tag{8.35}$$

where Y^k is an $N \times 1$ observed time-series vector at voxel k, X is an $N \times L$ design matrix, β^k is the parameter vector and ϵ^k is the error at voxel k. The error ϵ^k is normally distributed with $\epsilon \sim N(0, \sigma^{k^2} V)$. The critical difference, in relation to Eqn. **8.6**, is the distribution of the error; where the identity matrix I is replaced by the correlation matrix V. Note that V does not depend on the voxel position k, i.e. as mentioned above we assume that the correlation matrix V is the same for all voxels $k = 1, \ldots, K$. However, the variance σ^{k^2} is different for each voxel.

Since we assume that V is the same at each voxel, we can pool data from all voxels and then estimate V on this pooled data. The pooled data are given by summing the sampled covariance matrix of all interesting voxels k, i.e. $V_Y = 1/K \sum_k Y^k Y^{k^T}$. Note that the pooled V_Y is a mixture of two variance components; the experimentally induced variance and the error variance component:

$$V_Y = \sum_k X\beta^k \beta^{k^T} X^T + \epsilon^k \epsilon^{k^T} \tag{8.36}$$

The conventional way to estimate the components of the error covariance matrix $Cov(\epsilon^k) = \sigma^{k^2} V$ is to use restricted maximum likelihood (ReML) (Harville, 1997; Friston *et al.*, 2002). ReML returns an unbiased estimator of the covariance components, while accounting for uncertainty about the parameter estimates. ReML can work with precision or covariance components; in our case we need to estimate a mixture of covariance components. The concept of covariance components is a very general concept that can be used to model all kinds of non-sphericity (see Chapter 10). The model described in Appendix 8.1 (Eqn. **8.44**) is non-linear in the hyperparameters, so ReML cannot be used directly. But if we use a linear approximation:

$$V = \sum_l \lambda_l Q_l \tag{8.37}$$

where Q_l are $N \times N$ components and the λ_l are the hyperparameters, ReML can be applied. We want to specify

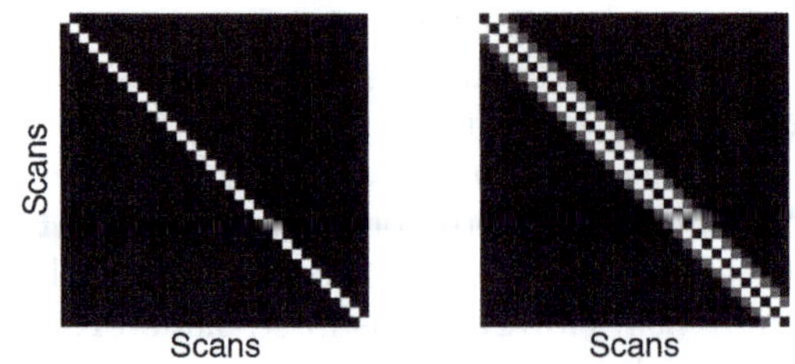

FIGURE 8.16 Graphical illustration of the two covariance components, which are used for estimating serial correlations. Left: component Q_1 that corresponds to a stationary white variance component; right: components Q_2 that implements the $AR(1)$ part with an autoregression coefficient of $1/e$.

Q_l such that they form an appropriate model for serial correlations in fMRI data. The default model in SPM is to use two components Q_1 and Q_2. These are $Q_1 = I_N$ and:

$$Q_{2_{ij}} = \begin{cases} e^{-|i-j|} & : \; i \neq j \\ 0 & : \; i = j \end{cases} \tag{8.38}$$

Figure 8.16 shows the shape of Q_1 and Q_2.

A voxel-wide estimate of V is then derived by re-scaling V such that V is a correlation matrix.

This method of estimating the covariance matrix at each voxel uses the two voxel-wide (global) hyperparameters λ_1 and λ_2. A third voxel-wise (local) hyperparameter (the *variance* σ^2) is estimated at each voxel using the usual estimator in a least squares mass-univariate scheme (Worsley and Friston, 1995):

$$\sigma^{2^k} = \frac{Y^{k^T} R Y^k}{trace(RV)} \tag{8.39}$$

where R is the residual forming matrix. This completes the estimation of the serial correlations at each voxel k. Before we can use these estimates to derive statistical tests, we will consider the highpass filter and the role it plays in modelling fMRI data.

Temporal filtering

Filtering is motivated by the observation that certain frequency bands in the data contain more noise than others. In an ideal world, our experimentally induced effects would lie in one frequency band and all of the noise in another. Applying a filter that removes the noise from the data would then give us increased sensitivity. However, the data are a mixture of activation and noise that share some frequency bands. The experimenter's task is therefore to make sure that the interesting effects do not lie in a frequency range which is especially exposed to noise. In fMRI, the low frequencies (say less than half a cycle per minute, i.e. 1/120 Hz) are known to contain scanner drifts and possibly cardiac/respiratory artefacts. Any activations that lie within this frequency range are virtually undistinguishable from noise. This is why: (i) fMRI data are highpass filtered to remove noise; and (ii) the experimenter constructs a design that puts the interesting effects into frequencies higher than 1/120 Hz. This issue is especially important for event-related designs and is described in more detail in Chapter 14. Here, we describe how the highpass filter is implemented.

The *highpass filter* is implemented using a discrete cosine transform (DCT) basis set. These are an implicit part of the design matrix. However, to the user of SPM, they are *invisible* in the sense that the DCT regressors are not actually estimated. This is simply to save space. In practice, the effects explained by the DCT or drift terms are removed from the data and design matrix using their residual forming matrix. This does not affect any of the remaining parameters estimates or any statistics, but is computationally more convenient.

Mathematically, for time points $t = 1, \ldots, N$, the discrete cosine set functions are $f_r(t) = \sqrt{2/N}\left(\cos\left(r\pi \frac{t}{N}\right)\right)$. (See Figure 8.17 for an example.) The integer index r ranges from 1 (giving half a cosine cycle over the N time points), to a user-specified maximum R. Note that SPM asks for a highpass cutoff d_{cut} in seconds. R is then chosen as $R = \lfloor 2NRT/d_{cut} + 1 \rfloor$.

To summarize, the regressors in the design matrix X account for all the components in the fMRI time series up to the level of residual noise. The highpass filter is effectively a part of the design matrix and removes unwanted low-frequency components. Therefore, long-range correlations are treated as fixed effects. The short range correlations are treated as random effects and are estimated using ReML and a linear approximation to the $AR(1) + \text{wn}$ model. In the next section, we describe how

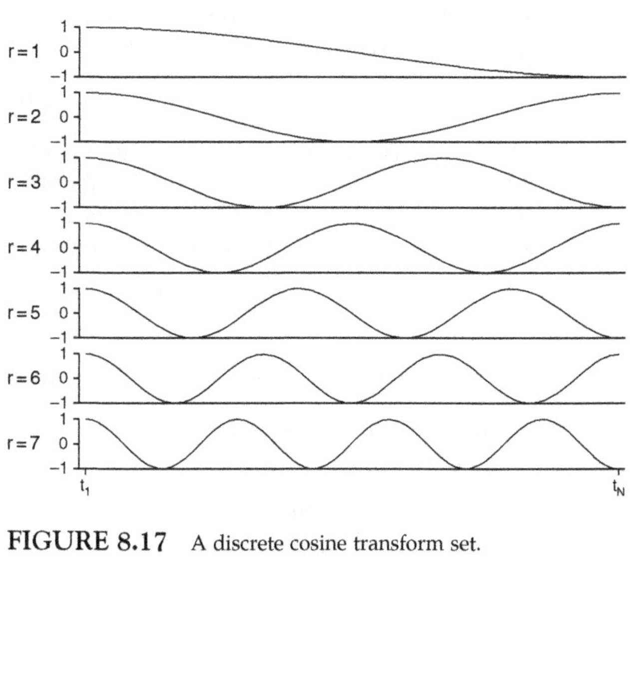

FIGURE 8.17 A discrete cosine transform set.

the model parameter estimates are used to form a t- or F-statistic at each voxel, in the context of serial correlations.

Parameter estimates and distributional results

The ordinary least-squares parameter estimates $\hat{\beta}$ are given by:

$$\hat{\beta} = (X^T X)^- X^T Y = X^- Y \qquad \textbf{8.40}$$

As described above, we estimate the error correlation matrix V using the ReML method. The error covariance matrix is then given by $\hat{\sigma}^2 V$ (Eqn. **8.39**). The covariance of the parameter estimates is:

$$Var(\hat{\beta}) = \sigma^2 X^- V X^{-T} \qquad \textbf{8.41}$$

A t-statistic can then be formed by dividing a contrast of the estimated parameters $c^T\hat{\beta}$ by its estimated standard deviation:

$$T = \frac{c^T \hat{\beta}}{\sqrt{\hat{\sigma}^2 c^T X^- V X^{-T} c}} \qquad \textbf{8.42}$$

where σ^2 is estimated using Eqn. **8.39**.

The key difference, in relation to the spherical case, i.e. when the error is IID, is that the correlation matrix V enters the denominator of the t-value. This gives us a more accurate t-statistic. However, because of V, the denominator of Eqn. **8.42** is not the square root of a χ^2-distribution. (The denominator would be exactly χ^2 distributed, when V describes a spherical distribution.) This means that Eqn. **8.42** is not t-distributed and we cannot simply make inferences by comparing with a null distribution with *trace(RV)* degrees of freedom.

Instead, one approximates the denominator with a χ^2-distribution (Eqn. **8.42**). T is then approximated by a t-distribution. The approximation proposed (Worsley and Friston, 1995) is the Satterthwaite approximation (see also Yandell, 1997), which is based on fitting the first two moments of the denominator distribution with a χ^2 distribution. The degrees of freedom of the approximating χ^2-distribution are called the *effective* degrees of freedom and are given by:

$$\nu = \frac{2E(\hat{\sigma}^2)^2}{Var(\hat{\sigma}^2)} = \frac{\text{trace}(RV)^2}{\text{trace}(RVRV)} \qquad \textbf{8.43}$$

See Appendix 8.2 for a derivation of this Satterthwaite approximation.

Similarly, the null distribution of an F-statistic in the presence of serial correlations can be approximated. In this case, both the numerator and denominator of the F-value are approximated by a χ^2-distribution.

Summary

After reconstruction, realignment, spatial normalization and smoothing, functional imaging data are ready for statistical analysis. This involves two steps: first, statistics indicating evidence against a null hypothesis of no effect at each voxel are computed to produce a statistical image (i.e. an SPM); second, this statistical image must be assessed, while limiting the possibility of false positives. These two steps are referred to as (1) estimation and (2) inference and they are covered separately in this book.

As models are designed with inference in mind it is often difficult to separate the two issues. However, the inference section, Part 4 of this book, is concerned largely with the multiple comparison problem, i.e. how to make inferences from large volumes of statistical images. A distinction can be made between such 'image-level' inference and statistical inference at a single voxel. This second sort of inference has been covered in this chapter and will be dealt with further in the remainder of Part 3.

We have shown how the general linear model, the workhorse of functional imaging analysis, provides a single framework for many statistical tests and models, giving great flexibility for experimental design and analysis. The use of such models will be further highlighted in the following chapters, especially Chapters 12, 14, and 16. In the next chapter, we take a closer look at contrasts and how they enable us to interrogate the parameter estimates described in this chapter.

APPENDIX 8.1 THE AUTOREGRESSIVE MODEL OF ORDER 1 PLUS WHITE NOISE

Mathematically, the $AR(1) + \text{wn}$ model at voxel k can be written in state-space form:

$$\epsilon(s) = z(s) + \delta_\epsilon(s)$$
$$z(s) = az(s-1) + \delta_z(s) \qquad 8.44$$

where $\delta_\epsilon(s) \sim N(0, {\sigma_\epsilon}^2)$, $\delta_z(s) \sim N(0, {\sigma_z}^2)$ and a is the $AR(1)$ coefficient. This model describes the error component $\epsilon(s)$ at time point s and at voxel k as the sum of an autoregressive component $z(s)$ plus white noise $\delta_\epsilon(s)$. We have three hyperparameters[15] at each voxel k, the variances of the two error components δ_ϵ and δ_z and the autoregressive coefficient a. The resulting error covariance matrix is then given by:

$$E(\epsilon\epsilon^T) = {\sigma_z}^2(I_N - A)^{-1}(I_N - A)^{-T} + {\sigma_\epsilon}^2 \qquad 8.45$$

where A is a matrix with all elements of the first lower off-diagonal set to $\boldsymbol{a}$ and zero elsewhere. I_N is the identity matrix of dimension N.

APPENDIX 8.2 THE SATTERTHWAITE APPROXIMATION

The unbiased estimator for σ^2 is given by dividing the sum of the squared residuals by its expectation (Worsley and Friston, 1995). Let e be the residuals $e = RY$, where R is the residual forming matrix. (Note that many steps in the following equations can be derived using properties of the trace operator; e.g. see the *Matrix Reference Manual* available under http://www.ee.ic.ac.uk/hp/staff/dmb/matrix/.)

$$\begin{aligned} E(e^Te) &= E(trace(ee^T)) \\ &= E(trace(RYY^TR^T)) \\ &= trace(R\sigma^2VR^T) \\ &= \sigma^2 trace(RV) \end{aligned}$$

An unbiased estimator of σ^2 is given by $\hat{\sigma}^2 = \frac{e^Te}{trace(RV)}$. If V is a diagonal matrix with identical non-zero elements, $trace(RV) = trace(R) = J - p$, where J is the number of observations and p the number of parameters.

In what follows, we derive the Satterthwaite approximation to a χ^2-distribution given a non-spherical error covariance matrix.

We approximate the distribution of the squared denominator of the t-value (Eqn. **8.42**) $d = \sigma^2 c^T(X^TX)^- X^TVX(X^TX)^- c\,\sigma^2$ with a scaled χ^2-variate, i.e.,

$$d \sim p(ay) \qquad 8.46$$

where $p(y) \sim \chi^2(\nu)$. We want to estimate the effective degrees of freedom ν. Note that, for a $\chi^2(\nu)$ distribution, $E(y) = \nu$ and $Var(y) = 2\nu$. The approximation is made by matching the first two moments of d to the first two moments of ay:

$$E(d) = a\nu \qquad 8.47$$

$$Var(d) = a^2 2\nu \qquad 8.48$$

If the correlation matrix V (Eqn. **8.42**) is assumed to be known, it follows that:

$$\nu = \frac{2E(\hat{\sigma}^2)^2}{Var(\hat{\sigma}^2)} \qquad 8.49$$

[15] We call these parameters hyperparameters to distinguish them from the parameter vector β.

With $E(\hat{\sigma}^2) = \sigma^2$ and:

$$\begin{aligned} E(e^T ee^T e) &= Var(e^T e) + (E(e^T e))^2 \\ &= 2trace(E(ee^T)^2) + trace(E(ee^T))^2 \\ &= \sigma^4(2trace(RV\,RV) + trace(RV)^2) \end{aligned}$$

we have:

$$\begin{aligned} Var(\hat{\sigma}^2) &= E(\hat{\sigma}^4) - E(\hat{\sigma}^2)^2 \\ &= \frac{\sigma^4(2trace(RV\,RV) + trace(RV)^2)}{trace(RV)^2} - \sigma^4 \\ &= \frac{2\sigma^4 trace(RV\,RV)}{trace(RV)^2} \end{aligned}$$

Using Eqn. **8.49**, we get:

$$\nu = \frac{trace(RV)^2}{trace(RV\,RV)} \tag{8.50}$$

REFERENCES

Andersson JL, Huttton C, Ashburner J *et al.* (2001) Modeling geometric deformations in EPI time series. *Neuroimage* **13**: 903–19

Bracewell R (1986) *The Fourier transform and its applications*, 2nd edn. McGraw-Hill International Editions, New York

Chatfield C (1983) *Statistics for technology*. Chapman & Hall, London

Christensen R (1996) *Plane answers to complex questions: the theory of linear models*. Springer-Verlag, Berlin

Draper N, Smith H (1981) *Applied regression analysis*, 2nd edn. John Wiley & Sons, New York

Friston K, Holmes A, Worsley K *et al.* (1995) Statistical parametric maps in functional imaging: a general linear approach. *Hum Brain Mapp* **2**: 189–210

Friston KJ, Penny WD, Phillips C *et al.* (2002) Classical and Bayesian inference in neuroimaging: theory. *NeuroImage* **16**: 465–83

Harville DA (1977) Maximum likelihood approaches to variance component estimation and to related problems. *Am Stat Assoc* **72**: 320–38

Healy M (1986) *Matrices for statistics*. Oxford University Press, Oxford

Mitra P, Ogawa S, Hu X *et al.* (1997) The nature of spatiotemporal changes in cerebral hemodynamics as manifested in functional magnetic resonance imaging. *Mag Res Imag Med* **37**: 511–18

Mould R (1989) *Introductory medical statistics*, 2nd edn. Institute of Physics Publishing, London

Purdon P, Weisskoff R (1998) Effect of temporal autocorrelation due to physiological noise and stimulus paradigm on voxel-level false-positive rates in fMRI. *Hum Brain Mapp* **6**: 239–49

Scheffé H (1959) *The analysis of variance*. Wiley, New York

Winer B, Brown D, Michels K (1991) *Statistical principles in experimental design*, 3rd edn. McGraw Hill, New York

Worsley K, Friston K (1995) Analysis of fMRI time-series revisited – again. *Neuroimage* **2**: 173–81

Yandell BS (1997) *Practical data analysis for designed experiments*. Chapman & Hall, London

CHAPTER

9

Contrasts and Classical Inference

J. Poline, F. Kherif, C. Pallier and W. Penny

INTRODUCTION

The general linear model (GLM) characterizes the relationship between our experimental manipulations and observed data. It allows us to ask questions like: does frontal lobe activity in a memory task depend on age? Is the activity greater for normal subjects than for patients? While many questions concern only one effect (e.g. age, group), often our questions speak to multiple effects. In 1926, John Russell wrote 'An experiment is simply a question put to nature... Even in the best planned experiment the answer can simply be yes or no... The chief requirement is simplicity: only one question should be asked at a time', but R.A. Fisher's answer in his 1935 *Design of experiments* was: 'I am convinced that this view is wholly mistaken. If we ask Nature a single question, she will often refuse to answer until some other topic has been discussed'. In other words, we model several effects that may or may not influence our measures and ask several questions by comparing the relative importance of and interactions among those effects. This chapter explains how one models and tests for effects through the use of 'contrasts'. These enable us to focus on specific questions that are put to the data.

There is no unique model of an experimental paradigm. For example, in a functional imaging experiment with three conditions 'A', 'B' and 'C', the 'C' condition (say a 'baseline'[1] or low level condition) can be modelled explicitly or implicitly. This issue generalizes to more complex designs. Contrast specification and the interpretation of the ensuing results depend on model specification, which, in turn, depends on the design of the experiment. The most important step is the specification of the experimental paradigm: if a design is clearly thought through, the questions asked of the data are generally formulated easily and contrasts are straightforward to interpret.

In general, it is not very useful simply to show that the measured signal in a specific brain area is higher under one condition relative to another. Rather, we want to know whether this difference is statistically significant. We will therefore review the aspects of hypothesis testing that relate directly to the specification of contrasts.

This chapter is organized as follows. First, we review the theoretical background behind the construction of contrasts. In the next section, we describe the rules for constructing contrasts that specify *t*-tests. We then discuss *F*-contrasts and the important issue of correlations between predictors and their impact on the interpretation of *t*- or *F*-tests. We conclude with some general remarks and a summary.

CONSTRUCTING MODELS

What should be included in the model?

Put simply, the model should include all factors (continuous or discrete) that might have an impact on the measurements. Deciding what should or should not be included is crucial (for instance, in a functional magnetic resonance imaging (fMRI) model, should the subjects' movement estimates be included?). The question 'should this factor be included in the model?' can be resolved with model selection, but *a-priori* knowledge is essential to limit the exploration of model space. With limited information about which factors influence the measured signal, the model will be larger and more complex.

[1] There is no absolute baseline condition. In fact, we generally only interpret the difference between two conditions, and therefore an activation pattern in neuroimaging is almost universally associated with at least two experimental conditions.

Statistical Parametric Mapping, by Karl Friston et al.
ISBN–13: 978-0-12-372560-8 ISBN–10: 0-12-372560-7

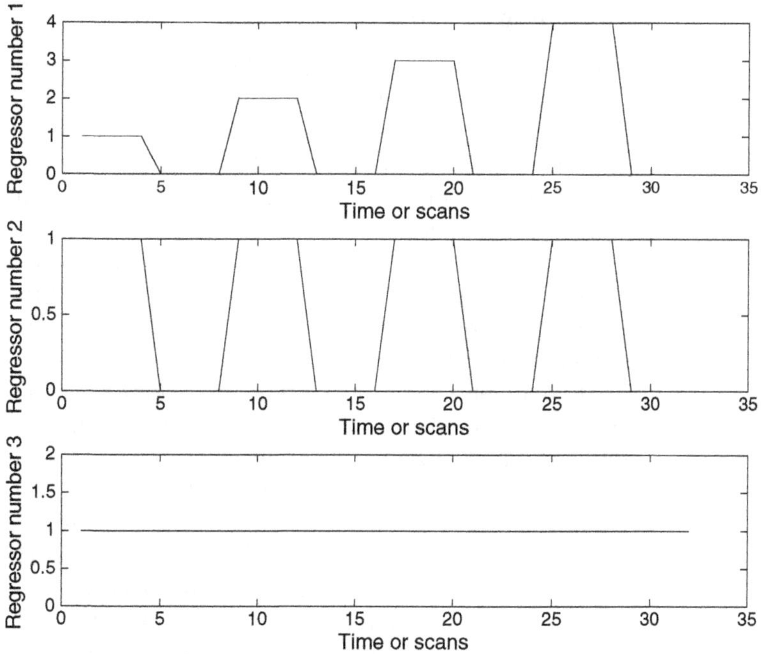

FIGURE 9.1 Model 1: design with simple linear increase. The regressors, from top to bottom, model (i) the effects of a linear increase in force, (ii) the effect of force itself and (iii) the baseline response.

To make this point clear, consider an fMRI experiment looking at motor cortex responses when a subject presses a device with four different force levels: 'press' conditions are interleaved with 'rest' periods. The conditions are ordered 'press force 1', 'rest', 'press force 2', 'rest',..., 'press force 4', etc.[2]

The first issue is how one models the 'press' and 'rest' conditions. One may have very specific prior assumptions, for example, that the response should be a *linear* function of the force. In this case, we construct a vector (a so-called *regressor, covariate,* or *predictor*) that represents this linear relationship. In the present example, this predictor could comprise 1s for all scans obtained during the first (lowest) force level, 2s for all scans acquired during the second force level, etc. If the 'rest' periods are represented by zeroes, the model assumes that the difference between rest and the first force level is the same as the difference between the first and the second force level (or between any two neighbouring force levels). To relax this assumption and construct a more flexible model, the difference between any 'press' condition and the rest period must be modelled explicitly in another predictor that takes value 1 during 'press' conditions and 0 during 'rest'.

Our model is then:

$$y_i = x_i^1\beta_1 + x_i^2\beta_2 + \epsilon_i \qquad 9.1$$

for which y_i is the ith measurement (scan), x^1 represents the predictor of the linear increase with force, and x^2 the difference between 'press' ($x_i^2 = 1$) and 'rest' ($x_i^2 = 0$). The parameters β_1 and β_2, which we need to estimate, are the coefficients of the linear functions encoded in our model. The error ϵ_i is the difference between the model prediction and the data y_i. If the signal is not zero during the rest condition (and this is always the case in neuroimaging), this offset has to be modelled by a constant term (i.e. a regressor consisting entirely of 1s). With this additional regressor, our model is written as:

$$y_i = x_i^1\beta_1 + x_i^2\beta_2 + 1\beta_3 + \epsilon_i \qquad 9.2$$

in which β_3 represents the absolute offset of the data. Figure 9.1 shows an example for the three regressors from this model[3] which, throughout this chapter, we refer to as a 'linear parametric model' or simply 'model 1'. Note that this model may or may not provide a good explanation for the measured data. It may lack important predictors, or the measured response may not be a linear function of force. Two things can be done with this model once its parameters have been estimated. One can

[2] This order would not be used in an actual experiment, where one would normally randomize the different force levels.

[3] For models of fMRI data, one needs to take into account the delay and dispersion of the haemodynamic signal. This is usually done by convolving the regressors with a haemodynamic response function (see Chapter 8). Here, we have omitted this convolution step to concentrate on the modelling aspect.

make statistical inferences about its parameters (the βs), i.e. specify a contrast, and one can compare it with an alternative model.

Modelling the 'baseline'

Should we add a predictor for the 'rest' periods to our model? This predictor could consist of 1 for scans during 'rest' and 0 for scans during all other conditions. This is not necessary because the difference between 'press' and 'rest' represented by predictor 2 (x^2) already encodes the difference between 'rest' and 'press'.

Given the model in Eqn. **9.2**, the following questions can be asked:

1 Does the measured response increase linearly with force, i.e. is β_1 significantly greater than zero?
2 Is there an additive offset for the 'press' condition that is not accounted for by the first predictor, i.e. is β_2 significantly greater than zero?
3 Is the signal during 'rest' above zero, i.e. is β_3 significantly greater than zero?

Note that the model in this example could be constructed differently, i.e. reparameterized, *while encoding exactly the same information*. For example, we could remove the average value of the first and second predictors (x^1 and x^2) so that their mean is zero. This operation is called 'mean centring'. This would not change the parameter estimates or interpretation of the first two predictors but would change the interpretation of the third predictor in this model (see below).

Extending the first model

The assumption that the response increases linearly with force is a rather strong one. There are at least two ways in which this assumption can be relaxed.

First, the first covariate can be expanded using a Taylor-like expansion, such that not only linear but higher-order (quadratic, cubic, etc.) increases are modelled. In this example, we restrict this expansion to second order, including a new regressor that is the square of the linear regressor. This results in a 'quadratic-parametric model' (model 2) which is shown in Figure 9.2.

Alternatively, one can choose a non-parametric form, enabling the model to capture any differences between the four force levels. This is achieved by representing each force level as a separate predictor. This 'non-parametric' model (model 3) is shown in Figure 9.3. Note that we would like to model two separate aspects of the data. First, the average activation over all force levels (the main effect of pressing). In model 3, this average can be computed from the sum of the different force levels. Second, we would like to model the differences between all pairs of neighbouring force levels, i.e. $(1-2)+(2-3)+(3-4)$. Modelling differences between levels is similar to modelling interactions in factorial designs (see Chapter 10). We therefore have the alternative choice to model the main effect and the interaction directly. This alternative model, model 4, is shown in Figure 9.4 (main effect and interactions). The questions that can be put to model 3 and model 4 are exactly the same; they just have to be 'rephrased' using appropriate contrasts.

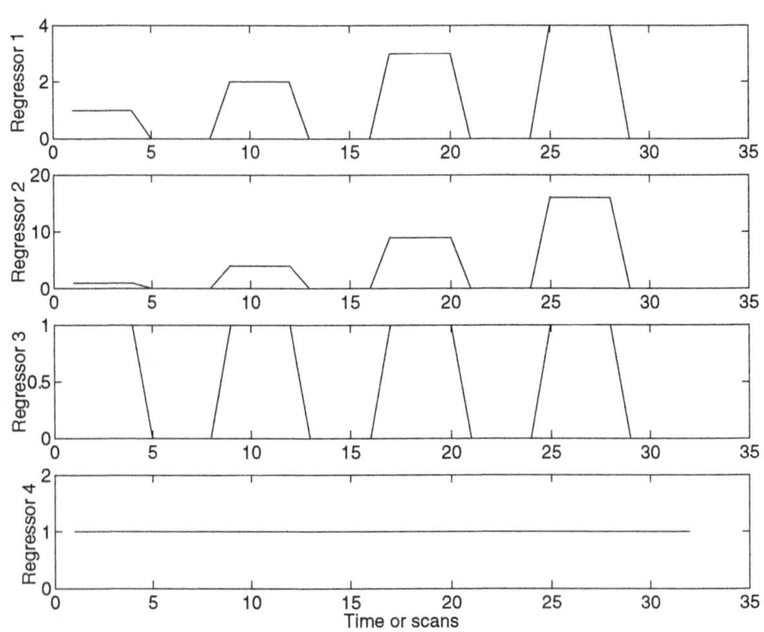

FIGURE 9.2 Model 2: linear and quadratic increase covariates. Note the scale of the second covariate.

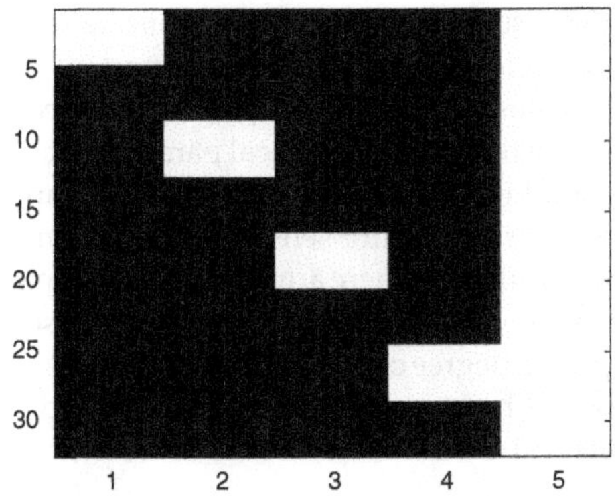

FIGURE 9.3 Model 3: different force levels are modelled using separate covariates. Black is 0 and white is 1 on this panel.

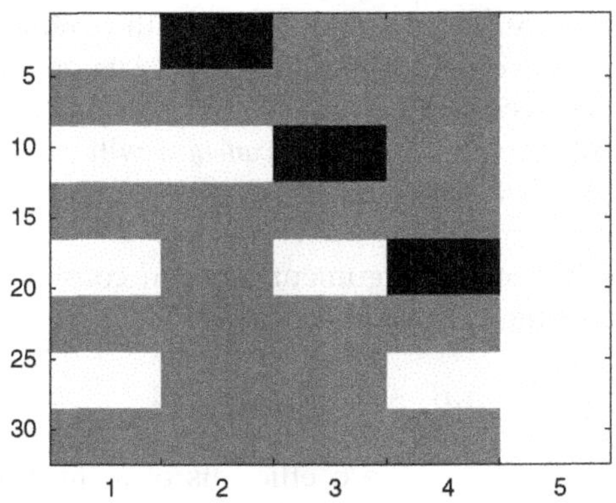

FIGURE 9.4 Model 4: the main effect of force is modelled with the first regressor and the interactions are modelled with regressors 2 to 4.

The choice between parametric and non-parametric models often depends on the number of parameters that are required. If this number is large, then parametric models might be preferred. Relatively few parameters (compared to the number of data points) and limited prior information would speak to using non-parametric models that are more flexible.

For the parametric models, we might be interested in the following questions:

- Is there a linear increase or decrease in activation with force level (modelled by the first covariate)?
- Is there a quadratic change in activation with force level *additionally* to the linear variation (modelled by the second covariate)?
- Is there any linear or quadratic dependency of the response on force (a joint test on the first and second covariate)?

Note that in the parametric model, the linear and quadratic regressors are not uncorrelated and therefore influence each other's parameter estimates and statistical inference. Issues concerning correlated regressors or contrasts are reviewed later in this chapter.

For the non-parametric models, interesting questions might be:

- Is there an overall difference between force levels and the rest condition? This question can be addressed by means of the first four regressors in model 3 and the first regressor in model 4, respectively.
- Are there any differences between different force levels? This can be addressed by looking jointly at all differences in force levels versus rest in model 3 and at regressors 2 to 4 in model 4.
- Would it be possible to test for a linear dependency of the measured signal on force level? Because any differences between force levels have been modelled, it is possible (but not easy) to test for a *specific* linear increase.

These model specification questions are often framed in the following form: should conditions A and B be modelled separately, or should the common part of A and B (A + B) be modelled together with the difference (A − B)? Note that if there is no third condition (or implicit baseline) only (A − B) can be estimated from the data.

CONSTRUCTING AND TESTING CONTRASTS

Parameter estimation

We now turn to the issue of parameter estimation. As reviewed in depth in Chapter 8, the general linear model[4] rests on the equation:

$$Y = X\beta + \epsilon \qquad 9.3$$

This equation models the data Y (comprising n measurements) as a linear combination of predictors which form the columns of the design matrix X. X is of dimension (n, p) and contains all effects $x^1, \ldots, x^p$ that are assumed to influence the measured data. The quantity ϵ is additive noise and has a normal distribution with zero mean and covariance $\sigma^2\Sigma_i$.

The model in Eqn. **9.3** states that the expectation of the data Y is equal to $X\beta$. If the data cannot be modelled by a linear combination of the predictors in X then the model is not appropriate and statistical results are difficult to interpret. This might occur if X does not contain all effects

[4] Most of the notation used in this chapter and Chapter 8 is identical but we also summarize notation in Appendix 9.1.

that influence the data, if it contains too many predictors that are unrelated to the data, or if the assumed linear relation between data and predictors does not hold.

A common method, used to solve the above equation, is called ordinary least squares (OLS).[5] OLS finds those parameter estimates β for which the sum of squared errors becomes minimal: $\| \epsilon \|^2 = \| Y - X\beta \|^2$.

This corresponds to finding a $\hat{\beta}$ such that $X\hat{\beta}$ is as close as possible to Y. This means that $X\hat{\beta}$ is the orthogonal projection of Y onto $C(X)$, the vector space spanned by the columns of X (see Figure 9.15 for an illustration). Therefore, if P_X is the orthogonal projection matrix (see Appendix 9.3) onto $C(X)$, $\hat{\beta}$ must satisfy:

$$P_X Y = X\hat{\beta}$$

This equation expresses the relationship between the parameters $\hat{\beta}$ and the data. For one-way analysis of variance ANOVA (Chapter 13), $P_X Y$ provides the means of the various groups, and the above equations describe the relationship between the $\hat{\beta}$ and these means (see below).

The matrix P_X depends only on the space spanned by Xs columns (i.e. C(X)). Therefore, two models with different design matrices X_1 and X_2 are equivalent if $C(X_1) = C(X_2)$: they explain the same aspects of the data ($X\beta$), have the same error components, and each contrast formulated for one model can be rephrased in the context of the other, such that it leads to the same statistical conclusions.

The parameters β are estimated from the data using:

$$\hat{\beta} = (X^T X)^- X^T Y \quad 9.4$$

where X^- denotes the (Moore-Penrose) pseudoinverse of X. The fitted data $\hat{Y}$ are defined as:

$$\hat{Y} = X\hat{\beta} \quad 9.5$$

and represent what is predicted by the model. The estimated noise (error) is:

$$Y - \hat{Y} = RY = \hat{\epsilon} \quad 9.6$$

where

$$R = I_n - P_X \quad 9.7$$

The noise variance is estimated with:

$$\hat{\sigma}^2 = Y^T RY / \mathrm{tr}[R\Sigma_i] \quad 9.8$$

Eqn. **9.4** has two important implications:

- Parameter estimates depend on the scaling of the regressors in X. This scaling is not important when a parameter estimate is compared to its standard deviation (see below). However, it is important if parameter estimates of different regressors are compared. When defined through statistical parametric mapping's (SPM) graphical user interface, regressors are appropriately scaled to ensure sensible comparisons.
- If X is not of full rank, there are infinitely many parameter vectors β which solve the equation. In this case, estimation of $\hat{\beta}$ has a degree of arbitrariness and only some compounds will be meaningful. These are called *estimable* contrasts and are the subject of the next section.

[5] If the properties of the noise are known, the most efficient way to estimate the parameters is a maximum likelihood procedure. This entails whitening the noise.

Estimability

One can appreciate that not all parameters may be estimable by looking at a model that contains the same regressor twice, say x_1 and $x_2 = x_1$ (with parameters β_1 and β_2). There is no information in the data on which to base the choice of $\hat{\beta}_1$ compared to $\hat{\beta}_2$. In this case, any solution of the form $\hat{\beta}_1 + \hat{\beta}_2 = constant$ will provide the same fitted data, the same residuals, but an infinity of solutions $\hat{\beta}_1$ and $\hat{\beta}_2$.

To generalize this argument, we can consider linear functions of the parameter estimates:

$$\lambda_1 \hat{\beta}_1 + \cdots + \lambda_p \hat{\beta}_p = \lambda^T \hat{\beta} \quad 9.9$$

The constants λ_i are the coefficients of a function that 'contrasts' the parameter estimates. The vector $\lambda^T = [\lambda_1, \ldots, \lambda_p]$, where p is the number of parameters in X, is referred to as the contrast vector. The word contrast is used for the result of the operation $\lambda^T \hat{\beta}$. A contrast is a random variable, because $\hat{\beta}$ is estimated from noisy data.

The matrix X is said to be rank deficient or degenerate when (some of) the parameter estimates are not unique and therefore do not convey any meaning on their own. At first sight, this situation seems unlikely. However, many designs for positron emission tomography (PET) data or population inference, are degenerate.

A contrast is estimable if (and only if) the contrast vector can be written as a linear combination of the rows of X. This is because the information about a contrast is obtained from combinations of the rows of Y. If no combination of rows of X is equal to λ^T, then the contrast is not estimable.[6]

In more technical terms, the contrast λ has to lie within the space of X^T, denoted by $\lambda \subset \mathcal{C}(X^T)$, or, equivalently, λ

[6] In Chapter 8, we define a contrast as an estimable function of the parameter estimates. If a linear combination of parameter estimates is not estimable then that linear combination is not a contrast. In this chapter, however, we often use the expression 'estimable contrast' for purposes of emphasis.

is unchanged when projected orthogonally onto the rows of X (i.e. $P_{X^T}\lambda = \lambda$ with P_{X^T} being the 'projector' onto X^T; see Appendix 9.3). The reason for this is as follows: if there is redundancy in X, for some linear combination q, we have $Xq = 0$. Therefore, $Y = X\beta + Xq + \epsilon = X(\beta + q) + \epsilon$. So, if we test $\lambda^T\beta$, we also test $\lambda^T(\beta + q)$, hence an estimable contrast λ will satisfy $\lambda^T q = 0$. A necessary and sufficient condition for this is that $\lambda^T = vX$.

The SPM interface ensures that any specified contrast is estimable, hence offering protection against contrasts that would not make sense in degenerate designs. However, a contrast may be estimable but misinterpreted. In this chapter, we hope to clarify the interpretation of contrasts.

Three design matrices for a two-sample *t*-test

The (unpaired) two-sample *t*-test, comparing the mean of two groups, can be implemented in the linear model framework as follows. Consider an experiment with two groups of 2 (group 1) and 3 (group 2) subjects. In imaging experiments, these numbers will be larger (at least 10 or so). We have:

$$X = \begin{bmatrix} 1 & 0 \\ 1 & 0 \\ 0 & 1 \\ 0 & 1 \\ 0 & 1 \end{bmatrix}$$

then

$$P_X Y = \begin{bmatrix} 1/2 & 1/2 & 0 & 0 & 0 \\ 1/2 & 1/2 & 0 & 0 & 0 \\ 0 & 0 & 1/3 & 1/3 & 1/3 \\ 0 & 0 & 1/3 & 1/3 & 1/3 \\ 0 & 0 & 1/3 & 1/3 & 1/3 \end{bmatrix} Y = X\beta = \begin{bmatrix} \bar{y}_1 \\ \bar{y}_1 \\ \bar{y}_2 \\ \bar{y}_2 \\ \bar{y}_2 \end{bmatrix}$$

where $\bar{y}_i$ is the mean observation in group i. We will now describe two other parameterizations of the same model (such that the matrix P_X is identical in all cases) and show how to specify meaningful contrasts.

Design matrix	Parameters	Contrasts
(1) $X = \begin{bmatrix} 1 & 0 \\ 1 & 0 \\ 0 & 1 \\ 0 & 1 \\ 0 & 1 \end{bmatrix}$	$\begin{cases} \hat{\beta}_1 = \bar{y}_1 \\ \hat{\beta}_2 = \bar{y}_2 \end{cases}$	$(1,0)\hat{\beta} = \bar{y}_1$ $(0,1)\hat{\beta} = \bar{y}_2$ $(1,-1)\hat{\beta} = \bar{y}_1 - \bar{y}_2$ $(.5,.5)\hat{\beta} = \text{mean}(\bar{y}_1, \bar{y}_2)$
(2) $X = \begin{bmatrix} 1 & 1 \\ 1 & 1 \\ 0 & 1 \\ 0 & 1 \\ 0 & 1 \end{bmatrix}$	$\begin{cases} \hat{\beta}_1 + \hat{\beta}_2 = \bar{y}_1 \\ \hat{\beta}_2 = \bar{y}_2 \end{cases}$	$(1,1)\hat{\beta} = \bar{y}_1$ $(0,1)\hat{\beta} = \bar{y}_2$ $(1,0)\hat{\beta} = \bar{y}_1 - \bar{y}_2$ $(.5,1)\hat{\beta} = \text{mean}(\bar{y}_1, \bar{y}_2)$
(3) $X = \begin{bmatrix} 1 & 0 & 1 \\ 1 & 0 & 1 \\ 0 & 1 & 1 \\ 0 & 1 & 1 \\ 0 & 1 & 1 \end{bmatrix}$	$\begin{cases} \hat{\beta}_1 + \hat{\beta}_3 = \bar{y}_1 \\ \hat{\beta}_2 + \hat{\beta}_3 = \bar{y}_2 \end{cases}$	$(1,0,1)\hat{\beta} = \bar{y}_1$ $(0,1,1)\hat{\beta} = \bar{y}_2$ $(1,-1,0)\hat{\beta} = \bar{y}_1 - \bar{y}_2$ $(.5,.5,1)\hat{\beta} = \text{mean}(\bar{y}_1, \bar{y}_2)$

The only intuitive case is the first parameterization. In the two other cases, the interpretation of the parameter estimates is not obvious and the contrasts are not intuitive. In case 3, parameters are not estimable and not all contrasts are meaningful. Estimable contrasts are orthogonal to $[1\ 1\ -1]$, because column 1 plus column 2 equals column 3.

Constructing and testing *t*-contrasts

If it is clear what the parameter estimates represent, then specification of contrasts is simple, especially in the case of *t*-contrasts. These contrasts are of the form described above, i.e. univariate linear combinations of parameter estimates. We return to our first model, which includes the four forces and 'rest' as regressors. For model 1, we can ask if there is a linear increase by testing β_1 using the combination $1\beta_1 + 0\beta_2 + 0\beta_3$ with the contrast vector $\lambda^T = [1\ 0\ 0]$. A linear decrease can be tested with $\lambda^T = [-1\ 0\ 0]$.

To test for the additive offset of the 'press' condition, not accounted for by the linear increase, we use $\lambda^T = [0\ 1\ 0]$. Note here that the linear increase is starting with a value of one for the first force level, and increases to 4 for the fourth level (see Figure 9.1).

When testing for the second regressor, *we are effectively removing that part of the signal that can be accounted for by the first regressor*. This means that the second parameter estimate is not the average of the difference between the 'press' conditions and the rest condition. To obtain the latter difference, we have to construct a re-parameterization of model 1 and replace the first regressor so that it models *only* differences of 'force levels' around an average difference between 'press' and 'rest'. This is achieved by orthogonalizing the first regressor with respect to the second. This new model, model 5, is shown in Figure 9.5. The parameter estimates of this new model are [10 30 100] as compared to [10 5 100] for model 1. This issue is detailed in Andrade *et al.* (1999) and an equivalent effect can be seen for *F*-tests. This emphasizes the principle that one should have in mind not only what is, but also what is *not*, tested by a contrast.

Another solution (useful in neuroimaging where estimating the parameters can be time consuming) is to

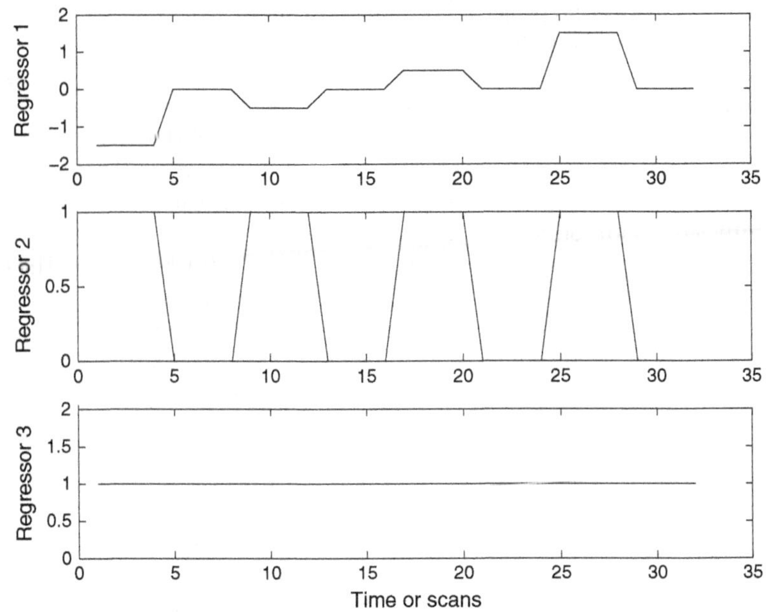

FIGURE 9.5 Model 5: this is the same as model 1 but the main effect of force has been removed from the first regressor. This changes the interpretation of the second regressor.

identify an equivalent contrast: the contrast vector $\lambda^T = [1\ 1\ 0]$ is valid but difficult to interpret. For example, the individual effects may be strong but, because they can have different signs, the overall effect may be weak.

For model 3 the average amplitude of the 'press' condition compared to 'rest' would be tested with $\lambda^T = [1\ 1\ 1\ 1\ 0]$. With model 4, the same effect can be tested with $\lambda^T = [1\ 0\ 0\ 0\ 0]$. The two contrasts give exactly the same t-maps. Note that in both cases the average over levels is tested, which could be significant just because of the effect of a single level.

An interesting question is whether we can test for the linearity of the response over the four levels. For model 3, the intuitive contrast to enter would be $\lambda^T = [1\ 2\ 3\ 4\ 0]$. This would indeed test for a linear increase with force level, but in a very unspecific manner: in the sense that the test might be significant in a situation where only the fourth condition has a greater signal than in the rest condition. This is because we are testing for the weighted sum of the corresponding parameters. The test is valid, but does not ensure that the signal changes linearly with force. In other words, the model is flexible and we are testing a very restricted hypothesis, such that the shape of the predicted signal may be distinct from the shape of the component tested.

Computing t-statistics

Whatever contrast is used, the contrast t-statistics are produced using (Friston *et al.*, 1995; Worsley and Friston, 1995):

$$t_{df} = \lambda^T \hat{\beta} / \mathrm{SD}(\lambda^T \hat{\beta}) \qquad \mathbf{9.10}$$

where SD(z) denotes the standard deviation of z and is computed as the square root of the variance:

$$\mathrm{var}[\lambda^T \hat{\beta}] = \hat{\sigma}^2 \lambda^T (X^T X)^- X^T \Sigma_i X (X^T X)^- \lambda \qquad \mathbf{9.11}$$

For Gaussian errors, t_{df} follows approximately a Student distribution with degrees of freedom given by $df = \mathrm{tr}[R\Sigma_i]^2 / \mathrm{tr}[R\Sigma_i R\Sigma_i]$. At the voxel level, the p-value of t_{df} is computed using its null distribution.

The important point is that the standard deviation of the contrast depends on the matrix X. More specifically, when regressors are correlated, the variance of the corresponding parameter estimates increases. In other words, the precision of the estimation for one component is greater when other components in the model are not correlated. The dependence of the covariance of the estimated effects and the correlation within the model can be used, for instance, to optimize event-related designs.

The test of t_{df} is one-tailed when testing exclusively for a positive (or negative) effect, and two-tailed when jointly testing for positive or negative effects.

CONSTRUCTING AND TESTING F-CONTRASTS

In this section, we consider an experiment with two event-related conditions using the simple case of right and left motor responses. The subject is asked to press a button with the right or left hand with a visual instruction. The events arrive pseudo-randomly but with a long

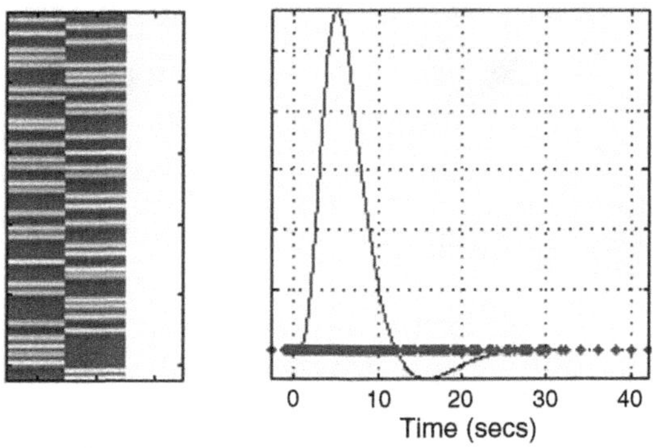

FIGURE 9.6 The left panel shows the design matrix for analysing two event-related conditions (left or right motor responses). The shape of the HRF is assumed to be known, up to a scaling factor. The two first regressors have been constructed by convolution of a series of Dirac functions with the 'canonical' HRF (right panel).

inter-stimulus interval. We are interested in brain regions that are more activated for right versus left movements.

Our first model assumes that the shape of the haemodynamic response function (HRF) can be modelled by a 'canonical HRF' (see Chapter 14). This model is shown in Figure 9.6. To find brain regions that are more active for left versus right motor responses we can use $\lambda^T = [1\ -1\ 0]$. Using Eqn. **9.10** we can compute the t-map shown in Figure 9.7. This shows activation of contralateral motor cortex plus other typical regions, such as ipsilateral cerebellum.

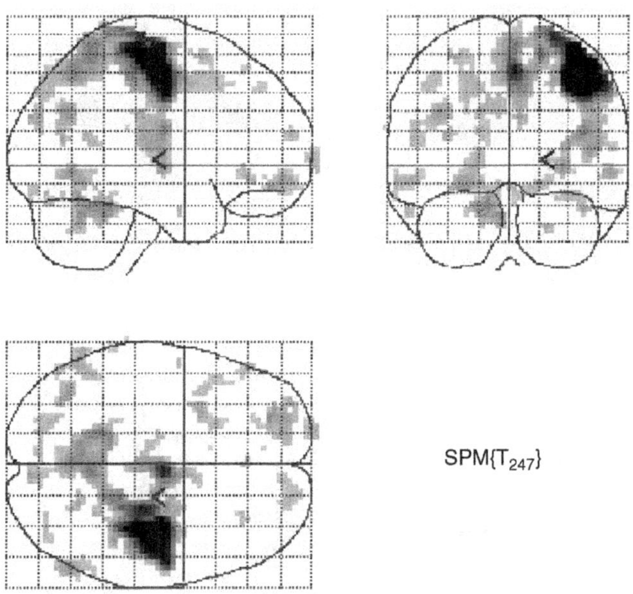

FIGURE 9.7 SPM-t image corresponding to the overall difference between the left and right responses. This map was produced using the [1 −1 0] contrast weights, using the model shown in Figure 9.6.

Because there is an implicit baseline, the parameters are also interpretable individually, and when tested (t-maps not shown) they reveal the appropriate visual and motor regions.[7] Instead of having the two regressors encoding the left and right responses separately, an equivalent model could have the first regressor modelling the response common to right and left and the second modelling the difference between them.

The fact that the HRF varies across brain regions and subjects can be accommodated as follows. A simple extension of the model of Figure 9.6 is presented in Figure 9.8, for which each response is modelled with three basis functions. These functions can model small variations in the delay and dispersion of the HRF, as described in Chapter 14. They are mean centred, so the mean parameter will represent the overall average of the data.

In this new model, how do we test for the effects of, for instance, the right motor response? The most obvious approach is to test for all regressors modelling this response. This does not entail the sum (or average) of the parameter estimates because the sign of those parameter estimates is not interpretable, but rather the (weighted) sum of squares of those parameter estimates. The appropriate F-contrast is shown in Figure 9.9.

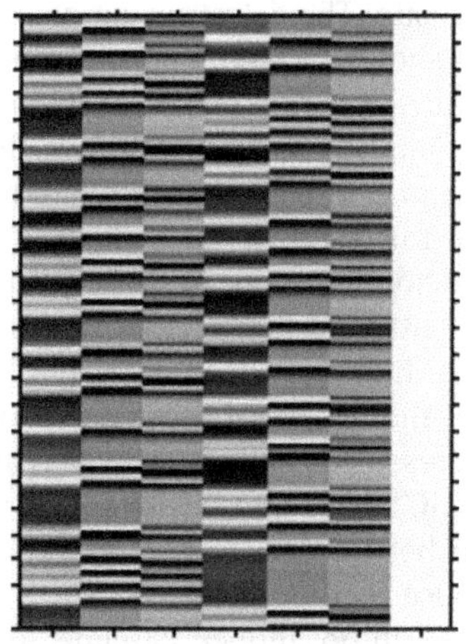

FIGURE 9.8 The same model as in Figure 9.6, but we use three regressors to model each condition. The first three columns model the first condition (left motor response) while columns 4 to 6 model the second condition (right motor response). Each set of three regressors is the result of the convolution of the stimulus onsets with the canonical HRF and its derivatives with respect to time and dispersion.

[7] Interestingly, there is some ipsilateral activation in the motor cortex such that the 'left-right' contrast is slightly less significant in the motor regions than the 'left' [1 0 0] contrast.

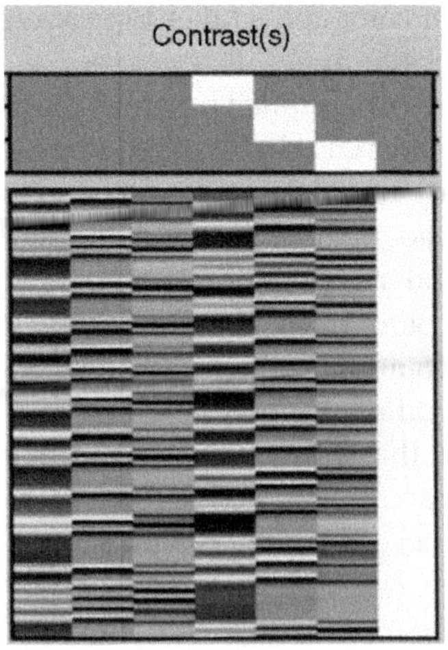

FIGURE 9.9 An '*F*-contrast' testing for the regressors modelling the right motor response. As described in the text, this corresponds to constructing the reduced model that does not contain the regressors that are 'marked' with the *F*-contrast.

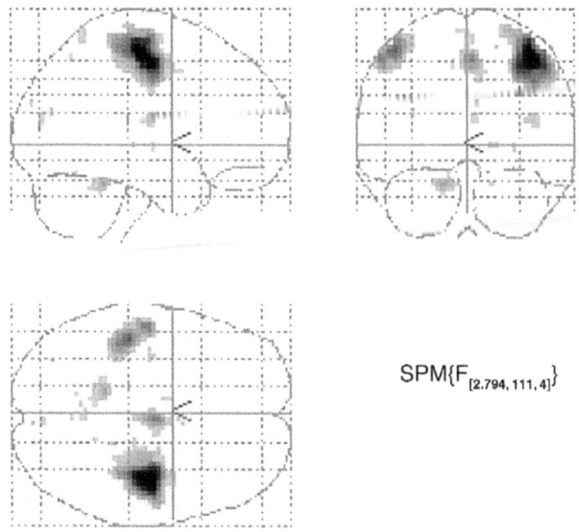

FIGURE 9.11 SPM-*F* image corresponding to the overall difference between the left and right responses. This map was produced using the *F*-contrast in Figure 9.10 and the design matrix in Figure 9.8.

One interpretation of the *F*-contrast is that it is a series of one-dimensional contrasts, each testing the null hypothesis that the relevant parameter is zero. To test for the *overall* difference between right and the left responses we use the contrast shown in Figure 9.10. Note that multiplying the *F*-contrast coefficients by -1 does not change the statistic. The *F*-test image corresponding to this contrast is shown in Figure 9.11. This image is very similar to the corresponding image for the simpler model (Figure 9.12). Finally, Figure 9.13 shows that the more complex model provides a better fit to the data.

To conclude this section, we look at another example; a 2 by 3 factorial design. In this experiment, words are presented either visually (V) or aurally (A) and belong to three different categories (C1, C2, C3). In the design matrix, the six event-types are ordered as follows: V-C1 (presented visually and in category one), V-C2, V-C3, A-C1, A-C2, A-C3. We can then test for the interaction between the modality and category factors. We suppose that the experiment is a rapid event-related design with no implicit baseline, such that only comparisons between different event-types are meaningful. In the first instance, we model each event using a single basis function. A test for the main effect of modality is presented in Figure 9.14(a). Figure 9.14(b) shows the test for the main effect of category. Note that because there is no implicit baseline here, the main effects of factors are given by differences between levels. Finally, the interaction term would be tested for as in Figure 9.14(c).

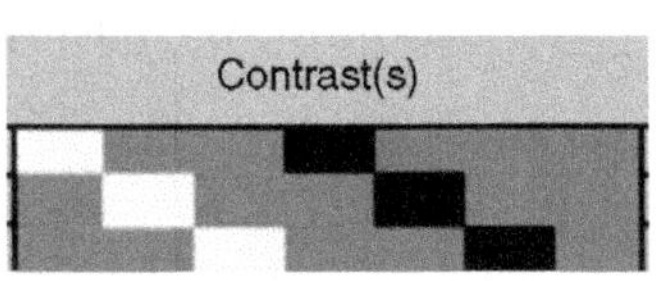

FIGURE 9.10 *F*-contrast used to test the overall difference (across basis functions) between the left and right responses.

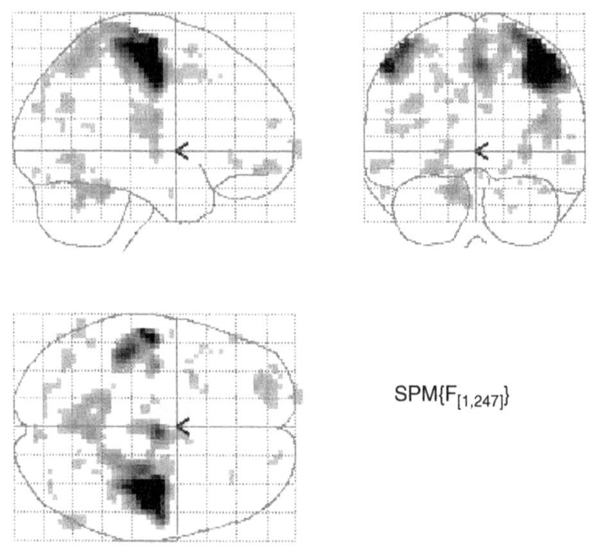

FIGURE 9.12 SPM-*F* image corresponding to the overall difference (positive or negative) from the left and right responses. This map was produced with an *F*-contrast [1 0 0;0 1 0] using the model shown in Figure 9.6.

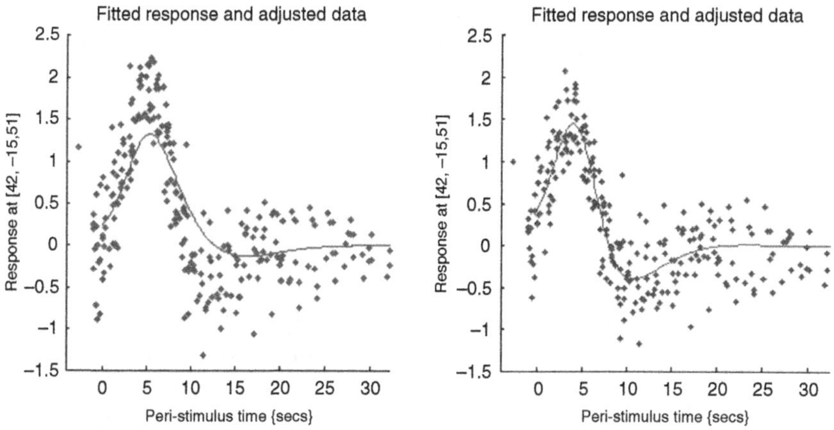

FIGURE 9.13 Haemodynamic responses at a single voxel (the maxima of the SPM-F map in Figure 9.11). The left plot shows the HRF as estimated using the simple model (Figure 9.6) and demonstrates a certain lack of fit. The fit based on a more flexible model (Figure 9.8) is better (right panel).

The number of rows in an interaction contrast (without implicit baseline) is given by:

$$N_{rows} = \prod_{i=1}^{N}(l_i - 1) \quad 9.12$$

where N is the number of factors and l_i the number of levels of factor i.

Interpretation of F-contrasts

There are two equivalent ways of thinking about F-contrasts. For example, we can think about the F-contrast in Figure 9.9 as fitting a reduced model that does not contain the 'right motor response'. This reduced model would have a design matrix X_0 with zero entries

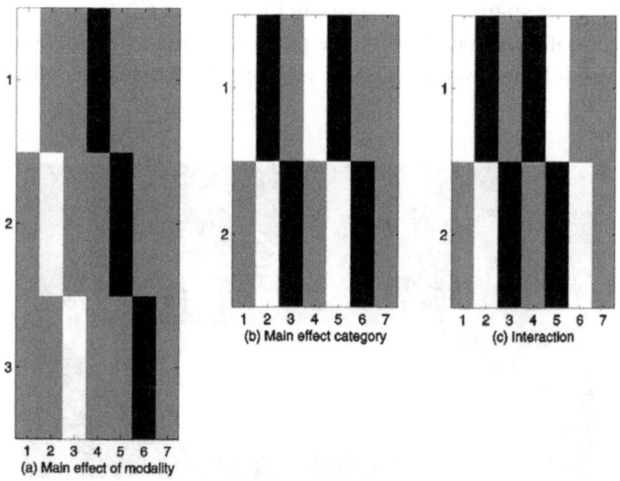

FIGURE 9.14 F-contrasts testing respectively for (a) the main effect of modality, (b) the main effect of categories, and (c) the interaction modality × category.

in place of the 'right motor response' regressors of the 'full' design matrix X. The test then compares the variance of the residuals as compared to that of the full model X. The F-test simply computes the extra sum of squares that can be accounted for by inclusion of the three 'right hand' regressors. Following any statistical textbook (e.g. Christensen, 1996) and the work of Friston *et al.* (1995) and Worsley and Friston (1995), this is expressed by testing the following quantity:

$$F_{df_1,df_2} = \frac{(Y^T(I-P_{X_0})Y - Y^T(I-P_X)Y)/\nu_1}{Y^T(I-P_X)Y/\nu_2} \quad 9.13$$

with

$$\nu_1 = \mathrm{tr}((R_0 - R)\Sigma_i)$$
$$\nu_2 = \mathrm{tr}(R\Sigma_i) \quad 9.14$$

and

$$df_1 = \mathrm{tr}((R_0 - R)\Sigma_i(R_0 - R)\Sigma_i)/tr((R_0 - R)\Sigma_i)^2 \quad 9.15$$
$$df_2 = tr(R\Sigma_i R\Sigma_i)/\mathrm{tr}(R\Sigma_i)^2 \quad 9.16$$

where R_0 is the projector onto the residual space of X_0 and P_X is the orthogonal projector onto X.

The second interpretation of the F-test is as a series of one-dimensional contrasts, each of them testing the null hypothesis that the respective contrast of parameters is zero.

We now show formally how these two interpretations are linked. The model in Eqn. **9.3**, $Y = X\beta + \epsilon$ is restricted by the test $c^T\beta = 0$ where c is now a 'contrast matrix'. If c yields an estimable function, then we can define a matrix H such that $c = H^T X$. Therefore, $H^T X\beta = 0$ which, together with Eqn. **9.3**, is equivalent to $Y \subset \mathcal{C}(X)$ and $Y \subset \mathcal{C}(H^\perp)$, the space orthogonal to H. It can be shown that the reduced model corresponding to this test is $X_0 =$

$P_X - P_H$. This is valid if, and only if, the space spanned by X_0 is the space defined by $\mathcal{C}(H)^{\perp} \cap \mathcal{C}(X)$: it is easy to show that this is indeed the case.

If $\mathcal{C}(H) \subset \mathcal{C}(X)$, the numerator of Eqn. **9.13** can be rewritten as:

$$Y^T(R_0 - R)Y = Y^T(X_0 - R)Y = Y^T(P_X - X_0)Y = Y^T(P_H)Y \quad \textbf{9.17}$$

We choose H such that it satisfies the condition above with $H = (X^T)^- c$, which yields:

$$\begin{aligned} Y^T(P_H)Y &= Y^T X(X^T X)^- X^T H(H^T H)^- H^T X(X^T X)^- X^T Y \\ &= \hat{\beta}^T c(H^T H)^- c^T \hat{\beta} \end{aligned} \quad \textbf{9.18}$$

This reformulation of the F-test is important for several reasons. First, it makes the specification and computation of F-tests feasible in the context of large data sets. Specifying a reduced model and computing the extra sum of squares using Eqn. **9.13** would be computationally too demanding. Second, it links the t-test and the test of a reduced model, and therefore makes it explicit that the 'extra' variability cannot be explained by the reduced model. Third, it makes the test of complex interactions using F-tests more intuitive.

The F-contrast that looks at the total contribution of all the 'right regressors' is, however, quite a non-specific test. One may have a specific hypothesis about the magnitude or the delay of the response and would like to test for this specifically. A reasonable test would be a t-test with contrast [0 0 0 1 0 0 0 0], testing for a positive value of the parameter that scales the standard HRF. This is perfectly valid, but it is not a test of the magnitude of the response. For instance, if the response has the shape implied by the standard model but is delayed significantly, the test might produce poor results, even if the delay is taken into account by the temporal derivative (Chapter 14). This may be important when comparing the magnitude of responses between two conditions: if the magnitudes are the same but the delays are different, across conditions, the test comparing the standard response regressors might be misinterpreted: a difference in delays might appear as a difference of magnitude *even if the basis functions are orthogonal to each other.*

Note that the simplest F-contrasts are unidimensional, in which case the F-statistic is simply the square of the corresponding t-statistic. To differentiate between unidimensional F-contrasts and t-contrasts in the SPM interface, the former are displayed in terms of images and the latter as bars.

An important point is that, generally, if we are confident about the shape of the expected response, F-tests are often less sensitive than t-tests. The reason is that, with increased model complexity, it becomes more likely that a signal of no interest could be captured by the F-contrast. The F-test implicitly corrects for this (Eqn. **9.13**), but this decreases sensitivity of the test, as compared to the more constrained t-test.

CORRELATION BETWEEN REGRESSORS

Correlations among regressors can make the interpretation of tests difficult. Unfortunately, such correlation is often imposed by the brain's dynamics, experimental design or the method of measurement. The risks of misinterpretation have been extensively discussed in Sen and Srivastava (1990) and Andrade *et al.*, (1999). To summarize, one could miss activations when testing for a given contrast if there is a substantial correlation with the rest of the design. A frequently encountered example is when the response to a stimulus is highly correlated with a motor response

If one believes that a region's activity will not be influenced by the motor response, then it is advisable to test this specific region by first removing, from the motor response regressor, all that can be explained by the stimulus. This can be seen as a 'dangerous' procedure because if, in fact, the motor response does influence the signal in this region, then an 'activation' could be wrongly attributed to a stimulus-induced effect.

Because the issue of what is and what is not tested in a model is so important, we use two complementary perspectives that might shed light on it. First, from a geometrical perspective, the model is understood as some low-dimensional space; for purposes of visualization we choose a two-dimensonal space. The data lie in a greater 3D space. The fitted data are an orthogonal projection of the data onto the model space (Figure 9.15). If the model space is spanned by two predictors C1 and C2, testing for C2 will, in effect, test for the part of C2 that is orthogonal to C1. If the two vectors are very similar (correlated), this

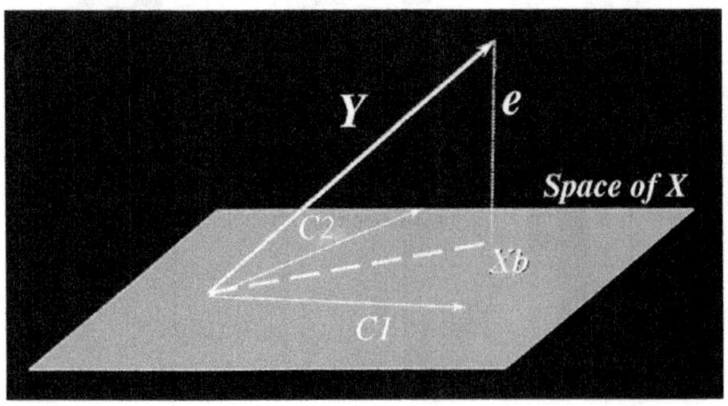

FIGURE 9.15 Geometrical perspective: estimation. The data Y are projected orthogonally onto the space of the design matrix (X) defined by two regressors C1 and C2. The error e is the distance between the data and the nearest point within the model space.

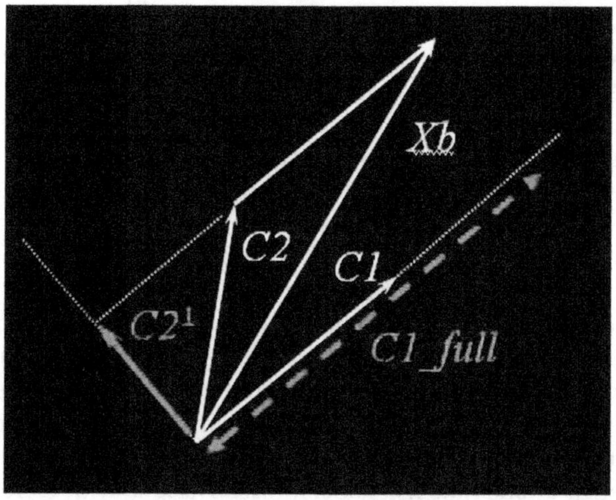

FIGURE 9.16 Hypothesis testing: the geometrical perspective. With a model defined by the two regressors C1 and C2, testing for C2 in effect measures its part orthogonal to C1. If the model is explicitly orthogonalized, (i.e. C2 is replaced by $C2^{orth}$), the test of C2 is unchanged, but the test of C1 is, and will capture more variability, as indicated by $C1_{full}$.

part can be very small. Explicit orthogonalization of C2 will make the effect tested by C1 appear much greater, while the effect tested by the $C2_{orth}$ is left unchanged (Figure 9.16).

A second perspective obtains from the following analogy. Let us consider a series of discs of different colours. Each disc represents a predictor, or more generally, a series of predictors in our model. Say we have two discs, a blue and a red one. The discs are placed on a table, where they might overlap. Testing for the effect of the first regressor would be analogous to measuring the surface of the blue disc that can be seen. If the two discs are non-overlapping (i.e. the regressors are not correlated), the two tests can be performed independently. But if the two discs do overlap (there is some correlation between the two regressors), testing for the blue disc corresponds to placing the red on top and measuring what remains of the blue. To put the blue on top amounts to orthogonalizing the red. Testing for the full surface of both discs corresponds to an F-test, and this does not depend on how the discs are placed on each other.

Moving the variance across correlated regressors

If one decides that regressors, or a combination of regressors, should be orthogonalized with respect to some part of the design matrix, it is not necessary to reparameterize and fit the model again. Once the model has been fitted, all the information needed can be found in the fitted parameter estimates. For instance, instead of testing for the *additional* variance explained by a regressor, one may wish to test for all the variance that can be explained by this regressor. If c is the contrast testing for the extra sum of squares, it is easy to show that the contrast matrix:

$$c_{Full\_space} = X^T Xc \qquad \textbf{9.19}$$

tests for all the variance explained by the subspace of X defined by Xc since we then have $H = Xc$.

Contrasts and reparameterized models

The above procedure can be generalized as follows: if the design matrix contains three subspaces (say S_1, S_2, S_3), one may wish to test for what is in S_1, having removed what could be explained by S_2 (but not by S_3). Other examples are conjunction analyses, in which a series of contrasts can be modified such that the effects they test are orthogonal. This involves orthogonalizing the subsequent subspaces tested. The results may therefore differ depending on the order in which these contrasts are entered.

The principle for computing the same contrast in two different model parameterizations, which span the same space, is simple. If X and X_p are two differently parameterized versions of the same model then we can define a matrix T such that $X_p = XT$. If c_p is a test expressed in X_p while the data have been fitted using X, the equivalent of c_p using the parameter estimates of X is

$$c = c_p(T^T X^T XT)^- T^T X^T X \qquad \textbf{9.20}$$

DESIGN COMPLEXITY

Before acquiring neuroimaging data one should think about how to model them and which contrasts are of interest. Most of the problems concerning contrast specification derive from poor design specification. Poor designs may be unclear about the objectives pursued, include factors that are confounded, or may try to answer too many questions in a single experiment. This often leads to compromises and it can become difficult to provide clear answers to the questions of interest.

This does not preclude the use of a complex paradigm, in the sense that many conditions can and (often should be) included in the design. The process of recruiting subjects and acquiring data is long and costly, and it is only natural that one would like to answer as many questions as possible with the same data. However, this requires careful thought about which contrasts will be specified and whether they actually answer the question of interest.

SUMMARY

In functional imaging experiments, one is often interested in many sorts of effects, e.g. the main effect of a factor and the possible interactions between factors. To analyse each of these effects one could fit several different GLMs and test hypotheses by looking at individual parameter estimates. However, this approach is impractical, because functional imaging data sets are very large. A more expedient approach is to fit larger models and test for effects using specific contrasts.

In this chapter, we have seen how the specification of the design matrix is intimately related to the specification of contrast weights. For example, it is often the case that main effects and interactions can be set up using parametric or non-parametric designs. These different designs lead to the use of different contrasts. Parametric approaches are favoured for factorial designs with many levels per factor. Contrasts must be estimable to be interpretable, and we have described the conditions for estimability.

In fMRI, one can model haemodynamic responses using the canonical HRF. This allows one to test for activations using t-contrasts. To account for the variability in the haemodynamic response, across subjects and brain regions, one can model the HRF using a canonical HRF plus its derivatives, with respect to time and dispersion. Inferences about differences in activation can then be made using F-contrasts. We have shown that there are two equivalent ways of interpreting F-contrasts, one employing the extra-sum-of-squares principle to compare the model and a reduced model, and one specifying a series of one-dimensional contrasts. Designs with correlations between regressors are less efficient and correlation can be removed by orthogonalizing one effect with respect to others. However, this may have a strong impact on the interpretation of subsequent tests. Finally, we have shown how such orthogonalization can be applied retrospectively, i.e. without having to refit the models.

In this chapter, we have focused on how to test for specific treatment effects encoded by the design matrix of the general linear model. However, the general linear model also entails assumptions about the random errors. In the next chapter, we examine these assumptions, in terms of covariance components and non-sphericity.

APPENDIX 9.1 NOTATION

Y:	Data	The $(n, 1)$ time series, where n is the number of time points or scans. y_i: one of those measures.
c or λ:	Contrast weights	Linear combination of the parameter estimates used to form the (numerator) of the statistics
X:	Design matrix or design model	the (n, p) matrix of regressors
β:	Model parameters	The true (unobservable) coefficients such that the weighted sum of the regressors is the expectation of our data (if X is correct)
$\hat{\beta}$:	Parameter estimates	The computed estimation of the β using the data Y: $\hat{\beta} = (X^TX)^-X^TY$
$C(X)$:	Vector space spanned by X	Given a model X, the vector space spanned by X are all vectors v that can be written as $v = X\lambda$
$P_X(X)$ or $M(X)$:	The orthogonal projector onto X	$P_X = X(X^TX)^-X^T$
R:	Residual forming matrix	Given a model X, the residual forming matrix $R = I_n - XX^-$ transforms the data Y into the residuals $r = RY$.
$\sigma^2\Sigma_i$:	scan (time) covariance	This (n, n) matrix describes the (noise) covariance between scans

APPENDIX 9.2 SUBSPACES

Let us consider a set of p vectors x_i of dimension $(n, 1)$ (with $p < n$), e.g. regressors in fMRI. The space spanned by this set of vectors is formed from all possible vectors (say u) that can be expressed as a linear combination of the x_i: $u = \alpha_1 x_1 + \alpha_2 x_2 + \ldots \alpha_p x_p$. If the matrix X is formed with the $x_i : X = [x_1 x_2 \ldots x_p]$, we denote this space as $\mathcal{C}(X)$.

Not all the x_i may be necessary to form $\mathcal{C}(X)$. The minimal number needed is called the rank of the matrix X. If only a subset of the x_i is selected, they form a smaller matrix X_0. The space spanned by X_0, $\mathcal{C}(X_0)$, is called a subspace of X. A contrast defines two subspaces of the design matrix X: one that is tested and one of 'no interest', corresponding to the reduced model.

APPENDIX 9.3 ORTHOGONAL PROJECTION

The orthogonal projection of a vector x onto the space of a matrix A is the vector (e.g a time-series) that is closest in the space $C(A)$, where distance is measured as the sum of squared errors. The projector onto A, denoted P_A, is unique and can be computed with $P_A = AA^-$, with A^- denoting the Moore-Penrose pseudoinverse[8] of A. For instance, the fitted data $\hat{Y}$ can be computed with

$$\hat{Y} = P_X Y = XX^- Y = X(X^T X)^- X^T Y = X\hat{\beta} \qquad 9.21$$

Most of the operations needed when working with linear models only involve computations in parameter space, as is shown in Eqn. **9.18**. For a further gain in computational expediency, one can work with an orthonormal basis of the space of X, if the design is degenerate. This is how the SPM code is implemented.

REFERENCES

Andrade A, Paradis A-L, Rouquette S *et al.* (1999) Ambiguous results in functional neuroimaging data analysis due to covariate correlation. *Neuroimage* **10**: 83–86

Christensen R (1996) *Plane answers to complex questions: the theory of linear models.* Springer, Berlin

Fisher RA (1935) *The Design of Experiments.* Oliver and Bayd, Edinburgh

Friston KJ, Holmes AP, Worsley KJ *et al.* (1995) Statistical parametric maps in functional imaging: a general linear approach. *Hum Brain Mapp* **2**: 189–210

Russell EJ (1926) Field experiments: how they are made and what they are. *Journal of the Ministry of Agriculture* **32**: 989–1001

Sen A, Srivastava M (1990) *Regression analysis – theory, methods, and applications.* Springer-Verlag, Berlin

Worsley KJ, Friston KJ (1995) Analysis of fMRI time-series revisited – again. *NeuroImage* **2**: 73–81

[8] Any generalized inverse could be used.

CHAPTER

10

Covariance Components

D. Glaser and K. Friston

INTRODUCTION

In this chapter, we take a closer look at covariance components and non-sphericity. This is an important aspect of the general linear model that we will encounter in different contexts in later chapters. The validity of F-statistics in classical inference depends on the sphericity assumption. This assumption states that the difference between two measurement sets (e.g. from two levels of a particular factor) has equal variance for all pairs of such sets. In practice, this assumption can be violated in several ways, for example, by differences in variance induced by different experimental conditions or by serial correlations within imaging time-series.

A considerable literature exists in applied statistics that describes and compares various techniques for dealing with sphericity violation in the context of repeated measurements (see e.g. Keselman *et al.*, 2001). The analysis techniques exploited by statistical parametrical mapping (SPM) also employ a range of strategies for dealing with the variance structure of random effects in imaging data. Here, we will compare them with conventional approaches.

Inference in imaging depends on a detailed model of what might arise by chance. If you do not know about the structure of random fluctuations in your signal, you will not know what features you should find 'surprising'. A key component of this structure is the covariance of the data. That is, the extent to which different sets of observations depend on each other. If this structure is specified incorrectly, one might obtain incorrect estimates of the variability of the parameters estimated from the data. This in turn can lead to false inferences.

Classical inference rests on the expected distribution of a test statistic under the null hypothesis. Both the statistic and its distribution depend on hyperparameters controlling different components of the error covariance (this can be just the variance, σ^2 in simple models). Estimates of variance components are used to compute statistics and variability in these estimates determines the statistic's degrees of freedom. Sensitivity depends, in part, upon precise estimates of the hyperparameters (i.e. high degrees of freedom).

In the early years of functional neuroimaging, there was debate about whether one could 'pool' (error variance) hyperparameter estimates over voxels. The motivation for this was an enormous increase in the precision of the hyperparameter estimates that rendered the ensuing t-statistics normally distributed with very high degrees of freedom. The disadvantage was that 'pooling' rested on the assumption that the error variance was the same at all voxels. Although this assumption was highly implausible, the small number of observations in positron emission tomography (PET) renders the voxel-specific hyperparameter estimates highly variable and it was not easy to show significant regional differences in error variance. With the advent of functional magnetic resonance imaging (fMRI) and more precise hyperparameter estimation, this regional heteroscedasticity was established and pooling was contraindicated. Consequently, most analyses of neuroimaging data use voxel-specific hyperparameter estimates. This is quite simple to implement, provided there is only one hyperparameter, because its restricted maximum likelihood (ReML) estimate (see Chapter 22) can be obtained non-iteratively and simultaneously through the sum of squared residuals at each voxel. However, in many situations, the errors have a number of variance components (e.g. serial correlations in fMRI or inhomogeneity of variance in hierarchical models). The ensuing non-sphericity presents a potential problem for mass-univariate tests of the sort implemented by SPM.

Two approaches to this problem can be adopted. First, departures from a simple distribution of the errors can be modelled using tricks borrowed from the classical statistical literature. This correction procedure is somewhat

Statistical Parametric Mapping, by Karl Friston et al.
ISBN–13: 978-0-12-372560-8 ISBN–10: 0-12-372560-7

crude, but can protect against the tendency towards liberal conclusions. This *post hoc* correction depends on an estimated or known non-sphericity. This estimation can be finessed by imposing a correlation structure on the data. Although this runs the risk of inefficient parameter estimation, it can condition the noise to ensure valid inference. Second, the non-sphericity, estimated in terms of different covariance components, can be used to whiten the data and effectively restore sphericity, enabling the use of conventional statistics without the need for a *post hoc* correction. However, this means we have to estimate the non-sphericity from the data.

In this chapter, we describe how the problem has been addressed in various implementations of SPM. We point first to a mathematical equivalence between the classical statistical literature and how SPM treats non-sphericity when using ordinary least squares parameter estimates. In earlier implementations of SPM, a temporal smoothing was employed to deal with non-sphericity in time-series models, as described in Worsley and Friston (1995). This smoothing 'swamps' any intrinsic autocorrelation and imposes a known temporal covariance structure. While this structure does not correspond to the assumptions underlying the classical analysis, it is known and can be used to provide *post hoc* adjustments to the degrees of freedom of the sort used in the classical literature. This correction is mathematically identical to that employed by the Greenhouse-Geisser univariate F-test.

In the second part of the chapter, we will describe a more principled approach to the problem of non-sphericity. Instead of imposing an arbitrarily covariance structure, we will show how iterative techniques can be used to estimate the actual nature of the errors, alongside the estimation of the model. While traditional multivariate techniques also estimate covariances, the iterative scheme allows the experimenter to 'build in' knowledge or assumptions about the covariance structure. This can reduce the number of hyperparameters which must be estimated and can restrict the solutions to plausible forms. These iterative estimates of non-sphericity use ReML. In fMRI time-series, for example, these variance components model the white noise component as well as the covariance induced by, for example, an AR(1) component. In a mixed-effects analysis, the components correspond to the within-subject variance (possibly different for each subject) and the between-subject variance. More generally, when the population of subjects consists of different groups, we may have different residual variance in each group. ReML partitions the overall degrees of freedom (e.g. total number of fMRI scans) in such a way as to ensure that the variance estimates are unbiased.

SOME MATHEMATICAL EQUIVALENCES

Assumptions underlying repeated-measures ANOVA

Inference on imaging data under SPM proceeds by the construction of an F-test based on the null distribution. Our inferences are vulnerable to violations of assumptions about the covariance structure of the data in just the same way as, for example, in the behavioural sciences:

Specifically, 'the conventional univariate method of analysis assumes that the data have been obtained from populations that have the well-known normal (multivariate) form, that the degree of variability (covariance) among the levels of the variable conforms to a spherical pattern, and that the data conform to independence assumptions. Since the data obtained in many areas of psychological inquiry are not likely to conform to these requirements... researchers using the conventional procedure will erroneously claim treatment effects when none are present, thus filling their literatures with false positive claims' (Keselman *et al.*, 2001).

It could be argued that limits on the computational power available to researchers led to a focus on models that can be estimated without recourse to iterative algorithms. In this account, sphericity and its associated literature could be considered a historically specific issue. Nevertheless, while the development of methods such as those described in Worsley and Friston (1995) and implemented in SPM do not refer explicitly to repeated measures designs they are, in fact, mathematically identical, as we will now show.

The assumptions required for both sorts of analysis can be seen easily by considering the variance-covariance matrix of the observation error. Consider a population variance-covariance matrix for a measurement error x under k treatments with n subjects. The errors on each subject can be viewed as a k-element vector with associated covariance matrix:

$$\Sigma_x = \begin{bmatrix} \sigma_{11} & \sigma_{12} & \dots & \sigma_{1k} \\ \sigma_{21} & \sigma_{22} & \dots & \sigma_{2k} \\ \vdots & & & \vdots \\ \sigma_{k1} & \sigma_{k2} & \dots & \sigma_{kk} \end{bmatrix} \qquad 10.1$$

This matrix can be estimated by the sample covariance matrix of the residuals:

$$\hat{\Sigma}_x = S_x = \begin{bmatrix} S_{11} & S_{12} & \dots & S_{1k} \\ S_{21} & S_{22} & \dots & S_{2k} \\ \vdots & & & \vdots \\ S_{k1} & S_{k2} & \dots & S_{kk} \end{bmatrix} \qquad 10.2$$

What is the most liberal criterion, which we can apply to this matrix, without violating the assumptions underlying repeated-measures analysis of variance (ANOVA)? By definition, the following equivalent properties are obeyed by the covariance matrix if the covariance structure is *spherical*:

$$\forall i \neq j$$
$$\sigma_{ii} + \sigma_{jj} - 2\sigma_{ij} = 2\lambda \quad \text{10.3}$$
$$\sigma^2_{X_i - X_j} = 2\lambda$$

In words, the statements in Eqn. **10.3** say that for any pair of levels, the sum of their variances minus twice their covariance is equal to a constant. Equivalently, the variance of the difference between a pair of levels is the same for all pairs. Intuitively, it is clear that this assumption is violated, for example, in the case of temporal autocorrelation. In such a case, by definition, pairs of nearby levels (in this case time points) are more highly correlated than those separated by longer times. Another example might be an analysis which took three activations from each member of two groups. Consider, for example, activation while reading, while writing, and while doing arithmetic. Imagine one wanted to test whether the populations from which two groups were drawn were significantly different, while considering the three types of task together. This would involve an F-test, and would assume that the covariation between the subject-specific residuals for reading and writing was the same as that between the writing and arithmetic. This may or may not be true. If it were not, sphericity would be violated, and the test would be overly liberal.

To illuminate the derivation of the term sphericity, we state without proof an equivalent condition to that in Eqn. **10.3**. The condition is that there can be found an orthonormal projection matrix M^* which can be used to transform the variables x of the original distribution to a new set of variables y. This new set of variables has a covariance matrix Σ_y which is spherical (i.e. is a scalar multiple of the identity matrix). This relation will be exploited in the next section.

$$M^* M^{*T} = I$$
$$M^* \Sigma_x M^{*T} = \Sigma_y = \lambda I = \begin{bmatrix} \lambda & 0 & \cdots \\ 0 & \lambda & \\ \vdots & & \ddots \end{bmatrix} \quad \text{10.4}$$
$$2\lambda = \sigma_{ii} + \sigma_{jj} - 2\sigma_{ij}$$

It is worth mentioning for completeness that while sphericity is necessary, it is not necessarily clear whether any particular dataset is spherical. Therefore, a more restricted sufficient condition has been adopted, namely compound symmetry. A matrix has compound symmetry if it has the following form:

$$\Sigma_x = \begin{bmatrix} \sigma^2 & \rho\sigma^2 & \cdots & \rho\sigma^2 \\ \rho\sigma^2 & \sigma^2 & \cdots & \rho\sigma^2 \\ \vdots & & & \vdots \\ \rho\sigma^2 & \rho\sigma^2 & \cdots & \sigma^2 \end{bmatrix} \quad \text{10.5}$$

To describe the relation in Eqn. **10.5** in words – all the within group variances are assumed equal, and separately all the covariances are assumed equal and this can be assessed directly from the data. There exist approaches to assess whether a dataset deviates from sphericity such as Mauchly's test (see e.g. Winer *et al.*, 1991), but these have low power.

A measure of departure from sphericity

Using the notation of the covariance matrix from Eqn. **10.1**, we can define a measure of departure from sphericity after Box (1954):

$$\varepsilon = \frac{k^2(\bar{\sigma}_{ii} - \sigma_{\bullet\bullet})^2}{(k-1)\sum\sum(\sigma_{ij} - \sigma_{i\bullet} - \sigma_{\bullet i} + \sigma_{\bullet\bullet})} \quad \text{10.6}$$

where $\bar{\sigma}_{ii}$ is the mean of diagonal entries, $\sigma_{\bullet\bullet}$ is the mean of all entries, $\sigma_{i\bullet}$ is the mean for row i, $\sigma_{\bullet i}$ is mean for column i. We can rewrite Eqn. **10.6** in terms of λ_i, the characteristic roots of the transformed matrix Σ_y from Eqn. **10.4**:

$$\varepsilon = \frac{(\sum \lambda_i)^2}{(k-1)\sum \lambda_i^2} \quad \text{10.7}$$

We now informally derive upper and lower bounds for our new measure. If Σ_y is spherical, i.e. of form λI then the roots are equal and since Σ_y is of size $(k-1)\times(k-1)$ then:

$$\varepsilon = \frac{(\sum \lambda)^2}{(k-1)\sum \lambda^2} = \frac{((k-1)\lambda)^2}{(k-1)(k-1)\lambda^2} = 1 \quad \text{10.8}$$

At the opposite extreme, it can be shown that for a maximum departure from sphericity:

$$\Sigma_x = \begin{bmatrix} c & c & \cdots & c \\ c & c & & c \\ \vdots & \vdots & & \vdots \\ c & c & \cdots & c \end{bmatrix} \quad \text{10.9}$$

for some constant c. Then the first characteristic root $\lambda_1 = (k-1)c$ and the rest are zeroes. From this we see that:

$$\varepsilon = \frac{(\sum \lambda_i)^2}{(k-1)\sum \lambda_i^2} = \frac{\lambda_1^2}{(k-1)\lambda_1^2} = \frac{1}{(k-1)} \quad \text{10.10}$$

Thus we have the following bounds:

$$\frac{1}{(k-1)} \le \varepsilon \le 1 \qquad 10.11$$

In summary, we have seen that the measure ϵ is well-defined using basic matrix algebra and expresses the degree to which the standard assumptions underlying the distribution are violated. In the next section, we employ this measure to protect ourselves against falsely positive inferences by correcting the parameters of the F-distribution.

Correcting degrees of freedom: the Satterthwaite approximation

Box's motivation for using this measure for the departure from sphericity was to harness an approximation due to Satterthwaite. This deals with the fact that the actual distribution of the variance estimator is not χ^2 if the errors are not spherical, and thus the F-statistic used for hypothesis testing is inaccurate. The solution adopted is to approximate the true distribution with a moment-matched scaled χ^2 distribution – matching the first and second moments. Under this approximation, in the context of repeated measures ANOVA with k measures and n subjects, the F-statistic is distributed as $F[(k-1)\varepsilon, (n-1)(k-1)\varepsilon]$. To understand the elegance of this approach, note that, as shown above, when the sphericity assumptions underlying the model are met, $\varepsilon = 1$ and the F-distribution is then just $F[(k-1), (n-1)(k-1)]$, the standard degrees of freedom. In short, the correction 'vanishes' when not needed.

Finally, we note that this approximation has been adopted for neuroimaging data in SPM. Consider the expression for the effective degrees of freedom from Worsley and Friston (1995):

$$\nu = \frac{tr(RV)^2}{tr(RVRV)} \qquad 10.12$$

Compare this with Eqn. **10.7** above, and see Chapter 8 for a derivation. Here R is the model's residual forming matrix and V are the serial correlations in the errors. In the present context: $\Sigma_x = RVR$. If we remember that the conventional degrees of freedom for the t-statistic are $k-1$ and consider ε as a correction for the degrees of freedom, then:

$$\nu = (k-1)\varepsilon = (k-1)\frac{(\sum\lambda_i)^2}{(k-1)\sum\lambda_i^2} = \frac{(\sum\lambda_i)^2}{\sum\lambda_i^2} = \frac{tr(RV)^2}{tr(RVRV)} \qquad 10.13$$

Thus, SPM applies the Satterthwaite approximation to correct the F-statistic, implicitly using a measure of sphericity violation. Next, we will see that this approach corresponds to that employed in conventional statistical packages.

But which covariance matrix is used to estimate the degrees of freedom?

Returning to the classical approach, in practice of course we do not know Σ_χ and so it is estimated by S_χ, the sample covariance matrix in Eqn. **10.2**. From this we can generate an $\hat{\varepsilon}$ by substituting s_{ij} for the σ_{ij} in Eqn. **10.6**. This correction, using the sample covariance, is often referred to as the 'Greenhouse-Geisser' correction (e.g. Winer *et al.*, 1991). An extensive literature treats the further steps in harnessing this correction and its variants. For example, the correction can be made more conservative by taking the lower bound on $\hat{\varepsilon}$ as derived in Eqn. **10.10**. This highly conservative test is [confusingly] also referred to as the 'Greenhouse-Geisser conservative correction'.

The important point to note, however, is that the construction of the F-statistic is predicated upon a model covariance structure, which satisfies the assumptions of sphericity as outlined above, but the degrees of freedom are adjusted based on the *sample* covariance structure. This contrasts with the approach taken in SPM which assumes either independent and identically distributed (IID) errors (a covariance matrix which is a scalar multiple of the identity matrix) or a simple autocorrelation structure, but corrects the degrees of freedom only on the basis of the *modelled* covariance structure. In the IID case, no correction is made. In the autocorrelation case, an appropriate correction was made, but ignoring the sample covariance matrix and assuming that the data structure was as modelled. These strategic differences are summarized in Table 10-1.

ESTIMATING COVARIANCE COMPONENTS

In earlier versions of SPM, the covariance structure V was imposed upon the data by smoothing them. Current implementations avoid this smoothing by modelling the non-sphericity. This is accomplished by defining a basis set of components for the covariance structure and then using an iterative ReML algorithm to estimate hyperparameters controlling these components. In this way, a wide range of sphericity violations can be modelled explicitly. Examples include temporal autocorrelation and more subtle effects of correlations induced by taking

TABLE 10-1 Different approaches to modelling non-sphericity

	Classical approach Greenhouse-Geisser	SPM99	SPM2/SPM5
Choice of model	Assume sphericity	Assume sphericity or AR(1)	Use ReML to estimate non-sphericity parameterized with a basis set
Corrected degrees of freedom based on covariance structure of:	Residuals	Model	Model
Estimation of degrees of freedom is voxel-wise or for whole brain	Single voxel	Many voxels	Many voxels

several measures on each of several subjects. In all cases, however, the modelled covariance structure is used to calculate the appropriate degrees of freedom using the moment-matching procedure described in Chapter 8. We do not discuss estimation in detail since this is covered in Part 4 (and Appendix 3). We will focus on the form of the covariance structure, and look at some typical examples. We model the covariance matrix as:

$$\Sigma_x = \sum \lambda_i Q_i \qquad 10.14$$

where λ_I are some hyperparameters and Q_i represent some basis set. The term Q_i embodies the form of the covariance components and could model different variances for different blocks of data or different forms of correlations within blocks. Estimation takes place using an ReML procedure where the model coefficients and variance estimates are re-estimated iteratively.

As will be discussed in the final section, what we in fact estimate is a correlation matrix or normalized covariance for many voxels at once. This can be multiplied by a scalar variance estimate calculated for each voxel separately. Since this scalar does not affect the correlation structure, the corrected degrees of freedom are the same for all voxels.

These components can be thought of as 'design matrices' for the second-order behaviour of the response variable and form a basis set for estimating the error covariance, where the hyperparameters scale the contribution of each constraint. Figure 10.1 illustrates two possible applications of this technique: one for first-level analyses and one for random effect analyses.

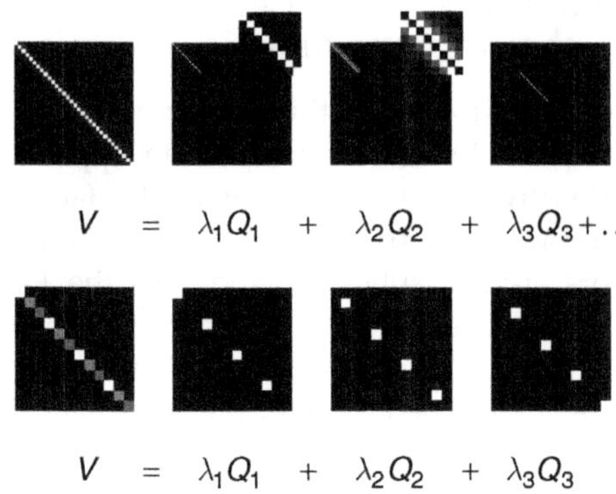

FIGURE 10.1 Examples of covariance components. Top row: here we imagine that we have a number of observations over time and a number of subjects. We decide to model the autocorrelation structure by a sum of a simple autocorrelation AR(1) component and a white noise component. A separately scaled combination of these two can approximate a wide range of actual structures, since the white noise component affects the 'peakiness' of the overall autocorrelation. For this purpose we generate two bases for each subject, and here we illustrate the first three. The first is an identity matrix (no correlation) restricted to the observations from the first subject; the second is the same but blurred in time and with the diagonal removed. The third illustrated component is the white noise for the second subject and so on. Second row: in this case we imagine that we have three measures for each of several subjects. For example, consider a second-level analysis in which we have a scan while reading, while writing and while doing arithmetic for several members of a population. We would like to make an inference about the population from which the subjects are drawn. We want to estimate what the covariation structure of the three measures is, but we assume that this structure is the same for each of the individuals. Here we generate three bases in total, one for all the reading scores, one for all the writing, and one for all the arithmetic. We then iteratively estimate the hyperparameters controlling each basis, and hence the covariance structure. After this has been normalized, so that $tr(V) = rank(V)$, it is the desired correlation.

Pooling non-sphericity estimates over voxels

So far we have discussed the covariance structure, drawing from a univariate approach. In this section, we ask whether we can harness the fact that our voxels come from the same brain. First, we will motivate the question by demonstrating that sphericity estimation involves a noisy measure, and that it might, therefore, be beneficial to pool over voxels. We will then show that under certain assumptions this strategy can be justified, and illustrate an implementation.

Simulating noise in sphericity measures

To assess the practicality of voxel-wise estimation of the covariance structure we simulated 10 000 voxels drawn from a known population with eight measures of three levels of a repeated measure. For each voxel we estimated the variance-covariance matrix using ReML and a basis set corresponding to the true distribution. We then calculated the ε correction factor and plotted a histogram for the distribution of this over the 10 000 voxels (Figure 10.2). Note the wide distribution, even for a uniform underlying covariance structure. The voxel-wide estimate was 0.65, which is higher (more spherical) than the average of the voxel-wise estimates (i.e. 0.56). In this case, the ε for the generating distribution was indeed 0.65. This highlights the utility of pooling the estimate over many voxels to generate a correlation matrix.

Degrees of freedom reprised

As the simulation shows, to make the estimate of effective degrees of freedom valid, we require very precise estimates of non-sphericity. However, as mentioned at the start of this chapter 'pooling' is problematic because the true error variance may change from voxel to voxel. We will now expand upon the form described in Eqn. **10.14** to describe in detail the strategy used by current fMRI analysis packages like SPM and *multistat* (Worsley *et al.*, 2002).

Critically, we factorize the error covariance at the i-th voxel $\Sigma_i = \sigma_i^2 V(\lambda)$ into a voxel-specific error variance and a voxel-wide correlation matrix. The correlation matrix is a function of hyperparameters controlling covariance components $V(\lambda) = \lambda_1 Q_1 +, \cdots, +\lambda_n Q_n$ that can be estimated with high precision over a large number of voxels. This allows one to use the reduced single-hyperparameter model and the effective degrees of freedom as in Eqn. **10.1**, while still allowing error variance to vary from voxel to voxel. Here the pooling is over 'similar' voxels (e.g. those that activate) and that are assumed to express various error variance components in the same proportion but not in the same amounts. In summary, we factorize the error covariance into voxel-specific variance and a correlation that is the same for all voxels in the subset. For time-series, this effectively factorizes the spatiotemporal covariance into non-stationary spatial variance and stationary temporal correlation. This enables pooling for, and only for, estimates of serial correlations.

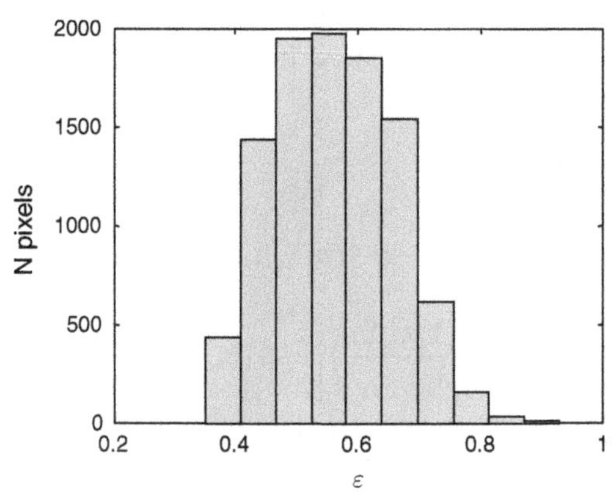

FIGURE 10.2 Histogram illustrating voxel-wise sphericity measure, ε, for 10 000 simulated voxels drawn from a known population with eight measures of three levels of a repeated measure. The average of the voxel-wise estimates is 0.56. The voxel-wide estimate was 0.65, and the ε for the generating distribution was indeed 0.65.

Once the serial correlations have been estimated, inference can then be based on the *post hoc* adjustment for the non-sphericity described above using the Satterthwaite approximation. The Satterthwaite approximation is exactly the same as that employed in the Greenhouse-Geisser (G-G) correction for non-sphericity in commercial packages. However, there is a fundamental distinction between the SPM adjustment and the G-G correction. This is because the non-sphericity V enters as a known constant (or as an estimate with very high precision) in SPM. In contradistinction, the non-sphericity in G-G uses the sample covariance matrix or multiple hyperparameter estimates, usually ReML, based on the data themselves to give $\hat{V} = \hat{\lambda}_1 Q_1 + \cdots + \hat{\lambda}_n Q_n$. This gives corrected degrees of freedom that are generally too high, leading to mildly capricious inferences. This is only a problem if the variance components interact (e.g. as with serial correlations in fMRI).

Compare the following with Eqn. **10.12**:

$$v_{GG} = tr(R)\varepsilon_{GG} = \frac{tr(R\hat{V})^2}{tr(R\hat{V}R\hat{V})} \qquad \textbf{10.15}$$

The reason the degrees of freedom are too high is that G-G fails to take into account the variability in the ReML hyperparameter estimates and ensuing variability in $\hat{V}$. There are solutions to this that involve abandoning the single variance component model and forming statistics using multiple hyperparameters directly (see Kiebel *et al.*, 2003 for details). However, in SPM this is unnecessary. The critical difference between conventional G-G corrections and the SPM adjustment lies in the fact that SPM is a mass-univariate approach that can pool non-sphericity estimates $\hat{V}$ over subsets of voxels to give a highly precise estimate V, which can be treated as a known quantity. Conventional univariate packages cannot do this because there is only one data sequence.

Non-sphericity, whitening and maximum likelihood

In this chapter, we have introduced the concept of non-sphericity and covariance component estimation through its impact on the distributional assumptions that underlie classical statistics. In this context, the non-sphericity is used to make a *post hoc* correction to the degrees of freedom of the statistics employed. Although current implementations of SPM can use non-sphericity to make *post hoc* adjustments to statistics based on ordinary least squares statistics, the default procedure is very different and much simpler; the non-sphericity is used to whiten the model. This renders the errors spherical and makes any correction redundant. Consider the general linear model (GLM) at voxel i, with non-spherical error:

$$\begin{aligned} y_i &= X\beta_i + e_i \\ e_i &\sim N(0, \sigma_i^2 V) \end{aligned} \quad \mathbf{10.16}$$

After V has been estimated using ReML (see below) it can be used to pre-multiply the model by a whitening matrix $W = V^{-1/2}$ giving:

$$\begin{aligned} Wy_i &= WX\beta_i + w_i \\ w_i &\sim N(0, \sigma_i^2 I) \\ w_i &= We_i \end{aligned} \quad \mathbf{10.17}$$

This new model now conforms to sphericity assumptions and can be inverted in the usual way at each voxel. Critically, the degrees of freedom revert to their classical values, without the need for any correction. Furthermore, the parameter estimates of this whitened model correspond to the maximum likelihood estimates of the parameters β_i and are the most efficient estimates for this model. This is used in current implementations of SPM and entails a two-pass procedure, in which the non-sphericity is estimated in a first pass and then the whitened model is inverted in a second pass to give maximum likelihood parameter and restricted maximum likelihood variance estimates. We will revisit this issue in later chapters that consider hierarchal models and the relationship of ReML to more general model inversion schemes. Note that both the *post hoc* and whitening procedures rest on knowing the non-sphericity. In the final section we take a closer look at this estimation.

Separable errors

The final issue we address is how the voxel-independent hyperparameters are estimated and how precise these estimates are. There are many situations in which the hyperparameters of mass-univariate observations factorize. In the present context, we can regard fMRI time-series as having both spatial and temporal correlations among the errors that factorize into a Kronecker tensor product. Consider the data matrix $Y = [y_1, \ldots, y_n]$ with one column, over time, for each of n voxels. The spatiotemporal correlations can be expressed as the error covariance matrix in a vectorized GLM:

$$\begin{aligned} Y &= X\beta + \varepsilon \\ vec(Y) &= \begin{bmatrix} y_1 \\ \vdots \\ y_n \end{bmatrix} = \begin{bmatrix} X & \cdots & 0 \\ \vdots & \ddots & \vdots \\ 0 & \cdots & X \end{bmatrix} \begin{bmatrix} \beta_1 \\ \vdots \\ \beta_n \end{bmatrix} + \begin{bmatrix} \varepsilon_1 \\ \vdots \\ \varepsilon_n \end{bmatrix} \\ cov(vec(\varepsilon)) &= \Sigma \otimes V = \begin{bmatrix} \Sigma_1 V & \cdots & \Sigma_{1n} V \\ \vdots & \ddots & \vdots \\ \Sigma_{n1} V & \cdots & \Sigma_{nn} V \end{bmatrix} \end{aligned} \quad \mathbf{10.18}$$

Note that Eqn. **10.16** assumes a separable form for the errors. This is the key assumption underlying the pooling procedure. Here V embodies the temporal non-sphericity and Σ the spatial non-sphericity. Notice that the elements of Σ are voxel-specific, whereas the elements of V are the same for all voxels. We could now enter the vectorized data into a ReML scheme, directly, to estimate the spatial and temporal hyperparameters. However, we can capitalize on the separable form of the non-sphericity over time and space by only estimating the hyperparameters of V and then use the usual estimator (Worsley and Friston, 1995) to compute a single hyperparameter $\hat{\Sigma}_i$ for each voxel.

The hyperparameters of V can be estimated with the algorithm presented in Friston *et al.* (2002, Appendix 3). This uses a Fisher scoring scheme to maximize the log likelihood $\ln p(Y|\lambda, \Sigma)$ (i.e. the ReML objective function) to find the ReML estimates. In the current context this scheme is:

$$\begin{aligned} \lambda &\leftarrow \lambda + H^{-1} g \\ g_i &= \frac{\partial \ln p(Y|\lambda, \Sigma)}{\partial \lambda_i} = \tfrac{n}{2} tr\{PQ_i\} + \tfrac{1}{2} tr(P^T Q_i P Y \Sigma^{-1} Y^T) \\ H_{ij} &= -\left\langle \frac{\partial^2 \ln p(Y|\lambda, \Sigma)}{\partial \lambda_{ij}^2} \right\rangle = \tfrac{n}{2} tr(PQ_i PQ_j) \\ P &= V^{-1} - V^{-1}X(X^T V^{-1} X)^{-1} X^T V^{-1} \\ V &= \lambda_1 Q_1 + \cdots + \lambda_n Q_n \end{aligned} \quad \mathbf{10.19}$$

This scheme is derived from basic principles in Appendix 4. Notice that the Kronecker tensor products and vectorized forms disappear. Critically H, the precision of the hyperparameter estimates, increases linearly with the number of voxels. With sufficient voxels this allows us to enter the resulting estimates, through V, into Eqn. **10.16** as known quantities, because they are so precise. The nice thing about Eqn. **10.19** is that the data enter only as $Y\Sigma^{-1}Y^T$ whose size is determined by the number of scans as opposed to the massive number of voxels. The term $Y\Sigma^{-1}Y^T$ is effectively the sample temporal covariance matrix, sampling over voxels (after spatial whitening) and can be assembled voxel by voxel in an efficient fashion. Eqn. **10.19** assumes that we know the spatial covariances. In practice, $Y\Sigma^{-1}Y^T$ is approximated by selecting voxels that are spatially dispersed (so that $\Sigma_{ij} = 0$) and scaling the data by an estimate of Σ_i^{-1} obtained non-iteratively, assuming temporal sphericity.

CONCLUSION

We have shown that classical approaches do not explicitly estimate the covariance structure of the noise in the data but instead assume it has a tractable form, and then correct for any deviations from the assumptions. This approximation can be based on the actual data, or on a defined structure which is imposed on the data. More principled approaches explicitly model those types of correlations which the experimenter expects to find in the data. This estimation can be noisy but, in the context of SPM, can be finessed by pooling over many voxels.

Estimating the non-sphericity allows the experimenter to perform types of analysis which were previously 'forbidden' under the less sophisticated approaches. These are of real interest to many researchers and include better estimation of the autocorrelation structure for fMRI data and the ability to take more than one scan per subject to a second level analysis and thus conduct F-tests. In event-related studies, where the exact form of the haemodynamic response can be critical, more than one aspect of this response can be analysed in a random-effects context. For example, a canonical form and a measure of latency or dispersion can cover a wide range of real responses. Alternatively, a more general basis set (e.g. Fourier or finite impulse response) can be used, allowing for non-sphericity among the different components of the set.

In this chapter, we have focused on the implications of non-sphericity for inference. In the next chapter we look at a family of models that have very distinct covariance components that can induce profound non-sphericity. These are hierarchical models.

REFERENCES

Box GEP (1954) Some theorems on quadratic forms applied in the study of analysis of variance problems. *Ann Math Stat* **25**: 290–302

Friston KJ, Glaser DE, Henson RN *et al.* (2002) Classical and Bayesian inference in neuroimaging: applications. *NeuroImage* **16**: 484–512[**10.5]

Keselman HJ, Algina J, Kowalchuk RK (2001) The analysis of repeated measures designs: a review. *Br J Math Stat Psychol* **54**: 1–20

Kiebel SJ, Glaser DE, Friston KJ (2003) A heuristic for the degrees of freedom of statistics based on multiple hyperparameters. *NeuroImage* **20**: 591–600

Winer BJ *et al.* (1991) *Statistical principles in experimental design*, McGraw-Hill 3rd edition, New York

Worsley KJ, Liao CH, Aston J *et al.* (2002) A general statistical analysis for fMRI data. *NeuroImage* **15**: 1–15

Worsley KJ, Friston KJ (1995) Analysis of fMRI time-series revisited – again. *NeuroImage* **2**: 173–81

C H A P T E R

11

Hierarchical Models

W. Penny and R. Henson

INTRODUCTION

Hierarchical models are central to many current analyses of functional imaging data including random effects analysis (Chapter 12), electroencephalographic (EEG) source localization (Chapters 28 to 30) and spatiotemporal models of imaging data (Chapters 25 and 26 and Friston *et al.*, 2002b). These hierarchical models posit linear relations among variables with error terms that are Gaussian. The general linear model (GLM), which to date has been so central to the analysis of functional imaging data, is a special case of these hierarchical models consisting of just a single layer.

Model fitting and statistical inference for hierarchical models can be implemented using a parametric empirical Bayes (PEB) algorithm described in Chapter 24 and in Friston *et al.* (2002a). The algorithm is sufficiently general to accommodate multiple hierarchical levels and allows for the error covariances to take on arbitrary form. This generality is particularly appealing as it renders the method applicable to a wide variety of modelling scenarios. Because of this generality, however, and the complexity of scenarios in which the method is applied, readers wishing to learn about PEB for the first time are advised to read this chapter first. Chapter 24 then goes on to discuss the more general case. It also shows that the variance components that are estimated using PEB, can also be estimated using an algorithm from classical statistics called restricted maximum likelihood (ReML).

In this chapter, we provide an introduction to hierarchical models and focus on some relatively simple examples. This chapter covers the relevant mathematics and numerical examples are presented in the following chapter. Each model and PEB algorithm we present is a special case of that described in Friston *et al.* (2002a). While there are a number of tutorials on hierarchical modelling (Lee, 1997; Carlin and Louis, 2000) what we describe here has been tailored for functional imaging applications. We also note that a tutorial on hierarchical models is, to our minds, also a tutorial on Bayesian inference, as higher levels act as priors for parameters in lower levels. Readers are therefore encouraged to also consult background texts on Bayesian inference, such as Gelman *et al.* (1995).

This chapter focuses on two-level models and shows how one computes the posterior distributions over the first- and second-level parameters. These are derived, initially, for completely general designs and error covariance matrices. We then consider two special cases: (i) models with equal error variances; and (ii) separable models. We assume initially that the covariance components are known, and then in the section on PEB, we show how they can be estimated. A numerical example is then given showing PEB in action. The chapter then describes how Bayesian inference can be implemented for hierarchical models with arbitrary probability distributions (e.g. non-Gaussian), using the belief propagation algorithm. We close with a discussion.

In what follows, the notation $\mathrm{N}(m, \Sigma)$ denotes a uni/multivariate normal distribution with mean m and variance/covariance Σ and lower-case ps denote probability densities. Upper case letters denote matrices, lower case denote column vectors and x^T denotes the transpose of x. We will also make extensive use of the normal density, i.e. if $p(x) = \mathrm{N}(m, \Sigma)$ then:

$$p(x) \propto \exp\left(-\frac{1}{2}(x-m)^T\Sigma^{-1}(x-m)\right) \qquad \textbf{11.1}$$

We also use Var[] to denote variance, $\otimes$ to denote the Kronecker product and X^+ to denote the pseudo-inverse.

Statistical Parametric Mapping, by Karl Friston et al.
ISBN–13: 978-0-12-372560-8 ISBN–10: 0-12-372560-7

TWO-LEVEL MODELS

We consider two-level linear Gaussian models of the form:

$$y = Xw + e$$
$$w = M\mu + z \qquad \textbf{11.2}$$

where the errors are zero mean Gaussian with covariances Cov $[e] = C$ and Cov$[z] = P$. The model is shown graphically in Figure 11.1. The column vectors y and w have K and N entries respectively. The vectors w and μ are the first- and second-level parameters and X and M are the first- and second-level design matrices. Models of this form have been used in functional imaging. For example, in random effects analysis, the second level models describe the variation of subject effect sizes about a population effect size, μ. In Bayesian inference with shrinkage priors, the second-level models variation of effect-size over voxels around a whole-brain mean effect size of $\mu - 0$ (i.e. for a given cognitive challenge, the response of a voxel chosen at random is, on average, zero). See, for example, Friston *et al.* (2002b).

The aim of Bayesian inference is to make inferences about w and μ based on the posterior distributions $p(w|y)$ and $p(\mu|y)$. These can be derived as follows. We first note that the above equations specify the likelihood and prior probability distributions:

$$p(y|w) \propto \exp\left(-\frac{1}{2}(y - Xw)^T C^{-1}(y - Xw)\right) \qquad \textbf{11.3}$$
$$p(w) \propto \exp\left(-\frac{1}{2}(w - M\mu)^T P^{-1}(w - M\mu)\right)$$

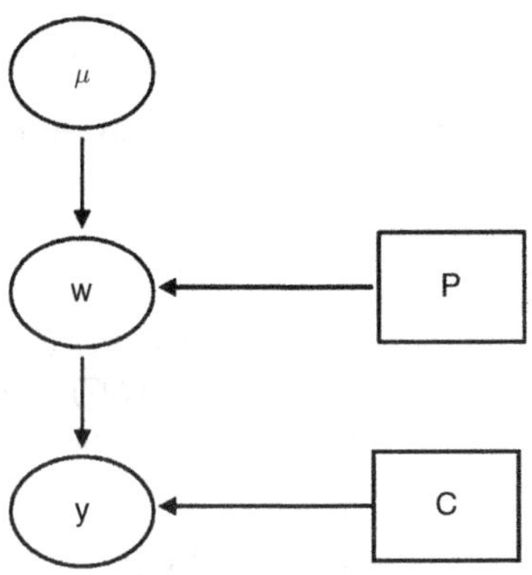

FIGURE 11.1 Two-level hierarchical model. The data y are explained as deriving from an effect w and a zero-mean Gaussian random variation with covariance C. The effects w in turn are random effects deriving from a superordinate effect μ and zero-mean Gaussian random variation with covariance P. The goal of Bayesian inference is to make inferences about μ and w from the posterior distributions $p(\mu|y)$ and $p(w|y)$.

The posterior distribution is then:

$$p(w|y) \propto p(y|w)p(w) \qquad \textbf{11.4}$$

Taking logs and keeping only those terms that depend on w gives:

$$\begin{aligned}\log p(w|y) &= -\frac{1}{2}(y - Xw)^T C^{-1}(y - Xw) \\ &\quad -\frac{1}{2}(w - M\mu)^T P^{-1}(w - M\mu) + .. \\ &= -\frac{1}{2}w^T(X^T C^{-1}X + P^{-1})w \\ &\quad + w^T(X^T C^{-1}y + P^{-1}M\mu) + ..\end{aligned} \qquad \textbf{11.5}$$

Taking logs of the Gaussian density $p(x)$ in Eqn. **11.1** and keeping only those terms that depend on x gives:

$$\log p(x) = -\frac{1}{2}x^T\Sigma^{-1}x + x^T\Sigma^{-1}m + .. \qquad \textbf{11.6}$$

Comparing Eqn. **11.5** with terms in the above equation shows that:

$$\begin{aligned}p(w|y) &= \mathrm{N}(m, \Sigma) \\ \Sigma^{-1} &= X^T C^{-1}X + P^{-1} \\ m &= \Sigma(X^T C^{-1}y + P^{-1}M\mu)\end{aligned} \qquad \textbf{11.7}$$

The posterior distribution over the second-level coefficient is given by Bayes' rule as:

$$p(\mu|y) = \frac{p(y|\mu)p(\mu)}{p(y)} \qquad \textbf{11.8}$$

However, because we do not have a prior $p(\mu)$, this posterior distribution becomes identical to the likelihood term, $p(y|\mu)$, which can be found by eliminating the first-level parameters from our two equations, i.e. by substituting the second-level equation into the first giving:

$$y = XM\mu + Xz + e \qquad \textbf{11.9}$$

which can be written as:

$$y = \tilde{X}\mu + \tilde{e} \qquad \textbf{11.10}$$

where $\tilde{X} = XM$ and $\tilde{e} = Xz + e$. The solution to Eqn. **11.10** then gives:

$$\begin{aligned}p(\mu|y) &= \mathrm{N}(\hat{\mu}, \Sigma_\mu) \\ \hat{\mu} &= (\tilde{X}^T\tilde{C}^{-1}\tilde{X})^{-1}\tilde{X}^T\tilde{C}^{-1}y \\ \Sigma_\mu &= (\tilde{X}^T\tilde{C}^{-1}\tilde{X})^{-1}\end{aligned} \qquad \textbf{11.11}$$

where the covariance term:

$$\tilde{C} = \mathrm{Cov}[\tilde{e}] \qquad \mathbf{11.12}$$
$$= XPX^T + C$$

We have now achieved our first goal, the posterior distributions of first- and second-level parameters being expressed in terms of the data, design and error-covariance matrices. We now consider the special cases of sensor fusion, equal variance models and separable models.

Sensor fusion

The first special case is the univariate model:

$$y = w + e \qquad \mathbf{11.13}$$
$$w = \mu + z$$

with a single scalar data point, y, and variances $C = 1/\beta$, $P = 1/\alpha$ specified in terms of the data precision β and the prior precision α (the 'precision' is the inverse variance). Plugging these values into Eqn. **11.7** gives

$$p(w|y) = \mathrm{N}(m, \lambda^{-1}) \qquad \mathbf{11.14}$$
$$\lambda = \beta + \alpha$$
$$m = \frac{\beta}{\lambda}y + \frac{\alpha}{\lambda}\mu$$

Despite its simplicity, this model possesses two important features of Bayesian learning in linear-Gaussian models. The first is that 'precisions add' – the posterior precision is the sum of the data precision and the prior precision. The second is that the posterior mean is the sum of the data mean and the prior mean, each weighted by their relative precisions. A numerical example is shown in Figure 11.2.

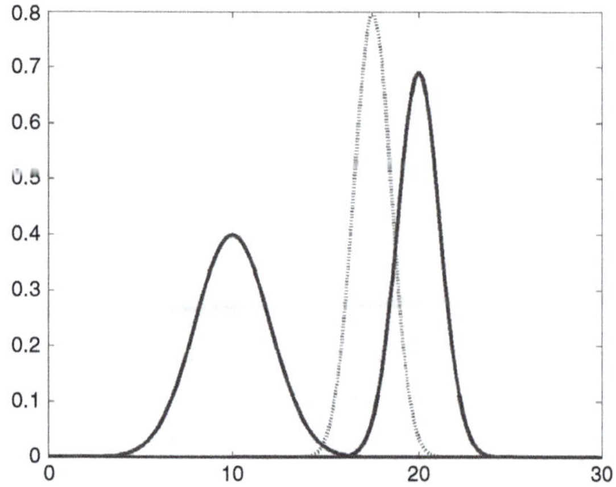

FIGURE 11.2 Bayes' rule for univariate Gaussians. The two solid curves show the probability densities for the prior $p(w) = \mathrm{N}(\mu, \alpha^{-1})$ with $\mu = 20$ and $\alpha = 1$ and the likelihood $p(y|w) = \mathrm{N}(w, \beta^{-1})$ with $w = 25$ and $\beta = 3$. The dotted curve shows the posterior distribution, $p(w|y) = \mathrm{N}(m, \lambda^{-1})$ with $m = 23.75$ and $\lambda = 4$, as computed from Eqn. **11.14**. The posterior distribution is closer to the likelihood because the likelihood has higher precision.

Equal variance

This special case is a two-level multivariate model as in Eqn. **11.2**, but with isotropic covariances at both the first and second levels. We have $C = \beta^{-1}I_K$ and $P = \alpha^{-1}I_N$. This means that observations are independent and have the same error variance. This is an example of the errors being independent and identically distributed (IID), where, in this case, the distribution is a zero-mean Gaussian having a particular variance. In this chapter, we will also use the term 'sphericity' for any model with IID errors. Models without IID errors will have 'non-sphericity' (as an aside we note that IID is not actually a requirement of 'sphericity' and readers looking for a precise definition are referred to Winer *et al.* (1991) and to Chapter 10).

On a further point of terminology, the unknown vectors w and μ will be referred to as 'parameters', whereas variables related to error covariances will be called 'hyperparameters'. The variables α and β are therefore hyperparameters. The posterior distribution over first level parameters is given by:

$$p(w|y) = \mathrm{N}(\hat{w}, \hat{\Sigma}) \qquad \mathbf{11.15}$$
$$\hat{\Sigma} = (\beta X^T X + \alpha I_N)^{-1}$$
$$\hat{w} = \hat{\Sigma}\,(\beta X^T y + \alpha M\mu)$$

Note that if $\alpha = 0$, we recover the maximum likelihood estimate:

$$\hat{w}_{ML} = (X^T X)^{-1} X^T y \qquad \mathbf{11.16}$$

This is the familiar ordinary least squares (OLS) estimate used in the GLM (Holmes *et al.*, 1997). The posterior distribution over the second level parameters is given by Eqn. **11.12** with:

$$\tilde{C} = \beta^{-1} I_K + \alpha^{-1} X X^T \qquad \mathbf{11.17}$$

Separable model

We now consider 'separable models' which can be used, for example, for random effects analysis. Figure 11.3

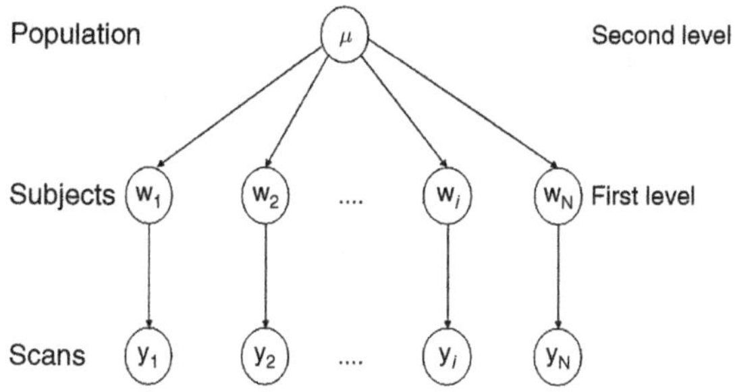

FIGURE 11.3 Generative model for random effects analysis.

shows the corresponding generative model. In these models, the first-level splits into N separate submodels. For each submodel, i, there are n_i observations. These form the n_i-element vector y_i giving information about the parameter w_i via the design vector x_i. For functional magnetic resonance imaging (fMRI) analysis, these design vectors comprise stimulus functions, e.g. boxcars or delta functions, convolved with an assumed haemodynamic response. The overall first-level design matrix X then has a block-diagonal form $X = \text{blkdiag}(x_1, \ldots, x_i, \ldots, x_N)$ and the covariance is given by $C = \text{diag}[\beta_1 1_{n_1}^T, \ldots, \beta_i 1_{n_i}^T, \ldots, \beta_N 1_{n_N}^T]$, where 1_n is a column vector of 1s with n entries. For example, for $N = 3$ groups with $n_1 = 2$, $n_2 = 3$ and $n_3 = 2$ observations in each group:

$$X = \begin{bmatrix} x_1(1) & 0 & 0 \\ x_1(2) & 0 & 0 \\ 0 & x_2(1) & 0 \\ 0 & x_2(2) & 0 \\ 0 & x_2(3) & 0 \\ 0 & 0 & x_3(1) \\ 0 & 0 & x_3(2) \end{bmatrix} \qquad 11.18$$

and $C^{-1} = \text{diag}[\beta_1, \beta_1, \beta_2, \beta_2, \beta_2, \beta_3, \beta_3]$. The covariance at the second level is $P = \alpha^{-1} I_N$, as before, and we also assume that the second level design matrix is a column of 1s, $M = 1_N$. The posterior distribution over first level parameters is found by substituting X and C into Eqn. **??**. This gives a distribution which factorizes over the different first level coefficients such that:

$$\begin{aligned} p(w|y) &= \prod_{i=1}^{N} p(w_i|y) \qquad 11.19 \\ p(w_i|y) &= N(\hat{w}_i, \hat{\Sigma}_{ii}) \\ \hat{\Sigma}_{ii}^{-1} &= \beta_i x_i^T x_i + \alpha \\ \hat{w}_i &= \hat{\Sigma}_{ii} \beta_i x_i^T y_i + \hat{\Sigma}_{ii} \alpha \mu \end{aligned}$$

The posterior distribution over second level parameters is, from Eqn. **11.12**, given by:

$$\begin{aligned} p(\mu|y) &= \mathrm{N}(\hat{\mu}, \sigma_\mu^2) \qquad 11.20 \\ \sigma_\mu^2 &= \frac{1}{\sum_{i=1}^{N} x_i^T (\alpha^{-1} x_i x_i^T + \beta_i^{-1})^{-1} x_i} \\ \hat{\mu} &= \sigma_\mu^2 \sum_{i=1}^{N} x_i^T (\alpha^{-1} x_i x_i^T + \beta_i^{-1})^{-1} y_i \end{aligned}$$

We note that, in the absence of any second level variability, i.e. $\alpha \to \infty$, the estimate $\hat{\mu}$ reduces to the mean of the first level coefficients weighted by their precision:

$$\hat{\mu} = \frac{\sum_i \beta_i x_i^T y_i}{\sum_i \beta_i x_i^T x_i} \qquad 11.21$$

PARAMETRIC EMPIRICAL BAYES

In the previous section, we have shown how to compute the posterior distributions $p(w|y)$ and $p(\mu|y)$. As can be seen from Eqns **11.7** and **11.11**, however, these equations depend on covariances P and C. In this section, we show how covariance components can be estimated for the special cases of equal variance models and separable models.

In Friston *et al.* (2002a), the covariances are decomposed using:

$$\begin{aligned} C &= \sum_j \lambda_j^1 Q_j^1 \qquad 11.22 \\ P &= \sum_j \lambda_j^2 Q_j^2 \end{aligned}$$

where Q_j^1 and Q_j^2 are basis functions that are specified by the modeller, depending on the application in mind. For example, for analysis of fMRI data from a single subject, two basis functions are used, the first relating to error variance and the second relating to temporal autocorrelation (Friston *et al.*, 2002b). The hyperparameters $\lambda = [\{\lambda_j^1\}, \{\lambda_j^2\}]$ are unknown, but can be estimated using the PEB algorithm described in Friston *et al.* (2002a). Variants of this algorithm are known as the *evidence framework* (Mackay, 1992) or *maximum likelihood II (ML-II)* (Berger, 1985). The PEB algorithm is also referred to as simply *empirical Bayes*, but we use the term PEB to differentiate it from the non-parametric empirical Bayes' methods described in Carlin and Louis (2000). The hyperparameters are set so as to maximize the evidence (also known as the marginal likelihood):

$$p(y|\lambda) = \int p(y|w, \lambda) p(w|\lambda) dw \qquad 11.23$$

This is the likelihood of the data after we have integrated out the first-level parameters. For the two multivariate special cases described above, by substituting in our expressions for the prior and likelihood, integrating, taking logs and then setting the derivatives to zero, we can derive a set of update rules for the hyperparameters. These derivations are provided in the following two sections.

Equal variance

For the equal variance model, the objective function is:

$$p(y|\alpha,\beta) = \int p(y|w,\beta)p(w|\alpha)dw \quad 11.24$$

Substituting in expressions for the likelihood and prior gives:

$$p(y|\alpha,\beta) = \left(\frac{\beta}{2\pi}\right)^{K/2}\left(\frac{\alpha}{2\pi}\right)^{N/2} \times \int \exp\left(-\frac{\beta}{2}e(w)^T e(w) - \frac{\alpha}{2}z(w)^T z(w)\right)dw$$

where $e(w) = y - Xw$ and $z(w) = w - M\mu$. By rearranging the terms in the exponent (and keeping all of them, unlike before where we were only interested in w-dependent terms) the integral can be written as:

$$I = \left[\int \exp\left(-\frac{1}{2}(w-\hat{w})^T\hat{\Sigma}^{-1}(w-\hat{w})\right)dw\right] \times \left[\exp\left(-\frac{\beta}{2}e(\hat{w})^T e(\hat{w}) - \frac{\alpha}{2}z(\hat{w})^T z(\hat{w})\right)\right] \quad 11.25$$

where the second term is not dependent on w. The first factor is then simply given by the normalizing constant of the multivariate Gaussian density:

$$(2\pi)^{N/2}|\hat{\Sigma}|^{1/2} \quad 11.26$$

Hence,

$$p(y|\alpha,\beta) = \left(\frac{\beta}{2\pi}\right)^{K/2}\alpha^{N/2}|\hat{\Sigma}|^{1/2} \times \exp\left(-\frac{\beta}{2}e(\hat{w})^T e(\hat{w}) - \frac{\alpha}{2}z(\hat{w})^T z(\hat{w})\right)$$

where $|\hat{\Sigma}|$ denotes the determinant of $\hat{\Sigma}$. Taking logs gives the 'log-evidence':

$$F = \frac{K}{2}\log\frac{\beta}{2\pi} + \frac{N}{2}\log\alpha + \frac{1}{2}\log|\hat{\Sigma}| - \frac{\beta}{2}e(\hat{w})^T e(\hat{w}) - \frac{\alpha}{2}z(\hat{w})^T z(\hat{w}) \quad 11.27$$

To find equations for updating the hyperparameters, we must differentiate F with respect to α and β and set the derivative to zero. The only possibly problematic term is the log-determinant, but this can be differentiated by first noting that the inverse covariance is given by:

$$\hat{\Sigma}^{-1} = \beta X^T X + \alpha I_N \quad 11.28$$

If λ_j are the eigenvalues of the first term, then the eigenvalues of $\hat{\Sigma}^{-1}$ are $\lambda_j + \alpha$. Hence,

$$|\hat{\Sigma}^{-1}| = \prod_j(\lambda_j+\alpha) \quad 11.29$$

$$|\hat{\Sigma}| = \frac{1}{\prod_j(\lambda_j+\alpha)}$$

$$\log|\hat{\Sigma}| = -\sum_j \log(\lambda_j+\alpha)$$

$$\frac{\partial}{\partial\alpha}\log|\hat{\Sigma}| = -\sum_j \frac{1}{\lambda_j+\alpha}$$

Setting the derivative $\partial F/\partial\alpha$ to zero then gives:

$$\alpha z(\hat{w})^T z(\hat{w}) = N - \sum_j \frac{\alpha}{\lambda_j+\alpha} = \sum_j\frac{\lambda_j+\alpha}{\lambda_j+\alpha} - \sum_j\frac{\alpha}{\lambda_j+\alpha} = \sum_j\frac{\lambda_j}{\lambda_j+\alpha} \quad 11.30$$

This is an implicit equation in α which leads to the following update rule. We first define the quantity γ which is computed from the 'old' value of α:

$$\gamma = \sum_{j=1}^{N}\frac{\lambda_j}{\lambda_j+\alpha} \quad 11.31$$

and then let:

$$\frac{1}{\alpha} = \frac{z(\hat{w})^T z(\hat{w})}{\gamma} \quad 11.32$$

The update for β is derived by first noting that the eigenvalues λ_j are linearly dependent on β. Hence,

$$\frac{\partial\lambda_j}{\partial\beta} = \frac{\lambda_j}{\beta} \quad 11.33$$

The derivative of the log-determinant is then given by:

$$\frac{\partial}{\partial\beta}\log|\hat{\Sigma}^{-1}| = \frac{1}{\beta}\sum_j\frac{\lambda_j}{\lambda_j+\alpha} \quad 11.34$$

which leads to the update:

$$\frac{1}{\beta} = \frac{e(\hat{w})^T e(\hat{w})}{K - \gamma} \tag{11.35}$$

The PEB algorithm consists of iterating the update rules in Eqn. **11.31**, Eqn. **11.32**, Eqn. **11.35** and the posterior estimates in Eqn. **11.15**, until convergence.

The update rules in Eqn. **11.31**, Eqn. **11.32** and Eqn. **11.35** can be interpreted as follows. For every j for which $\lambda_j >> \alpha$, the quantity γ increases by 1. As α is the prior precision and λ_j is the data precision (of the jth 'eigencoefficient'), γ therefore measures the number of parameters that are determined by the data. Given K data points, the quantity $K - \gamma$ therefore corresponds to the number of degrees of freedom in the data set. The variances α^{-1} and β^{-1} are then updated based on the sum of squares divided by the appropriate degrees of freedom.

Separable models

For separable models, the objective function is:

$$p(y|\alpha, \{\beta_i\}) = \int p(y|w, \{\beta_i\}) p(w|\alpha) dw \tag{11.36}$$

Because the second-level here is the same as for the equal variance case, so is the update for alpha. The updates for β_i are derived in a similar manner as before, but we also make use of the fact that the first-level posterior distribution factorizes (see Eqn. **11.19**). This decouples the updates for each β_i and results in the following PEB algorithm:

$$\begin{aligned}
\hat{e}_i &= y_i - \hat{w}_i x_i \\
\hat{z}_i &= \hat{w}_i - \hat{\mu} \\
\lambda_i &= \beta_i x_i^T x_i \\
\gamma_i &= \frac{\lambda_i}{\lambda_i + \alpha} \\
\gamma &= \sum_i \gamma_i \\
\beta_i &= (n_i - \gamma_i)/\hat{e}_i^T \hat{e}_i \\
\alpha &= \gamma / \hat{z}^T \hat{z} \\
\hat{w}_i &= (\beta_i x_i^T y_i + \alpha\mu)/(\lambda_i + \alpha) \\
d_i &= (\alpha^{-1} x_i x_i^T + \beta_i^{-1} I_{n_i})^{-1} \\
\sigma_{\hat{\mu}}^2 &= 1/(\sum_i x_i^T d_i x_i) \\
\hat{\mu} &= \sigma_{\hat{\mu}}^2 \sum_i x_i^T d_i y_i
\end{aligned} \tag{11.37}$$

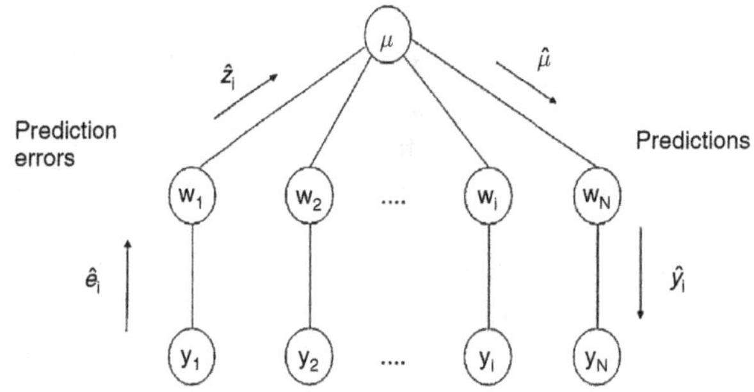

FIGURE 11.4 Part of the PEB algorithm for separable models requires the upwards propagation of prediction errors and downwards propagation of predictions. This passing of messages between nodes in the hierarchy is a special case of the more general belief propagation algorithm referred to in Figure 11.5.

Initial values for $\hat{w}_i$ and β_i are set using OLS, $\hat{\mu}$ is initially set to the mean of $\hat{w}_i$ and α is initially set to 0. The equations are then iterated until convergence (in our examples in Chapter 12, we never required more than ten iterations). While the above updates may seem somewhat complex, they can perhaps be better understood in terms of messages passing among nodes in a hierarchical network. This is shown in Figure 11.4 for the 'prediction' and 'prediction error' variables.

The PEB algorithms we have described show how Bayesian inference can take place when the variance components are unknown (in the previous section, we assumed the variance components were known). An application of this PEB algorithm to random effects analysis is provided in the next chapter. We now provide a brief numerical example demonstrating the iterations with PEB updates.

NUMERICAL EXAMPLE

This numerical example caricatures the use of PEB for estimating effect sizes from functional imaging data described in Chapter 23. The approach uses a 'global shrinkage prior' which embodies a prior belief that, across the brain: (i) the average effect is zero, $\mu = 0$; and (ii) the variability of responses follows a Gaussian distribution with precision α. Mathematically, we can write $p(w_i) = \text{N}(0, \alpha^{-1})$. Plate 5(a) (see colour plate section) shows effect sizes generated from this prior for an $N =$ 20-voxel brain and $\alpha = 1$.

Chapter 23 allows for multiple effects to be expressed at each voxel and for positron emission tomography (PET)/fMRI data to be related to effect sizes using the full flexibility of general linear models. Here, we just assume that data at each voxel are normally distributed about the effect size at that voxel. That is, $p(y_i|w_i) = \mathsf{N}(w_i, \beta_i^{-1})$. Plate 5(b) shows $n_i = 10$ data points at each voxel generated from this likelihood. We have allowed the observation noise precision β_i to be different at each voxel. Voxels 2, 15 and 18, for example, have noisier data than others.

Effect sizes were then estimated from these data using maximum likelihood (ML) and PEB. ML estimates are shown in Plate 5(c) and (d). These are simply computed as the mean value observed at each voxel. PEB was implemented using the updates in Eqn. **11.37** with $\mu = 0$ and $x_i = 1_{n_i}$ and initialized with $\alpha = 0$ and β_i and $\hat{w}_i$ set to ML-estimated values.

Eqn. **11.37** was then iterated, resulting in effect size estimates shown in Plate 6 before iterations one, three, five and seven. These estimates seem rather stable after only two or three iterations. Only the effects at voxels 5 and 15 seem markedly changed between iterations three and seven. The corresponding estimates of α were 0, 0.82, 0.91 and 0.95, showing convergence to the true prior response precision value of 1.

It is well known that PEB provides estimates that are, on average, more accurate than ML. Here, we quantify this using, σ_s, the standard deviation across voxels of the difference between the true and estimated effects. For ML, $\sigma_s = 0.71$ and for PEB, $\sigma_s = 0.34$. That PEB estimates are twice as accurate on average can be seen by comparing Plate 6(a) and (d). Of course, PEB is only better 'on average'. It does better at most voxels at the expense of being worse at a minority, for example, voxel 2. This trade-off is discussed further in Chapter 22.

PEB can do better than ML because it uses more information: here, the information that effects have a mean of zero across the brain and follow a Gaussian variability profile. This shows the power of Bayesian estimation, which combines prior information with data in an optimal way. In this example, a key parameter in this trade-off is the parameter γ_i which is computed as in Eqn. **11.37**. This quantity is the ratio of the data precision to the posterior precision. A value of 1 indicates that the estimated effect is determined solely by the data, as in ML. A value of 0 indicates the estimate is determined solely by the prior. For most voxels in our data set, we have $\gamma_i \approx 0.9$, but for the noisy voxels 2, 15 and 18, we have $\gamma_i \approx 0.5$. PEB thus relies more on prior information where data are unreliable.

PEB will only do better than ML if the prior is chosen appropriately. For functional imaging data, we will never know what the 'true prior' is, just as we will never know what the 'true model' is. But some priors and models are better than others, and there is a formal method for deciding between them. This is 'Bayesian model selection' and is described in Chapter 35.

Finally, we note that the prior used here does not use spatial information, i.e. there is no notion that voxel 5 is 'next to' voxel 6. It turns out that for functional imaging data, spatial information is important. In Chapter 25, we describe Bayesian fMRI inference with spatial priors. Bayesian model selection shows that models with spatial priors are preferred to those without (Penny *et al.*, 2006).

BELIEF PROPAGATION

This chapter has focused on the special case of two-level models and Gaussian distributions. It is worthwhile noting that the general solution to inference in tree-structured hierarchical models, which holds for all distributions, is provided by the 'sum-product' or 'belief propagation' algorithm (Pearl, 1988; Jordan and Weiss, 2002). This is a message passing algorithm which aims to deliver the marginal distributions[1] at each point in the hierarchy. It does this by propagating evidence up the hierarchy and marginal distributions down. If the downward messages are passed after the upward messages have reached the top, then this is equivalent to propagating the posterior beliefs down the hierarchy. This is shown schematically in Figure 11.5.

This general solution is important as it impacts on non-Gaussian and/or non-linear hierarchical models. Of particular relevance are the models of inference in cortical hierarchies (Friston, 2003) referred to in later chapters of the book. In these models, evidence flows up the hierarchy, in the form of prediction errors, and marginal distributions flow down, in the form of predictions. Completion of the downward pass explains late components of event-related potentials which are correlated with, e.g. extra-classical receptive field effects (Friston, 2003). This general solution also motivates a data analysis approach known as Bayesian model averaging (BMA), described further in Chapter 35, where, e.g. x_3 in Figure 11.5 embodies assumptions about model structure. The downward pass of belief propagation then renders our final inferences independent of these assumptions. See Chapter 16 of Mackay (2003) and Ghahramani (1998) for further discussion of these issues.

[1] The probability distribution over a set of variables is known as the joint distribution. The distribution over a subset is known as the marginal distribution.

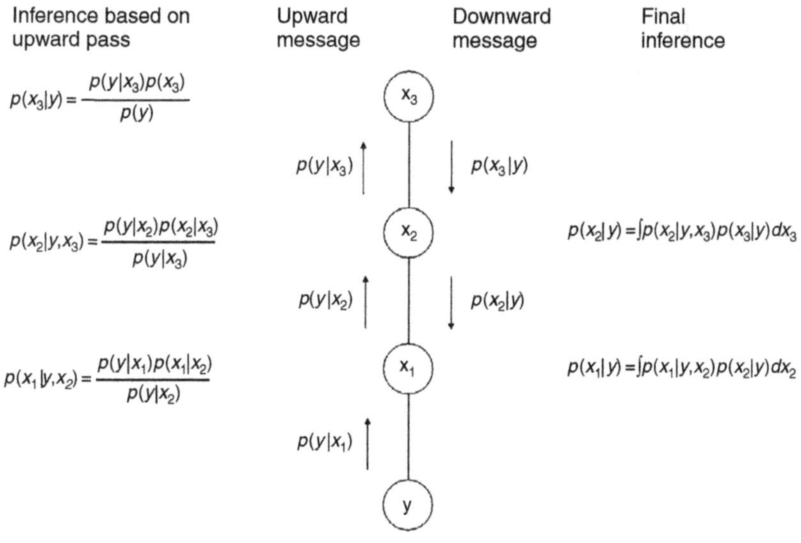

FIGURE 11.5 Belief propagation for inference in hierarchical models. This algorithm is used to update the marginal densities, i.e. to update $p(x_i)$ to $p(x_i|y)$. Inferences based on purely the upward pass are contingent on variables in the layer above, whereas inferences based on upward and downward passes are not. Completion of the downward pass delivers the marginal density. Application of this algorithm to the two-level Gaussian model will produce the update Eqn. **11.7** and Eqn. **11.11**. More generally, this algorithm can be used for Bayesian model averaging, where e.g. x_3 embodies assumptions about model structure, and as a model of inference in cortical hierarchies, where e.g. completion of the downward pass explains extra-classical receptive field effects (Friston, 2003).

DISCUSSION

We have described Bayesian inference for some particular two-level linear-Gaussian hierarchical models. A key feature of Bayesian inference in this context is that the posterior distributions are Gaussian with precisions that are the sum of the data and prior precisions. The posterior means are the sum of the data and prior means, but each weighted according to their relative precision. With zero prior precision, two-level models reduce to a single-level model (i.e. a GLM) and Bayesian inference reduces to the familiar maximum-likelihood estimation scheme. With non-zero and, in general unknown, prior means and precisions, these parameters can be estimated using PEB. These covariance components can also be estimated using the ReML algorithm from classical statistics. The relation between PEB and ReML is discussed further in Chapter 22.

We have described two special cases of the PEB algorithm, one for equal variances and one for separable models. Both algorithms are special cases of a general approach described in Friston *et al.* (2002a) and in Chapter 24. In these contexts, we have shown that PEB automatically partitions the total degrees of freedom (i.e. number of data points) into those to be used to estimate the hyperparameters of the prior distribution and those to be used to estimate hyperparameters of the likelihood distribution. The next chapter describes how PEB can be used in the context of random effects analysis.

REFERENCES

Berger JO, (1985) *Statistical decision theory and Bayesian analysis*. Springer-Verlag, New York

Carlin BP, Louis TA (2000) *Bayes and empirical Bayes methods for data analysis*. CRC Press, USA

Friston K (2003) Learning and inference in the brain. *Neural Netw* **16**: 1325–52

Friston KJ, Penny WD, Phillips C *et al.* (2002a) Classical and Bayesian inference in neuroimaging: theory. *NeuroImage* **16**: 465–83

Friston KJ, Glaser DE, Henson RNA *et al.* (2002b) Classical and Bayesian inference in neuroimaging: applications. *NeuroImage* **16**: 484–512

Gelman A, Carlin JB, Stern HS *et al.* (1995) *Bayesian data analysis*. Chapman and Hall, Boca Raton

Ghahramani Z (1998) Learning dynamic bayesian networks. In *Adaptive processing of temporal information*, Giles CL, Gori M (eds). Springer-Verlag, Berlin

Holmes AP, Poline JB, Friston KJ (1997) Characterizing brain images with the general linear model. In *Human brain function*, Frackowiak RSJ, Friston KJ, Frith C *et al.* (eds). Academic Press, London pp 59–84.

Jordan M, Weiss Y (2002) Graphical models: probabilistic inference. In *The handbook of brain theory and neural networks*, Arbib M (ed.). MIT Press, Cambridge

Lee PM (1997) *Bayesian statistics: an introduction*, 2nd edn. Arnold, London

Mackay DJC (1992) Bayesian interpolation. *Neural Comput* **4**: 415–47

Mackay DJC (2003) *Information theory, inference and learning algorithms*. Cambridge University Press, Cambridge

Pearl J (1988) *Probabilistic reasoning in intelligent systems: networks of plausible inference*. Morgan Kauffman

Penny WD, Flandin G, Trujillo-Barreto N (2006) Bayesian comparison of spatially regularised general linear models. *Hum Brain Mapp*, in press

Winer BJ, Brown DR, Michels KM (1991) *Statistical principles in experimental design*. McGraw-Hill

CHAPTER

12

Random Effects Analysis

W.D. Penny and A.J. Holmes

INTRODUCTION

In this chapter, we are concerned with making statistical inferences involving many subjects. One can envisage two main reasons for studying multiple subjects. The first is that one may be interested in individual differences, as in many areas of psychology. The second, which is the one that concerns us here, is that one is interested in what is common to the subjects. In other words, we are interested in the stereotypical effect in the population from which the subjects are drawn.

As every experimentalist knows, a subject's response will vary from trial to trial. Further, this response will vary from subject to subject. These two sources of variability, within-subject (also called between-scan) and between-subject, must both be taken into account when making inferences about the population.

In statistical terminology, if we wish to take the variability of an effect into account we must treat it as a 'random effect'. In a 12-subject functional magnetic resonance imaging (fMRI) study, for example, we can view those 12 subjects as being randomly drawn from the population at large. The subject variable is then a random effect and, in this way, we are able to take the sampling variability into account and make inferences about the population from which the subjects were drawn. Conversely, if we view the subject variable as a 'fixed effect' then our inferences will relate only to those 12 subjects chosen.

The majority of early studies in neuroimaging combined data from multiple subjects using a 'fixed-effects' (FFX) approach. This methodology only takes into account the within-subject variability. It is used to report results as case studies. It is not possible to make formal inferences about population effects using FFX. Random-effects (RFX) analysis, however, takes into account both sources of variation and makes it possible to make formal inferences about the population from which the subjects are drawn.

In this chapter, we describe FFX and RFX analyses of multiple-subject data. We first describe the mathematics behind RFX, for balanced designs, and show how RFX can be implemented using the computationally efficient 'summary-statistic' approach. We then describe the mathematics behind FFX and show that it only takes into account within-subject variance. The next section shows that RFX for unbalanced designs is optimally implemented using the PEB algorithm described in the previous chapter. This section includes a numerical example which shows that, although not optimal, the summary statistic approach performs well, even for unbalanced designs.

RANDOM EFFECTS ANALYSIS

Maximum likelihood

Underlying RFX analysis is a probability model defined as follows. We first envisage that the mean effect in the population (i.e. averaged across subjects) is of size w_{pop} and that the variability of this effect between subjects is σ_b^2. The mean effect for the ith subject (i.e. averaged across scans), w_i, is then assumed to be drawn from a Gaussian with mean w_{pop} and variance σ_b^2. This process reflects the fact that we are drawing subjects at random from a large population. We then take into account the within-subject (i.e. across scan) variability by modelling the jth observed effect in subject i as being drawn from a Gaussian with mean w_i and variance σ_w^2. Note that σ_w^2 is assumed to be the same for all subjects. This is a requirement of a balanced design. This two-stage process is shown graphically in Figure 12.1.

Statistical Parametric Mapping, by Karl Friston et al.
ISBN–13: 978-0-12-372560-8 ISBN–10: 0-12-372560-7

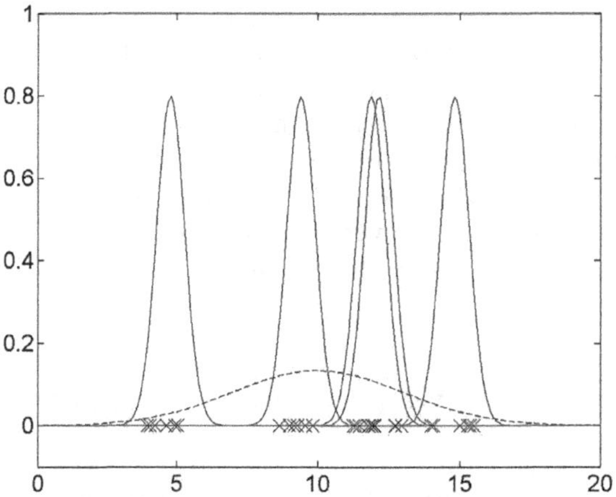

FIGURE 12.1 Synthetic data illustrating the probability model underlying random effects analysis. The dotted line is the Gaussian distribution underlying the second-level model with mean w_{pop}, the population effect, and variance σ_b^2, the between-subject variance. The mean subject effects, w_i, are drawn from this distribution. The solid lines are the Gaussians underlying the first level models with means w_i and variances σ_w^2. The crosses are the observed effects y_{ij} which are drawn from the solid Gaussians.

Given a data set of effects from N subjects with n replications of that effect per subject, the population effect is modelled by a two-level process:

$$y_{ij} = w_i + e_{ij} \qquad \textbf{12.1}$$

$$w_i = w_{pop} + z_i$$

where w_i is the true mean effect for subject i and y_{ij} is the jth observed effect for subject i, and z_i is the between-subject error for the ith subject. These Gaussian errors have the same variance, σ_b^2. For the positron emission tomography (PET) data considered below this is a differential effect, the difference in activation between word generation and word shadowing. The first equation captures the within-subject variability and the second equation the between-subject variability.

The within-subject Gaussian error e_{ij} has zero mean and variance $\text{Var}[e_{ij}] = \sigma_w^2$. This assumes that the errors are independent over subjects and over replications within subject. The between-subject Gaussian error z_i has zero mean and variance $\text{Var}[z_i] = \sigma_b^2$. Collapsing the two levels into one gives:

$$y_{ij} = w_{pop} + z_i + e_{ij} \qquad \textbf{12.2}$$

The maximum-likelihood estimate of the population mean is:

$$\hat{w}_{pop} = \frac{1}{Nn}\sum_{i=1}^{N}\sum_{j=1}^{n} y_{ij} \qquad \textbf{12.3}$$

We now make use of a number of statistical relations defined in Appendix 12.1 to show that this estimate has a mean $\text{E}[\hat{w}_{pop}] = w_{pop}$ and a variance given by:

$$\text{Var}[\hat{w}_{pop}] = \text{Var}\left[\sum_{i=1}^{N}\frac{1}{N}\sum_{j=1}^{n}\frac{1}{n}(w_{pop}+z_i+e_{ij})\right] \qquad \textbf{12.4}$$

$$= \text{Var}\left[\sum_{i=1}^{N}\frac{1}{N}z_i\right] + \text{Var}\left[\sum_{i=1}^{N}\frac{1}{N}\sum_{j=1}^{n}\frac{1}{n}e_{ij}\right]$$

$$= \frac{\sigma_b^2}{N} + \frac{\sigma_w^2}{Nn}$$

The variance of the population mean estimate contains contributions from both the within-subject and between-subject variance.

Summary statistics

Implicit in the summary-statistic RFX approach is the two-level model:

$$\bar{w}_i = w_i + e_i \qquad \textbf{12.5}$$

$$w_i = w_{pop} + z_i$$

where w_i is the true mean effect for subject i, $\bar{w}_i$ is the sample mean effect for subject i and w_{pop} is the true effect for the population.

The summary-statistic (SS) approach is of interest because it is computationally much simpler to implement than the full random effects model of Eqn. **12.1**. This is because it is based on the sample mean value, $\bar{w}_i$, rather than on all of the samples y_{ij}. This is important for neuroimaging as in a typical functional imaging group study there can be thousands of images, each containing tens of thousands of voxels.

In the first level, we consider the variation of the sample mean for each subject around the true mean for each subject. The corresponding variance is $\text{Var}[e_i] = \sigma_w^2/n$, where σ_w^2 is the within-subject variance. At the second level, we consider the variation of the true subject means about the population mean where $\text{Var}[z_i] = \sigma_b^2$, the between-subject variance. We also have $\text{E}[e_i] = \text{E}[z_i] = 0$. Consequently:

$$\bar{w}_i = w_{pop} + z_i + e_i \qquad \textbf{12.6}$$

The population mean is then estimated as:

$$\hat{w}_{pop} = \frac{1}{N}\sum_{i=1}^{N}\bar{w}_i \qquad \textbf{12.7}$$

This estimate has a mean $E[\hat{w}_{pop}] = w_{pop}$ and a variance given by:

$$\begin{aligned}\text{Var}[\hat{w}_{pop}] &= \text{Var}\left[\sum_{i=1}^{N}\frac{1}{N}\bar{w}_i\right] \\ &= \text{Var}\left[\sum_{i=1}^{N}\frac{1}{N}z_i\right] + \text{Var}\left[\sum_{i=1}^{N}\frac{1}{N}e_i\right] \\ &= \frac{\sigma_b^2}{N} + \frac{\sigma_w^2}{Nn}\end{aligned} \quad 12.8$$

Thus, the variance of the estimate of the population mean contains contributions from both the within-subject and between-subject variances. Importantly, both $E[\hat{w}_{pop}]$ and $\text{Var}[\hat{w}_{pop}]$ are identical to the maximum-likelihood estimates derived earlier. This validates the summary-statistic approach. Informally, the validity of the summary-statistic approach lies in the fact that what is brought forward to the second level is a *sample* mean. It contains an element of within-subject variability which, when operated on at the second level, produces just the right balance of within- and between-subject variance.

FIXED EFFECTS ANALYSIS

Implicit in FFX analysis is a single-level model:

$$y_{ij} = w_i + e_{ij} \quad 12.9$$

The parameter estimates for each subject are:

$$\hat{w}_i = \frac{1}{n}\sum_{j=1}^{n} y_{ij} \quad 12.10$$

which have a variance given by:

$$\begin{aligned}\text{Var}[\hat{w}_i] &= \text{Var}\left[\sum_{j=1}^{n}\frac{1}{n}y_{ij}\right] \\ &= \frac{\sigma_w^2}{n}\end{aligned} \quad 12.11$$

The estimate of the group mean is then:

$$\hat{w}_{pop} = \frac{1}{N}\sum_{i=1}^{N}\hat{w}_i \quad 12.12$$

which has a variance:

$$\begin{aligned}\text{Var}[\hat{w}_{pop}] &= \text{Var}\left[\sum_{i=1}^{N}\frac{1}{N}\hat{w}_i\right] \\ &= \frac{1}{N}\text{Var}[\hat{d}_i] \\ &= \frac{\sigma_w^2}{Nn}\end{aligned} \quad 12.13$$

The variance of the fixed-effects group mean estimate contains contributions from within-subject terms only. It is not sensitive to between-subject variance. We are not therefore able to make formal inferences about population effects using FFX. We are restricted to informal inferences based on separate case studies or summary images showing the average group effect. This will be demonstrated empirically in a later section.

PARAMETRIC EMPIRICAL BAYES

We now return to RFX analysis. We have previously shown how the SS approach can be used for the analysis of balanced designs, i.e. identical σ_w^2 for all subjects. This section starts by showing how parametric empirical Bayes (PEB) can also be used for balanced designs. It then shows how PEB can be used for unbalanced designs and provides a numerical comparison between PEB and SS on unbalanced data.

Before proceeding, we note that an algorithm from classical statistics, known as restricted maximum likelihood (ReML), can also be used for variance component estimation. Indeed, many of the papers on random effects analysis use ReML instead of PEB (Friston *et al.*, 2002, 2005).

The model described in this section is identical to the separable model in the previous chapter but with $x_i = 1_n$ and $\beta_i = \beta$. Given a data set of contrasts from N subjects with n scans per subject, the population effect can be modelled by the two-level process:

$$\begin{aligned}y_{ij} &= w_i + e_{ij} \\ w_i &= w_{pop} + z_i\end{aligned} \quad 12.14$$

where y_{ij} (a scalar) is the data from the ith subject and the jth scan at a particular voxel. These data points are accompanied by errors e_{ij} with w_i being the size of the effect for subject i, w_{pop} being the size of the effect in the population and z_i being the between-subject error. This may be viewed as a Bayesian model where the first equation acts as a likelihood and the second equation acts as a prior. That is:

$$\begin{aligned}p(y_{ij}|w_i) &= \text{N}(w_i, \sigma_w^2) \\ p(w_i) &= \text{N}(w_{pop}, \sigma_b^2)\end{aligned} \quad 12.15$$

where σ_b^2 is the between-subject variance and σ_w^2 is the within-subject variance. We can make contact with the hierarchical formalism of the previous chapter by making the following identities. We place the y_{ij} in the column vector y in the order – all from subject 1, all from subject 2, etc. (this is described mathematically by the *vec* operator and is implemented in MATLAB (Mathworks, Inc.) by the colon operator). We also let $X = I_N \otimes 1_n$ where $\otimes$ is the Kronecker product and let $w = [w_1, w_2, \ldots, w_N]^T$. With these values, the first level in Eqn. **11.2** of the previous chapter is then the matrix equivalent of the first level in Eqn. **12.14** (i.e. it holds for all i, j). For $y = Xw + e$ and e.g. $N = 3, n = 2$ we then have:

$$\begin{bmatrix} y_{11} \\ y_{12} \\ y_{21} \\ y_{22} \\ y_{31} \\ y_{32} \end{bmatrix} = \begin{bmatrix} 1 & 0 & 0 \\ 1 & 0 & 0 \\ 0 & 1 & 0 \\ 0 & 1 & 0 \\ 0 & 0 & 1 \\ 0 & 0 & 1 \end{bmatrix} \begin{bmatrix} w_1 \\ w_2 \\ w_3 \end{bmatrix} + \begin{bmatrix} e_{11} \\ e_{12} \\ e_{21} \\ e_{22} \\ e_{31} \\ e_{32} \end{bmatrix} \qquad 12.16$$

We then note that $X^T X = nI_N$, $\hat{\sum} = \text{diag}(\text{Var}[w_1], \text{Var}[w_2], \ldots, \text{Var}[w_N])$ and the ith element of $X^T y$ is equal to $\sum_{j=1}^{n} y_{ij}$.

If we let $M = 1_N$, then the second level in Eqn. **11.2** of the previous chapter is the matrix equivalent of the second-level in Eqn. **12.14** (i.e. it holds for all i). Plugging in our values for M and X and letting $\beta = 1/\sigma_w^2$ and $\alpha = 1/\sigma_b^2$ gives:

$$\text{Var}[\hat{w}_{pop}] = \frac{1}{N} \frac{\alpha + \beta n}{\alpha \beta n} \qquad 12.17$$

and

$$\hat{w}_{pop} = \frac{1}{N} \frac{\alpha + \beta n}{\alpha \beta n} \frac{\alpha \beta}{\alpha + \beta n} \sum_{i,j} y_{ij} \qquad 12.18$$

$$= \frac{1}{Nn} \sum_{i,j} y_{ij}$$

So the estimate of the population mean is simply the average value of y_{ij}. The variance can be re-written as:

$$\text{Var}[\hat{w}_{pop}] = \frac{\sigma_b^2}{N} + \frac{\sigma_w^2}{Nn} \qquad 12.19$$

This result is identical to the maximum-likelihood and summary-statistic results derived earlier. The equivalence between the Bayesian and ML results derives from the fact that there is no prior at the population level. Hence, $p(Y|\mu) = p(\mu|Y)$, as indicated in the previous chapter.

Unbalanced designs

The model described in this section is identical to the separable model in the previous chapter, but with $x_i = 1_{n_i}$. If the error covariance matrix is non-isotropic, i.e. $C \neq \sigma_w^2 I$, then the population estimates will change. This can occur, for example, if the design matrices are different for different subjects (so-called unbalanced-designs), or if the data from some of the subjects are particularly ill-fitting. In these cases, we consider the within-subject variances $\sigma_w^2(i)$ and the number of events n_i to be subject-specific. This will be the case in experimental paradigms where the number of events is not under experimental control, e.g. in memory paradigms where n_i may refer to the number of remembered items.

If we let $M = 1_N$, then the second level in Eqn. **11.2** in the previous chapter is the matrix equivalent of the second-level in Eqn. **12.14** (i.e. it holds for all i). Plugging in our values for M and X gives:

$$\text{Var}[\hat{w}_{pop}] = \left(\sum_{i=1}^{N} \frac{\alpha \beta_i n_i}{\alpha + n_i \beta_i} \right)^{-1} \qquad 12.20$$

and

$$\hat{w}_{pop} = \left(\sum_{i=1}^{N} \frac{\alpha \beta_i n_i}{\alpha + \beta_i n_i} \right)^{-1} \sum_{i=1}^{N} \frac{\alpha \beta_i}{\alpha + \beta_i n_i} \sum_{j=1}^{n_i} y_{ij} \qquad 12.21$$

This reduces to the earlier result if $\beta_i = \beta$ and $n_i = n$. Both of these results are different to the summary-statistic approach, which we note is therefore mathematically inexact for unbalanced designs. But as we shall see in the numerical example below, the summary-statistic approach is remarkably robust to departures from assumptions about balanced designs.

Estimation

To implement the PEB estimation scheme for the unequal variance case, we first compute the errors $\hat{e}_{ij} = y_{ij} - X\hat{w}_i$, $\hat{z}_i = \hat{w}_i - M\hat{w}_{pop}$. We then substitute $x_i = 1_{n_i}$ into the update rules derived in the PEB section of the previous chapter to obtain:

$$\sigma_b^2 \equiv \frac{1}{\alpha} = \frac{1}{\gamma} \sum_{i=1}^{N} \hat{z}_i^2 \qquad 12.22$$

$$\sigma_w^2(i) \equiv \frac{1}{\beta_i} = \frac{1}{n_i - \gamma_i} \sum_{j=1}^{n_i} \hat{e}_{ij}^2 \qquad 12.23$$

where:

$$\gamma = \sum_{i=1}^{N} \gamma_i \qquad 12.24$$

and

$$\gamma_i = \frac{n_i \beta_i}{\alpha + n_i \beta_i} \qquad 12.25$$

For balanced designs $\beta_i = \beta$ and $n_i = n$ we get:

$$\sigma_b^2 \equiv \frac{1}{\alpha} = \frac{1}{\gamma}\sum_{i=1}^{N} \hat{z}_i^2 \qquad 12.26$$

$$\sigma_w^2 \equiv \frac{1}{\beta} = \frac{1}{Nn-\gamma}\sum_{i=1}^{N}\sum_{j=1}^{n} \hat{e}_{ij}^2 \qquad 12.27$$

where:

$$\gamma = \frac{n\beta}{\alpha + n\beta} N \qquad 12.28$$

Effectively, the degrees of freedom in the data set (Nn) are partitioned into those that are used to estimate the between-subject variance, γ, and those that are used to estimate the within-subject variance, $Nn - \gamma$.

The posterior distribution of the first-level coefficients is:

$$p(w_i|y_{ij}) \equiv p(\hat{w}_i) = \mathrm{N}(\bar{w}_i, \mathrm{Var}[\hat{w}_i]) \qquad 12.29$$

where:

$$\mathrm{Var}[\hat{w}_i] = \frac{1}{\alpha + n_i\beta_i} \qquad 12.30$$

$$\hat{w}_i = \frac{\beta_i}{\alpha + n_i\beta_i}\sum_{j=1}^{n_i} y_{ij} + \frac{\alpha}{\alpha + n_i\beta_i}\hat{w}_{pop} \qquad 12.31$$

Overall, the PEB estimation scheme is implemented by first initializing $\hat{w}_i$, $\hat{w}_{pop}$ and α, β_i (e.g. to values given from the equal error-variance scheme). We then compute the errors $\hat{e}_{ij}$, $\hat{z}_i$ and re-estimate the α and β_is using the above equations. The coefficients $\hat{w}_i$ and $\hat{w}_{pop}$ are then re-estimated and the last two steps are iterated until convergence. This algorithm is identical to the PEB algorithm for the separable model in the previous chapter but with $x_i = 1_{n_i}$.

Numerical example

We now give an example of random effects analysis on simulated data. The purpose is to compare the PEB and SS algorithms. We generated data from a three-subject, two-level model with population mean $\mu = 2$, subject effect sizes $w = [2.2, 1.8, 0.0]^T$ and within subject variances $\sigma_w^2(1) = 1$, $\sigma_w^2(2) = 1$. For the third subject $\sigma_w^2(3)$ was varied from 1 to 10. The second-level design matrix was $M = [1, 1, 1]^T$ and the first-level design matrix was given by $X = \mathrm{blkdiag}(x_1, x_2, x_3)$ with x_i being a boxcar.

Figure 12.2 shows a realization of the three time-series for $\sigma_w^2(3) = 2$. The first two time-series contain stimulus-related activity but the third does not. We then applied the PEB algorithm, described in the previous section, to obtain estimates of the population mean $\hat{\mu}$ and estimated variances, σ_μ^2. For comparison, we also obtained equivalent estimates using the SS approach. We then computed the accuracy with which the population mean was estimated using the criterion $(\hat{\mu} - \mu)^2$. This was repeated for 1000 different data sets generated using the above parameter values, and for 10 different values of $\sigma_w^2(3)$. The results are shown in Figures 12.3 and 12.4.

First, we note that, as predicted by theory, both PEB and SS give identical results when the first-level error variances are equal. When the variance on the 'rogue' time-series approaches double that of the others we see different estimates of both $\hat{\mu}$ and σ_μ^2. With increasing

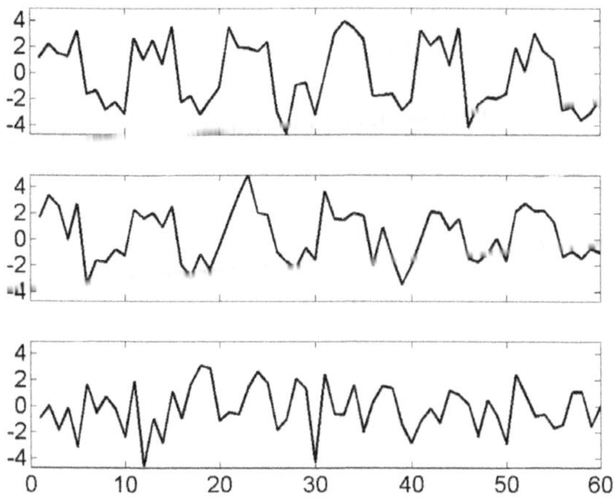

FIGURE 12.2 Simulated data for random effects analysis. Three representative time-series produced from the two-level hierarchical model. The first two time-series contain stimulus-related activity but the third does not.

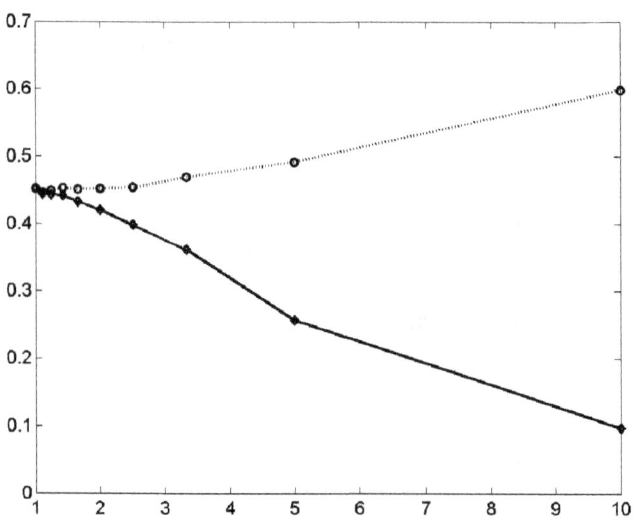

FIGURE 12.3 A plot of the error in estimating the population mean $E = \langle(\hat{\mu} - \mu)^2\rangle$ versus the observation noise level for the third subject, $\sigma_w^2(3)$, for the parametric empirical Bayes approach (solid line) and the summary-statistic approach (dotted line).

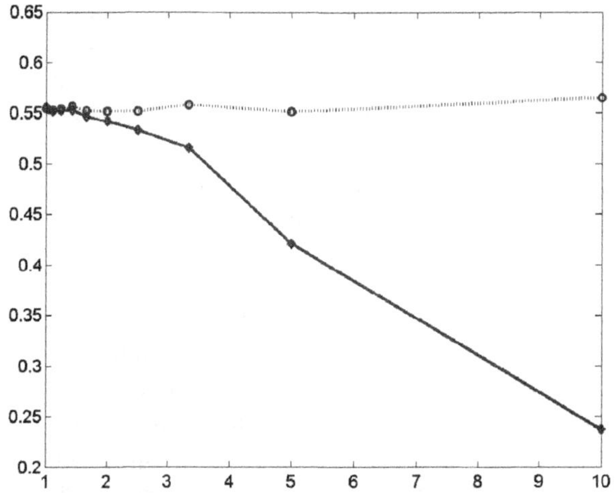

FIGURE 12.4 A plot of the estimated variance of the population mean, $\sigma_{\hat{\mu}}^2$, versus the observation noise level for the third subject, $\sigma_w^2(3)$, for the parametric empirical Bayes approach (solid line) and the summary-statistic approach (dotted line).

rogue error variance, the SS estimates get worse but the PEB estimates get better. There is an improvement with respect to the true values, as shown in Figure 12.3, and with respect to the variability of the estimate, as shown in Figure 12.4. This is because the third time-series is more readily recognized by PEB as containing less reliable information about the population mean and is increasingly ignored. This gives better estimates $\hat{\mu}$ and a reduced uncertainty, $\sigma_{\hat{\mu}}^2$.

We created the above example to reiterate a key point of this chapter, that SS gives identical results to PEB for equal within-subject error variances (homoscedasticity) and unbalanced designs, but not otherwise. In the numerical example, divergent behaviour is observed when the error variances differ by a factor of two. For studies with more subjects (12 being a typical number), however, this divergence requires a much greater disparity in error variances. In fact, we initially found it difficult to generate data sets where PEB showed a consistent improvement over SS! It is therefore our experience that the vanilla SS approach is particularly robust to departures from homoscedasticity. This conclusion is supported by what is known of the robustness of the t-test that is central to the SS approach. Lack of homoscedasticity only causes problems when the sample size (i.e. number of subjects) is small. As sample size increases so does the robustness (see e.g. Yandell, 1997).

PET DATA EXAMPLE

We now illustrate the difference between FFX and RFX analysis using data from a PET study of verbal fluency. These data come from five subjects and were recorded under two alternating conditions. Subjects were asked either to repeat a heard letter or to respond with a word that began with that letter. These tasks are referred to as word shadowing and word generation and were performed in alternation over 12 scans and the order randomized over subjects. Both conditions were identically paced with one word being generated every two seconds. PET images were re-aligned, normalized and smoothed with a 16 mm isotropic Gaussian kernel.[1]

Fixed-effects analysis

Analysis of multiple-subject data takes place within the machinery of the general linear model (GLM) as described in earlier chapters. However, instead of having data from a single-subject at each voxel, we now have data from multiple subjects. This is entered into a GLM by concatenating data from all subjects into the single column vector Y. Commensurate with this augmented data vector is an augmented multisubject design matrix,[2] X, which is shown in Figure 12.5. Columns 1 and 2 indicate scans taken during the word shadowing and word generation conditions respectively, for the first subject.

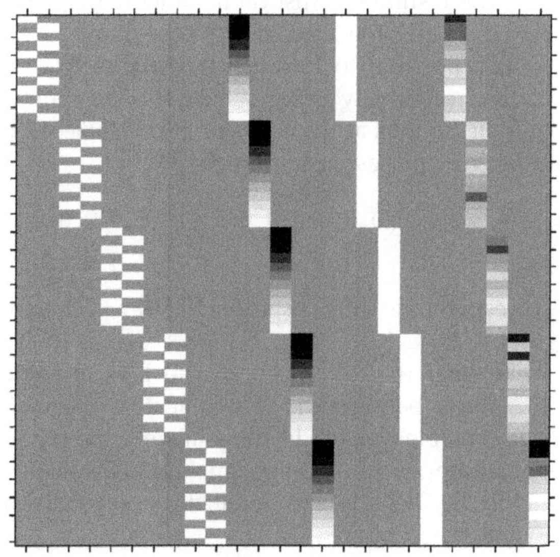

FIGURE 12.5 Design matrix for the five-subject FFX analysis of PET data. There are 60 rows, 12 for each subject. The first ten columns contain indicator variables showing which condition (word shadowing or word generation) relates to which scan. Columns 11 to 15 contain time variables, columns 16 to 20 subject-specific offsets and the last 5 columns the global effect at each scan.

[1] This data set and full details of the pre-processing are available from http://www.fil.ion.ucl.ac.uk/spm/data.

[2] This design was created using the 'Multisubject: condition by subject interaction and covariates' option in SPM-99.

Columns 3 to 10 indicate these conditions for the other subjects. The time variables in columns 11 to 15 are used to probe habituation effects. These variables are not of interest to us in this chapter but we include them to improve the fit of the model. The GLM can be written as:

$$Y = X\beta + E \tag{12.32}$$

where β are regression coefficients and E is a vector of errors. The effects of interest can then be examined using an augmented contrast vector, c. For example, for the verbal fluency data the contrast vector:

$$c = [-1, 1, -1, 1, -1, 1, -1, 1, -1, 1, 0, 0, 0, 0, 0, 0, 0, 0, 0, 0]^T \tag{12.33}$$

would be used to examine the differential effect of word generation versus word shadowing, averaged over the group of subjects. The corresponding t-statistic:

$$t = \frac{c^T\hat{\beta}}{\sqrt{\mathrm{Var}[c^T\hat{\beta}]}} \tag{12.34}$$

where Var[] denotes variance, highlights voxels with significantly non-zero differential activity. This shows the 'average effect in the group' and is a type of fixed-effects analysis. The resulting statistical parametric map (SPM) is shown in Plate 7(b) (see colour plate section).

It is also possible to look for differential effects in each subject separately using subject-specific contrasts. For example, to look at the activation from subject 2 one would use the contrast vector:

$$c_2 = [0, 0, -1, 1, 0, 0, 0, 0, 0, 0, 0, 0, 0, 0, 0, 0, 0, 0, 0, 0]^T \tag{12.35}$$

The corresponding subject-specific SPMs are shown in Plate 7(a).

We note that we have been able to look at subject-specific effects because the design matrix specified a 'subject-separable model'. In these models, the parameter estimates for each subject are unaffected by data from other subjects. This arises from the block-diagonal structure in the design matrix.

Random-effects analysis via summary statistics

An RFX analysis can be implemented using the 'summary-statistic (SS)' approach as follows (Frison and Pocock, 1992; Holmes and Friston, 1998).

1 Fit the model for each subject using different GLMs for each subject or by using a multiple-subject subject-separable GLM, as described above. The latter approach may be procedurally more convenient while the former is less computationally demanding. The two approaches are equivalent for the purposes of RFX analysis.
2 Define the effect of interest for each subject with a contrast vector. Each produces a contrast image containing the contrast of the parameter estimates at each voxel.
3 Feed the contrast images into a GLM that implements a one-sample t-test.

Modelling in step 1 is referred to as the 'first-level' of analysis, whereas modelling in step 3 is referred to as the 'second-level'. A balanced design is one in which all subjects have identical design matrices and error variances. Strictly, balanced designs are a requirement for the SS approach to be valid. But, as we have seen with the numerical example, the SS approach is remarkably robust to violations of this assumption.

If there are, say, two populations of interest and one is interested in making inferences about differences between populations, then a two-sample t-test is used at the second level. It is not necessary that the numbers of subjects in each population be the same, but it is necessary to have the same design matrices for subjects in the same population, i.e. balanced designs at the first-level.

In step 3, we have specified that only one contrast per subject be taken to the second level. This constraint may be relaxed if one takes into account the possibility that the contrasts may be correlated or be of unequal variance. This can be implemented using within-subject analyses of variance (ANOVAs) at the second level, a topic which is covered in Chapter 13.

An SPM of the RFX analysis is shown in Plate 7(c). We note that, as compared to the SPM from the average effect in the group, there are far fewer voxels deemed significantly active. This is because RFX analysis takes into account the between-subject variability. If, for example, we were to ask the question: 'Would a new subject drawn from this population show any significant posterior activity?', the answer would be uncertain. This is because three of the subjects in our sample show such activity but two subjects do not. Thus, based on such a small sample, we would say that our data do not show sufficient evidence against the null hypothesis that there is no population effect in posterior cortex. In contrast, the average effect in the group (in plate 7(b)) is significant over posterior cortex. But this inference is with respect to the group of five subjects, not the population.

We end this section with a disclaimer, which is that the results presented in this section have been presented for tutorial purposes only. This is because between-scan variance is so high in PET that results on single subjects are unreliable. For this reason, we have used uncorrected thresholds for the SPMs and, given that we have no prior anatomical hypothesis, this is not the correct thing to do (Frackowiak *et al.*, 1997) (see Chapter 14). But as our concern is merely to present a tutorial on the difference between RFX and FFX we have neglected these otherwise important points.

fMRI DATA EXAMPLE

This section compares RFX analysis as implemented using SS versus PEB. The dataset we chose to analyse comprised 1200 images that were acquired in 10 contiguous sessions of 120 scans. These data have been described elsewhere (Friston *et al.*, 1998).

The reason we chose these data was that each of the 10 sessions was slightly different in terms of design. The experimental design involved 30-second epochs of single word streams and a passive listening task. The words were concrete, monosyllabic nouns presented at a number of different rates. The word rate was varied pseudorandomly over epochs within each session.

We modelled responses using an event-related model where the occurrence of each word was modelled with a delta function. The ensuing stimulus function was convolved with a canonical haemodynamic response function and its temporal derivative to give two regressors of interest for each of the 10 sessions. These effects were supplemented with confounding and nuisance effects comprising a mean and the first few components of a discrete cosine transform, removing drifts lower than 1/128 Hz. Further details of the paradigm and analysis details are given in Friston *et al.* (2005).

The results of the SS and PEB analyses are presented in Figure 12.6 and have been thresholded at $p < 0.05$, corrected for the entire search volume. These results are taken from Friston *et al.* (2005) where PEB was implemented using the ReML formulation. It is evident that the inferences from these two procedures are almost identical, with PEB being slightly more sensitive. The results remain relatively unchanged despite the fact that the first-level designs were not balanced. This contributes to non-sphericity at the second level which is illustrated in Figure 12.7 for the SS and PEB approaches. This figure shows that heteroscedasticity can vary by up to a factor of 4.

DISCUSSION

We have shown how neuroimaging data from multiple subjects can be analysed using fixed-effects (FFX) or random-effects (RFX) analysis. FFX analysis is used for reporting case studies and RFX is used to make inferences

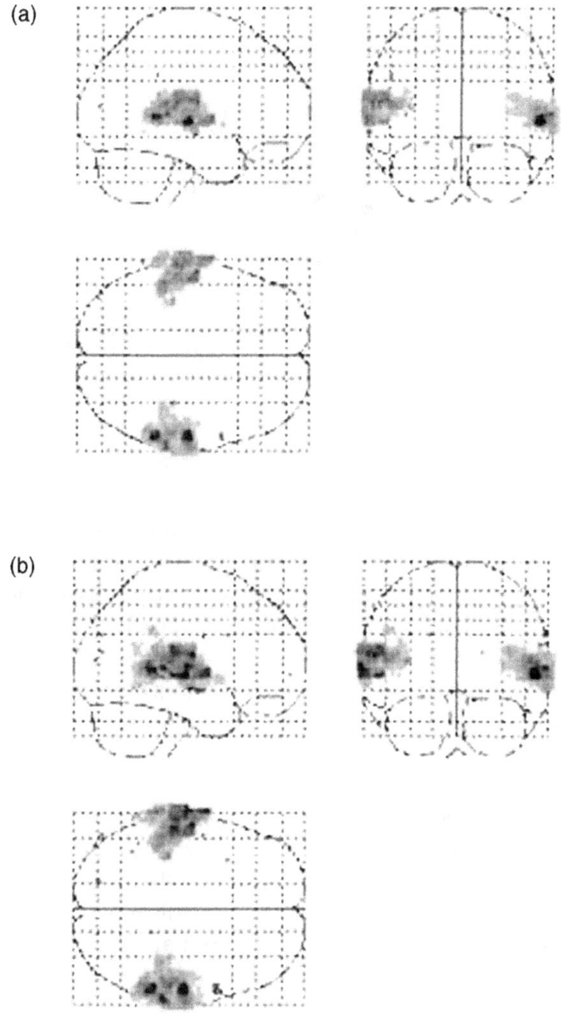

FIGURE 12.6 SPMs showing the effect of words in the population using (a) SS and (b) PEB approaches.

about the population from which subjects are drawn. For a comparison of these and other methods for combining data from multiple subjects see Lazar *et al.* (2002).

In neuroimaging, RFX is implemented using the computationally efficient summary-statistic approach. We have shown that this is mathematically equivalent to the more computationally demanding maximum likelihood procedure. For unbalanced designs, however, the summary-statistic approach is no longer equivalent. But we have shown, using a simulation study and fMRI data, that this lack of formal equivalence is not practically relevant.

For more advanced treatments of random effects analysis[3] see e.g. Yandell (1997). These allow, for example,

[3] Strictly, what in neuroimaging is known as random-effects analysis is known in statistics as mixed-effects analysis as the statistical models contain both fixed and random effects.

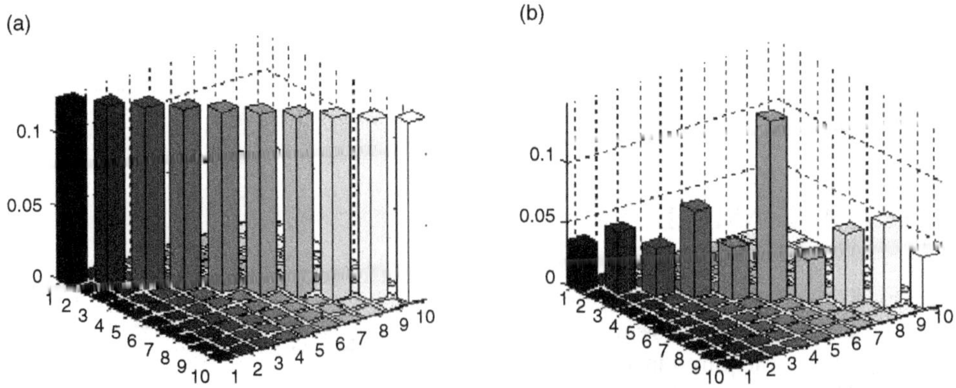

FIGURE 12.7 Within-session variance as (a) assumed by SS and (b) estimated using PEB. This shows that within-session variance can vary by up to a factor of 4, although this makes little difference to the final inference (see Figure 12.6).

for subject-specific within-subject variances, unbalanced designs and for Bayesian inference (Carlin and Louis, 2000). For a recent application of these ideas to neuroimaging, readers are referred to Chapter 17 in which hierarhical models are applied to single and multiple subject fMRI studies. As groundwork for this more advanced material readers are encouraged to first read the tutorial in Chapter 11.

A general point to note, especially for fMRI, is that because the between-subject variance is typically larger than the within-subject variance your scanning time is best used to scan more subjects rather than to scan individual subjects for longer. In practice, this must be traded off against the time required to recruit and train subjects (Worsley *et al.*, 2002).

Further points

We have so far described how to make inferences about univariate effects in a single population. This is achieved in the summary-statistic approach by taking forward a single contrast image per subject to the second level and then using a one sample t-test.

This methodology carries over naturally to more complex scenarios where we may have multiple populations or multivariate effects. For two populations, for example, we perform two-sample t-tests at the second level. An extreme example of this approach is the comparison of a single case study with a control group. While this may sound unfeasible, as one population has only a single member, a viable test can in fact be implemented by assuming that the two populations have the same variance.

For multivariate effects, we take forward multiple contrast images per subject to the second level and perform an analysis of variance. This can be implemented in the usual way with a GLM but, importantly, we must take into account the fact that we have repeated measures for each subject and that each characteristic of interest may have a different variability. This topic is covered in the next chapter.

As well as testing for whether univariate population effects are significantly different from hypothesized values (typically zero), it is possible to test whether they are correlated with other variables of interest. For example, one can test whether task-related activation in the motor system correlates with age (Ward and Frackowiak, 2003). It is also possible to look for conjunctions at the second level, e.g. to test for areas that are conjointly active for pleasant, unpleasant and neutral odour valences (Gottfried *et al.*, 2002). For a statistical test involving conjunctions of contrasts, it is necessary that the contrast effects be uncorrelated. This can be ensured by taking into account the covariance structure at the second level. This is also described in the next chapter on analysis of variance.

The validity of all of the above approaches relies on the same criteria that underpin the univariate single population summary-statistic approach. Namely, that the variance components and estimated parameter values are, on average, identical to those that would be obtained by the equivalent two-level maximum likelihood model.

APPENDIX 12.1 EXPECTATIONS AND TRANSFORMATIONS

We use E[] to denote the expectation operator and Var[] to denote variance and make use of the following results. Under a linear transform $y = ax + b$, the variance of x changes according to:

$$\mathrm{Var}[ax+b] = a^2\mathrm{Var}[x] \qquad \textbf{12.36}$$

Secondly, if $\text{Var}[x_i] = \text{Var}[x]$ for all i then:

$$\text{Var}\left[\frac{1}{N}\sum_{i=1}^{N} x_i\right] = \frac{1}{N}\text{Var}[x] \qquad \textbf{12.37}$$

For background reading on expectations, variance transformations and introductory mathematical statistics see Wackerley *et al.* (1996).

REFERENCES

Carlin BP, Louis TA (2000) *Bayes and empirical Bayes methods for data analysis*. Chapman and Hall, London

Frackowiak RSJ, Friston KJ, Frith C, *et al.* (1997) *Human brain function*. Academic Press, New York

Frison L, Pocock SJ (1992) Repeated measures in clinical trials: an analysis using mean summary statistics and its implications for design. *Stat Med*, **11**: 1685–1704

Friston KJ, Josephs O, Rees G, *et al.* (1998) Non-linear event-related responses in fMRI. *Magnet Res Med* **39**: 41–52

Friston KJ, Penny WD, Phillips C *et al.* (2002) Classical and Bayesian inference in neuroimaging: theory. *NeuroImage* **16**: 465–83

Friston KJ, Stephan KE, Lund TE *et al.* (2005) Mixed-effects and fMRI studies. *NeuroImage* **24**: 244–52

Gottfried JA, Deichmann R, Winston JS *et al.* (2002) Functional heterogeneity in human olfactory cortex: an event-related functional magnetic resonance imaging study. *Neurosci* **22**: 10819–28

Holmes AP, Friston KJ (1998) Generalisability, random effects and population inference. *NeuroImage* **7**: S754

Lazar NA, Luna B, Sweeney JA *et al.* (2002) Combining brains: a survey of methods for statistical pooling of information. *Neuroimage* **16**: 538–50

Wackerley DD, Mendenhall W, Scheaffer RL (1996) *Mathematical statistics with applications*. Duxbury Press, California

Ward NS, Frackowiak RSJ (2003) Age related changes in the neural correlates of motor performance. *Brain* **126**: 873–88

Worsley KJ, Liao CH, Aston J *et al.* (2002) A general statistical analysis for fMRI data. *NeuroImage* **15**(1): 1–15

Yandell BS (1997) *Practical data analysis for designed experiments*. Chapman and Hall, London

CHAPTER

13

Analysis of Variance

W. Penny and R. Henson

INTRODUCTION

The mainstay of many scientific experiments is the factorial design. These comprise a number of experimental factors which are each expressed over a number of levels. Data are collected for each factor/level combination and then analysed using analysis of variance (ANOVA). The ANOVA uses F-tests to examine a pre-specified set of standard effects, e.g. 'main effects' and 'interactions', as described in Winer *et al.* (1991).

ANOVAs are commonly used in the analysis of positron emission tomography (PET), electroencephalography (EEG), magnetoencephalography (MEG) and functional magnetic resonance imaging (fMRI) data. For PET, this analysis usually takes place at the 'first' level. This involves direct modelling of PET scans. For EEG, MEG and fMRI, ANOVAs are usually implemented at the 'second level'. As described in the previous chapter, first level models are used to create contrast images for each subject. These are then used as data for a second level or 'random-effects' analysis.

Some different types of ANOVA are tabulated in Table 13-1. A *two-way* ANOVA, for example, is an ANOVA with 2 factors; a K_1-*by*-K_2 ANOVA is a two-way ANOVA with K_1 levels of one factor and K_2 levels of the other. A *repeated measures* ANOVA is one in which the levels of one or more factors are measured from the same unit (e.g. subjects). Repeated measures ANOVAs are also sometimes called *within-subject* ANOVAs, whereas designs in which each level is measured from a different group of subjects are called *between-subject* ANOVAs. Designs in which some factors are within-subject, and others between-subject, are sometimes called *mixed* designs.

TABLE 13-1 Types of ANOVA

Factors	Levels	Simple	Repeated measures
1	2	Two-sample t-test	Paired t-test
1	K	One-way ANOVA	One-way ANOVA within-subject
M	$K_1, K_2, .., K_M$	M-way ANOVA	M-way ANOVA within-subject

This terminology arises because in a between-subject design the difference between levels of a factor is given by the difference between subject responses, e.g. the difference between levels 1 and 2 is given by the difference between those subjects assigned to level 1 and those assigned to level 2. In a within-subject design, the levels of a factor are expressed within each subject, e.g. the difference between levels 1 and 2 is given by the average difference of subject responses to levels 1 and 2. This is like the difference between two-sample t-tests and paired t-tests.

The benefit of repeated measures is that we can match the measurements better. However, we must allow for the possibility that the measurements are correlated (so-called 'non-sphericity' – see below).

The level of a factor is also sometimes referred to as a 'treatment' or a 'group' and each factor/level combination is referred to as a 'cell' or 'condition'. For each type of ANOVA, we describe the relevant statistical models and show how they can be implemented in a general linear model (GLM). We also give examples of how main effects and interactions can be tested for using F-contrasts.

The chapter is structured as follows: the first section describes one-way between-subject ANOVAs. The next section describes one-way within-subject ANOVAs and introduces the notion of non-sphericity. We then describe two-way within-subject ANOVAs and make a distinction

Statistical Parametric Mapping, by Karl Friston et al.
ISBN–13: 978-0-12-372560-8 ISBN–10: 0-12-372560-7

between models with pooled versus partitioned errors. The last section discusses issues particular to fMRI and we end with a discussion.

Notation

In the mathematical formulations below, $N(m, \Sigma)$ denotes a uni/multivariate Gaussian with mean m and variance/covariance Σ. I_K denotes the $K \times K$ identity matrix, X^T denotes transpose, X^{-T} the inverse transpose, X^- the generalized-inverse, 1_K is a $K \times 1$ vector of 1s, 0_K is a $K \times 1$ vector of zeroes and 0_{KN} is a $K \times N$ matrix of zeroes. We consider factorial designs with $n = 1..N$ subjects and $m = 1..M$ factors where the mth factor has $k = 1..K_m$ levels.

ONE-WAY BETWEEN-SUBJECT ANOVA

In a between-subject ANOVA, differences between levels of a factor are given by the differences between subject responses. We have one measurement per subject and different subjects are assigned to different levels/treatments/groups. The response from the nth subject (y_n) is modelled as:

$$y_n = \tau_k + \mu + e_n \tag{13.1}$$

where τ_k are the treatment effects, $k = 1..K$, $k = g(n)$ and $g(n)$ is an indicator function whereby $g(n) = k$ means the nth subject is assigned to the kth group, e.g. $g(13) = 2$ indicates the 13th subject being assigned to group 2. This is the single experimental factor that is expressed over K levels. The variable μ is sometimes called the *grand mean* or *intercept* or *constant term*. The random variable e_n is the residual error, assumed to be drawn from a zero mean Gaussian distribution.

If the factor is significant, then the above model is a significantly better model of the data than the simpler model:

$$y_n = \mu + e_n \tag{13.2}$$

where we just view all of the measurements as random variation about the grand mean. Figure 13.2 compares these two models on some simulated data.

In order to test whether one model is better than another, we can use an F-test based on the *extra sum of squares* principle (see Chapter 8). We refer to Eqn. **13.1** as the 'full' model and Eqn. **13.2** as the 'reduced' model. If RSS denotes the residual sum of squares (i.e. the sum of squares left after fitting a model) then:

$$F = \frac{(RSS_{reduced} - RSS_{full})/(K-1)}{RSS_{full}/(N-K)} \tag{13.3}$$

has an F-distribution with $K-1, N-K$ degrees of freedom. If F is significantly non-zero then the full model has a significantly smaller error variance than the reduced model. That is to say, the full model is a significantly better model, or the *main effect* of the factor is significant.

The above expression is also sometimes expressed in terms of sums of squares (SS) due to treatment and due to error:

$$F = \frac{SS_{treat}/DF_{treat}}{SS_{error}/DF_{error}} \tag{13.4}$$

where

$$\begin{aligned} SS_{treat} &= RSS_{reduced} - RSS_{full} \\ DF_{treat} &= K-1 \\ SS_{error} &= RSS_{full} \\ DF_{error} &= N-K \\ DF_{total} &= DF_{treat} + DF_{error} = N-1 \end{aligned} \tag{13.5}$$

Eqns **13.3** and **13.4** are therefore equivalent.

Numerical example

This subsection shows how an ANOVA can be implemented in a GLM. Consider a one-way ANOVA with $K = 4$ groups each having $n = 12$ subjects (i.e. $N = Kn = 48$ subjects/observations in total). The GLM for the full model in Eqn. **13.1** is:

$$y = X\beta + e \tag{13.6}$$

where the design matrix $X = [I_K \otimes 1_n, 1_N]$ is shown in Figure 13.1, where $\otimes$ denotes the Kronecker product (see Appendix 13.1). The vector of parameters is $\beta = [\tau_1, \tau_2, \tau_3, \tau_4, \mu]^T$.

Eqn. **13.3** can then be implemented using the *effects of interest* F-contrast, as introduced in Chapter 9:

$$C^T = \begin{bmatrix} 1 & -1/3 & -1/3 & -1/3 & 0 \\ -1/3 & 1 & -1/3 & -1/3 & 0 \\ -1/3 & -1/3 & 1 & -1/3 & 0 \\ -1/3 & -1/3 & -1/3 & 1 & 0 \end{bmatrix} \tag{13.7}$$

or equivalently:

$$C^T = \begin{bmatrix} 1 & -1 & 0 & 0 & 0 \\ 0 & 1 & -1 & 0 & 0 \\ 0 & 0 & 1 & -1 & 0 \end{bmatrix} \tag{13.8}$$

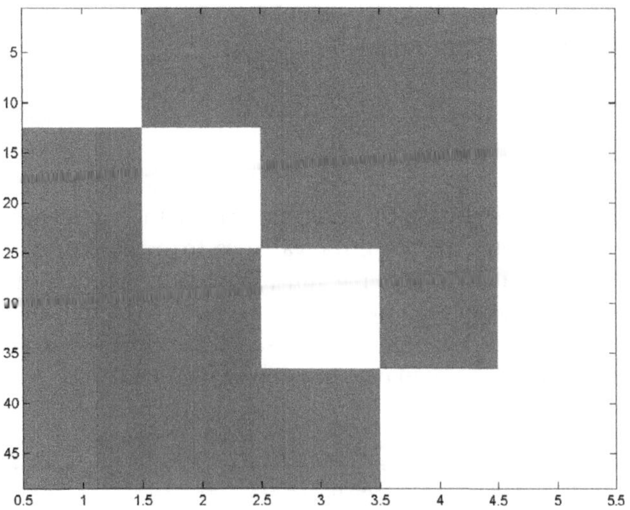

FIGURE 13.1 Design matrix for one-way (1 × 4) between-subject ANOVA. White and grey represent 1 and 0. There are 48 rows, one for each subject ordered by condition, and 5 columns, the first 4 for group effects and the 5th for the grand mean.

These contrasts can be thought of as testing the null hypothesis $\mathcal{H}_0$:

$$\mathcal{H}_0 : \tau_1 = \tau_2 = \tau_3 = \tau_4 \tag{13.9}$$

Note that a significant departure from $\mathcal{H}_0$ can arise from any pattern of these treatment means (parameter estimates) – they need not be monotonic across the four groups for example.

The correspondence between this F-contrast and the classical formulation in Eqn. **13.3** is detailed in Chapter 10. We now analyse the example data set shown in Figure 13.2. The results of a one-way between-subjects ANOVA are shown in Table 13-2. This shows that there is a significant main effect of treatment ($p < 0.02$).

Note that the design matrix in Figure 13.1 is rank-deficient (see Chapter 8) and the alternative design matrix $X = [I_K \otimes 1_n]$ could be used with appropriate F-contrasts (though the parameter estimates themselves would include a contribution of the grand mean, equivalent to the contrast $[1, 1, 1, 1]^T$). If β_1 is a vector of parameter estimates after the first four columns of X are mean-corrected (orthogonalized with respect to the fifth column), and β_0 is the parameter estimate for the corresponding fifth column, then:

$$\begin{aligned} SS_{treatment} &= n\beta_1^T\beta_1 = 51.6 \\ SS_{mean} &= nK\beta_0^2 = 224.1 \\ SS_{error} &= r^T r = 208.9 \\ SS_{total} &= y^T y = SS_{treatment} + SS_{mean} + SS_{error} = 484.5 \end{aligned} \tag{13.10}$$

where the residual errors are $r = y - XX^{-}y$.

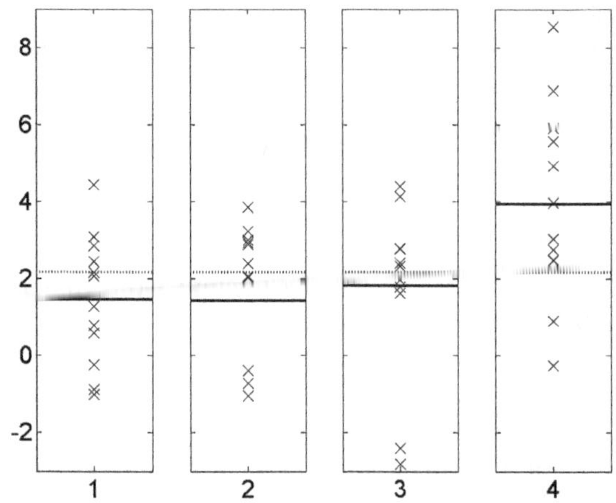

FIGURE 13.2 **One-way between-subject ANOVA.** 48 subjects are assigned to one of four groups. The plot shows the data points for each of the four conditions (crosses), the predictions from the 'one-way between-subjects model' or the 'full model' (solid lines) and the predicitons from the 'reduced model' (dotted lines). In the reduced model (Eqn. **13.2**), we view the data as random variation about a grand mean. In the full model (Eqn. **13.1**), we view the data as random variation about condition means. Is the full model significantly better than the reduced model? That responses are much higher in condition 4 suggests that this is indeed the case and this is confirmed by the results in Table 13-2.

TABLE 13-2 Results of one-way (1 × 4) between-subject ANOVA

Main effect of treatment	F = 3.62	DF = [3, 44]	$p = 0.02$

ONE-WAY WITHIN-SUBJECT ANOVA

In this model we have K measurements *per* subject. The treatment effects for subject $n = 1 \ldots N$ are measured relative to the average response made by subject n on all treatments. The kth response from the nth subject is modelled as:

$$y_{nk} = \tau_k + \pi_n + e_{nk} \tag{13.11}$$

where τ_k are the treatment effects (or *within-subject effects*), π_n are the *subject effects* and e_{nk} are the residual errors. We are not normally interested in π_n, but its explicit modelling allows us to remove variability due to differences in average responsiveness of each subject. See, for example, the data set in Figure 13.3. It is also possible to express the full model in terms of differences between treatments (see e.g. Eqn. **13.15** for the two-way case).

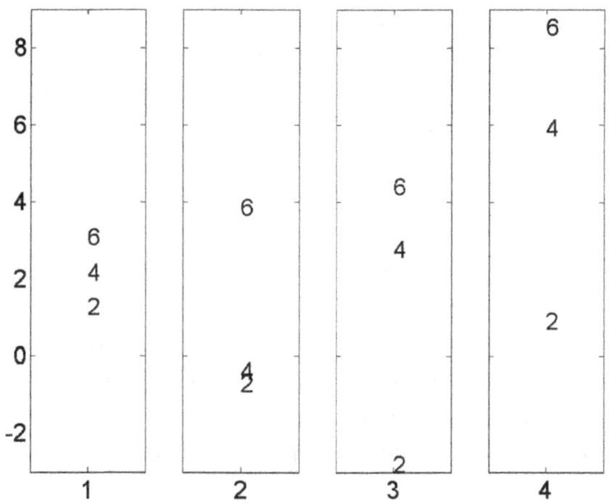

FIGURE 13.3 Portion of example data for one-way within-subject ANOVA. The plot shows the data points for 3 subjects in each of 4 conditions (in the whole data set there are 12 subjects). Notice how subject 6's responses are always high, and subject 2's are always low. This motivates modelling subject effects as in Eqns. **13.11** and **13.12**.

To test whether the experimental factor is significant, we compare the full model in Eqn. **13.11** with the reduced model:

$$y_{nk} = \pi_n + e_{nk} \tag{13.12}$$

An example of comparing these full and reduced models is shown in Figure 13.4. The equations for computing the relevant F-statistic and degrees of freedom are given, for example, in Chapter 14 of Howell (1992).

Numerical example

The design matrix $X = [I_K \otimes 1_N, 1_K \otimes I_N]$ for Eqn. **13.11**, with $K = 4$ and $N = 12$, is shown in Figure 13.5. The first 4 columns are treatment effects and the next 12 are subject effects. The main effect of the factor can be assessed using the same *effects of interest* F-contrast as in Eqn. **13.7**, but with additional zeroes for the columns corresponding to the subject effects.

We now analyse another example data set, a portion of which is shown in Figure 13.3. Measurements have been obtained from 12 subjects under each of $K = 4$ conditions.

Assuming sphericity (see below), we obtain the ANOVA results in Table 13-3. In fact this dataset contains exactly the same numerical values as the between-subjects example data. We have just relabelled the data as being measured from 12 subjects with 4 responses each instead of from 48 subjects with 1 response each. The reason that the p-value is less than in the between-subjects example (it has reduced from 0.02 to 0.001) is that the data were created to include subject effects. Thus, in repeated measures designs, the modelling of subject effects normally increases the sensitivity of the inference.

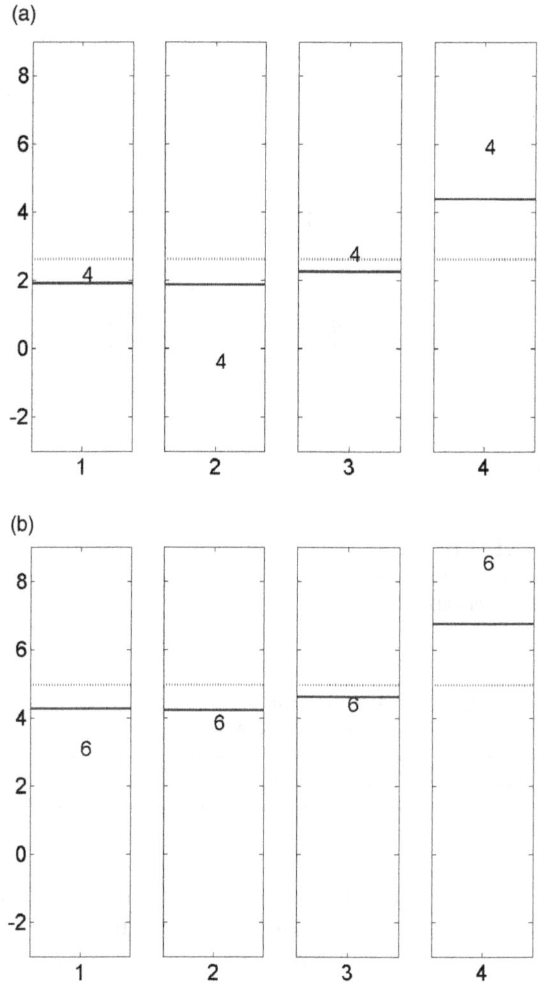

FIGURE 13.4 **One-way within-subject ANOVA.** The plot shows the data points for each of the four conditions for subjects (a) 4 and (b) 6, the predictions from the one-way within-subjects model (solid lines) and the reduced model (dotted lines).

Non-sphericity

Due to the nature of the levels in an experiment, it may be the case that if a subject responds strongly to level i, he may respond strongly to level j. In other words, there may be a correlation between responses. In Figure 13.6 we plot subject responses for level i against level j for the example data set. These show that for some pairs of conditions there does indeed seem to be a correlation. This correlation can be characterized graphically by fitting a Gaussian to each 2D data cloud and then plotting probability contours. If these contours form a sphere (a circle, in two dimensions) then the data are independent and identically distributed (IID), i.e. same

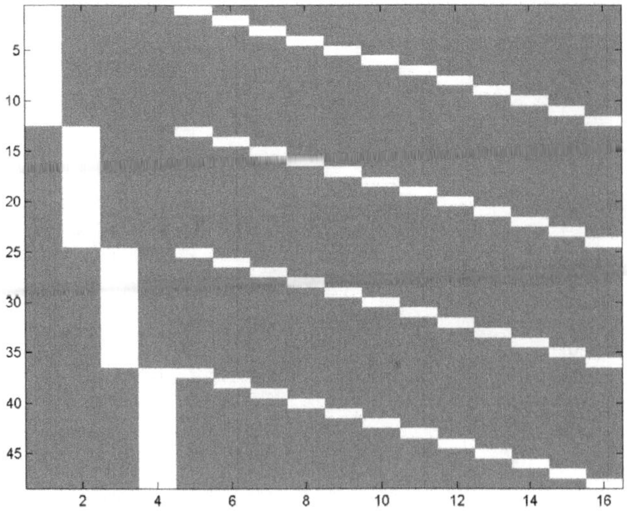

FIGURE 13.5 Design matrix for one-way (1×4) within-subject ANOVA. The first 4 columns are treatment effects and the last 12 are subject effects.

TABLE 13-3 Results of one-way (1×4) within-subjects ANOVA

Main effect of treatment	F = 6.89	DF = [3, 33]	$p = 0.001$

variance in all dimensions and there is no correlation. The more these contours look like ellipses, the more 'non-sphericity' there is in the data.

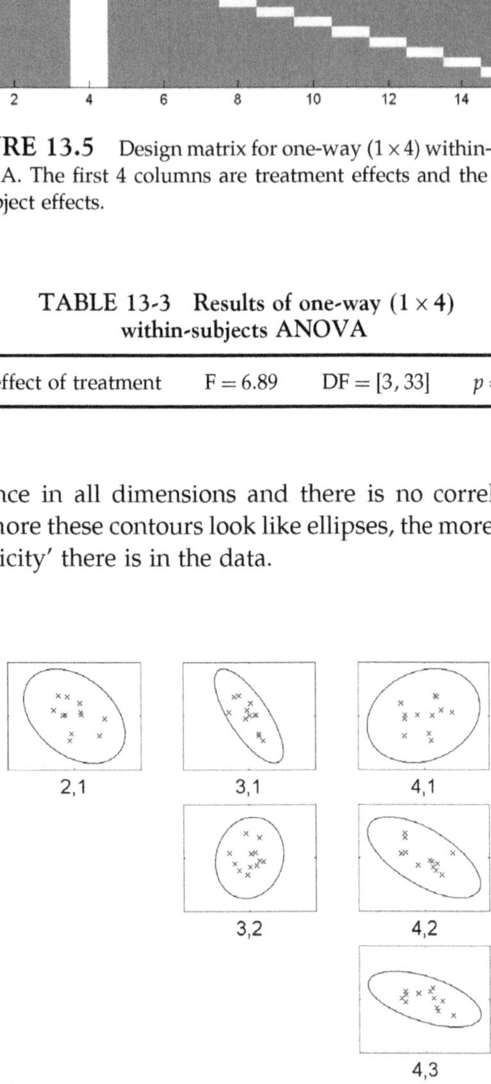

FIGURE 13.6 **One-way within-subjects ANOVA: Non-sphericity.** Each subgraph plots each subject's response to condition i versus condition j as a cross. There are twelve crosses, one from each subject. We also plot probability contours from the corresponding Gaussian densities. Subject responses, for example, to conditions 1 and 3 seem correlated – the sample correlation coefficient is −0.75. Overall, the more non-spherical the contours the greater the non-sphericity.

The possible non-sphericity can be taken into account in the analysis using a correction to the degrees of freedom (DFs). In the above example, a Greenhouse-Geisser (GG) correction (see Appendix 13.1 and Chapter 10) estimates $\varepsilon = .7$, giving DFs of $[2.1, 23.0]$ and a p-value (with GG we use the same F-statistic, i.e. $F = 6.89$) of $p = 0.004$. Assuming sphericity, as before, we computed $p = 0.001$. Thus the presence of non-sphericity in the data makes us less confident of the significance of the effect.

An alternative representation of the within-subjects model is given in Appendix 13.2. This shows how one can take into account non-sphericity. Various other relevant terminology is also defined in Appendices 13.1 and 13.2.

TWO-WAY WITHIN-SUBJECT ANOVAs

The full model for a two-way, K_1-*by*-K_2 repeated measures ANOVA, with $P = K_1K_2$ measurements taken from each of N subjects, can be written as:

$$y_{nkl} = \tau_{kl} + \pi_n + e_{nkl} \qquad \textbf{13.13}$$

where $k = 1 \ldots K_1$ and $l = 1 \ldots K_2$ index the levels of factor A and factor B respectively. Here we can think of indicator functions $k = g_k(i)$, $l = g_l(i)$ and $n = g_n(i)$ that return the levels of both factors and subject identity for the ith scan. Again, π_n are subject effects and e_{nkl} are residual errors. This equation can be written in matrix form:

$$y = X\beta + e \qquad \textbf{13.14}$$

where $X = [I_P \otimes 1_N, 1_N \otimes I_P]$ is the design matrix and $\beta = [\tau_{kl}, \pi_n]^T$ are the regression coefficients. This is identical to the one-way within-subject design but with P instead of K treatment effects.

However, rather than considering each factor/level combination separately, the key concept of ANOVA is to model the data in terms of a standard set of *experimental effects*. These consist of *main effects* and *interactions*. Each factor has an associated main effect, which is the difference between the levels of that factor, averaging over the levels of all other factors. Each pair of factors (and higher-order tuples; see below) has an associated interaction. Interactions represent the degree to which the effect of one factor depends on the levels of the other factor(s). A two-way ANOVA thus has two main effects and one interaction.

Eqn. **13.13** can be rewritten as:

$$\begin{aligned} y &= X\beta + e \qquad \textbf{13.15} \\ &= XC^{-T}C^T\beta + e \\ &= X_r\tilde{\beta} + e \end{aligned}$$

where $X_r = XC^{-T}$ is a rotated design matrix, the regression coefficients are $\tilde{\beta} = C^T\beta$, and C is a 'contrast matrix'. This equation is important as it says that the effects $\tilde{\beta}$ can be estimated by either (i) fitting the data using a GLM with design matrix X_r or (ii) fitting the original GLM, with design matrix X, and applying the contrast matrix $\tilde{\beta} = C^T\beta$.

For our two-way within-subjects ANOVA we choose C such that:

$$\tilde{\beta} = [\tau_q^A, \tau_r^B, \tau_{qr}^{AB}, m, \pi_n]^T \qquad 13.16$$

Here, τ_q^A represents the differences between each *successive* level $q = 1 \ldots (K_1 - 1)$ of factor A (e.g. the differences between levels 1 and 2, 2 and 3, 3 and 4 etc.), averaging over the levels of factor B. In other words, the main effect of A is modelled as $K_1 - 1$ differences among K_1 levels. The quantity τ_r^B represents the differences between each *successive* level $r = 1 \ldots (K_2 - 1)$ of factor B, averaging over the levels of factor A; and τ_{qr}^{AB} represents the differences between the differences of each level $q = 1 \ldots (K_1 - 1)$ of factor A across each level $r = 1 \ldots (K_2 - 1)$ of factor B. The quantity m is the mean treatment effect. Examples of contrast matrices and rotated design matrices are given below.

Pooled versus partitioned errors

In the above model, e is sometimes called a *pooled error*, since it does not distinguish between different sources of error for each experimental effect. This is in contrast to an alternative model in which the original residual error e is split into three terms e_{nq}^A, e_{nr}^B and e_{nqr}^{AB}, each specific to a main effect or interaction. This is a different form of *variance partitioning*. Each error term is a random variable and is equivalent to the interaction between that effect and the subject variable.

The F-test for, say, the main effect of factor A is then:

$$F = \frac{SS_k/DF_k}{SS_{nk}/DF_{nk}} \qquad 13.17$$

where SS_k is the sum of squares for the effect, SS_{nk} is the sum of squares for the interaction of that effect with subjects, $DF_k = K_1 - 1$ and $DF_{nk} = N(K_1 - 1)$.

Note that, if there are no more than two levels of every factor in an M-way repeated measures ANOVA (i.e., $K_m = 2$ for all $m = 1 \ldots M$), then the covariance of the errors Σ_e for each effect is a 2-by-2 matrix which necessarily has compound symmetry, and so there is no need for a non-sphericity correction.[1] A heuristic for this is that there is only one difference $q = 1$ between two levels $K_m = 2$. This is not necessarily the case if a pooled error is used, as in Eqn. **13.15**.

[1] Although one could model inhomegeneity of variance.

Models and null hypotheses

The difference between pooled and partitioned error models can be expressed by specifying the relevant models and null hypotheses.

Pooled errors

The pooled error model is given by Eqn. **13.15**. For the main effect of A we test the null hypothesis $\mathcal{H}_0 : \tau_q^A = 0$ for all q. Similarly, for the main effect of B. For an interaction we test the null hypothesis $\mathcal{H}_0 : \tau_{qr}^{AB} = 0$ for all q, r.

For example, for the 3-by-3 design shown in Figure 13.7 there are $q = 1..2$ differential effects for factor A and

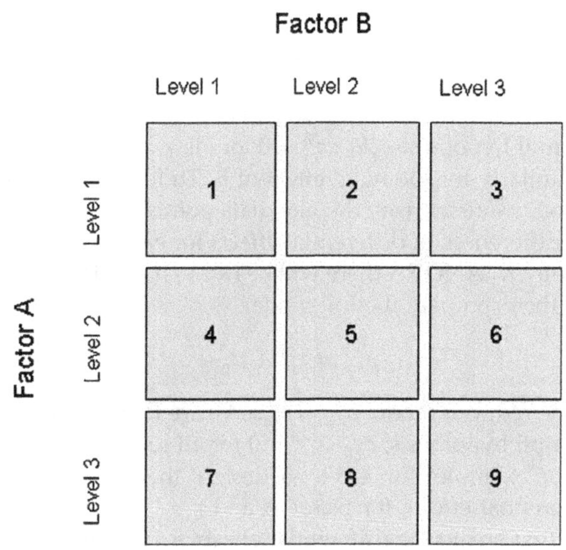

FIGURE 13.7 In a 3×3 ANOVA there are 9 cells or conditions. The numbers in the cells correspond to the ordering of the measurements when rearranged as a column vector y for a single-subject general linear model. For a repeated measures ANOVA there are 9 measurements per subject. The variable y_{nkl} is the measurement at the kth level of factor A, the lth level of factor B and for the nth subject. To implement the partitioned error models we use these original measurements to create differential effects for each subject. The differential effect τ_1^A is given by row 1 minus row 2 (or cells 1, 2, 3 minus cells 4, 5, 6 – this is reflected in the first row of the contrast matrix in Eqn. **13.28**). The differential effect τ_2^A is given by row 2 minus row 3. These are used to assess the main effect of A. Similarly, to assess the main effect of B we use the differential effects τ_1^B (column 1 minus column 2) and τ_2^B (column 2 minus column 3). To assess the interaction between A and B, we compute the four 'simple interaction' effects τ_{11}^{AB} (cells (1–4)-(2–5)), τ_{12}^{AB} (cells (2–5)-(3–6)), τ_{21}^{AB} (cells (4–7)-(5–8)) and τ_{22}^{AB} (cells (5–8)-(6–9)). These correspond to the rows of the interaction contrast matrix in Eqn. **13.30**.

$r = 1..2$ for factor B. The pooled error model therefore has regression coefficients:

$$\tilde{\beta} = [\tau_1^A, \tau_2^A, \tau_1^B, \tau_2^B, \tau_{11}^{AB}, \tau_{12}^{AB}, \tau_{21}^{AB}, \tau_{22}^{AB}, m, \pi_n]^T \qquad 13.18$$

For the main effect of A we test the null hypothesis $\mathcal{H}_0 : \tau_1^A = \tau_2^A = 0$. For the interaction we test the null hypothesis $\mathcal{H}_0 : \tau_{11}^{AB} = \tau_{12}^{AB} = \tau_{21}^{AB} = \tau_{22}^{AB} = 0$.

Partitioned errors

For partitioned errors, we first transform our data set y_{nkl} into a set of differential effects for each subject and then model these in a GLM. This set of differential effects for each subject is created using appropriate contrasts at the 'first-level'. The models that we describe below then correspond to a 'second-level' analysis. The difference between first and second level analyses are described in the previous chapter on random effects analysis.

To test for the main effect of A, we first create the new data points ρ_{nq} which are the differential effects between the levels in A for each subject n. We then compare the full model:

$$\rho_{nq} = \tau_q^A + e_{nq}$$

to the reduced model $\rho_{nq} = e_{nq}$. We are therefore testing the null hypothesis, $\mathcal{H}_0 : \tau_q^A = 0$ for all q.

Similarly for the main effect of B. To test for an interaction, we first create the new data points ρ_{nqr} which are the differences of differential effects for each subject. For a K_1 by K_2 ANOVA there will be $(K_1 - 1)(K_2 - 1)$ of these. We then compare the full model:

$$\rho_{nqr} = \tau_{qr}^{AB} + e_{nqr}$$

to the reduced model $\rho_{nqr} = e_{nqr}$. We are therefore testing the null hypothesis, $\mathcal{H}_0 : \tau_{qr}^{AB} = 0$ for all q, r.

For example, for a 3-by-3 design, there are $q = 1..2$ differential effects for factor A and $r = 1..2$ for factor B. We first create the differential effects ρ_{nq}. To test for the main effect of A we compare the full model:

$$\rho_{nq} = \tau_1^A + \tau_2^A + e_{nq}$$

to the reduced model $\rho_{nq} = e_{nq}$. We are therefore testing the null hypothesis, $\mathcal{H}_0 : \tau_1^A = \tau_2^A = 0$. Similarly for the main effect of B.

To test for an interaction we first create the differences of differential effects for each subject. There are $2 \times 2 = 4$ of these. We then compare the full model:

$$\rho_{nqr} = \tau_{11}^{AB} + \tau_{12}^{AB} + \tau_{21}^{AB} + \tau_{22}^{AB} + e_{nqr}$$

to the reduced model $\rho_{nqr} = e_{nqr}$. We are therefore testing the null hypothesis, $\mathcal{H}_0 : \tau_{11}^{AB} = \tau_{12}^{AB} = \tau_{21}^{AB} = \tau_{22}^{AB} = 0$, i.e. that all the 'simple' interactions are zero. See Figure 13.7 for an example with a 3-by-3 design.

Numerical example

Pooled error

Consider a 2×2 ANOVA of the same data used in the previous examples, with $K_1 = K_2 = 2$, $P = K_1 K_2 = 4$, $N = 12$, $J = PN = 48$. The design matrix for Eqn. **13.15** with a pooled error term is the same as that in Figure 13.5, assuming that the four columns/conditions are ordered:

$$\begin{matrix} 1 & 2 & 3 & 4 \\ A_1B_1 & A_1B_2 & A_2B_1 & A_2B_2 \end{matrix} \qquad 13.19$$

where A_1 represents the first level of factor A, B_2 represents the second level of factor B etc., and the rows are ordered; all subjects data for cell A_1B_1; all for A_1B_2 etc. The basic contrasts for the three experimental effects are shown in Table 13-4 with the contrast weights for the subject-effects in the remaining columns 5–16 set to 0.

Assuming sphericity, the resulting F-tests give the ANOVA results in Table 13-5. With a Greenhouse-Geisser correction for non-sphericity, on the other hand, ϵ is estimated as 0.7, giving the ANOVA results in Table 13-6.

Main effects are not really meaningful in the presence of a significant interaction. In the presence of an interaction, one does not normally report the main effects, but proceeds by testing the differences between the levels of one factor for each of the levels of the other factor in the interaction (so-called *simple effects*). In this case, the presence of a significant interaction could be used to justify further simple effect contrasts (see above), e.g. the effect of B at the first and second levels of A are given by the contrasts $c = [1, -1, 0, 0]^T$ and $c = [0, 0, 1, -1]^T$.

Equivalent results would be obtained if the design matrix were *rotated* so that the first three columns reflect the experimental effects plus a constant term in the fourth column (only the first four columns would be rotated). This is perhaps a better conception of the ANOVA approach, since it is closer to Eqn. **13.15**, reflecting the

TABLE 13-4 Contrasts for experimental effects in a two-way ANOVA

Main effect of A	[1	1	−1	−1]
Main effect of B	[1	−1	1	−1]
Interaction, A × B	[1	−1	−1	1]

TABLE 13-5 Results of 2 × 2 within-subject ANOVA with pooled error assuming sphericity

Main effect of A	F = 9.83	DF = [1, 33]	$p = 0.004$
Main effect of B	F = 5.21	DF = [1, 33]	$p = 0.029$
Interaction, A × B	F = 5.64	DF = [1, 33]	$p = 0.024$

TABLE 13-6 Results of 2 × 2 within-subject ANOVA with pooled error using Greenhouse-Geisser correction

Main effect of A	F = 9.83	DF = [0.7, 23.0]	$p = 0.009$
Main effect of B	F = 5.21	DF = [0.7, 23.0]	$p = 0.043$
Interaction, A × B	F = 5.64	DF = [0.7, 23.0]	$p = 0.036$

conception of factorial designs in terms of the experimental effects rather than the individual conditions. This rotation is achieved by setting the new design matrix:

$$X_r = X \begin{bmatrix} C^T & 0_{4,12} \\ 0_{12,4} & I_{12} \end{bmatrix} \quad \mathbf{13.20}$$

where

$$C^T = \begin{bmatrix} -1 & -1 & 1 & 1 \\ -1 & 1 & -1 & 1 \\ 1 & -1 & -1 & 1 \\ 1 & 1 & 1 & 1 \end{bmatrix} \quad \mathbf{13.21}$$

Notice that the rows of C^T are identical to the contrasts for the main effects and interactions plus a constant term (cf. Table 13-4). This rotated design matrix is shown in Figure 13.8. The three experimental effects can now be tested by the contrast weights $[1, 0, 0, 0]^T$, $[0, 1, 0, 0]^T$, $[0, 0, 1, 0]^T$ (again, padded with zeroes).

In this example, each factor only has two levels which results in one-dimensional contrasts for testing main

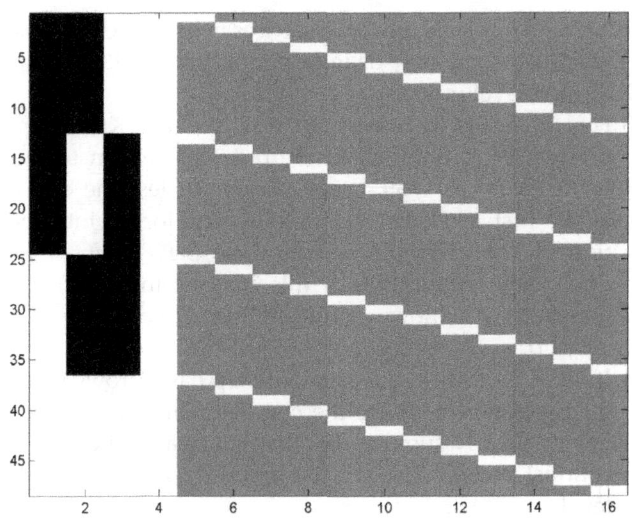

FIGURE 13.8 Design matrix for 2 × 2 within-subject ANOVA. This design is the same as in Figure 13.5 except that the first four columns are rotated. The rows are ordered all subjects for cell A_1B_1, all for A_1B_2 etc. White, grey and black represent 1, 0 and −1. The first four columns model the main effect of A, the main effect of B, the interaction between A and B and a constant term. The last 12 columns model subject effects. This model is a GLM instantiation of Eqn. **13.15**.

TABLE 13-7 Results of ANOVA using partitioned errors

Main effect of A	F = 12.17	DF = [1, 11]	$p = 0.005$
Main effect of B	F = 11.35	DF = [1, 11]	$p = 0.006$
Interaction, A × B	F = 3.25	DF = [1, 11]	$p = 0.099$

effects and interactions. The contrast weights form a *vector*. But factors with more than two levels require multi-dimensional contrasts. Main effects, for example, can be expressed as a linear combination of differences between successive levels (e.g. between levels 1 and 2, and 2 and 3). The contrast weights therefore form a *matrix*. An example using a 3-by-3 design is given later on.

Partitioned errors

Partitioned error models can be implemented by applying contrasts to the data, and then creating a separate model (i.e. separate GLM analysis) to test each effect. In other words, a two-stage approach can be taken, as described in the previous chapter on random effects analysis. The first stage is to create contrasts of the conditions for each subject, and the second stage is to put these contrasts or 'summary statistics' into a model with a block-diagonal design matrix.

Using the example dataset, and analogous contrasts for the main effect of B and for the interaction, we get the results in Table 13-7. Note how (1) the degrees of freedom have been reduced relative to Table 13-5, being split equally among the three effects; (2) there is no need for a non-sphericity correction in this case (since $K_1 = K_2 = 2$, see above); and (3) the p-values for some of the effects have decreased relative to Tables 13-5 and 13-6, while those for the other effects have increased. Whether p-values increase or decrease depends on the nature of the data (particularly correlations between conditions across subjects), but in many real data sets partitioned error comparisons yield more sensitive inferences. This is why, for repeated-measures analyses, the partitioning of the error into effect-specific terms is normally preferred over using a pooled error (Howell, 1992). But the partitioned error approach requires a new model to be specified for every effect we want to test.

GENERALIZATION TO M-WAY ANOVAs

The above examples can be generalized to M-way ANOVAs. For a K_1-by-K_2-..-by-K_M design, there are

$$P = \prod_{m=1}^{M} K_m \quad \mathbf{13.22}$$

conditions. An M-way ANOVA has $2^M - 1$ experimental effects in total, consisting of M main effects plus $M!/(M-r)!r!$ interactions of order $r = 2\ldots M$. A 3-way ANOVA for example has three main effects (A, B, C), three second-order interactions (A × B, B × C, A × C) and one third-order interaction (A × B × C). Or more generally, an M-way ANOVA has $2^M - 1$ interactions of order $r = 0\ldots M$, where a 0th-order interaction is equivalent to a main effect.

We consider models where every cell has its own coefficient (like Eqn. **13.13**). We will assume these conditions are ordered in a GLM so that the first factor *rotates* slowest, the second factor next slowest, etc., so that for a 3-way ANOVA with factors A, B, C:

$$\begin{array}{ccccccc} 1 & 2 & \ldots & K_3 & \ldots & P \\ A_1B_1C_1 & A_1B_1C_2 & \ldots & A_1B_1C_{K_3} & \ldots & A_{K_1}B_{K_2}C_{K_3} \end{array} \qquad 13.23$$

The data are ordered all subjects for cell $A_1B_1C_1$, all subjects for cell $A_1B_1C_2$ etc.

The F-contrasts for testing main effects and interactions can be constructed in an iterative fashion as follows. We define initial component contrasts.[2]

$$C_m = 1_{K_m} \quad D_m = -\mathrm{diff}(I_{K_m})^T \qquad 13.24$$

where $diff(A)$ is a matrix of column differences of A (as in the Matlab function *diff*). So for a 2-by-2 ANOVA:

$$C_1 = C_2 = [1, 1]^T \quad D_1 = D_2 = [1, -1]^T \qquad 13.25$$

The term C_m can be thought of as the *common effect* for the mth factor and D_m as the *differential effect*. Then contrasts for each experimental effect can be obtained by the Kronecker products of C_ms and D_ms for each factor $m = 1\ldots M$. For a 2-by-2 ANOVA, for example, the two main effects and interaction are respectively:

$$\begin{aligned} D_1 \otimes C_2 &= [1 \quad 1 \quad -1 \quad -1]^T \\ C_1 \otimes D_2 &= [1 \quad -1 \quad 1 \quad -1]^T \\ D_1 \otimes D_2 &= [1 \quad -1 \quad -1 \quad 1]^T \end{aligned} \qquad 13.26$$

This also illustrates why an interaction can be thought of as a *difference of differences*. The product $C_1 \otimes C_2$ represents the constant term.

[2] In fact, the contrasts presented here are incorrect. But we present them in this format for didactic reasons, because the rows of the resulting contrast matrices, which test for main effects and interactions, are then readily interpretable. The correct contrasts, which normalize row lengths, are given in Appendix 13.2. We also note that the minus sign is unnecessary. It makes no difference to the results but we have included it so that the contrast weights have the canonical form [1, −1, ...] etc. instead of [−1, 1, ...].

For a 3-by-3 ANOVA:

$$C_1 = C_2 = [1, 1, 1]^T \quad D_1 = D_2 = \begin{bmatrix} 1 & -1 & 0 \\ 0 & 1 & -1 \end{bmatrix}^T \qquad 13.27$$

and the two main effects and interaction are respectively:

$$D_1 \otimes C_2 = \begin{bmatrix} 1 & 1 & 1 & -1 & -1 & -1 & 0 & 0 & 0 \\ 0 & 0 & 0 & 1 & 1 & 1 & -1 & -1 & -1 \end{bmatrix}^T \qquad 13.28$$

$$C_1 \otimes D_2 = \begin{bmatrix} 1 & -1 & 0 & 1 & -1 & 0 & 1 & -1 & 0 \\ 0 & 1 & -1 & 0 & 1 & -1 & 0 & 1 & -1 \end{bmatrix}^T \qquad 13.29$$

$$D_1 \otimes D_2 = \begin{bmatrix} 1 & -1 & 0 & -1 & 1 & 0 & 0 & 0 & 0 \\ 0 & 1 & -1 & 0 & -1 & 1 & 0 & 0 & 0 \\ 0 & 0 & 0 & 1 & -1 & 0 & -1 & 1 & 0 \\ 0 & 0 & 0 & 0 & 1 & -1 & 0 & -1 & 1 \end{bmatrix}^T \qquad 13.30$$

The four rows of this interaction contrast correspond to the four 'simple interactions' τ_{11}^{AB}, τ_{12}^{AB}, τ_{21}^{AB}, and τ_{22}^{AB} depicted in Figure 13.7. This reflects the fact that an interaction can arise from the presence of one or more simple interactions.

Two-stage procedure for partitioned errors

Repeated measures M-way ANOVAs with partitioned errors can be implemented using the following summary-statistic approach.

1 Set up first-level design matrices where each cell is modelled separately as indicated in Eqn. **13.23**.
2 Fit first-level models.
3 For the effect you wish to test, use the Kronecker product rules outlined in the previous section to see what F-contrast you'd need to use to test the effect at the first level. For example, to test for an interaction in a 3×3 ANOVA you'd use the F-contrast in Eqn. **13.30** (application of this contrast to subject n's data tells you how significant that effect is in that subject).
4 If the F-contrast in the previous step has R_c rows then, for each subject, create the corresponding R_c contrast images. For N subjects this then gives a total of $N R_c$ contrast images that will be modelled at the second-level.
5 Set up a second-level design matrix, $X_2 = I_{R_c} \otimes 1_N$. The number of conditions is R_c. For example, in a 3×3 ANOVA, $X_2 = I_4 \otimes 1_N$ as shown in Figure 13.9.
6 Fit the second-level model.
7 Test for the effect using the F-contrast $C_2 = I_{R_c}$.

For each effect we wish to test we must get the appropriate contrast images from the first level (step 3) and

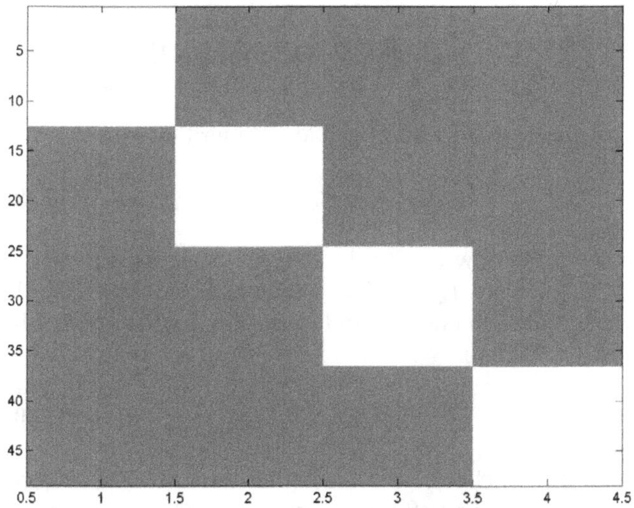

FIGURE 13.9 Second-stage design matrix for interaction in 3 × 3 ANOVA (partitioned errors).

implement a new second-level analysis (steps 4 to 7). Because we are taking *differential* effects to the second level we don't need to include subject effects at the second level.

fMRI BASIS FUNCTIONS

There are situations where one uses an 'ANOVA-type' model, but does not want to test a conventional main effect or interaction. One example is when one factor represents the basis functions used in an event-related fMRI analysis. So if one used three basis functions, such as a canonical haemodynamic response function (HRF) and two partial derivatives (see Chapter 14), to model a single event-type (versus baseline), one might want to test the reliability of this response over subjects. In this case, one would create for each subject the first-level contrasts: $[1, 0, 0]^T$, $[0, 1, 0]^T$ and $[0, 0, 1]^T$, and enter these as the data for a second-level 1-by-3 ANOVA, *without* a constant term.

In this model, we do not want to test for differences between the means of each basis function. For example, it is not meaningful to ask whether the parameter estimate for the canonical HRF differs from that for the temporal derivative. In other words, we do not want to test the null hypothesis for a conventional main effect, as described in Eqn. **13.9**. Rather, we want to test whether the sum of squares of the mean of each basis function explains significant variability relative to the total variability over subjects. This corresponds to the F-contrast:

$$c_2 = \begin{bmatrix} 1 & 0 & 0 \\ 0 & 1 & 0 \\ 0 & 0 & 1 \end{bmatrix} \quad \textbf{13.31}$$

This is quite different from the F-contrast:

$$c_2 = \begin{bmatrix} 1 & -0.5 & -0.5 \\ -0.5 & 1 & -0.5 \\ -0.5 & -0.5 & 1 \end{bmatrix} \quad \textbf{13.32}$$

which is the default 'effects of interest' contrast given for a model that includes a constant term (or subject effects) in statistical parametric mapping (SPM), and would be appropriate instead for testing the main effect of such a 3-level factor.

DISCUSSION

The mainstay of many neuroimaging experiments is the factorial design and data from these experiments can be analysed using an analysis of variance. This chapter has described ANOVAs in terms of model comparison. To test, for example for a main effect of a factor, one compares two models, a 'full model' in which all levels of the factor are modelled separately, versus a 'reduced model', in which they are modelled together. If the full model explains the data significantly better than the reduced model then there is a significant main effect. We have shown how these model comparisons can be implemented using F-tests and general linear models.

This chapter has also revisited the notion of non-sphericity, within the context of within-subject ANOVAs. Informally, if a subject's response to levels i and j of a factorial manipulation is correlated, then a plot of the bivariate responses will appear non-spherical. This can be handled at the inferential stage by making an adjustment to the degrees of freedom. In current implementations of SPM this is generally unnecessary, as global non-sphericty estimates are used which have very high precision. This non-sphericity is then implicitly removed during the formation of maximum-likelihood parameter estimates (see Chapter 10).

We have also described inference in multiway within-subject ANOVAs and made a distinction between models with pooled versus partitioned errors and noted that partitioning is normally the preferred approach. One can implement partitioning using the multistage summary-statistic procedure until, at the last level, there is only one contrast per subject. This is a simple way to implement inference based on partitioned errors using the pooled-errors machinery of SPM.

APPENDIX 13.1 THE KRONECKER PRODUCT

If A is an $m_1 \times m_2$ matrix and B is an $n_1 \times n_2$ matrix, then the Kronecker product of A and B is the $(m_1 n_1) \times (m_2 n_2)$ matrix:

$$A \otimes B = \begin{bmatrix} a_{11}B & \dots & a_{1m_2}B \\ \dots & & \\ a_{m_1 1}B & & a_{m_1 m_2}B \end{bmatrix} \qquad 13.33$$

Circularity

A covariance matrix Σ is circular if:

$$\Sigma_{ii} + \Sigma_{jj} - 2\Sigma_{ij} = 2\lambda \qquad 13.34$$

for all i, j.

Compound symmetry

If all the variances are equal to λ_1 and all the covariances are equal to λ_2 then we have *compound symmetry*.

Non-sphericity

If Σ is a $K \times K$ covariance matrix and the first $K-1$ eigenvalues are identically equal to:

$$\lambda = 0.5(\Sigma_{ii} + \Sigma_{jj} - 2\Sigma_{ij}) \qquad 13.35$$

then Σ is *spherical*. Every other matrix is non-spherical or has *non-sphericity*.

Greenhouse-Geisser correction

For a 1-way ANOVA between subjects with N subjects and K levels the overall F-statistic is approximately distributed as:

$$F[(K-1)\epsilon, (N-1)(K-1)\epsilon] \qquad 13.36$$

where

$$\epsilon = \frac{(\sum_{i=1}^{K-1} \lambda_i)^2}{(K-1)\sum_{i=1}^{K-1} \lambda_i^2} \qquad 13.37$$

and λ_i are the eigenvalues of the normalized matrix Σ_z where

$$\Sigma_z = M^T \Sigma_y M \qquad 13.38$$

and M is a K by $K-1$ matrix with orthogonal columns (e.g. the columns are the first $K-1$ eigenvectors of Σ_y).

APPENDIX 13.2 WITHIN-SUBJECT MODELS

The model in Eqn. 13.11 can also be written as:

$$y_n = 1_K \pi_n + \tau + e_n \qquad 13.39$$

where y_n is now the $K \times 1$ vector of measurements from the nth subject, 1_K is a $K \times 1$ vector of 1s, and τ is a $K \times 1$ vector with kth entry τ_k and e_n is a $K \times 1$ vector with kth entry e_{nk} where:

$$p(e_n) = \mathrm{N}(0, \Sigma_e) \qquad 13.40$$

We have a choice as to whether to treat the subject effects π_n as fixed-effects or random-effects. If we choose random-effects then:

$$p(\pi_n) = \mathrm{N}(\mu, \sigma_\pi^2) \qquad 13.41$$

and overall we have a mixed-effects model as the typical response for subject n, π_n, is viewed as a random variable whereas the typical response to treatment k, τ_k, is not a random variable. The reduced model is:

$$y_n = 1_K \pi_n + e_n \qquad 13.42$$

For the full model we can write:

$$p(y) = \prod_{n=1}^{N} p(y_n) \qquad 13.43$$

$$p(y_n) = \mathrm{N}(m_y, \Sigma_y)$$

and

$$m_y = 1_K \mu + \tau \qquad 13.44$$

$$\Sigma_y = 1_K \sigma_\pi^2 1_K^T + \Sigma_e$$

if the subject effects are random effects, and $\Sigma_y = \Sigma_e$ otherwise. If $\Sigma_e = \sigma_e^2 I_K$ then Σ_y has *compound symmetry*. It is also spherical (see Appendix 13.1). For $K = 4$ for example:

$$\Sigma_y = \begin{bmatrix} \sigma_\pi^2 + \sigma_e^2 & \sigma_\pi^2 & \sigma_\pi^2 & \sigma_\pi^2 \\ \sigma_\pi^2 & \sigma_\pi^2 + \sigma_e^2 & \sigma_\pi^2 & \sigma_\pi^2 \\ \sigma_\pi^2 & \sigma_\pi^2 & \sigma_\pi^2 + \sigma_e^2 & \sigma_\pi^2 \\ \sigma_\pi^2 & \sigma_\pi^2 & \sigma_\pi^2 & \sigma_\pi^2 + \sigma_e^2 \end{bmatrix} \qquad 13.45$$

If we let $\Sigma_y = (\sigma_\pi^2 + \sigma_e^2) R_y$ then:

$$R_y = \begin{bmatrix} 1 & \rho & \rho & \rho \\ \rho & 1 & \rho & \rho \\ \rho & \rho & 1 & \rho \\ \rho & \rho & \rho & 1 \end{bmatrix} \qquad 13.46$$

where

$$\rho = \frac{\sigma_\pi^2}{\sigma_\pi^2 + \sigma_e^2} \qquad 13.47$$

For a more general Σ_e, however, Σ_y will be non-spherical. In this case, we can attempt to correct for the non-sphericity. One approach is to reduce the degrees of freedom by a factor $\frac{1}{K-1} \leq \epsilon \leq 1$, which is an estimate of the degree of non-sphericity of Σ_y (the *Greenhouse-Geisser* correction; see Appendix 13.1). Various improvements of this correction (e.g. *Huhn-Feldt*) have also been suggested (Howell, 1992). Another approach is to parameterize explicitly the error covariance matrix Σ_e using a linear expansion and estimate the parameters using ReML, as described in Chapter 22.

Contrasts for M-way ANOVAs

The contrasts presented in the section 'Generalization to M-way ANOVAs' are actually incorrect. They were presented in a format that allowed the rows of the resulting contrast matrices, which test for main effects and interactions, to be readily interpretable. We now give the correct contrasts, which derive from speciying the initial differential component contrast as:

$$D_m = -\mathsf{orth}(\mathsf{diff}(I_{K_m})^T) \qquad 13.48$$

where $\mathsf{orth}(A)$ is the orthonormal basis of A (as in the Matlab function *orth*). This is identical to the expression in the main text but with the addition of an orth function which is necessary to ensure that the length of the contrast vector is unity.

This results in the following contrasts for the 2-by-2 ANOVA:

$$C_1 = C_2 = [1,1]^T \quad D_1 = D_2 = [0.71, -0.71]^T \qquad 13.49$$

$$\begin{aligned} D_1 \otimes C_2 &= [0.71 \quad 0.71 \quad -0.71 \quad -0.71]^T \\ C_1 \otimes D_2 &= [0.71 \quad -0.71 \quad 0.71 \quad -0.71]^T \\ D_1 \otimes D_2 &= [0.71 \quad -0.71 \quad -0.71 \quad 0.71]^T \end{aligned} \qquad 13.50$$

For the 3-by-3 ANOVA:

$$C_1 = C_2 = [1,1,1]^T D_1 = D_2 = \begin{bmatrix} 0.41 & -0.82 & 0.41 \\ 0.71 & 0.00 & -0.71 \end{bmatrix}^T \qquad 13.51$$

and the two main effects and interaction are respectively:

$$D_1 \otimes C_2 = \begin{bmatrix} 0.41 & 0.41 & 0.41 & -0.82 & -0.82 & -0.82 & 0.41 & 0.41 & 0.41 \\ 0.71 & 0.71 & 0.71 & 0 & 0 & 0 & -0.71 & -0.71 & -0.71 \end{bmatrix}^T \qquad 13.52$$

$$C_1 \otimes D_2 = \begin{bmatrix} 0.41 & -0.82 & 0.41 & 0.41 & -0.82 & 0.41 & 0.41 & -0.82 & 0.41 \\ 0.71 & 0 & -0.71 & 0.71 & 0 & -0.71 & 0.71 & 0 & -0.71 \end{bmatrix}^T \qquad 13.53$$

$$D_1 \otimes D_2 = \begin{bmatrix} 0.17 & -0.33 & 0.17 & -0.33 & 0.67 & -0.33 & 0.17 & -0.33 & 0.17 \\ 0.29 & 0 & -0.29 & -0.58 & 0 & 0.58 & 0.29 & 0 & -0.29 \\ 0.29 & -0.58 & 0.29 & 0 & 0 & 0 & -0.29 & 0.58 & -0.29 \\ 0.5 & 0 & -0.5 & 0 & 0 & 0 & -0.5 & 0 & 0.5 \end{bmatrix}^T \qquad 13.54$$

REFERENCES

Howell DC (1992) *Statistical methods for psychology*. Duxbury Press, Belmont California.

Winer BJ, Brown DR, Michels KM (1991) *Statistical principles in experimental design*. McGraw-Hill.

C H A P T E R

14

Convolution Models for fMRI

R. Henson and K. Friston

INTRODUCTION

This chapter reviews issues specific to the analysis of functional magnetic resonance imaging (fMRI) data. It extends the general linear model (GLM) introduced in Chapter 8 to convolution models, in which the blood oxygenation-level-dependent (BOLD) signal is modelled by neuronal causes that are expressed via a haemodynamic response function (HRF). We begin by considering linear convolution models and introduce the concept of temporal basis functions. We then consider the related issues of temporal filtering and temporal autocorrelation. Finally, we extend the convolution model to include non-linear terms and conclude with some example analyses of fMRI data.

THE HAEMODYNAMIC RESPONSE FUNCTION (HRF)

A typical BOLD response to a single, impulsive stimulation ('event') is shown in Figure 14.1. The response peaks approximately 5 s after stimulation, and is followed by an undershoot that lasts as long as 30 s (at high magnetic fields, an initial undershoot can sometimes be observed) (Malonek and Ginvald, 1996). Early event-related studies therefore used a long time between events (i.e. a long stimulus onset asynchrony (SOA)) to allow the response to return to baseline between stimulations. However, although the responses to successive events will overlap at shorter SOAs, this overlap can be modelled explicitly within the GLM via a convolution model and an HRF, as described below. Short SOAs of a few seconds are desirable because they are comparable to those typically used in behavioural and electrophysiological studies, and because they are generally more efficient from a statistical perspective, as we will see in the next chapter.

The shape of the BOLD impulse response appears similar across early sensory regions, such as V1 (Boynton *et al.*, 1996), A1 (Josephs *et al.*, 1997) and S1 (Zarahn *et al.*, 1997a). However, the precise shape has been shown to vary across the brain, particularly in higher cortical regions (Schacter *et al.*, 1997), presumably due mainly to variations in the vasculature of different regions (Lee *et al.*, 1995). Moreover, the BOLD response appears to vary considerably across people (Aguirre *et al.*, 1998).[1]

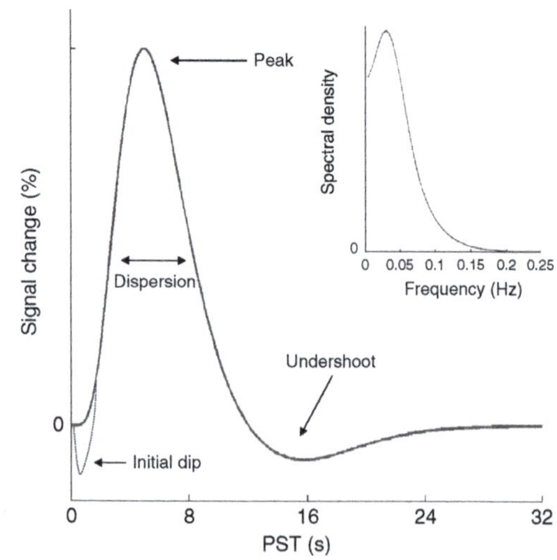

FIGURE 14.1 Typical (canonical) BOLD impulse response (power spectrum inset).

[1] This has prompted some to use subject-specific HRFs derived from a reference region known to respond to a specific task, (e.g. from central sulcus during a simple manual task performed during a pilot scan on each subject; Aguirre *et al.*, 1998). However, while this allows for inter-subject variability, it does not allow for inter-regional variability within subjects (or potential error in estimation of the reference response).

Statistical Parametric Mapping, by Karl Friston et al.
ISBN–13: 978-0-12-372560-8 ISBN–10: 0-12-372560-7

These types of variability can be accommodated by expressing the HRF in terms of a set of temporal basis functions.

A linear convolution model assumes that successive responses summate (superpose). However, there is good evidence for non-linearity in the amplitude of the BOLD response, as a function of the stimulus duration or stimulus magnitude (e.g. Vasquez and Noll, 1998), and as a function of SOA (Pollmann *et al.*, 1998; Friston *et al.*, 1998a; Miezin *et al.*, 2000). These non-linearities also appear to vary across different brain regions (Birn et al. 2001; Huettel and McCarthy, 2001). The non-linearity found as a function of SOA is typically a 'saturation' whereby the response to a run of events is smaller than would be predicted by the summation of responses to each event alone. This saturation is believed to arise in the mapping from blood flow to BOLD signal (Friston *et al.*, 2000a), though may also have a neuronal locus, particularly for very short SOAs or long stimulus durations (for biophysical models that incorporate such non-linearities, see Chapter 27). Saturation has been found for SOAs below approximately 8 s, and the degree of saturation increases as the SOA decreases. For typical SOAs of 2–4 s, however, its magnitude can be small (typically less than 20 per cent) (Miezin *et al.*, 2000). Later we will see how the linear convolution model is extended to handle such non-linearities via a Volterra expansion.

Linear time-invariant (convolution) models

It is useful to treat a session of fMRI scans as a time-series. This is because the data tend to be correlated across successive scans, given that the typical measurement interval, T_R, of 1–3 s, is less than the duration of the BOLD response. The GLM can be expressed as a function of time (Friston *et al.*, 1994):

$$y(t) = X(t)\beta + \varepsilon(t) \qquad 14.1$$
$$\varepsilon(t) \sim N(0, \sigma^2\Sigma)$$

where the data, $y(t)$, comprise the fMRI time-series, the explanatory variables, $X(t)$ are now functions of time, β are (time-invariant) parameters, and Σ is the noise autocorrelation. Though $y(t)$ and $X(t)$ are really discrete (sampled) time-series (normally represented by the vector y and design matrix X respectively), we will initially treat the data and model in terms of continuous time. For simplicity, we will consider the case of a single cause or regressor.

The explanatory variables $X(t)$ represents the predicted BOLD time course arising from neuronal activity, $u(t)$, up to some scaling factor. This neuronal activity (e.g. the mean synaptic activity of an ensemble of neurons – see Chapter 32) is assumed to be caused by a sequence of experimental manipulations and is usually referred to as the stimulus function. If we assume that the BOLD signal is the output of a linear time-invariant (LTI) system (Boynton *et al.*, 1996), i.e. that the BOLD response to a brief input has a finite duration and is independent of time, and that the responses to successive inputs superpose in a linear fashion, then we can express $X(t)$ as the convolution of the stimulus function with an impulse response, or HRF, $h(t)$:

$$X(t) = u(t) \otimes h(\tau) = \int_0^T u(t-\tau)h(\tau)d\tau \qquad 14.2$$

where τ indexes the peristimulus time (PST), over which the BOLD impulse response is expressed. The HRF is equivalent to the first-order Volterra kernel described below. The stimulus function $u(t)$ is usually a stick-function or boxcar function encoding the occurrence of an event or epoch. The result of convolving a random sequence of neuronal events with a 'canonical' HRF (see Figure 14.1) is shown in Figure 14.2(a). The smoothness of the resulting response is why the HRF is often viewed as a low-pass filter. The result of convolving more sustained periods of neuronal activity (called epochs in SPM) with the canonical HRF is shown in Figure 14.2(b). Note that the dominant effect of increasing the duration of neuronal activity, up to a few seconds, is to increase the peak amplitude of the BOLD response. In other words, the BOLD response integrates neuronal activity over a few seconds. The corollary is that a difference in the amplitude of the BOLD response (as tested conventionally) does not necessarily imply a difference in the mean level of neuronal activity: the difference could reflect different durations of neuronal activity at same mean level. One way to distinguish between these scenarios is to test for differences in the peak latency of the BOLD impulse response (Henson and Rugg, 2001).

In practice, the convolution must be performed in discrete time. Given that significant information may exist in the predicted BOLD time course beyond that captured by typical T_Rs of 1–3 s, SPM performs the convolution at a higher temporal resolution with N time points per scan (i.e. with resolution, $\Delta t = T_R/N$ seconds). This means, for example, that stimulus onsets do not need to be synchronized with scans (they can be specified in fractions of scans).[2] To create the explanatory variables, the predicted BOLD time course is then down-sampled every T_R with reference to a specified time point T_0 (Plate 8, see colour plate section).

[2] In SPM, an 'event' is defined as having a duration of 0, which in practice corresponds to a single non-zero value for one time bin of duration Δt, where the value of the stimulus function is $1/\Delta t$. For epochs, the stimulus function is scaled so that it sums to one over a second.

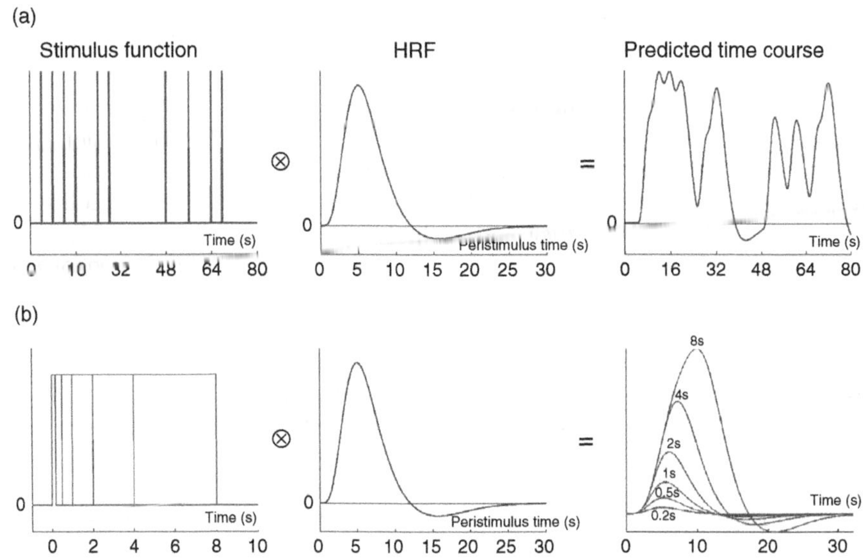

FIGURE 14.2 Linear convolution with a canonical HRF illustrated for (a) randomly presented events, and (b) epochs of neuronal activity with durations increasing from 200 ms to 16 s.

TEMPORAL BASIS FUNCTIONS

Given that the precise shape of the HRF may vary over brain regions and over individuals, variability in its shape needs to be accommodated. The simplest way to achieve this within the GLM is via an expansion in terms of K temporal basis functions, $f_k(\tau)$:

$$h(\tau) = \sum_{k=1}^{K} \beta_k f_k(\tau) \qquad \mathbf{14.3}$$

If the stimulation and resulting neuronal activity were a sequence of J impulses at times o_j, we can construct a stimulus stick-function:

$$u(t) = \sum_{j=1}^{J} \alpha_j \delta(t - o_j) \qquad \mathbf{14.4}$$

where $\delta(t)$ is the Dirac delta function. Note that variations in the stimulus – for example, its magnitude α_j on each trial – can be accommodated by modulating the delta-functions, prior to convolution with the HRF. These are called 'parametric modulations' in SPM. We will see an example of this parametric modulation in the last section. It is quite common to use a series of modulated stick-functions to model a single event type by using a polynomial or other expansions of α_j. For simplicity, we will assume that $\alpha_j = 1$ (i.e. a zeroth-order term).

Having specified the form of the stimulus and haemodynamic response functions in this way, the GLM equation in Eqn. **14.1** can be written:

$$y(t) = \sum_{j=1}^{J} \sum_{k=1}^{K} \beta_k f_k(t - o_j) + \varepsilon(t) \qquad \mathbf{14.5}$$

where β_k are the parameters to be estimated. Several temporal basis sets are offered in SPM, though not all are true 'basis' sets in the (mathematical) sense that they span the space of all possible impulse response shapes (over the finite duration of that response); the term 'basis' is used to reflect the user's assumption that the set of functions chosen capture BOLD impulse response shapes that occur in reality.

FIR and Fourier sets

The most flexible basis sets are the finite impulse response (FIR) and Fourier basis sets, which make the least assumptions about the shape of the response. The FIR set consists of contiguous boxcar functions of PST, each lasting T/K_{FIR} seconds (see Plate 9(a)), where T is duration of the HRF. The Fourier set (see Plate 9(b)) consists of a constant and K_F sine and K_F cosine functions of harmonic periods $T, T/2, \ldots, T/K_F$ seconds (i.e. $K = 2\,K_F + 1$ basis functions). Linear combinations of the (orthonormal) FIR or Fourier basis functions can capture any shape of response up to a specified timescale (T/K_{FIR} in the case of the FIR) or frequency (K_F/T in the case of the Fourier set).

Relationship between FIR and 'selective averaging'

In practice, there is little to choose between the FIR and Fourier sets. The Fourier set can be better suited when the sampling of peristimulus time (as determined by the relationship between the SOA and T_R) is non-uniform, whereas the parameter estimates for the FIR functions have a more direct interpretation in terms of the 'averaged' peristimulus time histogram (PSTH). Indeed, in the special case when $T/K_{FIR} = T_R = \Delta t$, the FIR functions can be considered as approximate delta-functions:

$$h(\tau) = \sum_{k=1}^{K} \beta_k \delta(\tau - k) \quad \textbf{14.6}$$

where τ ranges over post-stimulus scans. Then the model becomes (with some abuse of the delta-function):

$$y(t) = \sum_{j=1}^{J} \sum_{k=1}^{K} \beta_k \delta(t - k - o_j) + \varepsilon(t) \Rightarrow \quad \textbf{14.7}$$
$$y = X\beta + \varepsilon$$
$$X_{tk} = \sum_{j=1}^{J} \delta(t - k - o_j)$$

For the special case of non-overlapping responses and independent and identically distributed (IID) error (i.e. $\Sigma = 1$), the maximum likelihood estimates of the FIR parameters are equivalent to the simple trial-averaged data (much like with ERPs):

$$\hat{\beta}_k = \frac{1}{J} \sum_{j=1}^{J} y(o_j + k - 1) \Rightarrow \quad \textbf{14.8}$$
$$\beta = \frac{1}{J} X^T y$$

Estimating the haemodynamic response like this has been called 'selective averaging' (Dale and Buckner, 1997). However, in practice, this estimator is biased and suboptimal because it requires the information matrix (called the 'overlap correction matrix' (Dale, 1999)) to be a leading diagonal matrix, i.e. $X^T X = JI$, in which case the ordinary least-squares estimates become the selective average:

$$\hat{\beta} = (X^T X)^{-1} X^T y = \frac{1}{J} X^T y \quad \textbf{14.9}$$

With careful counterbalancing of different stimuli and the use of variable SOAs (e.g. via null events; see Chapter 15), this requirement can be met approximately. However, selective averaging as a procedure is redundant and represents a special case of the general deconvolution that obtains when simply inverting a linear convolution model. Selective averaging rests on undesirable and unnecessary constraints on the experimental design and is seldom used anymore.

Gamma functions

More parsimonious basis sets can be chosen that are informed about the shape of the HRF. For example, since the HRF is assumed to be bounded at zero for $\tau \leq 0$ and $\tau \geq T$, the Fourier basis functions can also be windowed (e.g. by a Hanning window) within this range. An alternative is based on the gamma function:

$$f(t) = \left(\frac{t - o}{d}\right)^{p-1} \left(\frac{\exp(-(t - o)/d)}{d(p-1)!}\right)^{p-1} \quad \textbf{14.10}$$

where o is the onset delay, d is the time-scaling, and p is an integer phase-delay (the peak delay is given by pd, and the dispersion by pd^2). This function is bounded and positively skewed (unlike a Gaussian for example). A single gamma function has been shown to provide a reasonably good fit to the BOLD impulse response (Boynton *et al.*, 1996), though it lacks an undershoot (Fransson *et al.*, 1999; Glover, 1999). A set of gamma functions of increasing dispersions can be obtained by increasing p (see Plate 9(c)). In SPM, these functions (as with all basis functions) are orthogonalized with respect to one another. This set is more parsimonious, in that fewer functions are required to capture the typical range of BOLD impulse responses than required by Fourier or FIR sets. This precludes overfitting and reduces the model's degrees of freedom, to provide more powerful tests.

The 'informed' basis set and the canonical HRF

Another more parsimonious basis set, suggested by Friston *et al.* (1998b), is based on a 'canonical HRF' and its partial derivatives (see Plate 9(d)). The canonical HRF is a 'typical' BOLD impulse response characterized by two gamma functions, one modelling the peak and one modelling the undershoot. The canonical HRF is parameterized by a peak delay of 6 s and an undershoot delay of 16 s, with a peak-undershoot amplitude ratio of six; these values were derived from a principal component analysis of the data reported in Friston *et al.* (1998a). To allow for variations about the canonical form, the partial derivatives of the canonical HRF with respect to its delay and dispersion can be added as further basis functions. For example, if the real BOLD impulse response is shifted by a small amount in time τ, then by the first-order Taylor expansion:

$$h(t + \tau) \approx h(t) + \tau h'(t) \quad \textbf{14.11}$$

This is the same as adding a small amount of the temporal derivative of $h'(t)$. Thus, if $h(t)$ and $h'(t)$ are used as two basis functions in the GLM to estimate the parameters β_1 and β_2 respectively, then small changes in the latency of the response can be captured by the parameter estimate for the temporal derivative (more precisely,

$\tau \approx \hat{\beta}_2/\hat{\beta}_1$; see Henson *et al.*, 2002a; Liao *et al.*, 2002, for a fuller treatment). In SPM, the temporal derivative is created from the finite difference between a canonical HRF and a canonical HRF shifted by one second. Thus, using the temporal derivative as a further response can capture differences in the latency of the BOLD response up to plus or minus a second (beyond this range, the first-order Taylor approximation breaks down). A similar logic applies to the use of dispersion derivative to capture [small] differences in the duration of the peak response. Together, these three functions comprise SPM's 'informed' basis set, in that they are informed by the range of typical BOLD impulse response shapes observed. Subsequent work, using more biophysically informed models of the haemodynamic response, revealed that the informed set is almost identical to the principal components of variation, with respect to the parameters of the Balloon model described in Chapter 27.

The temporal derivatives of an assumed HRF can also be used to allow for differences in the acquisition times of different slices with echo-planar imaging (EPI) sequences, in order to address the so-called slice-timing problem (see Chapter 15). The ability of the temporal derivative to capture these latency differences is appropriate for a T_R of up to 2 s (after synchronizing the model with the slice acquired half way though each scan), assuming that the true BOLD impulse responses match the canonical HRF in all brain regions (i.e. all slices; Henson *et al.*, 1999).

Other methods

Other methods for estimating the shape of the BOLD impulse response use non-linear (iterative) fitting techniques, beyond the GLM. These approaches are more powerful, but computationally more expensive. Various parameterizations have been used, such as a Gaussian function parameterized by amplitude, onset latency and dispersion (Rajapakse *et al.*, 1998), a gamma function parameterized by amplitude, onset latency and peak latency (Miezin *et al.*, 2000), or even SPM's canonical HRF, with the amplitude, onset latency and peak latency parameters free to vary (Henson and Rugg, 2001). A problem with unconstrained iterative fitting techniques is that the parameter estimates may not be optimal, because of local minima in the search space. Parameters that have correlated effects compound this problem (often requiring a re-parameterization into orthogonal components). One solution is to put priors on the parameters in a Bayesian estimation scheme (Chapter 34) in order to 'regularize' the solutions (see Gossl *et al.*, 2001, and Woolrich *et al.*, 2004, for other examples). Indeed, more recent Bayesian methods not only provide posterior densities for HRF parameters, but also provide metrics of the 'goodness' of different HRF models, using Bayesian model evidence (Friston, 2002; Penny *et al.*, in press).

Which temporal basis set?

Inferences using multiple basis functions are made with *F*-contrasts (see Chapter 9). An example *F*-contrast that tests for any difference in the event-related response to two trial-types modelled by SPM's informed basis set is shown in Plate13(c). If the real response matches an assumed HRF, models using just that HRF are statistically more powerful (Ollinger *et al.*, 2001). In such cases, *t*-tests on the parameter estimate for the HRF can be interpreted directly in terms of the 'amplitude' of the response. However, when the real response differs appreciably from the assumed form, tests on the HRF parameter estimates are biased (and unmodelled structure will exist in the residuals). In such cases, the parameter estimate for the canonical HRF, for example, can no longer necessarily be interpreted in terms of amplitude. The addition of partial derivatives of the HRF can ameliorate this problem: the inclusion of a temporal derivative, for example, can reduce the residual error by capturing systematic delays relative to the assumed HRF. Nonetheless, for responses that differ by more than a second in their latency (i.e. when the first-order Taylor approximation fails), different canonical HRF parameters will be estimated even when the responses have identical peak amplitudes (Henson *et al.*, 2002a).[3]

An important empirical question then arises: how much variability exists in the BOLD impulse response? Henson *et al.* (2001) addressed this question for a dataset involving rapid motor responses to the brief presentations of faces across twelve subjects. By modelling the event-related response with a canonical HRF, its partial derivatives *and* an FIR basis set, the authors assessed the contribution of the different basis functions using a series of *F*-contrasts (that collapsed over subjects within a single first-level design matrix). Significant additional variability was captured by both the temporal derivative and dispersion derivative, confirming that different regions exhibit variations around the canonical form (see

[3] Note that the inclusion of the partial derivatives of SPM's canonical HRF does not necessarily affect the parameter estimate for the HRF itself, since the basis functions are orthogonalized (unless correlations between the regressors arise due to under-sampling by the T_R, or by temporal correlations between the onsets of events of different types). Thus, their inclusion does not necessarily affect second-level *t*-tests on the canonical HRF parameter estimate alone.

Plate 10(a)). Little additional variability was captured by the FIR basis set. This suggests that the canonical HRF and its two partial derivatives are sufficient to capture the majority of experimental variability (at least in regions that were activated in this task). The same conclusion was reached using a second-level model and the twelve parameter estimates of a (pure) FIR model, by testing F-contrasts that specify the 'null-space' of either the canonical HRF or the canonical HRF plus its partial derivatives. Significant variability was not captured by the canonical HRF alone but there was little significant variability that could not be captured once the two partial derivatives were added (see Plate 10(b)). The latter data and analyses can be downloaded from the SPM website (http://www.fil.ion.ucl.ac.uk/spm/data).

This sufficiency of the informed basis set may be specific to this dataset and reflect the fact that neuronal activity was reasonably well approximated by a delta function. It is unlikely to hold for more complex experimental trials, such as working memory trials where information must be maintained for several seconds (e.g. Ollinger *et al.*, 2001). Nonetheless, such trials may be better accommodated by more complex neuronal models. This usually entails using multiple stimulus functions for different components of each trial (e.g. onset, delay-period, offset, etc.) while still using an informed model for the HRF. This allows more direct inferences about stimulus, response and delay components of a trial for example (Zarahn, 2000). More generally, the question of which basis set and how may components to use becomes a problem of model selection that can be addressed simply using F-contrasts or Bayesian techniques (Penny *et al.*, in press).

One issue arises when one wishes to use multiple basis functions to make inferences in second-level analyses (e.g. in 'random effects' analyses over subjects; see Chapter 12). Subject-specific contrast images created after fitting an FIR model in a first-level analysis could, for example, enter into a second-level model as a peristimulus time factor (differential F-contrasts which would correspond to a condition-by-time interaction in a conventional repeated-measures analysis of variance (ANOVA); Chapter 13). However, the parameter estimates are unlikely to be independent or identically distributed over subjects, violating the 'sphericity' assumption of univariate, parametric statistical tests (Chapter 10). This is one reason why researchers have tended to stick with t-tests on (contrasts of) the parameter estimate for a single canonical HRF at the second-level. This is at the expense of potentially missing response differences with a non-canonical form. One solution is to use multivariate tests (Henson *et al.*, 2000), though these are generally less sensitive (by virtue of making minimal assumptions about the data covariance) (Kiebel and Friston, 2004). Alternatively, restricted maximum likelihood (ReML) can be used to estimate the covariance components subtending any non-sphericity (Friston *et al.*, 2002; Chapter 22). In this case, one generally needs to model both unequal variances (given that different basis functions can have different scaling) and unequal covariances (given that parameter estimates for different basis functions are likely to be correlated across subjects). This allows one to make statistical inferences over multiple basis functions at the second-level, provided one is prepared to assume that the basic correlational structure of the error is invariant across 'activated' voxels (the 'pooling device'; see Chapter 10).

TEMPORAL FILTERING AND AUTOCORRELATION

We can also view our time-series in terms of frequency components via the Fourier transform. A schematic of the power spectrum, typical of a subject at rest in the scanner, is shown in Plate 11(a). This 'noise' spectrum is dominated by low frequencies and has been characterized by a $1/f$ form when expressed in amplitude (Zarahn *et al.*, 1997b). The noise arises from physical sources, sometimes referred to as 'scanner drift' (e.g. slowly varying changes in ambient temperature); from physiological sources (e.g. biorhythms, such as ~ 1 Hz respiratory or ~ 0.25 Hz cardiac cycles, which are aliased by the slower sampling rate); and from residual movement effects and their interaction with the static magnetic field (Turner *et al.*, 1998). When the subject is performing a task, signal components are added to this noise. For example, Plate 11(b) shows the approximate signal spectrum induced by a square-wave stimulation, with a duty cycle of 64 s. When averaging over all frequencies, this signal might be difficult to detect against background noise. However, by filtering the data with an appropriate highpass filter (see Plate 11(c)), we can remove most of the noise. Ideally, the remaining noise spectrum would be flat (i.e. 'white' noise, with equal power at all frequencies).

Highpass filtering

The choice of the highpass cut-off would ideally maximize the signal-to-noise ratio. However, we cannot distinguish signal from noise on the basis of the power spectrum alone. Usually, a cut-off period of approximately 128 s is used, based on observations that the noise becomes appreciable at frequencies below approximately 0.008 Hz (though this may vary considerably across scanners and subjects). In other words, some loss of signal may be necessary to minimize noise. Experimental designs therefore try to avoid significant power at low frequencies (i.e. conditions to be contrasted should be presented too far apart in time; see Chapter 15)

In the time domain, a highpass filter can be implemented by a discrete cosine transform (DCT) with harmonic periods up to the cut-off. These basis functions can be made explicit as confounds in the design matrix X_0 or they can be viewed as part of a filter matrix, S (as in current implementations of SPM).[4] This matrix is applied to both data and model:

$$y = X\beta + X_0\beta_0 + \varepsilon \qquad 14.12$$

$$\Leftrightarrow$$

$$Sy = SX\beta + S\varepsilon$$

$$S = I - X_0X_0^+$$

The effect of applying a highpass filter to real data (taken from a 42 s-epoch experiment; data available from the SPM website) is illustrated in Plate 11(d). Plate 11(e) shows the fitted responses after the filter S is applied to two boxcar models, one with and one without convolution with the HRF. The importance of convolving the neuronal model with an HRF is evident in the residuals (see Plate 11(f)); had the explanatory variables been directly equated with the stimulus function (or neuronal activity), significant temporal structure would remain in the residuals (e.g. as negative deviations at the start of each block, i.e. at higher frequency harmonics of the boxcar function).

Temporal autocorrelations

There are various reasons why the noise component may not be white even after highpass filtering. These include unmodelled neuronal sources that have their own haemodynamic correlates. Because these components live in the same frequency range as the effects of interest, they cannot be removed by the highpass filter. These noise sources induce temporal correlation between the residual errors. Such autocorrelation is a special case of non-sphericity, which is treated more generally in Chapter 10. Here, we review briefly the various (historical) solutions to the specific problem of temporal autocorrelation in fMRI time-series (see Friston *et al.*, 2000b, for a fuller treatment).

[4] Though the matrix form expedites mathematical analysis, in practice highpass filtering is implemented by the computationally efficient subtraction $Sy = y - X_0X_0^+y$, where X_0 is the matrix containing the DCT.

Pre-colouring

One solution proposed by Worsley and Friston (1995) is to apply temporal smoothing. This is equivalent to adding a lowpass filter component to S (such that S, together with the highpass filter, becomes a 'bandpass' filter). If the time-constants of the smoothing kernel are sufficiently large, the temporal autocorrelation induced by the smoothing can be assumed to swamp any intrinsic autocorrelation, Σ, such that:

$$V = S\Sigma S^T \approx SS^T \qquad 14.13$$

The effective degrees of freedom can then be calculated using the classical Satterthwaite correction (see Appendix 8.2):

$$v = \frac{tr(RV)^2}{tr(RVRV)} \qquad 14.14$$

$$R = I - SX(SX)^+$$

solely via knowledge of the filter matrix. Lowpass filters derived from a Gaussian smoothing kernel with full-width at half maximum (FWHM) of 4–6 s, or derived from the canonical HRF (see Figure 14.1, inset), have been suggested (Friston *et al.*, 2000b).

Pre-whitening

An alternative solution is to estimate the intrinsic autocorrelation directly, which can be used to create a filter to 'pre-whiten' the data before fitting the GLM. In other words, the smoothing matrix is set to $S = K^{-1}$, where $KK^T = \Sigma$ is the estimated autocorrelation matrix. If the estimation is exact, then:

$$V = S\Sigma S^T = I \qquad 14.15$$

All methods for estimating the autocorrelation rest on a model of its components. These include autoregressive (AR) models (Bullmore *et al.*, 1996) and 1/f models (Zarahn *et al.*, 1997b). An AR(p) is a pth-order autoregressive model, having the time domain form:

$$z_t = a_1z_{t-1} + a_2z_{t-2} + \ldots + a_pz_{t-p} + w_t \Rightarrow$$

$$z = Az + w$$

$$w_t \sim N(0, \lambda_w) \qquad 14.16$$

$$A = \begin{bmatrix} 0 & 0 & 0 & \cdots \\ a_1 & 0 & 0 & \\ a_2 & a_1 & 0 & \\ \vdots & & & \ddots \end{bmatrix}$$

where w_t is an IID innovation or Gaussian process and A is a lower-triangular matrix containing the coefficients in its lower leading diagonals. The regression coefficients a_i can be estimated by ordinary least-squares. Several authors (e.g. Bullmore *et al.*, 1996; Kruggel and von Cramon, 1999) use an AR(1) model, in which the autoregression parameters are estimated from the residuals after fitting the GLM. These estimates are then used to create the filter $S = (I - A)^{-1}$ that is applied to the data before re-fitting the GLM (a procedure that can be iterated until the residuals are white).

The 1/f model is a linear model with the frequency ω domain form:

$$s(\omega) = b_1/\omega + b_2$$
$$g(\omega) = |s(\omega)|^2 \qquad \textbf{14.17}$$

where $g(\omega)$ is the power spectrum, whose parameters, b_1 and b_2, can be estimated from the Fourier-transformed data. The advantage of these pre-whitening methods is that they produce the most efficient parameter estimates, under Gaussian assumptions (corresponding to Gauss-Markov or minimum variance estimators). Temporal smoothing is generally less efficient because it removes high-frequency components, which may contain signal. The disadvantage of the temporal autocorrelation models is that they can produce biased parameter estimates if the autocorrelation is not estimated accurately (i.e. they do not necessarily produce 'minimum bias estimators').

Friston *et al.* (2000b) argued that the AR(1) and 1/f models are not sufficient to estimate the typical temporal autocorrelation in fMRI data. This is illustrated in Plate 12(a), which shows the power spectra and 'autocorrelation functions'[5] for the residuals of an event-related dataset (used below). It can be seen that the AR(1) model underpredicts the intermediate-range correlations, whereas the 1/f model overpredicts the long-range correlations. Such a mismatch between the assumed and intrinsic autocorrelation will bias the statistics produced by pre-whitening the data.[6] This mismatch can be ameliorated by combining bandpass filtering (see Plate 12(b)) and modelling the autocorrelation, in which case both models provide a reasonable fit (see Plate12(c)). Indeed, highpass filtering alone (with an appropriate cut-off) is normally sufficient to allow either model to fit the remaining autocorrelation (Friston *et al.*, 2000b).

[5] An autocorrelation function plots the correlation, $\rho(t)$, as a function of 'lag', t=*0...n-1*, and is the Fourier transform of the power spectrum, $g(\omega)$.

[6] More complex models of the temporal autocorrelation have since been shown to minimize bias, such as Tukey tapers (Woolrich *et al.*, 2001) and autoregessive moving average (ARMA) models, a special case of the latter being an AR(1)+white noise model (Burock and Dale, 2000).

Estimating non-sphericity hyperparameters

The estimation of the autocovariance parameters or *hyperparameters* described so far is based on the residuals of the time-series and represents rather *ad hoc* procedures. They are *ad hoc* and biased because they do not allow for uncertainty about the fitted components that are removed from the data to produce the residuals. In other words, they fail to account for the loss of degrees of freedom due to parameter estimation *per se*. Current implementations of SPM avoid this shortcoming by partitioning the data covariance (rather than the residuals) using restricted maximum likelihood. This removes the bias resulting from correlations among the residuals induced by removing modelled effects (Friston *et al.*, 2002; though there are ways of reducing this bias, Worsley *et al.*, 2002).

Restricted maximum likelihood

Restricted maximum likelihood (ReML), allows simultaneous estimation of model parameters and hyperparameters, with proper partitioning of the effective degrees of freedom (see Chapter 22 for more details). ReML can be used with any temporal autocorrelation model. Friston *et al.* (2002) use an 'AR(1)+white noise' model (Purdon and Weisskoff, 1998) with an autoregressive error term, z_t and a white noise term e_t:

$$y_t = X_t\beta + z_t + e_t$$
$$z_t = a_1 z_{t-1} + w_t \qquad \textbf{14.18}$$
$$e_t \sim N(0, \lambda_e)$$
$$w_t \sim N(0, \lambda_w)$$

The autocorrelation coefficient $a_1 = \exp(-1)$ was fixed, leaving two unknown hyperparameters; λ_e and λ_w. The white-noise component contributes to the zero-lag autocorrelation, which allows the AR(1) model to capture better the shape of the autocorrelation at longer lags. Note that this approach still requires a highpass filter to provide accurate fits (see Plate 12(d)), though a subtle difference from the residual-based approaches is that the highpass filter is also treated as part of the complete model to be estimated, rather than a pre-whitening filter.

Pooled hyperparameter estimates

Iterative schemes like ReML are computationally expensive when performed at every voxel. Furthermore, the hyperparameter estimates from a single voxel can be quite imprecise. An expedient solution to both these issues is to assume that the relative values of the hyperparameters λ are stationary over voxels. This allows the data to be pooled over voxels in order to estimate the hyperparameters and implicitly $\Sigma(\lambda)$ for all voxels considered, in a single iterative procedure (see Chapter 22 for details). The ensuing autocorrelation matrix $\Sigma(\lambda)$ is extremely precise because thousands of voxel time-series have been used to estimate it. This means it can now be used to estimate the parameters in the usual way, assuming known non-sphericity. This ReML approach to modelling serial correlations or temporal non-sphericity retains the efficiency of pre-whitening approaches, properly accounts for the loss of degrees of freedom when estimating the parameters, and allows for spatial variability in the error variance. This obviates the need for temporal smoothing, a consequence particularly important for event-related designs, in which appreciable signal can exist at high frequencies.

It should be noted that if the temporal autocorrelation varies over voxels (Zarahn *et al.*, 1997) pooling may not be appropriate. For example, serial correlations are thought be higher in grey than white matter (Woolrich *et al.*, 2001). This can be accommodated by estimating voxel-specific hyperparameters with some spatial regularization (Worsley *et al.*, 2002). However, this means that different voxels can have different effective degrees of freedom, which complicates the application of random field theory (Chapter 17). The solution we prefer is to pool over a homogeneous subset of voxels that are likely to show the same serial correlations (e.g. all those that respond to the paradigm).

NON-LINEAR CONVOLUTION MODELS

The convolution model assumed thus far has been based on a linear approximation, for which there is counter-evidence, e.g. for events close together in time (see above). To allow non-linearities, a generalized convolution model can be used. This is based on the Volterra expansion (Friston *et al.*, 1998a; Josephs and Henson, 1999), which can be thought of as a generalization of the Taylor series approximation to dynamic systems and has the form:

$$y(t) = h_0 + \int_{-\infty}^{\infty} h_1(\tau_1) \cdot u(t-\tau_1) \cdot d\tau_1 + \int_{-\infty}^{\infty}\int_{-\infty}^{\infty} h_2(\tau_1, \tau_2) \cdot u(t-\tau_1) \cdot u(t-\tau_2) \cdot d\tau_1 d\tau_2 + \ldots \qquad 14.19$$

(with only terms up to second order shown here), where h_n is the n-th order Volterra kernel. This expansion can model any analytic time-invariant system, and is often used where the state equations (e.g. biophysical model) determining that system are unknown. In the present context, we assume we have a 'causal' system with finite memory (i.e. the integrals run from 0 to T) and that a second-order approximation is sufficient.

Basis functions and generalized HRFs

Again, temporal basis functions can be used to model the Volterra kernels:

$$\begin{aligned} h_0 &= f_0 \\ h(\tau_1) &= \sum_{k=1}^{K} \beta_k^{(1)} f_k(\tau_1) \\ h(\tau_1, \tau_2) &= \sum_{k=1}^{K}\sum_{l=1}^{K} \beta_{kl}^{(2)} f_k(\tau_1) f_l(\tau_2) \end{aligned} \qquad 14.20$$

This allows us to express (linearize) the Volterra expansion within the GLM, with each basis function coefficient associated with a column of the design matrix. The regressors for the first-order coefficients $\beta_k^{(1)}$ are simply the input convolved with each basis function in the usual way. The second-order coefficients $\beta_{kl}^{(2)}$ have regressors that are the [Hadamard] products of the first-order regressors. Friston *et al.* (1998a) used three gamma functions, leading to three columns for the first-order kernel plus a further nine columns for the second-order kernel (to model quadratic non-linearities). Using fMRI data from an experiment in which words were presented at different rates, F-tests on the non-linear partition showed reliable effects in bilateral superior temporal regions. The estimated first and second-order kernels are shown in Figure 14.3(a). The first-order kernel (a linear combination of the three gamma functions) closely resembles the canonical HRF. The second-order kernel shows evidence of under-additivity (e.g. saturation) at short SOAs below 5 s (the dark region in the lower left), consistent with other studies (see above). Interestingly, evidence of super-additivity was also found for SOAs of approximately 8 s (the light regions between 5 and 10 s; the kernel is necessarily symmetric).

Using these first- and second-order kernels, the response to any temporal pattern of word presentations can be simulated. Using only the first-order kernel (i.e. a linear convolution model), the response to two words presented one second apart is simply the sum of the BOLD responses to each word alone (Figure 14.3(b), top panel). However, adding the second-order kernel shows the expected effect of saturation, whereby the response to the pair of events is less than the sum of their responses when presented alone (Figure 14.3(b), bottom panel). In

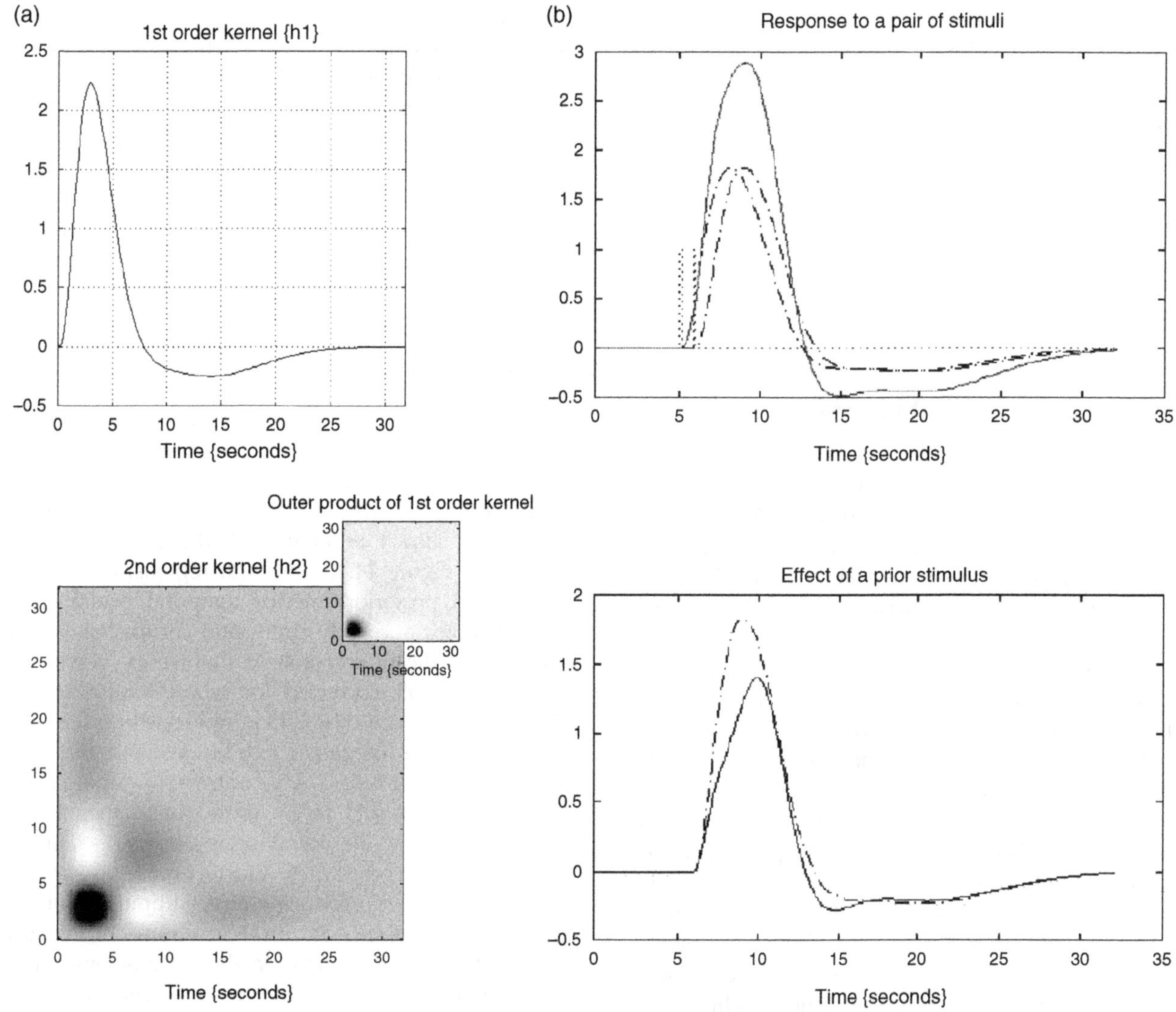

FIGURE 14.3 Volterra kernels. (a) shows the first-order (upper) and second-order (lower) Volterra kernels from superior temporal cortex in an experiment in which auditory words were presented at different rates (see Friston *et al.*, 1998a for more details). The second-order kernel shows non-linearities, resulting in both underadditivity (dark regions) and superadditivity (light regions). (b) shows the response predicted for two stimuli 1 s apart when using a linear convolution model – i.e. the first-order kernel only (upper) – and when adding the second-order kernel from (a), resulting in a predicted saturation of the response relative to the linear case.

principle, this saturation could be caused by neuronal factors, blood-flow factors or even blood-oxygenation factors. However, the fact that a PET experiment, using the same paradigm, showed that blood-flow increased linearly with word presentation rate suggests that the dominant source of saturation in these fMRI data arose in the mapping between perfusion and BOLD signal. Indeed, using a detailed biophysical, 'balloon' model of the BOLD response, Friston *et al.* (2000a) proposed that the reason the second stimulus is compromised, in terms of elaborating a BOLD signal, is because of the venous pooling, and consequent dilution of deoxyhaemoglobin, incurred by the first stimulus. This means that less deoxyhaemoglobin can be cleared for a given increase in flow. The second type of non-linearity – the superadditivity for events presented approximately 8 s apart – was attributed to the fact that, during the flow undershoot following a first stimulus, deoxyhaemoglobin concentration is greater than normal, thereby facilitating clearance of deoxyhaemoglobin following a second stimulus.

Although these non-linearities may be specific to this particular paradigm and auditory cortex, they do suggest caution in using event-related designs with very short SOAs. The saturation in particular provides important (and intuitive) limits on the statistical efficiency of event-related designs as a function of SOA (see next chapter). Even if the significant non-linearities are small enough that SOAs below 5 s (but above 1 s) are still more efficient from the statistical perspective, one could consider adding a second-order Volterra kernel (linearized via a number of basis functions) in order to capture systematic, event-related variability in the residuals.

A WORKED EXAMPLE

In this section, the concepts of this chapter are illustrated in a single-session event-related fMRI dataset from one of the 12 subjects reported in Henson *et al.* (2002b), and freely available from the SPM website http://www.fil.ion.ucl.ac.uk/spm/data. Events comprised 500 ms presentations of faces, to which the subject made a famous/non-famous decision with the index and middle fingers of their right hand. One half of the faces were famous, one half were novel (unfamiliar), and each face was presented twice during the session. This corresponds to a 2×2 factorial design consisting of first and second presentations of novel and famous faces (conditions N1, N2, F1 and F2 respectively, each containing $J = 26$ events). To these 104 events, 52 null events were added and the whole sequence permuted. This meant that the order of novel/famous faces was pseudo-randomized (given the finite sequence), though the order of first and second presentations, while intermixed, was constrained by the fact that second presentations were necessarily later than first presentations on average. The minimum SOA (SOA_{min}) was 4.5 s, but varied near-exponentially over multiples of SOA_{min} due to the null events (see next chapter). The time series comprised 351 images acquired continuously with a T_R of 2 s. The images were realigned spatially, slice-time corrected to the middle slice, normalized with a bilinear interpolation to $3 \times 3 \times 3$ mm voxels and smoothed with an isotropic Gaussian FWHM of 8 mm. The ratio of SOA_{min} to T_R ensured an effective peristimulus sampling rate of 2 Hz.

Events were modelled with $K = 3$ basis functions consisting of the canonical HRF, its temporal derivative and its dispersion derivative. The resolution of the simulated BOLD signal was set to 83 ms ($N = 24$) and the event onsets synchronized with the middle slice ($T_0 = 12$). Six user-specified regressors, derived from the rigid-body realignment parameters (3 translations and 3 rotations) were included to model residual (linear) movement effects.[7] A highpass filter with cut-off period of 120 s was applied to both model and data, with an AR(1) model for temporal autocorrelations. No global scaling was used. Two different models are considered below: a 'categorical' one and a 'parametric' one. In the categorical model, each event-type is modelled separately. In the parametric model, a single event-type representing all face-trials is modulated by their familiarity and the 'lag' since their last presentation.

Categorical model

The design matrix for the categorical model is shown in Figure 14.4(a). A (modified) effects-of-interest F-contrast, corresponding to a reduced F-test on the first 12 columns of the design matrix (i.e. removing linear movement effects), is shown in Figure 14.4(b) and the resulting SPM{F} in Figure 14.4(c). Several regions, most notably in bilateral posterior inferior temporal, lateral occipital, left motor and right prefrontal cortices, show some form of significant response to the events (versus baseline) at $p < 0.05$, corrected for whole brain. Note that these responses could be activations (positive amplitude) or deactivations (negative amplitude), and may differ across the event-types. A t-contrast like that inset in Figure 14.4(b) would test a more constrained hypothesis, namely that the response is positive when averaged across all event-types, and is a more powerful test for such responses (producing more suprathreshold voxels in this dataset). Also inset in Figure 14.4(c) is the SPM{F} from an F-contrast on the realignment parameters, in which movement effects can be seen at the edge of the brain.

The parameter estimates (plotting the modified effects-of-interest contrast) and best-fitting event-related responses for a right fusiform voxel (close to what has been called the 'Fusiform Face Area', Kanwisher *et al.*, 1997) are shown in Plate 13(a) and 13(b). First presentations of famous faces produced the greatest response (green fitted response). Furthermore, responses in this region appear to be slightly earlier and narrower than the canonical response (indicated by the positive

[7] One might also include the temporal derivatives of the realignment parameters, and higher-order interactions between them, in a Volterra approximation to residual movement effects (regardless of their cause). Note also that the (rare) events, for which the fame decision was erroneous, could be modelled as a separate event-type (since they may involve physiological changes that are not typical of face recognition). This was performed in the demonstration on the website, but is ignored here for simplicity.

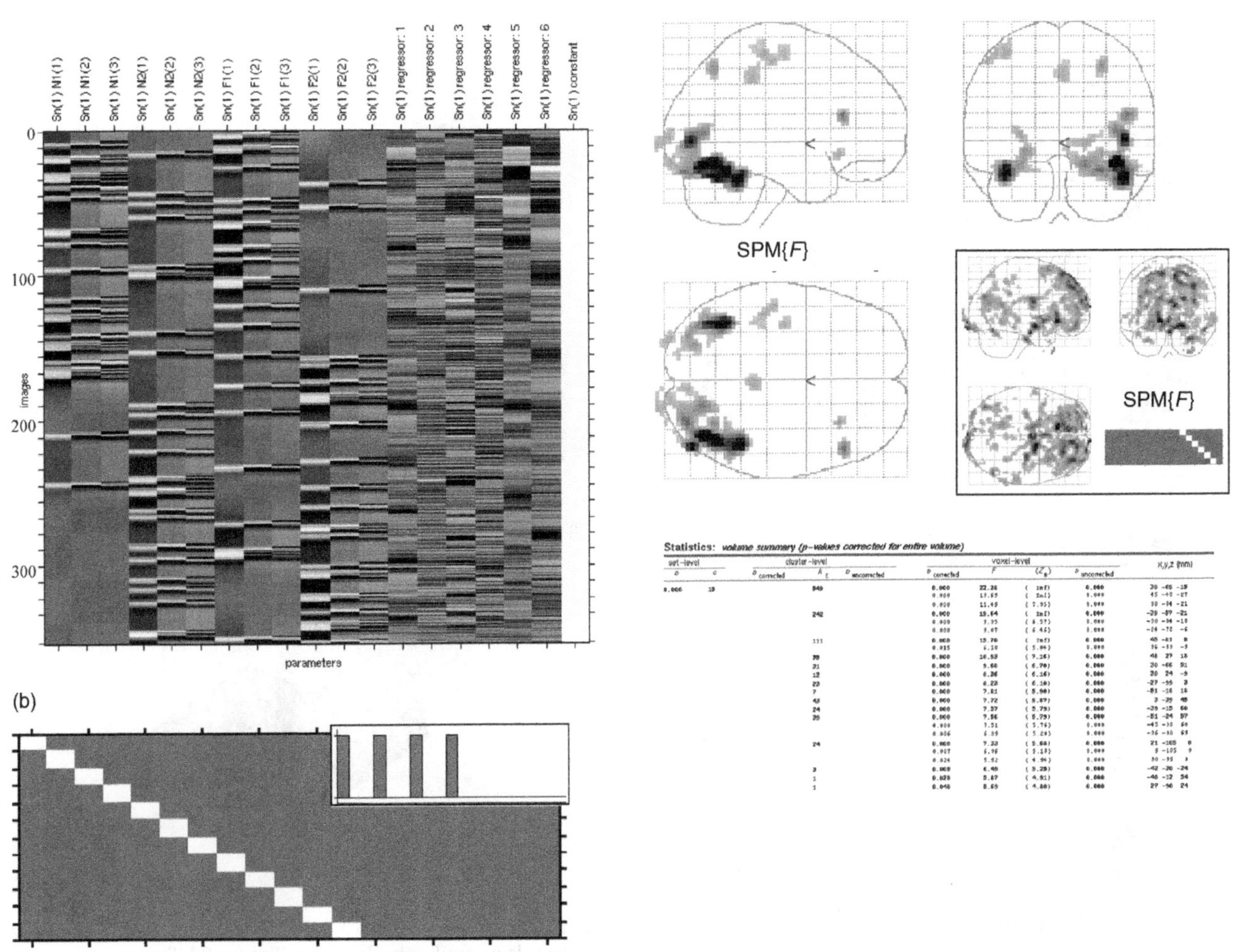

FIGURE 14.4 Categorical model: effects of interest. (a) Design matrix. (b) F-contrast for effects of interest (inset is t-contrast that tests for positive mean parameter estimate for canonical HRF). (c) SPM{F} MIP for effects of interest F-contrast, thresholded at $p < 0.05$ whole-brain corrected, together with SPM tabulated output (inset is SPM{F} for contrast on movement parameters, also at $p < 0.05$ corrected).

parameter estimates for the temporal and dispersion derivatives).

There are three obvious further effects of interest: the main effects of familiarity and repetition, and their interaction. The results from an F-contrast for the repetition effect are shown in Plate 13(c), after inclusive masking with the effects-of-interest F-contrast in Figure 14.4(c). This mask restricts analysis to regions that are generally responsive to faces (without needing a separate face-localizer scan, cf. Kanwisher *et al.*, 1997), and could be used for a small-volume correction (see Chapter 17). Note that this masking is facilitated by the inclusion of null events (otherwise the main effect of faces versus baseline could not be estimated efficiently, see Chapter 15). The contrast of parameter estimates and fitted responses for the single right posterior occipitotemporal region identified by the repetition contrast are shown in Plate 13(d). Differential effects were seen on all three basis functions, and represent decreased responses to repeated faces.[8]

Plate 14(a) shows the design matrix using a more general FIR basis set of $K = 16$, 2 s time bins. The effects-of-interest contrast (see Plate 14(b)) reveals a subset of the regions identified with the canonical basis set (cf. Plate 14(c) and Figure 14.4(c)). The absence of additional suprathreshold voxels when using the FIR model is likely to reflect the reduced statistical power for this F-test to detect BOLD responses with a canonical form (and the

[8] Note that this difference in the temporal derivative parameter estimates does not imply a difference in latency, given the concurrent difference in canonical parameter estimates: i.e. larger canonical responses require larger temporal derivatives to shift them in time (Henson *et al.*, 2002a); as mentioned previously, it is the ratio of the two parameter estimates that estimates latency.

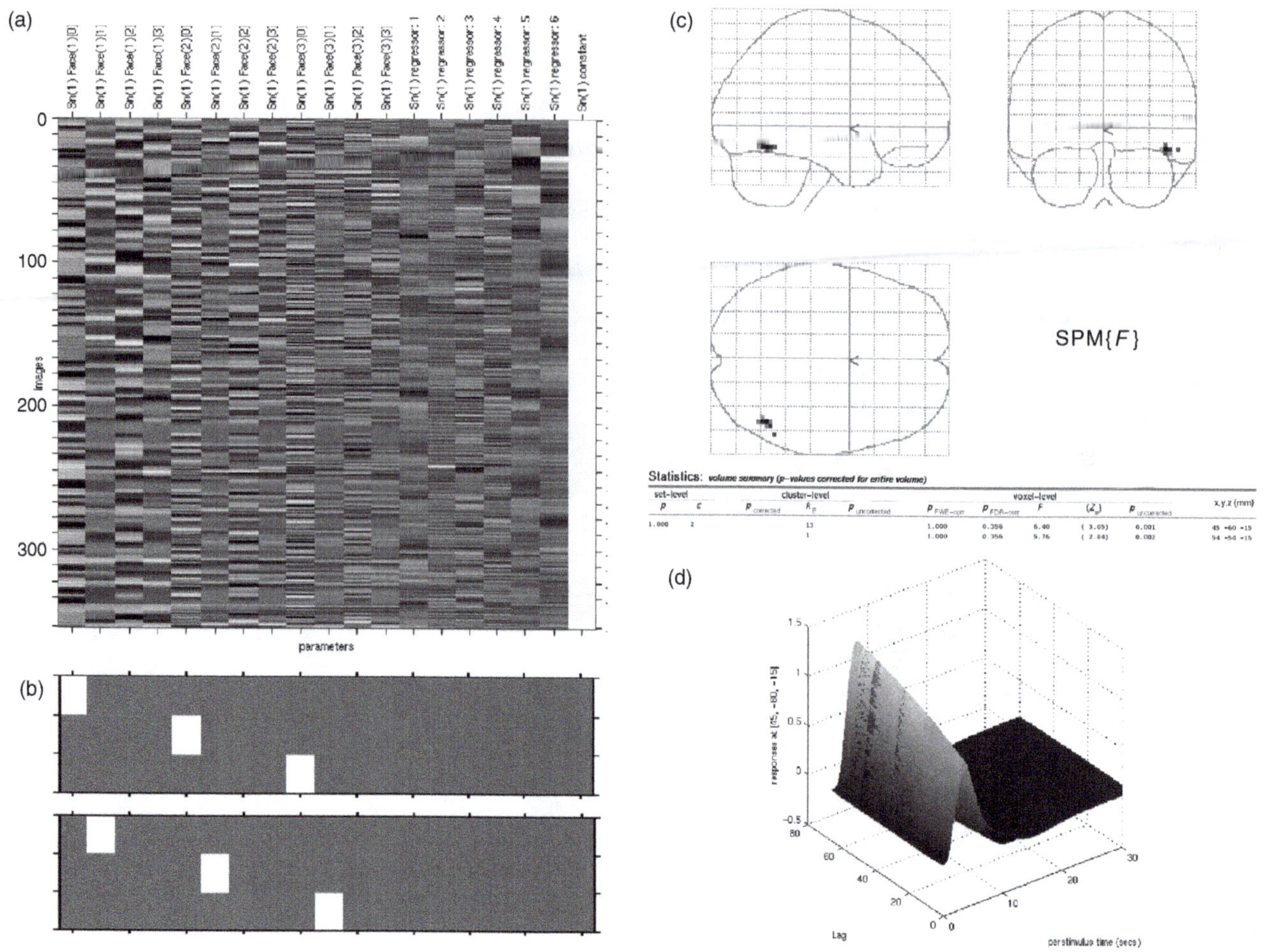

FIGURE 14.5 Parametric model (a) Design matrix, columns ordered by basis function – canonical HRF, temporal derivative, dispersion derivative – and within each basis function by parametric effect – main effect, lag, familiarity, lag-x-familiarity. (b) F-contrasts for main effect (top) and lag effect (bottom). (c) SPM{F} MIP for lag effect, together with SPM tabulated output, thresholded at $p < 0.005$ uncorrected, after inclusive masking with main effect at $p < 0.05$ corrected. (d) Parametric plot of fitted response from right occipitotemporal region (+45, −60, −15), close to that in Plate 14(c), in terms of percentage signal change versus PST and lag (infinite lag values for first presentations not shown).

likely absence of non-canonical responses). Plate 14(d) shows the parameter estimates from a right fusiform voxel for each of the event-types (concatenated), which clearly demonstrate canonical-like impulse responses in all four cases. No right occipitotemporal region was identified by an F-contrast testing for the repetition effect (inset in Plate 14(c)) when using the FIR basis set. This reflects the reduced power of this unconstrained contrast. Note that assumptions about the shape of the HRF can be imposed via appropriate contrasts within this FIR model, as illustrated by the t-contrast inset in Plate 14(b), which corresponds to a canonical HRF.

Parametric model

In this model, a single event-type was defined (collapsing the onsets for the four event-types above), which was modulated by three parametric modulations. The first modelled how the response varied according to the recency with which a face had been seen. This was achieved by an exponential parametric modulation of the form:

$$\alpha_j = \exp(-L_j/50) \qquad \textbf{14.21}$$

where L_j is the 'lag' for the j-th face presentation, defined as the number of stimuli between that presentation and the previous presentation of that face. The choice of an exponential function (rather than, say, a polynomial expansion) was based simply on the observation that many biological processes have exponential time-dependency, and the half-life of the function (50 scans) was somewhat arbitrary (ideally it would be derived empirically from separate data). Thus, as lag increases, the modulation decreases. For first presentations of faces,

$L_j = \infty$ and the modulation is zero (i.e. there is no possible adaptation or repetition suppression).

The second parametric modulation had a binary value of 1 or −1, indicating whether the face was famous or novel; the third modulation was the interaction between face familiarity and lag (i.e. the product of the first and second modulations, after mean-correction). Each modulation was applied to the three temporal basis functions, producing the design matrix in Figure 14.5(a). The F-contrast for the main effect of faces versus baseline (upper contrast in Figure 14.5(b)) identified regions similar to those identified by the effects-of-interest contrast in the categorical model above (since the models span similar spaces). As expected, the F-contrast for the lag effect (lower contrast in Figure 14.5(b)), after masking with the main effect, revealed the same right occipitotemporal region (Figure 14.5(c)) that showed a main effect of repetition in the categorical model. The best-fitting event-related parametric response in Figure 14.5(d) shows that the response increases with lag, suggesting that the repetition-related decrease observed in the categorical model may be transient.

These examples illustrate the use of basis functions and the convolution model for detecting non-stationary (adapting) haemodynamic responses of unknown form in the brain. The experimental design in this instance was as efficient as possible, under the psychological constraints imposed by our question. In the next chapter, we use the basic principles behind the convolution model to look at the design of efficient experiments.

REFERENCES

Aguirre GK, Zarahn E, D'Esposito M (1998) The variability of human, BOLD hemodynamic responses. *NeuroImage* **8**: 360–69

Birn RM, Saad ZS, Bandettini PA (2001) Spatial heterogeneity of the nonlinear dynamics in the fMRI bold response. *NeuroImage* **14**: 817–26

Boynton GM, Engel SA, Glover GH *et al.* (1996) Linear systems analysis of functional magnetic resonance imaging in human V1. *J Neurosci* **16**: 4207–21

Bullmore ET, Brammer MJ, Williams SCR *et al.* (1996) Statistical methods of estimation and inference for functional MR images. *Mag Res Med* **35**: 261–77

Burock MA, Dale AM (2000) Estimation and detection of event-related fMRI signals with temporally correlated noise: a statistically efficient and unbiased approach. *Hum Brain Mapp* **11**: 249–60

Dale AM (1999) Optimal experimental design for event-related fMRI. *Hum Brain Mapp* **8**: 109–14

Dale A, Buckner R (1997) Selective averaging of rapidly presented individual trials using fMRI. *Hum Brain Mapp* **5**: 329–40

Fransson P, Kruger G, Merboldt KD *et al.* (1999) MRI of functional deactivation: temporal and spatial characteristics of oxygenation-sensitive responses in human visual cortex. *NeuroImage* **9**: 611–18

Friston KJ (2002) Bayesian estimation of Dynamical systems: an application to fMRI. *NeuroImage* **16**: 513–30

Friston KJ, Jezzard PJ, Turner R (1994) Analysis of functional MRI time-series.*Hum Brain Mapp* **1**: 153–71

Friston KJ, Josephs O, Rees G *et al.* (1998a) Non-linear event-related responses in fMRI. *Mag Res Med* **39**: 41–52

Friston KJ, Fletcher P, Josephs O *et al.* (1998b) Event-related fMRI: characterizing differential responses. *NeuroImage* **7**: 30–40

Friston KJ, Mechelli A, Turner R *et al.* (2000a) Nonlinear responses in fMRI: the Balloon model, Volterra kernels, and other hemodynamics. *NeuroImage* **12**: 466–77

Friston KJ, Josephs O, Zarahn E *et al.* (2000b) To smooth or not to smooth? Bias and efficiency in fMRI time-series analysis. *NeuroImage* **12**: 196–208

Friston KJ, Glaser DE, Henson RNA *et al.* (2002) Classical and Bayesian inference in neuroimaging: applications. *NeuroImage* **16**: 484–512

Glover GH (1999) Deconvolution of impulse response in event-related BOLD fMRI. *NeuroImage* **9**: 416–29

Gossl C, Fahrmeir L, Auer DP (2001) Bayesian modeling of the hemodynamic response function in BOLD fMRI. *NeuroImage* **14**: 140–48

Henson R, Andersson J, Friston K (2000) Multivariate SPM: application to basis function characterisations of event-related fMRI responses. *NeuroImage* **11**: 468

Henson RNA, Rugg MD (2001) Effects of stimulus repetition on latency of the BOLD impulse response. *NeuroImage* **13**: 683

Henson RNA, Buechel C, Josephs O *et al.* (1999) The slice-timing problem in event-related fMRI. *NeuroImage* **9**: 125

Henson RNA, Rugg MD, Friston KJ (2001) The choice of basis functions in event-related fMRI. *NeuroImage* **13**: 149

Henson RNA, Price C, Rugg MD *et al.* (2002a) Detecting latency differences in event-related BOLD responses: application to words versus nonwords, and initial versus repeated face presentations. *NeuroImage* **15**: 83–97

Henson RNA, Shallice T, Gorno-Tempini M-L *et al.* (2002b) Face repetition effects in implicit and explicit memory tests as measured by fMRI. Cerebr Cortex **12**: 178–86

Huettel SA, McCarthy G (2001) Regional differences in the refractory period of the hemodynamic response: an event-related fMRI study. *NeuroImage* **14**: 967–76

Josephs O, Henson RNA (1999) Event-related fMRI: modelling, inference and optimisation. *Phil Trans Roy Soc London* **354**: 1215–28

Josephs O, Turner R, Friston KJ (1997) Event-related fMRI. *Hum Brain Mapp* **5**: 243–48

Kanwisher N, McDermott J, Chun MM (1997) The fusiform face area: a module in human extrastriate cortex specialised for face perception. *J Neurosci* **17**: 4302–11

Kiebel SJ, Friston KJ (2004) Statistical parametric mapping for event-related potentials. I: Generic considerations. *NeuroImage* **22**: 492–502

Kruggel F, von Cramon DY (1999) Temporal properties of the hemodynamic response in functional MRI. *Hum Brain Mapp* **8**: 259–71

Lee AT, Glover GH, Meyer CH (1995) Discrimination of large venous vessels in time-course spiral blood-oxygenation-level-dependent magnetic-resonance functional imaging. *Mag Res Med* **33**: 745–54

Liao CH, Worsley KJ, Poline J-B *et al.* (2002) Estimating the delay of the hemodynamic response in fMRI data. *NeuroImage* **16**: 593–606

Malonek D, Grinvald A (1996) Interactions between electrical activity and cortical microcirculation revealed by imaging spectroscopy: implications for functional brain mapping. *Science* **272**: 551–54

Miezin FM, Maccotta L, Ollinger JM *et al.* (2000) Characterizing the hemodynamic response: effects of presentation rate, sampling procedure, and the possibility of ordering brain activity based on relative timing. *NeuroImage* **11**: 735–59

Ollinger JM, Shulman GL, Corbetta M (2001) Separating processes within a trial in event-related functional MRI. *NeuroImage* **13**: 210–17

Penny WD, Flandin G, Trujillo-Barreto N (in press) Bayesian comparison of spatially regularised general linear models. *Hum Brain Mapp*

Pollmann S, Wiggins CJ, Norris DG *et al.* (1998) Use of short intertrial intervals in single-trial experiments: a 3T fMRI-study. *NeuroImage* **8**: 327–39

Purdon PL, Weisskoff RM (1998) Effect of temporal autocorrelation due to physiological noise and stimulus paradigm on voxel-level false-positive rates in fMRI. *Hum Brain Mapp* **6**: 239–95

Rajapakse JC, Kruggel F, Maisog JM *et al.* (1998) Modeling hemodynamic response for analysis of functional MRI time-series. *Hum Brain Mapp* **6**: 283–300

Schacter DL, Buckner RL, Koutstaal W *et al.* (1997) Late onset of anterior prefrontal activity during true and false recognition: an event-related fMRI study. *NeuroImage* **6**: 259–69

Turner R, Howseman A, Rees GE *et al.* (1998) Functional magnetic resonance imaging of the human brain: data acquisition and analysis. *Exp Brain Res* **123**: 5–12

Vasquez AL, Noll CD (1998) Nonlinear aspects of the BOLD response in functional MRI. *NeuroImage* **7**: 108–18

Woolrich MW, Behrens TE, Smith SM (2004) Constrained linear basis sets for HRF modelling using variational Bayes. *NeuroImage* **21**: 1748–61

Woolrich MW, Ripley BD, Brady M *et al.* (2001) Temporal autocorrelation in univariate linear modeling of fMRI data. *NeuroImage* **14**: 1370–86

Worsley KJ, Friston KJ (1995) Analysis of fMRI time-series revisited – again. *NeuroImage* **2**: 173–81

Worsley KJ, Liao, CH, Aston, J, Petre, V, Duncan, GH, Morales, F & Evans, AC (2002). A general statistical analysis for fMRI data. *NeuroImage* **15**: 1–15

Zarahn E (2000) Testing for neural responses during temporal components of trials with BOLD fMRI. *NeuroImage* **11**: 783–96

Zarahn E, Aguirre G, D'Esposito M (1997a) A trial-based experimental design for fMRI. *NeuroImage* **6**: 122–38

Zarahn E, Aguirre GK, D'Esposito M (1997b) Empirical analyses of BOLD fMRI statistics: I Spatially unsmoothed data collected under null-hypothesis conditions. *NeuroImage* **5**: 179–97

CHAPTER

15

Efficient Experimental Design for fMRI

R. Henson

INTRODUCTION

This chapter begins with an overview of the various types of experimental design, before proceeding to various modelling choices, such as the use of events versus epochs. It then covers some practical issues concerning the effective temporal sampling of blood oxygenation-level-dependent (BOLD) responses and the problem of different slice acquisition times. The final and main part of the chapter concerns the statistical efficiency of functional magnetic resonance imaging (fMRI) designs, as a function of stimulus onset asynchrony (SOA) and the ordering of different stimulus-types. These considerations allow researchers to optimize the efficiency of their fMRI designs.

TAXONOMY OF EXPERIMENTAL DESIGN

Most experiments involve the manipulation of a number of factors over a number of levels. For example, a factor of spatial attention might have two levels of left versus right covert attention (relative to fixation), while a second factor might be whether a stimulus is presented to the left or right visual hemi-field. Orthogonal manipulation of these two factors corresponds to a '2 × 2' 'factorial' design, in which each factor-level combination constitutes an experimental condition (i.e. four conditions in this case; see Chapter 13). Factors with a discrete number of levels, as in the above example, are often called 'categorical'. Other factors may have continuous values (the duration of the stimulus for example), and may have as many 'levels' as there are values. Such factors are called 'parametric'. Below, we discuss briefly different designs in the context of the general linear model (GLM) and some of the assumptions they entail.

Single-factor subtraction designs and 'pure insertion'

The easiest way to illustrate different types of design is with examples. Plate 15(a) (see colour plate section) shows an example design matrix with 12 conditions and 5 sessions (e.g. 5 subjects). The data could come from a positron emission tomography (PET) experiment or from a second-level analysis of contrast images from an fMRI experiment. We use this example to illustrate a number of designs and contrasts below. Initially, we will assume that there was only one factor of interest, with two levels (that happened to occur six times in alternation). These might be reading a visually presented cue word ('Read' condition) and generating a semantic associate of the cue ('Generate' condition). If one were interested in the brain regions involved in semantic association, then one might subtract the Read condition from the Generate condition, as shown by the *t*-contrast in Plate 15(a). The logic behind this subtraction is that brain regions involved in processes common to both conditions (such as visual processing of the cue word) will be equally active in both conditions, and therefore not appear in the resulting statistical parametric maps (SPM). In other words, the contrast should reveal activations related to those processes unique to generating semantic associations, relative to reading words.

A criticism often levelled at such 'cognitive subtractions' is that the conditions may differ in ways other than those assumed by the specific cognitive theory under investigation. For example, the Generate and Read conditions might differ in phonological processes, as well as semantic processes (i.e. the subtraction is 'confounded'). The assumption that tasks can be elaborated so that they call upon a single extra process is called the 'pure insertion' assumption, and has been the source of much debate in neuroimaging (Friston *et al.*, 1996). In fact, the debate goes back to the early days of experimental psychology, e.g. the 'Donders' method of subtraction and its

Statistical Parametric Mapping, by Karl Friston et al.
ISBN–13: 978-0-12-372560-8 ISBN–10: 0-12-372560-7

subsequent refinements (Sternberg, 1969). In short, the concerns about the pure insertion assumption are not unique to neuroimaging (Henson, 2005). Below we will consider some ways to ameliorate such concerns.

Cognitive conjunctions

One way to minimize the probability that interpretation of activation is confounded is to isolate the process of interest using multiple different subtractions. The probability of each subtraction being confounded by the same (uninteresting) differences is thus reduced. In other words, one only considers activation that is common to all subtractions: a method called 'cognitive conjunction' (Price and Friston, 1997). For example, consider an experiment with four conditions (Plate 16): passively viewing a colour-field (Viewing Colour), naming the colour of that field (Naming Colour), passively viewing an object (Viewing Object), and naming an object (Naming Object). One might try to isolate the neuronal correlates of visual object recognition by performing a conjunction of the two subtractions: (1) Object versus Colour Viewing and (2) Object versus Colour Naming. Both subtractions share a difference in the stimulus (the presence or absence of an object), but differ in the nature of the tasks (or 'contexts'). Thus a potential confound, such as number of possible names, which might confound the second subtraction, would not necessarily apply to the first subtraction, and thus would not apply to the conjunction as a whole.

The precise (statistical) definition of a conjunction has changed with the history of SPM, and different definitions may be appropriate for different contexts (the details are beyond the present remit, but for further discussion, see Friston *et al.*, 2005; Nichols *et al.*, 2005). In the present context of 'cognitive' conjunctions, a sufficient definition is that a region survives a statistical threshold in all component subtractions ('inclusive' masking), with the possible further constraint of no interaction between the subtractions ('exclusive' masking). A posterior temporal region shows this type of pattern in Plate 16 (upper panel) and might be associated with implicit object recognition.

A single parametric factor

To illustrate a parametric factor, let us return to the Generate and Read experiment in Plate 15. One might be interested whether there is any effect of time during the experiment (e.g. activation may decrease over the experiment as subjects acquire more practice). In this case, a time factor can be modelled with 12 discrete levels, over which the effects of time could be expressed in a number of different ways. For example, time may have a linear effect, or it may have a greater effect towards the start than towards the end of the experiment (e.g. an exponential effect). The t-contrast, testing the former linear effect – more specifically, for regions showing a decrease in activation over time – is shown in Plate 15(b) (in fact, the plot of activity in the highlighted region suggests an exponential decrease, but with a sufficiently linear component that it is identified with the linear contrast).

When the precise function relating a parametric experimental factor to neuronal activity is unknown, one option is to express the function in terms of a polynomial expansion, i.e.:

$$f(x) = \beta_0 + \beta_1 x + \beta_2 x^2 + \cdots \qquad 15.1$$

where β_i are the parameters to be estimated. For N levels of a factor, the expansion is complete when the terms run from 0th-order up to order $N-1$. In the latter case, the corresponding design matrix is simply a rotation of a design matrix where each level is modelled as a separate column. An example design matrix for an expansion up to second-order, over 12 images, is shown in Plate 17(a) (e.g. for a single subject in Plate 15): the first column models linear effects, the second column models quadratic effects, and the third column models the 0th-order (constant) effect. An F-contrast on the second column identifies a region that shows an inverted-U shape when activity is plotted as a function of the 12 levels of the factor. If this factor were rate of word generation, for example, one might conclude that activity in this region increases as the word rate increases to a certain level, but then decreases if that (optimal) level is surpassed. Parametric modulations that have only one level per value (i.e. are modelled as continuous rather than discrete values) can be modelled by a 'parametric modulation' in SPM. An example of a parametric modulation of event-related responses is shown in Chapter 14.

Factorial designs

Many experiments manipulate more than one factor concurrently. When each condition is associated with one level of every factor, it is called a 'factorial' design. These are common in experimental sciences because they allow tests of not only differences between the levels of each factor, collapsing over other factors ('main effects'), but also how the effect of one factor depends on the level of another factor ('interactions'). Let us return to the object–colour experiment in Plate 16. This experiment can be conceived as a '2 × 2' design, where one factor, Task, has two levels (viewing versus naming) and the other, Stimulus, also has two levels (colour-field or object). This '2-way' design, therefore, offers tests of two main effects and one interaction (see Chapter 13 for a generalization to 'M-way' factorial designs). The component subtractions considered for the cognitive conjunction above are sometimes called the 'simple effects'. The interaction in this design would test where the difference between objects and colour-fields varies between a naming task and a viewing task. If these conditions were ordered: Viewing Object, Viewing Colour, Naming Object, Naming Colour (i.e. with the Task factor 'rotating' slowest), then the interaction would have contrast weights$[1\ -1\ -1\ 1]$. This can be conceived as the difference of two differences, i.e. difference of the two simple effects, i.e. $[1\ -1\ 0\ 0]-[0\ 0\ 1\ -1]$, or as the 'product' of two differences, i.e. $[1\ -1]\otimes[1\ -1]$, where $\otimes$ is the Kronecker product.

When testing one tail of the interaction (i.e. with a t-rather than F-contrast), namely where objects produce greater activation relative to colour-fields when named, rather than when viewed, a region was found in temporal cortex (see Plate 16 – lower SPM), anterior to that in the conjunction (upper SPM). Given that the region showed little difference between objects and colour-fields under passive viewing (i.e. this simple effect was not significant), the pattern in Plate 16 might be termed 'naming-specific object-recognition'. Note also that, if one attempted to isolate visual object-recognition using only a naming task, this interaction could be used as evidence of a failure of pure insertion, i.e. that naming an object in the visual field involves more than simply visual recognition (Price and Friston, 1997).

An example of an interaction involving a parametric factor is shown in Plate 17(b). This contrast tests for a linear time-by-condition interaction in the Generate-Read experiment (when conceived as a 2×6 factorial design). Again, the contrast weights can be viewed as the Kronecker product of the Generate versus Read effect and the linear time effect, i.e. $[1\ -1]\otimes[5\ 3\ 1\ -1\ -3\ -5]$. This t-contrast asks where in the brain the process of semantic association decreases (linearly) over time (as might happen, for example, if subjects showed stronger practice effects on the generation task than the read task).

A final example of an interaction is shown in Figure 15.1. In this case, the effects Task (Generate versus Read), Time, and their interactions have been expressed in the design matrix (for a single subject), rather than in the contrast weights (cf. Plate 17(a)). This illustrates the general point that one can always re-represent contrasts by rotating both the design matrix and the contrast weights (see Chapter 13 for further discussion). More precisely, the columns of the design matrix in Figure 15.1 model (from left to right): effect of Task, linear then quadratic effects of Time, linear then quadratic interaction effects, and the constant. The F-contrast shown,

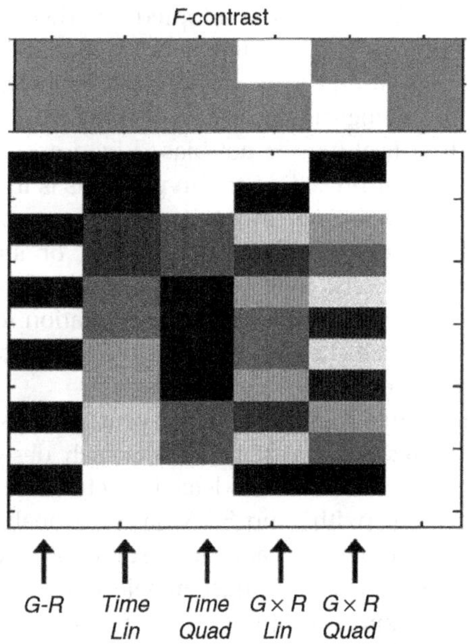

FIGURE 15.1 A single-subject design matrix and F-contrast showing non-linear (linear + quadratic) interactions in a 2×6 factorial design.

which picks out the fourth and fifth columns, would test for any type of time-by-condition interaction up to second order. Note that another common example of an interaction between a categorical factor and a parametric factor arises in psychophysiological interactions (PPIs; Chapter 38): in these cases, the psychological factor is often categorical (e.g. attended versus unattended) and the physiological factor is invariably parametric, since it reflects the continuous signal sampled by each scan from the source region of interest.

EVENT-RELATED fMRI, AND RANDOMIZED VERSUS BLOCKED DESIGNS

Event-related fMRI is simply the use of fMRI to detect responses to individual trials, in a manner analogous to the time-locked event-related potentials (ERPs) recorded with electroencephalography (EEG). The neuronal activity associated with each trial is normally (though not necessarily) modelled as a delta function – an 'event' – at the trial onset.

Historically, the advent of event-related methods (Dale and Buckner, 1997; Josephs *et al.*, 1997; Zarahn *et al.*, 1997), based on linear convolution models (see Chapter 14), offered several advantages. Foremost was the ability to intermix trials of different types (so-called randomized designs), rather than blocking them in the manner required for PET and initially adopted for fMRI (so-called blocked designs). The counterbalancing or randomizing of different trial-types, as is standard in behavioural or electrophysiological studies, ensures that the average response to a trial-type is not biased by a specific context or history of preceding trial-types. This is important because the blocking of trial-types might, for example, induce differences in the cognitive 'set' or strategies adopted by subjects. Johnson *et al.* (1997) for example, provided direct evidence that the presentation format – randomized or blocked – can affect the ERP associated with a trial-based memory effect.

Note that there are also disadvantages associated with randomized designs. Foremost, such designs are generally less efficient for detecting effects than are blocked designs (with short SOAs and reasonable block lengths; see below). In addition, some psychological manipulations, such as changes in selective attention or task, may be more appropriate when blocked.

Other advantages of event-related methods include:

1. the *post hoc* categorization of trial-types according to the subject's behaviour (e.g. Henson *et al.*, 1999b), or *post hoc* parametric modulation of neuronal activity by reaction time (RT) for each trial
2. modelling events whose occurrence is beyond experimental control, such as those that can only be indicated by the subject (e.g. perceptual transitions in the face-vase illusion, Kleinschmidt *et al.*, 1998)
3. the use of 'oddball' designs, in which the stimulus of interest is one that deviates from the prevailing context, and which therefore cannot be blocked (e.g. Strange *et al.*, 2000).

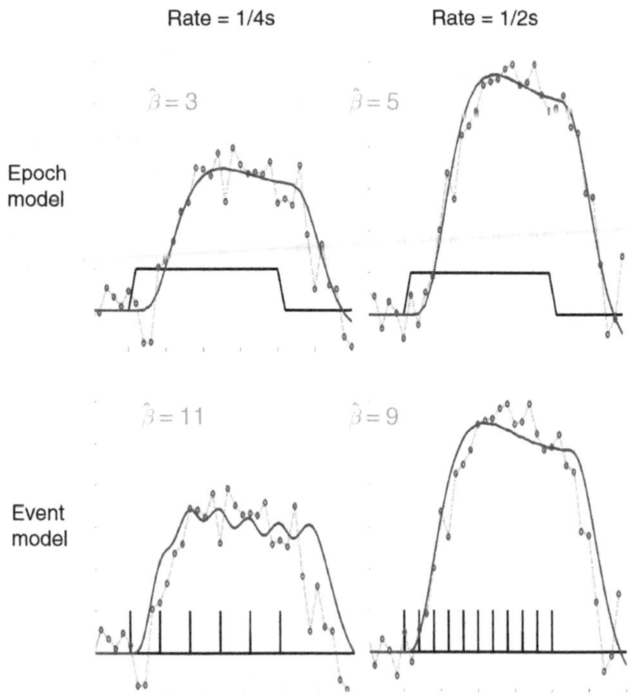

FIGURE 15.2 Effects of modelling the same data with events or epochs.

Epochs versus events and state- versus item-effects

It is important to distinguish between the experimental design (randomized versus blocked) and the neuronal model (events versus epochs). For example, a blocked design can be modelled as a series of events. Indeed, modelling the BOLD response to each stimulus within a block may capture variability that is not captured by a simple 'epoch' (or boxcar) model, particularly for SOAs of more than a few seconds, which will lead to small fluctuations of the BOLD response around the mean 'block' response (Price *et al.*, 1999; Mechelli *et al.*, 2003a; see, e.g. Figure 15.2 bottom left).

In SPM, the choice of events versus epochs can also have important conceptual consequences. Consider, for example, an experiment with two blocks of words presented at different rates (once every 4 s versus once every 2 s). The data may be such that mean activity during the block of words presented at the fast rate may be greater, but not twice as great, as that for the slow rate. When modelling both conditions as epochs (upper panels of Figure 15.2), the parameter estimates for the two rates may be, for example, 3 and 5 respectively. If individual words were modelled as events, however (lower panels of Figure 15.2), the *relative* size of the parameter estimates could be reversed, e.g. 11 and 9 respectively. This is simply because the parameter estimates have different interpretations for the two types of model: in the epoch model, they reflect the response *per block*, whereas in the event model, they reflect the response *per word*. Since there are twice as many words in the fast- relative to slow-rate blocks, and yet the mean block activity is not double, the response per word must be less (i.e. a non-linear saturation as a function of word rate).

Another situation where this issue arises concerns trials of different duration. If all trials are of the same duration (and that duration is below ~ 2 s), then they can be modelled effectively as events because, after convolution with the haemodynamic response function (HRF), a difference in the duration of a trial causes a difference in the scaling of the predicted response, but has little effect on its shape (see Chapter 14). Since it is the scaling of the predicted response that is estimated in the GLM, changing the duration of all trials (from approx. 0 to 2 s) simply changes the size of the resulting parameter estimates, but has no effect on statistics.[1] For longer duration trials, the response begins to plateau, meaning that an 'epoch model' can be a better model. More important, however, is the case of trials that vary in duration from trial to trial within a condition, or across conditions. Whether these are better modelled as events, or as epochs of different durations (e.g. with duration equal to the RT for each trial), is debatable. For example, if the stimulus duration were constant and only RTs varied, then the activity in V1 may not be expected to vary with RT, so an event model might fit better (and in this case, the parameter estimate can be interpreted as the response *per trial*). For activity in premotor cortex on the other hand, greater activity might be expected for trials with longer RTs, so a 'varying-duration' epoch model might fit better (and in this case, the parameter estimate can be interpreted as the response *per unit time*). So the choice of model depends on the assumptions about the duration of neuronal activity in the particular region of interest. If this is unknown, trials whose durations vary over a few seconds (as with typical RTs) are probably best modelled with two regressors: one modelling events, and a second modelling a parametric modulation of the response, by the RT on each trial.

Finally, note that one can combine both events and epochs within the same model. A common example of this is when trying to distinguish between sustained ('state') effects and transient ('item') effects. Chawla *et al.* (1999), for example, investigated the interaction between selective attention (a state-effect) and transient stimulus changes (an item-effect) in such a 'mixed epoch-event' design. Subjects viewed a visual stimulus that occasionally changed in either colour or motion. In some blocks, they detected the colour changes, in other blocks they detected the motion changes. By varying the interval between changes within a block, Chawla *et al.* were able to reduce the correlation between the corresponding epoch- and event-related regressors (which increases the statistical efficiency to detect either effect alone; see below). Tests of the epoch-related effect showed that attending to a specific visual attribute (e.g. colour) increased the baseline activity in regions selective for that attribute (e.g. V4). Tests of the event-related effect showed that the impulse response to the same change in visual attribute was augmented when subjects were attending to it (Plate 18). These combined effects of selective attention – raising endogenous baseline activity and increasing the gain of the exogenous response – could not be distinguished in a blocked or fully randomized design.

Timing issues

There are two practical issues concerning the timing within randomized designs (which also apply to blocked designs, but to a lesser extent): the effective sampling rate of the BOLD response, and the different acquisition times for different slices within a scan (i.e. volume) when using echo-planar imaging (EPI).

It is possible to sample the impulse response at post-stimulus intervals, T_S, shorter than the inter-scan interval, T_R, by dephasing event onsets with respect to scan onsets (Josephs *et al.*, 1997). This uncoupling can be effected by ensuring the SOA is not a simple multiple of the T_R, or by adding a random trial-by-trial delay in stimulus onsets relative to scan onsets (Figure 15.3). In both cases, responses at different peristimulus times (PST) are sampled over trials. The main difference between the two methods is simply whether the SOA is fixed or random, i.e. whether or not the stimulus onset is predictable. For example, an effective PST sampling of 0.5 Hz can be achieved with an SOA of 6 s and a T_R of 4 s; or by adding a delay of 0 or 2 s randomly to each trial (producing SOAs of 4–8 s, with a mean of 6 s). While effective sampling rates higher than the T_R do not necessarily improve response detection (since there is little power in the canonical response above 0.2 Hz), higher sampling rates are important for quantifying the response shape, such as its latency (Miezin *et al.*, 2000; Henson and Rugg, 2001).

Dephasing event onsets with respect to scan onsets does not help the second practical issue concerning different slice acquisition times. This 'slice-timing' problem (Henson *et al.*, 1999a) refers to the fact that, with a descending EPI sequence for example, the bottom slice is acquired T_R seconds later than the top slice. If a single basis function (such as a canonical HRF) were used to model the response, and onset times were specified relative to the start of each scan, the data in the bottom slice would be systematically delayed by T_R seconds relative to the model.[2] This would produce poor (and biased) parameter estimates for later slices, and mean

[1] This is despite the fact that the 'efficiency', as calculated by Eqn. **15.3**, increases with greater scaling of the regressors. This increase is correct, in the sense that a larger signal will be easier to detect in the presence of the same noise, but misleading in the sense that it is the size of the signal that we are estimating with our model (i.e. the data are unaffected by how we model the trials).

[2] One solution would be to allow different event onsets for different slices. However, slice-timing information is usually lost as soon as images are re-sliced relative to a different orientation (e.g. during spatial normalization).

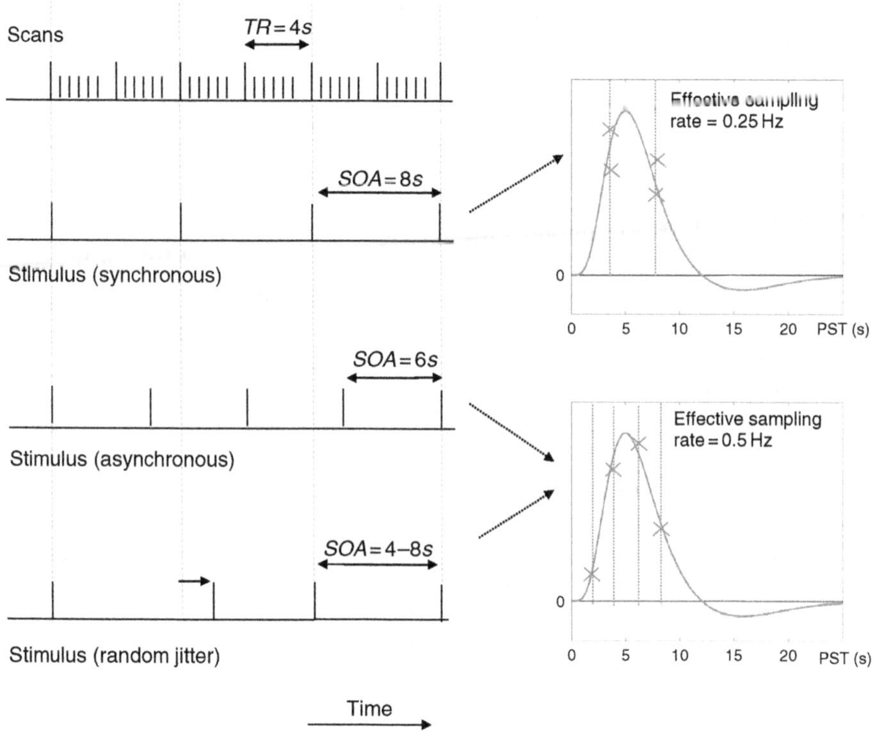

FIGURE 15.3 Effective sampling rate. Schematic (left) of event onsets relative to scan onsets (tall vertical lines represent first slice per scan; shorter lines represent subsequent slices) and resulting peristimulus sampling points (right).

that different sensitivities would apply to different slices (Figure 15.4(a)). There are two main solutions to this problem: to interpolate the data during pre-processing to make it seem as if the slices were acquired simultaneously; or use a temporal basis set that allows different response onset latencies.

Temporal interpolation of the data (using a full Fourier interpolation) is possible during pre-processing of images in SPM. One question that often arises is whether such temporal realignment should be performed before or after spatial realignment, given that movement often occurs. The answer depends on the order that slices are acquired within each scan. For sequential (contiguous) slice-acquisition, temporal interpolation is probably better performed after spatial realignment. This is because nearby voxels in space are sampled close in

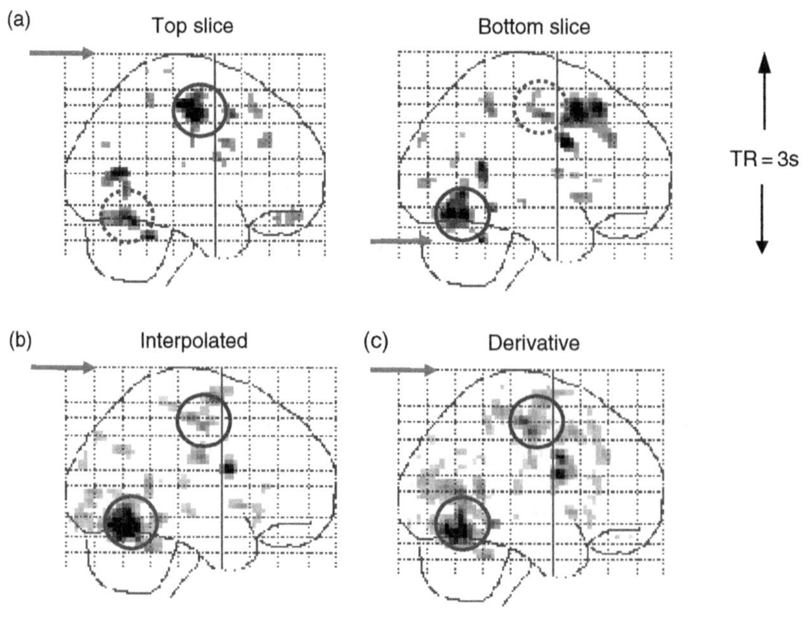

FIGURE 15.4 The slice-timing problem (from Henson *et al.*, 1999a) for a TR of 3 s. (a) SPM{t} for a [1] contrast on a canonical HRF synchronized with the top slice (left) or synchronized with the bottom slice (right). Note increased sensitivity to visual regions in the latter case, but reduced sensitivity to motor regions. (b) SPM{t} when the model is synchronized with the top slice, but the data have been interpolated as if all slices were acquired at the time of the top slice. Note sensitivity recovered in both motor and visual regions. (c) SPM{F} for the canonical HRF and its temporal derivative. Note sensitivity again recovered in both motor and visual regions.

time. Therefore, the temporal error for a voxel whose signal comes from different acquisition slices, due to re-slicing after correction for movement across scans, will be small (given that movement is rarely more than a few 'slices-worth'). The alternative, of performing temporal realignment before spatial realignment could cause greater error, particularly for voxels close to boundaries with large signal differences (e.g. the edge of the cortex): in such cases, rapid movement may cause the same voxel to sample quite different signal intensities across successive scans. Such high-frequency changes are difficult to interpolate (temporally in this case). The order of preprocessing is not so clear for interleaved slice-acquisition schemes, in which adjacent slices can be sampled $\frac{1}{2}T_R$ seconds apart. In this case, and when there is no rapid movement, it may be better to perform temporal realignment before spatial realignment.

During slice-time correction, the data are interpolated by an amount proportional to their sampling time relative to a reference slice (whose data are unchanged). The event onsets can then be synchronized with the acquisition of that reference slice. In SPM, this is equivalent to maintaining event onsets relative to scan onsets, but setting the time-point T_0 in the simulated time-space of N time bins, from which the regressors are sampled (see Chapter 14), to $T_0 = round(nN/S)$ where the reference slice is the nth slice acquired of the S slices per scan. This can ameliorate the slice-timing problem, if one wishes to use a single assumed response form (e.g. canonical HRF; see Figure 15.4(b)). A problem with slice-timing correction is that the interpolation will alias frequencies above the Nyquist limit $1/(2T_R)$. Ironically, this means that the interpolation accuracy decreases as the slice-timing problem (i.e. T_R) increases. For short $T_R < 2$–3 s, the interpolation error is likely to be small. For longer T_R, the severity of the interpolation error may depend on whether appreciable signal power exists above the Nyquist limit (which is more likely for rapid, randomized event-related designs).

An alternative solution to the slice-timing problem is to include additional temporal basis functions (see Chapter 14) to accommodate the timing errors within the GLM. The Fourier basis set, for example, does not have a slice-timing problem (i.e. it is phase-invariant). For more constrained sets, the addition of the temporal derivative of the response functions may be sufficient (see Figure 15.4(c)). The parameter estimates for the derivatives will vary across slices, to capture shifts in the data relative to the model, while those for the response functions can remain constant (up to a first-order Taylor approximation; Chapter 14). The temporal derivative of the canonical HRF, for example, can accommodate slice-timing differences of approximately plus or minus a second, or a T_R up to 2 s (when the model is synchronized to the middle slice in time). A potential problem with this approach occurs when the true impulse responses are also shifted in time relative to the assumed response functions: the combined latency shift may exceed the range afforded by the temporal derivatives.

EFFICIENCY AND OPTIMIZATION OF fMRI DESIGNS

This section is concerned with optimizing experimental fMRI designs for a specific contrast of interest. The properties of the BOLD signal measured by fMRI – particularly the 'sluggish' nature of the impulse response and the presence of low-frequency noise – can make the design of efficient experiments difficult to intuit. This section therefore starts with some general advice, before explaining the reasons for this advice from the perspectives of:

1 signal-processing
2 statistical 'efficiency'
3 correlations among regressors.

General points

Scan for as long as possible

This advice is of course conditional on the subject being able to perform the task satisfactorily in a sustained fashion. Longer is better because the power of a statistical inference depends primarily on the degrees of freedom (*df*), and the *df* depend on the number of scans. One might therefore think that reducing the T_R (inter-scan interval) will also increase your power. This is true to a certain extent, though the 'effective' *df* depend on the temporal autocorrelation of the sampled data (i.e. 100 scans rarely means 100 independent observations; Chapter 14), so there is a limit to the power increase afforded by a shorter T_R.

If you are only interested in group results (e.g. extrapolating from a random sample of subjects to a population), then the statistical power normally depends more heavily on the number of subjects than the number of scans per subject (Friston *et al.*, 2002). In other words, you are likely to have more power with 100 scans on 20 subjects, than with 400 scans on 5 subjects, particularly given that inter-subject variability tends to exceed inter-scan variability. Having said this, there are practical issues, like the preparation time necessary to position the subject in the scanner; that means that 100 scans on 20 subjects takes more time than 400 scans on 5 subjects. A common strategy is therefore to run several experiments on each subject while they are in the scanner.

Keep the subject as busy as possible

This refers to the idea that 'dead-time' – time during which the subject is not engaged in the task of interest – should be minimized. Again, of course, there may be psychological limits to the subject's performance (e.g. they may need rests), but apart from this, factors such as the SOA should be kept as short as possible (even within blocks of trials). The only situation where you might want longer SOAs (or blocks of rest) is if you want to measure 'baseline'. From a cognitive perspective though, baseline is rarely meaningful, since it is rarely under strong experimental control (see below).

Only stop the scanner – i.e. break your experiment into sessions – if it is strictly necessary. Breaks in scanning disrupt the spin equilibrium (i.e. require extra dummy scans), reduce the efficiency of any temporal filtering (since the data no longer constitute a single time-series), and introduce other potential 'session' effects (McGonigle *et al.*, 2000).

Do not contrast trials that are remote in time

One problem with fMRI is that there is a lot of low-frequency noise. This has various causes, from aliased biorhythms to gradual changes in physical parameters (e.g. ambient temperature). Thus, any low-frequency 'signal' (induced by your experiment) may be difficult to distinguish from background noise. This is why SPM recommends a highpass filter (see Chapter 14). Since contrasts between trials that are far apart in time correspond to low-frequency effects, they may be filtered out.

In SPM, for example, a typical highpass cut-off is $1/128\,\text{s} \sim 0.01\,\text{Hz}$, based on the observation that the amplitude as a function of frequency, f, for a subject at rest has a '$1/f+$ white noise' form (Plate 19), in which amplitude reaches a plateau for frequencies above approximately 0.01 Hz (the inflection point of the '1/f' and 'white' noise components). When summing over frequencies (in a statistical analysis), the removal of frequencies below this cut-off will increase the signal-to-noise ratio (SNR), provided that most of the signal is above this frequency.

In the context of blocked designs, the implication is not to use blocks that are too long. For two alternating conditions, for example, block lengths of more than 50 s would cause the majority of the signal (i.e. that at the fundamental frequency of the square-wave alternation) to be removed when using a highpass cut-off of 0.01 Hz. In fact, the optimal block length in an on-off design, regardless of any highpass filtering, is approximately 16 s (see below).

Randomize the order, or SOA, of trials close together in time

As will be explained below, in order to be sensitive to differences between trials close together in time (e.g. less than 20 s), one either uses a fixed SOA but varies the order of different trial-types (conditions), or constrains their order but varies the SOA. Thus, a design in which two trials alternate every 4 s is inefficient for detecting the difference between them. One could either randomize their order (keeping the SOA fixed at 4 s), or vary their SOA (keeping the alternating order).[3]

Signal-processing perspective

We begin by assuming that one has an event-related design, and the interest is in detecting the presence (i.e. measuring the amplitude) of a BOLD impulse response whose shape is well-characterized (i.e. a canonical HRF).[4] Given that we can treat fMRI scans as time-series, some intuition can be gained from adopting a signal-processing perspective, and by considering a number of simple examples.

To begin with, consider an event every 16 s. The result of convolving delta functions representing the events with the canonical HRF is shown in Figure 15.5(a) (see Chapter 14 for a discussion of linear convolution models). Maximizing the efficiency of a design is equivalent to maximizing the 'energy' of the predicted fMRI time-series, i.e. the sum of squared signal values at each scan (equal to the variance of the signal, after mean-correction). In other words, to be best able to detect the

[3] Note that, in this context, blocks can be viewed as runs of trials of the same type, and a blocked design corresponds to a varying-SOA design in which there is bimodal distribution of SOAs: a short SOA corresponding to the SOA within blocks, and a long SOA corresponding to the SOA between the last trial of one block and the first of the next.

[4] A distinction has been made between the ability to detect a response of known shape, 'detection efficiency' (as considered here), and the ability to estimate the shape of a response, 'estimation efficiency' [0] (Liu *et al.*, 2001; Birn *et al.*, 2002). This distinction actually reduces simply to the choice of temporal basis functions: The same efficiency equation (Eqn.**15.3** below) can be used to optimize either detection or estimation efficiency by using different response functions: e.g. either a canonical HRF or a FIR basis set respectively. A blocked design will optimize detection efficiency; whereas a randomized design with null events will optimize estimation efficiency (see Henson, 2004 for further details). Hagberg *et al.* (2001) considered a range of possible SOA distributions (bimodal in the case of blocked designs, exponential in the case of fully randomized designs) and showed that 'long-tail' distributions combine reasonable detection and estimation efficiency.

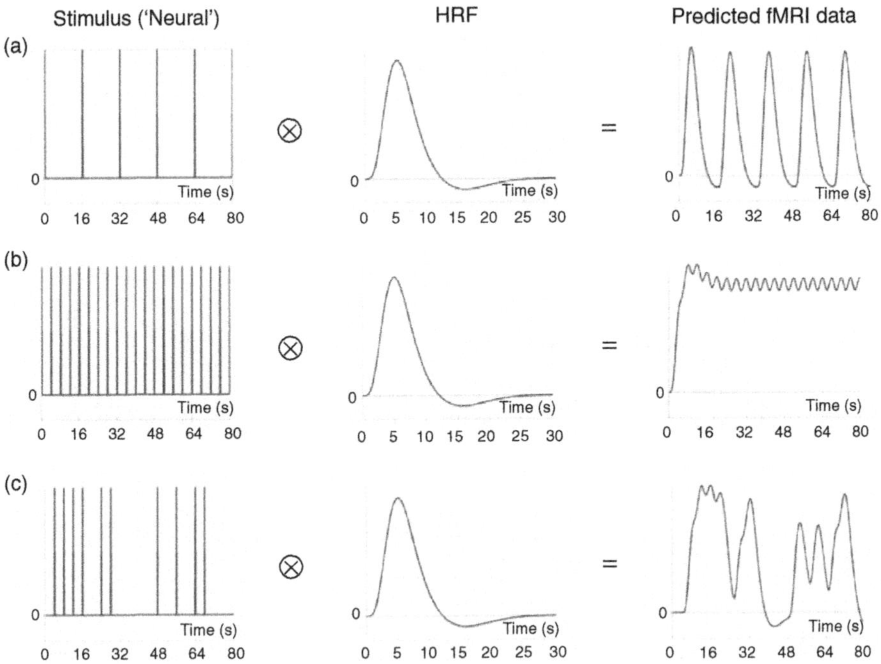

FIGURE 15.5 Effect of convolution by an assumed HRF on neuronal activity evoked by (a) events every 16 s, (b) events every 4 s and (c) events occurring with a 50 per cent probability every 4 s.

signal in the presence of background noise, we want to maximize the variability of that signal. A signal that varies little will be difficult to detect.

The above example (a fixed SOA of 16 s) is not particularly efficient, as we shall see later. What if we present the stimuli much faster, say every 4 s? The result is shown in Figure 15.5(b). Because the responses to successive events now overlap considerably, we see an initial build-up (transient) followed by small oscillations around a 'raised baseline'. Although the overall signal is high, its variance is low, and the majority of stimulus energy will be lost after highpass filtering (particularly after removal of the mean, i.e. lowest frequency). So this is an even less efficient design.

What if we vary the SOA randomly? Let's say we have a minimal SOA of 4 s, but only a 50 per cent probability of an event every 4 s. This is called a stochastic design (and one way to implement it is to intermix an equal number of 'null events' with 'true events'; see next section). This is shown in Figure 15.5(c). Though we only use half as many stimuli as in Figure 15.5(b), this is a more efficient design. This is because there is a much larger variability in the signal.

We could also vary the SOA in a more systematic fashion. We could have runs of events, followed by runs of no (null) events. This corresponds to a blocked design. For example, we could have blocks of 5 stimuli presented every 4 s, alternating with 20 s of rest, as shown in Figure 15.6(a). This is even more efficient than the previous stochastic design. To see why, we shall consider the Fourier transform of these time-series. First, however, note that, with short SOAs, the predicted fMRI time-series for a blocked design is similar to what would obtain if neuronal activity were sustained throughout the block (i.e. during the Interstimulus interval (ISI) as well) as in an epoch model (Figure 15.6(b)). Now, if we take the Fourier transform of each function in Figure 15.6(b), we can plot amplitude (magnitude) as a function of frequency (Figure 15.6(c)). The amplitude spectrum of the square-wave stimulus function has a dominant frequency corresponding to its 'fundamental' frequency ($F_0 = 1/(20\,\mathrm{s}+20\,\mathrm{s}) = 0.025\,\mathrm{Hz}$), plus a series of 'harmonics' ($3F_0$, $5F_0$, ... *etc.*) of progressively decreasing amplitude. The fundamental frequency corresponds to the frequency of a sinusoidal that best matches the basic on-off alternation; the harmonics can be thought of as capturing the 'sharper' edges of the square-wave function relative to this fundamental sinusoid.

The reason for performing the Fourier transform is that it offers a slightly different perspective. Foremost, a convolution in time is equivalent to a multiplication in frequency space. In this way, we can regard the stimulus function as our original data and the HRF as a 'filter'. One can see immediately from the shape of the Fourier transform of the HRF that this filter will 'pass' low frequencies, but attenuate higher frequencies (this is why it is sometimes called a 'lowpass filter' or 'temporal smoothing kernel'). This property is why, for example, much high-frequency information was lost with the fixed SOA of 4 s in Figure 15.5(b). In the present example, the result of multiplying the amplitude spectrum of the stimulus function by that of the filter is that some of

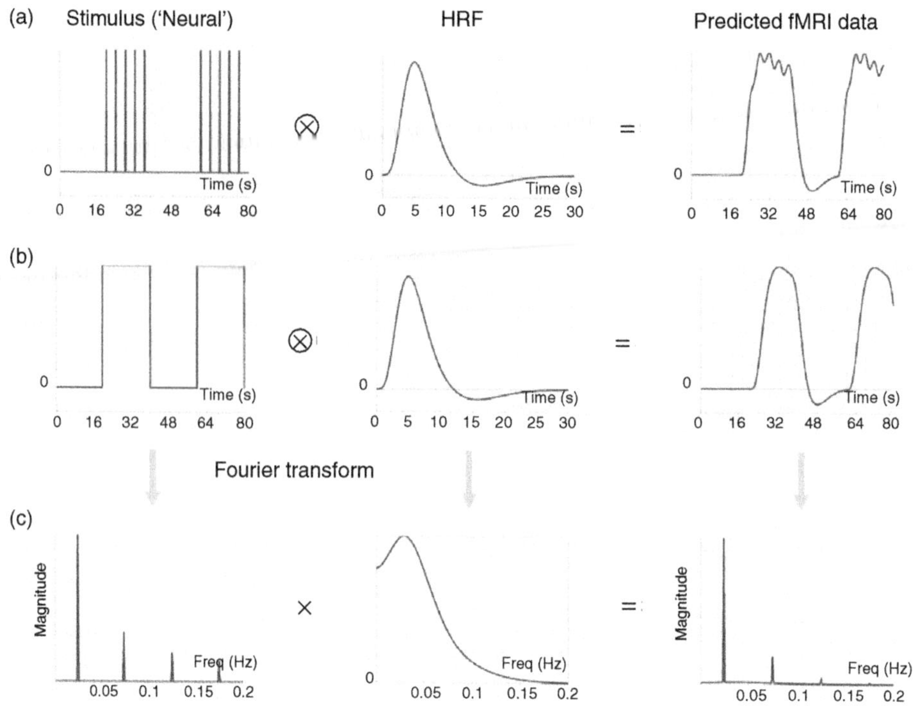

FIGURE 15.6 Effect of convolution by an assumed HRF on neuronal activity evoked by (a) blocks of events, (b) epochs of 20 s, and (c) the amplitude spectra after Fourier transform of (b).

the higher-frequency harmonics are attenuated, but the amplitude of the fundamental frequency is not. In other words, the majority of the signal is 'passed' by the HRF filter.

We are now in a position to answer the question: what is the most efficient design of all? Well, assuming we had a limited amount of total 'stimulus energy', the optimal design would be to modulate the neuronal activity in a sinusoidal fashion, with a frequency that matches the peak of the amplitude spectrum of the HRF filter. With the canonical HRF used here, this would be ~0.03 Hz (1/30 s). The sinusoidal modulation places all the stimulus energy at this single frequency, shown by the single line in frequency space in Figure 15.7.

We can now also turn to the question of highpass filtering. Because the filtering is commutative, we can apply the highpass filter to the lowpass filter inherent in the HRF to create a single bandpass filter (or 'effective HRF', Josephs and Henson, 1999). This is shown in Figure 15.8, in which the highpass filter reflects the 'chunk' of low frequencies that has been removed from the HRF filter (highpass cut-off here = 1/120 s ~0.008 Hz). The consequence of highpass filtering is shown for long blocks of 80 s (20 trials every 4 s). Because the fundamental

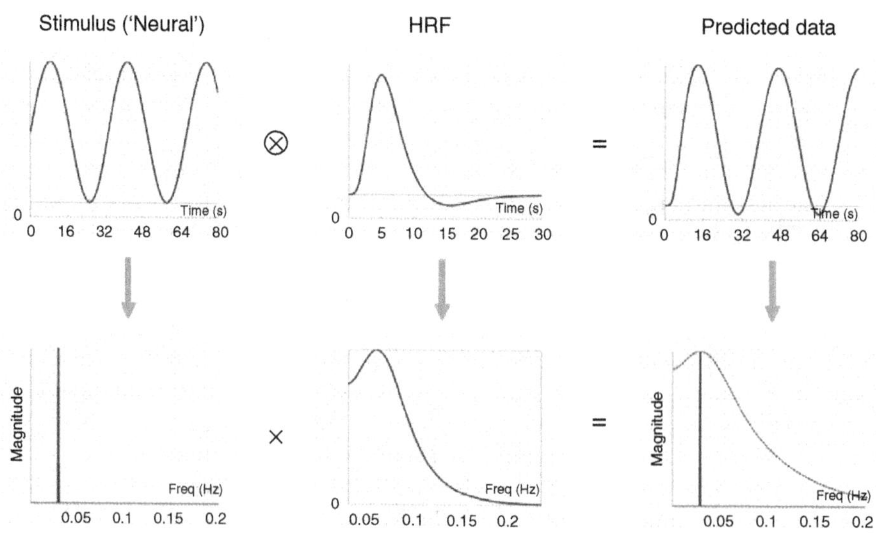

FIGURE 15.7 Effect of convolution by an assumed HRF on sinusoidal neuronal activity of 0.03 Hz.

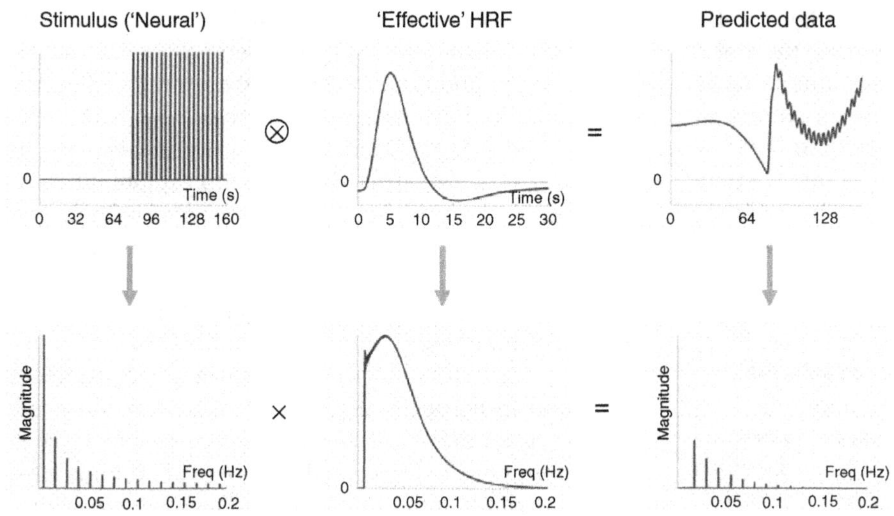

FIGURE 15.8 Effect of convolution by an 'effective' HRF (i.e. including a highpass filter) on 80-s blocks of trials.

frequency in this design (1/160 s ~0.006 Hz) is lower than the highpass cut-off, a large proportion of signal energy is lost (reflected by the rather strange shape of the predicted fMRI time-series, in which the lowest frequency has been removed). This is therefore not an efficient design (with this specific highpass cut-off). This illustrates the general point that blocked designs are only efficient when the block length is not too long: approx. 15 s-on, 15 s-off is optimal (see Figure 15.7). Block durations of up to 50 s-on, 50 s-off are also fine (given that the HRF filter does not attenuate low frequencies much), but block durations much longer than this (or contrasts between two of many different types of 50 s-blocks) may be in danger of being swamped by low-frequency noise.

Finally, we can return to consider what happens in a stochastic design like that in Figure 15.5(c). The effect of the randomized SOA is to 'spread' the signal energy across a range of frequencies, as shown in Figure 15.9. Some of the high- and low-frequency components are lost to the effective HRF filter, but much is passed, making it a reasonably efficient design.

Statistical perspective

From the statistical perspective, the aim is to minimize the standard error of a *t*-contrast, $c^T\hat{\beta}$ (i.e. the denominator of a *t*-statistic; Chapter 8). Given the specified contrast of interest, c, and parameter estimates, $\hat{\beta}$, the variance of $c^T\hat{\beta}$ is given by (Friston *et al.*, 2000):

$$\mathrm{var}(c^T\hat{\beta}) = \sigma^2 c^T (SX)^+ SVS^T (SX)^{+T} c \qquad \mathbf{15.2}$$

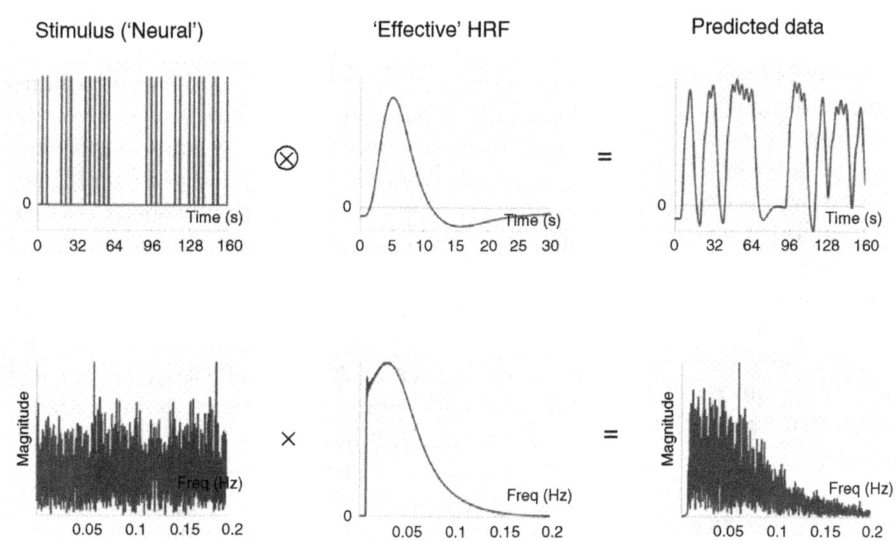

FIGURE 15.9 Effect of convolution by an 'effective' HRF on randomized SOA events (minimum = 4 s).

where S is a filter matrix incorporating the highpass filter and any temporal smoothing, and V is the noise autocorrelation matrix. We want to *minimize* this variance with respect to the design matrix, X. If we assume that the filter matrix S is specified appropriately to 'whiten' the residuals, such that $SVS^T = I$ (i.e. when $S = K^{-1}$, where $KK^T = V$; Chapter 14), and we incorporate S into X, then this is equivalent to *maximizing* the efficiency, ζ:

$$\xi(\sigma^2, c, X) = (\sigma^2 c^T (X^T X)^{-1} c)^{-1} \qquad 15.3$$

For a given contrast, c, this equation can be split into the 'noise variance', σ^2, and the 'design variance', $(X^T X)^{-1}$ (Mechelli *et al.*, 2003b). If one assumes that the noise variance is independent of the specific design used (which may not be the case, Mechelli *et al.*, 2003b; see later), then the efficiency of a contrast for a given design is proportional to:

$$\xi(c, X) = (c^T (X^T X)^{-1} c)^{-1} \qquad 15.4$$

(For F-contrasts, where c is a matrix, the *trace* operator can be used to reduce efficiency to a single number; Dale, 1999). Note that $\xi(c, X)$ has no units; it is a relative measure. It depends on the scaling of the design matrix and the scaling of the contrasts. Thus, all we can really say is that one design is more efficient than another (for a given contrast). In what follows, we use Eqn. **15.4** to evaluate the efficiency of different sorts of design and look at how designs can be characterized probabilistically.

Stochastic designs

For a single event-type, the space of possible experimental designs can be captured by two parameters: the minimal SOA (Δt) and the probability, p_t, of an event occurring at every Δt (Friston *et al.*, 1999). In 'deterministic' designs, $p_t = 1$ or $p_t = 0$, giving a series of events with fixed SOA, as in Figure 15.5(a). In 'stochastic' designs $0 \leq p_t \leq 1$, producing a range of SOAs (as in Figure 15.5(c)). For 'stationary' stochastic designs, p_t is constant, giving an exponential distribution of SOAs; for 'dynamic' stochastic designs, p_t changes with time. The extreme case of a dynamic stochastic design is one in which the temporal modulation of p_t conforms to a square-wave, corresponding to a blocked design. Notice that the quantities p_t and Δt parameterize a space of design matrices probabilistically. In other words, they specify the probability $p(X|p_t, \Delta t)$ of getting any particular design matrix. This allows one to compute the expected design efficiency for any class that can be parameterized in this way:

$$\langle \xi(c, p_t, \Delta t) \rangle = \int p(X|p_t, \Delta t) \xi(c, X) dX \qquad 15.5$$

This expected design efficiency can be evaluated numerically by generating large numbers of design matrices (using p_t and Δt) and taking the average efficiency according to Eqn. **15.4**. Alternatively, one can compute the expected efficiency analytically as described in Friston *et al.* (1999). This allows one to explore different sorts of designs by treating the design matrix itself as a random variable. For stochastic designs, efficiency is generally maximal when the Δt is minimal and the (mean) $p_t = 1/(L+1)$, where L is the number of trial types (see Friston *et al.*, 1999).

Figure 15.10 shows the expected efficiency for detecting canonical responses to a single event-type versus baseline, i.e. $L = 1$ and $c = 1$, for a range of possible designs. The deterministic design with $\Delta t = 8$ s (top row) is least efficient, whereas the dynamic stochastic design with a square-wave modulation with $\Delta t = 1$ s is the most efficient (corresponding, in this case, to a 32 s on-off blocked design). Intermediate between these extremes are the dynamic stochastic designs that use a sinusoidal modulation of p_t. In other words, these designs produce clumps of events close together in time, interspersed with periods in which events are rarer. Though such designs are less efficient than the blocked (square-wave) design, they are more efficient than the stationary stochastic design with $\Delta t = 1$ s (second row of Figure 15.10), and assuming that subjects are less likely to notice the 'clumping' of events (relative to a fully blocked design), may offer a good compromise between efficiency and subjective unpredictability.

Transition probabilities

The space of possible designs can also be characterized by Δt and a 'transition matrix' (Josephs and Henson, 1999). This is a generalization of the above formulation that introduces conditional dependencies over time. For $L > 1$ different event-types, a $L^m \times L$ transition matrix captures the probability of an event being of each type, given the history of the last m event-types. A fully randomized design with two event-types (A and B) has a simple first-order transition matrix in which each probability is a half. The efficiencies of two contrasts, the main effect of A and B (versus baseline), $c = [1\ 1]^T$, and the differential effect, $c = [1\ -1]^T$, are shown as a function of Δt in Plate 20(a). The optimal SOA for the main effect under these conditions is approximately 20 s. The efficiency of the main effect decreases for shorter SOAs, whereas the efficiency of the differential effect increases. Clearly, the efficiency for the differential contrast cannot increase indefinitely as the SOA decreases; at some point, the BOLD response must saturate (see below). Nonetheless, this graph clearly

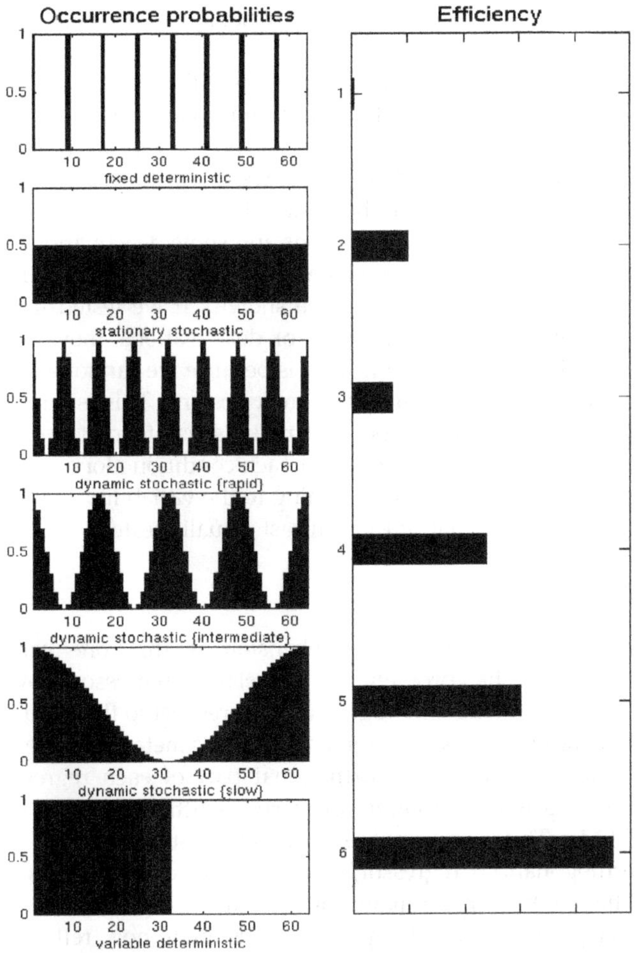

FIGURE 15.10 Efficiency for a single event-type (from Friston *et al.*, 1999). Probability of event each SOA (left column) and expected design efficiency (right column, increasing left-to-right) for a deterministic design with $\Delta t = 8$ s (1st row), a stationary stochastic (randomized) design with $p_t = 0.5$ (2nd row) and dynamic stochastic designs with modulations of p_t by different sinusoidal frequencies (3rd to 5th rows) and in a blocked manner every 32 s (6th row).

demonstrates how the optimal SOA depends on the specific contrast of interest.[5]

Various experimental constraints on multiple event-type designs can also be considered. In some situations, the order of event-types might be fixed, and the design question relates to the optimal SOA. For a design in which A and B must alternate (e.g. where A and B are transitions between two perceptual states), the optimal SOA for a differential effect is 10 s (Plate 20(b), i.e. half of that for the main effect). In other situations, experimental constraints may limit the SOA, to at least 10 s say, and the design question relates to the optimal stimulus ordering. An alternating design is more efficient than a randomized design for such intermediate SOAs. However, an alternating design may not be advisable for psychological reasons (subjects' behaviour might be influenced by the predictable pattern). In such cases, a permuted design (in which each of trial-types is presented successively in a randomly-permuted order) may be a more suitable choice (see Plate 20(b)).

A further design concept concerns 'null events'. These are not real events, in that they do not differ from the baseline and hence are not detectable by subjects (so are not generally modelled in the design matrix). They were introduced by Dale and Buckner (1997) as 'fixation trials', to allow 'selective averaging' (see Chapter 14). In fact, they are simply a convenient means of creating a stochastic design by shuffling a certain proportion of null events among the events of interest (producing an exponential distribution of SOAs). From the perspective of multiple event-type designs, the reason for null events is to buy efficiency for both the main effect and differential effect at short SOAs (at a slight cost to the efficiency for the differential effect; see Plate 20(c)).

The efficiencies shown in Plate 20 are unlikely to map simply (e.g. linearly) onto the size of the t-statistic. Nonetheless, if the noise variance, in Eqn. **15.3**, is independent of experimental design, the relationship should at least be monotonic. Mechelli *et al.* (2003b) showed that the noise variance can vary significantly between a blocked and a randomized design (both modelled with events). This suggests that the stimulus ordering did affect (un-modelled) psychological or physiological effects in this dataset, contributing to the residual error. When the data were highpass filtered however, the noise variance no longer differed significantly between the two designs. In this case, the statistical results were in agreement with the relative efficiencies predicted from the estimation variances.

Efficiency in terms of correlations

Another way of thinking about efficiency is in terms of the correlation between (contrasts of) regressors within the design matrix. In Eqn. **15.3** the term $X^T X$ is called the information matrix and reflects the orthogonality of the design matrix. High covariance between the columns of the design matrix introduces redundancy. This can increase the covariance of the parameter estimates $(X^T X)^{-1}$ and lead to low efficiency (depending on the particular contrast).

Consider the earlier example of two event-types, A and B, randomly intermixed, with a short SOA. If we plot

[5] The main effect, which does not distinguish A and B, is of course equivalent to a deterministic design, while the differential effect is equivalent to a stochastic design (from the perspective of any one event-type).

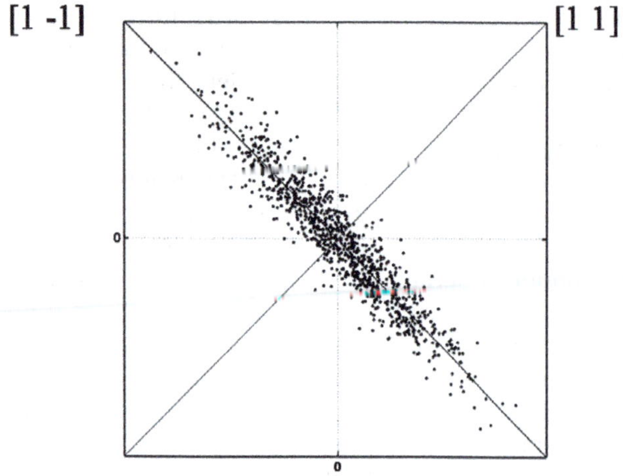

FIGURE 15.11 Scatter plot for two mean-corrected regressors (one point per scan) corresponding to two event-types randomly intermixed with a short SOA.

the resulting two regressors (after convolution with an HRF) against each other, we would end up with a scatter plot something like that in Figure 15.11, where each point reflects one scan. The high negative correlation between the regressors is because whenever there is high signal for A, there tends to be low signal for B, and vice versa. If we consider the projection of this distribution onto the $x = -y$ direction (corresponding to a [1 −1] contrast), it will have a large dispersion, i.e. high experimental variance, which means the difference between A and B will be detected efficiently in this design. However, if we project the distribution onto the $x = y$ direction (corresponding to a [1 1] contrast), it will have little spread, i.e. low experimental variance, which means that we will not detect the common effect of A and B versus baseline efficiently. This demonstrates the markedly different efficiency for these two contrasts at short SOAs that was shown in Plate 20(a).

Projection onto the x or y axes (i.e. [1 0] or [0 1] contrasts) will have less spread than if the two regressors were orthogonal and formed a spherical cloud of points. This shows how correlations can reduce efficiency and makes an important general point about correlations. High correlation between two regressors means that the parameter estimate for each one will be estimated inefficiently, i.e. the parameter estimate itself will have high variance. In other words, if we estimated each parameter many times we would get wildly different results. In the extreme case, that the regressors are perfectly correlated, the parameters would be inestimable (i.e. they would have infinite variance). Nonetheless, we could still estimate efficiently the difference between them. Thus, high correlations within the orthogonality matrix shown by SPM should not be a cause of concern for some contrasts: what is really relevant is the correlation between the *contrasts* of interest (i.e. linear combinations of columns of the design matrix) relative to the rest of the design matrix (i.e. null space of the contrast).

In short-SOA, randomized designs with no null events, for example, we might detect brain regions showing a reliable difference between event-types, yet when we plot the event-related response, we might find they are all 'activations' versus baseline, all 'deactivations' versus baseline or some activations and some deactivations. However, these impressions are more apparent than real (and should not really be shown). If we tested the reliability of these activations or deactivations, they are unlikely to be significant. This is because we cannot estimate the baseline reliably in such designs. This is why, for such designs, it does not make sense to plot error bars showing the variability of each condition alone: one should plot error bars pertaining to the variability of the difference (i.e. that of the contrast actually tested).

Orthogonalizing

Another common misapprehension is that one can overcome the problem of correlated regressors by 'orthogonalizing' one regressor with respect to the other. This rarely solves the problem. The parameter estimates always pertain to the orthogonal part of each regressor (this is an automatic property of fitting within the GLM). Thus, neither the parameter estimate for the orthogonalized regressor, nor its variance, will change. The parameter estimate for the other regressor will change. However, this parameter estimate now reflects the assumption that the common variance is uniquely attributed to this regressor. We must have an *a priori* reason for assuming this (i.e. without such prior knowledge, there is no way to determine which of the two correlated regressors caused the common effect). In the absence of such knowledge, there is no reason to orthogonalize.

The conception of efficiency in terms of correlations can help with the design of experiments where there is necessarily some degree of correlation among regressors. Two main experimental situations where this arises are:

1 when trials consist of two events, one of which must follow the other
2 blocks of events in which one wishes to distinguish 'item-' from 'state-' effects (see above).

A common example of the first type of experiment are 'working memory' designs, in which a trial consists of a stimulus, a short retention interval, and then a response. We shall ignore the retention interval and concentrate on how one can separate effects of the stimulus from those of the response. With short SOAs between each event-type (e.g. 4 s), the regressors for the stimulus and response will be negatively correlated, as shown in Figure 15.12(a). Two possible solutions to this problem are shown in

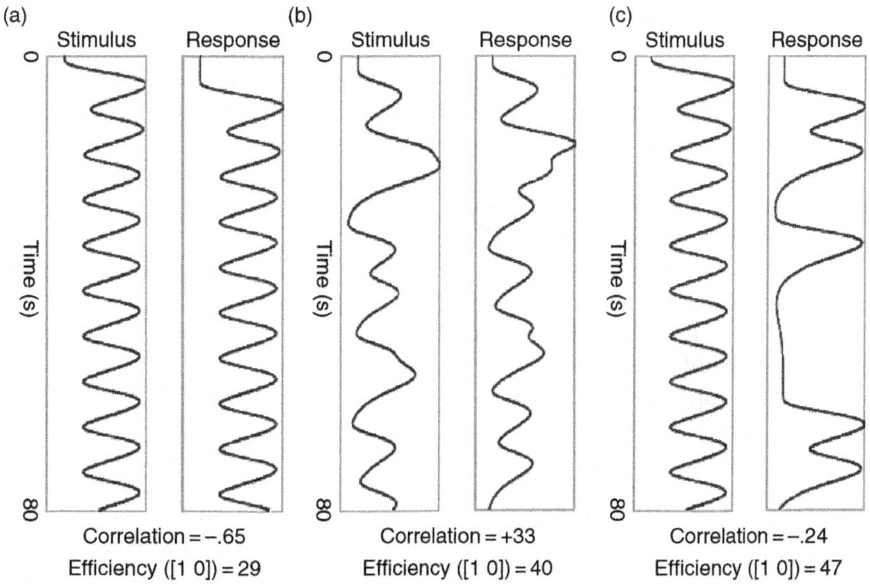

FIGURE 15.12 Regressors for 'working memory' trials presented every 8 s, consisting of (a) stimulus followed after 4 s by a response, (b) stimulus-response intervals varied from 0 to 8 s, and (c) responses following stimuli by 4 s, but only on 50 per cent of trials.

Figure 15.12(b) and 15.12(c). The first is to vary the time between successive stimuli and responses (assuming this is possible and that this variation is large; e.g. 1–8 s). The second is to keep the stimulus-response interval fixed at 4 s, but only cue a response on a random half of trials. The effect of both is to reduce the correlation between the regressors, which increases the efficiency of separate brain activity related to stimuli from that related to responses.

The second type of experiment tries to distinguish transient responses (item-effects) from sustained responses (state-effects). Such separation of transient and sustained effects requires modelling blocks of trials in terms of both individual events within blocks and sustained epochs throughout the blocks. An example with a fixed SOA of 4 s between events is shown in Figure 15.13(a). Here, the correlation between the event and epoch regressors is naturally high, and the efficiency for detecting either effect alone is low. Using the same total number of events per block, but with a pseudo-randomized design in which the events are randomly spread over the block with a minimal SOA of 2 s (Figure 15.13(b)), the correlation is reduced and efficiency increased. (Note that one perverse consequence of having to introduce some long SOAs between events within blocks in such 'mixed designs' is that subjects may be less able to maintain a specific cognitive 'state'.)

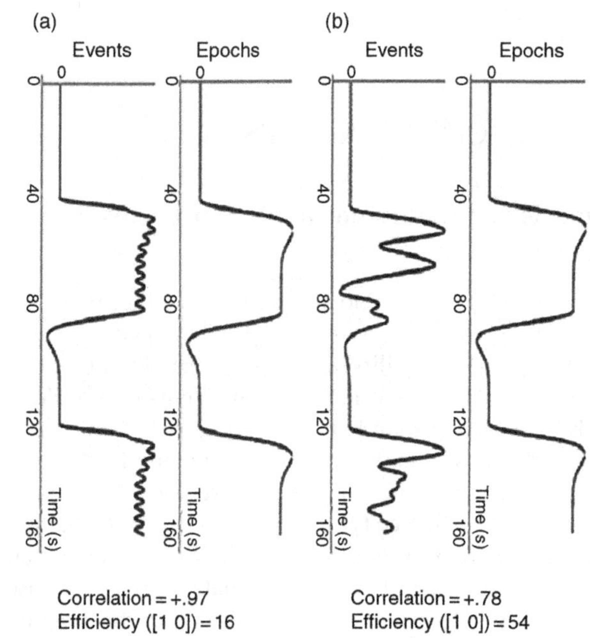

FIGURE 15.13 Regressors for 'mixed designs' that attempt to separate transient (item) from sustained (state) effects. (a) 10 events per block presented every SOA of 4 s, (b) 10 events per block distributed randomly over 2-s SOAs.

Effect of non-linearities on efficiency

The above efficiency arguments have assumed linearity, i.e. that the responses to successive trials summate linearly, no matter how close together in time they occur. In reality, we know there is a 'saturation', or under-additivity, even for SOAs of about 10 s (see Chapters 14 and 27). This means that the efficiency for stochastic designs does not increase indefinitely as the SOA decreases (e.g. for the differential effect in Plate 20(a)). By estimating non-linearity with a Volterra expansion (Chapter 14), Friston *et al.* (1998) calculated the impact of such non-linearity on evoked responses. The result is shown in the insert in Plate 20(a). The dotted line shows the average response to a train of stimuli under linear assumptions; the solid line shows the effects of saturation (using a second-order Volterra kernel). While the solid line is below the dotted line for all SOAs (below 10 s), the divergence is small until SOAs of 1–2 s. Indeed, the prediction of this calculation is that the optimal SOA can be as low as 1 s, i.e. the advantage of short SOAs can outweigh the saturation of responses until surprisingly short SOAs (though it should be noted that this prediction is based on a specific dataset, and may not generalize). Indeed, differential responses between randomized event-types have been detected with SOAs as short as 0.5 s (Burock *et al.*, 1998).

In this chapter, we have looked at how to detect evoked fMRI responses efficiently. Before turning to models of evoked responses in EEG in the next chapter, we will conclude with some common questions that exercise people designing fMRI studies

COMMON QUESTIONS

What is the minimum number of events I need?

Unfortunately, there is no answer to this, other than 'the more, the better'. The statistical power depends on the effect size and variability, and this is normally unknown. Heuristics like 'you cannot do an event-related fMRI analysis with less than N events' are fairly meaningless, unless one has a specific effect size in mind (which is likely to be a function of the brain region, the scanner strength, the sequence type, etc.). Note it is possible that fewer trials are required (for a given power) than for an equivalent contrast of behavioural data (e.g. if the noise level in, say, RTs exceeds that in a specific cortical region contributing to those RTs). Furthermore, it is not even the number of events *per se* that is relevant, it is also the SOA and event-ordering (see next question).

Do shorter SOAs mean more power simply because there are more trials?

It is not simply the number of trials: the temporal deployment of those trials is vital (as explained above). Thus 400 stimuli every 3 s is *less* efficient than 40 stimuli every 30 s for detecting a single event-related response (since a fixed SOA of 3 s produces little experimental variability after convolution by the HRF). Two hundred stimuli occurring with a 50 per cent probably every 3 s (i.e. pseudo-randomly mixed with 200 null events) is much more efficient than either.

What is the maximum number of conditions I can have?

A common interpretation of the rule – do not compare trials that are too far apart in time – is not to design experiments with too many experimental conditions. More conditions necessarily mean that replications of a particular condition will be further apart in time. However, the critical factor is not the number of conditions *per se*, but the specific contrasts performed over those conditions. For pair-wise comparisons of only two of, say, eight blocked conditions the above caveat would apply: if there were equal numbers of blocks of each condition, blocks longer than 12.5 s (100 s/8) are likely to entail a substantial loss of signal when using a highpass cut-off of 0.01 Hz. However, this caveat would not apply if the contrasts of interest included (i.e. 'spanned') all eight conditions. This would be the case if the experimenter were only interested in the two main effects and the interaction within a 2×4 factorial design (i.e. contrasts like $[1\ 1\ 1\ 1\ -1\ \ -1\ \ -1\ \ -1]$). If you must compare or plot only a subset of many such blocked conditions, you should consider presenting those blocks in a fixed order, rather than random or counterbalanced order, which will minimize the time between replications of each condition, i.e. maximize the frequency of the contrast.

Should I use null events?

Null events are simply a convenient means of achieving a stochastic distribution of SOAs, in order to allow estimation of the response versus inter-stimulus baseline, by randomly intermixing them with the events of interest. However, the 'baseline' may not always be meaningful. It may be well defined for V1, in terms of visual flashes versus a dark background. It becomes less well defined for 'higher' regions associated with cognition because it is unclear what these regions are 'doing' during the inter-stimulus interval. The experimenter normally has little control over this. Moreover, the baseline does not control for the fact that the events of interest are impulsive (rapid changes), whereas the baseline is sustained (and may entail adaptation or reduced attention). For this reason, it is often better to forget about baseline and add an extra low-level control event instead.

Another problem with null events is that, if they are too rare (e.g. less than approximately 33 per cent), they actually become 'true' events in the sense that subjects may be expecting an event at the next SOA and so be surprised when it does not occur (the so-called missing stimulus effect that is well-known in event-related potential (ERP) research). One solution is to replace randomly intermixed null events with periods of baseline between runs of events (i.e. 'block' the baseline periods). This will increase the efficiency for detecting the common effect versus baseline, at a slight cost in efficiency for detecting differences between the randomized event-types within each block. Yet another problem is that the unpredictability of the occurrence of true events (caused by the randomly intermixed null events) can cause delayed or even missed processing of the events of interest, if subjects cannot prepare for them.

In summary, null events are probably only worthwhile if:

1 you think the mean activity during the constant interstimulus interval is meaningful to contrast against
2 you do not mind null events being reasonably frequent (to avoid 'missing stimulus' effects)
3 you do not mind the stimulus occurrence being unpredictable (as far as the subject is concerned).

Having said this, some form of baseline can often serve as a useful 'safety net' (e.g. if you fail to detect differences between two visual event-types of interest, you can at least examine V1 responses and check that you are seeing a basic evoked response to both event-types – if not, you can question the quality of your data or accuracy of your model). Moreover, you may need randomly to inter-mix null events if you want to estimate more precisely the shape of the BOLD impulse response (see footnote 4). It is often the case that people include a low-level baseline or null event to use as reference for a localizing contrast on tests for differences among true events. In other words, the contrast testing for all events versus baseline can serve as a useful constraint on the search volume for interesting comparisons among events.

Should I generate multiple random designs and choose the most efficient?

This is certainly possible, though be wary that such designs are likely to converge on designs with some structure (e.g. blocked designs, given that they tend to be optimal, as explained above). This may be problematic if such structure affects subjects' behaviour (particularly if they notice the structure). Note, however, that there are software tools available that optimize designs at the same time as allowing users to specify a certain level of counterbalancing (to avoid fully blocked designs, e.g. Wager and Nichols, 2003).

REFERENCES

Birn RM, Cox RW, Bandettini PA (2002) Detection versus estimation in event-related fMRI: choosing the optimal stimulus timing. *NeuroImage* **15**: 252–64

Burock MA, Buckner RL, Woldorff MG *et al.* (1998) Randomized event-related experimental designs allow for extremely rapid presentation rates using functional MRI. *NeuroReport* **9**: 3735–39

Chawla D, Rees G, Friston KJ (1999) The physiological basis of attentional modulation in extrastriate visual areas. *Nat Neurosci* **2**: 671–76

Dale AM (1999) Optimal experimental design for event-related fMRI. *Hum Brain Mapp* **8**: 109–14

Dale A, Buckner R (1997) Selective averaging of rapidly presented individual trials using fMRI. *Hum Brain Mapp* **5**: 329–40

Friston KJ, Price CJ, Fletcher P *et al.* (1996) The trouble with cognitive subtraction. *NeuroImage* **4**: 97–104

Friston KJ, Josephs O, Rees G *et al.* (1998) Non-linear event-related responses in fMRI. *Mag Res Med* **39**: 41–52

Friston KJ, Zarahn E, Josephs O *et al.* (1999) Stochastic designs in event-related fMRI. *NeuroImage* **10**: 607–19

Friston KJ, Josephs O, Zarahn E *et al.* (2000) To smooth or not to smooth? Bias and efficiency in fMRI time-series analysis. *NeuroImage* **12**: 196–208

Friston KJ, Glaser DE, Henson RNA *et al.* (2002) Classical and Bayesian inference in neuroimaging: applications. *NeuroImage* **16**: 484–512

Friston KJ, Penny WD, Glaser DE (2005) Conjunction revisited. *NeuroImage* [15.3]

Hagberg GE, Zito G, Patria F *et al.* (2001) Improved detection of event-related functional MRI signals using probability functions. *NeuroImage* **14**: 193–205

Henson RNA (2004) Analysis of fMRI time-series: linear time-invariant models, event-related fMRI and optimal experimental design. In *Human Brain Function*, 2nd edn, Frackowiak RS, Friston KJ, Frith CD *et al.* (eds). Elsevier, London, pp 793–822

Henson RNA (2005) What can functional neuroimaging tell the experimental psychologist? *Quart J Exp Psych A* **58**: 193–234

Henson RNA, Büchel C, Josephs O *et al.* (1999a) The slice-timing problem in event-related fMRI. *NeuroImage* **9**: 125

Henson RNA, Rugg MD, Shallice T *et al.* (1999b) Recollection and familiarity in recognition memory: an event-related fMRI study. *J Neurosci* **19**: 3962–72

Henson RNA, Rugg MD (2001) Effects of stimulus repetition on latency of the BOLD impulse response. *NeuroImage* **13**: 683

Johnson MK, Nolde SF, Mather M *et al.* (1997) Test format can affect the similarity of brain activity associated with true and false recognition memory. *Psychol Sci* **8**: 250–57

Josephs O, Henson RNA (1999) Event-related fMRI: modelling, inference and optimisation. *Phil Trans Roy Soc Lond* **354**: 1215–28

Josephs O, Turner R, Friston KJ (1997) Event-related fMRI. *Hum Brain Mapp* **5**: 243–48

Kleinschmidt A, Büchel C, Zeki S *et al.* (1998) Human brain activity during spontaneously reversing perception of ambiguous figures. *Proc R Soc Lond B Biol Sci* **265**: 2427–33

Liu TT, Frank LR, Wong EC *et al.* (2001) Detection power, estimation efficiency and predictability in event-related fMRI. *NeuroImage* **13**: 759–73

McGonigle DJ, Howseman AM, Athwal BS *et al.* (2000) Variability in fMRI: an examination of intersession differences. *NeuroImage* **11**: 708–34

Mechelli A, Henson RNA, Price CJ *et al.* (2003a) Comparing event-related and epoch analysis in blocked design fMRI. *NeuroImage* **18**: 806–10

Mechelli A, Price CJ, Henson RNA *et al.* (2003b) The effect of high-pass filtering on the efficiency of response estimation: a comparison between blocked and randomised designs. *NeuroImage* **18**: 798–805

Miezin FM, Maccotta L, Ollinger JM *et al.* (2000) Characterizing the hemodynamic response: effects of presentation rate, sampling procedure, and the possibility of ordering brain activity based on relative timing. *NeuroImage* **11**: 735–59

Nichols TE, Brett M, Andersson J *et al.* (2005) Valid conjunction inference with the minimum statistic. *NeuroImage* **25**: 653–60

Price CJ, Friston KJ (1997) Cognitive conjunction: a new approach to brain activation experiments. *NeuroImage* **5**: 261–70

Price CJ, Veltman DJ, Ashburner J *et al.* (1999) The critical relationship between the timing of stimulus presentation and data acquisition in blocked designs with fMRI. *NeuroImage* **10**: 36–44

Sternberg S (1969) The discovery of processing stages: extensions of Donders method. *Acta Psychol* **30**: 276–315

Strange BA, Henson RN, Friston KJ *et al.* (2000) Brain mechanisms for detecting perceptual, semantic, and emotional deviance. *NeuroImage* **12**: 425–33

Wager TD, Nichols TE (2003) Optimization of experimental design in fMRI: a general framework using a genetic algorithm. *NeuroImage* **18**: 293–309

Zarahn E, Aguirre G, D'Esposito M (1997) A trial-based experimental design for fMRI. *NeuroImage* **6**: 122–38

CHAPTER

16

Hierarchical models for EEG and MEG

S. Kiebel, J. Kilner and K. Friston

INTRODUCTION

In this chapter, we will look at hierarchical linear models for magneto- and electroencephalographic (M/EEG) data. Traditionally, the analysis of evoked responses is based on analysis of variance (ANOVA) (see Chapter 13). Statistical parametric mapping (SPM) follows the same route, but we motivate the model within a hierarchical framework, paying special attention to the distinction between time as a factor and time as a dimension of a multivariate response variable. We describe the analysis of both event-related responses (ERR)[1] and power data.

The chapter is divided into two parts. First, we discuss some fundamental modelling issues pertaining to the analysis of M/EEG data over space and time. Equivalent considerations emerged in positron emission tomography/functional magnetic resonance imaging (PET/fMRI) more than a decade ago where one can treat imaging data either in a mass-univariate (Friston *et al.*, 1991) or multivariate fashion (Friston *et al.*, 1995; Worsley *et al.*, 1997). The same considerations also apply to time bins in M/EEG data. However, the situation is different for event-related responses, because one might want to make inferences about the temporal form of responses. This means time has to be treated as an experiential factor, as opposed to another dimension of the response space. Finely resolved temporal features in M/EEG are important, because they may contain important information about neuronal dynamics. The implications for models of M/EEG are that the data can be analysed in one of two ways: they can be regarded as high-dimensional responses in space and time, or they can be treated as a time-series at each point in space. The first section discloses this distinction and their relative merits. In the second part, we describe the analysis of M/EEG data using linear models in which time becomes a factor. The models are essentially the same as those presented in Chapters 8 and 13 and we recommend these chapters are read first.

We will focus on the analysis of averaged (ERR) data (Figure 16.1) and then extend the ideas to cover other forms of M/EEG data analysis (e.g. single-trial and power data). Source reconstruction, using informed basis functions and restricted maximum likelihood (ReML) covariance estimates, artefact removal or correction and averaging are considered here to be pre-processing issues (see Chapters 28 and 29). After pre-processing, the ERR data constitute a time-series of three-dimensional images over peristimulus time bins. These images may be scalar images corresponding to current source density or three-variate images retaining information about source orientation. Here, we assume that we are dealing with univariate or scalar response variables, at each voxel and time bin.

The approaches we consider can also be applied to ERR data which have been projected onto the scalp surface. Of course, this two-dimensional representation does not allow for a full integration with other neuroimaging data (e.g. fMRI) but might be an appropriate way to proceed when source reconstruction is not feasible (Plate 21; see colour plate section).

Some key issues

After projection to voxel-space during interpolation or source reconstruction, the data can be represented as an array of several dimensions. These dimensions include: (i) space (in two or three dimensions); (ii) peristimulus

[1] By ERR, we mean averaged event-related time courses (Rugg and Coles, 1995), where each of these time courses has been averaged within subject and trial-type (condition) to provide one peristimulus time-series for each trial-type and each subject. The modality can be either EEG or MEG.

Statistical Parametric Mapping, by Karl Friston et al.
ISBN–13: 978-0-12-372560-8 ISBN–10: 0-12-372560-7

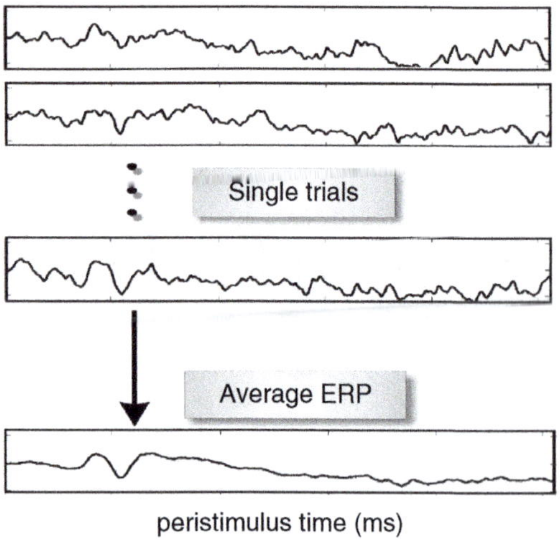

FIGURE 16.1 The average ERR is estimated over single trials for each trial type and each subject.

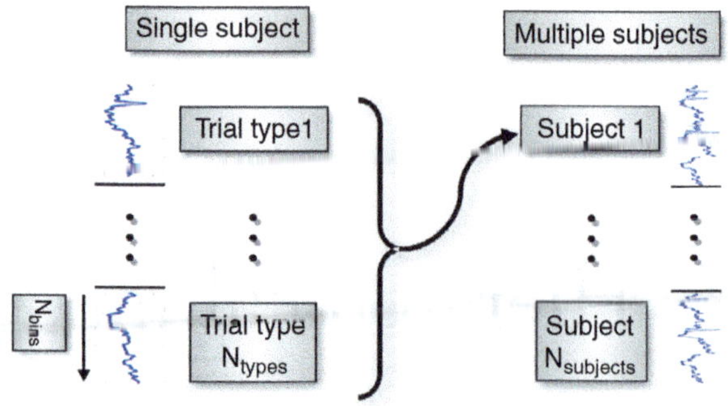

FIGURE 16.2 Data vectorization. For each voxel, one has $N_{subjects}N_{types}$ ERRs, giving $N = N_{subjects}N_{types}N_{bins}$ data points.

time or peristimulus time-frequency; and (iii) design factors (e.g. group, subjects, experimental conditions, single trials). For a given experiment, our task is to specify a model that captures not only experimental effects, but also the correlations between space, time and design. These correlations are important, because they allow us to assess the significance of experimentally induced effects. A failure to capture correlations properly will cause inexact inference, which will lead to either lenient or conservative tests.

The three key questions addressed in the first part of this chapter are:

1 Shall we treat space as an experimental factor or as a dimension of the data?
2 Shall we treat time as an experimental factor or as a dimension of the data?
3 If time is an experimental factor, what is a good (linear) model?

In the following sections we explain what these questions mean and why they are important.

Notation

We assume that the source reconstructed data consists of three-dimensional images with M voxels. Each image contains data for one time bin. In other words, each voxel in three-dimensional space has one time-series over peristimulus time. We assume that we have measured the same trial types in each subject and all ERR data have the same number of time bins.[2] The number of subjects is $N_{subjects}$, the number of trial types is N_{types}, and the number of time bins per ERR is N_{bins}. The total number of images is given by:

$$N = N_{subjects}N_{types}N_{bins}$$

(see also Figure 16.2).

SPATIAL MODELS

In this section, we discuss the different ways one can model spatial correlations in the error (spatial non-sphericity). As described in Chapter 10, accounting for these correlations is important for valid inference. To avoid confusing this issue with temporal correlations, we will assume that the data comprise one point in peristimulus time. The difference between treating space (i.e. voxel) as an experimental factor and a treating it as a dimension of the data array is closely related to the difference between a multivariate and a mass-univariate approach. This is the difference between treating the data as a single M-dimensional response or M univariate observations. In other words, we consider each image as a single observation or as a family of single-voxel observations.

Multivariate models

Statistical parametric mapping (SPM) represents a mass-univariate approach, while something like a multivariate analysis of variance (MANOVA) would constitute a multivariate approach (cf. Yandell, 1997). These are fundamentally different because the mass-univariate approach ignores correlations between voxels during estimation and precludes inference about effects that are expressed in more than one voxel. Conversely, multivariate approaches model spatial correlations explicitly.

[2] These assumptions are not strictly necessary but simplify our notation.

Furthermore, we can rearrange multivariate models and test for voxel-times-treatment interactions. We will look first at this rearrangement because it shows that conventional multivariate models can be treated as large univariate models, in which the components of the multivariate response become levels of an extra factor.

We can convert a multivariate observation model into a univariate observation model by simply rearranging the matrix formulation. Consider the multivariate linear model with a response variable y comprising $N_d = N_{subjects}\ N_{types}$ images (e.g. images of current source density at 100 ms after presentation of a visual stimulus) over M voxels:

$$y = X\beta + \varepsilon \qquad 16.1$$

where y is the $N_d \times M$ data matrix, X is an $N_d \times P$ design matrix, β is a $P \times M$ parameter matrix and ϵ is an $N_d \times M$ error matrix, where each row of ϵ is assumed to be sampled independently from the same multivariate normal distribution with zero mean. The classical analysis of this model, the MANOVA, proceeds by computing sample covariance matrices of the predicted or treatment effects and the residuals. Wilk's lambda (Chatfield and Collins, 1980) is then used to test for treatment effects, relative to the covariance of the residuals and, after transformation, compared with an F distribution. The important point about MANOVA is that the errors are assumed to be correlated over all pairs of voxels and that this correlation is taken into account when deriving the statistic.

Vectorized forms

It is helpful to understand the implicit assumptions about spatio-temporal non-sphericity in MANOVA by considering a more general univariate formulation: Eqn. **16.1** can be rearranged into a univariate model by stacking the columns of the response matrix on top of each other to form a response vector and forming an augmented design matrix using a Kronecker tensor product. The parameter and error matrices are similarly vectorized:

$$vec(y) = (I_M \otimes X)vec(\beta) + vec(\varepsilon) \qquad 16.2$$

where $\otimes$ denotes the Kronecker tensor product and $vec(.)$ is the operator that stacks a matrix column-wise to produce one column vector. The matrix I_M is the $M \times M$ identity matrix. The essential difference between Eqn. **16.1** and Eqn. **16.2**, lies in the, hitherto, unspecified assumptions about the error terms on the right hand side. Generally, when using MANOVA, the covariance matrix of the error has $M \times M$ elements. A covariance matrix is symmetrical and therefore contains $M(M+1)/2$ unknown elements or, in our case, variance parameters. Each variance parameter controls the (co)variance between the error at voxels i and j. These variance parameters must be estimated. In MANOVA, this is done using the residuals of the fitted model.

Similarly, in Eqn. **16.2**, the error covariance matrix has dimensions $N_dM \times N_dM$ and is fully specified by $M(M+1)/2$ variance parameters (remember that we assume that each row of ϵ is sampled from the same distribution):

$$\begin{aligned} Cov(vec(\varepsilon)) &= \sum_{i,j=1}^{M} \lambda_{ij} Q_{ij} \\ Q_{ij} &= \tilde{Q}_{ij} \otimes I_{N_d} \end{aligned} \qquad 16.3$$

where $\tilde{Q}_{ij}$ is an $M \times M$ matrix with $\tilde{Q}_{ij}(k,l) = \tilde{Q}_{ij}(k,l) = 1$ and zeroes elsewhere. The quantities λ_{ij} are variance parameters (i.e. hyperparameters) that can be estimated using restricted maximum likelihood (ReML). However, one does not need to estimate all the variance parameters in an unconstrained way. The point made by Eqn. **16.3** is that it can accept constraints on the variance parameters. Such constraints allow us to use (and estimate) much fewer variance components. The use of constraints is critical in neuroimaging, because the number of images N is typically much smaller than the number of voxels M. It would be impossible to estimate all the variance parameters (Eqn. **16.3**) from the data without using constraints. This is the reason why one cannot apply a MANOVA to neuroimaging data directly. Instead, one reduces its dimensionality by using a principal component analysis (PCA) or a similar device (Friston *et al.*, 1996; Worsley *et al.*, 1997).

In summary, multivariate models have an equivalent vectorized form. In both forms, the number of covariance components scales quadratically with the number of components, in our case voxels. However, the vectorized form offers an opportunity to specify these components explicitly and any constraints upon them. There is another important opportunity that is afforded by the vectorized form; this is the specification of contrasts that span the different components of the multivariate response variable. One can specify these contrasts because the Kronecker product $I_M \otimes X$ in Eqn. **16.2** treats the different components (e.g. voxels) as different levels of an additional factor. This will be an important consideration below when we consider whether time bins should be treated as components of a response or as difference of time factor.

Mass-univariate models

In contrast to multivariate approaches and their vectorized forms, mass-univariate approaches consider the data at each voxel i in isolation:

$$y_i = X\beta_i + \varepsilon_i \qquad 16.4$$

by ignoring the spatial correlations (at this stage). Note that ordinary least squares (OLS) estimates of β are identical for Eqns **16.1**, **16.2** and **16.4**. This enables us to estimate, for each voxel i, P parameters β_I and one variance parameter λ_i independently of other voxels.

The critical issue for mass-univariate approaches is how to deal with the spatial covariances that have been ignored in Eqn. **16.4**. The impact of spatial covariance is accommodated at the inference stage through adjusting the p-values associated with the SPM. This adjustment or correction uses random field theory (RFT) and assumes that the error terms conform to a good lattice approximation to an underlying continuous spatially extended process (Chapters 17 and 18). In other words, it assumes that the errors are continuous in space. The RFT correction plays the same role as a Bonferroni correction (Yandell, 1997) for discrete data. The power of the RFT approach is that valid inference needs only one spatial covariance parameter for each voxel. This is the smoothness, which is the determinant of the covariance matrix of the spatial first partial derivatives of the error fields (Worsley *et al.*, 1999). As with the MANOVA, these are estimated using the residuals about the fitted model. The RFT correction does not assume spatial stationarity of the errors or that the spatial autocovariance function is Gaussian. All it assumes is that the error fields are continuous (i.e. smooth). The important distinction between the SPM mass-univariate approach with RFT correction and the equivalent MANOVA approach, with a full covariance matrix, is that the former only requires $2M$ (M spatial and M temporal) variance parameters, whereas the latter requires $M(M+1)/2$ variance parameters.

A further difference between SPM and multivariate approaches is that SPM inferences are based on regionally specific effects as opposed to spatially distributed modes. In SPM, classical inference proceeds using the voxel-specific t- or F-value, whereas in multivariate statistics inference is made about effects over all voxels. Rejection of the null hypothesis in MANOVA allows one to infer that there is a treatment effect in some voxel(s) but it does not tell one where. In principle, if the treatment effect was truly spatially distributed, SPM would be much less sensitive than MANOVA. However, the aim of functional neuroimaging is to establish regionally specific responses. By definition, diffuse spatially distributed responses are not useful in trying to characterize functional specialization. Furthermore, the goal of fMRI or EEG integration is to endow electrophysiological measures with a spatial precision. This goal is met sufficiently by mass-univariate approaches.

In conclusion, the special nature of neuroimaging data and the nature of the regionally specific questions that are asked of them, point clearly to the adoption of mass-univariate approaches and the use of RFT to accommodate spatial non-sphericity. This conclusion is based upon the fact that, for spatially continuous data, we only need the covariances of the first partial derivatives of the error at each point in space, as opposed to the spatial error covariances among all pairs of points. Secondly, the nature of the hypotheses we wish to test is inherently region-specific. The price we pay is that there is no opportunity to specify contrasts over different voxels as in a vectorized multivariate approach that treats voxels as a factor.

TEMPORAL MODELS

Having motivated a mass-univariate approach for the analysis of each voxel time-series, we now have to consider whether time (and/or frequency) is an extra dimension of the response variable (mass-univariate) or an experimental factor (vectorized-multivariate). One could simply treat time as another dimension of the response variable to produce four-dimensional SPMs that span anatomical space and peristimulus time. These SPMs would have activations or regions above the threshold (excursion sets) that covered a cortical region and a temporal domain following the stimulus. This would allow both for anatomical and temporal specificity of inferences using adjusted p-values. We will see an example of this later.

The appeal of this approach echoes the points made in the previous section. The nice thing about creating four-dimensional (over space and time) SPMs is that temporal correlations or non-sphericity among the errors over time can be dealt with in a parsimonious way, at the inference stage, using random field corrections. This means that one only needs to estimate the temporal smoothness at each time bin as opposed to the temporal correlations over all time bins. The assumption underpinning the RFT is clearly tenable because the M/EEG data are continuous in time. An example of this approach is shown in Figure 16.3 (see also Kilner *et al.*, 2006).

The alternative to treating time as a dimension is to assume that it is an experimental factor with the same number of levels as there are bins in peristimulus time. This is simply a time-series model at each voxel, of the sort used by fMRI. In this instance, one has to estimate the temporal variance parameters by analogy with the spatial variance parameters in Eqn. **16.3**. In other words, one has to estimate the temporal correlations of the error to make an appropriate non-sphericity adjustment to ensure valid inference.[3] ReML estimation procedures

[3] This applies if one uses OLS parameter estimates. For maximum likelihood (ML) estimates, temporal correlations have to be estimated to whiten the data.

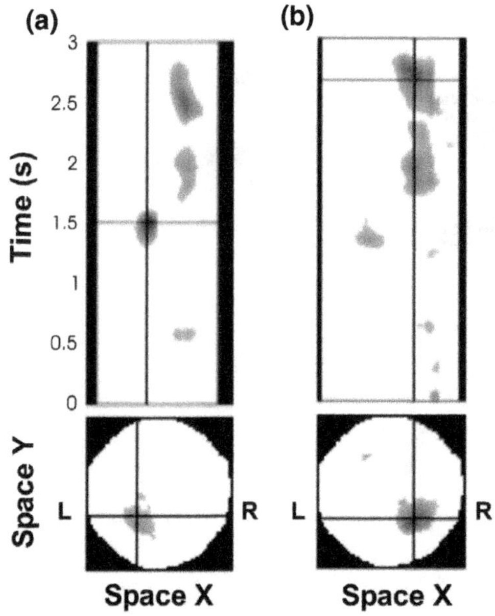

FIGURE 16.3 Analysis of ERP data based on a spatio-temporal SPM, at the sensor level (see (Kilner *et al.*, 2006 for details). The two plots show orthogonal slices through a 3-dimensional statistical image SPM{t}, thresholded at $p < 0.001$ (uncorrected). In (a), the display is centred on a left parietal sensor, in (b) on a right parietal sensor. In the top row, the temporal profiles are over the spatial x-axis at the two selected y-locations.

based upon expectation maximization (EM) allow the temporal variance parameters to be specified in terms of separable covariance components (see Chapters 10 and 22, and Friston *et al.*, 2002). This technique allows for a flexible model of serial correlations, where the estimated non-sphericity is used either to whiten the data (to furnish maximum likelihood estimates) or in the estimation of the effective degrees of freedom (for ordinary least-squares estimates).

Both the mass-univariate (spatio-temporal SPM) and time-series SPM approaches are available to us. Which is the most suitable for a given study? The answer is that both approaches have their pros and cons, i.e. it depends on the data and the question.

Time-series SPM

The advantage of treating time as an experimental effect is that one can specify contrasts that cover different peristimulus times. It is not possible to make inferences about the spatial extent of activation foci in SPMs (this is because mass-univariate approaches do not need any spatial parameters on which they could infer). Similarly, in the context of space-time SPMs, inferences about the temporal extent of evoked responses (e.g. differential latencies among trial-types) are precluded. To enable inferences about the temporal form of an effect, it is necessary to specify contrasts that encompass many time bins. This means time has to enter as an experimental effect or factor.

In the spatial domain, we are interested in region- or voxel-specific inferences because activation in one part of the brain does not have any quantitative meaning in relation to activation in a different structure. Conversely, the relative responses over time, a given voxel, are meaningful because they define the form of the evoked transient. Therefore, in some instances, it is useful to compare responses over time explicitly. This means time is an experimental factor. An example of a question requiring a time-series SPM would be a contrast testing for a decreased latency of the N170 under a particular level of attention (e.g. by using a temporal derivative in a time-series model). Because the N170 is defined by its temporal form and deployment (a component that peaks around 170 ms that is typically evoked by face stimuli) this contrast would necessarily cover many time bins.

Spatio-temporal SPMs

Conversely, when the question cannot be framed in terms effects with an explicit temporal form, a spatio-temporal SPM may be more appropriate. These SPMs will identify differences in evoked responses where and whenever they occur. In this context, the search for effects with a specific temporal form or scale can be implemented by temporally convolving the data with the desired form in accord with the match filter theorem (see Figure 16.3 for an example and Kilner *et al.*, 2006). This allows one to search for effects, over peristimulus time, in an exploratory way. Importantly, inference using random field theory adjusts p-values to accommodate the fact that one was searching over serially correlated time bins. We have found that treating time as a dimension is especially useful for time-frequency power analyses (Kilner *et al.*, 2005). See Figure 16.4 for an example of this approach using time-frequency SPMs.

Summary

We have addressed some key issues about how one models responses in the spatial and temporal (or time-frequency) domains. These issues touch upon the underlying question of how to model the error covariance of spatio-temporal data. By assuming a factorization of the spatial and temporal domain, we can separate the modelling of the spatial and temporal correlations.

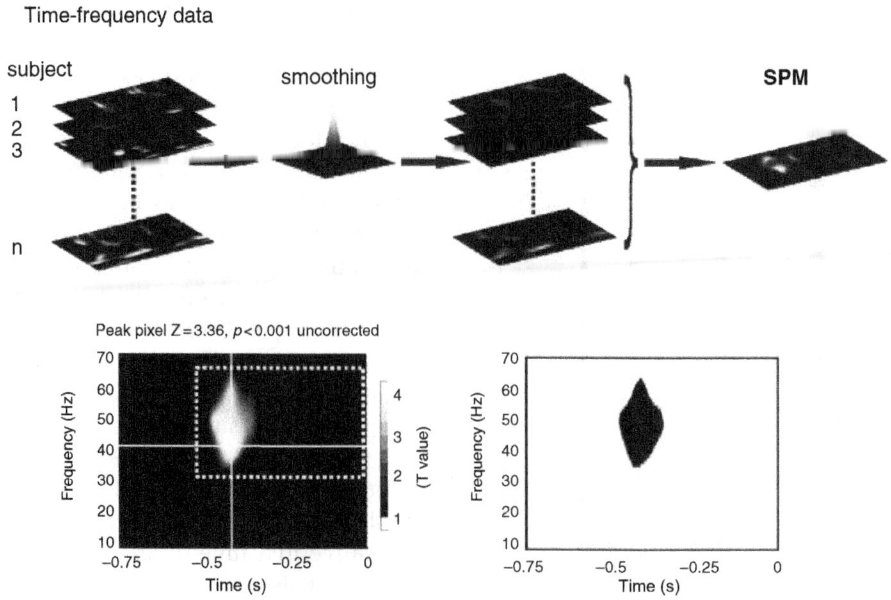

FIGURE 16.4 Upper panels: the pipeline for time-frequency SPMs. Time-frequency contrasts were calculated for each subject and smoothed by convolution with a Gaussian kernel. These data were then analysed in SPM (Wellcome Department of Imaging Neuroscience, London, UK). Lower left: the SPM calculated from the smoothed time-frequency images and thresholded at $p < 0.01$ uncorrected. The location of the peak time-frequency bin is shown. The white dotted box indicates our illustrative *a priori* window of interest. Lower right: the significant cluster of bins at $p < 0.01$ uncorrected. With this analysis, both the cluster level and bin level results are significant at $p < 0.05$ corrected for family-wise error (FWE). Although the data were smoothed with a Gaussian kernel of FWHM 96 ms and 12 Hz, the smoothness calculated from the data was greater with FWHM of 107.8 ms and 16.8 Hz. This difference reflects the correlation in the underlying data between adjacent bins in both the time and frequency dimensions.

(See Figure 16.5 for a schematic summary of these issues.) The discussion about multivariate versus mass-univariate approaches concerns the modelling of spatial correlations among voxels. In space, we chose the mass-univariate approach. Using this approach, at the estimation stage, the spatial correlations are ignored. To account for spatial correlations, we use random field corrections at the subsequent inference stage. The same arguments can be applied to the temporal aspect of the response. Generally, one searches for differences in space-time or time-frequency in the usual way using established SPM protocols. These include the use of summary statistics to emulate random-effect analyses (as described in Chapter 12) to look at between-trial, between-subject or between-group effects. However, in some cases, one might want to treat time as an experimental factor so that questions about the temporal deployment or form of response differences can be addressed. In the next section, we focus on the ensuing time-series models.

Hierarchical models

When electing to treat time as a factor, we create an interesting distinction between explanatory variables that model time effects and experimental design variables that model the treatment effects of other experimental factors (e.g. trial differences). As mentioned above, we assume that each peristimulus time series represents one particular trial-type within an M/EEG session. The temporal explanatory variables model responses in each subject- and trial-type-specific ERR. Further explanatory variables model treatment effects among trial types and/or sessions or subjects. We will refer to the temporal explanatory variables as temporal effects and to the experimental design variables as experimental effects. These are encoded by the design matrices X^t and X^d, respectively. This natural distinction points to a hierarchical modelling framework (see Chapter 11). A hierarchical model can be used to decompose the data into within-ERR (temporal effects) and between-ERR components (experimental effects). There is an important difference between the sorts of inferences that can be made with ERR data. This distinction rests upon the form of the hierarchical observation model and the level in the hierarchy at which the inferences are made. Usually, these hierarchical observation models have two levels, engendering the distinction between fixed- and random-effects analyses. In two-level hierarchical observation models, the response at the first level is caused by first-level

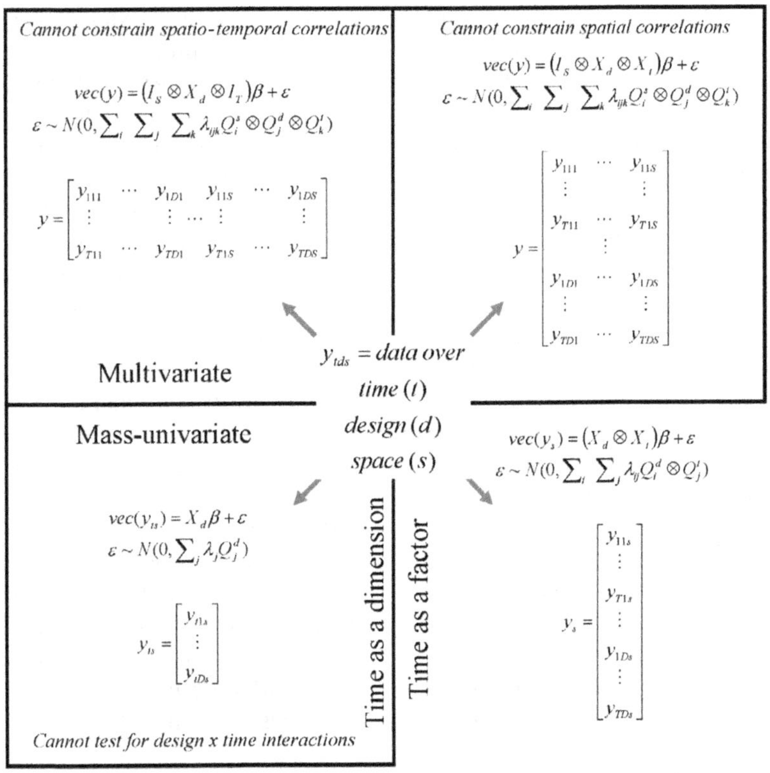

FIGURE 16.5 Schematic demonstrating the different formulations of observation models implied by the multivariate versus mass-univariate and temporal dimension versus factor distinctions. The upper panels represent multivariate formulations in which there is no opportunity to place continuity constraints on the spatial or (upper left) spatio-temporal correlations. By treating each time bin or brain location as a separate component of the observed response, one effectively creates a model which is not informed about the continuity of these responses over space and time. The mass-univariate model in the lower left panel can be used to embody continuity constraints in both space and time through the application of random field theory. However, there would be no opportunity to assess design by time interactions. The lower right panel represents the formulation in which time is treated as a factor but space is not.

parameters that themselves are modelled as random or stochastic variables at the second level, i.e.:

$$\begin{aligned} y &= X^{(1)}\beta^{(1)} + \varepsilon^{(1)} \\ \beta^{(1)} &= X^{(2)}\beta^{(2)} + \varepsilon^{(2)} \\ X^{(1)} &= I_{N_{subjects}N_{types}} \otimes X^t \\ X^{(2)} &= X^d \otimes I_{N_p} \end{aligned} \qquad 16.5$$

where the data y consists of one column vector of length $N_{subjects}N_{types}N_{bins}$ (see Figure 16.2). The ERR data are ordered such that trial type-specific ERRs of an individual subject are next to each other. This means the first-level design is $X^{(1)} = I_{N_{subjects}} \otimes I_{N_{types}} \otimes X^t$, where X^t is some first-level design matrix, embodying temporal effects, for a single ERR. The second-level design matrix is given by $X^{(2)} = X^d \otimes I_{N_p}$, where, for example, $X^d = 1_{N_{subjects}} \otimes I_{N_{types}}$ would be a simple averaging matrix. N_p is the number of columns in X^t and 1_N denotes a column vector of ones of length N.

The model in Eqn. **16.5** reflects the natural hierarchy of observed data. At the first level, the observations are modelled in a subject- and trial-type-specific fashion within peristimulus time, where $\varepsilon^{(1)}$ is the observation error. At the second level, we model the parameters $\beta^{(1)}$ over subjects and trial types. In other words, the ensuing hierarchy models temporal within-trial effects at the first level and between-trial effects at the second. At this level, the error $\varepsilon^{(2)}$ represents between-subject variability not modelled in X^d.

There are several reasons why hierarchical linear models are useful for characterizing ERR data. To start with, they afford substantial latitude for modelling and hypothesis testing. Note that the first-level design matrix $X^{(1)}$ defines a projection onto some subspace of the data. Each parameter is associated with a specific dimension of this subspace. One can consider many subspaces for ERR data, where the Fourier transform or the wavelet transform are just two examples. The Fourier transform is useful for making inferences about power in some frequency range. The wavelet transform is appropriate when making inferences about localized time-frequency effects. By specifying contrasts on the first- or second-level parameter estimates, we can test for effects localized in peristimulus time and within certain frequency ranges (Kiebel and Friston, 2004). The important point about the two-level hierarchical model is that it enables different (linear) transforms at the first level.

A further motivation for hierarchical models is that they finesse the parameterization of non-sphericity. In this context, the error is decomposed into level-specific components. From Eqn. **16.5**, we see that the first-level error $\varepsilon^{(1)}$ is the error about the fitted response, i.e. the observation error. The second-level error $\varepsilon^{(2)}$, with an averaging design component X^d, is the deviation of each first-level parameter from the average value for a particular trial-type in a specific subject. This error arises from between-subject variability. The distinction between these two components allows the decomposition of the overall error into two partitions, the within- and between-subject variability. These partitions have distinct non-sphericity structures, which can be modelled using level-specific variance components. These facilitate robust and accurate error estimation, which is necessary for valid inference.

An advantage of level-specific error components is that one can make inferences at either level. This relates to the distinction between fixed- and random-effects analyses (see Chapter 12). For example, in Eqn. **16.5**, $\beta^{(2)}$ corresponds to the average response over subjects for a particular trial type. The variability of $\beta^{(2)}$ (its covariance), can be derived by first collapsing the two-level model to one level:

$$y = X^{(1)}X^{(2)}\beta^{(2)} + X^{(1)}\varepsilon^{(2)} + \varepsilon^{(1)} \qquad \mathbf{16.6}$$

Using an ordinary least-squares estimator, the covariance matrix of the parameter estimates is:

$$Cov(\hat{\beta}^{(2)}) = (X^{(1)}X^{(2)})^{-}Cov(X^{(1)}\varepsilon^{(2)} + \varepsilon^{(1)})(X^{(1)}X^{(2)})^{-T} \qquad \mathbf{16.7}$$

where X^{-} denotes the generalized inverse. Eqn. **16.7** says that the covariance of the estimated second-level parameters is given by the projected error covariance of the collapsed model. This is a mixture of the variability of the errors from both levels. Therefore, $\hat{\beta}^{(2)}$ not only varies because of inter-subject, but also because of within-subject variability.

In summary, the two-level model is useful if there is a meaningful linear projection to some low-dimensional space. For instance, in fMRI, the blood oxygenation-level-dependent (BOLD) response can be linearly modelled with three basis functions. These capture a large part of the stimulus-induced variability. Similarly, in M/EEG, temporal basis functions in X^t allow us to summarize a response with a few parameters. At the second level, we can then test hypotheses about these parameters (i.e. differences and interactions). This dimension reduction is an efficient way of testing for effects. Note that these basis functions can extend over peristimulus time. They can describe simple averages or other more complicated shapes like damped oscillations.

An alternative is to forgo modelling the first level completely and employ the identity matrix as temporal design matrix $X^t = I$. This effectively renders the two-level model (Eqn. **16.5**) a one-level model, because there is no error $\epsilon^{(1)}$ at the first level. This is the traditional way of analysing evoked responses. One simply forms contrasts at the first level and models them at the second. Traditionally, researchers test for averages over a particular window of peristimulus time, but one can actually use any linear combination over peristimulus time. The advantage of this analysis is that it is simple and straightforward and appeals to exactly the same arguments as the summary-statistic approach to hierarchical models of fMRI data.

This concludes our discussion about how we model the spatio-temporal and experimental dimensions of M/EEG data. Next we focus on modelling the experimental factors, e.g. groups or trial-types. To simplify things we will use the identity matrix (i.e. a complete basis set) as temporal design matrix X^t.

HYPOTHESIS TESTING WITH HIERARCHICAL MODELS

Hierarchical models for M/EEG are the same as those used for other imaging modalities (see Chapter 13) and inference can proceed using the same multistage protocol, by taking contrasts from one level to the next. This summary statistic approach can be used for the analysis of either evoked responses or time-frequency power data. In this section, we describe how contrasts can be specified to ask a wide range of questions about evoked responses in multisubject studies, using time-series SPMs.

For ERR data, we follow the approach outlined above. First, one projects data from channel to voxel-space. This can be either a two-dimensional approximation to the scalp surface or a three-dimensional source reconstruction (Phillips *et al.*, 2002; Mattout *et al.*, 2005). These time-series of images form the input to the first level of a hierarchical model to provide estimates of the response for each temporal basis function (in this case each time bin), each trial-type and each subject. The inputs to the second level are contrasts over time bins, for each subject and trial type. These can be simple averages over several time points in peristimulus time, e.g. the average between 150 and 190 ms to measure the N170 component. Note that one can choose any temporal shape as contrast, e.g. a Gaussian or damped sinusoid.

For time-frequency analyses, one first computes power in the time-frequency domain (Kiebel *et al.*, 2005). In current software implementations of SPM, we use the Morlet wavelet transform. This renders the data 4- or 5-dimensional (2 or 3 spatial dimensions, time, and frequency). As outlined above, there are two different approaches to these data. One could treat space, time, and frequency as dimensions of a random field. This leaves only dimensions like subject or trial-types as experimental factors (see Figure 16.4 and Kilner *et al.*, 2005). An alternative is to treat both time and frequency as factors. In this case, the input to the second level is formed by subject and trial type-specific contrasts over time and frequency (e.g. averages in windows of time-frequency space). In the following, we describe some common second-level models for M/EEG.

Evoked responses

A typical analysis of ERRs is the one-way (repeated measures) analysis of variance. For each of the $n = 1, \dots, N$ subjects there are K measurements (i.e. trial-types). The second-level summary statistics are contrasts over peristimulus time, for each subject and trial type. The design matrix in this case is $X^{(2)} = [I_K \otimes 1_N, 1_K \otimes I_N]$ (see also Chapter 13). In terms of the model's covariance components, one could assume that the between-subject errors $\varepsilon^{(2)}$ are uncorrelated and have unequal variances for each trial-type. This results in K covariance components. After parameter estimation, one tests for main effects or interactions among the trial-types at the between-subject level, with the appropriate contrast. Note that this model uses a 'pooled' estimate of non-sphericity over voxels (see Chapter 13). An alternative is to compute the relevant contrasts (e.g. a main effect or interaction) at the first level and use a series of one-sample t-tests at the second.

Induced responses

In SPM, the time-frequency decomposition of data uses the Morlet wavelet transform. Note that other approaches using short-term Fourier transform, or the Hilbert transform on bandpassed filtered data are largely equivalent to the wavelet decomposition (Kiebel *et al.*, 2005). For frequency $\omega = 2\pi f$ and peristimulus time t, the Morlet wavelet kernel is:

$$h(t, \omega) = c_\omega \exp(-\frac{t^2}{\sigma_t^2}) \exp(i\omega t) \qquad 16.8$$

where c_ω is some normalization factor and σ_t^2 is the temporal width of the kernel. The transform itself is the convolution:

$$z(t, \omega)_{ij} = h(\omega) * y_{ij} \qquad 16.9$$

where y_{ij} is the i-th (single) trial measured at the j-th channel (or voxel) and * denotes convolution. The power is:

$$P(t, \omega)_{ij} = z(t, \omega)_{ij} z(t, \omega)^*_{ij} \qquad 16.10$$

where $z(t, \omega)^*_{ij}$ is the complex conjugate. Induced activity is computed by averaging P_{ij} over trials and subtracting the power of the evoked response. When time and frequency are considered experimental factors, we compute contrasts at the first level and pass them to the second. This might be an average in the time-frequency plane (e.g. Kilner *et al.*, 2005). Alternatively, one can compute, per trial type and subject, several averages in the time-frequency plane and take them up to the second level. These contrasts can be modelled using repeated-measures ANOVA, where time and frequency are both factors with multiple levels.

In general, we assume that the second-level error for contrasts of power is normally distributed, whereas power data *per se* have a χ^2-distribution. However, in most cases, the contrasts have a near-normal distribution because of averaging over time, frequency and trials and, more importantly, taking differences between peristimulus times or trial-types. These averaging operations render the contrasts or summary statistics normally distributed, by central limit theorem. When there is still doubt about the normality assumption, one can apply a log or square-root transform (Kiebel *et al.*, 2005).

SUMMARY

In this chapter, we have covered various ways of analysing M/EEG data, with a special focus on the distinction between treating time as a factor and treating it as a dimension of the response variable. This corresponds to the distinction between inference based on time-series models (of the sort using in fMRI) and spatio-temporal SPMs that span space and time. In the case of time-series models, the hierarchical nature of our observation models calls for a multistage summary-statistic approach, in which contrasts at each level are passed to higher levels to enable between-trial, between-subject and between-group inferences. The central role of hierarchal models will be taken up again in Section 4, in the context of empirical Bayes. In the next section we consider, in greater depth, the nature of inference on SPMs.

REFERENCES

Chatfield C, Collins A (1980) *Introduction to multivariate analysis*. Chapman & Hall, London

Friston KJ, Frith CD, Frackowiak RS *et al.* (1995) Characterizing dynamic brain responses with fMRI: a multivariate approach. *NeuroImage* **2**: 166–72

Friston KJ, Frith CD, Liddle PF *et al.* (1991) Comparing functional (PET) images: the assessment of significant change. *J Cereb Blood Flow Metab* **11**: 690–99

Friston KJ, Penny W, Phillips C *et al.* (2002) Classical and Bayesian inference in neuroimaging: theory. *NeuroImage* **16**: 465–83

Friston KJ, Stephan KM, Heather JD *et al.* (1996) A multivariate analysis of evoked responses in EEG and MEG data. *NeuroImage* **3**: 167–74

Kiebel SJ, Friston KJ (2004) Statistical parametric mapping for event-related potentials (II): a hierarchical temporal model. *NeuroImage* **22**: 503–20

Kiebel SJ, Tallon-Baudry C, Friston KJ (2005) Parametric analysis of oscillatory activity as measured with M/EEG. *Hum Brain Mapp* **26**: 170–77

Kilner JM, Kiebel SJ, Friston KJ (2005) Applications of random field theory to electrophysiology. *Neurosci Lett* **374**: 174–78

Kilner JM, Otten LJ, Friston KJ (2006) Application of random field theory to EEG data in space and time. O Human Brain Mapping, Florence

Mattout J, Phillips C, Penny WD *et al.* (2005) MEG source localization under multiple constraints: an extended Bayesian framework. *Neuroimage* **3**: 753–67

Phillips C, Rugg MD, Friston KJ (2002) Anatomically informed basis functions for EEG source localization: combining functional and anatomical constraints. *Neuroimage* **16**: 678–95

Rugg MD, Coles MG (1995) *Electrophysiology and mind*. Oxford University Press, Oxford

Worsley KJ, Andermann M, Koulis T *et al.* (1999) Detecting changes in nonisotropic images. *Hum Brain Mapp* **8**: 98–101

Worsley KJ, Poline JB, Friston KJ *et al.* (1997) Characterizing the response of PET and fMRI data using multivariate linear models. *Neuroimage* **6**: 305–19

Yandell BS (1997) Practical data analysis for designed experiments. Chapman & Hall, London

PART 4

Classical inference

C H A P T E R

17

Parametric procedures

M. Brett, W. Penny and S. Kiebel

INTRODUCTION

This chapter is an introduction to inference using parametric procedures with focus on the multiple comparison problem in functional imaging, and the way it can be solved using random field theory (RFT).

In a standard functional imaging analysis, we fit a statistical model to the data, to give us model parameters. We then use the model parameters to look for an effect we are interested in, such as the difference between a task and baseline. To do this, we usually calculate a statistic for each brain voxel that tests for the effect of interest in that voxel. The result is a large volume of statistic values.

We now need to decide if this volume shows any evidence of the effect. To do this, we have to take into account that there are many thousands of voxels and therefore many thousands of statistic values. This is the multiple comparison problem in functional imaging. Random field theory is a recent branch of mathematics that can be used to solve this problem.

To explain the use of RFT, we will first go back to the basics of hypothesis testing in statistics. We describe the multiple comparison problem and the usual solution, which is the Bonferroni correction. We explain why spatial correlation in imaging data causes problems for the Bonferroni correction and introduce RFT as a solution. Finally, we discuss the assumptions underlying RFT and the problems that arise when these assumptions do not hold. We hope this chapter will be accessible to those with no specific expertise in mathematics or statistics. Those more interested in mathematical details and recent developments are referred to Chapters 18 and 19.

Rejecting the null hypothesis

When we calculate a statistic, we often want to decide whether the statistic represents convincing evidence of the effect we are interested in. Usually, we test the statistic against the null hypothesis, which is the hypothesis that there is no effect. If the statistic is not compatible with the null hypothesis, we may conclude that there is an effect. To test against the null hypothesis, we can compare our statistic value to a *null distribution*, which is the distribution of statistic values we would expect if there is no effect. Using the null distribution, we can estimate how likely it is that our statistic could have come about by chance. We may find that the result we found has a 5 per cent chance of resulting from a null distribution. We therefore decide to reject the null hypothesis, and accept the alternative hypothesis that there is an effect. In rejecting the null hypothesis, we must accept a 5 per cent chance that the result has in fact arisen when there is in fact no effect, i.e. the null hypothesis is true. Five per cent is our expected *type I* error rate, or the chance that we take that we are wrong when we reject the null hypothesis.

For example, when we do a single *t*-test, we compare the *t*-value we have found to the null distribution for the *t*-statistic. Let us say we have found a *t*-value of 2.42, and have 40 degrees of freedom. The null distribution of *t*-statistics with 40 degrees of freedom tells us that the probability of observing a value greater than or equal to 2.42, if there is no effect, is only 0.01. In our case, we can reject the null hypothesis with a 1 per cent risk of type I error.

The situation is more complicated in functional imaging because we have many voxels and therefore many statistic values. If we do not know where in the brain our effect will occur, our hypothesis refers to the whole volume of statistics in the brain. Evidence against the null hypothesis would be that the whole observed *volume* of values is unlikely to have arisen from a null distribution. The question we are asking is now a question about the volume, or *family* of voxel statistics, and the risk of error that we are prepared to accept is the family-wise error

Statistical Parametric Mapping, by Karl Friston et al.
ISBN–13: 978-0-12-372560-8 ISBN–10: 0-12-372560-7

rate (FWE) – which is the likelihood that this family of voxel values could have arisen by chance.

We can test a family-wise null hypothesis in a variety of ways, but one useful method is to look for any statistic values that are larger than we would expect, if they all had come from a null distribution. The method requires that we find a threshold to apply to every statistic value, so that any values above the threshold are unlikely to have arisen by chance. This is often referred to as 'height thresholding', and it has the advantage that if we find voxels above threshold, we can conclude that there is an effect at these voxel locations, i.e. the test has localizing power. Alternative procedures based on cluster- and set-level inferences are discussed in the next chapter.

A height threshold that can control family-wise error must take into account the number of tests. We saw above that a single t-statistic value from a null distribution with 40 degrees of freedom has a 1 per cent probability of being greater than 2.42. Now imagine our experiment has generated 1000 t values with 40 degrees of freedom. If we look at any single statistic, then by chance it will have a 1 per cent probability of being greater than 2.42. This means that we would expect 10 t values in our sample of 1000 to be greater than 2.42. So, if we see one or more t values above 2.42 in this family of tests, this is not good evidence against the *family-wise* null hypothesis, which is that all these values have been drawn from a null distribution. We need to find a new threshold, such that, in a family of 1000 t statistic values, there is a 1 per cent probability of there being *one or more t* values above that threshold. The Bonferroni correction is a simple method of setting this threshold.

THE BONFERRONI CORRECTION

The Bonferroni correction is based on simple probability rules. Imagine we have taken our t values and used the null t distribution to convert them to probability values. We then apply a probability threshold α to each of our n probability values; in our previous example α was 0.01, and n was 1000. If all the test values are drawn from a null distribution, then each of our n probability values has a probability α of being greater than threshold. The probability of *all* the tests being less than α is therefore $(1-\alpha)^n$. The family-wise error rate (P^{FWE}) is the probability that one or more values will be greater than α, which is simply:

$$P^{FWE} = 1 - (1-\alpha)^n \tag{17.1}$$

Because α is small this can be approximated by the simpler expression:

$$P^{FWE} < n\alpha \tag{17.2}$$

Using Eqn. **17.2**, we can find a single-voxel probability threshold α that will give us our required family-wise error rate, P^{FWE}, such that we have a P^{FWE} probability of seeing any voxel above threshold in all of the n values. We simply solve Eqn. **17.2** for α:

$$\alpha = P^{FWE}/n \tag{17.3}$$

If we have a brain volume of 100 000 t-statistics, all with 40 degrees of freedom, and we want an FWE rate of 0.05, then the required probability threshold for a single voxel, using the Bonferroni correction, would be $0.05/100,000 = 0.0000005$. The corresponding t-statistic is 5.77. If any voxel t-statistic is above 5.77, then we can conclude that a voxel statistic of this magnitude has only a 5 per cent chance of arising anywhere in a volume of 100 000 t-statistics drawn from the null distribution.

The Bonferroni procedure gives a *corrected* p-value; in the case above, the uncorrected p-value for a voxel with a t-statistic of 5.77 was 0.0000005; the p-value corrected for the number of comparisons is 0.05.

The Bonferroni correction is used for calculating FWE rates for some functional imaging analyses. However, in many cases, the Bonferroni correction is too conservative because most functional imaging data have some degree of spatial correlation, i.e. there is correlation between neighbouring statistic values. In this case, there are fewer *independent* values in the statistic volume than there are voxels.

Spatial correlation

Some degree of spatial correlation is almost universally present in functional imaging data. In general, data from any one voxel in the functional image will tend to be similar to data from nearby voxels, even after the modelled effects have been removed. Thus the errors from the statistical model will tend to be correlated for nearby voxels. The reasons for this include factors inherent in collecting and reconstructing the image, physiological signal that has not been modelled, and spatial preprocessing applied to the data before statistical analysis.

For positron emission tomography (PET) data, much more than for functional magnetic resonance imaging (fMRI), nearby voxels are related because of the way that the scanner collects and reconstructs the image. Thus, data that do in fact arise from a single voxel location in the brain will also cause some degree of signal change in neighbouring voxels in the resulting image. The extent to which this occurs is a measure of the performance of the PET scanner, and is referred to as the point spread function.

Spatial pre-processing of functional data introduces spatial correlation. Typically, we will realign images for an individual subject to correct for motion during the scanning session (see Chapter 4), and may also spatially normalize a subject's brain to a template to compare data between subjects (see Chapter 5). These transformations will require the creation of new resampled images, which have voxel centres that are very unlikely to be the same as those in the original images. The resampling requires that we estimate the signal for these new voxel locations from the values in the original image, and typical resampling methods require some degree of averaging of neighbouring voxels to derive the new voxel value (see Chapter 4).

It is very common to smooth the functional images before statistical analysis. A proportion of the noise in functional images is independent from voxel to voxel, whereas the signal of interest usually extends over several voxels. This is due both to the distributed nature of neuronal sources and to the spatially extended nature of the haemodynamic response. According to the matched filter theorem, smoothing will therefore improve the signal-to-noise ratio. For multiple subject analyses, smoothing may also be useful for blurring the residual differences in location between corresponding areas of functional activation. Smoothing involves averaging over voxels, which will by definition increase spatial correlation.

The Bonferroni correction and independent observations

Spatial correlation means that there are fewer independent observations in the data than there are voxels. This means that the Bonferroni correction will be too conservative because the family-wise probability from Eqn. **17.1** relies on the individual probability values being independent, so that we can use multiplication to calculate the probability of combined events. For Eqn. **17.1**, we used multiplication to calculate the probability that all tests will be below threshold with $(1-\alpha)^n$. Thus, the n in the equation must be the number of *independent* observations. If we have n voxels in our data, but there are only n_i independent observations, then Eqn. **17.1** becomes $P^{FWE} = 1-(1-\alpha)^{n_i}$, and the corresponding α from Eqn. **17.3** is given by $\alpha = P^{FWE}/n_i$. This is best illustrated by example.

Let us take a single image slice, of 100 by 100 voxels, with a t-statistic value for each voxel. For the sake of simplicity, let the t-statistics have very high degrees of freedom, so that we can consider the t-statistic values as being from the normal distribution, i.e. that they are Z scores. We can simulate this slice from a null distribution by filling the voxel values with independent random numbers from the normal distribution, which results in an image such as that in Figure 17.1.

If this image had come from the analysis of real data, we might want to test if any of the numbers in the image are more positive than is likely by chance. The values are independent, so the Bonferroni correction will give an accurate threshold. There are 10 000 Z scores, so the Bonferroni threshold, α, for an FWE rate of 0.05, is 0.05/10 000=0.000005. This corresponds to a Z-score of 4.42. Given the null hypothesis (which is true in this case) we would expect only 5 out of 100 such images to have one or more Z scores more positive than 4.42.

The situation changes if we add spatial correlation. Let us perform the following procedure on the image: break up the image into squares of 10 by 10 pixels; for each square, calculate the mean of the 100 values contained; replace the 100 random numbers in the square by the mean value.[1] The image that results is shown in Figure 17.2.

We still have 10 000 numbers in our image, but there are only 10 by $10 = 100$ numbers that are independent. The appropriate Bonferroni correction is now $0.05/100 = 0.0005$, which corresponds to a Z-score of 3.29. We would expect only 5 of 100 of such images to have a square block of values greater than 3.29 by chance. If we had assumed

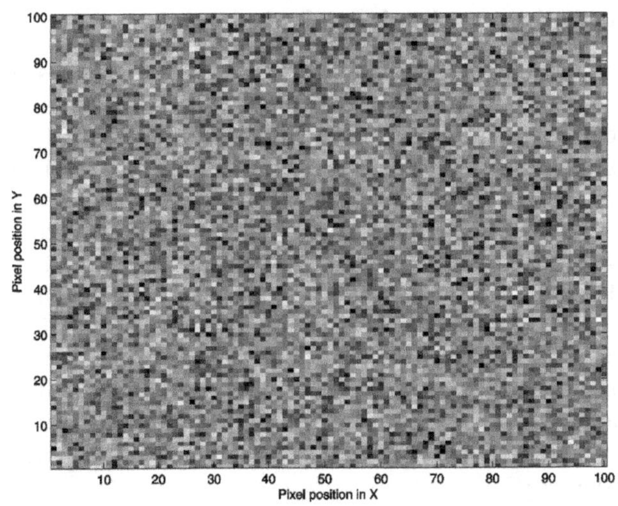

FIGURE 17.1 Simulated image slice using independent random numbers from the normal distribution. Whiter pixels are more positive.

[1] Averaging the random numbers will make them tend to zero; to return the image to a variance of 1, we need to multiply the numbers in the image by 10; this is $\sqrt{n}$, where n is the number of values we have averaged.

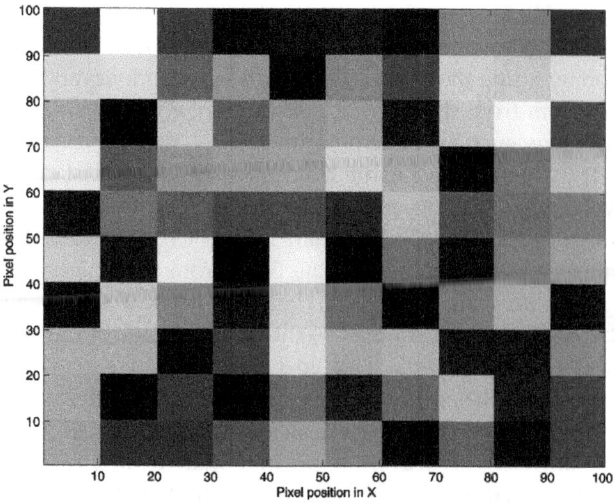

FIGURE 17.2 Random number image from Figure 17.1 after replacing values in the 10 by 10 squares by the value of the mean within each square.

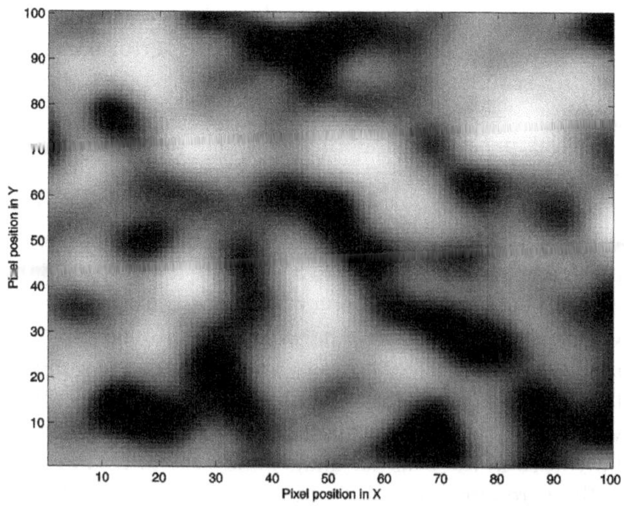

FIGURE 17.3 Random number image from Figure 17.1 after smoothing with a Gaussian smoothing kernel of full width at half maximum of 10 pixels.

all the values were independent, then we would have used the correction for 10 000 values, of $\alpha = 0.000005$. Because we actually have only 100 independent observations, Eqn. **17.2**, with $n = 100$ and $\alpha = 0.000005$, tells us that we expect an FWE rate of 0.0005, which is one hundred times lower (i.e. more conservative) than the rate that we wanted.

Smoothing and independent observations

In the preceding section, we replaced a block of values in the image with their mean in order to show the effect of reducing the number of independent observations. This procedure is a very simple form of smoothing. When we smooth an image with a smoothing kernel, such as a Gaussian, each value in the image is replaced with a weighted average of itself and its neighbours. Figure 17.3 shows the image from Figure 17.1 after smoothing with a Gaussian kernel of full width at half maximum (FWHM) of 10 pixels.[2] An FWHM of 10 pixels means that, at five pixels from the centre, the value of the kernel is half its peak value. Smoothing has the effect of blurring the image, and reduces the number of independent observations.

[2] As for the procedure where we took the mean of the 100 observations in each square, the smoothed values will no longer have a variance of one, because the averaging involved in smoothing will make the values tend to zero. As for the square example, we need to multiply the values in the smoothed image by a scale factor to return the variance to one; the derivation of the scale factor is rather technical, and not relevant to our current discussion.

The smoothed image contains spatial correlation, which is typical of the output from the analysis of functional imaging data. We now have a problem, because there is no simple way of calculating the number of independent observations in the smoothed data, so we cannot use the Bonferroni correction. This problem can be addressed using random field theory.

RANDOM FIELD THEORY

Random field theory (RFT) is a recent body of mathematics defining theoretical results for smooth statistical maps. The theory has been versatile in dealing with many of the thresholding problems that we encounter in functional imaging. Among many other applications, it can be used to solve our problem of finding the height threshold for a smooth statistical map which gives the required family-wise error rate.

The way that RFT solves this problem is by using results that give the expected *Euler characteristic* (EC) for a smooth statistical map that has been thresholded. We will discuss the EC in more detail below; for now it is only necessary to note that the expected EC leads directly to the expected number of clusters above a given threshold, and that this in turn gives the height threshold that we need.

The application of RFT proceeds in stages. First, we estimate the smoothness (spatial correlation) of our statistical map. Then we use the smoothness values in the appropriate RFT equation, to give the expected EC at different thresholds. This allows us to calculate the threshold at which we would expect 5 per cent of equivalent statistical maps arising under the null hypothesis to contain at least one area above threshold.

Smoothness and resels

Usually we do not know the smoothness of our statistical map. This is so even if the map resulted from smoothed data, because we usually do not know the extent of spatial correlation in the underlying data before smoothing. If we do not know the smoothness, it can be calculated using the observed spatial correlation in the images. For our example (see Figure 17.3), however, we *know* the smoothness, because the data were independent before smoothing. In this case, the smoothness results entirely from the smoothing we have applied. The smoothness can be expressed as the width of the smoothing kernel, which was 10 pixels FWHM in the X and Y direction. We can use the FWHM to calculate the number of *resels* in the image. 'Resel' was a term introduced by Worsley (Worsley *et al.*, 1992) and allows us to express the search volume in terms of the number of 'resolution elements' in the statistical map. This can be thought of as *similar to* the number of independent observations, but it is not the same, as we will see below. A resel is defined as a volume (in our case, of pixels) that has the same size as the FWHM. For the image in Figure 17.3, the FWHMs were 10 by 10 pixels, so that a resel is a block of 100 pixels. As there are 10 000 pixels in our image, there are 100 resels. Note that the number of resels depends only on the smoothness (FWHM) and the number of pixels.

The Euler characteristic

The Euler characteristic is a property of an image after it has been thresholded. For our purposes, the EC can be thought of as the number of blobs in an image after thresholding. For example, we can threshold our smoothed image (Figure 17.3) at $Z = 2.5$; all pixels with Z scores less than 2.5 are set to zero, and the rest are set to one. This results in the image in Figure 17.4.

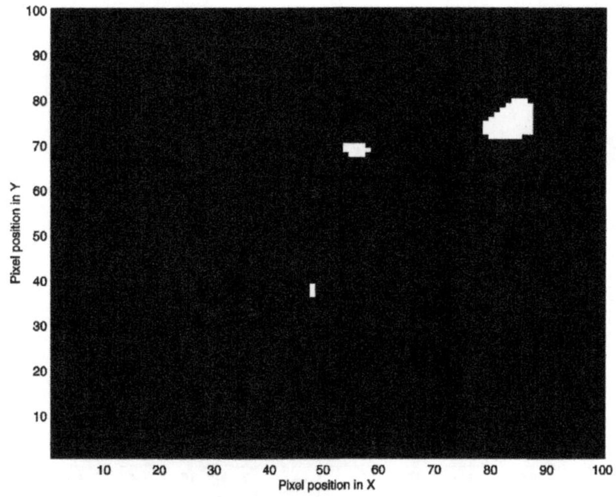

FIGURE 17.4 Smoothed random number image from Figure 17.3 after thresholding at $Z = 2.5$. Values less than 2.5 have been set to zero (displayed as black). The remaining values have been set to one (displayed as white).

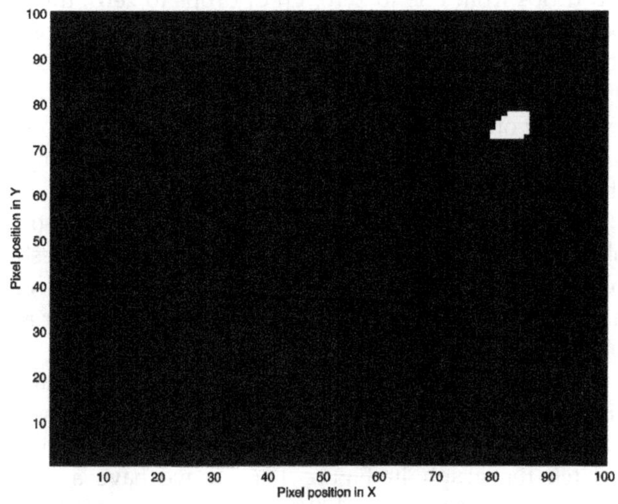

FIGURE 17.5 Smoothed random number image from Figure 17.3 after thresholding at $Z = 2.75$. Values less than 2.75 have been set to zero (displayed as black). The remaining values have been set to one (displayed as white).

There are three white blobs in Figure 17.4, corresponding to three areas with Z scores higher than 2.5. The EC of this image is therefore 3. If we increase the Z-score threshold to 2.75, we find that the two central blobs disappear – because the Z scores were less than 2.75 (Figure 17.5).

The area in the upper right of the image remains; the EC of the image in Figure 17.5 is therefore one. At high thresholds the EC is either one or zero. Hence, the average or expected EC, written $\mathsf{E}[EC]$, corresponds (approximately) to the probability of finding an above threshold blob in our statistic image. That is, the probability of a family-wise error is approximately equivalent to the expected Euler characteristic, $P^{FWE} \approx \mathsf{E}[EC]$.

It turns out that if we know the number of resels in our image, it is possible to calculate $\mathsf{E}[EC]$ at any given threshold. For an image of two dimensions $\mathsf{E}[EC]$ is given by Worsley (Worsley *et al.*, 1992). If R is the number of resels, Z_t is the Z-score threshold, then:

$$\mathsf{E}\,[EC] = R(4 log_e 2)(2\pi)^{-\frac{3}{2}} Z_t e^{-\frac{1}{2}Z_t^2}; \qquad \mathbf{17.4}$$

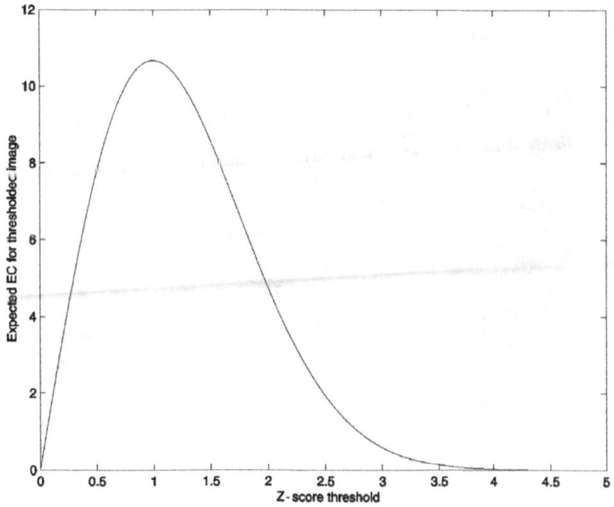

FIGURE 17.6 Expected EC values for an image of 100 resels.

Figure 17.6 shows E[*EC*] for an image of 100 resels, for *Z*-score thresholds between zero and five. As the threshold drops from one to zero, E[*EC*] drops to zero; this is because the precise definition of the EC is more complex than simply the number of blobs (Worsley *et al.*, 1994). This makes a difference at low thresholds but is not relevant for our purposes because, as explained above, we are only interested in the properties of E[*EC*] at high thresholds, i.e. when it approximates P^{FWE}.

Note also that the graph in Figure 17.6 does a reasonable job of predicting the EC in our image; at a *Z* threshold of 2.5 it predicted an EC of 1.9, when we observed a value of 3; at $Z = 2.75$ it predicted an EC of 1.1, for an observed EC of 1.

We can now apply RFT to our smoothed image (Figure 17.3) which has 100 resels. For 100 resels, Eqn. **17.4** gives an E[*EC*] of 0.049 for a *Z* threshold of 3.8 (cf. the graph in Figure 17.6). If we have a two-dimensional image with 100 resels, then the probability of getting one or more blobs where *Z* is greater than 3.8, is 0.049. We can use this for thresholding. Let x be the *Z*-score threshold that gives an E[*EC*] of 0.05. If we threshold our image at x, we can conclude that any blobs that remain have a probability of less than or equal to 0.05 that they have occurred by chance. From Eqn. **17.4**, the threshold, x, depends only on the number of resels in our image.

Random field thresholds and the Bonferroni correction

The random field correction derived using the EC is not the same as a Bonferroni correction for the number of resels. We stated above that the resel count in an image is not exactly the same as the number of independent observations. If it were the same, we could use a Bonferroni correction based on the number of resels, instead of using RFT. However, these two corrections give different answers. For $\alpha = 0.05$, the *Z* threshold according to RFT, for our 100 resel image, is $Z = 3.8$. The Bonferroni threshold for 100 independent tests is 0.05/100, which equates to a *Z*-score of 3.3. Although the RFT maths gives us a correction that is similar in principle to a Bonferroni correction, it is not the same. If the assumptions of RFT are met (see Section 4) then the RFT threshold is more accurate than the Bonferroni.

Random fields and functional imaging

Analyses of functional imaging data usually lead to three-dimensional statistical images. So far we have discussed the application of RFT to an image of two dimensions, but the same principles apply in three dimensions. The EC is the number of 3D blobs of *Z* scores above a certain threshold and a resel is a cube of voxels of size (FWHM in x) by (FWHM in y) by (FWHM in z). The equation for E[*EC*] is different in the 3D case, but still depends only on the resels in the image.

For the sake of simplicity, we have only considered a random field of *Z*-scores, i.e. numbers drawn from the normal distribution. There are now equivalent results for t, F and χ^2 random fields (Worsley, 1994). For example, the statistical parametric mapping (SPM) software uses formulae for t and F random fields to calculate corrected thresholds for height.

As noted above, we usually do not know the smoothness of a statistic volume from a functional imaging analysis, because we do not know the extent of spatial correlation before smoothing. We cannot assume that the smoothness is the same as any explicit smoothing that we have applied and will need to calculate smoothness from the images themselves. In practice, smoothness is calculated using the residual values from the statistical analysis as described in Kiebel *et al.* (1999).

Small volume correction

We noted above that the results for the expected Euler characteristic depend only on the number of resels contained in the volume of voxels we are analysing. This is not strictly accurate, although it is a very close approximation when the voxel volume is large compared to the size of a resel (Worsley *et al.* 1996). In fact, E[*EC*] also depends on the shape and size of the volume. The shape of the volume becomes important when we have a small or oddly shaped region. This is unusual if we are analysing statistics from the whole brain, but there are often situations where we wish to restrict our search to a smaller subset of the volume, for example where we have a specific hypothesis as to where our signal is likely to occur.

The reason that the shape of the volume may influence the correction is best explained by example. Let us return to the 2D image of smoothed random numbers (Figure 17.3). We could imagine that we had reason to believe that signal change will occur only in the centre of the image. Our search region will not be the whole image, but might be a box at the image centre, with size 30 by 30 pixels (Figure 17.7).

The box contains 9 resels. The figure shows a grid of X-shaped markers; these are spaced at the width of a resel, i.e. 10 pixels. The box can contain a maximum of 16 of these markers. Now let us imagine we had a more unusually shaped search region. For some reason, we might expect that our signal of interest will occur within a frame 2.5 pixels wide around the outside of the image. The frame contains the same number of voxels as the box, and therefore has the same volume in terms of resels. However, the frame contains many more markers (32), so the frame is sampling from the data of more resels than the box. Multiple comparison correction for the frame must therefore be more stringent than for the box.

In fact, E[*EC*] depends on the volume, surface area, and diameter of the search region (Worsley *et al.*, 1996). These parameters can be calculated for continuous shapes for which formulae are available for volume, surface area and diameter, such as spheres or boxes (see Appendix 6).

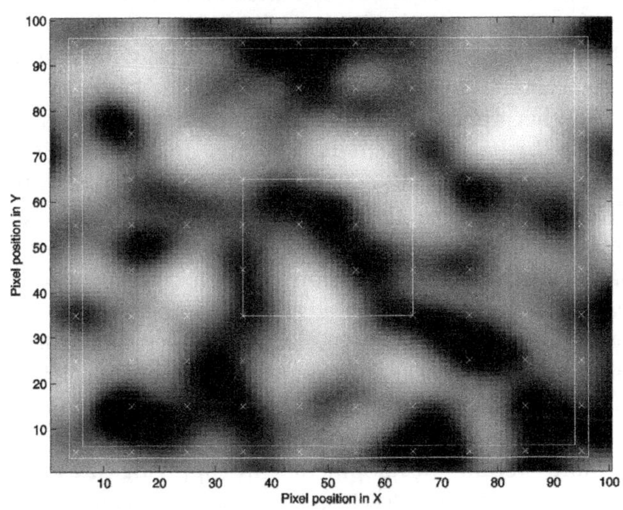

FIGURE 17.7 Smoothed random number image from Figure 17.3 with two example search regions: a box (centre) and a frame (outer border of image). X-shaped markers are spaced at one-resel widths across the image.

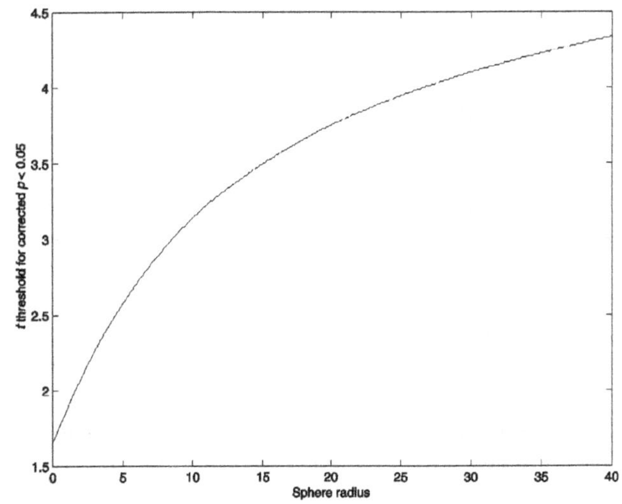

FIGURE 17.8 t threshold giving an FWE rate of 0.05, for spheres of increasing radius. Smoothness was 8 mm in X Y and Z directions, and the example analysis had 200 degrees of freedom.

Otherwise, the parameters can be estimated from any shape that has been defined in an image. Restricting the search region to a small volume within a statistical map can lead to greatly reduced thresholds for given FWE rates. For Figure 17.8, we assumed a statistical analysis that resulted in a t-statistic map with 8 mm smoothness in the X, Y and Z directions. The t-statistic has 200 degrees of freedom, and we have a spherical search region. The graph shows the t-statistic value that gives a corrected p-value of 0.05 for spheres of increasing radius.

For a sphere of zero radius, the threshold is simply that for a single t-statistic (uncorrected = corrected $p = 0.05$ for $t = 1.65$ with 200 degrees of freedom). The corrected t threshold increases sharply as the radius increases to ≈ 10 mm, and less steeply thereafter.

Uncorrected *p* values and regional hypotheses

When making inferences about regional effects (e.g. activations) in SPMs, one often has some idea about where the activation should be. In this instance a correction for the entire search volume is inappropriate.

If the hypothesized region contained a single voxel, then inference could be made using an uncorrected p-value (as there is no extra search volume to correct for). In practice, however, the hypothesized region will usually contain many voxels and can be characterized, for example, using spheres or boxes centred on the region of interest, and we must therefore use a p-value that has been appropriately corrected. As described in the previous section, this corrected p-value will depend on the size and shape of the hypothesized region, and the smoothness of the statistic image.

Some research groups have used uncorrected p-value thresholds, such as $p < 0.001$, in order to control FWE when there is a regional hypothesis of where the activation will occur. This approach gives unquantified error control, however, because any hypothesized region is likely to contain more than a single voxel. For example, for 8 mm smoothness, a spherical region with a radius greater than 6.7 mm will require an uncorrected p-value threshold of less than 0.001 for a FWE rate ≤ 0.05. For a sphere of radius 15 mm, an uncorrected p-value threshold of 0.001 gives an $\mathrm{E}[EC]$ of 0.36, so there is approximately a 36 per cent chance of seeing one or more voxels above threshold even if the null hypothesis is true.

DISCUSSION

In this chapter, we have focused on voxel-level inference based on height thresholds to ask the question: is activation at a given voxel significantly non-zero? More generally, however, voxel-level inference can be placed in a larger framework involving cluster-level and set-level inference. These require height and spatial extent thresholds to be specified by the user. Corrected p-values can then be derived that pertain to: (i) the number of activated regions (i.e. number of clusters above the height and volume threshold) – set-level inferences; (ii) the number of activated voxels (i.e. volume) comprising a particular region – cluster-level inferences; and (iii) the p-value for each peak within that cluster – peak-level inferences. Typically, people use peak-level inferences and a spatial extent threshold of zero. This reflects the fact that characterizations of functional anatomy are generally more useful when specified with a high degree of anatomical precision (see Chapter 19 for more details)

There are two assumptions underlying RFT. The first is that the error fields are a reasonable lattice approximation to an underlying random field with a multivariate Gaussian distribution. The second is that these fields are continuous, with a twice-differentiable autocorrelation function. A common misconception is that the autocorrelation function has to be Gaussian, but this is not the case.

If the data have been sufficiently smoothed and the general linear models correctly specified (so that the errors are indeed Gaussian) then the RFT assumptions will be met. If the data are not smooth; one solution is to reduce the voxel size by subsampling. Alternatively, one can turn to different inference procedures. One such alternative is the non-parametric framework described in Chapter 21.

Other inference frameworks are the false discovery rate (FDR) approach and Bayesian inference. While RFT controls the family-wise error, the probability of reporting a false positive anywhere in the volume, FDR controls the proportion of false positive voxels, among those that are declared positive. This very different approach is discussed in Chapter 20. Finally, Chapter 22 introduces Bayesian inference where, instead of focusing on how unlikely the data are under a null hypothesis, inferences are made on the basis of a posterior distribution which characterizes our uncertainty about the parameter estimates, without reference to a null distribution.

Bibliography

The mathematical basis of RFT is described in a series of peer-reviewed articles in statistical journals (Siegmund and Worsley, 1994; Worsley, 1994; Cao, 1999). The core paper for RFT as applied to functional imaging is Worsley *et al.*, (1996) (see also Worsley *et al.*, 2004). This provides estimates of p-values for local maxima of Gaussian, t, χ^2 and F fields over search regions of any shape or size in any number of dimensions. This unifies earlier results on 2D (Friston *et al.*, 1991) and 3D (Worsley *et al.*, 1992) images.

The above analysis requires an estimate of the smoothness of the images. Poline *et al.* (1995) estimate the dependence of the resulting SPMs on the estimate of this parameter. While the applied smoothness is usually fixed, Worsley *et al.*, (1995) propose a scale-space procedure for assessing significance of activations over a range of proposed smoothings. In Kiebel *et al.* (1999), the authors implement an unbiased smoothness estimator for Gaussianized t-fields and t-fields. Worsley *et al.* (1999) derive a further improved estimator, which takes into account non-stationarity of the statistic field.

Another approach to assessing significance is based, not on the height of activity, but on spatial extent (Friston *et al.*, 1994), as described in the previous section. In Friston *et al.* (1996), the authors consider a hierarchy of tests that are regarded as peak-level, cluster-level and set-level inferences. If the approximate location of an activation can be specified in advance then the significance of the activation can be assessed using the spatial extent or volume of the nearest activated region (Friston, 1997). This test is particularly elegant as it does not require a correction for multiple comparisons.

More recent developments in applying RFT to neuroimaging are described in the following chapters. Finally, we refer readers to an online resource, http://www.mrc-cbu.cam.ac.uk/Imaging/Common/randomfields.shtml, from which much of the material in this chapter was collated.

REFERENCES

Cao J (1999) The size of the connected components of excursion sets of chi-squared, t and F fields. *Adv Appl Prob* **31**: 577–93

Friston KJ, Frith CD, Liddle PF *et al.* (1991) Comparing functional (PET) images: the assessment of significant change. *J Cereb Blood Flow Metab* **11**: 690–99

Friston KJ, Worsley KJ, Frackowiak RSJ *et al.* (1994) Assessing the significance of focal activations using their spatial extent. *Hum Brain Mapp* **1**: 214–20

Friston KJ, Holmes A, Poline J-B *et al.* (1996) Detecting activations in PET and fMRI: levels of inference and power. *NeuroImage* **4**: 223–35

Friston K (1997) Testing for anatomically specified regional effects. *Hum Brain Mapp* **5**: 133–36

Kiebel SJ, Poline JB, Friston KJ *et al.* (1999) Robust smoothness estimation in statistical parametric maps using standardized residuals from the general linear model. *NeuroImage* **10**: 756–66

Poline JB, Worsley KJ, Holmes AP *et al.* (1995) Estimating smoothness in statistical parametric maps: variability of p values. linear model. *J Comput Assist Tomogr* **19**: 788–96

Siegmund DO, Worsley KJ (1994) Testing for a signal with unknown location and scale in a stationary Gaussian random field. *Ann Stat* **23**: 608–39

Worsley KJ, Evans AC, Marrett S *et al.* (1992) A three-dimensional statistical analysis for rCBF activation studies in human brain. *J Cereb Blood Flow Metab* **12**: 900–18

Worsley KJ (1994) Local maxima and the expected Euler characteristic of excursion sets of x2, F and t fields. *Adv Appl Prob* **26**: 13–42

Worsley KJ, Marrett S, Neelin P *et al.* (1995) Searching scale space for activation in PET images. *Hum Brain Mapp* **4**: 74–90

Worsley KJ, Marrett S, Neelin P *et al.* (1996) A unified statistical approach for determining significant signals in images of cerebral activation. *Hum Brain Mapp* **4**: 58–73

Worsley KJ, Andermann M, Koulis T *et al.* (1999) Detecting changes in nonisotropic images. *Hum Brain Mapp* **8**: 98–101

Worsley KJ, Taylor JE, Tomaiuolo F *et al.* (2004) Unified univariate and multivariate random field theory. *NeuroImage* **23**: 189–95

CHAPTER

18

Random Field Theory

K. Worsley

INTRODUCTION

Random field theory is used in the statistical analysis of statistical parametric maps (SPMs) whenever there is a spatial component to the inference. Most important is the question of detecting an effect or activation at an unknown spatial location. Very often we do not know in advance where to look for an effect, and we are interested in searching the whole brain, or part of it. This chapter presents special statistical problems related to the problem of multiple comparisons, or multiple tests. Two methods have been proposed, the first based on the maximum of the T or F statistic, the second based on the spatial extent of the region where these statistics exceed some threshold value. Both involve results from random field theory (Adler, 1981).

THE MAXIMUM TEST STATISTIC

An obvious method is to select those locations where a test statistic Z (which could be a T, χ^2, F or Hotelling's T^2 statistic) is large, i.e. to threshold the image of Z at a height z. The problem is then to choose the threshold z to exclude false positives with a high probability, say 0.95. Setting z to the usual (uncorrected) $p = 0.05$ critical value of Z (1.64 in the Gaussian case) means that 5 per cent of the unactivated parts of the brain will show false positives. We need to raise z so that the probability of finding any activation in the non-activated regions is 0.05. This is a type of multiple comparison problem, since we are testing the hypothesis of no activation at a very large number of voxels.

A simple solution is to apply a Bonferroni correction. The probability of detecting any activation in the unactivated locations is bounded by assuming that the unactivated locations cover the entire search region. By the Bonferroni inequality, the probability of detecting any activation is further bounded by:

$$P(\max Z > z) \leq N\, P(Z > z) \qquad \textbf{18.1}$$

where the maximum is taken over all N voxels in the search region. For a $p = 0.05$ test of Gaussian statistics, critical thresholds of 4–5 are common. This procedure is conservative if the image is smooth, although for functional magnetic resonance imaging (fMRI) data it can give accurate thresholds in single-subject analyses in which no smoothing has been applied.

Random field theory gives a less conservative (lower) p-value if the image is smooth:

$$P(\max Z > z) \approx \sum_{d=0}^{D} \text{Resels}_d \text{EC}_d(z) \qquad \textbf{18.2}$$

where D is the number of dimensions of the search region, Resels_d is the number of d-dimensional resels (resolution elements) in the search region, and $\text{EC}_d(z)$ is the d-dimensional Euler characteristic density. The approximation Eqn. **18.2** is based on the fact that the left hand side is the exact expectation of the Euler characteristic of the region above the threshold z. The Euler characteristic (EC) counts the number of clusters if the region has no holes, which is likely to be the case if z is large. Details can be found in Appendix 6 and Worsley *et al.* (1996a).

The approximation Eqn. **18.2** is accurate for search regions of any size or shape, even a single point, but it is best for search regions that are not too concave. Sometimes it is better to surround a highly convoluted search region, such as grey matter, by a convex hull with slightly higher volume but less surface area, to get a lower and more accurate p-value. This is because the Euler characteristic includes terms that depend on both the volume and surface area of the search volume.

Statistical Parametric Mapping, by Karl Friston et al.
ISBN–13: 978-0-12-372560-8 ISBN–10: 0-12-372560-7

For large search regions, the last term ($d = D$) is the most important. The number of resels is:

$$\text{Resels}_D = V/\text{FWHM}^D$$

where V is the volume of the search region and FWHM is the effective full width at half maximum of a Gaussian kernel that encodes the smoothness (i.e. the kernel that would be applied to an unsmooth image to produce the same smoothness). The corresponding EC density for a T-statistic image with ν degrees of freedom is:

$$\text{EC}_3(z) = \frac{(4\log_e 2)^{\frac{3}{2}}}{(2\pi)^2}\left(\frac{\nu-1}{\nu}z^2-1\right)\left(1+\frac{z^2}{\nu}\right)^{-\frac{1}{2}(\nu-1)}$$

For small search regions, the lower dimensional terms $d < D$ become important. However the p-value (Eqn. **18.2**) is not very sensitive to the shape of the search region, so that assuming a spherical search region gives a very good approximation.

Figure 18.1 shows the threshold z for a $p = 0.05$ test calculated by the two methods. If the FWHM is small relative to the voxel size, then the Bonferroni threshold is actually less than the random field one (Eqn. **18.2**). In practice, it is better to take the minimum of the the two thresholds (Eqn. **18.1** and Eqn. **18.2**).

EC densities for F fields can be found in Worsley *et al.* (1996a), and for Hotelling's T^2, see Cao and Worsley (1999a). Similar results are also available for correlation random fields, useful for detecting functional connectivity (see Cao and Worsley, 1999b).

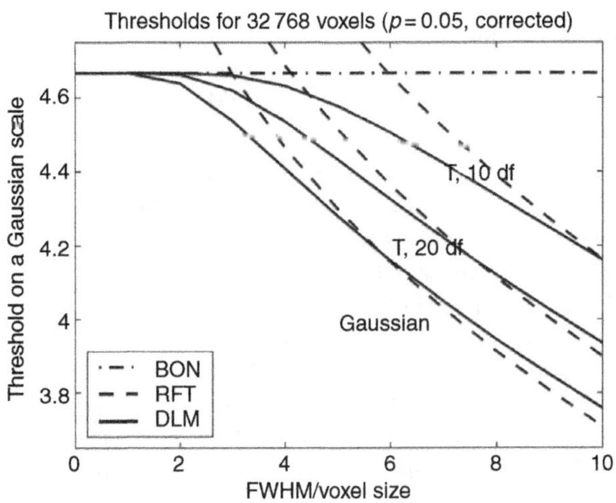

FIGURE 18.1 Thresholds for a volume with $N = 32^3 = 32\,768$ voxels ($p = 0.05$, corrected). Note that if the full width half maxim (FWHM) is less than ~ 3 voxels, then the Bonferroni (BON) method is better than the random field (RFT) method for a Gaussian statistic. For T statistics with $\nu = 20$ df, this limit is higher (~ 4), and much higher (~ 6) with $\nu = 10$ df. The discrete maxim (DLM) method bridges the gap between the two, and gives the best results.

Extensions of the result in Eqn. **18.2** to scale space random fields are given in Worsley *et al.* (1996b). Here, the search is over all spatial filter widths as well over location, so that the width of the signal is estimated as well as its location. The price paid is an increase in critical threshold of about 0.5.

A recent improvement fills in the gap between low FWHM (when Bonferroni is accurate) and high FWHM (when random field theory is accurate). This new method, based on discrete local maxima (DLM), uses the correlation between adjacent voxels. The formula is somewhat complicated, and can be found in Worsley (2005). Like Bonferroni, DLM is a lower bound, so we suggest taking the minimum of Bonferroni, random field theory, and DLM. Thresholds found using DLM are also shown in Figure 18.1.

THE MAXIMUM SPATIAL EXTENT OF THE TEST STATISTIC

An alternative test can be based on the spatial extent of clusters of connected components of suprathreshold voxels where $Z > z$ (Friston *et al.*, 1994). Typically z is chosen to be about 3 for a Gaussian random field. Once again, the image must be a smooth stationary random field. The idea is to approximate the shape of the image by a quadratic with a peak at the local maximum. For a Gaussian random field, the spatial extent S is then approximated by the volume where the quadratic of height H above z cuts the threshold z:

$$S \approx cH^{D/2} \qquad \textbf{18.3}$$

where

$$c = \text{FWHM}^D(2\pi/z)^{D/2}(4\log 2)^{-D/2}/\Gamma(D/2+1)$$

For large z, the upper tail probability of H is well approximated by:

$$\text{P}(H > h) = \text{P}(\max Z > z+h)/\text{P}(\max Z > z) \approx \exp(-zh) \qquad \textbf{18.4}$$

from which we conclude that H has an approximate exponential distribution with mean $1/z$. From this we can find the approximate p-value of the spatial extent S of a single cluster:

$$\text{P}(S > s) \approx \exp(-z(s/c)^{2/D}) \qquad \textbf{18.5}$$

The p-value for the largest spatial extent can be obtained by a simple Bonferroni correction for the expected number of clusters K:

$$\mathrm{P}(\max S > s) \approx \mathrm{E}(K)\ \mathrm{P}(S > s), \text{ where } \mathrm{E}(K) \approx \mathrm{P}(\max Z > z) \tag{18.6}$$

from Eqn. **18.2**. See Chapter 19 for a fuller discussion.

We can substantially improve the value of the constant c by equating the expected total spatial extent, given by $V\ \mathrm{P}(Z > z)$, to that obtained by summing up the spatial extents of all the clusters $S_1, \ldots, S_K$:

$$V\ \mathrm{P}(Z > z) = \mathrm{E}(S_1 + \cdots + S_K) = \mathrm{E}(K)\ \mathrm{E}(S)$$

Using the fact that:

$$\mathrm{E}(S) \approx c\Gamma(D/2+1)/z^{D/2}$$

from Eqn. **18.3**, and the expected number of clusters from Eqn. **18.2**, it follows that:

$$c \approx \mathrm{FWHM}^D z^{D/2} \mathrm{P}(Z > z)/\{\mathrm{EC}_D(z)\ \Gamma(D/2+1)\}$$

Cao (1999) has extended these results to T, χ^2 and F fields, but unfortunately there are no theoretical results for non-smooth fields such as raw fMRI data.

SEARCHING IN SMALL REGIONS

For small prespecified search regions such as the cingulate, the p-values for the maximum test statistic are very well estimated by Eqn. **18.2**, but the results in the previous section only apply to large search regions. Friston (1997) has proposed a fascinating method that avoids the awkward problem of prespecifying a small search region altogether. We threshold the image of test statistics at z, then simply pick the nearest peak to a point or region of interest. The clever part is this. Since we have identified this peak based only on its spatial location and not based on its height or extent, there is now no need to correct for searching over all peaks. Hence, the p-value for its spatial extent S is simply $\mathrm{P}(S > s)$ from Eqn. **18.5**, and the p-value for its peak height H above z is simply $\mathrm{P}(H > h)$ from Eqn. **18.4**.

ESTIMATING THE FWHM

The only data-dependent component required for setting the random field threshold is Resels_D, and indirectly, the FWHM. The FWHM often depends on the location: raw fMRI data are considerably smoother in cortex than white matter (Figure 18.2). This means that the random field is not isotropic, so the above random field theory is not valid. Fortunately, there is a simple way of allowing for this by estimating the FWHM separately at each voxel.

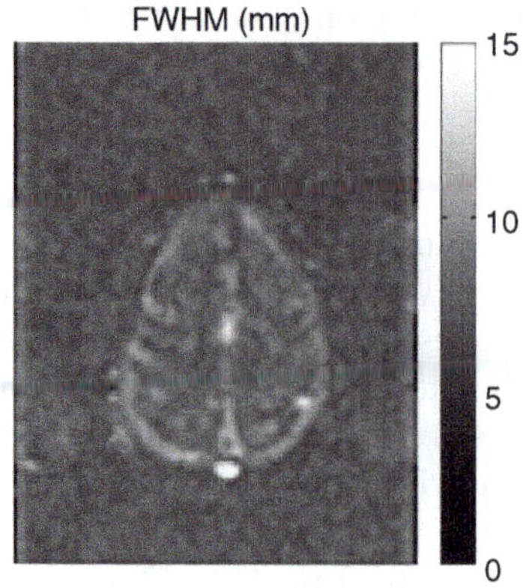

FIGURE 18.2 The estimated FWHM for one slice of raw fMRI data. Note the ~6 mm FWHM outside the brain due to smoothing imposed by motion correction. The FWHM in cortex is much higher, ~10 mm, while white matter is lower ~6 mm.

Let $\mathbf{r}$ be the n-vector of least-squares residuals from the (possibly whitened) linear model fitted at each voxel, and let $\mathbf{u}$ be the vector of normalized residuals $\mathbf{u} = \mathbf{r}/(\mathbf{r}'\mathbf{r})^{1/2}$. Let $\dot{\mathbf{u}}$ be the $n \times 3$ spatial derivative of $\mathbf{u}$ in the three orthogonal directions of the voxel lattice. The estimated FWHM is:

$$\widehat{\mathrm{FWHM}} = (4\log 2)^{1/2}|\dot{\mathbf{u}}'\dot{\mathbf{u}}|^{-1/(2D)} \tag{18.7}$$

and the estimated Resels_D is:

$$\widehat{\mathrm{Resels}}_D = \sum_{\text{volume}} \widehat{\mathrm{FWHM}}^{-D} v$$

where summation is over all voxels in the search region and v is the volume of a single voxel (Worsley *et al.*, 1999). The extra randomness added by estimating Resels_D can be ignored if the search region is large.

However, spatially varying FWHM can have a strong effect on the validity of the p-value for spatial extent. If the cluster is in a region where FWHM is large, then its extent will be larger by chance alone, and so its p-value will be too small. In other words, clusters will look more significant in smooth regions than in rough regions of the image. To correct for this, we simply replace cluster volume by cluster resels, defined as:

$$\tilde{S} = \sum_{\text{cluster}} v\widehat{\mathrm{FWHM}}^{-D}$$

where summation is over all voxels in the cluster (Hayasaka *et al.*, 2004).

There is one remaining problem: since the above summation is over a small cluster, rather than a large search region, the randomness in estimating FWHM now makes a significant contribution to the randomness of $\tilde{S}$, and hence its p-value. Hayasaka *et al.* (2004) suggest allowing for this by the approximation:

$$\tilde{S} \approx \tilde{c} H^{D/2} \prod_{k=1}^{D+1} X_k^{p_k} \qquad \textbf{18.8}$$

where $X_1, \ldots, X_{D+1}$ are independent χ^2 random variables. The degrees of freedom of X_k is $\nu - k + 1$ where $\nu = n - p$ and p is the number of regressors in the linear model, raised to the power $p_k = -D/2$ if $k = 1$ and $p_k = 1/2$ if $k > 1$. Again the constant $\tilde{c}$ is chosen so that the expected total resels of all clusters matches the probability of exceeding the threshold times the volume of the search region:

$$\tilde{c} \approx z^{D/2} \mathrm{P}(Z > z)/\{\mathrm{EC}_D(z)\ \Gamma(D/2+1)\}$$

Combining this with the approximate distributions of spatial extents for T, χ^2 and F fields from Cao (1999) requires no extra computational effort. H is replaced by a beta random variable in Eqn. **18.8**, multiplied by powers of yet more χ^2 random variables, with appropriate adjustments to $\tilde{c}$.

In practice, the distribution function of $\tilde{S}$ is best calculated by first taking logarithms, so that $\log \tilde{S}$ is then a sum of independent random variables. The density of a sum is the convolution of the densities whose Fourier transform is the sum of the Fourier transforms. It is easier to find the upper tail probability of $\log \tilde{S}$ by replacing the density of one of the random variables by its upper tail probability *before* doing the convolution. The obvious choice is the exponential or beta random variable, since its upper tail probability has a simple closed form expression. This method has been implemented in the stat_threshold.m function of fmristat, available from http://www. math.mcgill.ca/keith/fmristat.

FALSE DISCOVERY RATE

A remarkable breakthrough in multiple testing was made by Benjamini and Hochberg, in 1995, who took a completely different approach. Instead of controlling the probability of ever reporting a false positive, they devised a procedure for controlling the *false discovery rate* (FDR), the expected proportion of false positives amongst those voxels declared positive (the *discoveries*) (Figure 18.3). The procedure is extremely simple to implement. Simply calculate the uncorrected p-value for each voxel and order them so that the ordered p-values are $P_1 \leq P_2 \leq \ldots \leq P_N$. To control the FDR at α, find the largest value k so that $P_k < \alpha k/N$. This procedure is conservative if the voxels are positively dependent, which is a reasonable assumption for most unsmoothed or smoothed imaging data. See Chapter 20 and Genovese *et al.* (2002) for an application of this method to fMRI data, and for further references.

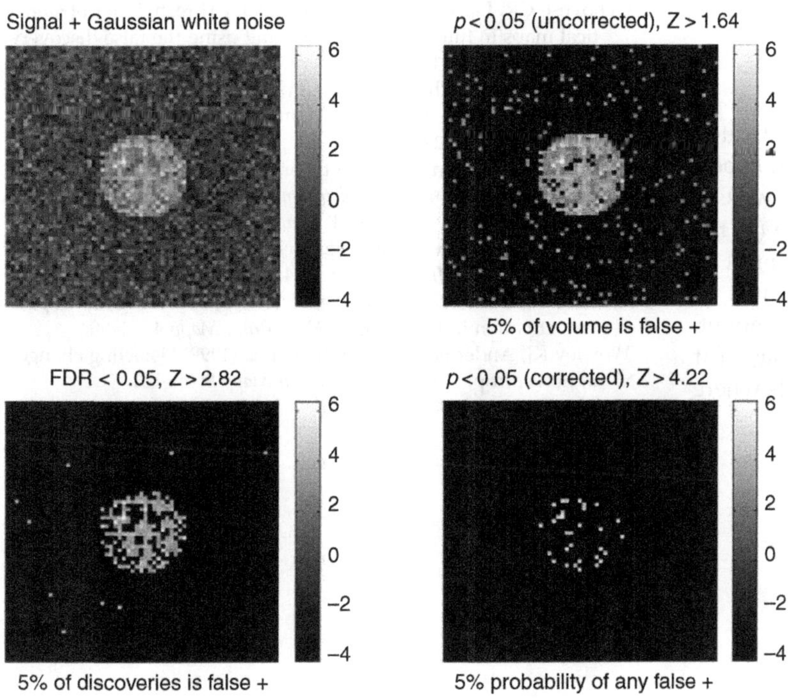

FIGURE 18.3 Illustration of the difference between false discovery rate and Bonferroni/random field methods for thresholding an image.

TABLE 18-1 Examples of the thresholds of false discovery rate, Bonferroni and random field methods for thresholding a Gaussian image

Proportion of true + in image	1	0.1	0.01	0.001	0.0001
FDR threshold	1.64	2.56	3.28	3.88	4.41
Number of voxels in image	1	10	100	1000	10 000
Bonferroni threshold	1.64	2.58	3.29	3.89	4.42
Number of resels in image	0	1	10	100	1000
Random fields threshold	1.64	2.82	3.46	4.09	4.65

The resulting threshold, corresponding to the value of Z for P_k, depends on the amount of signal in the data, not on the number of voxels or the smoothness. Table 18-1 compares thresholds for the FDR, Bonferroni and random field methods. Thresholds of 2–3 are typical for brain mapping data with a reasonably strong signal, quite a bit lower than the Bonferroni or random field thresholds.

But we must remember that the interpretation of the FDR is quite different. False positives will be detected; we are simply controlling them so that they make up no more than α of our discoveries. On the other hand, the Bonferroni and random field methods control the probability of *ever* reporting a false discovery (see Figure 18.3). Furthermore, FDR controls the expected false discovery rate of voxels or volume (i.e. the proportion of the volume that is false positive), whereas RFT controls the false positive rate of regions or maxima (note that in Figure 18.3 (lower left panel), there are 9 false maxima but there is only one regional effect).

CONCLUSION

The idea of using a hypothesis test to detect activated regions does contain a fundamental flaw that all experimenters should be aware of. Think of it this way: if we had enough data, T statistics would increase (as the square root of the number of scans or subjects) until *all* voxels were 'activated'! In reality, *every* voxel must be affected by the stimulus, perhaps by a very tiny amount; it is impossible to believe that there is never any signal at all. So thresholding simply excludes those voxels where we don't yet have enough evidence to distinguish their effects from zero. If we had more evidence, perhaps with better scanners, or simply more subjects, we surely would be able to do so. But then we would probably not want to detect activated regions. As for satellite images, the job for statisticians would then be signal *enhancement* rather than signal detection (see also Chapter 23 for a Bayesian perspective on this issue). The distinguishing feature of most brain mapping data is that there is so little signal to enhance. Even with the advent of better scanners this is still likely to be the case, because neuroscientists will surely devise yet more subtle experiments that are always pushing the signal to the limits of detectability.

REFERENCES

Adler RJ (1981) *The Geometry of random fields*. Wiley, New York

Benjamini Y, Hochberg Y (1995) Controlling the false discovery rate: a practical and powerful approach to multiple testing. *J Roy Stat Soc Series B, Methodological* **57**: 289–300

Cao J (1999) The size of the connected components of excursion sets of χ^2, t and F fields. *Adv Appl Prob* **31**: 577–93

Cao J, Worsley KJ (1999a) The detection of local shape changes via the geometry of Hotelling's T^2 fields. *Ann Stat* **27**: 925–42

Cao J, Worsley KJ (1999b) The geometry of correlation fields, with an application to functional connectivity of the brain. *Ann Appl Prob*, **9**: 1021–57

Friston KJ (1997) Testing for anatomically specified regional effects. *Hum Brain Mapp* **5**: 133–36

Friston KJ, Worsley KJ, Frackowiak RSJ *et al.* (1994) Assessing the significance of focal activations using their spatial extent. *Hum Brain Mapp* **1**: 214–20

Genovese CR, Lazar NA, Nichols TE (2002) Thresholding of statistical maps in functional neuroimaging using the false discovery rate. *NeuroImage* **15**: 772–86

Hayasaka S, Luan-Phan K, Liberzon I *et al.* (2004) Non-stationary cluster-size inference with random field and permutation methods. *NeuroImage* **22**: 676–87

Worsley KJ (2005) An improved theoretical P-value for SPMs based on discrete local maxima. *Neuroimage* **28**: 1056–62

Worsley KJ, Marrett S, Neelin P *et al.* (1996a) A unified statistical approach for determining significant signals in images of cerebral activation. *Hum Brain Mapp* **4**: 58–73

Worsley KJ, Marrett S, Neelin P *et al.* (1996b) Searching scale space for activation in PET images. *Hum Brain Mapp* **4**: 74–90

Worsley KJ, Andermann M, Koulis T *et al.* (1999) Detecting changes in nonisotropic images. *Hum Brain Mapp* **8**: 98–101

CHAPTER

19

Topological Inference

K. Friston

INTRODUCTION

The previous two chapters have established the heuristics behind inference on statistical maps and how this inference rests upon random field theory. In this chapter, we revisit these themes by looking at the nature of topological inference and the different ways it can be employed. The key difference between statistical parametric mapping (SPM) and conventional statistics lies in the thing one is making an inference about. In conventional statistics, this is usually a scalar quantity (i.e. a model parameter) that generates measurements, such as reaction times. This inference is based on the size of a statistic that is a function of scalar measurements. In contrast, in SPM one makes inferences about the topological features of a statistical process that is a function of space or time. The SPM is a function of the data, which is a function of position in the image. This renders the SPM a functional (a function of a function). Put succinctly, conventional inference is based on a statistic, which is a *function* of the data. Topological inference is based on an SPM which is a *functional* of the data. This is important because a statistic can only have one attribute – its size. Conversely, an SPM can have many topological attributes, such as the height of a peak, the number of peaks, the volume or extent of an excursion set (part of the SPM above some height). All these topological features are quantities that have a distribution under the null hypothesis and can be used for inference. Critically, the nature of this inference is determined by the topological quantity used. This chapter introduces some common topological inferences and their relative sensitivities to different sorts of treatments effects or activations.

TOPOLOGICAL INFERENCE

This chapter is about detecting activations in statistical parametric maps and considers the relative sensitivity of a nested hierarchy of tests that are framed in terms of the level of inference (peak-level, cluster-level, and set-level). These tests are based on the probability of obtaining c, or more, clusters with k, or more, voxels, above a threshold u. This probability has a reasonably simple form and is derived using distributional approximations from differential topology (i.e. the theory of random fields; see below and Chapter 18). The different levels of inference pertain to different topological quantities, which vary in their spatial specificity: set-level inference refers to the inference that the number of clusters comprising an observed activation profile is unlikely to have occurred by chance. This inference pertains to the set of clusters (connected excursion sets) reaching criteria and represents an inference about distributed effects. Cluster-level inferences are a special case of set-level inferences, which obtain when the number of clusters c is one. Similarly, peak-level inferences are special cases of cluster-level inferences that result when the cluster is very small (i.e. $k < 0$). Set-level inferences are generally more powerful than cluster-level inferences and cluster-level inferences are generally more powerful than peak-level inferences. The price paid for this increased sensitivity is reduced localizing power: peak-level tests permit individual maxima to be identified as significant, whereas cluster and set-level inferences only allow clusters or sets of clusters to be so identified.

Peaks, clusters and sets

In what follows, we consider tests based on three topological features: a peak, a cluster, and a set of clusters. We then consider the relative sensitivity of the ensuing tests in terms of power and how that power varies as a function of resolution and the nature of the underlying signal. This treatment is concerned primarily with distributed signals that have no *a priori* anatomical specification. Activations in positron emission

Statistical Parametric Mapping, by Karl Friston et al.
ISBN–13: 978-0-12-372560-8 ISBN–10: 0-12-372560-7

tomography (PET) and functional magnetic resonance imaging (fMRI) are almost universally detected using some form of statistical mapping. The statistical processes that ensue (i.e. statistical parametric maps) are characterized in terms of regional excursions above some threshold and p-values are assigned to these excursions. These p-values reflect the probability of false positives or type 1 error. There are two forms of control over family-wise error (FWE), weak and strong, which determine the level at which departures from the null hypothesis can be reported. A test procedure controls FWE in the weak sense if the probability of false rejection is less than α. A procedure with only weak control has no localizing power. If the null hypothesis is rejected, then all that can be said is that there is a departure from the null hypothesis somewhere in the SPM. Here, the level of inference is the whole volume analysed. A procedure controls FWE in the strong sense if the probability of a false positive peak or cluster, for which the null hypothesis is true, is less than α, regardless of the truth of the null hypothesis elsewhere. This more stringent criterion gives localizing power. Peaks or clusters identified by such a procedure may be declared individually significant. Another way of looking at this is in terms of the attributes of the topological features whose false positive rate is controlled. Controlling the family-wise false positive rate of peaks means that inference is about a peak, which has the attribute 'location'. Controlling the false positive rate of clusters enables one to infer the cluster is significant and implicitly its spatial support. However, inferring a set of clusters is significant controls FWE in a weaker sense because it is the ensemble that is significant, not any single cluster.

As noted in Chapter 17, the simplest multiple comparisons procedure which maintains strong control over the error rate in a family of discrete tests is based on the Bonferroni inequality. Here, the p-values belong, not to peaks, but to voxels and are adjusted for the number of voxels. However, for even mild dependencies between the voxels, this method is excessively conservative and inappropriate. It is inappropriate because we are not interested in controlling the false positive rate of voxels; we want to control the false positive rate of peaks. The peak is a topological feature, a voxel is not. Critically, for any given threshold, there will be more suprathreshold voxels than peaks. This means the false positive rate of voxels is always greater than the false positive rate of peaks. This is why SPM uses a lower threshold than required by a Bonferroni correction and why SPM is more powerful. It is easy to see that controlling the false positive rate of voxels is meaningless; imagine we simply halved the size of each voxel by interpolating the SPM. This would increase the false positive rate of voxels by a factor of eight. But nothing has changed. On the other hand, the false positive rate of peaks remains constant and the inference furnished by SPM remains exact. This simple example illustrates that SPM and the topological inference it entails, is central to the analysis of data that are a function of some position in space, time, frequency etc.

Random field theory

The most successful approach to statistical inference on analytic (continuous) statistical processes is predicated on the theory of random fields. Early work was based on the theory of level crossings (Friston *et al.*, 1991) and differential topology (Worsley *et al.*, 1992). These approaches control FWE strongly, allowing for inference at the peak-level: a corrected p-value is assigned to a peak using the probability that its height, or a higher one, could have occurred by chance in the volume analysed. There have been a number of interesting elaborations at this level of inference (e.g. searching scale-space and other high-dimensional SPMs (e.g. Siegmund and Worsley 1994)) and results for many statistics exist (e.g. Worsley, 1994). The next development, using the random field theory, was to use the spatial extent of a cluster of voxels defined by a height threshold (Friston *et al.*, 1994; see also Poline and Mazoyer, 1993, and Roland *et al.*, 1993). These procedures control FWE strongly at the cluster-level, permitting inference about each cluster, and are based on the probability of getting a cluster of the extent observed (defined by a height threshold), or a larger one, in the volume analysed. In Friston *et al.* (1996) set-level inference was introduced. This is based on the probability of getting the observed number of clusters (defined by a height and an extent threshold), or more, in the volume analysed. This inference is about the set of clusters (contiguous regions above some height and size thresholds) or more simply about the excursion *set*. In this chapter, we compare the relative power of these different levels of inference, under different conditions. In the sense that all these inferences are based on adjusted p-values, we consider only the case where no *a priori* knowledge about the deployment of activation is available, within the volume considered. This volume may be the entire cerebrum, or could be a smaller volume encompassing a region in which one wants to focus statistical power. The underlying theory is exactly the same for whole brain and small volume corrections and all levels of inference apply.

The results in the previous two chapters and used below, derive from random field theory. The assumptions implicit in this approach are: that the SPMs are reasonable lattice representations of underlying continuous fields; that the components of the fields have a multivariate Gaussian distribution; and that the height thresholds employed are high. These are reasonable assumptions in neuroimaging as long as the voxel-size or bin-size is small relative to the smoothness. There has been some interest in revising cluster-level approaches in the context of fMRI (where the voxel sizes are larger in relation to resolution) using Monte Carlo simulations and adjustments to the smoothness estimators (e.g. Forman *et al.*, 1995). Usual estimates of smoothness (e.g. Friston *et al.*, 1991; Worsley *et al.*, 1992; Kiebel *et al.*, 1999) fail when the reasonable lattice assumption is violated. In our work, we sidestep this issue by interpolating the data to reduce voxel size or smoothing the data to increase smoothness. It is generally accepted that the voxel size should be less than half the full width at half maximum (FWHM) of the smoothness. The good lattice and Gaussian assumptions can be further ensured by slight spatial smoothing of the data, which usually increases the sensitivity of the ensuing analysis.

Control, levels and regional specificity

There is a fundamental difference between rejecting the null hypothesis of no activation at a particular peak and rejecting the null hypothesis over the entire volume analysed. As noted above, the former requires the strongest control over FWE and the latter the weakest. One way of thinking about this difference is to note that if activation is confirmed at a particular point in the brain then, implicitly, the hypothesis of activation somewhere is also confirmed (but the converse is not true). The distinction between weak and strong control, in the context of statistical parametric mapping, relates to the level at which the inference is made. The stronger the control, the more regional specificity it confers. For example, a peak-level inference is stronger than a cluster-level inference because the latter disallows inferences about any peak within the cluster. In other words, cluster level inferences maintain strong control at the cluster level but only weak control at the peak level. Similarly, set-level inferences are weaker than cluster-level inferences because they refer to the set of regional effects but not any individual peak or cluster in that set. Procedures with the weakest control have been referred to as 'omnibus' tests (e.g. Fox and Mintun, 1989) and frame the alternative hypothesis in terms of effects anywhere in the brain. These hypotheses are usually tested using the volume above some threshold (e.g. exceedence proportion tests, Friston *et al.*, 1991) or use all the SPM values (e.g. quadratic tests, Worsley *et al.*, 1995). A weaker control over FWE, or high-level inference, has less regional specificity but remains a valid way of establishing the significance of an activation profile. Intuitively, one might guess that the weaker procedures provide more powerful tests because there is a trade-off between sensitivity and regional specificity (see below).

Here we focus on the weaker hypotheses and consider peak-level and cluster-level inferences subordinate to set-level inferences. This allows us to ask: which is the most powerful approach for detecting brain activations? The remainder of this chapter is divided into two sections. The first section reprises the distributional approximations used to make statistical inferences about an SPM and frames the results to show that all levels of inference can be regarded as special cases of a single probability (namely, the probability of getting c, or more, clusters with k, or more, voxels above height u). The final section deals with the relative power of voxel-level, cluster-level, and set-level inferences and its dependency on signal characteristics, namely, the spatial extent of the underlying haemodynamics and the signal-to-noise ratio.

THEORY AND DISTRIBUTIONAL APPROXIMATIONS

In this section, we review the basic results from the random field theory that are used to provide a general expression for the probability of getting any excursion set defined by three quantities: a height-threshold; a spatial extent threshold; and a threshold on the number of clusters. We then show that peak-level, cluster-level and set-level inferences are all special cases of this general formulation and introduce its special cases.

A general formulation

We assume that a D-dimensional SPM conforms to a reasonable lattice representation of a statistical functional of volume $R(\theta)$ expressed in resolution elements, or resels. This volume is a function of the volume's size θ, for example radius. This measure is statistically flattened or normalized by the smoothness W. The smoothness is, as before, $W = |\Lambda|^{1/2D} = (4\ln 2)^{-\frac{1}{2}} FWHM$, where Λ is the covariance matrix of the first partial derivatives of the underlying component fields and $FWHM$ is the full width at half maximum. An excursion set is defined as the set of voxels that exceeds some threshold u. This excursion set comprises m clusters each with a volume of n voxels. At high thresholds, m approximates the number of maxima and has been shown to have a Poisson distribution (Adler and Hasofer, 1981, Theorem 6.9.3):

$$p(m = c) = \lambda(c, \psi_0) = \frac{1}{c!}\psi_0{}^c e^{-\psi_0} \quad \mathbf{19.1}$$

where ψ_0 is the expected number of maxima (i.e. clusters). This depends on the search volume and height threshold. The expected number of maxima (Hasofer, 1978) would generally be approximated with the expected Euler characteristic $\psi_0 = EC_D$, for the SPM in question (see Chapter 18, Appendix 6 and Figure 19.1). The volume of a cluster n is distributed according to (e.g. Friston *et al.*, 1994):

$$p(n \geq k) = \exp(-\beta k^{2/D})$$
$$\beta = \left(\frac{\Gamma(\frac{D}{2}+1)}{\eta}\right)^{2/D} \qquad \textbf{19.2}$$

where η is the expected volume of each cluster. This is simply the total number of voxels expected by chance divided by expected number of maxima (see Figure 19.1). This specific form for the distribution of cluster volume is for SPM{Z} and should be replaced with more accurate expressions for SPM{t} or SPM{F} (see Cao, 1999 and Chapter 18), but serves here to illustrate the general form of topological inference. With these results it is possible to construct an expression for the probability of getting c, or more, clusters of volume k, or more, above a threshold u (Friston *et al.*, 1996)

$$P(u,k,c,) = 1 - \sum_{i=0}^{c-1} \lambda(i, \psi_0 p(n \geq k)) \qquad \textbf{19.3}$$

Eqn. **19.3** can be interpreted in the following way: consider clusters as 'rare events' that occur in a volume according to the Poisson distribution with expectation ψ_0. The proportion of these rare events that meets the spatial extent criterion will be $p(n \geq k)$. These criterion events will themselves occur according to a Poisson distribution with expectation $\psi_0 p(n \geq k)$. The probability that the number of events will be c or more is simply one minus the probability that the number of events lies between 0 and c minus one (i.e. the sum in Eqn. **19.3**).

We now consider various ways in which $P(u,c,k)$ can be used to make inferences about brain activations. In brief, if the number of clusters $c = 1$, the probability reduces to that of getting one, or more, clusters with k, or more, voxels. This is the p-value for a single cluster of volume k. This corresponds to a cluster-level inference. Similarly if $c = 1$ and the number of suprathreshold voxels $k = 0$, the resulting cluster-level probability (i.e. the probability of getting one or more excursions of any volume above u) is the p-value of any peak of height u. In other words, cluster and peak-level inferences are special cases of set-level inferences.

Peak-level inferences

Consider the situation in which the threshold u is the height of a peak. The probability of this happening by chance is the probability of getting one or more clusters (i.e. $c = 1$) with non-zero volume (i.e. $k - 0$). The p-value is therefore:

$$P(u,0,1) = 1 - \exp(-\psi_0) \qquad \textbf{19.4}$$

This is simply the corrected probability based on the expected number of maxima or Euler characteristic.

Cluster-level inferences

Consider now the case in which we base our inference on spatial extent k, which is defined by specifying a height threshold u. The probability of getting more than one cluster of volume k or more is:

$$P(u,k,1) = 1 - \exp(-\psi_0 p(n \geq k) \qquad \textbf{19.5}$$

This is the corrected p-value based on spatial extent (Friston *et al.*, 1994) and has proved to be more powerful than peak-based inference when applied to high-resolution data (see below).

Set-level inferences

Now consider the instance where inference is based on cluster number c. In this case, both height u and extent k threshold need to be specified before the statistic c is defined. The corresponding probability is given by Eqn. **19.3** and is the corrected p-value for the set of activation foci surviving these joint criteria. There is a conceptual relationship between set-level inferences and non-localizing tests based on the exceedence proportion (i.e. the total number of voxels above a threshold u). Exceedence proportion tests (e.g. Friston *et al.*, 1991) and threshold-less quadratic tests (Worsley *et al.*, 1995) have been proposed to test for activation effects over the volume analysed in an omnibus sense. These tests have not been widely used because they have no localizing power and do not pertain to a set of well-defined activations. In this sense, these tests differ from set-level tests because the latter do refer to a well-defined set of activation foci. However, in the limiting case of a small spatial extent threshold k the set-level inference approaches an omnibus test:

$$P(u,0,c) = 1 - \sum_{i=0}^{c-1} \lambda(i, \psi_0) \qquad \textbf{19.6}$$

This test simply compares the expected and observed number of maxima in an SPM using the Poisson distribution under the null hypothesis. These set-level inferences are seldom employed in conventional analyses because they have no localizing power. However, they form a reference for the sensitivity of alternative (e.g. multivariate) analyses that focus on more distributed responses.

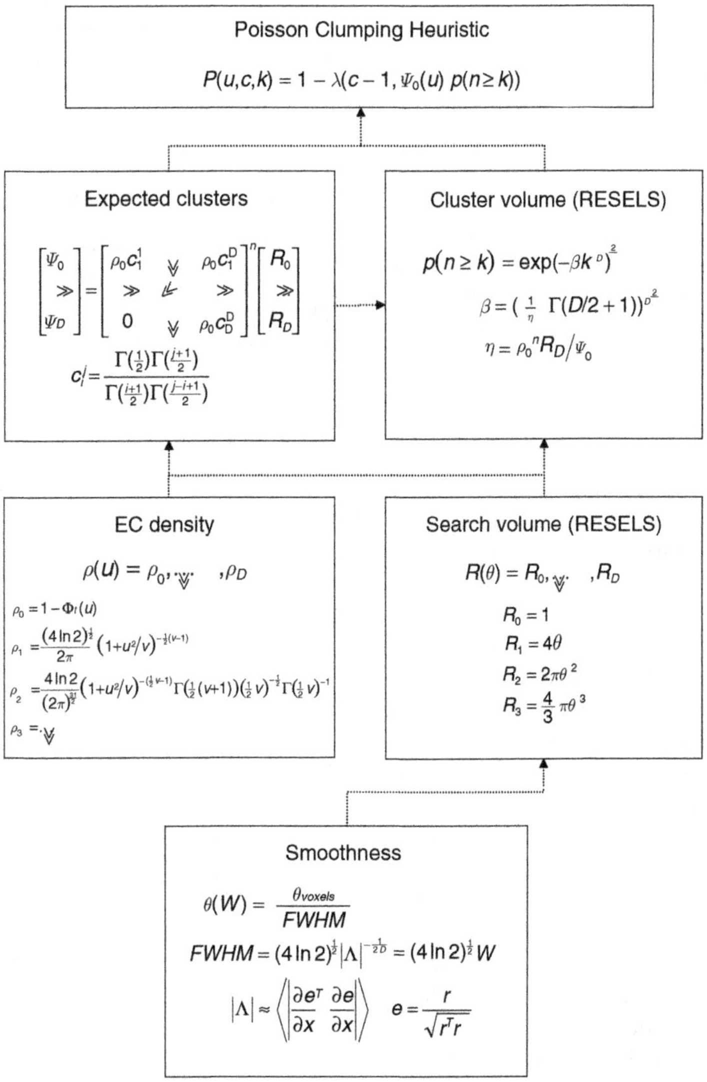

FIGURE 19.1 Schematic illustrating the use of random field theory in making inferences about SPMs. A fuller treatment is provided in Appendix 6. If one knew where to look exactly, then inference can be based on the value of the statistic at a specified location in the SPM, without correction. However, if one did not have an anatomical constraint *a priori*, then an adjustment for multiple dependent comparisons has to be made. These corrections are usually made using distributional approximations from Gaussian random field (GRF) theory. This schematic deals with a general case of n SPM$\{t\}$ whose voxels all survive a common threshold u (i.e. a conjunction of n component SPMs). The central probability, upon which all peak-, cluster- or set-level inferences are made, is the probability $P(u, c, k)$ of getting c or more clusters with k or more resels (resolution elements) above this threshold. By assuming that clusters behave like a multidimensional Poisson point process (i.e. the Poisson clumping heuristic) $P(u, c, k)$ is simply determined. The distribution of c is Poisson with an expectation that corresponds to the product of the expected number of clusters, of any size, and the probability that any cluster will be bigger than k resels. The latter probability is shown using a form for a single Z-variate field constrained by the expected number of resels per cluster η. The expected number of resels per cluster is simply the expected number of resels in total divided by the expected number of clusters. The expected number of clusters ψ_0 is estimated with the Euler characteristic (EC) (effectively the number of blobs minus the number of holes). This estimate is, in turn, a function of the EC density for the statistic in question (with degrees of freedom v) and the resel counts. The EC density is the expected EC per unit of D-dimensional volume of the SPM where the D-dimensional volume of the search space is given by the corresponding element in the vector of resel counts. Resel counts can be thought of as a volume metric that has been normalized by the smoothness of the SPM's component fields expressed in terms of the full width at half maximum (*FWHM*). This is estimated from the determinant of the variance-covariance matrix of the first spatial derivatives of e, the normalized residual fields r (from Plate 2, see colour plate section). In this example, equations for a sphere of radius θ are given. Φ denotes the cumulative density function for the statistic in question.

POWER ANALYSES

In this section, we describe a model which enables analytic power analyses and use it to compare various levels of inference and thresholds in the final section. The specificity of a test is the probability of correctly rejecting the null hypothesis. The sensitivity of a test is the probability of correctly accepting the alternative hypothesis. A plot of specificity against sensitivity is called a receiver operator characteristic (ROC) curve. Examples of these curves will be provided below. In order to determine the power of a test analytically, it is necessary to define the nature of the signal implied by the alternative hypothesis. In this chapter, we consider a simple model (Friston *et al.*, 1996) which assumes that the activations are spatially distributed with no predilection for a particular anatomical area. Although this model is used for mathematical convenience, it is not physiologically unreasonable and embodies an ignorance of where activations will be found. More specifically, it models activations that are distributed throughout the volume and the power analysis below applies to this, and only this, model. Different models (i.e. a single activation focus) would yield different results. Here we focus on a 'distributed' model, where we expect set-level inferences to be more sensitive. Suppose the signal comprises Gaussian foci, of random height, distributed continuously throughout the volume. The shape of the signal is characterized by the width w of these foci expressed as a proportion of W. This signal can be modelled by a continuous ensemble of kernels with randomly distributed heights or equivalently by convolving an uncorrelated Gaussian random process with a kernel of the same height. Let the signal (following convolution with the point spread function) have a standard deviation s, where s corresponds to the amplitude of the measured signal. Following Friston *et al.* (1996) the specificity and power at a given specificity are:

$$
\begin{aligned}
1-\alpha &= P(u, c, k)_W \\
\gamma &= P(\tilde{u}, c, k)_{\tilde{W}} \\
\tilde{u} &= \frac{u}{\sqrt{1+s^2}} \\
\tilde{W} &= W\sqrt{\frac{1+s^2}{1+s^2/(1+f^2)}}
\end{aligned} \tag{19.7}
$$

This simple approach assumes that the SPM under the alternate hypothesis is smoother and has larger amplitude than the SPM under the null hypothesis. The sensitivity is simply the probability of an outcome under the alternate hypothesis. Applying the threshold u to the alternate SPM is the same as applying the threshold $\tilde{u}$ to a null SPM of smoothness $\tilde{W}$ under the null hypothesis, which means $\gamma = P(\tilde{u}, c, k)_{\tilde{W}}$ can be taken as a measure of sensitivity.

Eqn. **19.7** allows us to compute the specificity and sensitivity as function of thresholds, amplitude or smoothness of the signal. In what follows we will compare the power of voxel-, cluster-, and set-level inferences for signals of different sorts to identify the most powerful sorts of inference.

Peak-level inferences

In this instance, the only parameter that can be varied is the threshold u. An example of an ROC curve for peak-level inferences is seen in Figure 19.2 (top). The influence of signal parameters on power is shown in the

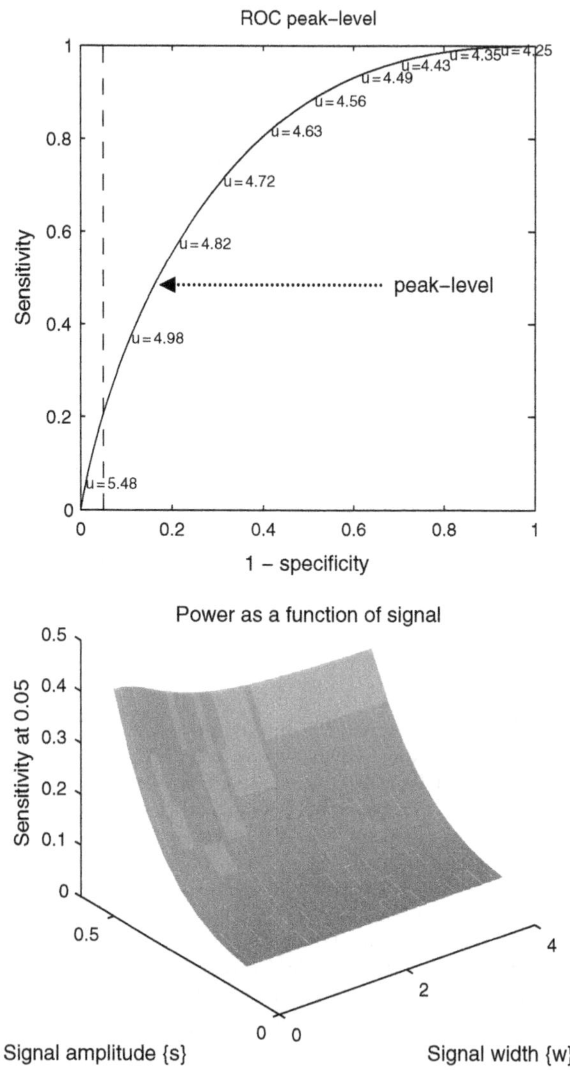

FIGURE 19.2 (Top) ROC curve for peak-level inference, where W corresponds to an FWHM of 3 voxels and a volume of 64 voxels cubed. Signal amplitude $s = 0.3$ and width $w = 2$. The dotted line corresponds to 95 per cent specificity. Three-dimensional plot of power (at 95 per cent specificity) as a function of signal amplitude s and width.

lower panel by plotting the sensitivity (at 95 per cent specificity) as a function of amplitude s and size w. It can be seen that power is a strong function of signal amplitude for all sizes of signal. It is also evident that higher thresholds are slightly more powerful, when the signals are smaller than the resolution ($w < 1$). High-resolution fMRI and optical imaging data suggest that haemodynamic changes are typically expressed on a spatial scale of 2–4 mm. This is around or below the resolution of PET (especially when the data are smoothed), however, it is greater than the resolution of fMRI data which, before any interpolation or smoothing, can be equated with voxel size (e.g. 3 mm). This suggests that peak-level tests might be more powerful for PET than for fMRI, in the absence of any smoothing. This dependency of power on signal smoothness is consistent with the matched filter theorem, which says that the best smoothing to apply to a signal matches the smoothness of the signal. In the present context, smoothing in accord with the matched filter theorem will increase the smoothness of noise (i.e. decrease w) without markedly suppressing signal. It should be noted that this link with the matched filter theorem is very heuristic.

Cluster-level inferences

In this context, we reiterate previous observations that cluster-level inferences are generally more powerful than peak-level inferences (although they have weaker control). Figure 19.3 (top) shows an ROC curve for cluster-level inferences at threshold $u = 2.8$ (solid line). This curve was calculated by varying the cluster threshold k (Eqn. **19.7**). The equivalent ROC curve from the previous analysis is also shown (broken line). The lower panel of Figure 19.3 demonstrates the effect of different signal sizes (for signal amplitude of 0.3). This represents a plot of sensitivity (at 95 per cent specificity) as a function of u and w. This shows a number of interesting features: it is immediately obvious that, for small signals (i.e. low resolution), the most powerful tests obtain when the height threshold is high and k is small (cf. peak-tests). Conversely, when the signal is smooth, relative to noise, the more powerful tests are associated with a low height threshold and a high extent threshold. In practical terms, this is consistent with the experience of people searching for smooth, non-focal effects in SPMs, such as in voxel-based-morphometry, in which cluster-level tests are often used.

Set-level inferences

Here we observe that set-level inferences are generally more powerful than cluster-level inferences and that this

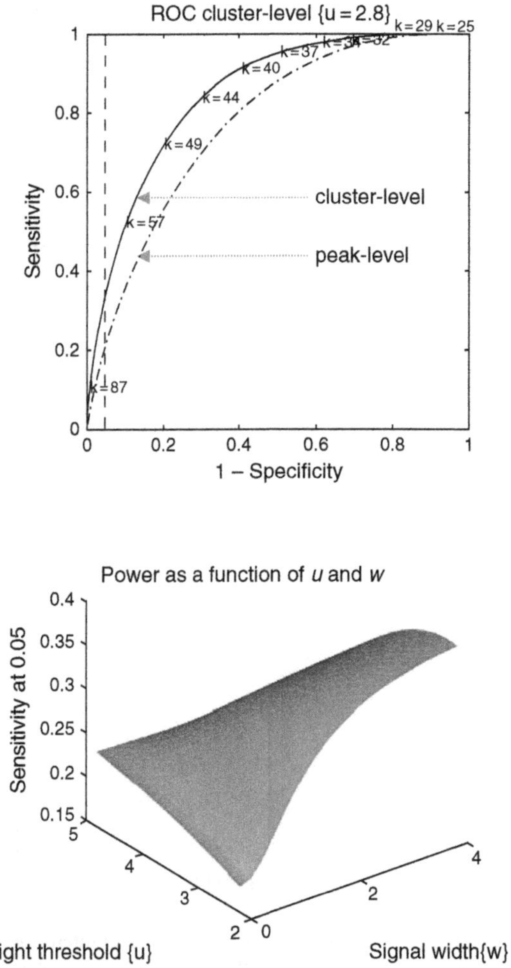

FIGURE 19.3 (Top) ROC curve for cluster-level inference, where W corresponds to an FWHM of 3 voxels and a volume of 64 voxels cubed. Signal amplitude $s = 0.3$ and width $w = 2$. Here $u = 2.8$. The broken line corresponds to the equivalent peak-level ROC curve of the previous figure. (Bottom) Three-dimensional plot of power (at 95 per cent specificity) as a function of signal width w and threshold u.

holds irrespective of signal characteristics. Figure 19.4 shows an ROC curve that obtains by varying the number of clusters c for a fixed height and extent threshold. It can be seen that the set-level inference (solid line) is much more powerful than either the cluster-level (dashed line) or peak-level (broken line) tests. To determine whether there are any special cases of the set-level test (i.e. cluster or voxel level) that are more powerful than the general case, we computed sensitivity (at 95 per cent specificity) by allowing k to vary for different values of u and c. The lower panels of Figure 19.4 show the results of this analysis and demonstrate that the most powerful tests result when $c > 1$ (i.e. set level). This is the case for both low- and high-resolution data (left and right lower panels, respectively).

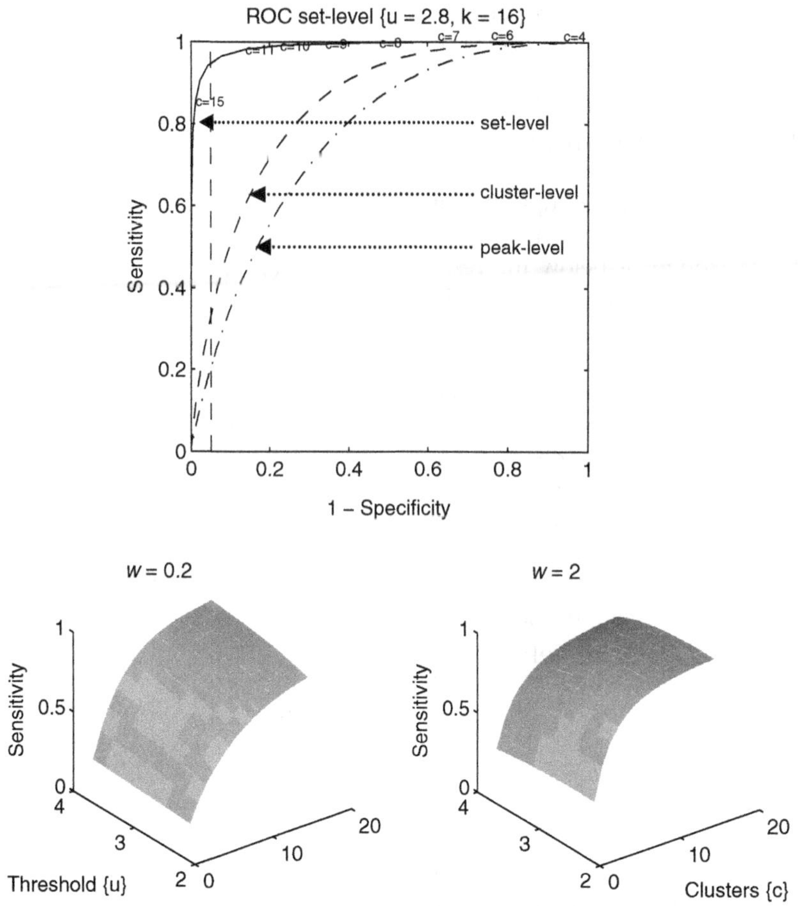

FIGURE 19.4 (Top) ROC curve for set-level inference, where W corresponds to an FWHM of 3 voxels and a volume of 64 voxels cubed. Signal amplitude $s = 0.3$ and width $w = 2$. Here $u = 2.8$ and $k = 16$. The dashed and broken lines correspond to the equivalent cluster- and peak-level ROC curves of the previous figures, respectively. (Bottom). Three-dimensional plot of sensitivity (at 95 per cent specificity) as a function of cluster number c and threshold u. Left $w = 0.2$ and right $w = 2$.

SUMMARY

We have addressed the sensitivity of various tests for activation foci in the brain considering levels of inference (voxel-level, cluster-level and set-level) in a topological context. All these tests inferences are based on a single probability of obtaining c, or more, clusters with k, or more, voxels above a threshold u. High levels have weaker regional specificity (cf. control over family-wise error). The weakest case of set-level inference is based on the total number of maxima above a height threshold and corresponds to omnibus tests. Cluster-level inferences are a special case of set-level inferences that obtain when the number of clusters is one. Similarly, peak-level inferences are special cases of cluster-level inferences that result when the cluster has an unspecified volume. On the basis of an analytical power analysis, we concluded that set-level inferences are generally more powerful than cluster-level inferences and cluster-level inferences are generally more powerful than peak-level inferences, for distributed signals.

Generally speaking, people use peak-level inferences because of their local specificity. This is appropriate for carefully designed experiments that elicit responses in one region (or a small number of foci), where set-level inferences may not be appropriate. However, set-level tests are sometimes useful in studies of cognitive function with many separable cognitive components that are instantiated in distributed neuronal systems. In this context, the set of activation foci that ensue are probably more comprehensive descriptors of evoked responses. An important point that can be made here is that set-level inferences do not preclude lower-level inferences. When confronted with the task of characterizing an unknown and probably distributed activation profile, set-level inferences should clearly be considered, provided the implicit loss of regional specificity is acceptable. However voxel-, cluster-, and set-level inferences can be made concurrently. For example, using thresholds of $u = 2.4$

and $k = 4$ allows for a set-level inference in terms of the clusters reaching criteria. At the same time, each cluster in that set has a corrected p-value based on its size and the cluster-level inference. Similarly, each peak in that cluster has a corrected p-value based on the voxel-level inference (i.e. its value). The nested taxonomy presented here allows for all levels to be reported, each providing protection for the lower level. As long as the levels are clearly specified, there is no reason why different levels cannot be employed, in a step-down fashion, in characterizing an SPM.

REFERENCES

Adler RJ, Hasofer AM (1981) *The geometry of random fields*. Wiley, New York

Cao J (1999) The size of the connected components of excursion sets of $\chi 2$, t and F fields. *Adv Appl Prob* **31**: 577–93

Forman SD, Cohen JD, Fitzgerald M *et al.* (1995) Improved assessment of significant activation in functional magnetic resonance imaging (fMRI): use of a cluster-size threshold. *Mag Res Med* **33**: 636–47

Fox PT, Mintun MA (1989) Non-invasive functional brain mapping by change distribution analysis of averaged PET images of H15O2 tissue activity. *J Nucl Med* **30**: 141–49

Friston KJ, Frith CD, Liddle PF *et al.* (1991) Comparing functional (PET) images: the assessment of significant change. *J Cereb Blood Flow Metab* **11**: 690–99

Friston KJ, Worsley KJ, Frackowiak RSJ *et al.* (1994) Assessing the significance of focal activations using their spatial extent. *Hum Brain Mapp* **1**: 214–20

Friston KJ, Holmes A, Poline JB *et al.* (1996) Detecting activations in PET and fMRI: levels of inference and power. *NeuroImage* **4**: 223–35

Hasofer AM (1978) Upcrossings of random fields. *Suppl Adv Appl Prob* **10**: 14–21

Kiebel SJ, Poline JB, Friston KJ *et al.* (1999) Robust smoothness estimation in statistical parametric maps using standardized residuals from the general linear model. *NeuroImage* **10**: 756–66

Poline J-B, Mazoyer BM (1993) Analysis of individual positron emission tomography activation maps by detection of high signal-to-noise-ratio pixel clusters. *J Cereb Blood Flow Metab* **13**: 425–37

Roland PE, Levin B, Kawashima R *et al.* (1993) Three dimensional analysis of clustered voxels in 15O-Butanol brain activation images. *Hum Brain Mapp* **1**: 3–19

Siegmund DO, Worsley KJ (1994) Testing for a signal with unknown location and scale in a stationary Gaussian random field. *Ann Stat* **23**: 608–39

Worsley KJ (1994) Local maxima and the expected Euler characteristic of excursion sets of x2, F and t fields. *Adv Appl Prob* **26**: 13–42

Worsley KJ, Evans AC, Marrett S *et al.* (1992) A three-dimensional statistical analysis for rCBF activation studies in human brain. *J Cereb Blood Flow Metab* **12**: 900–18

Worsley KJ, Poline J-B, Vandal AC *et al.* (1995) Tests for distributed, nonfocal brain activations. *NeuroImage* **2**: 183–94

CHAPTER

20

False Discovery Rate procedures

T. Nichols

INTRODUCTION

In the previous three chapters, we have seen how inferences can be made to control false positives while searching the brain for activations. While those chapters consider inferences on different types of features of a statistic image (e.g. cluster-level versus peak-level), they all focus on a single measure of false positives, the family-wise error rate (FWE). Methods that control FWE are very specific: if one were to use a level 0.05 FWE threshold throughout one's career, one is guaranteed that, on average, no more than 1 out of 20 of the examined statistical parametric maps (SPMs) will have any false positives. In statistical terms, we say that a 0.05 FWE method has 95 per cent confidence of producing results totally free of type I errors. This remarkable control of false positives, however, comes with relatively poor sensitivity, or risk of false negatives.

A new perspective on the multiple testing problem is the consideration of a different, more lenient measure of false positives. Instead of controlling the chance of one or more false positives, one could instead control the fraction of false positives present. More precisely, the false discovery rate (FDR) is the expected proportion of false positives among all detected voxels. A level 0.05 FDR procedure allows *some* false positives but, on average, the false positives are controlled to be no more than 5 per cent of the number of voxels above the threshold used.

This chapter introduces FDR and related false positive measures and describes methods that control FDR. We illustrate FDR control with two synthetic examples and one real dataset. Throughout, we assume that peak or voxel-level inferences are of interest, though cluster-level inferences are briefly mentioned in the first illustration.

MULTIPLE TESTING DEFINITIONS

Our starting point is a completed SPM analysis, with a statistic image that assesses evidence of an experimental or group effect. Let a statistic image comprised of v voxels be denoted $\{T_i\} = \{T_1, T_2, \ldots, T_v\}$, and their corresponding p-values be $\{P_i\} = \{P_1, P_2, \ldots, P_v\}$. Consider applying a threshold u to the image, classifying V_P suprathreshold voxels as 'positives' and $V_N = v - V_P$ subthreshold voxels as 'negatives'. As shown in Table 20-1, each voxel can be cross-classified according to whether or not there is truly any signal, and whether or not the voxel is classified as signal (possibly incorrectly). When there is truly no signal at a voxel, we say that the voxel's null hypothesis is true, and here we will refer to such voxels as 'null voxels'. When there is a signal present the null is false, and we call these 'non-null voxels'. Based on threshold u, we either reject (for $T_i \geq u$) or fail to reject the null hypothesis (for $T_i < u$). We say the 'complete null' is true when every voxel's null hypothesis is true ($v_0 = v$).

Table 20-1 defines all of the quantities needed to define a range of false positive measures. Note that these measures are not generally observable. For example,

TABLE 20-1 Cross-classification of all v voxels in an image. For some threshold applied, V_P positives are found, while v_1 true signal voxels actually exist in the image. Among the V_P detections, V_{0P} are false positives, while V_{1N} false negatives exist

	Negatives Do not reject H_0 'Subthreshold'	Positives Reject H_0 'Suprathreshold'	
H_0 True, 'True noise'	V_{0N}	V_{0P}	v_0
H_0 False, 'True signal'	V_{1N}	V_{1P}	v_1
	V_N	V_P	v

Statistical Parametric Mapping, by Karl Friston et al.
ISBN–13: 978-0-12-372560-8 ISBN–10: 0-12-372560-7

while we can count the number of detected positives V_P, we can never know the number of true positive voxels v_1. (Following the statistical convention, the quantities in lower case are fixed, while upper case variables are random and will vary from realization to realization.) Among the V_P detected positives, V_{0P} are false positives.

Family-wise error rate

The presence of any false positives ($V_{0P} \geq 1$) is referred to as a family-wise (type I) error, and the family-wise error rate (FWE) is defined as:

$$\text{FWE} = P\{V_{0P} \geq 1\} \quad \textbf{20.1}$$

There are two types of FWE control, strong and weak. A procedure that controls the chance of a family-wise error when there is no signal ($v_0 = v$) has weak FWE control. When there is some signal present ($v_1 > 0$), a procedure that controls the chance of a family-wise error over any collection of null voxels has strong FWE control. A test with strong FWE can localize an activation, asserting that any group of voxels are falsely detected with probability at most α (Holmes, 1996). A test with weak FWE is an omnibus test, and can only assert that there is some signal somewhere. Hence, for imaging, strong FWE methods are generally sought, and all the methods described in Chapters 17 through 19 (save set-level inference) control FWE strongly.

False discovery proportion and false discovery rate

By measuring false positives as a fraction, we now define the false discovery proportion (FDP) as:

$$\text{FDP} = \frac{V_{0P}}{V_P} \quad \textbf{20.2}$$

Throughout this chapter, we use the convention that zero divided by zero is defined to be zero ($0/0 \triangleq 0$). While for any given model of a statistic image FWE is a fixed number, FDP is, in contrast, a random quantity; with each new realized statistic image, V_P and V_{0P} will vary, even for well-behaved data. One way of summarizing FDP is through its expectation, which defines false discovery rate (FDR) as:

$$\text{FDR} = E\{\text{FDP}\} \quad \textbf{20.3}$$

Per-comparison error rate

Finally, for completeness, we define the per-comparison error rate (PCE), the nominal false positive rate for any voxel:

$$\text{PCE} = P\{T_i > u | H_o\} \quad \textbf{20.4}$$

For a single test and a given threshold u, PCE is the α-level of the test.

Note that we use α generically, i.e. as the tolerable limit of some measure of false positives. We can thus discuss a level α PCE, a level α FDR and a level α FWE threshold, often referring to PCE procedures as 'uncorrected' and FDR and FWE methods as 'corrected', as they account for the multiplicity of v tests.

False discovery rate versus family-wise error rate

An important connection between FDR and FWE is in the case of a totally null image ($v_0 = v$). In this setting, there can be no true positives ($V_{1P} = 0$), and so the FDP takes on only two values, 0 when there are no detections ($V_P = 0$) or 1 when there are one or more (all false) detections. For this particular case, FDR is equivalent to FWE:

$$\text{FDR} = E(\text{FDP}) = P(V_{0P} > 0) = \text{FWE} \quad \textbf{20.5}$$

showing that FDR has weak control of FWE. This is an important feature of FDR and contributes to the interpretability of FDR results: when there is *no* signal in a statistic image, FDR and FWE methods control false positives in exactly the same manner.

False discovery exceedance and the control of the number of false positives

A common misconception about FDR is that it controls the number of false positives present (V_{0P}). Consider a dataset where an $\alpha = 0.05$ FDR procedure is used and yields $V_P = 260$ detections. One is tempted to conclude that there are $V_{0P} = \alpha \times V_P = 0.05 \times 260 = 13$ false positives present. However, this is incorrect, as shown by:

$$E(V_{0P}/V_P) \leq \alpha \nRightarrow E(V_{0P}) \leq \alpha \times V_P \quad \textbf{20.6}$$

In words, FDR controls the random fraction FDP in expectation, which does not imply that the expected false positive count V_{0P} is controlled by a fraction of the random V_P.

A recent direction in FDR methodology is to control the chance that FDP exceeds some quantity, or to control the false discovery exceedance (FDX). While this is a more stringent measure of false positives, it allows control on V_{0P}. A level α, $(1-\gamma)$ confidence FDX procedure guarantees that the FDP is controlled at α with probability $(1-\gamma)$:

$$P(\text{FDP} \leq \alpha) \geq 1-\gamma \qquad \textbf{20.7}$$

Inference on V_{0P} follows:

$$P(\text{FDP} \leq \alpha) \geq 1-\gamma \Leftrightarrow P(V_{0P} \leq \alpha \times V_P) \geq 1-\gamma \qquad \textbf{20.8}$$

In words, control of the random FDP implies the false positive count is controlled as a fraction of the number of detections V_P.

Now consider a dataset where a level 0.05, 90 per cent confidence FDX procedure is used that detects 200 voxels; we could conclude with 90 per cent confidence that no more than $200 \times 0.05 = 10$ voxels were false detections. While a few authors have proposed FDX methods (e.g. Pacifico *et al.*, 2004; Farcomeni *et al.*, 2005), they are not widely used in neuroimaging, though they are mentioned to highlight a limitation of FDR.

FDR METHODS

Like FWE, FDR is simply a measure of false positives and, as with FWE, there are many different proposed methods which produce thresholds that control FDR. The first authors who defined FDR, however, proposed a method which is straightforward and consequently is the most widely used FDR method.

Benjamini and Hochberg (BH) FDR method

Benjamini and Hochberg (1995) introduced the FDR metric and proved that the Simes method, an existing weak FWE procedure, controlled their newly defined measure for independent tests. The level α FDR threshold method starts by finding the p-values $\{P_i\}$ for each of the test statistics $\{T_i\}$ and by ranking the p-values from smallest to largest $\{P_{(i)}\} = \{P_{(i)} : P_{(1)} \leq P_{(2)} \leq \cdots \leq P_{(v)}\}$. Next, the following expression is considered for different i:

$$P_{(i)} \leq \frac{i}{v}\alpha \qquad \textbf{20.9}$$

and largest index i' is found such that the inequality holds. The value $P_{(i')}$ can then used as a statistic threshold, and all p-values less than or equal to it can have their null hypotheses rejected. Benjamini and Hochberg show this procedure to control FDR conservatively:

$$FDR \leq \frac{v_0}{v}\alpha \leq \alpha \qquad \textbf{20.10}$$

with the inequality becoming equality for continuous $\{T_i\}$.

P-P plot interpretation

The defining inequality (Eqn. **20.9**) can be plotted versus i/v, giving insight to how the FDR threshold is found (see plots in Figure 20.2). When the left side is plotted it produces a 'P-P plot', ordered p-values $\{P_{(i)}\}$ plotted versus i/v, a scale of their index. Plotting the right-hand-side produces a line of slope α. The FDR threshold $P_{(i)}$ is found as the largest p-value below the line.

The value of this plot stems from two observations. First is the fundamental property of p-values: under the null hypothesis, p-values follow a uniform distribution, $P_i \sim \text{uniform}(0, 1)$. Thus when $v_0 = v$ the p-values should be uniformly spread between 0 and 1, each with $E(P_{(i)}) = i/(v+1) \approx i/v$. This shows the P-P plot to be the ordered p-values plotted against their null-hypothesis expected value (almost), and under the complete null the p-values should roughly follow the identity.

The second observation is simply that, when there is signal present, we expect an excess of small p-values, causing the P-P plot to bend down at the left and below the slope-α line. This also shows that the exact threshold found will depend on the distribution of p-values observed, something considered in the second illustration in the next section.

Conservativeness of the BH FDR method

It may seem that the conservativeness of the method, by a factor of v_0/v, would reduce its utility. However, in the context of brain imaging this is not a problem.

An aspect of most functional imaging experiments is that the number of tests considered (v) is quite large and the fraction of active voxels is quite small ($v_1 \ll v$). Consequently, the fraction of null voxels is large ($v_0/v \approx 1$) and as a result the conservativeness implicit in the BH FDR method is not severe. Some authors (e.g. Benjamini *et al.*, 2003) propose to estimate v_0 with $\hat{v}_0$ and replace α with $\alpha \times (v/\hat{v}_0)$ in the BH FDR method (Eqn. **20.9**), though that same work suggests these methods do not work well under dependence and when v_0/v is close to 1. Hence, estimation of v_0 probably will not aid most neuroimaging applications.

BH FDR method under-dependence

An assumption of independence between voxels is untenable in brain imaging, and so it would seem that the BH

FDR method would not be applicable. Fortunately, this assumption was relaxed in a subsequent paper, in which Benjamini and Yekutieli (2001) proved that the BH FDR procedure is valid under positive regression dependency on subsets (PRDS). PRDS is a very technical, generalized notion of dependence, but can be concisely stated for Gaussian data:

$$Corr(T_i, T_j) \geq 0, \text{ for } i = 1, \ldots, v, \; j \in \mathcal{H}_0 \qquad \textbf{20.11}$$

where $\mathcal{H}_0$ is the set of indices for all null tests. This condition requires that there must be zero or positive correlation between all pairs of null voxels, and between all pairs of null and signal voxels. No constraint is made on the correlation between pairs of signal voxels. For smooth imaging data, this assumption seems reasonable, though strong physiological artefacts could induce structured noise with negative correlations.

If the PRDS assumption is regarded as untenable, Benjamini and Yekutieli (2001) provide a version of the BH FDR method that is valid for *any* correlation structure. In Eqn. **20.9** replace α with $\alpha/c(v)$, where $c(v) = \sum_{i=1,\ldots,v} 1/i \approx \log(v) + 0.5772$. This modified method is, however, much more stringent and is much less sensitive than the original BH FDR method.

EXAMPLES AND DEMONSTRATIONS

In the following section, we demonstrate a number of important features of FDR with simulated examples and data.

Comparison of PCE, FWE and FDR

Figure 20.1 demonstrates the application of three different measures of false positive control. The top row shows ten statistic images, each of which is a realization of smooth background noise with a deterministic signal in the centre. In the bottom two rows, signal voxels are indicated by a grey line. Think of the ten images as the results of one's next ten experiments.

Per-comparison error rate

The second row shows the use of an uncorrected $\alpha = 0.1$ threshold, i.e., control of the per comparison error rate at 10 per cent. There are many false positives: 11.3 per cent of the first image's null voxels are marked as significant, 12.5 per cent of the second image, etc. Theory dictates 10 per cent of the null voxels are falsely detected *on average*, but for any particular experiment it may be either higher or lower. While the magnitude of false positives

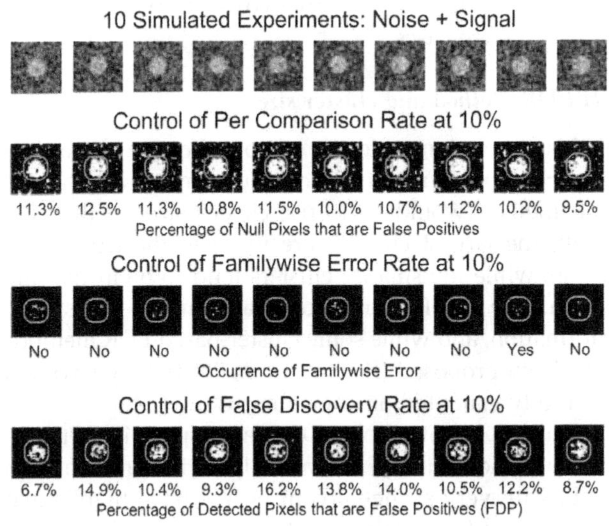

FIGURE 20.1 Monte Carlo demonstration of the control of different false positive metrics.

is unacceptable, observe that such an approach is very sensitive, and most of the non-null voxels are correctly identified as signal.

Family-wise error rate

The third row shows the use of a 0.1-level FWE threshold. In nine out of ten images no false positives occur; in one image a family-wise error is visible due to the two false positive voxels. This illustrates how a valid FWE procedure controls the *long-run* chance of any false positives: over many, many experiments, no more than 1 out of 10 will have *any* false positives. That is, there is 90 per cent confidence of the FWE thresholded image being false-positive-free. This precise specificity sacrifices sensitivity: only a small fraction of signal voxels have been detected in each case.

False discovery rate

The fourth row depicts the use of a 0.1-level FDR threshold. There are fewer false positives than with PCE; measured as the false discovery proportion (FDP), false positives vary from image to image: 6.7 per cent of the first image's significant voxels are false positives; 14.9 per cent of the second image's detections are false positives. Note that the FDP can be large at times, here as much as 16.2 per cent over these ten realizations and could be even larger in other realizations. As an extreme, recall that for complete null data FDP can only be either 0 or 100 per cent. A valid FDR procedure merely guarantees that *on average* FDP will not exceed the nominal 10 per cent. The use of this more liberal false positive measure, though, results in more true positives than with FWE, with most of the true positive voxels being detected in each of the ten cases.

BH FDR method and cluster size

In this simulation, where we know the true signal is a large contiguous region, an obvious way to improve the FDR method would be to use cluster size. In the FDR result, the largest clusters are those in the true signal region, while the smaller clusters tend to be in the null region. The BH FDR method makes no use of spatial information, and while some cluster-based FDR methods have been proposed (Pacifico *et al.*, 2004), they have not been fully validated or implemented in SPM.

Figure 20.1 shows how FDR is a compromise between PCE, with no control of the multiple testing problem, and FWE, with very stringent control. We have highlighted the variation from realization to realization to emphasize that, for any one dataset, one cannot know whether a family-wise error has occurred or if the FDP is below α. Rather, valid FWE and FDR procedures guarantee the long-run behaviour of the method.

Adaptiveness of FDR

An important feature of FDR-controlling procedures is how they adapt to the signal in the data. One can build intuition with the following artificial case: consider a situation where all v_1 non-null voxels have extremely strong signal, to such an extent that the p-values are essentially zero, and the remaining null $v_0 = v - v_1$ voxels have p-values that follow a uniform distribution. Specifically, let us assume that $P_{(i)} \approx 0$ for $i = 1, \ldots, v_1$, and $P_{(i)} \sim$ Uniform (0,1) for $i = v_1 + 1, \ldots, v$, with $E(P_{(i)}) = (i - v_1 + 1)/(v_0)$.

Figure 20.2 shows three cases: the complete null, an intermediate case where half of the nulls are true, and an extreme case where exactly one test is null and all others have signal. In each case, we have plotted the assumed p-values (0 or expected value) as open circles, a fine dotted line showing expected p-values under the complete null, a solid line of slope $\alpha = 0.1$, and a dot on the solid line showing the likely p-value threshold that will be obtained.

In the first case, there is no signal at all ($v_1 = 0$) and the p-values fall nearly on the identity. Since there are no unusually small p-values, we could guess that the p-value threshold will come from $i' = 1$ in Eqn. **20.9**

$$P_{(1)} \leq \frac{1}{v}\alpha \qquad \mathbf{20.12}$$

A p-value threshold of α/v is of course the threshold that the Bonferroni method specifies.

Now consider the second case, where half of the tests have strong signal. Here the smallest $v_1 = v/2$ p-values are zero and the remaining are uniformly spread between zero and one. For the smallest non-null p-value, Eqn. **20.9** implies a p-value threshold of:

$$P_{(v_1+1)} \leq \frac{v_1+1}{v}\alpha \approx \frac{1}{2}\alpha \qquad \mathbf{20.13}$$

a much less stringent threshold than with Bonferroni.

The third case shows a most extreme situation, in which all but one of the tests have strong signal ($v_1 = v - 1$). All but one of the p-values are zero, so the sole non-null test has a p-value threshold of:

$$P_{(v)} \leq \frac{v}{v}\alpha = \alpha \qquad \mathbf{20.14}$$

In other words, here the FDR procedure will use an uncorrected α-level threshold. This result may seem startling, as we have to search v tests for activation, but actually it is quite sensible. In this setting, with exactly one null test, there is only one opportunity for a false positive to occur, namely when the test corresponding to $P_{(v)}$ is falsely detected. Hence, in this extreme case, the multiple comparisons problem has vanished and the ordinary α threshold is appropriate.

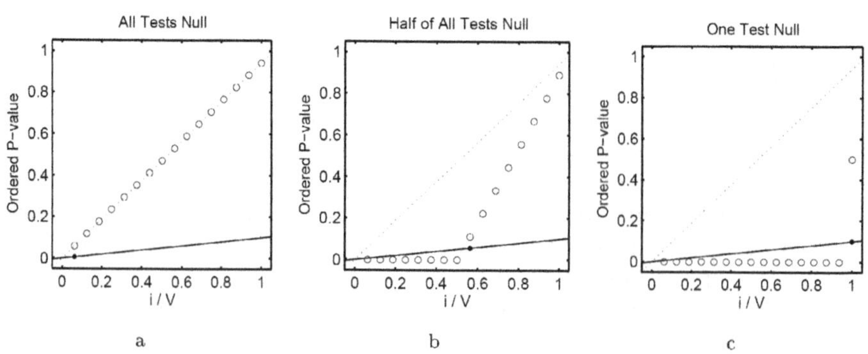

FIGURE 20.2 Demonstration of adaptiveness of the Benjamini and Hochberg FDR procedure.

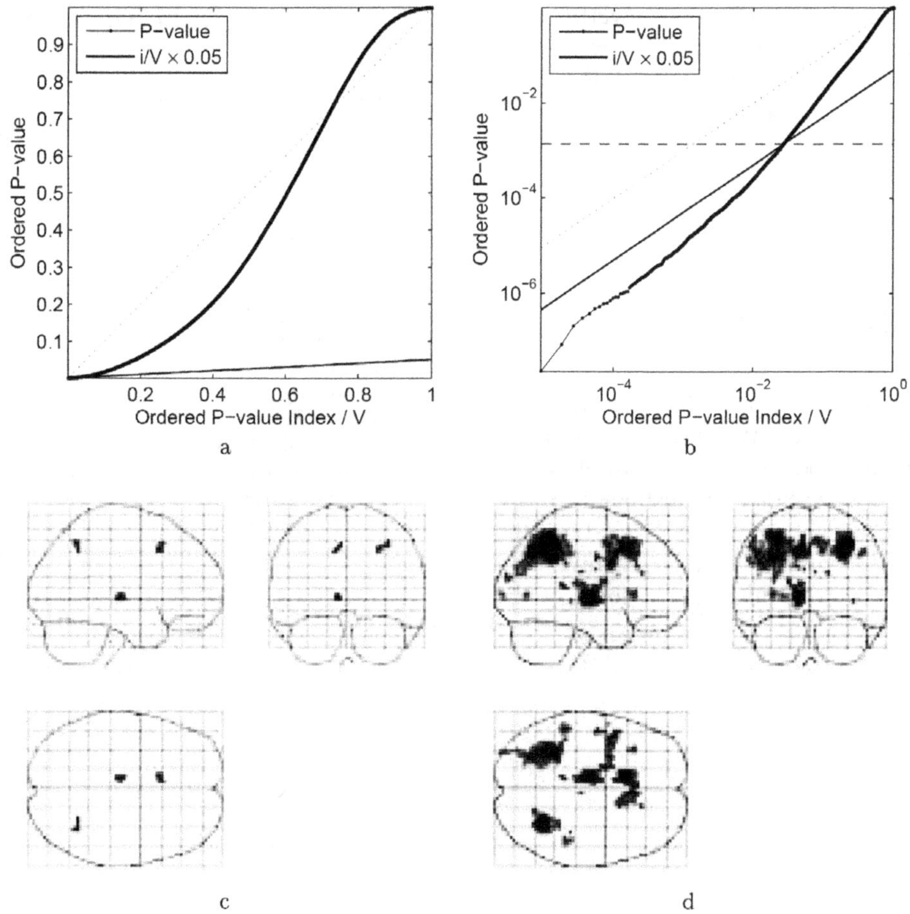

FIGURE 20.3 Application of the FDR method to real data, with comparison to a FWE result.

This illustration demonstrates the adaptiveness of the BH FDR procedure. As the number of signal voxels (v_1) increases, the number of opportunities for a false positive to occur (v_0) shrinks, and hence we have a less severe multiple testing problem. At the same time, as the number of detections V_P increases, a greater absolute number of false positives can be tolerated. Both of these factors contribute to the BH FDR method finding a threshold that is somewhere between Bonferroni and uncorrected, depending on the relative abundance of signal voxels.

Application to real data

Figure 20.3 illustrates the application of the FDR method to a real data set. Marshuetz *et al.* (2000) studied order-effects in working memory using functional magnetic resonance imaging (fMRI). They analysed 12 subjects using a random-effects analysis (see Chapter 12 on random effects, and Marshuetz *et al.* (2000) for complete details on the study). The contrast considered compares item recognition with a control condition.

On the top left of Figure 20.3 is the P-P plot and on the top right plot is the data rootogram[1] showing both the null-hypothesis distribution and the empirical FDR rate for every possible threshold.

The bottom right of the figure shows the results from using a permutation method (see Chapter 21) to find a level 0.05 FWE threshold (permutation can be more sensitive than RFT thresholds, and hence is used here). The bottom right depicts the results with a level-0.05 FDR threshold. The FDR threshold 3.83 detects 3073 voxels, while the FWE threshold of 7.67 detects 58 voxels. The FDR finds the same three regions as the FWE result (left anterior cingulate, thalamus, right parietal), as well as several others (right anterior cingulate, left pre-motor, and left parietal).

[1] A rootogram is similar to a histogram, except that the bars plot the square root counts of observations that fall in each bin. This type of plot allows better visualization of the low-count tails of a distribution.

While the FDR result is clearly more powerful, it is important to remember the limitations of the result. For the FWE result, we have 95 per cent confidence that there are no false positives in the image, while in the FDR result we expect (on average) 5 per cent of the detected voxels to be false.

CONCLUSION

This chapter has described the false discovery rate and related false positive metrics. We have reviewed the Benjamini-Hochberg method implemented in SPM and used demonstrations and real data to illustrate the properties of the method. We have emphasized how the method is adaptive and generally more sensitive than FWE methods, but also have highlighted some of the limitations of FDR.

REFERENCES

Benjamini Y, Hochberg Y (1995) Controlling the false discovery rate: a practical and powerful approach to multiple testing. *J Roy Stat Soc, Series B, Methodological* **57**: 289–300

Benjamini Y, Krieger AM, Yekutieli D (2003) Adaptive linear step-up procedures that control the false discovery rate. Research paper 01–03, Department of Statistics and OR, Tel Aviv University, Tel Aviv, Israel

Benjamini Y, Yekutieli D (2001) The control of the false discovery rate in multiple testing under dependency. *Ann Stat* **29**: 1165–88

Farcomeni A, Lasinio GJ, Alisi C, *et al.* (2005) A new multiple testing procedure with applications to quantitative biology and wavelet thresholding. In *The Proceedings of the 24th LASR*, Leeds Annual Statistical Research Workshop, Leeds University, pp 123–26

Holmes AP, Blair RC, Watson JD *et al.* (1996) Nonparametric analysis of statistic images from functional mapping experiments. *J Cereb Blood Flow Metab* **16**: 7–22

Marshuetz C, Smith EE, Jonides J, *et al.* (2000) Order information in working memory: fMRI evidence for parietal and prefrontal mechanisms. *J Cogn Neurosci* **12**/S2: 130–44

Pacifico P, Genovese CR, Verdinelli I, *et al.* (2004) False discovery rates for random fields. *J Am Stat Assoc* **99**: 1002–14

CHAPTER

21

Non-parametric procedures

T. Nichols and A. Holmes

INTRODUCTION

The statistical analyses of functional mapping experiments usually proceed at the voxel level, involving the formation and assessment of a *statistic image*: at each voxel, a statistic indicating evidence of the experimental effect of interest, at that voxel, is computed, giving an image of statistics, a *statistic image* or *statistical parametric map* (SPM). In the absence of *a priori* anatomical hypotheses, the entire statistic image must be assessed for significant experimental effects, using a method that accounts for the inherent multiplicity incurred by testing all voxels simultaneously.

Traditionally, this has been accomplished in a classical *parametric* statistical framework. In the methods discussed in Chapters 8 and 9 of this book, the data are assumed to be normally distributed, with a mean parameterized by a general linear model. This flexible framework encompasses t-tests, F-tests, paired t-tests, analysis of variance (ANOVA), correlation, linear regression, multiple regression, and analysis of covariance (ANCOVA), among others. The estimated parameters of this model are contrasted to produce a test statistic at each voxel, which has a Student's t-distribution under the null hypothesis. The resulting t-statistic image is then assessed for statistical significance, using distributional results for continuous random fields to identify voxels or regions where there is significant evidence against the null hypothesis (Friston *et al.*, 1994, 1996; Worsley *et al.*, 1995; Worsley, 1995; Poline *et al.*, 1997).

Holmes *et al.* (1996) introduced a non-parametric alternative based on permutation test theory. This method is conceptually simple, relies only on minimal assumptions, deals with the multiple testing issue, and can be applied when the assumptions of a parametric approach are untenable. Furthermore, in some circumstances, the permutation method outperforms parametric approaches. Arndt *et al.*, (1996), working independently, also discussed the advantages of similar approaches. Subsequently, Grabrowski *et al.* (1996) demonstrated empirically the potential power of the approach in comparison with other methods. Halber *et al.* (1997), discussed further by Holmes *et al.* (1998), also favour the permutation approach. Methods to improve the computational efficiency of the method have been proposed by Heckel *et al.* (1998) and Belmonte and Yurgelun-Todd (2001).

Nichols and Holmes (2001) review the non-parametric theory and demonstrate how multisubject functional magnetic resonance images (fMRI) can be analysed. One use of permutations is to evaluate methods with more assumptions. Nichols and Hayasaka (2003; 2004) compare voxels-wise parametric to non-parametric performance on several multisubject datsets, as well as cluster-wise performance under stationarity and non-stationarity.

Applications of permutation testing methods to *single* subject fMRI require models of the temporal autocorrelation in the fMRI time series. Bullmore *et al.* (1996) have developed permutation based procedures for periodic fMRI activation designs using a simple autoregression moving average (ARMA) model for temporal autocorrelations, though they eschew the problem of multiple testing; later they generalized their method to multiple subjects (Brammer *et al.*, 1997). In later work, Bullmore and colleagues used a wavelet transformation to account for more general forms of fMRI correlation (Bullmore *et al.*, 2001; Fadili and Bullmore, 2001), though Friman and Westin (2005) have criticized this approach. Locascio *et al.* (1997) describe an application to fMRI combining the general linear model (Friston *et al.*, 1995), ARMA modelling (Bullmore *et al.*, 1996), and a multiple testing permutation procedure (Holmes *et al.*, 1996). In a randomized experiment, Raz *et al.* (2003) proposed permuting the experimental labels instead of the data. Bullmore *et al.* (1999) apply non-parametric methods to compare groups of structural MR images.

Statistical Parametric Mapping, by Karl Friston et al.
ISBN–13: 978-0-12-372560-8 ISBN–10: 0-12-372560-7

The aim of this chapter is to present the theory of multiple testing using non-parametric permutation for independent data (e.g. positron emission tomography (PET) or intersubject fMRI), including detailed examples. While the traditional approach to multiple testing controls the family-wise error rate, the chance of any false positives, another perspective has been introduced recently, the false discovery rate. Chapter 20 covers this new false positive metric, though we note that a permutation approach to FDR has been proposed (Yekutieli and Benjamini, 1999) and evaluated (Logan and Rowe, 2004).

We begin with an introduction to non-parametric permutation testing, reviewing experimental design and hypothesis testing issues, and illustrating the theory by considering testing a functional neuroimaging dataset at a single voxel. The problem of searching the brain volume for significant activations is then considered, and the extension of the permutation methods to the *multiple testing problem* of simultaneously testing at all voxels is described. With appropriate methodology in place, we conclude with three annotated examples illustrating the approach. Software implementing the approach, called statistical non-parametric mapping, is available as an extension of the MATLAB based SPM package.

PERMUTATION TESTS

Permutation tests are one type of non-parametric test. They were proposed in the early twentieth century, but have only recently become popular with the availability of inexpensive, powerful computers to perform the computations involved.

The essential concept of a permutation test is relatively intuitive. Consider a simple single-subject PET activation experiment, where a subject is scanned repeatedly under 'rest' and 'activation' conditions. Considering the data at a particular voxel, if there is really no difference between the two conditions, then we would be fairly surprised if most of the 'activation' observations were larger than the 'rest' observations, and would be inclined to conclude that there was evidence of some activation at that voxel. Permutation tests simply provide a formal mechanism for quantifying this 'surprise' in terms of probability, thereby leading to significance tests and p-values.

If there is no experimental effect, then the labelling of observations by the corresponding experimental condition is arbitrary, since the same data would have arisen whatever the condition. These *labels* can be any relevant attribute: condition 'tags', such as 'rest' or 'active'; a covariate, such as task difficulty or response time; or a label, indicating group membership. Given the null hypothesis that the labellings are arbitrary, the significance of a statistic expressing the experimental effect can then be assessed by comparison with the distribution of values obtained when the labels are permuted.

The justification for exchanging the labels comes from either weak distributional assumptions, or by appeal to the randomization scheme used in designing the experiment. Tests justified by the initial randomization of conditions to experimental units (e.g. subjects or scans), are sometimes referred to as *randomization tests*, or *re-randomization tests*. Whatever the theoretical justification, the mechanics of the tests are the same. Many authors refer to both generically as permutation tests, a policy we shall adopt unless a distinction is necessary.

In this section, we describe the theoretical underpinning for randomization and permutation tests. Beginning with simple univariate tests at a single voxel, we first present randomization tests, describing the key concepts at length, before turning to permutation tests. These two approaches lead to exactly the same test, which we illustrate with a simple worked example, before describing how the theory can be applied to assess an entire statistic image. For simplicity of exposition, the methodology is developed using the example of a simple single-subject PET activation experiment. However, the approach is not limited to PET nor single-subject datasets.

Randomization test

We first consider randomization tests, using a single-subject activation experiment to illustrate the thinking: suppose we are to conduct a simple single-subject PET activation experiment, with the regional cerebral blood flow (rCBF) in 'active' (A) condition scans to be compared with that in scans acquired under an appropriate 'baseline' (B) condition. The fundamental concepts are of experimental *randomization*, the *null hypothesis*, *exchangeability*, and the *randomization distribution*.

Randomization

To avoid unexpected confounding effects, suppose we randomize the allocation of conditions to scans prior to conducting the experiment. Using an appropriate scheme, we label the scans as A or B according to the conditions under which they will be acquired, and hence specify the *condition presentation order*. This allocation of condition labels to scans is chosen randomly according to the randomization scheme, and any other possible labelling of this scheme is equally likely to have been chosen.

Null hypothesis

In the randomization test, the null hypothesis is explicitly about the acquired data. For example: $\mathcal{H}_0$: 'Each scan would have been the same whatever the condition, A or B'. The hypothesis is that the experimental conditions did not affect the data differentially, such that had we run the experiment with a different condition presentation order, we would have observed exactly the same data. In this sense we regard the data as fixed, and the experimental design as random. (In contrast to regarding the design as fixed, and the data as a realization of a random process.) Under this null hypothesis, the labelling of the scans as A or B is arbitrary; since these labellings arose from the initial random allocation of conditions to scans, and any initial allocation would have given the same data. Thus, we may re-randomize the labels on the data, effectively permuting the labels, subject to the restriction that each permutation could have arisen from the initial randomization scheme. The observed data are equally likely to have arisen from any of these permuted labellings.

Exchangeability

This leads to the notion of *exchangeability*. Consider the situation before the data are collected, but after the condition labels have been assigned to scans. Formally, a set of labels on the data (still to be collected) are *exchangeable* if the distribution of the statistic (still to be evaluated) is the same whatever the labelling (Good, 1994). For our activation example, we would use a statistic expressing the difference between the 'active' and 'baseline' scans. Thus, under the null hypothesis of no difference between the A and B conditions, the labels are exchangeable, provided the permuted labelling could have arisen from the initial randomization scheme. The initial randomization scheme gives us the probabilistic justification for permuting the labels; the null hypothesis asserts that the data would have been the same.

Randomization distribution

Consider now some statistic expressing the experimental effect of interest at a particular voxel. For the current example of a PET single-subject activation, this could be the mean difference between the A and the B condition scans, a two-sample t-statistic, a t-statistic from an ANCOVA, or any appropriate statistic. We are not restricted to the common statistics of classical parametric hypothesis whose null distributions are known under specific assumptions, because the appropriate distribution will be derived from the data.

The computation of the statistic depends on the labelling of the data. For example, with a two-sample t-statistic, the labels A and B specify the groupings. Thus, permuting the labels leads to an alternative value of the statistic.

Given exchangeability under the null hypothesis, the observed data are equally likely to have arisen from any of the possible labellings. Hence, the statistics associated with each of the possible labellings are also equally likely. Thus, we have the permutation (or randomization) distribution of our statistic: the *permutation distribution* is the *sampling distribution* of the statistic under the null hypothesis, given the data observed. Under the null hypothesis, the observed statistic is randomly chosen from the set of statistics corresponding to all possible relabellings. This gives us a way to formalize our 'surprise' at an outcome: the probability of an outcome as or more extreme than the one observed, the p-value, is the proportion of statistic values in the permutation distribution greater or equal to that observed. The actual labelling used in the experiment is one of the possible labellings, so if the observed statistic is the largest of the permutation distribution, the p-value is $1/N$, where N is the number of possible labellings of the initial randomization scheme. Since we are considering a test at a single voxel, these would be *uncorrected* p-values in the language of multiple testing (see below).

Randomization test: summary

To summarize, the null hypothesis asserts that the scans would have been the same whatever the experimental condition, A or B. Under this null hypothesis, the initial randomization scheme can be regarded as arbitrarily labelling scans as A or B, under which the experiment would have given the same data, and the labels are exchangeable. The statistic corresponding to any labelling from the initial randomization scheme is as likely as any other, since the permuted labelling could equally well have arisen in the initial randomization. The sampling distribution of the statistic (given the data) is the set of statistic values corresponding to all the possible labellings of the initial randomization scheme, each value being equally likely.

Randomization test: mechanics

Let N denote the number of possible relabellings, t_i the statistic corresponding to relabelling i. (After having performed the experiment, we refer to *re*labellings for the data, identical to the labellings of the randomization scheme.) The set of t_i for all possible relabellings constitutes the *permutation distribution*. Let T denote the value of the statistic for the actual labelling of the experiment. As usual in statistics, we use a capital letter for a *random variable*. T is random, since under $\mathcal{H}_0$ it is chosen from the permutation distribution according to the initial randomization.

Under $\mathcal{H}_0$, all of the t_i are equally likely, so we determine the significance of our observed statistic T by counting the proportion of the permutation distribution as or more extreme than T, giving us our p-value. We reject the null hypothesis at significance level α if the p-value is less than α. Equivalently, T must be greater than or equal to the $100(1-\alpha)$ percentile of the permutation distribution. Thus, the *critical value* is the $(c+1)$th largest member of the permutation distribution, where $c = \lfloor \alpha N \rfloor$, αN rounded down. If T exceeds this critical value then the test is significant at level α.

Permutation test

In many situations, it is impractical randomly to allocate experimental conditions, or perhaps we are presented with data from an experiment that was not randomized. For instance, we cannot randomly assign subjects to be patients or normal controls. Or, for example, consider a multisubject fMRI second-level analysis where a covariate is measured for each subject, and we seek brain regions whose activation appears to be related to the covariate value.

In the absence of an explicit randomization of conditions to scans, we must make weak distributional assumptions to justify permuting the labels on the data. Typically, all that is required is that distributions have the same shape, or are symmetric. The actual permutations that are performed again depend on the degree of exchangeability which, in turn, depend on the actual assumptions made. With the randomization test, the experimenter designs the initial randomization scheme carefully to avoid confounds. The randomization scheme reflects an implicitly assumed degree of exchangeability. With the permutation test, the degree of exchangeability must be assumed *post hoc*. Usually, the reasoning that would have led to a particular randomization scheme can be applied *post hoc* to an experiment, leading to a permutation test with the same degree of exchangeability. Given exchangeability, computation proceeds as for the randomization test.

Permutation test: summary

Weak distributional assumptions are made, which embody the degree of exchangeability. The exact form of these assumptions depends on the experiment at hand, as illustrated in the following section and in the examples section.

For a simple single-subject activation experiment, we might typically assume the following: for a particular voxel, 'active' and 'baseline' scans within a given block have a distribution with the same shape, though possibly different means. The null hypothesis asserts that the distributions for the 'baseline' and 'active' scans have the same mean, and hence are the same. Then the labels are arbitrary within the chosen blocks, which are thus the exchangeability blocks. Any permutation of the labels within the exchangeability blocks leads to an equally likely statistic. (Note, if this were a multisubject dataset, the exchangeability block would be the entire dataset, as subjects are regarded as independent.)

The mechanics are then the same as with the randomization test: for each of the possible relabellings, compute the statistic of interest; for relabelling i, call this statistic t_i. Under the null hypothesis each of the t_i are equally likely, so the p-value is the proportion of the t_is greater than or equal to the statistic T corresponding to the correctly labelled data.

Single voxel example

To make these concepts concrete, consider assessing the evidence of an activation effect at a single voxel of a single-subject PET activation experiment consisting of six scans, three in each of the 'active' (A) and 'baseline' (B) conditions. Suppose that the conditions were presented alternately, starting with rest, and that the observed data at this voxel are {90.48, 103.00, 87.83, 99.93, 96.06, 99.76} to 2 decimal places. (These data are from a voxel in the primary visual cortex of the second subject in the PET visual activation experiment presented in the examples section.)

As mentioned before, any statistic can be used, so for simplicity of illustration we use the 'mean difference', i.e. $T = \frac{1}{3}\sum_{j=1}^{3}(A_j - B_j)$ where B_j and A_j indicate the value of the jth scan at the particular voxel of interest, under the baseline and active conditions respectively. Thus, we observe statistic $T = 9.45$.

Randomization test

Suppose that the condition presentation order was randomized, the actual ordering of BABABA having been selected randomly from all allocations of three As and three Bs to the six available scans, a simple balanced randomization within a single randomization block of size six. By combinatorics, or some counting, we find that this randomization scheme has twenty (${}_6C_3 = 20$) possible outcomes.

Then we can justify permuting the labels on the basis of this initial randomization. Under the null hypothesis $\mathcal{H}_0$: 'The scans would have been the same whatever the experimental condition, A or B', the labels are exchangeable, and the statistics corresponding to the twenty possible labellings are equally likely. The twenty possible labellings are:

1: AAABBB 6: ABABAB 11: BAAABB 16: BABBAA
2: AABABB 7: ABABBA 12: BAABAB 17: BBAAAB
3: AABBAB 8: ABBAAB 13: BAABBA 18: BBAABA
4: AABBBA 9: ABBABA 14: BABAAB 19: BBABAA
5: ABAABB 10: ABBBAA 15: BABABA 20: BBBAAA

Permutation test

Suppose there was no initial randomization of conditions to scans, and that the condition presentation order ABABAB was chosen. With no randomization, we must make weak distributional assumptions to justify permuting the labels, effectively prescribing the degree of exchangeability.

For this example, consider permuting the labels freely among the six scans. This corresponds to *full exchangeability*, a single exchangeability block of size six. For this to be tenable, we must either assume the absence of any temporal or similar confounds, or model their effect such that they do not affect the statistic under permutations of the labels. Consider the former. This gives twenty possible permutations of the labels, precisely those enumerated for the randomization justification above. Formally, we are assuming that the voxel values for the 'baseline' and 'active' scans come from distributions that are the same except for a possible difference in location, or mean. Our null hypothesis is that these distributions have the same mean, and therefore are the same.

Clearly, the mean difference statistic under consideration in the current example is confounded with time for labellings such as AAABBB (#1) and BBBAAA (#20), where a time effect will result in a large mean difference between the A and the B labelled scans. The test is still valid, but possibly conservative. The actual condition presentation order of BABABA is relatively unconfounded with time, but the contribution of confounds to the statistics for alternative labellings such as #1 and #20 will potentially increase the number of statistics greater than the observed statistic.

Computation

Let t_i be the mean difference for labelling i, as enumerated above. Computing for each of the twenty relabellings:

$t_1 = +4.82$ $t_6 = +9.45$ $t_{11} = -1.48$ $t_{16} = -6.86$
$t_2 = -3.25$ $t_7 = +6.97$ $t_{12} = +1.10$ $t_{17} = +3.15$
$t_3 = -0.67$ $t_8 = +1.38$ $t_{13} = -1.38$ $t_{18} = +0.67$
$t_4 = -3.15$ $t_9 = -1.10$ $t_{14} = -6.97$ $t_{19} = +3.25$
$t_5 = +6.86$ $t_{10} = +1.48$ $t_{15} = -9.45$ $t_{20} = -4.82$

This is our permutation distribution for this analysis, summarized as a histogram in Figure 21.1. Each of the

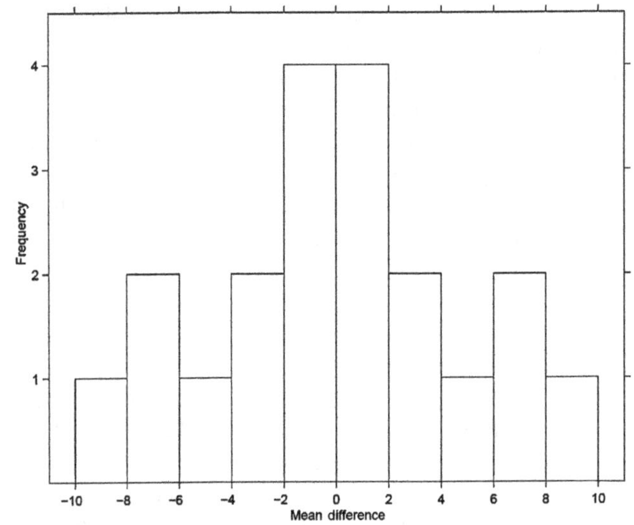

FIGURE 21.1 Histogram of permutation distribution for the single voxel example, using a mean difference statistic. Note the symmetry of the histogram about the y-axis. This occurs because, for each possible labelling, the opposite labelling is also possible, and yields the same mean difference but in the opposite direction. This trick can be used in many cases to halve the computational burden.

possible labellings was equally likely. Under the null hypothesis the statistics corresponding to these labellings are equally likely. The p-value is the proportion of the permutation distribution greater than or equal to T. Here the actual labelling #6 with $t_6 = +9.4$ gives the largest mean difference of all the possible labellings, so the p-value is $1/20 = 0.05$. For a test at given α level, we reject the null hypothesis if the p-value is less than α, so we conclude that there is significant evidence against the null hypothesis of no activation at this voxel at level $\alpha = 0.05$.

Permutation tests accounting for the multiple testing problem

Thus far we have considered using a permutation test at a single voxel: for each voxel we can produce a p-value, p^k, for the null hypothesis $\mathcal{H}_0^k$, where the superscript k indexes the voxel. If we have an *a priori* anatomical hypothesis concerning the experimentally induced effect at a single voxel, then we can simply test at that voxel using an appropriate α level test. If we do not have such precise anatomical hypotheses, evidence for an experimental effect must be assessed at each and every voxel. We must take account of the multiplicity of testing. Clearly 5 per cent of voxels are expected to have p-values less than $\alpha = 0.05$. This is the essence of the *multiple testing problem*. In the language of multiple testing, these p-values are *uncorrected* p-values. Type I errors must be controlled overall, such that the probability of falsely declaring any region as significant is less than the nominal test level α. This is known as controlling the family-wise error rate, the family being the collection of tests performed over the entire brain. Formally, we require a test procedure maintaining strong control over *family-wise* type I error, giving *adjusted* p-values, p-values *corrected* for the multiplicity of tests examined.

The construction of suitable multiple testing procedures for the problem of assessing statistic images from functional mapping experiments within parametric frameworks has occupied many authors e.g. Friston *et al.*, 1991, 1994, 1996; Worsley *et al.*, 1992, 1995; Poline and Mazoyer, 1993; Roland *et al.* 1993; Worsley, 1994; Forman *et al.* 1995; Poline *et al.* 1997; Cao, 1999). In contrast to these parametric and simulation based methods, a non-parametric resampling based approach provides an intuitive and easily implemented solution (Westfall and Young, 1993). The key realization is that the reasoning presented above for permutation tests at a single voxel relies on relabelling entire *images*, so the arguments can be extended to image-level inference by considering an appropriate *maximal statistic*. If, under the omnibus null hypothesis, the labels are exchangeable with respect to the voxel statistic under consideration, then the labels are exchangeable with respect to any statistic summarizing the voxel statistics, such as their maxima.

We consider two popular types of test, *single threshold* and *suprathreshold cluster* size tests, but note again the ability of these methods to use any statistic.

Single threshold test

With a single threshold test, the statistic image is thresholded at a given *critical threshold*, and voxels with statistic values exceeding this threshold have their null hypotheses rejected. Rejection of the *omnibus hypothesis* (that all the voxel hypotheses are true) occurs if any voxel value exceeds the threshold, a situation clearly determined by the value of the maximum value of the statistic image. Thus, consideration of the maximum voxel statistic deals with the multiple testing problem. For a valid omnibus test, the critical threshold is such that the probability that it is exceeded by the maximal statistic is less than α; therefore, we require the distribution of the maxima of the null statistic image. Approximate parametric derivations based on the theory of strictly stationary continuous random fields are given by Friston *et al.* (1991) and Worsley *et al.* (1992, 1995); Worsley, 1994.

The permutation approach can furnish the distribution of the maximal statistic in a straightforward manner: Rather than compute the permutation distribution of the statistic at a particular voxel, we compute the permutation distribution of the maximal voxel statistic over the volume of interest. We reject the omnibus hypothesis at level α if the maximal statistic for the actual labelling of the experiment is in the top 100α per cent of the permutation distribution for the maximal statistic. The critical value is the $(c+1)$th largest member of the permutation distribution, where $c = \lfloor \alpha N \rfloor$, αN rounded down. Furthermore, we can reject the null hypothesis at any voxel with a statistic value exceeding this threshold: the critical value for the maximal statistic is the critical threshold for a single threshold test over the same volume of interest. This test can be shown to have *strong* control over *experiment-wise* type I error. A formal proof is given by Holmes *et al.* (1996).

The mechanics of the test are as follows: for each possible relabelling $i = 1, \ldots, N$, note the maximal statistic $t_i^{\max}$, the maximum of the voxel statistics for relabelling i: $t_i^{\max} = \max\{t_i\}_{i=1}^{N}$. This gives the permutation distribution for $T^{\max}$, the maximal statistic. The critical threshold is the $c+1$ largest member of the permutation distribution for $T^{\max}$, where $c = \lfloor \alpha N \rfloor$, αN rounded down. Voxels with statistics exceeding this threshold exhibit evidence against the corresponding voxel hypotheses at level α. The corresponding corrected p-value for each voxel is the proportion of the permutation distribution for the maximal statistic that is greater than or equal to the voxel statistic.

Suprathreshold cluster tests

Suprathreshold cluster tests start by thresholding the statistic image at a predetermined *primary* threshold, and then assess the resulting pattern of suprathreshold activity. Suprathreshold cluster size tests assess the size of connected suprathreshold regions for significance, declaring regions greater than a critical size as activated. Thus, the distribution of the maximal suprathreshold cluster size (for the given primary threshold) is required. Simulation approaches have been presented by Poline and Mazoyer (1993) and Roland *et al.* (1993) for PET, Forman *et al.* (1995) for fMRI. Friston *et al.* (1994) give a theoretical parametric derivation for Gaussian statistic images based on the theory of continuous Gaussian random fields; Cao (1999) gives results for χ^2, t and F fields.

Again, as noted by Holmes *et al.* (1996), a non-parametric permutation approach is simple to derive. Simply construct the permutation distribution of the maximal suprathreshold cluster size. For the statistic image corresponding to each possible relabelling, note the size of the largest suprathreshold cluster above the primary threshold. The critical suprathreshold cluster size for this primary threshold is the $(\lfloor \alpha N \rfloor + 1)$th largest member of this permutation distribution. Corrected p-values for each suprathreshold cluster in the observed statistic image are obtained by comparing their size to the permutation distribution.

In general, such suprathreshold cluster tests are more powerful for functional neuroimaging data than the single threshold approach (see Chapter 19 and Friston *et al.*, 1995 for a fuller discussion). However, it must be remembered that this additional power comes at the price of reduced localizing power: the null hypotheses for voxels within a significant cluster are not tested, so individual voxels cannot be declared significant. Only the omnibus null hypothesis for the cluster can be rejected. Further, the choice of primary threshold dictates the power of the test in detecting different types of deviation from the omnibus null hypothesis. With a low threshold, large suprathreshold clusters are to be expected, so intense focal 'signals' will be missed. At higher thresholds these focal activations will be detected, but lower intensity diffuse 'signals' may go undetected below the primary threshold.

Poline *et al.* (1997) addressed these issues within a parametric framework by considering the suprathreshold cluster size and height jointly. A non-parametric variation could be to consider the *exceedance mass*, the excess mass of the suprathreshold cluster, defined as the integral of the statistic image above the primary threshold within the suprathreshold cluster (Holmes, 1994; Bullmore *et al.*, 1999). Calculation of the permutation distribution and p-values proceeds exactly as before.

Considerations

Before turning to example applications of the non-parametric permutation tests described above, we note some relevant theoretical issues. The statistical literature (referenced below) should be consulted for additional theoretical discussion. For issues related to the current application to functional neuroimaging, see also Holmes (1994), Holmes *et al.* (1996), and Arndt *et al.* (1996).

Non-parametric statistics

First, it should be noted that these methods are neither new nor contentious: originally expounded by Fisher (1935), Pitman (1937a,b,c), and later Edgington (1964, 1969a,b), these approaches are enjoying a renaissance as computing technology makes the requisite computations feasible for practical applications. Had R.A. Fisher and his peers had access to similar resources, it is possible that large areas of parametric statistics would have gone undeveloped! Modern texts on the subject include Good's *Permutation Tests* (1994), Edgington's *Randomization Tests* (1995), and Manly's *Randomization, Bootstrap and Monte-Carlo Methods in Biology* (1997). Recent interest in more general resampling methods, such as the bootstrap, has further contributed to the field. For a treatise on resampling based multiple testing procedures, see Westfall and Young (1993).

Many standard statistical tests are essentially permutation tests: The 'classic' non-parametric tests, such as the Wilcoxon and Mann-Whitney tests, are permutation tests with the data replaced by appropriate ranks, such that the critical values are only a function of sample size and can therefore be tabulated. Fisher's exact test (1990), and tests of Spearman and Kendall correlations (Kendall and Gibbons 1990), are all permutation/randomization based.

Assumptions

The only assumptions required for a valid permutation test are those to justify permuting the labels. Clearly, the experimental design, model, statistic and permutations must also be appropriate for the question of interest. For a randomization test, the probabilistic justification follows directly from the initial randomization of condition labels to scans. In the absence of an initial randomization, permutation of the labels can be justified via weak distributional assumptions. Thus, only minimal assumptions are required for a valid test. (The notable case when exchangeability under the null hypothesis is not tenable is fMRI time-series, due to temporal autocorrelation.)

In contrast to parametric approaches where the statistic must have a known null distributional form, the permutation approach is free to consider any statistic summarizing evidence for the effect of interest at each voxel. The consideration of the maximal statistic over the volume of interest then deals with the multiple testing problem.

However, there are additional considerations when using the non-parametric approach with a maximal statistic to account for multiple testing. In order for the single threshold test to be equally sensitive at all voxels, the (null) sampling distribution of the chosen statistic should be similar across voxels. For instance, the simple mean difference statistic used in the single voxel example could be considered as a voxel statistic, but areas where the mean difference is highly variable will dominate the permutation distribution for the maximal statistic. The test will still be valid, but will be less sensitive at those voxels with lower variability. So, although for an individual voxel, a permutation test on group mean differences is equivalent to one using a two-sample t-statistic (Edgington, 1995), this is not true in the multiple testing setting using a maximal statistic.

One approach to this problem is to consider multistep tests, which iteratively identify activated areas, cut them out, and continue assessing the remaining volume. These are described below, but are additionally computationally intensive. A preferable solution is to use a voxel statistic with approximately homogeneous null permutation distribution across the volume of interest, such as an appropriate t-statistic. A t-statistic is essentially a mean difference normalized by a variance estimate, effectively measuring the reliability of an effect. Thus, we consider the same voxel statistics for a non-parametric approach as we would for a comparable parametric approach.

Pseudo t-statistics

Nonetheless, we can still do a little better than a straight t-statistic, particularly at low degrees of freedom. A t-statistic is a change divided by the square root of the estimated variance of that change. When there are few degrees of freedom available for variance estimation, say, less than 20, this variance is estimated poorly. Errors in estimation of the variance from voxel to voxel appear as high (spatial) frequency noise in images of the estimated variance or near-zero variance estimates, which in either case cause noisy t-statistic images. Given that PET and fMRI measure (or reflect) blood flow, physiological considerations would suggest that the variance be roughly constant over small localities. This suggests pooling the variance estimate at a voxel with those of its neighbours to give a locally pooled variance estimate as a better estimate of the actual variance. Since the model is of the same form at all voxels, the voxel variance estimates have the same degrees of freedom, and the locally pooled variance estimate is simply the average of the variance estimates in the neighbourhood of the voxel in question. More generally, weighted locally pooled voxel variance estimates can be obtained by smoothing the raw variance image. The filter kernel then specifies the weights and neighbourhood for the local pooling. The *pseudo* t-statistic images formed with smoothed variance estimators are smooth. In essence the noise (from the variance image) has been smoothed, but not the signal. A derivation of the parametric distribution of the pseudo t requires knowledge of the variance-covariances of the voxel-level variances, and has so far proved elusive. This precludes parametric analyses using a pseudo t-statistic, but poses no problems for a non-parametric approach.

Number of relabellings and test size

A constraint on the permutation test is the number of possible relabellings. Since the observed labelling is always one of the N possible labellings, the smallest p-value attainable is $1/N$. Thus, for a level $\alpha = 0.05$ test potentially to reject the null hypothesis, there must be at least twenty possible relabellings.

More generally, the permutation distribution is *discrete*, consisting of a finite set of possibilities corresponding to the N possible relabellings. Hence, any p-values produced will be multiples of $1/N$. Further, the $100(1-\alpha)^{\text{th}}$ percentile of the permutation distribution, the critical threshold for a level α test, may lie between two values. Equivalently, α may not be a multiple of $1/N$, such that a p-value of exactly α cannot be attained. In these cases, an exact test with size exactly α is not possible. It is for this reason that the critical threshold is computed as the $(c+1)$th largest member of the permutation distribution, where $c = \lfloor \alpha N \rfloor$, αN rounded down. The test can be described as *almost* exact, since the size is at most $1/N$ less than α.

Monte Carlo tests

A large number of possible relabellings is also problematic, due to the computations involved. In situations where it is not feasible to compute the statistic images for all the relabellings, a random subsample of relabellings can be used (Dwass, 1957) (see also Edgington, 1969a for a less mathematical description). The set of N possible relabellings is reduced to a more manageable N' consisting of the true labelling and $N'-1$ randomly chosen from the set of $N-1$ possible relabellings. The test then proceeds as before.

Such a test is sometimes known as an approximate or *Monte Carlo permutation test*, since the permutation distribution is approximated by a random selection of all possible values. The p-values found are random, and will vary between Monte Carlo realizations. Despite the name, the resulting test is still exact. However, as might be expected from the previous section, using an approximate permutation distribution results in a test that is less powerful than one using the full permutation distribution.

Fortunately, as few as 1000 permutations can yield an effective approximate permutation test (Edgington, 1969a). However, for a Monte Carlo test with minimal loss of power in comparison to the full test (i.e. with high efficiency), one should consider rather more (Jöckel, 1986). A margin of error can characterize the degree to which the p-values vary. When N is large and $N' \ll N$ the approximate 95 per cent margin of error is $2\sqrt{p(1-p)/N'}$ where p is the true p-value found with all N permutations. This result shows that, if one is to limit the margin of error at 10 per cent of a nominal 0.05 p-value, an N' of approximately 7500 is required.

Power

Generally, non-parametric approaches are less powerful than equivalent parametric approaches when the assumptions of the latter are true. The assumptions provide the parametric approach with additional information, which the non-parametric approach must 'discover'. The more relabellings, the better the power of the non-parametric approach relative to the parametric approach. In a sense, the method has more information from more relabellings, and 'discovers' the null distribution assumed in the parametric approach. However, if the assumptions required for a parametric analysis are not credible, a non-parametric approach provides the only valid method of analysis.

In the current context of assessing statistic images from functional neuroimaging experiments, the prevalent statistical parametric mapping techniques require a number of assumptions and involve some approximations. Experience suggests that the permutation methods described here do at least as well as the parametric methods, at least on real (PET) data (Arndt *et al.*, 1996). For noisy statistic images, such as t-statistic images with low degrees of freedom, the ability to consider pseudo t-statistics constructed with locally pooled (smoothed) variance estimates affords the permutation approach additional power (Holmes, 1994; Holmes *et al.*, 1996, and examples below).

Multistep tests

The potential for confounds to affect the permutation distribution via the consideration of unsuitable relabellings has already been considered. Recall also the above comments regarding the potential for the maximum-based permutation test to be differentially sensitive across the volume of interest if the null permutation distribution varies dramatically from voxel to voxel. There is also the prospect that departures from the null hypothesis influence the permutation distribution. Thus far, our non-parametric multiple testing permutation testing technique has consisted of a *single-step*: the null sampling distribution (given the data), is the permutation distribution of the maximal statistic computed over all voxels in the volume of interest, potentially including voxels where the null hypothesis is not true. A large departure from the null hypothesis will give a large statistic, not only in the actual labelling of the experiment, but also in other relabellings, similar to the true labelling. This does not affect the overall validity of the test, but may make it more conservative for voxels other than that with the maximum observed statistic.

One possibility is to consider *step-down* tests, where significant regions are iteratively identified and cut out, and the remaining volume is reassessed. The resulting procedure still maintains strong control over family-wise type I error, our criteria for a test with localizing power, but will be more powerful (at voxels other than that with the maximal statistic). However, the iterative nature of the procedure multiplies the computational burden of an already intensive procedure. Holmes *et al.* (1996), give a discussion and efficient algorithms, developed further in Holmes (1998), but find that the additional power gained was negligible for the cases studied.

Recall also the motivations for using a normalized voxel statistic, such as the t-statistic: an inappropriately normalized voxel statistic will yield a test that is differentially sensitive across the image. In these situations the step-down procedures may be more beneficial.

Generalizability

Questions often arise about the scope of inference, or generalizability of non-parametric procedures. For parametric tests, when a collection of subjects have been randomly selected from a population of interest and intersubject variability is considered, the inference is on the sampled population and not just the sampled subjects. The randomization test, in contrast, only makes inference on the data at hand: a randomization test regards the data as fixed and uses the randomness of the experimental design to justify exchangeability. A permutation test, while operationally identical to the randomization test, *can* make inference on a sampled population: a permutation test also regards the data as fixed but it additionally assumes the presence of a population distribution to justify exchangeability, and hence can be used for population inference. The randomization test is truly assumption free, but has a limited scope of inference.

In practice, since subjects rarely constitute a random sample of the population of interest, we find the issue of little practical concern. Scientists routinely generalize results, integrating prior experience, other findings, existing theories, and common sense in a way that a simple hypothesis test does not admit.

WORKED EXAMPLES

The following sections illustrate the application of the techniques described above to three common experimental designs: single-subject PET 'parametric', multisubject PET activation, and multisubject fMRI activation. In each example we will illustrate the key steps in performing a permutation analysis:

1 **Null hypothesis**
Specify the null hypothesis
2 **Exchangeability**
Specify exchangeability of observations under the null hypothesis
3 **Statistic**
Specify the statistic of interest, usually broken down into specifying a voxel-level statistic and a summary statistic
4 **Relabellings**
Determine all possible relabellings given the exchangeability scheme under the null hypothesis

5 **Permutation distribution**
Calculate the value of the statistic for each relabelling, building the permutation distribution

6 **Significance**
Use the permutation distribution to determine significance of correct labelling and threshold for statistic image.

The first three items follow from the experimental design and must be specified by the user; the last three are computed by the software, though we will still address them here. When comparable parametric analyses are available (within SPM), we will compare the permutation results to the parametric results.

Single-subject PET: parametric design

The first study will illustrate how covariate analyses are implemented and how the suprathreshold cluster size statistic is used. This example also shows how randomization in the experimental design dictates the exchangeability of the observations.

Study description

The data come from a study of Silbersweig *et al.* (1994). The aim of the study was to validate a novel PET methodology for imaging transient, randomly occurring events, specifically events that were shorter than the duration of a PET scan. This work was the foundation for later work imaging hallucinations in schizophrenics (Silbersweig *et al.*, 1995). We consider one subject from the study, who was scanned 12 times. During each scan the subject was presented with brief auditory stimuli. The proportion of each scan over which stimuli were delivered was chosen randomly, within three randomization blocks of size four. A score was computed for each scan, indicating the proportion of activity infused into the brain during stimulation. This scan activity score is our covariate of interest, which we shall refer to as Duration. This is a type of parametric design, though in this context parametric refers not to a set of distributional assumptions, but rather an experimental design where an experimental parameter is varied continuously. This is in contradistinction to a factorial design where the experimental probe is varied over a small number of discrete levels.

We also have to consider the global cerebral blood flow (gCBF), which we account for here by including it as a nuisance covariate in our model. This gives a multiple regression, with the slope of the Duration effect being of interest. Note that regressing out gCBF like this requires an assumption that there is no linear dependence between the score and global activity; examination of a scatter plot and a correlation coefficient of 0.09 confirmed this as a tenable assumption (see Chapter 8 for further discussion of global effects in PET).

Null hypothesis

Since this is a randomized experiment, the test will be a randomization test, and the null hypothesis pertains directly to the data, and no assumptions are required:

$\mathcal{H}_0$: The data would be the same whatever the Duration.

Exchangeability

Since this experiment was randomized, our choice of EB (i.e. exchangeability block) matches the randomization blocks of the experimental design, which was chosen with temporal effects in mind. The values of Duration were grouped into three blocks of four, such that each block had the same mean and similar variability, and then randomized within block. Thus we have three EBs of size four.

Statistic

We decompose our statistic of interest into two statistics: one voxel-level statistic that generates a statistic image, and a maximal statistic that summarizes that statistic image in a single number. An important consideration will be the degrees of freedom. We have one parameter for the grand mean, one parameter for the slope with Duration, and one parameter for confounding covariate gCBF. Hence 12 observations less three parameters leaves just nine degrees of freedom to estimate the error variance at each voxel.

Voxel-level statistic With only nine degrees of freedom, this study shows the characteristic noisy variance image (Figure 21.2). The high frequency noise from poor variance estimates propagates into the t-statistic image, when one would expect an image of evidence against $\mathcal{H}_0$ to be smooth (as is the case for studies with greater degrees of freedom) since the raw images are smooth.

We address this situation by smoothing the variance images (see pseudo t-statistics), replacing the variance estimate at each voxel with a weighted average of its neighbours. We use weights from an 8 mm spherical Gaussian smoothing kernel. The statistic image consisting of the ratio of the slope and the square root of the smoothed variance estimate is smoother than that computed with the raw variance. At the voxel level, the resulting statistic does not have a Student's t-distribution under the null hypothesis, so we refer to it as a *pseudo* t-statistic.

Figure 21.3 shows the effect of variance smoothing. The smoothed variance image creates a smoother statistic image, the pseudo t-statistic image. The key here

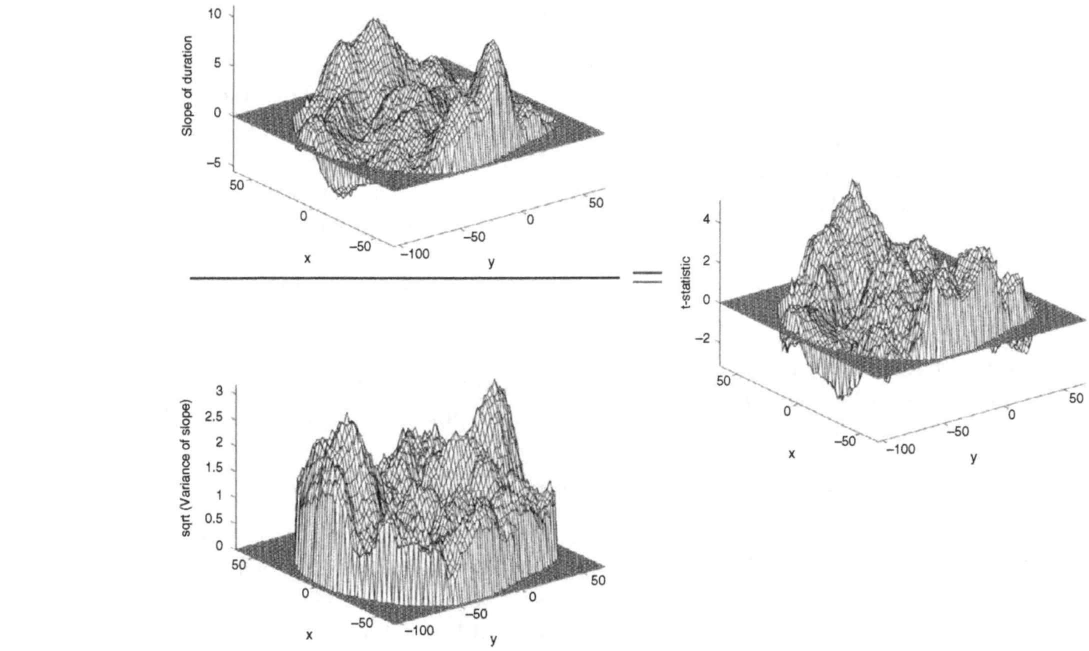

FIGURE 21.2 Mesh plots of parametric analysis, $z = 0$ mm. Upper left: slope estimate. Lower left: standard deviation of slope estimate. Right: t image for Duration. Note how the standard deviation image is much less smooth than the slope image, and how the t image is correspondingly less smooth than the slope image.

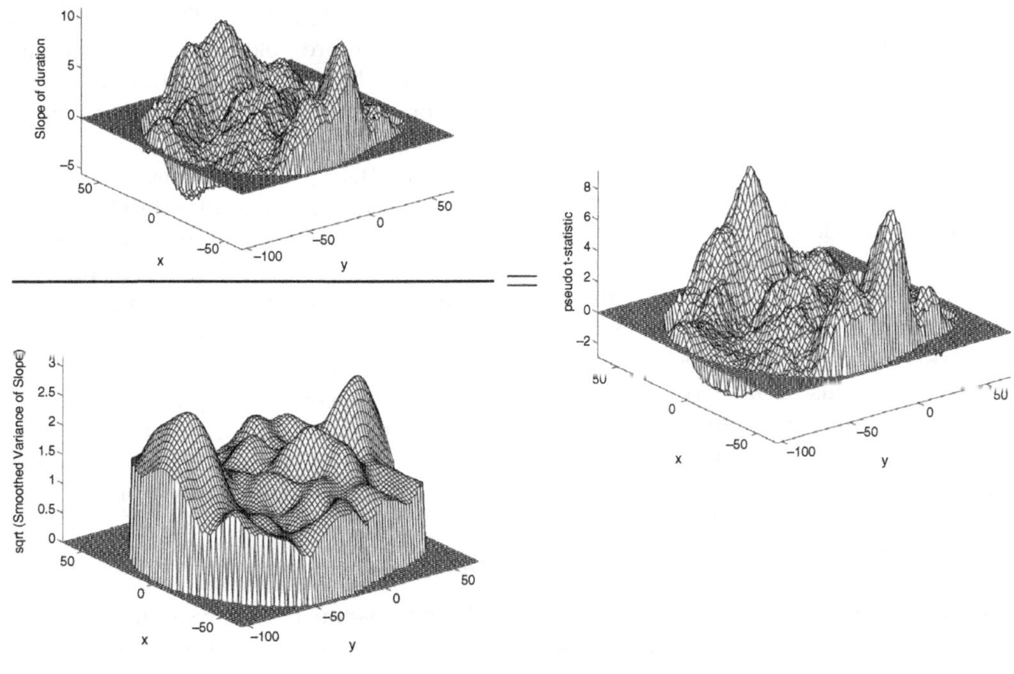

FIGURE 21.3 Mesh plots of permutation analysis, $z = 0$ mm. Upper left: slope estimate. Lower left: square root of smoothed variance of slope estimate. Right: pseudo t image for Duration. Note that the smoothness of the pseudo t image is similar to that of the slope image (cf. Figure 21.2).

is that the parametric t-statistic introduces high spatial frequency noise via the poorly estimated standard deviation – by smoothing the variance image we are making the statistic image more like the 'signal'.

Summary statistic We summarize evidence against $\mathcal{H}_0$ for each relabelling with the maximum statistic and, in this example, consider the maximum suprathreshold cluster size (max STCS).

Clusters are defined by connected suprathreshold voxels. Under $\mathcal{H}_0$, the statistic image should be random with no features or structure, hence large clusters are unusual and indicate the presence of an activation. A primary threshold is used to define the clusters. The selection of the primary threshold is crucial. If set too high there will be no clusters of any size; if set too low the clusters will be too large to be useful.

Relabelling enumeration

Each of the three previous sections corresponds to a choice that a user of the permutation test has to make. Those choices and the data are sufficient for an algorithm to complete the permutation test. This and the next two sections describe the ensuing computational steps.

To create the labelling used in the experiment, the labels were divided into three blocks of four, and randomly ordered within blocks. There are $4! = 4 \times 3 \times 2 \times 1 = 24$ ways to permute 4 labels, and since each block is independently randomized, there are a total of $4!^3 =$ 13 824 permutations of the labels.

Computations for 13 824 permutations would be burdensome, so we use a Monte Carlo test. We randomly select 999 relabellings to compute the statistic, giving 1000 relabellings including the actual labelling used in the experiment. Recall that while the p-values are random approximations, the test is still exact.

Permutation distribution

For each of the 1000 relabellings, the statistic image is computed and thresholded, and the maximal suprathreshold cluster size is recorded. For each relabelling this involves model fitting at each voxel, smoothing the variance image, and creating the pseudo t-statistic image. This is the most computationally intensive part of the analysis, but is not onerous on modern computing hardware. (See below for computing times.)

Selection of the primary threshold is a quandary. For the results to be valid we need to pick the threshold before the analysis is performed. With a parametric voxel-level statistic we could use its null distribution to specify a threshold from the uncorrected p-value (e.g. by using a t table). Here we cannot take this approach since we are using a non-parametric voxel-level statistic whose null distribution is not known *a priori*. Picking several thresholds is not valid, as this introduces a new multiple testing problem. We suggest gaining experience with similar datasets from *post hoc* analyses: apply different thresholds to get a feel for an appropriate range and then apply such a threshold to the data on hand. Using data from other subjects in this study we found 3.0 to be a reasonable primary threshold.

Significance threshold

The distribution of max STCS is used to assess the overall significance of the experiment and the significance of individual clusters: The significance is the proportion of relabellings that had max STCS greater than or equal to the maximum STCS of the correct labelling. Put another way, if max STCS of the correct labelling is at or above the 95th percentile of the max STCS permutation distribution, the experiment is significant at $\alpha = 0.05$. Also, any cluster in the observed image with size greater than the 95th percentile is significant at $\alpha = 0.05$. Since we have 1000 relabellings, $1000 \times 0.95 = 950$, so the 950th largest max STCS will be our significance threshold.

Results

The permutation distribution of max STCS under $\mathcal{H}_0$ is shown in Figure 21.4(a). Most relabellings have max STCS less than 250 voxels. The vertical dotted line indicates the 95th percentile: the top 5 per cent are spread from about 500 to 3000 voxels.

For the correctly labelled data the max STCS was 3101 voxels. This is unusually large in comparison to the permutation distribution. Only five relabellings yield max STCS equal to or larger than 3101, so the p-value for the experiment is $5/1000 = 0.005$. The 95th percentile is 462, so any suprathreshold clusters with size greater than 462 voxels can be declared significant at level 0.05, accounting for the multiple testing implicit in searching over the brain.

Figure 21.4(b), is a *maximum intensity projection* (MIP) of the significant suprathreshold clusters. Only these two clusters are significant, i.e. there are no other suprathreshold clusters larger than 462 voxels. These two clusters cover the bilateral auditory (primary and associative) and language cortices. They are 3101 and 1716 voxels in size, with p-values of 0.005 and 0.015 respectively. Since the test concerns suprathreshold clusters, it has no localizing power: significantly large suprathreshold clusters contain voxels with a significant experimental effect, but the test does not identify them.

Discussion

The non-parametric analysis presented here uses maximum STCS on a pseudo t-statistic image. Since the distribution of the pseudo t-statistic is not known, the corresponding primary threshold for a parametric analysis using a standard t-statistic cannot be computed. This precludes a straightforward comparison of this non-parametric analysis with a corresponding parametric analysis such as that of Friston *et al.* (1994).

While the necessity to choose the primary threshold for suprathreshold cluster identification is a problem, the

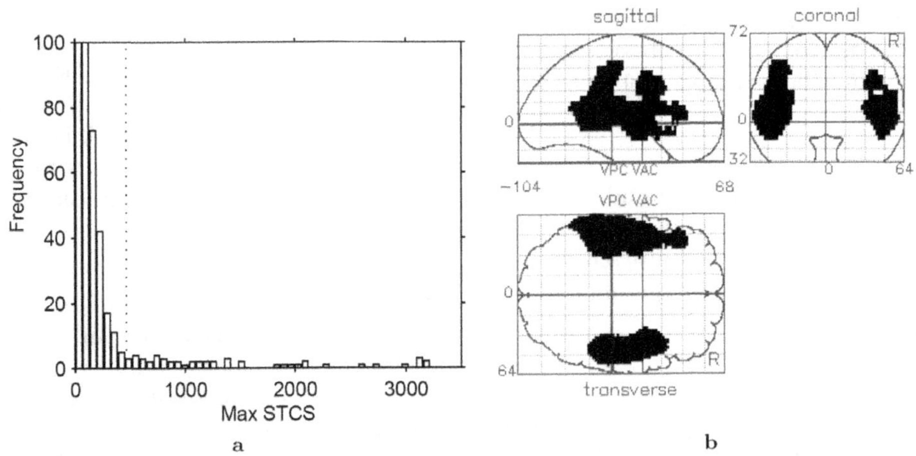

FIGURE 21.4 (a) Distribution of maximum suprathreshold cluster size with a primary threshold of 3. Dotted line shows 95th percentile. The count axis is truncated at 100 to show low-count tail; first two bars have counts 579 and 221. (b) Maximum intensity projection image of significantly large clusters.

same is true for parametric approaches. The only additional difficulty occurs with pseudo t-statistic images, when specification of primary thresholds in terms of upper tail probabilities from a Students' t-distribution is impossible. Further, parametric suprathreshold cluster size methods (Friston *et al.*, 1994; Poline *et al.*, 1997) utilize asymptotic distributional results, and therefore require high primary thresholds. The non-parametric technique is free of this constraint, giving exact p-values for any primary threshold (although very low thresholds are undesirable due to the large suprathreshold clusters expected and consequent poor localization of an effect).

Although only suprathreshold cluster size has been considered, any statistic summarizing a suprathreshold cluster could be considered. In particular, an exceedance mass statistic could be employed.

Multisubject PET: activation

For the second example, we consider a multisubject, two condition activation experiment. Here we will use a standard t-statistic with a single threshold test, enabling a direct comparison with the standard parametric random field approach.

Study description

Watson *et al.* (1993) localized the region of visual cortex sensitive to motion, area MT/V5, using high resolution 3D PET imaging of twelve subjects. These data were analysed by Holmes *et al.* (1996), using proportional scaling global flow normalization and a repeated measures pseudo t-statistic. Here, we consider the same data, but use a standard repeated measures t-statistic, allowing direct comparison of parametric and non-parametric approaches.

The visual stimulus consisted of randomly placed squares. During the baseline condition the pattern was stationary, whereas during the active condition the squares smoothly moved in independent directions. Prior to the experiment, the twelve subjects were randomly allocated to one of two scan condition presentation orders in a balanced randomization. Thus six subjects had scan conditions ABABABABABAB, the remaining six having BABABABABABA, which we will refer to as AB and BA orders respectively.

Null hypothesis

In this example, the labels of the scans as A and B are allocated by the initial randomization, so we have a randomization test, and the null hypothesis concerns the data directly:

> $\mathcal{H}_0$: For each subject, the experiment would have yielded the same data were the conditions reversed.

Exchangeability

Given the null hypothesis, exchangeability follows directly from the initial randomization scheme: the experiment was randomized at the subject level, with six AB and six BA labels randomly assigned to the twelve subjects. Correspondingly, the labels are exchangeable subject to the constraint that they could have arisen from the initial randomization scheme. Thus we consider all permutations of the labels that result in six subjects having scans labelled AB, and the remaining six BA. The initial randomization could have resulted in any six subjects having the AB condition presentation order (the remainder being BA), and under the null hypothesis the data would have been the same, hence exchangeability.

Statistic

We are interested in the activation magnitude relative to the intersubject variability in activation, hence we use the statistic associated with a *random-effects* model which incorporates a random subject by condition interaction term.

Voxel-level statistic A random-effects analysis is easily effected by collapsing the data within subject and computing the statistic across subjects (Worsley et al., 1991; Holmes and Friston, 1999). In this case, the result is a repeated measures t-statistic, after proportional scaling for global normalization: each scan is proportionally scaled to a common global mean of 50; each subject's data are collapsed into two average images, one for each condition; a paired t-statistic is computed across the subjects' 'rest'-'active' pairs of average images. By computing this paired t-statistic on the collapsed data, both the inter-subject and intra-subject (error) components of variance are accounted for appropriately. Since there are twelve subjects, there are twelve pairs of average condition images, and the t-statistic has 11 degrees of freedom. With just 11 degrees of freedom we anticipate the same problems with noisy variance images as in the previous examples, but in order to make direct comparisons with a parametric approach, we will not consider variance smoothing and pseudo t-statistics for this example.

Summary statistic To consider a single threshold test over the entire brain, the appropriate summary statistic is the maximum t-statistic.

Relabelling enumeration

This example is different from the previous one in that we permute across subjects instead of across replications of conditions. Here our EB is not in units of scans, but subjects. The EB size here is twelve subjects, since the six AB and six BA labels can be permuted freely among the twelve subjects. There are $\binom{12}{6} = \frac{12!}{6!(12-6)!} = 924$ ways of choosing six of the twelve subjects to have the AB labelling. This is a sufficiently small number of permutations to consider a complete enumeration.

One may consider permuting labels within subjects, particularly in the permutation setting when there is no initial randomization dictating the exchangeability. However, the bulk of the permutation distribution is specified by these between-subject permutations, and any within-subject permutations just flesh out this framework, yielding little practical improvement in the test.

Permutation distribution

For each of 924 relabellings, we calculate the maximum repeated measures t-statistic, resulting in the permutation distribution shown in Figure 21.5(a). Note that for each possible relabelling and t-statistic image, the opposite relabelling is also possible, and gives the negative of the t-statistic image. Thus, it is only necessary to compute t-statistic images for half of the relabellings, and retain their maxima and minima. The permutation distribution is then that of the maxima for half the relabellings concatenated with the negative of the corresponding minima.

Significance threshold

As before, the 95th percentile of the maximum t distribution provides both a threshold for omnibus experimental

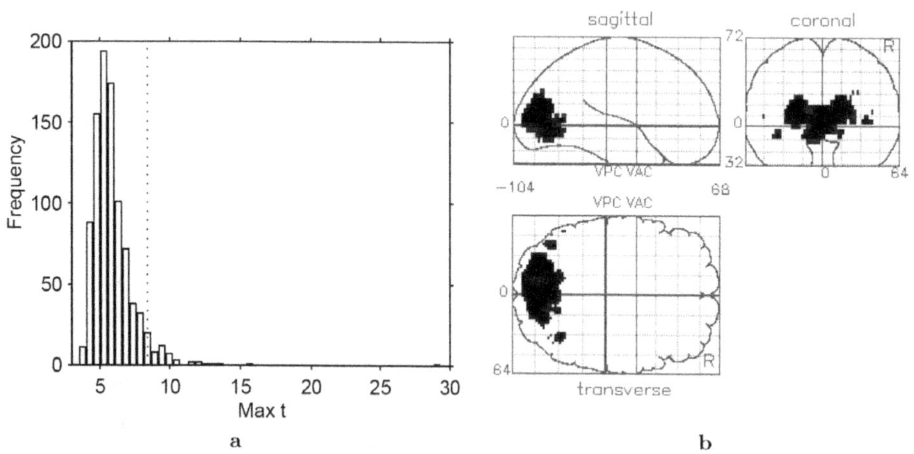

FIGURE 21.5 (a) Permutation distribution of maximum repeated measures t-statistic. Dotted line indicates the 5 per cent level corrected threshold. (b) Maximum intensity projection of t-statistic image, thresholded at critical threshold for a 5 per cent level permutation test analysis of 8.401.

significance and a voxel-level significance threshold appropriate for the multiple testing problem. With 924 permutations, the 95th percentile is at $924 \times 0.05 = 46.2$, so the critical threshold is the 47th largest member of the permutation distribution. Any voxel with intensity greater than this threshold can be declared significant at the 0.05 level.

Results

Figure 21.5(a) shows the permutation distribution of the maximum repeated measures *t*-statistic. Most maxima lie between about 4 and 9, though the distribution is skewed in the positive direction.

The outlier at 29.30 corresponds to the observed *t*-statistic, computed with correctly labelled data. Since no other relabellings are higher, the *p*-value is $1/924 = 0.0011$. The 47th largest member of the permutation distribution is 8.40, the critical threshold (marked with a dotted vertical line on the permutation distribution). The *t*-statistic image thresholded at this critical value is shown in Figure 21.5(b). There is a primary region of 1424 significant voxels covering the V1/V2 region, flanked by two secondary regions of 23 and 25 voxels corresponding to area V5, plus six other regions of 1 or 2 voxels.

For a *t*-statistic image of 43 724 voxels of size $2 \times 2 \times 4$ mm, with an estimated smoothness of $7.8 \times 8.7 \times 8.7$ mm FWHM, the parametric theory gives a 5 per cent level critical threshold of 11.07, substantially higher than the corresponding 4.61 of the non-parametric result. The thresholded image is shown in Figure 21.6(b); the image is very similar to the non-parametric image (Figure 21.5(b)), with the primary region having 617 voxels, with two secondary regions of 7 and 2 voxels. Another parametric result is the well-known, but conservative Bonferroni correction; here it specifies a 5 per cent threshold of 8.92 which yields a primary region of 1212 voxels and five secondary regions with a total of 48 voxels. In Figure 21.6(a) we compare these three approaches by plotting the significance level versus the threshold. The critical threshold based on the expected Euler characteristic (Worsley *et al.*, 1995) for a *t*-statistic image is shown as a dashed line and the critical values for the permutation test is shown as a solid line. For a given test level (a horizontal line), the test with the smaller threshold has the greater power. At all thresholds in this plot the non-parametric threshold is below the random field threshold, though it closely tracks the Bonferroni threshold below the 0.05 level. Thus random field theory (see Chapters 17, 18 and 19) appears to be quite conservative here.

Discussion

This example again demonstrates the role of the permutation test as a reference for evaluating other procedures, here the parametric analysis of Friston *et al.* (1995). The *t* field results are conservative for low degrees of freedom and low smoothness; the striking difference between the non-parametric and random field thresholds makes this clear.

Figure 21.6(a) provides an informative comparison between the two methods. For all typical test sizes ($\alpha \leq 0.05$), the non-parametric method specifies a lower threshold than the parametric method: for these data, this is exposing the conservativeness of the *t* field results. For lower thresholds the difference between the methods is even greater, though this is anticipated since the parametric results are based on high threshold approximations.

A randomization test applied to a random-effects statistic presents an interesting contradiction. While we use a statistic corresponding to a model with a random subject by condition interaction, we are performing a

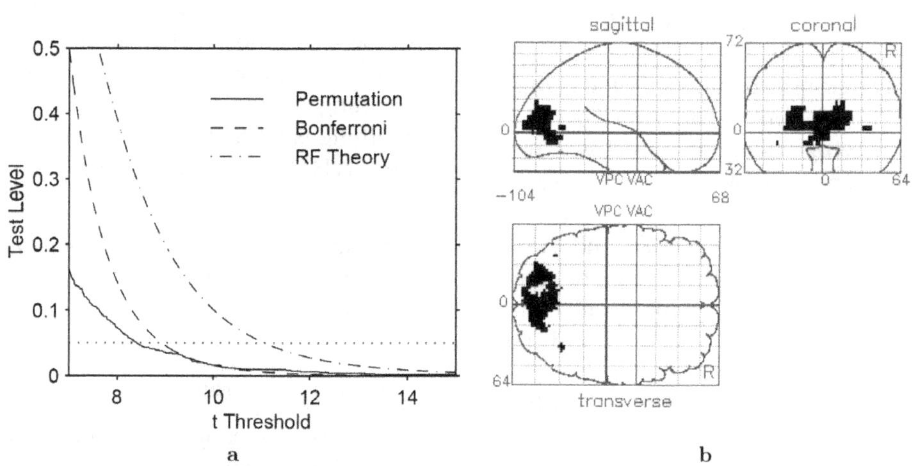

FIGURE 21.6 (a) Test significance (α) levels plotted against critical thresholds, for non-parametric and parametric analyses. (b) Maximum intensity projection of *t* image, thresholded at parametric 5 per cent level critical threshold of 11.07.

randomization test that technically excludes inference on a population. However, if we assume that the subjects of this study constitute a random sample of the population of interest, we can ignore the experimental randomization and perform a permutation test, as we do in the next example.

Multisubject fMRI: activation

For this third and final example, consider a multisubject fMRI activation experiment. Here we will perform a permutation test so that we can make inference on a population. We will use a smoothed variance t-statistic with a single threshold test and will make qualitative and quantitative comparisons with the parametric results.

Before discussing the details of this example, we note that fMRI data present a special challenge for non-parametric methods. Since fMRI data exhibit temporal autocorrelation (Smith *et al.*, 1999), an assumption of exchangeability of scans within subjects is not tenable. However, to analyse a group of subjects for population inference, we need only assume exchangeability of subjects. Hence, while intrasubject fMRI analyses are not straightforward with the permutation test, multisubject analyses are.

Study description

Marshuetz *et al.* (2000) studied order effects in working memory using fMRI. The data were analysed using a random-effects procedure (Holmes and Friston, 1999), as in the last example. For fMRI, this procedure amounts to a generalization of the repeated measures t-statistic.

There were 12 subjects, each participating in eight fMRI acquisitions. There were two possible presentation orders for each block, and there was randomization across blocks and subjects. The TR was two seconds, a total of 528 scans collected per condition. Of the study's three conditions we consider only two, item recognition and control. For item recognition, the subject was presented with five letters and, after a two second interval, presented with a probe letter. They were to respond 'yes' if the probe letter was among the five letters and 'no' if it was not. In the control condition, they were presented with five Xs and, two seconds later, presented with either a 'y' or an 'n'; they were to press 'yes' for y and 'no' for n.

Each subject's data were analysed, creating a difference image between the item recognition and control effects. These images were analysed with a one-sample t-test, yielding a random-effects analysis that accounts for intersubject differences.

Null hypothesis

While this study used randomization within and across subject and hence permits the use of a randomization test, we will use a permutation approach to generalize the results to a population (see above).

Again using a random-effects statistic, we only analyse each subject's item versus control difference image. We make the weak distributional assumption that the values of a subject's difference images at any given voxel (across subjects) are drawn from a symmetric distribution. (The distribution may be different at different voxels, so long as it is symmetric.) The null hypothesis is that these distributions are centred on zero:

$\mathcal{H}_0$: The symmetric distributions of the (voxel values of the) subjects' difference images have zero mean.

Exchangeability

The conventional assumption of independent subjects implies exchangeability, and hence a single EB consisting of all subjects.

Exchanging the item and control labels has exactly the effect of flipping the sign of the difference image. So we consider subject labels of '+1' and '−1', indicating an unflipped or flipped sign of the data. Under the null hypothesis, we have data symmetric about zero, and hence can randomly flip the signs of a subject's difference images.

Statistic

In this example we focus on statistic magnitude.

Voxel-level statistic As noted above, this analysis amounts to a one-sample t-test on the first level difference images, testing for a zero-mean effect across subjects. We use a pseudo t-test, with a variance smoothing of 4 mm FWHM, comparable to the original within-subject smoothing. In our experience, the use of *any* variance smoothing is more important than the particular magnitude (FWHM) of the smoothing.

Summary statistic Again we are interested in searching over the whole brain for significant changes, hence we use the maximum pseudo t.

Relabelling enumeration

Based on our exchangeability under the null hypothesis, we can flip the sign on some or all of our subjects' data. There are $2^{12} = 4096$ possible ways of assigning either '+1' or '−1' to each subject. We consider all 4096 relabellings.

Permutation distribution

For each relabelling we found the maximum pseudo t-statistic, yielding the distribution in Figure 21.7(a). As in the last example, we have a symmetry in these labels; we need only compute 2048 statistic images and save both the maxima and minima.

Significance threshold

With 4096 permutations, the 95th percentile is $4096 \times 0.05 = 452.3$, and hence the 453rd largest maxima defines the 0.05 level corrected significance threshold.

Results

The permutation distribution of the maximum pseudo t-statistic under $\mathcal{H}_0$ is shown in Figure 21.7(a). It is centred around 4.5 and is slightly positively skewed; all maxima found were between about 3 and 8.

The correctly labelled data yielded the largest maximum, 8.471. Hence the overall significance of the experiment is $1/4096 = 0.0002$. The dotted line indicates the 0.05 corrected threshold, 5.763. Figure 21.7(b) shows the thresholded maximum intensity projection (MIP) of significant voxels. There are 312 voxels in eight distinct regions; in particular there are bilateral posterior parietal regions, a left thalamic region and an anterior cingulate region; these are typical of working memory studies (Marshuetz *et al.*, 2000).

It is informative to compare this result to the traditional t-statistic, using both a non-parametric and parametric approach to obtain corrected thresholds. We reran this non-parametric analysis using no variance smoothing. The resulting thresholded data is shown in Figure 21.7(c);

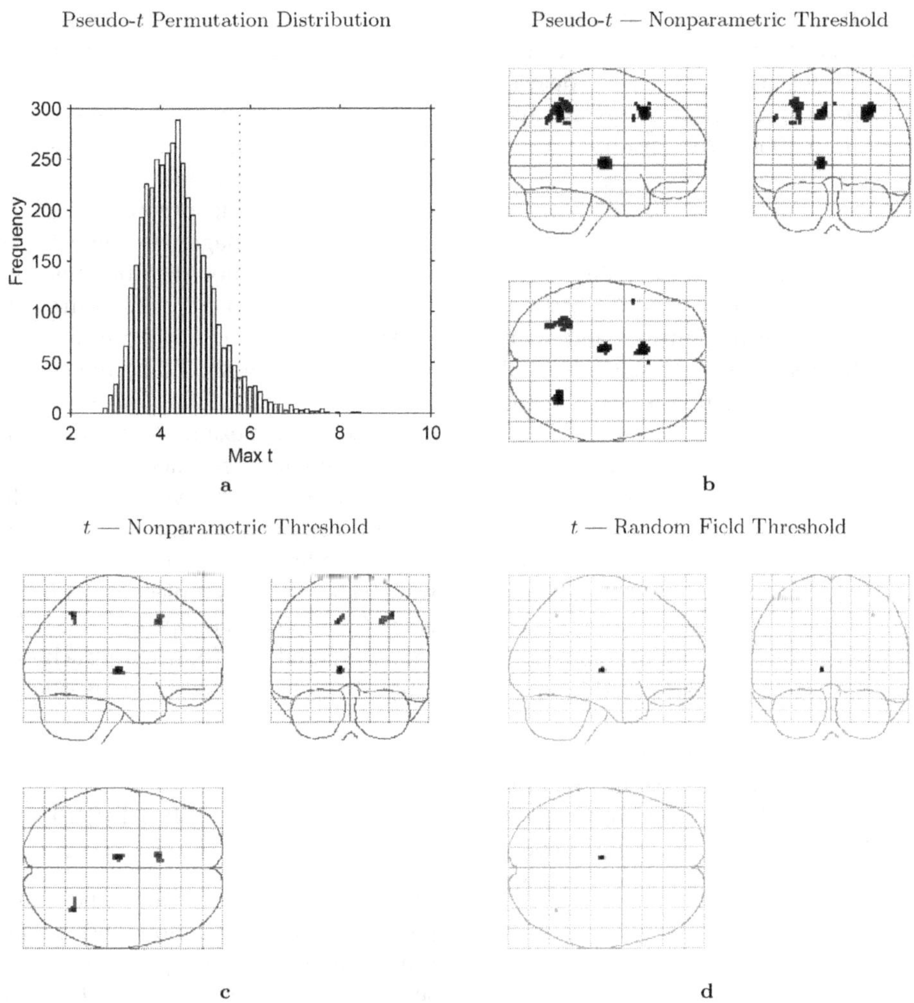

FIGURE 21.7 (a) Permutation distribution of maximum repeated-measures t-statistic. Dotted line indicates the 5 per cent level corrected threshold. (b) Maximum intensity projection of pseudo t-statistic image threshold at 5 per cent level, as determined by permutation distribution. (c) Maximum intensity projection of t-statistic image threshold at 5 per cent level as determined by permutation distribution. (d) Maximum intensity projection of t-statistic image threshold at 5 per cent level as determined by random field theory.

TABLE 21-1 Comparison of four inference methods for the item recognition fMRI data. The minimum corrected *p*-value and number of significant voxels give an overall measure of sensitivity; corrected thresholds can only be compared within statistic type. For these data, the Bonferroni and random field results are very similar, and the non-parametric methods are more powerful; the non-parametric *t* method detects 10 times as many voxels as the parametric method, and the non-parametric pseudo *t* detects 60 times as many

Statistic	Inference method	Corrected threshold		Minimum corrected *p*-value	Number of significant voxels
		t	Pseudo *t*		
t	Random field	9.870		0.0062	5
t	Bonferroni	9.802		0.0025	5
t	Permutation	7.667		0.0002	58
Pseudo *t*	Permutation		5.763	0.0002	312

there are only 58 voxels in three regions that exceeded the corrected threshold of 7.667. Using standard parametric random field methods produced the result in Figure 21.7(d). For 110 776 voxels of size $2 \times 2 \times 2$ mm, with an estimated smoothness of $5.1 \times 5.8 \times 6.9$ mm FWHM, the parametric theory finds a threshold of 9.870; there are only five voxels in three regions above this threshold. Note that only the pseudo *t*-statistic detects the bilateral parietal regions. Table 21-1 summarizes the three analyses along with the Bonferroni result.

Discussion

In this example, we have demonstrated the utility of the non-parametric method for intersubject fMRI analyses. Based only on independence of the subjects and symmetric distribution of difference images under the null hypothesis, we can create a permutation test that yields inferences on a population.

Multiple subject fMRI studies often have few subjects, many fewer than 20 subjects. By using the smoothed variance *t*-statistic we have gained sensitivity, relative to the standard *t*-statistic. Even with the standard *t*-statistic, the non-parametric test proved more powerful, detecting five times as many voxels as active. Although the smoothed variance *t* is statistically valid, it does not overcome any limitations of *face* validity of an analysis based on only 12 subjects.

We note that this relative ranking of sensitivity (non-parametric pseudo *t*, non-parametric *t*, parametric *t*) is consistent with the other second-level data sets we have analysed. We believe this is due to a conservativeness of the random *t*-field results under low degrees of freedom.

Discussion of examples

These examples have demonstrated the non-parametric permutation test for PET and fMRI with a variety of experimental designs and analyses. We have addressed each of the steps in sufficient detail to follow the algorithmic steps that the SPM software performs. We have shown that the ability to utilize smoothed variances via a pseudo *t*-statistic can offer increased power over a corresponding standard *t*-statistic image. Using standard *t*-statistics, we have seen how the permutation test can be used as a reference against which parametric results can be validated.

However, note that the comparison between parametric and non-parametric results must be made very carefully. Comparable models and statistics must be used, and multiple testing procedures with the same degree of control over image-wise type I error used. Further, since the permutation distributions are derived from the data, critical thresholds are specific to the data set under consideration. Although the examples presented above are compelling, it should be remembered that these are only a few specific examples. However, the points noted for these specific examples are indicative of our general experience with these methods.

Finally, while we have noted that the non-parametric method has greater computational demands than parametric methods, they are reasonable on modern hardware. The PET examples presented here would take about 5 minutes on a typical desktop PC, while the fMRI example could take longer, as much as 20 minutes due to more permutations (2048 versus. 500) and larger images.

CONCLUSIONS

In this chapter, the theory and practicalities of non-parametric randomization and permutation tests for functional neuroimaging experiments have been presented and illustrated with worked examples.

As has been demonstrated, the permutation approach offers various advantages. The methodology is intuitive and accessible. With suitable maximal summary statistics, the multiple testing problem can be accounted for easily; only minimal assumptions are required for valid inference, and the resulting tests are almost exact, with size at most $1/N$ less than the nominal test level α, where N is the number of relabellings.

The non-parametric permutation approaches described give results similar to those obtained from a comparable statistical parametric mapping approach using a general linear model with multiple testing corrections derived from random field theory. In this respect, these non-parametric techniques can be used to verify the validity of less computationally expensive parametric approaches. When the assumptions required for a parametric approach are not met, the non-parametric approach described provides a viable alternative.

In addition, the approach is flexible. Choice of voxel and summary statistic are not limited to those whose null distributions can be derived from parametric assumptions. This is particularly advantageous at low degrees of freedom, when noisy variance images lead to noisy statistic images and multiple testing procedures based on the theory of continuous random fields are conservative. By assuming a smooth variance structure, and using a pseudo t-statistic computed with smoothed variances, the permutation approach gains considerable power.

Therefore, the non-parametric permutation approach should be considered for experimental designs with low degrees of freedom. These include small-sample size problems, such as single-subject PET/SPECT (single photon emission computed tomography), but also PET/SPECT and fMRI multisubject and between-group analyses involving small numbers of subjects, where analysis must be conducted at the subject level to account for intersubject variability. It is our hope that this chapter, and the accompanying software, will encourage appropriate application of these non-parametric techniques.

REFERENCES

Arndt S, Cizadlo T, Andreasen NC *et al.* (1996) Tests for comparing images based on randomization and permutation methods. *J Cereb Blood Flow Metab* **16**: 1271–79

Belmonte M, Yurgelun-Todd D (2001) Permutation testing made practical for functional magnetic resonance image analysis. *IEEE Trans Med Imag* **20**: 243–48

Brammer MJ, Bullmore ET, Simmons A *et al.* (1997) Generic brain activation mapping in functional magnetic resonance imaging: a nonparametric approach. *Mag Res Med* **15**: 763–70

Bullmore E, Brammer M, Williams SCR *et al.* (1996) Statistical methods of estimation and inference for functional MR image analysis. *Mag Res Med* **35**: 261–77

Bullmore E, Long C, Suckling J (2001) Colored noise and computational inference in neurophysiological (fMRI) time series analysis: resampling methods in time and wavelet domains. *Hum Brain Mapp* **12**: 61–78

Bullmore ET, Suckling J, Overmeyer S *et al.* (1999) Global, voxel, and cluster tests, by theory and permutation, for a difference between two groups of structural MR images of the brain. *IEEE Trans Med Imag* **18**: 32–42

Cao J (1999) The size of the connected components of excursion sets of χ^2, t and f fields. *Adv Appl Prob* **31**: 577–93

Dwass M (1957) Modified randomization tests for nonparametric hypotheses. *Ann Math Stat* **28**: 181–87

Edgington ES (1964) Randomization tests. *J Psychol* **57**: 445–49

Edgington ES (1969a) Approximate randomization tests. *J Psychol* **72**: 143–49

Edgington ES (1969b) *Statistical inference: the distribution free approach.* McGraw-Hill, New York

Edgington ES (1995) *Randomization tests*, 3rd edn. Marcel Dekker, New York

Fadili J, Bullmore ET (2001) Wavelet-generalised least squares: a new blu estimator of regression models with long-memory errors. *NeuroImage* **15**: 217–32

Fisher RA (1935) *The design of experiments.* Oliver, Boyd, Edinburgh

Fisher RA, Bennett JH (1990) *Statistical methods, experimental design, and scientific inference.* Oxford University Press, Oxford

Forman SD, Cohen JD, Fitzgerald M *et al.* (1995) Improved assessment of significant activation in functional magnetic resonance imaging (fMRI): use of a cluster-size threshold. *Mag Res Med* **33**: 636–47

Friman O, Westin CF (2005) Resampling FMRI time series. *NeuroImage* **25**: 859–67

Friston KJ, Frith CD, Liddle PF *et al.* (1991) Comparing functional (PET) images: the assessment of significant change. *J Cereb Blood Flow Metab* **11**: 690–99

Friston KJ, Holmes AP, Poline J-B *et al.* (1996) Detecting activations in PET and fMRI: levels of inference and power. *NeuroImage* **4**: 223–35

Friston KJ, Holmes AP, Worsley KJ *et al.* (1995) Statistical parametric maps in functional imaging: a general linear approach. *Hum Brain Mapp* **2**: 189–210

Friston KJ, Worsley KJ, Frackowiak RSJ *et al.* (1994) Assessing the significance of focal activations using their spatial extent. *Hum Brain Mapp* **1**: 214–20

Good P (1994) *Permutation tests. A practical guide to resampling methods for testing hypotheses.* Springer-Verlag, New York

Grabowski TJ, Frank RJ, Brown CK *et al.* (1996) Reliability of PET activation across statistical methods, subject groups, and sample sizes. *Hum Brain Mapp*, **4**: 23–46

Halber M, Herholz K, Wienhard K *et al.* (1997) Performance of a randomization test for single-subject 15-O-water PET activation studies. *J Cereb Blood Flow Metab* **17**: 1033–39

Heckel D, Arndt S, Cizadlo T *et al.* (1998) An efficient procedure for permutation tests in imaging research. *Comput Biomed Res* **31**: 164–71

Holmes AP (1994) *Statistical issues in functional brain mapping.* PhD thesis, University of Glasgow, Glasgow. Available from http://www.fil.ion.ucl.ac.uk/spm/papers/APH_thesis

Holmes AP, Blair RC, Watson JDG *et al.* (1996) Nonparametric analysis of statistic images from functional mapping experiments. *J Cereb Blood Flow Metab* **16**: 7–22

Holmes AP, Friston KJ (1999) Generalisability, random effects & population inference. *NeuroImage* **7**: S754

Holmes AP, Watson JDG, Nichols TE (1998) Holmes & Watson, on 'Sherlock'. *Cereb Blood Flow Metab* **18**: S697 Letter to the editor, with reply.

Jöckel K-H (1986) Finite sample properties and asymptotic efficiency of monte-carlo tests. *Ann Stat* **14**: 336–47

Kendall M, Gibbons JD (1990) *Rank correlation methods*, 5th edn. Edward Arnold, London

Locascio JJ, Jennings PJ, Moore CI *et al.* (1997) Time series analysis in the time domain and resampling methods for studies of functional magnetic resonance brain imaging. *Hum Brain Mapp* **5**: 168–93

Logan BR, Rowe DB (2004) An evaluation of thresholding techniques in FMRI analysis. *NeuroImage* **22**: 95–108

Manly BFJ (1997) *Randomization, bootstrap and Monte-Carlo methods in biology*. Chapman & Hall, London

Marshuetz C, Smith EE, Jonides J *et al.* (2000) Order information in working memory: fMRI evidence for parietal and prefrontal mechanisms. *J Cogn Neurosci* **12/S2**: 130–44

Nichols TE, Hayasaka S (2003) Controlling the familywise error rate in functional neuroimaging: a comparative review. *Stat Meth Med Res* **12**: 419–46

Nichols TE, Holmes AP (2001) Nonparametric permutation tests for functional neuroimaging: a primer with examples. *Hum Brain Mapp* **15**: 1–25

Pitman EJG (1937a) Signficance tests which may be applied to samples from any population. *J R Stat Soc (Suppl)* **4**: 119–30

Pitman EJG (1937b) Signficance tests which may be applied to samples from any population. II. The correlation coefficient test. *J R Stat Soc (Suppl)* **4**: 224–32

Pitman EJG (1937c) Signficance tests which may be applied to samples from any population. III. The analysis of variance test. *Biometrika* **29**: 322–35

Poline JB, Mazoyer BM (1993) Analysis of individual positron emission tomography activation maps by detection of high signal-to-noise-ratio pixel clusters. *J Cereb Blood Flow Metab* **13**: 425–37

Poline JB, Worsley KJ, Evans AC *et al.* (1997) Combining spatial extent and peak intensity to test for activations in functional imaging. *NeuroImage* **5**: 83–96

Raz J, Zheng H, Ombao H *et al.* (2003) Statistical tests for FMRI based on experimental randomization. *NeuroImage* **19**: 226–32

Roland PE, Levin B, Kawashima R *et al.* (1993) Three-dimensional analysis of clustered voxels in 15-o-butanol brain activation images. *Hum Brain Mapp* **1**: 3–19

Silbersweig DA, Stern E, Frith C *et al.* (1995) A functional neuroanatomy of hallucinations in schizophrenia. *Nature* **378**: 167–69

Silbersweig DA, Stern E, Schnorr L *et al.* (1994) Imaging transient, randomly occurring neuropsychological events in single-subjects with positron emission tomography: an event-related count rate correlational analysis. *J Cereb Blood Flow Metab* **14**: 771–82

Smith AM, Lewis BK, Ruttimann UE *et al.* (1999) Investigation of low frequency drift in FMRI signal. *NeuroImage* **9**: 526–33

Watson JDG, Myers R, Frackowiak RSJ *et al.* (1993) Area v5 of the human brain: evidence from a combined study using positron emission tomography and magnetic resonance imaging. *Cereb Cort* **3**: 79–94

Westfall PH, Young SS (1993) *Resampling-based multiple testing: examples and methods for p-value adjustment*. John Wiley & Sons Inc, New York

Worsley KJ (1994) Local maxima and the expected Euler characteristic of excursion sets of χ^2, f and t fields. *Adv Appl Prob* **26**: 13–42

Worsley KJ (1996) The geometry of random images. *Chance* **9**: 27–40

Worsley KJ, Evans AC, Marrett S *et al.* (1992) A three-dimensional statistical analysis for CBF activation studies in human brain. *J Cereb Blood Flow Metab* **12**: 1040–42. See comment in *J Cereb Blood Flow Metab* **13**: 1040–42.

Worsley KJ, Evans AC, Strother SC *et al.* (1991) A linear spatial correlation model, with applications to positron emission tomography. *J Am Stat Assoc* **86**: 55–67

Worsley KJ, Marrett S, Neelin P *et al.* (1995) A unified statistical approach for determining significant signals in images of cerebral activation. *Hum Brain Mapp* **4**: 58–73

Yekutieli D, Benjamini Y (1999) Resampling-based false discovery rate controlling multiple test procedures for correlated test statistics. *J Stat Plann Inference* **82**: 171–96

Part 5

Bayesian inference

CHAPTER

22

Empirical Bayes and hierarchical models

K. Friston and W. Penny

INTRODUCTION

Bayesian inference plays a central role in many imaging neuroscience applications. Empirical Bayes is the estimation and inference framework associated with a Bayesian treatment of hierarchical models. Empirical Bayes and the inversion of hierarchical models can almost be regarded as the same thing. Empirical Bayes is important because nearly all the models we use in neuroimaging have some explicit or implicit hierarchical structure. Bayesian inference will play an increasing role in subsequent chapters and rests on the equality:

$$p(\theta, y) = p(y|\theta)p(\theta) \tag{22.1}$$

The first term specifies a generative model and an associated energy function, which we usually try to optimize with respect to the parameters of the model θ, given some data y. The model is specified in terms of the second two terms; the likelihood $p(y|\theta)$ and prior $p(\theta)$. However, in hierarchical models there are conditional independences among the model parameters that mean the probabilities on the right factorize:

$$p(\theta, y) = p(y|\theta^{(1)})p(\theta^{(1)}|\theta^{(2)}), \ldots, p(\theta^{(n)}) \tag{22.2}$$

There are many examples of these models, ranging from Markov models of hidden states that evolve over time to complicated graphical models, in which the parameters at one level control the distribution of parameters at lower levels, sometimes in a non-linear and possibly time-varying fashion. In this chapter, we will be looking at relatively simple hierarchical linear models. Irrespective of the exact form of hierarchical models, they all have intermediate terms $p(\theta^{(i)}|\theta^{(i+1)})$. These are *empirical* priors. They have an ambiguous role that is intermediate between the likelihood and priors (i.e. the first and last terms). On the one hand, like a full prior, they provide constraints on $\theta^{(i)}$, on the other, they specify the likelihood of $\theta^{(i)}$, given $\theta^{(i+1)}$. This means the empirical priors depend on $\theta^{(i+1)}$, which has to be inferred from the data, hence empirical Bayes. Empirical Bayes is critical to model specification and inversion because empirical priors embody formal constraints on the generation of observed data and these constraints can be used in a powerful way. The chapters in this section try to illustrate this point, starting with simple hierarchies of linear models and ending with complicated and informed spatio-temporal models of functional magnetic resonance imaging (fMRI) time-series.

This chapter revisits hierarchical observation models (see Chapter 11) used in functional neuroimaging. It emphasizes the common ground shared by classical and Bayesian methods to show that conventional analyses of neuroimaging data can be usefully extended within an *empirical* Bayesian framework. In particular, we formulate the procedures used in conventional data analysis in terms of hierarchical linear models and establish a connection between classical inference and parametric empirical Bayes (PEB) through covariance component estimation. This estimation is based on *expectation maximization* or EM (see Appendix 3). Hierarchical models not only provide for inference at the highest level, but allow one to revisit lower levels, to make Bayesian inferences. Bayesian inferences eschew many of the difficulties encountered with classical inference and characterize brain responses in a way that is more directly related to what one is interested in.

We start with a theoretical summary and then deal with applications of the theory to a range of issues in neuroimaging. These include: estimating non-sphericity or variance components in fMRI time-series that can arise from serial correlations within subject, or are induced by multisubject (i.e. hierarchical) studies; Bayesian models for imaging data, in which effects at one voxel are constrained by responses in others (see Chapters 23 and 25); and Bayesian estimation of dynamic models of brain

Statistical Parametric Mapping, by Karl Friston et al.
ISBN–13: 978-0-12-372560-8 ISBN–10: 0-12-372560-7

responses (see Chapter 34). Although diverse, all these estimation and inference problems are accommodated by the framework described next.

Classical and Bayesian inference

Since its inception, statistical parametric mapping (SPM) has proved useful for characterizing neuroimaging data sequences. However, SPM is limited because it is based on classical inference procedures. In this chapter, we introduce a more general framework, which places SPM in a broader context and points to alternative ways of characterizing and making inferences about regionally specific effects in neuroimaging. In particular, we formulate the procedures used in conventional data analysis in terms of hierarchical linear models and establish the connection between classical inference and *empirical* Bayesian inference through covariance component estimation. This estimation is based on the expectation maximization (or EM) algorithm.

Statistical parametric mapping entails the use of the general linear model and classical statistics, under parametric assumptions, to create a statistic (usually the t-statistic) at each voxel. Inferences about regionally specific effects are based on the ensuing image of t-statistics, the SPM$\{t\}$. The requisite distributional approximations for the peak height, or spatial extent, of voxel clusters, surviving a specified threshold, are derived using Gaussian random field theory (see Part 3). Random field theory enables the use of classical inference procedures, and the latitude afforded by the general linear model, to give a powerful and flexible approach to continuous, spatially extended data. It does so by protecting against family-wise false positives over all the voxels that constitute a search volume, i.e. it provides a way of adjusting the p-values, in the same way that a Bonferroni correction does for discrete data (Worsley, 1994; Friston *et al.*, 1995).

Despite its success, statistical parametric mapping has a number of limitations: the p-value, ascribed to a topological feature of the SPM, does not reflect the likelihood that the effect is present but simply the probability of getting the feature in the effect's absence. There are several shortcomings to this classical approach. First, one can never reject the alternate hypothesis (i.e. say that an activation has not occurred) because the probability that an effect is exactly zero is itself zero. This is problematic, for example, in trying to establish double dissociations or indeed functional segregation; one can never say one area responds to colour *but not motion* and another responds to motion *but not colour*. Secondly, because the probability of an effect being zero is vanishingly small, given enough scans or subjects one can always demonstrate a significant effect at every voxel. This fallacy of classical inference is relevant practically, with the thousands of scans that enter some fixed-effect analyses of fMRI data. The issue here is that trivially small activations can be declared significant if there are sufficient degrees of freedom to render their estimated variability small enough. A third problem, which is specific to SPM, is the correction or adjustment applied to the p-values to resolve the multiple comparison problem. This has the somewhat nonsensical effect of changing the inference about one part of the brain in a way that depends on whether another part is examined. Put simply, the threshold increases with search volume, rendering inference very sensitive to what it encompasses. Clearly, the probability that any voxel has activated does not change with the search volume and yet the p-value does.

All these problems would be eschewed by using the probability that a voxel had activated, or indeed its activation was greater than some threshold. This sort of inference is precluded by classical approaches, which simply give the likelihood of getting *the data, given no effect*. What one would like is the probability of *the effect given the data*. This is the *posterior* probability used in Bayesian inference. The Bayesian approach to significance testing in functional neuroimaging was introduced by Andrew Holmes and Ian Ford (Holmes and Ford, 1993) four years after SPM was invented. The Bayesian approach requires both the *likelihood*, afforded by assumptions about the distribution of errors, and the *prior* probability of activation. These priors can enter as known values or can be estimated from the data, provided we have observed multiple instances of the effect we are interested in. The latter is referred to as *empirical* Bayes and rests upon a hierarchical observation model. In many situations we do assess the same effect over different subjects, or indeed different voxels, and are in a position to adopt empirical Bayes. This chapter describes one such approach. In contrast to other proposals, this approach is not a novel way of analysing neuroimaging data. The use of Bayes for fMRI data has been usefully explored elsewhere (e.g. spatio-temporal Markov field models, Descombes *et al.*, 1998; and mixture models, Everitt and Bullmore, 1999). See also Hartvig and Jensen (2000) who combine both these approaches and Højen-Sørensen *et al.* (2000) who focus on temporal aspects with hidden Markov models. Generally, these approaches assume that voxels are either active or not and use the data to infer their status. Because of this underlying assumption, there is little connection with conventional models that allow for continuous or parametric responses. Our aim here is to highlight the fact that conventional models we use routinely conform to hierarchical observation models that can be treated in a Bayesian fashion. The importance of this rests on the connection between classical and Bayesian inference and the use of Bayesian procedures that are overlooked from a classical perspective. For example, random-effect analyses of fMRI data (Holmes and Friston, 1998, and see Chapter 12) adopt two-level hierarchical models. In this context, people generally focus on classical inference at the second level, unaware that the same model can support Bayesian inference at the first. Revisiting the first level, within a Bayesian framework, provides a much better characterization of single-subject responses, both in terms of the estimated effects and the nature of the inference. We will see an example of this later.

Overview

The aim of the first section below is to describe hierarchical observation models and establish the relationship between classical *maximum likelihood* (ML) and empirical Bayes estimators. Parametric empirical Bayes can be formulated in terms of covariance component estimation (e.g. within-subject versus between-subject contributions to error). The covariance component formulation is important because it is ubiquitous in fMRI and electroencephalography (EEG). Different sources of variability in the data induce non-sphericity that has to be estimated before any inferences about an effect can be made. Important sources of non-sphericity include serial or temporal correlations among errors in single-subject studies, or in multisubject studies, the differences between within- and between-subject variability. These issues are used in the second section to emphasize both the covariance component estimation and Bayesian perspectives, in terms of the difference between response estimates based on classical maximum likelihood estimators and the conditional expectations or modes, using a Bayesian approach.

In the next chapter, we use the same theory to elaborate hierarchical models that allow the construction of posterior probability maps. Again, this employs two-level models but focuses on Bayesian inference at the first level. It complements the preceding fMRI application by showing how global shrinkage priors can be estimated using observations *over voxels* at the second level. This is a special case of more sophisticated hierarchical models of fMRI data presented in Chapter 25, which use local shrinkage priors to enforce spatial constraints on the inversion.

THEORETICAL BACKGROUND

In this section, we focus on theory and procedures. The key points are reprised in subsequent sections where they are illustrated using real and simulated data. This section describes how the parameters and hyperparameters of a hierarchical model can be estimated given data. The distinction between a *parameter* and a *hyperparameter* depends on the context established by the inference. Here, parameters are quantities that determine the expected response, which is observed. Hyperparameters pertain to the probabilistic behaviour of the parameters. Perhaps the simplest example is provided by a single-sample t-test. The parameter of interest is the true effect causing the observations to differ from zero. The hyperparameter corresponds to the variance of the observation error (usually denoted by σ^2). Note that one can *estimate* the parameter, with the sample mean, without knowing the hyperparameter. However, if one wanted to make an *inference* about the estimate, it is necessary to know (or estimate using the residual sum of squares) the hyperparameter. In this chapter, all the hyperparameters are simply variances of different quantities that cause the measured response (e.g. within-subject variance and between-subject variance).

The aim of this section is to show the close relationship between Bayesian and maximum likelihood estimation implicit in conventional analyses of imaging data, using the general linear model. Furthermore, we want to place classical and Bayesian inference within the same framework. In this way, we show that conventional analyses are special cases of a more general parametric empirical Bayes (PEB) approach. First, we reprise hierarchical linear observation models that form the cornerstone of the ensuing estimation procedures. These models are then reviewed from the classical perspective of estimating the model parameters using maximum likelihood (ML) and statistical inference using the t-statistic. The same model is then considered in a Bayesian light to make an important point: the estimated error variances, at any level, play the role of priors on the variability of the parameters in the level below. At the highest level, the ML and Bayes estimators are the same.

Both classical and Bayesian approaches rest upon covariance component estimation using EM. This is described briefly and presented in detail in Appendix 3. The EM algorithm is related to that described in Dempster *et al.* (1981), but extended to cover hierarchical models with any number of levels. For an introduction to EM in generalized linear models, see Fahrmeir and Tutz (1994). This text provides an exposition of EM and PEB in linear models, usefully relating EM to classical methods (e.g. restricted maximum likelihood (ReML)). For an introduction to Bayesian statistics see Lee (1997). This text adopts a more explicit Bayesian perspective and again usefully connects empirical Bayes with classical approaches, e.g. the Stein 'shrinkage' estimator and empirical Bayes estimators used below (Lee, 1997). In many standard texts the hierarchical models considered here are referred to as random-effects models.

Hierarchical linear models

We will deal with hierarchical linear observation models of the form:

$$\begin{aligned} y &= X^{(1)}\theta^{(1)} + \varepsilon^{(1)} \\ \theta^{(1)} &= X^{(2)}\theta^{(2)} + \varepsilon^{(2)} \\ &\vdots \\ \theta^{(n-1)} &= X^{(n)}\theta^{(n)} + \varepsilon^{(n)} \end{aligned} \qquad 22.3$$

under Gaussian assumptions about the errors $\varepsilon^{(i)} \sim N(0, C_\varepsilon^{(i)})$. The response variable, y, is usually observed both within units over time and over several units (e.g. subject or voxels). $X^{(i)}$ are specified [design] matrices containing explanatory variables or constraints on the parameters $\theta^{(i-1)}$ of the level below. If the hierarchical model has only one level, it reduces to the familiar general linear model employed in conventional data analysis (see Chapter 8). Two-level models will be familiar to readers who use mixed- or random-effects analyses. In this instance, the first-level design matrix models the activation effects, over scans within subjects, in a subject-separable fashion (i.e. in partitions constituting the blocks of a block diagonal matrix). The second-level design matrix models the subject-specific effects over subjects. Usually, but not necessarily, the design matrices are block diagonal matrices where each block models the observations in each unit at that level (e.g. session, subject or group).

$$X^{(i)} = \begin{bmatrix} X_1^{(i)} & 0 & \cdots & 0 \\ 0 & X_2^{(i)} & & \\ \vdots & & \ddots & \\ 0 & & & X_J^{(i)} \end{bmatrix} \qquad 22.4$$

Some examples are shown in Figure 22.1 (these examples are used in the next section). The design matrix at any level has as many rows as the number of columns in the design matrix of the level below. One can envisage three-level models, which embody activation effects in scans modelled for each session, effects expressed in each session modelled for each subject and, finally, effects over subjects.

The Gaussian or parametric assumptions, implicit in these models, imply that all the random sources of variability, in the observed response, have a Gaussian distribution. This is appropriate for most models in neuroimaging and makes the relationship between classical approaches and Bayesian treatments (that can be generalized to non-Gaussian densities) much more transparent.

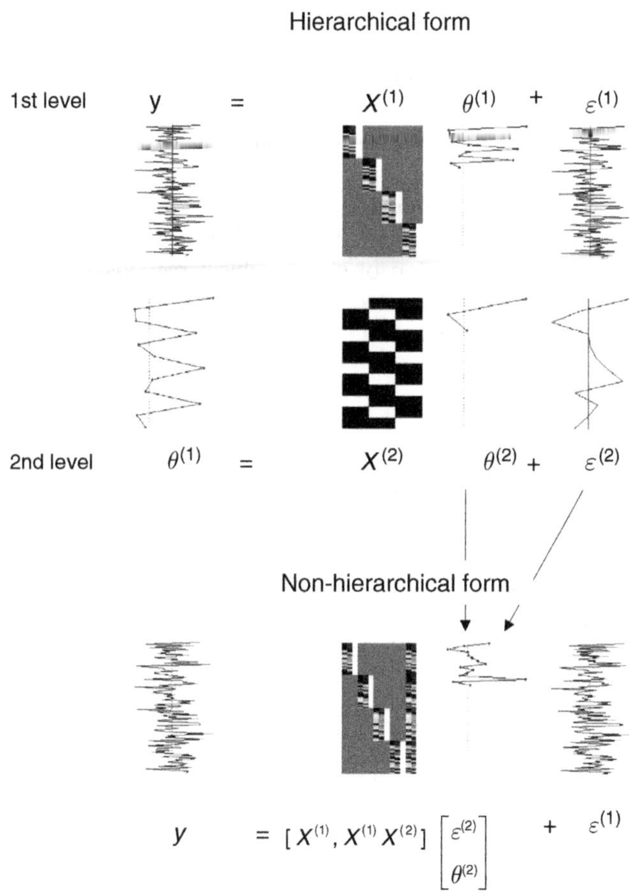

FIGURE 22.1 Schematic showing the form of the design matrices in a two-level model and how the hierarchical form (upper panel) can be reduced to a non-hierarchical form (lower panel). The design matrices are shown in image format with an arbitrary colour scale. The response variable, parameters and error terms are depicted as plots. In this example there are four subjects or units observed at the first level. Each subject's response is modelled with the same three effects, one of these being a constant term. These design matrices are part of those used in Friston *et al.* (2002b) to generate simulated fMRI data and are based on the design matrices used in the subsequent empirical event-related fMRI analyses.

Technically, models that conform to Eqn. **22.3** fall into the class of conditionally independent hierarchical models when the response variables and parameters are independent across units, conditionally on the hyperparameters controlling the error terms (Kass and Steffey, 1989). These models are also called *parametric empirical Bayes* (PEB) models because the obvious interpretation of the higher-level densities as priors led to the development of PEB methodology (Efron and Morris, 1973). Although the procedures considered in this chapter accommodate general models that are not conditionally independent, we refer to the Bayesian procedures below as PEB because the motivation is identical and most of the examples assume conditional independence. Having posited a model with a hierarchical form, the aim is to estimate its parameters and make some inferences about these estimates using their estimated variability or, more generally, find the posterior or conditional density of its parameters. In classical inference, one is usually only interested in inference about the parameters at the highest level to which the model is specified. In a Bayesian context, the highest level is regarded as providing constraints or empirical priors that enable posterior inferences about the parameters in lower levels. Identifying the system of equations in Eqn. **22.3** can proceed under two perspectives that are formally identical: a classical statistical perspective and a Bayesian one.

After recursive substitution, to eliminate all but the final level parameters, Eqn. **22.3** can be written as:

$$y = \varepsilon^{(1)} + X^{(1)}\varepsilon^{(2)} + \cdots + X^{(1)} \dots X^{(n-1)}\varepsilon^{(n)} + X^{(1)} \dots X^{(n)}\theta^{(n)} \tag{22.5}$$

In this non-hierarchical form the components of the response comprise linearly separable contributions from all levels. Those components are referred to as *random effects* $\tilde{C}_\varepsilon$ where the last-level parameters enter as *fixed effects*. The covariance partitioning implied by Eqn. **22.5** is:

$$\begin{aligned} E(yy^T) &= \tilde{C}_\varepsilon + X^{(1)} \dots X^{(n)}\theta^{(n)}\theta^{(n)T}X^{(n)T} \dots X^{(1)T} \\ \tilde{C}_\varepsilon &= C_\varepsilon^{(1)} + \cdots + X^{(1)} \dots X^{(n-1)}C_\varepsilon^{(i)}X^{(n-1)T} \dots X^{(1)T} \end{aligned} \tag{22.6}$$

where $C_\varepsilon^{(i)}$ is the covariance of $\epsilon^{(i)}$. If only one level is specified, the random effects vanish and a fixed-effect analysis ensues. If n is greater than one, the analysis corresponds to a random-effect analysis (or more exactly a *mixed-effect analysis* that includes random terms). Eqn. **22.5** can be interpreted in two ways that form respectively the basis for a classical:

$$\begin{aligned} y &= \tilde{X}\theta^{(n)} + \tilde{\varepsilon} \\ \tilde{X} &= X^{(1)}X^{(2)} \dots X^{(n)} \\ \tilde{\varepsilon} &= \varepsilon^{(1)} + X^{(1)}\varepsilon^{(2)} + \cdots + X^{(1)}X^{(2)} \dots X^{(n-1)}\varepsilon^{(n)} \end{aligned} \tag{22.7}$$

and Bayesian estimation:

$$\begin{aligned} y &= X\theta + \varepsilon^{(1)} \\ X &= \left[X^{(1)}, \dots, X^{(1)}X^{(2)} \dots X^{(n-1)}, X^{(1)}X^{(2)} \dots X^{(n)}\right] \\ \theta &= \begin{bmatrix} \varepsilon^{(2)} \\ \vdots \\ \varepsilon^{(n)} \\ \theta^{(n)} \end{bmatrix} \end{aligned} \tag{22.8}$$

In the first formulation, the random effects are lumped together and treated as a compound error, rendering the last-level parameters the only parameters to appear explicitly. Inferences about n-th level parameters are obtained by simply specifying the model to the order required. In contradistinction, the second formulation treats the error terms as parameters, so that θ comprises the errors at all levels and the final-level parameters. Here we have effectively collapsed the hierarchical model into a single level by treating the error terms as parameters (see Figure 22.1 for a graphical depiction).

A classical perspective

From a classical perceptive, Eqn. **22.7** represents an observation model with response variable y, design matrix $\tilde{X}$ and parameters $\theta^{(n)}$. Classically, estimation proceeds using the maximum likelihood (ML) estimator of the final-level parameters. Under our model assumptions, this is the Gauss-Markov estimator:

$$\begin{aligned} \eta_{ML} &= My \\ M &= (\tilde{X}^T C_{\tilde{\varepsilon}}^{-1}\tilde{X})^{-1}\tilde{X}^T C_{\tilde{\varepsilon}}^{-1} \end{aligned} \tag{22.9}$$

where M is a matrix that projects the data onto the estimate. Inferences about this estimate are based upon its covariance, against which any contrast (i.e. linear mixture specified by the contrast weight vector c) can be compared using the t-statistic:

$$\begin{aligned} t &= \frac{c^T\eta_{ML}}{\sqrt{c^T Cov\{\eta_{ML}\}c}} \\ Cov\{\eta_{ML}\} &= M\tilde{C}_\varepsilon M^T = (\tilde{X}^T\tilde{C}_\varepsilon^{-1}\tilde{X})^{-1} \end{aligned} \tag{22.10}$$

The covariance of the ML estimator represents a mixture of covariances of lower-level errors projected to the highest level. To implement this classical procedure we need the random effects $\tilde{C}_\varepsilon = Cov\{\tilde{\varepsilon}\}$ projected down the hierarchy onto the response or observation space. In other words, we need the error covariance components of the model, $C_\varepsilon^{(i)}$ from Eqn. **22.6**. To estimate these one has to turn to the second formulation, Eqn. **22.8**, and some iterative procedure (i.e. EM). This covariance component estimation reflects the underlying equivalence between classical and empirical Bayes methods. There are special cases where one does not need to resort to iterative covariance component estimation, e.g. single-level models. With balanced designs, where $X_1^{(i)} = X_j^{(i)}$ for all i and j, one can replace the response variable with the ML estimates at the penultimate level and proceed as if one had a single-level model. This is used in summary-statistic implementation of random-effect analyses (Holmes and Friston, 1998, see Chapter 12).

In summary, parameter estimation and inference, in hierarchical models, can proceed given estimates of the hierarchical covariance components. The reason for introducing inference based on ML is to establish the central role of covariance component estimation. In the next section, we take a Bayesian approach to the same issue.

A Bayesian perspective

Bayesian inference is based on the conditional probability of the parameters given the data $p(\theta^{(i)}|y)$. Under the assumptions above, this conditional density is Gaussian and the problem reduces to finding its first two moments, the conditional mean $\eta_{\theta|y}^{(i)}$ and covariance $C_{\theta|y}^{(i)}$. These densities are determined for all levels, enabling inference at any level, using the same hierarchical model. Given the posterior density we can work out the maximum *a posteriori* (MAP) estimate of the parameters (a point estimator equivalent to $\eta_{\theta|y}^{(i)}$ for the linear systems considered here) or the probability that the parameters exceed some specified value. Consider Eqn. **22.3** from a Bayesian point of view. Here, level i can be thought of as providing *prior* constraints on the expectation and covariances of the parameters below:

$$E\{\theta^{(i-1)}\} = \eta_\theta^{(i-1)} = X^{(i)}\theta^{(i)}$$
$$Cov\{\theta^{(i-1)}\} = C_\theta^{(i-1)} = C_\varepsilon^{(i)} \qquad \mathbf{22.11}$$

In other words, the parameters at any level play the role of hyperparameters for the subordinate level that control the prior expectation under the constraints specified by $X^{(i)}$. Similarly, the prior covariances are specified by the error covariances of the level above. For example, given several subjects, we can use information about the distribution of activations, over subjects, to inform an estimate pertaining to any single subject. In this case, the between-subject variability, from the second level, enters as a *prior* on the parameters of the first level. In many instances, we measure the same effect repeatedly in different contexts. The fact that we have some handle on this effect's inherent variability means that the estimate for a single instance can be constrained by knowledge about others. At the final level we can treat the parameters as unknown, in which case their priors are flat[1] (cf. fixed effects) giving an empirical Bayesian approach, or known. In the latter case, the connection with the classical formulation is lost because there is nothing to make an inference about at the final level.

The objective is to estimate the conditional means and covariances of lower-level parameters in a way that is consistent with information from higher levels. All the information we require is contained in the conditional mean and covariance of θ from Eqn. **22.8**. From Bayes' rule, the posterior probability is proportional to the likelihood of obtaining the data, conditional on θ, times the prior probability of θ;

$$p(\theta|y) \propto p(y|\theta)p(\theta) \qquad \mathbf{22.12}$$

where the Gaussian priors $p(\theta)$ are specified in terms of their expectation and covariance:

$$\eta_\theta = E\{\theta\} = \begin{bmatrix} 0 \\ \vdots \\ 0 \\ \eta_\theta^{(n)} \end{bmatrix} \quad C_\theta = Cov\{\theta\} = \begin{bmatrix} C_\varepsilon^{(2)} & \cdots & 0 & 0 \\ \vdots & \ddots & \vdots & \vdots \\ 0 & \cdots & C_\varepsilon^{(n)} & 0 \\ 0 & \cdots & 0 & C_\theta^{(n)} \end{bmatrix} \qquad \mathbf{22.13}$$

Under Gaussian assumptions the likelihood and priors are given by:

$$p(y|\theta) \propto \exp(-\frac{1}{2}(X\theta - y)^T C_\varepsilon^{(1)-1}(X\theta - y)) \qquad \mathbf{22.14}$$
$$p(\theta) \propto \exp(-\frac{1}{2}(\theta - \eta_\theta)^T C_\theta^{-1}(\theta - \eta_\theta))$$

Substituting Eqn. **22.14** into Eqn. **22.12** gives a posterior density with a Gaussian form:

$$p(\theta|y) \propto \exp(-\frac{1}{2}(\theta - \eta_{\theta|y})^T C_{\theta|y}^{-1}(\theta - \eta_{\theta|y})) \qquad \mathbf{22.15}$$
$$C_{\theta|y} = (X^T C_\varepsilon^{(1)-1} X + C_\theta^{-1})^{-1}$$
$$\eta_{\theta|y} = C_{\theta|y}(X^T C_\varepsilon^{(1)-1} y + C_\theta^{-1}\eta_\theta)$$

Note that when we adopt an empirical Bayesian scheme $C_\theta^{(n)} = \infty$ and $C_\theta^{-1}\eta_\theta = 0$. This means we never have to specify the prior expectation at the last level because it never appears explicitly in Eqn. **22.15**.

The solution Eqn. **22.15** is ubiquitous in the estimation literature and is presented under various guises in different contexts. If the priors are flat, i.e. $C_\theta^{-1} = 0$, the expression for the conditional mean reduces to the minimum variance linear estimator, referred to as the *Gauss-Markov* estimator. The Gauss-Markov estimator is identical to the ordinary least square (OLS) estimator that obtains after pre-whitening. If the errors are assumed to be independently and identically distributed, i.e. $C_\varepsilon^{(1)} = I$, then Eqn. **22.15** reduces to the ordinary least square estimator. With non-flat priors, the form of Eqn. **22.15** is identical to that employed by *ridge regression* and [weighted] *minimum norm* solutions (e.g. Tikhonov and Arsenin, 1977) commonly found in the inverse solution literature. The Bayesian perspective is useful for minimum norm formulations because it motivates plausible forms for the constraints that can be interpreted in terms of priors.

[1] Flat or uniform priors denote a probability distribution that is the same everywhere, reflecting a lack of any predilection for specific value. In the limit of very high variance, a Gaussian distribution becomes flat.

Equation **22.15** can be expressed in an equivalent but more compact (Gauss-Markov) form by augmenting the design matrix with an identity matrix and augmenting the data matrix with the prior expectations such that:

$$
\begin{aligned}
C_{\theta|y} &= (\overline{X}^T C_\varepsilon^{-1} \overline{X})^{-1} \\
\eta_{\theta|y} &= C_{\theta|y}(\overline{X}^T C_\varepsilon^{-1} \overline{y}) \\
\overline{y} = \begin{bmatrix} y \\ \eta_\theta \end{bmatrix} \overline{X} &= \begin{bmatrix} X \\ I \end{bmatrix} C_\varepsilon = \begin{bmatrix} C_\varepsilon^{(1)} & 0 \\ 0 & C_\theta \end{bmatrix}
\end{aligned}
\qquad \textbf{22.16}
$$

Figure 22.2 shows a schematic illustration of the linear model implied by this augmentation. If the last-level priors are flat, the last-level prior expectation can be set to zero. This augmented form is computationally more efficient to deal with and simplifies the exposition of the EM algorithm. Furthermore, it highlights the fact that a Bayesian scheme of this sort can be reformulated as a simple weighted least square or ML problem. The problem now reduces to estimating the error covariances C_ε that determine the weighting. This is exactly where we ended up in the classical approach, namely reduction to a covariance component estimation problem.

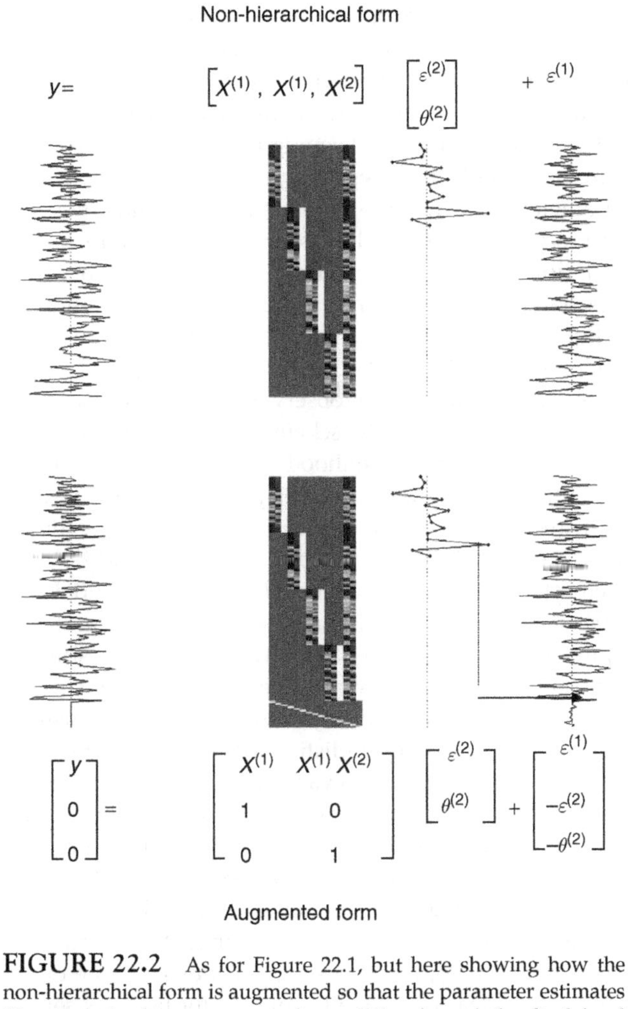

FIGURE 22.2 As for Figure 22.1, but here showing how the non-hierarchical form is augmented so that the parameter estimates (that include the error terms from all levels and the final level parameters) now appear in the model's residuals. A Gauss-Markov estimator will minimize these residuals in proportion to their prior precision.

Covariance component estimation

The classical approach was portrayed above as using the error covariances to construct an appropriate statistic. The PEB approach was described as using the error covariances as priors to estimate the conditional means and covariances; note from Eqn. **22.13** that $C_\theta^{(i-1)} = C_\varepsilon^{(i)}$. Both approaches rest on estimating the covariance components. This estimation depends upon some parameterization of these components; in this chapter, we use $C_\varepsilon^{(i)} = \sum \lambda_j^{(i)} Q_j^{(i)}$ where $\lambda_j^{(i)}$ are hyperparameters and $Q_j^{(i)}$ represent components of the covariance matrices. The components can be construed as constraints on the prior covariance structures in the same way as the design matrices $X^{(i)}$ specify constraints on the prior expectations. $Q_j^{(i)}$ embodies the form of the j-th covariance component at the i-th level and can model different variances for different levels and different forms of correlations within levels. The components Q_j are chosen to model the sort of non-sphericity anticipated. For example, they could specify serial correlations within-subject or correlations among the errors induced hierarchically, by repeated measures over subjects (Figure 22.3 illustrates both these examples). We will illustrate a number of forms for Q_j in the subsequent sections.

One way of thinking about these covariance components is in terms of the Taylor expansion of any function of hyperparameters that produced the covariance structure:

$$
C(\lambda)_\varepsilon^{(i)} = \sum \lambda_j^{(i)} \frac{\partial C_\varepsilon^{(i)}}{\partial \lambda_j^{(i)}} + \cdots \qquad \textbf{22.17}
$$

where the components correspond to the partial derivatives of the covariances with respect to the hyperparameters. In variance component estimation, the high-order terms are generally zero. In this context, a linear decomposition of $C_\varepsilon^{(i)}$ is a natural parameterization because the different sources of conditionally independent variance add linearly and the constraints can be specified directly in terms of these components. There are other situations where a different parameterization may be employed. For example, if the constraints were implementing several independent priors in a non-hierarchical model, a more natural expansion might be in terms of the precision $C_\theta^{-1} = \sum \lambda_j Q_j$. The precision is simply the inverse of the covariance matrix. Here Q_j correspond to precisions specifying the form of independent prior densities. However,

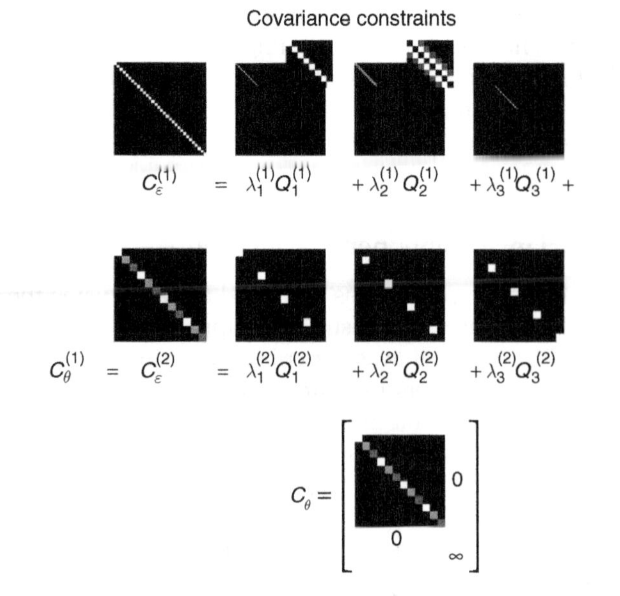

FIGURE 22.3 Schematic illustrating the form of the covariance components. These can be thought of as 'design matrices' for the second-order behaviour of the response variable and form a basis set for estimating the error covariance and implicitly the empirical prior covariances. The hyperparameters scale the contribution to the covariances of each component. These components correspond to the model in Figure 22.1. The top row depicts the constraints on the errors. For each subject there are two components, one modelling white (i.e. independent) errors and another serial correlation with an AR(1) form. The second level components simply reflect the fact that each of the three parameters estimated on the basis of repeated measures at the first level has its own variance. The estimated priors at each level are assembled with the prior for the last level (here a flat prior) to specify completely the model's empirical priors (lower panel). Components of this form are used in Friston *et al.* (2002b) during the simulation of serially correlated fMRI data-sequences and covariance component estimation using real data.

in this chapter, we deal with linearly mixed variance components that are induced by the hierarchical model. See Harville (1977) for comments on the usefulness of making the covariances linear in the hyperparameters and Appendix 4 for the usefulness of making the precisions linear in the hyperparameters.

The augmented form of the covariance components obtains by placing them in the appropriate partition in relation to the augmented error covariance matrix:

$$C_\varepsilon = Q_\theta + \sum \lambda_k Q_k$$

$$Q_\theta = \begin{bmatrix} 0 & \dots & 0 & 0 \\ \vdots & \ddots & \vdots & \vdots \\ 0 & \cdots & 0 & 0 \\ 0 & \cdots & 0 & C_\theta^{(n)} \end{bmatrix} \quad Q_k = \begin{bmatrix} 0 & & \dots & & 0 & 0 \\ & \ddots & & & & \\ \vdots & & Q_j^{(i)} & & \vdots & \vdots \\ & & & \ddots & & \\ 0 & & \dots & & 0 & 0 \\ 0 & & \dots & & 0 & 0 \end{bmatrix} \qquad 22.18$$

where the subscript k runs over both levels and components within each level. Having framed the covariance estimation in terms of estimating hyperparameters, we can now use EM for their estimation.

Expectation maximization

Expectation maximization is a generic, iterative parameter re-estimation procedure that encompasses many iterative schemes devised to estimate the parameters and hyperparameters of a model (Dempster *et al.*, 1977, 1981). It was originally introduced as an iterative method to obtain maximum likelihood estimators in incomplete data situations (Hartley, 1958) and was generalized by Dempster *et al.* (1977). More recently, it has been formulated (e.g. Neal and Hinton, 1998) in a way that highlights its connection to statistical mechanics (see Appendix 4). This formulation considers EM as a coordinate descent on the *free energy* of a system. The descent comprises an *E*-step, which finds the conditional *E*xpectation of the parameters, holding the hyperparameters fixed and an *M*-step, which updates the *M*aximum likelihood estimate of the hyperparameters, keeping the parameters fixed.

In brief, EM provides a way to estimate both the parameters and hyperparameters from the data. For linear models under Gaussian assumptions, EM returns the posterior density of the parameters, in terms of their expectation and covariance and restricted ML estimates of the hyperparameters. The EM algorithm described in Appendix 3 is depicted schematically in Figure 22.4. In the context of the linear observation models discussed in this chapter, this EM scheme is the same as using restricted maximum likelihood (ReML) estimates of the hyperparameters. ReML properly accounts for the loss of degrees of freedom incurred by parameter estimation. The formal equivalence between ReML and EM has been established for many years (see Fahrmeir and Tutz, 1994, p. 226). However, it is useful to understand the equivalence because EM is usually employed to estimate the conditional densities of model parameters when the hyperparameters of the likelihood and prior densities are not known. In contradistinction, ReML is generally used to estimate unknown variance components without explicit reference to the parameters. In the hierarchical linear observation model considered here, the unknown hyperparameters are variance components, which can be estimated using ReML. It should be noted that EM is not restricted to linear observation models or Gaussian priors, and has found diverse applications. On the other hand, ReML was developed explicitly for linear observation models, under Gaussian assumptions.

In Appendices 3 and 4, we have made an effort to reconcile the free-energy formulation based on statistical

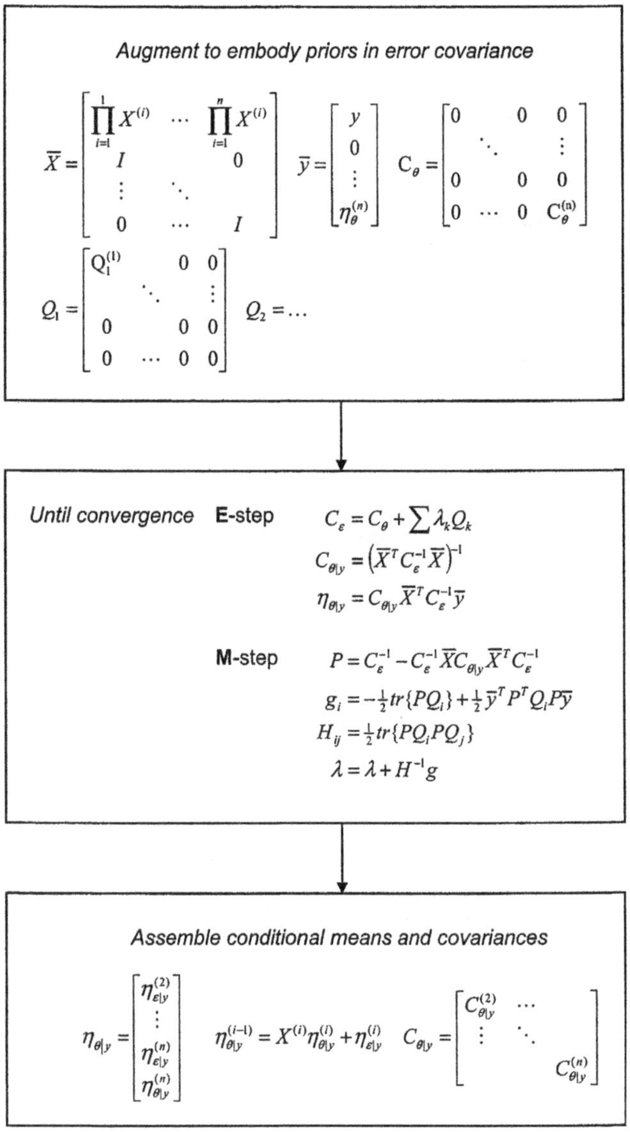

FIGURE 22.4 Pseudo-code schematic showing the recursive structure of the EM algorithm (described in Appendix 3) as applied in the context of conditionally independent hierarchical models. See main text for a full explanation. This formulation follows Harville (1977).

mechanics with classical ReML (Harville, 1977). This helps understand ReML in the context of extensions within the free-energy formulation, afforded by the use of hyperpriors (priors on the hyperparameters). One key insight into EM is that the M-step returns, not simply the ML estimate of the hyperparameters, but the *restricted* ML that is properly restricted from a classical perspective.

Having computed the conditional mean and covariances of the parameters using EM, one can make inferences about the effects at any level using their posterior density.

The conditional density

Given an estimate of the error covariance of the augmented form C_ε and implicitly the priors it entails, one can compute the conditional mean and covariance at each level where the conditional means for each level obtain recursively from:

$$\eta_{\theta|y} = E(\theta|y) = \begin{bmatrix} \eta_{\varepsilon|y}^{(2)} \\ \vdots \\ \eta_{\varepsilon|y}^{(n)} \\ \eta_{\theta|y}^{(n)} \end{bmatrix} \qquad 22.19$$

$$\eta_{\theta|y}^{(i-1)} = E(\theta^{(i-1)}|y) = X^{(i)} \eta_{\theta|y}^{(i)} + \eta_{\varepsilon|y}^{(i)}$$

These conditional expectations represent a better characterization of the model parameters than the equivalent ML estimates because they are constrained by higher levels (see summary). However, the conditional mean and ML estimators are the same at the last level. This convergence of classical and Bayesian inference rests on adopting an empirical Bayesian approach and establishes a close connection between classical random effect analyses and hierarchical Bayesian models. The two approaches diverge if we consider that the real power of Bayesian inference lies in coping with incomplete data or unbalanced designs and inferring on the parameters of lower levels. These are the issues considered in the next section.

Summary

This section has introduced three key components that play a role in the estimation of the linear models: Bayesian estimation; hierarchical models; and EM. The summary points below attempt to clarify the relationships among these components. It is worth while keeping in mind there are essentially three sorts of estimation: fully Bayesian, when the priors are known; empirical Bayes, when the priors are unknown but they can be parameterized in terms of some hyperparameters estimated from the data; and maximum likelihood estimation, when the priors are assumed to be flat. In the final instance, the ML estimators correspond to weighted least square or minimum norm solutions. All these procedures can be implemented with EM (Figure 22.5).

- Model estimation and inference are enhanced by being able to make probabilistic statements about model parameters given the data, as opposed to probabilistic statements about the data under some arbitrary assumptions about the parameters (e.g. the null hypothesis), as in classical statistics. The former is predicated on the posterior or conditional distribution of the parameters that is derived using Bayes' rule.

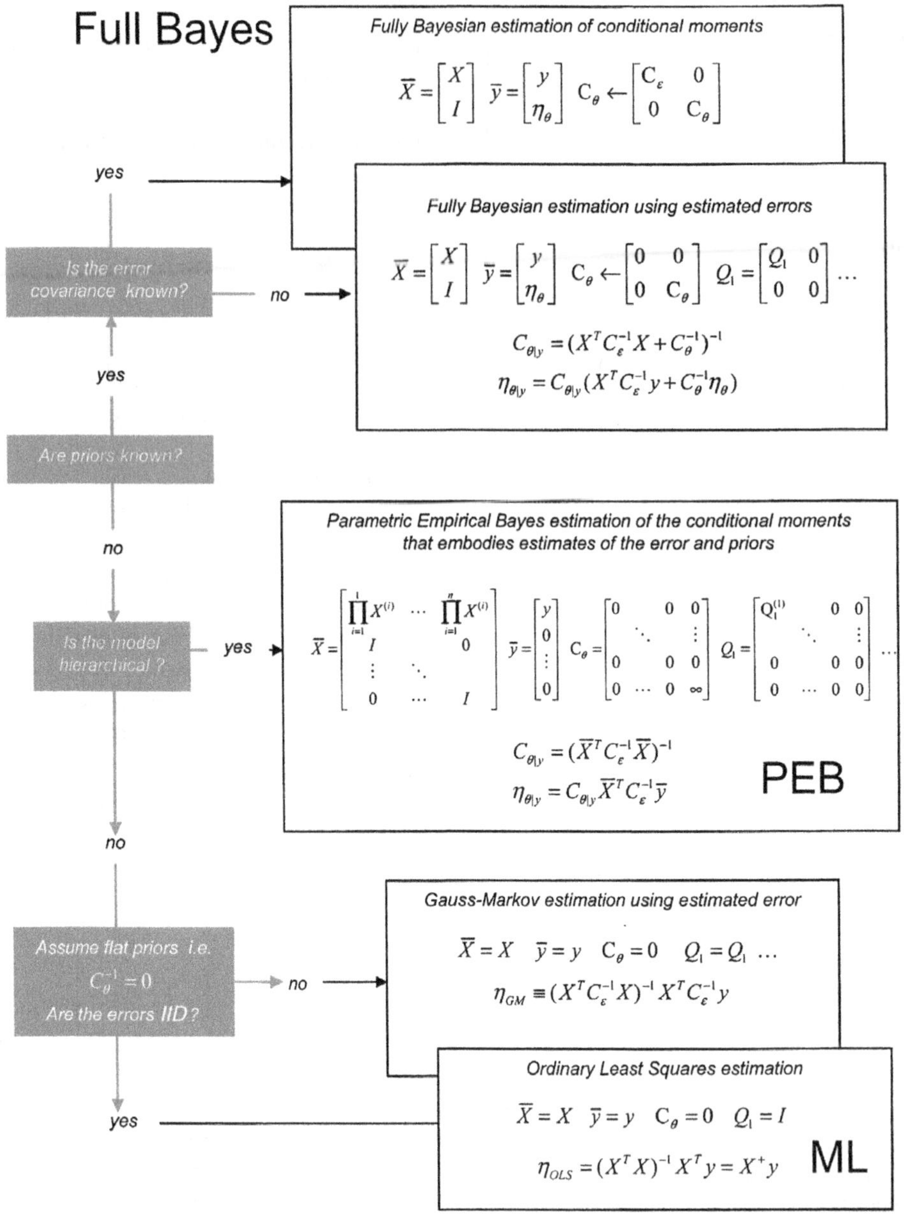

FIGURE 22.5 Schematic showing the relationship among estimation schemes for linear observation models under parametric assumptions. This figure highlights the universal role of EM, showing that all conventional estimators can be cast in terms of, or implemented with, the EM algorithm in Figure 22.4.

- Bayesian estimation and inference require priors. If the priors are known, then a fully Bayesian estimation can proceed. In the absence of known priors, there may be constraints on the form of the model that can be harnessed using *empirical* Bayes estimates of the associated hyperparameters.
- A model with a hierarchical form embodies implicit constraints on the form of the prior distributions. Hyperparameters that, in conjunction with these constraints, specify empirical priors can then be estimated, using EM to invert the model. The inversion of linear models under Gaussian assumptions corresponds to parametric empirical Bayes (PEB). In short, a hierarchical form for the observation model enables an empirical Bayesian approach.
- If the observation model does not have a hierarchical structure then one knows nothing about the form of the priors, and they are assumed to be flat. Bayesian estimation with flat priors reduces to maximum likelihood estimation.
- In the context of empirical Bayes, the priors at the last level are generally unknown and enter as flat priors.

This is equivalent to treating the parameters at the last level as fixed effects (i.e. effects with no intrinsic or random variability). One consequence of this is that the conditional mean and the ML estimate, at the last level, are identical.

- At the last level, PEB and classical approaches are formally identical. At subordinate levels PEB can use the posterior densities for Bayesian inference about the effects of interest. This is precluded in classical treatments because there are no empirical priors.
- EM provides a generic framework in which full Bayes, PEB or ML estimation can proceed. Its critical utility in this context is the estimation of covariance components, given some data, through the ReML estimation of hyperparameters mixing these covariance components. EM can be applied to hierarchical models by augmenting the design matrix and data (see Figures 22.2 and 22.4) to convert a hierarchical inverse problem into a non-hierarchical form. We will see later (Appendix 4) that EM is a special case of variational Bayes and that ReML is a special case of EM.
- In the absence of priors, or hierarchical constraints, EM can be used in an ML setting to estimate the error covariance for Gauss-Markov estimates (see Figure 22.5). These estimators are the optimum weighted least square estimates in the sense they have the minimum variance of all unbiased linear estimators. In the limiting case that the covariance constraints reduce to a single basis (synonymous with known correlations or a single hyperparameter), the EM scheme converges in a single iteration and emulates a classical sum of square estimation of error variance. When this single basis is the identity matrix, EM simply implements ordinary least squares.

In this section, we have reviewed hierarchical observation models of the sort commonly encountered in neuroimaging. Their hierarchical nature induces different sources of variability in the observations at different levels (i.e. variance components) that can be estimated using EM. The use of EM, for variance component estimation, is not limited to hierarchical models, but finds a useful application whenever non-sphericity of the errors is specified with more than one hyperparameter (e.g. serial correlations in fMRI). This application will be illustrated next. The critical thing about hierarchical models is that they enable empirical Bayes, where variance estimates at higher levels can be used as constraints on the estimation of effects at lower levels. This perspective rests upon exactly the same mathematics that pertains to variance component estimation in non-hierarchical models, but allows one to frame the estimators in conditional or Bayesian terms. An intuitive understanding of the conditional estimators, at a given level, is that they 'shrink' towards their average, in proportion to the error variance at that level, relative to their intrinsic variability (error variance at the supraordinate level). See Lee (1997: 232) for a discussion of PEB and Stein 'shrinkage' estimators.

In what sense are these Bayes predictors a better characterization of the model parameters than the equivalent ML estimates? In other words, what is gained in using a shrinkage estimator? This is a topic that has been debated at great length in the statistics literature and even in the popular press (see the *Scientific American* article 'Stein's paradox in statistics', Efron and Morris, 1977). The answer depends on one's definition of 'better', or in technical terms, the *loss function*. If the aim is to find the best predictor for a specific subject, then one can do no better than the ML estimator for that subject. Here the loss function is simply the squared difference between the estimated and real effects for the subject in question. Conversely, if the loss function is averaged over subjects then the shrinkage estimator is best. This has been neatly summarized in a discussion chapter read before the Royal Statistical Society entitled 'Regression, prediction and shrinkage' by Copas (1983). The vote of thanks was given by Dunsmore, who said:

> Suppose I go to the doctor with some complaint and ask him to predict the time, y, to remission. He will take some explanatory measurements $\mathbf{x}$ and provide some prediction for y. What I am interested in is a prediction for my $\mathbf{x}$, not for any other $\mathbf{x}$ that I might have had – but did not. Nor am I really interested in his necessarily using a predictor which is 'best' over all possible $\mathbf{x}$s. Perhaps rather selfishly, but I believe justifiably, I want the best predictor for my $\mathbf{x}$. Does it necessarily follow that the best predictor for my $\mathbf{x}$ should take the same form as for some other $\mathbf{x}$? Of course, this can cause problems for the esteem of the doctor or his friendly statistician. Because we are concerned with actual observations, the goodness or otherwise of the prediction will eventually become apparent. In this case, the statistician will not be able to hide behind the screen provided by averaging over all possible future $\mathbf{x}$s.

Copas then replied:

> Dr Dunsmore raises two general points that repay careful thought. First, he questions the assumption made at the very start of the chapter that predictions are to be judged in the context of a population of future $\mathbf{x}$s and not just at some specific $\mathbf{x}$. To pursue the analogy of the doctor and the patient, all I can say is that the chapter is written from the doctor's point of view and not from the patient's! No doubt the doctor will feel he is doing a better job if he cures 95 per cent of patients rather than only 90 per cent, even though

> a particular patient (Dr Dunsmore) might do better in the latter situation than the former. As explained in the chapter, pre-shrunk predictors do better than least squares for most **x**s at the expense of doing worse at a minority of **x**s. Perhaps if we think our symptoms are unusual we should seek a consultant who is prepared to view our complaint as an individual research problem rather than rely on the blunt instrument of conventional wisdom.

The implication for Bayesian estimators, in the context of neuroimaging, is that they are the best for each subject [or voxel] *on average over subjects [or voxels]*. In this sense, Bayesian or conditional estimates of individual effects are only better, on average, over the individual effects estimated. The issues, framed by Keith Worsley above, speak to the important consideration that Bayesian estimates, of the sort discussed in this chapter, are only 'better' in a collective sense. One example of this collective context is presented in the next chapter, where between-voxel effects are used to 'shrink' within-voxel estimates that are then reported together in a posterior probability map (PPM).

The estimators and inference from a PEB approach do not inherently increase the sensitivity or specificity of the analysis. The most appropriate way to do this would be simply to increase sample size. PEB methodology can be better regarded as providing a set of estimates or predictors that are internally consistent within and over hierarchies of the observation model. Furthermore, they enable Bayesian inference (comments about the likelihood of an effect given the data) that complements classical inference (comments about the likelihood of the data). Bayesian inference does not necessarily decide whether activation is present or not, it simply estimates the probability of activation, specified in terms of the size of the effect. Conversely, classical inference is predicated on a decision (is the null hypothesis true or is the size of the effect different from zero?). The product of classical inference is a decision or declaration, which induces a sensitivity and specificity of the inference. One motivation, behind Bayesian treatments, is to circumvent the difficult compromise between sensitivity and specificity engendered by classical inference in neuroimaging.

EM AND COVARIANCE COMPONENT ESTIMATION

In this section we present a series of models that exemplify the diverse issues that can be addressed with EM. In hierarchical linear observation models, both classical and empirical Bayesian approaches can be framed in terms of *covariance component estimation* (e.g. variance partitioning). To illustrate the use of EM in covariance component estimation, we focus on two important problems in fMRI: non-sphericity induced by serial or temporal correlations among errors and variance components caused by the hierarchical nature of multisubject studies. In hierarchical observation models, variance components at higher levels can be used as constraints on the parameter estimates of lower levels. This enables the use of PEB estimators, as distinct from classical ML estimates. We develop this distinction to address the difference between response estimates based on ML and conditional estimators.

As established in the previous section, empirical Bayes enables the estimation of a model's parameters (e.g. activations) and hyperparameters that specify the model's variance components (e.g. within- and between-subject-variability). The estimation procedures conform to EM, which, considering just the hyperparameters in linear models, is formally identical to ReML. If there is only one variance component, these iterative schemes simplify to conventional, non-iterative sum of squares variance estimates. However, there are many situations when several hyperparameters have to be estimated: for example, when the correlations among errors are unknown but can be parameterized with a small number of hyperparameters (cf. serial correlations in fMRI time-series). Another important example, in fMRI, is the multisubject design, in which the hierarchical nature of the observation induces different variance components at each level. The aim of this section is to illustrate how variance component estimation can proceed in both single-level and hierarchical models. In particular, the examples emphasize that, although the mechanisms inducing non-sphericity can be very different, the variance component estimation problems they represent, and the analytic approaches called for, are identical.

We will use two fMRI examples. In the first, we deal with the issue of variance component estimation using serial correlations in single-subject fMRI studies. Because there is no hierarchical structure to this problem there is no Bayesian aspect. However, in the second example, we add a second level to the observation model for the first to address inter-subject variability. Endowing the model with a second level invokes empirical Bayes. This enables a quantitative comparison of classical and conditional single-subject response estimates.

Variance component estimation in a single-level model

In this section, we review serial correlations in fMRI and use simulated data to compare ReML estimates to estimates of correlations based simply on the model residuals. The importance of modelling temporal correlations for classical inference based on the t-statistic is discussed in terms of correcting for non-sphericity in fMRI time-series. This section concludes with a quantitative assessment of serial correlations within and between subjects.

Serial correlations in fMRI

In this section, we restrict ourselves to a single-level model and focus on the covariance component estimation afforded by EM. We have elected to use a simple but important covariance estimation problem to illustrate one of the potential uses of the scheme described in the previous section. Namely, serial correlations in fMRI embodied in the error covariance matrix for the first (and only) level of this model $C_\varepsilon^{(1)}$. Serial correlations have a long history in the analysis of fMRI time-series. fMRI time-series can be viewed as a linear admixture of signal and noise. Noise has many contributions that render it rather complicated in relation to other neurophysiological measurements. These include neuronal and non-neuronal sources. Neuronal noise refers to neurogenic signal not modelled by the explanatory variables and has the same frequency structure as the signal itself. Non-neuronal components have both white (e.g. RF noise) and coloured components (e.g. pulsatile motion of the brain caused by cardiac cycles and local modulation of the static magnetic field B_0 by respiratory movement). These effects are typically low frequency or wide-band and induce long-range correlations in the errors over time. These serial correlations can be used to provide ML estimates of the parameters (see previous section), whiten the data (Bullmore *et al.*, 1996; Purdon and Weisskoff, 1998) or enter the non-sphericity corrections described in Worsley and Friston (1995). These approaches depend upon an accurate estimation of the serial correlations. To estimate correlations $C(\lambda)_\varepsilon$, in terms of some hyperparameters, λ, one needs both the residuals of the model, r, and the conditional covariance of the parameter estimates that produced those residuals. These combine to give the required error covariance (cf. Eqn.**A3.5** in Appendix 3).

$$C(\lambda)_\varepsilon = rr^T + XC_{\theta|y}X^T \qquad \mathbf{22.20}$$

The term $XC_{\theta|y}X^T$ represents the conditional covariance of the parameter estimates $C_{\theta|y}$ 'projected' onto the measurement space, by the design matrix X. The problem is that the covariance of the parameter estimates *is itself a function of the error covariance*. This circular problem is solved by the recursive parameter re-estimation implicit in EM. It is worth noting that estimators of serial correlations based solely on the residuals (produced by any estimator) will be biased. This bias results from ignoring the second term in Eqn. **22.20**, which accounts for the component of error covariance due to uncertainty about the parameter estimates themselves. It is likely that any valid recursive scheme for estimating serial correlations in fMRI time-series conforms to EM (or ReML), even if the connection is not made explicit. See Worsley *et al.* (2002) for a non-iterative approach to autoregressive models.

In summary, the covariance estimation afforded by EM can be used to estimate serial correlations in fMRI time-series that coincidentally provide the most efficient (i.e. Gauss-Markov) estimators of the effect one is interested in. In this section, we apply EM as described in Friston *et al.* (2002a) to simulated fMRI data sequences and take the opportunity to establish the connections among some commonly employed inference procedures based upon the t-statistic. This example concludes with an application to empirical data to demonstrate quantitatively the relative variability in serial correlations over voxels and subjects.

Estimating serial correlations

For each fMRI session, we have a single-level observation model that is specified by the design matrix $X^{(1)}$ and constraints on the observation's covariance structure $Q_i^{(1)}$, in this case serial correlations among the errors.

$$\begin{aligned} y &= X^{(1)}\theta^{(1)} + \varepsilon^{(1)} \\ Q_1^{(1)} &= I \\ Q_2^{(1)} &= KK^T, \qquad k_{ij} = \begin{cases} e^{j-i} & i > j \\ 0 & i \le j \end{cases} \end{aligned} \qquad \mathbf{22.21}$$

The measured response y has errors $\varepsilon^{(1)} \sim N\{0, C_\varepsilon^{(1)}\}$. I is the identity matrix. Here $Q_1^{(1)}$ and $Q_2^{(1)}$ represent covariance components of $C_\varepsilon^{(1)}$ that model a white noise and an autoregressive AR(1) process with an AR coefficient of $1/e = 0.3679$. Notice that this is a very simple model of autocorrelations; by fixing the AR coefficient there are just two hyperparameters that allow for different mixtures of an AR(1) process and white-noise (cf. the 3 hyperparameters needed for a full AR(1) plus white noise model). The AR(1) component is modelled as an exponential decay of correlations over non-zero lag.

These components were chosen given the popularity of AR plus white-noise models in fMRI (Purdon and Weisskoff, 1998). Clearly, this basis set can be extended in any fashion using Taylor expansions to model deviations of the AR coefficient from $1/e$ or, indeed, model any other form of serial correlations. Non-stationary autocorrelations can be modelled by using non-Toeplitz forms for the bases that allow the elements in the diagonals of $Q_i^{(1)}$ to vary over observations. This might be useful, for example, in the analysis of event-related potentials, where the structure of errors may change with peristimulus time.

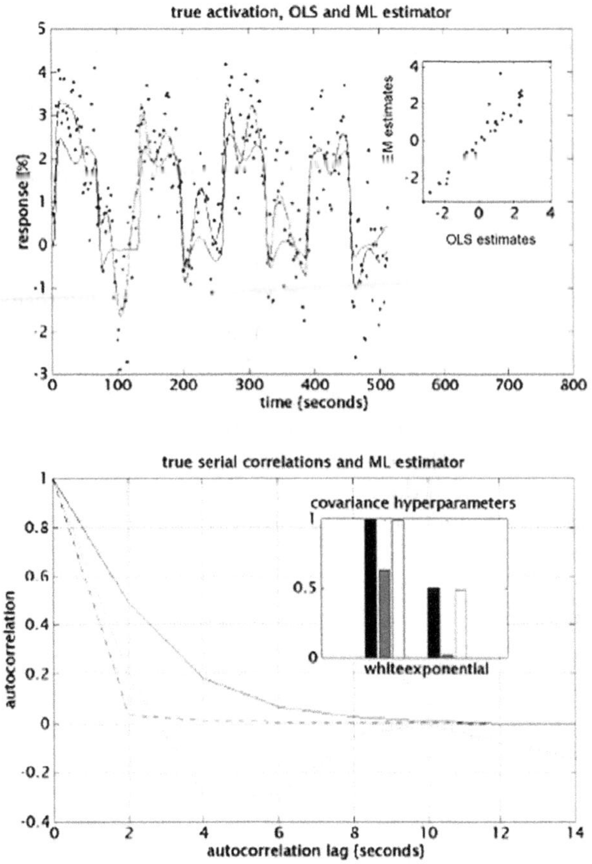

FIGURE 22.6 Top panel: true response (activation plus random low frequency components) and that based on the OLS and ML estimators for a simulated fMRI experiment. The insert shows the similarity between the OLS and ML predictions. Lower panel: true (dashed) and estimated (solid) autocorrelation functions. The sample autocorrelation function of the residuals (dotted line) and the best fit in terms of the covariance components (dot-dashed) are also shown. The insert shows the true covariance hyperparameters (black), those obtained using just the residuals (grey) and those estimated by EM (white). Note, in relation to the EM estimates, those based directly on the residuals severely underestimate the actual correlations. The simulated data comprised 128 observations with an inter-scan interval of 2 s. The activations were modelled with a boxcar (duty cycle 64 s) convolved with a canonical haemodynamic response function and scaled to a peak height of two. The constant terms and low frequency components were simulated with a linear combination of the first sixteen components of a discrete cosine set, each scaled by a random unit Gaussian variate. Serially correlated noise was formed by filtering unit Gaussian noise with a convolution kernel based on covariance hyperparameters of 1.0 [uncorrelated or white component] and 0.5 [AR(1) component].

In the examples below, the covariance constraints were scaled to a maximum of one. This means that the second hyperparameter can be interpreted as the correlation between one scan and the next. The components enter, along with the data, into the EM algorithm in Figure 22.4 to provide ML estimates of the parameters $\theta^{(1)}$ and ReML estimates of the hyperparameters $\lambda^{(1)}$. An example, based on simulated data, is shown in Figure 22.6. In this example, the design matrix comprised a boxcar regressor and the first sixteen components of a discrete cosine set. The simulated data corresponded to a compound of this design matrix (see figure legend) plus noise, coloured using hyperparameters of one and a half for the white and AR(1) components respectively. The top panel shows the data (dots), the true and fitted effects (broken and solid lines). For comparison, fitted responses based on both ML and OLS (ordinary least squares) are provided. The insert in the upper panel shows these estimators are similar but not identical. The lower panel shows the true (dashed) and estimated (solid) autocorrelation function based on $C_\varepsilon^{(1)} = \lambda_1^{(1)} Q_1^{(1)} + \lambda_2^{(1)} Q_2^{(1)}$. They are nearly identical. For comparison the sample autocorrelation function (dotted line) and an estimate based directly on the residuals, i.e. ignoring the second term of Eqn. **22.20** (dot-dash line) are provided. The underestimation that ensues using the residuals is evident in the insert that shows the true hyperparameters (black), those estimated properly using ReML (white) and those based on the residuals alone (grey). By failing to account for the uncertainty about the parameters, the hyperparameters based only on the residuals are severe underestimates. The sample autocorrelation function even shows negative correlations. This is a result of fitting the low frequency components of the design matrix. One way of understanding this is to note that the autocorrelations among the residuals are not unbiased estimators of $C_\varepsilon^{(1)}$ but of $RC_\varepsilon^{(1)}R^T$, where R is the residual-forming matrix. In other words, the residuals are not the true errors but what is left after projecting them onto the null space of the design matrix. The full details of this simulated single-session, boxcar design fMRI study are provided in the figure legend.

Inference in the context of non-sphericity[2]

This subsection reprises why covariance component estimation is so important for inference. In short, although the parameter estimates may not depend on sphericity, the standard error, and ensuing inference, does. The impact of serial correlations on inference was noted early in the fMRI analysis literature (Friston *et al.*, 1994) and led to the generalized least squares (GLS) scheme described in Worsley and Friston (1995). In this scheme one starts with any observation model that is pre-multiplied by some weighting or convolution matrix S to give:

$$Sy = SX^{(1)}\theta^{(1)} + S\varepsilon^{(1)} \qquad \textbf{22.22}$$

[2] An IID process is identically and independently distributed and has a probability distribution whose iso-contours conform to a ***sphere***. Any departure from this is referred to as non-sphericity.

The GLS parameter estimates and their covariance are:

$$\begin{aligned} \eta_{LS} &= Ly \\ Cov\{\eta_{LS}\} &= LC_{\varepsilon}^{(1)}L^T \\ L &= (SX^{(1)})^+Sy \end{aligned} \qquad 22.23$$

These estimators minimize the generalized least square index $(y - X^{(1)}\eta_{LS})^T SS^T (y - X^{(1)}\eta_{LS})$. This family of estimators is unbiased but they are not necessarily ML estimates. The Gauss-Markov estimator is the minimum variance and ML estimator that obtains as a special case when $(SS^T)^{-1} = C_{\varepsilon}^{(1)}$. The t-statistic corresponding to the GLS estimator is distributed with v degrees of freedom where (Worsley and Friston, 1995):

$$\begin{aligned} t &= \frac{c^T\eta_{LS}}{\sqrt{c^T Cov\{\eta_{LS}\}c}} \\ v &= \frac{tr\{RSC_{\varepsilon}^{(1)}S\}^2}{tr\{RSC_{\varepsilon}^{(1)}SRSC_{\varepsilon}^{(1)}S\}} \\ R &= 1 - X^{(1)}L \end{aligned} \qquad 22.24$$

The effective degrees of freedom are based on an approximation due to Satterthwaite (1941). This formulation is formally identical to the non-sphericity correction elaborated by Box (1954), which is commonly known as the Geisser-Greenhouse correction in classical analysis of variance, ANOVA (Geisser and Greenhouse, 1958).

The point here is that EM can be employed to give ReML estimates of correlations among the errors that enter into Eqn. **22.24** to enable classical inference, properly adjusted for non-sphericity, *about any GLS estimator*. EM finds a special role in enabling inferences about GLS estimators in statistical parametric mapping; when the relative amounts of different covariance components can be assumed to be the same over a subset of voxels, ReML estimates can be obtained using the sample covariance of the data over these voxels, in a single EM (see Appendix 4). After re-normalization, the ensuing estimate specifies non-sphericity with a single component. Voxel-specific hyperparameters can now be estimated non-iteratively, in the usual way, because there is only one hyperparameter to estimate.

An application to empirical data

In this subsection, we address the variability of serial correlations over voxels within subject and over subjects within the same voxel. Here we are concerned only with the form of the correlations. The next subsection addresses between-subject error variance *per se*. Using the model specification in Eqn. **22.21**, serial correlations were estimated using EM in twelve randomly selected voxels from the same slice, from a single subject. The results are shown in Figure 22.7 (left panel) and show that the correlations from one scan to the next can vary between about 0.1 and 0.4. The data sequences and experimental paradigm are described in the figure legend. Briefly, these data came from an event-related study of visual word processing in which *new* and *old* words (i.e. encoded during a pre-scanning session) were presented in a random order with a stimulus onset asynchrony (SOA) of about 4 s. Although the serial correlations within subject vary somewhat there is an even greater variability from subject to subject at the same voxel. The right-hand panel of Figure 22.7 shows the autocorrelation functions estimated separately for twelve subjects at a single voxel. In this instance, the correlations between one scan and the next range from about -0.1 to 0.3 with a greater dispersion relative to the within-subject autocorrelations.

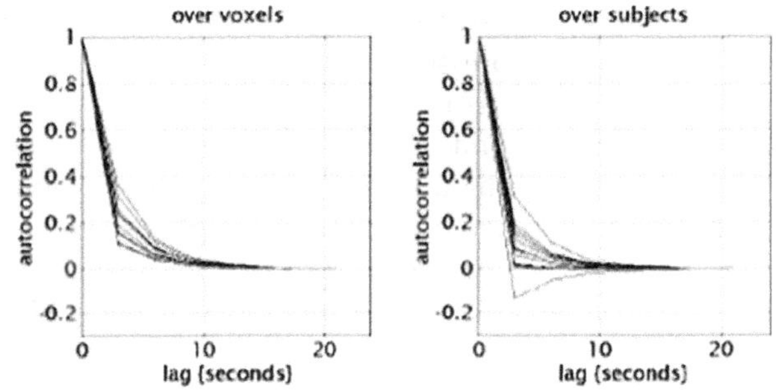

FIGURE 22.7 Estimates of serial correlations expressed as autocorrelation functions based on empirical data. Left panel: estimates from twelve randomly selected voxels from a single subject. Right panel: estimates from the same voxel over twelve different subjects. The voxel was in the cingulate gyrus. The empirical data are described in Henson *et al.* (2000). They comprised 300 volumes, acquired with EPI at two tesla and a TR of 3 s. The experimental design was stochastic and event-related looking for differential response evoked by *new* relative to *old* (studied prior to the scanning session) words. Either a new or old word was presented visually with a mean stimulus onset asynchrony (SOA) of 4 s (SOA varied randomly between 2.5 and 5.5 s). Subjects were required to make an old versus new judgement for each word. The design matrix for these data comprised two regressors (early and late) for each of the four trial types (old versus new and correct versus incorrect) and the first sixteen components of a discrete cosine set (as in the simulations).

Summary

We have chosen to focus on a covariance estimation example that requires an iterative parameter re-estimation procedure in which the hyperparameters controlling the covariances depend on the variance of the parameter estimates and vice versa. There are other important applications of covariance component estimation we could have considered (although not all require an iterative scheme). One example is the estimation of condition-specific error variances in PET and fMRI. In conventional SPM analyses, one generally assumes that the error variance expressed in one condition is the same as that in another. This represents a sphericity assumption over conditions and allows one to pool several conditions when estimating the error variance. Assumptions of this sort, and related sphericity assumptions in multisubject studies, can be easily addressed in unbalanced designs, or even in the context of missing data, using EM.

Variance component estimation in fMRI in two-level models

In this subsection, we augment the model above with a second level. This engenders a number of issues, including the distinction between fixed- and random-effect models of subject responses and the opportunity to make Bayesian inferences about single-subject responses. As above, we start with model specification, proceed to simulated data and conclude with an empirical example. In this example, the second level represents observations over subjects. Analyses of simulated data are used to illustrate the distinction between fixed- and random-effect inferences by looking at how their respective t-values depend on the variance components and design factors. The fMRI data are the same as used above and comprise event-related time-series from twelve subjects. We chose a dataset that would be difficult to analyse rigorously using routine software. These data not only evidence serial correlations but also the number of trial-specific events varied from subject to subject, giving an unbalanced design.

Model specification

The observation model here comprises two levels with the opportunity for subject-specific differences in error variance and serial correlations at the first level and parameter-specific variance at the second. The estimation model here is simply an extension of that used in the previous subsection to estimate serial correlations. Here it embodies a second level that accommodates observations over subjects.

Level one $y = X^{(1)}\theta^{(1)} + \varepsilon^{(1)}$

$$\begin{bmatrix} y_1 \\ \vdots \\ y_s \end{bmatrix} = \begin{bmatrix} X_1^{(1)} & \cdots & 0 \\ \vdots & \ddots & \vdots \\ 0 & \cdots & X_s^{(1)} \end{bmatrix} \begin{bmatrix} \theta_1^{(1)} \\ \vdots \\ \theta_s^{(1)} \end{bmatrix} + \varepsilon^{(1)}$$

$$Q_1^{(1)} = \begin{bmatrix} I_t & \cdots & 0 \\ \vdots & \ddots & \vdots \\ 0 & \cdots & 0 \end{bmatrix}, \ldots, Q_s^{(1)} = \begin{bmatrix} 0 & \cdots & 0 \\ \vdots & \ddots & \vdots \\ 0 & \cdots & I_t \end{bmatrix}$$

$$Q_{s+1}^{(1)} = \begin{bmatrix} KK^T & \cdots & 0 \\ \vdots & \ddots & \vdots \\ 0 & \cdots & 0 \end{bmatrix}, \ldots, Q_{2s}^{(1)} = \begin{bmatrix} 0 & \cdots & 0 \\ \vdots & \ddots & \vdots \\ 0 & \cdots & KK^T \end{bmatrix}$$

Level two $\theta^{(1)} = X^{(2)}\theta^{(2)} + \varepsilon^{(2)}$

$$X^{(2)} = 1_s \otimes I_p$$

$$Q_1^{(2)} = I_s \otimes \begin{bmatrix} 1 & \cdots & 0 \\ \vdots & \ddots & \vdots \\ 0 & \cdots & 0 \end{bmatrix}, \ldots, Q_p^{(2)} = I_s \otimes \begin{bmatrix} 0 & \cdots & 0 \\ \vdots & \ddots & \vdots \\ 0 & \cdots & 1 \end{bmatrix} \quad \mathbf{22.25}$$

for s subjects each scanned on t occasions and p parameters. The Kronecker tensor product $A \otimes B$ simply replaces the element of A with $A_{ij}B$. An example of these design matrices and covariance constraints are shown, respectively, in Figures 22.1 and 22.3. Note that there are $2 \times s$ error covariance constraints, one set for the white noise components and one for AR(1) components. Similarly, there are as many prior covariance constraints as there are parameters at the second level.

Simulations

In the simulations we used 128 scans for each of twelve subjects. The design matrix comprised three effects, modelling an event-related haemodynamic response to frequent but sporadic trials (in fact the instances of correctly identified 'old' words from the empirical example below) and a constant term. Activations were modelled with two regressors, constructed by convolving a series of delta functions with a canonical haemodynamic response function (HRF)[3] and the same function delayed by three seconds. The delta functions indexed the occurrence of each event. These regressors model event-related responses with two temporal components, which we will refer to as 'early' and 'late' (cf. Henson *et al.*, 2000). Each subject-specific design matrix therefore comprised three columns giving a total of thirty-six parameters at the first level and three at the second (the third being a constant term). The HRF basis functions were scaled so that a parameter estimate of one corresponds to a peak response of unity. After division by the grand mean, and multiplication by 100, the units of the response variable and parameter estimates were rendered adimensional and correspond to per cent whole brain mean over all scans. The simulated data were generated using Eqn. **22.25** with unit Gaussian

[3] The canonical HRF was the same as that employed by SPM. It comprises a mixture of two gamma variates modelling peak and undershoot components and is based on a principal component analysis of empirically determined haemodynamic responses, over voxels, as described in Friston *et al.* (1998)

noise coloured using a temporal, convolution matrix with first-level hyperparameters 0.5 and −0.1 for each subject's white and AR(1) error covariance components respectively. The second-level parameters and hyperparameters were $\theta^{(2)} = [0.5, 0, 0]^T$, $\lambda^{(2)} = [0.02, 0.006, 0]^T$. These values model substantial early responses, with an expected value of 0.5 per cent and a standard deviation over subjects of 0.14 per cent (i.e. square root of 0.02). The late component was trivial with zero expectation and a standard deviation of 0.077 per cent. The third or constant terms were discounted with zero mean and variance. These values were chosen because they are typical of real data (see below).

Figures 22.8 and 22.9 show the results after subjecting the simulated data to EM to estimate the conditional

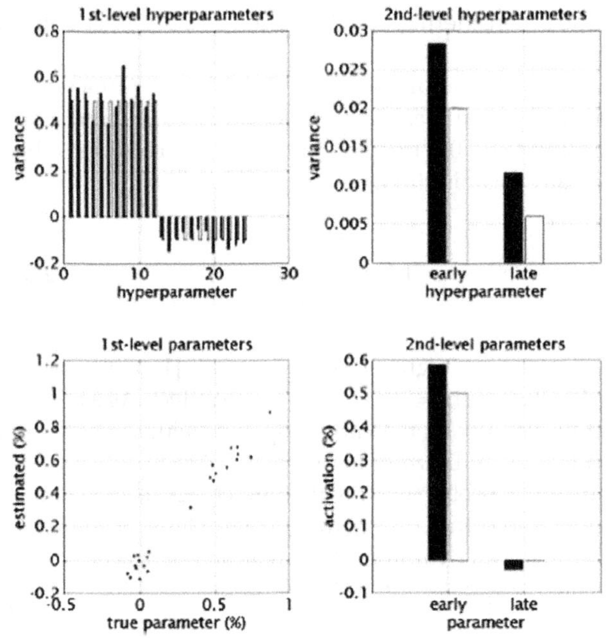

FIGURE 22.8 The results of an analysis of simulated event-related responses in a single voxel. Parameter and hyperparameter estimates based on a simulated fMRI study are shown in relation to their true values. The simulated data comprised 128 scans for each of twelve subjects with a mean peak response over subjects of 0.5 per cent. The construction of these data is described in the main text. Stimulus presentation conformed to the presentation of 'old' words in the empirical analysis described in the main text. Serial correlations were modelled as in the main text. Upper left: first-level hyperparameters. The estimated subject-specific values (black) are shown alongside the true values (white). The first twelve correspond to the 'white' term or variance. The second twelve control the degree of autocorrelation and can be interpreted as the correlation between one scan and the next. Upper right: hyperparameters for the early and late components of the evoked response. Lower-left: the estimated subject-specific parameters pertaining to the early- and late-response components are plotted against their true values. Lower right: the estimated and true parameters at the second level, representing the conditional mean of the distribution from which the subject-specific effects are drawn.

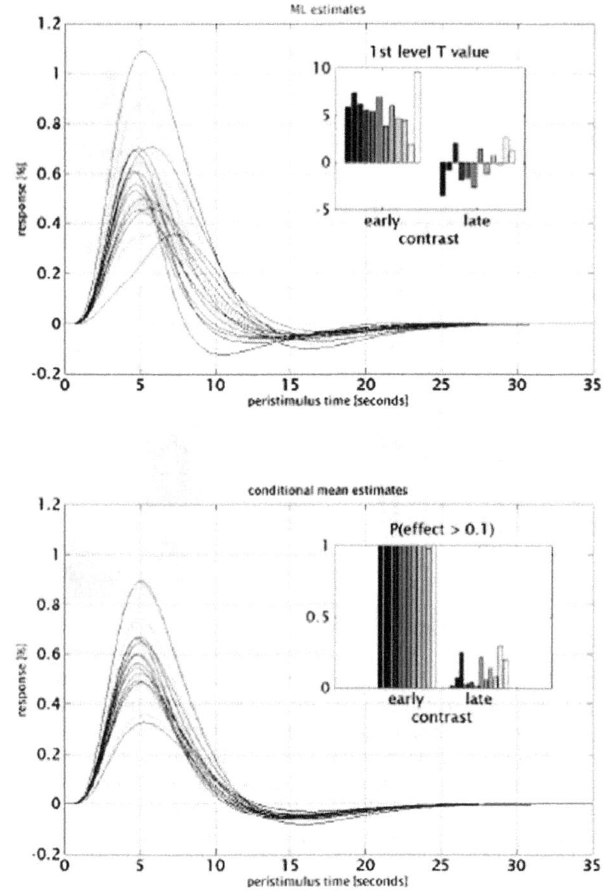

FIGURE 22.9 Response estimates and inferences about the estimates presented in Figure 22.8. Upper panel: true (dotted) and ML (solid) estimates of event-related responses to a stimulus over 12 subjects. The units of activation are adimensional and correspond to per cent of whole-brain mean. The insert shows the corresponding subject-specific t-values for contrasts testing for early and late responses. Lower panel: the equivalent estimates based on the conditional means. It can be seen that the conditional estimates are much 'tighter' and reflect better the inter-subject variability in responses. The insert shows the posterior probability that the activation was greater than 0.1 per cent. Because the responses were modelled with early and late components (basis functions corresponding to canonical haemodynamic response functions, separated by 3 s) separate posterior probabilities could be computed for each. The simulated data comprised only early responses as reflected in the posterior probabilities.

mean and covariances of the subject-specific evoked responses. Figure 22.8 shows the estimated hyperparameters and parameters (black) alongside the true values (white). The first-level hyperparameters controlling within-subject error (i.e. scan to scan variability) are estimated in a reasonably reliable fashion, but note that these estimates show a degree of variation about the veridical values (see Conclusion). In this example, the second-level hyperparameters are over-estimated but remarkably good, given only twelve subjects. The parameter estimates at the first and second levels are again very

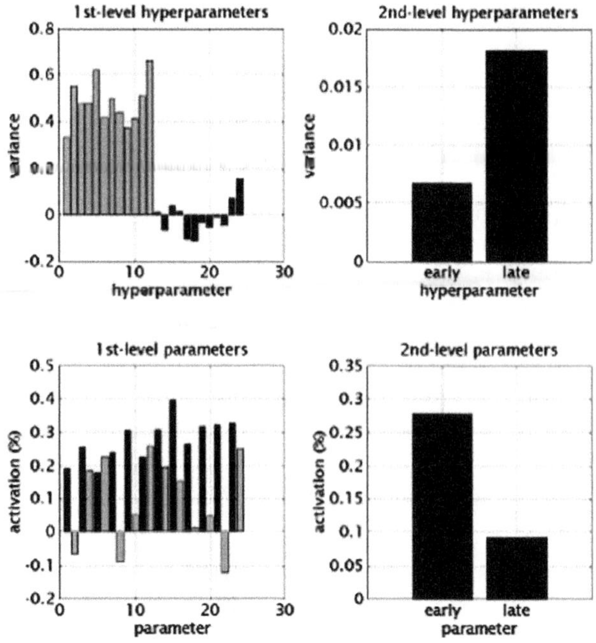

FIGURE 22.10 Estimation of differential event-related responses using real data. The format of this figure is identical to that of Figure 22.8. The only differences are that these results are based on real data and the response is due to the difference between studied or familiar (old) words and novel (new) words. In this example we used the first 128 scans from twelve subjects. Clearly, in this figure, we cannot include true effects.

reasonable, correctly attributing the majority of the experimental variance to an early effect. Figure 22.8 should be compared with Figure 22.10, which shows the equivalent estimates for real data.

The top panel in Figure 22.9 shows the ML estimates that would have been obtained if we had used a single-level model. These correspond to response estimates from a conventional fixed-effects analysis. The insert shows the classical fixed-effect t-values, for each subject, for contrasts testing early and late response components. Although these t-values properly reflect the prominence of early effects, their variability precludes any threshold that could render the early components significant and yet exclude false positives pertaining to the late component. The lower panel highlights the potential of revisiting the first level in the context of a hierarchical model. It shows the equivalent responses based on the conditional mean and the posterior inference (insert) based on the conditional covariance. This allows us to reiterate some points made in the previous section. First, the parameter estimates and ensuing response estimates are informed by information abstracted from higher levels. Second, these empirical priors enable Bayesian inference about the probability of an activation that is specified in neurobiological terms.

In Figure 22.9 the estimated responses are shown (solid lines) with the actual responses (broken lines). Note how the conditional estimates show a regression or 'shrinkage' to the conditional mean. In other words, their variance shrinks to reflect, more accurately, the variability in real responses. In particular, the spurious variability in the apparent latency of the peak response in the ML estimates disappears when using the conditional estimates. This is because the contribution of the late component, which causes latency differences, is suppressed in the conditional estimates. This, in turn, reflects the fact that the variability in its expression over subjects is small relative to that induced by the observation error. Simulations like these suggest that characterizations of inter-subject variability using ML approaches can severely overestimate the true variability. This is because the ML estimates are unconstrained and simply minimize observation error without considering how likely the ensuing inter-subject variability is.

The posterior probabilities (insert) are a function of the conditional mean $\eta^{(1)}_{\theta|y}$ and covariance $C^{(1)}_{\theta|y}$ and a size threshold $\gamma = 0.1$ that specifies what we consider a biologically meaningful effect.

$$1 - \Phi\left(\frac{\gamma - c^T \eta^{(1)}_{\theta|y}}{\sqrt{c^T C^{(1)}_{\theta|y} c}}\right) \quad 22.26$$

The contrast weight vectors were $c = [1, 0, 0]^T$ for the early effect and $c = [0, 1, 0]^T$ for a late effect. As expected, the probability of the early response being greater than $\gamma = 0.1$ was uniformly high for all subjects, whereas the equivalent probability for the late component was negligible. Note that, in contrast to the classical inference, there is now a clear indication that each subject expressed an early response but no late response.

An empirical analysis

Here the analysis is repeated using real data and the results are compared to those obtained using simulated data. The empirical data are described in Henson *et al.* (2000). Briefly, they comprised 128+ scans in twelve subjects. Only the first 128 scans were used below. The experimental design was stochastic and event-related, looking for differential responses evoked by *new* relative to *old* (studied prior to the scanning session) words. Either a new or old word was presented every 4 seconds or so (SOA varied between 2.5 and 5.5 s). In this design one is interested only in the differences between responses evoked by the two stimulus types. This is because the efficiency of the design to detect the effect of stimuli *per se* is negligible with such a short SOA. Subjects were required to make an old versus new judgement for each word. Drift (the first 8 components of a discrete cosine set) and the effects of incorrect trials were treated as confounds and removed using linear regression.[4] The first-level subject-specific design matrix partitions comprised four regressors with early and late effects for both old and new words.

The analysis proceeded in exactly the same way as above. The only difference was that the contrast tested for *differences* between the two word types, i.e. $c = [1, 0, -1, 0]^T$ for an old minus new early effect. The hyperparameter and parameter estimates, for a voxel in the cingulate gyrus (BA 31; −3, −33, 39 mm), are shown in Figure 22.10, adopting the same format as in Figure 22.8. Here we see that the within-subject error varies much more in the empirical data, with the last subject showing almost twice the error variance of the first. As above, serial correlations vary considerably from subject to subject and are not consistently positive or negative. The second-level hyperparameters showed the early component of the differential response to be more reliable over subjects than the late component (0.007 and 0.19, respectively). All but two subjects had a greater early response, relative to late, which on average was about 0.28 per cent. In other words, activation differentials, in the order of 0.3 per cent, occurred in the context of an observation error with a standard deviation of 0.5 per cent (see Figure 22.10). The inter-subject variability was about 30 per cent of the mean response amplitude. A component of the variability in within-subject error is due to uncertainty in the ReML estimates of the hyperparameters (see below), but this degree of inhomogeneity is substantially more than in the simulated data (where subjects had equal error variances). It is interesting to note that, despite the fact that the regressors for the early and late components had exactly the same form, the between-subject error for one was less than half that of the other. Results of this sort speak of the prevalence of non-sphericity (in this instance heteroscedasticity or unequal variances) and a role for the analyses illustrated here.

The response estimation and inference are shown in Figure 22.11. Again we see the characteristic 'shrinkage' when comparing the ML to the conditional estimates. It can be seen that all subjects, apart from the first and third, had over a 95 per cent chance of expressing an early differential of 0.1 per cent or more. The late differential response was much less consistent, although one subject expressed a difference with about 84 per cent confidence.

[4] Strictly speaking the projection matrix implementing this adjustment should also be applied to the covariance constraints but this would render the components singular and ruin their sparsity structure. We therefore omitted this and ensured, in simulations, that the adjustment had a negligible effect on the hyperparameter estimates.

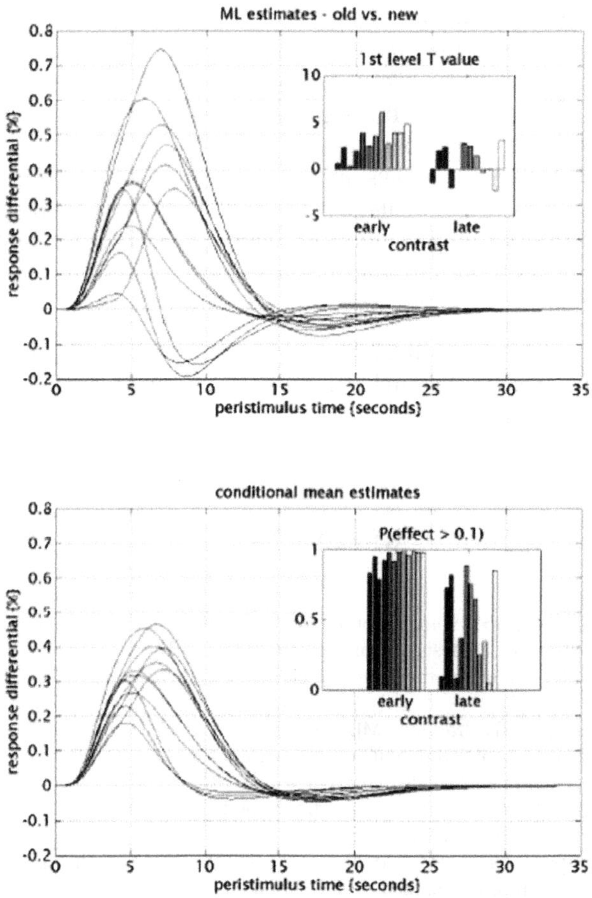

FIGURE 22.11 The format of this figure is identical to that of Figure 22.9. The only differences are that these results are based on real data and the response is due to the difference between studied or familiar (old) words and novel (new) words. The same regression of conditional responses to the conditional mean is seen on comparing the ML and conditional estimates. In relation to the simulated data, there is more evidence for a late component but no late activation could be inferred for any subject with any degree of confidence. The voxel from which these data were taken was in the cingulate gyrus (BA 31) at −3, −33, 39 mm.

Summary

The examples presented above allow us to reprise a number of important points made in the previous section (see also Friston *et al.*, 2002a). In conclusion, the main points are:

- There are many instances when an iterative parameter re-estimation scheme is required (e.g. dealing with serial correlations or missing data). These schemes are generally variants of EM.
- Even before considering the central role of covariance component estimation in hierarchical models or empirical Bayes, it is an important aspect of model estimation in its own right, particularly in estimating non-sphericity. Parameter estimates can either be obtained

directly from an EM algorithm, in which case they correspond to the ML or Gauss-Markov estimates, or the hyperparameters can be used to determine the error correlations which re-enter a generalized least-square scheme, as a non-sphericity correction.

- Hierarchical models enable a collective improvement in response estimates by using conditional, as opposed to maximum-likelihood, estimators. This improvement ensues from the constraints derived from higher levels that enter as empirical priors on lower levels.

In the next chapter, we revisit two-level models but consider hierarchical observations over voxels as opposed to subjects.

REFERENCES

Box GEP (1954) Some theorems on quadratic forms applied in the study of analysis of variance problems, I. Effect of inequality of variance in the one-way classification. *Ann Math Stats* **25**: 290–302

Bullmore ET, Brammer MJ, Williams SCR *et al.* (1996) Statistical methods of estimation and inference for functional MR images. *Mag Res Med* **35**: 261–77

Copas JB (1983) Regression prediction and shrinkage. *J Roy Stat Soc Series B* **45**: 311–54

Dempster AP, Laird NM, Rubin DB (1977) Maximum likelihood from incomplete data via the EM algorithm. *J Roy Stat Soc Series* B **39**: 1–38

Dempster AP, Rubin DB, Tsutakawa RK (1981) Estimation in covariance component models. *J Am Stat Assoc* **76**: 341–53

Descombes X, Kruggel F, von Cramon DY (1998) fMRI signal restoration using a spatio-temporal Markov random field preserving transitions. *NeuroImage* **8**: 340–49

Efron B, Morris C (1973) Stein's estimation rule and its competitors – an empirical Bayes approach. *J Am Stat Assoc* **68**: 117–30

Efron B, Morris C (1977) Stein's paradox in statistics. *Sci Am* **May**: 119–27

Everitt BS, Bullmore ET (1999) Mixture model mapping of brain activation in functional magnetic resonance images. *Hum Brain Mapp* **7**: 1–14

Fahrmeir L, Tutz G (1994) *Multivariate statistical modelling based on generalised linear models.* Springer-Verlag Inc., New York, pp 355–56

Friston KJ, Josephs O, Rees G *et al.* (1998) Nonlinear event-related responses in fMRI. *Magn Reson Med* **39**: 41–52

Friston KJ, Jezzard PJ, Turner R (1994) Analysis of functional MRI time-series. *Hum Brain Mapp* **1**: 153–71

Friston KJ, Holmes AP, Worsley KJ *et al.* (1995) Statistical parametric maps in functional imaging: a general linear approach. *Hum Brain Map* **2**: 189–210

Friston KJ, Penny W, Phillips C *et al.* (2002a) Classical and Bayesian inference in neuroimaging: theory. *NeuroImage* **16**: 465–83

Friston KJ, Glaser DE, Henson RNA *et al.* (2002b) Classical and Bayesian inference in neuroimaging: applications. *NeuroImage* **16**: 484–512

Geisser S, Greenhouse SW (1958) An extension of Box's results on the use of the F distribution in multivariate analysis. *Ann Math Stats* **29**: 885–91

Hartley H (1958) Maximum likelihood estimation from incomplete data. *Biometrics* **14**: 174–94

Hartvig NV, Jensen JL (2000) Spatial mixture modelling of fMRI data. *Hum Brain Mapp* **11**: 233–48

Harville DA (1977) Maximum likelihood approaches to variance component estimation and to related problems. *J Am Stat Assoc* **72**: 320–38

Henson RNA, Rugg MD, Shallice T *et al.* (2000) Confidence in recognition memory for words: dissociating right prefrontal roles in episodic retrieval. *J Cog Neurosci* **12**: 913–23

Højen-Sørensen P, Hansen LK, Rasmussen CE (2000) Bayesian modelling of fMRI time-series. In *Advances in neural information processing systems* **12**, Solla SA, Leen TK, Muller KR (eds). MIT Press, pp 754–60

Holmes A, Ford I (1993) A Bayesian approach to significance testing for statistic images from PET. In *Quantification of brain function, tracer kinetics and image analysis in brain PET*, Uemura K, Lassen NA, Jones T *et al.* (eds). Excerpta Medica, Int. Cong. Series No. 1030: 521–34

Holmes AP, Friston KJ (1998) Generalisability, random effects and population inference. *NeuroImage* S754

Kass RE, Steffey D (1989) Approximate Bayesian inference in conditionally independent hierarchical models (parametric empirical Bayes models). *J Am Stat Assoc* **407**: 717–26

Lee PM (1997) *Bayesian Statistics an Introduction.* John Wiley and Sons Inc., New York

Neal RM, Hinton GE (1998) A view of the EM algorithm that justifies incremental, sparse and other variants. In *Learning in graphical models*, Jordan MI (ed.). Kluwer Academic Press, Dordrecht, pp 355–68

Purdon PL, Weisskoff R (1998) Effect of temporal autocorrelations due to physiological noise stimulus paradigm on voxel-level false positive rates in fMRI. *Hum Brain Mapp* **6**: 239–49

Satterthwaite EF (1941) Synthesis of variance. *Psychometrika* **6**: 309–16

Tikhonov AN, Arsenin VY (1977) *Solution of ill posed problems.* Washington and Sons, Washington

Worsley KJ (1994) Local maxima and the expected Euler characteristic of excursion sets of chi squared, F and t fields. *Adv Appl Prob* **26**: 13–42

Worsley KJ, Friston. KJ (1995) Analysis of fMRI time-series revisited – again. *NeuroImage* **2**: 173–81

Worsley KJ, Liao C, Aston J *et al.* (2002) A general statistical analysis for fMRI data. *NeuroImage* **15**: 1–14

C H A P T E R

23

Posterior probability maps

K. Friston and W. Penny

INTRODUCTION

This chapter describes the construction of *posterior probability maps* that enable conditional or Bayesian inferences about regionally specific effects in neuroimaging. Posterior probability maps are images of the probability or confidence that an activation exceeds some specified threshold, given the data. Posterior probability maps (PPMs) represent a complementary alternative to statistical parametric maps (SPMs) that are used to make classical inferences. However, a key problem in Bayesian inference is the specification of appropriate priors. This problem can be finessed using *empirical Bayes* in which prior variances are estimated from the data, under some simple assumptions about their form. Empirical Bayes requires a hierarchical observation model, in which higher levels can be regarded as providing prior constraints on lower levels. In neuroimaging, observations of the same effect over voxels provide a natural, two-level hierarchy that enables an empirical Bayesian approach. In this section, we present the motivation and the operational details of a simple empirical Bayesian method for computing posterior probability maps. We then compare Bayesian and classical inference through the equivalent PPMs and SPMs testing for the same effect in the same data. The approach adopted here is a natural extension of parametric empirical Bayes described in the previous chapter. The resulting model entails global shrinkage priors to inform the estimation of effects at each voxel or bin in the image. These global priors can be regarded as a special case of spatial priors in the more general spatiotemporal models for functional magnetic resonance imaging (fMRI) introduced in Chapter 25.

To date, inference in neuroimaging has been restricted largely to classical inferences based upon statistical parametric maps (SPMs). The alternative approach is to use Bayesian or conditional inference based upon the posterior distribution of the activation given the data (Holmes and Ford, 1993). This necessitates the specification of priors (i.e. the probability distribution of the activation). Bayesian inference requires the posterior distribution and therefore rests upon a posterior density analysis. A useful way to summarize this posterior density is to compute the probability that the activation exceeds some threshold. This computation represents a Bayesian inference about the effect, in relation to the specified threshold. We now describe an approach to computing posterior probability maps for activation effects or, more generally, treatment effects in imaging data sequences. This approach represents the simplest and most computationally expedient way of constructing PPMs.

As established in the previous chapter, the motivation for using conditional or Bayesian inference is that it has high face-validity. This is because the inference is about an effect, or activation, being greater than some specified size that has some meaning in relation to underlying neurophysiology. This contrasts with classical inference, in which the inference is about the effect being significantly different from zero. The problem for classical inference is that trivial departures from the null hypothesis can be declared significant, with sufficient data or sensitivity. Furthermore, from the point of view of neuroimaging, posterior inference is especially useful because it eschews the multiple-comparison problem. Posterior inference does not have to contend with the multiple-comparison problem because there are no false-positives. The probability that activation has occurred, given the data, at any particular voxel is the same, irrespective of whether one has analysed that voxel or the entire brain. For this reason, posterior inference using PPMs may represent a relatively more powerful approach than classical inference in neuroimaging. The reason that there is no need to adjust the p-values is that we assume independent prior distributions for the activations over voxels. In this simple Bayesian model, the Bayesian perspective is similar to that of the frequentist who makes inferences on a

Statistical Parametric Mapping, by Karl Friston et al.
ISBN–13: 978-0-12-372560-8 ISBN–10: 0-12-372560-7

per-comparison basis (see Berry and Hochberg, 1999 for a detailed discussion).

Priors and Bayesian inference

PPMs require the posterior distribution or conditional distribution of the activation (a contrast of conditional parameter estimates) given the data. This posterior density can be computed, under Gaussian assumptions, using Bayes' rule. Bayes' rule requires the specification of a likelihood function and the prior density of the model's parameters. The models used to form PPMs, and the likelihood functions, are exactly the same as in classical SPM analyses. The only extra bit of information that is required is the prior probability distribution of the parameters of the general linear model employed. Although it would be possible to specify these in terms of their means and variances using independent data, or some plausible physiological constraints, there is an alternative to this fully Bayesian approach. The alternative is empirical Bayes, in which the variances of the prior distributions are estimated directly from the data. Empirical Bayes requires a hierarchical observation model, where the parameters and hyperparameters at any particular level can be treated as priors on the level below. There are numerous examples of hierarchical observation models. For example, the distinction between fixed- and mixed-effects analyses of multisubject studies relies upon a two-level hierarchical model. However, in neuroimaging, there is a natural hierarchical observation model that is common to all brain mapping experiments. This is the hierarchy induced by looking for the same effects at every voxel within the brain (or grey matter). The first level of the hierarchy corresponds to the experimental effects at any particular voxel and the second level of the hierarchy comprises the effects over voxels. Put simply, the variation in a particular contrast, over voxels, can be used as the prior variance of that contrast at any particular voxel. A caricature of the approach presented in this chapter appears as a numerical example in Chapter 11 on hierarchical models.

The model used here is one in which the spatial relationship among voxels is discounted. The advantage of treating an image like a 'gas' of unconnected voxels is that the estimation of between-voxel variance in activation can be finessed to a considerable degree (see below and Appendix 4). This renders the estimation of posterior densities tractable because the between-voxel variance can then be used as a prior variance at each voxel. We therefore focus on this simple and special case and on the 'pooling' of voxels to give precise [ReML] estimates of the variance components required for Bayesian inference. The main focus of this chapter is the pooling procedure that affords a computational saving necessary to produce PPMs of the whole brain. In what follows, we describe how this approach is implemented and provide some examples of its application.

THEORY

Conditional estimators and the posterior density

Here we describe how the posterior distribution of the parameters of any general linear model can be estimated at each voxel from imaging data sequences. Under Gaussian assumptions about the errors $\varepsilon \sim N\{0, C_\varepsilon\}$ of a general linear model with design matrix X, the responses are modelled as:

$$y = X\theta + \varepsilon \qquad 23.1$$

The conditional or posterior covariances and mean of the parameters θ are given by (Friston *et al.*, 2002a):

$$\begin{aligned} C_{\theta|y} &= (X^T C_\varepsilon^{-1} X + C_\theta^{-1})^{-1} \\ \eta_{\theta|y} &= C_{\theta|y} X^T C_\varepsilon^{-1} y \end{aligned} \qquad 23.2$$

where C_θ is the prior covariance (assuming a prior expectation of zero). Once these moments are known, the posterior probability that a particular effect or contrast specified by a contrast weight vector c exceeds some threshold γ is computed easily:

$$p = 1 - \Phi\left(\frac{\gamma - c^T \eta_{\theta|y}}{\sqrt{c^T C_{\theta|y} c}}\right) \qquad 23.3$$

$\Phi(\cdot)$ is the cumulative density function of the unit normal distribution. An image of these posterior probabilities constitutes a PPM.

Estimating the error covariance

Clearly, to compute the conditional moments above one needs to know the error and prior covariances C_ε and C_θ. In the next section, we will describe how the prior covariance C_θ can be estimated. In this section, we describe how the error covariance can be estimated in terms of a hyperparameter λ_ε where $C_\varepsilon = \lambda_\varepsilon V$, and V is the correlation or non-sphericity matrix of the errors (see below). This hyperparameter is estimated simply using restricted maximum likelihood (ReML) or EM as described in the previous chapter.[1]

Until convergence {**E**-step

$$C_\varepsilon = \lambda_\varepsilon V$$
$$C_{\theta|y} = (X^T C_\varepsilon^{-1} X + C_\theta^{-1})^{-1}$$

M-step

$$P = C_\varepsilon^{-1} - C_\varepsilon^{-1} X C_{\theta|y} X^T C_\varepsilon^{-1}$$
$$g = \frac{1}{2} tr(P^T VPyy^T) - \frac{1}{2} tr(PV)$$
$$H = \frac{1}{2} tr(PVPV)$$
$$\lambda_\varepsilon \leftarrow \lambda_\varepsilon + H^{-1} g$$

} **23.4**

In brief, P represents the residual forming matrix, pre-multiplied by the error precision. It is this projector matrix that 'restricts' the estimation of variance components to the null space of the design matrix. g and H are the first- and expected second-order derivatives (i.e. gradients and expected negative curvature) of the ReML objective function (a special case of the variational free energy). The **M**-step can be regarded as a Fisher-Scoring scheme that maximizes the ReML objective function. Given that there is only one hyperparameter to estimate, this scheme converges very quickly (2 to 3 iterations for a tolerance of 10^{-6}).

Estimating the prior density

Simply computing the conditional moments using Eqn. **23.4** corresponds to a fully Bayesian analysis at each and every voxel. However, there is an outstanding problem in the sense that we do not know the prior covariances of the parameters. It is at this point that we introduce the hierarchical perspective that enables empirical Bayes. If we consider Eqn. **23.1** as the first level of a two-level hierarchy, where the second level corresponds to observations over voxels, we have a hierarchical observation model for all voxels that treats some parameters as random effects and others as fixed. The random effects θ_1 are those that we are interested in and the fixed effects θ_0 are nuisance variables or confounds (e.g. drifts or the constant term) modelled by the regressors in X_0 where $X = [X_1, X_0]$ and

$$y = [X_1, X_0]\begin{bmatrix}\theta_1\\ \theta_0\end{bmatrix} + \varepsilon^{(1)}$$
$$\theta_1 = 0 + \varepsilon^{(2)} \qquad \mathbf{23.5}$$

This model posits that there is a voxel-wide prior distribution for the parameters θ_1 with zero mean and unknown covariance $E(\varepsilon^{(2)}\varepsilon^{(2)^T}) = \sum_i \lambda_i Q_i^{(2)}$. The components $Q_i^{(2)}$ specify the prior covariance structure of the interesting effects and would usually comprise a component for each parameter whose i-th leading diagonal element was one and zero elsewhere. This implies that, if we selected a voxel at random from the search volume, the i-th parameter at that voxel would conform to a sample from a Gaussian distribution of zero expectation and variance λ_i. The reason this distribution can be assumed to have zero mean is that parameters of interest reflect region-specific effects that, by definition sum to zero over the search volume. By concatenating the data from all voxels and using Kronecker tensor products of the design matrices and covariance components, it is possible to create a very large hierarchical observation model that could be subject to EM (see, for example, Friston *et al.*, 2002b: Section 3.2). However, given the enormous number of voxels in neuroimaging, this is computationally prohibitive. A mathematically equivalent but more tractable approach is to consider the estimation of the prior hyperparameters as a variance component estimation problem after collapsing Eqn. **23.5** to a single-level model:

$$y = X_0\theta_0 + \xi$$
$$\xi = X_1\varepsilon^{(2)} + \varepsilon^{(1)} \qquad \mathbf{23.6}$$

This is simply a rearrangement of Eqn. **23.5** to give a linear model with a compound error covariance that includes the observation error covariance and m components for each parameter in θ_1. These components are induced by variation of the parameters over voxels:

$$C_\xi = E(\xi\xi^T) = \sum \lambda_i Q_i^{(1)}$$
$$Q^{(1)} = X_1 Q_1^{(2)} X_1^T, \ldots, X_1 Q_m^{(2)} X_1^T, V \qquad \mathbf{23.7}$$
$$\lambda = [\lambda_1, \ldots, \lambda_m, \lambda_\varepsilon]^T$$

This equation says that the covariance of the compound error can be linearly decomposed into m components (usually one for each parameter) and the error variance. The form of the observed covariances, due to variation in the parameters, is determined by the design matrix X and $Q_i^{(2)}$ that model variance components in parameter space.

Equation **23.7** furnishes a computationally expedient way to estimate the prior covariances for the parameters that then enter into Eqn. **23.4** to provide for voxel-specific error hyperparameter estimates and conditional moments. In brief, the hyperparameters are estimated by pooling the data from all voxels to provide ReML estimates of the variance components of C_ξ according to Eqn. **23.7**. The nice thing about this pooling is that the hyperparameters of the parameter covariances are, of course,

[1] Note that the augmentation step shown in Figure 22.4 of Chapter 22 is unnecessary because the prior covariance enters explicitly into the conditional covariance.

the same for all voxels. This is not the case for the error covariance that may change from voxel to voxel. The pooled estimate of λ_ϵ can be treated as an estimate of the average λ_ϵ over voxels. These global hyperparameters are estimated by iterating:

Until convergence { **E**-step

$$C_\xi = \sum \lambda_i Q_i^{(1)}$$
$$C_{\theta_0|y} = (X_0^T C_\xi^{-1} X_0)^{-1}$$

M-step

$$P = C_\xi^{-1} - C_\xi^{-1} X_0 C_{\theta_0|y} X_0^T C_\xi^{-1}$$
$$g_i = \frac{1}{2} tr(P^T Q_i P \tfrac{1}{n} YY^T) - \frac{1}{2} tr(PQ_i)$$
$$H_{ij} = \frac{1}{2} tr(PQ_i PQ_j)$$
$$\lambda \leftarrow \lambda + H^{-1} g$$

} **23.8**

It can be seen that this has exactly the form as Eqn. **23.4** used for the analysis at each voxel. The differences are yy^T has been replaced by its sample mean over voxels $\frac{1}{n}YY^T$ and there are no priors because the parameters controlling the expression of confounding effects or nuisance variables are treated as fixed effects. This is equivalent to setting their prior variance to infinity (i.e. flat priors) so that $C_{\theta_0}^{-1} \to 0$. Finally, the regressors in X_1 have disappeared from the design matrix because they are embodied in the covariance components of the compound error. As above, the inclusion of confounds restricts the hyperparameter estimation to the null space of X_0, hence *restricted* maximum likelihood (ReML). In the absence of confounds, the hyperparameters would simply be maximum likelihood (ML) estimates that minimize the difference between the estimated and observed covariance of the data, averaged over voxels. The ensuing ReML estimates are very high precision estimators. Their precision increases linearly with the number of voxels n and is in fact equal to nH. These hyperparameters now enter as priors into the voxel-specific estimation along with the flat priors for the nuisance variables:

$$C_\theta = \begin{bmatrix} \sum \lambda_i Q_i^{(2)} \cdots & & 0 \\ \vdots & \infty & \\ & & \ddots & \\ 0 & & \infty \end{bmatrix} \qquad \mathbf{23.9}$$

We now have a very precise estimate of the prior covariance that can be used to re-visit each voxel to compute the conditional or posterior density. Finally, the conditional moments enter Eqn. **23.3** to give the posterior probability for each voxel. Figure 23.1 is a schematic illustration of this scheme.

Summary

A natural hierarchy characterizes all neuroimaging experiments, where the second level is provided by variation over voxels. Although it would be possible to form a very large two-level observation model and estimate the conditional means and covariances of the parameters at the first level, this would involve dealing with matrices of size $(ns) \times (ns)$ (number of voxels n times the number of scans s). The same conditional estimators can be computed using the two-step approach described above. First, the data covariance components induced by parameter variation over voxels and observation error are computed using ReML estimates of the associated covariance hyperparameters. Second, each voxel is revisited to compute voxel-specific error variance hyperparameters and the conditional moments of the parameters, using the empirical priors from the first step (see Figure 23.1). Both these steps deal only with matrices of size $n \times n$. The voxel-specific estimation sacrifices the simplicity of a single large iterative scheme for lots of quicker iterative schemes at each voxel. This exploits the fact that the same first-level design matrix is employed for all voxels. We have presented this two-stage scheme in some detail because it is used in subsequent chapters on the inversion of source models for electroencephalography (EEG). The general idea is to compute hyperparameters in measurement space and then use them to construct empirical priors for inference at higher levels (see Appendix 4 for a summary of this two-stage procedure and its different incarnations).

EMPIRICAL DEMONSTRATIONS

In this section, we compare and contrast Bayesian and classical inference using PPMs and SPMs based on real data. The first dataset is the positron emission tomography (PET) verbal fluency data that has been used to illustrate methodological advances in SPM over the years. In brief, these data were required from five subjects each scanned twelve times during the performance of one of two word generation tasks. The subjects were asked either to repeat a heard letter or to respond with a word that began with the heard letter. These tasks were performed in alternation over the twelve scans and the order randomized over subjects. The second dataset comprised data from a study of attention to visual motion (Büchel and Friston, 1997). The data used in this note came from the first subject studied. This subject was scanned at 2T to give a time series of 360 images comprising ten block epochs of different visual motion conditions. These

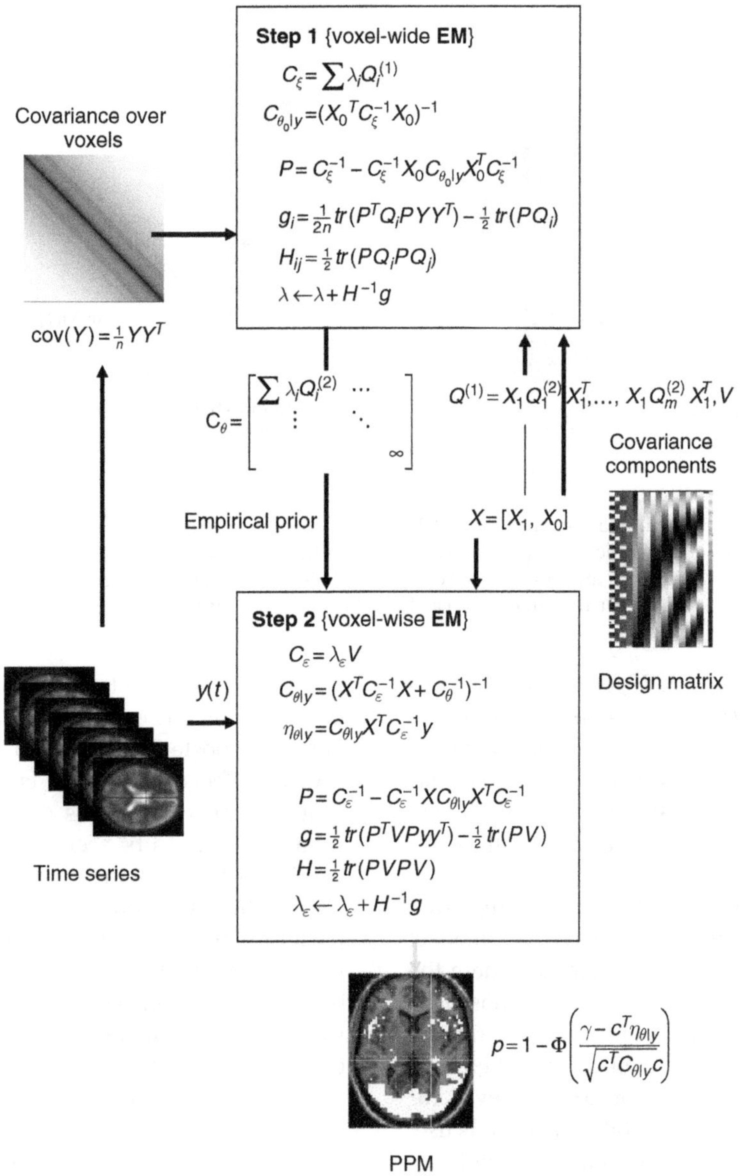

FIGURE 23.1 Schematic summarizing the two-step procedure for (1) ReML estimation of the empirical prior covariance based on the sample covariance of the data, pooled over voxels and (2) a voxel-by-voxel estimation of the conditional expectation and covariance of the parameters, required for inference. See the main text for a detailed explanation of the equations.

conditions included a fixation condition, visual presentation of static dots, and visual presentation of radially moving dots under attention and no-attention conditions. In the attention condition, subjects were asked to attend to changes in speed (which did not actually occur). These data were re-analysed using a conventional SPM procedure and using the empirical Bayesian approach described in the previous section. The ensuing SPMs and PPMs are presented below for the PET and fMRI data respectively. The contrast for the PET data compared the word generation with the word shadowing condition and the contrast for the fMRI data tested for the effect of visual motion above and beyond that due to photic stimulation with stationary dots.

Inference for the PET data

The right panel of Figure 23.2 shows the PPM for a deactivating effect of verbal fluency. There are two thresholds for the PPM. The first and more important is γ in Eqn. **23.3**. This defines what we mean by 'activation' and, by default, is set at one standard deviation of the prior variance of the contrast, in this instance 2.2. This corresponds to a change

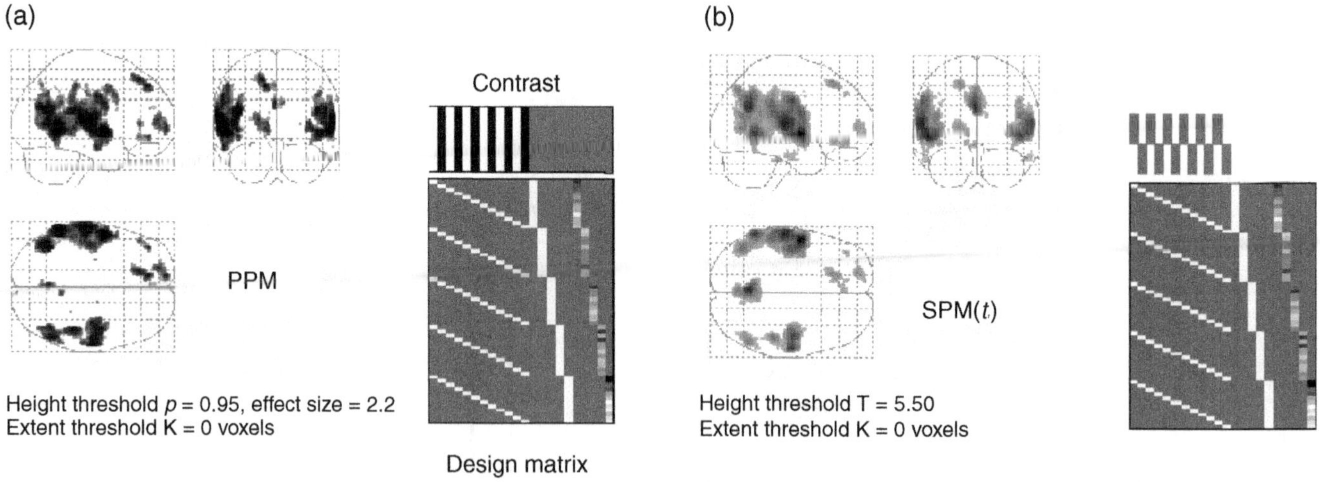

FIGURE 23.2 Bayesian and classical and inference for a PET study of word generation. (a) PPM for a contrast reflecting the difference between word-shadowing and word-generation, using an activation threshold of 2.2 and a confidence of 95 per cent. The design matrix and contrast for this model are shown (right) in image format. We have modelled each scan as a specific effect that has been replicated over subjects. (b) Classical SPM of the t-statistic for the same contrast. This SPM has been thresholded at $p = 0.05$, corrected using a random-field adjustment.

in regional cerebral blood flow (rCBF) of 2.2 adimensional units (equivalent to ml/dl/min). The second threshold is more trivial and simply enables the use of maximum intensity projections. This is the probability the voxel has to exceed in order to be displayed. In the PPM shown, this was set at 95 per cent. This means that all voxels shown have greater than 95 per cent probability of being deactivated by 2.2 or more. The PPM can be regarded as a way of summarizing one's confidence that an effect is present (cf. the use of confidence intervals where the lower bound on the interval is set at γ). It should be noted that posterior inference would normally require the reporting of the conditional probability whether it exceeded some arbitrary threshold or not. However, for the visual display of posterior probability maps, it is useful to remove voxels that fall below some threshold.

Figure 23.3 provides a quantitative representation of Bayesian inference afforded by PPMs. In the left-hand panel, the posterior expectation for the twelve condition-specific effects are shown, encompassed by the 95 per cent confidence intervals (bars) based on the posterior covariance. It can be seen that in the fifth condition (the third word-shadowing condition) one could be almost certain the activation is greater than zero. The prior and posterior densities for this activation are shown in the right-hand panel. These are the probability distributions before and after observing the data. Note that the posterior variance is always smaller than the prior variance, depending on how noisy the data are.

The corresponding SPM is shown in the right-hand panel (see Figure 23.2). The SPM has been thresholded at 0.05 adjusted for the search volume using a

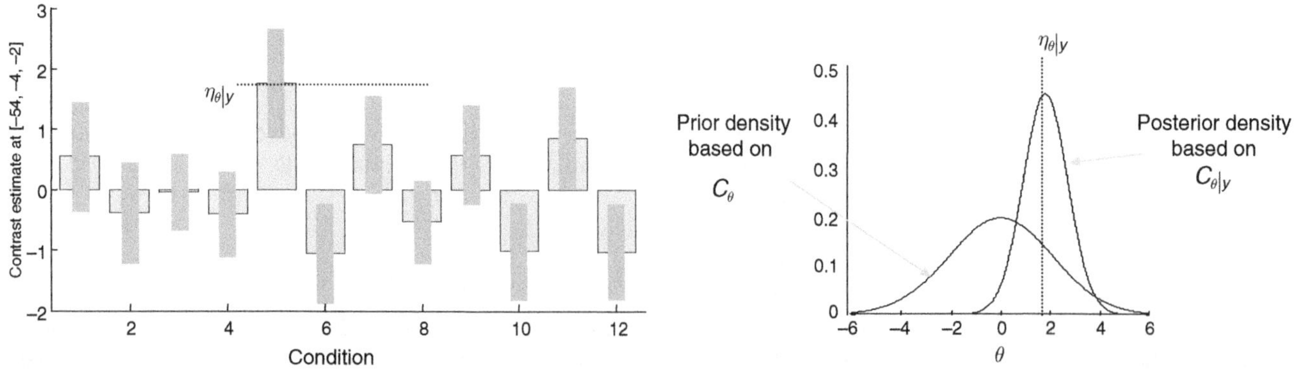

FIGURE 23.3 Illustrative results for a single voxel – the maximum in the left temporal region of the PPM in Figure 23.3 (−54, −4, −2 mm). Right panel: these are the conditional or posterior expectations and 95 per cent confidence intervals for the activation effect associated with each of the 12 conditions. Note that the odd conditions (word shadowing) are generally higher. In condition 5, one would be more than 95 per cent certain the activation exceeded 2.2. Left panel: the prior and posterior densities for the parameter estimate for condition 5.

random-field correction. There is a remarkable correspondence between the activation profiles inferred by the PPM and the SPM. The similarity between the PPM and the SPM for these data should not be taken as characteristic. The key difference between Bayesian inference, based on the confidence we have about activation, and classical inference, based on rejecting the null hypothesis, is that the latter depends on the search volume. The classical approach, when applied in a mass-univariate setting (i.e. over a family of voxels) induces a multiple comparison problem that calls for a procedure to control for family-wise false positives. In the context of imaging data, this procedure is a random-field adjustment to the threshold. This adjustment depends on the search volume. This means that, if we increased the search volume, the threshold would rise and some of the voxels seen in the SPM would disappear. Because the PPM does not label any voxel as 'activated', there is no multiple comparison problem and the 95 per cent confidence threshold is the same irrespective of search volume. This difference between PPMs and SPMs is highlighted in the analysis of the fMRI data. Here, the search volume is effectively increased by reducing the smoothness of the data. We do this by switching from PET to fMRI. Smoothness controls the 'statistical' search volume, which is generally much greater for fMRI than for PET.

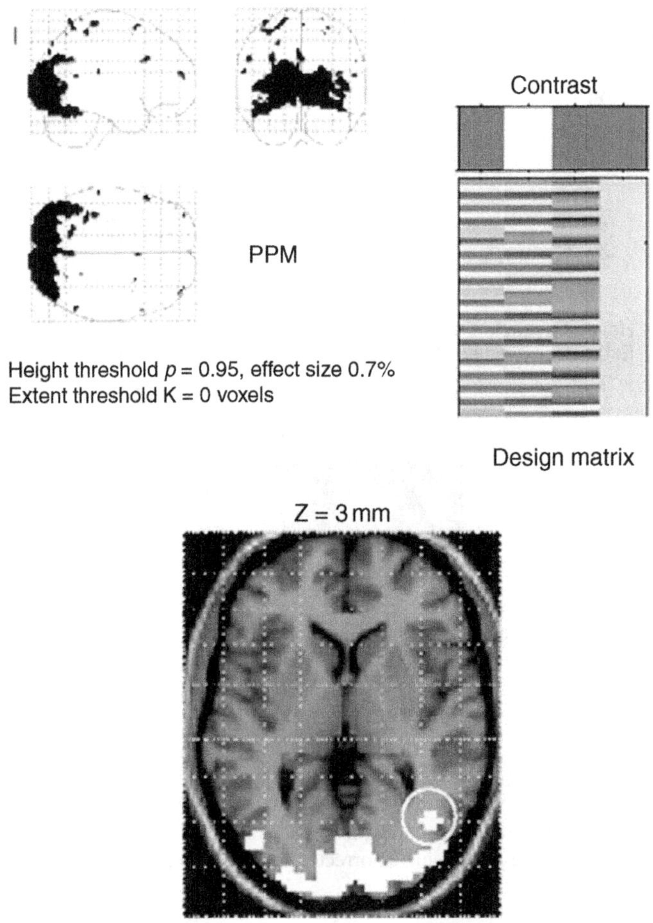

FIGURE 23.4 PPM for the fMRI study of attention to visual motion. The display format in the lower panel uses an axial slice through extra-striate regions but the thresholds are the same as employed in maximum intensity projections (upper panels). The activation threshold for the PPM was 0.7. As can be imputed from the design matrix, the statistical model of evoked responses comprised boxcar regressors convolved with a canonical haemodynamic response function.

Inference for the fMRI data

The difference between the PPM and SPM for the fMRI analysis is immediately apparent on inspection of Figures 23.4 and 23.5. Here the default threshold for the PPM was 0.7 per cent (equivalent to percentage whole-brain mean signal). Again, only voxels that exceed 95 per cent confidence are shown. These are restricted to visual and extra-striate cortex involved in motion processing. The thing to note here is that the corresponding SPM identifies a smaller number of voxels than the PPM. Indeed, the SPM appears to have missed a critical and bilaterally represented part of the V5 complex (circled cluster on the PPM in the lower panel of Figure 23.4). The SPM is more conservative because the correction for multiple comparisons in these data is very severe, rendering classical inference relatively insensitive. It is interesting to note that dynamic motion in the visual field has such widespread (if small) effects at a haemodynamic level.

PPMs and false discovery rate

There is an interesting connection between false discovery rate (FDR – see Chapter 20) control and thresholded PPMs. Subjecting PPMs to a 95 per cent threshold means that surviving voxels have, at most, a 5 per cent probability of not exceeding the default threshold γ. In other words, if we declared these voxels as 'activated', 5 per cent of the voxels could be false activations. This is exactly the same as FDR in the sense that the FDR is the proportion of voxels that are declared significant but are not. It should be noted that many voxels will have a posterior probability that is more than 95 per cent. Therefore, the 5 per cent is an upper bound on the FDR. This interpretation rests explicitly on thresholding the PPM and labelling the excursion set as 'activated'. It is reiterated that this declaration is unnecessary and only has any meaning in relation to classical inference. However, thresholded PPMs do have this interesting connection to SPMs in which false discovery rate has been controlled.

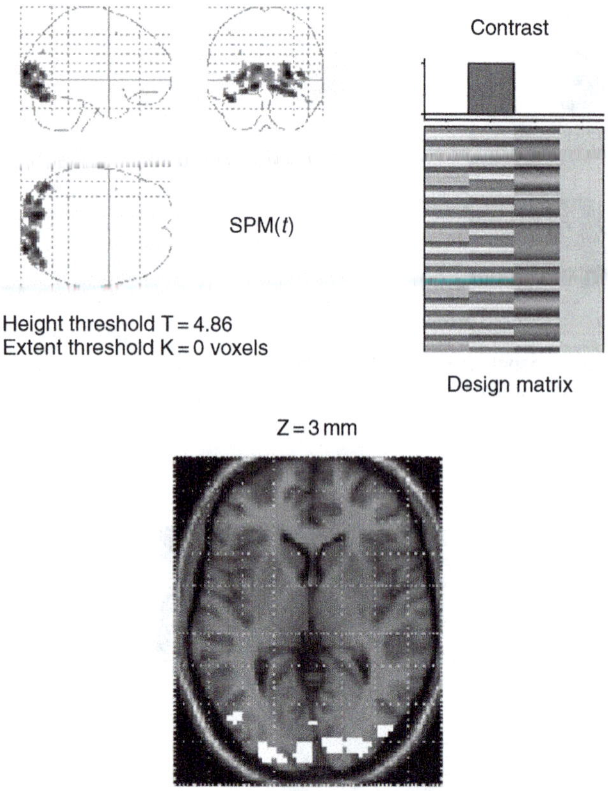

FIGURE 23.5 As for Figure 23.4, but this time showing the corresponding SPM using a corrected threshold of $p = 0.05$.

Conclusion

In this section, we looked at a simple way to construct posterior probability maps using empirical Bayes. Empirical Bayes can be used because of the natural hierarchy in neuroimaging engendered by looking for the same thing over multiple voxels. The approach provides simple shrinkage priors based on between-voxel variation in parameters controlling effects of interest. A computationally expedient way of computing these priors using ReML was presented that pools over voxels. This pooling device offers an enormous computational saving through simplifying the matrix algebra and enabling the construction of whole-brain PPMs. The same device has found an interesting application in the ReML estimation of prior variance components in space, by pooling over time bins, in the EEG source reconstruction problem (see Chapters 29 and 30).

A key consideration in the use of empirical Bayes in this setting is 'which voxels to include in the hierarchy?' There is no right or wrong answer here (cf. the search volume in classical inference with SPMs). The most important thing to bear in mind is that the conditional estimators of an activation or effect are those which minimize some cost function. This cost function can be regarded as the ability to predict the observed response with minimum error, on average, over the voxels included in the hierarchical model. In other words, the voxels over which the priors are computed define the space one wants, on average, the best estimates for. In this work we have simply used potentially responsive voxels within the brain as defined by thresholding the original images (to exclude extra-cranial regions).

In the next chapter, we turn to variational techniques for Bayesian inversion, of which EM and ReML can be regarded as special cases. These variational approaches can accommodate a wide range of biophysically informed generative models, as we will see.

REFERENCES

Berry DA, Hochberg Y (1999) Bayesian perspectives on multiple comparisons. *J Stat Plann Inference* **82**: 215–27

Büchel C, Friston KJ (1997) Modulation of connectivity in visual pathways by attention: cortical interactions evaluated with structural equation modelling and fMRI. *Cereb Cortex* **7**: 768–78

Friston KJ, Penny W, Phillips C *et al.* (2002a) Classical and Bayesian inference in neuroimaging: theory. *NeuroImage* **16**: 465–83

Friston KJ, Glaser DE, Henson RNA *et al.* (2002b) Classical and Bayesian inference in neuroimaging: applications. *NeuroImage* **16**: 484–512

Holmes A, Ford I (1993) A Bayesian approach to significance testing for statistic images from PET. In *Quantification of brain function, tracer kinetics and image analysis in brain PET*, Uemura K, Lassen NA, Jones T *et al.* (eds). Excerpta Medica, Int. Cong. Series No. 1030: 521–34

CHAPTER

24

Variational Bayes

W. Penny, S. Kiebel and K. Friston

INTRODUCTION

Bayesian inference can be implemented for arbitrary probabilistic models using Markov chain Monte Carlo (MCMC) (Gelman *et al.*, 1995). But MCMC is computationally intensive and so not practical for most brain imaging applications. This chapter describes an alternative framework called 'variational Bayes (VB)' which is computationally efficient and can be applied to a large class of probabilistic models (Winn and Bishop, 2005).

The VB approach, also known as 'ensemble learning', takes its name from Feynmann's variational free energy method developed in statistical physics. VB is a development from the machine learning community (Peterson and Anderson, 1987; Hinton and van Camp, 1993) and has been applied in a variety of statistical and signal processing domains (Bishop *et al.*, 1998; Jaakola *et al.*, 1998; Ghahramani and Beal, 2001; Winn and Bishop, 2005). It is now also widely used in the analysis of neuroimaging data (Penny *et al.*, 2003; Sahani and Nagarajan, 2004; Sato *et al.*, 2004; Woolrich, 2004; Penny *et al.*, 2006; Friston *et al.*, 2006).

This chapter is structured as follows. We describe the fundamental relationship between model evidence, free energy and Kullback-Liebler (KL) divergence that lies at the heart of VB. Before this, we review the salient properties of the KL-divergence. We then describe how VB learning delivers a factorized, minimum KL-divergence approximation to the true posterior density in which learning is driven by an explicit minimization of the free energy. The theoretical section is completed by relating VB to Laplace approximations and describing how the free energy can also be used as a surrogate for the model evidence, allowing for Bayesian model comparison. Numerical examples are then given showing how VB differs from Laplace and providing simulation studies using models of functional magnetic resonance imaging (fMRI) data (Penny *et al.*, 2003). These are based on a general linear model with autoregressive errors, or GLM-AR model.

THEORY

In what follows we use upper-case letters to denote matrices and lower-case to denote vectors. $\mathsf{N}(m, \Sigma)$ denotes a uni/multivariate Gaussian with mean m and variance/covariance Σ. X^T denotes the matrix transpose and $\log x$ denotes the natural logarithm.

Kullback-Liebler divergence

For densities $q(\theta)$ and $p(\theta)$ the relative entropy or Kullback-Liebler (KL) divergence from q to p is (Cover and Thomas, 1991):

$$KL[q\|p] = \int q(\theta) \log \frac{q(\theta)}{p(\theta)} d\theta \qquad \textbf{24.1}$$

The KL-divergence satisfies the Gibb's inequality (Mackay, 2003):

$$KL[q\|p] \geq 0 \qquad \textbf{24.2}$$

with equality only if $q = p$. In general $KL[q\|p] \neq KL[p\|q]$, so KL is not a distance measure. Formulae for computing KL, for both Gaussian and gamma densities, are given in Appendix, 24.1.

Model evidence and free energy

Given a probabilistic model of some data, the log of the 'evidence' or 'marginal likelihood' can be written as:

$$\begin{aligned}\log p(Y) &= \int q(\theta) \log p(Y) d\theta \\ &= \int q(\theta) \log \frac{p(Y, \theta)}{p(\theta|Y)} d\theta\end{aligned}$$

Statistical Parametric Mapping, by Karl Friston et al.
ISBN–13: 978-0-12-372560-8 ISBN–10: 0-12-372560-7

$$= \int q(\theta) \log \left[\frac{p(Y,\theta) q(\theta)}{q(\theta) p(\theta|Y)} \right] d\theta$$
$$= F + KL(q(\theta) \| p(\theta|Y)) \quad \textbf{24.3}$$

where $q(\theta)$ is considered, for the moment, as an arbitrary density. We have:

$$F = \int q(\theta) \log \frac{p(Y,\theta)}{q(\theta)} d\theta \quad \textbf{24.4}$$

which, in statistical physics is known as the *negative* free energy. The second term in Eqn. **24.3** is the KL-divergence between the density $q(\theta)$ and the true posterior $p(\theta|Y)$. Eqn. **24.3** is the fundamental equation of the VB-framework and is shown graphically in Figure 24.1.

Because KL is always positive, due to the Gibbs inequality, F provides a lower bound on the model evidence. Moreover, because KL is zero when two densities are the same, F will become equal to the model evidence when $q(\theta)$ is equal to the true posterior. For this reason $q(\theta)$ can be viewed as an *approximate posterior*.

The aim of VB-learning is to maximize F and so make the approximate posterior as close as possible to the true posterior. This approximate posterior will be the one that best approximates the true posterior in the sense of minimizing KL-divergence. We should point out that this divergence cannot be minimized explicitly because $p(\theta|y)$ is only known up to a constant. Instead, it is minimized implicitly by maximizing F and by virtue of Eqn. **24.3**. Of course, maximizing F, the negative free energy, is the same as minimizing $-F$, the free energy.

Factorized approximations

To obtain a practical learning algorithm we must also ensure that the integrals in F are tractable. One generic procedure for attaining this goal is to assume that the

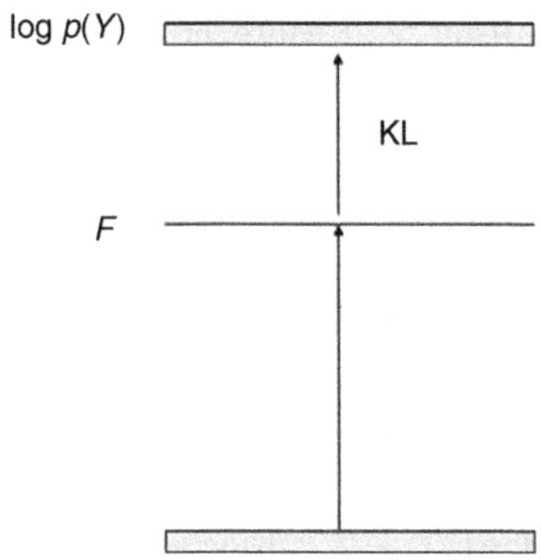

FIGURE 24.1 The negative free energy, F, provides a lower bound on the log-evidence of the model with equality when the approximate posterior equals the true posterior.

approximating density factorizes over groups of parameters. In physics, this is known as the mean field approximation. Thus, we consider:

$$q(\theta) = \prod_i q(\theta_i) \quad \textbf{24.5}$$

where θ_i is the ith group of parameters. We can also write this as:

$$q(\theta) = q(\theta_i) q(\theta_{\backslash i}) \quad \textbf{24.6}$$

where $\theta_{\backslash i}$ denotes all parameters *not* in the ith group. The distributions $q(\theta_i)$ which maximize F can then be derived as follows:

$$F = \int q(\theta) \log \left[\frac{p(Y,\theta)}{q(\theta)} \right] d\theta \quad \textbf{24.7}$$
$$= \int \int q(\theta_i) q(\theta_{\backslash i}) \log \left[\frac{p(Y,\theta)}{q(\theta_i) q(\theta_{\backslash i})} \right] d\theta_{\backslash i} d\theta_i$$
$$= \int q(\theta_i) \left[\int q(\theta_{\backslash i}) \log p(Y,\theta) d\theta_{\backslash i} \right] d\theta_i$$
$$- \int q(\theta_i) \log q(\theta_i) d\theta_i + C$$
$$= \int q(\theta_i) I(\theta_i) d\theta_i - \int q(\theta_i) \log q(\theta_i) d\theta_i + C$$

where the constant C contains terms not dependent on $q(\theta_i)$ and:

$$I(\theta_i) = \int q(\theta_{\backslash i}) \log p(Y,\theta) d\theta_{\backslash i} \quad \textbf{24.8}$$

Writing $I(\theta_i) = \log \exp I(\theta_i)$ gives:

$$F = \int q(\theta_i) \log \left[\frac{\exp(I(\theta_i))}{q(\theta_i)} \right] d\theta_i + C \quad \textbf{24.9}$$
$$= KL\left[q(\theta_i) \| \exp(I(\theta_i)) \right] + C$$

This is minimized when:

$$q(\theta_i) = \frac{\exp[I(\theta_i)]}{Z} \quad \textbf{24.10}$$

where Z is the normalization factor needed to make $q(\theta_i)$ a valid probability distribution. Importantly, this means we are often able to determine the optimal analytic *form* of the component posteriors. This results in what is known as a 'free-form' approximation.

For example, Mackay (1995) considers the case of linear regression models with gamma priors over error precisions, λ, and Gaussian priors over regression coefficients β, with a factorized approximation $q(\beta, \lambda) = q(\beta) q(\lambda)$. Application of Eqn. **24.10** then leads to an expression in which $I(\lambda)$ has terms in λ and $\log \lambda$ only. From this we can surmise that the optimal form for $q(\lambda)$ is a gamma density (see Appendix 24.1).

More generally, free-form approximations can be derived for models from the 'conjugate-exponential' family (Attias, 1999; Ghahramani and Beal, 2001; Winn and Bishop, 2005). Exponential family distributions include Gaussians and discrete multinomials and conjugacy requires the posterior (over a factor) to have the same functional form as the prior.

This allows free-form VB to be applied to arbitrary directed acyclic graphs comprising discrete multinomial variables with arbitrary subgraphs of univariate and multivariate Gaussian variables. Special cases include hidden Markov models, linear dynamical systems, principal component analysers, as well as mixtures and hierarchical mixtures of these. Moreover, by introducing additional variational parameters, free-form VB can be applied to models containing non-conjugate distributions. This includes independent component analysis (Attias, 1999) and logistic regression (Jaakola and Jordan, 1997).

Application of Eqn. **24.10** also leads to a set of update equations for the *parameters* of the component posteriors. This is implemented for the linear regression example by equating the coefficients of λ and $\log \lambda$ with the relevant terms in the gamma density (see Appendix 24.1). In the general case, these update equations are coupled as the solution for each $q(\theta_i)$ depends on expectations with respect to the other factors $q(\theta_{\backslash i})$. Optimization proceeds by initializing each factor and then cycling through each factor in turn and replacing the current distribution with the estimate from Eqn. **24.10**. Examples of these update equations are provided in the following chapter, which applies VB to spatio-temporal models of fMRI data.

Laplace approximations

Laplace's method approximates the integral of a function $\int f(\theta)d\theta$ by fitting a Gaussian at the maximum $\hat{\theta}$ of $f(\theta)$, and computing the volume of the Gaussian. The covariance of the Gaussian is determined by the Hessian matrix of $\log f(\theta)$ at the maximum point $\hat{\theta}$ (Mackay, 1998).

The term 'Laplace approximation' is used for the method of approximating a posterior distribution with a Gaussian centred at the maximum *a posteriori* (MAP) estimate. This is the application of Laplace's method with $f(\theta) = p(Y|\theta)p(\theta)$. This can be justified by the fact that under certain regularity conditions, the posterior distribution approaches a Gaussian as the number of samples grows (Gelman *et al.*, 1995). This approximation is derived in detail in Chapter 35.

Despite using a full distribution to approximate the posterior, instead of a point estimate, the Laplace approximation still suffers from most of the problems of MAP estimation. Estimating the variances at the end of iterated learning does not help if the procedure has already led to an area of low probability mass. This point will be illustrated in the results section.

This motivates a different approach where, for nonlinear models, the Laplace approximation is used at each step of an iterative approximation process. This is described in Chapters 22 and 35. In fact, this method is an expectation maximization (EM) algorithm, which is known to be a special case of VB (Beal, 2003). This is clear from the fact that, at each step of the approximation, we have an ensemble instead of a point estimate.

The relations between VB, EM, iterative Laplace approximations, and an algorithm from classical statistics called restricted maximum likelihood (ReML) are discussed in Appendix 4 and Friston *et al.* (2006). This algorithm uses a 'fixed-form' for the approximating ensemble, in this case being a full-covariance Gaussian. This is to be contrasted with the 'free-form' VB algorithms described in the previous section, where the optimal form for $q(\theta)$ is derived from $p(Y, \theta)$ and the assumed factorization.

Model inference

As we have seen earlier, the negative free energy, F, is a lower bound on the model evidence. If this bound is tight then F can be used as a surrogate for the model evidence and so allow for Bayesian model selection and averaging.[1] This provides a mechanism for fine-tuning models. In neuroimaging, F has been used to optimize the choice of haemodynamic basis set (Penny *et al.*, 2006), the order of autoregressive models (Penny *et al.*, 2003) (see also Chapter 40), and the spatial diffusivity of EEG sources (see Chapter 26).

Earlier, the negative free energy was written:

$$F = \int q(\theta) \log \frac{p(Y, \theta)}{q(\theta)} d\theta \qquad \textbf{24.11}$$

By using $p(Y, \theta) = p(Y|\theta)p(\theta)$ we can express it as the sum of two terms:

$$F(\theta) = \int q(\theta) \log p(Y|\theta) d\theta - KL[q(\theta)\|p(\theta)] \qquad \textbf{24.12}$$

where the first term is the average likelihood of the data and the second term is the KL between the approximating posterior and the *prior*. This is not to be confused with the KL in Eqn. **24.3** which was between the approximate posterior and the true posterior. In Eqn. **24.12**, the KL

[1] Throughout this chapter our notation has, for brevity, omitted explicit dependence on the choice of model, m. But strictly, e.g. $p(Y)$, F, $p(\theta|Y)$ and $q(\theta)$ should be written as $p(Y|m)$, $F(m)$, $p(\theta|Y, m)$ and $q(\theta|m)$.

term grows with the number of model parameters and so penalizes more complex models. Thus, F contains both accuracy and complexity terms, reflecting the two conflicting requirements of a good model, that it fit the data yet be as simple as possible. Model selection principles are also discussed in Chapter 35.

In the very general context of probabilistic graphical models, Beal and Ghahramani (2003) have shown that the above VB approximation of model evidence is considerably more accurate than the Bayesian information criterion (BIC), while incurring little extra computational cost. Chapter 35 discusses the utility of the BIC in the context of fixed priors. Moreover, it is of comparable accuracy to a much more computationally demanding method based on annealed importance sampling (AIS) (Beal and Ghahramani, 2003).

EXAMPLES

This section first provides an idealized example which illustrates the difference between Laplace and VB approximations. We then present some simulation results showing VB applied to a model of fMRI data. In what follows 'Laplace approximation' refers to a Gaussian centred at the MAP estimate and VB uses either a fixed-form approximation (see 'univariate densities' below) or a free-form approximation (see 'factorized approximation' below).

Univariate densities

Figure 24.2 and Plate 22 (see colour plate section) provide an example showing what it means to minimize KL for univariate densities. The solid lines in Figure 24.2 show a posterior distribution p which is a Gaussian mixture density comprising two modes. The first contains the maximimum *a posteriori* (MAP) value and the second contains the majority of the probability mass.

The Laplace approximation to p is therefore given by a Gaussian centred around the first, MAP mode. This is shown in Figure 24.2(a). This approximation does not have high probability mass, so the model evidence will be underestimated.

Figure 24.2(b) shows a Laplace approximation to the second mode, which could arise if MAP estimation found a local, rather than a global, maximum. Finally, Figure 24.2(c) shows the minimum KL-divergence approximation, assuming that q is a Gaussian. This is a fixed-form VB approximation, as we have fixed the form of the approximating density (i.e. q is a Gaussian). This VB solution corresponds to a density q which is moment-matched to p.

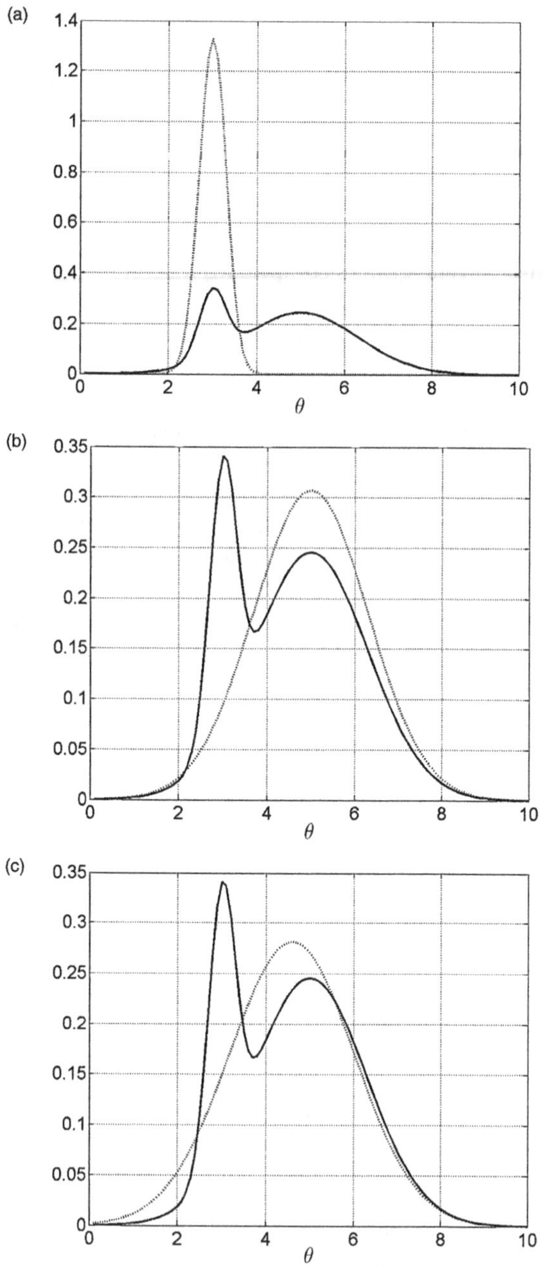

FIGURE 24.2 Probability densities $p(\theta)$ (solid lines) and $q(\theta)$ (dashed lines) for a Gaussian mixture $p(\theta) = 0.2 \times \mathrm{N}(m_1, \sigma_1^2) + 0.8 \times \mathrm{N}(m_2, \sigma_2^2)$ with $m_1 = 3, m_2 = 5, \sigma_1 = 0.3, \sigma_2 = 1.3$, and a single Gaussian $q(\theta) = \mathrm{N}(\mu, \sigma^2)$ with (a) $\mu = \mu_1, \sigma = \sigma_1$ which fits the first mode, (b) $\mu = \mu_2, \sigma = \sigma_2$ which fits the second mode and (c) $\mu = 4.6, \sigma = 1.4$ which is moment-matched to $p(\theta)$.

Plate 22 plots $KL[q\|p]$ as a function of the mean and standard deviation of q, showing a minimum around the moment-matched values. These KL values were computed by discretizing p and q and approximating Eqn. **24.1** by a discrete sum. The MAP mode, maximum mass mode and moment-matched solutions have $KL[q\|p]$

values of 11.7, 0.93 and 0.71 respectively. This shows that low KL is achieved when q captures most of the probability mass of p and, minimum KL when q is moment-matched to p. Plate 22 also shows that, for reasonable values of the mean and standard deviation, there are no local minima. This is to be contrasted with the posterior distribution itself which has two maxima, one local and one global.

Capturing probability mass is particularly important if one is interested in non-linear functions of parameter values, such as model predictions. This is the case for the dynamic causal models described in later chapters. Figures 24.3 and 24.4 show histograms of model predictions for squared and logistic-map functions indicating that VB predictions are qualitatively better than those from the Laplace approximation.

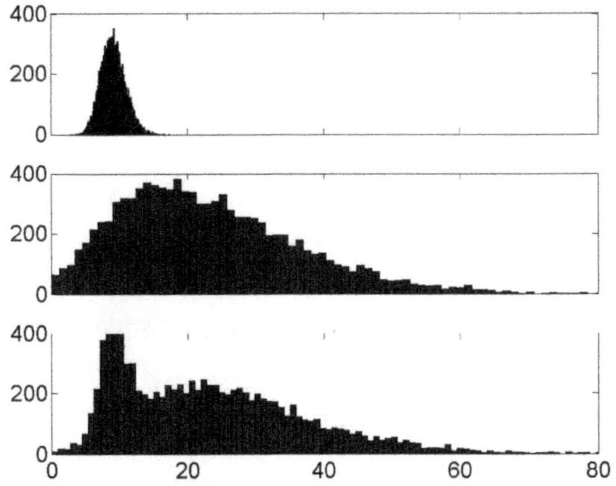

FIGURE 24.3 Histograms of 10 000 samples drawn from $g(\theta)$ where the distribution over θ is from the Laplace approximation (top), VB approximation (middle) and true distribution, p (bottom) for $g(\theta) = \theta^2$.

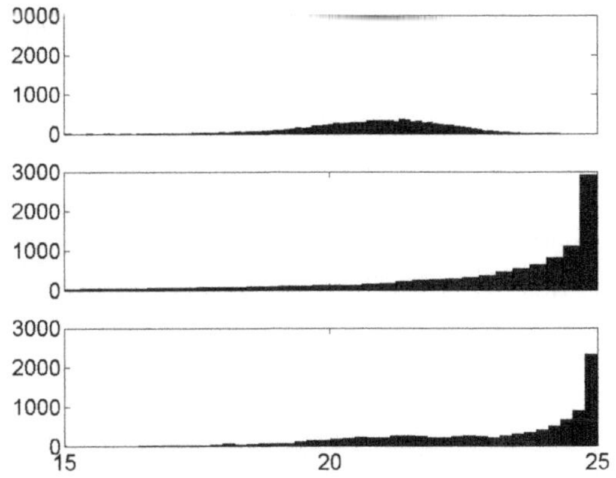

FIGURE 24.4 Histograms of 10000 samples drawn from $g(\theta)$ where the distribution over θ is from the Laplace approximation (top), VB approximation (middle) and true distribution, p (bottom) for $g(\theta) = \theta * (10 - \theta)$. This is akin to a logistic map function encountered in dynamical systems (Mullin, 1993).

Often in Bayesian inference, one quotes posterior exceedance probabilities. Examples of this are the posterior probability maps described in Chapter 23 and dynamic causal models in Chapter 41. For the squared function, Laplace says 5 per cent of samples are above $g = 12.2$. But in the true density, 71 per cent of samples are. For the logisitic function 62 per cent are above Laplace's 5 per cent point. The percentage of samples above VB's 5 per cent points are 5.1 per cent for the squared function and 4.2 per cent for the logistic-map function. So, for this example, Laplace can tell you the posterior exceedance probability is 5 per cent when, in reality, it is an order of magnitude greater. This is not the case for VB.

As we shall see later on, the VB solution depends crucially on our assumptions about q, either in terms of the factorization assumed (this is of course, irrelevant for univariate densities) or the family of approximating densities assumed for q. For example, if q were a mixture density, as in Bishop *et al.* (1998), then VB would provide an exact approximation of p. It is also important to note that the differences between VB and Laplace depend on the nature of p. For unimodal p, these differences are likely to be less significant than those in the above example.

Factorized approximation

We now present results of a simulation study using a general linear model with autoregressive errors, or GLM-AR model. The GLM-AR model can describe both the signal and noise characteristics of fMRI data. This model is used in the rest of the results section. For simplicity, we describe application to data at a single voxel. But the next chapter augments the model with a spatial prior and shows it can be applied to whole slices of data.

We first illustrate VB's factorized approximation to the posterior and compare the marginal distributions obtained with VB to those from exact evalution. We generated data from a known GLM-AR model:

$$y_t = x_t w + e_t \qquad 24.13$$

$$e_t = a e_{t-1} + z_t \qquad 24.14$$

where $x_t = 1$ for all t, $w = 2.7$, $a = 0.3$ and $1/\lambda = \mathrm{Var}(z) = \sigma^2 = 4$. We generated $N = 128$ samples. Given any particular values of parameters $\theta = \{w, a, \lambda\}$ it is possible to compute the exact posterior distribution up to a normalization factor, as:

$$p(w, a, \lambda | Y) \propto p(Y | w, a, \lambda) p(w|\alpha) p(a|\beta) p(\lambda) \qquad 24.15$$

where α is the prior precision of regression coefficients and β is the prior precision of AR coefficients (see next chapter for more details). If we evaluate the above quantity over a grid of values w, a, λ, we can then normalize it so it sums to one and so make plots of the exact posterior density. We then assumed an approximate posterior $q(w, a, \lambda) = q(w)q(a)q(\lambda)$ and used VB to fit it to the data. Update equations are available in Penny *et al.* (2003).

Figure 24.5 compares the exact and approximate posterior joint densities for w, a. In the true posterior, it is clear that there is a dependence between w and a but VB's approximate posterior ignores this dependence. Figure 24.6 compares the exact and approximate posterior marginal densities for w, a and σ^2. In this example, VB has accurately estimated the marginal distributions.

Model inference

We generated data from a larger GLM-AR model having two regression coefficients and three autoregressive coefficients:

$$y_t = x_t w + e_t \tag{24.16}$$

$$e_t = \sum_{j=1}^{m} a_j e_{t-j} + z_t \tag{24.17}$$

where x_t is a two-element row vector, the first element flipping between a '−1' and '1' with a period of 40 scans (i.e. 20 − 1s followed by 20 1s) and the second element being '1' for all t. The two corresponding entries in w reflect the size of the activation, $w_1 = 2$, and the mean signal level, $w_2 = 3$. We used an AR(3) model for the errors with parameters $a_1 = 0.8$, $a_2 = -0.6$ and $a_3 = 0.4$. The noise precision was set to $1/\lambda = \mathrm{Var}(z) = \sigma^2 = 1$ and

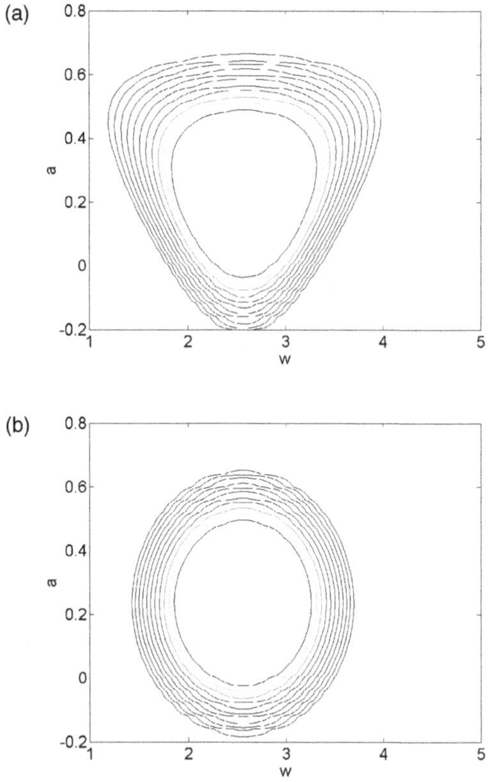

FIGURE 24.5 The figures show contour lines of constant probability density from (a) the exact posterior $p(a, w|Y)$ and (b) the approximate posterior used in VB, $q(a, w)$ for the GLM-AR model. This clearly shows the effect of the factorization, $q(a, w) = q(a)q(w)$.

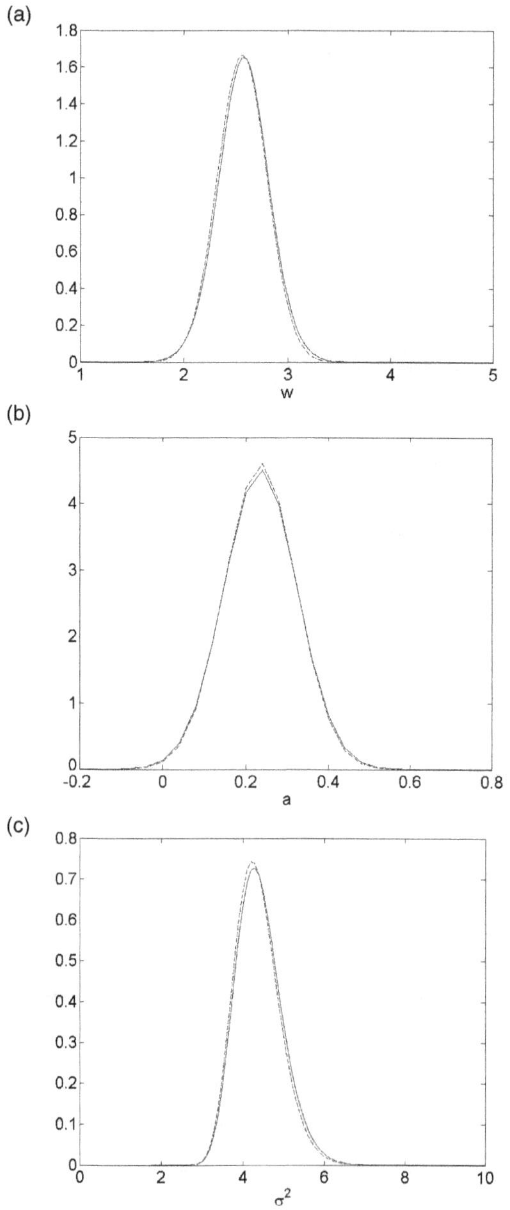

FIGURE 24.6 The figures compare the exact (solid lines) and approximate (dashed lines) marginal posteriors (a) $p(w|Y)$ and $q(w)$, (b) $p(a|Y)$ and $q(a)$, (c) $p(\sigma^2|Y)$ and $q(\sigma^2)$ (where $\sigma^2 = 1/\lambda$).

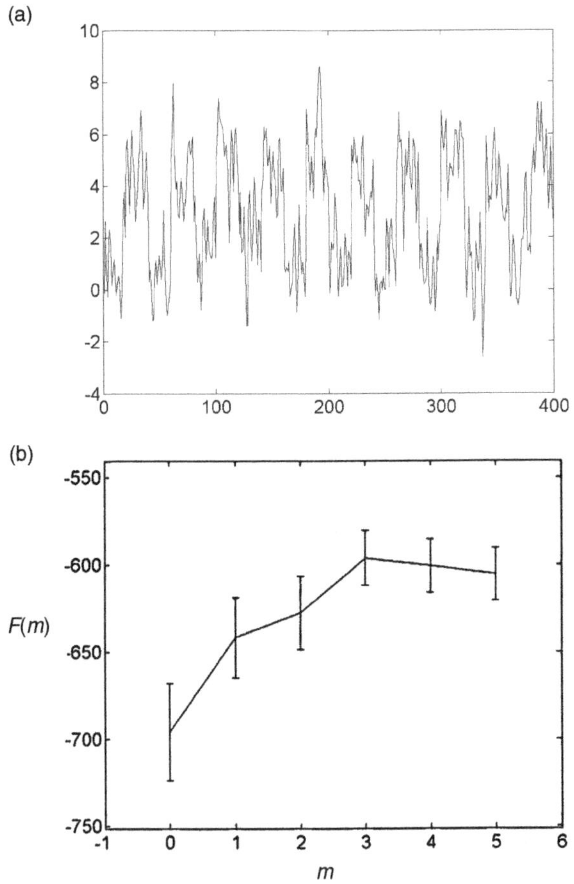

FIGURE 24.7 The figures show (a) an example time-series from a GLM-AR model with AR model order $m = 3$ and (b) a plot of the average negative free energy $F(m)$, with error bars, versus m. This shows that $F(m)$ picks out the correct model order.

we initially generated $N = 400$ samples. This is a larger model than in the previous example as we have more AR and regression coefficients. An example time series produced by this process is shown in Figure 24.7(a).

We then generated 10 such time-series and fitted GLM-AR(p) models to each using the VB algorithm. In each case, the putative model order was varied between $m = 0$ and $m = 5$ and we estimated the model evidence for each. Formulae for the model evidence approximation are available in Penny *et al.* (2003). Figure 24.7(b) shows a plot of the average value of the negative free energy, $F(m)$ as a function of m, indicating that the maximum occurs at the true model order.

Gibbs sampling

While it is possible, in principle, to plot the exact posteriors using the method described previously, this would require a prohibitive amount of computer time for this larger model. We therefore validated VB by comparing it with Gibbs sampling (Gelman *et al.*, 1995; Penny *et al.*, 2003).

We generated a number of datasets containing either $N = 40$, $N = 160$ or $N = 400$ scans. At each dataset size we compared Gibbs and VB posteriors for each of the regression coefficients. For the purpose of these comparisons the model order was kept fixed at $m = 3$. Figure 24.8 shows representative results indicating a better agreement with increasing number of scans. We also note that VB requires more iterations for fewer scans (typically 4 iterations for $N = 400$, 5 iterations for $N = 160$ and 7 iterations for $N = 40$). This is because the algorithm was initialized with an ordinary least squares (OLS) solution which is closer to the VB estimate if there is a large number of scans.

Estimation of effect size

Finally, we generated a number of data sets of various sizes to compare VB and OLS estimates of activation size with the true value of $w_1 = 2$. This comparison was made using a paired t-test on the absolute estimation error. For $N > 100$, the VB estimation error was significantly smaller for VB than for OLS ($p < 0.05$). For $N = 160$, for example, the VB estimation error was 15 per cent smaller than the OLS error ($p < 0.02$).

DISCUSSION

Variational Bayes delivers a factorized, minimum KL-divergence approximation to the true posterior density and model evidence. This provides a computationally efficient implementation of Bayesian inference for a large class of probabilistic models (Winn and Bishop, 2005). It allows for parameter inference, based on the approximating density $q(\theta|m)$ and model inference based on a free energy approximation, $F(m)$ to the model evidence, $p(y|m)$.

The quality of inference provided by VB depends on the nature of the approximating distribution. There are two distinct approaches here. Fixed-form approximations fix the form of q to be, for example, a diagonal (Hinton and van Camp, 1993) or full-covariance Gaussian ensemble (Friston *et al.*, 2006). Free-form approximations choose a factorization that depends on $p(Y, \theta)$. These range from fully factorized approximations, where there are no dependencies in q, to structured approximations. These identify substructures in $p(Y, \theta)$, such as trees or mixtures of trees, in which exact inference is possible. Variational methods are then used to handle interactions between them (Ghahramani, 2002).

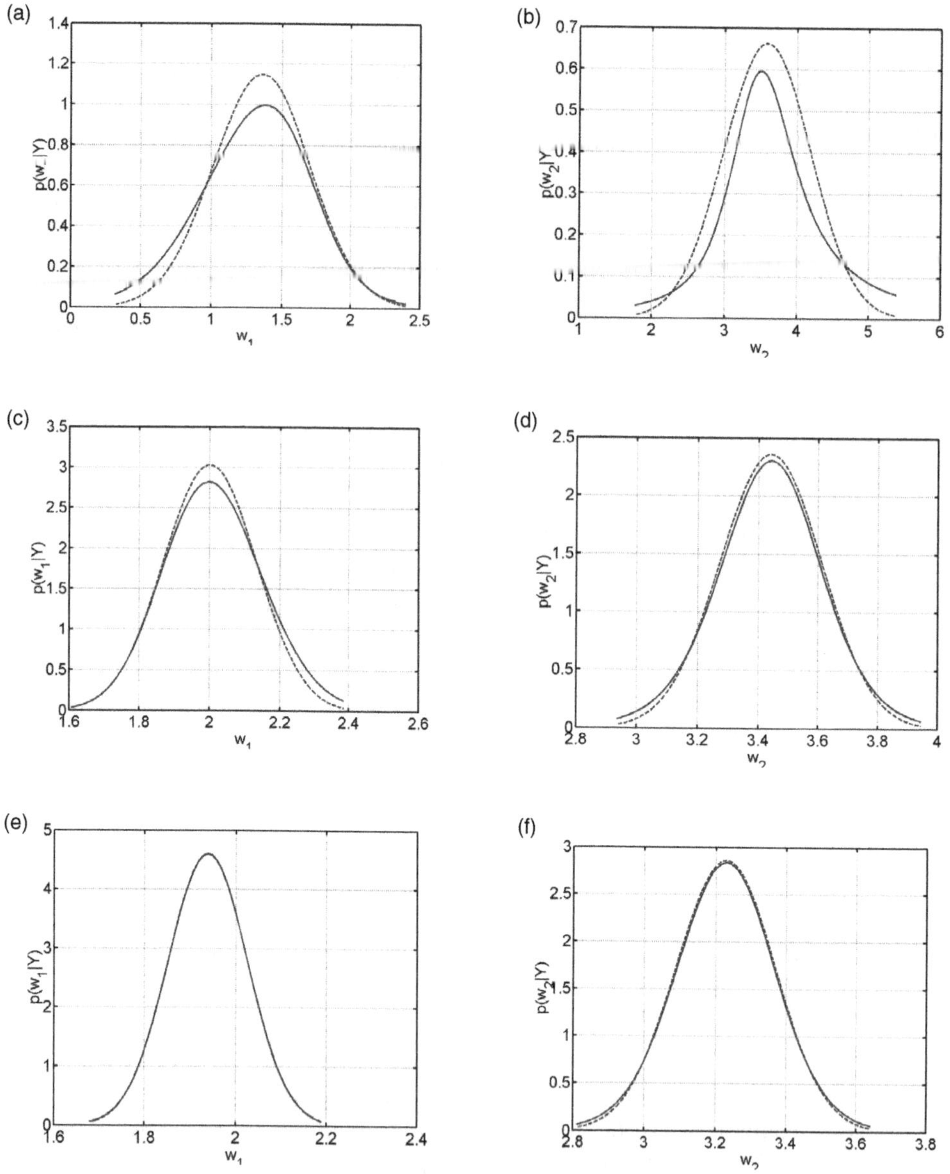

FIGURE 24.8 The figures show the posterior distributions from Gibbs sampling (solid lines) and variational Bayes (dashed lines) for data sets containing 40 scans (top row), 160 scans (middle row) and 400 scans (bottom row). The distributions in the left column are for the first regression coefficient (size of activation) and in the right column for the second regression coefficient (offset). The fidelity of the VB approximation increases with number of scans.

VB also delivers an approximation to the model evidence, allowing for Bayesian model comparison. However, it turns out that model selections based on VB are systematically biased towards simpler models (Beal and Ghahramani, 2003). Nevertheless, they have been shown empirically to be more accurate than BIC approximations and faster than sampling approximations (Beal and Ghahramani, 2003). Bayesian model selection is discussed further in Chapter 35.

Chapter 22 described a parametric empirical Bayes (PEB) algorithm for inference in hierarchical linear Gaussian models. This algorithm may be viewed as a special case of VB with a fixed-form full-covariance Gaussian ensemble (Friston *et al.*, 2006). More generally, however, VB can be applied to models with discrete as well as continuous variables.

A classic example here is the Gaussian mixture model. This has been applied to an analysis of intersubject variability in fMRI data. Model comparisons based on VB identified two overlapping degenerate neuronal systems in subjects performing a cross-modal priming task (Noppeney *et al.*, 2006).

In the dynamic realm, VB has been used to fit and select hidden Markov Models (HMMs) for the analysis of electroencephalographic (EEG) data (Cassidy and Brown, 2002). These HMMs use discrete variables to enumerate the hidden states and continuous variables to parameterize the activity in each. Here, VB identifies the number of stationary dynamic regimes, when they occur, and describes activity in each with a multivariate autoregressive (MAR) model. The application of VB to MAR models is described further in Chapter 40.

The following chapter uses a spatio-temporal model for the analysis of fMRI. This includes spatial regularization of the autoregressive processes which characterize fMRI noise. This regularization requires a prior over error terms which is precluded in Chapter 22's PEB framework but is readily accommodated using free-form VB.

APPENDIX 24.1

For univariate normal densities $q(x) = \mathsf{N}(\mu_q, \sigma_q^2)$ and $p(x) = \mathsf{N}(\mu_p, \sigma_p^2)$ the KL-divergence is:

$$KL_{N_1}(\mu_q, \sigma_q; \mu_p, \sigma_p) = 0.5 \log \frac{\sigma_p^2}{\sigma_q^2} + \frac{\mu_q^2 + \mu_p^2 + \sigma_q^2 - 2\mu_q\mu_p}{2\sigma_p^2} - 0.5 \quad 24.18$$

The multivariate normal density is given by:

$$\mathsf{N}(\mu, \Sigma) = (2\pi)^{-d/2}|\Sigma|^{-1/2} \exp\left(-\frac{1}{2}(x-\mu)^T\Sigma^{-1}(x-\mu)\right) \quad 24.19$$

The KL divergence for normal densities $q(x) = \mathsf{N}(\mu_q, \Sigma_q)$ and $p(x) = \mathsf{N}(\mu_p, \Sigma_p)$ is:

$$KL_N(\mu_q, \Sigma_q; \mu_p, \Sigma_p) = 0.5 \log \frac{|\Sigma_p|}{|\Sigma_q|} + 0.5Tr(\Sigma_p^{-1}\Sigma_q) + 0.5(\mu_q - \mu_p)^T\Sigma_p^{-1}(\mu_q - \mu_p) - \frac{d}{2} \quad 24.20$$

where $|\Sigma_p|$ denotes the determinant of the matrix Σ_p.

The gamma density is defined as:

$$Ga(b, c) = \frac{1}{\Gamma(c)} \frac{x^{c-1}}{b^c} \exp\left(\frac{-x}{b}\right) \quad 24.21$$

The log of the gamma density:

$$\log Ga(b, c) = -\log\Gamma(c) - c\log b + (c-1)\log x - \frac{x}{b} \quad 24.22$$

In Mackay (1995), application of Eqn. **24.10** for the approximate posterior over the error precision $q(\lambda)$ leads to an expression containing terms in λ and $\log\lambda$ only. This identifies $q(\lambda)$ as a gamma density. The coefficients of these terms are then equated with those in the above equation to identify the parameters of $q(\lambda)$.

For gamma densities $q(x) = Ga(x; b_q, c_q)$ and $p(x) = Ga(x; b_p, c_p)$ the KL-divergence is:

$$KL_{Ga}(b_q, c_q; b_p, c_p) = (c_q - 1)\Psi(c_q) - \log b_q - c_q - \log\Gamma(c_q) + \log\Gamma(c_p) + c_p \log b_p - (c_p - 1)(\Psi(c_q) + \log b_q) + \frac{b_q c_q}{b_p} \quad 24.23$$

where $\Gamma()$ is the gamma function and $\Psi()$ the digamma function (Press *et al.*, 1992). Similar equations for multinomial and Wishart densities are given in Penny (2001).

REFERENCES

Attias H (1999) Inferring parameters and structure of latent variable models by variational Bayes. In: *Proceedings of the Fifteenth Conference on Uncertainty in Artificial Intelligence* Morgan Kaufmann, California

Beal M (2003) *Variational algorithms for approximate Bayesian inference.* PhD thesis, Gatsby Computational Neuroscience Unit, University College London

Beal M, Ghahramani Z (2003) The variational Bayesian EM algorithms for incomplete data: with application to scoring graphical model structures. In *Bayesian statistics 7*, Bernardo J, Bayarri M, Berger J *et al.* (eds). Cambridge University Press, Cambridge

Bishop CM, Lawrence N, Jaakkola TS *et al.* (1998) Approximating posterior distributions in belief networks using mixtures. In *Advances in neural information processing systems 10*, Kearns MJ, Jordan MI, Solla SA (eds) MIT Press, Cambridge, MA

Cassidy MJ, Brown P (2002) Hidden Markov based autoregressive analysis of stationary and non-stationary electrophysiological signals for functional coupling studies. *J Neurosci Meth* **116**: 35–53

Cover TM, Thomas JA (1991) *Elements of information theory.* John Wiley, New York

Friston K, Mattout J, Trujillo-Barreto N *et al.* (2006) Variational free energy and the Laplace approximation *NeuroImage* (in press)

Gelman A, Carlin JB, Stern HS *et al.* (1995) *Bayesian data analysis.* Chapman and Hall, Boca Raton

Ghahramani Z (2002) On structured variational approximations. Technical report, Gatsby Computational Neuroscience Unit, UCL, London

Ghahramani Z, Beal MJ (2001) Propagation algorithms for variational Bayesian learning. In *NIPS 13*, Leen T *et al.* (eds). MIT Press, Cambridge, MA

Hinton GE, van Camp D (1993) Keeping neural networks simple by minimizing the description length of the weights. In *Proceedings of the Sixth Annual Conference on Computational Learning Theory*, pp 5–13 ACM Press, New York

Jaakkola TS, Jordan MI (1997) A variational approach to Bayesian logistic regression models and their extensions. Technical Report 9702, MIT Computational Cognitive Science, Cambridge, MA

Jaakola TS, Jordan MI, Ghahramani Z *et al.* (1998) An introduction to variational methods for graphical models. In *Learning in graphical models*, Jordan MI (ed.). Kluwer Academic Press, Dordrecht

Mackay DJC (1995) Ensemble learning and evidence maximization. Technical report, Cavendish Laboratory, University of Cambridge, Cambridge

MacKay DJC (1998) Choice of basis for Laplace approximations. *Machine Learning* **33**: 77–86

Mackay DJC (2003) *Information theory, inference and learning algorithms*. Cambridge University Press, Cambridge

Mullin T (1993) *The nature of chaos*. Oxford Science Publications, Oxford

Noppeney U, Penny WD, Price CJ *et al.* (2006) Identification of degenerate neuronal systems based on intersubject variability. *NeuroImage* **30**: 885–90

Penny WD (2001) Kullback-Liebler divergences of normal, gamma, Dirichlet and Wishart densities. Technical report, Wellcome Department of Imaging Neuroscience, London

Penny WD, Kiebel SJ, Friston KJ (2003) Variational Bayesian inference for fMRI time series. *NeuroImage* **19**: 727–41

Penny WD, Flandin G, Trujillo-Barreto N (2006) Bayesian comparison of spatially regularised general linear models. *Hum Brain Mapp* in press

Peterson C, Anderson J (1987) A mean field theory learning algorithm for neural networks. *Complex Syst* **1**: 995–1019

Press WH, Teukolsky SA, Vetterling WT *et al.* (1992) *Numerical Recipes in C*. Cambridge University Press

Sahani M, Nagarajan SS (2004) Reconstructing MEG sources with unknown correlations. In *Advances in neural information processing systems*, Saul L, Thrun S, Schoelkopf B (eds), volume 16. MIT, Cambridge, MA

Sato M, Yoshioka T, Kajihara S *et al.* (2004) Hierarchical Bayesian estimation for MEG inverse problem. *NeuroImage* **23**: 806–26

Winn J, Bishop C (2005) Variational message passing. *J Machine Learning Res* **6**: 661–94

Woolrich MW, Behrens TE, Smith SM (2004) Constrained linear basis sets for HRF modelling using variational Bayes. *NeuroImage* **21**: 1748–61

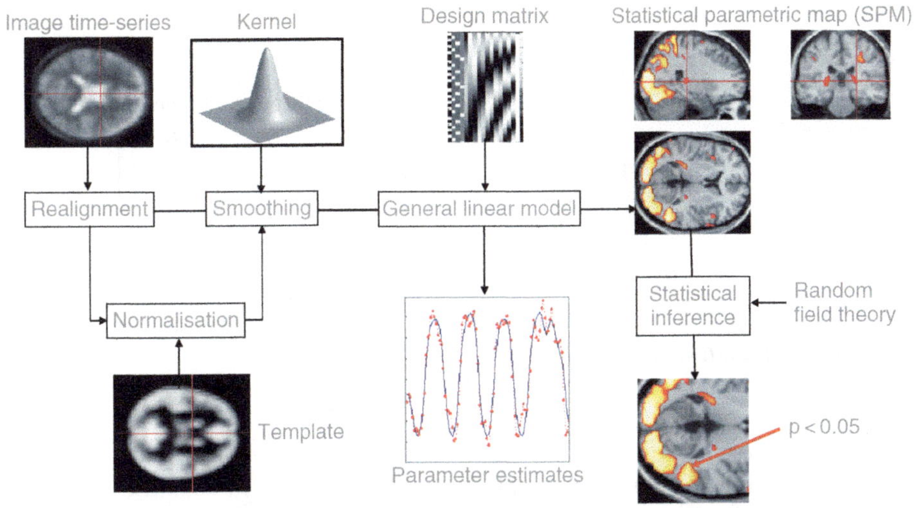

Plate 1 This schematic depicts the transformations that start with an imaging data sequence and end with a statistical parametric map (SPM). An SPM can be regarded as an 'X-ray' of the significance of regional effects. Voxel-based analyses require the data to be in the same anatomical space: this is effected by realigning the data. After realignment, the images are subject to non-linear warping so that they match a spatial model or template that already conforms to a standard anatomical space. After smoothing, the general linear model is employed to estimate the parameters of a temporal model (encoded by a design matrix) and derive the appropriate univariate test statistic at every voxel (see Figure 2.3). The test statistics (usually t- or F-statistics) constitute the SPM. The final stage is to make statistical inferences on the basis of the SPM and random field theory (see Figure 2.4) and characterize the responses observed using the fitted responses or parameter estimates.

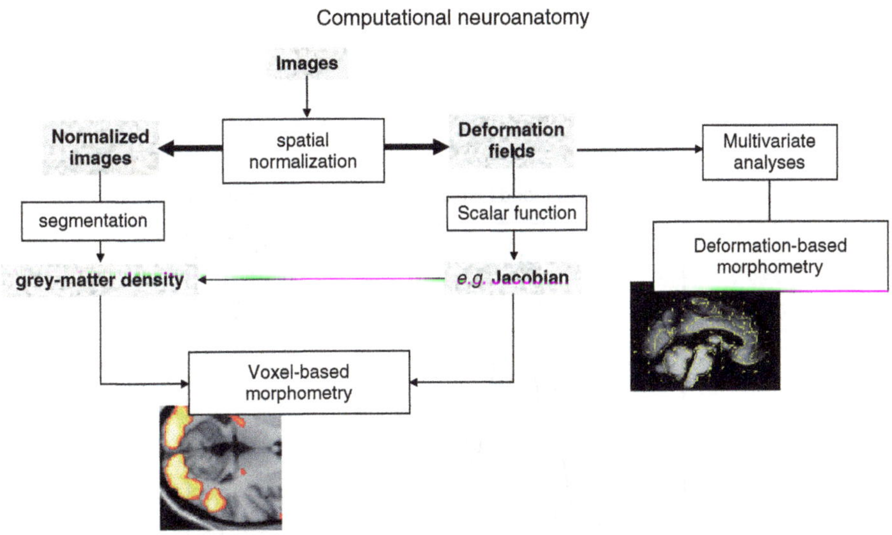

Plate 2 Schematic illustrating different procedures in computational anatomy. After spatial normalization, one has access to the normalized image and the deformation field implementing the normalization. The deformation or tensor field can be analysed directly (deformation-based morphometry) or can be used to derive maps of formal attributes (e.g. compression, dilatation, shear, etc.). These maps can then be subject to conventional voxel-based analyses (tensor-based morphometry). Alternatively, the normalized images can be processed (e.g. segmented) to reveal some interesting aspect of anatomy (e.g. the tissue composition) and analysed in a similar way (voxel-based morphometry). Tensor-based morphometry can be absorbed into voxel-based morphometry. For example, before statistical analysis, Jacobian, or voxel-compression maps can be multiplied by grey-matter density maps. This endows volumetric changes, derived from the tensor, with tissue specificity, based on the segmentation.

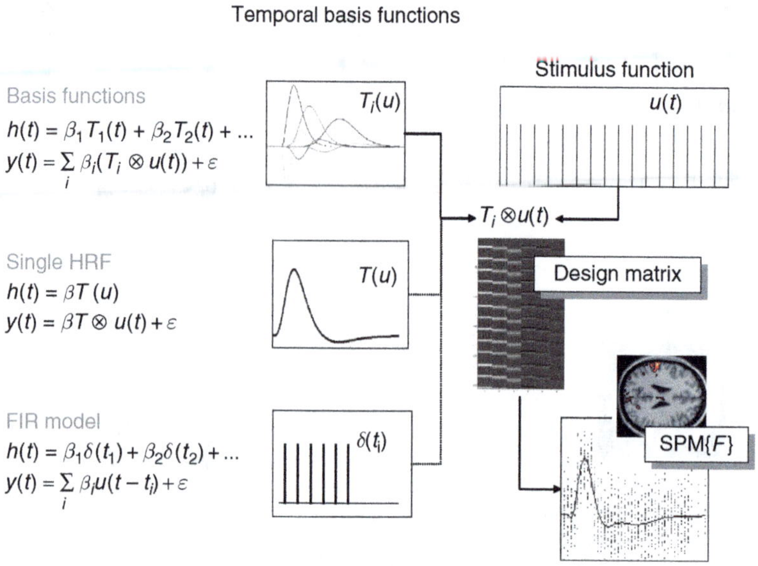

Plate 3 Temporal basis functions offer useful constraints on the form of the estimated response that retain the flexibility of FIR models and the efficiency of single regressor models. The specification of these constrained FIR models involves setting up stimulus functions $u(t)$ that model expected neuronal changes, e.g. boxcar-functions of epoch-related responses or spike-(delta)-functions at the onset of specific events or trials. These stimulus functions are then convolved with a set of basis functions $T_i(t)$ of peristimulus time that, in some linear combination, model the HRF. The ensuing regressors are assembled into the design matrix. The basis functions can be as simple as a single canonical HRF (middle), through to a series of top-hat-functions $\delta_i(t)$ (bottom). The latter case corresponds to an FIR model and the coefficients constitute estimates of the impulse response function at a finite number of discrete sampling times. Selective averaging in event-related fMRI (Buckner *et al.*, 1998) is mathematically equivalent to this limiting case.

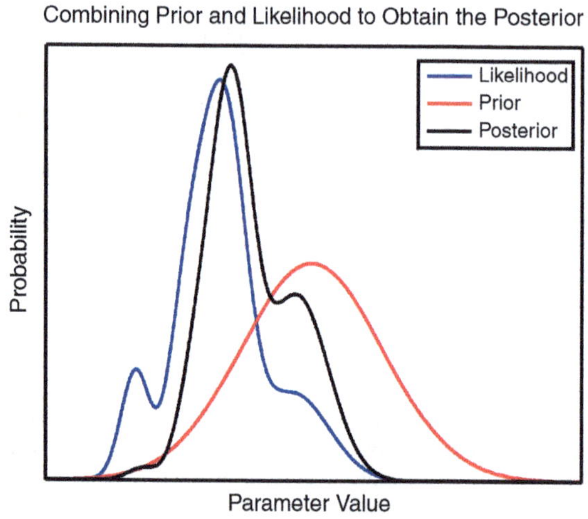

Plate 4 This figure illustrates a hypothetical example of Bayes' theory with a single parameter. By combining the likelihood and prior probability density, it is possible to obtain a tighter posterior probability density. Note that the area under each of the curves is one.

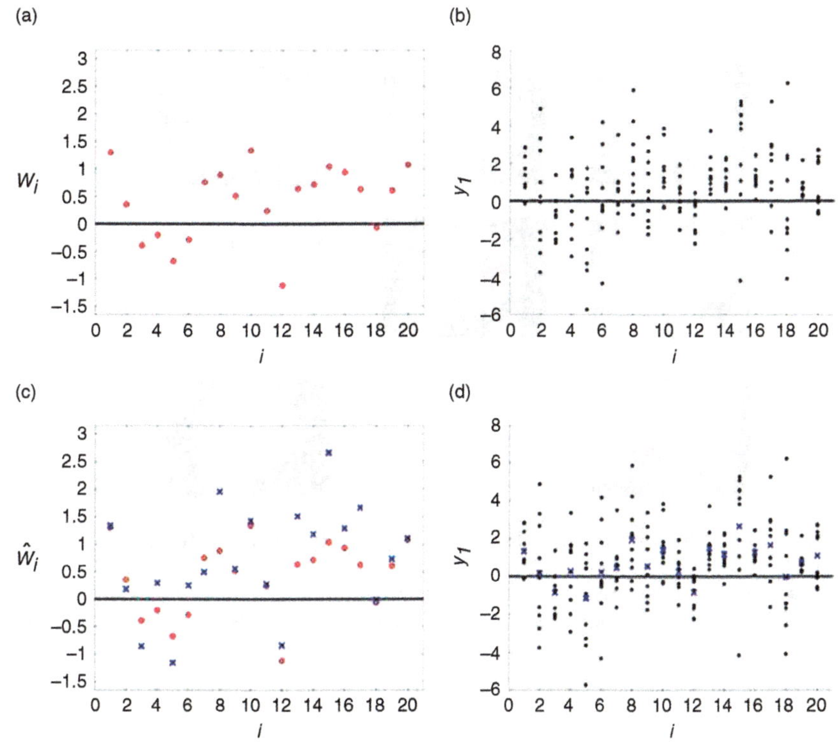

Plate 5 Data for PEB example. (a) Red circles denote 'true' effect sizes, w_i, for each voxel i, generated from the prior $p(w_i|\alpha) = \mathsf{N}(0, \alpha^{-1})$ with $\alpha = 1$. (b) The black dots denote $n_i = 10$ data points at each voxel generated from the likelihood $p(y_i|w_i) = \mathsf{N}(w_i, \beta_i^{-1})$ with β_i drawn from a uniform distribution between 0.1 and 1. Thus some voxels, e.g. voxels 2, 15 and 18, have noisier data than others. Plots (c) and (d) are identical to (a) and (b) but with blue crosses indicating maximum likelihood (ML) estimates of the effect size, $\hat{w}_i$. These are simply computed as the mean of the data at each voxel, and are used to initialize PEB – see Plate 6.

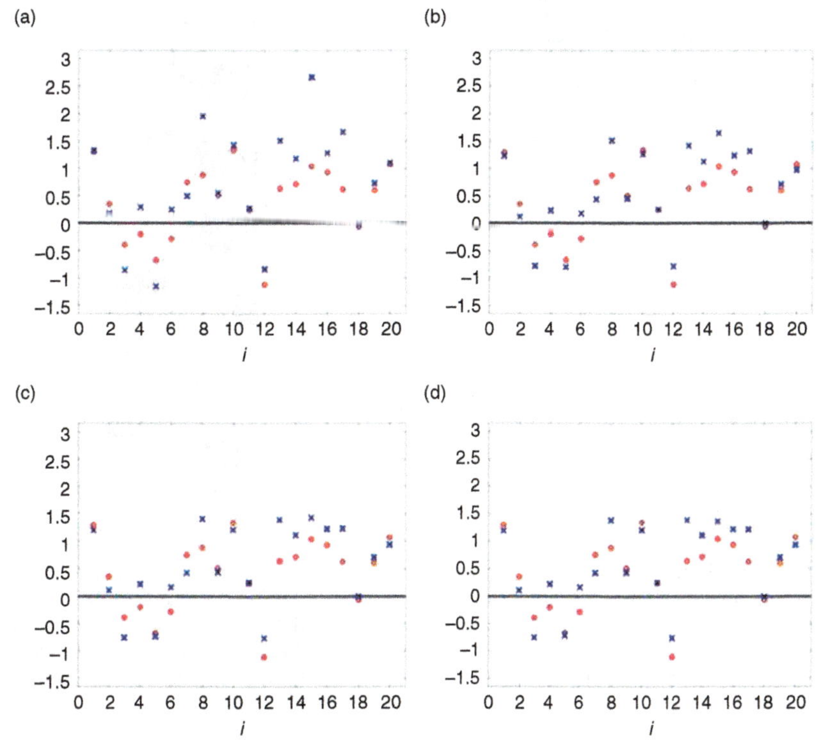

Plate 6 The plots show the true effect sizes, w_i (red circles) and estimated effect sizes, $\hat{w}_i$ (blue crosses), before PEB iteration number (a) one, (b) three, (c) five and (d) seven. Plot (a) here is the same as plot (c) in Plate 5, as the estimates were initialized using ML.

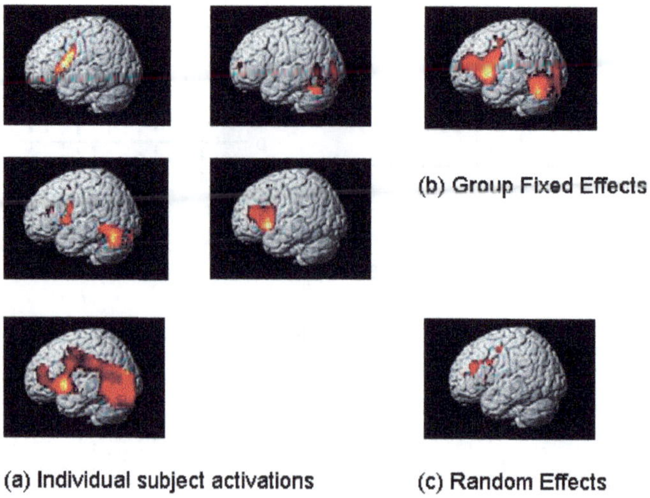

Plate 7 Analysis of PET data showing active voxels ($p < 0.001$ uncorrected). The maps in (a) show the significance of subject-specific effects whereas map (b) shows the significance of the average effect over the group. Map (c) shows the significance of the population effect from an RFX analysis

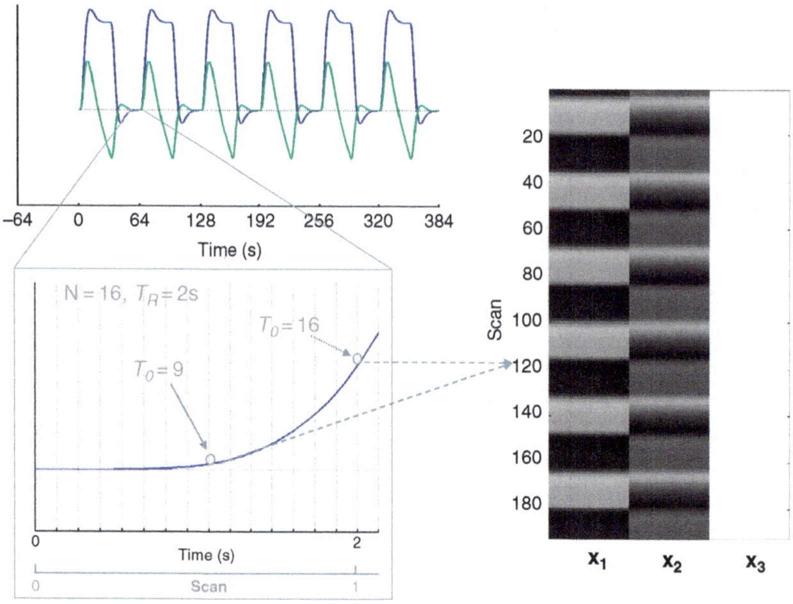

Plate 8 Creation of regressors for design matrix. The predicted BOLD signal for sustained (blue) and decaying (green) activity, during boxcar stimulation after convolution with an HRF in a time-space with resolution $\Delta t = T_R/N$ seconds (upper left). This predicted time course is down-sampled every scan ($T_R = 2\,\text{s}$) at time point T_0 to create the columns x_1 (boxcar) and x_2 (exponential decay) of the design matrix (together with the constant term x_3). Two possible sample points (T_0) are shown: at the middle and end of the scan (the specific choice should match the relative position of the reference slice within the slice order, if any slice-timing correction is performed).

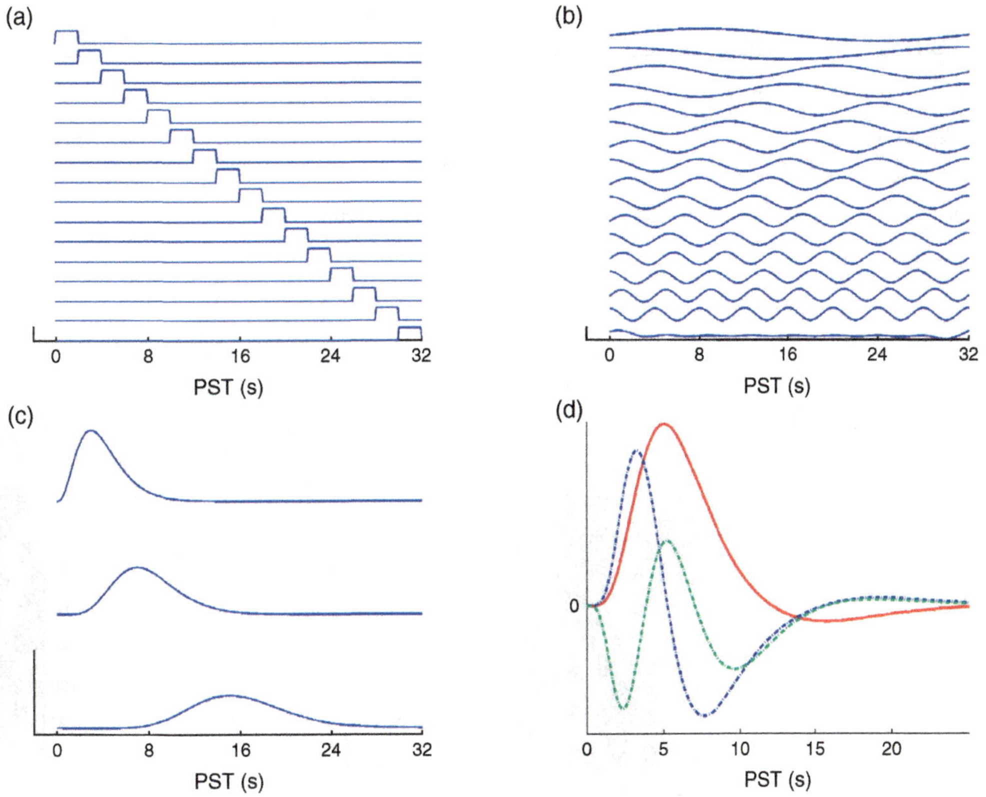

Plate 9 Temporal basis functions offered by SPM, $T = 32\,s$: (a) FIR basis set, $K_{FIR} = 16$; (b) Fourier basis set, $K_F = 8$; (c) Gamma functions, $K = 3$; (d) Canonical HRF (red) and its temporal (blue) and dispersion (green) derivatives. The temporal derivative is approximated by the orthogonalized finite difference between canonical HRFs with peak delay of 7 s versus 6 s; the dispersion derivative is approximated by the orthogonalized finite difference between canonical HRFs with peak dispersions of 1 versus 1.01.

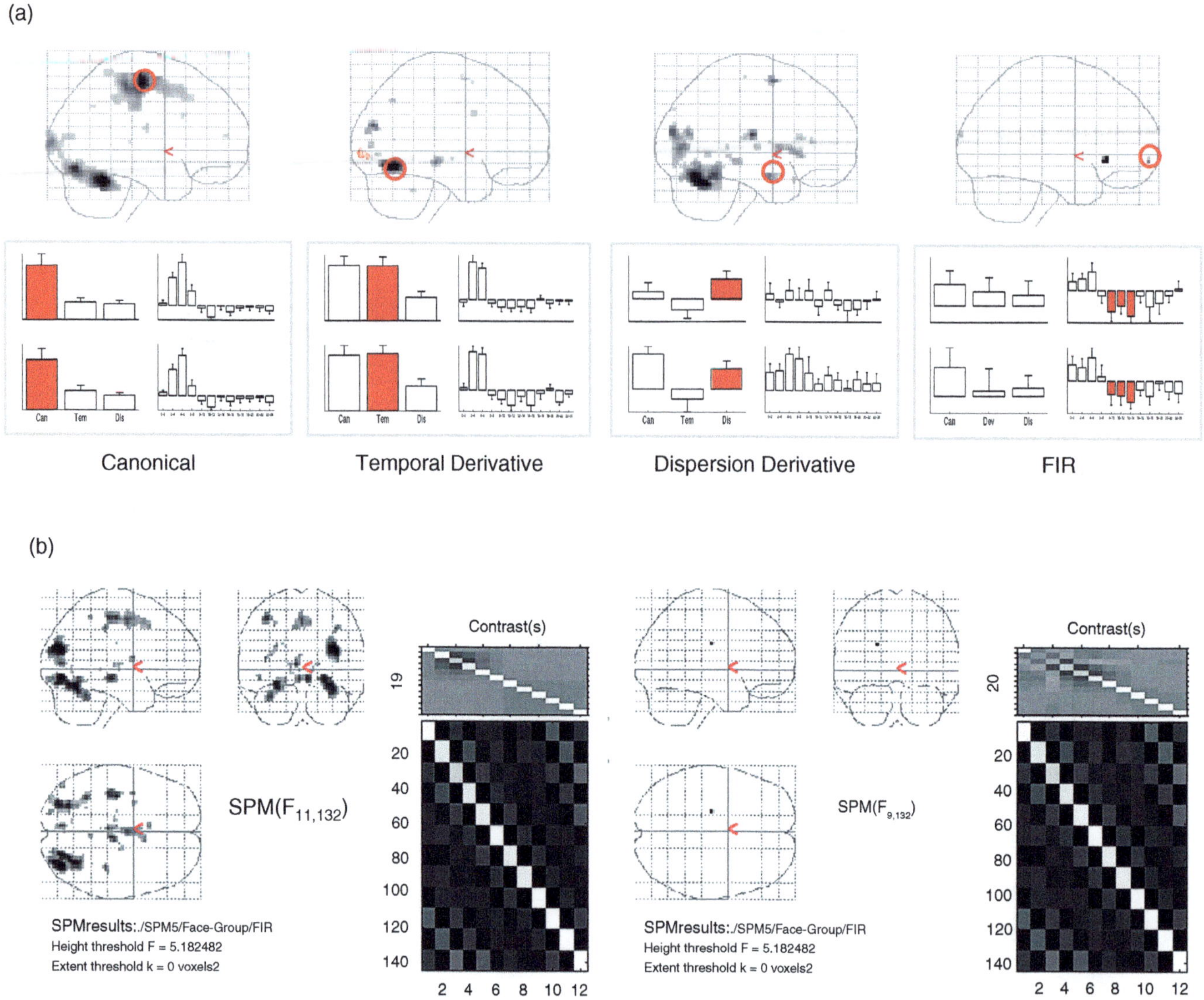

Plate 10 Sufficiency of the informed basis set. In (a), two event-types were modelled with both the informed basis set and an FIR (12, 2 s time bins) for 12 subjects within a single, first-level (fixed-effects) design matrix. The event-types were novel or famous faces presented for 0.5 s and requiring a response with the right hand (mean reaction times less than second; see Henson *et al.*, 2002, for more details). The maximum intensity projections (MIP) show F-contrasts thresholded at $p < 0.05$ corrected for (from left to right): the canonical HRF only, its temporal derivative, its dispersion derivative and all time bins of the FIR. The plots below the MIPs show, for the region circled in each MIP, the parameter estimates of the three 'informed' response functions (left) and for a re-fitted FIR (right), for both event-types (upper and low plots). The canonical HRF and its two derivatives explain a lot of variability. A left motor region shows a canonical response (i.e. loading mainly on the canonical HRF); an occipital region shows a response earlier than the canonical; an anterior temporal region shows a response more dispersed than the canonical, but only for the second event-type (famous faces). Little additional variability is, however, picked up by the FIR model: only a few voxels in anterior prefrontal cortex, which show a sustained undershoot (which could reflect a non-haemodynamic artefact). In (b), contrast images of the average of the two event-types for each of 12, 2 s FIR time bins were taken to a second-level ('random effects') analysis. The F-contrast $I - h^+h$, where h is the canonical HRF (sampled every 2 s) and $^+$ is the pseudoinverse, shows some regions in which significant variability across subjects is not captured by the canonical HRF (the 'null space' of the canonical HRF; left). The F-contrast $I - H^+H$, on the other hand, where H is now a matrix including the canonical HRF and its two derivatives, shows little that cannot be captured by these three functions (right).

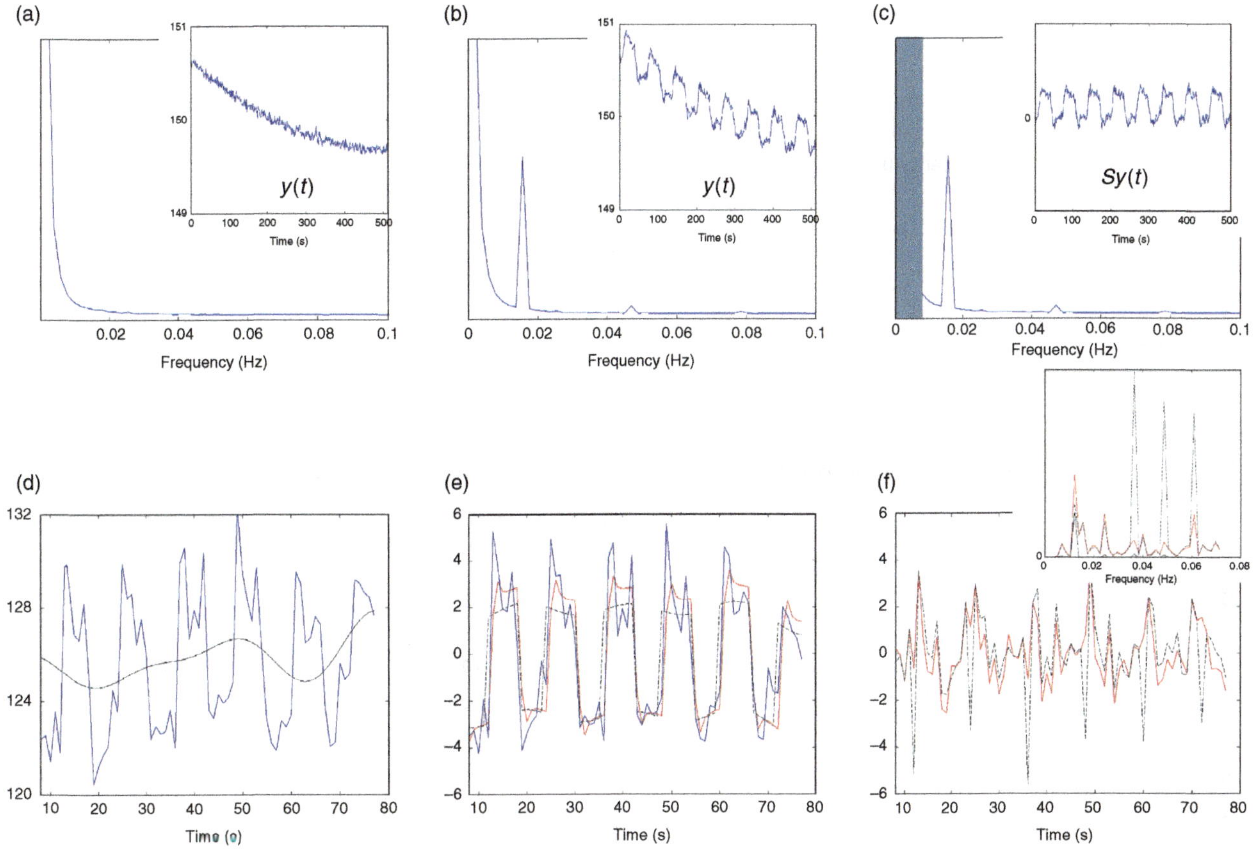

Plate 11 Power spectra, highpass filtering and HRF convolution. Schematic power spectrum and time series (inset) for (a) subject at rest, (b) after square-wave stimulation at 32 s on, 32 s off, (c) after highpass filtering with cut-off 128 s. (d) Real data (blue) and low-frequency drift (black) fitted by DCT highpass filter matrix S (cut-off 168 s) derived from the global maximum in a 42 s on; 42 s off auditory blocked design ($T_R = 7$ s). (e) Fits of a boxcar epoch model with (red) and without (black) convolution by a canonical HRF, together with the data (blue), after application of the highpass filter. (f) Residuals after fits of models with and without HRF convolution: note large systematic errors for model *without* HRF convolution (black) at onset of each block, corresponding to (non-white) harmonics of the stimulation frequency in the residual power spectrum (inset).

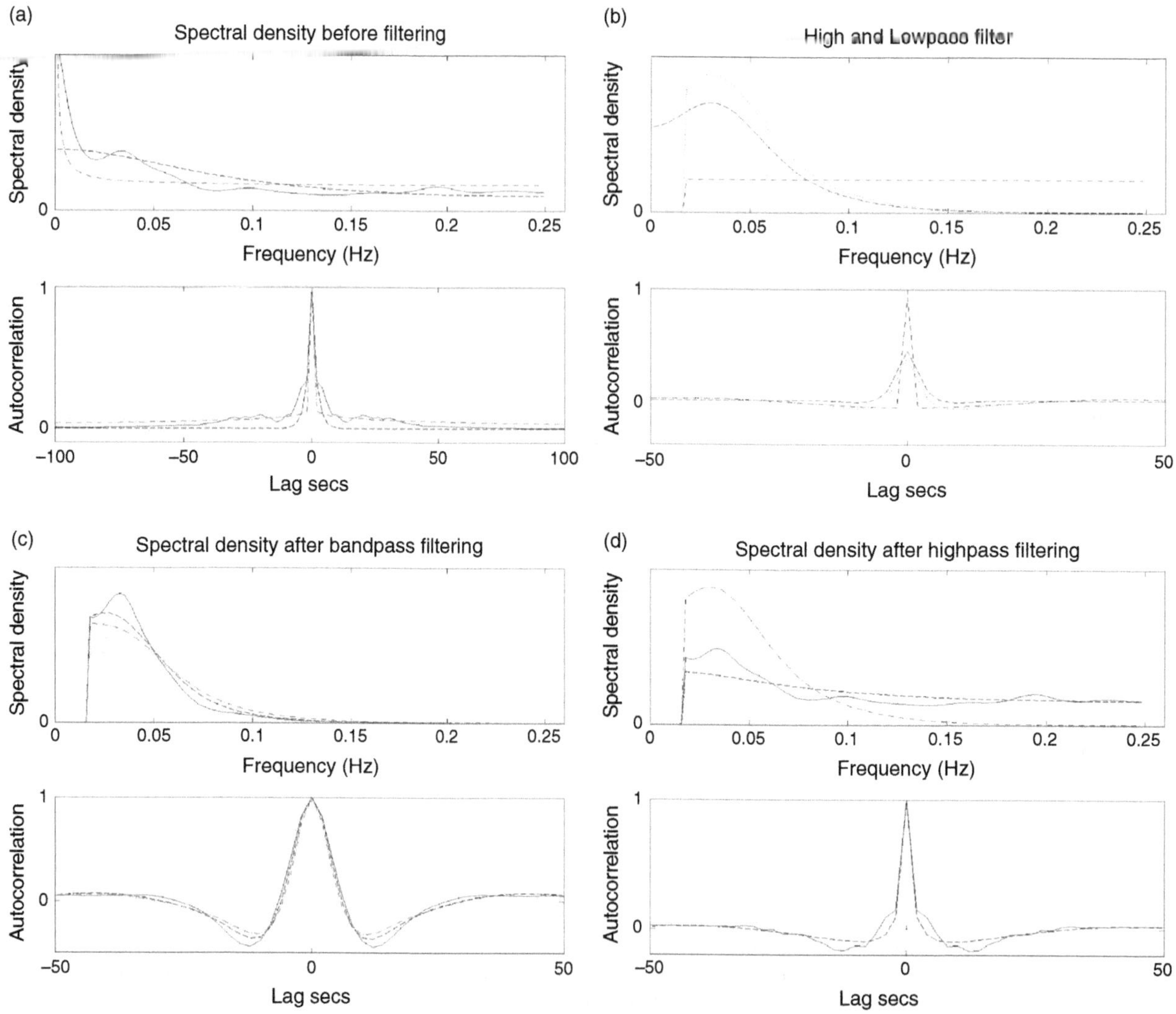

Plate 12 Models of fMRI temporal autocorrelation. Power spectra and autocorrelation functions for: (a) data (solid black), derived from an AR(16) estimation of the mean, globally-normalized residuals from one slice of an event-related dataset; together with fits of an AR(1) model (dashed blue) and 1/f amplitude model (dashed red); (b) high- (dot-dash) and low- (dotted) pass filters, comprising a bandpass filter (dashed); (c) data and both models after bandpass filtering (note that bandpass filter characteristics in (b) would also provide a reasonable approximation to residual autocorrelation); (d) data (solid black) and ReML fit of AR(1)+white noise model (dashed blue) after highpass filtering (also shown is the bandpass filter power spectrum, demonstrating the high-frequency information that would be lost by lowpass smoothing).

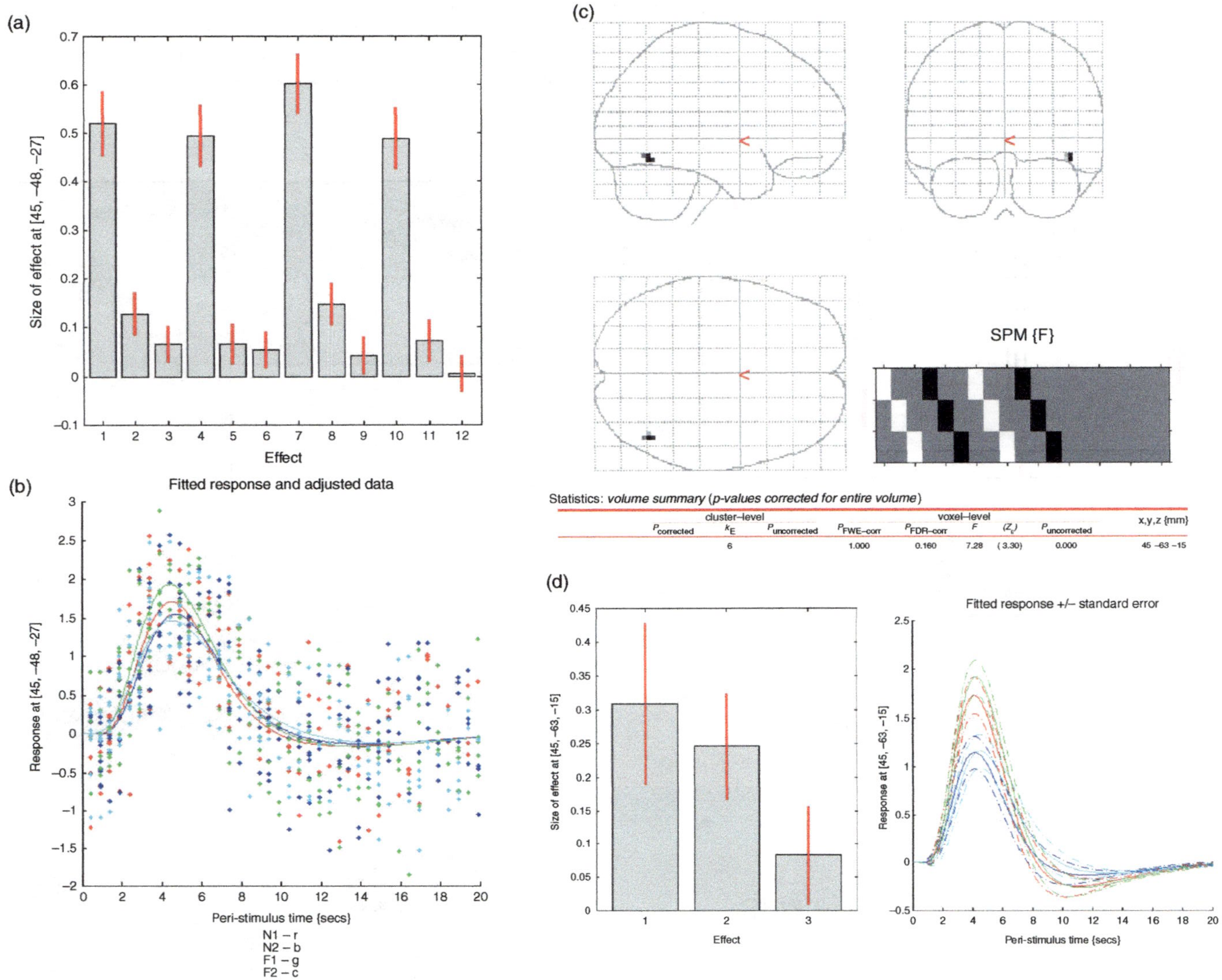

Plate 13 Categorical model: repetition effect. (a) Parameter estimates (scale arbitrary) from local maximum in right fusiform (+45, −48, −27), ordered by condition – N1, N2, F1, F2 – and within each condition by basis function – canonical HRF, temporal derivative and dispersion derivative. (b) Fitted event-related responses (solid) and adjusted data (dots) in terms of percentage signal change (relative to grand mean over space and time) against PST for N1 (red), N2 (blue), F1 (green) and F2 (cyan). (c) SPM{F} MIP for repetition effect contrast (inset), thresholded at $p < 0.001$ uncorrected, after inclusive masking with effects of interest (Figure 14.4) at $p < 0.05$ corrected. (d) Contrast of parameter estimates for repetition effect (difference between first and second presentations) in right occipitotemporal region (+45, −63, −15) for canonical HRF, temporal derivative and dispersion derivative, together with fitted responses (solid) ± one standard error (dashed).

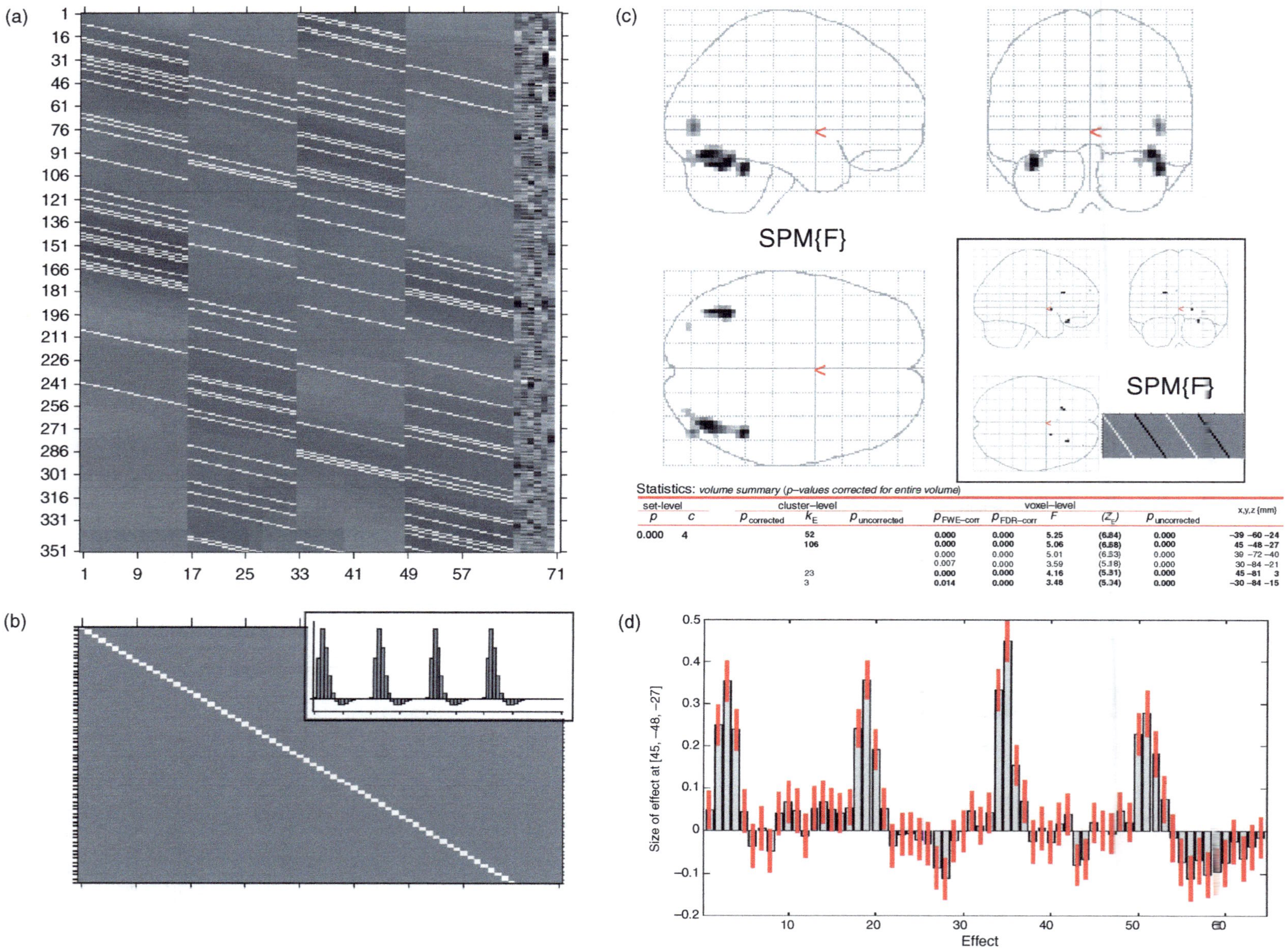

Plate 14 Categorical model: FIR basis set. (a) Design matrix. (b) Effects of interest F-contrast (canonical HRF weighted t-contrast inset). (c) SPM{F} MIP for effects of interest, thresholded at $p < 0.05$ whole-brain corrected, together with SPM tabulated output (inset is SPM{F} for unconstrained repetition effect F-contrast, thresholded at $p < 0.005$ uncorrected). (d) Parameter estimates for effects of interest from right fusiform region (+45, −48, −27), as in Plate 13(a), ordered by condition – N1, N2, F1, F2 – and within each condition by the 16 basis functions (i.e. mean response every 2 s from 0 to 32 s PST).

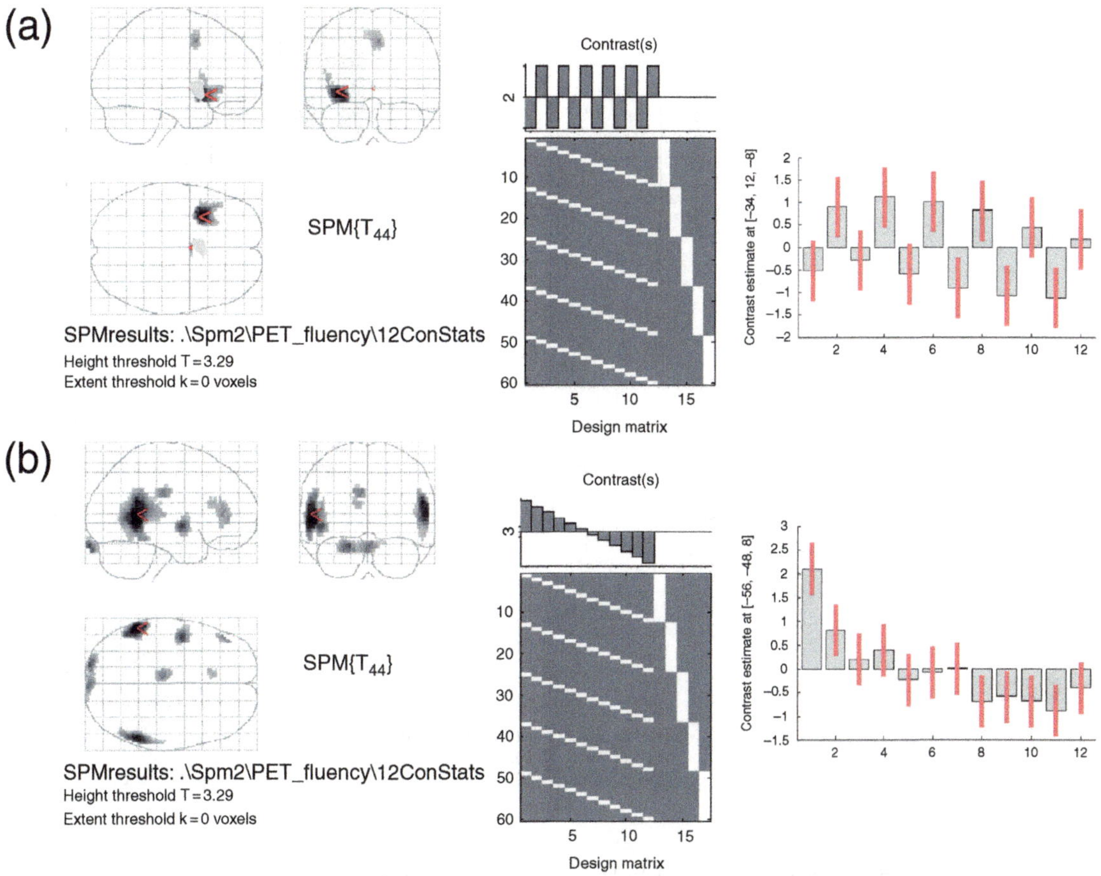

Plate 15 A simple design matrix with 5 subjects (sessions) and 12 conditions. (a) A simple subtraction of two levels of a categorical factor (Generate minus Read) that alternate six times. (b) A linear contrast across a parametric factor, time during experiment, with 12 levels.

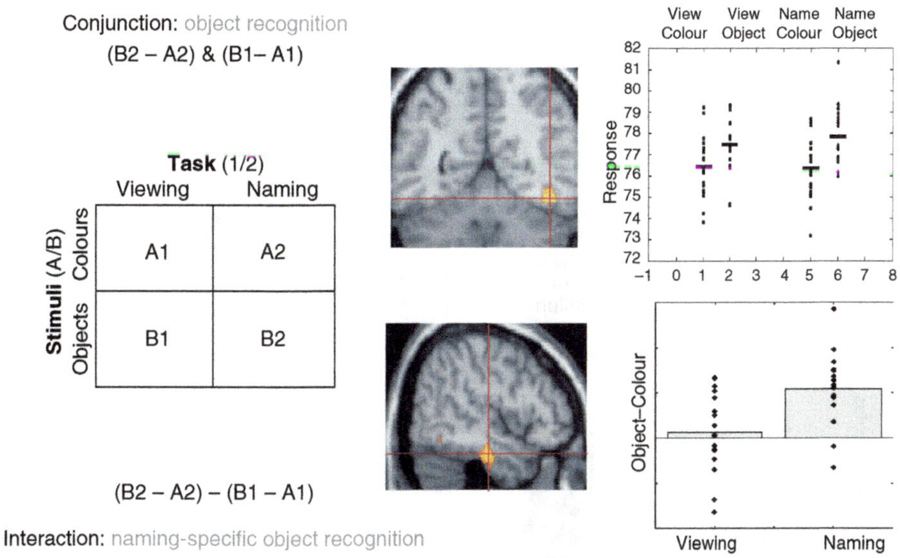

Plate 16 Cognitive conjunction and interactions (from Price and Friston, 1997). (Upper panels) The conjunction of two contrasts shows a posterior temporal region that is activated when an object is present, regardless of whether the task was passive viewing or naming. (Lower panels) When the same conditions are treated as a 2×2 factorial design (left), an interaction contrast reveals a more anterior temporal region that is only active when an object is present and subjects are naming that object.

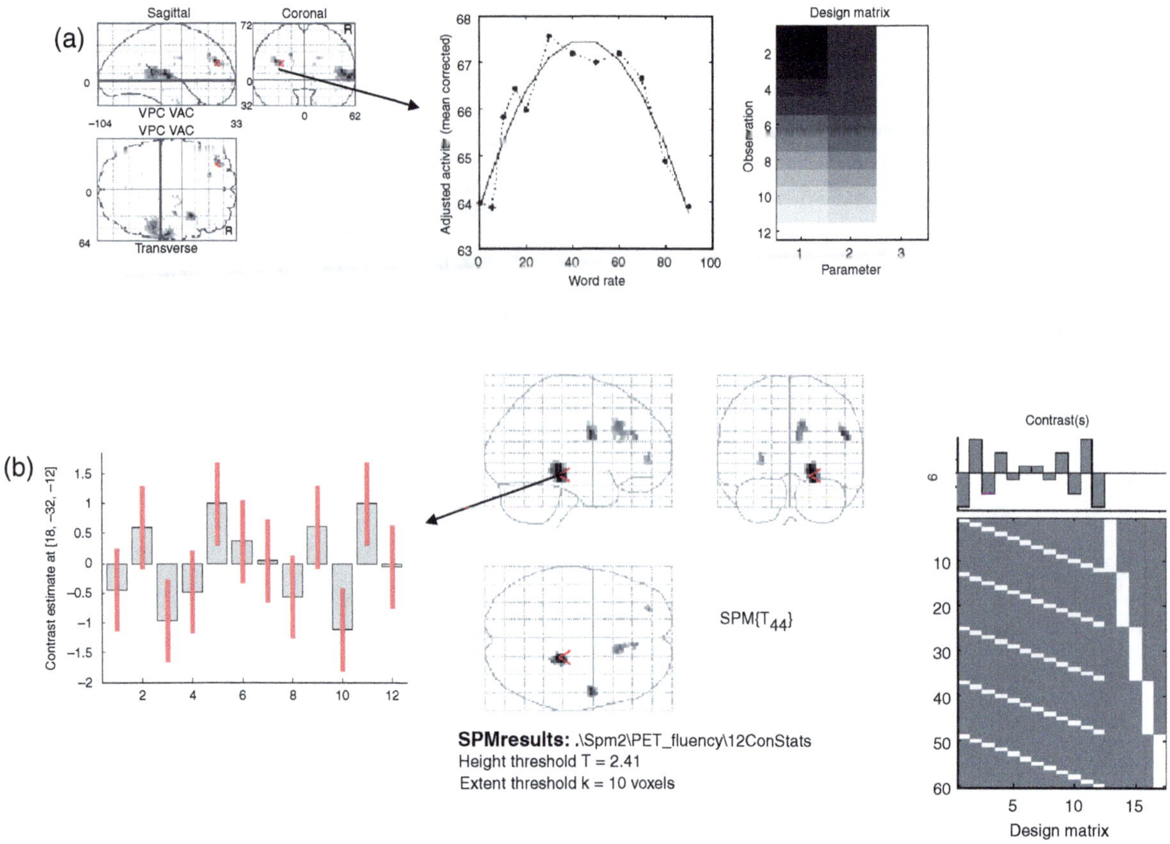

Plate 17 (a) Non-linear effects of a parametric factor modelled in a single-subject design matrix using a second-order polynomial expansion. (b) A linear time-by-condition interaction contrast in a 2 × 6 factorial design across 5 subjects.

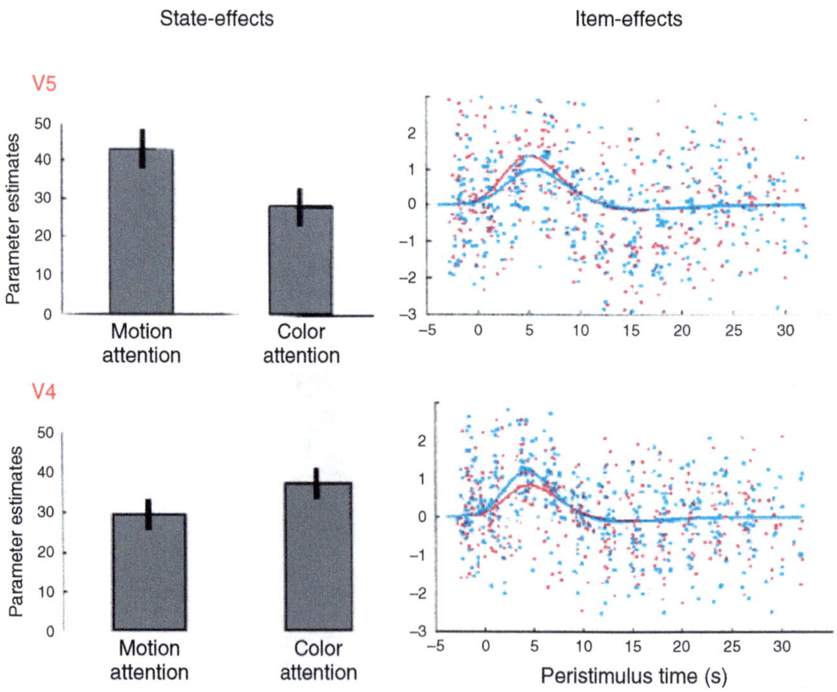

Plate 18 State- and item-effects (from Chawla *et al.*, 1999). Attention to colour of radially-moving coloured dots increased both baseline activity (top left) and evoked responses (top right) – i.e. both offset and gain – in V4 relative to attention to motion. The opposite pattern was found in V5 (bottom row).

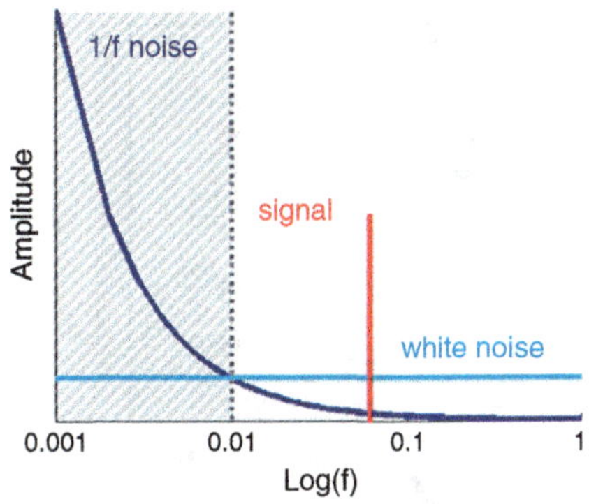

Plate 19 Schematic form of $1/f$ and white noise (dark and light blue respectively) typical of fMRI data, together with experimentally-induced signal at 0.03 Hz (red) and highpass filtering (hatched area).

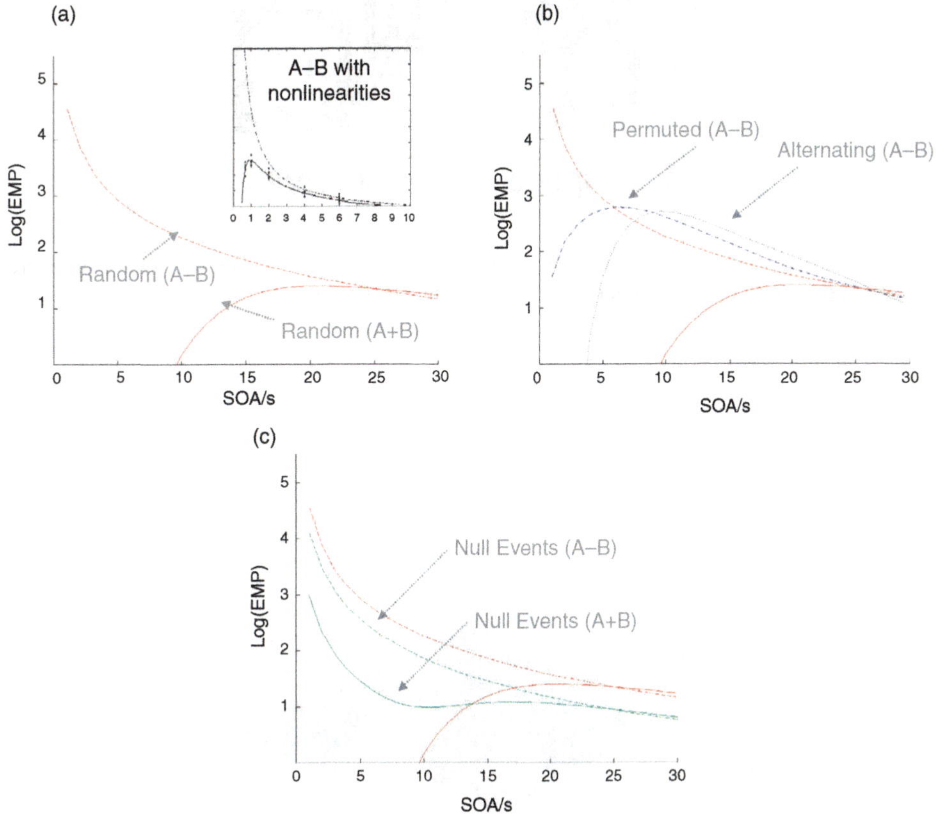

Plate 20 Efficiency for two event-types (from Josephs and Henson, 1999). Efficiency is expressed in terms of 'estimated measurable power' (EMP) passed by an effective HRF, characterized by a canonical HRF, highpass filter with cut-off period of 60 s and lowpass smoothing by a Gaussian 4 s full-width at half maximum (FWHM), as a function of Δt for main (solid) effect ([1 1] contrast) and differential (dashed) effect ([1 −1] contrast). (a) Randomized design. (b) Alternating (black) and permuted (blue) designs. (c) With (green) and without (red) null events. Insert: effect of non-linearity (saturation) on average response as a function of SOA within a 32 s blocked design. Solid line: average response to a train of stimuli predicted using a second-order Volterra model of haemodynamic responses. The broken line shows the predicted response in the absence of non-linear or second-order effects (Friston *et al.*, 2000).

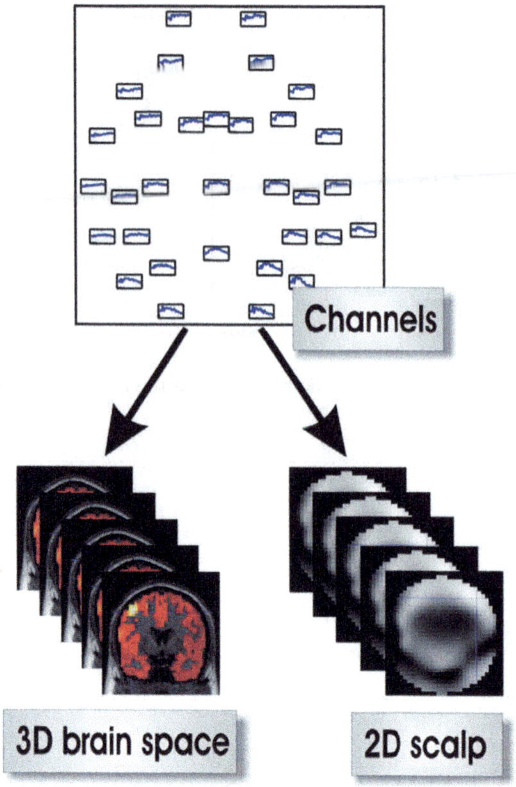

Plate 21 For each trial type and subject, the ERR, for each channel, is projected to either three-dimensional brain space (source reconstruction) or interpolated on the scalp surface. This results in either 3-D or 2-D image time-series.

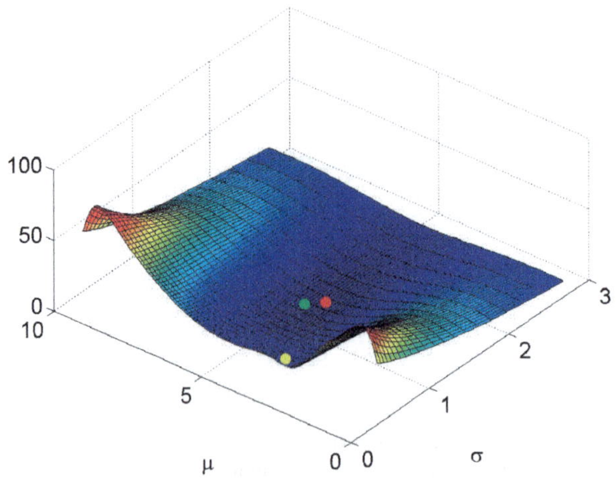

Plate 22 KL-divergence, $KL(q||p)$ for p as defined in Figure 24.2 and q being a Gaussian with mean μ and standard deviation σ. The KL-divergences of the approximations in Figure 24.2 are (a) 11.73 for the first mode (yellow ball), (b) 0.93 for the second mode (green ball) and (c) 0.71 for the moment-matched solution (red ball).

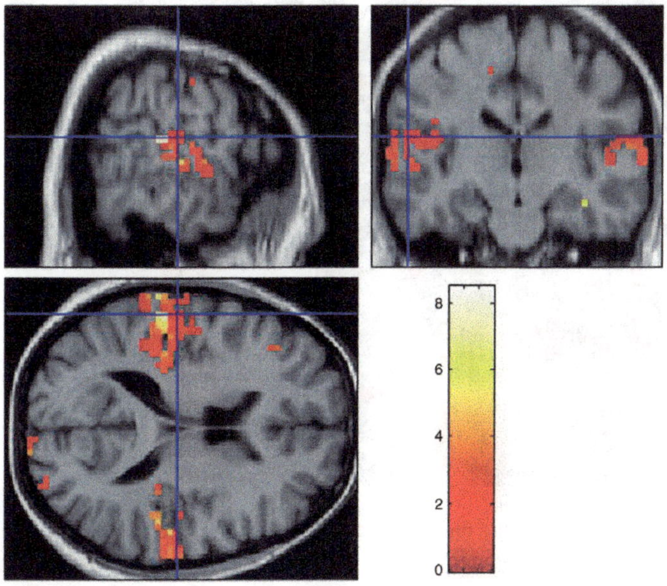

Plate 23 PPM for positive auditory activation. Overlay of effect-size, in units of percentage of global mean, on subjects' MRI for above-threshold voxels. The default thresholds were used, i.e., we plot c_n for voxels which satisfy $p(c_n > 0) > 1 - 1/N$.

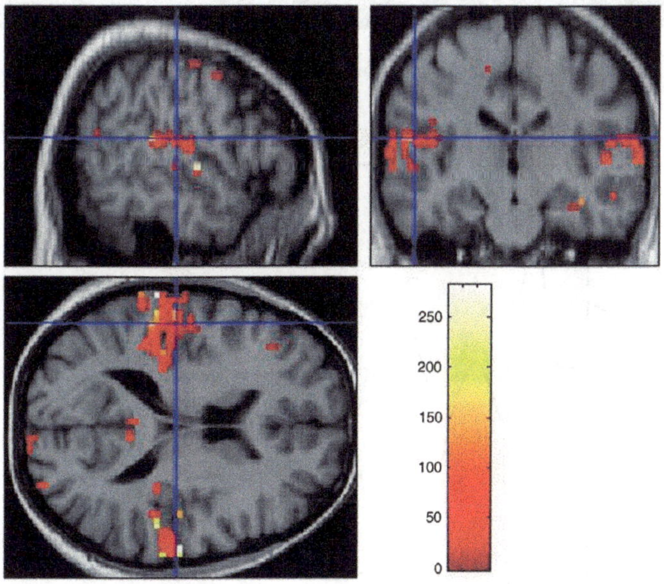

Plate 24 PPM for positive or negative auditory activation. Overlay of χ^2 statistic on subjects' MRI for above-threshold voxels. The default thresholds were used, that is, we plot χ_n^2 for voxels which satisfy $p(c_n > 0) > 1 - 1/N$.

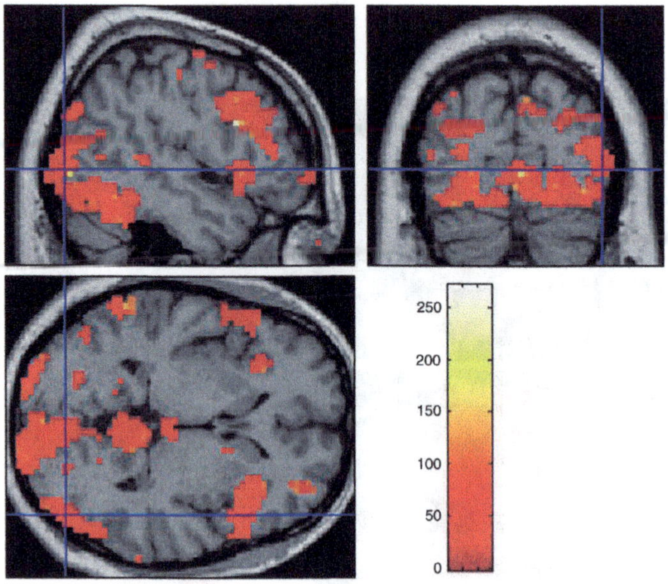

Plate 25 PPM showing above-threshold χ^2 statistics for any effect of faces.

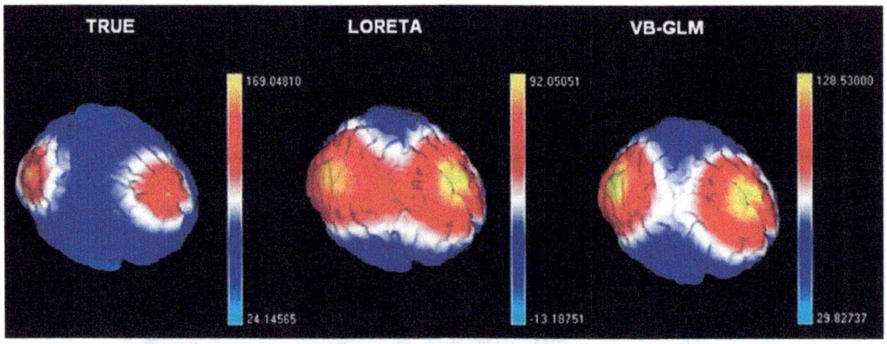

Plate 26 True and estimated source distributions at time $t = 20$ ms. Note the scaling in the figures. The VB-GLM approach is better both in terms of spatial localization and the scaling of source estimates.

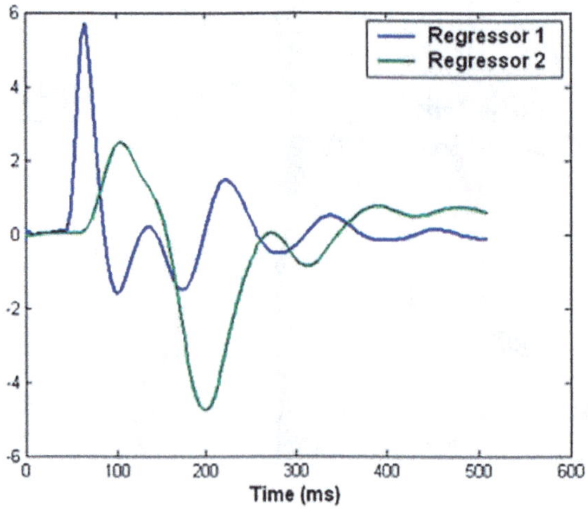

Plate 27 Two ERP components, derived from a biophysical model, used to generate simulated ERP data. These mimic an early component and a late component.

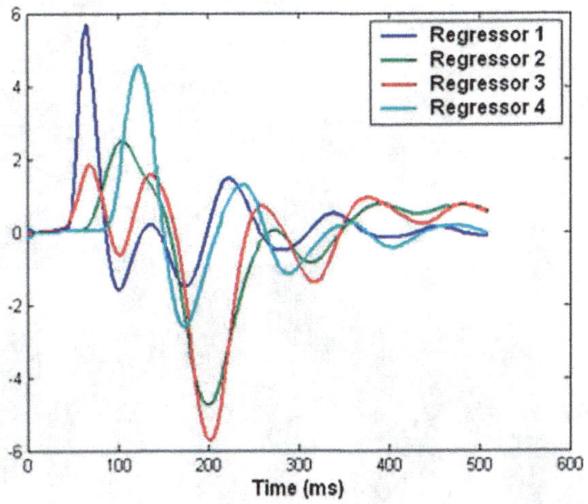

Plate 28 Four components, derived from a biophysical model, used in an over-specified ERP model.

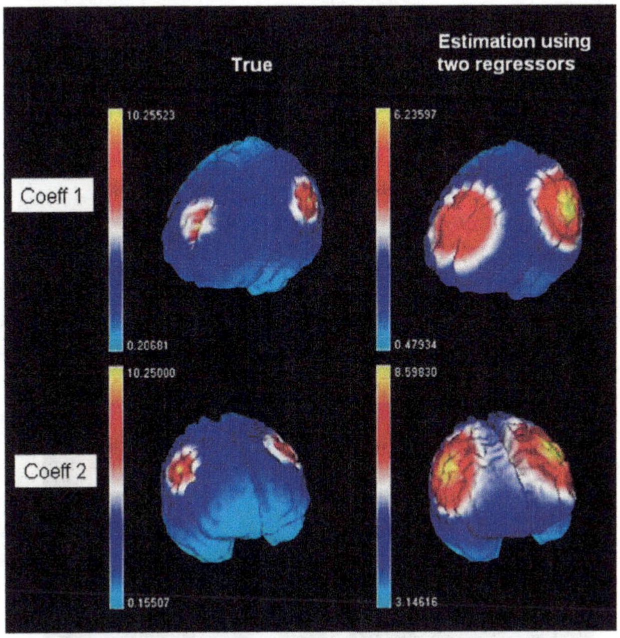

Plate 29 Regression coefficients, w_g, from ERP simulation. 'Coeff 1' and 'Coeff 2' denote the first and second entries in the regression coefficient vector w_g. True model (left) and estimates from correctly specified model (right).

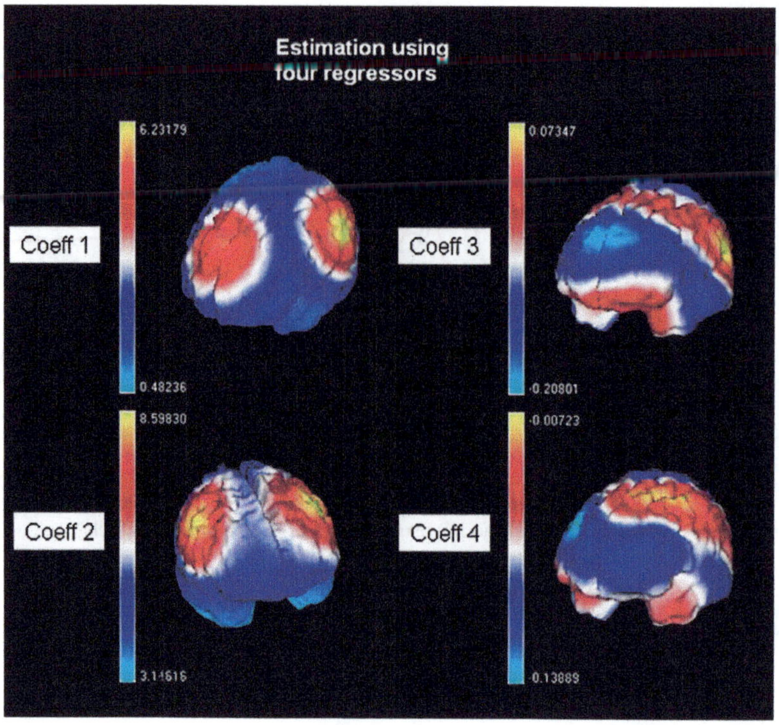

Plate 30 Estimated regression coefficients, $\hat{w}_g$, from over-specified model. The true coefficients are shown in Plate 29. Note the scaling of coefficients 3 and 4 (the true values are zero). Despite the high temporal correlation between regressors 2 and 3, the coefficients for regressor 3 have been correctly shrunk towards zero. This is a consequence of the spatial prior and the iterative nature of the spatio-temporal deconvolution (see Figure 26.6).

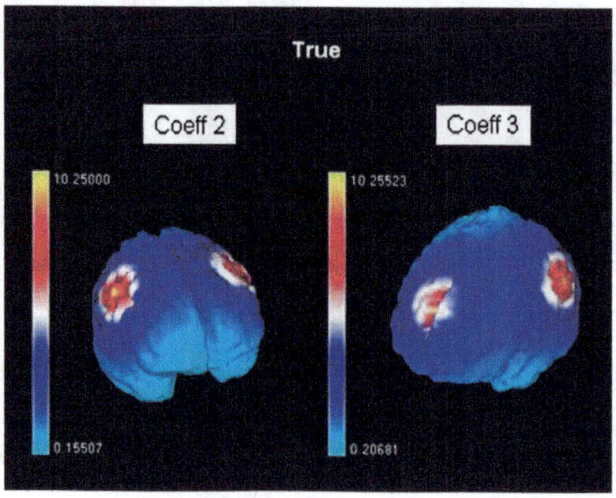

Plate 31 True regression coefficients for ERP simulation with correlated sources. This simulation used a design matrix comprising the regressors shown in Plate 28, with the first and fourth coefficients set to zero and the second and third set as shown in this figure.

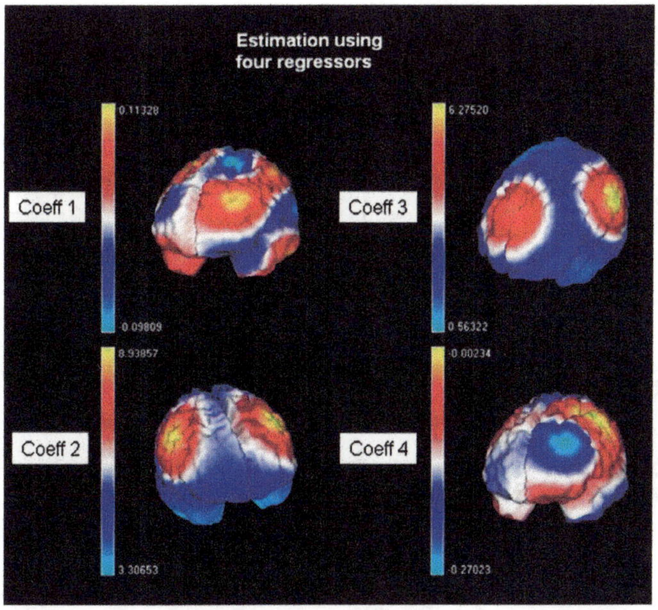

Plate 32 Estimated regression coefficents, $\hat{w}_g$, for ERP simulation with correlated sources. Coefficients 2 and 3 resemble the true values shown in Plate 31, whereas regressors 1 and 4 have been correctly shrunk towards zero by the spatio-temporal deconvolution algorithm.

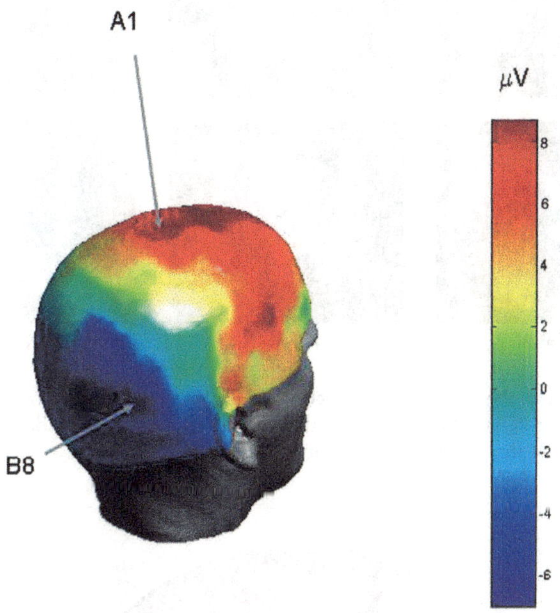

Plate 33 The figure shows differential EEG topography for faces minus scrambled faces at $t = 160$ ms poststimulus.

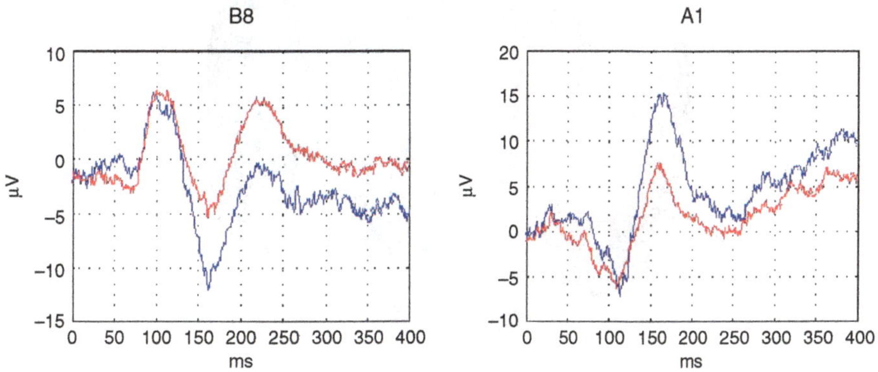

Plate 34 Sensor time courses for face data at occipito-temporal electrode B8 (left) and vertex A1 (right) for faces (blue) and scrambled faces (red).

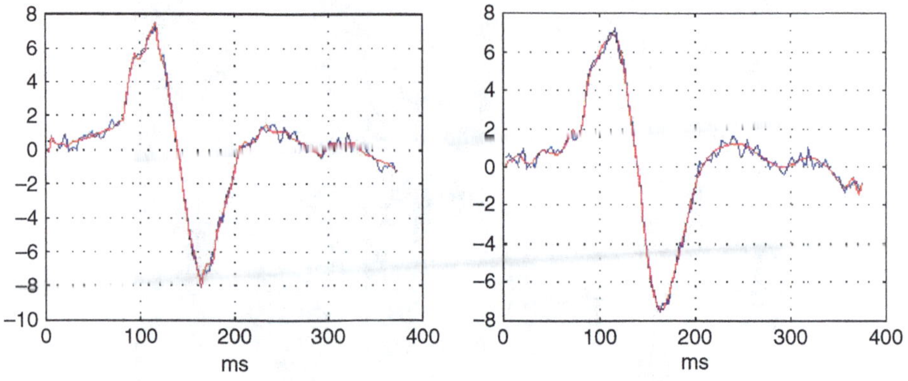

Plate 35 First eigen-time-series of downsampled ERP for unfamiliar faces (blue lines in both plots) with wavelet shrinkage approximations using Daubechies basis (left) and Battle-Lemarie basis (right).

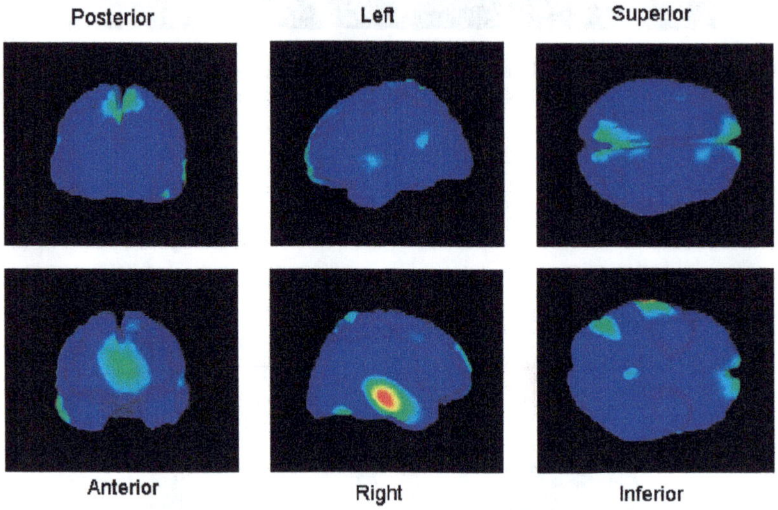

Plate 36 These images are derived from the source reconstruction of ERPs in response to faces and scrambled faces. The plots show absolute differences between faces and scrambled faces at t = 160 ms post-stimulus. The maps have been thresholded such that the largest difference appears in red and 50 per cent of the largest difference appears in blue.

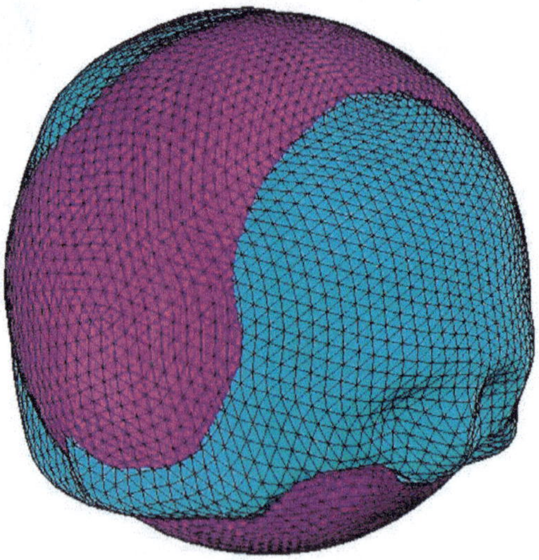

Plate 37 Best fitting sphere on a scalp surface extracted from a subject's structural MRI.

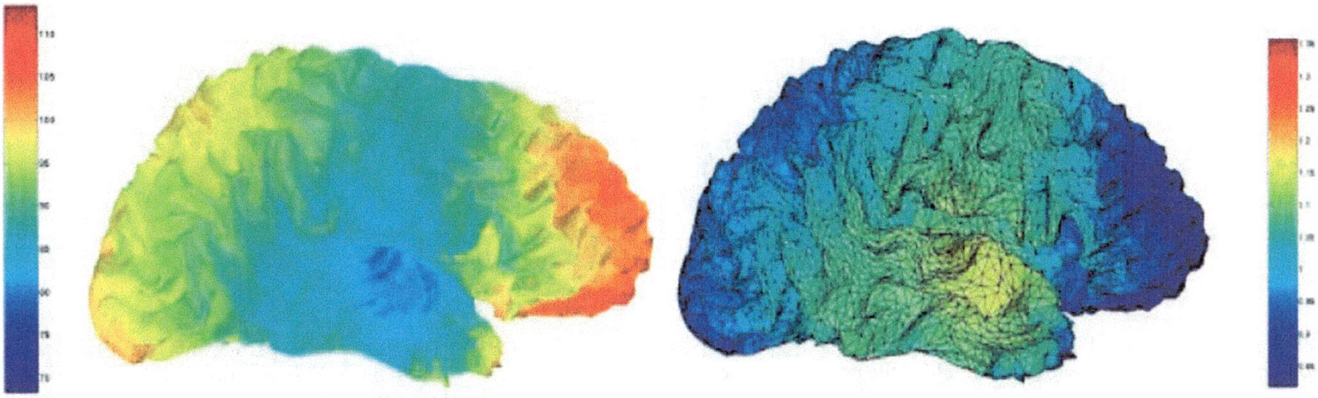

Plate 38 The left figure shows the original source location, defined by the cortical sheet. The colour scale indicates the radius of the scalp $R_{scalp}(\theta, \varphi)$ in the direction (θ, φ) of the dipole locations $(R_{sb}, \theta, \varphi)$. The right figure is the transformed source space obtained after applying Eqn. **28.60**. The colour scale shows the scaling of the source radii, i.e. $R_{sphere}/R_{scalp}\ (\theta, \varphi)$.

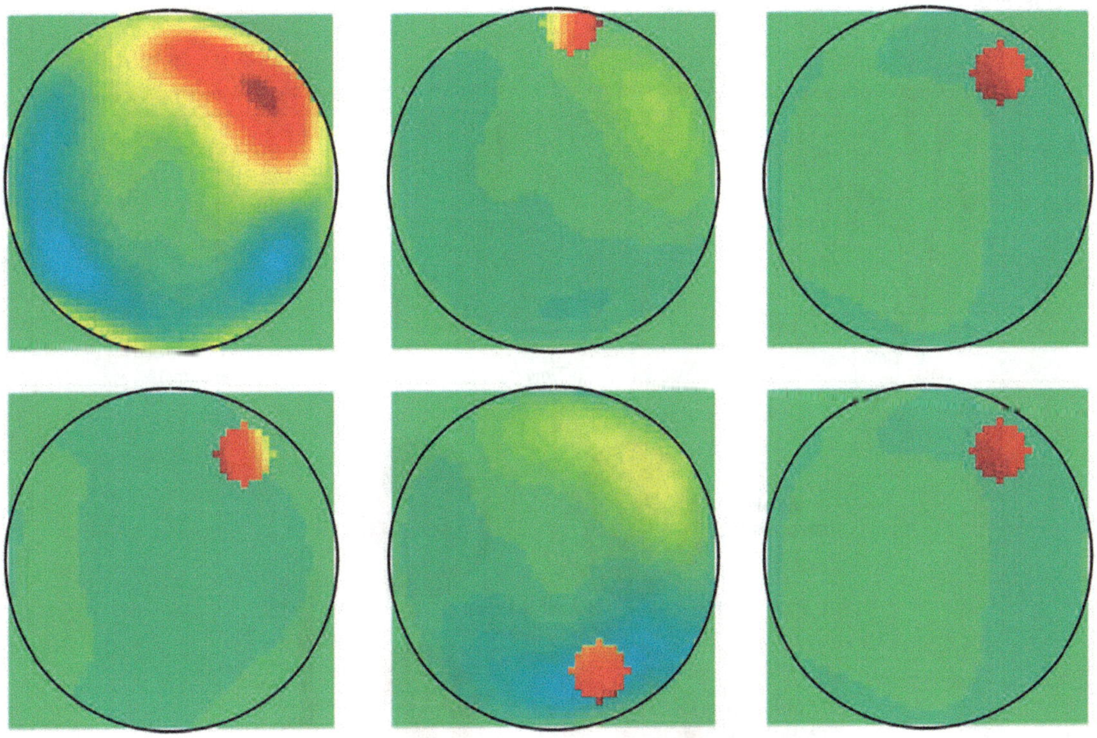

Plate 39 Reconstructed sources obtained with ReML using the example shown in Figure 29.2 (SNR = 12): no location priors (top left), with accurate location priors (bottom left), with close inaccurate location priors (top middle), with distant inaccurate location priors (bottom middle), with both accurate and close inaccurate location priors (top right) and with both accurate and distant inaccurate location priors (bottom right).

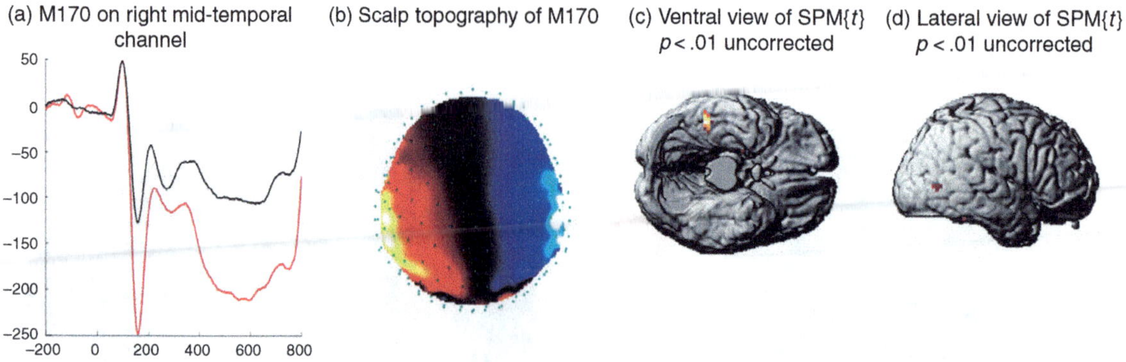

Plate 40 Multisubject analysis of face-selective response based on ReML analysis. (a) The mean M170 across participants and (b) its scalp topography. An identity matrix was used for the noise covariance in sensor space. When using only the MSP source prior, subtraction of the absolute values of the separate source reconstructions for faces versus scrambled faces revealed (c) activation of right fusiform (+51 −39 −15), T(7) = 6.61, (d) right middle temporal gyrus (+63 −69 +3), T(7) = 3.48, and right parahippocampal gyrus (+27 −6 −18), T(7) = 3.32, when thresholded at $p < 0.01$ uncorrected. MEG data from a 151-channel CTF Omega system were acquired while 9 participants made symmetry judgements to faces and scrambled faces. The MEG epochs were baseline-corrected from −100 to 0 ms, averaged over trials (approx. 70 face and 80 scrambled trials) and low-pass filtered to 20 Hz. A time-window around the peak in the global field power of the difference between the event-related field (ERF) for faces and scrambled faces that corresponded to the M170 was selected for each participant (mean window = 120–200 ms). Segmented cortical meshes of approximately 7200 dipoles oriented normal to the grey matter were created using Anatomist, and single-shell spherical forward models were constructed using Brainstorm. Multivariate source prelocalization (MSP) was used to reduce the number of dipoles to 1500. The localizations on the mesh were converted into 3D images, warped to MNI space using normalization parameters determined from participants' MRIs using SPM2, and smoothed with a 20 mm full width half maximum (FWHM) isotropic Gaussian kernel. These smoothed, normalized images were used to create an SPM of the t-statistic over participants (final smoothness approx 12 × 12 × 12 mm) (c) and (d).

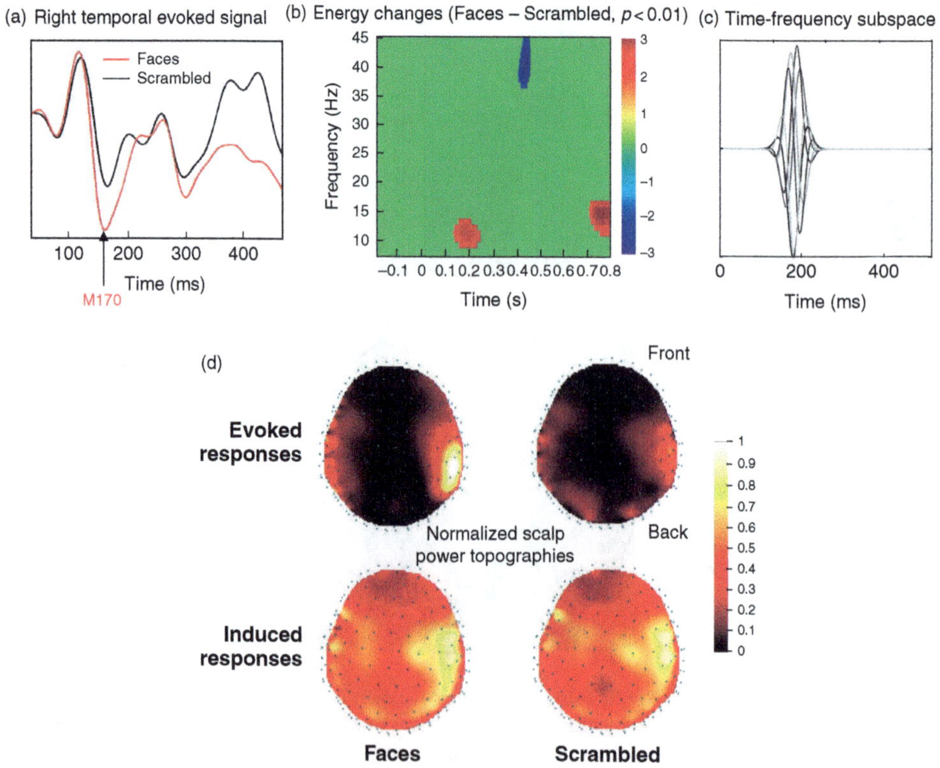

Plate 41 Real data analysis – sensor level: (a) and (b) show the differences, measured on the scalp, between faces and scrambled faces, in terms of the event-related field (ERF) from a single sensor (a), and the global energy over sensors (b) using standard time-frequency analysis and statistical parametric mapping (Kilner *et al.*, 2005). The time-frequency subspace W we tested is shown in (c) by plotting each column as a function of time. This uses the same representation as the first panel of the previous figure. This subspace tests for responses in the alpha range, around 200 ms (see corresponding time-frequency effect in (a)). The corresponding induced and evoked energy distributions over the scalp are shown in (d), for two conditions (faces and scrambled faces).

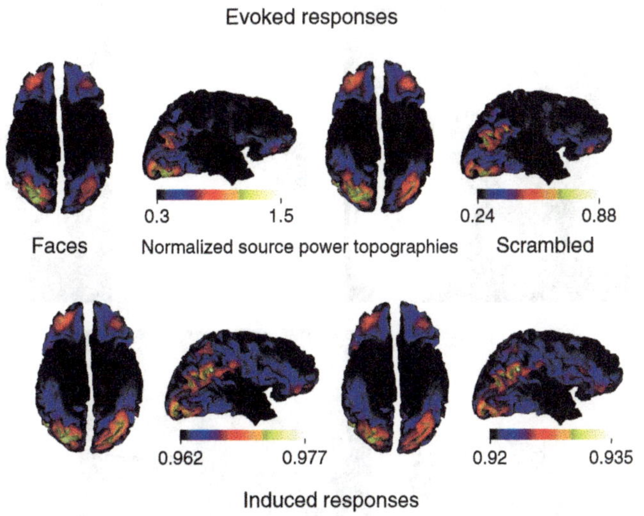

Plate 42 Real data analysis – source level: reconstructed evoked and induced responses are shown for both faces and scrambled face trials. These data correspond to conditional expectations, rendered onto a cortical surface. Note that these views of the cortical surface are from below (i.e. left is on the right). Evoked power was normalized to the maximum over cortical sources.

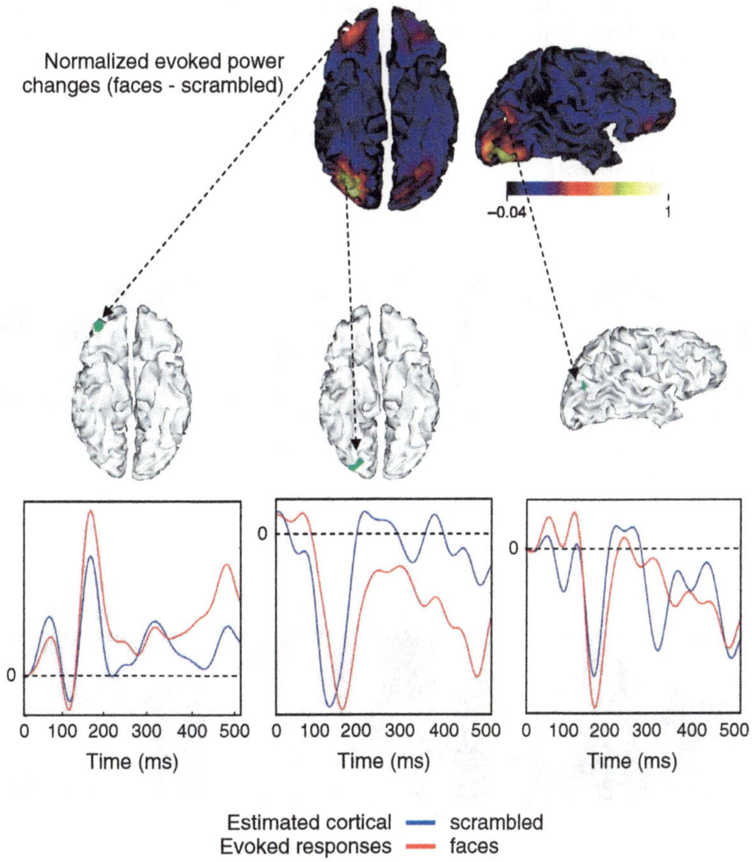

Plate 43 Real data analysis – evoked responses: the upper panels show the reconstructed evoked power changes between faces and scrambled faces. The lower panels show the reconstructed evoked responses associated with three regions where the greatest energy change was elicited.

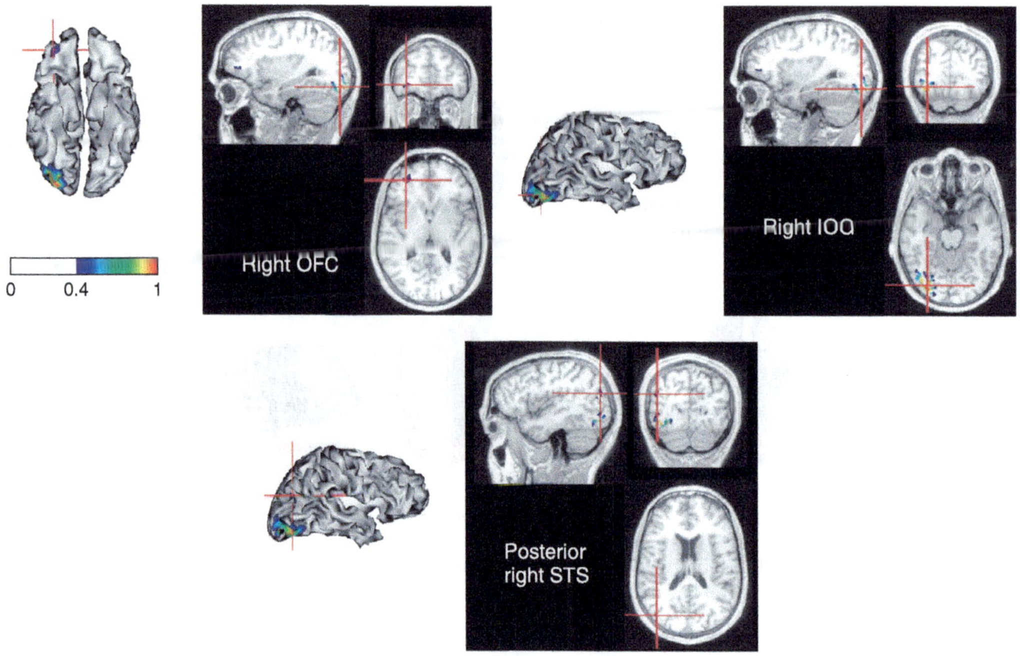

Plate 44 Visualization on the subjects' MRI: the regions identified as showing energy changes for faces versus scrambled faces in Plate 43 are shown, co-registered with a MRI scan: the right OFC (upper left panel), the right IOG (upper right panel) and the posterior right STS (lower panel). These source estimates are shown both as cortical renderings (from below) and on orthogonal sections through a structural MRI, using the radiological convention (right is left).

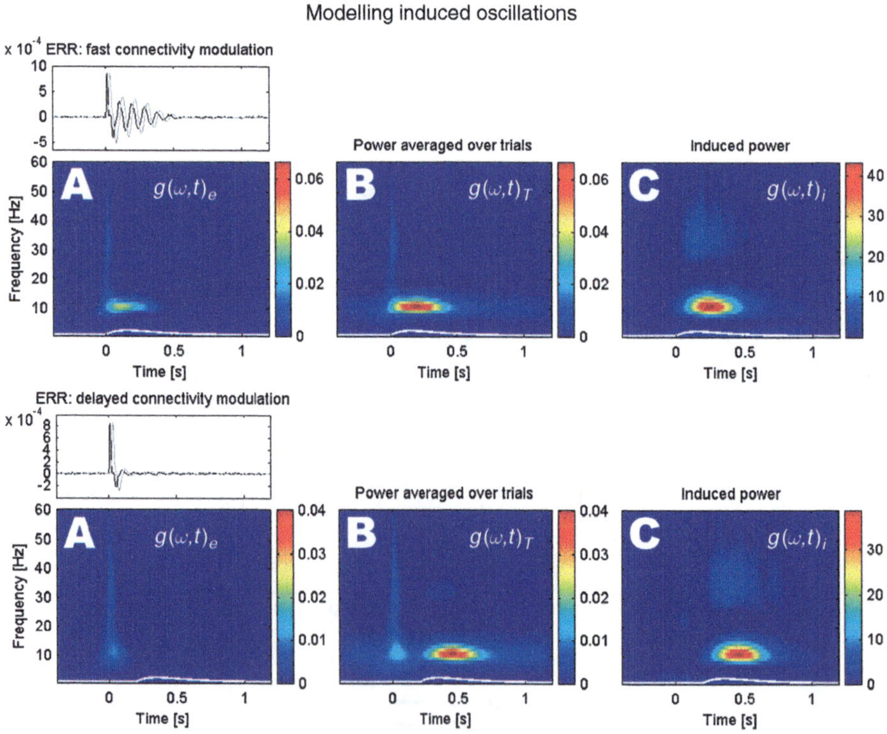

Plate 45 Upper panel: simulation of fast stimulus-related modulation of backward connectivity, using the model depicted in Figure 33.18. Black curves are the responses of area 1; grey curves correspond to area 2. Time-frequency responses are shown for area 1 only. The white line, superimposed on these spectral profiles, shows the time course of the modulatory input. (a) Evoked power, after averaging over trials, showing late oscillations that have been augmented by modulatory input. (b) Total power, averaged over trials. (c) Induced power, normalized over frequency. Lower panel: as for the upper panel, but here the modulatory effect has been delayed. The main difference is that low-frequency evoked components have disappeared because dynamic and structural perturbations are now separated in time and cannot interact. See main text for further details.

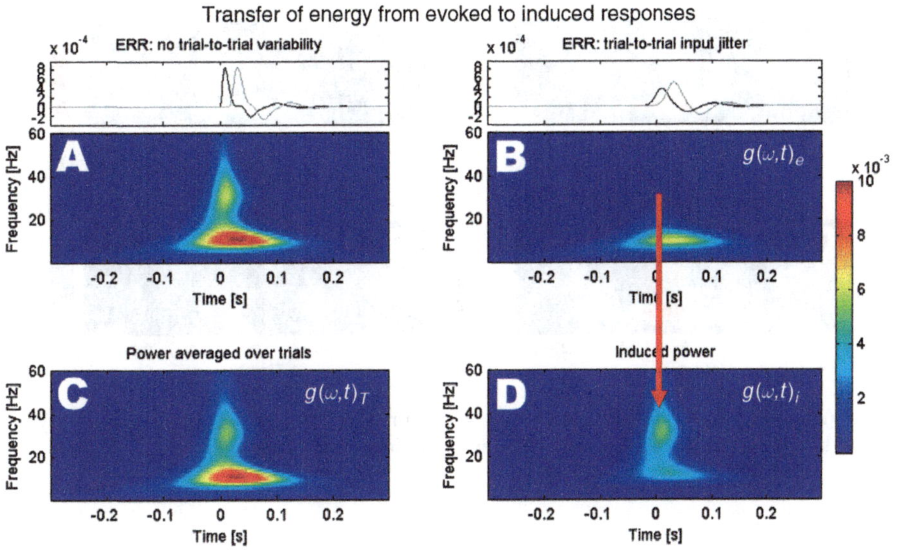

Plate 46 Simulation of trial-to-trial latency jitter (2000 trials), using the model depicted in Figure 33.18. Black curves are the responses of area 1; grey curves correspond to area 2. Time-frequency responses are shown for area 1 only. (a) Canonical response to a stimulus at time zero. (b) Evoked responses, after averaging over trials. (c) Total power, averaged over trials. (d) Induced power. As predicted, high-frequency induced oscillations emerge with latency jittering. This is due to the fact that trial-averaging removes high frequencies from the evoked power; as a result, they appear in the induced response.

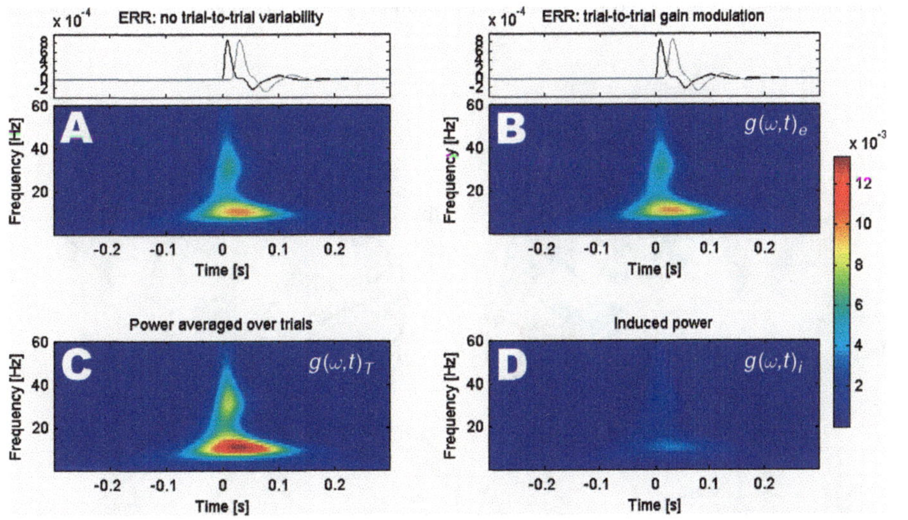

Plate 47 Simulation of gain variations over trials (2000 trials). The format is the same as in Plate 46. As predicted, although gain variation has no effect on evoked power it does affect induced power, rendering it a 'ghost' of the evoked power. See main text for details.

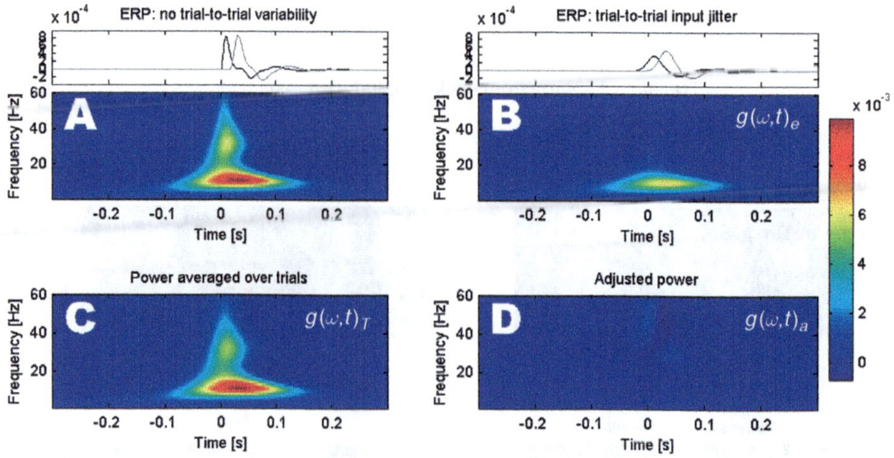

Plate 48 Adjusted power (d). The format is the same as in Plate 46. As predicted, the adjusted power is largely immune to the effects of latency variation, despite the fact that evoked responses still lose their high-frequency components.

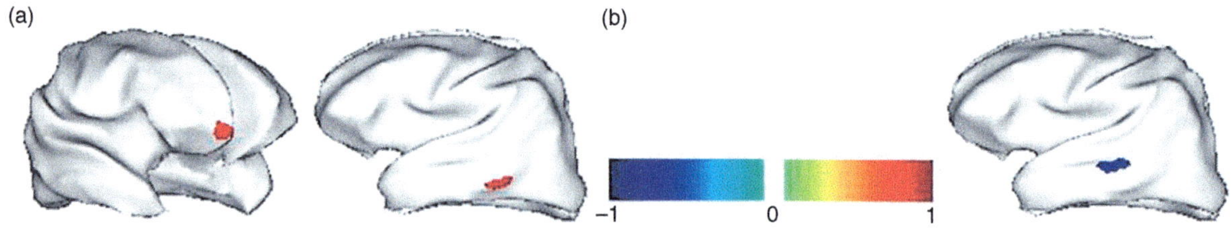

Plate 49 Inflated cortical representation of (a) two simulated source locations ('valid' prior) and (b) 'invalid' prior location.

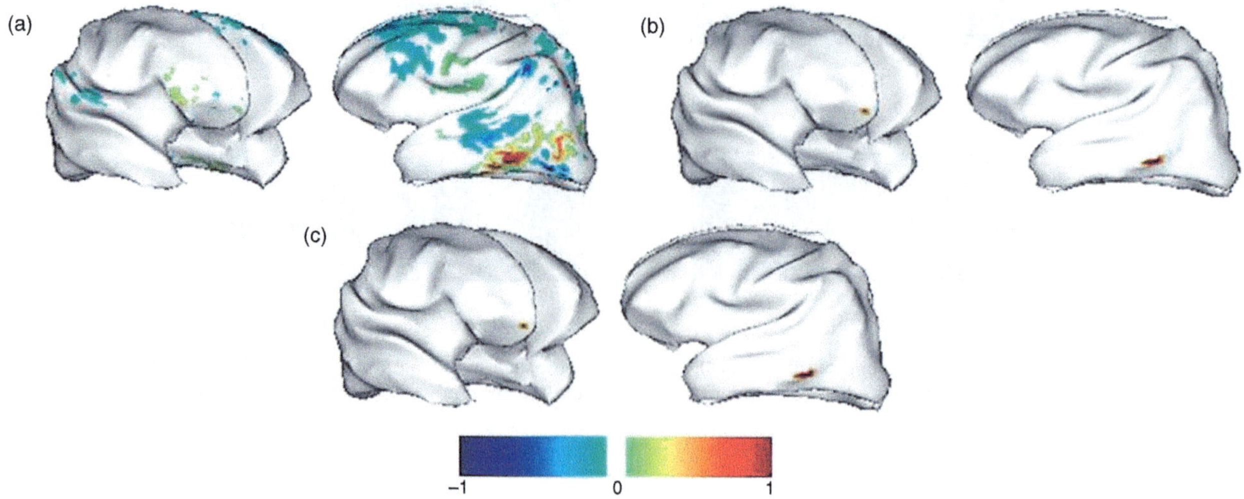

Plate 50 Inflated cortical representation of representative source reconstructions using (a) smoothness prior, (b) smoothness and valid priors and (c) smoothness, valid and invalid priors. The reconstructed values have been normalized between −1 and 1.

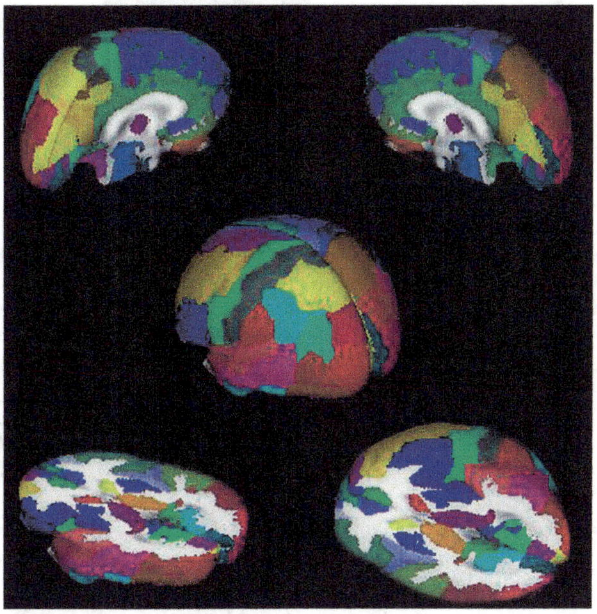

Plate 51 3D segmentation of 71 structures of the probabilistic MRI atlas developed at the Montreal Neurological Institute. As shown in the colour scale, brain areas belonging to different hemispheres were segmented separately.

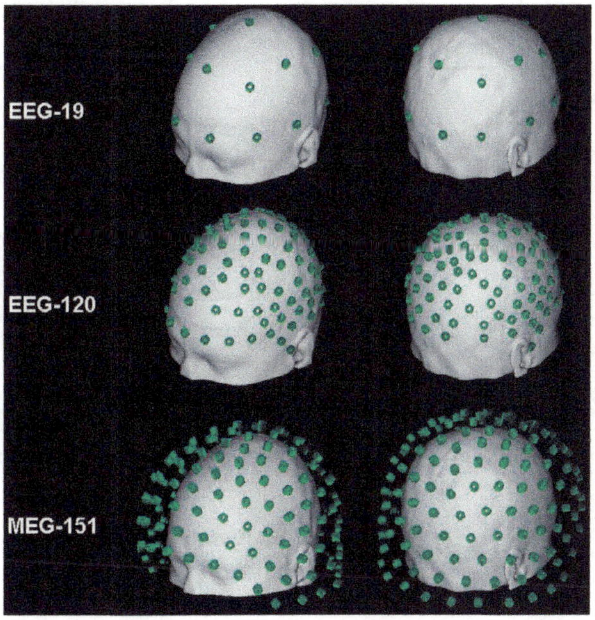

Plate 52 Different arrays of sensors used in the simulations. EEG-19 represents the 10/20 electrode system; EEG-120 is obtained by extending and refining the 10/20 system; and MEG-151 corresponds to the spatial configuration of MEG sensors in the helmet of the CTF System Inc.

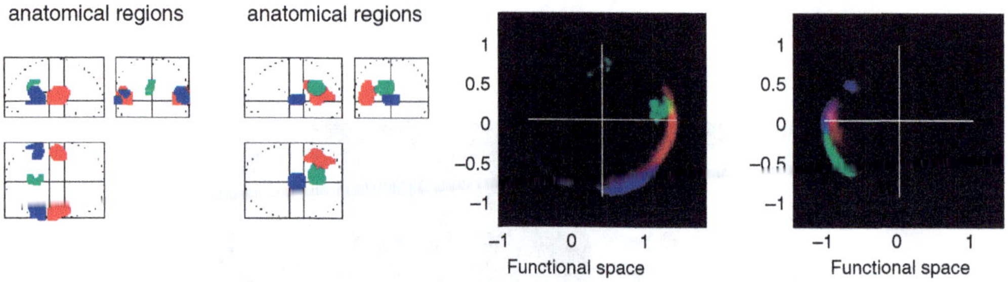

Plate 53 Classical or metric scaling analysis of the functional topography of intrinsic word generation in normal subjects. Left: anatomical regions categorized according to their colour. The designation was by reference to the atlas of Talairach and Tournoux (1988). Right: regions plotted in a functional space, following the scaling transformation. In this space the proximity relationships reflect the functional connectivity among regions. The colour of each voxel corresponds to the anatomical region it belongs to. The brightness reflects the local density of points corresponding to voxels in anatomical space. This density was estimated by binning the number of voxels in 0.02 'boxes' and smoothing with a Gaussian kernel of full width at half maximum of 3 boxes. Each colour was scaled to its maximum brightness.

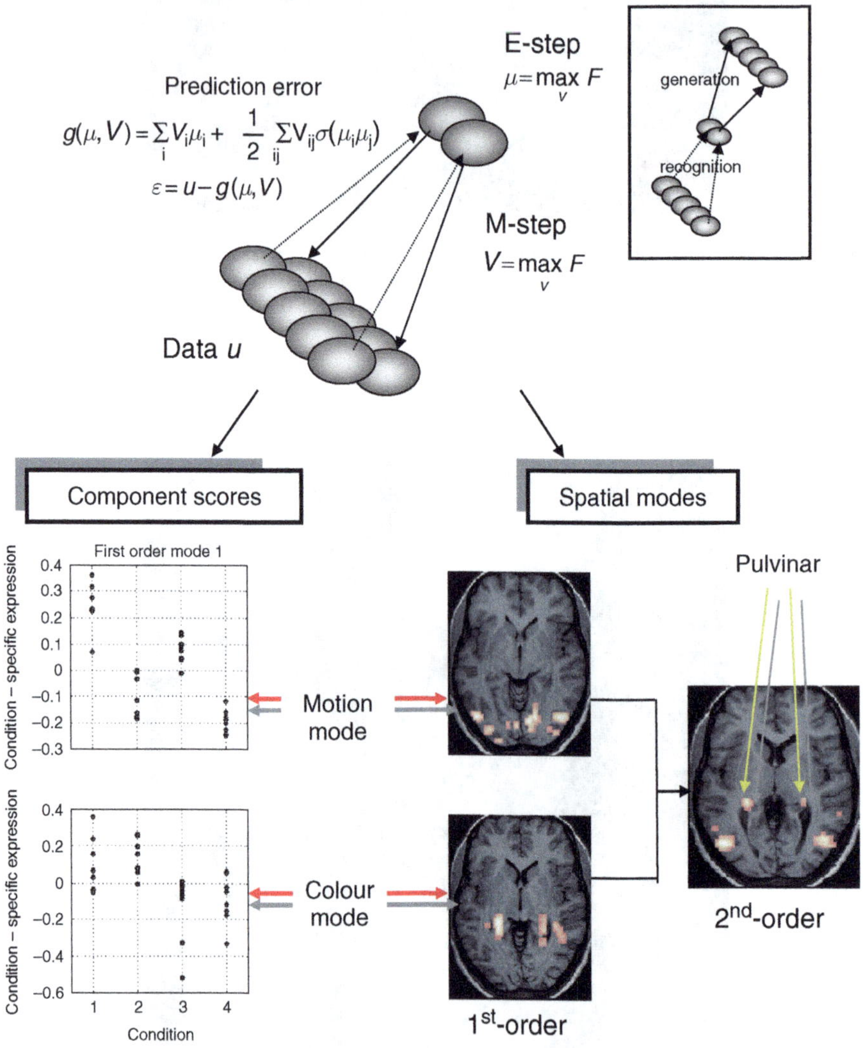

Plate 54 Upper panel: schematic of the neural net architecture used to estimate causes and modes. Feed-forward connections from the input layer to the hidden layer provide an estimate of the causes using some recognition model (the **E**-step). This estimate minimizes prediction error under the constraints imposed by prior assumption about the causes. The modes or parameters are updated in an **M**-step. The architecture is quite ubiquitous and when 'unwrapped' discloses the hidden layer as a 'bottleneck' (see insert). These bottleneck-architectures are characteristic of manifold learning algorithms, like non-linear PCA. Lower panel (left): condition-specific expression of the two first-order modes from the visual processing fMRI study. These data represent the degree to which the first principal component of epoch-related responses over the 32 photic stimulation-baseline pairs was expressed. These condition-specific responses are plotted in terms of the four conditions for the two modes. **Motion** – motion present. **Stat.** – stationary dots. **Colour** – isoluminant, chromatic contrast stimuli. **Isochr.** – isochromatic, luminance contrast stimuli. Lower panels (right): the axial slices have been selected to include the maxima of the corresponding spatial modes. In this display format, the modes have been thresholded at 1.64 of each mode's standard deviation over all voxels. The resulting excursion set has been superimposed onto a structural T1-weighted MRI image.

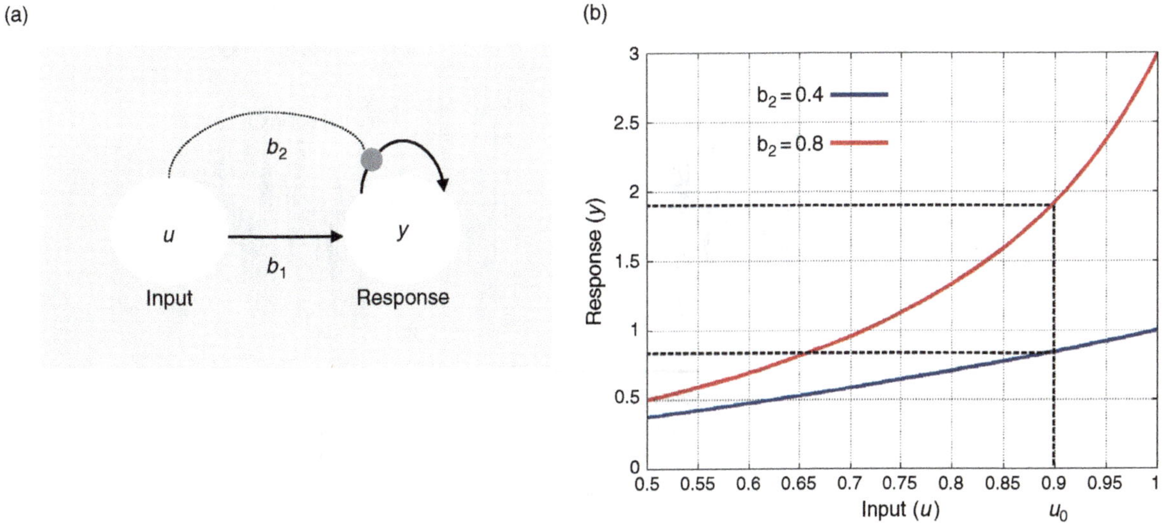

Plate 55 A simple non-linear model involving u (input) and y (response). (a) Non-linearity in the response is generated by a bilinear term uy, which models a non-additive interaction between input and intrinsic activity. The model is noise free for simplicity. The interaction term is scaled by b_2, effectively quantifying the model's sensitivity to input at different levels of intrinsic activity. (b) Plots of input and output at different values of b_2 disclose the model's sensitivity to b_2. At a fixed input, $u = u_0$, the response varies depending on its value.

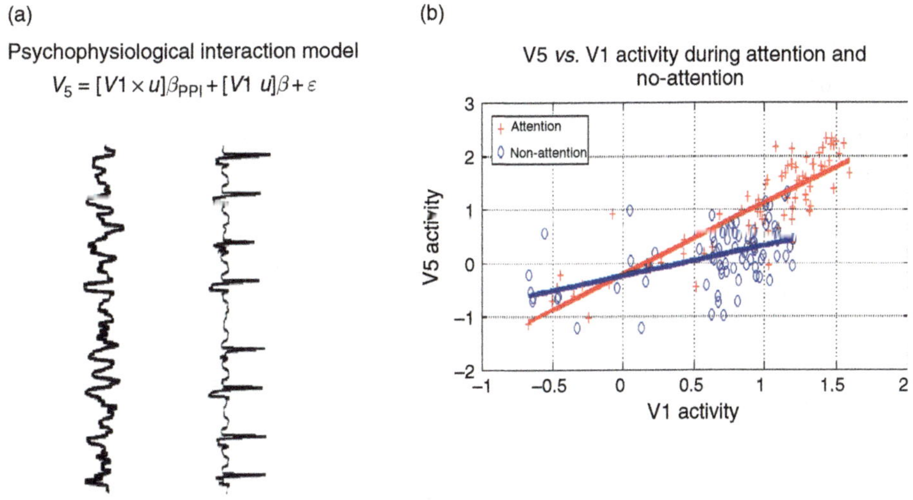

Plate 56 (a) Models for psychophysiological interaction (PPI): subjects were asked to detect changes in velocity of a radially moving stimulus or just to observe the stimulus. The velocity of the actual stimulus remained constant, so that only the attentional set changed. An analysis based on the PPI model in (a) identified a significant response in V5 that was consistent with an attentional modulation of input from V1. The PPI term is basically an interaction between attentional set, u, and V1 activity as measured with fMRI. (b) The change in sensitivity of V5 to V1 input, depending on attentional set. This is a simple comparative regression analysis, after partitioning the data according to the level of the attention factor.

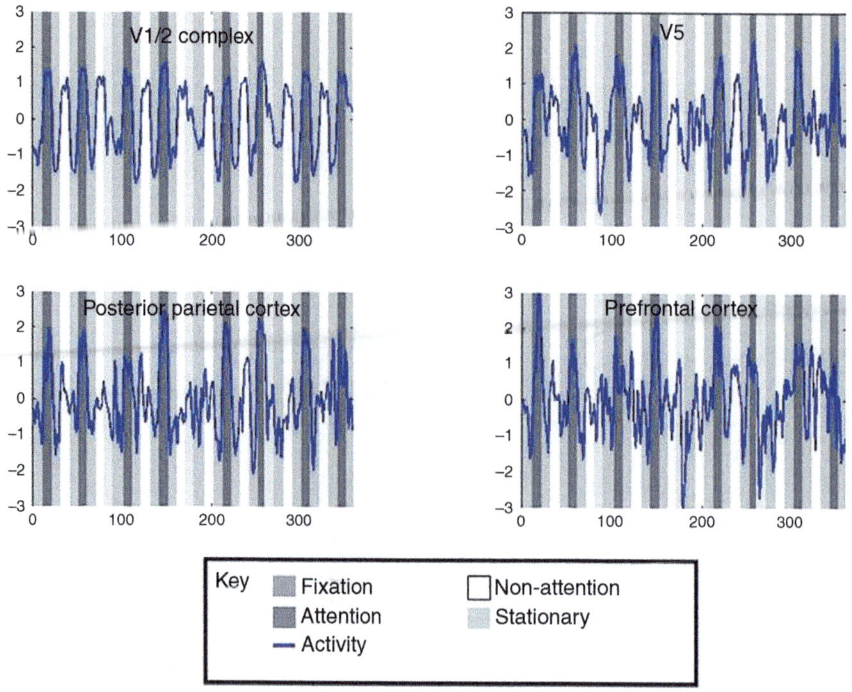

Plate 57 These are the time-series of regions V1/2 complex, V5, PPC and PFC from subject 1, in the right hemisphere. All plots have the same axes of activity (adjusted to zero mean and unit variance) versus scan number (360 in total). The experiment consisted of four conditions in four blocks of 90 scans. Periods of 'attention' and 'non-attention' were separated by a 'fixation' interval where the screen was dark and the subject fixated on a central cross. Each block ended with a 'stationary' condition where the screen contained a freeze frame of the previously moving dots. Epochs of each task are indicated by the background greyscale (see key) of each series. Visually evoked activity is dominant in the lower regions of the V1/2 complex, whereas attentional set becomes the prevalent influence in higher PPC and PFC regions.

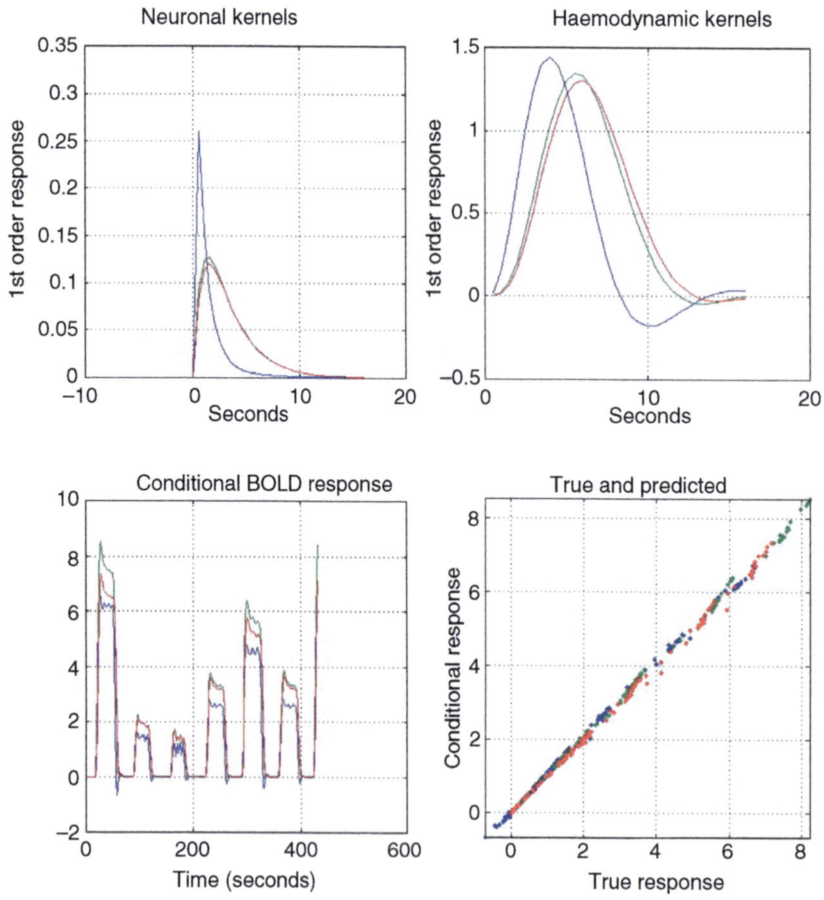

Plate 58 These results are based upon the conditional or MAP estimates of Figure 41.7. The upper panels show the implied first-order kernels for neuronal responses (upper-left) and equivalent haemodynamic responses (upper-right) as a function of peristimulus time for each of the three regions. The lower panels show the predicted response based upon the MAP estimators and a comparison of this response to the true response. The agreement is self-evident.

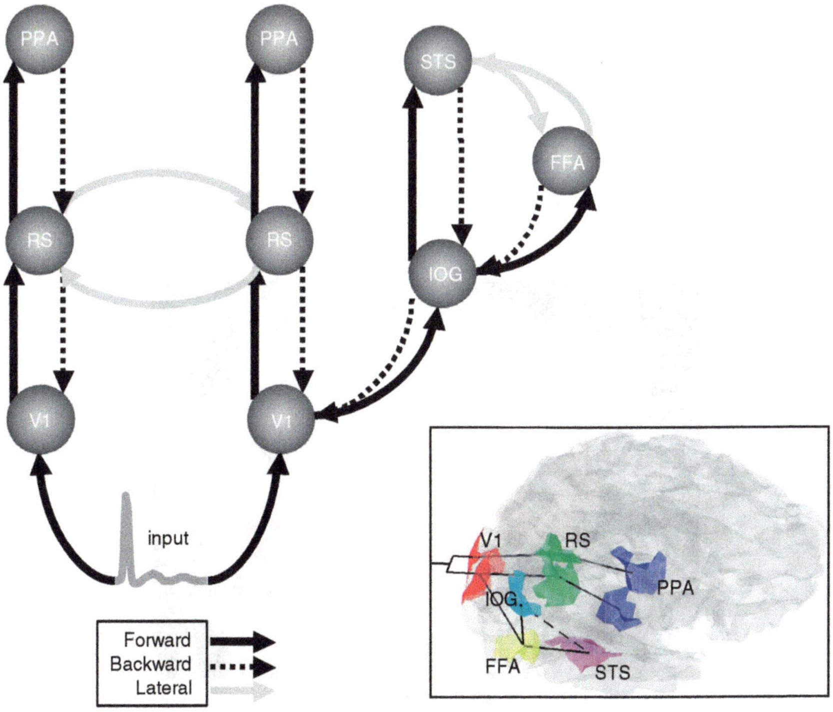

Plate 59 Model definition for the category-selectivity paradigm: the sources comprising the DCM are connected with forward (solid), backward (broken) or lateral (grey) connections as shown. V1: primary visual cortex, RS: retrosplenial cortex, PPA: parahippocampal place area, IOG: inferior occipital gyrus, STS: superior temporal sulcus, FFA: fusiform face area (left is on the left). Insert: transparent views of the subject's cortical mesh from the top-right, showing the sources that defined the lead-field for the DCM: a bilateral extrinsic input acts on the primary visual cortex (red). Two pathways are considered: (i) bilaterally from occipital regions to the parahippocampal place area (blue) through the retrosplenial cortex (green, laterally interconnected); (ii) in the right hemisphere, from primary visual areas to inferior occipital gyrus (yellow) which projects to the superior temporal sulcus (cyan) and the lateral fusiform gyrus (magenta). The superior temporal sulcus and lateral fusiform gyrus are laterally connected.

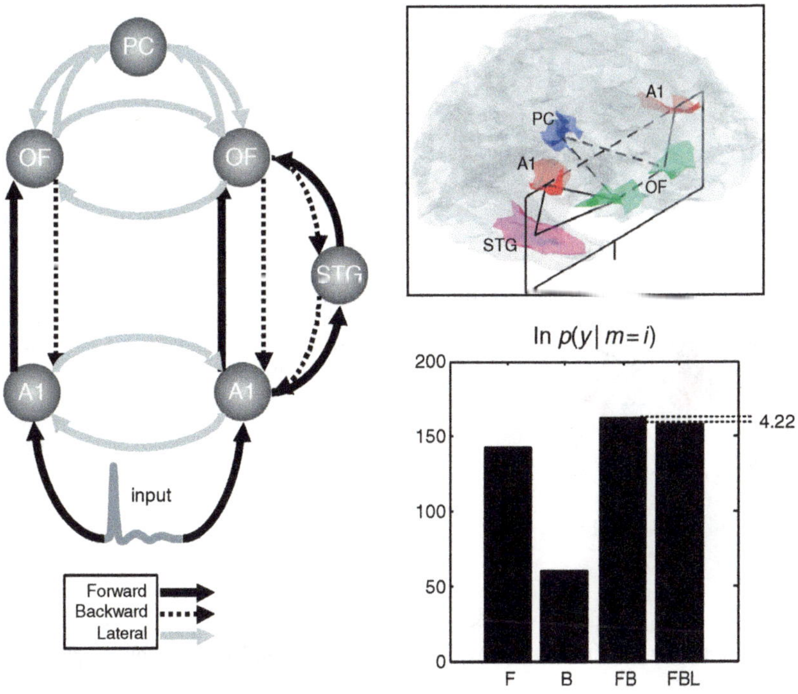

Plate 60 DCM specification for the auditory oddball paradigm: Left: graph depicting the sources and connections of the DCM using the same format as Plate 59: A1: primary auditory cortex, OF: orbitofrontal cortex, PC: posterior cingulate cortex, STG: superior temporal gyrus. Insert: localized sources corresponding to the lead fields that entered the DCM: a bilateral extrinsic input acts on primary auditory cortex (red) which project to orbitofrontal regions (green). In the right hemisphere, an indirect pathway was specified, via a relay in the superior temporal gyrus (magenta). At the highest level in the hierarchy, orbitofrontal and left posterior cingulate (blue) cortices were assumed to be laterally and reciprocally connected. Lower right: results of the Bayesian model selection among DCMs allowing for learning-related changes in forward F, backward B, forward and backward FB and all connections FBL. The graph shows the Laplace approximation to the log-evidence and demonstrates that the FB model supervenes. The log-evidence is expressed relative to a DCM in which no connections were allowed to change.

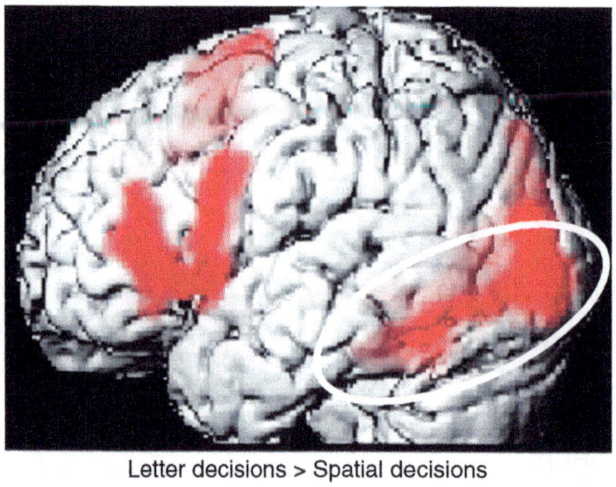

Plate 61 Results from an SPM analysis of the fMRI data from Stephan *et al.* (2003). Comparing letter decisions to visuo-spatial decisions about identical stimuli showed strongly left-lateralized responses, including classical language areas in the left inferior frontal gyrus and visual areas in the left ventral visual stream (white ellipse), e.g. in the fusiform gyrus, middle occipital gyrus and lingual gyrus. Results are shown at $p < 0.05$, corrected at the cluster level for multiple comparisons across the whole brain. Adapted, with permission, from Figure 1 in Stephan *et al.*, 2003.

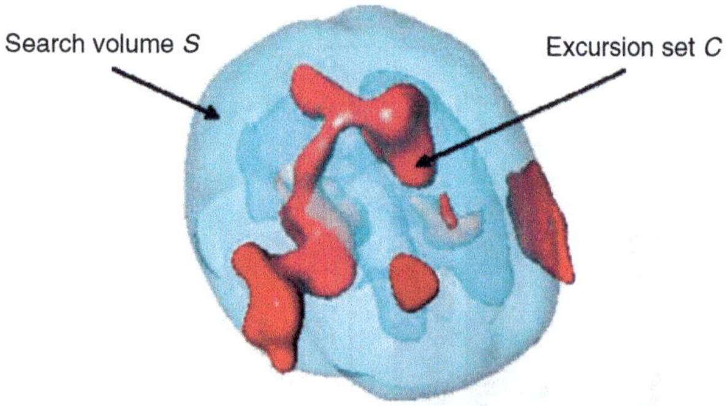

Intrinsic volumes of C

C	$\mu_0(C)$	$\mu_1(C)$	$\mu_2(C)$	$\mu_3(C)$
Sphere, radius r	1	$4r$	$2\pi r^2$	$(4/3)\pi r^3$
Hemisphere, radius r	1	$(2+\pi/2)r$	$(3/2)\pi r^2$	$(2/3)\pi r^3$
Disk, radius r	1	πr	πr^2	0
Sphere surface, radius r	2	0	$4\pi r^2$	0
Hemisphere surface, radius r	1	πr	$2\pi r^2$	0
Box, $a \times b \times c$	1	$a+b+c$	$ab+bc+ac$	abc
Rectangle, $a \times b$	1	$a+b$	ab	0
Line, length a	1	a	0	0

Plate 62 Left: picture of a search volume $S \subset \Re^D$ and its excursion set $C = \{t \in S : X(t) > x\}$, defined by a height threshold x. Right: Minkowski functionals for some common search volumes. After statistical flattening (i.e. with unit roughness) these correspond to the resel counts.

C H A P T E R

25

Spatio-temporal models for fMRI

W. Penny, G. Flandin and N. Trujillo-Barreto

INTRODUCTION

Functional magnetic resonance imaging (fMRI) using blood oxygen-level-dependent (BOLD) contrast is an established method for making inferences about regionally specific activations in the human brain (Frackowiak *et al.*, 2003). From measurements of changes in blood oxygenation one can use various statistical models, such as the general linear model (GLM) (Friston *et al.*, 1995), to make inferences about task-specific changes in underlying neuronal activity.

This chapter reviews previous work (Penny *et al.*, 2003, 2005, 2006; Penny and Flandin, 2005a) on the development of spatially regularized general linear models (SRGLMs) for the analysis of fMRI data. These models allow for the characterization of subject and regionally specific effects using spatially regularized posterior probability maps (PPMs). This spatial regularization has been shown (Penny *et al.*, 2005) to increase the sensitivity of inferences one can make.

The chapter is structured as follows. The theoretical section describes the generative model for SRGLM. This is split into descriptions of the prior and likelihood. We show how the variational Bayes (VB) algorithm, described in the previous chapter, can be used for approximate inference. We describe how these inferences are implemented for uni- and multivariate contrasts, and discuss the rationale for thresholding the resulting PPMs. We also discuss the spatio-temporal nature of the model and compare it with standard approaches. The results section looks at null fMRI data, synthetic data and fMRI from functional activation studies of auditory and face processing. The chapter finishes with a discussion.

Notation

Lower case variable names denote vectors and scalars. Whether the variable is a vector or scalar should be clear from the context. Upper case names denote matrices or dimensions of matrices. In what follows $\mathsf{N}(x;\mu,\Sigma)$ denotes a multivariate normal density over x, having mean μ and covariance Σ. The precision of a Gaussian variate is the inverse (co)variance. A gamma density over the scalar random variable x is written as $\mathsf{Ga}(x;a,b)$. Normal and gamma densities are defined in Chapter 24. We also use $||x||^2 = x^T x$, denote the trace operator as $\mathsf{Tr}(X)$, X^+ for the pseudo-inverse, and use $\mathsf{diag}(x)$ to denote a diagonal matrix with diagonal entries given by the vector x.

THEORY

We denote an fMRI data set consisting of T time points at N voxels as the $T \times N$ matrix Y. In mass-univariate models (Friston *et al.*, 1995), these data are explained in terms of a $T \times K$ design matrix X, containing the values of K regressors at T time points, and a $K \times N$ matrix of regression coefficients W, containing K regression coefficients at each of the N voxels. The model is written:

$$Y = XW + E \qquad \textbf{25.1}$$

where E is a $T \times N$ error matrix.

It is well known that fMRI data are contaminated with artefacts. These stem primarily from low-frequency drifts due to hardware instabilities, aliased cardiac pulsation and respiratory sources, unmodelled neuronal activity and residual motion artefacts not accounted for by rigid body registration methods (Woolrich *et al.*, 2001). This

Statistical Parametric Mapping, by Karl Friston et al.
ISBN–13: 978-0-12-372560-8 ISBN–10: 0-12-372560-7

results in the residuals of an fMRI analysis being temporally autocorrelated.

In previous work, we have shown that, after removal of low-frequency drifts using discrete cosine transform (DCT) basis sets, low-order voxel-wise autoregressive (AR) models are sufficient for modelling this autocorrelation (Penny *et al.*, 2003). It is important to model these noise processes because parameter estimation then becomes less biased (Gautama and Van Hulle, 2004) and more accurate (Penny *et al.*, 2003).

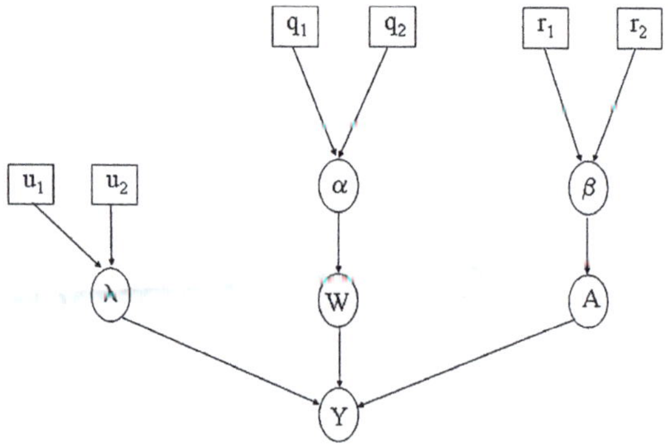

FIGURE 25.1 The figure shows the probabilistic dependencies underlying the SRGLM generative model for fMRI data. The quantities in squares are constants and those in circles are random variables. The spatial regularization coefficients α constrain the regression coefficients W. The parameters λ and A define the autoregressive error processes which contribute to the measurements. The spatial regularization coefficients β constrain the AR coefficients A.

Model likelihood

We now describe the approach taken in our previous work. For a Pth-order AR model, the likelihood of the data is given by:

$$p(Y|W, A, \lambda) = \prod_{t=P+1}^{T} \prod_{n=1}^{N} \mathrm{N}(y_{tn} - x_t w_n; (d_{tn} - X_t w_n)^T a_n, \lambda_n^{-1}) \quad 25.2$$

where n indexes the nth voxel, a_n is a $P \times 1$ vector of autoregressive coefficients, w_n is a $K \times 1$ vector of regression coefficients and λ_n is the observation noise precision. The vector x_t is the tth row of the design matrix and X_t is a $P \times K$ matrix containing the previous P rows of X prior to time point t. The scalar y_{tn} is the fMRI scan at the tth time point and nth voxel and $d_{tn} = [y_{t-1,n}, y_{t-2,n}, \ldots, y_{t-P,n}]^T$. Because d_{tn} depends on data P time steps before, the likelihood is evaluated starting at time point $P+1$, thus ignoring the GLM fit at the first P time points.

Eqn. **25.2** shows that higher model likelihoods are obtained when the prediction error $y_{tn} - x_t w_n$ is closer to what is expected from the AR estimate of prediction error.

The voxel-wise parameters w_n and a_n are contained in the nth columns of matrices W and A, and the voxel-wise precision λ_n is the nth entry in λ. The next section describes the prior distributions over these parameters. Together, the likelihood and prior define the generative model, which is shown in Figure 25.1.

Priors

The graph in Figure 25.1 shows that the joint probability of parameters and data can be written:

$$\begin{aligned} p(Y, W, A, \lambda, \alpha, \beta) = \; & p(Y|W, A, \lambda)p(W|\alpha) \\ & p(A|\beta)p(\lambda|u_1, u_2) \\ & p(\alpha|q_1, q_2)p(\beta|r_1, r_2) \end{aligned} \quad 25.3$$

where the first term is the likelihood and the other terms are the priors. The likelihood is given in Eqn. **25.2** and the priors are described below.

Regression coefficients

For the regressions coefficients we have:

$$\begin{aligned} p(W|\alpha) &= \prod_{k=1}^{K} p(w_k^T|\alpha_k) \\ p(w_k^T|\alpha_k) &= N(w_k^T; 0, \alpha_k^{-1} D_w^{-1}) \end{aligned} \quad 25.4$$

where D_w is a spatial precision matrix. This can be set to correspond to, for example, a low resolution tomography (LORETA) prior, a Gaussian Markov random field (GMRF) prior or a minimum norm prior ($D_w = I$) (Friston and Penny, 2003) as described in earlier work (Penny *et al.*, 2005). These priors implement the spatial regularization and are specified separately for each slice of data. Specification of 3-dimensional spatial priors (i.e. over multiple slices) is desirable from a modelling perspective, but is computationally too demanding for current computer technology.

We can also write $w_v = vec(W)$, $w_r = \mathrm{vec}(W^T)$, $w_v = H_w w_r$ where H_w is a permutation matrix. This leads to:

$$\begin{aligned} p(W|\alpha) &= p(w_v|\alpha) \\ &= N(w_v; 0, B^{-1}) \end{aligned} \quad 25.5$$

where B is an augmented spatial precision matrix given by:

$$B = H_w(\mathrm{diag}(\alpha) \otimes D_w)H_w^T \quad 25.6$$

where $\otimes$ is the Kronecker product. This form of the prior is useful as our specification of approximate posteriors is based on similar quantities. It can be seen that α encodes the spatial precision of the regression coefficients.

The above Gaussian priors underly GMRFs and LORETA and have been used previously in fMRI (Woolrich *et al.*, 2004) and electroencephalography (EEG) (Pascal Marqui *et al.*, 1994). They are by no means, however, the optimal choice for imaging data. In EEG, for example, much interest has focused on the use of L^p-norm priors (Auranen *et al.* 2005) instead of the L^2-norm implicit in the Gaussian assumption. Additionally, we are currently investigating the use of wavelet priors. This is an active area of research and will be the topic of future publications.

AR coefficients

We also define a spatial prior for the AR coefficients so that they too can be spatially regularized. We have:

$$p(A|\beta) = \prod_{p=1}^{P} p(a_p|\beta_p) \tag{25.7}$$

$$p(a_p|\beta_p) = N(a_p; 0, \beta_p^{-1} D_a^{-1})$$

Again, D_a is a user-defined spatial precision matrix, $a_v = \text{vec}(A)$, $a_r = \text{vec}(A^T)$ and $a_v = H_a a_r$, where H_a is a permutation matrix. This prior is used to implement the spatial regularization of the AR coefficients. We can write:

$$\begin{aligned} p(A|\beta) &= p(a_v|\beta) \\ &= N(a_v; 0, J^{-1}) \end{aligned} \tag{25.8}$$

where J is an augmented spatial precision matrix:

$$J = H_a (\text{diag}(\beta) \otimes D_a) H_a^T \tag{25.9}$$

This form of the prior is useful as our specification of approximate posteriors is based on similar quantities. The parameter β plays a similar role to α and controls the spatial regularization of the temporal autoregression coefficients.

We have also investigated 'tissue-type' priors which constrain AR estimates to be similar for voxels in the same tissue-type, e.g. grey matter, white matter or cerebrospinal fluid. Bayesian model selection (Penny *et al.*, 2006), however, favours the smoothly varying priors defined in Eqn. **25.7**.

Precisions

We use gamma priors on the precisions α, β and λ:

$$p(\lambda|u_1, u_2) = \prod_{n=1}^{N} \text{Ga}(\lambda_n; u_1, u_2) \tag{25.10}$$

$$p(\alpha|q_1, q_2) = \prod_{k=1}^{K} \text{Ga}(\alpha_k; q_1, q_2)$$

$$p(\beta|r_1, r_2) = \prod_{p=1}^{P} \text{Ga}(\beta_p; r_1, r_2)$$

where the gamma density is defined in Chapter 24. Gamma priors were chosen as they are the conjugate priors for Gaussian error models. The parameters are set to $q_1 = r_1 = u_1 = 10$ and $q_2 = r_2 = u_2 = 0.1$. These parameters produce gamma densities with a mean of 1 and a variance of 10. The robustness of, for example, model selection to the choice of these parameters is discussed in Penny *et al.* (2003).

Approximate posteriors

Inference for SRGLMs has been implemented using the variational Bayes (VB) approach described in the previous chapter. In this section, we describe the algorithm developed in previous work (Penny *et al.*, 2005) where we assumed that the approximate posterior factorizes over voxels and subsets of parameters.

Because of the spatial priors, the regression coefficients in the true posterior $p(W|Y)$ will clearly be correlated. Our perspective, however, is that this is too computationally burdensome for current personal computers to take account of. Moreover, as we shall see later, updates for the approximate factorized densities $q(w_n)$ do encourage the approximate posterior means to be similar at nearby voxels, thereby achieving the desired effect of the prior.

Our approximate posterior is given by:

$$\begin{aligned} q(W, A, \lambda, \alpha, \beta) = &\prod_n q(w_n) q(a_n) q(\lambda_n) \\ &\prod_k q(\alpha_k) \prod_p q(\beta_p) \end{aligned} \tag{25.11}$$

and each component of the approximate posterior is described below. These update equations are self-contained except for a number of quantities that are marked out using the 'tilde' notation. These are $\tilde{A}_n$, $\tilde{b}_n$, $\tilde{C}_n$, $\tilde{d}_n$ and $\tilde{G}_n$, which are all defined in Appendix 25.1.

Regression coefficients

We have:

$$\begin{aligned} q(w_n) &= \text{N}(w_n; \hat{w}_n, \hat{\Sigma}_n) \\ \hat{\Sigma}_n &= \left(\bar{\lambda}_n \tilde{A}_n + \bar{B}_{nn}\right)^{-1} \\ \hat{w}_n &= \hat{\Sigma}_n \left(\bar{\lambda}_n \tilde{b}_n^T + r_n\right) \\ r_n &= - \sum_{i=1, i \neq n}^{N} \bar{B}_{ni} \hat{w}_i \end{aligned} \tag{25.12}$$

where $\hat{w}_n$ is the estimated posterior mean and $\hat{\Sigma}_n$ is the estimated posterior covariance. The quantity $\bar{B}$ is defined as in Eqn. **25.6** but uses $\bar{\alpha}$ instead of α. The quantities $\tilde{A}_n$ and $\tilde{b}_n$ are expectations related to autoregressive processes and are defined in Appendix 25.1. In the absence of temporal autocorrelation we have $\tilde{A}_n = X^T X$ and $\tilde{b}_n^T = X^T y_n$.

AR coefficients

We have:

$$q(a_n) = \mathsf{N}(a_n; m_n, V_n)$$

where

$$V_n = \left(\bar{\lambda}_n \tilde{C}_n + \bar{J}_{nn}\right)^{-1} \qquad 25.13$$

$$m_n = V_n(\bar{\lambda}_n \tilde{d}_n + j_n)$$

$$j_n = -\sum_{i=1, i\neq n}^{N} \bar{J}_{ni} m_i$$

and m_n is the estimated posterior mean and V_n is the estimated posterior covariance. The quantity $\bar{J}$ is defined as in Eqn. **25.9** but $\bar{\beta}$ is used instead of β. The subscripts in $\bar{J}_{ni}$ denote that part of $\bar{J}$ relevant to the nth and ith voxels. The quantities $\tilde{C}_n$ and $\tilde{d}_n$ are expectations that are defined in Appendix 25.1.

Precisions

The approximate posteriors over the precision variables are gamma densities. For the precisions on the observation noise we have:

$$q(\lambda_n) = \mathsf{Ga}(\lambda_n; b_n, c_n) \qquad 25.14$$

$$\frac{1}{b_n} = \frac{\tilde{G}_n}{2} + \frac{1}{u_1}$$

$$c_n = \frac{T}{2} + u_2$$

$$\bar{\lambda}_n = bc$$

where $\tilde{G}_n$ is the expected prediction error defined in Appendix 25.1. For the precisions of the regression coefficients we have:

$$q(\alpha_k) = \mathsf{Ga}(\alpha_k; g_k, h_k) \qquad 25.15$$

$$\frac{1}{g_k} = \frac{1}{2}\left(\mathrm{Tr}(\hat{\Sigma}_k D_w) + \hat{w}_k^T D_w \hat{w}_k\right) + \frac{1}{q_1}$$

$$h_k = \frac{N}{2} + q_2$$

$$\bar{\alpha}_k = g_k h_k$$

For the precisions of the AR coefficients we have:

$$q(\beta_p) = \mathsf{Ga}(\beta_p; r_{1p}, r_{2p}) \qquad 25.16$$

$$\frac{1}{r_{1p}} = \frac{1}{2}\left(\mathrm{Tr}(V_p D_a) + m_p^T D_a m_p\right) + \frac{1}{r_1}$$

$$r_{2p} = \frac{N}{2} + r_2$$

$$\bar{\beta}_p = r_{1p} r_{2p}$$

Practicalities

Our empirical applications use spatial precision matrices D_a and D_w, defined above, which produce GMRF priors. Also, we use AR models of order $P = 3$. Model selection using VB showed that this model order was sufficient for all voxels in a previous analysis of fMRI (Penny *et al.*, 2003).

The VB algorithm is initialized using ordinary least square (OLS) estimates for regression and autoregressive parameters as described in Penny *et al.* (2003). Quantities are then updated using Eqns. **25.12, 25.13, 25.14, 25.15, 25.16.**

As described in the previous chapter, the aim of VB is to match an approximate posterior to the true posterior density in the sense of minimizing Kullback-Liebler (KL) divergence. This is implemented implicitly by maximizing the quantity F, known in statistical physics as the negative free energy. In the implemention of VB for SRGLMs, F is monitored during optimization. Convergence is then defined as less than a 1 per cent increase in F.

Expressions for computing F are given in Penny *et al.* (2006). This is an important quantity as it can also be used for model comparison. This is described at length in Penny *et al.* (2006) and reviewed in Chapters 24 and 35.

The algorithm we have described is implemented in SPM version 5 and can be downloaded from SPM Software (2006). Computation of a number of quantites (e.g. $\tilde{C}_n$, $\tilde{d}_n$ and $\tilde{G}_n$) is now much more efficient than in previous versions (Penny *et al.*, 2005). These improvements are described in a separate document (Penny and Flandin, 2005b). To analyse a single session of data (e.g. the face fMRI data) takes about 30 minutes on a typical personal computer.

Spatio-temporal deconvolution

The central quantity of interest in fMRI analysis is our estimate of effect sizes, embodied in contrasts of regression coefficients. A key update equation in our VB scheme is, therefore, the approximate posterior for the regression coefficients. This is given by Eqn. **25.12**. For the special case of temporally uncorrelated data we have:

$$\hat{\Sigma}_n = \left(\bar{\lambda}_n X^T X + \bar{B}_{nn}\right)^{-1} \qquad \textbf{25.17}$$
$$\hat{w}_n = \hat{\Sigma}_n \left(\bar{\lambda}_n X^T y_n + r_n\right)$$

where $\bar{B}$ is a spatial precision matrix and r_n is the weighted sum of neighbouring regression coefficient estimates.

This update indicates that the estimate at a given voxel regresses towards those at nearby voxels. This is the desired effect of the spatial prior and is preserved despite the factorization over voxels in the approximate posterior (see Eqn. **25.11**). Eqn. **25.17** can be thought of as the combination of a temporal prediction $X^T y_n$ and a spatial prediction from r_n. Each prediction is weighted by its relative precision to produce the optimal estimate $\hat{w}_n$. In this sense, VB provides a spatio-temporal deconvolution of fMRI data. Moreover, the parameters controlling the relative precisions, $\bar{\lambda}_n$ and $\bar{\alpha}$, are estimated from the data. The effect size estimates therefore derive from an automatically regularized spatio-temporal deconvolution.

Contrasts

After having estimated a model, we will be interested in characterizing a particular effect, c, which can usually be expressed as a linear function or 'contrast' of parameters, w. This is described at length in Chapter 9. That is,

$$c_n = C^T w_n \qquad \textbf{25.18}$$

where C is a contrast vector or matrix. For example, the contrast vector $C^T = [1, -1]$ computes the difference between two experimental conditions.

Our statistical inferences are based on the approximate distribution $q(W)$, which implies a distribution on c, $q(c)$. Because $q(W)$ factorizes over voxels we can write:

$$q(c) = \prod_{n=1}^{N} q(c_n) \qquad \textbf{25.19}$$

where c_n is the effect size at voxel n. Given a contrast matrix C we have:

$$q(c_n) = \mathsf{N}(c_n; \mu_n, S_n) \qquad \textbf{25.20}$$

with mean and covariance:

$$\mu_n = C^T \hat{w}_n \qquad \textbf{25.21}$$
$$S_n = C^T \hat{\Sigma}_n C$$

Bayesian inference based on this posterior can then take place using confidence intervals (Box and Tiao, 1992). For univariate contrasts we have suggested the use of posterior probability maps (PPMs), as described in Chapter 23.

If c_n is a vector, then we have a multivariate contrast. Inference can then proceed as follows. The probability α that the zero vector lies on the $1-\alpha$ confidence region of the posterior distribution at each voxel must then be computed. We first note that this probability is the same as the probability that the vector μ_n lies on the edge of the $1-\alpha$ confidence region for the distribution $N(\mu_n; 0, S_n)$. This latter probability can be computed by forming the test statistic:

$$d_n = \mu_n^T S_n^{-1} \mu_n \qquad \textbf{25.22}$$

which will be the sum of $r_n = rank(S_n)$ independent, squared Gaussian variables. As such, it has a χ^2 distribution:

$$p(d_n) = \chi^2(r_n) \qquad \textbf{25.23}$$

This procedure is identical to that used for making inferences in Bayesian multivariate autoregressive models (Harrison *et al.*, 2003). We can also use this procedure to test for two-sided effects, i.e., activations or deactivations. Though, strictly, these contrasts are univariate we will use the term 'multivariate contrasts' to cover these two-sided effects.

Thresholding

In previous work (Friston and Penny, 2003), we have suggested deriving PPMs by applying two thresholds to the posterior distributions: (i) an effect size threshold, γ; and (ii) a probability threshold p_T. Voxel n is then included in the PPM if $q(c_n > \gamma) > p_T$. This approach was described in Chapter 23.

If voxel n is to be included, then the posterior exceedance probability $q(c_n > \gamma)$ is plotted. It is also possible to plot the effect size itself, c_n. The following exploratory procedure can be used for exploring the posterior distribution of effect sizes. First, plot a map of effect sizes using the thresholds $\gamma = 0$ and $p_T = 1 - 1/N$ where N is the number of voxels. We refer to these values as the 'default thresholds'. Then, after visual inspection of the resulting map use a non-zero γ, the value of which reflects effect sizes in areas of interest. It will then be possible to reduce p_T to a value such as 0.95. Of course, if previous imaging analyses have indicated what effect sizes are physiologically relevant then this exploratory procedure is unnecessary.

False positive rates

If we partition effect-size values into two hypothesis spaces $H_0 : c \leq \gamma$ and $H_1 : c > \gamma$, then we can characterize the sensitivity and specificity of our algorithm. This is different to classical inference which uses $H_0 : c = 0$. A false positive (FP) occurs if we accept H_1 when H_0 is true.

If we use the default threshold, and the approximate posterior were exact, then the distribution of FPs is binomial with rate $1/N$. The expected number of false positives in each PPM is therefore $N \times 1/N = 1$. The variance is $N \times 1/N \times (1 - 1/N)$ which is approximately 1. We would therefore expect 0, 1 or 2 false positives per PPM.

Of course, the above result only holds if the approximate posterior is equal to the true posterior. But, given that all of our computational effort is aimed at this goal, it would not be surprising if the above analysis were indicative of actual FP rates. This issue will be investigated using null fMRI data in the next section.

RESULTS

Null data

This section describes the analysis of a null dataset to find out how many false positives are obtained using PPMs with default thresholds.

Images were acquired from a 1.5T Sonata (Siemens, Erlangen, Germany) which produced T2*-weighted transverse echo-planar images (EPIs) with BOLD contrast, while a subject was lying in the scanner, asked to rest and was not provided with any experimental stimulus. These data are thus collected under the null hypothesis, H_0, that experimental effects are zero. This should hold whatever the design matrix and contrast we conceive. Any observed activations will be false positives.

Whole brain EPIs consisting of 48 transverse slices were acquired every $TR = 4.32\,s$ resulting in a total of $T = 98$ scans. The voxel size is $3 \times 3 \times 3\,mm$. All images were realigned to the first image using a six-parameter rigid-body transformation to account for subject movement. These data were not spatially smoothed. While spatial smoothing is necessary for the standard application of classical inference (see e.g. Chapter 2), it is not necessary for the spatio-temporal models described in this chapter. Indeed, the whole point of SRGLM is that the optimal smoothness can be inferred from the data.

We then implemented a standard whole volume analysis on images comprising $N = 59\,945$ voxels. We used the design matrix shown in the left panel of Figure 25.2. Use of the default thresholds resulted in no spurious activations in the PPM.

We then repeated the above analysis but with a number of different design matrices. First, we created a number of epoch designs. These were based on the design in Figure 25.2, but epoch onsets were jittered by a number between plus or minus 9 scans. This number was drawn from a uniform distribution, and the epoch durations were drawn from a uniform distribution between 4 and 10 scans. Five such design matrices were created and VB models fitted with each to the null data. For every analysis, the number of false positives was 0.

Secondly, we created a number of event-related designs by sampling event onsets with inter-stimulus intervals drawn from a poisson distribution with rate five scans. These event streams were then convolved with a canonical HRF (Henson, 2003). Again, five such design matrices were created and VB models fitted with each to the null data. Over the five analyses, the average number of false positives was 9.4. The higher false positive rate for event-related designs is thought to occur because event-related regressors are more similar than epoch regressors to fMRI noise.

Synthetic data

We then added three synthetic activations to a slice of null data ($z = -13\,mm$). These were created using the design matrix and regression coefficient image shown in Figure 25.2 (the two regression coefficient images, i.e. for the activation and the mean, were identical). These images were formed by placing delta functions at three locations and then smoothing with Gaussian kernels having full width at half maxima (FWHMs) of 2, 3 and 4 pixels (going clockwise from the top-left blob). Images were then rescaled to make the peaks unity.

In principle, smoothing with a Gaussian kernel renders the true effect size greater than zero everywhere because a Gaussian has infinite spatial support. In practice, however, when implemented on a digital computer with finite numerical precision, most voxels will be numerically zero. Indeed, our simulated data contained 299 'activated' voxels, i.e. voxels with effect sizes numerically greater than zero.

This slice of data was then analysed using VB. The contrast $C^T = [1, 0]$ was then used to look at the estimated activation effect, which is shown in the left panel of Figure 25.3. For comparison, we also show the effect as estimated using OLS. Clearly, OLS estimates are much noisier than VB estimates.

Figure 25.4 shows plots of the exceedance probabilities for two different effect-size thresholds, $\gamma = 0$ and $\gamma = 0.3$. Figure 25.5 shows thresholded versions of these images. These are PPMs. Neither of these PPMs contains any false positives. That is, the true effect size is greater than

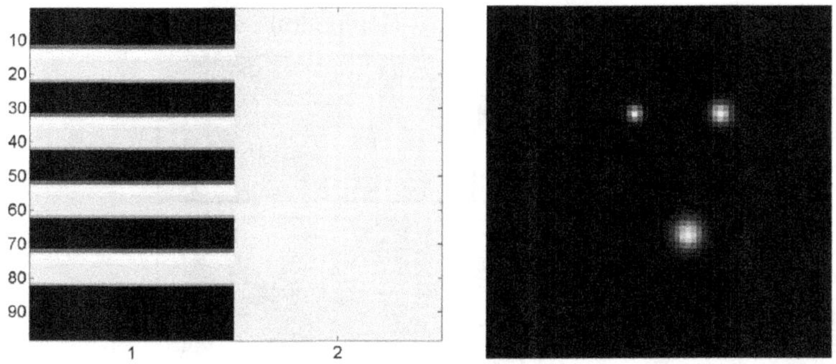

FIGURE 25.2 Left: design matrix for null fMRI data. The first column models a boxcar activation and the second column models the mean. There are $n = 1..98$ rows corresponding to the 98 scans. Right: image of regression coefficients corresponding to a synthetic activation. This image is added to the null data. In this image, and others that follow, black is 0 and white is 1.

zero wherever a white voxel occurs. This shows, informally, that use of the default thresholds provides good specificity while retaining reasonable sensitivity. Also, a combination of non-zero effect-size thresholds and more liberal probability thresholds can do the same.

Auditory data

This section describes the use of multivariate contrasts for an auditory fMRI data set comprising whole brain BOLD/EPI images acquired on a modified 2T Siemens Vision system. Each acquisition consisted of 64 contiguous slices ($64 \times 64 \times 64$, $3\,\text{mm} \times 3\,\text{mm} \times 3\,\text{mm}$ voxels) and a time-series of 84 images was acquired with $TR = 7\,\text{s}$ from a single subject.

This was an epoch fMRI experiment in which the condition for successive epochs alternated between rest and auditory stimulation, starting with rest. Auditory stimulation was bi-syllabic words presented binaurally at a rate of 60 per minute.

These data were analysed using VB with the design matrix shown in Figure 25.6. To look for voxels that increase activity in response to auditory stimulation, we used the univariate contrast $C^T = [1, 0]$. Plate 23 (see colour plate section) shows a PPM that maps effect-sizes of above-threshold voxels.

To look for either increases or decreases in activity, we use the multivariate contrast $C^T = [1, 0]$. This inference uses the χ^2 approach described earlier. Plate 24 shows the PPM obtained using default thresholds.

Face data

This is an event-related fMRI data set acquired by Henson *et al.* (2002) during an experiment concerned with the processing of faces. Greyscale images of faces were presented for 500 ms, replacing a baseline of an oval chequerboard which was present throughout the interstimulus interval. Some faces were of famous people, and were therefore familiar to the subject, and others were not. Each face

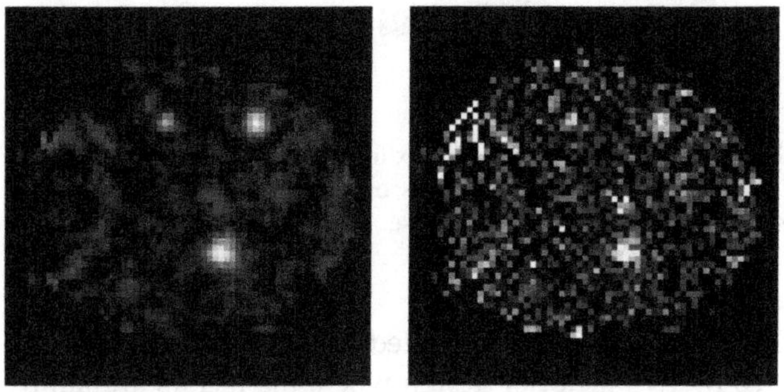

FIGURE 25.3 Left: estimated effect using VB (the true effect is shown in the right section in Figure 25.2). Right: estimated effect using OLS.

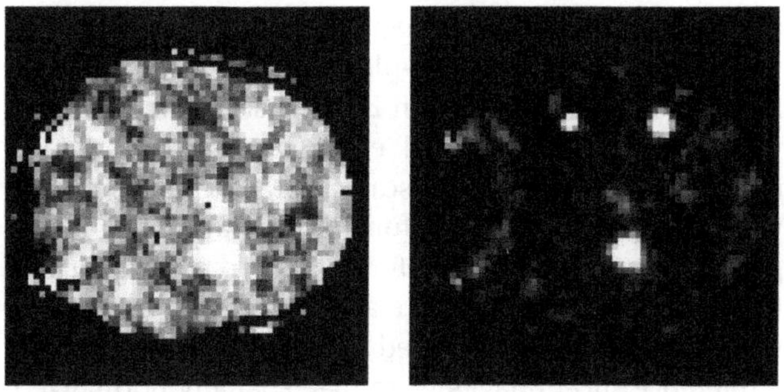

FIGURE 25.4 Plots of exceedance probabilities for two γ thresholds. Left: a plot of $p(c_n > 0)$. Right: a plot of $p(c_n > 0.3)$.

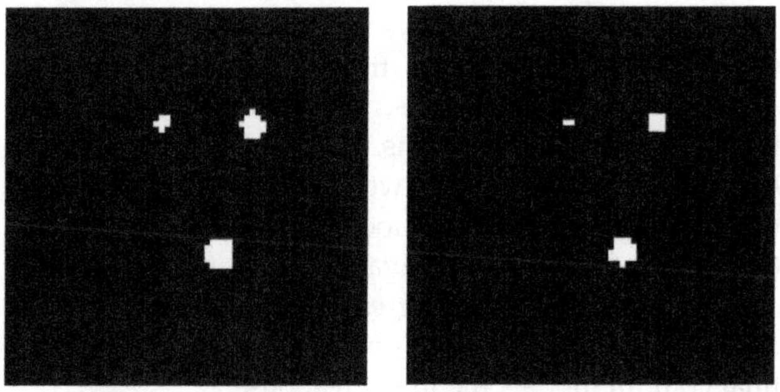

FIGURE 25.5 PPMs for two thresholds. Left: the default thresholds ($\gamma = 0$, $p_T = 1 - 1/N$) Right: the thresholds $\gamma = 0.3$, $p_T = 0.95$.

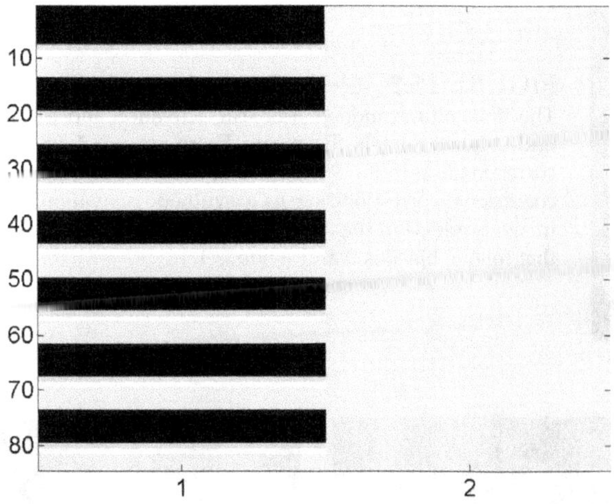

FIGURE 25.6 Design matrix for analysis of the auditory data. The first column models epochs of auditory stimulation and the second models the mean response.

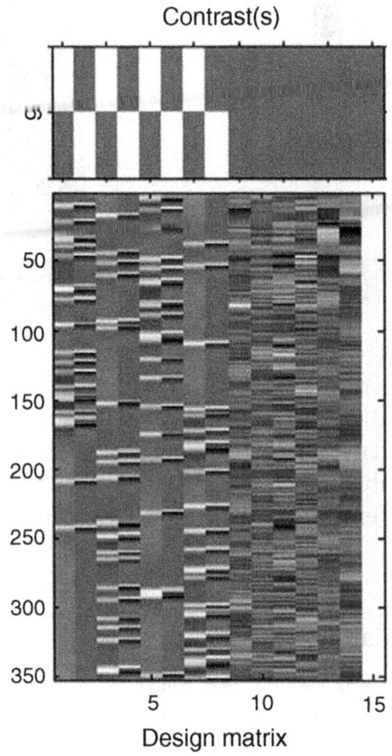

FIGURE 25.7 Lower part: design matrix for analysis of face data. Upper part: multivariate contrast used to test for any effect of faces.

in the database was presented twice. This paradigm is a two-by-two factorial design where the factors are familiarity and repetition. The four experimental conditions are 'U1', 'U2', 'F1' and 'F2', which are the first or second (1/2) presentations of images of familiar 'F' or unfamiliar 'U' faces.

Images were acquired from a 2T VISION system (Siemens, Erlangen, Germany) which produced T2*-weighted transverse echo-planar images (EPIs) with BOLD contrast. Whole brain EPIs consisting of 24 transverse slices were acquired every two seconds resulting in a total of $T = 351$ scans. All functional images were realigned to the first functional image using a six-parameter rigid-body transformation. To correct for the fact that different slices were acquired at different times, time-series were interpolated to the acquisition time of a reference slice. Images were then spatially normalized to a standard EPI template using a non-linear warping method (Ashburner and Friston, 2003). Each time-series was then highpass filtered using a set of discrete cosine basis functions with a filter cut-off of 128 seconds.

The data were then analysed using the design matrix shown in Figure 25.7. The first eight columns contain stimulus related regressors. These correspond to the four experimental conditions, where each stimulus train has been convolved with two different haemodynamic bases: (i) the canonical haemodynamic response function (HRF); and (ii) the time derivative of the canonical (Henson, 2003). The next six regressors in the design matrix describe movement of the subject in the scanner and the final column models the mean response.

Figure 25.8 plots a map of the first autoregressive component as estimated using VB. This shows a good deal of heterogeneity and justifies our assumption that that AR coefficients are spatially varying. The estimated spatial variation is smooth, however, due to the spatial prior. Plate 25 shows a PPM for 'any effect of faces' which was obtained using the multivariate contrast matrix shown in Figure 25.7.

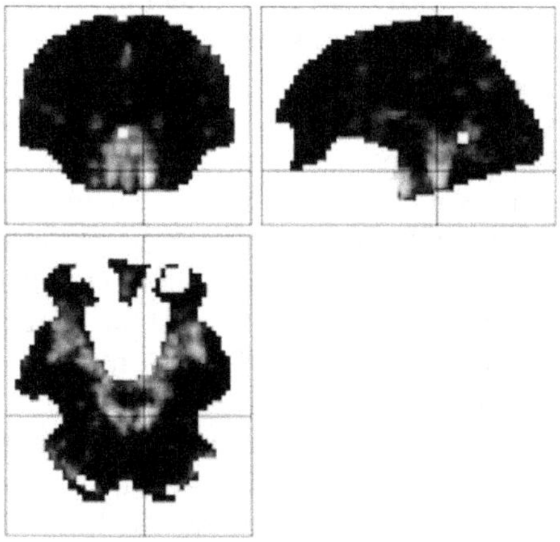

FIGURE 25.8 Image of the first autoregressive coefficient estimated from the face fMRI data (in all, there were $P = 3$ AR coefficients per voxel). Black denotes 0 and white 1. Large values can be seen around the circle of Willis and middle cerebral artery. This makes sense as cardiac-induced pulsatile motion is likely to be stronger in these regions.

DISCUSSION

We have reviewed a framework for the analysis of fMRI data based on spatially regularized GLMs. This model embodies prior knowledge that evoked responses are spatially homogeneous and locally contiguous.

As compared to standard approaches based on spatially smoothing the imaging data itself, the spatial regularization procedure has been shown to result in inferences with higher sensitivity (Penny *et al.*, 2005). The approach may be viewed as an automatically regularized spatio-temporal deconvolution scheme.

Use of PPMs with default thresholds resulted in low false positive rates for null fMRI data, and physiologically plausible activations for auditory and face fMRI data sets. We have recently developed a similar approach for source localization of EEG/MEG, which is described in the following chapter.

APPENDIX 25.1

This appendix provides a number of formulae required for updating the approximate posteriors. These have been derived in Penny *et al.* (2003). First, auxiliary quantities for updating $q(w_n)$ are:

$$\tilde{A}_n = \sum_t x_t^T x_t + X_t^T (m_n^T m_n + V_n) X_t$$

$$-x_t^T m_n X_t - X_t^T m_n^T x_t \qquad \textbf{25.24}$$

$$\tilde{b}_n = \sum_t y_{tn} x_t - m_n d_{tn} x_t - y_{tn} m_n X_t$$

$$+d_{tn}^T (m_n^T m_n + V_n) X_t \qquad \textbf{25.25}$$

For the special case in which the errors are uncorrelated, i.e. $P = 0$, we have $\tilde{A}_n = X^T X$ and $\tilde{b}_n = X^T y_n$. If we also have no spatial prior on the regression coefficients, i.e. $\alpha = 0$, we then recover the least squares update:

$$\hat{w}_n = (X^T X)^{-1} X^T y_n \qquad \textbf{25.26}$$

Secondly, auxiliary quantities for updating $q(a_n)$ are:

$$\tilde{C}_n = \sum_t d_{tn} d_{tn}^T + X_t (\hat{w}_n \hat{w}_n^T + \hat{\Sigma}_n) X_t^T \qquad \textbf{25.27}$$

$$-d_{tn} \hat{w}_n^T X_t^T - X_t \hat{w}_n d_{tn}^T$$

$$\tilde{d}_n = \sum_t y_{tn} d_{tn}^T - x_t \hat{w}_n d_{tn}^T - y_{tn} \hat{w}_n^T \tilde{X}^T + x_t (\hat{w}_n \hat{w}_n^T + \hat{\Sigma}_n) X_t^T$$

Thirdly, the auxiliary quantity for updating $q(\lambda_n)$ is:

$$\tilde{G}_n = \tilde{G}_{n1} + \tilde{G}_{n2} + \tilde{G}_{n3} \qquad \textbf{25.28}$$

where

$$\tilde{G}_{n1} = \sum_t y_{tn}^2 + d_{tn}^T (m_n^T m_n + V_n) d_{tn} - 2 y_{tn} d_{tn}^T m_n \qquad \textbf{25.29}$$

$$\tilde{G}_{n2} = \sum_t x_t (\hat{w}_n \hat{w}_n^T + \hat{\Sigma}) x_t^T + \mathrm{Tr}(X_t^T (m_n^T m_n + V_n) X_t \hat{\Sigma}_n)$$

$$+\hat{w}_n^T X_t^T (m_n^T m_n + V_n) X_t \hat{w}_n - 2 x_t (\hat{w}_n \hat{w}_n^T + \hat{\Sigma}_n) X_t m_n^T$$

$$\tilde{G}_{n3} = \sum_t -2 y_{tn} x_t \hat{w}_n + 2 m_n d_{tn} x_t \hat{w}_n$$

$$+2 y_{tn} m_n X_t \hat{w}_n - 2 d_{tn}^T (m_n^T m_n + V_n) X_t \hat{w}_n$$

Many terms in the above equations do not depend on model parameters and so can be pre-computed for efficient implementation. See Penny and Flandin (2005b) for more details.

REFERENCES

Ashburner J, Friston KJ (2003) Spatial normalization using basis functions. In *Human brain function,* 2nd edn, Frackowiak RSJ, Friston KJ, Frith C *et al.* (eds). Academic Press, London

Auranen T, Nummenmaa A, Hammalainen M *et al.* (2005) Bayesian analysis of the neuromagnetic inverse problem with l^p norm priors. *NeuroImage* **26**: 70–84

Box GEP, Tiao GC (1992) *Bayesian inference in statistical analysis.* John Wiley, New York

Frackowiak RSJ, Friston KJ, Frith C *et al.* (2003) *Human brain function,* 2nd edn. Academic Press, London

Friston KJ, Holmes AP, Worsley KJ *et al.* (1995). Statistical parametric maps in functional imaging: a general linear approach. *Hum Brain Mapp* **2**: 189–210

Friston KJ, Penny WD (2003) Posterior probability maps and SPMs. *NeuroImage* **19**: 1240–49

Gautama T, Van Hulle MM (2004) Optimal spatial regularisation of autocorrelation estimates in fMRI analysis. *NeuroImage* **23**: 1203–16

Harrison L, Penny WD, Friston KJ (2003) Multivariate autoregressive modelling of fMRI time series. *NeuroImage* **19**: 1477–91

Henson RNA. (2003) Analysis of fMRI time series. In *Human brain function,* 2nd edn, Frackowiak RSS, Friston KJ, Frith C *et al.* (eds). Academic Press, London

Henson RNA, Shallice T, Gorno-Tempini ML *et al.* (2002) Face repetition effects in implicit and explicit memory tests as measured by fMRI. *Cereb Cortex* **12**: 178–86

Pascual Marqui R, Michel C, Lehman D (1994) Low resolution electromagnetic tomography: a new method for localizing electrical activity of the brain. *Int J Psychophysiol* **18**: 49–65

Penny WD, Kiebel SJ, Friston KJ (2003) Variational Bayesian inference for fMRI time series. *NeuroImage* **19**: 727–41

Penny WD, Flandin G (2005a) Bayesian analysis of fMRI data with spatial priors. In *Proceedings of the Joint Statistical Meeting (JSM), Section on Statistical Graphics (CDROM),* 2005. American Statistical Association, Alexandria, VA

Penny WD, Flandin G (2005b) Bayesian analysis of single-subject fMRI: SPM implementation. Technical report, Wellcome Department of Imaging Neuroscience, London

Penny WD, Trujillo-Barreto N, Friston KJ (2005) Bayesian fMRI time series analysis with spatial priors. *NeuroImage,* **24**: 350–62

Penny WD, Flandin G, Trujillo-Barreto N (2006) Bayesian comparison of spatially regularised general linear models. *Hum Brain Mapp* (in press)

SPM Software (2006) Wellcome Department of Imaging Neuroscience. Available from http://www.fil.ion.ucl.ac.uk/spm/software/

Woolrich MW, Ripley BD, Brady M *et al.* (2001) Temporal autocorrelation in univariate linear modelling of fMRI data. *NeuroImage* **14**: 1370–86

Woolrich MW, Behrens TE, Smith SM (2004) Constrained linear basis sets for HRF modelling using variational Bayes. *NeuroImage* **21**: 1748–61

CHAPTER

26

Spatio-temporal models for EEG

W. Penny, N. Trujillo-Barreto and E. Aubert

INTRODUCTION

Imaging neuroscientists have at their disposal a variety of imaging techniques for investigating human brain function (Frackowiak *et al.*, 2003). Among these, the electroencephalogram (EEG) records electrical voltages from electrodes placed on the scalp, the magnetoencephalogram (MEG) records the magnetic field from sensors placed just above the head and functional magnetic resonance imaging (fMRI) records magnetization changes due to variations in blood oxygenation.

However, as the goal of brain imaging is to obtain information about the neuronal networks that support human brain function, one must first transform measurements from imaging devices into estimates of intracerebral electrical activity. Brain imaging methodologists are therefore faced with an inverse problem, 'How can one make inferences about intracerebral neuronal processes given extracerebral or vascular measurements?'

We argue that this problem is best formulated as a model-based spatio-temporal deconvolution problem. For EEG and MEG, the required deconvolution is primarily spatial, and for fMRI it is primarily temporal. Although one can make minimal assumptions about the source signals by applying 'blind' deconvolution methods (McKeown *et al.*, 1998; Makeig *et al.*, 2002), knowledge of the underlying physical processes can be used to great effect. This information can be implemented in a forward model that is inverted during deconvolution. In M/EEG, forward models make use of Maxwell's equations governing propagation of electromagnetic fields (Baillet *et al.*, 2001) and in fMRI haemodynamic models link neural activity to 'balloon' models of vascular dynamics (Friston *et al.*, 2000).

To implement a fully spatio-temporal deconvolution, time-domain fMRI models must be augmented with a spatial component and spatial-domain M/EEG models with a temporal component. The previous chapter showed how this could be implemented for fMRI. This chapter describes a model-based spatio-temporal deconvolution method for M/EEG.

The underlying forward or 'generative' model incorporates two mappings. The first specifies a time-domain general linear model (GLM) at each point in source space. This relates effects of interest at each voxel to source activity at that voxel. This is identical to the 'mass-univariate' approach that is widely used in the analysis of fMRI (Frackowiak *et al.*, 2003). The second mapping relates source activity to sensor activity at each time point using the usual spatial-domain lead-field matrix (see Chapter 28).

Our model therefore differs from the standard generative model implicit in source reconstruction by having an additional level that embodies temporal priors. There are two potential benefits of this approach. First, the use of temporal priors can result in more sensitive source reconstructions. This may allow signals to be detected that cannot be detected otherwise. Second, it provides an analysis framework for M/EEG that is very similar to that used in fMRI. The experimental design can be coded in a design matrix, the model fitted to data, and various effects of interest can be characterized using 'contrasts' (Frackowiak *et al.*, 2003). These effects can then be tested for statistically using posterior probability maps (PPMs), as described in previous chapters. Importantly, the model does not need to be refitted to test for multiple experimental effects that are potentially present in any single data set. Sources are estimated once only using a spatio-temporal deconvolution rather than separately for each temporal component of interest.

The chapter is organized as follows. In the theory section, we describe the model and relate it to existing distributed solutions. The success of the approach rests on our ability to characterize neuronal responses, and task-related differences in them, using GLMs.

Statistical Parametric Mapping, by Karl Friston et al.
ISBN–13: 978-0-12-372560-8 ISBN–10: 0-12-372560-7

We describe how this can be implemented for the analysis of event-related potentials (ERPs) and show how the model can be inverted to produce source estimates using variational Bayes (VB). The framework is applied to simulated data and data from an EEG study of face processing.

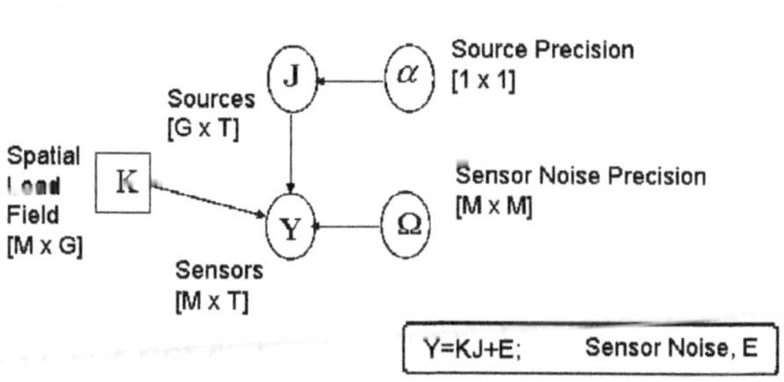

FIGURE 26.1 Generative model for source reconstruction. This is a graphical representation of the probabilistic model implicit in many distributed source solutions.

THEORY

Notation

Lower case variable names denote vectors and scalars. Whether the variable is a vector or scalar should be clear from the context. Upper case names denote matrices or dimensions of matrices. In what follows $\mathsf{N}(x; \mu, \Sigma)$ denotes a multivariate normal density over x, having mean μ and covariance Σ. The precision of a Gaussian variate is the inverse (co)variance. A gamma density over the scalar random variable x is written as $\mathsf{Ga}(x; a, b)$. Normal and gamma densities are defined in Chapter 26. We also use $\|x\|^2 = x^T x$, denote the trace operator as $\mathsf{Tr}(X)$, X^+ for the pseudo-inverse, and use $\mathsf{diag}(x)$ to denote a diagonal matrix with diagonal entries given by the vector x.

Generative model

The aim of source reconstruction is to estimate primary current density (PCD) J from M/EEG measurements Y. If we have $m = 1..M$ sensors, $g = 1..G$ sources and $t = 1..T$ time points then J is of dimension $G \times T$ and Y is of dimension $M \times T$. The applications in this chapter use a cortical source space in which dipole orientations are constrained to be perpendicular to the cortical surface. Each entry in J therefore corresponds to the scalar current density at particular locations and time points. Sensor measurements are related to current sources via Maxwell's equations governing electromagnetic fields (Baillet *et al.*, 2001) (see Chapter 28).

Most established distributed source reconstruction or 'imaging' methods (Darvas *et al.*, 2004) implicitly rely on the following two-level generative model:

$$p(Y|J, \Omega) = \prod_{t=1}^{T} \mathsf{N}(y_t; Kj_t, \Omega) \qquad \textbf{26.1}$$
$$p(J|\alpha) = \prod_{t=1}^{T} \mathsf{N}(j_t; 0, \alpha^{-1}D^{-1})$$

where j_t and y_t are the source and sensor vectors at time t, K is the $[M \times G]$ lead-field matrix and Ω is the sensor noise covariance. The matrix D reflects the choice of spatial prior and α is a spatial precision variable. This generative model is shown schematically in Figure 26.1 and can be written as a hierarchical model:

$$Y = KJ + E \qquad \textbf{26.2}$$
$$J = Z$$

in which random fluctuations E correspond to sensor noise and the source activity is generated by random innovations Z. Critically, these assumptions provide empirical priors on the spatial deployment of source activity (see Chapter 29).

Because the number of putative sources is much greater than the number of sensors, $G >> M$, the source reconstruction problem is ill-posed. Distributed solutions therefore depend on the specification of a spatial prior for estimation to proceed. A common choice is the Laplacian prior used, e.g. in low resolution electromagnetic tomography (LORETA) (Pascual Marqui *et al.*, 1994). This can be implemented in the above generative model by setting D to compute local differences as measured by an L^2-norm, which embodies a belief that sources are diffuse and highly distributed. Other spatial priors, such as those based on L^1-norms (Fuchs *et al.*, 1999), L^p-norms (Auranen *et al.*, 2005), or variable resolution electromagnetic tomography (VARETA) (Valdes-Sosa *et al.*, 2000) can provide more focal source estimates. These are all examples of schemes that use a single spatial prior and are special cases of a more general model (Mattout *et al.*, 2006) that covers multiple priors. In this model, the sensor noise and spatial prior covariances are modelled as mixtures of components Ω_i and Q_i respectively:

$$p(Y|J, \Omega) = \prod_{t=1}^{T} \mathsf{N}(y_t; Kj_t, \rho_1\Omega_1 + \rho_2\Omega_2 + \ldots) \qquad \textbf{26.3}$$
$$p(J|\alpha) = \prod_{t=1}^{T} \mathsf{N}(j_t; 0, \gamma_1 Q_1 + \gamma_2 Q_2 + \ldots)$$

The advantage of this model is that multiple priors can be specified and are mixed adaptively by adjusting the covariance parameters ρ_i and γ_i, as described in

Chapter 29. One can also use Bayesian model selection to compare different combinations of priors, as described in Chapter 35. For simplicity, we will deal with a single spatial prior component because we want to focus on temporal priors. However, it would be relatively simple to extend the approach to cover multiple prior covariance (or precision) components.

Also, in Chapter 35 we will describe a prior over a model class that, when used with Bayesian model averaging (BMA), can automatically provide either focal or distributed solutions depending on the reconstruction at hand. The applications in this chapter use Laplacian priors.

Whatever the choice of spatial prior, the majority of source reconstruction applications follow a single-pass serial processing strategy. Either (i) spatial processing first proceeds to create source estimates at each time point and then (ii) temporal models are applied at these 'virtual depth electrodes' (Brookes *et al.*, 2004; Darvas *et al.*, 2004; Kiebel and Friston, 2004). Or (ii) time series methods are applied in sensor space to identify components of interest using, e.g. time windowing (Rugg and Coles, 1995) or time-frequency estimation and then (iii) source reconstructions are based on these components.

In this chapter, we review a multiple-pass strategy in which temporal and spatial parameter estimates are improved iteratively to provide an optimized and mutually constrained solution. It is based on the following three-level generative model:

$$p(Y|J,\Omega) = \prod_{t=1}^{T} \mathrm{N}(y_t; Kj_t, \Omega) \qquad 26.4$$

$$p(J|W,\lambda) = \prod_{t=1}^{T} \mathrm{N}(j_t^T; x_t W, \lambda^{-1} I_G) \qquad 26.5$$

$$p(W|\alpha) = \prod_{k=1}^{K} \mathrm{N}(w_k; 0, \alpha^{-1} D^{-1}) \qquad 26.6$$

The first level, Eqn. **26.4**, is identical to the standard model. In the second level, however, source activity at each voxel is constrained using a $[T \times K]$ matrix of temporal basis functions, X. The tth row of X is x_t. The generative model is shown schematically in Figure 26.2.

The precision of the source noise is given by λ. In this chapter, λ is a scalar; and we will apply the framework to analyse event-related potentials (ERPs) (Rugg and Coles, 1995). Event-related source activity is described by the time domain GLM and remaining source activity will correspond to spontaneous activity. The quantity λ^{-1} can therefore be thought of as the variance of spontaneous activity in source space.

The regression coefficients W determine the weighting of the temporal basis functions. The third level of the model is a spatial prior that reflects our prior uncertainty about W. Each regression coefficient map, w_k (row of W), is constrained by setting D to correspond to the usual L^2-norm spatial prior. The spatial prior that is usually on the sources now, therefore, appears at a superordinate level.

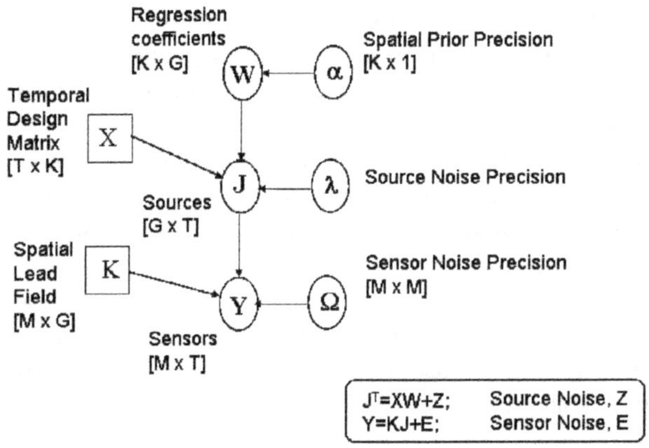

FIGURE 26.2 Generative model for source reconstruction with temporal priors. This is a hierarchical model with regression coefficients at the 'top' and M/EEG data at the 'bottom'.

Different choices of D result in different weights and different neighbourhood relations. The applications in this paper use $D = L^T L$, where L is the surface Laplacian defined as:

$$L_{ij} = \begin{cases} 1, & if\ i = j \\ -\frac{1}{N_{ij}}, & if\ i\ and\ j\ are\ geodesic\ neighbours \\ 0, & otherwise. \end{cases}$$

where N_{ij} is the geometric mean of the number of neighbors of i and j. This prior has been used before in the context of fMRI with Euclidean neighbours (Woolrich *et al.*, 2001, Penny and Flandin, 2005).

The first level of the model assumes that there is Gaussian sensor noise, e_t, with zero mean and covariance Ω. This covariance can be estimated from pre-stimulus or baseline periods when such data are available (Sahani and Nagarajan, 2004). Alternatively, we assume that $\Omega = \mathrm{diag}(\sigma^{-1})$ where the mth element of σ^{-1} is the noise variance on the mth sensor. We provide a scheme for estimating σ_m, should this be necessary.

We also place gamma priors on the precision variables σ, λ and α:

$$p(\sigma) = \prod_{m=1}^{M} \mathrm{Ga}(\sigma_m; b_{\sigma_{prior}}, c_{\sigma_{prior}}) \qquad 26.7$$

$$p(\lambda) = \mathrm{Ga}(\lambda; b_{\lambda_{prior}}, c_{\lambda_{prior}})$$

$$p(\alpha) = \prod_{k=1}^{K} \mathrm{Ga}(\alpha_k; b_{\alpha_{prior}}, c_{\alpha_{prior}})$$

This allows the inclusion of further prior information into the source localization. For example, instead of using baseline periods to estimate a full covariance matrix Ω, we could use these data to estimate the noise variance at each sensor. This information could then be used to set $b_{\sigma_{prior}}$ and $c_{\sigma_{prior}}$, allowing noise estimates during periods of interest to be constrained softly by those from baseline periods. Similarly, we may wish to enforce stronger or weaker spatial regularization on w_k by setting $b_{\alpha_{prior}}$ and $c_{\alpha_{prior}}$ appropriately. The applications in this chapter, however, use uninformative gamma priors. This means that σ, λ and α will be estimated solely from the data Y.

In summary, the addition of the supraordinate level to our generative model induces a partitioning of source activity into signal and noise. We can see this clearly by reformulating the probabilistic model as before:

$$Y = KJ + E \qquad \textbf{26.8}$$
$$J^T = XW + Z$$
$$W = P$$

Here we have random innovations Z which are 'temporal errors', i.e. lack of fit of the temporal model, and P which are 'spatial errors', i.e. lack of fit of a spatial model. Here the spatial model is simply a zero mean Gaussian with covariance $\alpha^{-1}D^{-1}$. We can regard XW as an empirical prior on the expectation of source activity. This empirical Bayes perspective means that the conditional estimates of source activity J are subject to bottom-up constraints, provided by the data, and top-down predictions from the third level of our model. We will use this heuristic later to understand the update equations used to estimate source activity.

Temporal priors

The usefulness of the spatio-temporal approach rests on our ability to characterize neuronal responses using GLMs. Fortunately, there is a large literature that suggests this is possible. The type of temporal model necessary will depend on the M/EEG response one is interested in. These components could be (i) single trials, (ii) evoked components (steady-state or ERPs (Rugg and Coles, 1995)) or (iii) induced components (Tallon Baudry *et al.*, 1996).

In this chapter, we focus on ERPs. We briefly review three different approaches for selecting an appropriate ERP basis set. These basis functions will form columns in the GLM design matrix, X (see Eqn. **26.5** and Figure 26.2).

Damped sinusoids

An ERP basis set can be derived from the fitting of damped sinusoidal (DS) components (Demiralp *et al.*, 1998). These are given by:

$$j = \sum_{k=1}^{K} w_k x_k \qquad \textbf{26.9}$$
$$x_k = \exp(i\phi_k)\exp(\alpha_k + i2\pi f_k)\delta_t$$

where $i = \sqrt{-1}$, δ_t is the sampling interval and w_k, ϕ_k, α_k and f_k are the amplitude, phase, damping and frequency of the kth component. The $[T \times 1]$ vector x_k will form the kth column in the design matrix. Figure 26.3 shows how damped sinusoids can generate an ERP.

Fitting DS models to ERPs from different conditions allows one to make inferences about task related changes in constituent rhythmic components. For example, in Demiralp *et al.* (1998), responses to rare auditory events elicited higher amplitude, slower delta and slower damped theta components than did responses to frequent events. Fitting damped sinusoids, however, requires a non-linear estimation procedure. But approximate solutions can also be found using the Prony and related methods (Osborne and Smyth, 1991) which require two stages of linear estimation.

Once a DS model has been fitted, e.g. to the principal component of the sensor data, the components x_k provide a minimal basis set. Including extra regressors from a first-order Taylor expansion about phase, damping and frequency ($\frac{\partial x_k}{\partial \phi_k}, \frac{\partial x_k}{\partial \alpha_k}, \frac{\partial x_k}{\partial f_k}$) provides additional flexibility. Use of this expanded basis in our model would allow these attributes to vary with source location. Such

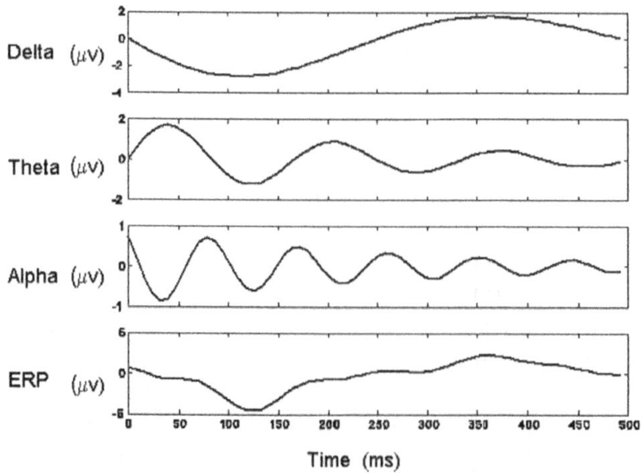

FIGURE 26.3 The figure shows how damped sinusoids can model ERPs. In this example, damped delta, theta and alpha sinusoids, of particular phase, amplitude and damping, add together to form an ERP with an early negative component and a late positive component.

Taylor series expansions have been particularly useful in GLM characterizations of haemodynamic responses in fMRI (Frackowiak *et al.*, 2003).

Wavelets

ERPs can also be modelled using wavelets:

$$j = \sum_{k=1}^{K} w_k x_k \qquad \textbf{26.10}$$

where x_k are wavelet basis functions and w_k are wavelet coefficients. Wavelets provide a tiling of time-frequency space that gives a balance between time and frequency resolution. The Q-factor of a filter or basis function is defined as the central frequency to bandwidth ratio. Wavelet bases are chosen to provide constant Q (Unser and Aldroubi, 1996). This makes them good models of non-stationary signals, such as ERPs and induced EEG components (Tallon Baudry *et al.*, 1996). Wavelet basis sets are derived by translating and dilating a mother wavelet. Figure 26.4 shows wavelets from two different basis sets, one based on Daubechies wavelets and one based on Battle-Lemarie (BL) wavelets. These basis sets are orthogonal. Indeed the BL wavelets have been designed from an orthogonalization of cubic B-splines (Unser and Aldroubi, 1996).

If $K = T$, then the mapping $j \rightarrow w$ is referred to as a wavelet transform, and for $K > T$ we have an overcomplete basis set. More typically, we have $K \leq T$. In the ERP literature, the particular subset of basis functions

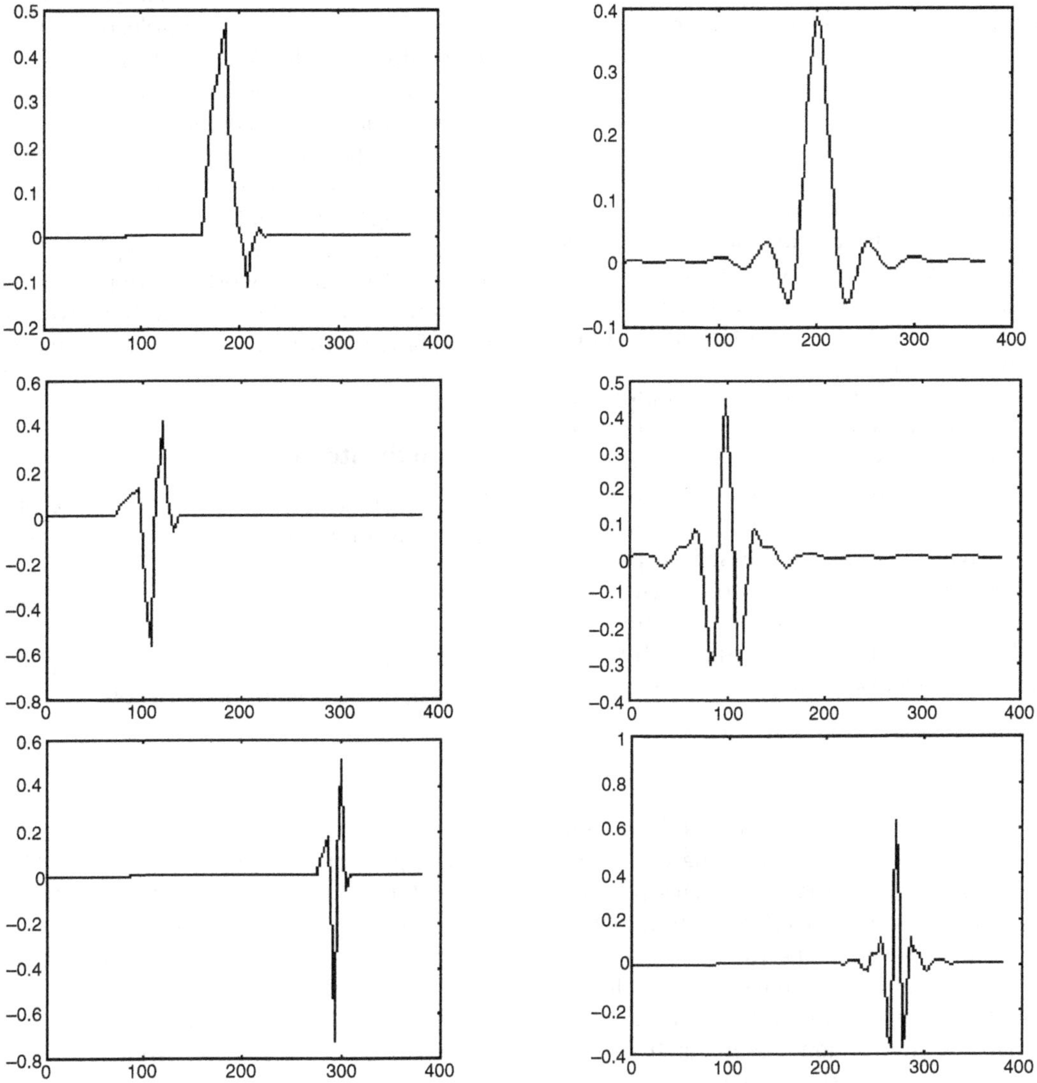

FIGURE 26.4 The graphs show wavelets from a Daubechies set of order 4 (left) and a Battle-Lemarie basis set of order 3 (right). The wavelets in the lower panels are higher frequency translations of the wavelets in the top panels. Each full basis set comprises multiple frequencies and translations covering the entire time domain.

used is chosen according to the type of ERP component one wishes to model. Popular choices are wavelets based on B-splines (Unser and Aldroubi, 1996).

In statistics, however, it is well known that an appropriate subset of basis functions can be automatically selected using a procedure known as 'wavelet shrinkage' or 'wavelet denoising'. This relies on the property that natural signals, such as images, speech or neuronal activity, can be represented using a sparse code comprising just a few large wavelets coefficients. Gaussian noise signals, however, produce Gaussian noise in wavelet space. This comprises a full set of wavelet coefficients whose size depends on the noise variance. By 'shrinking' these noise coefficients to zero using a thresholding procedure (Donoho and Johnstone, 1994; Clyde *et al.*, 1998), and transforming back into signal space, one can denoise data. This amounts to defining a temporal model. We will use this approach for the empirical work reported later on.

PCA

A suitable basis can also be derived from principal components analysis (PCA). Trejo and Shensa (1999), for example, applied PCA and varimax rotation to the classification of ERPs in a signal detection task. They found, however, that better classification was more accurate with a Daubechies wavelet basis.

PCA decompositions are also used in the multiple signal classification (MUSIC) approach (Mosher and Leahy, 1998). The dimension of the basis set is chosen to separate the signal and noise subspaces. Source reconstruction is then based on the signal, with information about the noise used to derive statistical maps based on pseudo-z scores. In Friston *et al.* (2006), a temporal basis set is defined using the principal eigenvectors of a full-rank prior temporal covariance matrix. This approach makes the link between signal subspace and prior assumptions transparent.

Dimensionality

Whatever the choice of basis, it is crucial that the dimension of the signal subspace is less than the dimension of the original time series. That is, $K < T$. This is necessary for the temporal priors to be effective, from both a statistical and computational perspective.

Theoretically, one might expect the dimensionality of ERP generators to be quite small. This is because of the low-dimensional synchronization manifolds that arise when non-linear dynamical systems are coupled into an ensemble (Breakspear and Terry, 2002).

In practice, the optimal reduced dimensionality can be found automatically using a number of methods. For wavelets this can be achieved using shrinkage methods (Donoho and Johnstone 1994; Clyde *et al.*, 1998) and, for PCA, using various model order selection criteria (Minka, 2000); for damped sinusoids, Prony-based methods can use AR model order criteria (Roberts and Penny, 2002). Moreover, it is also possible to compute the model evidence of the source reconstruction model we have proposed, as shown in the following section. This can then be used to optimize the basis set.

Bayesian inference

To make inferences about the sources underling M/EEG, we need to invert our probabilistic model to produce the posterior density $p(J|Y)$. This is straightforward in principle and can be achieved using standard Bayesian methods (Gelman *et al.*, 1995). For example, one could use Markov chain Monte Carlo (MCMC) to produce samples from the posterior. This has been implemented efficiently for dipole-like inverse solutions (Schmidt *et al.*, 1999) in which sources are parameterized as spheres of unknown number, extent and location. It is, however, computationally demanding for distributed source solutions, taking several hours for source spaces comprising $G > 1000$ voxels (Auranen *et al.*, 2005). In this work, we adopt the computationally efficient approximate inference framework called variational Bayes (VB), which was reviewed in Chapter 24.

Approximate posteriors

For our source reconstruction model we assume the following factorization of the approximate posterior:

$$q(J, W, \alpha, \sigma, \lambda) = q(J)q(W)q(\alpha)q(\sigma)q(\lambda) \qquad \textbf{26.11}$$

We also assume that the approximate posterior for the regression coefficients factorizes over voxels:

$$q(W) = \prod_{g=1}^{G} q(w_g) \qquad \textbf{26.12}$$

This approximation was used in the spatio-temporal model for fMRI described in the previous chapter.

Because of the spatial prior (Eqn. **26.6**), the regression coefficients in the true posterior $p(W|Y)$ will clearly be correlated. Our perspective, however, is that this is too computationally burdensome for current personal computers to take account of. Moreover, as we shall see below, updates for our approximate factorized densities $q(w_g)$ do encourage the approximate posterior means to be similar at nearby voxels, thereby achieving the desired effect of the prior.

Now that we have defined the probabilistic model and our factorization of the approximate posterior, we can use the procedure described in Chapter 24 to derive expressions for each component of the approximate posterior. We do not present details of these derivations in this chapter. Similar derivations have been published elsewhere (Penny *et al.*, 2005). The following sections describe each distribution and the updates of its sufficient statistics required to maximize the lower bound on the model evidence, F.

Sources

Updates for the sources are given by:

$$q(J) = \prod_{t=1}^{T} q(j_t) \quad \textbf{26.13}$$

$$q(j_t) = \mathsf{N}(j_t; \hat{j}_t, \hat{\Sigma}_{j_t}) \quad \textbf{26.14}$$

$$\hat{\Sigma}_{j_t} = \left(K^T \hat{\Omega} K + \hat{\lambda} I_G\right)^{-1} \quad \textbf{26.15}$$

$$\hat{j}_t = \hat{\Sigma}_{j_t} \left(K^T \hat{\Omega} y_t + \hat{\lambda} \hat{W}^T x_t^T\right) \quad \textbf{26.16}$$

where $\hat{j}_t$ is the tth column of $\hat{J}$ and $\hat{\Omega}$, $\hat{\lambda}$ and $\hat{W}$ are estimated parameters defined in the following sections. Eqn. **26.16** shows that our source estimates are the result of a spatio-temporal deconvolution. The spatial contribution to the estimate is $K^T y_t$ and the temporal contribution is $\hat{W}^T x_t^T$. From the perspective of the hierarchical model, shown in Figure 26.5, these are the 'bottom-up' and 'top-down' predictions. Importantly, each prediction is weighted by its relative precision. Moreover, the parameters controlling the relative precisions, $\hat{\lambda}$ and $\hat{\Omega}$, are estimated from the data. This means that our source estimates derive from an automatically regularized spatio-temporal deconvolution. This property is shared by the spatio-temporal model for fMRI, described in the previous chapter.

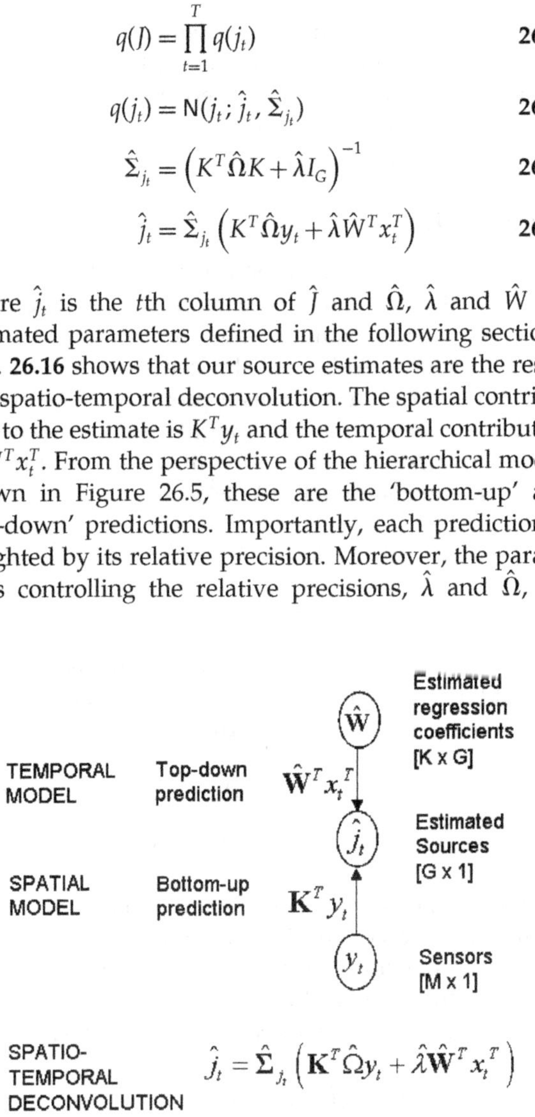

FIGURE 26.5 Probabilistic inversion of the generative model leads to a source reconstruction based on a spatio-temporal deconvolution in which bottom-up and top-down predictions, from sensor data and temporal priors, are optimally combined using Bayesian inference.

An alternative perspective on this computation is given by ignoring the regularization term in Eqn. **26.16**. We then see that $\hat{\Sigma}_{j_t} K^T \hat{\Omega} = (K^T \hat{\Omega} K)^{+} K^T \hat{\Omega} = B_w^T$, which is equivalent to a beamformer (Darvas *et al.*, 2004). Eqn. **26.16** then shows that our source estimates use beamformer predictions $B_w^T y_t$ that are modified using a temporal model. Beamformers cannot localize temporally correlated sources. But, as we shall see later, the spatio-temporal model can.

We end this section by noting that statistical inferences about current sources are more robust than point predictions. This property has been used to great effect with pseudo-z beamformer statistics (Robinson and Vrba, 1999), sLORETA (Pascual Marqui, 2002) and VARETA (Valdes-Sosa *et al.*, 2000) source reconstructions, which divide current source estimates by their standard deviations. This approach can be adopted in the current framework as the standard deviations are readily computed from the diagonal elements of $\hat{\Sigma}_{j_t}$ using Eqn. **26.23**. Moreover, we can threshold these statistic images to create posterior probability maps (PPMs), as introduced in Chapter 25.

Regression coefficients

Updates for the regression coefficients are given by:

$$q(w_g) = \mathsf{N}(w_g; \hat{w}_g, \hat{\Sigma}_{w_g}) \quad \textbf{26.17}$$

$$\hat{\Sigma}_{w_g} = \left(\hat{\lambda} X^T X + d_{gg} \mathsf{diag}(\hat{\alpha})\right)^{-1}$$

$$\hat{w}_g = \hat{\Sigma}_{w_g} \left(\hat{\lambda} X^T \hat{j}_g^T + \mathsf{diag}(\hat{\alpha}) r_g\right)$$

where $\hat{\alpha}$ is defined in Eqn. **26.21**, d_{ij} is the i, jth element of D and r_g is given by:

$$r_g = \sum_{g'=1, g' \neq g}^{G} d_{gg'} \hat{w}_{g'} \quad \textbf{26.18}$$

As shown in the previous chapter, r_g is the weighted sum of neighbouring regression coefficient estimators.

The update for $\hat{w}_g$ in Eqn. **26.17** therefore indicates that the regression coefficient estimates at a given voxel regress towards those at nearby voxels. This is the desired effect of the spatial prior and it is preserved despite the factorization in the approximate posterior. This equation can again be thought of in terms of the hierarchical model where the regression coefficient estimate is a combination of a bottom-up prediction from the level below, $X^T \hat{j}_g^T$, and a top-down prediction from the prior, r_g. Again, each contribution is weighted by its relative precision.

The update for the covariance in Eqn **26.17** shows that the only off-diagonal contributions are due to the design matrix. If the temporal basis functions are therefore chosen to be orthogonal then this posterior covariance will be diagonal, thus making a potentially large saving of computer memory. One benefit of the proposed framework, however, is that non-orthogonal bases can be accommodated. This may allow for a more natural and compact description of the data.

Precision of temporal models

Updates for the precision of the temporal model are given by:

$$q(\lambda) = \mathrm{Ga}(\lambda; b_{\lambda_{post}}, c_{\lambda_{post}}) \qquad \textbf{26.19}$$

$$\frac{1}{b_{\lambda_{post}}} = \frac{1}{b_{\lambda_{prior}}} + \frac{1}{2}\sum_t \left(\|\hat{j}_t - \hat{W}^T x_t^T\|^2 + \mathrm{Tr}\left(\hat{\Sigma}_{j_t}\right) + \sum_{g=1}^{G} x_t \hat{\Sigma}_{w_g} x_t^T \right)$$

$$c_{\lambda_{post}} = c_{\lambda_{prior}} + \frac{GT}{2}$$

$$\hat{\lambda} = b_{\lambda_{post}} c_{\lambda_{post}}$$

In the context of ERP analysis, these expressions amount to an estimate of the variance of spontaneous activity in source space, $\hat{\lambda}^{-1}$, given by the squared error between the ERP estimate, $\hat{W}^T x_t^T$, and source estimate, $\hat{j}_t$, averaged over time and space and the other approximate posteriors.

Precision of forward model

Updates for the precision of the sensor noise are given by:

$$q(\sigma) == \prod_{m=1}^{M} q(\sigma_m) \qquad \textbf{26.20}$$

$$q(\sigma_m) = \mathrm{Ga}(\sigma_m; b_{\sigma_{post}}, c_{\sigma_{post}})$$

$$\frac{1}{b_m} = \frac{1}{b_{\sigma_{prior}}} + \frac{1}{2}\sum_t \left(y_{mt} - k_m^T \hat{j}_t\right)^2 + \frac{1}{2} k_m^T \hat{\Sigma}_{j_t} k_m$$

$$c_m = c_{\sigma_{prior}} + \frac{T}{2}$$

$$\hat{\sigma}_m = b_m c_m$$

$$\hat{\Omega}^{-1} = \mathrm{diag}(\hat{\sigma})$$

These expressions amount to an estimate of observation noise at the mth sensor, $\hat{\sigma}_m^{-1}$, given by the squared error between the forward model and sensor data, averaged over time and the other approximate posteriors.

Precision of spatial prior

Updates for the precision of the spatial prior are given by:

$$q(\alpha) = \prod_{k=1}^{K} q(\alpha_k) \qquad \textbf{26.21}$$

$$q(\alpha_k) = \mathrm{Ga}(\alpha_k; b_{\alpha_{post}}, c_{\alpha_{post}})$$

$$\frac{1}{b_{\alpha_{post}}} = \frac{1}{b_{\alpha_{prior}}} + \|D\hat{w}_k^T\|^2 + \sum_{g=1}^{G} d_g s_{gk}$$

$$c_{\alpha_{post}} = c_{\alpha_{prior}} + \frac{G}{2}$$

$$\hat{\alpha}_k = b_{\alpha_{post}} c_{\alpha_{post}}$$

where s_{gk} is the kth diagonal element of $\hat{\Sigma}_{w_g}$. These expressions amount to an estimate of the 'spatial noise variance', $\hat{\alpha}_k^{-1}$, given by the discrepancy between neighbouring regression coefficients, averaged over space and the other approximate posteriors.

Implementation details

A practical difficulty with the update equations for the sources is that the covariance matrix $\hat{\Sigma}_{j_t}$ is of dimension $G \times G$ where G is the number of sources. Even low resolution source grids typically contain $G > 1000$ elements. This therefore presents a problem. A solution is found, however, with use of a singular value decomposition (SVD). First, we define a modified lead field matrix $\bar{K} = \hat{\Omega}^{1/2} K$ and compute its SVD:

$$\bar{K} = USV^T \qquad \textbf{26.22}$$
$$= U\bar{V}$$

where $\bar{V}$ is an $M \times G$ matrix, the same dimension as the lead field, K. It can then be shown using the matrix inversion lemma (Golub and Van Loan, 1996) that:

$$\hat{\Sigma}_{j_t} = \hat{\lambda}^{-1}(I_G - R_G) \qquad \textbf{26.23}$$

$$R_G = \bar{V}^T(\hat{\lambda} I_M + SS^T)^{-1}\bar{V}$$

which is simple to implement computationally, as it only requires inversion of an $M \times M$ matrix.

Source estimates can be computed as shown in Eqn. **26.16**. In principle, this means the estimated sources over all time points and source locations are given by:

$$\hat{J} = \hat{\Sigma}_{j_t} K^T \hat{\Omega} Y + \hat{\lambda}\hat{\Sigma}_{j_t} \hat{W}^T X^T$$

In practice, however, it is inefficient to work with such a large matrix during estimation. We therefore do not implement Eqns **26.15** and **26.16** but, instead, work in the reduced space $\hat{J}_X = \hat{J}X$ which are the sources projected onto the design matrix. These projected source estimates are given by:

$$\begin{aligned}\hat{J}_X &= \hat{J}X \\ &= \hat{\Sigma}_{j_t} K^T \hat{\Omega} YX + \hat{\lambda}\hat{\Sigma}_{j_t} \hat{W}^T X^T X \\ &= A_{K\Omega} YX + \hat{\lambda} A_W X^T X\end{aligned} \quad 26.24$$

where YX and X^TX can be pre-computed and the intermediate quantities are given by:

$$\begin{aligned}A_{K\Omega} &= \hat{\Sigma}_{j_t} K^T \hat{\Omega} \\ &= \hat{\lambda}^{-1}(K^T - R_G K^T) \\ A_W &= \hat{\Sigma}_{j_t} \hat{W}^T \\ &= \hat{\lambda}^{-1}\left(\hat{W}^T - R_G \hat{W}^T\right)\end{aligned} \quad 26.25$$

Because these matrices are only of dimension $G \times M$ and $G \times K$ respectively, $\hat{J}_X$ can be efficiently computed. The term $X^T \hat{j}_g^T$ in Eqn. **26.17** is then given by the gth row of $\hat{J}_X$.

The intermediate quantities can also be used to compute model predictions as:

$$\begin{aligned}\hat{Y} &= K\hat{J} \\ &= KA_{K\Omega} Y + \hat{\lambda} K A_W X^T\end{aligned} \quad 26.26$$

The m, tth entry in $\hat{Y}$ then corresponds to the $k_m^T \hat{j}_t$ term in Eqn. **26.20**. Other computational savings are as follows. For Eqn. **26.20** we use the result:

$$k_m^T \hat{\Sigma}_{j_t} k_m = \frac{1}{\hat{\sigma}_m} \sum_{m'=1}^{M} \frac{s_{m'm'}^2 u_{mm'}^2}{s_{m'm'}^2 + \hat{\lambda}} \quad 26.27$$

where s_{ij} and u_{ij} are the i, jth entries in S and U respectively. For Eqn. **26.19** we use the result:

$$\mathrm{Tr}(\hat{\Sigma}_{j_t}) = \sum_{i=1}^{M} \frac{1}{s_{ii}^2 + \hat{\lambda}} + \frac{G - M}{\hat{\lambda}} \quad 26.28$$

To summarize, our source reconstruction model is fitted to data by iteratively applying the update equations until the change in the negative free energy (see Chapter 24), F, is less than some user-specified tolerance. This procedure is summarized in the pseudo-code in Figure 26.6. This amounts to a process in which sensor data are spatially deconvolved, time-series models are fitted in source space, and then the precisions (accuracy) of the temporal and spatial models are estimated. This process is then iterated and results in a spatio-temporal deconvolution in which all aspects of the model are optimized to maximize a lower bound on the model evidence.

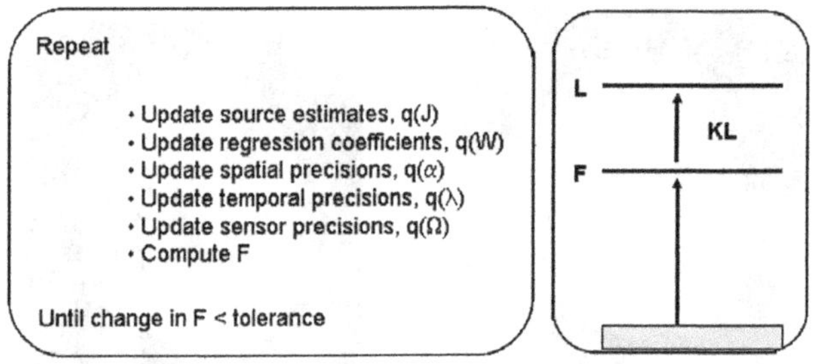

FIGURE 26.6 Pseudo-code for spatio-temporal deconvolution of M/EEG. The parameters of the model $\theta = \{J, W, \Omega, \lambda, \alpha\}$ are estimated by updating the approximate posteriors until the negative free energy is maximized to within a certain tolerance (left panel). At this point, because the log evidence $L = \log p(Y)$ is fixed, the approximate posteriors will best approximate the true posteriors in the sense of KL-divergence (right panel), as described in Chapter 24. The equations for updating the approximate posteriors are given in the theory section.

RESULTS

This section presents some preliminary qualitative results. In what follows we refer to the spatio-temporal approach as 'VB-GLM'.

Comparison with LORETA

We generated data from our model as follows. First, we created two regressors consisting of a 10 Hz and 20 Hz sinewave with amplitudes of 10 and 8 respectively. These formed the two columns of a design matrix shown in Figure 26.7. We generated 600 ms of activity with a sample period of 5 ms, giving 120 time points.

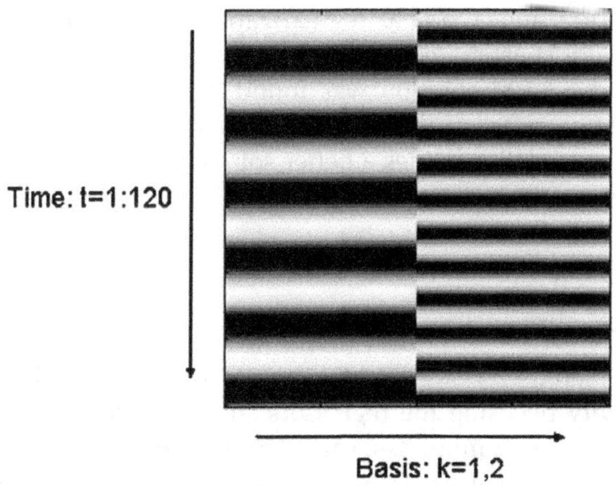

FIGURE 26.7 Simulations that compare VB-GLM with LORETA use the above design matrix, X. The columns in this matrix comprise a 10 Hz and a 20 Hz sinewave.

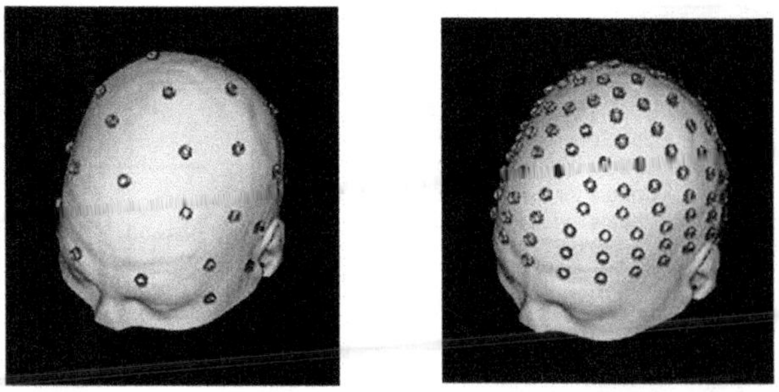

FIGURE 26.8 Electrode positions for the 32-sensor BESA system (left) and 128-sensor BioSemi system (right).

The sensor space was defined using $M = 32$ electrodes from the brain electrical source activity (BESA) system (Scherg and von Cramon, 1986) shown in Figure 26.8. We used the three concentric sphere model to calculate the electric lead field (Rush and Driscoll, 1969). The centre and radius of the spheres were fitted to the scalp, skull and cerebral tissue of a 'typical' brain from the Montreal Neurological Institute (MNI) data base (Evans *et al.*, 1993). The source space consisted of a mesh of points corresponding to the vertices of the triangles obtained by tessellation of the cortical surface of the same brain. A medium resolution spatial grid was used containing $G = 10\,242$ points.

We define the signal-to-noise ratio (SNR) as the ratio of the signal standard deviation to noise standard deviation and used sensor and source SNRs of 10 and 40 respectively. The spatial distributions of the two regression coefficients were identical, each of them consisting of two Gaussian blobs with a maximum amplitude of 10, and a full width at half maximum (FWHM) of 20 mm.

Plate 26 (see colour plate section) shows the true and estimated sources at time point $t = 20$ ms. The LORETA solution was found from an instantaneous reconstruction of the sensor data at that time point, using an L^2-norm and a spatial regularization parameter $\hat{\alpha}$ (see Eqn. **26.1**) estimated using generalized cross-validation. The VB-GLM solution was found by applying the VB update equations described in the Theory section. As expected, VB provides a better solution both in terms of localization accuracy and scaling.

ERP simulation

We then used our generative model to simulate ERP-like activity by using the regressors shown in Plate 27. The first regressor mimics an early component and the second a later component. These regressors were derived from a neural-mass model describing activity in a distributed network of cortical areas (David and Friston, 2003), which lends these simulations a degree of biological plausibility. These neural-mass models are described at length in Chapter 32.

We then specified two source activations with the same amplitude and FWHM as in the previous example. The source space, sensor space and forward model were also identical to the previous example. Ten trials of sensor data were then generated using the same SNR as in the previous set of simulations. Signal epochs of 512 ms were produced with a sampling period of 4 ms giving a total of 5120 ms of EEG. The data were then averaged over trials to calculate the sample ERP shown in Figure 26.9.

We then estimated the sources underlying the sample ERP with (i) a correctly specified model using the same two regressors used for generating the data and (ii) an over-specified model that also incorporated two additional spurious regressors shown in Plate 28. The design matrices for each of these models are shown in Figure 26.10. In the over-specified model, regressors 2 and 3 are highly correlated ($r = 0.86$). This can be seen most clearly in Figure 26.10.

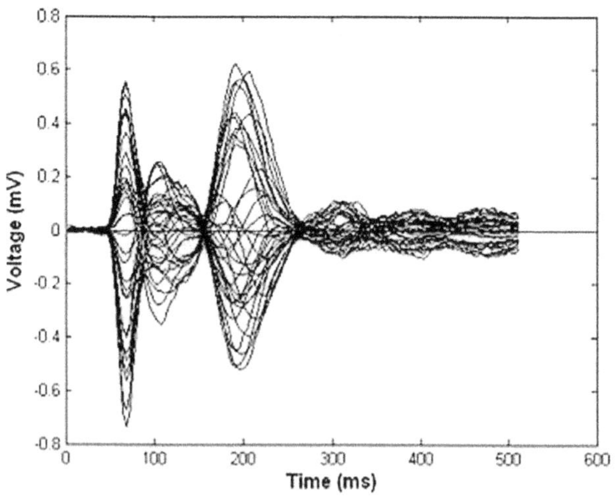

FIGURE 26.9 A butterfly plot of simulated ERPs at 32 sensors.

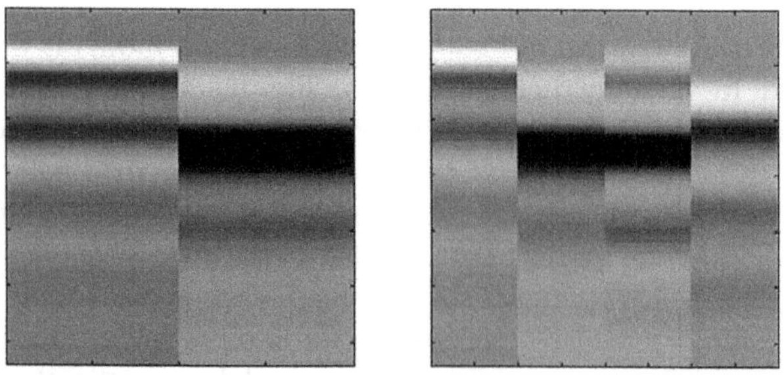

FIGURE 26.10 Design matrices, X, used for localization of biophysical components. Model 1 (left) contains the regressors used to generate the data and Model 2 (right) contains two additional spurious regressors. These regressors have been plotted as time-series in Plates 27 and 28.

The models were then fitted to the data using the VB update rules. As shown in Plates 29 and 30, the true effects (regression coefficients) are accurately recovered even for the over-specified model. The spurious regression coefficients are shrunk towards zero. This is a consequence of the spatial prior and the iterative spatio-temporal deconvolution. This also shows that source reconstruction with temporal priors is robust to model mis-specification.

We then performed a second simulation with the set of regressors shown in Plate 28, and identical specifications of source space, sensor space, forward model and SNR. But in this example we generated data from the regression coefficients shown in Plate 31, regression coefficients 1 and 4 being set to zero. These data therefore comprise three distributed sources: (i) a right-lateralized source having time-series given by a scaled, noise-corrupted regressor 2; (ii) a frontal source given by a scaled, noise-corrupted regressor 3; and (iii) a left-lateralized source comprising a noisy, scaled mixture of regressors 2 and 3. These sources are therefore highly correlated.

The VB-GLM model, using a full design matrix comprising all four regressors, was then fitted to these data. The estimated regression coefficients are shown in Plate 32. Regressors 1 and 4 have been correctly estimated to be close to zero, whereas regressors 2 and 3 bear a close resemblance to the true values. This shows that VB-GLM, in contrast to, for example beamforming approaches, is capable of localizing temporally correlated sources.

Face ERPs

This section presents an analysis of a face processing EEG data set from Henson *et al.* (2003). The experiment involved presentation of images of faces and scrambled faces, as described in Figure 26.11.

The EEG data were acquired on a 128-channel BioSemi ActiveTwo system, sampled at 1024 Hz. The data were referenced to the average of left and right earlobe electrodes and epoched from −200 ms to +600 ms. These epochs were then examined for artefacts, defined as time-points that exceeded an absolute threshold of 120 microvolts. A total of 29 of the 172 trials were rejected. The epochs were then averaged to produce condition specific ERPs at each electrode.

The first clear difference between faces and scrambled faces is maximal around 160 ms, appearing as an enhancement of a negative component (peak 'N160') at occipito-temporal channels (e.g. channel 'B8'), or enhancement of a positive peak at Cz (e.g. channel 'A1').

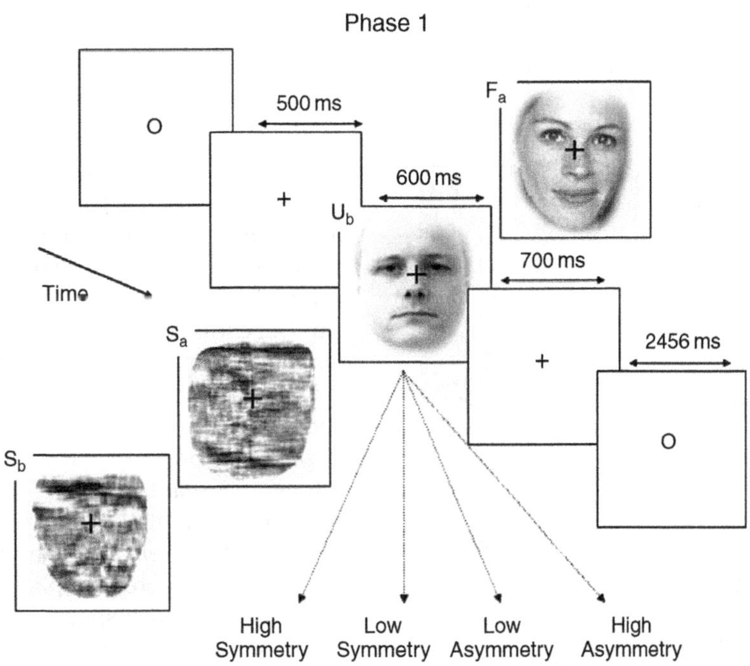

FIGURE 26.11 Face paradigm. The experiment involved randomized presentation of images of 86 faces and 86 scrambled faces. Half of the faces belong to famous people, half are novel, creating three conditions in total. In this chapter, we consider just two conditions: (i) faces (famous or not); and (ii) scrambled faces. The scrambled faces were created by 2D Fourier transformation, random phase permutation, inverse transformation and outline-masking. Thus faces and scrambled faces are closely matched for low-level visual properties such as spatial frequency. The subject judged the left-right symmetry of each image around an imaginary vertical line through the centre of the image. Faces were presented for 600 ms, every 3600 ms.

These effects are shown as a differential topography in Plate 33 and as time-series in Plate 34.

A temporal model was then fitted using wavelet shrinkage (Donoho and Johnstone, 1994). Before applying the model, the data were first downsampled and the 128 samples following stimulus onset were extracted. These steps were taken as we used WaveLab[1] to generate the wavelet bases. This uses a pyramid algorithm to compute coefficients, so requiring the number of samples to be a power of two.

We then extracted the first eigenvector of the sensor data using a singular value decomposition (SVD) and fitted wavelet models to this time-series. A number of wavelet bases were examined, two samples of which are shown in Figure 26.4. These are the Daubechies-4 and Battle-Lemarie-3 wavelets. Plate 35 shows the corresponding time-series estimates. These employed $K = 28$ and $K = 23$ basis functions respectively, as determined by application of the wavelet shrinkage algorithm (Donoho and Johnstone, 1994). We used the smaller Battle-Lemarie basis set in the source reconstruction that follows.

ERPs for faces and scrambled faces were then concatenated to form a vector of 256 elements at each electrode. The overall sensor matrix Y was then of dimension 256×128. The design matrix, of dimension 256×46, was created by having identical block diagonal elements each comprising the Battle-Lemarie basis. This is shown in Figure 26.12. The source space was then defined using a medium resolution cortical grid defined using the typical MNI brain, as in the previous sections. Current source

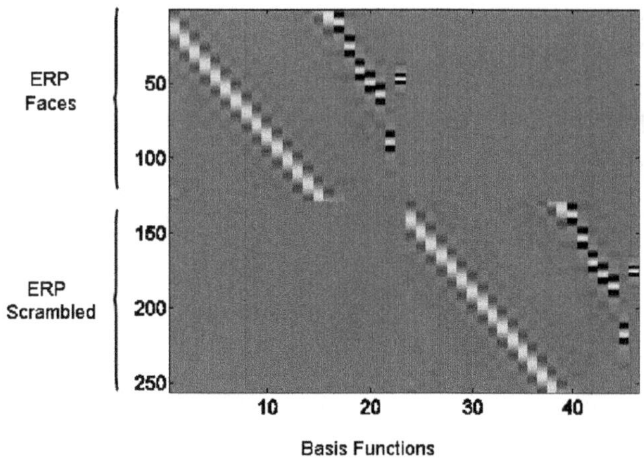

FIGURE 26.12 Design matrix for source reconstruction of ERPs from face data. Each block contains a 23-element Batte-Lemarie basis set. The first components, forming diagonals in the picture, are low frequency wavelets. The high frequency wavelets are concentrated around the N160, where the signal is changing most quickly.

[1] WaveLab is available from http://www-stat.stanford.edu/wavelab.

orientations were assumed perpendicular to the cortical surface. Constraining current sources based on a different individual anatomy is clearly suboptimal, but nevertheless allows us to report some qualitative results.

We then applied the source reconstruction algorithm and obtained a solution after 20 minutes of processing. Plate 36 shows differences in the source estimates for faces minus scrambled faces at time $t = 160$ ms. The images show differences in absolute current at each voxel. They have been thresholded at 50 per cent of the maximum difference at this time point. The maximum difference is plotted in red and 50 per cent of the maximum difference in blue. At this threshold four main clusters of activation appear at (i) right fusiform, (ii) right anterior temporal, (iii) frontal and (iv) superior centroparietal.

These activations are consistent with previous fMRI (Henson *et al.*, 2003) and MEG analyses of faces minus scrambled faces in that face processing is lateralized to the right hemisphere and, in particular, to fusiform cortex. Additionally, the activations in temporal and frontal regions, although not significant in group random effects analyses, are nonetheless compatible with observed between-subject variability on this task.

DISCUSSION

This chapter has described a model-based spatio-temporal deconvolution approach to source reconstruction. Sources are reconstructed by inverting a forward model comprising a temporal process as well as a spatial process. This approach relies on the fact that EEG and MEG signals are extended in time as well as in space.

It rests on the notion that MEG and EEG reflect the neuronal activity of a spatially distributed dynamical system. Depending on the nature of the experimental task, this activity can be highly localized or highly distributed and the dynamics can be more, or less, complex. At one extreme, listening, for example to simple auditory stimuli produces brain activations that are highly localized in time and space. This activity is well described by a single dipole located in brainstem and reflecting a single burst of neuronal activity at, for example $t = 20$ ms post-stimulus. More complicated tasks, such as oddball paradigms, elicit spatially distributed responses and more complicated dynamics that can appear in the ERP as damped sinusoidal responses. In this chapter, we have taken the view that, by explictly modelling these dynamics, one can obtain better source reconstructions.

This view is not unique within the source reconstruction community. Indeed, there have been a number of approaches that also make use of temporal priors. Baillet and Garnero (1997), in addition to considering edge-preserving spatial priors, have proposed temporal priors that penalize quadratic differences between neighbouring time points. Schmidt *et al.* (2000) have extended their dipole-like modelling approach using a temporal correlation prior which encourages activity at neighbouring latencies to be correlated. Galka *et al.* (2004) have proposed a spatio-temporal Kalman filtering approach which is implemented using linear autoregressive models with neighbourhood relations. This work has been extended by Yamashita *et al.* (2004), who have developed a 'Dynamic LORETA' algorithm in which the Kalman filtering step is approximated using a recursive penalized least squares solution. The algorithm is, however, computationally costly, taking several hours to estimate sources in even low-resolution source spaces.

Compared to these approaches, our algorithm perhaps embodies stronger dynamic constraints. But the computational simplicity of fitting GLMs, allied to the efficiency of variational inference, results in a relatively fast algorithm. Also, the GLM can accommodate damped sinusoidal and wavelet approaches that are ideal for modelling transient and non-stationary responses.

The dynamic constraints implicit in our model help to regularize the solution. Indeed, with M sensors, G sources, T time points and K temporal regressors used to model an ERP, if $K < MT/G$, the inverse problem is no longer underdetermined. In practice, however, spatial regularization will still be required to improve estimation accuracy.

This chapter has described a spatio-temporal source reconstruction method embodying well known phenomenological descriptions of ERPs. A similar method has recently been proposed in Friston *et al.* (2006) (see also Chapter 30), but the approaches are different in a number of respects. First, in Friston *et al.* (2006) scalp data Y are (effectively) projected onto a temporal basis set X and source reconstructions are made in this reduced space. This results in a computationally efficient procedure based on restricted maximum likelihood (ReML), but one in which the fit of the temporal model is not taken into account. This will result in inferences about W and J which are overconfident. If one is simply interested in population inferences based on summary statistics (i.e. $\hat{W}$) from a group of subjects, then this does not matter. If, however, one wishes to make within-subject inferences then the procedure described in this chapter is the preferred approach. Secondly, in Friston *et al.* (2006), the model has been augmented to account for trial-specific responses. This treats each trial as a 'random effect' and provides a method for making inferences about induced responses. The algorithm described in this chapter, however, is restricted to treating trials as fixed effects. This mirrors standard first-level analyses of fMRI in which multiple trials are treated by forming concatenated data and design matrices.

A further exciting recent development in source reconstruction is the application of dynamic causal models (DCMs) to M/EEG. DCMs can also be viewed as providing spatio-temporal reconstructions, but ones where the temporal priors are imposed by biologically informed neural-mass models. This offers the possibility of making inferences about task-specific changes in the synaptic efficacy of long range connections in cortical hierarchies, directly from imaging data. These developments are described in Chapter 43.

REFERENCES

Auranen T, Nummenmaa A, Hammalainen M *et al.* (2005) Bayesian analysis of the neuromagnetic inverse problem with l^p norm priors. *NeuroImage* **26**: 870–84

Baillet S, Garnero L (1997) A Bayesian approach to introducing anatomo-functional priors in the EEG/MEG inverse problem. *IEEE Trans Biomed Eng*, **44**: 374–85

Baillet S, Mosher JC, Leahy RM. (2001) Electromagnetic brain mapping. *IEEE Signal Process Mag*, **18**: 14–30

Breakspear M, Terry JR (2002) Nonlinear interdependence in neural systems: motivation, theory and relevance. *Int J Neurosci* **112**: 1263–84

Brookes M, Gibson A, Hall S *et al.* (2004) A general linear model for MEG beamformer imaging. *NeuroImage* **23**: 936–46

Clyde M, Parmigiani G, Vidakovic B (1998) Multiple shrinkage and subset selection in wavelets. *Biometrika* **85**: 391–402

Darvas F, Pantazis D, Kucukaltun Yildirim E *et al.* (2004) Mapping human brain function with MEG and EEG: methods and validation. *NeuroImage* **25**: 383–94

David O, Friston KJ (2003) A neural mass model for MEG/EEG: coupling and neuronal dynamics. *NeuroImage* **20**: 1743–55

Demiralp T, Ademoglu A, Istefanopoulos Y *et al.* (1998) Analysis of event-related potentials (ERP) by damped sinusoids. *Biol Cybernet* **78**: 487–93

Donoho DL, Johnstone IM (1994) Ideal spatial adaptation by wavelet shrinkage. *Biometrika* **81**: 425–55

Evans A, Collins D, Mills S *et al.* (1993) 3d statistical neuroanatomical models from 305 mri volumes. *Proc. IEEE Nuclear Science Symposium and Medical Imaging Conference* **95**: 1813–17

Frackowiak RSJ, Friston KJ, Frith C *et al.* (2003) *Human brain function*, 2nd edn. Academic Press, London

Friston KJ, Mechelli A, Turner R *et al.* (2000) Nonlinear responses in fMRI: the Balloon model, Volterra kernels and other hemodynamics. *NeuroImage* **12**: 466–77

Friston KJ, Henson RNA, Phillips C *et al.* (2006) Bayesian estimation of evoked and induced responses. *NeuroImage* (in press)

Fuchs M, Wagner M, Kohler T *et al.* (1999) Linear and nonlinear current density reconstructions. *J Clin Neurophysiol* **16**: 267–95

Galka A, Yamashita O, Ozaki T *et al.* (2004) A solution to the dynamical inverse problem of EEG generation using spatiotemporal Kalman filtering. *NeuroImage* **23**: 435–53

Gelman A, Carlin JB, Stern HS *et al.* (1995) *Bayesian data analysis*. Chapman and Hall, Boca Raton

Golub GH, Van Loan CF (1996) *Matrix computations*, 3rd edn. Johns Hopkins University Press, Baltimore

Henson RNA, Goshen-Gottstein Y, Ganel T *et al.* (2003) Electrophysiological and hemodynamic correlates of face perception, recognition and priming. *Cerebr Cort* **13**: 793–805

Kiebel SJ, Friston KJ (2004) Statistical parametric mapping for event-related potentials II: a hierarchical temporal model. *NeuroImage* **22**: 503–20

Makeig S, Westerfield M, Jung T-P *et al.* (2002) Dynamic brain sources of visual evoked responses. *Science* **295**: 690–94

Mattout J, Phillips C, Penny WD (2006) MEG source localisation under multiple constraints: an extended Bayesian framework. *NeuroImage* **30**: 753–67

McKeown MJ, Makeig S, Brown GG *et al.* (1998) Analysis of fMRI data by blind separation into independent spatial components. *Hum Brain Mapp* **6**: 160–88

Minka TP (2000) Automatic choice of dimensionality for PCA. Technical Report 514, MIT Media Laboratory, Perceptual Computing Section, Cambridge, MA

Mosher JC, Leahy RM (1998) Recursive MUSIC: a framework for eeg and meg source localization. *IEEE Trans Biomed Eng* **47**: 332–40

Osborne MR, Smyth GK (1991) A modified Prony algorithm for fitting functions defined by difference equations. *J Sci Stat Comput* **12**: 362–82

Pascual Marqui R (2002) Standardized low resolution electromagnetic tomography (sLORETA): technical details. *Meth Find Exp Clin Pharmacol* **24**: 5–12

Pascual Marqui R, Michel C, Lehman D (1994) Low resolution electromagnetic tomography: a new method for localizing electrical activity of the brain. *Int J Psychophysiol* **18**: 49–65

Penny WD, Flandin G (2005) Bayesian analysis of single-subject fMRI: SPM implementation. Technical report, Wellcome Department of Imaging Neuroscience, London

Penny WD, Trujillo-Barreto N, Friston KJ (2005) Bayesian fMRI time series analysis with spatial priors. *NeuroImage* **24**: 350–62

Roberts SJ, Penny WD (2002) Variational Bayes for generalised autoregressive models. *IEEE Trans Signal Process* **50**: 2245–57

Robinson S, Vrba J (1999) Functional neuroimaging by synthetic aperture magnetometry (SAM). In *Recent advances in biomagnetism*. Tohoku University Press, Sendai

Rugg MD, Coles MGH (1995) *Electrophysiology of mind: event-related potentials and cognition*. Oxford University Press, Oxford

Rush S, Driscoll D (1969) EEG electrode sensitivity – an application of reciprocity. *IEEE Trans Biomed Eng* **16**: 15–22

Sahani M, Nagarajan SS (2004) Reconstructing MEG sources with unknown correlations. In *Advances in neural information processing systems*, Saul L, Thrun S, Schoelkopf B (eds), volume 16. MIT, Cambridge, MA

Scherg M, von Cramon D (1986) Evoked dipole source potentials of the human auditory cortex. *Electroencephalogr Clin Neurophysiol* **65**: 344–60

Schmidt DM, George JS, Wood CC (1999) Bayesian inference applied to the electromagnetic inverse problem. *Hum Brain Mapp* **7**: 195–212

Schmidt DM, Ranken DM, George JS (2000) Spatial-temporal Bayesian inference for MEG/EEG. In *12th International Conference on Biomagnetism*, Helsinki, Finland, August

Tallon Baudry C, Bertrand O, Delpuech C *et al.* (1996) Stimulus specificity of phase-locked and non phase-locked 40 Hz visual responses in human. *J Neurosci* **16**: 4240–49

Trejo L, Shensa MJ (1999) Feature extraction of event-related potentials using wavelets: an application to human performance monitoring. *Brain Lang* **66**: 89–107

Unser M, Aldroubi A (1996) A review of wavelets in biomedical applications. *Proc IEEE* **84**: 626–38

Valdes-Sosa P, Marti F, Garcia F *et al.* (2000) Variable resolution electric-magnetic tomography. In *Proceedings of the 10th International Conference on Biomagnetism*, volume 2, 373–76, Springer-Verlag, New York

Woolrich MW, Ripley BD, brady M *et al.* (2001) Temporal autocorrelation in univariate linear modelling of fMRI data. *NeuroImage* **14**: 1370–86

Yamashita O, Galka A, Ozaki T *et al.* (2004) Recursive penalised least squares solution for dynamical inverse problems of EEG generation. *Hum Brain Mapp* **21**: 221–35

PART 6

Biophysical models

C H A P T E R

27

Forward models for fMRI

K. Friston and D. Glaser

INTRODUCTION

Part 6 is concerned with biophysical models of neuronal responses and the inversion of these models to make inferences about their parameters. In relation to the previous chapters, there is a greater focus on how signals observed with neuroimaging are generated and the underlying physical and physiological mechanisms. In this chapter, we look at haemodynamics. In the next chapter, we turn to electrical and magnetic sources that generate electroencephalography (EEG) and magnetoencephalography (MEG) signals. These chapters start with conventional electromagnetic forward models. In the subsequent chapters, we consider physiology in more depth, through mean-field models of neuronal populations or ensembles and how these provide a motivation for neural-mass models of event-related potentials. In the final chapters, we consider some general issues encountered during model inversion and selection.

There is a growing appreciation of the importance of non-linearities in evoked responses in functional magnetic resonance imaging (fMRI), particularly with the advent of event-related fMRI. These non-linearities are commonly expressed as interactions among stimuli that can lead to the suppression and increased latency of responses to stimuli that are incurred by a preceding stimulus. We presented previously a model-free characterization of these effects using generic techniques from non-linear system identification, namely a Volterra series formulation. At the same time, Buxton *et al.* (1998) described a plausible and compelling dynamical model of haemodynamic signal transduction in fMRI. Subsequent work by Mandeville *et al.* (1999) provided important theoretical and empirical constraints on the form of the dynamic relationship between blood flow and volume that underpins the evolution of the fMRI signal. In this chapter, we combine these system identification and model-based approaches and ask whether the Balloon model is sufficient to account for the non-linear behaviours observed in real time-series. We conclude that it can and, furthermore, the model parameters that ensue are biologically plausible. This conclusion is based on the observation that the Balloon model can produce Volterra kernels that emulate empirical kernels.

To enable this evaluation we have had to embed the Balloon model in a haemodynamic input-state-output model that included the dynamics of perfusion changes that are contingent on underlying synaptic activation. This chapter presents the haemodynamic model, its associated Volterra kernels and addresses the model's validity in relation to empirical characterizations of evoked responses in fMRI and other neurophysiological constraints.

Background

This chapter is about modelling the relationship between neural activity and the BOLD (blood oxygenation-level-dependent) fMRI signal. The link between blood supply and brain activity has been established for over a hundred years. In their seminal paper, Roy and Sherrington (1890) concluded that functional activity increased blood flow and inferred that there was a coupling that increased blood flow in response to increased metabolic demand. Interestingly, their observation of the consequences of metabolic demand came before the demonstration of the increase in demand itself. It was more than seventy years later that the regional measurement of metabolic changes was convincingly achieved using autoradiographic techniques. These used a substitute for glucose, called deoxyglucose (2DG) radioactively labelled with C^{14}. 2DG enters the cells by the same route as glucose but is not metabolized and thus accumulates inside the cells at a rate that depends on their metabolic activity. By examining the density of labelled 2DG in brain slices,

Statistical Parametric Mapping, by Karl Friston et al.
ISBN–13: 978-0-12-372560-8 ISBN–10: 0-12-372560-7

Sokoloff and colleagues (Kennedy *et al.*, 1976) obtained functional maps of the activity during the period in which 2DG was injected. This activity period was generally around 45 minutes, which limited the time-resolution of the technique. In addition, only one measurement per subject could be made since the technique involved the sacrifice of the animal (further developments allowed the injection of two tracers, but this was still very restrictive). However, the spatial resolution could be microscopic, since the label is contained in the cells themselves, rather than being limited to the blood vessels surrounding them. Through modelling of the enzyme kinetics for the uptake of 2DG and practical experiments, the relationships between neural function and glucose metabolism have been established and underpin the development of 'metabolic encephalography'.

Positron emission tomography (PET) measures an intermediate stage in the chain linking neural activity, via metabolism, to the BOLD signal. By using a tracer such as O^{15}-labelled water, one can measure changes in regional cerebral blood flow (rCBF), which accompany changes in neural activity. This was originally thought of as an autoradiographic technique, but has many advantages over 2DG and is clearly much less invasive, making it suitable for human studies. Also, substantially shorter times are required for measurements, typically well below a minute. As suggested above, the elucidation of the mechanisms underlying the coupling of neural activity and blood flow lags behind its exploitation. There are several candidate signals, including rapidly diffusible second messengers such as nitric oxide. This remains an active area of research irrespective of its consequences for models of functional brain imaging.

Below, we follow Miller *et al.* (2000), among others, and assume that blood flow and neural activity are related linearly over normal ranges. However, there are ongoing arguments about the nature of the linkage between neural activity, the rate of metabolism of oxygen and cerebral blood flow. Some PET studies have suggested that, while an increase in neural activity produces a proportionate increase in glucose metabolism and cerebral blood flow, oxygen consumption does not increase proportionately (Fox and Raichle, 1986). This decoupling between blood flow and oxidative metabolism is known as the 'anaerobic brain' hypothesis by analogy with muscle physiology. Arguing against this position, other groups have adopted an even more radical interpretation. They suggest that immediately following neural stimulation there is a transient decoupling between neural activity and blood flow (Vanzetta and Grinvald, 1999). By this argument, there is an immediate increase in oxidative metabolism which produces a transient localized increase in deoxyhaemoglobin. Only later do the mechanisms regulating blood flow kick in, causing the observed increase in rCBF and hence blood volume. Evidence for this position comes from optical imaging studies and depends on modelling the absorption and light-scattering properties of cortical tissue and the relevant chromophores, principally (de)oxyhaemoglobin. Other groups have questioned these aspects of the work, and the issue remains controversial (Lindauer *et al.*, 2001). One possible consequence of this position is that better spatial resolution would be obtained by focusing on the early phase of the haemodynamic response.

As this chapter will demonstrate, the situation is even more complicated with regard to fMRI using a BOLD contrast. As its name suggests, the technique uses the amount of oxygen in the blood as a marker for neural activity, exploiting the fact that deoxyhaemoglobin is less diamagnetic than oxyhaemoglobin. Blood oxygenation level refers to the proportion of oxygenated blood, but the signal depends on the total amount of deoxyhaemoglobin and so the total volume of blood is also a factor. Another factor is the change in the amount of oxygen leaving the blood to enter the tissue and meet changes in metabolic demand. Since the blood which flows into the capillary bed is fully oxygenated, changes in blood flow also change the blood oxygenation level. Finally, the elasticity of the veins and venules means that an increase in blood flow causes an increase in blood volume. All these factors are modelled and discussed in this chapter. Of course, even more factors can be considered; for example, Mayhew and colleagues (Zheng *et al.*, 2002) have extended the treatment described here to include (among others) the dynamics of oxygen buffering in the tissue.

Notwithstanding these complications, it is a standard assumption that 'the fMRI signal is approximately proportional to a measure of local neural activity' (reviewed in Heeger and Ress, 2002), and this linear model is still used in many studies, particularly where inter-stimulus intervals are more than a second or two. Empirical evidence against this hypothesis is outlined below, but note that there are now theoretical objections too. In particular, the models which have been developed to account for observed non-linearities point to quite specific ranges in simulation parameters outside which the linearity assumption fails.

The last link in the chain concerns the relation between a complete description of the relevant aspects of blood supply and the physics underlying the BOLD signal. While this is not the principal focus of this chapter, a couple of points are worth emphasizing. First, different sized blood vessels will give different changes in BOLD signal for the same changes in blood flow, volume and oxygenation. This is because of differences in the inhomogeneity of the magnetic fields in their vicinity. Second, and for related reasons, heuristic equations, as employed in this and other models, depend on the strength of the magnet. In particular, the equations used here may be relevant only for 1.5 T scanners, although other versions for different field strength have been developed.

Finally, a word about 'neural activity'. So far, we have deliberately not specified what type of neural activity we are considering. First, it is worth remembering that different electrophysiological measures can emphasize different aspects of neural activity. In particular, recording of multiple single units, with an intracortical microelectrode, can tend to sample action potentials from large pyramidal output neurons. Such studies are frequently referred to when characterizing the response properties of a primate cortical area. However, the metabolic demands of various cellular processes suggest that spiking is not the major drain on the resources of a cell but rather synaptic transmission, restoration of postsynaptic potentials and cytoskeletal turnover dominate (Attwell and Laughlin, 2001). The processes are just as important as spiking, whether in interneurons and whether excitatory or inhibitory. An example that shows the importance of this distinction is feed-forward (spike-dependent) versus feed-back (modulatory) activity in low-level visual cortex. BOLD fMRI experiments in humans have shown a good agreement with studies of spiking in V1 in response to modulating the contrast of a visual stimulus, however, attentional (top-down) modulation effects in V1 have proved elusive in monkey electrophysiology but robust with BOLD studies in humans. Aside from their neurobiological significance, these issues should be borne in mind when considering the neural activity disclosed by BOLD.

A further subtlety relates to the modelled time course of neural activity; a typical set of spike- or block-functions used to model neural activity will fail to capture adaptation and response transients which are well known from the neurophysiological literature. Note that the adaptation paradigm deliberately exploits these effects (Grill-Spector and Malach, 2001). Studies using simultaneous fMRI and intracortical electrical recording in monkeys have empirically validated many of the theoretical points considered above (Logothetis *et al.*, 2001). In particular, the closeness of the BOLD signal to local field potentials (LFP) and multiunit activity (MUA) rather than spiking activity was emphasized. These studies also demonstrated that the linear assumption can predict up to 90 per cent of the variance in BOLD responses in some cortical regions. However, there was considerable variability in the accuracy of prediction. In Chapter 32, we will revisit some of these themes in a more abstract way using a simple dimensional analysis of the energetics entailed by neuronal dynamics and the implications for electrical and haemodynamic responses.

Having surveyed the general issues surrounding the coupling of neural activity and the BOLD signal, we will now consider a specific and detailed model. This could be considered as an instantiation of current knowledge and further extensions of this model have already been proposed, which incorporate new data. What follows is largely a reprise of Friston *et al.* (2000), and contains some mathematical material.

NON-LINEAR EVOKED RESPONSES

In this section, we focus on the non-linear aspects of evoked responses in functional neuroimaging and present a dynamical approach to modelling and characterizing event-related signals in fMRI. We will show that the Balloon-Windkessel model (Buxton and Frank, 1997; Buxton *et al.*, 1998; Mandeville *et al.*, 1999) can account for non-linearities in event-related responses that are seen empirically and describe a non-linear dynamical model that couples changes in synaptic activity to fMRI signals. This haemodynamic model obtains by combining the Balloon-Windkessel model (henceforth Balloon model) with a model of how synaptic activity causes changes in regional flow.

In Friston *et al.* (1994), we presented a linear model of haemodynamic responses in fMRI time-series, wherein underlying neuronal activity (inferred on the basis of changing stimulus or task conditions) is convolved, or smoothed with a haemodynamic response function. In Friston *et al.* (1998) we extended this model to cover non-linear responses using a Volterra series expansion. At the same time, Buxton and colleagues developed a mechanistic model of how evoked changes in blood flow were transformed into a BOLD signal (Buxton *et al.*, 1998). A component of the Balloon model, namely the relationship between blood flow and volume, was then elaborated in the context of standard Windkessel theory by Mandeville *et al.* (1999). The Volterra approach, in contradistinction to other non-linear characterization of haemodynamic responses (cf. Vazquez and Noll, 1996), is model-independent, in the sense that Volterra series can model the behaviour of any non-linear time-invariant dynamical system.[1] The principal aim of the work presented below was to see if the Balloon model would be

[1] In principle Volterra series can represent any dynamical input-state-output system and in this sense a characterization in terms of Volterra kernels is model-independent. However, by using basis functions to constrain the solution space, constraints are imposed on the form of the kernels and, implicitly, the underlying dynamical system (i.e. state-space representation). The characterization is therefore only assumption free to the extent the basis set is sufficiently comprehensive.

sufficient to explain the non-linearities embodied in a purely empirical Volterra characterization.

Volterra series

Volterra series express the output of a system, in this case the BOLD signal from a particular voxel, as a function of some input, here the assumed synaptic activity that is changed experimentally. This series is a function of the input over its recent history and is expressed in terms of generalized convolution kernels. Volterra series are often referred to as non-linear convolutions or polynomial expansions with memory. They are simply Taylor expansions extended to cover dynamical input-state-output systems by considering the effect of the input now and in the recent past. The zeroth order kernel is simply a constant about which the response varies. The first-order kernel represents the weighting applied to a sum of inputs over the recent past (cf. the haemodynamic response function) and can be thought of as the change in output for a change in the input at each time point. Similarly, the second-order coefficients represent interactions that are simply the effect of the input at one point in time on its contribution at another. The second-order kernel comprises coefficients that are applied to interactions among (i.e. products of) inputs, at different times in the past, to predict the response. (See Appendix 2 for more mathematical details.)

In short, the output can be considered a non-linear convolution of the input where non-linear behaviours are captured by high-order kernels. For example, the presence of a stimulus can be shown to attenuate the magnitude of, and induce a longer latency in, the response to a second stimulus that occurs within a second or so. The example shown in Figure 27.1 comes from our previous analysis (Friston *et al.*, 1998) and shows how a preceding stimulus can modify the response to a subsequent stimulus. This sort of effect led to the notion of haemodynamic refractoriness and is an important example of non-linearity in fMRI time-series.

The important thing about Volterra series is that they do not refer to hidden state variables that mediate between the input and output (e.g. blood flow, venous volume, oxygenation, the dynamics of endothelium derived relaxing factor, kinetics of cerebral metabolism etc.). This renders them powerful because they provide for a complete specification of the dynamical behaviour of a system without ever having to measure the state variables or make any assumptions about how these variables interact to produce a response. On the other hand, the Volterra formulation is impoverished because it yields no mechanistic insight into how the response is mediated. The alternative is to posit some model of interacting

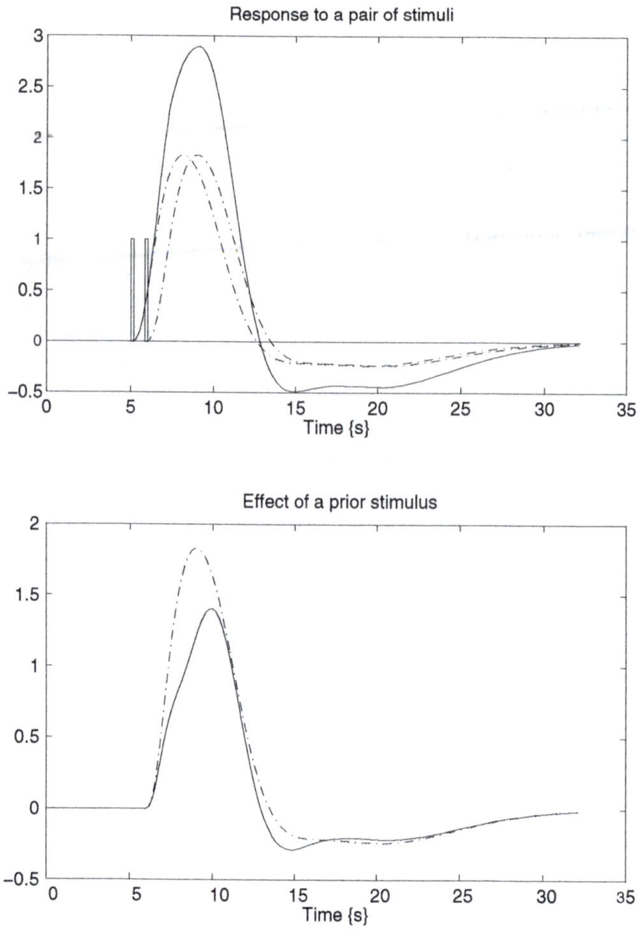

FIGURE 27.1 Top panel: simulated responses to a pair of words (bars) one second apart, presented together (solid line) and separately (broken lines) based on the kernels shown in Figure 27.5. Lower panel: the response to the second word when presented alone (broken line as above) and when preceded by the first (solid line). The latter obtains by subtracting the response to the first word from the response to both. The difference reflects the effect of the first word on the response to the second.

state variables and establish the validity of that model in relation to observed input-output behaviours and the dynamics of the states themselves. This involves specifying a series of differential equations that express the change in one state variable as a function of the others and the input. Once these equations are specified, the equivalent Volterra representation can be derived analytically (see Appendix 2 for details). The Balloon model is an example of such a model.

The Balloon model

The Balloon model (Buxton and Frank, 1997; Buxton *et al.*, 1998) is an input-state-output model with two state variables: volume, v and deoxyhaemoglobin content, q. The input to the system is blood flow, f and the output is the BOLD signal, y. The BOLD signal is partitioned into an extra- and intravascular component, weighted by their respective volumes. These signal components depend on the deoxyhaemoglobin content and render the signal a non-linear function of v and q. The effect of flow on v and q (see below) determines the output and it is these effects that are the essence of the Balloon model: increases in flow effectively inflate a venous 'balloon' such that deoxygenated blood is diluted and expelled at a greater rate. The clearance of deoxyhaemoglobin reduces intra-voxel dephasing and engenders an increase in signal. Before the balloon has inflated sufficiently the expulsion and dilution may be insufficient to counteract the increased delivery of deoxygenated blood to the venous compartment and an 'early dip' in signal may be expressed. After the flow has peaked, and the balloon has relaxed again, reduced clearance and dilution contribute to the post-stimulus undershoot commonly observed. This is a simple and plausible model that is predicated on a minimal set of assumptions and relates closely to the Windkessel formulation of Mandeville *et al.* (1999). Furthermore, the predictions of the Balloon model concur with the steady-state models of Hoge and colleagues, and their elegant studies of the relationship between blood flow and oxygen consumption in human visual cortex (e.g. Hoge *et al.*, 1999).

The Balloon model is inherently non-linear and may account for the sorts of non-linear interactions revealed by the Volterra formulation. One simple test of this hypothesis is to see if the Volterra kernels associated with the Balloon model compare with those derived empirically. The Volterra kernels estimated in Friston *et al.* (1998) clearly did not use flow as input because flow is not measurable with BOLD fMRI. The input comprised a stimulus function as an index of synaptic activity. In order to evaluate the Balloon model in terms of Volterra kernels, it has to be extended to accommodate the dynamics of how flow is coupled to synaptic activity encoded in the stimulus function. This chapter presents one such extension.

In summary, the Balloon model deals with the link between flow and BOLD signal. By extending the model to cover the coupling of synaptic activity and flow, a complete model, relating experimentally induced changes in neuronal activity to BOLD signal, obtains. The input-output behaviour of this model can be compared to the real brain in terms of their respective Volterra kernels.

The remainder of this chapter is divided into three sections. In the next section, we present a haemodynamic model of the coupling between synaptic activity and BOLD response that builds upon the Balloon model. The second section presents an empirical evaluation of this model by comparing its Volterra kernels with those obtained using real fMRI data. This is not a trivial exercise because there is no guarantee that the Balloon model could produce the complicated forms of the kernels seen empirically and, even if it could, the parameters needed to do so may be biologically implausible. This section provides estimates of these parameters, which allow some comment on the face validity of the model, in relation to known physiology. The final section presents a discussion of the results in relation to known biophysics and neurophysiology. This chapter is concerned with the validation and evaluation of the Balloon model in relation to the Volterra characterizations, and the haemodynamic model presented below in relation to real haemodynamics. In Chapter 34, we will use the haemodynamic model to illustrate Bayesian inversion of dynamic models and subsequent chapters will use the model as part of larger dynamic causal models for fMRI (see Chapter 41).

THE HAEMODYNAMIC MODEL

In this section, we describe a haemodynamic model that mediates between synaptic activity and measured BOLD responses. This model essentially combines the Balloon model and a simple linear dynamical model of changes in regional cerebral blood flow (rCBF) caused by neuronal activity. The model architecture is summarized in Figure 27.2. To motivate the model components more clearly we will start at the output and work towards the input.

The Balloon component

This component links rCBF and the BOLD signal as described in Buxton *et al.* (1998). All variables are expressed in normalized form, relative to resting values. The BOLD signal is taken to be a static non-linear function of normalized venous volume v, normalized total deoxyhaemoglobin voxel content q and resting net oxygen extraction fraction by the capillary bed E_0:

$$\begin{aligned} y &= g(v,q) = V_0(k_1(1-q)+k_2(1-q/v)+k_3(1-v)) \\ k_1 &= 7E_0 \\ k_1 &= 2 \\ k_1 &= 2E_0-0.2 \end{aligned} \qquad 27.1$$

where V_0 is resting blood volume fraction. This signal comprises a volume-weighted sum of extra- and intravascular signals that are functions of volume and deoxyhaemoglobin content. The latter are the state variables

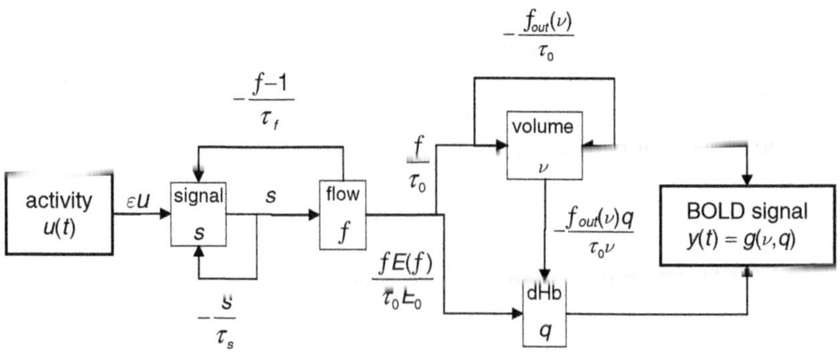

FIGURE 27.2 Schematic illustrating the organization of the haemodynamic model. This is a fully non-linear single-input-single-output state model with four state variables. The form and motivation for the changes in each state variable, as functions of the others, are described in the main text.

whose dynamics need specifying. The rate of change of volume is simply:

$$\tau_0 \dot{v} = f - f_{out}(v) \qquad 27.2$$

(See Mandeville *et al.* (1999) for an excellent discussion of this equation in relation to Windkessel theory.) Eqn. **27.2** says that volume changes reflect the difference between inflow and outflow from the venous compartment with a time constant τ_0. This constant represents the mean transit time (i.e. the average time it takes to traverse the venous compartment or for that compartment to be replenished) and is V_0/f_0 where f_0 is resting flow. The physiology of the relationship between flow and volume is determined by the evolution of the transit time. Mandeville *et al.* (1999) reformulated the temporal evolution of transit time into a description of the dynamics of resistance and capacitance of the balloon using Windkessel theory; 'Windkessel' means leather bag. This enabled them to posit a form for the temporal evolution of a downstream elastic response to arteriolar vasomotor changes and estimate mean transit times using measurements of volume and flow in rats, using fMRI and laser-Doppler flowmetry. We will compare these estimates to our empirical estimates in the next section.

Note that outflow is a function of volume. This function models the balloon-like capacity of the venous compartment to expel blood at a greater rate when distended. We model it with a single parameter α based on the Windkessel model:

$$f_{out}(v) = v^{\frac{1}{\alpha}} \qquad 27.3$$

where $1/\alpha = \gamma + \beta$ (cf. Eqn. **27.6** in Mandeville *et al.*, 1999). $\gamma = 2$ represents laminar flow. $\beta > 1$ models diminished volume reserve at high pressures and can be thought of as the ratio of the balloon's capacitance to its compliance. At steady state, empirical results from PET suggest $\alpha \approx 0.38$ (Grubb *et al.*, 1974). However, when flow and volume are changing dynamically, this value is smaller. Mandeville *et al.* (1999) were the first to measure the dynamic flow-volume relationship and estimated $\alpha \approx 0.18$, after 6 s of stimulation, with a projected asymptotic [steady-state] value of 0.36.

The change in deoxyhaemoglobin reflects the delivery of deoxyhaemoglobin into the venous compartment minus that expelled (outflow times concentration):

$$\tau_0 \dot{q} = f\frac{E(f)}{E_0} - \frac{f_{out}(v)q}{v} \qquad 27.4$$

where $E(f)$ is the fraction of oxygen extracted from the inflowing blood. This is assumed to depend on oxygen delivery and is consequently flow-dependent. A reasonable approximation for a wide range of transport conditions is (Buxton *et al.*, 1998):

$$E(f) = 1 - (1 - E_0)^{1/f} \qquad 27.5$$

The second term in Eqn. **27.4** represents an important non-linearity: the effect of flow on signal is largely determined by the inflation of the balloon, resulting in an increase of outflow and clearance of deoxyhaemoglobin. This effect depends upon the concentration of deoxyhaemoglobin such that the clearance attained by the outflow will be severely attenuated when the concentration is low (e.g. during the peak response to a prior stimulus). The implications of this will be illustrated in the next section.

This concludes the Balloon model component, where there are only three unknown parameters that determine the dynamics, namely resting oxygen extraction fraction, mean transit time and Grubb's exponent E_0, τ_0, α. The only thing required, to specify the BOLD response, is flow.

rCBF component

It is generally accepted that, over normal ranges, blood flow and synaptic activity are linearly related. A recent empirical verification of this assumption can be found in Miller *et al.* (2000), who used MRI perfusion imaging to address this issue in visual and motor cortices. After modelling neuronal adaptation they were able to conclude: 'Both rCBF responses are consistent with a linear transformation of a simple non-linear neural response model'. Furthermore, our own work using PET and fMRI replications of the same experiments suggested that the observed non-linearities enter into the translation of rCBF into a BOLD response (as opposed to a non-linear relationship between synaptic activity and rCBF) in the auditory cortices (see Friston *et al.*, 1998). Under the constraint that the dynamical system linking synaptic activity and rCBF is linear we will use the most parsimonious model:

$$\dot{f} = s \qquad \textbf{27.6}$$

where s is some flow inducing signal. Although it may seem more natural to express the effect of this signal directly on vascular resistance (r), e.g. $\dot{r} = -s$, Eqn. **27.6** has the more plausible form. This is because the effect of signal is much smaller when r is small (when the arterioles are fully dilated, signals such as endothelium-derived relaxing factor or nitric oxide will cause relatively small decrements in resistance). This can be seen by noting Eqn. **27.6** is equivalent to $\dot{r} = -r^2 s$, where $f = 1/r$.

The signal is assumed to subsume many neurogenic and diffusive signal sub-components and is generated by neuronal activity $u(t)$:

$$\dot{s} = \varepsilon u - s/\tau_s - (f-1)/\tau_f \qquad \textbf{27.7}$$

The unknown parameters here represent the efficacy with which neuronal activity causes an increase in signal ε, the time-constant for signal decay or elimination τ_s and the time-constant for autoregulatory feedback τ_f from blood flow. The existence of this feedback term can be inferred from post-stimulus undershoots in rCBF (e.g. Irikura *et al.*, 1994) and the well-characterized vasomotor signal in optical imaging (Mayhew *et al.*, 1998). The critical aspect of the latter oscillatory (~ 0.1 Hz) component of intrinsic signals is that it shows variable phase relationships from region to region, supporting strongly the notion of local closed-loop feedback mechanisms as modelled in Eqn. **27.6** and Eqn. **27.7**.

There are three unknown parameters for each of the two components of the haemodynamic model above (see also Figure 27.2 for a schematic summary). Figure 27.3 illustrates the behaviour of the haemodynamic model for typical values of the six parameters; $\varepsilon = 0.5$, $\tau_s = 0.8$, $\tau_f = 0.4$, $\tau_0 = 1$, $\alpha = 0.2$, $E_0 = 0.8$ and assuming $V_0 = 0.02$ here and throughout. We have used a very high value for oxygen extraction to accentuate the early dip (see discussion).

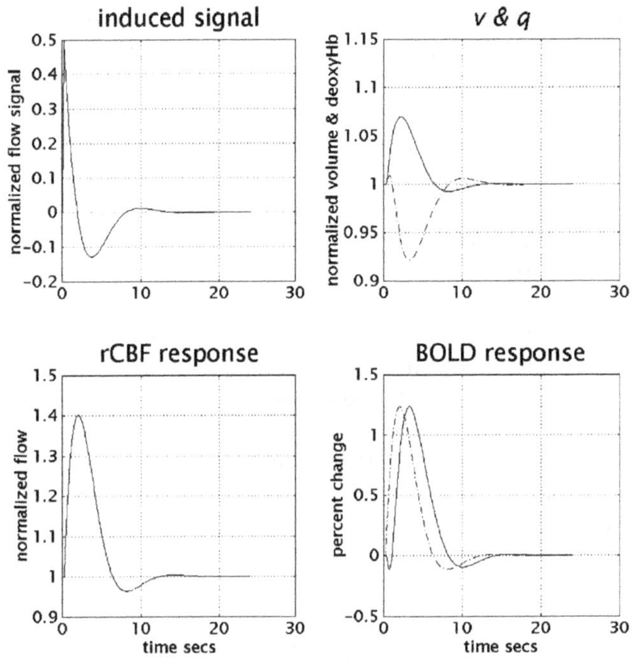

FIGURE 27.3 Dynamics of the haemodynamic model. Upper left panel: the time-dependent changes in the neuronally induced perfusion signal that causes an increase in blood flow. Lower left panel: the resulting changes in normalized blood flow f. Upper right panel: the concomitant changes in normalized venous volume v (solid line) and normalized deoxyhaemoglobin content q (broken line). Lower right panel: the per cent change in BOLD signal that is contingent on v and q. The broken line is inflow normalized to the same maximum as the BOLD signal. This highlights the fact that BOLD signal lags the rCBF signal by about a second.

Following a short-lived neuronal transient, signal is generated and starts to decay immediately. This signal induces an increase in flow that itself augments signal decay, to the extent the signal is suppressed below resting levels (see the upper left panel in Figure 27.3). This behaviour corresponds to a dampened oscillator. Increases in flow (lower-left panel) dilate the venous balloon, which responds by ejecting deoxyhaemoglobin. In the first few hundred milliseconds the net deoxyhaemoglobin q increases with an accelerating flow-dependent delivery. It is then cleared by volume-dependent outflow expressing a negative peak a second or so after the positive volume v peak (the broken and solid lines in the upper right panel correspond to q and v respectively). This results in an early dip in the BOLD signal followed by a pronounced positive peak at about 4 s (lower right panel) that reflects the combined effects of reduced net deoxyhaemoglobin, increased venous volume and consequent dilution of deoxyhaemoglobin. Note that the rise and peak in volume (solid line in the upper right panel) lags flow by about a second. This is very similar to the predictions of the Windkessel formulation and the empirical results presented in Mandeville *et al.*

(1999) (see their Figure 2). After about 8 s, flow experiences a rebound due to its suppression of the perfusion signal. The reduced venous volume and ensuing outflow permit a re-accumulation of deoxyhaemoglobin and a consequent undershoot in the BOLD signal.

The rCBF component of the haemodynamic model is a linear dynamical system and, as such, has only zeroth and first-order kernels. This means it cannot account for the haemodynamic refractoriness and non-linearities observed in BOLD responses. Although the rCBF component may facilitate the Balloon component's capacity to model non-linearities (by providing appropriate input), the rCBF component alone cannot generate second-order kernels. The question addressed in this chapter is whether the Balloon component can produce second-order kernels that are realistic and do so with physiologically plausible parameters.

KERNEL ESTIMATION

In what follows, we describe the data used to estimate Volterra kernels. The six free parameters of the haemodynamic model that best reproduce these empirical kernels are then identified by minimizing the difference between the model kernels and the empirical kernels. The critical questions this section addresses are: 'can the haemodynamic model account for the form of empirical kernels up to second-order?'; and 'are the model parameters required to do this physiologically plausible?'

Empirical analyses

The data and Volterra kernel estimation are described in detail in Friston *et al.* (1998). In brief, we obtained fMRI time-series from a single subject at 2-tesla using a Magnetom VISION (Siemens, Erlangen) whole body MRI system, equipped with a head volume coil. Contiguous multislice T2*-weighted fMRI images were obtained with a gradient echo-planar sequence using an axial slice orientation (TE = 40 ms, TR = 1.7 s, 64 × 64 × 16 voxels). After discarding initial scans (to allow for magnetic saturation effects) each time-series comprised 1200 volume images with 3 mm isotropic voxels. The subject listened to monosyllabic or bi-syllabic concrete nouns (i.e. 'dog', 'radio', 'mountain', 'gate') presented at five different rates (10, 15, 30, 60 and 90 words per minute) for epochs of 34 s, intercalated with periods of rest. The presentation rates were successively repeated according to a Latin Square design.

Volterra kernels were estimated by expanding the kernels in terms of temporal basis functions and estimating the kernel coefficients up to second-order using maximum likelihood estimates with a general linear model (Worsley and Friston, 1995). The basis set comprised three gamma varieties of increasing dispersion and their temporal derivatives (as described in Friston *et al.* 1998). The stimulus function $u(t)$, the supposed neuronal activity, was simply the word presentation rate at which the scan was acquired. We selected voxels that showed a robust response to stimulation from two superior temporal regions in both hemispheres (Figure 27.4). These were the 128 voxels showing the most significant response when testing for the null hypothesis that the first- and second-order kernels were zero. Selecting these voxels ensured that the kernel estimates had minimal variance.

The haemodynamic parameters

For each voxel we identified the six parameters of the haemodynamic model of the previous section whose kernels corresponded, in a least squares sense, to the empirical kernels for that voxel. The model's kernels were computed, for a given parameter vector, as described in Appendix 2 and entered, with the corresponding

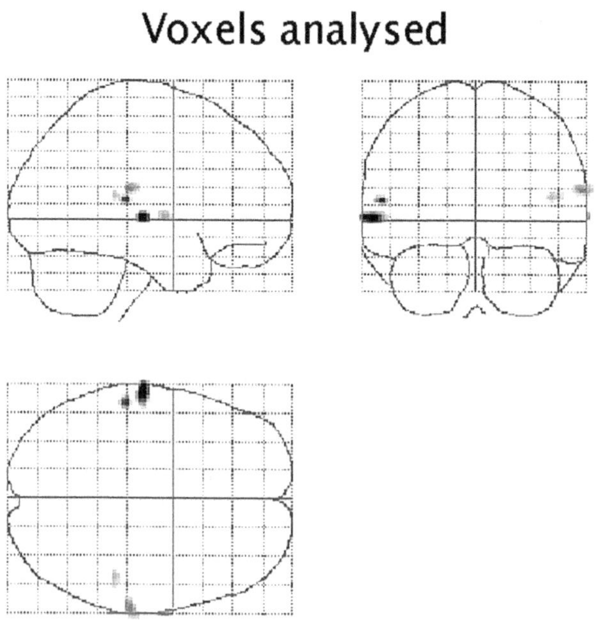

FIGURE 27.4 Voxels used to estimate the parameters of the haemodynamic model shown in Figure 27.2. This is an SPM{F} testing for the significance of the first- and second-order kernel coefficients in the empirical analysis and represents a maximum intensity projection of a statistical process of the F-ratio, following a multiple regression analysis at each voxel. This regression analysis estimated the kernel coefficients after expanding them in terms of a small number of temporal basis functions (see Friston *et al.*, 1998 for details). The format is standard and provides three orthogonal projections in the standard space conforming to that described in Talairach and Tournoux (1988). The grey scale is arbitrary and the SPM has been thresholded to show the 128 most significant voxels.

empirical estimates, into the objective function that was minimized.

RESULTS AND DISCUSSION

The model-based and empirical kernels for the first voxel are shown in Figure 27.5. It can be seen that there is a remarkable agreement in terms of the first- and second-order kernels. This is important because it suggests that the non-linearities inherent in the Balloon component of the haemodynamic model are sufficient to account for the non-linear responses observed in real time-series. The first-order kernel corresponds to the conventional haemodynamic response function and shows the characteristic peak at about 4 s and the post-stimulus undershoot. The empirical undershoot appears more protracted than the model's prediction, suggesting that the model is not perfect in every respect. The second-order kernel has a pronounced negativity on the upper left, flanked by two smaller positivities. This negativity accounts for the refractoriness seen when two stimuli are temporally proximate; from the perspective of the Balloon model, the second stimulus is compromised in terms of elaborating a BOLD signal, because of the venous pooling, and consequent dilution of deoxyhaemoglobin incurred by the first stimulus. This means that less deoxyhaemoglobin can be cleared for a given increase in flow. More interesting are the positive regions, which suggest stimuli separated by about 8 s should show super-additive effects. This can be attributed to the fact that, during the flow undershoot following the first stimulus, deoxyhaemoglobin concentration is greater than normal (see the upper right panel in Figure 27.3), thereby facilitating clearance of deoxyhaemoglobin following the second stimulus.

Figure 27.6 shows the various functions implied by the haemodynamic model parameters averaged over all

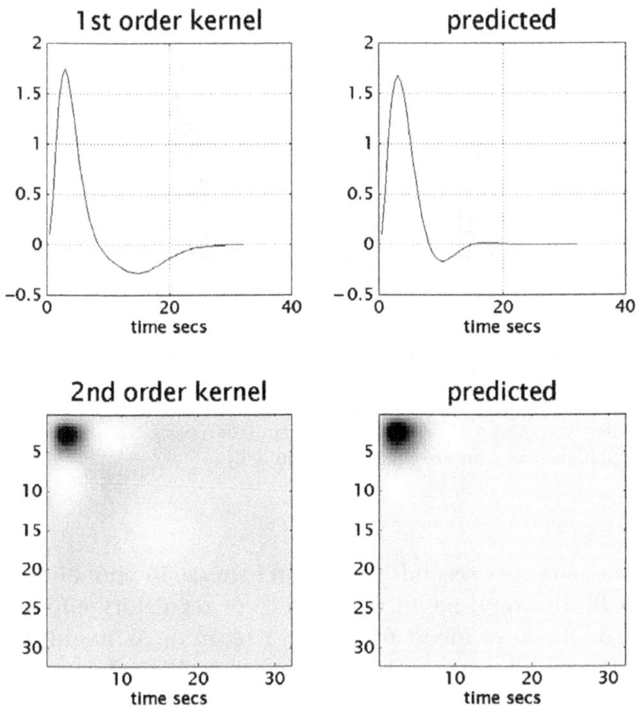

FIGURE 27.5 The first- and second-order Volterra kernels based on parameter estimates from a voxel in the left superior temporal gyrus at −56, −28, 12 mm. These kernels can be thought of as a second-order haemodynamic response function. The first-order kernels (upper panels) represent the (first-order) component usually presented in linear analyses. The second-order kernels (lower panels) are presented in image format. The colour scale is arbitrary; white is positive and black is negative. The left-hand panels are kernels based on parameter estimates from the analysis described in Figure 27.4. The right-hand panels are the kernels associated with the haemodynamic model using parameter estimates that best match the empirical kernels.

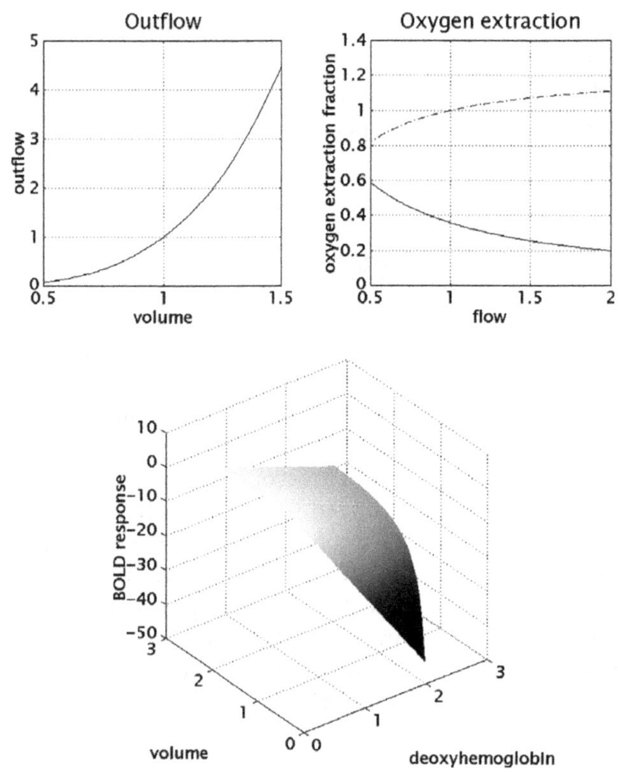

FIGURE 27.6 Functions implied by the (mean) haemodynamic model parameters over the voxels shown in Figure 27.4. Upper left panel: outflow as a function of venous volume $f_{out}(v)$. Upper right panel: oxygen extraction as a function of inflow. The solid line is extraction *per se* $E(f)$ and the broken line is the net normalized delivery of deoxyhaemoglobin to the venous compartment $fE(f)/E_0$. Lower panel: this is a plot of the non-linear function of volume and deoxyhaemoglobin that represents BOLD signal $y(t) = g(v, q)$.

voxels. These include outflow as a function of venous volume $f_{out}(v)$ and oxygen extraction fraction as a function of inflow. The solid line in the upper right panel is extraction $E(f)$ and the broken line is the net normalized delivery of deoxyhaemoglobin to the venous compartment $fE(f)/E_0$. Note that, although the fraction of oxygen extracted decreases with flow, the net delivery of deoxygenated haemoglobin increases with flow. In other words, flow increases *per se* actually reduce signal. It is only the secondary effects of flow on dilution and volume-dependent outflow that cause an increase in BOLD signal. The lower panel depicts the non-linear function of volume and deoxyhaemoglobin that represents BOLD signal $y(t) = g(v, q)$. Here one observes that positive BOLD signals are expressed only when deoxyhaemoglobin is low. The effect of volume is much less marked and tends to affect signal predominantly through dilution.

The distributions of the parameters over voxels are shown in Figure 27.7 with their mean in brackets at the top of each panel. It should be noted that the data from which these estimates came were not independent. However, given they came from four different brain regions they are remarkably consistent. In the next section we will discuss each of these parameters and the effect it exerts on the BOLD response.

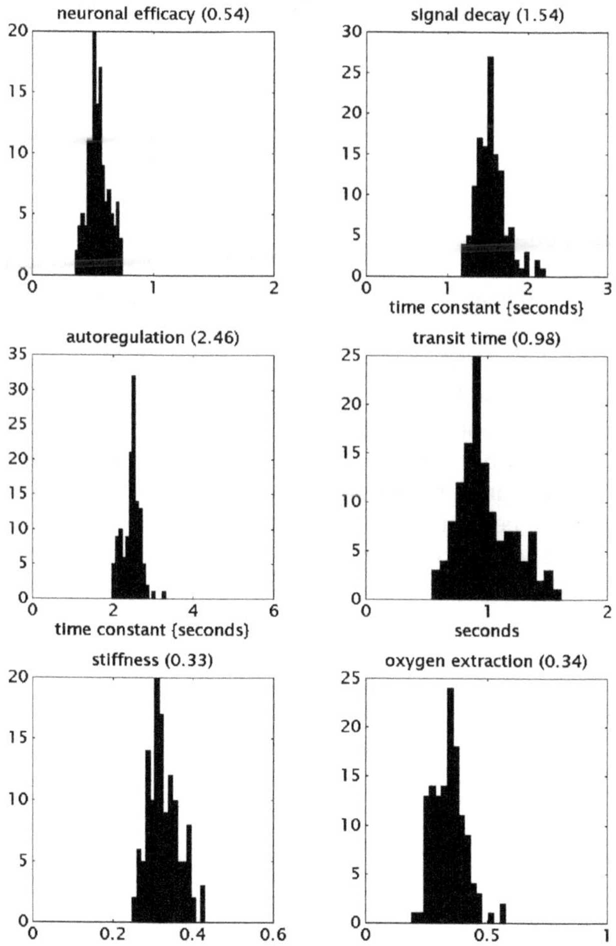

FIGURE 27.7 Histograms of the distribution of the six free parameters of the haemodynamic model estimated over the voxels shown in Figure 27.4. The number in brackets at the top of each histogram is the mean value for the parameters in question: neuronal efficacy is ε, signal decay is τ_s, autoregulation is τ_f, transit time is τ_0, stiffness is α and oxygen extraction is E_0.

DISCUSSION

The main point is that the Balloon model, suitably extended to incorporate the dynamics of rCBF induction by synaptic activity, can reproduce the same form of Volterra kernels that are seen empirically. As such, the Balloon model is sufficient to account for the more important non-linearities observed in evoked fMRI responses. The remainder of this section deals with the validity of the haemodynamic model in terms of the plausibility of the parameter estimates from the previous section. The role of each parameter in shaping the haemodynamic response is illustrated in the associated panel in Figure 27.8 and is discussed in the following subsections.

Neuronal efficacy

This represents the increase in signal elicited by neuronal activity, expressed in terms of event density (i.e. number of evoked transients per second). From a biophysical perspective it is not very interesting because it reflects both the potency of the stimulus in eliciting a neuronal response and the efficacy of the ensuing synaptic activity to induce the signal. It is interesting to note, however, that one word per second invokes an increase in normalized rCBF of unity (i.e. in the absence of regulatory effects, a doubling of blood flow over a second). As might be expected, changes in this parameter simply modulate the evoked haemodynamic responses (see the first panel in Figure 27.8).

Signal decay

This parameter reflects signal decay or elimination. Transduction of neuronal activity into perfusion changes, over a few $100\,\mu$, has a substantial neurogenic component (that may be augmented by electrical conduction along the vascular endothelium). However, at spatial scales of several millimetres it is likely that rapidly diffusing spatial signals mediate increases in rCBF through relaxation

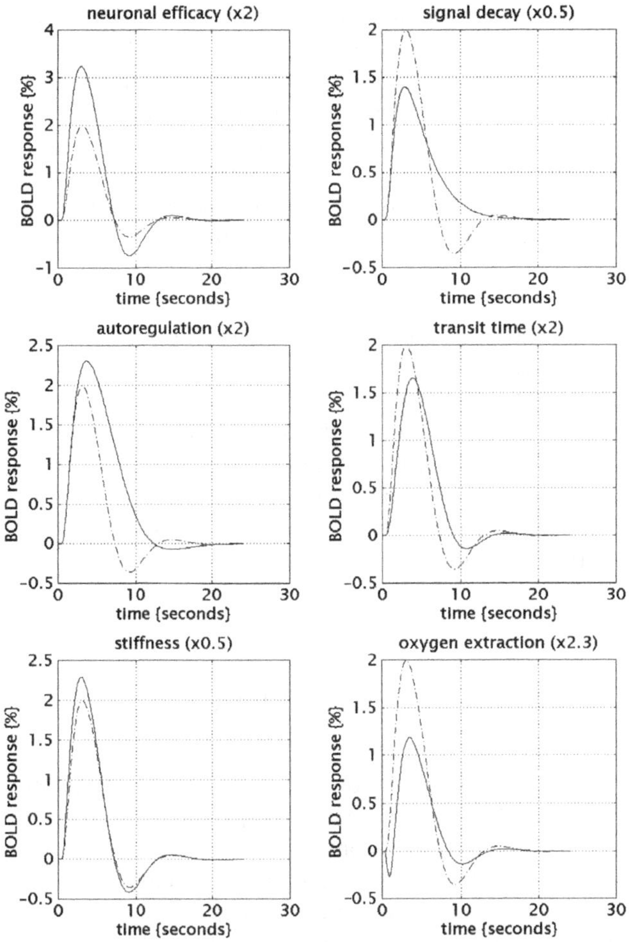

FIGURE 27.8 The effects of changing the model parameters on the evoked BOLD response. The number in brackets at the top of each graph is the factor applied to the parameter in question. Solid lines correspond to the response after changing the parameter and the broken line is the response for the original parameter values (the mean values given in Figure 27.7): neuronal efficacy is ε, signal decay is τ_s, autoregulation is τ_f, transit time is τ_0, stiffness is α and oxygen extraction is E_0.

of arteriolar smooth muscle. There are a number of candidates for this signal, nitric oxide (NO) being the primary one. It has been shown that the rate of elimination is critical in determining the effective time-constants of haemodynamic transduction (Friston, 1995). Our decay parameter had a mean of about 1.54 s giving a half-life of 1067 ms. The half-life of NO is between 100 and 1000 ms (Paulson and Newman, 1987), whereas that of potassium ions is about 5 s. Our results are therefore consistent with spatial signalling with NO. It should be remembered that the model signal subsumes all the actual signalling mechanisms employed by the real brain. Increases in this parameter dampen the rCBF response to any input and will also suppress the undershoot (see next subsection) because the feedback mechanisms, that are largely responsible for the undershoot, are selectively suppressed (relative to just reducing neuronal efficacy during signal induction).

Autoregulation

This parameter is the time-constant of the feedback autoregulatory mechanisms, whose physiological nature remain unspecified (but see Irikura *et al.*, 1994). The coupled differential equations Eqn. **27.6** and Eqn. **27.7** represent a damped oscillator with a resonance frequency of $1/(2\pi\sqrt{\tau_f} = 0.101$ per second. This is exactly the frequency of the vasomotor signal that typically has a period of about 10 s. This is a pleasing result that emerges spontaneously from the parameter estimation. The nature of these oscillations can be revealed by increasing the signal decay time-constant (i.e. reducing the dampening) and presenting the model with low-level random neuronal input (uncorrelated Gaussian noise with a standard deviation of 1/64) as shown in Figure 27.9. The characteristic oscillatory dynamics are readily expressed. The effect of increasing the feedback time-constant is to decrease the resonance frequency and render the BOLD (and rCBF) response more enduring with a reduction or elimination of the undershoot. The third panel in Figure 27.8 shows the effect of doubling it.

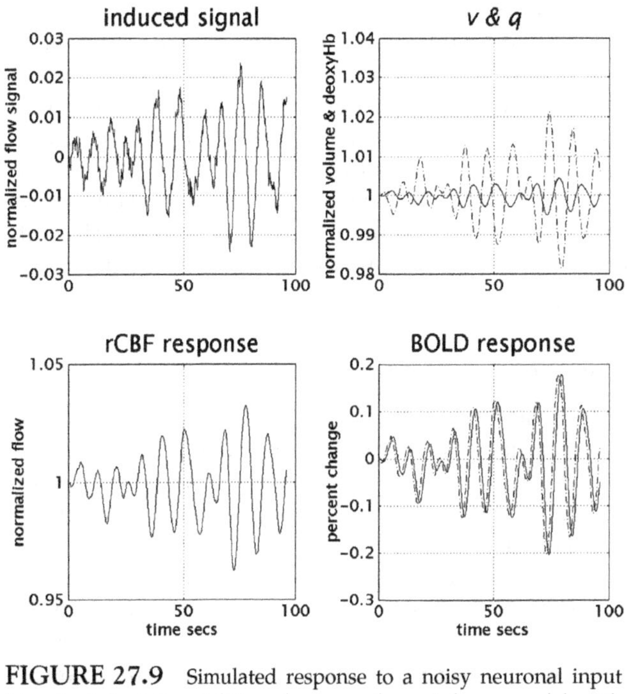

FIGURE 27.9 Simulated response to a noisy neuronal input (standard deviation 1/64 and mean of zero) for a model with decreased signal decay (i.e. less dampening). The model parameters were the same as in Figure 27.3 with the exception of τ_s which was increased by a factor of four. The characteristic 0.1 Hz oscillations are very similar to the oscillatory vasomotor signal seen in optical imaging experiments and low-frequency fluctuations studied in fMRI (e.g. Stanberry *et al.*, 2006).

Transit time

This is an important parameter that determines the dynamics of the signal. It is effectively resting venous volume divided by resting flow and, in our data, is estimated at about 1 s (0.98 s). The transit time through the rat brain is roughly 1.4 s at rest and, according to the asymptotic projections for rCBF and volume, falls to 0.73 s during stimulation (Mandeville *et al.*, 1999). In other words, it takes about a second for a blood cell to traverse the venous compartment. The effect of increasing mean transit time is to slow down the dynamics of the BOLD signal with respect to the flow changes. The shape of the response remains the same but it is expressed more slowly. In the fourth panel of Figure 27.8 a doubling of the mean transit time is seen to retard the peak BOLD response by about a second and the undershoot by about 2 s.

Stiffness parameter

Under steady state conditions this would be about 0.38. The mean over voxels considered above was about 0.33. This discrepancy, in relation to steady state levels, is anticipated by the Windkessel formulation and is attributable to the fact that volume and flow are in a state of continuous flux during the evoked responses. Recall from Eqn. **27.3** that $1/\alpha = \gamma + \beta = 3.03$, in our data. Under the assumption of laminar flow, $\gamma = 2 \Rightarrow \beta \approx 1$, which is less than Mandeville *et al.* (1999) found for rats during forepaw stimulation, but is certainly in a plausible range. Increasing this parameter increases the degree of non-linearity in the flow-volume behaviour of the venous balloon that underpins the non-linear behaviour we are trying to account for. However, its direct effect on evoked responses to single stimuli is not very marked. The fifth panel of Figure 27.8 shows the effects when it is decreased by 50 per cent.

Resting oxygen extraction

This is about 34 per cent and the range observed in our data fits exactly with known values for resting oxygen extraction fraction (between 20 and 55 per cent). Oxygen extraction fraction is a potentially important factor in determining the nature of evoked fMRI responses because it may be sensitive to the nature of the baseline that defines the resting state. Increases in this parameter can have quite profound effects on the shape of the response that bias it towards an early dip. In the example shown (last panel in Figure 27.8) the resting extraction has been increased to 78 per cent. This is a potentially important observation that may explain why the initial dip has been difficult to observe in all studies. According to the results presented in Figure 27.8, the initial dip is very sensitive to resting oxygen extraction fraction, which should be high before the dip is expressed. Extraction fraction will be high in regions with very low blood flow, or in tissue with endogenously high extraction. It may be that cytochrome oxidase rich cortex, like the visual cortices, has a higher fraction and is more likely to evidence early dips.

In summary, the parameters of the haemodynamic model that best reproduce empirically derived Volterra kernels are all biologically plausible and lend the model a construct validity (in relation to the Volterra formulation) and face validity (in relation to other physiological characterizations of the cerebral haemodynamics reviewed here). In this extended haemodynamic model, non-linearities, inherent in the Balloon model, and output non-linearity have been related directly to non-linearities in responses. Their role in mediating the post-stimulus undershoot is emphasized less here because the rCBF component cannot model undershoots.

CONCLUSION

In conclusion, we have described an input-state-output model of the haemodynamic response to changes in synaptic activity that combines the Balloon model of flow to BOLD signal coupling and a dynamical model of the transduction of neuronal activity into perfusion changes. This model has been characterized in terms of its Volterra kernels and easily reproduces empirical kernels with parameters that are biologically plausible. This means that the non-linearities inherent in the Balloon model are sufficient to account for haemodynamic refractoriness and other non-linear aspects of evoked responses in fMRI.

In the next chapter, we consider the mapping from neuronal activity to measurements made with EEG and MEG. Here the dynamics are less important, but the spatial aspect of the forward model becomes much more complicated.

REFERENCES

Attwell D, Laughlin SB (2001) An energy budget for signaling in the grey matter of the brain. *J Cereb Blood Flow Metab* **21**: 1133–45

Buxton RB, Frank LR (1997) A model for the coupling between cerebral blood flow and oxygen metabolism during neural stimulation. *J Cereb Blood Flow Metab* **17**: 64–72

Buxton RB, Wong EC, Frank LR (1998) Dynamics of blood flow and oxygenation changes during brain activation: the Balloon model. *Mag Res Med* **39**: 855–64

Fox PT, Raichle ME (1986) Focal physiological uncoupling of cerebral blood flow and oxidative metabolism during somatosensory stimulation in human subjects. *Proc Natl Acad Sci USA* **83**: 1140–44

Friston KJ (1995) Regulation of rCBF by diffusible signals: an analysis of constraints on diffusion and elimination. *Hum Brain Mapp* **3**: 56–65

Friston KJ, Jezzard P, Turner R (1994) Analysis of functional MRI time series. *Hum Brain Mapp* **1**: 153–71

Friston KJ, Josephs O, Rees G *et al.* (1998) Nonlinear event-related responses in fMRI. *Mag Res Med* **39**: 41–52

Friston KJ, Mechelli A, Turner R *et al.* (2000) Nonlinear responses in fMRI: the Balloon model, Volterra kernels, and other hemodynamics. *NeuroImage* **12**: 466–77

Grill-Spector K, Malach R (2001) fMR-adaptation: a tool for studying the functional properties of human cortical neurons. *Acta Psychol (Amst)* **107**: 293–321

Grubb RL, Rachael ME, Euchring JO *et al.* (1974) The effects of changes in PCO_2 on cerebral blood volume, blood flow and vascular mean transit time. *Stroke* **5**: 630–39

Heeger DJ, Ress D (2002) What does fMRI tell us about neuronal activity? *Nat Rev Neurosci* **3**: 142–51

Hoge RD, Atkinson J, Gill B *et al.* (1999) Linear coupling between cerebral blood flow and oxygen consumption in activated human cortex. *Proc Natl Acad Sci USA* **96**: 9403–08

Irikura K, Maynard KI, Moskowitz MA (1994) Importance of nitric oxide synthase inhibition to the attenuated vascular responses induced by topical l-nitro-arginine during vibrissal stimulation. *J Cereb Blood Flow Metab* **14**: 45–48

Kennedy C, Des Rosiers MH, Sakurada O *et al.* (1976) Metabolic mapping of the primary visual system of the monkey by means of the autoradiographic [14C]deoxyglucose technique. *Proc Natl Acad Sci USA* **73**: 4230–34

Lindauer U, Royl G, Leithner C *et al.* (2001) No evidence for early decrease in blood oxygenation in rat whisker cortex in response to functional activation. *NeuroImage* **13**: 988–1001

Logothetis NK, Pauls J, Augath MT *et al.* (2001) Neurophysiological investigation of the basis of the fMRI signal. *Nature* **412**: 150–57

Mandeville JB, Marota JJ, Ayata C *et al.* (1999) Evidence of a cerebrovascular postarteriole Windkessel with delayed compliance. *J Cereb Blood Flow Metab* **19**: 679–89

Mayhew J, Hu D, Zheng Y *et al.* (1998) An evaluation of linear models analysis techniques for processing images of microcirculation activity. *NeuroImage* **7**: 49–71

Miller KL, Luh WM, Liu TT *et al.* (2000) Characterizing the dynamic perfusion response to stimuli of short duration. *Proc ISRM* **8**: 580

Paulson OB, Newman EA (1987) Does the release of potassium from astrocyte endfeet regulate cerebral blood? *Science* **237**: 896–98

Roy CS, Sherrington CS (1890) On the regulation of the blood supply of the brain. *J Physiol Lond* **11**: 85–108

Stanberry LI, Richards TL, Berninger VW *et al.* (2006). Low-frequency signal changes reflect differences in functional connectivity between good readers and dyslexics during continuous phoneme mapping. *Mag Res Imag* **24**: 217–29

Talairach J, Tournoux P (1988) *A co-planar stereotaxic atlas of a human brain*. Thieme, Stuttgart

Vanzetta I, Grinvald A (1999) Increased cortical oxidative metabolism due to sensory stimulation: implications for functional brain imaging. *Science* **286**: 1555–58

Vazquez AL, Noll DC (1996) Non-linear temporal aspects of the BOLD response in fMRI. *Proc Int Soc Mag Res Med* **3**: S1765

Worsley KJ, Friston KJ (1995) Analysis of fMRI time-series revisited – again. *NeuroImage* **2**: 173–81

Zheng Y, Martindale J, Johnston D *et al.* (2002) A model of the hemodynamic response and oxygen delivery to brain. *NeuroImage* **16**: 617–37

C H A P T E R

28

Forward models for EEG

C. Phillips, J. Mattout and K. Friston

INTRODUCTION

In this chapter, we turn to forward models for electrical and magnetic measurements **v** of neuronal activity **j**. The form of such models is very simple,

$$\mathbf{v} = \mathbf{L}\mathbf{j} \quad 28.1$$

where **L** is called a lead field. We will refer to the lead field repeatedly in subsequent chapters, where it is treated as a simple linear mapping. However, as we will see below, it entails a substantial amount of theory and forward modelling: electroencephalography (EEG) and magnetoencephalography (MEG) involves recording the electric potential (or magnetic field) produced by neuronal activity, at the surface of the scalp. The advantage of M/EEG lies in its excellent temporal resolution, as activity is seen in 'real time'. However, non-invasive measurements of the electromagnetic field can only be obtained from a limited number of sensors placed on, or around, the scalp.

In EEG (and MEG), the *forward problem* entails computing the electromagnetic field at the scalp given a source configuration and volume conductor. The *inverse problem* describes the opposite problem: given the volume conductor and the electromagnetic field at the scalp, what is the location and time course of the sources? The solution of the forward problem, even if approximate, is required to solve the inverse problem. The forward problem itself is a 'simple electromagnetic problem' that can be expressed and solved with Maxwell's equations. The difficult aspect lies in modelling the volume conductor, a human head, and the sources, the neuronal activity of the brain. To do this, it is necessary to understand cerebral anatomy and the nature of neuronal activity.

To be detected, neural activity must sum coherently. Because of their very short time course, synchronous firing of action potentials, APs, is unlikely but the longer time course of postsynaptic potentials (PSPs) allows them to superpose temporally. Moreover the electromagnetic field generated by a dipolar source (like a PSP) decreases with distance – approximately as $1/r^2$, more slowly than the $1/r^3$-dependent field of a quadrupolar source (like an AP). Therefore, even though APs are much larger in amplitude than PSPs, it is generally accepted that postsynaptic potentials are the generators of scalp fields usually recorded in EEG and MEG (Nunez, 1981; Hämäläinen *et al.*, 1993). If the dendrites supporting PSPs are oriented randomly or radially on a complete spherical surface (or small closed surface), no net electromagnetic field can be detected outside the immediate vicinity of the active neurons; this is called a 'closed field' configuration. Because of the uniform spatial organization of their dendrites (perpendicular to the cortical surface), the pyramidal cells are the only neurons that can generate a net macroscopic current dipole over the cortical surface, whose field is detectable on the scalp. This is named an 'open field' configuration. The brain's electrical activity is thus generally modelled as small current dipoles, located in the grey matter.

The overall structure of the head is rather complicated. The brain, skull, scalp and other parts of the head (eyes, vessels, nerves, cerebrospinal fluid, etc.) comprise various tissues and cavities of different electrical conductivity. Moreover, the electrical conductivity of brain tissue is highly anisotropic: in the white matter, conduction is 10 times greater along an axon fibre than in the transverse direction. These complications are generally ignored; at the present time, it is impossible to measure accurately *in vivo* the detailed conductivity of all tissues and to account for them in the solution of the forward problem (Marin *et al.*, 1998; Huiskamp *et al.*, 1999). Therefore, the head is usually modelled as a set of concentric homogeneous volume conductors: the brain (comprising the white and grey matter), the skull and the scalp. Once the relationship between brain electrical activity and electromagnetic

Statistical Parametric Mapping, by Karl Friston et al.
ISBN–13: 978-0-12-372560-8 ISBN–10: 0-12-372560-7

scalp fields has been established, the inverse problem can be addressed. Non-invasive measurements of the electromagnetic field can only be obtained from the scalp surface, and the spatial configuration of neuronal activity *cannot* be determined uniquely if based on EEG and/or MEG recordings alone (von Helmholtz, 1853; Nunez, 1981).

In summary, estimates of electromagnetic activity depend on cortical anatomy: cellular-level structures determine how neuronal electrical activity produces macroscopic current sources detectable outside the head, the local electrical conductivity influences the solution of the forward problem, and the anatomy of the brain at the sulcal level can be used to constrain the inverse solution.

ANALYTICAL FORMULATION

Maxwell's equations

For the forward problem, the electromagnetic properties of the head (or at least its simplified model) and the location of the electric current generators in the brain are assumed to be known. Maxwell's equations can then be used to calculate the electric (and magnetic) field on the surface of the scalp or inside the head volume if necessary. In differential equation form, Maxwell's equations can be expressed as (Ramo *et al.*, 1984):

$$\vec{\nabla}\vec{E} = \frac{\rho}{\epsilon} \quad \textbf{28.2a}$$

$$\vec{\nabla}\vec{B} = 0 \quad \textbf{28.2b}$$

$$\vec{\nabla} \times \vec{E} = -\frac{\partial \vec{B}}{\partial t} \quad \textbf{28.2c}$$

$$\vec{\nabla} \times \vec{B} = \mu\vec{j} + \mu\epsilon\frac{\partial \vec{E}}{\partial t} \quad \textbf{28.2d}$$

where $\vec{E}$ is the electric field, $\vec{B}$ is the magnetic field, $\vec{j}$ is the current density, ρ is the charge density, ϵ is the electric permittivity, and μ is the magnetic permeability.

These are the basic set of equations of classical electromagnetism. Together with two auxiliary relations, Ohm's law $\vec{j} = \sigma\vec{E}$ and the continuity equation $\vec{\nabla}\vec{j} = \partial\rho/\partial t$, they describe all electromagnetic phenomena. In the case of bioelectric phenomena such as electro- and magnetoencephalography (M/EEG), we are only interested in the electric (and magnetic) field $\vec{E}$ (and $\vec{B}$). The treatment of Maxwell's equations can be simplified significantly by noting that the media comprising the head have no significant capacitance: they are either purely resistive or the frequency of activity is sufficiently low that the capacitance can be neglected. This means there are no electromagnetic wave propagation phenomena (Hämäläinen *et al.*, 1993; He, 1998).

This allows us to adopt a quasistatic approximation of Maxwell's equations, which means that, in the calculation of $\vec{E}$ (and $\vec{B}$), $\partial\vec{E}/\partial t$ and $\partial\vec{B}/\partial t$ can be ignored as source terms. Physically, these assumptions mean that the instantaneous current density depends only on the instantaneous current sources and conforms to the superposition theorem. Eqn. **28.2c** then becomes $\vec{\nabla} \times \vec{E} = 0$ and the electric field $\vec{E}$ can be expressed as the negative gradient of a scalar field, the electric potential V:

$$\vec{E} = -\vec{\nabla}V \quad \textbf{28.3}$$

'Current sources' are, by definition, a distribution of forced current density $\vec{j}_f$. The current sources $\vec{j}_f$ can be seen as the summed coherent electric activity of activated cell membranes, i.e. the current density produced directly by neural activity, with arbitrarily close sink and source currents. The total current density $\vec{j}_{tot}$ flowing through the media is equal to the sum of the imposed sources $\vec{j}_f$ and the return current $\vec{j}_r$. The latter is the result of the macroscopic electric field on charge carriers in the conducting medium, as expressed by Ohm's law. With Ohm's law and **28.3**, we have:

$$\vec{j}_{tot} = \vec{j}_r + \vec{j}_f = \sigma\vec{E} + \vec{j}_f = -\sigma\vec{\nabla}V + \vec{j}_f \quad \textbf{28.4}$$

By neglecting the capacitance of head tissues, charge does not accumulate in the volume or on tissue interfaces, i.e. the charges are redistributed in negligible time. This translates mathematically into zero divergence of the current density $\vec{\nabla}\vec{j}_{tot} = 0$. And, by taking the divergence of **28.4**, we obtain the simplified Maxwell's equation:

$$\vec{\nabla}(\sigma\vec{\nabla}V) = \vec{\nabla}\vec{j}_f \quad \textbf{28.5}$$

This equation is at the heart of the forward problem in EEG: it links explicitly the current sources $\vec{j}_f$ and the electric potential V.

Equation **28.5** can be solved for V in various ways, depending on the geometry of the model, the form of the conductivity σ and the location of the sources $\vec{j}_f$. An analytical solution is possible only for particular cases: a highly symmetrical geometry (e.g. concentric spheres or cubes) and homogeneous isotropic conductivity. For other more general cases, numerical methods are required. Common methods include the 'finite element method' (FEM) (Chari and Silvester, 1980; van den Broek *et al.*, 1996; Awada *et al.*, 1997; Buchner *et al.*, 1997; Klepfer *et al.*, 1997), the 'finite difference method' (FDM) (Rosenfeld *et al.*, 1996; Saleheen and Ng, 1997, 1998) and the 'boundary element method' (BEM) (Becker, 1992; Ferguson and Stroink, 1997; Mosher *et al.*, 1999).

The FEM and FDM make no assumptions about the shape of the volume conductivity and allow the estimation of V at any location in the volume. The volume is tessellated into small volume elements in which Maxwell's equations are solved locally. As each volume element is characterized by its own conductivity (isotropic or not), any configuration of conductive volume can be modelled. With the FEM, the volume elements are of arbitrary shape (usually tetrahedron or regular polyhedron), while the volume elements are cubic with the FDM. In contrast, the BEM is based on the hypothesis that the volume is divided into subvolumes of homogeneous and isotropic conductivity, and the potential V is estimated only on the surfaces separating these subvolumes.

The FEM and FDM offer a more general solution of the forward problem but the complexity of the numerical problem and the computing time needed to solve it are increased greatly, compared with the BEM. Moreover, the conductivity throughout the volume, necessary to make full use of the FEM and FDM, cannot be estimated for individual subjects. Therefore, a simple 'three sphere shell' model with an analytical solution is still used generally for M/EEG source localization. A more anatomically realistic but still tractable approach uses the BEM to solve the forward problem. We will focus on the BEM from now on.

The BEM approach for EEG

One way of solving Eqn. **28.5** is the boundary element method, which relies on one major assumption: the volume is divided into subvolumes of *homogeneous* and *isotropic* conductivity. The solution of the forward problem then can be obtained with the help of some boundary conditions and Green's theorem.

Boundary conditions

Consider S_l the surface separating two volumes, vol^- and vol^+, of conductivity σ^- and σ^+. Let's define dS_l an infinitesimal element of this surface and let $\vec{n}$ be the unit vector normal to the surface oriented from the inside towards the outside or, by convention, from vol^- to vol^+, as shown in Figure 28.1.

There are no sources located on the surfaces between homogeneous volumes and the current normal to the surfaces is continuous, so on surface S_l, we have $\vec{j}_{tot} = -\sigma\vec{\nabla}V$ and

$$\sigma^-\vec{\nabla}V^-\,\vec{dS_l} = \sigma^+\vec{\nabla}V^+\,\vec{dS_l} \qquad \textbf{28.6}$$

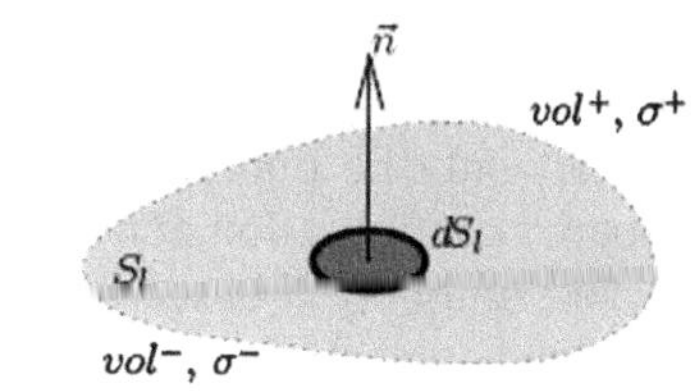

FIGURE 28.1 The surface S_l separates the two homogeneous volumes vol^- and vol^+ of isotropic conductivity σ^- and σ^+. dS_l is an infinitesimal element of the surface S_l and $\vec{n}$ is the unit vector normal to S_l, oriented from vol^- to vol^+.

where $\vec{ds_l} = \vec{n}\,dS_l$ is the *oriented* infinitesimal element of this surface. Moreover, the potential V must also be continuous on S_l:

$$V^-(S_l) = V^+(S_l) \qquad \textbf{28.7}$$

The Eqns **28.6** and **28.7** provide the two boundary conditions necessary to solve the quasistatic form of Maxwell's equations as expressed in Eqn. **28.5**.

Green's theorem

Let dv_k be an element of the homogeneous regional volume vol_k (where $k = 1, \ldots, N_v$ and N_v is the number of homogeneous volumes) and let $\vec{dS_l}$ be an oriented element, $\vec{ds_l} = \vec{n}\,dS_l$, of the surface S_l separating two regions of homogeneous conductivity (where $l = 1, \ldots, N_S$ and N_S is the number of such surfaces). Take two well-behaved functions ψ and ϕ in each region vol_k; then Green's theorem states (Smythe, 1950):

$$\sum_l^{N_S}\int_{S_l}\left[\sigma_l^-(\psi^-\vec{\nabla}\phi^- - \phi^-\vec{\nabla}\psi^-) - [\sigma_l^+(\psi^+\vec{\nabla}\phi^+ - \phi^+\vec{\nabla}\psi^+)\right]\vec{dS_l} \qquad \textbf{28.8}$$

$$= \sum_k^{N_v}\int_{vol_k}\left[\psi\vec{\nabla}(\sigma_k\vec{\nabla}\phi) - \phi\vec{\nabla}(\sigma_k\vec{\nabla}\psi)\right]dv_k$$

where the sums run over the N_v volumes and the N_S surfaces, and the symbols $-$ and $+$ refer to the volumes inside and outside surface S_l.

The forward problem entails evaluating the electric potential V from the current sources distribution $\vec{j}_f$ with the quasistatic form of Maxwell's equation, Eqn. **28.5** and the boundary conditions Eqn. **28.6** and Eqn. **28.7**, using Green's theorem, Eqn. **28.8**.

Analytical BEM equation

In Eqn. **28.8**, if we take $\psi = 1/r$, where r is the distance between an arbitrary point $\vec{r}$ and the origin $\vec{o}$, then, for smooth surfaces, Eqn. **28.8** becomes, as shown by Geselowitz (1967):

$$\sum_{l}^{N_S} \int_{S_l} \left[(\sigma_l^- \vec{\nabla}\phi^- - \sigma_l^+ \vec{\nabla}\phi^+) \frac{1}{r} - (\sigma_l^- \phi^- - \sigma_l^+ \phi^+) \vec{\nabla}\left(\frac{1}{r}\right) \right] \vec{d}S_l \quad \textbf{28.9}$$

$$= 4\pi\sigma\phi + \sum_{k}^{N_v} \int_{vol_k} \frac{1}{r} \vec{\nabla}\left(\sigma_k \vec{\nabla}\phi\right) dv_k$$

where, σ and ϕ in the first term of the right hand side, are evaluated at the origin $\vec{o}$, i.e. for $r = 0$. Consider $\phi = V$ and suppose that $\vec{j}_f$ is distributed in only one homogeneous volume, the brain volume vol_{br}; then, thanks to the simplified Maxwell's equation, Eqn. **28.5** and the boundary conditions, Eqn. **28.6** and Eqn. **28.7**, Eqn. **28.9** becomes:

$$-\sum_{l}^{N_S} (\sigma_l^- - \sigma_l^+) \int_{S_l} V \vec{\nabla}\left(\frac{1}{r}\right) \vec{d}S_l = 4\pi\sigma V + \int_{vol_{br}} \frac{1}{r} \vec{\nabla}\vec{j}_f \, dv_{br} \quad \textbf{28.10}$$

By the divergence theorem and the definition of the divergence operator $\vec{\nabla}$, we have: $\int_{vol} \vec{\nabla}\left(\frac{\vec{j}_f}{r}\right) dv = \int_S \frac{\vec{j}_f}{r} \vec{d}S = \int_{vol} \left[\vec{j}_f \vec{\nabla}\left(\frac{1}{r}\right) + \frac{1}{r}\vec{\nabla}\vec{j}_f\right] dv$. And, as there are no sources $\vec{j}_f$ on any surface S_l, $\vec{j}_f(S_l) = 0$, then $\int_{vol} \frac{1}{r}\vec{\nabla}\vec{j}_f \, dv = -\int_{vol} \vec{j}_f \vec{\nabla}\left(\frac{1}{r}\right) dv$. Therefore, Eqn. **28.10** becomes:

$$4\pi\sigma V = \int_{vol_{br}} \vec{j}_f \vec{\nabla}\left(\frac{1}{r}\right) dv_{br} - \sum_{l}^{N_S} (\sigma_l^- - \sigma_l^+) \int_{S_l} V \vec{\nabla}\left(\frac{1}{r}\right) \vec{d}S_l \quad \textbf{28.11}$$

On the left hand side of Eqn. **28.11**, V is still evaluated at the origin $\vec{o}$ of space (an arbitrary point) and r is the distance from that origin to a point on the surface S_l (in the surface integrals) or in the volume vol_{br} (in the volume integral).

Let us consider $\vec{x}$ the point where V is evaluated, $\vec{s}'$ a point on the surface S_l and $\vec{r}'$ a point in the volume vol_{br}. The distance r between the point $\vec{x}$ where V is evaluated and any point $\vec{s}'$ on the surface S_l' (or $\vec{r}'$ in the volume vol_{br}) will be expressed by $|\vec{x} - \vec{s}'|$ (or $|\vec{x} - \vec{r}'|$). Thus Eqn. **28.11** is rewritten (Sarvas, 1987) like this:

$$4\pi\sigma(\vec{x})V(\vec{x}) = \int_{Vol_{br}} \vec{j}_f(\vec{r}')\vec{\nabla}'\left(\frac{1}{|\vec{x} - \vec{r}'|}\right) dv_{br}$$
$$- \sum_{l}^{N_S} (\sigma_l^- - \sigma_l^+) \int_{S_l'} V(\vec{s}')\vec{\nabla}'\left(\frac{1}{|\vec{x} - \vec{s}'|}\right) \vec{d}S_l' \quad \textbf{28.12}$$

where $\vec{\nabla}'$ means that the gradient is with respect to $\vec{r}'$ or $\vec{s}'$. The potential V should be evaluated on the surfaces S_l but, if $\vec{x}$ approaches the point $\vec{s}$ on a surface S_k, the kth surface integral becomes singular in Eqn. **28.12**.

Consider now $F_k(\vec{x})$ the integral on the smooth surface S_k in Eqn. **28.12**, $F_k(\vec{x}) = \int_{S_k'} V(\vec{s}')\vec{\nabla}(1/|\vec{x} - \vec{s}'|)\vec{d}S_k'$; then it follows from Vladimirov (1971):

$$\lim_{\vec{x} \to \vec{s}} F_k(\vec{x}) = -2\pi V(\vec{s}) + F_k(\vec{s}) \quad \textbf{28.13}$$

where $\vec{x}$ approaches the point $\vec{s}$ on the surface S_k from the inside, so in the volume of conductivity σ_k^-. With Eqn. **28.13**, the point $\vec{x}$ can be placed on any surface S_k and, with $\vec{x} \to \vec{s}$, Eqn. **28.12** becomes:

$$4\pi\sigma_k^- V(\vec{s}) = \int_{Vol_{br}} \vec{j}_f(\vec{r}')\vec{\nabla}'\left(\frac{1}{|\vec{s} - \vec{r}'|}\right) dv_{br}$$
$$- \sum_{l}^{N_S} (\sigma_l^- - \sigma_l^+) \int_{S_l'} V(\vec{s}')\vec{\nabla}'\left(\frac{1}{|\vec{s} - \vec{s}'|}\right) \vec{d}S_l'$$
$$+ 2\pi(\sigma_k^- - \sigma_k^+)V(\vec{s}) \quad \textbf{28.14}$$

and eventually:

$$V(\vec{s}) = V_\infty(\vec{s}) - \frac{1}{2\pi} \sum_{l}^{N_S} \frac{\sigma_l^- - \sigma_l^+}{\sigma_k^- + \sigma_k^+} \int_{S_l'} V(\vec{s}')\vec{\nabla}'\left(\frac{1}{|\vec{s} - \vec{s}'|}\right) \vec{n}(\vec{s}')\, dS_l' \quad \textbf{28.15}$$

where $V_\infty(\vec{s})$ is the potential due to $\vec{j}_f$ in a conductor of infinite extent and homogeneous conductivity $(\sigma_k^- + \sigma_k^+)/2$:

$$V_\infty(\vec{s}) = \frac{1}{2\pi(\sigma_k^- + \sigma_k^+)} \int_v \vec{j}_f(\vec{r}')\vec{\nabla}'\left(\frac{1}{|\vec{s} - \vec{r}'|}\right) dv \quad \textbf{28.16}$$

and

- the sum $\sum_l^{N_S}$ runs over all the surfaces separating volumes of homogeneous isotropic conductivity.
- σ_l^- and σ_l^+ are the conductivity inside and outside the surface S_l.
- $\vec{s}$ and $\vec{s}'$ are points on the surfaces S_k and S_l respectively.
- $\vec{n}(\vec{s}')$ is a unit vector normal to the surface S_l at the point $\vec{s}'$ and oriented from the inside towards the outside of S_l.

This is an explicit relationship between the current sources $\vec{j}_f$ and the surface potential V. As V is present on both sides of this integral equation, V can only be evaluated in closed form for particular geometries, such as concentric spheres. For more general cases, i.e. realistic head models, numerical methods are required to solve this integral equation.

The BEM approach for MEG

Magnetoencephalographic (MEG) signals are not very different from EEG data because they represent complementary effects generated by neuronal activity. Both EEG and MEG are related to the imposed current sources $\vec{j}_f$ by Maxwell's equations, through $\vec{E}$ and V for EEG and $\vec{B}$ for MEG.

As with the derivations above for V (Eqn. **28.15**), an integral form of Maxwell's equation can be derived for the magnetic field $\vec{B}$ (Hämäläinen *et al.*, 1993):

$$\vec{B}(\vec{r}) = \vec{B}_\infty(\vec{r}) + \frac{\mu_0}{4\pi}\sum_l^{N_S}(\sigma_l^- - \sigma_l^+)\int_{S_l'} V(\vec{s}')\frac{\vec{r}-\vec{s}'}{|\vec{r}-\vec{s}'|^3} \times \vec{n}(\vec{s}')\,dS_l' \quad \mathbf{28.17}$$

where

$$\vec{B}_\infty(\vec{r}) = \frac{\mu_0}{4\pi}\int_{vol_{br}} \vec{j}_f(\vec{r}') \times \frac{\vec{r}-\vec{r}'}{|\vec{r}-\vec{r}'|^3}\,dv \quad \mathbf{28.18}$$

Importantly, the potential distribution V over the surfaces is assumed to be known. As with EEG, an analytical solution exists only for highly symmetric models; otherwise, numerical methods are used to solve this equation.

NUMERICAL SOLUTION OF THE BEM EQUATION

The main task now, in solving the BEM forward problem, is to evaluate accurately the integrals on the right hand side of Eqn. **28.15**.

Approximation of the BEM analytical equation

The volume integral over the continuous sources distribution $\vec{j}_f$ can be calculated easily by approximating $\vec{j}_f$ with a superposition of independent point sources of known location and orientation. On the other hand, the surface integrals are more difficult to calculate: they run on different and irregular surfaces and, moreover, they involve the potential $V(\vec{s})$ that we want to solve for. Therefore, it is necessary to express the surface integrals in terms of the value of the unknown function V at some discrete set of points on the surfaces, and to tessellate the surfaces into sets of regular patches. The most obvious approximation for the surfaces is to model each of them by a set of plane triangles. With this surface tessellation, the surface integrals of Eqn. **28.15** can be expressed as a sum of integrals over triangles:

$$V(\vec{s}) = V_\infty(\vec{s}) - \frac{1}{2\pi}\sum_{l=1}^{N_S}\frac{\sigma_l^- - \sigma_l^+}{\sigma_k^- + \sigma_k^+}$$

$$\sum_{m=1}^{N_{tr}^{(l)}}\int_{\Delta_m^{(l)}} V(\vec{s}')\vec{\nabla}'\left(\frac{1}{|\vec{s}-\vec{s}'|}\right)\vec{n}(\vec{s}')\,dS' \quad \mathbf{28.19}$$

where the surface S_l has been modelled by a set of $N_{tr}^{(l)}$ triangles $\Delta_m^{(l)}$. The function V is rendered discrete by choosing on which nodal points V is evaluated and how the function V behaves on each individual plane triangle. This would allow an explicit calculation of the integrals over the triangles and Eqn. **28.19** could eventually be simplified to a sum of known or, at least, easily evaluated functions.

Three different approximations of V over a triangle are usually considered. First, one could evaluate V at the centre of gravity of each triangle and consider this value constant over the triangle: one value is thus obtained for each triangle. This approximation is referred as the 'centre of gravity' (or CoG) method (Hämäläinen and Sarvas, 1989; Meijs *et al.*, 1989). The function V could also be evaluated on the vertices of the triangles; this is generally called a 'vertex' approximation (one value per vertex). If the potential over the triangle is supposed to be constant and equal to the mean of the potential at its vertices, this approximation will be called the 'constant potential at vertices' (or CPV) method (Schlitt *et al.*, 1995). A better approximation would be to consider that the potential varies linearly over the triangle. We will call this approximation the 'linear potential at vertices' (or LPV) method (Schlitt *et al.*, 1995). In what follows, we will consider only the CoG and LPV approximations.

The CoG and LPV approaches differ mainly in the choice of the nodal points where the unknown potential function V is calculated. It is important to note that for a closed tessellated surface there are about twice as many triangles as vertices. The number and arrangement of the triangles determine how well the true surface is approximated spatially. The choice of the potential approximation method determines the number of equations to be solved (one per triangle or vertex) and how well the true potential is modelled over each triangle (constant or linear approximation).

Current source model

In Eqn. **28.16**, the source function $\vec{j}_f(\vec{r})$ is a continuous function throughout the volume. The source function $\vec{j}_f(\vec{r})$ can be approximated by a distribution of N_j independent dipole sources of known location $\vec{r}_i$:

$$\vec{j}_f(\vec{r}) = \sum_{i=1}^{N_j}\vec{j}_f(\vec{r}_i)\delta(\vec{r}-\vec{r}_i) \quad \mathbf{28.20}$$

where $\vec{j}_f(\vec{r}_i) = \int_{v_i}\vec{j}_f(\vec{r}')\,dv_i$ is the summed activity, in the small volume v_i around the location $\vec{r}_i$, modelled as a punctate current source, and $\delta(\vec{r})$ is the discrete Dirac delta function. Now, with this source model, Eqn. **28.16** becomes:

$$V_\infty(\vec{s}) = \frac{1}{2\pi(\sigma_k^- + \sigma_k^+)} \int_v \frac{\vec{s}-\vec{r}'}{|\vec{s}-\vec{r}'|^3} \sum_{i=1}^{N_j} \vec{j}_f(\vec{r}_i)\delta(\vec{r}' - \vec{r}_i)\, dv$$

$$= \frac{1}{2\pi(\sigma_k^- + \sigma_k^+)} \sum_{i=1}^{N_j} \vec{j}_f(\vec{r}_i) \int_v \frac{\vec{s}-\vec{r}'}{|\vec{s}-\vec{r}'|^3} \delta(\vec{r}' - \vec{r}_i)\, dv$$

$$= \frac{1}{2\pi(\sigma_k^- + \sigma_k^+)} \sum_{i=1}^{N_j} \frac{\vec{s}-\vec{r}_i}{|\vec{s}-\vec{r}_i|^3} \vec{j}_f(\vec{r}_i) \qquad \textbf{28.21}$$

Potential function model

The centre of gravity approximation With this approximation, the unknown function V is calculated on nodal points located at the centre of gravity of each triangle. The potential over the triangle is assumed to be constant and equal to the potential at the centre of gravity $V = V(\vec{s}_{cog})$, as shown in Figure 28.2. With this approximation, the integral over each triangle in Eqn. **28.19** can be simplified:

$$\int_{\Delta_m^{(l)}} V(\vec{s}')\vec{\nabla}' \left(\frac{1}{|\vec{s}-\vec{s}'|} \right) \vec{n}(\vec{s}')\, dS' = -V(\vec{s}'_{cog})\Omega^{(l,m)}(\vec{s}) \qquad \textbf{28.22}$$

where $\Omega^{(l,m)}(\vec{s})$ is the solid angle at $\vec{s}$ subtended by the triangle $\Delta_m^{(l)}$:

$$\Omega^{(l,m)}(\vec{s}) = -\int_{\Delta_m^{(l)}} \vec{\nabla}' \left(\frac{1}{|\vec{s}-\vec{s}'|} \right) \vec{n}(\vec{s}')\, dS' = \int_{\Delta_m^{(l)}} \frac{\vec{s}'-\vec{s}}{|\vec{s}'-\vec{s}|^3} \vec{n}(\vec{s}')\, dS' \qquad \textbf{28.23}$$

This integral depends only on the three vector differences between $\vec{s}$ (the 'point of view') and the three vertices $\vec{s}'_1$, $\vec{s}'_2$ and $\vec{s}'_3$ (the 'points of support') determining the triangle $\Delta_m^{(l)}$. There exists an explicit analytic formula to calculate $\Omega^{(l,m)}(\vec{s})$.

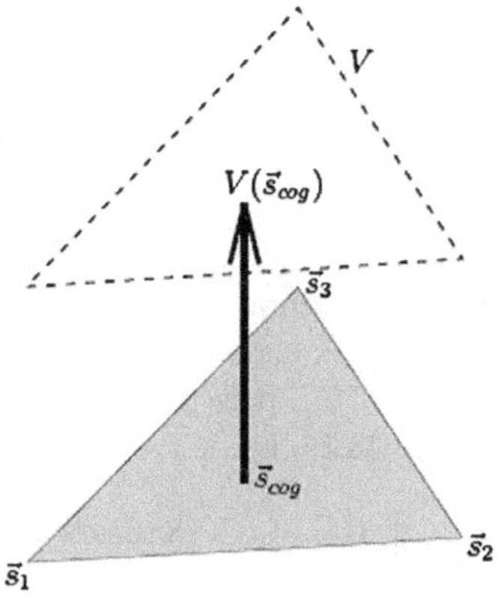

FIGURE 28.2 The centre of gravity (CoG) potential approximation: the potential V over the triangle is assumed to be constant and equal to the potential at the centre of gravity $\vec{s}_{cog}$ of the triangle, $V = V(\vec{s}_{cog})$.

The BEM Eqn. **28.19** eventually becomes a 'simple sum of known analytical functions':

$$V(\vec{s}_{cog,p}) = V_\infty(\vec{s}_{cog,p}) + \frac{1}{2\pi} \sum_{l=1}^{N_S} \frac{\sigma_l^- - \sigma_l^+}{\sigma_k^- + \sigma_k^+} \sum_{m=1}^{N_{tr}^{(l)}} V(\vec{s}_{cog,m})\Omega^{(l,m)}(\vec{s}_{cog,p}) \qquad \textbf{28.24}$$

where $\vec{s}_{cog,m}$ (resp. $\vec{s}_{cog,p}$) is the 'centre of gravity' of the mth (resp. p^{th}) triangle $\Delta_m^{(l)}$ (resp. $\Delta_p^{(k)}$) of the lth (resp. k^{th}) surface S_l (resp. S_k). The BEM problem now has the form of a set of linear equations.

The linear potential at vertices approximation Here the potential is evaluated on the vertices of the triangles but a better approximation of the potential over the triangles is used: the potential is assumed to vary linearly over each triangle, as shown in Figure 28.3. As only three values are needed to specify a linear function on a plane surface, the value of the potential V at the three vertices of the triangle can be used. Moreover, this ensures that the potential varies continuously from one triangle to the next which was not the case with the CoG approximation.

The integral over each triangle in Eqn. **28.19** can be simplified to give a weighted sum of the potential at the vertices:

$$\int_{\Delta_m^{(l)}} V(\vec{s}')\vec{\nabla}' \left(\frac{1}{|\vec{s}-\vec{s}'|} \right) \vec{n}(\vec{s}')dS' \qquad \textbf{28.25}$$

$$= -\left(V(\vec{s}'_1)\Omega_1^{(l,m)}(\vec{s}) + V(\vec{s}'_2)\Omega_2^{(l,m)}(\vec{s}) + V(\vec{s}'_3)\Omega_3^{(l,m)}(\vec{s}) \right)$$

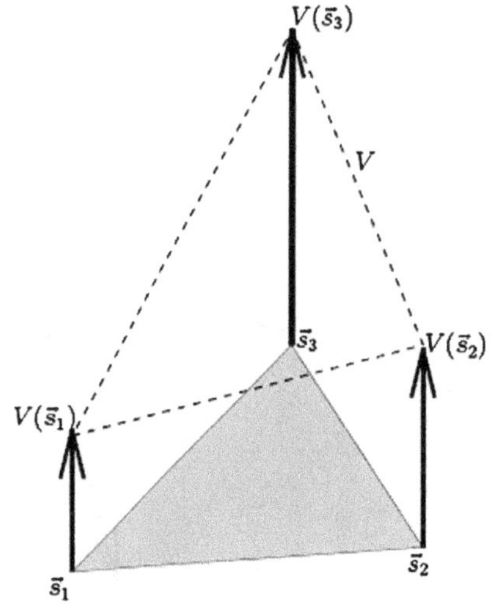

FIGURE 28.3 The linear potential at vertices (LPV) potential approximation: the potential V over the triangle is assumed to be varying linearly between the potential calculated at each vertex $\vec{s}_1$, $\vec{s}_2$ and $\vec{s}_3$ of the triangle.

The three $\Omega_{\bullet}^{(l,m)}(\vec{s})$ are also purely geometric quantities depending on the vector differences between the 'point of view' $\vec{s}$ and the vertices $\vec{s}'_{\bullet}$ of the triangle. An analytic formula to calculate the $\Omega_{\bullet}^{(l,m)}(\vec{s})$ from $\vec{s}$ and $\vec{s}'_{\bullet}$ exists. With this approximation the BEM Eqn. **28.19** simplifies to:

$$V(\vec{s}_{\bullet}) = V_{\infty}(\vec{s}_{\bullet}) + \frac{1}{2\pi}\sum_{l=1}^{N_S}\frac{\sigma_l^- - \sigma_l^+}{\sigma_k^- + \sigma_k^+} \quad \mathbf{28.26}$$

$$\sum_{m=1}^{N_{tr}^{(l)}}\left(V(\vec{s}'_{1,m})\Omega_1^{(l,m)}(\vec{s}_{\bullet}) + V(\vec{s}'_{2,m})\Omega_2^{(l,m)}(\vec{s}_{\bullet}) + V(\vec{s}'_{3,m})\Omega_3^{(l,m)}(\vec{s}_{\bullet})\right)$$

where $\vec{s}_{\bullet}$ is one of the three vertices of a triangle of the kth surface S_k and $\vec{s}'_{i,m}$ is the ith vertex of the mth triangle $\Delta_m^{(l)}$ of the lth surface S_l. Again, the BEM problem is reduced to a set of linear equations.

Solid angles and potential approximation

As shown in the previous section, to solve the BEM equations, it is necessary to calculate solid angles, i.e. purely geometrical quantities.

Solid angle calculation

For the CoG approximation, the solid angle $\Omega^{(l,m)}(\vec{s})$ subtended by a plane triangle $\Delta_m^{(l)}$ at some point $\vec{s}$ has to be calculated. Without loss of generality, the observation point $\vec{s}$ can be placed at the origin $\vec{o}^*$. The three vertices $\vec{s}_1$, $\vec{s}_2$ and $\vec{s}_3$ of the plane triangle are then specified by the vectors $\vec{v}_1 = \vec{s}_1 - \vec{o}^*$, $\vec{v}_2 = \vec{s}_2 - \vec{o}^*$ and $\vec{v}_3 = \vec{s}_3 - \vec{o}^*$ relative to this origin $\vec{o}^*$, as shown in Figure 28.4. The solid angle Ω can be expressed analytically as a function of $\vec{v}_1$, $\vec{v}_2$ and $\vec{v}_3$ by the formula taken from van Oosterom and Strackee (1983):

$$\tan\left(\frac{1}{2}\Omega\right) = \frac{\vec{v}_1(\vec{v}_2 \times \vec{v}_3)}{|\vec{v}_1||\vec{v}_2||\vec{v}_3| + (\vec{v}_1\vec{v}_2)|\vec{v}_3| + (\vec{v}_1\vec{v}_3)|\vec{v}_2| + (\vec{v}_2\vec{v}_3)|\vec{v}_1|} \quad \mathbf{28.27}$$

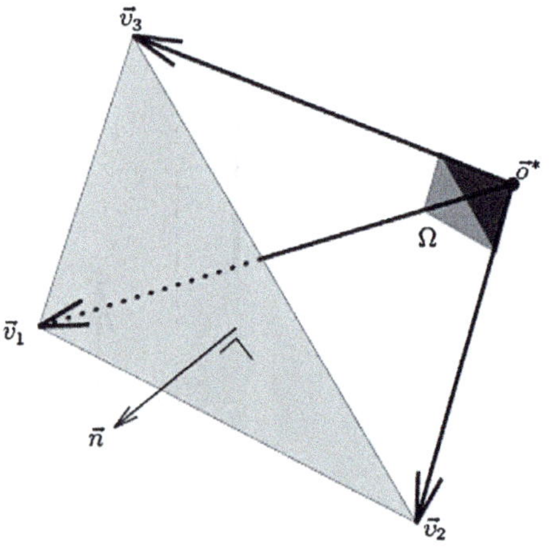

FIGURE 28.4 Solid angle supported by a plane triangle: the solid angle Ω supported at the point $\vec{o}^*$ by the plane triangle (grey shade) depends only on the three vectors $\vec{v}_1$, $\vec{v}_2$ and $\vec{v}_3$ and can be easily calculated by Eqn. **28.27**.

For the LPV approximation, three geometric quantities $\Omega_i (i = 1, 2, 3)$ have to be calculated for each triangle, under the assumption that the potential V varies linearly over this triangle. There also exists an analytic formula for Ω_i (de Munck, 1992; Schlitt *et al.*, 1995):

$$\Omega_i = \frac{1}{2A}\left(\vec{z}_i\vec{n}\Omega + \beta(\vec{v}_j - \vec{v}_k)\vec{\Omega}\right) \quad \mathbf{28.28}$$

where

- A is the surface of the plane triangle
- $\vec{z}_i = \vec{v}_j \times \vec{v}_k$ with (i, j, k) a cyclic permutation of $(1, 2, 3)$
- $\vec{n}$ is a unit vector normal to the triangle
- Ω is the solid angle subtended by the plane triangle at the origin as expressed in Eqn. **28.27**
- $\beta = \vec{n}\vec{v}_i$ is equal to the perpendicular distance from the origin to the triangle
- $\vec{\Omega}$ is a vector defined by $\vec{\Omega} = \sum_{i=1}^{3}(\gamma_j - \gamma_i)\vec{v}_j$ with $\gamma_i = \frac{1}{|\vec{v}_j - \vec{v}_i|}\ln\left(\frac{|\vec{v}_j - \vec{v}_i||\vec{v}_j| + (\vec{v}_j - \vec{v}_i)\vec{v}_j}{|\vec{v}_j - \vec{v}_i||\vec{v}_i| + (\vec{v}_j - \vec{v}_i)\vec{v}_i}\right)$.

The Ω_i also satisfy the equality: $\Omega_1 + \Omega_2 + \Omega_3 = \Omega$.

Solid angle properties

An important property of solid angles concerns their integral over a single closed surface. We know from Eqn. **28.23** that the infinitesimal solid angle $d\Omega'$ subtended by the infinitesimal surface dS' around the point $\vec{s}'$ at the point of view $\vec{s}$ is expressed by: $d\Omega'(\vec{s}, \vec{s}') = \vec{n}(\vec{s}')\frac{\vec{s}' - \vec{s}}{|\vec{s}' - \vec{s}|^3}dS'$. Then the integral of $d\Omega'(\vec{s}, \vec{s}')$ over a smooth closed surface is:

$$\Omega_S(\vec{s}) = \int_S d\Omega'(\vec{s}, \vec{s}') = \begin{bmatrix} 0 \\ 2\pi \\ 4\pi \end{bmatrix}, \text{ for } \vec{s} \begin{bmatrix} \text{outside} \\ \text{on} \\ \text{inside} \end{bmatrix} \text{ the surface.} \quad \mathbf{28.29}$$

The BEM Eqn. **28.15** contains an integral of the form: $\int_S d\Omega'(\vec{s}, \vec{s}')V(\vec{s}')$ which is converted into a discrete sum, of the form: $\sum_{m=1}^{M}\Omega_{nm}V_m$. It is therefore important that the Ω_{nm} satisfy Eqn. **28.29**, i.e.

$$\Omega_{n,S} = \sum_{m=1}^{M}\Omega_{nm} = \begin{bmatrix} 0 \\ 2\pi \\ 4\pi \end{bmatrix}, \text{ for } \vec{s} \begin{bmatrix} \text{outside} \\ \text{on} \\ \text{inside} \end{bmatrix} \text{ the surface.} \quad \mathbf{28.30}$$

When $\vec{s}$ is outside or inside the surface, these equalities are satisfied: all the Ω_{nm} can be calculated unambiguously with Eqn. **28.27** or Eqn. **28.28**. But when $\vec{s}$ is on the surface itself we meet the 'auto-solid angle' (ASA) problem: the solid angle subtended by a triangle which contains the point of view is zero and the second equality of Eqn. **28.30** may not be satisfied automatically.

The ASA problem for the CoG approximation

In the CoG approximation, as the potential is evaluated at the 'centre of gravity' of each triangle, there will be only one null solid angle: $\Omega_{mm} = 0$, as can be seen in Figure 28.5. Since the rest of the solid angle subtended by the closed surface is already 2π, there is no missing angle and the second equality of Eqn. **28.30** is satisfied.

Nevertheless, in reality the surface modelled by the triangle is not plane, and it should thus support some non-zero solid angle. There is no obvious way to improve the solution, but to use a finer meshing of the surface.

The ASA problem for the LPV approximation

With the LPV approximation, all the adjacent triangles (grey triangles in Figure 28.6) containing the 'point of view' are supporting an *estimated* null solid angle. The solid angle subtended by the rest of the surface will *not* be equal to 2π because the adjacent triangles do not represent a flat surface.

$$\Omega_{miss} = 2\pi - \sum_{m=1}^{M} \Omega_{nm} \neq 0 \qquad \textbf{28.31}$$

There are two main problems here: (1) how do we divide up the missing solid angle Ω_{miss} between the adjacent triangles, and (2) within each of them, how do we share between its vertices the missing part, as illustrated

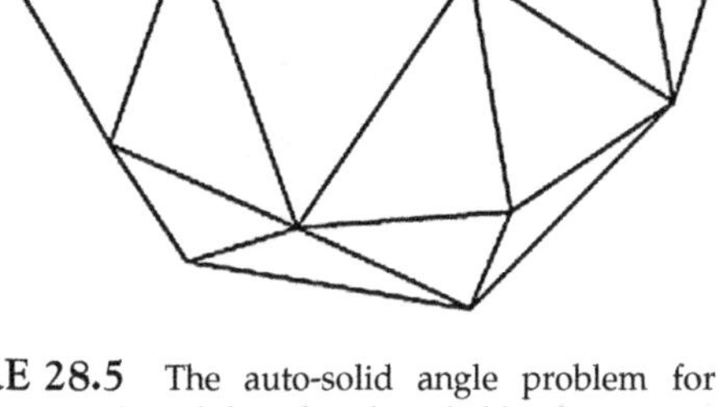

FIGURE 28.5 The auto-solid angle problem for the CoG approximation: the solid angle subtended by the grey triangle from its centre of gravity (the black dot) is zero and the total solid angle subtended by the rest of the surface (white triangles) is equal to 2π.

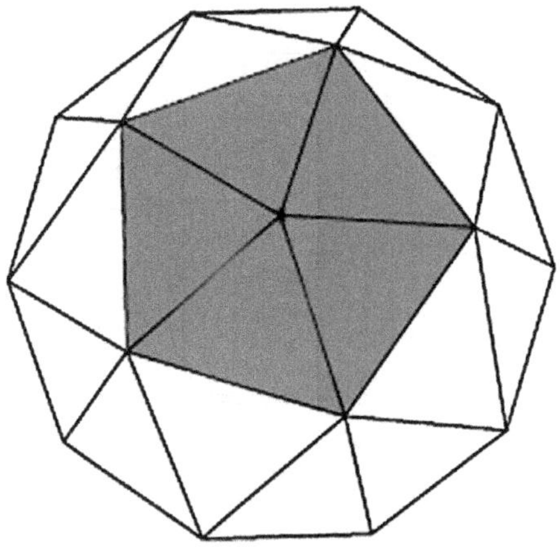

FIGURE 28.6 The auto-solid angle problem for the LPV approximation: the estimated solid angle supported by the 'central' point and the adjacent grey triangles is zero but the total solid angle supported by the remaining white triangles is less than 2π.

in Figure 28.7. An analytic solution exists (Heller, 1990) but it requires each triangle around the point of view $\vec{s}_0$ be approximated by a portion of sphere of centre $\vec{r}_c$ and radius R. If the surface is regular and smooth compared to the density of the mesh, this local spherical approximation will hold as R will be much larger than the length of the edges of the triangles.

Since three points do not determine a sphere, a fourth point must be chosen. A suitable point would be the next adjacent vertex, e.g. the sphere that passes through the triplet $[\vec{s}_0\vec{s}_1\vec{s}_2]$ could be required to pass through $\vec{s}_3$ as well. A better and anatomically more tenable approximation can be obtained, if, at the tessellation stage, the centre of gravity $\vec{s}_{cog}$ of each triangle is projected perpendicular to the triangular plane onto the actual surface of the volume $\vec{s}^{\perp}_{cog}$. A sphere can now be fitted easily to four points: the three vertices defining the triangle and its projected centre of gravity, as shown in Figure 28.8.

Once the spheres have been fitted for $\vec{s}_0$ and its adjacent vertices $\vec{s}_1, \vec{s}_2, \ldots, \vec{s}_{N_{adj}}$, an approximate value for the solid angle subtended by each triplet $[\vec{s}_0\vec{s}_1\vec{s}_2]$, $[\vec{s}_0\vec{s}_2\vec{s}_3]$, ..., $[\vec{s}_0\vec{s}_{N_{adj}}\vec{s}_1]$ at $\vec{s}_0$ can be calculated. Using spherical coordinates for the vertices, as shown in

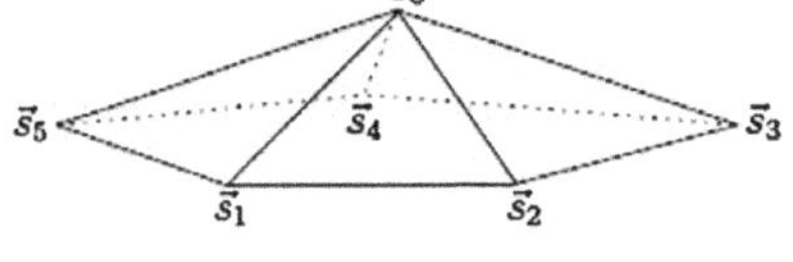

FIGURE 28.7 The surface defined by $\vec{s}_1, \vec{s}_2, \ldots, \vec{s}_5$ around $\vec{s}_0$ supports a non-zero solid angle. Each triangle, defined by a triplet of vertices ($[\vec{s}_0\vec{s}_1\vec{s}_2]$, $[\vec{s}_0\vec{s}_2\vec{s}_3]$, ..., $[\vec{s}_0\vec{s}_5\vec{s}_1]$), supports a part of the missing solid angle, which in turn must be shared between its vertices. Each triangle can be locally approximated by a portion of sphere.

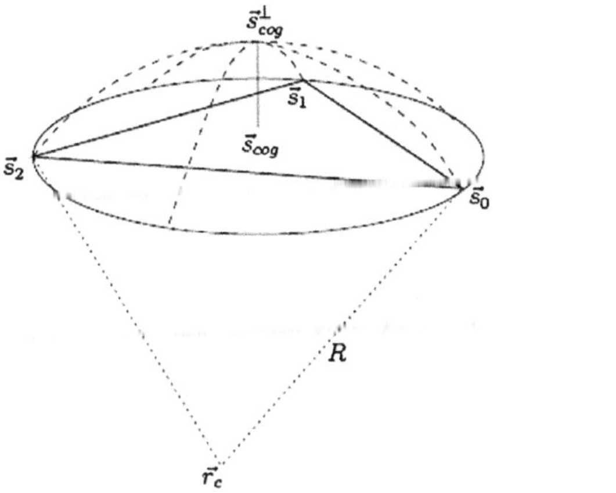

FIGURE 28.8 Spherical approximation of an adjacent plane triangle: the triplet of vertices $[\vec{s}_0\vec{s}_1\vec{s}_2]$ and the projection $\vec{s}_{cog}^{\perp}$ of the centre of gravity $\vec{s}_{cog}$ of the triangle onto the actual surface determine a sphere, with centre $\vec{r}_c$ and radius R, that approximates the actual surface.

Figure 28.9, the solid angle $\Omega_{[\vec{s}_0\vec{s}_1\vec{s}_2]}$ subtended at $\vec{s}_0$ by the spherical region bounded by $\vec{s}_0$, $\vec{s}_1$ and $\vec{s}_2$, is approximated by: $\Omega_{[\vec{s}_0\vec{s}_1\vec{s}_2]} = \frac{\psi_1+\psi_2}{4}\phi_{12}$ where ψ_1 and ψ_2 are easily obtained and $\sin\frac{\phi_{12}}{2} = \frac{|\vec{s}_b-\vec{s}_2|}{2R\sin\psi_2}$ with $\vec{s}_b = \vec{r}_c + \frac{1}{\sin\psi_1}\left[\sin(\psi_1-\psi_2)(\vec{s}_0-\vec{r}_c)+\sin\psi_2(\vec{s}_1-\vec{r}_c)\right]$.

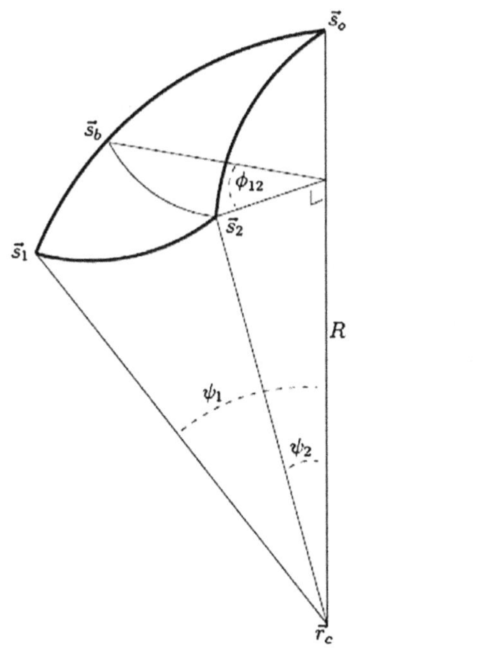

FIGURE 28.9 Auto-solid angle approximation: the triangle defined by the triplet $[\vec{s}_0\vec{s}_1\vec{s}_2]$ is approximated by a portion of a sphere. The solid angle subtended at $\vec{s}_0$ by the curved surface (bold line) can be calculated using the spherical coordinates of $\vec{s}_1$ and $\vec{s}_2$: ψ_1, ψ_2 and ϕ_{12}.

The fraction $f_{[\bullet]}$ of missing solid angle Ω_{miss} to be assigned to each triangle $[\bullet]$, e.g., $[\vec{s}_0\vec{s}_1\vec{s}_2]$, is obtained by:

$$f_{[\vec{s}_0\vec{s}_1\vec{s}_2]} = \frac{\Omega_{[\vec{s}_0\vec{s}_1\vec{s}_2]}}{\Omega_{[\vec{s}_0\vec{s}_1\vec{s}_2]}+\Omega_{[\vec{s}_0\vec{s}_2\vec{s}_3]}+\cdots+\Omega_{[\vec{s}_0\vec{s}_{N_{adj}}\vec{s}_1]}} \qquad 28.32$$

Note that, even though these equations entail approximations, since they only involve ratios, the total solid angle subtended by the region around $\vec{s}_0$ will sum to Ω_{miss}, and the total solid angle subtended by the entire surface at $\vec{s}_0$ will be exactly 2π. Now, it is necessary to share further this portion of missing solid angle between the vertices of the adjacent triangles.

Assuming that ψ_1, ψ_2 and ϕ_{12} are small and that the potential V varies linearly with distance on the sphere, Heller (1990) showed that it is possible to share the solid angle $f_{[\vec{s}_0\vec{s}_1\vec{s}_2]}\Omega_{miss}$ between the three vertices $\vec{s}_0$, $\vec{s}_1$ and $\vec{s}_2$ such that $\Omega_{\vec{s}_0,[\vec{s}_0\vec{s}_1\vec{s}_2]}+\Omega_{\vec{s}_1,[\vec{s}_0\vec{s}_1\vec{s}_2]}+\Omega_{\vec{s}_2,[\vec{s}_0\vec{s}_1\vec{s}_2]} = f_{[\vec{s}_0\vec{s}_1\vec{s}_2]}\Omega_{miss}$, where:

$$\Omega_{\vec{s}_0,[\vec{s}_0\vec{s}_1\vec{s}_2]} = \frac{1}{12(\psi_1+\psi_2)}\left(7\psi_1+7\psi_2-\frac{\psi_1^2}{\psi_2}-\frac{\psi_2^2}{\psi_1}\right)f_{[\vec{s}_0\vec{s}_1\vec{s}_2]}\Omega_{miss} \qquad 28.33a$$

$$\Omega_{\vec{s}_1,[\vec{s}_0\vec{s}_1\vec{s}_2]} = \frac{1}{12(\psi_1+\psi_2)}\left(3\psi_1+2\psi_2+\frac{\psi_2^2}{\psi_1}\right)f_{[\vec{s}_0\vec{s}_1\vec{s}_2]}\Omega_{miss} \qquad 28.33b$$

$$\Omega_{\vec{s}_2,[\vec{s}_0\vec{s}_1\vec{s}_2]} = \frac{1}{12(\psi_1+\psi_2)}\left(2\psi_1+3\psi_2+\frac{\psi_1^2}{\psi_2}\right)f_{[\vec{s}_0\vec{s}_1\vec{s}_2]}\Omega_{miss} \qquad 28.33c$$

Matrix form of the BEM equation

A simple realistic head model can be obtained by considering three concentric volumes of homogeneous conductivity: the brain, skull and scalp volumes, of conductivity σ_{br}, σ_{sk} and σ_{sc} respectively, as depicted in Figure 28.10. The three interfaces: 'brain-skull', 'skull-scalp' and 'scalp-air' separating the three volumes are numbered 1, 2 and 3. With this convention, the conductivity inside and outside each surface is defined by:

$$\begin{aligned} \sigma_1^- &= \sigma_{br} \\ \sigma_1^+ &= \sigma_2^- = \sigma_{sk} \\ \sigma_2^+ &= \sigma_3^- = \sigma_{sc} \\ \sigma_3^+ &= \quad 0 \end{aligned} \qquad 28.34$$

With the discrete approximation of the source term (Eqn. **28.21**) and the approximation of the boundary element equation: CoG approximation (Eqn. **28.24**) or LPV

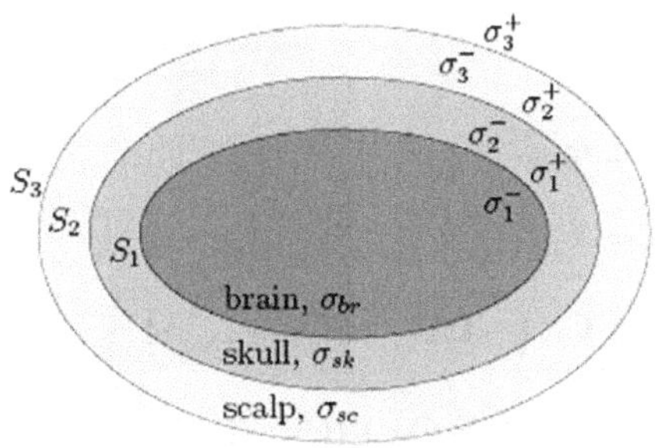

FIGURE 28.10 Simple realistic head model: three concentric volumes of homogeneous conductivity, brain (σ_{br}), skull (σ_{sk}) and scalp (σ_{sc}), separated by the three surfaces S_1, S_2 and S_3, comprise the head model.

approximation (Eqn. **28.27**), the BEM problem can be expressed in matrix form:

$$\begin{bmatrix} \mathbf{v}_1 \\ \mathbf{v}_2 \\ \mathbf{v}_3 \end{bmatrix} = \begin{bmatrix} \mathbf{B}_{11}\ \mathbf{B}_{12}\ \mathbf{B}_{13} \\ \mathbf{B}_{21}\ \mathbf{B}_{22}\ \mathbf{B}_{23} \\ \mathbf{B}_{31}\ \mathbf{B}_{32}\ \mathbf{B}_{33} \end{bmatrix} \begin{bmatrix} \mathbf{v}_1 \\ \mathbf{v}_2 \\ \mathbf{v}_3 \end{bmatrix} + \begin{bmatrix} \mathbf{G}_1 \\ \mathbf{G}_2 \\ \mathbf{G}_3 \end{bmatrix} [\mathbf{j}] \quad \Leftrightarrow \quad \mathbf{v} = \mathbf{B}\,\mathbf{v} + \mathbf{G}\,\mathbf{j} \tag{28.35}$$

where:

- $\mathbf{v}_k$, an $N_{v_k} \times 1$ vector, contains the potential at the N_{v_k} nodal points of surface S_k: centre of gravity of each triangle for the CoG approximation or vertices of the triangles for the LPV approximation. $\mathbf{v}$ is $N_v \times 1$ vector with $N_v = N_{v_1} + N_{v_2} + N_{v_3}$.
- $\mathbf{B}_{kl}$, an $N_{v_k} \times N_{v_l}$ matrix, represents the influence of the potential of surface S_l on the potential of surface S_k. Its elements depend on the conductivity inside and outside the surfaces S_k and S_l, and on the solid angles used in the BEM Eqn. **28.24**. $\mathbf{B}$ is an $N_v \times N_v$ matrix.
- $\mathbf{j} = [\vec{j}_1^{\,t}\ \vec{j}_2^{\,t} \ldots \vec{j}_{N_j}^{\,t}]^t$, a $3N_j \times 1$ vector, is the source distribution vector, where each $\vec{j}_n = [j_{n,x}\ j_{n,y}\ j_{n,z}]^t$ is an orientation-free source vector.
- $\mathbf{G}_k$, an $N_{v_k} \times 3N_j$ matrix, is the free space potential matrix depending on the location $\vec{r}_n$ of the sources $\vec{j}_n$, the nodal points on surface S_k and the conductivity inside and outside surface S_k (σ_k^- and σ_k^+). $\mathbf{G}$ is an $N_v \times 3N_j$ matrix.

Constructing the matrices B and G

The matrices $\mathbf{B}$ and $\mathbf{G}$ can be constructed directly as follows.

Matrix B with the CoG approximation The element (p, q) of the matrix $\mathbf{B}_{kl}$ is:

$$\mathbf{B}_{kl}^{(p,q)} = \frac{1}{2\pi}\left(\frac{\sigma_l^- - \sigma_l^+}{\sigma_k^- + \sigma_k^+}\right)\Omega_{pq} \tag{28.36}$$

where p (resp. q) is the index of the nodal point on the surface S_k (resp. S_l), and Ω_{pq} is the solid angle at the centre of gravity of the pth triangle of S_k subtended by the qth triangle of S_l.

Matrix B with the LPV approximation The element (p, q) of the matrix $\mathbf{B}_{kl}$ is:

$$\mathbf{B}_{kl}^{(p,q)} = \frac{1}{2\pi}\left(\frac{\sigma_l^- - \sigma_l^+}{\sigma_k^- + \sigma_k^+}\right)\sum_n^{N_q}\Omega_{pn}^q \tag{28.37}$$

where p (resp. q) is the index of the nodal point on the surface S_k (resp. S_l), N_q is the number of triangles comprising the qth vertex, and Ω_{pn}^q is the portion of solid angle attributed to the qth vertex and subtended by the nth triangle (containing the qth vertex) of S_l at the pth vertex of the surface S_k.

Free potential matrix G The elements $(p, 3q-2)$, $(p, 3q-1)$ and $(p, 3q)$ of the matrix $\mathbf{G}_k$ is:

$$\left[\mathbf{G}_k^{(p,3q-2)}\mathbf{G}_k^{(p,3q-1)}\mathbf{G}_k^{(p,3q)}\right] = \frac{(\vec{s}_p - \vec{r}_q)^t}{2\pi(\sigma_k^- + \sigma_k^+)|\vec{s}_p - \vec{r}_q|^3} \tag{28.38}$$

where $\vec{s}_p$ is the pth nodal point of the surface S_k, and $\vec{r}_q$ is the location of the qth current source $\vec{j}_q$.

Solving the numerical BEM equation

The solution of the forward problem now rests on establishing a linear relationship between the source distribution $\mathbf{j}$ and the potential on the surfaces $\mathbf{v}$ (or at least on the scalp $\mathbf{v}_3$) of the form:

$$\mathbf{v} = \mathbf{L}\,\mathbf{j} \tag{28.39}$$

An obvious solution of Eqn. **28.35** would be simply to solve the system of equations:

$$(\mathbf{I}_{N_v} - \mathbf{B})\mathbf{v} = \mathbf{G}\,\mathbf{j} \tag{28.40}$$

by inverting the matrix $(\mathbf{I}_{N_v} - \mathbf{B})$. However, we are dealing here with a problem of electric potential and a potential function can only be measured relative to some reference point, i.e. calculated to within a constant. The systems of Eqn. **28.40** therefore rank deficient and the matrix $(\mathbf{I}_{N_v} - \mathbf{B})$ ***cannot*** be inverted.[1] The only way to

[1] As $\mathbf{v}_a = \mathbf{v}$ and $\mathbf{v}_b = \mathbf{v} + c\,\mathbf{1}_{N_v}$ (with $c \neq 0$) must both satisfy Eqn. **28.35** and Eqn. **28.40**, it follows:

$$\left.\begin{aligned} \mathbf{v} &= \mathbf{B}\,\mathbf{v} + \mathbf{G}\,\mathbf{j} \\ (\mathbf{v} + c\mathbf{1}_{N_v}) &= \mathbf{B}(\mathbf{v} + c\mathbf{1}_{N_v}) + \mathbf{G}\,\mathbf{j}, \quad c \neq 0 \end{aligned}\right\} \Rightarrow c\mathbf{1}_{N_v} = \mathbf{B}\,c\,\mathbf{1}_{N_v}, \quad c \neq 0$$

$$\Rightarrow \mathbf{B}\,\mathbf{1}_{N_v} = \mathbf{1}_{N_v}$$

$$\Rightarrow (\mathbf{I}_{N_v} - \mathbf{B})\mathbf{1}_{N_v} = 0$$

The matrix $(\mathbf{I}_{N_v} - \mathbf{B})$ has a **null** eigenvalue associated with the eigenvector $\mathbf{1}_{N_v}$, i.e. $\mathbf{B}$ has a **unit** eigenvalue associated with the eigenvector $\mathbf{1}_{N_v}$.

solve Eqn. **28.40** is to use a 'deflation technique' (Lynn and Timlake, 1968a, b; Chan, 1984).

Deflation technique

By assuming that the unit eigenvalue of **B** is simple, it can easily be shown that any other solution will only differ by an additive constant, i.e., a scalar multiple of $\mathbf{1}_{N_v}$. Let **p** be any vector such that $\mathbf{1}_{N_v}^t\,\mathbf{p} = 1$ and suppose that we seek the solution of Eqn. **28.35** such that $\mathbf{p}^t\,\mathbf{v} = 0$. Then looking for this particular solution, Eqn. **28.35** becomes:

$$\mathbf{v} = (\mathbf{B} - \mathbf{1}_{N_v}\mathbf{p}^t)\mathbf{v} + \mathbf{G}\,\mathbf{j} \qquad \mathbf{28.41}$$

Under the assumption that $\mathbf{p}^t\,\mathbf{v} = 0$, the matrix $\mathbf{C} = (\mathbf{B} - \mathbf{1}_{N_v}\mathbf{p}^t)$ is a deflation of **B** and has no unit eigenvalue, so that $(\mathbf{I}_{N_v} - \mathbf{C})^{-1} = (\mathbf{I}_{N_v} - \mathbf{B} - \mathbf{1}_{N_v}\mathbf{p}^t)^{-1}$ exists. Eqn. **28.35** can be rewritten as:

$$\begin{bmatrix}\mathbf{v}_1\\ \mathbf{v}_2\\ \mathbf{v}_3\end{bmatrix} = \begin{bmatrix}\mathbf{C}_{11}\,\mathbf{C}_{12}\,\mathbf{C}_{13}\\ \mathbf{C}_{21}\,\mathbf{C}_{22}\,\mathbf{C}_{23}\\ \mathbf{C}_{31}\,\mathbf{C}_{32}\,\mathbf{C}_{33}\end{bmatrix}\begin{bmatrix}\mathbf{v}_1\\ \mathbf{v}_2\\ \mathbf{v}_3\end{bmatrix} + \begin{bmatrix}\mathbf{G}_1\\ \mathbf{G}_2\\ \mathbf{G}_3\end{bmatrix}[\mathbf{j}] \qquad \mathbf{28.42}$$

and this system of equations can be solved by calculating:

$$\mathbf{v} = (\mathbf{I}_{N_v} - \mathbf{C})^{-1}\mathbf{G}\,\mathbf{j} = (\mathbf{I}_{N_v} - \mathbf{B} - \mathbf{1}_{N_v}\mathbf{p}^t)^{-1}\mathbf{G}\,\mathbf{j} \qquad \mathbf{28.43}$$

where **v** satisfies $\mathbf{p}^t\,\mathbf{v} = 0$.

Each vector $\mathbf{v}_\bullet$ is of size $N_{v_\bullet} \times 1$, so if, for example, **p** is defined by:

$$\mathbf{p} = [\underbrace{0\,0\ldots0}_{N_{v_1}}\,\underbrace{0\,0\ldots0}_{N_{v_2}}\,\underbrace{p\,p\ldots p}_{N_{v_3}}]^t \qquad \mathbf{28.44}$$

with $p = 1/N_{v_3}$, then $\mathbf{p}^t\,\mathbf{v} = 0$ simply means that the mean of $\mathbf{v}_3$ is zero. Therefore Eqn. **28.43** provides us with the solution that is mean corrected over the scalp surface.

Partial solution for the scalp

The number of equations (N_v) to solve in Eqn. **28.42** is rather large, but only the direct relationship between the source distribution **j** and the potential on the scalp $\mathbf{v}_3$, i.e. the lead field between the sources and the scalp **L**, is of interest in the EEG problem. After some algebraic manipulations, Eqn. **28.42** can be rewritten as:

$$\Gamma_1\mathbf{v}_3 = \Gamma_2\mathbf{j} \qquad \mathbf{28.45}$$

where

$$\Gamma_1 = -\left((\mathbf{C}_{33} - \mathbf{I}_{N_{v_3}}) + \Upsilon_5\mathbf{C}_{13} + \Upsilon_6\mathbf{C}_{23}\right) \qquad \mathbf{28.46a}$$

$$\Gamma_2 = \mathbf{G}_3 + \Upsilon_5\mathbf{G}_1 + \Upsilon_6\mathbf{G}_2 \qquad \mathbf{28.46b}$$

and

$$\Upsilon_6 = \Upsilon_4\Upsilon_2 - \Upsilon_3 \qquad \mathbf{28.47a}$$

$$\Upsilon_5 = \Upsilon_3\Upsilon_1 - \Upsilon_4 \qquad \mathbf{28.47b}$$

$$\Upsilon_4 = \mathbf{C}_{31}\left(-\Upsilon_2\mathbf{C}_{21} + (\mathbf{C}_{11} - \mathbf{I}_{N_{v_1}})\right)^{-1} \qquad \mathbf{28.47c}$$

$$\Upsilon_3 = \mathbf{C}_{32}\left(-\Upsilon_1\mathbf{C}_{12} + (\mathbf{C}_{22} - \mathbf{I}_{N_{v_2}})\right)^{-1} \qquad \mathbf{28.47d}$$

$$\Upsilon_2 = \mathbf{C}_{12}\left(\mathbf{C}_{22} - \mathbf{I}_{N_{v_2}}\right)^{-1} \qquad \mathbf{28.47e}$$

$$\Upsilon_1 = \mathbf{C}_{21}\left(\mathbf{C}_{11} - \mathbf{I}_{N_{v_1}}\right)^{-1} \qquad \mathbf{28.47f}$$

By proceeding carefully, one has only to solve four systems of equations (**28.47f, 28.47e, 28.47d** and **28.47c**) to obtain Γ_1 and Γ_2. The matrices to invert are only of size N_{v_1} and N_{v_2}, thus the calculation of Γ_1 and Γ_2 require much less computational effort than inverting **C**, directly, which is of size $N_v = N_{v_1} + N_{v_2} + N_{v_3}$. The lead field for all nodal points on the scalp surface can then be obtained from Eqn. **28.45** by calculating:

$$\mathbf{v}_3 = \Gamma_1^{-1}\Gamma_2\mathbf{j} = \mathbf{L}\,\mathbf{j} \qquad \mathbf{28.48}$$

where Γ_1 is only of size N_{v_3}. It would of course be possible to obtain a relation such as Eqn. **28.45** for the other two surfaces.

It is important to note that Γ_1 depends ***only*** on the matrix **C**, i.e. on the geometry and the conductivity of the volumes, but not on the source distribution **j**. By pre-calculating and saving Γ_1^{-1}, Υ_5 and Υ_6, the lead field matrix **L** can be calculated very easily for any source distribution, using Eqns **28.46b** and **28.48**.

Partial solution for the electrode sites

In general, it is not necessary to calculate the potential V over the entire scalp surface as the EEG is recorded from a limited number of electrodes. Therefore, only the lead field for the electrode sites, $\mathbf{L}_{el}$ is required:

$$\mathbf{v}_{3,el} = \mathbf{L}_{el}\mathbf{j} \qquad \mathbf{28.49}$$

In a realistic head model, the location of the electrodes is defined relative to the triangular mesh of the scalp. As the electrodes typically have a diameter of a few millimetres, their location of can be approximated with the nodal point directly underneath.

A partial solution of Eqn. **28.45** for a few nodal points is possible thanks to the Frobenius-Schur formula that allows the partial calculation of the inverse of a matrix:

$$\begin{bmatrix} \mathbf{M} & \mathbf{N} \\ \mathbf{P} & \mathbf{Q} \end{bmatrix}^{-1} = \begin{bmatrix} \mathbf{M}^{-1} + \mathbf{M}^{-1}\mathbf{N}\,\mathbf{F}^{-1}\mathbf{P}\,\mathbf{M}^{-1} & -\mathbf{M}^{-1}\mathbf{N}\,\mathbf{F}^{-1} \\ -\mathbf{F}^{-1}\mathbf{P}\,\mathbf{M}^{-1} & \mathbf{F}^{-1} \end{bmatrix} \quad \mathbf{28.50}$$

where $\mathbf{M}$ and $\mathbf{Q}$ must be square, and $\mathbf{M}$ and $\mathbf{F} = \mathbf{Q} - \mathbf{P}\,\mathbf{M}^{-1}\mathbf{N}$ must be invertible.

Considering that the N_{el} interesting (respectively, N_{ot} other) nodal points are the last N_{el} (respectively, first N_{ot}) elements of $\mathbf{v}_3 : \mathbf{v}_3 = [\mathbf{v}_{3,ot}^t \mathbf{v}_{3,el}^t]^t$, then Eqn. **28.45** can be rewritten as:

$$\underbrace{\left[\begin{array}{c|c} \mathbf{M} & \mathbf{N} \\ \hline \mathbf{P} & \mathbf{Q} \end{array}\right]}_{\Gamma_1} \underbrace{\left[\begin{array}{c} \mathbf{v}_{3,ot} \\ \hline \mathbf{v}_{3,et} \end{array}\right]}_{\mathbf{v}_3} = \underbrace{\left[\begin{array}{c} \mathbf{R} \\ \hline \mathbf{S} \end{array}\right]}_{\Gamma_2} [\mathbf{j}] \quad \mathbf{28.51}$$

This partitioning of the vertices is not natural (as the electrodes are spread over the scalp surface), but such ordering may be obtained easily by permuting the rows and columns in Γ_1 and Γ_2. The lead field for the electrode sites can be obtained from the sub-matrices of Γ_1 and Γ_2 by:

$$\mathbf{L}_{el} = [-\mathbf{F}^{-1}\mathbf{P}\,\mathbf{M}^{-1}\;\mathbf{F}^{-1}]\,\Gamma_2 \quad \mathbf{28.52}$$

and only the two matrices $\mathbf{F}$ and $\mathbf{M}$ have to be inverted.

By using the simplified Eqn. **28.46b** for Γ_2, Eqn. **28.52** becomes:

$$\mathbf{L}_{el} = [-\mathbf{F}^{-1}\mathbf{P}\,\mathbf{M}^{-1}\;\mathbf{F}^{-1}](\Upsilon_5\mathbf{G}_1 + \Upsilon_6\mathbf{G}_2 + \mathbf{G}_3) \quad \mathbf{28.53a}$$

$$= \Xi_1\mathbf{G}_1 + \Xi_2\mathbf{G}_2 + \Xi_3\mathbf{G}_3 \quad \mathbf{28.53b}$$

where

$$\Xi_1 = [-\mathbf{F}^{-1}\mathbf{P}\,\mathbf{M}^{-1}\;\mathbf{F}^{-1}]\,\Upsilon_5 \quad \mathbf{28.54a}$$

$$\Xi_2 = [-\mathbf{F}^{-1}\mathbf{P}\,\mathbf{M}^{-1}\;\mathbf{F}^{-1}]\,\Upsilon_6 \quad \mathbf{28.54b}$$

$$\Xi_3 = [-\mathbf{F}^{-1}\mathbf{P}\,\mathbf{M}^{-1}\;\mathbf{F}^{-1}] \quad \mathbf{28.54c}$$

The three matrices Ξ_1, Ξ_2 and Ξ_3 depend ***only*** on the geometry and conductivity of the head model. If they are pre-calculated (and saved), the lead field $\mathbf{L}_{el}$ can be calculated rapidly for any source distribution $\mathbf{j}$ using Eqn. **28.53** (as only the matrices $\mathbf{G}$ have to be computed before calculating $\mathbf{L}_{el}$). This is of particular interest if the location of the dipoles has to be modified; e.g. if a denser mesh of dipoles is required in a linear distributed solution, or if an iterative procedure is used to optimize the location of the 'equivalent current dipoles' (ECDs) in an ECD-based solution of the inverse problem.

ANALYTIC SOLUTION OF THE BEM EQUATION

The three sphere shell model in EEG

The analytic solution of the BEM form of Maxwell's Eqn. **28.12** is possible for particular volume models. One such model, commonly used for EEG source localization, is the 'three sphere shell model' (Figure 28.11). It comprises three concentric spheres of radius r_1, r_2 and r_3, with $r_1 < r_2 < r_3$. The innermost spherical volume represents the brain volume. The volume between the spheres of radius r_1 and r_2 models the skull layer. The outer layer volume, between radii r_2 and r_3, corresponds to the scalp.

We will assume that the brain and scalp volumes have the same conductivity σ, and that the skull volume has a conductivity $\sigma_{sk} \ll \sigma$. A current source dipole $\vec{m}$, located at a height z on axis $\vec{e}_z$, generates a potential distribution $V(\vec{s})$ on the surface of the outer sphere. In spherical coordinates, i.e. $\vec{s} = \vec{s}(\theta, \varphi)$ as shown in Figure 28.12, this potential is calculated by the following, from Ary *et al.* (1981):

$$V(\vec{s}(\theta,\varphi)) = \frac{1}{4\pi\sigma}\sum_{n=1}^{\infty}\frac{2n+1}{n}b^{n-1}\left[\frac{\xi(2n+1)^2}{d_n(n+1)}\right] \quad \mathbf{28.55}$$
$$\left[n m_z P_n(\cos\theta) + (m_x\cos\varphi + m_y\sin\varphi)P_n^1(\cos\theta)\right]$$

where

- $b = z/r_3$ is the eccentricity of the dipole
- m_x, m_y and m_z are the components of the dipole $\vec{m} = [m_x m_y m_z]^t$ along the main axes
- $\xi = \sigma_{sk}/\sigma$ is the relative conductivity of the skull volume to the conductivity of the brain and scalp volumes
- $P_n(\cos\theta)$ and $P_n^1(\cos\theta)$ are Legendre and associated Legendre polynomials

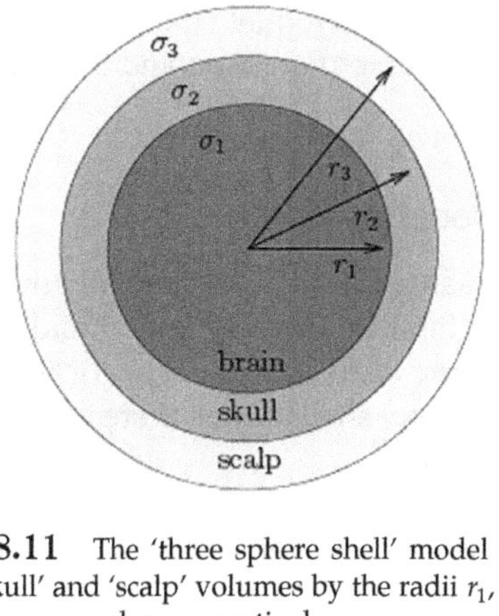

FIGURE 28.11 The 'three sphere shell' model is defined for the 'brain', 'skull' and 'scalp' volumes by the radii r_1, r_2 and r_3, and conductivity σ_1, σ_2, and σ_3 respectively.

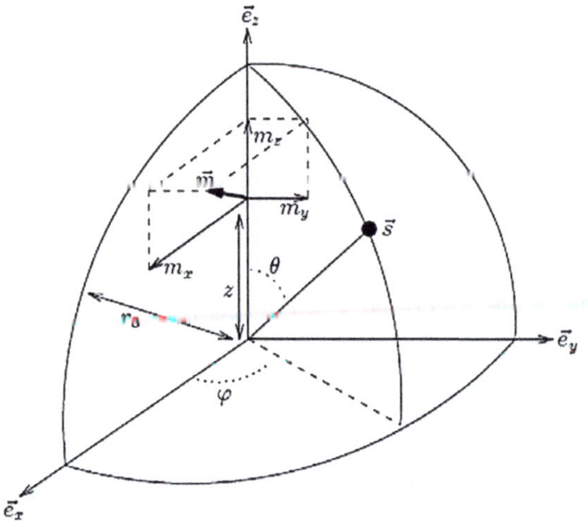

FIGURE 28.12 Current source dipole in a three sphere shell model: the dipole $\vec{m} = [m_x m_y m_z]^t$, located at a height z on axis $\vec{e}_z$, generates a potential distribution $V(\vec{s})$, with $\vec{s} = \vec{s}(\theta, \varphi)$, on the surface of the outer sphere.

- d_n is defined by:

$$\begin{aligned} d_n &= [(n+1)\xi + n]\left[\frac{n\xi}{n+1} + 1\right] \\ &+ (1-\xi)[(n+1)\xi + n](f_1^{2n+1} - f_2^{2n+1}) \\ &- n(1-\xi)^2\left(\frac{f_1}{f_2}\right)^{2n+1} \end{aligned} \tag{28.56}$$

with $f_1 = r_1/r_3$ and $f_2 = r_2/r_3$.

Any source location can be reduced to this configuration, i.e. current dipole located on the z axis, by a couple of rotations Even though the potential $V(\vec{s})$ is expressed as an infinite sum of terms, it is only necessary to calculate a few tens of them, as the terms converge rapidly to zero.[2]

The analytic expression of the potential for more complicated models can be found in the literature: four spheres with different conductivity (Arthur and Geselowitz, 1970; Cuffin and Cohen, 1979), cylindrical volume (Lambin and Troquet, 1983; Kleinermann *et al.*, 2000) or cubic volume (Ferguson and Stroink, 1994).

Spherical model in MEG

The signal measured in MEG is generally the radial field component B_r because of the position and geometry of the sensors. If the head model is spherically symmetric, like the three sphere shell model, there is no contribution of the return (or volume) current $\vec{j}_r$ to the radial field component B_r. Therefore, the radial field B_r outside the head model can be obtained analytically by:

$$B_r(\vec{r}) = \frac{\mu_0}{4\pi}\int_{vol_{br}} \vec{j}_f(\vec{r}') \times \frac{\vec{r} - \vec{r}'}{|\vec{r} - \vec{r}'|^3} . \vec{e}_r dv' \tag{28.57}$$

where vol_{br} is the brain volume containing the imposed current sources $\vec{j}_f(\vec{r}')$ and $\vec{e}_r$ is the unit vector in the radial direction.

If the current sources are modelled by discrete dipoles as in Eqn. **28.20**, then the magnetic lead field becomes:

$$B_r(\vec{r}) = -\frac{\mu_0}{4\pi}\sum_{i=1}^{N_j} \frac{\vec{j}_f(\vec{r}_i') \times \vec{r}_i' . \vec{e}_r}{|\vec{r} - \vec{r}_i'|^3} \tag{28.58}$$

where the origin $\vec{o}$ of space is placed at the centre of the sphere.

This expression highlights three important features of the MEG lead field calculated for a spherical model: (1) a source at the centre of the sphere will produce no magnetic field outside; (2) MEG will only be sensitive to the tangential component of the current sources; and (3) the lead field is not affected by the layers surrounding the brain volume, especially the poorly conducting skull.

DISCUSSION

Limitations of the BEM numerical solution

The lead field obtained with the BEM approach is affected by numerical errors due to the approximations employed. The main sources of errors are the limited number of nodal points, where the potential is actually estimated, and the way the potential is interpolated between those points. The CoG and LPV approximations can be interpreted as zeroth and first-order approximations respectively.

For sources far away from any surface, i.e. relatively deep sources, the constant or linear approximation will be sufficient to model the potential distribution over the tessellated surfaces. Errors appear when these approximations do not hold anymore, i.e. when the distance between the source and the surface becomes small compared to the (mean) distance between nodal points (Meijs *et al.*, 1989; Schlitt *et al.*, 1995; Ferguson and Stroink, 1997; Fuchs *et al.*, 1998). Figure 28.13 shows the difference (mean relative difference measure, RDM) between the lead field estimated analytically, $\mathbf{v}_a$, and with the BEM approach, $\mathbf{v}_{BEM}$ as a function of the eccentricity of the dipoles in a three sphere shell model:

$$\text{RDM} = \sqrt{\frac{\sum_{i=1}^{N}(v_{a,i} - v_{BEM,i})^2}{\sum_{i=1}^{N} {v_{a,i}}^2}} \tag{28.59}$$

[2] There also exists a closed-form approximation that requires less computational effort (at the cost of some minor error) as shown by Sun (1997).

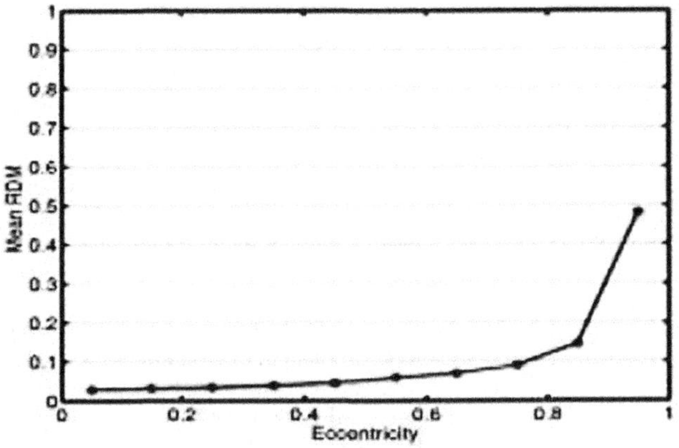

FIGURE 28.13 In a three sphere shell model, both analytical and BEM solutions are calculated for sources placed at different eccentricities. The lead fields obtained are compared using a relative difference measure (RDM). The mean RDM is plotted as a function of the sources eccentricity.

The difference between the exact analytic solution and the approximated BEM solution increases abruptly when the source is placed close to the inner skull surface. Note that this problem is also the same with other numerical approaches such as the FEM (Marin *et al.*, 1998).

Accuracy could be improved by higher order approximation of the potential on the tessellated surface, at the cost of increased computational load and complexity. Other improvements have been suggested, such as the 'linear Galerkin' approximation (Mosher *et al.*, 1999), or the 'isolated skull approach' (Hämäläinen and Sarvas, 1989). Still, whatever the complexity of the mesh elements, there will always be errors caused by replacing a smooth surface with a polyhedron.

Anatomically constrained spherical head model

The three sphere shell model can be used directly to model a subject's head. If images of the subject's head and/or the electrode locations are not available, then a standard spherical model can be used. The electrode locations are fixed according to the system used, for example the classic 10-20 system. If anatomical information and the electrode locations are available, then the spherical model can be adjusted to the subject's head. The spherical model can be either fitted to the overall head shape (or to all the electrodes simultaneously), or fitted locally to the area with the largest EEG activity. The former ensures a global fit of the model to the subject's anatomy, while the latter offers a better fit just for the area of interest.

Whatever the approach adopted, the human head is clearly not spherical and the brain volume, i.e. the source space, will not fit in some areas. For example (see Plate 37, colour plate section), when globally fitting the three sphere shell model to the scalp surface, the frontal and occipital lobes protrude beyond the sphere model, and the temporal lobes are too deep within. To overcome these modelling errors, an anatomically constrained spherical head model was proposed by Spinelli *et al.* (2000). In order to retain the simplicity of the three sphere shell model, the anatomy of the head is itself transformed to a best fitting sphere. The spherical transformation and the lead field of the sources are obtained as follows:

1 From an anatomical image, the scalp surface (it could also be the inner skull surface or both) is extracted and tessellated with a regular mesh.
2 The best fitting sphere is estimated with this scalp mesh. The spherical model is defined by its centre $\vec{c}_{sphere}$ and radius R_{sphere}.
3 The source locations inside the brain volume can be expressed in spherical coordinates $(R_{sb}, \theta, \varphi)$ around the best fitting sphere centre. The radius of each source depends on its longitude and latitude, i.e. $R_{sb} = R_{sb}(\theta, \varphi)$. Similarly, the radius of the scalp surface in the direction (θ, φ) of any source is given by $R_{scalp} = R_{scalp}(\theta, \varphi)$.
4 Then the brain volume, i.e. source space, can be rendered spherical by scaling the radius of all the source locations (Plate 38) like this:

$$R_{sb,sphere}(\theta, \varphi) = R_{sb}(\theta, \varphi)\frac{R_{sphere}}{R_{scalp}(\theta, \varphi)} \qquad \mathbf{28.60}$$

5 The lead field for the re-located sources $(R_{sb,sphere}, \theta, \varphi)$, can then be estimated in the spherical model in the usual way.

This process clearly modifies the solution space, introducing discrepancies between the anatomy of the subject and the model. However, compared to the simple best fitting sphere model and the BEM numerical approach, the anatomically constrained spherical head model offers two main advantages. First, even though the solution space is warped, the relative depth of sources is preserved: superficial (resp. deep) sources remain close to (resp. far from) the scalp surface and the electrodes. This is crucial, as the strength of the lead field depends on the distance between the source and the electrodes. Second, the analytic solution of the forward problem can be used to calculate the lead field. This provides a fast and accurate estimation, free from numerical errors that attend the BEM equations. The anatomically constrained spherical head model may be a good compromise between anatomical accuracy and computational efficiency.

This concludes our treatment of the forward model for M/EEG. In the next chapter, we look at the inversion of this model to estimate distributed sources in the brain.

REFERENCES

Arthur RM, Geselowitz DB (1970) Effect of inhomogeneities on the apparent location and magnitude of a cardiac current dipole source. *IEEE Trans Biomed Eng* **17**: 141–46

Ary JP, Klein SA, Fender DH (1981) Location of sources of evoked scalp potentials: Corrections for skull and scalp thickness. *IEEE Trans Biomed Eng* **28**: 447–52

Awada KA, Jackson DR, Williams JT *et al.* (1997) Computational aspects of finite element modeling in EEG source localization. *IEEE Trans Biomed Eng* **44**: 736–52

Becker A (1992) *The boundary element method in engineering. A complete course.* McGraw-Hill Book Company, London

Buchner H, Knoll G, Fuchs M *et al.* (1997) Inverse localization of electric dipole current sources in finite element models of the human head. *Electroencephalogr Clin Neurophysiol* **102**: 267–78

Chan TF (1984) Deflated decomposition of solutions of nearly singular systems. *SIAM J Numeric Anal* **21**: 738–54

Chari M, Silvester P (1980) *Finite elements in electrical and magnetic field problems. Chichester* John Wiley & Sons

Cuffin BN, Cohen D (1979) Comparison of the magnetoencephalogram and electroencephalogram. *Electroencephalogr Clin Neurophysiol* **47**: 132–46

de Munck JC (1992) A linear discretization of the volume conductor boundary integral equation using analytically integrated elements. *IEEE Trans Biomed Eng* **39**: 986–90

Ferguson A, Stroink G (1997) Factors affecting the accuracy of the boundary element method in the forward problem – I: calculating surface potential. *IEEE Trans Biomed Eng* **44**: 1139–55

Ferguson AS, Stroink G (1994) The potential generated by current sources located in an insulated rectangular volume conductor. *J Appl Phys* **76**: 7671–76

Fuchs M, Wagner M, Wischmann HA *et al.* (1998) Improving source reconstructions by combining bioelectric and biomagnetic data. *Electroencephalogr Clin Neurophysiol* **107**: 93–111

Geselowitz DB (1967) On bioelectric potentials in an inhomogeneous volume conductor. *Biophys J* **7**: 1–11

Hämäläinen MS, Hari R, Ilmoniemi RJ *et al.* (1993) Magnetoencephalography – theory, instrumentation, and applications to noninvasive studies of the working human brain. *Rev Mod Phys* **65**: 413–97

Hämäläinen MS, Sarvas J (1989) Realistic conductivity geometry model of the human head for interpretation of neuromagnetic data. *IEEE Trans Biomed Eng* **36**: 165–71

He S (1998) Frequency series expansion of an explicit solution for a dipole inside a conducting sphere at low frequency. *IEEE Trans Biomed Eng* **45**: 1249–58

Heller L (1990) Computation of the return current in encephalography: the auto solid angle. *SPIE Digit Image Synth Inverse Optics* **1351**: 376–90

Huiskamp G, Vroeijenstijn M, van Dijk R *et al.* (1999) The need for correct realistic geometry in the inverse EEG problem. *IEEE Trans Biomed Eng* **46**: 1281–87

Kleinermann F, Avis N, Alhargan F (2000) Analytical solution to the three-dimensional electrical forward problem for a circular cylinder. *Inverse Probl* **16**: 461–68

Klepfer RN, Johnson CR, Macleod RS (1997) The effect of inhomogeneities and anisotropies on electrocardiographic fields: a 3-D finite-element study. *IEEE Trans Biomed Eng* **44**: 706–19

Lambin P, Troquet J (1983) Complete calculation of the electric potential produced by a pair of current source and sink energizing a circular finite-length cylinder. *J Appl Phys* **54**: 4174–84

Lynn MS, Timlake WP (1968a) The numerical solution of singular integral equation of potential theory. *Numer Math* **11**: 77–98

Lynn MS, Timlake WP (1968b) The use of multiple deflations in the numerical solution of singular systems of equations, with application to potential theory. *SIAM J Numer Anal* **5**: 303–22

Marin G, Guerin C, Baillet S *et al.* (1998) Influence of skull anisotropy for the forward and inverse problem in EEG: simulations studies using fem on realistic head models. *Hum Brain Mapp* **6**: 250–69

Meijs JWH, Weier OW, Peters MJ *et al.* (1989) On the numerical accuracy of the boundary element method. *IEEE Trans Biomed Eng* **36**: 1038–49

Mosher JC, Leahy RM, Lewis PS (1999) EEG and MEG: forward solutions for inverse methods. *IEEE Trans Biomed Eng* **46**: 245–59

Nunez PL (1981) *Electric fields of the brain: the neurophysics of EEG.* Oxford University Press, New York

Ramo S, Whinnery JR, van Duze T (1984) *Fields and Waves in Communication Electronics.* John Wiley & Sons, New York

Rosenfeld M, Tanami R, Abboud S (1996) Numerical solution of the potential due to dipole sources in volume conductors with arbitrary geometry and conductivity. *IEEE Trans Biomed Eng* **43**: 679–89

Saleheen HI, Ng KT (1997) New finite difference formulations for general inhomogeneous anisotropic bioelectric problems. *IEEE Trans Biomed Eng* **44**: 800–09

Saleheen HI, Ng KT (1998) A new three-dimensional finite-difference bidomain formulation for inhomogeneous anisotropic cardiac tissues. *IEEE Trans Biomed Eng* **45**: 15–25

Sarvas J (1987) Basic mathematical and electromagnetic concepts of the biomagnetic inverse problem. *Phys Med Biol* **32**: 11–22

Schlitt HA, Heller L, Aaron R *et al.* (1995) Evaluation of boundary element methods for the eeg forward problem: effect of linear interpolation. *IEEE Trans Biomed Eng* **42**: 52–58

Smythe WR (1950) *Static and dynamic electricity.* MacGraw Hill, New York

Spinelli L, Andino SG, Lantz G *et al.* (2000) Electromagnetic inverse solutions in anatomically constrained spherical heads models. *Brain Topogr* **13**: 115–25

Sun M (1997) An efficient algorithm for computing multishell spherical volume conductor models in eeg dipole source localization. *IEEE Trans Biomed Eng* **44**: 1243–52

van den Broek S, Zhou H, Peters M (1996) Computation of neuromagnetic fields using finite-element method and Biot-Savart law. *Med Biol Eng Comput* **34**: 21–26

van Oosterom A, Strackee J (1983) The solid angle of a plane triangle. *IEEE Trans Biomed Eng* **30**: 125–26

Vladimirov VS (1971) *Equations of mathematical physics.* Marcel Dekker, New York

Von Helmholtz HL (1853) Ueber einige Gesetze der Vertheilung elektrischer Ströme in Koperlichen Leitern mit Anwendung auf die thierisch-elektrischen Versuche. *Ann Phys Chem* **89**: 211–33, 354–77.

C H A P T E R

29

Bayesian inversion of EEG models

J. Mattout, C. Phillips, J. Daunizeau and K. Friston

INTRODUCTION

In this chapter, we consider a generative model for evoked neuronal responses as observed with electroencephalography (EEG) and magnetoencephalography (MEG). Because of its linear and hierarchical nature, this model can be estimated efficiently using empirical Bayes. Importantly, multiple constraints on the source distribution can be incorporated in terms of variance components that are estimated from the data. A dual estimation is obtained via an expectation maximization (EM) scheme to give the restricted maximum likelihood (ReML) estimate of the prior covariance components (in terms of hyperparameters) and the maximum *a posteriori* (MAP) estimate of the sources. The Bayesian formalism yields a generic approach to source reconstruction under multiple constraints, which is extended to cover spatio-temporal models for induced responses in the next chapter.

Background

The problem of recovering volume current sources from superficial electromagnetic measurement is intrinsically ill-posed. This means the spatial configuration of neuronal activity cannot be determined uniquely, based on EEG and/or MEG recordings alone (Nunez, 1981). To resolve the non-uniqueness of this inverse problem, assumptions about the solution are necessary for a unique solution. There are two approaches to this:

- Equivalent current dipole (ECD) approaches, where the M/EEG signals are assumed to be generated by a relatively small number of focal sources (Miltner *et al.*, 1994; Scherg and Ebersole, 1994; Scherg *et al.*, 1999; Aine *et al.*, 2000).
- Distributed linear (DL) approaches, where all possible source locations are considered simultaneously (Backus and Gilbert, 1970; Sarvas, 1987; Hamalainen and Ilmoniemi, 1994; Grave de Peralta Menendez and Gonzalez Andino, 1999; Pascual-Marqui, 1999; Uutela *et al.*, 1999).

In the context of distributed models, constraints or priors can be introduced probabilistically using Bayes. The major problem here is the handling of multiple constraints and their appropriate weighting, in relation to each other and observation noise (Gonzalez Andino *et al.*, 2001). In Phillips *et al.* (2002a), we introduced a simple restricted maximum likelihood (ReML) procedure to estimate a single hyperparameter, i.e. the balance between fitting the data and conforming to the priors. Here, we reformulate the implicit weighted minimum norm (WMN) solution in terms of a hierarchical linear model. With this approach, any number of constraints (or priors) on the source or noise covariance matrices can be introduced. An expectation maximization (EM) algorithm is used to obtain an ReML estimate of the hyperparameters associated with each constraint. This enables the MAP solution for the sources to be calculated uniquely and efficiently.

This chapter is divided into five sections. In the first three sections, the theoretical background and operational details of the approach are described. The first section introduces the WMN solution in a Bayesian framework, while the second and third sections introduce the hierarchical parametric empirical Bayes (PEB) framework and ReML approach, respectively. The two last sections give some examples of the application of ReML in comparison with classical approaches, using EEG and MEG data.

Statistical Parametric Mapping, by Karl Friston et al.
ISBN–13: 978-0-12-372560-8 ISBN–10: 0-12-372560-7

THE BAYESIAN FORMULATION OF CLASSICAL REGULARIZATION

The instantaneous source localization problem in EEG can be summarized as:

$$v = F(r, j) + \varepsilon \tag{29.1}$$

where v, a vector of size $N_e \times 1$, is the potential at N_e electrodes; r and j are the 3×1 vectors of source location and moment; ε is the additive noise. F is a function linking the source (r, j) and the potential v. The function F is the solution of the forward problem (i.e. the forward model) and depends only on the head model (conductivities and spatial configuration).

For N_d sources defined by r_i and $j_i (I = 1, K, N_d)$, the source localization problem Eqn. **29.1** can be rewritten, by appeal to the superposition theorem, as:

$$v = \sum_{i=1}^{N_d} F(r_i, j_i) + \varepsilon \tag{29.2}$$

In this chapter, sources of the M/EEG signal are modelled using current dipoles, either on a three-dimensional grid throughout the brain (see fourth section) or on a discrete three-dimensional manifold that corresponds to the cortical surface (see fifth section). This represents a distributed linear approach where N_d is much smaller than N_e. Because the location r_i of each current source i is now fixed, Eqn. **29.1** becomes an underdetermined but linear problem:

$$v = Lj + \varepsilon \tag{29.3}$$

where j now indicates a vector of current dipoles at N_d locations and L is the lead field matrix linking the source amplitudes j to the electrical potential v. If the source orientation is free, then $j = [j_1, j_2, \ldots, j_{N_d}]^T$, where $j_i = [j_{x,i}, j_{y,i}, j_{z,i}]^T$ encodes both orientation and amplitude of the i-th dipole. Otherwise, for oriented sources, $j = [j_1, j_2, \ldots, j_{N_d}]^t$, where each j_i specifies only the amplitude. For discrete data, with N_t time bins, Eqn. **29.3** can be expressed as a multivariate linear model:

$$V = LJ + \varepsilon \tag{29.4}$$

with $V = [v_1, v_2, \ldots, v_{N_t}]$, $J = [j_1, j_2, \ldots, j_{N_t}]$ and $\varepsilon = [\varepsilon_1, \varepsilon_2, \ldots, \varepsilon_{N_t}]$ where v_i, j_i and ε_i are the potential, source parameters and additive noise at the ith time sample.

Weighted minimum norm and Bayesian solutions

As mentioned above, the source localization problem is intrinsically ill posed. With the DL approach, we face the linear but under-determined problem expressed in Eqn. **29.4**. One common approach to this problem is the weighted minimum norm (WMN) solution or Tikhonov regularization method (Tikhonov and Arsenin, 1977), in which the *a priori* regularization constraints have a Bayesian interpretation.

The WMN solution constrains the reconstructed source distribution by minimizing a linear mixture of some weighted norm $\|Hj\|$ of the source amplitudes j and the residuals of the fit. Assuming noise is Gaussian $\varepsilon \sim N(0, C_\varepsilon)$ with a known covariance matrix C_ε, the regularized problem is expressed as:

$$\begin{aligned}\hat{j} &= \min_j \left(\left\| C_\varepsilon^{-1/2}(Lj - v) \right\|^2 + \lambda \left\| Hj \right\|^2 \right) \\ &= \min_j \left((Lj - v)^T C_\varepsilon^{-1} (Lj - v) + \lambda j^T H^T H j \right)\end{aligned} \tag{29.5}$$

where the hyperparameter λ expresses the balance between fitting the model $\left\| C_\varepsilon^{-1/2}(Lj - v) \right\|$ and minimizing the *a priori* constraint $\|Hj\|$. The solution of Eqn. **29.5** for a given λ is:

$$\hat{j} = Tv \tag{29.6}$$

where, using the matrix inversion Lemma:

$$\begin{aligned}T &= [L^T C_\varepsilon^{-1} L + \lambda H^T H]^{-1} L^T C_\varepsilon^{-1} \\ &= (H^T H)^{-1} L^T [L(H^T H)^{-1} L^T + \lambda C_\varepsilon]^{-1}\end{aligned} \tag{29.7}$$

The important connection with Bayesian estimates rests on Gaussian assumptions, under which the conditional expectation of the source amplitudes j is:

$$\begin{aligned}\hat{j} &= [L^T C_\varepsilon^{-1} L + C_j^{-1}]^{-1} L^T C_\varepsilon^{-1} v \\ &= C_j L^T [L C_j L^T + C_\varepsilon]^{-1} v\end{aligned} \tag{29.8}$$

where C_j is the prior covariance of the sources. Comparing Eqn. **29.8** with Eqn. **29.7** provides the motivation for choosing forms of H, where:

$$\lambda H^T H = C_j^{-1} \tag{29.9}$$

In other words, H specifies the form of the precision or our prior belief on where sources are expressed (precision is the inverse of the variance).

In summary, the WMN solution depends on the hyperparameter λ that balances the relative contribution of fitting the model (or likelihood of the data) and the constraint on the solution (or prior). As λ varies, the regularized solution $\hat{j}_\lambda$ also changes. Therefore, the choice of λ is crucial. A heuristic way to understand the properties of $\hat{j}_\lambda$ is to plot the weighted norm of the regularized solution $\left\|H\hat{j}_\lambda\right\|$, against the norm of the residuals $\left\|C_\varepsilon^{-1/2}(L\hat{j}_\lambda - v)\right\|$ for different values of λ. The ensuing curve has an L shape (in ordinary or double logarithmic scale), hence its name *L-curve* (see Appendix 29.1). A satisfactory λ would lie close to the inflection of the *L-curve* (Hansen, 1992). A major disadvantage of the L-curve approach is that solutions must be calculated for a large number of values of λ to find an appropriate level of regularization. Moreover, the *L-curve* approach cannot be extended to estimate multiple hyperparameters. This would require an extensive search in hyperparameter space to determine the inflection in a hyperplane (Brooks *et al.*, 1999).

In Phillips *et al.* (2002a) we introduced an iterative restricted ML (ReML) procedure to estimate λ while calculating $\hat{j}$ for the simple case of one hyperparameter. In the following section, we extend the approach to solve the generalized WMN problem with multiple hyperparameters, thereby determining the relative contribution of different priors to the solution.

A HIERARCHICAL OR PARAMETRIC EMPIRICAL BAYES APPROACH

The source localization problem can be expressed in the context of a two-level hierarchical parametric empirical Bayes (PEB) model:

$$\begin{aligned} v &= Lj + \varepsilon^{(1)} \\ j &= 0 + \varepsilon^{(2)} \end{aligned} \qquad \textbf{29.10}$$

where both random terms have a Gaussian distribution with zero mean:

$$\begin{aligned} \varepsilon^{(1)} &\sim N(0, C_\varepsilon) \\ \varepsilon^{(2)} &\sim N(0, C_j) \end{aligned} \qquad \textbf{29.11}$$

Within this framework, the covariance matrices C_ϵ and C_j, which are equivalent to those in Eqn. **29.5**, can be modelled as a linear combination of covariance components:

$$\begin{aligned} C_\varepsilon &= \lambda_1^{(1)} Q_1^{(1)} + \lambda_2^{(1)} Q_2^{(1)} + \ldots \\ C_j &= \lambda_1^{(2)} Q_1^{(2)}{}_1 + \lambda_2^{(2)} Q_2^{(2)} + \ldots \end{aligned} \qquad \textbf{29.12}$$

where the [restricted] maximum likelihood estimates of $\lambda = \lambda_1^{(1)}, \lambda_2^{(1)}, \ldots, \lambda_1^{(2)}, \lambda_2^{(3)}$ can be identified using expectation maximization (EM) (see Appendix 3 for the operational details). This allows one to compute $C(\lambda)_\varepsilon$ and $C(\lambda)_j$, and the conditional expectation of the sources $\hat{j}$ using Eqn. **29.8**.

In this hierarchical model, the unknown parameters j are assumed to be Gaussian variables with zero mean (i.e. have shrinkage priors). Regional variance can be increased to render sources at some locations more likely. A source with a larger prior variance is less constrained and is more likely to be different from zero. Eqn. **29.12** furnishes a better approximation of the sources' covariance C_j when its exact form is not known *a priori*. A variety of components can be used simultaneously, but the relative weight of each constraint is not fixed. This is particularly important when, for instance, spatial coherence and explicit location priors are used together. For example, some sources can be specified as *a priori* more active: this can be based on prior knowledge or derived from functional magnetic resonance imaging (fMRI) measures of brain activity. Introducing priors from fMRI in terms of covariance components is natural because fMRI provides only spatial information at the temporal scale of M/EEG. Note that specifying priors in terms of covariance components means that sources can be expressed preferentially in different parts of the brain. This should be contrasted with a prior specification in terms of precision components, where sources are precluded from being expressed in areas of high precision. The former is more appropriate because it enables permissive priors as opposed to restrictive priors. For example, each fMRI activation blob could be used as a prior covariance component, which means sources can be expressed in each blob.

Similarly, the noise covariance matrix C_ε can be modelled more accurately with a mixture of components. For example, an independent and uniform noise component over electrodes can be introduced by defining $Q_1^{(1)}$ as the identity matrix. Covariances among electrodes can be introduced in $Q_2^{(1)}$, which is generally estimated from the data (e.g. within the prestimulus interval of the event-related potential (ERP)). If a subset of electrodes picks up more noise than the others, this can also be modelled in $Q_2^{(1)}$.

By selecting the EEG episode carefully, during which the hyperparameters are stationary, the EM algorithm favours the relevant priors by increasing their hyperparameters and suppresses the others by rendering their hyperparameters very small.

RESTRICTED MAXIMUM LIKELIHOOD

In practice, the hierarchical model is inverted by minimizing the ReML objective function with respect to the hyperparameters (see Appendix 3). The critical aspect of this inversion is that it can proceed in channel space. This means the size of the matrices are relatively small and the inversion is extremely quick. The two-level model in Eqn. **29.10** can be collapsed into a single equation:

$$v = L\varepsilon^{(2)} + \varepsilon^{(1)} \quad \textbf{29.13}$$

This means the covariance C_v of the channel data[1] has the following components:

$$\begin{aligned} C_v &= LC_jL^T + C_\varepsilon \\ &= \lambda_1^{(1)}Q_1^{(1)} + \lambda_2^{(1)}Q_2^{(1)} + \cdots + \lambda_1^{(2)}LQ_1^{(2)}L^T + \lambda_2^{(2)}LQ_2^{(2)}L^T + \cdots \end{aligned} \quad \textbf{29.14}$$

where only the hyperparameters are unknown and can be estimated using restricted maximum likelihood, as described in Appendix 4. This gives the same [restricted] maximum likelihood estimate as the **M**-step of the EM algorithm, but does so in a computationally expedient way, by variance partitioning in sensor-space. To get very precise estimates of the hyperparameters one can use multiple observations to calculate the covariance matrix $C_v \approx vv^T$. This sample covariance can be based on successive time bins, assuming that the noise and prior covariances are locally stationary. Although not pursued here, it is also possible to use instantaneous estimates of C_v sampled from the same time bin of multiple trials.

APPLICATION TO SYNTHETIC MEG DATA

In this section, we illustrate the ReML scheme using simulated MEG data. Further details about these simulations and results can be found in Mattout *et al.* (2006) and Chapter 35.

The simulated data

To simulate MEG data, a 3D high resolution (voxel size: 0.9375 mm × 0.9375 mm × 1.5 mm) MRI volume from a healthy volunteer was segmented. The boundary between white and grey matter was approximated with small triangles whose vertices provided about 7000 dipole locations spread uniformly over the cortex. We computed the forward operator L for this dipole mesh, using a single-shell spherical head model (Sarvas, 1987).

MEG data were simulated over 130 sensors, by activating two extended sources. Each source was a cluster comprising one randomly chosen dipole and its four nearest neighbours. The extent of each simulated source was about 5 mm in radius. The activation was modelled with a half-period sine function (over 15 time bins). A delay of two time bins was applied to waveforms of the two sources. After projection onto sensor space, white Gaussian noise was added ($SNR = 20$ dB).[2] Five hundred data sets were simulated to compare the ReML approach with the classical WMN estimation based on the *L-curve* approach and to study the performance of the ReML scheme under various combinations of priors.

The priors

At the sensor level, we consider a single measurement noise component defined by $Q_1^{(1)} = I_{N_e}$, i.e. independent measurement noise on each sensor with identical variance. I_{N_e} is the $N_e \times N_e$ identity matrix. At the source level, three types of priors are considered, either individually or together.

Smoothness constraint

This is defined by the covariance component:

$$Q_S^{(1)}(i, j) = \exp(-d_{ij}^2/2s^2) \quad \textbf{29.15}$$

where d_{ij} is the distance[3] between dipoles i and j. The spatial smoothness parameter was $s = 8$ mm. Like LORETA (Pascual-Marqui *et al.*, 1994), this prior enforces correlation among neighbouring sources.

Intrinsic functional constraint

Multivariate source prelocalization (MSP) provides, from the normalized MEG data itself, an estimate α_i of the likelihood of activation at each source location (Mattout *et al.*, 2005). These estimates can be incorporated as quantitative priors. They can also provide a substantial reduction of the inverse solution space by only considering the dipoles that are most likely to be active. For each simulated source configuration we restrict the solution space

[1] We refer to C_v as a covariance matrix but, strictly speaking, it is just a second order matrix. Covariance matrices are second order matrices of mean corrected variables, whereas our variables are usually baseline corrected.

[2] *SNR* stands for *signal-to-noise ratio* and is here expressed in decibels, i.e. $SNR = \log_{10}(A_s/A_n)$, where A_s (resp. A_n) refers to the maximum absolute signal (resp. noise) value. An *SNR* of 20 dB thus corresponds to a 10 per cent noise level.

[3] The Euclidian distance was used for these simulations. However, the smoothness constraint as implemented in SPM5 uses the geodesic distance over the cortical surface.

to the 1500 dipoles with the highest likelihood of activation. Within this subset, the intrinsic prior covariance component is the leading diagonal matrix:

$$Q_i^{(2)}(i, i) = \alpha_i \qquad \textbf{29.16}$$

Extrinsic functional constraints

The third constraint we consider is based on data from other imaging modalities, typically fMRI, and enters simply as a binary mask. This mask distinguishes qualitatively between *a priori* active and non-active cortical areas. The corresponding prior source variance component is defined by the leading diagonal matrix:

$$Q_e^{(2)}(i, i) = 1 \qquad \textbf{29.17}$$

when the source is part of an active area, and zero otherwise. We modelled two sorts of extrinsic priors – valid $Q_e^{(2)}$ and invalid $\tilde{Q}_e^{(2)}$ (Figure 29.1) – because we were interested in the impact of invalid priors, particularly in the context of multiple priors. In this instance, there is an opportunity to discount invalid priors in favour of valid priors.

Results

Since the L-curve approach can only accommodate a single constraint, the classical WMN provides only four solutions for each data set, one per source prior, whereas ReML provides nineteen, corresponding to all possible mixtures of priors. Here we use the receiver operating characteristic (ROC) to evaluate and compare the various source estimates. The ROC characterizes inverse methods in terms of correctly classifying each dipole, as either active or not. For each estimate of the source distribution, a ROC curve represents the true positive rate (*sensitivity*) versus the false positive rate (1−*specificity*). The area under the curve (AUC) quantifies the overall power. The AUC ranges between 0 and 1, indicating the probability of correct separation of an active source from a non-active one. Comparing the AUC of different inverse models allows one to assess the relative performance of methods and sets of priors.

Table 29-1 shows the averaged AUC value over our simulations. We analysed the AUC using analysis of variance (ANOVA). The main effect of method (ReML versus

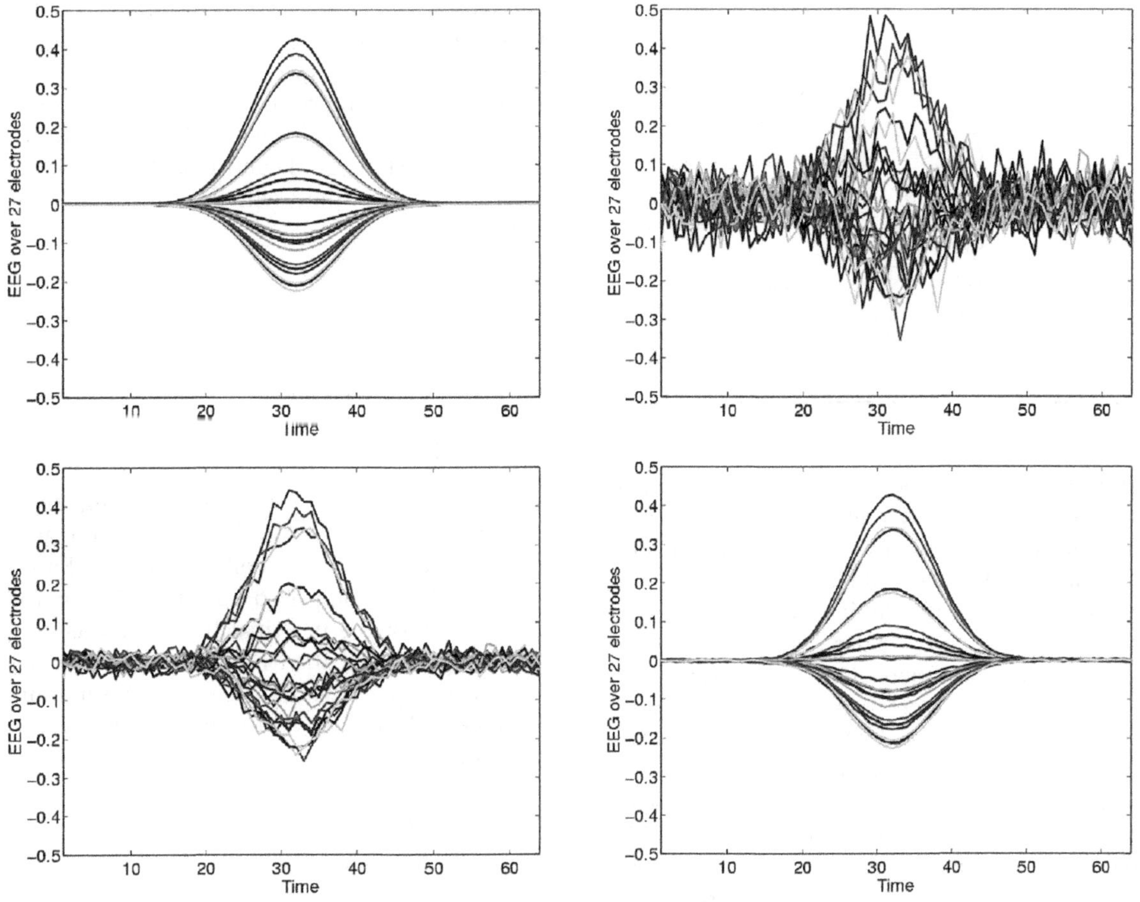

FIGURE 29.1 Synthetic EEG data: data without noise (top left). Same data with added noise, at different SNRs: 4 (top right), 12 (bottom left), 100 (bottom right).

TABLE 29-1 Mean values of the AUC for the WMN and ReML approaches and different priors

AUC	Prior models	ReML	WMN
1 constraint	$Q_s^{(2)}$	0.7833	0.777
	$Q_i^{(2)}$	0.7944	0.7746
	$Q_e^{(2)}$	0.8560	0.8560
	$\tilde{Q}_e^{(2)}$	0.4994	0.4994
2 constraints	$Q_s^{(2)}$	0.7999	
	$Q_i^{(2)}$		
	$Q_s^{(2)}$	0.8211	
	$Q_e^{(2)}$		
	$Q_s^{(2)}$	0.7931	
	$\tilde{Q}_e^{(2)}$		
	$Q_i^{(2)}$	0.8211	
	$Q_e^{(2)}$		
	$Q_i^{(2)}$	0.7962	
	$\tilde{Q}_e^{(2)}$		
	$Q_i^{(2)}$	0.8536	
	$\tilde{Q}_e^{(2)}$		
3 constraints	$Q_s^{(2)}$	0.8211	
	$Q_i^{(2)}$		
	$Q_e^{(2)}$		
	$Q_s^{(2)}$	0.7972	
	$Q_i^{(2)}$		
	$\tilde{Q}_e^{(2)}$		
	$Q_s^{(2)}$	0.8211	
	$Q_e^{(2)}$		
	$\tilde{Q}_e^{(2)}$		
	$Q_i^{(2)}$	0.8211	
	$Q_e^{(2)}$		
	$\tilde{Q}_e^{(2)}$		
4 constraints	$Q_s^{(2)}$	0.8206	
	$Q_i^{(2)}$		
	$Q_e^{(2)}$		
	$\tilde{Q}_e^{(2)}$		

WMN) proves highly significant ($F(1,499) = 1.01$; $p < 0.001$), implying a much better source detection profile with ReML. Since the ReML and WMN approaches differ only in the way they estimate the hyperparameters, these results suggest that the ReML estimates of the balance between the priors and data fit are significantly better than obtained with the traditional *L-curve* approach.

When using ReML, the effect of the valid extrinsic prior can be assessed by a two-by-seven ANOVA, whose factors are the inclusion or not of the valid extrinsic prior and the seven possible prior models. The main effect on the valid extrinsic prior was highly significant ($F(1,499) = 2565.272$; $p < 0.001$). This shows that relevant priors are properly assigned a high weight, even in the context of invalid or irrelevant priors. Conversely, any deterioration in the reconstruction, due to the inclusion of the invalid prior, was insignificant ($F(1,499) = 0.140$; $p < 0.708$). Again, this is important because it shows that irrelevant priors are properly discounted by the inversion.

APPLICATION TO SYNTHETIC EEG DATA

The simulated data

We used a simplified head model that comprises 1716 dipoles distributed uniformly on a horizontal grid (with a maximum of 24 sources along a radius), within a three-sphere shell model. Twenty-seven electrodes were placed on the upper hemisphere according to a pseudo 10–20 electrode setup. The orientation of each source was fixed and the lead-field for all sources was calculated analytically (Ary *et al.*, 1981). Two hundred locations were selected randomly to assess the efficiency of the ReML approach. At each of these locations an instantaneous distributed source set j_0 was generated as a set of connected dipoles within a 1.5 grid-size radius of a 'central' dipole. On average, each source comprised about nine dipoles. Each source vector j_0 was modulated over time to generate a time series J_0. Data with different SNRs were obtained by adding scaled white noise $\varepsilon^{(1)}$ to the noise-free data $V = LJ_0 + \varepsilon$. Three levels of noise were used by adopting an SNR of 4, 12 and 100. The electrical potential at electrodes, over time, with these three SNRs are shown in Figure 29.1.

The priors

We considered three kinds of priors on the sources.

Spatial coherence

This component was constructed using anatomically informed basis functions (IBFs) based on grey matter density and a spatial smoothness kernel (Phillips *et al.*, 2002b, 2005). It comprised a leading diagonal matrix of eigenvalues of the principal IBFs that was rotated into source space (and then projected onto channel space). This prior implies that the basis functions in source space, with the highest eigenvalues, are more likely to represent the source activity, ensuring a smooth solution.

Depth constraint

To ensure that sources contribute to the solution equally, irrespective of their depth (Ioannanides *et al.*, 1990; Gorodnitsky *et al.*, 1995; Grave de Peralta Menendez and Gonzales Andino, 1998), deeper sources are given a larger prior variance than superficial sources. The depth is indexed by the norm of the gain vector for each source. This covariance component is defined by the leading diagonal matrix $diag(L^T L)^{-1}$.

Location priors

These were introduced as leading diagonal matrices whose elements encode the prior probability of whether the source is active or not. As above, these extrinsic functional priors take values of zero (the prior variance is not affected), or one (the prior variance increases with the corresponding hyperparameter). Three different types of location priors are considered, either separately or together:

1 accurate priors, centred on the active source set
2 close inaccurate priors, located between 6 and 16 grid-size from the truly active source
3 distant inaccurate priors, located between 24 and 48 grid-size from the truly active source.

Each simulation condition (SNR, priors) was assessed for the 200 different source configurations.

Results

Figure 29.2 and Plate 39 (see colour plate section) show the reconstructions obtained with the various constraints

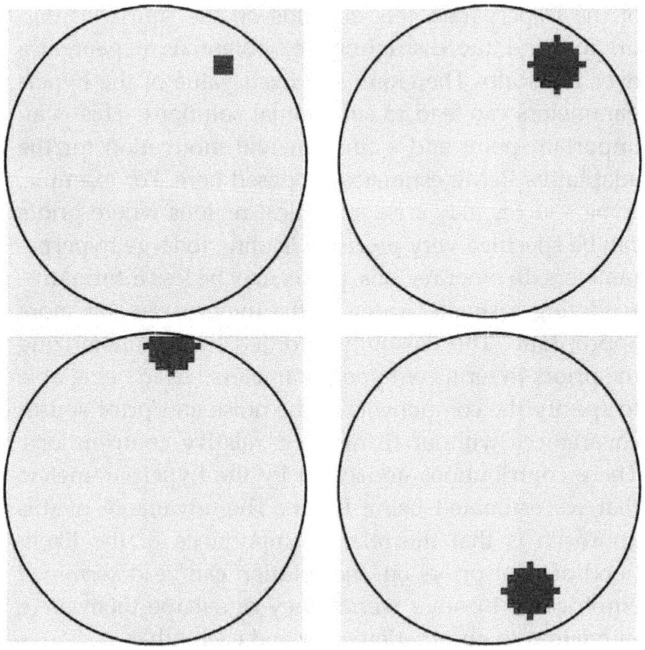

FIGURE 29.2 Example of a source configuration used in the simulations (top left) and the corresponding priors; accurate location priors (top right), and inaccurate location priors (close, bottom left, and distant, bottom right).

TABLE 29-2 Localization error (mm) for the three SNRs and the four location priors (none, accurate, inaccurate and both accurate and inaccurate)

Localization error	Low	Medium	High
No priors	6	4	10
Accurate priors	4	4	4
Close inaccurate priors	12	15	13
Distant inaccurate priors	34	32	32
Accurate and close inaccurate priors	4	4	4
Accurate and distant inaccurate priors	4	4	4

described above. To evaluate the performance of the inversion we used a lower bound on the localization error such that 80 per cent of the sources are recovered within this bound (Table 29-2). Table 29-3 provides the mean and standard deviation of the hyperparameter of the noise component. This estimate is relatively stable over simulations. It reflects accurately the actual variance of noise, i.e. 2.3 (low SNR), 0.26 (medium SNR) and 0.0037 (high SNR). Although the estimated noise variance is accurate, the standard deviation of its estimate is much smaller when accurate location priors are used. The inclusion of accurate location priors seems to help the partitioning of signal variance into its components, namely, source activity and additive noise. For some sources, the noise variance was greatly over-estimated.

Finally, Table 29-4 gives the mean and standard deviation of the hyperparameters pertaining to the source variance components. Note that some hyperparameters are negative. This simply means that the corresponding covariance component is used to reduce the (co)variance of the sources. The combination of all the covariance components should always lead to a positive definite (co)variance matrix estimate. In general, the hyperparameters decrease as the SNR increases, except when inaccurate location priors are used in isolation. The use of accurate location priors has a large influence on the other priors. Indeed, the hyperparameter corresponding to the accurate location priors is several orders larger than any other hyperparameter (i.e. the spatial coherence, depth or inaccurate location priors). Apart from the hyperparameter corresponding to the accurate location priors, the hyperparameters vary over a relatively large range and can have positive or negative values depending on the source, the noise and the priors. Generally, the standard deviation of the hyperparameters estimates decreases as the SNR increases.

TABLE 29-3 Mean and standard deviation of the hyperparameters of the noise component*

Hyperparameters	Low SNR	Medium SNR	High SNR
Without accurate location priors	2.3 ± 0.15	0.26 ± 0.030	0.0082 ± 0.047
With accurate location priors	2.3 ± 0.04	0.26 ± 0.0065	0.0037 ± 0.00012

*The precision of this estimate depends mainly on the presence or absence of accurate location priors. Therefore, the simulations without any location priors and with inaccurate location priors are pooled in the category 'Without accurate location priors'. Similarly, results obtained 'With accurate location priors' cover simulations with accurate location priors, with and without inaccurate location priors.

TABLE 29-4 Mean and standard deviation of the hyperparameters pertaining to the sources' prior variance components

Hyperparameters	SNR	Spatial coherence	Depth constraint	Accurate location	Inaccurate location
No location priors	Low	−0.5 ± 1.9	6.8 ± 14.0	–	–
	Medium	−0.35 ± 1.7	5.6 ± 12.0	–	–
	High	−0.058 ± 1.3	3.1 ± 8.8	–	–
Accurate location priors	Low	0.016 ± 0.043	−0.2 ± 0.32	16 ± 4.1	–
	Medium	0.0073 ± 0.0097	−0.72 ± 0.076	9.1 ± 1.9	–
	High	0.00016 ± 0.00032	−0.0016 ± 0.0026	6.9 ± 1.6	–
Inaccurate location priors	Low	−0.39 ± 1.7	5.6 ± 1.3	–	4.7 ± 1.9
	Medium	−0.41 ± 1.5	5.5 ± 11	–	18 ± 72
	High	−0.092 ± 1.2	2.9 ± 7.9	–	70 ± 300
Accurate and inaccurate location priors	Low	0.021 ± 0.064	−0.23 ± 0.49	16 ± 4.4	−0.15 ± 0.36
	Medium	0.088 ± 0.014	−0.082 ± 0.11	9.4 ± 2.1	−0.076 ± 0.14
	High	0.00017 ± 0.00034	−0.0016 ± 00.0029	6.9 ± 1.6	−0.0021 ± 0.013

CONCLUSION

The hierarchical PEB-ReML approach presented here provides efficient and optimal estimators of M/EEG evoked responses. Our analyses of synthetic EEG and MEG data show that: the noise variance estimate is accurate and consistent; localization and detection error is greatly reduced by the introduction of valid location priors; and the further introduction of invalid priors had no effect on the reconstruction.

Combining data obtained from different modalities within the same framework can overcome the intrinsic limitations (on temporal or spatial resolution) of any one modality. In this chapter, we have outlined a way in which structural and functional MRI data can be used as priors in the estimation of M/EEG sources. Crucially, we have illustrated the role of ReML hyperparameter estimates in modelling the relative contributions of M/EEG residuals and fMRI-based priors to the estimation.

Classical approaches to noise regularization of distributed linear solutions are usually empirical and proceed on a trial-and-error basis: the level of regularization is adapted manually such that the ensuing solution and assumed noise component seem reasonable. In contrast, the ReML procedure successfully controls the noise regularization by a hyperparameter in a principled and unique way. Even at constant SNR, the values of the hyperparameters vary over a wide range. The value of the hyperparameters depends on the source configuration and the distribution of potentials it generates over the scalp. Therefore, any fixed value of the hyperparameters can lead to suboptimal solutions. This is an important point and a fundamental motivation for the adaptative ReML estimates proposed here. For example, some sources may arise in cortical regions where priors can be specified very precisely leading to large hyperparameters. In other regions, priors may be less informative rendering a smaller value of the hyperparameter more appropriate. The flexibility afforded by parameterizing the priors in terms of hyperparameters lies in being able to specify the components of the noise and prior source covariances without fixing their relative contributions. These contributions are scaled by the hyperparameters that we estimated using ReML. The advantage of this approach is that the relative importance of the likelihood of, and priors on, the solution can be determined empirically. In other words, they can shape themselves in relation to observation error and each other.

In this chapter, we have focused on the face validity of Bayesian inversion (i.e. showing that the scheme does what it is supposed to do). It is also fairly easy to establish the construct validity in terms of neurobiological plausibility. The example in Plate 40 shows that the ReML scheme localizes face-selective responses to the appropriate part of the fusiform gyrus. In this instance, intrinsic functional priors produced the best results. Clearly, one cannot assess the intrinsic quality of the priors based on the solution. However, the ReML objective function (which is also the variational free energy – see Chapter 24) can be used for model comparison and selection, i.e. to evaluate different reconstructions obtained from different sets of priors (Mattout *et al.*, 2006). This will be the subject of Chapter 35. In the next chapter, we extend the instantaneous model described above to cover peristimulus time and generalize the inversion to provide conditional estimates of both evoked and induced responses.

APPENDIX 29.1 THE L-CURVE APPROACH

The L-curve heuristic involves estimating the WMN solution for various values of hyperparameter λ (see Eqn. **29.5** and Eqn. **29.7**). A plot of the norm of the prior term against the norm of the data fit leads to an *L-shape* curve whose inflection point indicates an optimal hyperparameter. This amounts to maximizing the following:

$$-\frac{1}{2}(v - L\hat{j}_\lambda)^T C_\varepsilon^{-1}(v - L\hat{j}_\lambda) - \frac{1}{2}\hat{j}_\lambda^T C_j^{-1} \hat{j}_\lambda \qquad \textbf{29.A1}$$

A precise estimation entails an exhaustive scanning of hyperparameter space. This is one drawback of the approach, namely, the need for a large number of estimations to find an appropriate level of regularization. We use L-curve analysis as a reference for the estimation of single hyperparameters in the WMN simulations. Practically, we use:

$$\lambda = \beta \frac{\|LL^T\|}{N_e} \qquad \textbf{29.A2}$$

with thirty values of β. This ensures an *L-curve* with a relatively fine sampling in the vicinity of its inflection. Note that the *L-curve* minimization criterion is a poor approximation to the ReML (EM) objective function (see Appendix 3):

$$\begin{aligned} F &= p(v|\lambda) \\ &= -\frac{1}{2}(v - L\hat{j})^T C_\varepsilon^{-1}(v - L\hat{j}) - \frac{1}{2}\hat{j}^T C_j^{-1} j - \frac{1}{2}\log|C_\varepsilon| \\ &\quad -\frac{1}{2}\ln|C_j| + \frac{1}{2}\ln|C_{v|y}| + \ldots \end{aligned} \qquad \textbf{29.A3}$$

which accounts properly for uncertainty in the source estimates.

REFERENCES

Aine C, Huang M, Stephen J *et al.* (2000) Multistart algorithms for MEG empirical data analysis reliably characterize locations and time courses of multiple sources. *NeuroImage* **12**: 159–72

Ary JP, Klein SA, Fender DH (1981) Location of sources of evoked scalp potentials: corrections for skull and scalp thickness. *IEEE Trans Biomed Eng* **28**: 447–52

Backus GE, Gilbert JF (1970) Uniqueness in the inversion of inaccurate gross earth data. *Philos Trans R Soc* **266**: 123–92

Brooks DH, Ahmad GF, MacLeod RS *et al.* (1999) Inverse electrocardiography by simultaneous imposition of multiple constraints. *IEEE Trans Biomed Eng* **46**: 3–17

Gorodnitsky L, George JS, Rao BD (1995) Neuromagnetic source imaging with FOCUSS: a recursive weighted minimum norm algorithm. *Electroencephalogr Clin Neurophysiol* **95**: 231–51

Gonzales Andino SL, Blanke O, Lantz G *et al.* (2001) The use of functional constraints for the neuroelectromagnetic inverse problem: Alternatives and caveats. *Int J Bioelectromagnetism* **3**: 1–17

Grave de Peralta Menendez R, Gonzales Andino SL (1998) A critical analysis of linear inverse solutions to the neuroelectromagnetic inverse problem. *IEEE Trans Biomed Eng* **45**: 440–48

Grave de Peralta Menendez R, Gonzales Andino SL (1999) Backus and Gilbert method for vector fields. *Hum Brain Mapp* **7**: 161–65

Hamalainen MS, Ilmoniemi RJ (1994) Interpreting magnetic fields of the brain – minimum norm estimates. *Med Biol Eng Comput* **32**: 35–42

Hansen PC (1992) Analysis of discrete ill-posed problems by means of the L-curve. *SIAM Rev* **34**: 561–80

Ionnanides AA, Bolton JPR, Clarke CJS (1990) Continuous probabilistic solutions to the biomagnetic inverse problem. *Inverse Probl* **6**: 523–43

Mattout J, Pélégrini-Issac M, Garnero L *et al.* (2005) Multivariate source prelocalization (MSP): use of functionally informed basis functions for better conditioning the MEG inverse problem. *NeuroImage* **26**: 356–73

Mattout J, Phillips C, Penny WD *et al.* (2006) MEG source localization under multiple constraints: an extended Bayesian framework. *NeuroImage* **30**: 753–67

Miltner W, Braun C, Johnson R *et al.* (1994) A test of brain electrical source analysis (BESA): a simulation study. *Electroencephalogr Clin Neurophysiol* **91**: 295–310

Nunez PL (1981) *Electric fields of the brain: the neurophysics of EEG*. Oxford University Press, New York

Pascual-Marqui RD (1999) Review of methods for solving the EEG inverse problem. *Int J Bioelectromag* **1**: 75–86

Pascual-Marqui RD, Michel CM, Lehmann D (1994) Low resolution electromagnetic tomography: a new method for localizing electrical activity in the brain. *IEEE Trans Biomed Eng* **418**: 49–65

Phillips C, Rugg MD, Friston KJ (2002a) Anatomically informed basis functions for EEG source localization: combining functional and anatomical constraints. *NeuroImage* **16**: 678–95

Phillips C, Rugg MD, Friston KJ (2002b) Systematic regularisation of linear inverse solution of the EEG source localisation problem. *NeuroImage* **17**: 287–301

Phillips C, Mattout J, Rugg MD *et al.* (2005) An empirical Bayesian solution to the source reconstruction problem in EEG. *NeuroImage* **24**: 997–1011

Sarvas J (1987) Basic mathematical and electromagnetic concepts of the biomagnetic inverse problem. *Phys Med Biol* **32**: 11–22

Scherg M, Ebersole JS (1994) Brain source imaging of focal and multifocal epileptiform EEG activity. *Clin Neurophysiol* **24**: 51–60

Scherg M, Bast T, Berg P (1999) Multiple source analysis of interictal spikes: goals, requirements, and clinical value. *J Clin Neurophysiol* **16**: 214–24

Tikhonov AN, Arsenin VY (1977) *Solutions of ill-posed problems*. John Wiley, New York

Uutela K, Hamalainen MS, Somersolo E (1999) Visualization of magnetoencephalographic data using minimum current estimates. *NeuroImage* **10**: 173–80

C H A P T E R

30

Bayesian inversion for induced responses

J. Mattout, C. Phillips, J. Daunizeau and K. Friston

INTRODUCTION

Electroencephalography (EEG) and magnetoencephalography (MEG) provide a non-invasive and instantaneous measure of the whole brain activity. These measures reflect synchronous postsynaptic potentials of cortical populations of neurons (Nunez and Silberstein, 2000). Unfortunately, localizing these electromagnetic sources represents an ill-posed inverse problem that, in the absence of constraints, does not admit a unique solution. Consequently, deriving a realistic and unique solution rests on prior knowledge, in addition to the observed measurements.

Any source reconstruction approach comprises three components. The first relates to the definition of the solution space and a parametric representation of the sources. The second embodies information about the physical and geometrical properties of the head; this models the propagation of brain electromagnetic fields through the various tissues (i.e. a forward model). Together, these two components constitute a generative model of the EEG/MEG data. By nature, a generative model can be used for both simulating synthetic datasets and estimating the parameters of local neuronal activity from experimental data. Finally, the generative model is inverted to provide conditional estimates of the sources (Baillet and Garnero, 1997; Schmidt *et al.*, 1999; Phillips *et al.*, 2002; Amblard *et al.*, 2004; Daunizeau *et al.*, 2006).

Two types of inverse methods can be distinguished by their respective source model: the equivalent current dipole (ECD) and distributed modelling (DM). Although other source models have been used, such as multipoles (Jerbi *et al.*, 2004) or continuous current densities (Riera *et al.*, 1998), both approaches usually rely upon a dipolar representation of cortical sources, which are parameterized in terms of location, orientation and intensity. An ECD models the activity of a large cortical area. MEG and EEG data are then explained by few ECDs (usually less than five). Distributed models consider a large number (a few thousands) of dipoles deployed at fixed locations over the cortical surface. Although the underlying parametric models are the same, the parameterization of the solution space is very different, calling for different forward calculations as well as different inverse operators and solutions. In this chapter, we focus on distributed models.

In contradistinction to most ECD approaches, DM uses the subject's anatomy, usually derived from high resolution anatomical magnetic resonance imaging (MRI) (Dale and Sereno, 1993). The solution space and associated forward models can then be made as realistic as allowed by computational constraints and the precision of head tissue conductivity measures. Moreover, due to the use of fixed dipole locations, the forward computation only need be computed once, prior to any inverse operation. DM yields a highly under-determined but linear system which is formally similar to those encountered in signal and image processing. These problems can be treated in a Bayesian way, using priors to furnish a unique solution. Prior constraints are needed due to the under-determinacy of the system.

In the context of DM, priors based on mathematical, anatomical, physiological and functional heuristics have been considered (Hamalainen and Ilmoniemi, 1994; Pascual-Marqui *et al.*, 1994; Gorodnitsky *et al.*, 1995; Baillet and Garnero, 1997; Dale *et al.*, 2000; Phillips *et al.*, 2002; Babiloni *et al.*, 2004; Mattout *et al.*, 2005). Although these approaches involve different constraints and inverse criteria, they all obtain a unique solution by optimizing a goodness of fit term and a prior term in a carefully balanced way. Most can be framed in terms of a weighted minimum norm (WMN) criterion, which represents the classical and most popular distributed approach (Hauk, 2004).

Statistical Parametric Mapping, by Karl Friston et al.
ISBN–13: 978-0-12-372560-8 ISBN–10: 0-12-372560-7

However, a critical outstanding issue lies in the relative weighting of the accuracy and regularization criteria upon which the solution depends. Usually, in the context of Tikhonov regularization or WMN solutions, this weighting is fixed arbitrarily, or by using the L-curve heuristic. The latter case, which we will refer to as the (classical) WMN, is limited because it can only accommodate a single constraint on the source parameters. This means that multiple constraints (e.g. spatial and temporal) (Baillet and Garnero, 1997) have to be mixed into a single prior term, using *ad hoc* criteria.

The inverse approach considered in this chapter is a generalization of the approach described in the previous chapter. This inversion uses a hierarchical (general) linear model that embraces, under the assumption of Gaussian errors, multiple constraints specified in terms of variance components. These constraints can be formulated in sensor or source space. The optimal weight associated with each constraint is estimated from the data using empirical Bayes and is computed iteratively using expectation–maximization (EM) (Friston *et al.*, 2002). These weights are equivalent to restricted maximum likelihood (ReML) estimates of the prior covariance components.

Using temporal basis functions, the same approach can be extended to estimate both evoked and induced responses. This chapter shows how one can estimate evoked responses which are phase-locked to the stimulus, and induced responses that are not. For a single trial, the model is exactly the same. However, in the context of multiple trials, the inherent distinction between evoked and induced responses calls for different treatments of a hierarchical multitrial model. This is because there is a high correlation between the response evoked in one trial and that of the next. Conversely, induced responses have a random phase-relationship over trials and are, *a priori*, independent. In what follows, we derive the respective models and show how they can be estimated efficiently using ReML. This enables the Bayesian estimation of evoked and induced changes in power.

This chapter comprises four sections. The first section describes the ReML identification operators based on covariances, over time, for a single trial. We then consider an extension of this scheme that accommodates constraints on the temporal expression of responses using temporal basis functions. In the third section, we show how the same conditional operator can be used to estimate response energy or power. In the fourth section, we consider extensions to the model that cover multiple trials and show that evoked responses are based on the covariance of the average response over trials, whereas induced responses are based on the average covariance.

THE BASIC ReML APPROACH TO DISTRIBUTED SOURCE RECONSTRUCTION

Hierarchical linear models

Inversion of hierarchical models for M/EEG was covered in the previous chapter, so we focus here on the structure of the problem and on the nature of the variables that enter the ReML scheme. The empirical Bayes approach to multiple priors, in the context of unknown observation noise, rests on the hierarchical observation model:

$$\begin{aligned} y &= Lj + \varepsilon^{(1)} \\ j &= \varepsilon^{(2)} \\ Cov(vec(\varepsilon^{(1)})) &= V^{(1)} \otimes C^{(1)} \\ Cov(vec(\varepsilon^{(2)})) &= V^{(2)} \otimes C^{(2)} \\ C^{(1)} &= \sum \lambda_i^{(1)} Q_i^{(1)} \\ C^{(2)} &= \sum \lambda_i^{(2)} Q_i^{(2)} \end{aligned} \qquad 30.1$$

where y represents a $c \times t$ data matrix of channels $\times$ time bins. L is a $c \times s$ lead-field matrix, linking the channels to the s sources, and j is an $s \times t$ matrix of source activity over peristimulus time. $\varepsilon^{(1)}$ and $\varepsilon^{(2)}$ are random effects, representing observation error or noise and unknown source activity respectively. $V^{(1)}$ and $V^{(2)}$ are the temporal correlation matrices of these random effects. Most approaches, including our previous work, have assumed them to be the identity matrix. However, they could easily model serial correlations and, indeed, non-stationary components. $C^{(1)}$ and $C^{(2)}$ are the spatial covariances for noise and sources respectively; they are linear mixtures of covariance components $Q_i^{(1)}$ and $Q_i^{(2)}$, which embody spatial constraints on the solution. $V^{(1)} \otimes C^{(1)}$ represents a parametric noise covariance model (Huizenga *et al.*, 2002) in which the temporal and spatial components factorize. Here the spatial component can have multiple components estimated through $\lambda_i^{(1)}$, whereas the temporal form is fixed. At the second level, $V^{(2)} \otimes C^{(2)}$ can be regarded as spatio-temporal priors on the sources $p(j) = N(0, V^{(2)} \otimes C^{(2)})$, whose spatial components are estimated empirically in terms of $\lambda_i^{(2)}$.

The ReML scheme described here is based on the two main results for *vec* operators and Kronecker tensor products:

$$\begin{aligned} vec(ABC) &= (C^T \otimes A)vec(B) \\ tr(A^T B) &= vec(A)^T vec(B) \end{aligned} \qquad 30.2$$

The *vec* operator stacks the columns of a matrix on top of each other to produce a long column vector. The trace operator $tr(A)$ sums the leading diagonal elements of a matrix A and the Kronecker tensor product $A \otimes B$ replaces each element A_{ij} of A with $A_{ij}B$ to produce a larger matrix. These equalities enable us to express the forward model in Eqn. **30.1** as:

$$\begin{aligned} vec(y) &= (I \otimes L)vec(j) + vec(\varepsilon^{(1)}) \\ vec(j) &= vec(\varepsilon^{(2)}) \end{aligned} \qquad \mathbf{30.3}$$

and express the conditional mean $\hat{j}$ and covariance $\hat{\Sigma}$ as:

$$\begin{aligned} vec(\hat{j}) &= \hat{\Sigma}(V^{(1)-1} \otimes L^T C^{(1)-1})vec(y) \\ &= (V^{(2)} \otimes C^{(2)} L^T)(V^{(2)} \otimes LC^{(2)} L^T + V^{(1)} \otimes C^{(1)})^{-1} vec(y) \\ \hat{\Sigma} &= (V^{(2)-1} \otimes C^{(2)-1} + V^{(1)-1} \otimes L^T C^{(1)-1} L)^{-1} \end{aligned} \qquad \mathbf{30.4}$$

The first and second lines of Eqn. **30.4** are equivalent by the matrix inversion lemma. The conditional mean $vec(\hat{j})$ or maximum *a posteriori* (MAP) estimate is the most likely source, given the data. The conditional covariance $\hat{\Sigma}$ encodes uncertainty about $vec(j)$ and can be regarded as the dispersion of a distribution over an ensemble of solutions centred on the conditional mean.

In principle, Eqn. **30.4** provides a way of computing conditional means of the sources and their conditional covariances. However, it entails pre-multiplying a very long vector with an enormous $(st \times ct)$ matrix. Things are much simpler when the temporal correlations of noise and signal are the same, i.e. $V^{(1)} = V^{(2)} = V$. In this special case, we can compute the conditional mean and covariances using much smaller matrices:

$$\begin{aligned} \hat{j} &= My \\ \hat{\Sigma} &= V \otimes \hat{C} \\ M &= C^{(2)} L^T C^{-1} \\ C &= LC^{(2)} L^T + C^{(1)} \\ \hat{C} &= (L^T C^{(1)-1} L + C^{(2)-1})^{-1} \end{aligned} \qquad \mathbf{30.5}$$

Here M is an $(s \times c)$ matrix that corresponds to a MAP operator that maps the data to the conditional mean. This compact form depends on assuming the temporal correlations V of the observation error and the sources are the same. This ensures the covariance of the data $\mathrm{cov}(vec(y)) = \Sigma$, and of the sources conditioned on the data $\hat{\Sigma}$, factorize into separable spatial and temporal components:

$$\begin{aligned} \Sigma &= V \otimes C \\ \hat{\Sigma} &= V \otimes \hat{C} \end{aligned} \qquad \mathbf{30.6}$$

This is an important point because Eqn. **30.6** is not generally true if the temporal correlations of the error and sources are different, i.e. $V^{(1)} \neq V^{(2)}$. Even if, *a priori*, there is no interaction between the temporal and spatial responses, a difference in the temporal correlations, from the two levels, induces conditional spatio-temporal dependencies. This means that the conditional estimate of the spatial distribution changes with time. This dependency precludes the factorization implicit in Eqn. **30.5** and enforces a full-vectorized spatio-temporal analysis (Eqn. **30.4**), which is computationally expensive.

For the moment, we will assume the temporal correlations are the same and then generalize the approach in the next section for some special cases of $V^{(1)} \neq V^{(2)}$.

Estimating the covariances

Under the assumption that $V^{(1)} = V^{(2)} = V$, the only quantities that need to be estimated are the covariance components in Eqn. **30.1**. This proceeds using an iterative ReML scheme in which the covariance parameters maximize the log-likelihood or log-evidence:

$$\begin{aligned} \lambda &= \max_{\lambda} \ln p(y|\lambda, Q) \\ &= REML(vec(y)vec(y)^T, V \otimes Q) \end{aligned} \qquad \mathbf{30.7}$$

In brief, the $\lambda = REML(A, B)$ operator decomposes a sample covariance matrix A into a number of specified components $B = B_1, \ldots$ so that $A \approx \sum_i \lambda_i B_i$ (see previous chapter and Appendix 4). The ensuing covariance parameters $\lambda = \lambda_i, \ldots$ render the sample covariance the most likely. In our application, the sample covariance is simply the outer product of the vectorized data $vec(y)vec(y)^T$ and the components are $V \otimes Q_i$. Here, $Q = Q_1^{(1)}, \ldots LQ_1^{(2)} L^T, \ldots$ are the spatial covariance components from the first level of the model and the second level, after projection onto channel space through the lead-field.

ReML was originally formulated in terms of covariance component analysis, but is now appreciated as a special case of expectation maximization (EM). The use of the ReML estimate properly accounts for the degrees of freedom lost in estimating the model parameters (i.e. sources), when estimating the covariance components. The 'restriction' means that the covariance component estimated is restricted to the null space of the model. This ensures that uncertainty about the source estimates is accommodated in the covariance estimates. The key thing is how the data enter the log-likelihood that is maximized by ReML:[1]

$$\begin{aligned}\ln p(y\,|\,\lambda, Q) &= -\frac{1}{2}tr\,(\Sigma^{-1}vec(y)vec(y)^T) - \frac{1}{2}\ln|\Sigma| \\ &= -\frac{1}{2}tr\,(C^{-1}yV^{-1}y^T) - \frac{1}{2}\ln|C|rank(V)\end{aligned} \qquad \mathbf{30.8}$$

The second line uses the results in Eqn. **30.2** and shows that the substitutions $vec(y)vec(y)^T \to yV^{-1}y^T/rank(V)$ and $V \otimes Q \to Q$ do not change the maximum of the objective function. This means we can replace the ReML arguments in Eqn. **30.7** with much smaller $(c \times c)$ matrices:

$$\begin{aligned}\lambda &= REML(vec(y)vec(y)^T, V \otimes Q) \\ &= REML(yV^{-1}y^T/rank(V), Q)\end{aligned} \qquad \mathbf{30.9}$$

Assuming the data are zero mean, this second-order matrix $yV^{-1}y^T/rank(V)$ is simply the sample covariance matrix of the whitened data over the t time bins, where $rank(V) = t$. The greater the number of time bins, the more precise the ReML covariance component estimators.

This reformulation of the ReML scheme requires the temporal correlations of the observation error and the sources to be the same. This ensures $\Sigma = V \otimes C$ can be factorized and affords the computational saving implicit in Eqn. **30.9**. However, there is no reason to assume that the processes generating the signal and noise have the same temporal correlations. In the next section, we finesse this unlikely assumption by restricting the estimation to a subspace defined by temporal basis functions.

A TEMPORALLY INFORMED SCHEME

In this section, we describe a simple extension to the basic ReML approach that enables some constraints to be placed on the form of evoked or induced responses. This involves relaxing the assumption that $V^{(1)} = V^{(2)}$. The basic idea is to project the data onto a subspace (via a matrix S) in which the temporal correlation of signal and noise are formally equivalent. This falls short of a full spatio-temporal model, but retains the efficiency of ReML scheme above and allows for differences between $V^{(1)}$ and $V^{(2)}$ subject to the constraint that $S^TV^{(2)}S = S^TV^{(1)}S$.

In brief, we have already established a principled and efficient Bayesian inversion of the inverse problem for M/EEG using ReML. To extend this approach to multiple time bins we need to assume that the temporal correlations of channel noise and underlying sources are the same. In reality, sources are generally smoother than noise because of the generalized convolution implicit in synaptic and population dynamics at the neuronal level (Friston, 2000). However, by projecting the time-series onto a carefully chosen subspace we can make the temporal correlations of noise and signal the same. This enables us to solve a spatio-temporal inverse problem, using the re-formulation of the previous section. Heuristically, this projection removes high-frequency noise components so that the remaining smooth components exhibit the same correlations as signal. We now go through the maths that this entails.

Consider the forward model, where, for notational simplicity $V^{(1)} = V$:

$$\begin{aligned}y &= LkS^T + \varepsilon^{(1)} \\ k &= \varepsilon^{(2)} \\ Cov(vec(\varepsilon^{(1)})) &= V \otimes C^{(1)} \\ Cov(vec(\varepsilon^{(2)})) &= S^TVS \otimes C^{(2)}\end{aligned} \qquad \mathbf{30.10}$$

This is the same as Eqn. **30.1** with the substitution $j = kS^T$. The only difference is that the sources are estimated in terms of the activity k of temporal modes. The orthonormal columns of the temporal basis set S define these modes, where $S^TS = I_r$. When S has fewer columns than rows $r < t$, it defines an r-subspace in which the sources lie. In other words, the basis set allows us to preclude temporal response components that are, *a priori*, unlikely (e.g. very high frequency responses or responses before stimulus onset). This restriction enables one to define a signal that lies in the subspace of the errors.

In short, the subspace S encodes prior beliefs about when and how signal will be evoked. It specifies temporal priors on the sources through $V^{(2)} = SS^TV^{(1)}SS^T$. This ensures that $S^TV^{(2)}S = S^TV^{(1)}S$ because $S^TS = I_r$ and renders the restricted temporal correlations formally equivalent. We will see later that the temporal priors on sources are also their posteriors, $V^{(2)} = \hat{V}$, because the temporal correlations are treated as fixed and known.

The restricted model can be transformed into a spatio-temporally separable form by post-multiplying the first line of Eqn. **30.10** by S to give:

$$\begin{aligned}yS &= Lk + \varepsilon^{(S)} \\ k &= \varepsilon^{(2)} \\ Cov(vec(\varepsilon^{(S)})) &= S^TVS \otimes C^{(1)} \\ Cov(vec(\varepsilon^{(2)})) &= S^TVS \otimes C^{(2)}\end{aligned} \qquad \mathbf{30.11}$$

In this model, the temporal correlations of signal and noise are now the same. This restricted model has exactly

[1] Ignoring constant terms. The rank of a matrix corresponds to the number of dimensions it spans. For full-rank matrices, the rank is the same as the number of columns (or rows).

the same form as Eqn. **30.1** and can be used to provide ReML estimates of the covariance components in the usual way, using equation Eqn. **30.9**:

$$\lambda = REML(\frac{1}{r}yS(S^TVS)^{-1}S^Ty^T, Q) \qquad \textbf{30.12}$$

These are then used to compute the conditional moments of the sources as a function of time:

$$\begin{aligned} \hat{j} &= \hat{k}S^T = MySS^T \\ \hat{\Sigma} &= \hat{V} \otimes \hat{C} \\ \hat{V} &= SS^TVSS^T \end{aligned} \qquad \textbf{30.13}$$

Note that the temporal correlations $\hat{V}$ are rank deficient and non-stationary, because the conditional responses do not span the null space of S. This scheme does not represent a full spatio-temporal analysis; it is simply a device to incorporate constraints on the temporal component of the solution. A full analysis would require covariances that could not be factorized into spatial and temporal factors. This would preclude the efficient use of ReML covariance estimation described above. However, in most applications, a full temporal analysis would proceed, using the above estimates from different trial types and subjects (see for example, Kiebel *et al.*, 2004b).

In the later examples, we specify S as the principal eigenvectors of a temporal prior source covariance matrix based on a windowed autocorrelation (i.e. Toeplitz) matrix. In other words, we took a Gaussian autocorrelation matrix and multiplied the rows and columns with a window-function to embody our *a priori* assumption that responses are concentrated early in peristimulus time. The use of prior constraints in this way is very similar to the use of anatomically informed basis functions to restrict the solution space anatomically (see Phillips *et al.*, 2002). Here, S can be regarded as a temporally informed basis set that defines a signal subspace.

ESTIMATING RESPONSE ENERGY

In this section, we consider the estimation of evoked and induced responses in terms of their energy or power. The energy is simply the squared norm (i.e. squared length) of the response projected onto some time-frequency subspace defined by W. The columns of W correspond to the columns of a [wavelet] basis set that encompasses time-frequencies of interest, e.g. a sine-cosine pair of windowed sinusoids of a particular frequency. We deal first with estimating the energy of a single trial and then turn to multiple trials. The partitioning of energy into evoked and induced components pertains only to multiple trials.

For a single trial the energy expressed by the i-th source is:

$$j_{i,\bullet}WW^Tj_{i,\bullet}^T \qquad \textbf{30.14}$$

$j_{i,\bullet}$ is the i-th row of the source matrix, over all time bins. The conditional expectation of this energy obtains by averaging over the conditional density of the sources. The conditional density for the i-th source, over time, is:

$$\begin{aligned} p(j_{i,\bullet}|y,\lambda) &= N(\hat{j}_{i,\bullet}, \hat{C}_{ii}\hat{V}) \\ \hat{j}_{i,\bullet} &= M_{i,\bullet}ySS^T \end{aligned} \qquad \textbf{30.15}$$

and the conditional expectation of the energy is:

$$\begin{aligned} \langle j_{i,\bullet}WW^Tj_{i,\bullet}^T\rangle_p &= tr(WW^T\langle j_{i,\bullet}^Tj_{i,\bullet}\rangle_p) \\ &= tr(WW^T(\hat{j}_{i,\bullet}^T\hat{j}_{i,\bullet} + \hat{C}_{ii}\hat{V})) \\ &= M_{i,\bullet}yGy^TM_{i,\bullet}^T + \hat{C}_{ii}tr(GV) \\ G &= SS^TWW^TSS^T \end{aligned} \qquad \textbf{30.16}$$

Note that this is a function of yGy^T, the corresponding energy E_y in channel space. The expression in Eqn. **30.16** can be generalized to cover all sources, although this would be a rather large matrix to interpret:

$$\begin{aligned} \hat{E} &= \langle jWW^Tj^T\rangle_p = ME_yM^T + \hat{C}tr(GV) \\ E_y &= yGy^T \end{aligned} \qquad \textbf{30.17}$$

The matrix $\hat{E}$ is the conditional expectation of the energy over sources. The diagonal terms correspond to energy at the corresponding source (e.g. spectral density if W comprised sine and cosine functions). The off-diagonal terms represent cross energy (e.g. cross-spectral density or coherence).

Eqn. **30.17** means that the conditional energy has two components, one attributable to the energy in the conditional mean (the first term) and one related to conditional covariance (the second). The second component may seem a little counterintuitive: it suggests that the conditional expectation of the energy increases with conditional uncertainty about the sources. In fact, this is appropriate; when conditional uncertainty is high, the priors shrink the conditional mean of the sources towards zero. This results in an underestimate of energy based solely on the conditional expectations of the sources. By including the second term, the energy estimator becomes unbiased. It would be possible to drop the second term if conditional uncertainty was small. This would be equivalent to approximating the conditional density of the sources with a point mass over its mean. The advantage of this is that one does not have to compute the $s \times s$ conditional covariance of the sources. However, we will assume the number of sources is sufficiently small to use Eqn. **30.17**.

In this section, we have derived expressions for the conditional energy of a single trial. In the next section, we revisit the estimation of response energy over multiple trials. In this context, there is a distinction between induced and evoked energy.

AVERAGING OVER TRIALS

With multiple trials we have to consider trial-to-trial variability in responses. Conventionally, the energy associated with between-trial variations, around the average or evoked response, is referred to as induced. Induced responses are normally characterized in terms of the energy of oscillations within a particular time-frequency window. Because, by definition, they do not show a systematic phase-relationship with the stimulus, they are expressed in the average energy over trials, but not in the energy of the average. In this chapter, we use the term global response in reference to the total energy expressed over trials and partition this into evoked and induced components. In some formulations, a third component due to stationary, ongoing activity is considered. Here, we will subsume this component under induced energy. This is perfectly sensible, provided induced responses are compared between trial types, when ongoing or baseline power cancels.

Multitrial models

Hitherto we have dealt with single trials. When dealing with multiple trials, the same procedures can be adopted, but there is a key difference for evoked and induced responses. The model for n trials is:

$$
\begin{aligned}
Y &= Lk^{(1)}(I_n \otimes S)^T + \varepsilon^{(1)} \\
k^{(1)} &= (1_n \otimes k^{(2)}) + \varepsilon^{(2)} \\
k^{(2)} &= \varepsilon^{(3)} \\
Cov(vec(\varepsilon^{(1)})) &= I_n \otimes V \otimes C^{(1)} \\
Cov(vec(\varepsilon^{(2)})) &= I_n \otimes S^T VS \otimes C^{(2)} \\
Cov(vec(\varepsilon^{(3)})) &= S^T VS \otimes C^{(3)}
\end{aligned}
\qquad \textbf{30.18}
$$

where $1_n = [1, \ldots, 1]$ is a $1 \times n$ vector and $Y = [y_1, \ldots, y_n]$ represents data concatenated over trials. Note that multiple trials induce a third level in the hierarchical model. In this three-level model, sources have two components: a component that is common to all trials $k^{(2)}$ and a trial-specific component $\varepsilon^{(2)}$. These are related to evoked and induced response components as follows.

Operationally, we can partition the responses $k^{(1)}$ in source space into a component that corresponds to the average response over trials, the evoked response and an orthogonal component, the induced response:

$$
\begin{aligned}
k^{(e)} &= k^{(1)}(1_n^- \otimes I_r) \\
&= k^{(2)} + \varepsilon^{(2)}(1_n^- \otimes I_r) \\
k^{(i)} &= k^{(1)}((I_n - 1_n^- 1_n) \otimes I_r)
\end{aligned}
\qquad \textbf{30.19}
$$

$1_n^- = [\frac{1}{n}, \ldots, \frac{1}{n}]^T$ is the generalized inverse of 1_n and is simply an averaging vector. As the number of trials n increases, the random terms at the second level are averaged away and the evoked response $k^{(e)} \rightarrow k^{(2)}$ approximates the common component. Similarly, the induced response $k^{(i)} \rightarrow \varepsilon^{(2)}$ becomes the trial specific component. With the definition of evoked and induced components in place we can now turn to their estimation.

Evoked responses

The multitrial model can be transformed into a spatiotemporally separable form by simply averaging the data $\overline{Y} = Y(1_n^- \otimes I_t)$ and projecting onto the signal subspace. This is exactly the same restriction device used above to accommodate temporal basis functions but applied here to the trial-average. This corresponds to post-multiplying the first level by the trial-averaging and projection operator $1_n^- \otimes S$ to give:

$$
\begin{aligned}
\overline{Y}S &= Lk^{(e)} + \overline{\varepsilon}^{(1)} \\
k^{(e)} &= \varepsilon^{(e)} \\
Cov(vec(\overline{\varepsilon}^{(1)})) &= S^T VS \otimes \overline{C}^{(1)} \\
Cov(vec(\varepsilon^{(e)})) &= S^T VS \otimes C^{(e)}
\end{aligned}
\qquad \textbf{30.20}
$$

Here, $\overline{C}^{(1)} = \frac{1}{n}C^{(1)}$ and $C^{(e)} = \frac{1}{n}C^{(2)} + C^{(3)}$ is a mixture of trial-specific and non-specific spatial covariances. This model has exactly the same form as the single-trial model, enabling ReML estimation of $\overline{C}^{(1)}$ and $C^{(e)}$ that are needed to form the conditional estimator M (see Eqn. **30.4**):

$$
\lambda = REML(\frac{1}{r}\overline{Y}S(S^T VS)^{-1}S^T\overline{Y}^T, Q) \qquad \textbf{30.21}
$$

The conditional expectation of the evoked response amplitude (e.g. event-related potential, ERP, or event-related field, ERF) is simply:

$$
\begin{aligned}
\hat{j}^{(e)} &= M\overline{Y}SS^T \\
M &= C^{(e)}L^T(LC^{(e)}L^T + \overline{C}^{(1)})^{-1}
\end{aligned}
\qquad \textbf{30.22}
$$

$$\overline{C}^{(1)} = \sum \lambda_i^{(1)} Q_i^{(1)}$$
$$C^{(e)} = \sum \lambda_i^{(e)} Q_i^{(e)}$$

The conditional expectation of evoked power is then:

$$\begin{aligned}\hat{E}^{(e)} &= ME_y^{(e)} M^T + \hat{C}tr(GV)\\ E_y^{(e)} &= \overline{Y} G \overline{Y}^T \\ \hat{C} &= (L^T \overline{C}^{(1)-1} L + C^{(e)-1})^{-1}\end{aligned} \qquad 30.23$$

where $E_y^{(e)}$ is the evoked cross-energy in channel space. In short, this is exactly the same as a single-trial analysis but using the channel-data averaged over trials. However, this averaging is not appropriate for the induced responses considered next.

Induced responses

To isolate and characterize induced responses, we effectively subtract the evoked response from all trials to give $\tilde{Y} = Y((I_n - 1_n^- 1_n) \otimes I_t)$, and project this mean-corrected data onto the signal subspace. The average covariance of the ensuing data is then decomposed using ReML. This entails post-multiplying the first level of the multi-trial model by $(I_n - 1_n^- 1_n) \otimes S$ to give:

$$\begin{aligned}\tilde{Y}(I_n \otimes S) &= Lk^{(i)} + \tilde{\varepsilon}^{(1)}\\ k^{(i)} &= \varepsilon^{(i)}\\ Cov(vec(\tilde{\varepsilon}^{(1)})) &= I_n \otimes S^T VS \otimes \tilde{C}^{(1)}\\ Cov(vec(\varepsilon^{(i)})) &= I_n \otimes S^T VS \otimes C^{(i)}\end{aligned} \qquad 30.24$$

In this transformation $k^{(i)}$ is a large $s \times nr$ matrix that covers all trials. Again, this model has the same spatio-temporally separable form as the previous models, enabling an efficient ReML estimation of the covariance components of $\tilde{C}^{(1)}$ and $C^{(i)}$:

$$\lambda = REML(\frac{1}{nr} \tilde{Y}(I_n \otimes S(S^T VS)^{-1} S^T) \tilde{Y}^T, Q) \qquad 30.25$$

The first argument of the ReML function is just the covariance of the whitened, mean-corrected data averaged over trials. The conditional expectation of induced energy, per trial, is then:

$$\begin{aligned}\hat{E}^{(i)} &= \frac{1}{n} M \tilde{Y}(I_n \otimes G) \tilde{Y}^T M^T + \frac{1}{n} \hat{C}tr(I_n \otimes GV)\\ &= ME_y^{(i)} M^T + \hat{C}tr(GV)\\ E_y^{(i)} &= \frac{1}{n} \tilde{Y}(I_n \otimes G) \tilde{Y}^T\end{aligned} \qquad 30.26$$

where $E_y^{(i)}$ is the induced cross-energy per trial, in channel space. The spatial conditional projector M and covariance $\hat{C}$ are defined as above (Eqn. **30.22** and Eqn. **30.23**). Although it would be possible to estimate the amplitude of induced responses for each trial, this is seldom interesting.

Summary

The key thing to take from this section is that the estimation of evoked responses involves averaging over trials and estimating the covariance components. Conversely, the analysis of induced responses involves estimating covariance components and then averaging. In both cases, the iterative ReML scheme operates on small $c \times c$ matrices.

The various uses of the ReML scheme and conditional estimators are shown schematically in Figure 30.1. Note that all applications, be they single-trial or trial-average, estimates of evoked responses or induced energy, rest on a two-stage procedure in which ReML covariance component estimators are used to form conditional estimators of the sources. The second thing to take from this figure is that the distinction between evoked and induced responses only has meaning in the context of multiple trials. This distinction rests on an operational definition, established in the decomposition of response energy in channel space. The corresponding decomposition in source space affords the simple and efficient estimation of evoked and induced power described in this section. However, it is interesting to note that conditional estimators of evoked and induced components are not estimates of the fixed $k^{(2)}$ and random $\epsilon^{(2)}$ effects in the hierarchical model. These estimates would require a full mixed-effects analysis. Another interesting issue is that evoked and induced responses in channel space (where there is no estimation *per se*) represent a bi-partitioning of global responses. This is not the case for their conditional estimates in source space. In other words, the conditional estimate of global power is not necessarily the sum of the conditional estimates of evoked and induced power.

SOME EXAMPLES

In this section, we illustrate the above procedures using toy and real data. The objective of the toy example is to clarify the nature of the operators and matrices, to highlight the usefulness of restricting the signal space and to show, algorithmically, how evoked and induced responses are recovered. The real data are presented to establish a degree of face validity, given that face-related responses have been fully characterized in terms of their functional anatomy. The toy example deals with the

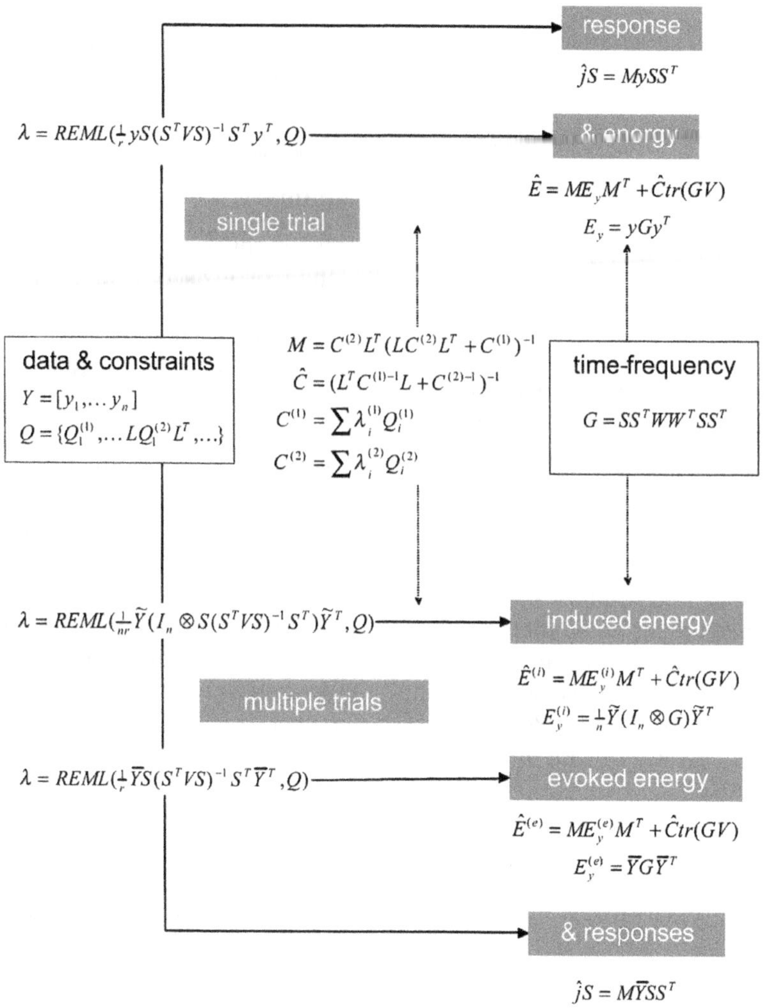

FIGURE 30.1 ReML scheme: schematic showing the various applications of the ReML scheme in estimating evoked and induced responses in multiple trials. See main text for an explanation of the variables.

single-trial case and the real data, looking for face-specific responses, illustrates the multi-trial case.

Toy example

We generated data according to the model in Eqn. **30.10** using $s = 128$ sources, $c = 16$ channels and $t = 64$ time bins. The lead field L was a matrix of random Gaussian variables. The spatial covariance components comprised:

$$Q_1^{(1)} = I_c$$
$$Q_1^{(2)} = DD^T$$
$$Q_2^{(2)} = DFD^T \qquad \textbf{30.27}$$

where D was a spatial convolution or dispersion operator, using a Gaussian kernel with a standard deviation of four voxels. This can be considered a smoothness or spatial coherence constraint. F represents structural or functional MRI constraints and was a leading diagonal matrix encoding the prior probability of a source at each voxel. This was chosen randomly by smoothing a random Gaussian sequence raised to the power four. The noise was assumed to be identically and independently distributed, $V^{(1)} = V = I_t$. The signal subspace in time S was specified by the first $r = 8$ principal eigenvectors of a Gaussian autocorrelation matrix of standard deviation two, windowed with a function of peristimulus time $t^2 \exp(-t/8)$. This constrains the prior temporal correlation structure of the sources $V^{(2)} = SS^TVSS^T$, which are smooth and restricted to earlier time bins by the window-function.

The hyperparameters were chosen to emphasize the MRI priors $\lambda = [\lambda_1^{(1)}, \lambda_1^{(2)}, \lambda_2^{(2)}] = [1, 0, 8]$ and provide a signal to noise of about one, measured as the ratio of the standard deviation of signal divided by noise, averaged over channels. The signal to noise in the example shown in Figure 30.2 was 88 per cent. The spatial coherence and MRI priors are shown at the top of Figure 30.2. The resulting spatial priors are shown below and are simply $\lambda_1^{(2)}Q_1^{(2)} + \lambda_2^{(2)}Q_2^{(2)}$. The temporal priors $SS^T VSS^T$ are shown on the middle right. Data (middle panel) were generated in source space using random Gaussian variates and the spatial and temporal priors above, according to the forward model in Eqn. **30.10**. These were passed through the lead-field matrix and added to observation noise to simulate channel data. The lower left panels show the channel data with and without noise.

ReML solution

The simulated channel data were used to estimate the covariance components and implicitly the spatial priors using Eqn. **30.12**. The resulting estimates of $\lambda = [\lambda_1^{(1)}, \lambda_1^{(2)}, \lambda_2^{(2)}]$ are shown in Figure 30.3 (upper panel). The small bars represent 90 per cent confidence intervals, about the ReML estimates, based on the curvature

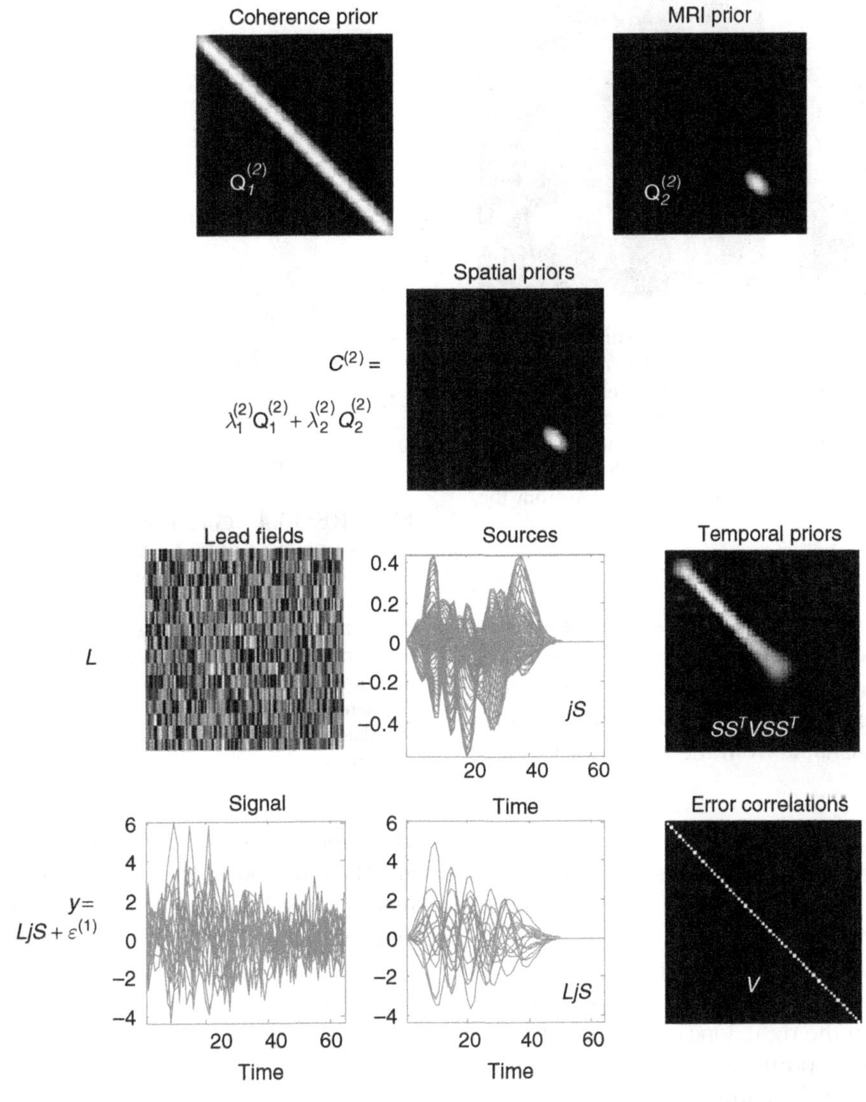

FIGURE 30.2 Simulated data: the spatial smoothness or coherence and MRI priors are shown in the top panels. These are prior covariance components, over sources, shown in image format. Note that the smoothness component is stationary (i.e. does not change along the diagonals), whereas the fMRI prior changes with source location. The resulting spatial priors $\lambda_1^{(2)}Q_1^{(2)} + \lambda_2^{(2)}Q_2^{(2)}$ are shown below. The temporal priors on the sources $SS^T VSS^T$ are shown on the middle right. Again, these are depicted as a covariance matrix, over time bins. Notice how this prior concentrates signal variance in the first forty time bins. Data (middle panel) were generated in source space, using random Gaussian variates according to the forward model in Eqn. **30.10** and the spatial and temporal priors above. These were passed through the lead-field matrix to simulate channel data. In this example, the lead-field matrix was simply a matrix of independent Gaussian variates. The lower left panels show the channel data after (left) and before (right) adding noise, over time bins.

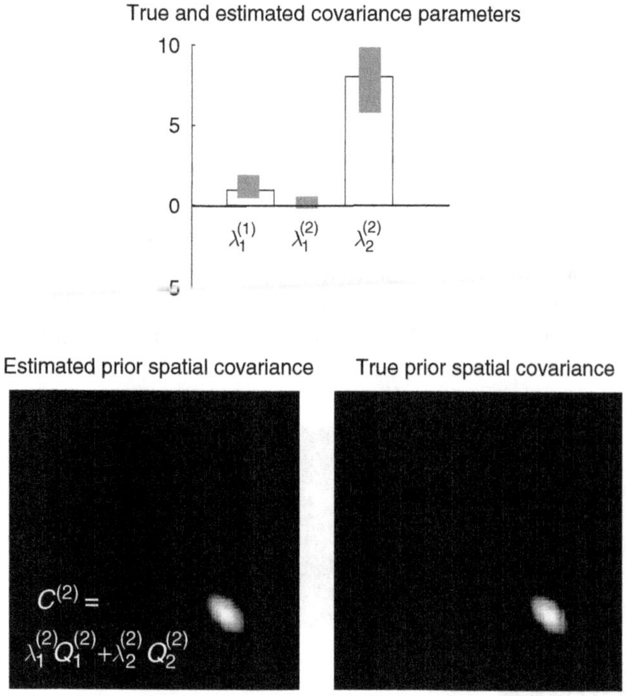

FIGURE 30.3 ReML solution: the ReML estimates of $\lambda = [\lambda_1^{(1)}, \lambda_1^{(2)}, \lambda_2^{(2)}]$ are shown in the upper panel. The small bars represent 90 per cent confidence intervals about the ReML estimates, based on the curvature of the log likelihood. The large bars are the true values. The ReML scheme correctly assigns more weight to the MRI priors to provide the empirical prior in the lower panel (left). This ReML estimate is virtually indistinguishable from the true prior (right).

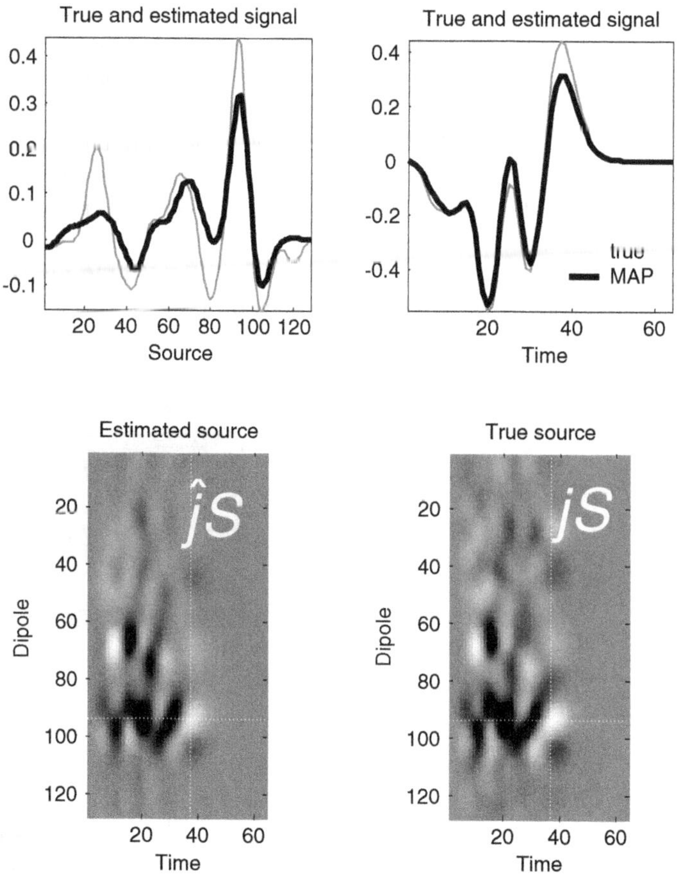

FIGURE 30.4 Conditional estimates of responses: the upper panel shows the true and estimated spatial profile at the time bin expressing the largest activity (upper left). The equivalent profile, over time, is shown on the upper right for the source expressing the greatest response. These graphs correspond to sections (dotted lines) though the full spatio-temporal profiles shown in image format (lower panels). Note the characteristic shrinkage of the MAP estimates, relative to the true values, that follows from the use of shrinkage priors (that shrink the conditional expectations to the prior mean of zero).

of the log-likelihood in Eqn. **30.7**. The large bars are the true values. The ReML scheme correctly assigns much more weight to the MRI priors to provide the empirical prior in the lower panel (left). This ReML estimate (left) is virtually indistinguishable from the true prior (right).

Conditional estimates of responses

The conditional expectations of sources, over time, are shown in Figure 30.4 using the expression in Eqn. **30.13**. The upper left panel shows the true and estimated spatial profile at the time bin expressing the largest activity (maximal deflection). The equivalent source estimate, over time, is shown on the right. One can see the characteristic shrinkage of the conditional estimators, in relation to the true values. The full spatio-temporal profiles are shown in the lower panels.

Conditional estimates of response energy

To illustrate the estimation of energy, we defined a time-frequency window $W = [w(t)\sin(\omega t), w(t)\cos(\omega t)]$ for one frequency, ω, over a Gaussian time window, $w(t)$. This time-frequency subspace is shown in the upper panels of Figure 30.5. The corresponding energy was estimated using Eqn. **30.17** and is shown, with the true values, in the lower panels. The agreement is evident.

Analysis of real data

We used MEG data from a single subject while they made symmetry judgements on faces and scrambled faces (for a detailed description of the paradigm see Henson *et al.*, 2003). MEG data were sampled at 625 Hz from a 151-channel CTF Omega system at the Wellcome Trust Laboratory for MEG Studies, Aston University, UK. The epochs (80 face trials, collapsing across familiar and unfamiliar faces, and 84 scrambled trials) were

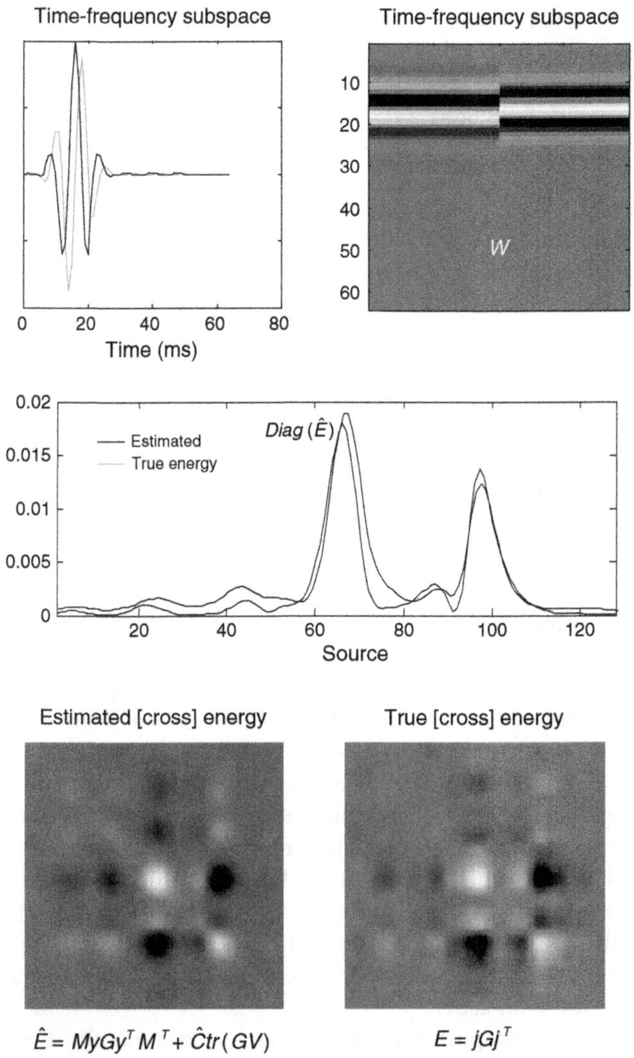

FIGURE 30.5 Conditional estimates of response energy: a time-frequency subspace W is shown in the upper panels as functions of time (left) and in image format (right). This subspace defines the time-frequency response of interest. In this example, we are testing for frequency-specific responses between 10 and 20 time bins. The corresponding energy *estimates* are shown over sources, with the true values, in the middle panel. Note the remarkable correspondence. The lower panels show the cross-energy over sources, with estimates on the left and true values on the right. The energies in the middle panel are the leading diagonals of the cross-energy matrices, shown as images below. Again, note the formal similarity between the true and estimated cross-energies.

baseline-corrected from −100 ms to 0 ms. The 500 ms, after stimulus onset, of each trial entered the analysis. A T1-weighted MRI was also obtained with a resolution $1 \times 1 \times 1\,\text{mm}^3$. Head-shape was digitized with a 3-D Polhemus Isotrak and used to co-register the MRI and MEG data. A segmented cortical mesh was created using Anatomist (Mangin *et al.*, 2004), with approximately 4000 dipoles oriented normal to the grey matter. Finally, a single-shell spherical head model was constructed using BrainStorm (Baillet *et al.*, 2004) to compute the forward operator L.

The spatial covariance components comprised:

$$\begin{aligned} Q_1^{(1)} &= I_c \\ Q_1^{(2)} &= DD^T \end{aligned} \qquad \textbf{30.28}$$

where spatial smoothness operator D was defined on the cortical mesh, using a Gaussian kernel with a standard deviation of 8 mm. Note we used only one, rather smooth, spatial component in this analysis. This was for simplicity. A more thorough analysis would use multiple components and Bayesian model selection to choose the optimum number of components (Mattout *et al.*, 2006). As with the simulated data analysis, the noise correlations were assumed to be identical and independently distributed, $V^{(1)} = I_t$, and the signal subspace in time S was specified by the first $r = 50$ principal eigenvectors of the windowed autocorrelation matrix used above.

We focused our analysis on the earliest reliable difference between faces and scrambled faces, as characterized by the M170 component in the ERF (Plate 41(c) – see colour plate section). A frequency band of 10–15 Hz was chosen on the basis of reliable differences ($p < 0.01$; corrected) in a statistical parametric map (time-frequency SPM) of the global energy differences (Plate 41(b)) around the M170 latency (Henson *et al.*, 2005a). The ensuing time-frequency subspace was centred at 170 ms (Plate 41(c)).

Results

Plate 41(d) shows evoked and induced power in channel space as defined in Eqn. **30.23** and Eqn. **30.26** respectively. Power maps were normalized to the same maximum for display. In both conditions, maximum power is located over the right temporal regions. However, the range of power values is much wider for the evoked response. Moreover, whereas scalp topographies of induced responses are similar between conditions, the evoked energy is clearly higher for faces, relative to scrambled faces. This suggests that the M170 is mediated by differences in phase-locked activity.

This is confirmed by the power analysis in source space (Plate 42), using Eqn. **30.23** and Eqn. **30.26**. Evoked and induced responses are generated by the same set of cortical regions. However, the differences between faces and scrambled faces, in terms of induced power, are weak compared to the equivalent differences in evoked power (see the scale bars in Plate 42). Furthermore, the variation in induced energy over channels and conditions is small, relative to evoked power. This non-specific profile suggests that ongoing activity may contribute substantially to the induced component. As mentioned above, the interesting aspect of induced power usually resides in trial-specific differences. A full analysis of induced differences will be presented elsewhere. Here we focus on the functional anatomy implied by evoked differentials. The functional anatomy of evoked responses, in this context, is sufficiently well known to establish the face validity of our conditional estimators.

The upper panel of Plate 43 shows the cortical projection of the difference between the conditional expectations of evoked energy, for faces versus scrambled faces. The largest changes were expressed in the right inferior occipital gyrus (IOG), the right orbitofrontal cortex (OFC) and the horizontal posterior segment of the right superior temporal sulcus (STS). Plate 44 shows the co-registration of these energy changes with the subject's structural MRI. Happily, the 'activation' of these regions is consistent with the equivalent comparison of fMRI responses (Henson *et al.*, 2003).

The activation of the ventral and lateral occipitotemporal regions is also consistent with recent localizations of the evoked M170 (Henson *et al.*, 2005b; Tanskanen *et al.*, 2005). This is to be expected, given that most of the energy change appears to be phase-locked (Henson *et al.*, 2005a). Indeed, the conditional estimates of evoked responses at the location of the maximum of energy change in the right IOG and right posterior STS show a deflection around 170 ms that is greater for faces than scrambled faces (Plate 43, lower panel).

Note we have not made any inferences about these effects. SPMs of energy differences would normally be constructed using conditional estimates of power changes over subjects (see Plate 40).

DISCUSSION

We have described an empirical Bayes approach to M/EEG source reconstruction that covers both evoked and induced responses. The estimation scheme is based on classical covariance component estimation using restricted maximum likelihood (ReML). We used temporal basis functions to place constraints on the temporal form of the responses and showed how one can estimate evoked responses, which are phase-locked to the stimulus, and induced responses that are not. This inherent distinction calls for different transformations of a hierarchical model of multiple trial responses to provide Bayesian estimates of power.

Oscillatory activity is well known to be related to neural coding and information processing in the brain (Hari *et al.*, 1997; Tallon-Baudry *et al.*, 1999; Fries *et al.*, 2001). Oscillatory activity refers to signals generated in a particular frequency band time-locked but not necessary phase-locked to the stimulus. Classical data averaging approaches may not capture this activity, which calls for trial-to-trial analyses. However, localizing the sources of oscillatory activity on a trial-by-trial basis is computationally demanding and requires data with low SNR. This is why early approaches were limited to channel space (e.g. Tallon-Baudry *et al.*, 1997). Recently, several inverse algorithms have been proposed to estimate the sources of induced oscillations. Most are distributed (or imaging) methods, since equivalent current dipole models are not suitable for explaining a few hundreds of milliseconds of non-averaged activity. Among distributed approaches, two main types can be distinguished: the beam-former (Gross *et al.*, 2001; Sekihara *et al.*, 2001; Cheyne *et al.*, 2003) and minimum-norm-based techniques (David *et al.*, 2002; Jensen and Vanni, 2002), although both can be formulated as (weighted) minimum norm estimators (Hauk, 2004). A strict minimum norm solution obtains when no weighting matrix is involved (Hamalainen *et al.*, 1993), but constraints such as fMRI-derived priors have been shown to condition the inverse solution (Lin *et al.*, 2004). Beam-former approaches implement a constrained inverse using a set of spatial filters (see Huang *et al.*, 2004 for an overview). The basic principle employed by beam-formers is to estimate the activity at each putative source location while suppressing the contribution of other sources. This means that beam-formers look explicitly for uncorrelated sources. Although some robustness has been reported in the context of partially correlated sources (Van Veen *et al.*, 1997), this aspect of beam-forming can be annoying when trying to characterize coherent or synchronized cortical sources (Gross *et al.*, 2001).

In this chapter, we have looked at a generalization of the weighted minimum norm approach based on hierarchical linear models and empirical Bayes, which can accommodate multiple priors in an optimal fashion (Phillips *et al.*, 2005). The approach involves a partitioning of the data covariance matrix into noise and prior source variance components, whose relative contributions are estimated using ReML. Each model (i.e. set of partitions or components) can be evaluated using Bayesian model selection (Mattout *et al.*, 2006). Moreover, the ReML scheme is computationally efficient, requiring only the inversion of small matrices. With temporal constraints, the scheme offers a general Bayesian framework that can incorporate all kind of spatial priors such as beam-former-like spatial filters and/or fMRI-derived constraints (Friston et al., 2006). Furthermore, basis functions enable both the use of computationally efficient ReML-based variance component estimation and the definition of priors on the temporal form of the response. This implies a separation of the temporal and spatial dependencies, at both the sensor and source levels, using a Kronecker formulation (Huizenga *et al.*, 2002).

Thanks to this spatio-temporal approximation, the estimation of induced responses, from multi-trial data, does not require a computationally demanding trial-by-trial approach (Jensen *et al.*, 2002) or drastic dimension reduction of the solution space (David *et al.*, 2002).

The approach described in this chapter allows for spatio-temporal modelling of evoked and induced responses under the assumption that there is a subspace S, in which temporal correlations among the data and signal have the same form. Clearly, this subspace should encompass as much of the signal as possible. In this work, we used the principal eigenvariates of a prior covariance based on smooth signals, concentrated early in peristimulus time. This subspace is therefore informed by prior assumptions about how and when signal is expressed. A key issue here is what would happen if the prior subspace did not coincide with the true signal subspace. In this instance, there may be a loss of efficiency as experimental variance is lost to the null space of S. However, there will be no bias in the [projected] response estimates. Similarly, the estimate of the error covariance components will be unbiased but lose efficiency as high frequency noise components are lost in the restriction. Put simply, this means the variability in the covariance parameter estimates will increase, leading to a slight overconfidence in conditional inference. The overconfidence problem is not an issue here because we are only interested in the conditional expectations, which would normally be taken to a further (between-subject) level for inference.

Importantly, statistical parametric mapping of the estimated power changes in a particular time-frequency window, over conditions and/or over subjects, can now be achieved at the cortical level (Brookes *et al.*, 2004; Kiebel *et al.*, 2004a). Finally, with the appropriate wavelet transformation, instantaneous power and phase could also be estimated in order to study cortical synchrony.

In the previous three chapters, we considered the inversion of electromagnetic models to estimate current sources. In the next three chapters, we turn to models of how neuronal dynamics generate the activity in these sources. The neuronal and electromagnetic models of this section are combined later (Chapter 42) in the dynamic causal modelling of measured M/EEG responses.

REFERENCES

Amblard C, Lapalme E, Lina JM (2004) Biomagnetic source detection by maximum entropy and graphical models. *IEEE Trans Biomed Eng* **51**: 427–42

Babiloni F, Babiloni C, Carducci F *et al.* (2004) Multimodal integration of EEG and MEG data: a simulation study with variable signal-to-noise ratio and number of sensors. *Hum Brain Mapp* **14**: 197–209

Baillet S, Garnero L (1997) A Bayesian approach to introducing anatomo-functional priors in the EEG/MEG inverse problem. *IEEE Trans Biomed Eng* **44**: 374–85

Baillet S, Mosher JC, Leahy RM (2004) Electromagnetic brain imaging using Brainstorm. Proceedings of the IEEE International Symposium on Biomedical Imaging (ISBI) **1**: 652–5

Brookes MJ, Gibson AM, Hall SD *et al.* (2005) A general linear model for MEG beam former imaging. *NeuroImage* **23**: 936-46

Cheyne D, Gaetz W, Garnero L *et al.* (2003) Neuromagnetic imaging of cortical oscillations accompanying tactile stimulation. *Cogn Brain Res* **17**: 599–611

Dale AM, Sereno M (1993) Improved localization of cortical activity by combining EEG and MEG with MRI surface reconstruction: a linear approach. *J Cogn Neurosci* **5**: 162–76

Dale AM, Liu AK, Fischl BR *et al.* (2000) Dynamic statistical parametric mapping: combining fMRI and MEG for high-resolution imaging of cortical activity. *Neuron* **26**: 55–67

Daunizeau J, Mattout J, Clonda D, *et al.* (2006) Bayesian spatio-temporal approach for EEG-source reconstruction: conciliating ECD and distributed models. *IEEE Trans Biomed Eng* **53**: 503–16

David O, Garnero L, Cosmelli D *et al.* (2002) Estimation of neural dynamics from MEG/EEG cortical current density maps: application to the reconstruction of large-scale cortical synchrony. *IEEE Trans Biomed Eng* **49**: 975–87

Dietterich TG (2000) Ensemble methods in machine learning. *Proceedings of the First International Workshop on Multiple Classifier Systems*, Lecture Notes in Computer Science, Springer-Verlag

Fries P, Reynolds JH, Rorie AE *et al.* (2001) Modulation of oscillatory neuronal synchronization by selective visual attention. *Science* **291**: 1506–07

Friston KJ (2000) The labile brain I. Neuronal transients and nonlinear coupling. *Phil Trans R Soc Lond B* **355**: 215–36

Friston KJ, Penny W, Phillips C *et al.* (2002) Classical and Bayesian inference in neuroimaging: theory. *NeuroImage* **16**: 465–83

Friston KJ, Mattout J, Trujillo-Barreto N *et al.* (2006) Variational free energy and the Laplace approximation, *NeuroImage* (in Press)

Gorodnitsky LF, George JS, Rao BD (1995) Neuromagnetic source imaging with FOCUSS: a recursive weighted minimum norm algorithm. *Electroencephalogr Clin Neurophysiol* **95**: 231–51

Gross J, Kujala J, Hamalainen M *et al.* (2001) Dynamic imaging of coherent sources: studying neural interactions in the human brain. *Proc Natl Acad Sci* **98**: 694–99

Hamalainen MS, Hari R, Ilmoniemi R *et al.* (1993) Magnetoencephalography: theory, instrumentation and applications to non-invasive study of human brain functions. *Rev Mod Phys* **65**: 413–97

Hamalainen MS, Ilmoniemi RJ (1994) Interpreting magnetic fields of the brain – minimum norm estimates. *Med Biol Eng Comput* **32**: 35–42

Hari R, Salmelin R (1997) Human cortical oscillations: a neuromagnetic view through the skull. *Trends Neurosci* **20**: 44–49

Hauk O (2004) Keep it simple: a case for using classical minimum norm estimation in the analysis of EEG and MEG data. *NeuroImage* **21**: 1612–21

Henson RN, Goshen-Gottstein Y, Ganel T *et al.* (2003) Electrophysiological and haemodynamic correlates of face perception, recognition and priming. *Cereb Cortex* **13**: 793–805

Henson R, Kiebel S, Kilner J *et al.* (2005a) Time-frequency SPMs for MEG data on face perception: power changes and phase-locking. Human Brain Mapping Conference (HBM'05)

Henson R, Mattout J, Friston K *et al.* (2005b) Distributed source localization of the M170 using multiple constraints. Human Brain Mapping Conference (HBM'05)

Huang M-X, Shih JJ, Lee DL *et al.* (2004) Commonalities and differences among vectorized beamformers in electromagnetic source imaging. *Brain Topogr* **16**: 139–58

Huizenga HM, de Munk JC, Waldorp LJ *et al.* (2002) Spatio-temporal EEG/MEG source analysis based on a parametric noise covariance model. *IEEE Trans Biomed Eng* **49**: 533–39

Jensen O, Vanni S (2002) A new method to identify sources of oscillatory activity from magnetoencephalographic data. *NeuroImage* **15**: 568–74

Jerbi K, Baillet S, Mosher JC *et al.* (2004) Localization of realistic cortical activity in MEG using current multipoles. *NeuroImage* **22**: 779–93

Kiebel SJ, Friston KJ (2004a) Statistical parametric mapping for event-related potentials: I. Generic considerations. *NeuroImage* **22**: 492–502

Kiebel SJ, Friston KJ (2004b) Statistical parametric mapping for event-related potentials: II. A hierarchical temporal model. *NeuroImage* **22**: 503–20

Kilner JM, Kiebel SJ, Friston KJ (2005) Applications of random field theory to electrophysiology. *Neurosci Lett* **374**: 174–78

Lin F-H, Witzel T, Hamalainen MS *et al.* (2004) Spectral spatiotemporal imaging of cortical oscillations and interactions in the human brain. *NeuroImage* **23**: 582–95

Mangin JF, Riviere D, Cachia A *et al.* (2004) A framework to study the cortical folding patterns. *NeuroImage* **23**: 129–38

Mattout J, Pélégrini-Issac M, Garnero L *et al.* (2005) Multivariate source prelocalization (MSP): use of functionally informed basis functions for better conditioning the MEG inverse problem. *NeuroImage* **26**: 356–73

Mattout J, Phillips C, Penny WD *et al.* (2006) MEG source localization under multiple constraints: an extended Bayesian framework. *NeuroImage* **30**: 753–67

Nunez PL, Silberstein RB (2000) On the relationship of synaptic activity to macroscopic measurements: does co-registration of EEG with fMRI make sense? *Brain Topogr* **13**: 79–96

Pascual-Marqui RD, Michel CM, Lehmann D (1994) Low resolution electromagnetic tomography: a new method for localizing electrical activity in the brain. *IEEE Trans Biomed Eng* **418**: 49–65

Phillips C, Rugg MD, Friston KJ (2002) Anatomically informed basis functions for EEG source localization: combining functional and anatomical constraints. *NeuroImage* **16**: 678–95

Phillips C, Mattout J, Rugg MD *et al.* (2005) An empirical Bayesian solution to the source reconstruction problem in EEG. *NeuroImage* **24**: 997–1011

Riera JJ, Fuentes ME, Valdes PA *et al.* (1998) EEG distributed inverse solutions for a spherical head model. *Inverse Problems* **14**: 1009–19

Schmidt D, George J, Wood C (1999) Bayesian inference applied to the electromagnetic inverse problem. *Hum Brain Mapp* **7**: 195–212

Sekihara K, Nagarajan SS, Poeppel D *et al.* (2001) Reconstructing spatio-temporal activities of neural sources using an MEG vector beamformer technique. *IEEE Trans Biomed Eng* **48**: 760–71

Tallon-Baudry C, Bertrand O (1999) Oscillatory gamma activity in humans and its role in object representation. *Trends Cogn Sci* **3**: 151–62

Tallon-Baudry C, Bertrand O, Delpuech C *et al.* (1997) Oscillatory γ-band (30–70 Hz) activity induced by a visual search task in humans. *J Neurosci* **17**: 722–34

Tanskanen T, Nasanen R, Montez T *et al.* (2005) Face recognition and cortical responses show similar sensitivity to noise spatial frequency. *Cereb Cortex* **15**: 526–34

Van Veen BD, van Drongelen W, Yuchtman M *et al.* (1997) Localization of brain electrical activity via linearly constrained minimum variance spatial filtering. *IEEE Trans Biomed Eng* **44**: 867–80

C H A P T E R

31

Neuronal models of ensemble dynamics

L. Harrison, O. David and K. Friston

INTRODUCTION

In this chapter, we introduce some of the basis principles behind modelling ensemble or population dynamics. In particular, we focus on mean-field approximations that allow one to model the evolution of states averaged over large numbers of neurons that are assumed to respond similarly to external influences. This is the basis of the neural mass models used in subsequent chapters to model interactions among subpopulations that constitute sources of measured brain signals. These models are important because they are parameterized in biological terms. This means their inversion allows one to ask questions that are framed in terms of biological processes, rather than at a purely phenomenological level.

Neuronal responses are the product of coupling among neuronal populations. Sensory information is encoded and distributed among neuronal ensembles in a way that depends on biophysical parameters that control this coupling. Because coupling can be modulated by experimental factors, estimates of coupling parameters provide a systematic way to parameterize experimentally induced responses. This chapter is about biologically informed models that embed this coupling.

The electrical properties of nervous tissue derive from the electrochemical activity of coupled neurons that generate current sources within the cortex, which can be estimated from multiple scalp electrode electroencephalography (EEG) recordings. These electrical traces express large-scale coordinated patterns of electrical activity. There are two commonly used methods to characterize event-related changes in these signals: averaging over many traces to form event-related potentials (ERP) and calculating the spectral profile of ongoing oscillatory activity (cf. evoked and induced responses). The assumption implicit in the averaging procedure is that the evoked signal has a fixed temporal relationship to the stimulus, whereas the latter procedure relaxes this assumption (Pfurtscheller and Lopes da Silva, 1999). We will return to this distinction, using neural-mass models, in Chapter 33.

Particular characteristics of ERPs are associated with cognitive states, e.g. the mismatch negativity in auditory oddball paradigms (Winkler *et al.*, 2001). The changes in ERP evoked by an 'event' are assumed to reflect event-dependent changes in cortical processing. In a similar way, spectral peaks of ongoing oscillations are generally thought to reflect the degree of synchronization among oscillating neuronal populations, with specific changes in the spectral profile being associated with various cognitive states. These changes have been called event-related desynchronization (ERD) and event-related synchronization (ERS). ERD is associated with an increase in processing information, e.g. voluntary hand movement (Pfurtscheller, 2001), whereas ERS is associated with reduced processing, e.g. during little or no motor behaviour. These observations led to the thesis that ERD represents increased cortical excitability and conversely that ERS reflects deactivation. We will look at this from the point of view neuronal energetics and relationship to metabolic activation in the next chapter.

The conventional approach to interpreting the EEG in terms of computational processes (Churchland and Sejnowski, 1994), is to correlate task-dependent changes in the ERP or time-frequency profiles of ongoing activity with cognitive or pathological states. A complementary strategy is to invert a generative model of how data are caused and estimate its parameters. This approach goes beyond associating particular activities with cognitive states to model the self-organization of neuronal systems during functional processing. Candidate models, developed in theoretical neuroscience, can be divided into mathematical and computational (Dayan, 1994). Mathematical models entail the biophysical mechanisms behind neuronal activity, such as the Hodgkin-Huxley model neuron of action potential generation (Dayan and Abbott, 2001). Computational models are

Statistical Parametric Mapping, by Karl Friston et al.
ISBN–13: 978-0-12-372560-8 ISBN–10: 0-12-372560-7

concerned with how a computational device could implement a particular task, e.g. representing saliency in a hazardous environment. Both approaches have produced compelling models, which speaks to the use of biologically and computationally informed forward or generative models in neuroimaging. We focus here on mathematical models.

The two broad classes of generative models are neural-mass models (NMM) (Jansen and Rit, 1995; Valdes *et al.*, 1999, David and Friston, 2003) and population density models (Knight, 2000; Gerstner and Kistler, 2002). NMMs were developed as parsimonious models of the mean activity (firing rate or membrane potential) of neuronal populations and have been used to generate a wide range of oscillatory behaviours associated with the ERP/EEG. The model equations for a population are a set of non-linear differential equations forming a closed loop between the influence neuronal firing has on mean membrane potential and how this potential changes the consequent firing rate of a population. Usually, two operators are required – linking membrane responses to input from afferent neurons (pulse-to-wave) and the dependence of action potential density on membrane potential (wave-to-pulse) (see Jirsa, 2004 for a review). They are divided into lumped models, where populations of neurons are modelled as discrete nodes that interact through cortico-cortical connections, or as continuous neuronal fields (Jirsa and Haken, 1996; Rennie *et al.*, 2002), where the cortical sheet is modelled as a continuum on which the cortical dynamics unfold. Frank (2005) has extended the continuum model to include stochastic effects with an application to magnetoencephalography (MEG) data. David *et al.* (2006) have recently extended an NMM proposed by Jansen and Rit (1995) and implemented it as a forward model to analyse ERP data. In doing this, they were able to infer changes in effective connectivity, defined as the influence one region exerts on another (Friston *et al.*, 2003). We will deal with this NMM model in the next chapter and its inversion in Chapter 41.

Analyses of effective connectivity in the neuroimaging community were first used with positron emission tomography (PET) and later with functional magnetic resonance imaging (fMRI) data. The latter applications led to the development of dynamic causal modelling (DCM). DCM for neuroimaging data (see Chapter 41) embodies organizational principles of cortical hierarchies and neurophysiological knowledge (e.g. time constants of biophysical processes) to constrain a parameterized non-linear dynamic model of observed responses. A principled way of incorporating these constraints is in the context of Bayesian estimation (Friston *et al.*, 2002). Furthermore, established Bayesian model comparison and selection techniques can be used to disambiguate different models and their implicit assumptions. The development of this methodology by David *et al.* (2006) for NMMs of ERP/EEG was an obvious extension. In this chapter, we apply the same treatment to population density models of interacting neuronal subpopulations.

An alternative to NMMs are population density models. These model the effect of stochastic influences (e.g. variability of presynaptic spike-time arrivals) by considering how the probability density of neuronal states evolves over time. In contradistinction, NMMs consider only the evolution of the density's mode or mass. Stochastic effects are important for many phenomena, e.g. stochastic resonance (Wiesenfeld and Moss, 1995). The probability density is over trajectories through state-space. The ensuing densities can be used to generate measurements, such as the mean firing rate or membrane potential of an average neuron within a population. A key tool for modelling population densities is the Fokker-Planck equation (FPE) (Risken, 1996). This equation has a long history in the physics of transport processes and has been applied to a wide range of physical phenomena, e.g. Brownian motion, chemical oscillations, laser physics and biological self-organization (Kuramoto, 1984; Haken, 1996). The beauty of the FPE is that, given constraints on the smoothness of stochastic forces (Kloeden and Platen, 1999), stochastic effects are equivalent to a diffusive process. This can be modelled by a deterministic equation in the form of a partial differential equation. The Fokker-Planck formalism uses notions from mean-field theory, but is dynamic and can model transitions from non-equilibrium to equilibrium states. The key point here is that all the random fluctuations and forces that shape neuronal dynamics at a microscopic level can be summarized in terms of a deterministic (i.e. non-random) evolution of probability densities using the FPE.

Local field potentials (LFPs) and ERPs represent the average response over millions of neurons, which means it is sufficient to model their population density to generate responses. This means the FPE is a good candidate for a forward or generative model of LFPs and ERPs. Because the population dynamics entailed by the FPE are deterministic, established Bayesian techniques for inverting deterministic dynamical systems (Chapter 34) can be applied directly.

Overview

In the first section, we review the theory of integrate-and-fire neurons with synaptic dynamics and its formulation as an FPE of interacting populations mediated through mean-field quantities (see Risken, 1996; Dayan and Abbott, 2001 for further details). The model encompasses four basic characteristics of neuronal activity and organization – neurons are: dynamic, driven by stochastic forces, organized into populations with similar biophysical properties, and have multiple populations that interact to form functional networks. In the second section, we discuss features of the model and demonstrate its face-validity using simulated data. This involves inverting a population density model to estimate model parameters given synthetic data. The discussion focuses on outstanding issues with this approach in the context of generative models for LFP/ERP data.

THEORY

A deterministic model neuron

The response of a model neuron to input, $s(t)$, has a generic form, which can be represented by the differential equation:

$$\dot{x} = f(x(t), s(t), \theta) \quad \textbf{31.1}$$

where $\dot{x} = dx/dt$. The state vector, x, (e.g. including variables representing membrane potential and proportion of open ionic channels) defines a space within which its dynamics unfold. The number of elements in x defines the dimension of this space and specific values identify a coordinate within it. The temporal derivative of x quantifies the motion of a point in state-space and the solution of the differential equation is its trajectory. The right-hand term is a function of the states, $x(t)$, and input, $s(t)$, where input can be exogenous or internal, i.e. mediated by coupling with other neurons. The model parameters, namely, the characteristic time-constants of the system, are represented by θ. As states are not generally observed directly, an observation equation is needed to link them to measurements, y:

$$y = g(x, \theta) + \varepsilon \quad \textbf{31.2}$$

where ε is observation noise (usually modelled as a Gaussian random variable). An example of an observation equation is an operator that returns the mean firing rate or membrane potential of a neuron. These equations form the basis of a forward or generative model to estimate the conditional density $p(\theta|y)$ given real data.

Neurons are electrical units. A simple expression for the rate of change of membrane potential, $V(t)$, in terms of membrane currents, $I_i(t)$, and capacitance is:

$$C\dot{V}(t) = \sum_i I_i(t) \quad \textbf{31.3}$$

Figure 31.1 shows a schematic of a model neuron and its resistance-capacitance circuit equivalent. Models of action potential generation (e.g. the Hodgkin-Huxley neuron) are based on specifying the currents in Eqn. **31.3** as functions of voltage or other quantities; typically, currents are categorized as voltage-, calcium- or neurotransmitter-dependent. The dynamic repertoire of a specific model depends on the nature of the different currents. This repertoire can include fixed-point attractors, limit cycles and chaotic dynamics.

A caricature of a spiking neuron is the simple integrate-and-fire (SIF) model. It is one-dimensional as all voltage

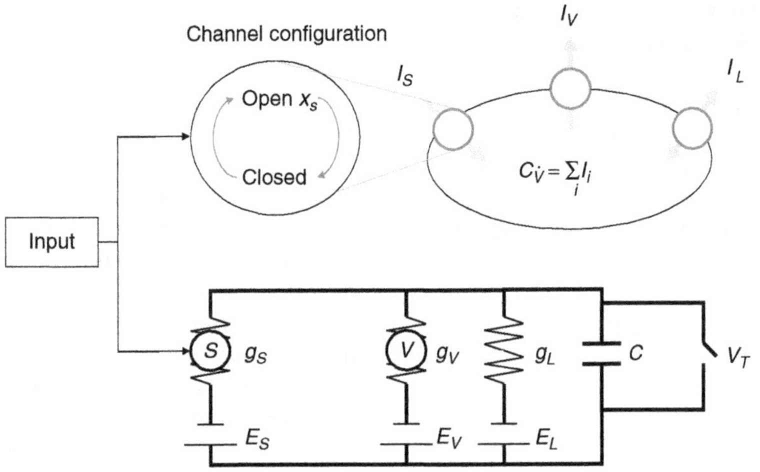

FIGURE 31.1 Schematic of a single-compartment model neuron including synaptic dynamics and its RC circuit analogue. Synaptic channels and voltage-dependent ion channels are shown as a circle containing S or V respectively. There may be several species of channels, indicated by the dots. Equilibrium potentials, conductance and current due to neurotransmitter (synaptic), voltage-dependent and passive (i.e. leaky) channels are E_S, E_V, E_L, g_S, g_V, g_L, I_S, I_V and I_L respectively. Depolarization occurs when membrane potential exceeds threshold, V_T. Input increases the opening rate of synaptic channels. Note that if synaptic channels are dropped from a model (e.g. as in a simple integrate-and-fire neuron) then input is directly into the circuit.

and synaptic channels are ignored. Instead, current is modelled as a constant passive leak of charge, thereby reducing the right-hand side of Eqn. **31.3** to 'leakage' and input currents:

$$C\dot{V} = g_L(E_L - V) + s(t) \qquad \textbf{31.4}$$

where g_L and E_L are the conductance and equilibrium potential of the leaky channel respectively. This model does not incorporate the biophysics needed to generate action potentials. Instead, spiking is modelled as a threshold process, i.e. once membrane potential exceeds a threshold value, V_T, a spike is generated and membrane potential is reset to V_R, where $V_R \leq E_L < V_T$. No spike is actually emitted; only subthreshold dynamics are modelled.

Modelling suprathreshold dynamics

First, we will augment a simple integrate-and-fire model with an additional variable T, inter-spike time (IST). Typically, as described above, the threshold potential is modelled as an absorbing boundary condition with re-entry at the reset voltage. However, we will model the IST as a state variable for a number of reasons. First, it constrains the neuronal trajectory to a finite region of state-space, which only requires natural boundary conditions, i.e. the probability mass decays to zero as state-space extends to infinity. This makes our later treatment generic, because we do not have to consider model-specific boundary conditions when, for example, formulating the FPE or deriving its eigensystem. Another advantage is that the time between spikes can be calculated directly from the density on IST. Finally, having an explicit representation of the time since the last spike allows us to model time-dependent changes in the system's parameters (e.g. relative refractoriness), which would be much more difficult in conventional formulations. The resulting model is two-dimensional and automates renewal to reset voltage once threshold has been exceeded. We will refer to this as a temporally augmented (TIF) model. With this additional state-variable, we have:

$$\begin{aligned}
\dot{V} &= \frac{1}{C}(g_L(E_L - V) + s(t)) + \alpha(V_R - V)\beta \\
\dot{T} &= 1 - \alpha T H(V) \\
\beta &= \exp(-T^2/2\gamma^2) \\
H(V) &= \begin{cases} 1 & V \geq V_T \\ 0 & V < V_T \end{cases}
\end{aligned} \qquad \textbf{31.5}$$

A characteristic feature of this deterministic model is that the input has to reach a threshold before spikes are generated, after which firing rate increases monotonically. This is in contrast to a stochastic model that has non-zero probability of firing, even with low input. We will see an example of this later. Given a suprathreshold input to the TIF neuron, membrane voltage is reset to V_R using the Heaviside function (last term in Eqn. **31.5**). This ensures that once $V > V_T$ the rate of change of T with respect to time is large and negative ($\alpha = 10^4$), reversing the progression of inter-spike time and returning it to zero, after which it increases constantly for $V_R < V < V_T$. Membrane potential is coupled to T, via an expression involving β, which is a Gaussian function, centred at $T = 0$ with a small dispersion ($\gamma = 1\,\text{ms}$). During the first few milliseconds following a spike, this term provides a brief impulse to clamp membrane potential near to V_R (cf. the refractory period).

Modelling spike-rate adaptation and synaptic dynamics

The TIF model can be extended to include ion-channel dynamics, i.e. to model spike-rate adaptation and synaptic transmission. We will call this channel model a CIF model.

$$\begin{aligned}
\dot{V} =& \frac{1}{C}(g_L(E_L - V) + g_1 x_1(E_1 - V) \\
&+ g_2 x_2(E_2 - V) + g_3 x_3(E_3 - V) \\
&+ g_4 x_4(E_4 - V)/(1 + \exp(-(V - a)/b)) + \alpha(V_R - V)\beta \\
\dot{T} =& 1 - \alpha T H(V) \\
\tau_1 \dot{x}_1 =& (1 - x_1)4\beta - x_1 \\
\tau_2 \dot{x}_2 =& (1 - x_2)(p_2 + s(t)) - x_2 \\
\tau_3 \dot{x}_3 =& (1 - x_3)p_3 - x_3 \\
\tau_4 \dot{x}_4 =& (1 - x_4)p_4 - x_4
\end{aligned} \qquad \textbf{31.6}$$

(Table 31-1 gives a list of variables used throughout this chapter.) These equations model spike-rate adaptation and synaptic dynamics (fast excitatory AMPA, slow excitatory NMDA and inhibitory GABA channels) by a generic synaptic channel mechanism, which is illustrated in Figure 31.1. The proportion of open channels is modelled by $x_1, \ldots, x_4$; these correspond to K (slow potassium), AMPA, GABA or NMDA channels, where $0 \leq x_i \leq 1$. Given no input, the proportion of open channels relaxes to an equilibrium state, e.g. $p_i/(1+p_i)$ for GABA and NMDA channels. The rate at which channels close is proportional to x_i. Conversely, the rate of opening is proportional to $1 - x_i$. External input now enters by increasing the opening rate of AMPA channels (see the RC circuit of Figure 31.1).

TABLE 31-1 Variables and symbols

Description	Symbol
Membrane potential	V
Inter-spike time	T
Proportion of open channels: K, AMPA, GABA and NMDA	$x_1, \ldots, x_4$
State vector	$x = [x_1, \ldots, x_4, V, T]^T$
Probability density	ρ
Probability mode coefficients	μ
Dynamic operator	Q
Right eigenvector matrix	R
Left eigenvector matrix	L
Eigenvalue matrix	D
Observations	y
Observation operator	M

Stochastic dynamics

The response of a deterministic system to input is known from its dynamics and initial conditions as it follows a well-defined trajectory in state-space. The addition of system noise, i.e. random input, to the deterministic equation turns it into a stochastic differential equation. If the random input is Langevin, it is referred to as a Langevin equation (Frank, 2005). In contrast to deterministic systems, the Langevin equation has an ensemble of solutions. The effect of stochastic terms, e.g. variable spike-time arrival, is to disperse trajectories through state-space. A simple example of this is shown in Figure 31.2. Three trajectories are shown, each with the same initial condition. The influence of stochastic input is to disperse the trajectories.

Under smoothness constraints on the random input, the ensemble of solutions are described exactly by the Fokker-Planck equation (Risken, 1996; Kloeden and Platen, 1999). The FPE frames the evolution of the ensemble of solutions as a deterministic process, which models the dispersive effect of stochastic input as a diffusive process. This leads to a parsimonious description in terms of a probability density over state space, represented by $\rho(x, t)$. This is important, as stochastic effects can have a substantial influence on dynamic behaviour, as we will see in the next section.

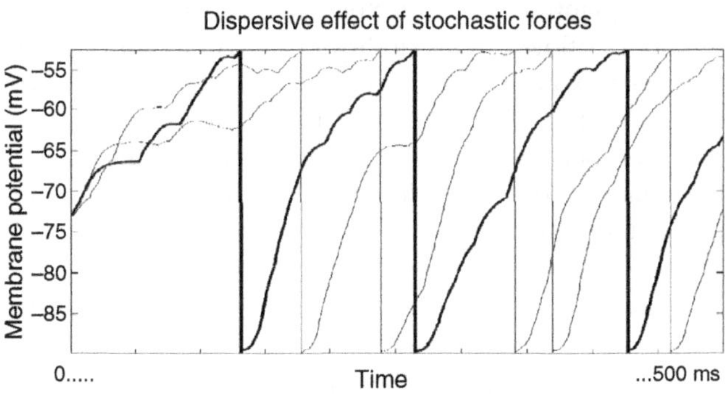

FIGURE 31.2 Dispersive effect of stochastic input on trajectories. Three trajectories of the CIF model (Eqn. **31.6**) are shown, one in bold to emphasize the dispersion of trajectories despite the same initial position. Once threshold is exceeded V is reset to V_R.

Population density methods have received much attention over the past few decades as a means of modelling efficiently the activity of thousands of similar neurons. Knight and Sirovich (Sirovich, 2003) describe an eigenfunction approach to solving these equations and extend the method to a time-dependent perturbation solution. Comparative studies by (Omurtag *et al.*, 2000; Haskell *et al.*, 2001) have demonstrated the efficiency and accuracy of population density methods using Monte Carlo simulations of neuronal populations. The effects of synaptic dynamics have been explored and applied to orientation tuning (Nykamp and Tranchina, 2000, 2001). Furthermore, Casti *et al.* (2002) have modelled bursting activity in the lateral geniculate nucleus. Below, we introduce the FPE and its eigensolution. We have adopted some terminology of Knight and others for consistency. We start with an intuitive derivation of the FPE for random inputs that are Poisson in nature. This is a good model for input that comprises random spike trains.

The Fokker-Planck formalism

To simplify the description, we will deal with a simple integrate-and-fire model with one excitatory input. Consider a one-dimensional system with equations of motion described by Eqn. **31.1**, but now with $s(t)$ as a random variable encoding the stochastic arrival of spikes:

$$\begin{aligned} \dot{x} &= f(x) + s(t) \\ s(t) &= h\sum_n \delta(t - t_n) \end{aligned} \qquad \textbf{31.7}$$

where h represents a discrete change in postsynaptic membrane potential due to each spike and t_n represents the time of the nth spike. If we consider the input over a short time interval the mean spike-rate and the associated input are:

$$\begin{aligned} r(t) &= \frac{1}{T}\int_0^T \sum_n \delta(\tau - t_n)d\tau \\ s(t) &= hr(t) \end{aligned} \qquad \textbf{31.8}$$

The input is now a Poisson process whose expectation and variance scales with firing rate. How does this variability affect the evolution of the density? The master equation (Risken, 1996), detailing the rate of change of

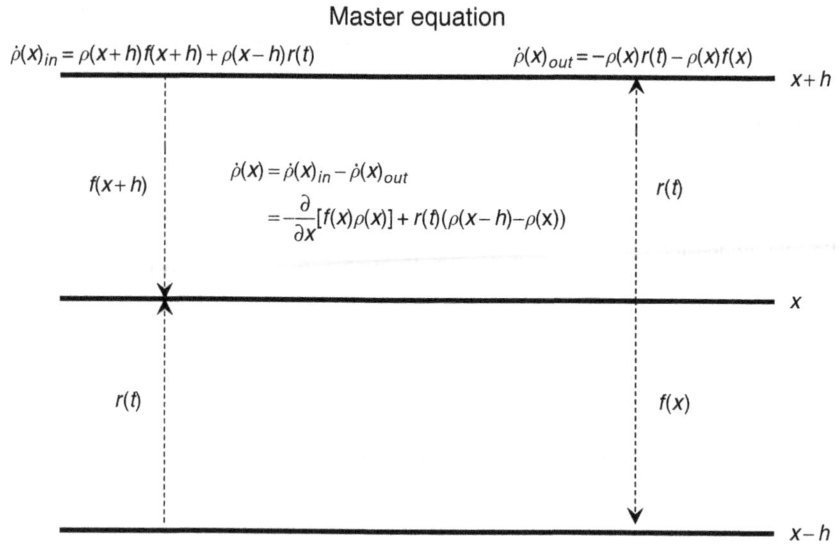

FIGURE 31.3 The master equation of a simple integrate-and-fire model with excitatory input (no synaptic dynamics). Three values of the state x (x and $x \pm h$) and transition rates (f(x), $f(x+h)$ and $r(t)$) are shown. The rate of change of $\rho(x, t)$ with time is attained by considering the rates in and out of x (formulas at the top left and right respectively).

$\rho(x, t)$ with time, can be intuited from the ladder diagram of Figure 31.3:

$$\dot{\rho}(x) = -\frac{\partial}{\partial x}[f(x)\rho(x)] + r(t)(\rho(x-h) - \rho(x)) \qquad \textbf{31.9}$$

The first term is due to leakage current and generates a steady flow towards E_L. The second expression is due to input and is composed of two terms, which can be considered as replenishing and depleting terms respectively. Given an impulse of input the probability mass between $x - h$ and x will flow to x, whereas the probability mass at x flows away from it. The replenishing term in Eqn. **33.9** can be approximated using the second-order Taylor expansion:

$$\rho(x-h) \approx \rho(x) - h\frac{\partial \rho}{\partial x} + \frac{h^2}{2}\frac{\partial^2 \rho}{\partial x^2} \qquad \textbf{31.10}$$

Substituting into Eqn. **31.9** and using Eqn. **31.8** we get:

$$\dot{\rho} = -\frac{\partial}{\partial x}[(f+s)\rho] + \frac{w^2}{2}\frac{\partial^2 \rho}{\partial x^2} \qquad \textbf{31.11}$$
$$w^2 = rh^2$$

where w^2 is the strength or variance of the stochastic fluctuations. This is also known as the diffusion coefficient $c = w^2/2$. This equation can be written more simply by:

$$\dot{\rho} = Q(x, s)\rho \qquad \textbf{31.12}$$

where $Q(x, s)$ contains all the dynamic information entailed by the differential equations of the model. This is the Fokker-Planck or dynamic operator (Knight, 2000). The first term of Eqn. **31.11**, known as the advection term, describes movement of the probability density due to the systems' deterministic dynamics. The second describes dispersion of density brought about by stochastic variations in the input. Inherent in this approximation is the assumption that h is small, i.e. the accuracy of the Taylor series increases as $h \to 0$. Comparisons between the diffusion approximation and direct simulations have demonstrated its accuracy in the context of neuronal models. The FPE above generalizes easily to cover models with multiple states, such as the TIF and CIF models of the previous section. We can see this with an alternative derivation of Eqn. **31.11** in terms of scalar and vector fields.

Derivation in terms of vector fields

For an alternative perspective on Eqn. **31.11**, we can think of the density dynamics in terms of scalar and vector fields. The probability density, $\rho(x, t)$, is a scalar function which specifies the probability mass at x, corresponding to the number of neurons with state x. This quantity will change if the states are acted on by a force whose influence is quantified by a vector field $J(x)$. This field represents the flow of probability mass within state-space. The net flux at any point is given by the divergence of the vector field $\nabla \cdot J$ (where ∇ is the divergence operator). The net flux is a scalar field and contains all the information needed to determine the rate of change of probability with time:

$$\dot{\rho} = -\nabla \cdot J \qquad \textbf{31.13}$$

is the continuity equation. The negative sign ensures that probability mass flows from high to low densities. The two forces in our simplified model are the leakage current and excitatory input. The former moves a neuron towards its equilibrium potential, E_L, while excitatory input drives voltage towards V_T. Each force generates its flux, which are in opposite directions. The overall flux is:

$$J(x) = f\rho(x) + r(t)\int_{x-h}^{x} \rho(x', t)dx'$$

$$\int_{x-h}^{x} \rho(x', t)dx' \approx h(1 - \frac{h}{2}\nabla)\rho \Rightarrow \qquad \mathbf{31.14}$$

$$\dot{\rho} = -\nabla \cdot J$$
$$= -\nabla \cdot ((f+s) - \frac{w^2}{2}\nabla)\rho$$

which is the same as Eqn. **31.11**, but has been formulated for a vector of states as opposed to a single state.

A general formulation

In the derivations so far, we have dealt with excitatory input that is Poisson in nature, where the expectation and variance of the process scale with each other. A more general formulation of the FPE considers deterministic input $s(t)$ and stochastic input $w\Gamma(t)$ as separate quantities. This gives the Langevin and non-linear Fokker-Planck equations (ensemble dynamics):

$$\dot{x} = f(x, s) + w(x, s)\Gamma(t)$$
$$\dot{\rho} = \nabla \cdot (-f(x, s) + \frac{w^2}{2}\nabla)\rho \qquad \mathbf{31.15}$$
$$= -\sum_{i=1}^{n} \frac{\partial}{\partial x_i}(f+s)\rho + \frac{1}{2}\sum_{i,k=1}^{n} \frac{\partial^2}{\partial x_i \partial x_k} w_{ik}^2(x, s)\rho$$

Zero mean, Langevin fluctuation, $\Gamma(t)$ (cf. a Weiner process) of unit variance, is scaled by $w(x, s)$. The last expression in Eqn. **31.15** has been written in a way that separates the two parts of the advection-diffusion equation. The deterministic input contributes to flow, which confers structure on the density. An example of $f(x, s)$ is Eqn. **31.6**, where $x = [x_1, \ldots, x_4, V, T]^T$ and $n = 6$. The advection or flow changes the local density in proportion to its gradient; clearly, if the density is flat, flow will have no effect. The second term is a diffusion term that tends to smooth the density; in other words, the density will decrease when it is peaked and has a high negative curvature (the curvature is the second partial derivative). This dispersion or diffusion reflects the stochastic fluctuations that dispel states from areas of high density. In the example above, $w(x, s) = \sqrt{hs}$; however, if $w(x, s)$ is fixed and diagonal the dispersion or diffusion of each state is isotropic. In what follows, we will use input in reference to deterministic inputs that shape density dynamics in the context of random fluctuations where $w(x, s)$ is diagonal and constant (see Table 31-2 for values).

TABLE 31-2 Parameter values used in simulations

Parameter description	Symbol	Value/units
Firing threshold	V_T	−53 mV
Reset voltage	V_R	−90 mV
Equilibrium potential	E_L	−73 mV
Equilibrium potential: K, AMPA, GABA and NMDA	$E_1, \ldots, E_4$	−90, 0, −70 and 0 mV
Passive conductance	g_L	25 nS
Active conductance: K, AMPA, GABA and NMDA	$g_1, \ldots, g_4$	128, 24, 64 and 8 nS
Membrane capacitance	C	0.375 nF
Time constant: K, AMPA, GABA and NMDA	$\tau_1, \ldots, \tau_4$	80, 2.4, 7 and 100 ms
Background opening coefficient: AMPA, GABA and NMDA	$p_1, \ldots, p_3$	0.875, 0.0625 and 0.0625 a.u.
Diffusion coefficient: V, T, K, AMPA, GABA and NMDA	$w_1^2, \ldots, w_6^2$	4, 0, 0.125, 0.125, 0.125 and 0.125 ms^{-1}

Solving the Fokker-Planck equation

Generally, the FPE is difficult to solve using analytic techniques. Exact solutions exist for only a limited number of models (Risken, 1996). However, approximate analytic and numerical techniques offer ways of solving a general equation. We have chosen a solution based on projection onto a bi-orthogonal set. This results in a system of uncoupled equations that approximate the original system and enables an important dimension reduction of the original set of equations.

The dynamic operator, $Q(s)$, is generally input-dependent and non-symmetric. By diagonalizing $Q(s)$, we implicitly reformulate the density dynamics in terms of probability modes. Two sets of eigenvectors (right and left) are associated with the dynamic operator, forming a bi-orthogonal set. This set encodes modes or patterns of probability over state-space. The right eigenvectors are column vectors of the matrix R, where $QR = RD$ and left eigenvectors are row vectors of matrix L, where $LQ = DL$. Both sets of eigenvectors share the same eigenvalues in the diagonal matrix D, which are sorted so that $\lambda_0 > \lambda_1 > \lambda_2 \ldots$. The left-eigenvector matrix is simply the generalized inverse of the right-eigenvector matrix. The number of eigenvectors and values, n, is equal to the dimensionality of Q. After normalization Q can be diagonalized:

$$LQR = D \qquad \mathbf{31.16}$$

Assume for the moment that input is constant, i.e. $s = 0$. Projecting the probability density $\rho(x, t)$ onto the space L generates an equivalent representation, but within a

TABLE 31-3 State variable ranges and number of bins used to grid state-space

State variable	Number of bins	Range of values
V	16	$[-92, -48]$ mV
T	8	$[0, 0.1]$ s
x_1	4	$[0, 1]$ a.u.
x_2	2	$[0, 1]$ a.u.
x_3	2	$[0, 1]$ a.u.
x_4	2	$[0, 1]$ a.u.

different coordinate system. Conversely, R projects back to the original coordinate system:

$$\mu = L\rho, \quad \rho = R\mu \qquad \mathbf{31.17}$$

Substituting $Q = RDL$ and the right expression of Eqn. **31.17** into Eqn. **31.12** gives a diagonalized system, $\dot{\mu} = D\mu$, which has the solution:

$$\mu(t) = \exp(Dt)\mu(0) \qquad \mathbf{31.18}$$

where $\mu(0)$ is a vector of initial conditions. This solution can be framed in terms of independent modes, i.e. columns of R, each of which contributes linearly to the evolution of the density. The expression of each mode is encoded by the elements of μ. The modes are independent in the sense that the changes in one mode do not affect the changes in another. This is because we have uncoupled the differential equations when projecting onto the bi-orthogonal set.

The rate of exponential decay of the i-th mode is characterized by its eigenvalue according to $\tau_i = -1/\lambda_i$, where τ_i is its characteristic time-constant. The key thing here is that, in many situations, most modes decay rapidly to zero, i.e. they have large negative eigenvalues. Heuristically, these modes dissipate very quickly because dispersion smoothes them away. The contribution of these unstable modes, to the dynamics at longer time-scales, is negligible. This is the rationale for reducing the dimension of the solution by ignoring them. Another advantage is that the equilibrium solution, i.e. the probability density that the system relaxes to (given constant input), is given by the principal mode, whose eigenvalue is zero. An approximate solution can then be written as:

$$\rho(x, t) = R_m \exp(D_m t) L_m \rho(x, 0) \qquad \mathbf{31.19}$$

where $\rho(x, 0)$ is the initial density profile, R_m and L_m are the principal m modes and D_m contains the first m eigenvalues, where $m \leq n$. The benefit of an approximate solution is that computational demand is greatly reduced when modelling the population dynamics. From now on, we will drop the subscript m and assume the eigensystem has been reduced.

Time-dependent solutions

The dynamic operator, $Q(s)$, is input-dependent and ideally needs calculating for each new input vector. This is time consuming and can be circumvented by using a perturbation expansion around a solution we already know, i.e. for $s = 0$. Approximating $Q(s)$ with a Taylor expansion about $s = 0$:

$$Q(s) \approx Q(0) + \sum_i s_i \frac{\partial Q}{\partial s_i} \qquad \mathbf{31.20}$$

where $Q(0)$ is evaluated at zero input and $\partial Q / \partial s_i$ is a measure of its dependency on s_i. Substituting this into the above equations and using $D(0) = LQ(0)R$ gives the dynamic equation for the coefficients μ, in terms of the bi-orthogonal set of $Q(0)$:

$$\dot{\mu} = \left(D(0) + \sum_i s_i L \frac{\partial Q}{\partial s_i} R \right) \mu = \hat{D}(s)\mu \qquad \mathbf{31.21}$$

We are assuming here that the inputs can be treated as constant during the small time interval over which the system is integrated. Eqn. **31.21** provides a simple set of locally linear equations that can be integrated to emulate the density dynamics of any ensemble.

So far, density dynamics have been presented as describing a statistical ensemble of solutions of a single neuron's response to input. An alternative interpretation is that $\rho(x, t)$ represents an ensemble of trajectories describing a population of neurons. The shift from a single neuron to an ensemble interpretation entails additional constraints on the population dynamics. An ensemble or mean-field type population equation assumes that neurons within an ensemble are indistinguishable. This means that each neuron 'feels' the same influence from internal interactions and external input. We will assume that this approximation holds for a subpopulation of neurons. However, for this to be a useful assumption, we need some mechanism for coupling different subpopulations that are acted upon by different inputs.

Multiple populations and their coupling

Interactions within a network of ensembles are modelled by coupling activities among populations with mean-field terms, such as average activity or spike rate. Coupled populations have been considered by Nykamp and Tranchina (2000) in modelling orientation tuning in the visual cortex. Coupling among populations, each described by standard Fokker-Planck dynamics, via mean-field quantities, induces non-linearities and thereby extends the networks' dynamic repertoire. This can result in a powerful set of equations that have been used to model physical, biological and social phenomena.

To model influences among populations, we simply treat an average state of an ensemble $s^{(i)} = M\rho^{(i)}$ as an input. The linear operator M acts on the density of the i-th population to provide an expected state (or some other moment) such as mean firing rate. A simple example is self-feedback, where the mean of an ensemble couples to its own dynamics as an input. The mean-field now depends on the population's own activity and has been described as a dynamic mean-field. This notion of mean-field coupling can be extended to cover multiple populations. A schematic of two interacting populations of the CIF model (Eqn. **31.6**) is shown in Figure 31.4.

Generally, coupling is modelled as a modulation of the parameters of the target population i, by inputs from the source k. The effect of a small perturbation of these parameters is used to approximate the time-dependent operator, which leads to a further term in Eqn. **31.21**. To a first-order approximation, for the i-th ensemble with j external inputs:

$$\begin{aligned} \dot{\mu}^{(i)} &= \hat{D}^{(i)}\mu^{(i)} \\ \hat{D}^{(i)}(s, \mu^{(1)}, \mu^{(2)}, \ldots,) &= D^{(i)}(0) + \sum_j s_j L \frac{\partial Q^{(i)}}{\partial s_j} R \\ &+ \sum_{k,l} \mu_l^{(k)} L \frac{\partial Q^{(i)}}{\partial \mu_l^{(k)}} R \end{aligned} \qquad \textbf{31.22}$$

The effect of the l-th mode of the k-th ensemble $\mu_l^{(k)}$ on the i-th ensemble follows from the chain rule:

$$\frac{\partial Q^{(i)}}{\partial \mu_l^{(k)}} = -\nabla \cdot \left(\frac{\partial \dot{\mu}^{(i)}}{\partial \theta^{(i)}} \frac{\partial \theta^{(i)}}{\partial s^{(k)}} \frac{\partial s^{(k)}}{\partial \mu_l^{(k)}} \right) \qquad \textbf{31.23}$$

where $\partial\theta^{(i)}/\partial s^{(k)}$ specifies the mechanism whereby the input from the k-th ensemble affects the i-th by changing its parameters. The derivative $\partial s^{(k)}/\partial \mu_l^{(k)}$ encodes how the input changes with the l-th mode of the source. For example, evolution of the probability modes in the source region causes a change in its average firing rate, which modulates synaptic AMPA channel opening dynamics in the target region, parameterized by $\theta^{(i)} = p_2^{(i)}$ in Eqn. **31.6**. This leads to a change in the flux of the target region $\dot{\mu}^{(i)}$ and a change in its density.

Once Eqn. **31.22** has been integrated, the measured output of the i-th region can be calculated in the same way that mean-field effects are calculated:

$$y^{(i)} = MR^{(i)}\mu^{(i)} \qquad \textbf{31.24}$$

Estimation and inference of mean field models

The formulation above furnishes a relatively simple dynamic model for measured electrophysiological responses reflecting population dynamics. This model can be summarized using Eqn. **31.22** and Eqn. **31.24**:

$$\begin{aligned} \dot{\mu}^{(i)} &= \hat{D}^{(i)}\mu^{(i)} \\ y^{(i)} &= MR^{(i)}\mu^{(i)} + \varepsilon^{(i)} \end{aligned} \qquad \textbf{31.25}$$

This is a deterministic input-state-output system with hidden states $\mu^{(i)}$ controlling the expression of probability density modes of the i-th population. Notice that the states no longer refer to biophysical or neuronal states (e.g. depolarization) but to the densities over states. The inputs are known deterministic perturbations $s(t)$ and the outputs are $y^{(i)}$. The architecture and mechanisms of this system are encoded in its parameters. In Chapter 34, we will see how this class of model can be inverted to provide conditional estimates of the parameters. In the remainder of this chapter, we focus on the sorts of systems that can be modelled and how inversion of these

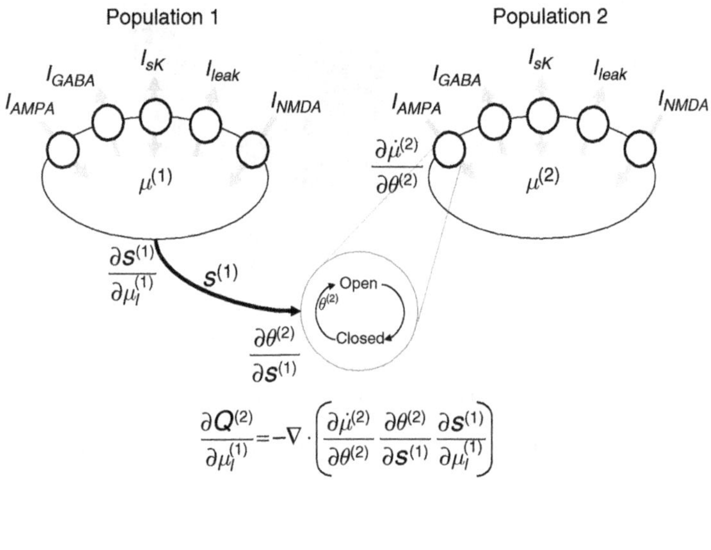

FIGURE 31.4 Schematic of interacting subpopulations. The rate of change of density over time, $\dot{\mu}^{(2)}$, depends on $\mu^{(1)}$ due to coupling. The density in population 1 leads to a firing rate $s^{(1)}$ which modulates the synaptic channel opening rate of population 2, parameterized by $\theta^{(2)}$, which in turn modulates $\dot{\mu}^{(2)}$. The coupling is calculated using the chain rule.

models can be used to address questions about coupling among neuronal ensembles.

ILLUSTRATIVE APPLICATIONS

In this section, we illustrate the nature of the generative model of the previous sections and its use in making inferences about the functional architecture of neuronal networks. We focus first on a single population and the approximations entailed by dimension reduction. We then consider the coupling between two populations that comprise a simple network.

Dynamics of a single population

Figure 31.5 shows the response of a population of TIF neurons (Eqn. **31.5**) to an external input. The top figure shows the mean firing rate over time. The black bar indicates the duration of sustained input. The rate oscillates briefly before being damped, after which it remains constant at a new equilibrium firing rate that is determined by the magnitude of the input. Once the input is removed, the rate decays to its background level. Below are two 3-D plots of the evolution of marginal distributions over V and T with time. The results were obtained by integrating Eqn. **31.21** for a single population and single (boxcar) input, using a dynamic operator based on Eqn. **31.5**. See Appendix 31.1 for numerical details.

Just prior to input there is very little probability mass at inter-spike times below 0.1 ms. This is seen as the large peak in $\rho(T, t)$ at $t = 0$ at the far right corner of the lower right figure. The inverse of the expected inter-spike interval corresponds to baseline-firing rate. After input, both distributions change dramatically. Density over the shorter inter-spike times increases, as the population is driven to fire more frequently. This is also seen in $\rho(V, t)$ (lower left), where density accumulates close to V_T and V_R, indicating a higher firing rate. These distributions

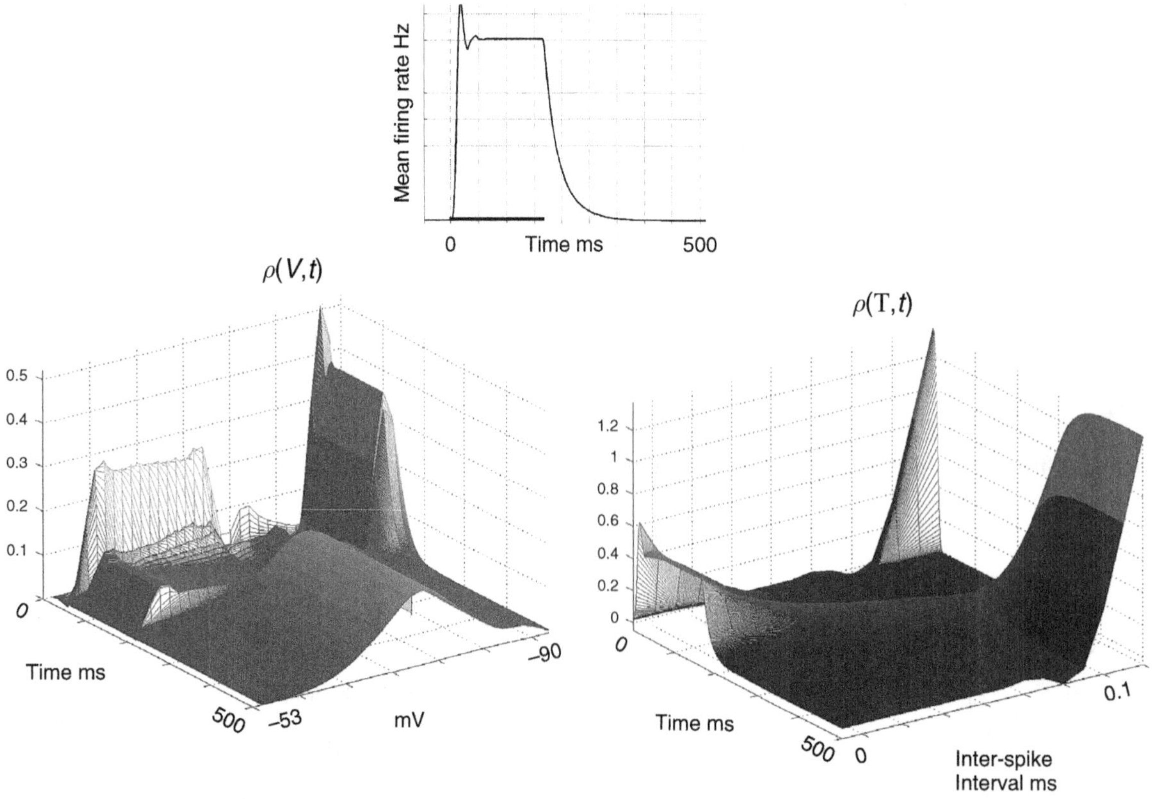

FIGURE 31.5 Response of a single population of neurons (FPE based on the TIF model) to a step rise in input, i.e. a boxcar input. The top figure shows the mean firing rate per neuron within the ensemble. The horizontal bar indicates the duration of input. The population responds with an increase in firing that oscillates briefly before settling to a new equilibrium and returns to its original firing rate after the input is removed. Below are two 3-D images of the marginal distributions over V and T (left and right respectively). Before input, the majority of probability over $\rho(T, 0)$ is peaked close to 0.1 ms. However, input causes a shift in density towards shorter time intervals and an increase in mean firing rate. This is also seen in the left figure, where input forces the density towards the firing threshold and reset potential. After input is removed, both densities return to their prior distributions.

return to their original disposition after the input returns to baseline.

Dimension reduction

The question now is how many probability modes are required to retain the salient aspects of these dynamics? An indication comes from the characteristic time-constants of each mode. These are shown for all modes, excluding the principal mode, which is stationary, in Figure 31.6. The time-constants decrease rapidly with mode number. A comparison of approximations in terms of response to different levels of input is shown in Figure 31.7. We considered approximations truncated at 16, 64 and 128 ($\hat{D}_{16}(s)$, $\hat{D}_{64}(s)$ and $\hat{D}_{128}(s)$ respectively) and the response curve for the deterministic model.

First, the stochastic models (full and approximate solutions) exhibit a key difference in relation to the deterministic model, i.e. firing rate does not have an abrupt start at an input threshold. Instead, there is a finite probability of firing below threshold and the response curve tapers off with lower input. Second, the solution using all probability modes of the approximation compares well with $D(s)$ computed explicitly at each input value. The approximation is less accurate as input increases, however, it remains close to and retains the character of the true response curve. Third, the truncated approximation using 64 modes is almost indistinguishable from the full approximation. The solution using only 16 modes, despite losing accuracy with larger inputs, still maintains some of the character of the response curve and, at low input levels, is a reasonable approximation. Given that this approximation represents an eightfold decrease in the number of modes, this degree of approximation is worth considering when optimizing the balance between computational efficiency and accuracy.

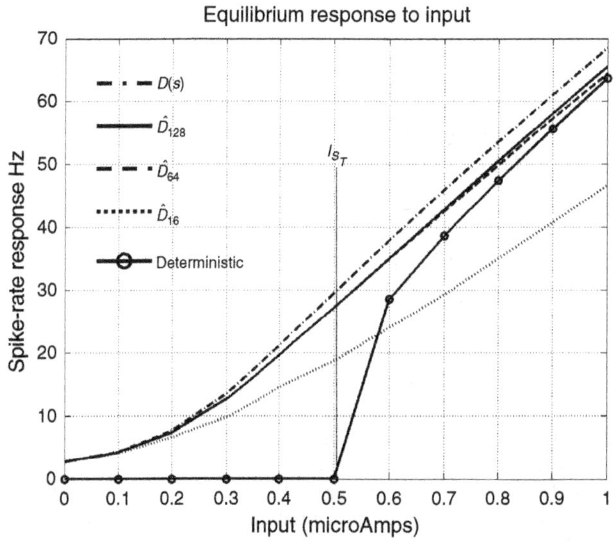

FIGURE 31.7 Comparison of approximation of equilibrium response rates of a stochastic and deterministic model-neuron (Eqn. 31.5). All rates from diffusion equations taper off gradually as input falls below threshold, s_T, in contrast to the deterministic model. The curve labelled $D(s)$ is from an explicit re-calculation of the dynamic operator at each input, whereas $\hat{D}_{128}$, $\hat{D}_{64}$ and $\hat{D}_{16}$ are first-order approximations using 128, 64 or 16 modes (out of 128). $\hat{D}_{128}$ is in good agreement with D, losing some accuracy as input increases. $\hat{D}_{64}$ is almost indistinguishable from $\hat{D}_{128}$. $\hat{D}_{16}$ is less accurate; however, at low inputs, is still in reasonable agreement with $\hat{D}_{128}$ and maintains the characteristic profile of the response curve.

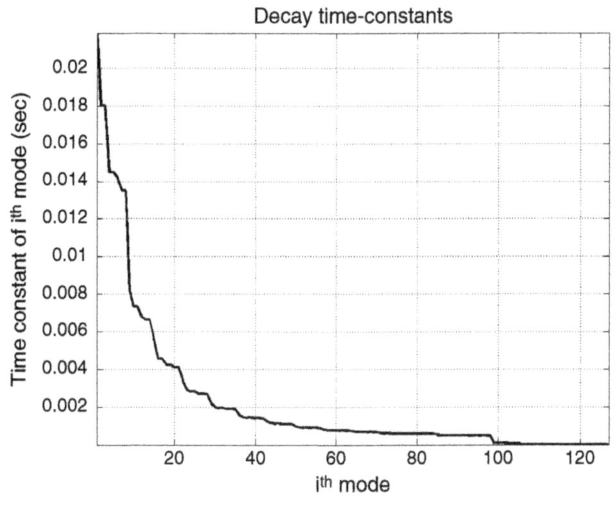

FIGURE 31.6 Decay time-constants versus mode number. The model used in Figure 31.5 was approximated using a system of 128 coupled ordinary differential equations (see text). This can be transformed into 128 uncoupled equations, where each equation describes the dynamics of a probability mode. The characteristic time to decay (i.e. negative inverse eigenvalue) for each mode is shown. Right-most modes, i.e. short time-constants, decay very rapidly and do not contribute significantly to dynamics over a relatively longer time period. This is the rationale for approximating a solution by excluding these unstable modes.

Coupling among populations

We next simulated a small network of populations. A schematic of the network is shown in Figure 31.8. The model consisted of two identical regions, each containing sub-populations of excitatory and inhibitory neurons. The excitatory sub-population exerted its effect through AMPA synaptic channels, while GABA channels mediated inhibition of excitatory neurons. The regions were reciprocally connected, with the second region driven by inputs from the first that targeted fast excitatory AMPA channels. Conversely, the first region was modulated by feedback from the second that was mediated by slow excitatory NMDA channels. Only the first

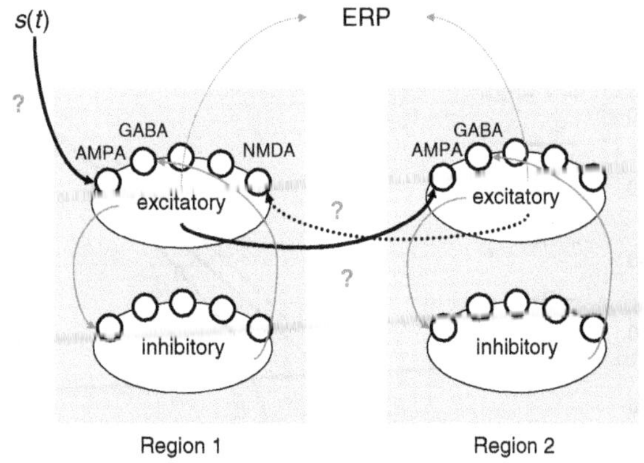

FIGURE 31.8 Schematic of the simulated network. Two regions were coupled via excitatory synaptic channels and Region 1 was driven by an external input. Each region was identical and composed of an excitatory and inhibitory population coupled via AMPA and GABA synaptic channels. Region 2 was driven by 1 via fast excitatory AMPA channels, while Region 1 received excitatory feedback from 2 mediated by slow NMDA channels. Local field potentials were modelled as mean membrane potential from the excitatory populations of both regions. Results from the simulation were used in an estimation scheme to identify the three coupling parameters indicated by question marks.

region received external input. This conforms to a simple cortical hierarchy with the second region being supraordinate. These receptor-specific effects were specified by making $\partial\theta^{(i)}/\partial s^{(k)}$ non-zero for the rate of the appropriate receptor-specific channel opening (see Eqn. **31.6** and Figure 31.8) and using the mean spike rate as the output $s^{(k)}$ from the source population.

Event-related signals ($y^{(i)} = MR^{(i)}\mu^{(i)}$, mean depolarization of excitatory populations), generated by the network in response to an impulse of exogenous input are shown in Figure 31.9. These responses have early and late components around 150 and 350 ms respectively, which are characteristic of real evoked response potentials.

Inverting the model to recover coupling parameters

Gaussian observation noise (~10 per cent) was added to the mean potentials from both regions to simulate data. The model was then inverted to estimate the known parameters, using EM as described in Chapter 34. The predicted response (solid line) of the generative model is compared to the synthetic data (broken line) in Figure 31.9. Three coupling parameters $\partial\theta^{(i)}/\partial s^{(k)}$

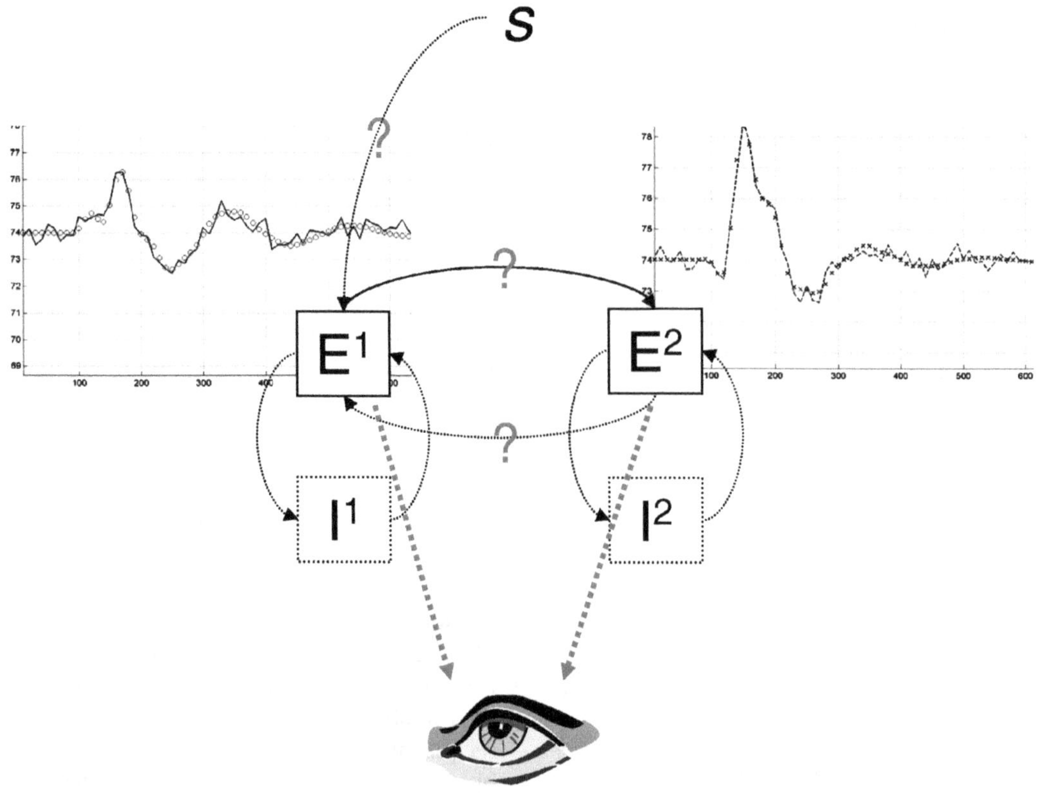

FIGURE 31.9 Comparison of simulated data (plus observation noise) and predicted responses after estimation (superimposed) of coupling parameters (see Figure 31.10) from the network simulation. Mean membrane potential from the excitatory populations of each region, in response to a brief input, is shown.

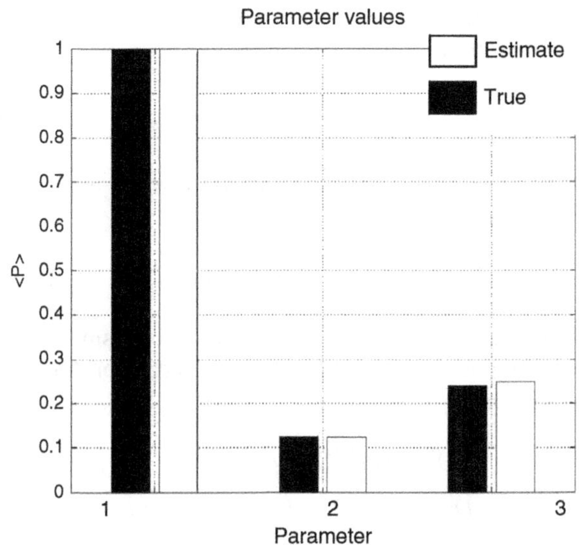

FIGURE 31.10 The conditional expectations of the three coupling parameters between input to Region 1, Region 1 to 2 and backward coupling from 2 to 1 (parameter indices 1–3 respectively) are shown next to the known parameters of the simulation.

mediating exogenous input and the extrinsic connections between regions (bold connections with question marks in Figures 31.8 and 31.9) were given uninformative priors. These represent unknown parameters that the estimation scheme was trying to identify. Their conditional expectations are shown with the true values in Figure 31.10. They show good agreement and speak to the possibility of inverting mean-field models using real data.

CONCLUSION

The aim of this chapter was to demonstrate the feasibility of using the FPE in a generative model for LFP/ERPs. The ensuing model embodies salient features of real neuronal systems: neurons are dynamic units, driven by stochastic forces and organized into populations with similar response characteristics; and multiple populations interact to form functional networks. Despite the stochastic nature of neuronal dynamics, the FPE formulates the solution in terms of a deterministic process, where the dispersive effect of noise is modelled as a diffusive process. The motivation for using such a model is that its associated parameters have a clear biological meaning, enabling unambiguous and mechanistic interpretations.

We have reviewed well-known material on integrate-and-fire model neurons with synaptic dynamics, which included fast excitatory AMPA, slow excitatory NMDA and inhibitory GABA mediated currents. The FPE was used to model the effect of stochastic input, or system noise, on population dynamics. Its time-dependent solution was approximated using a perturbation expansion about zero input. Decomposition into a bi-orthogonal set enabled a dimension reduction of the system of coupled equations, due to the rapid dissipation of unstable probability modes. Interactions among populations were modelled as a change in the parameters of a target population that depended on the average state of source populations.

To show that the model produces realistic responses and, furthermore, it could be used as an estimation or forward model, separate ensembles were coupled to form a small network of two regions. The coupled model was used to simulate ERP data, i.e. mean potentials from excitatory subpopulations in each region. Signals were corrupted by Gaussian noise and subject to expectation maximization (EM). Three parameters were estimated – input, forward and backward connection strengths – and were shown to compare well to the known values. It is pleasing to note that the stochastic model produces signals that exhibit salient features of real ERP data and the estimation scheme was able to recover its parameters.

The key aspect of the approach presented here is the use of population density dynamics as a forward model of observed data. These models have been used to explore the cortical dynamics underlying orientation tuning in the visual cortex. These models may also find a place in LFP/ERP data analysis. In these models, random effects are absorbed into the FPE and the population dynamics become deterministic. This is a critical point because it means system identification has only to deal with observation noise. Heuristically, the deterministic noise induced by stochastic effects is effectively 'averaged away' by measures like ERPs. However, the effect of stochastic influence is still expressed and modelled, deterministically, at the level of population dynamics.

There are many issues invoked by this modelling approach. The dimensionality of solutions for large systems can become extremely large in probability space. Given an N-dimensional dynamic system, dividing each dimension into M bins results in an approximation to the FPE with a total of M^N ordinary differential equations. The model used to simulate a network of populations used 4096 equations to approximate the dynamics of one population. Dimension reduction, by using a truncated bi-orthogonal set, is possible. However, as was demonstrated in Figure 31.7, there is a trade-off between accuracy and dimension reduction. Generally, a more realistic model requires more variables, so there is a balance between biological realism and what we can expect from current computational capabilities.

The model neuron used in this chapter is just one of many candidates. Much can be learnt from comparing models. For instance, is modelling the inter-spike time as a state variable an efficient use of dimensions? This may eschew the need for detailed boundary conditions; however, it may well be an extravagant use of dimensions given computational limitations. The current modelling approach is not limited to electrophysiological data. Any measurement which is coupled to electrical activity of a neuronal population could, in principle, also be included in the generative model, which could be used to combine measurements of electrical and metabolic origin, e.g. fMRI.

The use of Bayesian system identification enables the formal inclusion of additional information when estimating model parameters from data. We have not given the topic of priors much consideration in this chapter, but it is an important issue. Analytic priors derived from stability or bifurcation analyses could be used to ensure parameter values which engender dynamics characteristic of the signals measured, i.e. stable fixed point, limit cycle or chaotic attractors. Empirical priors derived from data also have great potential in constraining system identification. A Bayesian framework also facilitates model comparison through quantifying the 'evidence' that a data set has for a number of different models (Penny *et al.*, 2004).

In the next two chapters, we turn to neural-mass models of EEG data. Neural-mass models are a special case of mean-field models in which the ensemble densities are approximated with a single mode or point mass. In other words, the density is summarized with a state vector that encodes the most likely or expected state of each ensemble. This can be thought of as a dimension reduction to the smallest number of modes (i.e. one) or can be regarded as approximating the full density with a point mass over its expectation. Clearly, neural-mass models lose the ability properly to model random fluctuations, however, the computational saving enables a much greater number of biophysical states to be modelled with a large repertoire of dynamical behaviours.

APPENDIX 31.1 NUMERICAL SOLUTION OF FOKKER-PLANCK EQUATION

The equation we wish to integrate is:

$$\dot{\rho}(x, t) = Q\rho \qquad \textbf{31.A1}$$

First, vectorize and grid up the n-D state-space $x_i = ih$, where $i = 1, \ldots, N$ and evaluate $\dot{x} = f(x, s)$ at all grid points. Calculate the operators $-\nabla \cdot f$ and ∇^2, where $\nabla^2 = \nabla \cdot \nabla$ is the Laplace operator, required to construct an approximation to the dynamic operator:

$$Q = -\nabla \cdot (f + s) + \frac{w^2}{2}\nabla^2 \qquad \textbf{31.A2}$$

Eqn. 31.A1 is a system of coupled differential equations with the solution:

$$\rho(x, t+\Delta t) = \exp(\Delta t Q)\rho(x, t) \qquad \textbf{31.A3}$$

This system is reduced and integrated over small time steps using the eigenvectors of Q, where for the i-th ensemble or population:

$$\begin{aligned} \mu^{(i)}(t+\Delta t) &= \exp(D^{(i)}\Delta t)\mu^{(i)}(t) \\ \hat{D}^{(i)} &= D_0^{(i)} + \sum_j s_j D_j^{(i)} + \sum_{k,l} \mu_l^{(k)} D_{kl}^{(i)} \\ y^{(i)}(t) &= MR^{(i)}\mu^{(i)}(t) \end{aligned} \qquad \textbf{31.A4}$$

The Jacobian matrices are pre-computed using $Q^{(i)}$ from 31.A2:

$$D_0^{(i)} = LQ^{(i)}R \quad D_j^{(i)} = L\frac{\partial Q^{(i)}}{\partial s_j}R \quad D_{kl}^{(i)} = L\frac{\partial Q^{(i)}}{\partial \mu_l^{(k)}}R \qquad \textbf{31.A5}$$

After discretizing state-space and approximating $Q^{(i)}$ for one population, the eigenvectors and values can be calculated using the Matlab function 'eig.m' (or 'eigs.m' for specific eigenvector/values) and saved. Given these, a reduced or full model can be used to model a network of populations by specifying the connectivity among ensembles. The exponential matrix of the reduced or full model can be calculated using the Matlab function 'expm' (Moler and Van Loan, 2003). This uses a (6,6) Pade approximation. Explicit and implicit numerical integration schemes can be reformulated into a Pade approximation, e.g. (0,1) approximation of order 1 is a forward Euler scheme, whereas (1,1) approximation of order 2 is a Crank-Nicolson implicit scheme. As the 'expm' function uses an implicit approximation the scheme is accurate and unconditionally stable (Smith, 1985).

REFERENCES

Casti AR, Omurtag A, Sornborger A *et al.* (2002) A population study of integrate-and-fire-or-burst neurons. *Neural Comput* **14**: 957–86

Churchland PS and Sejnowski TJ (1994) *The computational brain.* MIT Press, Cambridge, MA

David O, Friston KJ (2003) A neural mass model for MEG/EEG: coupling and neuronal dynamics. *NeuroImage* **20**: 1743–55

David O, Kiebel SJ, Harrison LM *et al.* (2006) Dynamic causal modeling of evoked responses in EEG and MEG. *NeuroImage* **30**: 1255–72

Dayan P (1994) Computational modelling. *Curr Opin Neurobiol* **4**: 212–17

Dayan P, Abbott L (2001) *Theoretical neuroscience: computational and mathematical modeling of neural systems*. MIT Press, Cambridge, MA

Frank T (2005) *Nonlinear Fokker-Planck equations: fundamentals and applications*. Springer, Heidelberg

Friston KJ, Glaser DE, Henson RN *et al.* (2002) Classical and Bayesian inference in neuroimaging: applications. *NeuroImage* **16**: 484–512

Friston KJ, Harrison L, Penny W (2003) Dynamic causal modelling. *NeuroImage* **19**: 1273–302

Gerstner W, Kistler W (2002) *Spiking neuron models. Single neurons, populations, plasticity*. Cambridge University Press, Cambridge

Haken H (1996) *Principles of brain function*. Springer, Heidelberg

Haskell E, Nykamp DQ, Tranchina D (2001) Population density methods for large-scale modelling of neuronal networks with realistic synaptic kinetics: cutting the dimension down to size. *Network* **12**: 141–74

Jansen BH, Rit VG (1995) Electroencephalogram and visual evoked potential generation in a mathematical model of coupled cortical columns. *Biol Cybern* **73**: 357–66

Jirsa VK (2004) Connectivity and dynamics of neural information processing. *Neuroinformatics* **2**: 183–204

Jirsa VK, Haken H (1996) Field theory of electromagnetic brain activity. *Phys Rev Lett* 77: 960–63

Kloeden and Platen (1999) *Numerical solution of stochastic differential equations*. Springer, Berlin

Knight BW (2000) Dynamics of encoding in neuron populations: some general mathematical features. *Neural Comput* **12**: 473–518

Kuramoto (1984) *Chemical oscillations, waves and turbulence*. Springer, Berlin

Moler C, Van Loan C (2003) Nineteen dubious ways to compute the exponential of a matrix, twenty five years later. *SIAM Review* **45**: 3–49

Nykamp DQ, Tranchina D (2000) A population density approach that facilitates large-scale modeling of neural networks: analysis and an application to orientation tuning. *J Comput Neurosci* **8**: 19–50

Nykamp DQ, Tranchina D (2001) A population density approach that facilitates large-scale modeling of neural networks: extension to slow inhibitory synapses. *Neural Comput* **13**: 511–46

Omurtag A, Knight BW, Sirovich L (2000) On the simulation of large populations of neurons. *J Comput Neurosci* **8**: 51–63

Penny WD, Stephan KE, Mechelli A *et al.* (2004) Comparing dynamic causal models. *NeuroImage* **22**: 1157–72

Pfurtscheller G (2001) Functional brain imaging based on ERD/ERS. *Vision Res* **41**: 1257–60

Pfurtscheller G, Lopes da Silva FH (1999) Event-related EEG/MEG synchronization and desynchronization: basic principles. *Clin Neurophysiol* **110**: 1842–57

Rennie CJ, Robinson PA, Wright JJ (2002) Unified neurophysical model of EEG spectra and evoked potentials. *Biol Cybern* **86**: 457–71

Risken H (1996) *The Fokker-Planck equation*, 3rd edn. Springer-Verlag, New York, Berlin, Heidelberg

Sirovich L (2003) Dynamics of neuronal populations: eigenfunction theory; some solvable cases. *Network* **14**: 249–72

Smith G (1985) *Numerical solutions of partial differential equations: finite difference methods*. Clarendon Press, Oxford

Valdes PA, Jimenez JC, Riera J *et al.* (1999) Nonlinear EEG analysis based on a neural mass model. *Biol Cybern* **81**: 415–24

Wiesenfeld K, Moss F (1995) Stochastic resonance and the benefits of noise: from ice ages to crayfish and SQUIDs. *Nature* **373**: 33–36

Winkler I, Schroger E, Cowan N (2001) The role of large-scale memory organization in the mismatch negativity event-related brain potential. *J Cogn Neurosci* **13**: 59–71

CHAPTER

32

Neuronal models of energetics

J. Kilner, O. David and K. Friston

INTRODUCTION

In this chapter, we will describe a simple biophysical model that links neuronal dynamics to functional magnetic resonance imaging (fMRI) and electroencephalography (EEG) measurements. The previous chapter described mean-field models and the next chapter will cover neural-mass approaches to modelling EEG signals. Here, we look at a simple mean-field model that was developed with the specific aim of relating neuronal activity recorded with EEG and the neuronal activity that leads to a modulation of the blood oxygenation-level-dependent (BOLD) signal recorded with fMRI.

The chapter is divided into three sections. First, we will describe the motivation for relating the different measures of brain activity afforded by EEG and fMRI and review different approaches that have been adopted for multimodal fusion. Second, we will outline the neuronal model, starting with a dimensional analysis. In brief, this model suggests that neuronal activity causes an acceleration of temporal dynamics leading to: increased energy dissipation; decreased effective membrane time-constants; increased coupling among neuronal ensembles; and a shift in the EEG spectral profile to higher frequencies. Finally, we will show that these predictions are consistent with empirical observations of how changes in the EEG spectrum are expressed haemodynamically.

EEG AND fMRI INTEGRATION

It is now generally accepted that the integration of fMRI and electromagnetic measures of brain activity has an important role in characterizing evoked brain responses. These measures are both related to the underlying neural activity. However, electromagnetic measures are direct and capture neuronal activity with millisecond temporal resolution, while fMRI provides an indirect measure with poor temporal resolution, in the order of seconds. Conversely, fMRI has excellent spatial resolution, in the order of millimetres, compared to EEG-MEG (electroencephalography-magnetoencephalography). Therefore, the obvious motivation for integrating these two measures is to provide a composite recording that has high temporal and spatial resolution. The possibility of integrating these two measures in humans is supported by the study of Logothetis *et al.* (2001). In this study, the authors demonstrated that within the macaque monkey visual cortex, intracortical recordings of the local field potential (LFP) and the BOLD signal were linearly related.

Approaches to integration can be classified into three sorts: integration through prediction; integration through constraints; and integration through fusion. These are depicted schematically in Figure 32.1. Integration through prediction (dotted line) uses temporally resolved EEG signals as a predictor of changes in concurrently recorded fMRI. The ensuing region-specific haemodynamic correlates can then be characterized with high spatial resolution with conventional imaging methodology (Lovblad *et al.*, 1999; Lemieux *et al.*, 2001; Czisch *et al.*, 2004). Several studies of this type (Goldman *et al.*, 2002; Laufs *et al.*, 2003a, b; Martinez-Montes, 2004) have focused on correlating modulations in ongoing oscillatory activity measured by EEG with the haemodynamic signal. They have demonstrated that modulations in alpha rhythms (oscillations at $\sim$10Hz) are negatively correlated with modulations in the BOLD signal, i.e. an increase in alpha power is associated with a decrease in BOLD. Studies employing integration through constraints (dashed line), have used the spatial resolution of focal fMRI activations to constrain equivalent dipole or distributed estimates of EEG-MEG sources (Dale et al., 2000; Phillips *et al.*, 2002). However, neither of these

Statistical Parametric Mapping, by Karl Friston et al.
ISBN–13: 978-0-12-372560-8 ISBN–10: 0-12-372560-7

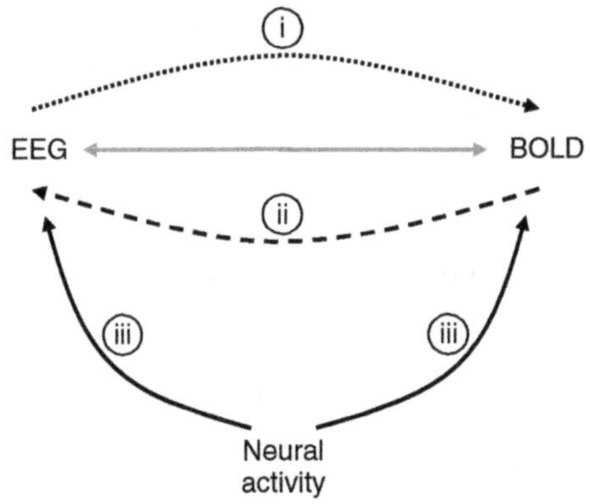

FIGURE 32.1 Schematic showing the approaches to EEG/fMRI integration. (i) Integration through prediction. (ii) Integration through constraints. (iii) Integration through fusion with forward models.

schemes can be considered as true integration of multimodal data in the sense that there is no common forward model that links the underlying neuronal dynamics of interest to measured haemodynamic and electrical responses (solid black lines).

The characterization of the relationship between the electrophysiology of neuronal systems and their slower haemodynamics is crucial from a number of perspectives: not only for forward models of electrical and haemodynamic data, but also for the utility of spatial priors, derived from fMRI, in the inverse EEG/MEG source reconstruction problem, and for the disambiguation of induced, relative to evoked, brain responses using both modalities.

In the next section, we will describe perhaps the simplest of all models which helps to explain some empirical observations from EEG-fMRI integration, using a dimensional analysis and a biophysical model.

A HEURISTIC FOR EEG-fMRI INTEGRATION

A dimensional analysis

Given the assumption that haemodynamics reflect the energetics of underlying neuronal changes (Jueptner and Weiller, 1995; Shulman and Rothman, 1998; Hoge *et al.*, 1999; Magistretti *et al.*, 1999; Magistretti and Pellerin, 1999; Shin, 2000), we assume here that the BOLD signal b, at any point in time, is proportional to the rate of energy dissipation, induced by transmembrane currents. It is important to note that we are not assuming the increase in blood flow, which is the major contributor to the BOLD signal, is a direct consequence of the rate of energy dissipation, but rather that these two measures are proportional (see Hoge *et al.*, 1999). Although recent work has suggested that the neurovascular coupling is driven by glutamate release (see Lauritzen, 2001; Attwell and Iadecola, 2002), glutamate release, BOLD signal and energy usage are correlated and therefore the assumption here that the BOLD signal is proportional to the rate of energy dissipation is tenable. This dissipation is expressed in terms of Joules per second and corresponds to the product of transmembrane potential (V, joules per coulomb) and transmembrane current (I, coulombs per second):

$$b \propto \langle V^T I \rangle \tag{32.1}$$

where $V = [V_1, \ldots, V_k]^T$ corresponds to a [large] column vector of potentials for each neuronal compartment k within a voxel, similarly for the currents. Clearly, Eqn. **32.1** will not be true instantaneously, because it may take some time for the energy cost to be expressed in terms of increased oxygen delivery, extraction and perfusion. However, over a suitable timescale, of order seconds, Eqn. **32.1** will be approximately correct. Assuming a single-compartment model, currents are related to changes in membrane potential though their capacitance C, which we will assume is constant (see Dayan and Abbott, 2001: 156). By convention, the membrane current is defined as positive when positive ions leave the neuron and negative when positive ions enter the neuron:

$$I = -C\dot{V} \tag{32.2}$$

then

$$b \propto V \langle V^T \dot{V} \rangle \tag{32.3}$$

To relate changes in membrane potential to the BOLD signal, we need to adopt some model of a neuronal system and how it activates. A simple integrate-and-fire model of autonomous neuronal dynamics can be expressed in the form:

$$\begin{aligned} \dot{V}_k &= -V_k/\tau_k + u_k \\ &= f_k(V) \end{aligned} \tag{32.4}$$

for the k-th compartment or unit (See Eqn. **31.1** of the previous chapter). We have assumed here that synaptic currents are caused by some non-linear function of the depolarization status of all units in the population (cf. a mean-field effect as described in the previous chapter): i.e. $u_k = g_k(V)$. For a system of this form we can approximate the dynamics of perturbations with the first-order system:

$$\begin{aligned} \dot{V}(t) &= -JV \\ J &= \frac{\partial f}{\partial v} \end{aligned} \tag{32.5}$$

The Jacobian J summarizes the functional or causal architecture of the neuronal system. The leading diagonal elements of J correspond to self-inhibition and play the role of effective membrane rates or conductances. In the absence of influences from any other units, the k-th potential will decay exponentially to the equilibrium or resting potential ($V = 0$):

$$\dot{V}_k = -J_{kk} V_k \quad \mathbf{32.6}$$

It can be seen that $J_{kk} = 1/\tau_k$ has units of per second and is the inverse of the effective membrane time constant. In fact, in most biophysical models of neuronal dynamics, this 'rate' is usually considered as the ratio of the membrane's conductance to its capacitance. Conductivity will reflect the configuration of various ion channels and the ongoing postsynaptic receptor occupancy. In a similar way, the off-diagonal elements of the Jacobian characterize the intrinsic coupling among units, where $J_{kj} = \partial f_k/\partial V_j = \partial \dot{V}_k/\partial V_j$. It is interesting to note that plausible neuronal models of ensemble dynamics suggest a tight coupling between average spiking rate and decreases in effective membrane time constants (e.g. Chawla *et al.*, 2000). However, as we will see below, we do not need to consider spikes to close the link between BOLD and frequency profiles of ongoing EEG or MEG dynamics.

From Eqn. **32.3** and Eqn. **32.5**, we have:

$$\begin{aligned} b &\propto C < V^T J V > \\ &\propto Ctr(J < VV^T >) \\ &\propto Ctr(JCov\{V\}) \end{aligned} \quad \mathbf{32.7}$$

This means that the metabolic response is proportional to the trace of the product of the Jacobian (i.e. coupling matrix) and the temporal covariance of the transmembrane potentials.

Modelling activations

At this point, we have to consider how 'activation' is mediated. In other words, how the dynamics over an extended period of time could change. If we treat the units within any voxel as an autonomous system then any extrinsic influence must be mediated by changes in the Jacobian, e.g. changes in conductance or coupling among neurons induced by afferent input. The attending changes in potential are secondary to these changes in the functional architecture of the system and may, or may not, change their covariances $Cov\{V\}$. According to Eqn. **32.7**, a metabolic cost is induced through changes in J, even in the absence of changes in the covariance.

To link BOLD and EEG responses we need to model the underlying changes in the Jacobian that generate them. This is accomplished in a parsimonious way by introducing an activation variable, α, that changes J. Here, α is a parameter that changes the coupling (i.e. synaptic efficacies) and, implicitly, the dynamics of neuronal activity. In this model, different modes of brain activity are associated with the Jacobian:

$$J(\alpha) = J(0) + \alpha \partial J/\partial \alpha \quad \mathbf{32.8}$$

We will assume that $\partial J/\partial \alpha = J(0)$. In other words, the change in intrinsic coupling (including self-inhibition), induced by activation, is proportional to the coupling in the 'resting' state when $\alpha = 0$. The motivations for this assumption include:

- its simplicity
- guaranteed stability, in the sense that if $J(0)$ has no unstable modes (positive eigenvalues) then neither will $J(\alpha)$. For example, it ensures that activation does not violate the 'no-strong-loops hypothesis' (Crick and Koch, 1998)
- it ensures the intrinsic balance between inhibitory and excitatory influences that underpins 'cortical gain control' (Abbott *et al.*, 1998)
- it models the increases in membrane conductance associated with short-term increases in synaptic efficacy.

Effect of neuronal activation on BOLD

These considerations suggest that the coupling J_{kj} among neurons (positive and negative) will scale in a similar way and that these changes will be reflected by changes in effective membrane time constants J_{kk}. Under this model for activation, the effect of α is to accelerate the dynamics and increase the system's energy dissipation. This acceleration can be seen most easily by considering the responses to perturbations around v_0 under $J = J(0)$ and $\tilde{J} = J(\alpha) = (1+\alpha)J$:

$$\begin{aligned} V(t) &= e^{-Jt} V_0 \\ \tilde{V}(t) &= e^{-\tilde{J}t} V_0 = V((1+\alpha)t) \end{aligned} \quad \mathbf{32.9}$$

In other words, the perturbation under $J(\alpha)$ at time t is exactly the same as that under $J(0)$ at $(1+\alpha)t$. This acceleration will only make dynamics faster; it will not change their form. Consequently, there will be no change in the covariation of membrane potentials and the impact on the fMRI responses is mediated by, and only by, changes in J:

$$\frac{\tilde{b}}{b} \propto \frac{tr(\tilde{J}Cov\{V\})}{tr(JCov\{V\})} = (1+\alpha) \quad \mathbf{32.10}$$

In other words, the activation α is proportional to the relative increase in metabolic demands. This is intuitive from the perspective of fMRI, but what does activation look like in terms of the EEG?

Effect of neuronal activation on EEG

From the point of view of the fast temporal activity reflected in the EEG, activation will cause an acceleration of the dynamics, leading to a 'rougher' looking signal with loss of lower frequencies, relative to higher frequencies. A simple way to measure this effect is in terms of the roughness r, which is the normalized variance of the first temporal derivative of the EEG. From the theory of stationary processes (Cox and Miller, 1965), this is mathematically the same as the negative curvature of the EEGs autocorrelation function evaluated at zero lag. Thus, for an EEG signal, e:

$$r = \frac{Var(\dot{e})}{Var(e)} = -\rho(0)''$$

Assuming e (measured at a single site) is a linear mixture of potentials, i.e. $e = lV$, where l is a lead-field row vector, its autocorrelation at lag h is:

$$\rho(h) = < V(t)^T l^T l V(t+h) > \qquad 32.11$$

From Eqn. **32.9** and Eqn. **32.11**, we have:

$$\begin{aligned} \tilde{\rho}(h) &= \rho((1+\alpha)h) \\ \tilde{\rho}(h)'' &= (1+\alpha)^2 \rho(h)'' \end{aligned} \qquad 32.12$$

It follows that the change in r is related to neuronal activation by:

$$\frac{\tilde{r}}{r} = \frac{\tilde{\rho}(0)''}{\rho(0)''} = (1+\alpha)^2 \qquad 32.13$$

As the spectral density of a random process is the Fourier transform of its autocorrelation function, $g(\omega) = \int \rho(h) e^{-i\omega h} dh$, the equivalent relationship in the frequency domain that obtains from the 'roughness' expressed in terms of spectral density $g(\omega)$ is:

$$r = \frac{\int \omega^2 g(\omega) d\omega}{\int g(\omega) d\omega}$$

From Eqn. **32.12**, the equivalent of the activated case, in terms of the spectral density is:

$$\tilde{g}(\omega) = \frac{g((1+\alpha)\omega)}{(1+\alpha)} \qquad 32.14$$

Here, the effect of activation is to shift the spectral profile toward higher frequencies with a reduction in amplitude

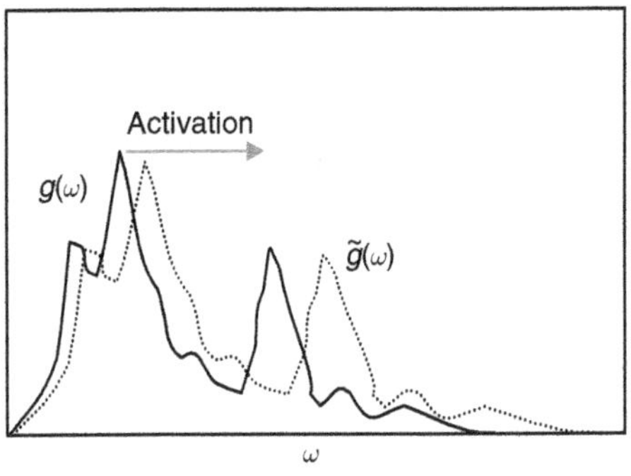

FIGURE 32.2 Schematic showing the effect of activation on the spectral profile.

(Figure 32.2). The activation can be expressed in terms of the 'normalized' spectral density:

$$\begin{aligned} \frac{\tilde{r}}{r} &= \frac{\int \omega^2 \tilde{p}(\omega) d\omega}{\int \omega^2 p(\omega) d\omega} = (1+\alpha)^2 \\ p(\omega) &= \frac{g(\omega)}{\int g(\omega) d\omega} \end{aligned} \qquad 32.15$$

$p(\omega)$ could be treated as an empirical estimate of probability, rendering roughness equivalent to the expected or mean square frequency. Gathering the above equalities together, we can express relative values of fMRI and spectral measures in terms of each other:

$$\left[\frac{\tilde{b}}{b}\right]^2 \propto (1+\alpha)^2 \propto \frac{\int \omega^2 \tilde{p}(\omega) d\omega}{\int \omega^2 p(\omega) d\omega} \qquad 32.16$$

Eqn. **32.16** means that as neuronal activation increases, there is a concomitant increase in BOLD signal and a shift in the spectral profile to higher frequencies. High-frequency dynamics are associated with small effective membrane time constants and high [leaky] transmembrane conductances. The ensuing currents and fast changes in potential incur an energy cost to which the BOLD signal is sensitive. Such high-frequency dynamics have also been shown to be dependent upon the firing patterns of inhibitory interneurons (Traub *et al.*, 1996; Whittington and Traub, 2003). The conjoint effect of inhibitory and excitatory synaptic input is to open ion channels, rendering the postsynaptic membrane leaky with high rate constants. The effect is captured in the model by the scaling of the leading diagonal elements of the Jacobian. This suggests that the changes in the temporal dynamics to which the BOLD signal is sensitive are mediated by changes in the firing patterns of both excitatory and inhibitory subpopulations.

Critically, however, the predicted BOLD signal is a function of the frequency profile as opposed to any particular frequency. For example, an increase in alpha

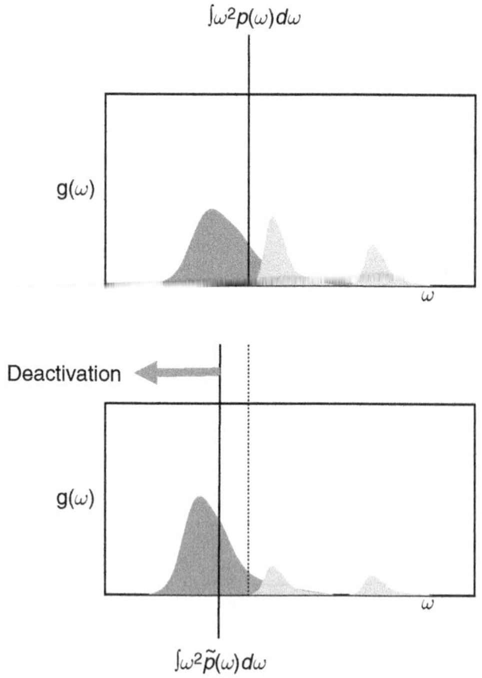

FIGURE 32.3 Schematic showing the effect of deactivation on mean square frequency.

(low-frequency), without a change in total power, would reduce the mean square frequency and suggest deactivation. Conversely, an increase in gamma (high-frequency) would increase the mean square frequency and speak to activation (Figure 32.3).

EMPIRICAL EVIDENCE

The model described in this chapter ties together expected changes in BOLD and EEG measures and makes a clear prediction about the relationship between the different frequency components of ongoing EEG or MEG activity and the expected BOLD response. According to the model, any modulations in the degree of low-frequency relative to the high-frequency components in the EEG signal will be inversely correlated with the BOLD signal. This is largely in agreement with the empirical data available. It is now generally accepted that modulations in the ongoing alpha rhythm, 8–12 Hz, when the eyes are shut, are inversely correlated with the BOLD signal at voxels within the parietal, parieto-occipital and frontal cortices (Goldman *et al.*, 2002; Laufs *et al.*, 2003a, b; Moosmann *et al.*, 2003; Martinez-Montes, 2004). Furthermore, during low-frequency visual entrainment, using a periodic checkerboard stimulus, the BOLD signal is reduced compared to an aperiodic stimulus (Parkes *et al.*, 2004). The model presented here also predicts that a shift in the frequency profile of the EEG to high-frequency components should be correlated with an increase in the BOLD signal. Although there is a much smaller literature on high frequency EEG-BOLD correlations, what has been published is broadly in agreement with this prediction. Laufs *et al.* (2003b) report predominantly positive correlations between the BOLD signal and the EEG power in the 17–23 Hz and the 24–30 Hz bandwidth and Parkes *et al.* (2004) demonstrate that an aperiodic checkerboard stimulus induces a greater BOLD signal than a low-frequency periodic stimulus. However, perhaps the most convincing empirical data that support the prediction of the model described here come not from a human study but from a study on anaesthetized adult cats by Niessing *et al.* (2005). In this study, Niessing and colleagues recorded a measure of the electrical activity, the local field potential (LFP), which is analogous to the EEG signal, directly from primary visual cortex using implanted electrodes. They recorded the BOLD signal simultaneously, using optical imaging, while the cats were shown moving whole field gratings of different orientations. Niessing *et al.* showed that, in trials in which the optical signal was strong, the neural response tended to oscillate at higher frequencies. In other words, an increase in the BOLD signal was associated with a shift in the spectral mass of the electrical signal to higher frequencies, in agreement with the analysis described here.

However, there are a number of observations that are not captured by the model. First, the model does not address very slow changes in potentials, < 0.1 Hz, that are unlikely to contribute to the event-related response. As such it does not capture the very slow modulations in LFP that have been shown previously to be related to the BOLD response (Leopold *et al.*, 2003). Secondly, and most notably, it does not explain the positive correlation of the BOLD signal with alpha oscillations in the thalamus (Goldman *et al.*, 2002; Martinez-Montes, 2004). This discrepancy could reflect the unique neuronal dynamics of the thalamus. Thalamic neurons are characterized by complex intrinsic firing properties, which may range from the genesis of high-frequency bursts of action potentials to tonic firing (Steriade *et al.*, 1993). However, it could also reflect the fact that the model is based on several assumptions that are wrong:

- The first assumption is that the dynamics of transmembrane potentials conform to an autonomous ordinary differential equation (ODE). The shortcomings of this assumption are that there is no opportunity for deterministic noise. However, this is a fairly mild restriction in relation to the autonomy, which precludes extrinsic input. This enforces afferent inputs outside the voxel or source to exert their influence vicariously through changes in the systems' parameters, encoded by the Jacobian. This can be seen as a limitation, given the driving nature of [forward] extrinsic connections in sensory cortex, but does fit comfortably with other perspectives on functional integration in the brain (see below and Chapter 36).

- The use of an autonomous ODE precludes hidden states that mediate changes in conductance and limits the model to a simple integrate-and-fire-like summary of neuronal behaviour. A more general form for Eqn. **32.2** would require:

$$\begin{aligned}\dot{V} &= f_V(V, x)\\ \dot{x} &= f_x(V, x)\end{aligned} \qquad 32.17$$

where, other hidden states x might correspond to conductances and channel gating terms as in more detailed biophysical models of neurons (see Dayan and Abbott, 2001 and Eqn. **31.6** in the previous chapter). The only argument that can be offered in defence of Eqn. **32.2** is that it may be sufficient to capture important behaviours by appeal to mean field approximations (see Chapter 31).

- The activation is modelled effectively by a scaling of the Jacobian. Functionally, this is a severe assumption because it precludes the selective enabling of particular connections. The consequence of this assumption is that any neuronal system can vary its rate or speed of computation, but can only do one thing. In reality, a better approximation would be bilinear with a multidimensional activation denoted by the vector $\alpha = [\alpha_1, \ldots]$:

$$\begin{aligned}\tilde{J} &= J + \sum \alpha_k B_k\\ B_k &= \frac{\partial J}{\partial \alpha_k}\end{aligned} \qquad 32.18$$

However, this model is too unconstrained to make any generic comments, without assuming a particular form for the bilinear terms B_k.

Having briefly deconstructed the model, it is worth noting that it highlights some important conceptual issues. These include:

- First, it reframes the notion of 'activation' in dynamic terms, suggesting that activation is not simply an excess of spikes, or greater power in any particular EEG frequency band; activation may correspond to an acceleration of dynamics, subserving more rapid computations. This sort of activation can manifest with no overall change in power but a change in the frequencies at which power is expressed. Because more rapid or dissipative dynamics are energetically more expensive, it may be that they are reserved for 'functionally' adaptive or necessary neuronal processing.
- Secondly, under the generative model of activation, a speeding up of the dynamics corresponds to a decrease in the width of the cross-correlation functions between all pairs of units in the population. At the macroscopic level of EEG recordings, considered here, the synchronization between pairs of units, as measured by the cross-correlation function, is captured in the width of the autocorrelation of the EEG signal, because the EEG signal is a measure of synchronous neural activity. This width is a ubiquitous measure of synchronization that transcends any frequency-specific changes in coherence. In short, activation as measured by fMRI is caused, in this model, by increased synchronization and, implicitly, a change in the temporal structure of neuronal dynamics.
- Thirdly, our analysis suggests that the underlying 'activation' status of neuronal systems is not expressed in any single frequency, but is exhibited across the spectral profile. This has implications for use of classical terms like 'event-related desynchronization'. If one only considered modulations in spectral density at one frequency, as in the classical use of the term desynchronization, one would conclude that the effect of activation was a desynchronization of low-frequency components. According to the model, however, the effect of activation is a shift in the entire spectral profile to higher frequencies with a concomitant attenuation in amplitude of all frequencies. This general conclusion does not preclude the selective expression of certain frequencies during specific cognitive operations (e.g. increases in theta oscillations during mental arithmetic (Mizuhara *et al.*, 2004)). However, the model suggests that the context in which these frequencies are expressed is an important determinant of the BOLD response. In other words, it is not the absolute power of any frequency but the profile which determines expected metabolic cost.
- Fourthly, as introduced above, one of the assumptions treats neuronal systems as autonomous, such that the evolution of their states is determined in an autonomous fashion. This translates into the assumption that the presynaptic influence of intrinsic connections completely overshadows extrinsic inputs. This may seem an odd assumption. However, it is the basis of non-linear coupling in the brain and may represent, quantitatively, a much more important form of integration than simple linear coupling. We have addressed this issue both empirically and theoretically in Friston (2000) and in Chapter 39; in brief, if extrinsic inputs affect the excitability of neurons, as opposed to simply driving a response, the coupling can be understood in terms of changes

in the system parameters, namely the Jacobian. This means the response to input will be non-linear. Quantitative analyses (a simple form of bi-coherence analysis) of MEG data suggest this form of non-linear coupling can account for much more variation in power than linear coupling, i.e. coherence (see Chapter 39).

SUMMARY

The integration of EEG and fMRI data is likely to play an important role in our understanding of brain function. Through multimodal fusion it should be possible to harness the temporal resolution of EEG and the spatial resolution of fMRI in characterizing neural activity. The majority of the studies integrating EEG and fMRI to date have focused on directly correlating the two measures, after first transforming the data so that they are on the same temporal scale (usually by convolution of the EEG time-series with a canonical haemodynamic response function). This approach has proved successful in demonstrating that multimodal fusion is feasible and that regionally specific dependencies between the two measures exist. However, this approach to characterizing the relationship between the EEG and the fMRI is limited as it does not characterize the integration in terms of the underlying neuronal causes. Full EEG-fMRI integration rests on understanding the relationship between the underlying neuronal activity and the BOLD and EEG signals through their respective forward models. Although the model discussed in this chapter falls a long way short of this, it can explain the nature of some of the previously reported correlations between EEG and fMRI by considering the integration in terms of the underlying neuronal dynamics.

We have proposed a simple model that relates BOLD changes to the relative spectral density of an EEG trace and the roughness of the EEG time-series. Neuronal activation affects the relative contribution of high and low EEG frequencies. This model accommodates the observations that BOLD signal correlates negatively with the expression of alpha power and positively with the expression of higher frequencies. Clearly, many of the assumptions are not correct in detail, but the overall picture afforded may provide a new perspective on some important issues in neuroimaging. In the next chapter, we go beyond the quantitative heuristics entailed by the dimensional analysis of this chapter and look at neural-mass models with a much more detailed form. These models embed constraints on synaptic physiology and connectivity and can reproduce many of the phenomena seen in EEG recordings. In Chapter 39, we will again use detailed neural-mass models to help understand the mechanistic basis of non-linear coupling in the brain.

REFERENCES

Abbott LF, Varela JA, Karmel Sen S *et al.* (1997) Synaptic depression and cortical gain control. *Science* **275**: 220–23

Attwell D, Iadecola C (2002) The neural basis of functional imaging brain signals. *Trends Neurosci* **25**: 621–25

Chawla D, Lumer ED, Friston KJ (2000) Relating macroscopic measures of brain activity to fast, dynamic neuronal interactions. *Neural Comput* **12**: 2805–21

Cox DR, Miller HD (1965) *The theory of stochastic processes*. Chapman and Hall, London

Crick F, Koch C (1998) Constraints on cortical and thalamic projections: the no-strong-loops hypothesis. *Nature* **39**: 245–50

Czisch M, Wehrle R, Kaufmann C *et al.* (2004) Functional MRI during sleep: BOLD signal decreases and their electrophysiological correlates. *Eur J Neurosci* **20**: 566–74

Dale AM, Liu AK, Fischl BR *et al.* (2000) Dynamic statistical parametric mapping: combining fMRI and MEG for high-resolution imaging of cortical activity. *Neuron* **26**: 55–67

Dayan P, Abbott LF (2001). *Theoretical neuroscience. computational and mathematical modelling of neural systems*. MIT Press, Cambridge, MA

Friston KJ (2000) The labile brain I. Neuronal transients and nonlinear coupling. *Phil Trans R Soc Lond B* **355**: 215–36

Goldman R, Stern JM, Engel J Jr *et al.* (2002) Simultaneous EEG and fMRI of the alpha rhythm. *Neuroreport* **13**: 2487–92

Hoge RD, Atkinson J, Gill B *et al.* (1999) Linear coupling between cerebral blood flow and oxygen consumption in activated human cortex. *Proc Natl Acad Sci USA* **96**: 9403–08

Jueptner M, Weiller C (1995) Review: does measurement of regional cerebral blood flow reflect synaptic activity? Implications for PET and fMRI. *NeuroImage* **2**: 148–56

Laufs H, Krakow K, Sterzer P *et al.* (2003a) Electroencephalographic signatures of attentional and cognitive default modes in spontaneous brain activity fluctuations at rest. *Proc Natl Acad Sci* **100**: 11053–58

Laufs H, Kleinschmidt A, Beyerle A *et al.* (2003b) EEG-correlated fMRI of human alpha activity. *NeuroImage* **19**: 1463–76

Lauritzen M (2001) Relationship of spikes, synaptic activity, and local changes of cerebral blood flow. *J Cereb Blood Flow Metab* **21**: 1367–83

Lemieux L, Krakow K, Fish DR (2001) Comparison of spike-triggered functional MRI BOLD activation and EEG dipole model localization. *NeuroImage* **14**: 1097–104

Leopold DA, Murayama Y, Logothetis NK (2003) Very slow activity fluctuations in monkey visual cortex: implications for functional brain imaging. *Cereb Cortex* **13**: 422–33

Logothetis NK, Pauls J, Augath M *et al.* (2001) Neurophysiological investigation of the basis of the fMRI signal. *Nature* **412**: 150–57

Lovblad KO, Thomas R, Jakob PM *et al.* (1999) Silent functional magnetic resonance imaging demonstrates focal activation in rapid eye movement sleep. *Neurology* **53**: 2193–95

Magistretti PJ, Pellerin L (1999) Cellular mechanisms of brain energy metabolism and their relevance to functional brain imaging. *Philos Trans R Soc Lond B Biol Sci* **354**: 1155–63

Magistretti PJ, Pellerin L, Rothman DL *et al.* (1999) Energy on demand. *Science* **283**: 496–97

Martinez-Montes E, Valdes-Sosa PA, Miwakeichi F *et al.* (2004) Concurrent EEG/fMRI analysis by multiway Partial Least Squares. *NeuroImage* **22**: 1023–34

Mizuhara H, Wang LQ, Kobayashi K *et al.* (2004) A long-range cortical network emerging with theta oscillation in a mental task. *Neuroreport* **15**: 1233–38

Moosmann M, Ritter P, Krastel I *et al.* (2003) Correlates of alpha rhythm in functional magnetic resonance imaging and near infrared spectroscopy. *NeuroImage* **20**: 145–58

Niessing J, Ebisch B, Schmidt KE *et al.* (2005) Hemodynamic signals correlate tightly with synchronized gamma oscillations. *Science* **309**: 948–51

Parkes LM, Fries P, Kerskens CM *et al.* (2004) Reduced BOLD response to periodic visual stimulation. *NeuroImage* **21**: 236–43

Phillips C, Rugg MD, Friston KJ (2002) Anatomically informed basis functions for EEG source localization: combining functional and anatomical constraints. *NeuroImage* **16**: 678–95

Shin C (2000) Neurophysiologic basis of functional neuroimaging: animal studies. *J Clin Neurophysiol* **17**: 2–9

Shulman RG, Rothman DL (1998) Interpreting functional imaging studies in terms of neurotransmitter cycling. *Proc Natl Acad Sci USA* **95**: 11993–98

Steriade M, McCormick DA, Sejnowski T (1993) Thalamocortical oscillations in the sleeping and aroused brain. *Science* **262**: 679–85

Traub RD, Whittington MA, Stanford IM *et al.* (1996) A mechanism for generation of long-range synchronous fast oscillations in the cortex. *Nature* **383**: 621–24

Whittington MA, Traub RD (2003) Interneuron diversity series: inhibitory interneurons and network oscillations in vitro. *Trends Neurosci* **26**: 676–82

CHAPTER

33

Neuronal models of EEG and MEG

O. David, L. Harrison and K. Friston

INTRODUCTION

This chapter considers the mechanisms that shape evoked electroencephalographic (EEG) and magnetoencephalographic (MEG) responses. We introduce neural-mass models and focus on a particular model of hierarchically arranged areas, defined with three kinds of inter-area connections (forward, backward and lateral). Using this model, we will investigate the role of connections or coupling on the generation of oscillations and event-related activity. Neural-mass models can reproduce nearly all the characteristics of event-related activity observed with M/EEG. Critically, they enforce a neurobiological perspective on ERPs (even-related potentials). This chapter uses neural-mass models to emulate common M/EEG phenomena and, in doing so, addresses their underlying mechanisms. In Chapter 42, we will show how the parameters of these models can be estimated from real data. This is the goal of dynamic causal modelling, where differences in ERP components, between conditions, are explained by connectivity changes within the brain.

Overview

This chapter gathers together the ideas and results presented in a series of papers (David and Friston, 2003; David *et al.*, 2004, 2006b) describing the development of neural-mass models for dynamic causal modelling (DCM). These neural-mass models are used in statistical parametric mapping (SPM) as the basis of DCM for event-related potentials (ERP) and event-related fields (ERFs) (see Chapter 42). In this chapter, we describe what a neural-mass model is; we then construct, *a posteriori*, the model used in later chapters, starting with a single neuronal population and ending with hierarchical cortical models. Finally, we illustrate how this model can be used to understand the basis of key phenomena observed with EEG and MEG.

NEURAL-MASS MODELS

M/EEG signals result mainly from extracellular current flow, associated with summed postsynaptic potentials in synchronously activated and vertically oriented neurons (i.e. the dendritic activity of macrocolumns of pyramidal cells in the cortical sheet) (see Chapter 28). Often, signals measured in MEG and EEG are decomposed into distinct frequency bands (delta: 1–4 Hz; theta: 4–8 Hz; alpha: 8–12 Hz; beta: 12–30 Hz; gamma: 30–70 Hz) (Nunez and Srinivasan, 2005). These rhythms exhibit robust correlates of behavioural states but often with no obvious functional role. The exact neurophysiological mechanisms, which constrain synchronization to a given frequency band, remain obscure, however, the generation of oscillations appears to depend on interactions between inhibitory and excitatory populations, whose kinetics determine their oscillation frequency.

ERPs and ERFs are obtained by averaging EEG and MEG signals in reference to some change in stimulus or task (Coles and Rugg, 1995). They show transient activity that lasts a second or so and which is correlated with changes in the state of the subject. ERPs and ERFs have been used for decades as putative electrophysiological correlates of perceptual and cognitive operations. However, like M/EEG oscillations, the exact neurobiological mechanisms underlying their generation are largely unknown. Recently, there has been a growing interest in the distinction between evoked and induced responses (see Chapter 30). Evoked responses are disclosed by conventional averaging procedures (classical

Statistical Parametric Mapping, by Karl Friston et al.
ISBN–13: 978-0-12-372560-8 ISBN–10: 0-12-372560-7

ERPs/ERFs), whereas the latter usually call for single-trial analyses of induced oscillations (Tallon-Baudry and Bertrand, 1999). Understanding the mechanistic relationship between evoked and induced responses could be a key for revealing how the electrical activity reacts to experimental changes. In other words, generative models of both M/EEG and ERP/ERF could be important for revealing how information is coded and processed in neural networks.

What is a neural-mass model?

There are several ways to model neural signals (Whittington *et al.*, 2000): by using either a detailed model, in which it is difficult to determine the influence of each model parameter, or a simplified one, in which realism is sacrificed for a more parsimonious description of key mechanisms. The complexity of neural networks generating M/EEG signals (Thomson and Deuchars, 1997; DeFelipe *et al.*, 2002) is considerable and usually makes the second approach more viable. Neural-mass models (Lopes da Silva *et al.*, 1974; Freeman, 1978; Van Rotterdam *et al.*, 1982; Stam *et al.*, 1999; Valdes *et al.*, 1999; Wendling *et al.*, 2000; Robinson *et al.*, 2001; David and Friston, 2003) are examples of simplified models, which usually model cortical macrocolumns as surrogates for cortical areas and, sometimes, thalamic nuclei. They use only one or two state variables to represent the mean activity of neuronal populations. These states summarize the behaviour of millions of interacting neurons. This procedure, sometimes referred to loosely as a *mean-field approximation*, is very efficient for determining the steady-state behaviour of neuronal systems, but its utility in a dynamic or non-stationary context is less established (Haskell *et al.*, 2001) (see Chapter 31 for a detailed discussion). In what follows, we will assume that the mean-field approximation is sufficient for our purposes. Figure 33.1 shows a schematic that tries to convey the intuition behind mean-field approximations.

As we saw in Chapter 31, the mean-field approximation involves partitioning the neuronal system into separable ensembles. Each ensemble or population is then coupled with mean-field quantities like average firing rate. In neural-mass models, we make the further approximation that the density of each ensemble can be described as a point mass. In other words, we ignore the variability in the states of neurons in an ensemble and assume that their collective behaviour can be approximated with a single value (i.e. the density's mode) for each state variable (e.g. voltage, current, conductance etc.).

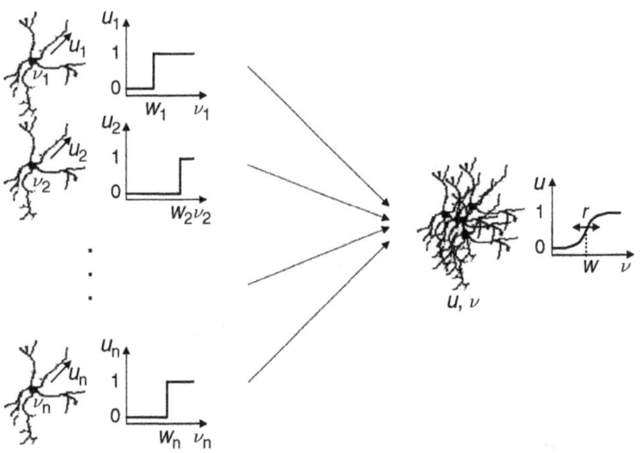

FIGURE 33.1 Mean-field approximation (MFA). Left: consider a neuronal population comprising n neurons. v_i, m_i and w_i denote the membrane potential, the normalized firing rate and the firing threshold of the i-th neuron, respectively. The input-output step-functions show that, in this example, the neurons fire at 1 or do not fire, depending on the threshold of firing which may vary between neurons. Right: the MFA of the neural-mass models stipulates that the dynamics of the neuronal ensemble, or neuronal mass, is described sufficiently by the mean of the state variables (v and m), using the mean relationship between v and m (the step function is transformed into a sigmoid function by averaging). Thus the effect of the MFA is that it reduces a huge system into a small one.

Neural-mass models of M/EEG

M/EEG signals are generated by the massively synchronous dendritic activity of pyramidal cells (Nunez and Srinivasan, 2005), but modelling M/EEG signals is seldom tractable using realistic models because of the complexity of real neural networks. Since the 1970s, the preferred approach has been neural-mass models, i.e. models which describe the average activity with a small number of state variables (see Figure 33.1). Basically, these models use two conversion operations (Jirsa and Haken, 1997; Robinson *et al.*, 2001): a wave-to pulse operator at the soma of neurons, which is generally a static sigmoid function; and a linear pulse-to-wave conversion implemented at a synaptic level. The first operator relates the mean firing rate to average postsynaptic depolarization. This is assumed to be instantaneous. The second operator depends on synaptic kinetics and models the average postsynaptic response as a linear convolution of incoming spike rate. The shape of the convolution kernels embodies the synaptic and dendritic kinetics of the population (Figures 33.2 and 33.3).

The majority of neural-mass models of M/EEG have been designed to generate alpha rhythms (Lopes da Silva *et al.*, 1974; Van Rotterdam *et al.*, 1982; Jansen and Rit, 1995; Stam *et al.*, 1999). Recent studies have shown that it is possible to reproduce the whole spectrum of M/EEG oscillations, using appropriate model parameters (Robinson

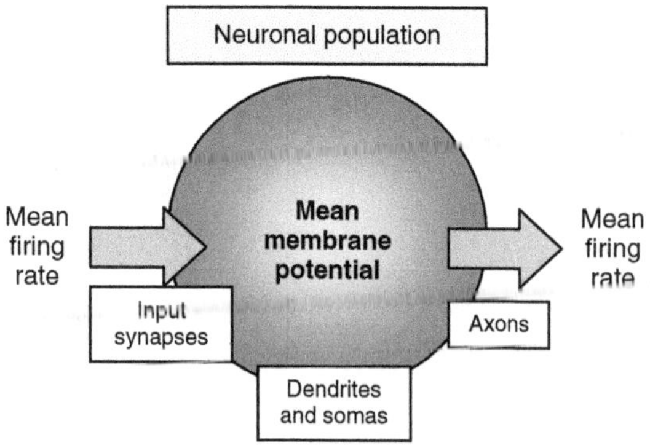

FIGURE 33.2 In neural-mass models, a neuronal population is usually described by its mean membrane potential. Each mass receives input (usually interpreted as the mean firing rate of all afferent axons). These inputs enter via synapses whose kinetics are usually modelled with a linear lowpass filter. The output of the neuronal population is modelled as the mean firing rate of the neurons. It is generally assumed that the mean firing rate is an instantaneous non-linear function (often a sigmoid as in Figure 33.1) of the mean membrane potential.

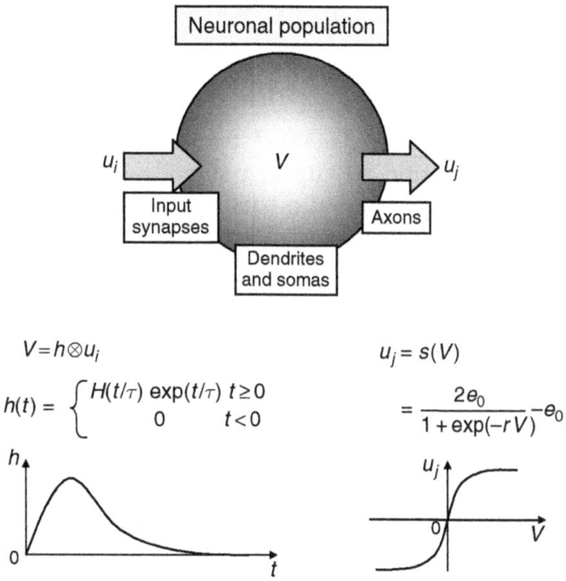

FIGURE 33.3 Model of a neuronal population as used in the Jansen model. This model rests on two operators: the first transforms u, the average density of presynaptic input arriving at the population, into V, the average postsynaptic membrane potential (PSP). This is modelled by a linear transformation. The kernel h models specific properties of synapses. The parameter H tunes the maximum amplitude of PSPs and τ is a lumped representation of the sum of the rate constants of passive membrane and other spatially distributed delays in the dendritic tree. The second operator transforms the average membrane potential of the population into an average rate of action potentials fired by the neurons. This transformation is assumed to be instantaneous and is a sigmoid function parameterized with e_0 and r.

et al., 2001; David and Friston, 2003). In addition, these models have been used to test specific hypotheses about brain function, e.g. focal attention (Suffczynski *et al.*, 2001). Pathological activity such as epilepsy can also be emulated. This means, in principle, generative models of the sort employed above could be used to characterize the pathophysiological mechanisms underlying seizure activity (Robinson *et al.*, 2002; Wendling *et al.*, 2002).

To date, modelling event-related activity using neural-mass models has received much less attention (Jansen and Rit, 1995; Suffczynski *et al.*, 2001; Rennie *et al.*, 2002 David *et al.*, 2005). An early attempt, in the context of visual ERPs, showed that it was possible to emulate ERP-like damped oscillations (Jansen and Rit, 1995). A more sophisticated thalamo-cortical model has been used to simulate event-related synchronization (ERS) and event-related desynchronization (ERD), commonly found in the alpha band (Suffczynski *et al.*, 2001). Finally, it has been shown that model parameters can be adjusted to fit real ERPs (Rennie *et al.*, 2002). These studies (Suffczynski *et al.*, 2001; Rennie *et al.*, 2002), emphasize the role of the thalamo-cortical interactions by modelling the cortex as a single compartment. Our work has focused on cortico-cortical interactions. Although the framework below is very general, for the sake of simplicity, we will concentrate on modelling interactions that are restricted to the cortex. As mentioned in Rennie *et al.* (2002), neural-mass models offer a unified view of M/EEG oscillations and event-related activity. This is an important point, which we will reiterate throughout this chapter.

MODELLING CORTICAL SOURCES

In this section, we introduce the Jansen model (Jansen and Rit, 1995) and its generalization to hierarchal networks (David and Friston, 2003). Before introducing the mathematical model of the basic neuronal population used in the Jansen model, we describe briefly the neuronal microcircuitry found in most cortical areas. Our goal is to motivate the Jansen model for modelling M/EEG activity anywhere in the cortex. However, strictly speaking, the Jansen model was developed to model the visual cortex. Again, for simplicity, we will ignore area-specific differences in laminar organization at the expense of neurobiological accuracy.

Basic cytoarchitecture of the cortical macrocolumn

The neocortex is commonly described as a six-layered structure (DeFelipe et al., 2002). Spiny neurons (pyramidal cells and spiny stellate cells) and smooth neurons comprise the two major groups of cortical neurons. The majority of cortical neurons are pyramidal cells that are found in layers 2 to 6. Most spiny stellate cells are interneurons that are located in the middle cortical layers. Smooth neurons are essentially GABA (γ-aminobutyric acid)-ergic interneurons that are distributed in all layers. In general, cortical neurons are thought to be organized into multiple, repeating microcircuits. In spite of cortical heterogeneity, a basic microcircuit has emerged: its skeleton is formed by a pyramidal cell, which receives excitatory inputs from extrinsic afferents and spiny cells. Inhibitory inputs originate mostly from GABAergic interneurons. These microanatomical characteristics have been found in all cortical areas and species examined so far and can be considered as fundamental aspects of cortical organization (DeFelipe *et al.*, 2002).

These canonical microcircuits are commonly referred to as cortical minicolumns ($\sim 50\,\mu$m diameter, containing about 400 principal cells), which are themselves grouped into cortical macrocolumns ($\sim 900\,\mu$m diameter) (Jones, 2000). A cortical area ($\sim$ 1–2 cm diameter) is composed of many cortical macrocolumns. Depending upon the level of integration one is interested in, each of these structures (cortical minicolumn, macrocolumn or area) can be considered as the functional unit, whose behavioural is approximated by neural-mass models. As we are interested here in cognitive neuroscience, using macroscopic measurements, we will consider the highest level of organization and build a neural-mass model (Generalized Jansen Model) reflecting the activity of cortical areas.

Modelling a neuronal population

A cortical area, macrocolumn or minicolumn, comprises several neuronal subpopulations. In this section, we describe the mathematical model of one area, which specifies the evolution of the dynamics of each subpopulation. This evolution rests on two operators (see Figure 33.3): the first transforms $u(t)$, the average density of presynaptic input arriving at the population, into $V(t)$, the average postsynaptic membrane potential (PSP). This is modelled by the linear transformation:

$$V = h \otimes u \qquad \mathbf{33.1}$$

where $\otimes$ denotes the convolution operator in the time domain and h is the impulse response or first-order kernel:

$$h(t) = \begin{cases} H(t/\tau)\exp(-t/\tau) & t \geq 0 \\ 0 & t < 0 \end{cases} \qquad \mathbf{33.2}$$

The kernel h is parameterized by H and τ that model specific properties of synapses: the parameter H tunes the maximum amplitude of PSPs and τ is a lumped representation of the sum of the rate constants of passive membrane and other spatially distributed delays in the dendritic tree. Eqn. **33.1** and Eqn. **33.2** are mathematically equivalent to the following state equations:

$$\begin{aligned} \dot{g} &= \frac{H}{\tau}u - \frac{2}{\tau}g - \frac{1}{\tau^2}V \\ \dot{V} &= g \\ \Rightarrow \ddot{V} &= \frac{H}{\tau}u - \frac{2}{\tau}\dot{V} - \frac{1}{\tau^2}V \end{aligned} \qquad \mathbf{33.3}$$

Note that this has the same form as Eqn. **32.2** in the previous chapter; $I = -C\dot{V}$, if we assume $g = -I/C$ is proportional to transmembrane current. When the external input to the population $u(t)$ is known, it is possible to obtain the membrane potential $v(t)$ by integrating Eqn. **33.3** (see Kloeden and Platen, 1999 for a description of various numerical solutions of differential equations). Eqn. **33.3** is effectively a second-order differential equation governing the evolution of postsynaptic potential in response to input and is formally similar to the equations of motion in Chapter 31 (e.g. Eqn. **31.6**) for integrate-and-fire neurons.

The second operator transforms the average membrane potential of the population into the average rate of action potentials fired by the neurons. This transformation is assumed to be instantaneous and is described by the sigmoid function:

$$s(V) = \frac{2e_0}{1+\exp(-rV)} - e_0 \qquad \mathbf{33.4}$$

where e_0 and r are parameters that determine its shape (e.g. voltage sensitivity). It is this function that endows the model with non-linear behaviours that are critical for phenomena like phase-resetting of the M/EEG (see below). Note that the form of $s(V)$ specifies implicitly that the resting state for the mean membrane potential v and the mean firing rate $u = s(V)$ is zero. This means that we are modelling variations of v around its resting value, which is usually negative (tens of mV) but unknown. In other words, states are treated as small perturbations around their resting values, which are set to zero.

Neural-mass versus mean-field models

The sigmoid function in Eqn. **33.4** is quite important for neural-mass models because it summarizes the effects of variability, over the population, considered in full mean-field treatments. In Chapter 31, we coupled populations with the mean-field quantities $s^{(i)} = M\rho^{(i)}$, where M was a linear operator that returned the mean firing rate and $\rho^{(i)}$ was the population density of the i-th population. In neural-mass models we only use the mode of this density, i.e. $V^{(i)} = \max_V \rho^{(i)}(V)$. This means we have to replace the linear mapping between the density and firing rate with a non-linear function of its mode $s^{(i)} = M\rho^{(i)} \rightarrow s^{(i)} = s(V^{(i)})$. The sigmoid form for this function is based on two premises. The first is that most neurons, most of the time, have membrane potentials that fluctuate just below the threshold for firing. In this subthreshold range, the probability of firing increases exponentially with depolarization. This property contributes the concave upward lower part of the curve. The second premise is that population firing rate approaches an upper limit, determined by the hyperpolarizing after-potentials of spikes. This property forms the convex upward part of the curve (Freeman, 1979). See Figures 33.1, 33.2 and 33.3 for schematics of this effect and Figure 31.4 in Chapter 31 for an example of spike rate responses of a population with increasing depolarization, using a full mean-field model.

Jansen's model

The Jansen model (Jansen and Rit, 1995) uses the microcircuitry described above to emulate a cortical area. It is based upon an earlier lumped parameter model (Lopes da Silva *et al.*, 1974). The basic idea behind these models is to make excitatory and inhibitory populations interact, such that oscillations emerge. A cortical area, taken here to be an ensemble of macrocolumns, is modelled by a population of excitatory pyramidal cells, receiving inhibitory and excitatory feedback from local (i.e. intrinsic) interneurons and excitatory input from neighbouring or remote (i.e. extrinsic) areas. It is composed of three subpopulations: a population of excitatory pyramidal (output) cells receives inputs from inhibitory and excitatory populations of interneurons, via intrinsic connections (intrinsic connections are confined to the cortical sheet). Within this model, excitatory interneurons can be regarded as spiny stellate cells found predominantly in layer 4 and in receipt of forward connections (Miller, 2003). Excitatory pyramidal cells and inhibitory interneurons will be considered to occupy agranular layers and receive backward and lateral inputs.

Interactions among the different subpopulations depend on the constants γ_i, which control the strength of intrinsic connections and the total number of synapses expressed by each subpopulation. The relative values of these constants are fixed, using anatomical information from the literature, as described in Jansen and Rit (1995): $\gamma_2 = 0.8\gamma_1, \gamma_3 = \gamma_4 = 0.25\gamma_1$. The model is summarized in Figure 33.4 using a state-space representation (as opposed to a kernel or convolution representation). We integrate the differential equations of the state-space form to simulate dynamics *per se*. These state equations cover the dynamics of PSPs and have the same form as Eqn. **33.3**, where we have used the second-order form for clarity. In this example, we have only allowed exogenous input to affect the excitatory interneurons.

The M/EEG signal is assumed to reflect $V_0(t)$, the average depolarization of pyramidal cells (Figure 33.4). For the sake of simplicity, we ignore the observation equation, i.e. how $V_0(t)$ is measured. This observer would include the effects of amplifiers (which are an additional bandpass filter), the lead fields (see Chapter 28 and Baillet *et al.*, 2001). The Jansen model can produce a large variety of M/EEG-like waveforms (broad-band noise, epileptic-like activity) and alpha rhythms (Jansen and Rit, 1995; Wendling *et al.*, 2000; David and Friston, 2003) when extrinsic inputs u are random (Gaussian) processes. When extrinsic inputs comprise transient inputs, event-related activity like ERP/ERF can be generated (Jansen and Rit, 1995; David *et al.*, 2005). This is illustrated in Figure 33.5.

Coupling cortical areas

Neurophysiological studies have shown that extrinsic cortico-cortical connections are exclusively excitatory. Moreover, experimental evidence suggests that M/EEG activity is generated by strongly coupled but remote cortical areas (Rodriguez *et al.*, 1999; Engel *et al.*, 2001; Varela *et al.*, 2001; David *et al.*, 2002). Fortunately, modelling excitatory coupling is straightforward using the Jansen model and some consequences of excitatory coupling have been described already (Jansen and Rit, 1995; Wendling *et al.*, 2000). In this section, we describe how coupling between two cortical areas, each one modelled as above, is implemented. In addition, we illustrate the effects of such coupling, both on M/EEG oscillations and on ERP/ERF.

Coupling in state equations

Let us first summarize the state equations describing the activity of one cortical area with

$$\dot{x} = f(x, u, \theta) \qquad x = \{V, \dot{V}\} \qquad 33.5$$

Eqn. **33.5** embeds the differential equations shown in Figure 33.4 where $x = \{V, \dot{V}\}$ are the states of the area (remember that the M/EEG signal $V_0 = V_2 - V_3$ represents the mean depolarization of pyramidal cells that comprise excitatory and inhibitory components), u are the extrinsic inputs and θ are the various parameters of the model. Coupling several cortical areas is implemented

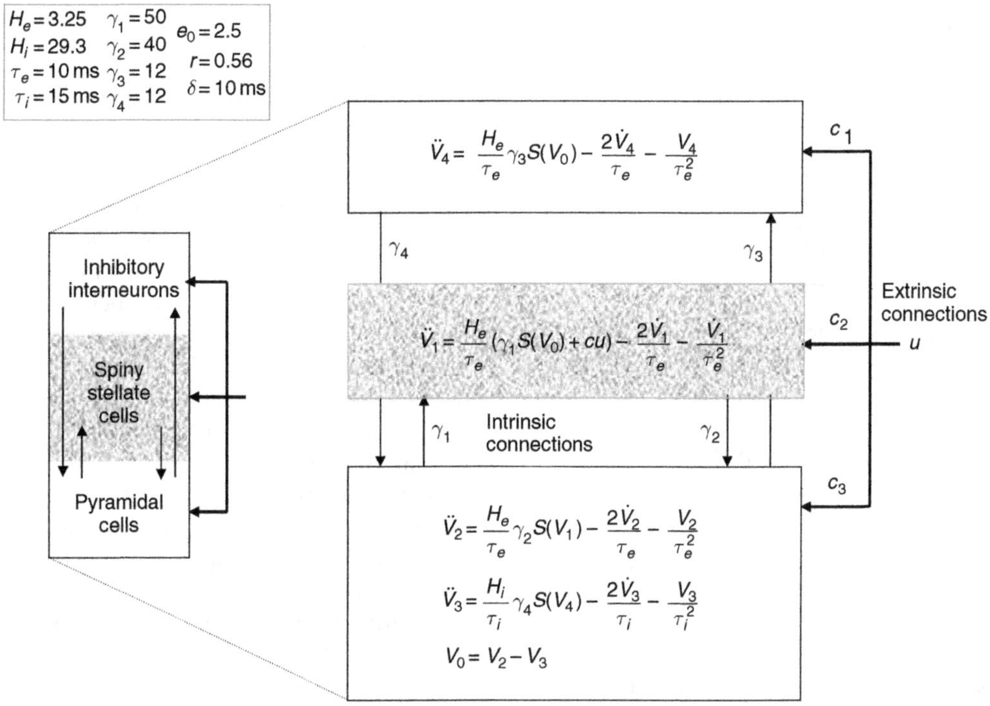

FIGURE 33.4 State-space representation of Jansen's model of a cortical area. Three neuronal subpopulations are considered to model a cortical area. Pyramidal cells interact with both excitatory and inhibitory interneurons with the connectivity constants $\gamma_2 = 0.8\gamma_1$, $\gamma_3 = \gamma_4 = 0.25\gamma_1$. The parameters H and τ control the expression of postsynaptic potentials (see previous figure). Subscripts e and i denote excitatory and inhibitory, respectively. We assume the average depolarization of pyramidal cells V_0 is proportional to cortical current densities obtained with source reconstruction algorithms using M/EEG scalp data. The insert provides the values of parameters used for all the simulations in this chapter: δ refers to propagation delays on extrinsic connections.

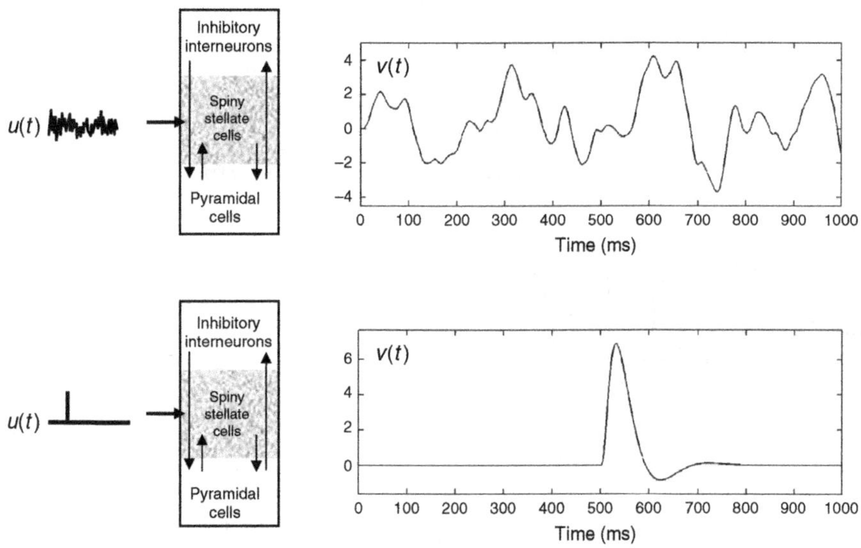

FIGURE 33.5 Effect of the nature of extrinsic inputs u on the responses of a cortical area (Jansen model). Upper: M/EEG-like oscillations are obtained when u is stochastic (Gaussian in this simulation). Lower: ERP/ERF-like waveforms are obtained when u contains a fast transient (a delta function in this simulation). This simulation used the model described in Figure 33.4.

by treating the outputs of one cortical area as the extrinsic input to another. The input $a_{ij}s(V^{(j)})$ to the i-th is the mean firing of pyramidal cells in the j-th, multiplied by the strength of the connection or coupling a_{ij}.

For several areas with states $x^{(I)}$, we can express the dynamics as a set of coupling differential equations where, ignoring propagation delays for simplicity:

$$\dot{x}^{(1)} = f(x^{(1)}, a_{12}S(V_0^{(2)}) + a_{13}S(V_0^{(3)}) + \cdots + c_1 u, \theta^{(1)})$$
$$\dot{x}^{(2)} = f(x^{(2)}, a_{21}S(V_0^{(1)}) + a_{23}S(V_0^{(3)}) + \cdots + c_2 u, \theta^{(2)})$$
$$\dot{x}^{(2)} = \ldots \qquad 33.6$$

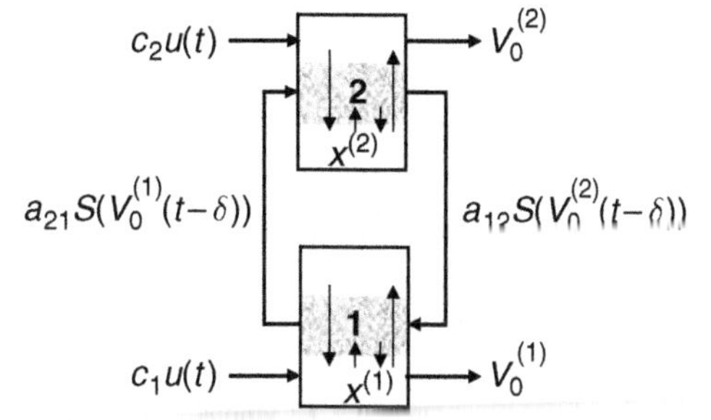

FIGURE 33.6 Graphical representation of the connection between two cortical areas. Coupling coefficients are assembled into matrices A and C (see main text).

In matrix form, Eqn. **33.6** is equivalent to:

$$\dot{x} = f(x, AS(x_0) + Cu, \theta)$$

$$x = \begin{bmatrix} x^{(1)} \\ x^{(2)} \\ \vdots \end{bmatrix} \quad x_0 = \begin{bmatrix} V_0^{(1)} \\ V_0^{(2)} \\ \vdots \end{bmatrix} \quad A = \begin{bmatrix} 0 & a_{12} & \cdots \\ a_{21} & 0 & \\ \vdots & & \ddots \end{bmatrix} \quad C = \begin{bmatrix} c_1 \\ c_2 \\ \vdots \end{bmatrix} \qquad \mathbf{33.7}$$

In other words, coupling is specified using connectivity matrices: the matrix A, which specifies the coupling among cortical areas, and the matrix C which specifies where exogenous inputs, such as stimuli, enter the model (Figure 33.6).

Coupling and functional connectivity

Coupling between areas models effective connectivity, i.e. the influence of one system on another (Friston, 2001). Using a model composed of two areas, it is easy to look at the effects of coupling on functional connectivity, i.e. the correlation between the signals that are generated. The simulations shown in Figure 33.7 used the same parameters as in Figure 33.5 for each area. The connections between areas were asymmetrical (forward from 1 to 2 and backward from 2 to 1). The distinction between forward and backward connections will be explained in the next section. Extrinsic inputs u were applied to spiny stellate cells and a propagation delay of 10 ms between the areas was assumed.

The upper panels of Figure 33.7 show that synchronous oscillations appear when neuronal coupling between cortical areas is increased. This suggests that coupling and the synchrony of the M/EEG (Varela *et al.*, 2001) are closely related. The lower panels show that coupling can

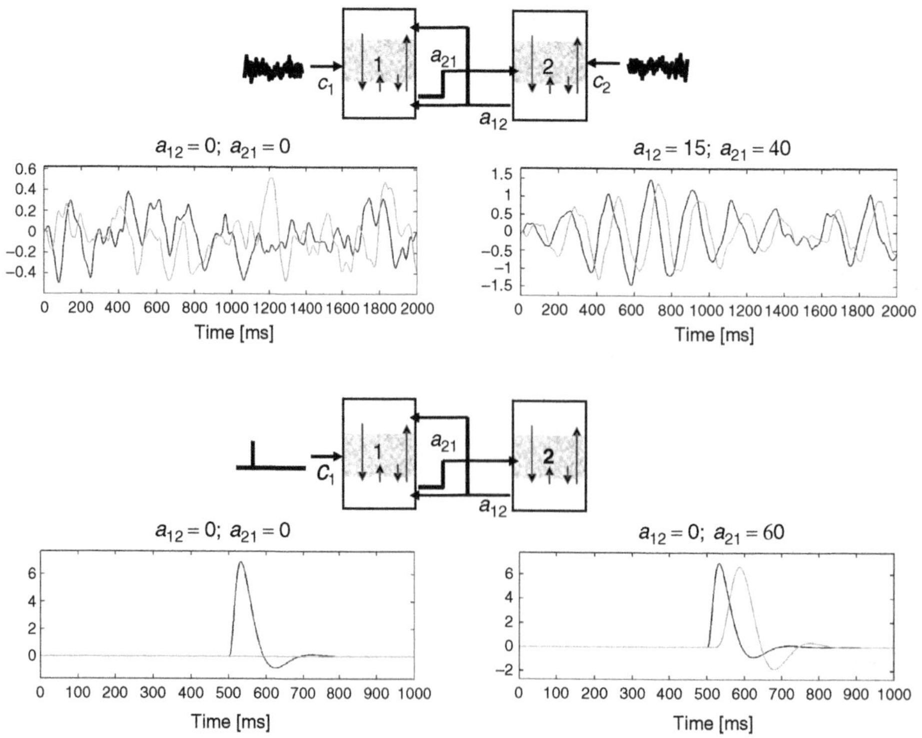

FIGURE 33.7 Relationship between coupling and functional connectivity (correlations between observed signals). The model was composed of two areas, coupled asymmetrically. Extrinsic inputs u were applied to spiny stellate cells only. Top: ongoing activity was simulated using stochastic inputs that entered each area. One can see that synchronous oscillations appear when increasing neuronal coupling between cortical areas (despite the fact that the inputs were independent). Bottom: a fast transient (delta function) was the extrinsic input to area 1. When areas are not coupled (left hand side) the ERP does not propagate to area 2, whereas one can observe a late response in area 2 when coupling is present.

propagate neuronal transients such as ERP/ERF. This topic will be discussed in more depth later. The important point here is that changes in coupling can explain many aspects of synchronization and ERP/ERFs and are in a position to bridge the study of evoked and induced or ongoing M/EEG activity. We will come back to this issue later.

So far, we have been introducing the different components and key ideas behind neural-mass models. In the next section, we describe the model which we have used to simulate several aspects of cortical responses and use as the generative model for DCM in later chapters. The key feature of this model is its hierarchal architecture that arises from the distinction between forward, backward and lateral extrinsic connections.

HIERARCHICAL MODELS OF CORTICAL NETWORKS

It is well known that the cortex has a hierarchical organization (Felleman and Van Essen, 1991; Crick and Koch, 1998), comprising bottom-up, top-down and lateral processes that can be understood from an anatomical and cognitive perspective (Engel *et al.*, 2001). We have discussed previously the importance of hierarchical processes, in relation to perceptual inference in the brain, using the intimate relationship between hierarchical models and empirical Bayes (Friston, 2002). Using a hierarchical neural-mass model, the work described in David *et al.* (2005) and reprised here, was more physiologically motivated. We were primarily interested in the effects, on event-related M/EEG activity, of connections strengths, and how these effects were expressed at different hierarchical levels. In addition, we were interested in how non-linearities in these connections might be expressed in observed responses. In this section, we describe the architecture and the state equations of the neural-mass model which is used in DCM (David *et al.*, 2006a; Kiebel *et al.*, 2006; and Chapter 42). The subsequent sections in this chapter explore the emergent properties of the model and the mechanistic insights provided.

Forward, backward and lateral connections

Although neural-mass models originated in the early 1970s (Wilson and Cowan, 1972; Lopes da Silva *et al.*, 1974; Freeman, 1978), none has addressed the hierarchical nature of cortical organization. The minimal model we propose, which accounts for directed extrinsic connections, uses the rules described in Felleman and Van Essen (1991). Extrinsic connections are connections that traverse white matter and connect cortical regions (and subcortical structures). These rules, based upon a tri-partitioning of the cortical sheet (into supra-, infra-granular layers and granular layer 4), have been derived from experimental studies of cat visual cortex. We will assume that they can be generalized to the whole cortex. The ensuing model is general, and can be used to model various cortical networks (David *et al.*, 2006a; Kiebel *et al.*, 2006), where variability among different cytoarchitectonic regions is modelled by different area-specific parameters, under the same microcircuitry. Under this simplifying assumption, the connections can be defined as in Figure 33.8: bottom-up or forward connections originate in agranular layers and terminate in layer 4. Top-down or backward connections only connect agranular layers. Lateral connections originate in agranular layers and target all layers. All these long-range or extrinsic cortico-cortical connections are excitatory and are mediated through the axons of pyramidal cells. For schematic reasons, we lump the superficial and deep pyramidal layers into the infra-granular layer in our model.

Although the thalamo-cortical connections have been the focus of several modelling studies, they represent a minority of extrinsic connections: in contrast, it is thought that at least 99 per cent of axons in white matter link cortical areas of the same hemisphere (Abeles, 1991). For this reason, and for simplicity, we do not include the thalamic

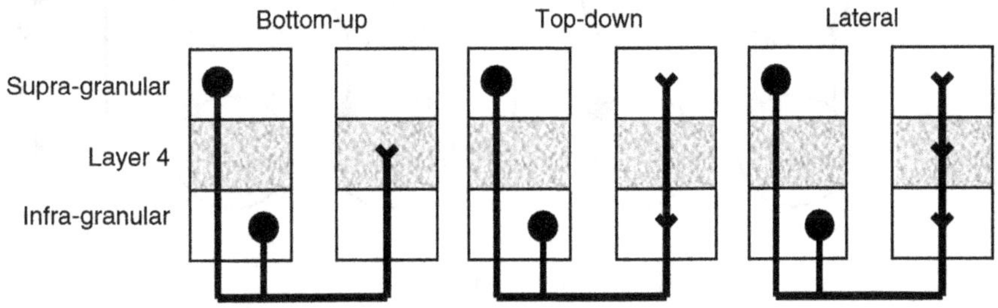

FIGURE 33.8 Connection rules adopted for the construction of hierarchical models. These rules are a simplified version of those proposed by Felleman and Van Essen (1991). The cortical sheet is divided into two components: the granular layer (layer 4) and the agranular layers (supra- and infra-granular layers). Bottom-up connections originate in agranular layers and terminate in layer 4. Top-down connections only engage agranular layers. Lateral connections originate in agranular layers and target all layers.

nuclei in our model. However, they can be included if the role of the thalamus (or other subcortical structure) is thought important.

State equations

Using the connection rules above, it is straightforward to construct hierarchical cortico-cortical networks using Jansen models of cortical areas. The different types of connections are shown in Figure 33.9, in terms of connections among the three subpopulations. To model event-related responses, the network receives inputs via exogenous input connections. These connections are exactly the same as forward connections delivering fixed or stochastic inputs u to the spiny stellate cells in layer 4. In the present context, they can be regarded as connections from thalamic or geniculate nuclei. Inputs u can model incoming stimuli and stochastic background activity.

Connections among areas are mediated by long-range excitatory (glutaminergic) pathways. As discussed above, we consider three types of extrinsic connections (Figure 33.9): forward, backward, and lateral. The strength of each type of connection is controlled by a coupling parameter $a : a^F$ for forward, a^B for backward and a^L for lateral. We model propagation delays for these connections. The state equations for a single area are shown in Figure 33.10. It can be seen that the distinction between forward, backward and lateral connections is modelled in terms of which subpopulation is affected by input from the pyramidal cells of another area. The insert shows the values of the intrinsic parameters $\theta^{(i)}$ used in all the simulations of this chapter.

Using these connections, hierarchical cortical models for M/EEG can be constructed to test various hypotheses, and represent examples of dynamic causal models (Friston *et al.*, 2003). The causal model here is a multiple-input multiple-output system that comprises m inputs and l outputs with one output per region. The m inputs correspond to designed causes (e.g. stimulus functions encoding the occurrence of events) or stochastic processes modelling background activity. In principle, each input could have direct access to every region. However, in practice the effects of inputs are usually restricted to a single input region, usually the lowest in the hierarchy. Each of the l regions produces a measured output that corresponds to the M/EEG signal. Each region has five $(H_{e,i}, \tau_{e,i}, \gamma_1)$ intrinsic parameters that correspond to the time constants described above. These play a crucial role in generating regional responses. However, we will consider them fixed and focus on the extrinsic coupling parameters or effective connectivity. These are the matrices C, A^F, A^B and A^L that contain the coupling parameters c, a^F, a^B and a^L. The values of these parameters, used in the following simulations, are provided in the figure legends. These are the parameters that are estimated from the data in DCM using the expectation and maximization (EM) (Friston *et al.*, 2002).

The neuronal model described above embodies many neuroanatomical and physiological constraints which lend it a neuronal plausibility. It has been designed to explore emergent behaviours that may help understand empirical phenomena and, critically, as the basis of dynamic observation models. Although the model comprises coupled systems, the coupling is highly asymmetric and heterogeneous. This contrasts with homogeneous

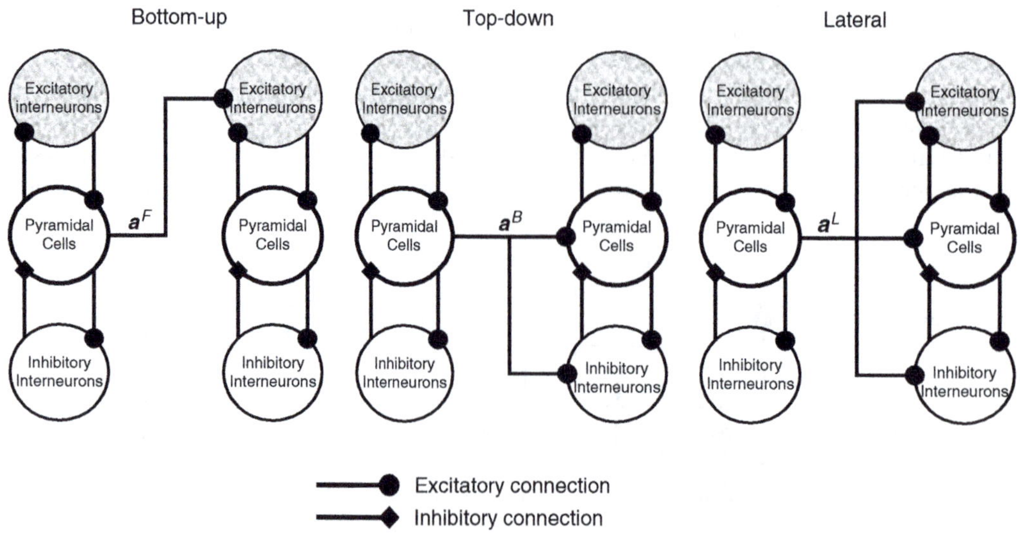

FIGURE 33.9 Hierarchical connections among Jansen units based on simplified Felleman and van Essen rules (Figure 33.8). Long range connectivity is mediated by pyramidal cells axons. Their targets depend upon the type of connections. Coupling or connectivity parameters control the strength of each type of connection: a^F for forward, a^B for backward, and a^L for lateral.

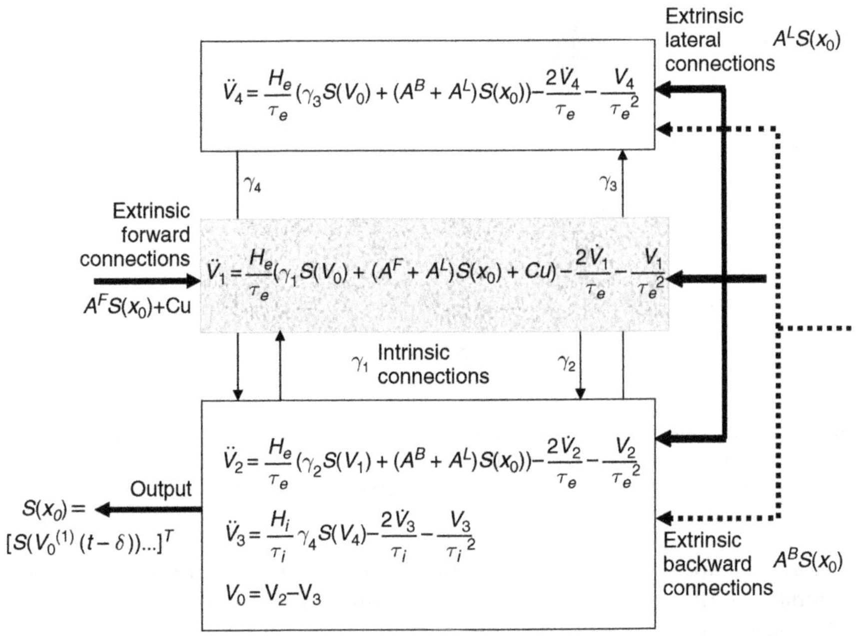

FIGURE 33.10 Schematic of the model of a single source with its extrinsic connections. This schematic includes the (simplified) differential equations describing the dynamics of the source or regions states. Each source is modelled with three subpopulations (pyramidal, spiny-stellate and inhibitory interneurons) as described in Jansen and Rit (1995). These have been assigned to granular and agranular cortical layers, which receive forward and backward connection respectively.

and symmetrically coupled map lattices (CML) and globally coupled maps (GCMs) encountered in more analytic treatments. Using the concepts of chaotic dynamical systems, GCMs have motivated a view of neuronal dynamics that is cast in terms of high-dimensional transitory dynamics among 'exotic' attractors (Tsuda, 2001). Much of this work rests on uniform coupling, which induces a synchronization manifold, around which the dynamics play. The ensuing chaotic itinerancy has many intriguing aspects that can be related to neuronal systems (Breakspear *et al.*, 2003; Kaneko and Tsuda, 2003). However, the focus of the work presented below is not chaotic *itinerancy* but chaotic *transience* (the transient dynamics evoked by perturbations to the systems state), in systems with asymmetric coupling. This focus precludes much of the analytic treatment available for GCMs; but see Jirsa and Kelso (2000) for an analytical description of coherent pattern formation in a spatially continuous neural system with a heterogeneous connection topology. However, as we hope to show, simply integrating the model, to simulate responses, can be a revealing exercise. This is what we will pursue in subsequent sections.

MECHANISMS OF ERP GENERATION

In this section, we characterize the input-output behaviour of a series of canonical networks in terms of their impulse response functions. This is effectively the response (mean depolarization of pyramidal subpopulations) to a delta-function-input or impulse. The simulations of this section can be regarded as modelling event-related responses to events of short duration, in the absence of spontaneous activity or stochastic input. In the next section, we will use more realistic inputs that comprise both stimulus-related and stochastic components.

The effects of inputs

Inputs u act directly on the spiny stellate neurons of layer 4. Their influence is mediated by the forward connections parameterized by the matrix C. When these connections are sufficiently strong, the output of the spiny stellate subpopulation saturates, due to the non-linear sigmoid function in Eqn. **33.4**. This non-linearity has important consequences for event-related responses and the ensuing dynamics. In brief, the form of the impulse response function changes qualitatively with input strength. To illustrate this point, we modelled a single area, which received an impulse at time zero, and calculated the corresponding response for different values of c (Figure 33.11). With weak inputs, the response is linear, leading to a linear relationship between c and peak M/EEG responses. However, with large values of c, neuronal activity leaves the linear domain of the sigmoid function, the spiking saturates and the shape of the evoked response changes.

This behaviour is not surprising and simply reflects the non-linear relationship between firing rates and post-synaptic depolarization, modelled by the non-linearity. This non-linearity causes saturation in the responses of units to intrinsic and extrinsic inputs. For example, when the input is strong enough to saturate spiny stellate

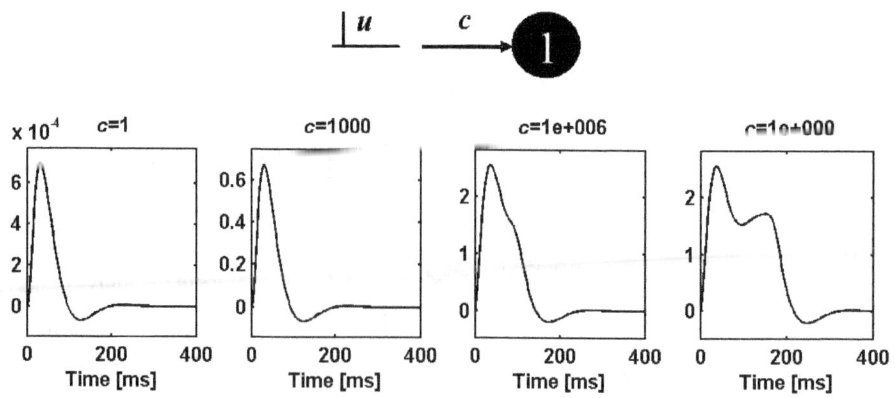

FIGURE 33.11 The strength of input modulates the shape of the M/EEG signal. The output of one area has been calculated for different values of c, the strength of forward connections mediating input u (delta function). When c is small ($c = 1, c = 1000$), the output is not saturated and the M/EEG amplitude is linearly related to c. For large values of $c(c = 10^6, c = 10^9)$, spiny stellate cells saturate and the shape of event-related M/EEG response changes substantially.

spiking, the pyramidal response exhibits a short plateau (right panel in Figure 33.11). This saturation persists until the membrane potential of spiny stellate cells returns to its resting state. The sigmoid function models phenomena at the single unit level, like refractoriness and spike rate adaptation and aspects of neuronal ensembles at the population level, like the distribution of thresholds involved in the generation of action potentials. The ensuing behaviour confers an inherent stability on dynamics because it is recapitulated in the response to all bottom-up influences, as shown next.

Bottom-up effects

The targets of forward connections and extrinsic inputs are identical. Therefore, the effects of c and a^F on event-related responses are exactly the same. Figure 33.12 shows the simplest case of two areas (area 1 drives area 2). The difference, in relation to the previous configuration, is that area 1 has a gating effect. This is basically a lowpass filter, which leads to greater variation of the response in area 2, relative to responses elicited by direct input to area 2 (cf. Figure 33.11). For instance, the small negative response component in area 1, which follows the first positive deflection, is dramatically enhanced in area 2 for strong forward couplings. Again, this reflects the non-linear behaviour of subpopulations responding to synaptic inputs.

As mentioned above, activity is subject to lowpass filtering, by synaptic processes, each time it encounters a cortical region. A simple and intuitive consequence of this is that the form of event-related responses changes with each successive convolution in the hierarchy. To illustrate this point, consider a feed-forward configuration composed of five regions (Figure 33.13). We see in Figure 33.13 that, in addition to the propagation lag

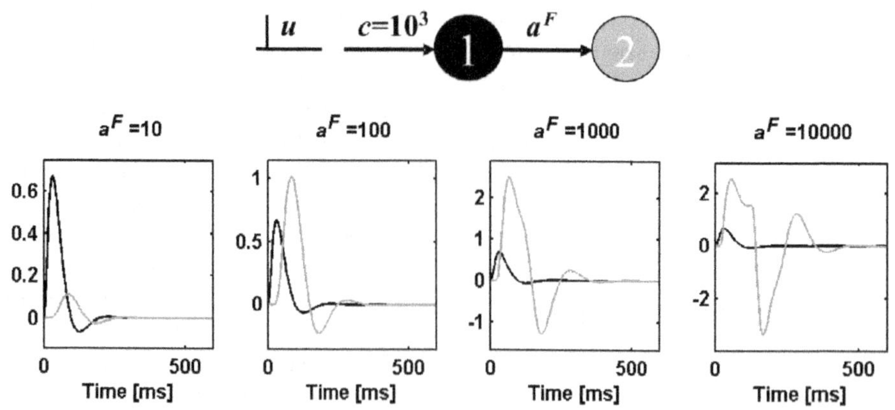

FIGURE 33.12 The M/EEG signal of area 1 (black) and area 2 (grey) is plotted as a function of the forward connectivity a^F. Bottom-up connectivity has the same effect as input connectivity c: high values cause a saturation of spiny stellate cells (input cells), with a dramatic effect on M/EEG event-related responses. Non-linear effects are particularly strong for the largest value of a^F (right panel) as the small negative component of area 1 (seen best in the left panel) induces a huge negative response in area 2.

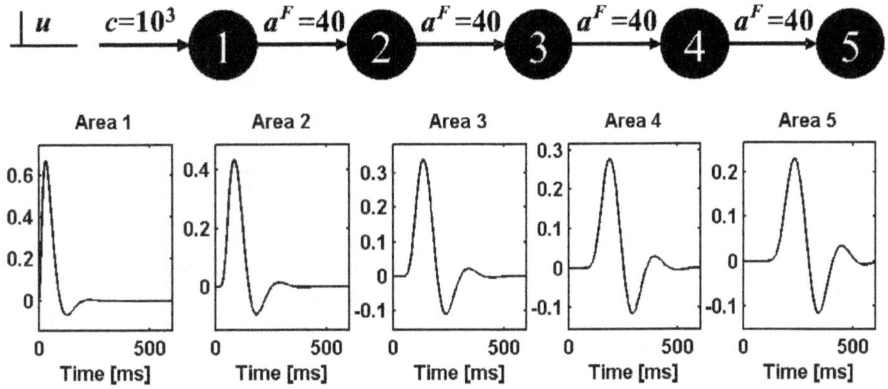

FIGURE 33.13 A feed-forward system composed of five areas. The M/EEG signal of each area elicited by a single pulse to area 1 is plotted in successive panels from left to right. Event-related activity lasts longer in high-level cortical areas in feed-forward architectures. At each level in the hierarchy, the event-related response of pyramidal cells experiences successive lowpass filtering, embodied by synaptic processes that transform the input signals to output.

that delays the waveform at each level, the event-related response is more enduring and dispersed in higher-level areas. A useful heuristic here is that late components of evoked responses may reflect hierarchical processing at a higher level. This effect is independent of synaptic time constants and connectivity parameters.

This simple delay and dispersion is not necessarily seen with more realistic configurations that involve top-down effects. In this context, late response components in higher cortical areas can re-enter (Edelman, 1993) lower levels, engendering complicated and realistic impulse response functions. In the reminder of this section, we look at the effects of adding backward and then lateral connections to the forward architecture considered above.

Loops and late components

Interactions in the brain are mostly reciprocal. This means that re-entrant or recurrent loops are ubiquitous and play a major role in brain dynamics. In our model, two types of re-entrant loop can be imagined: between different levels of the hierarchy using forward and backward connections; and between two areas at the same level, using lateral connections.

Top-down effects

Top-down connections mediate influences from high- to low-level regions. Incoming sensory information is promulgated through the hierarchy via forward, and possibly lateral, connections to high-level areas. To demonstrate the effect of backward connections on M/EEG, we will consider a minimal configuration composed of two areas (Figure 33.14). Although asymmetric, the presence of forward and backward connections creates loops. This induces stability issues as shown in Figure 33.14; when backward connections are made stronger, damped oscillations ($a^B = 1$; $a^B = 10$) are transformed into oscillations, which ultimately stabilize ($a^B = 50$) because of the saturation described in the previous subsection. Therefore, with $a^B = 50$, the attractor is a limit cycle and the resting state point attractor loses its dynamic stability. The dependence of oscillations on layers, loops and propagation delays has been the subject of much study in computational models (e.g. Lumer *et al.*, 1997).

From a neurobiological perceptive, the most interesting behaviours are shown just prior to this phase-transition,[1] when damped oscillations are evident. Note that the peaks of the evoked response, in this domain, occur every 100 milliseconds or so. This emulates the expression of late components seen empirically, such as the N300 or P400. The key point here is that late components, in the EEG/MEG, may reflect re-entrant effects mediated by backward connections in hierarchical architectures. This observation fits comfortably with the notion that late M/EEG components reflect endogenous processing

[1] A phase-transition refers to the qualitative change in the system's attractor caused by changes in the system's parameters, here the coupling parameters. In the present context, increasing the backward coupling causes the point attractor to lose its dynamic stability (stability under small perturbations) and the emergence of a limit-cycle attractor. The nature of the phase-transition is usually assessed in terms of Lyapunov exponents (eigenvalues of the system's Jacobian $\partial f/\partial x$). When the system has a point attractor the imaginary part of the principal or largest exponent is zero. A limit cycle has non-zero imaginary parts and chaotic attractors have at least one real positive exponent. We do not present a stability analysis or the Lyapunov exponents in this work, because the phase-transitions are self-evident in the trajectories of the system.

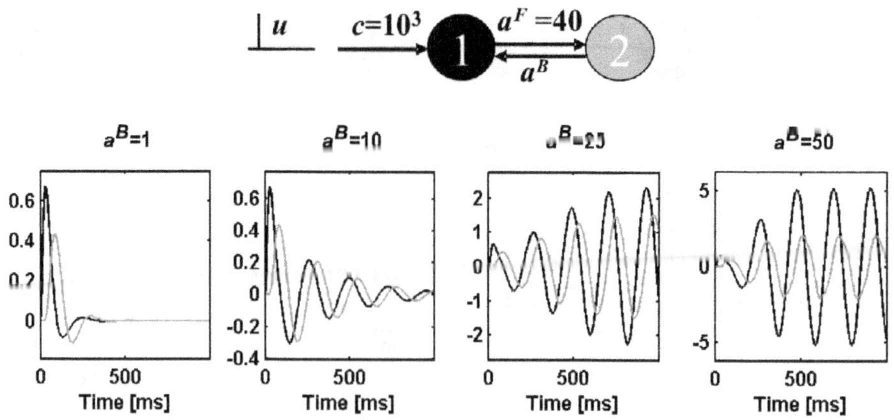

FIGURE 33.14 Backward connections have a key influence on the stability of M/EEG event-related activity as demonstrated by this simple model composed of two areas (area 1 coded in black and area 2 coded in grey). The forward connectivity a^F has been fixed at 40 and backward connectivity a^B varies between 1 and 50 from left to right. When top-down effects are small, their re-entry leads to longer lasting event-related responses characterized by damped oscillations ($a^B = 1$; $a^B = 10$). However, over a critical threshold of a^B (which depends upon a^F), the system undergoes a phase-transition, loses its point attractor and expresses oscillatory dynamics ($a^B = 25$; $a^B = 50$).

and depend explicitly on top-down effects. In short, late components may depend on backward connections and reflect a re-entry of dynamics to hierarchically lower processing areas. This dependency can be seen clearly by comparing the two left-hand panels in Figure 33.14 that show the emergence of late components on increasing the backward connection from one to ten.

The phase transition from damped late components to oscillations is critical. Before the transition the system is controllable. This means that the response can be determined analytically given the input. As discussed in Friston (2000a), long impulse responses endow the brain with a 'memory' of past inputs that enables perceptual processing of temporally extended events. In Friston (2000b), this was demonstrated using a Volterra kernel formulation and the simulation of spatiotemporal receptive fields in the visual system (see also Chapter 39). However, after the transition, it is no longer possible to determine when the input occurred given the output. This violates the principle of maximum information transfer (Linsker, 1990) and precludes this sort of response in the brain. In short, it is likely that re-entrant dynamics prolong neuronal transients but will stop short of incurring a phase transition. If this phase transition occurs it is likely to be short-lived or pathological (e.g. photosensitive seizure activity).

It should be noted that the oscillations in the right hand panels of Figure 33.14 do *not* represent a mechanism for induced oscillations. The oscillations here are deterministic components of the system's impulse response function and are time-locked to the stimulus. Induced oscillations, by definition, are not time-locked to the stimulus and probably arise from a stimulus-related change in the system's control parameters (i.e. short-term changes in connectivity). We will return to this point later.

Lateral connections

Lateral connections link different regions at the same level in the hierarchy. They can be unidirectional or bidirectional as shown for the model in Figure 33.15 with two areas. The main difference between forward and unidirectional lateral connections is that the latter target pyramidal cells. This means that the M/EEG signal is not so constrained by non-linear saturation in layer 4 units. Therefore, as shown in Figure 33.15(a), the event-related response does not saturate for strong lateral connectivity values a^L. On the other hand, bilateral connections are completely symmetric, which enables them to create a synchronization manifold (Breakspear, 2002; Breakspear and Terry, 2002b). A comparison of Figure 33.15(b) and Figure 33.14 shows that a special aspect of bilateral connections is their ability to support dynamics that are in phase. This sort of zero-lag phase-synchronization is commonplace in the brain. Its mediation by lateral connections in this model concurs with previous modelling studies of zero-lag coupling in triplets of cortical areas that involve at least one set of bilateral or reciprocal connections (Chawla *et al.*, 2001). For very large values of a^L, architectures with bilateral connections are highly non-linear and eventually undergo a second phase transition (see Figure 33.15(b)).

In this section, we have provided a deterministic characterization of simple hierarchical models in terms of their impulse responses. We have tried to show that the model exhibits a degree of face validity in relation to real evoked responses and have related certain mechanistic aspects to previous modelling work to provide some construct validity. We now turn to another biological issue, namely the plausibility of non-linear mechanisms that might explain ERP/ERF components.

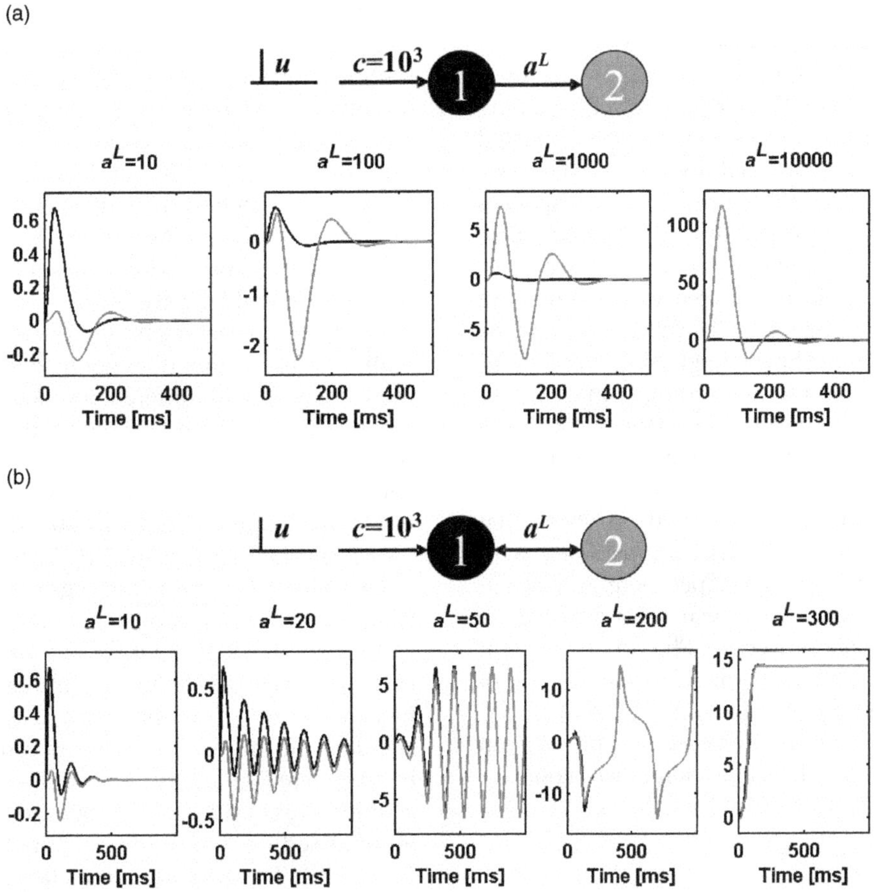

FIGURE 33.15 The effects of lateral connections are shown with a simple model composed of two areas (black: area 1, grey: area 2). The depolarization of pyramidal cells is plotted for several values of lateral connectivity a^L. (a) Unilateral connections support transients that differ from those elicited by forward connections (Figure 33.9). In particular, the saturation of layer 4 is not so important and the signal exhibits less saturation for large a^L. (b) Increasing bi-directional lateral connections has a similar effect to increasing backward connections. The main difference is the relative phase of evoked oscillations, which are synchronized at zero-lag. For very large values of a^L, the model is highly non-linear and eventually exits the oscillatory domain of parameter space.

PHASE-RESETTING AND THE ERP

It is generally held that an ERP/ERF is the result of averaging a set of discrete stimulus-evoked brain transients (Coles and Rugg, 1995). However, several groups (Kolev and Yordanova, 1997; Makeig *et al.*, 2002; Jansen *et al.*, 2003; Klimesch *et al.*, 2004; Fuentemilla *et al.*, 2005) have suggested that some ERP/ERF components might be generated by stimulus-induced changes in ongoing brain dynamics. This is consistent with views emerging from several fields suggesting that phase-synchronization of ongoing rhythms, across different spatiotemporal scales, mediates the functional integration necessary to perform higher cognitive tasks (Varela *et al.*, 2001; Penny *et al.*, 2002). In brief, a key issue is the distinction between processes that do and do not rely on phase-resetting of ongoing spontaneous activity. Both can lead to the expression of ERP/ERF components but their mechanisms are very different.

EEG and MEG signals are effectively ergodic and cancel when averaged over a sufficient number of randomly chosen epochs. The fact that ERPs/ERFs exhibit systematic waveforms, when the epochs are stimulus locked, suggests either a reproducible stimulus-dependent modulation of amplitude or phase-locking of ongoing M/EEG activity (Tass, 2003). The key distinction between these two explanations is whether the stimulus-related component interacts with ongoing or spontaneous activity. If there is no interaction, the spontaneous component will be averaged away because it has no consistent phase-relationship with stimulus onset. Conversely, if there is an interaction, dominant frequencies of the spontaneous activity must experience a phase-change, so that they acquire a degree of phase-locking to the stimulus. Note that phase-resetting is a stronger-requirement than induced oscillations. It requires any induced dynamics to be phase-locked in peristimulus time. In short, phase-resetting is explicitly non-linear and implies an interaction between stimulus-related response and ongoing activity. Put simply, this means that the event-related response depends on ongoing activity. This dependency can be assessed with the difference between responses elicited with and without the stimulus (if we could reproduce exactly the same ongoing activity). In the absence of interactions there will be no difference. Any difference implies non-linear interactions. Clearly, this cannot be done empirically but it can be pursued using simulations.

We will show next that phase-resetting is an emergent phenomenon and a plausible candidate for causing ERPs/ERFs. Phase-resetting is used in this chapter as an interesting example of non-linear responses that have been observed empirically. We use it to show that non-linear mechanisms can be usefully explored with neuronal models of the sort developed here. In particular, static non-linearities, in neuronal mass models, are sufficient to explain phase-resetting. Phase-resetting represents non-linear behaviour because, in the absence of amplitude changes, phase-changes can only be mediated in a non-linear way. This is why phase-synchronization plays a central role in detecting non-linear coupling among sources (Breakspear, 2002; Tass, 2003).

Simulations

In this section, we investigate the effect of ongoing activity on stimulus-dependent responses to reconcile apparently contradictory conclusions from studies of event-related potentials. On one hand, classical studies have shown that event-related potentials are associated with amplitude changes in the M/EEG signal that represent a linear summation of an impulse response and ongoing activity (Arieli *et al.*, 1996; Shah *et al.*, 2004). In this scheme, the variability at the single-trial level is due to, and only to, ongoing activity, which is removed after averaging to estimate the impulse response. On the other hand, it has been hypothesized that event-related waveforms, obtained after averaging, could be due to a phase-resetting of ongoing activity with no necessary change in the amplitude (i.e. power) of any stimulus-locked transient (Makeig *et al.*, 2002; Jansen *et al.*, 2003). Although mathematically well defined, the neural mechanisms that could instantiate phase-resetting of ongoing activity are unknown.

We will take phase-resetting to imply a non-linear interaction between ongoing activity and stimulus-related input that results in phase-locking to stimulus onset. Although phase-*locking* can be produced by evoking oscillatory transients (i.e. amplitude modulation), this mechanism involves no change or *resetting* of the ongoing dynamics. To assess the contribution of phase-resetting in our simulations, we therefore need to look for interactions between ongoing and stimulus-related inputs that produced phase-locking in the outputs. We can address this, in a simple way, by subtracting the response to ongoing activity alone from the response to a mixture of ongoing activity and stimulus input. In the absence of interactions, this difference (the evoked response) should be the same. On the other hand, if interactions are prevalent, the difference should change with each realization of ongoing activity. We performed these analyses with different levels of input and assessed the degree of phase-locking in the outputs with the phase-locking value (PLV) (Tallon-Baudry *et al.*, 1996; Lachaux *et al.*, 1999): $PLV(t) = \left|\langle \exp(j\phi(t)) \rangle_{trials}\right|$ where the instantaneous phase $\phi(t)$ was obtained from the Hilbert transform (Le Van Quyen *et al.*, 2001).

To evaluate the effect of background activity on single-trial event-related responses, we used the two area hierarchical model above, with $a^B = 1$ (see Figure 33.14). The first area was driven by an impulse function (stimulus) and Gaussian random noise (background activity) of standard deviation of 0.05. The output of this region can be considered a mixture of evoked response and ongoing activity. We considered two conditions: one with low levels of mixed input ($c = 10^2$) and another with high levels ($c = 2.10^4$). These values were chosen to emphasize the system's nonlinear properties; with the smaller value of c, neuronal responses remain largely in the linear regime. The larger value of c was chosen so that excursions of the states encroached on the non-linear regime, to produce neuronal saturation in some trials. In both cases, the stimulus was a delta-function. The simulated responses, for 100 trials, are shown in Figure 33.16.

When input levels are low (left hand side of Figure 33.16), event-related activity, at the single-trial level, shows a relatively reproducible waveform after stimulus onset (Figure 33.16(b)). This transient is reflected in the ERP/ERF after averaging (Figure 33.16(c)). To confirm the experimental results of Arieli *et al.* (1996), we decomposed each event-related response into two components. First, the stochastic component (the response to ongoing activity alone – Figure 33.16(d)) and second, an extra component elicited by adding the stimulus (Figure 33.16(e)). This is the difference between the response elicited by the stochastic component alone (Figure 33.16(d)) and the response to the mixed input (Figure 33.16(b)). If the system was linear, these differences should not exhibit any variability over trials, and thus define the 'reproducible response' (Arieli *et al.*, 1996). Effectively, the stimulus-dependent component shows no variability and we can conclude that the response components due to stimulus and ongoing activity are linearly

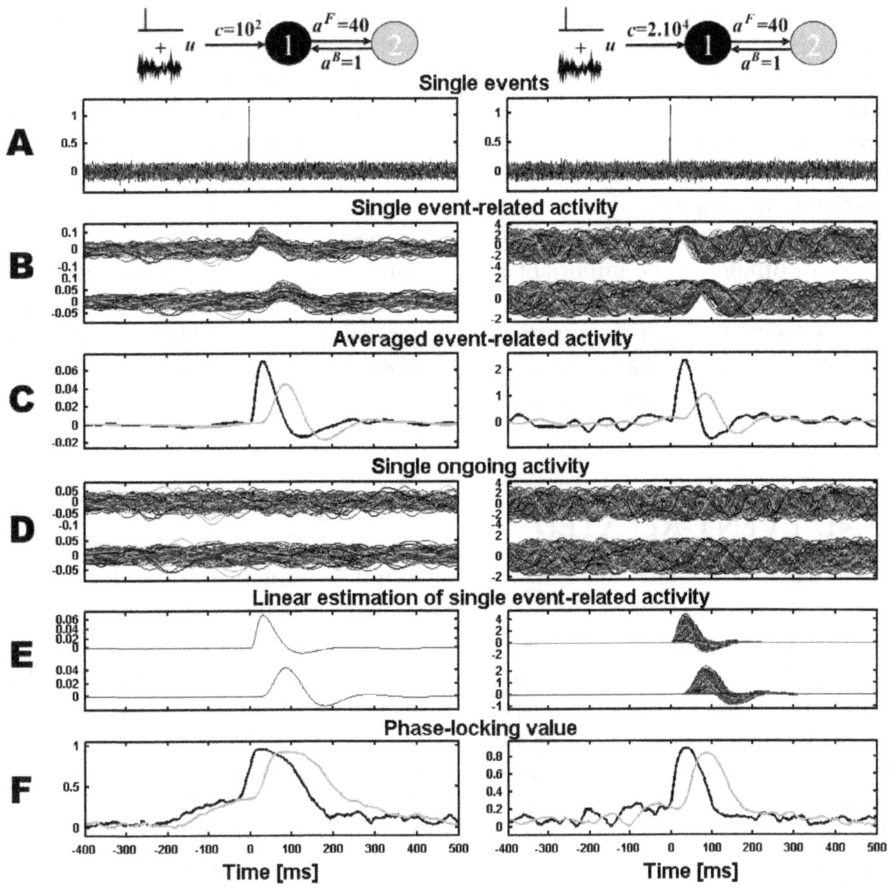

FIGURE 33.16 Event-related responses in the context of ongoing activity (100 trials). Two hierarchically connected regions are considered. Input (ongoing and stimulus-related) enters into the system through region 1. Two levels of input are considered: weak on the left hand side ($c = 10^2$), strong on the right hand side ($c = 2.10^4$). The successive horizontal panels show different types of activity. The time scale is identical for each panel and shown at the bottom. (a) Inputs, comprising a delta function and Gaussian noise of standard deviation 0.05 (stimulus onset at time zero). (b) Event-related activity at the single-trial level. The time series over trials is shown (area 1 is above area 2). (c) Averaged event-related response estimated by averaging over epochs shown in (b) (area 1 in black, area 2 in grey). (d) Responses to the noisy input without the delta function, shown in the same format as in (b). (e). Stimulus-dependent component obtained from subtracting (d) from (b). (f). Phase-locking value computed from time series in (b), which exhibits a transient phase-synchronization to peristimulus time (Black: area 1, grey: area 2).

separable. In other words, there are no interactions that could mediate phase-resetting. Despite this, there is ample evidence for phase-*locking*. This is shown in Figure 33.16(f), using the phase-locking index.

However, the situation is very different when we repeat the simulations with high input levels (right hand side of Figure 33.16). In this context, the event-related responses do not show any obvious increase in amplitude after the stimulus (Figure 33.16(b)). However, the averaged event-related activity (Figure 33.16(c)) is very similar to that above (left hand side of Figure 33.16(c)). The fact that one obtains an ERP by averaging in this way suggests that the stimulus input induced phase-resetting of the ongoing oscillations. This is confirmed by the large variation in stimulus-dependent components from trial to trial. This variation reflects non-linear interactions between the stimulus and ongoing activity (Figure 33.16(e)). These interactions are associated with phase-locking as shown in Figure 33.16(f).

In summary, the fact that the difference in evoked responses with and without background noise (panel (e), Figure 33.16) shows so much variability, suggests that background activity interacts with the stimulus: when ongoing activity is high, cells saturate and the stimulus-related response is attenuated. Conversely, when ongoing activity is low the evoked-response is expressed fully. This dependency on ongoing activity is revealed by variation in the evoked responses with high input levels. In conclusion, the apparently contradictory results presented in Arieli *et al.* (1996), Makeig *et al.* (2002), Jansen *et al.* (2003) and Shah *et al.* (2004) can be reproduced for the most part and reconciled within the same framework.

With high activity levels, the ongoing and stimulus-dependent components interact, through non-linearities in the population dynamics, to produce phase-resetting and a classical ERP on averaging. When activity is lower, the stimulus and endogenous dynamics do not interact and the ERP simply reflects the transient evoked by stimuli that is linearly separable from ongoing dynamics.

We have shown that non-linear mechanisms due to the saturation of neuronal outputs can be important in measured EPR/ERF. In the next section, we take another perspective on event-related activity (ERPs are a particular type of event-related activity) and focus on the modulation of ongoing activity by the experimental context.

ONGOING AND EVENT-RELATED ACTIVITY

Ongoing activity, i.e. oscillations in the M/EEG signal that share no phase relationship with the stimulus, refers to dynamics that are regarded as random fluctuations, or autonomous dynamics with a high complexity. Ongoing activity is shaped by the same non-linear convolution experienced by deterministic inputs. In the context of stationary inputs, the outputs can be characterized in terms of their spectral properties, which are determined by the generalized transfer functions of the Volterra kernels (see Chapter 39) associated with any controllable analytic system. The impulse response function is the first-order kernel. As soon as the connectivity parameters of a hierarchical network change, the principal modes of this network, defined by the principal frequencies of oscillations, are modulated (David and Friston, 2003). As an illustration, consider the simple hierarchical model of two cortical areas established in the previous sections with two configurations, which differ in the strength of backward connections ($a^B = 1$ and $a^B = 10$) (Figure 33.17). The corresponding frequency spectra, of pyramidal cell depolarization of the two areas, show that the change in connectivity induces a profound modulation of their spectral profile. As one might intuit, strong backward connections induce a peak at the same frequency of the damped oscillations in the impulse response function. This is an important aspect of ongoing activity in the sense that its spectral behaviour may be very close to that of evoked transients as shown in (Makeig *et al.*, 2002).

Induced versus evoked responses

This modulation of oscillatory dynamics, by the system's coupling parameters, provides a natural model for event-related changes in rhythmic activity. This phenomenon is known as event-related synchronization (ERS) in frequency bands showing an evoked increase in power, or conversely, event-related desynchronization (ERD) for decreases (Basar, 1980; Pfurtscheller and Lopes da Silva, 1999). In light of the above connectivity-dependent changes in power, ERD and ERS may reflect the dynamics *induced* by *evoked* changes in short-term plasticity. The key difference between evoked and induced transients relates to the presence or absence of changes in the system's control parameters, here coupling or synaptic efficacy. Evoked changes are not necessarily associated with parameter changes and any complicated response can be ascribed to transients that arise as the systems trajectory returns to its attractor. Conversely, induced responses arise from perturbation of the attractor manifold itself, by changes in the parameters and ensuing changes in

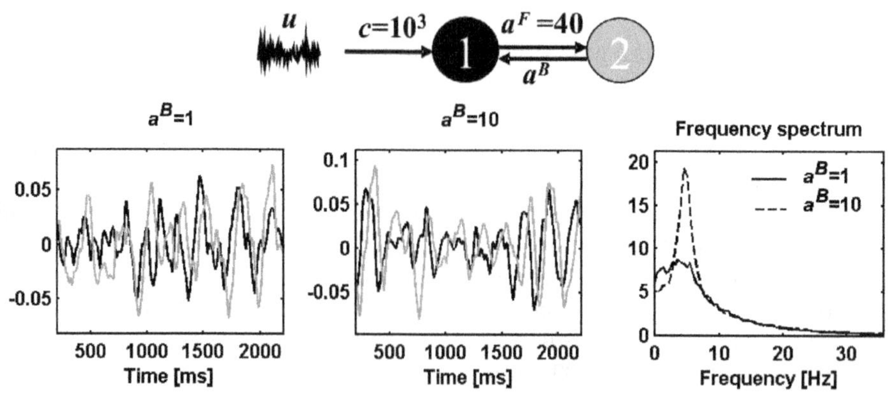

FIGURE 33.17 The modulation of backward connectivity ($a^B = 1$ or $a^B = 10$) has huge effect on the power spectrum of ongoing M/EEG dynamics (responses are plotted in black for area 1 and in grey for area 2). When a^B increases from 1 to 10, there is loss of power below 3 Hz, and an excess between 3 and 7 Hz. The amplitude spectra in the right panel were obtained by averaging the modulus of the fast Fourier transform of pyramidal cell depolarization, over 100 epochs of 2.5 s (examples are shown in the two left-hand panels, black: first area, grey: second area).

the dynamics. This distinction was discussed in Friston (1997a) in relation to MEG dynamics and modelled using asymmetric connections between two areas in Friston (2000a) and Chapter 39.

Empirically, the ERS/ERD approach is used to look for M/EEG power changes of rhythmic activity induced by external events. This phenomenon has been modelled, in the case of alpha rhythms, by a computational model of thalamocortical networks (Suffczynski *et al.*, 2001). It has been shown that a key mechanism is the modulation of functional interaction between populations of thalamocortical cells and the reticular nucleus. In the last application of neural-mass models for ERPs in this chapter, we consider the differences between induced and evoked responses in more depth. Our goal is to look at the generative mechanisms behind induced responses. The next section is a summary of David *et al.* (2006b).

INDUCED RESPONSES AND ERPs

Cortical oscillatory activity, as disclosed by local field potentials (LFPs), EEG and MEG recordings, can be categorized as ongoing, evoked or induced (Tallon-Baudry and Bertrand, 1999). Evoked and induced oscillations differ in their phase-relationships to the stimulus. Evoked oscillations are phase-locked to the stimulus, whereas induced oscillations are not. Operationally, these two phenomena are revealed by the order of trial-averaging and spectral analysis (see Chapter 30). To estimate evoked power, the M/EEG signal is first averaged over trials and then subject to time-frequency analysis to give an event-related response (ERR). To estimate induced oscillations, the time-frequency decomposition is applied to each trial and the ensuing power is averaged across trials. The power of evoked and background components is subtracted from this total power to reveal induced power. In short, evoked responses can be characterized as the power of the average, while induced responses are the average power that cannot be explained by the power of the average.

A common conception is that evoked oscillations reflect a stimulus-locked event-related response in time-frequency space and that induced oscillations are generated by some distinct high-order process. Following Singer and Gray (1995), this process is often described in terms of 'binding' and/or neuronal synchronization. The tenet of the binding hypothesis is that coherent firing patterns can induce large fluctuations in the membrane potential of neighbouring neurons which, in turn, facilitate synchronous firing and information transfer (as defined operationally in Varela, 1995). Oscillations are induced because their self-organized emergence is not evoked directly by the stimulus, but induced vicariously through non-linear and possibly autonomous mechanisms.

Our treatment of induced responses is divided into three parts. In the first, we establish a key distinction between *dynamic* mechanisms, normally associated with classical evoked responses like the ERP and *structural* mechanisms, implicit in the genesis of induced responses. Dynamic effects are simply the effect of inputs on a system's response. Conversely, structural mechanisms entail a transient change in the system's causal structure, i.e. its parameters (e.g. synaptic coupling). These changes could be mediated by non-linear effects of input. We relate the distinction between dynamic and structural mechanisms to series of dichotomies in dynamical system theory and neurophysiology. These include the distinction between driving and modulatory effects in the brain. This part concludes with a review of how neuronal responses are characterized operationally, in terms of evoked and induced power, and how these characterizations relate to dynamic and structural perturbations. In the second part, we show that structural mechanisms can indeed produce induced oscillations. In the example provided, responses are induced by a stimulus-locked modulation of the backward connections from one source to another. However, we show that this structural effect is also expressed in evoked oscillations when dynamic and structural effects interact. In the final part, we show the converse, namely that dynamic mechanisms can produce induced oscillations, even in the absence of structural effects. This can occur when trial-to-trial variations in input suppress high-frequency responses after averaging. Our discussion focuses on the rather complicated relationship between the two types of mechanisms that can cause responses in M/EEG and the ways in which evoked and induced responses are measured. We introduce *adjusted* power as a complement to induced power that resolves some of these ambiguities.

Dynamic and structural mechanisms

From Eqn. **33.5**, it is immediately clear that the states and implicitly the system's response, can only be changed by perturbing $u(t)$ or θ. We will refer to these as *dynamic* and *structural* effects respectively. This distinction arises in a number of different contexts. From a purely dynamical point of view, transients elicited by dynamic effects are the system's response to input changes, e.g. the presentations of a stimulus in an ERP study. If the system is dissipative and has a stable fixed point, then the response is a generalized convolution of the input with associated kernels. The duration and form of the resulting dynamics effect depends on the *dynamical stability* of the system to perturbations of its states (i.e. how the system's trajectories change with the state). Structural effects depend on *structural stability* (i.e. how the system's trajectories change with the parameters). Systematic changes in the parameters can produce systematic changes in the response, even in the absence of input. For systems that show autonomous (i.e. periodic or chaotic) dynamics, changing the parameters is equivalent to changing the attractor manifold, which induces a change in the system's states. We have discussed this in the context of non-linear coupling and classical neuromodulation (Friston, 1997b; Breakspear *et al.*, 2003). For systems with fixed points and Volterra kernels, changing the parameters is equivalent to changing the kernels and transfer functions. This changes the spectral density relationships between the inputs and outputs. As such, structural effects are clearly important in the genesis of induced oscillations because they can produce frequency modulation of ongoing activity that does not entail phase-locking to any event. In summary, dynamic effects are expressed directly on the states and conform to a convolution of inputs to form responses. Structural effects are expressed indirectly, through the Jacobian (see Appendix 2), and are inherently non-linear, inducing high-order kernels and associated transfer functions.

Drivers and modulators

The distinction between dynamic and structural inputs speaks immediately of the difference between 'drivers' and 'modulators' (Sherman and Guillery, 1998). In sensory systems, a driver ensemble can be identified as the transmitter of receptive field properties. For instance, neurons in the lateral geniculate nuclei drive primary visual area responses in the cortex, so that retinotopic mapping is conserved. Modulatory effects are expressed as changes in certain aspects of information transfer, by the changing responsiveness of neuronal ensembles in a context-sensitive fashion. A common example is attentional gain. Other examples involve extra-classical receptive field effects that are expressed beyond the classical receptive field. Generally, these are thought to be mediated by backward and lateral connections. In terms of synaptic processes, it has been proposed that the postsynaptic effects of drivers are fast (ionotropic receptors), whereas those of modulators are slower and more enduring (e.g. metabotropic receptors). The mechanisms of action of drivers refer to classical neuronal transmission, either biochemical or electrical, and are well understood. Conversely, modulatory effects can engage a complex cascade of highly non-linear cellular mechanisms (Turrigiano and Nelson, 2004). Modulatory effects can be understood as transient departures from homeostatic states, lasting hundreds of milliseconds, due to synaptic changes in the expression and function of receptors and intracellular messaging systems.

Classical examples of modularity mechanisms involve voltage-dependent receptors, such as NMDA receptors. These receptors do not cause depolarization directly (cf. a dynamic effect) but change the unit's sensitivity to depolarization (i.e. a structural effect). It is interesting to note that backward connections, usually associated with modulatory influences, target supragranular layers in the cortex where NMDA receptors are expressed in greater proportion. Having established the difference between dynamics and structural effects and their relationship to driving and modulatory afferents in the brain, we now turn to the characterization of evoked and induced responses in terms of time-frequency analyses.

Evoked and induced power

The criterion that differentiates induced and evoked responses is the degree to which oscillatory activity is phase-locked to the stimulus over trials. An ERR is the waveform that is expressed in the EEG signal after every repetition of the same stimulus. Due to physiological and measurement noise, the ERR is often only evident after averaging over trials. More formally, the *evoked* response $y(t)_e$ to a stimulus is defined as the average of measured responses in each trial $y(t)$:

$$y(t)_e = \langle y(t) \rangle \tag{33.8}$$

where t is peristimulus time. A time-frequency representation $s(\omega, t)$ of a response $y(t)$ obtains by successively filtering $y(t)$ using a kernel or filter-bank parameterized by frequencies $\omega_j = 2\pi v_j$, over the frequency range of interest:

$$s(\omega, t) = \begin{bmatrix} k(\omega_1, t) \otimes y(t) \\ \vdots \\ k(\omega_J, t) \otimes y(t) \end{bmatrix} \tag{33.9}$$

$k(\omega_j, t)$ can take several forms (Kiebel *et al.*, 2005). The total power, averaged over trials and the power of the average are respectively:

$$\begin{aligned} g(\omega, t)_T &= \langle s(\omega, t) s(\omega, t)* \rangle \\ g(\omega, t)_e &= \langle s(\omega, t) \rangle \, \langle s(\omega, t)* \rangle \end{aligned} \tag{33.10}$$

where * denotes the complex conjugate. $g(\omega, t)_e$ is evoked power and is simply the power of $y(t)_e$. Induced power $g(\omega, t)_i$ is defined as the component of total power that cannot be explained by baseline and evoked power.[2] This implicitly partitions total power into three orthogonal components (induced, baseline and evoked):

$$g(\omega, t)_T = g(\omega, t)_i + g(\omega, t)_e + g(\omega)_b \qquad \textbf{33.11}$$

Baseline power $g(\omega)_b$ is a frequency-specific constant due to ongoing activity and experimental noise, both of which are assumed to be stationary. This component is usually calculated over a period of time preceding stimulus presentation.

Mechanisms of generation

In this subsection, we establish how dynamic and structural mechanisms are expressed in terms of evoked and induced power. As illustrated in Figure 33.18, the inputs for the i-th trial $u^{(i)}$ can be decomposed into a deterministic stimulus-related component α and trial-specific background activity $\beta^{(i)}$, which is stochastic and unrelated to the stimulus:

$$u^{(i)} = \alpha + \beta^{(i)} \qquad \textbf{33.12}$$

For simplicity, we will assume that the state-space defined by Eqn. **33.4** operates largely in its linear regime,

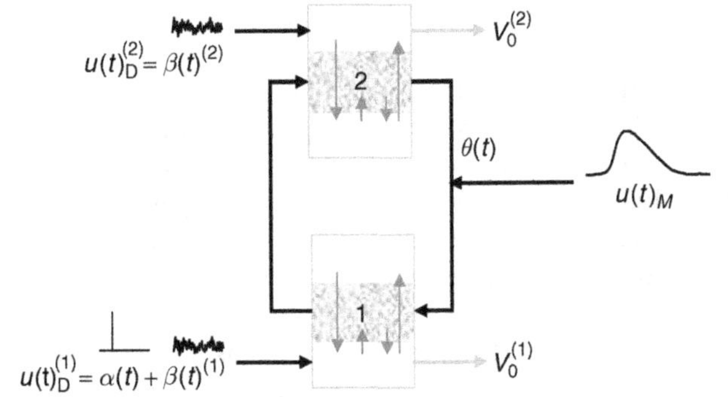

FIGURE 33.18 Neuronal model used in the simulations. Two cortical areas interact with forward and backward connections. Both areas receive a stochastic input, which simulates ongoing activity from other brain areas. In addition, area 1 receives a stimulus, modelled as a delta function. A single modulatory effect is considered. It simulates a stimulus-related slow modulation of extrinsic backward connectivity. The outputs of the neuronal system are the pyramidal depolarizations of both areas.

as suggested by studies which have found only weak non-linearities in EEG oscillations (Stam *et al.*, 1999; Breakspear and Terry, 2002a). This allows us to focus on the first-order kernels and transfer functions. We will also assume the background activity is stationary. In this instance, the total power is:

$$\begin{aligned} g(\omega, t)_T &= |\Gamma(\omega, t)|^2 g(\omega, t)_u \\ g(\omega, t)_u &= g(\omega, t)_\alpha + g(\omega)_\beta \end{aligned} \qquad \textbf{33.13}$$

In words, the total power is the power of the input, modulated by the transfer function $|\Gamma(\omega, t)|^2$ (see Appendix 2). The power of the input is simply the power of the deterministic component, at time t, plus the power of ongoing activity. The evoked power is simply the power of the input, because the noise and background terms are suppressed by averaging:

$$\begin{aligned} g(\omega, t)_e &= |\Gamma(\omega, t)|^2 \langle s(\omega, t)_\alpha\rangle \langle s(\omega, t)^*_\alpha\rangle \\ &= |\Gamma(\omega, t)|^2 g(\omega, t)_\alpha \end{aligned} \qquad \textbf{33.14}$$

The baseline power at $t = t_0$ is:

$$g(\omega)_b = |\Gamma(\omega, t_0)|^2 g(\omega)_\beta \qquad \textbf{33.15}$$

This means that induced power is:

$$g(\omega, t)_i = (|\Gamma(\omega, t)|^2 - |\Gamma(\omega, t_0)|^2) g(\omega)_\beta \qquad \textbf{33.16}$$

This is an important result. It means that the only way induced power can be expressed is if the transfer function $\Gamma(\omega, t, \theta)$ changes at time t. This can only happen if the parameters of the neuronal system change. In other words, only structural effects can mediate induced power. However, this does not mean to say that structural effects are expressed only in induced power. They can also be expressed in the evoked power: Eqn. **33.14** shows clearly that evoked power at a particular point in peristimulus time depends on both $g(\omega, t)_\alpha$ and $\Gamma(\omega, t, \theta)$. This means that structural effects mediated by changes in the transfer function can be expressed in evoked power, provided $g(\omega, t)_\alpha > 0$. In other words, structural effects can modulate the expression of stationary components due to ongoing activity and also deterministic components elicited dynamically. To summarize so far:

- Dynamic effects (of driving inputs) conform to a generalized convolution of inputs to form the system's response.
- Structural effects can be formulated as a time-dependent change in the parameters (that may be mediated by modulatory inputs). This translates into time-dependent change in the convolution kernels and ensuing response.

[2] A different definition is sometimes used, where induced responses are based on the difference in amplitude between single-trials and the ERR: $y(t) - y(t)_e$ (Truccolo *et al.*, 2002). The arguments in this work apply to both formulations. However, it is simpler for us to use Eqn. **33.11** because it discounts ongoing activity. This allows us to develop the arguments by considering just one trial-type (opposed to differences between trial-types)

- If the ongoing activity is non-zero and stationary, only structural effects can mediate induced power.
- If stimulus-related input is non-zero, structural effects can also mediate evoked power, i.e. dynamic and structural effects can conspire to produce evoked power.

In the next subsection, we demonstrate these theoretical considerations in a practical setting, using the neural-mass model of event-related responses. In this section and in the simulations below, we only consider a single trial-type. In practice, one would normally compare the responses evoked and induced by two trial types. However, the conclusions are exactly the same in both contexts. One can regard the simulations below as a comparison of one trial-type to a baseline that caused no response (and had no baseline power).

Modelling induced oscillations

We consider a simple model composed of two sources, inter-connected with forward and backward connections (Figure 33.18). The sources receive two types of inputs. The first $u(t)_D$, models afferent activity that delivers dynamic perturbations to the system's states (by changing postsynaptic currents). This dynamic perturbation had stochastic and deterministic components: background inputs $\beta^{(i)}$ comprised Gaussian noise that was delivered to both sources. The deterministic part modelled a stimulus with an impulse $\alpha(t) = \delta(0)$, delivered to the first source at the beginning of each trial. The second sort of input $u(t)_M$ induced a structural change by modulating extrinsic connections. As one might expect, the effects of these two input classes differ considerably. On the one hand, synaptic inputs perturb the system nearly instantaneously and the deterministic part evokes responses that are phase-locked to the stimulus. On the other hand, modulatory inputs modify the manifold that attracts ongoing activity, without necessarily resetting its phase. For simplicity, we restrict our modulatory effects to a modulation of the extrinsic backward connection, thus encompassing various synaptic mechanisms which modify the gain of excitatory synapses (Salinas and Thier, 2000). We chose the backward connection because backward connections are associated with modulatory effects, both in terms of physiology (e.g. the mediation of extra-classical receptive field effects) (see also Allman *et al.*, 1985; Murphy *et al.*, 1999) and anatomy (e.g. they terminate in supragranular layers that expressed large number of voltage-dependent NMDA receptors) (see also Maunsell and Van, 1983; Angelucci *et al.*, 2002). There may be many other modulatory mechanisms that will produce the same pattern of oscillatory activity and it will be an interesting endeavour to disambiguate the locus of structural changes using these sorts of models and empirical data.

Structural perturbation and induced oscillations

To illustrate the points of the previous section, we will consider two scenarios in which the modulatory effect arrives at the same time as the driving input and one in which it arrives after the dynamic perturbation has dissipated. Let us assume that the modulatory input has a slow time-constant $\tau = 150$ ms compared to the main frequency of ongoing oscillations (10 Hz). The modulatory effects can be expressed with stimulus onset, or after some delay. In the first case, evoked oscillations will be modulated and these effects will be visible in the ERR. In the second case, phase-locking with the stimulus will have been lost and no effect will be seen in the ERR. However, in both cases, structural changes will appear as induced oscillations.

This is illustrated in Plate 45 (see colour plate section) using 500 trial-averages. In the upper panel we consider a modulatory input immediately after stimulus onset. As expected, evoked responses are much more pronounced relative to delayed modulation (lower panel). The induced power (c) shows that increases in the backward connection induce oscillations in the alpha and gamma band. The induced power in Plate 45 has been frequency normalized (by removing the mean and dividing by the standard deviation at $t = 0$) to show increased power in the gamma band more clearly. These simulations provide a nice model for induced responses using a structural perturbation, in this instance a slow modulation of the efficacy of backward connections in a simple hierarchy of neuronal populations. Critically, these simulations also show that responses can be evoked structurally by a modulation of dynamic perturbations. This dual mechanism depends on driving and modulatory effects occurring at the same time, causing evoked and induced responses in the same time-frequency window. Having established that evoked responses can also be mediated by structural mechanisms we now show that induced responses can be mediated by dynamic mechanisms.

Induced oscillations and trial-to-trial variability

Above we considered the stimulus as a deterministic input. Here we consider what would happen if the stimulus-related input was stochastic. This randomness is most easily understood in terms of trial-to-trial variability in the inputs. As suggested in Truccolo *et al.* (2002), we consider two types of variability in the input. The first relates to a trial-to-trial gain, or amplitude variations. For an identical stimulus, early processing may introduce variations in the amplitude of driving inputs to a source. Gain modulation is a ubiquitous phenomenon in the central nervous system (Salinas and Thier, 2000), but its causes are not completely understood. Two neurophysiological mechanisms that may mediate gain modulation include fluctuations of extracellular calcium concentration (Smith *et al.*, 2002) and/or of the overall level of synaptic input to a neuron (Chance *et al.*, 2002). These may act as a gain control signal that modulates responsiveness to excitatory drive. A common example of gain effects, in a psychophysiological context, is the effect of attention (McAdams and Maunsell, 1999; Treue and Martinez Trujillo, 1999). The second commonly observed source of variability is in the latency of input onset, i.e. the time between the presentation of the stimulus and the peak response of early processing. Azouz and Gray (1999) have investigated the sources of such latency variations at the neuronal level. Basically, they describe two major phenomena: coherent fluctuations in cortical activity preceding the onset of a stimulus have an impact on the latency of neuronal responses (spikes). This indicates that the time needed to integrate activity to reach action potential threshold varies between trials. The other source of latency variability is fluctuations in the action potential threshold itself.

Both types of trial-to-trial variability, gain modulation and latency, can be modelled by introducing appropriate random variables. The theoretical analysis described in David *et al.* (2006b) shows that:

- High frequency components are lost in the evoked responses when latency varies randomly over trials. This means that ERR will be estimated badly at high frequencies. This variation effectively blurs or smoothes the average, and suppresses fast oscillations in the evoked response. However, the total power remains unchanged, because the power expressed in each trial does not depend on latency. Therefore, the high frequencies lost from the evoked responses now appear in the induced response. In summary, the induced power has now acquired a stimulus-locked component. Note that this dynamically induced power can only be expressed in frequencies that show evoked responses. This is illustrated in Plate 46 as a transfer of power from the evoked component to the induced component.
- Gain variations also allow non-structural mechanisms to induce power. Here the time-dependent changes in stimulus-dependent power again contribute to induced responses. In this instance, the contribution is not frequency specific, as with latency variations, but proportional to the variance in gain. This is illustrated in Plate 47.

Summary

In summary, we have made a distinction between dynamic and structural mechanisms that underlie transient responses to perturbations. We then considered how responses are measured in time-frequency in terms of evoked and induced responses. Theoretical predictions (see David *et al.*, 2006b), confirmed by simulations, show that there is no simple relationship between the two mechanisms causing responses and the two ways in which they are characterized. Specifically, evoked responses can be mediated both structurally and dynamically. Similarly, if there is trial-to-trial variability, induced responses can be mediated by both mechanisms (see Figure 33.19 for a schematic summary).

For evoked responses this is not really an issue. The fact that evoked responses reflect both dynamic and structural perturbations is sensible, if one allows for the fact that any input can have dynamic and structural effects. In other words, the input perturbs the states of the neuronal system and, at the same time, modulates interactions among the states. The structural component here can be viewed as a non-linear (e.g. bilinear) effect that simply involves interactions between the input and parameters (e.g. synaptic status). Generally, the structurally mediated component of evoked responses will occur at the same time and frequency as the dynamically mediated components. This precludes ambiguity when interpreting evoked responses, if one allows for both dynamic and structural causes.

The situation is more problematic for induced responses. In the absence of trial-to-trial variability, induced responses must be caused by structural perturbations. Furthermore, there is no necessary co-localization of evoked and induced responses in time-frequency, because induced responses are disclosed by ongoing activity. However, if trial-to-trial variability is sufficient, induced responses with no structural component will be expressed. This means that induced

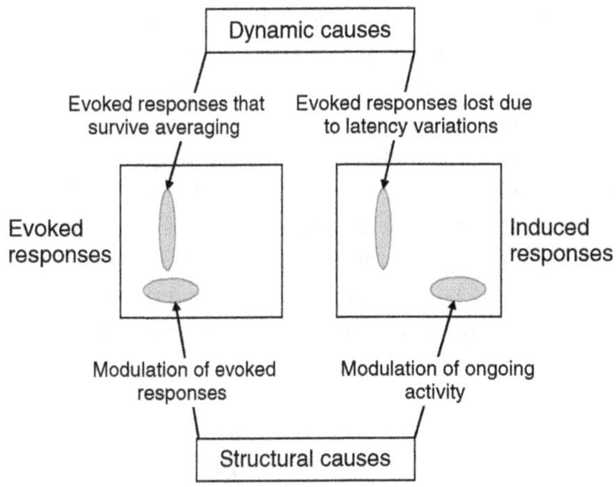

FIGURE 33.19 Schematic illustrating the many-to-many mapping between dynamic versus structural causes and evoked versus induced responses.

responses that occur at the same time as evoked responses have an ambiguity in relation to their cause. Happily, this can be addressed at two levels. First, induced responses that do not overlap in peristimulus time cannot be attributed to dynamic mechanisms and are therefore structural in nature. Second, one can re-visit the operational definition of induced responses to derive a measure that is immune to the effects of trial-to-trial variability.

Adjusted power

Here we introduce the notion of adjusted power as a complementary characterization of structurally mediated responses. Adjusted power derives from a slightly more explicit formulation of induced responses as that component of total power that cannot be explained by evoked or ongoing activity. The adjusted response is simply the total power, orthogonalized, at each frequency, with respect to baseline and evoked power:

$$
\begin{aligned}
g(\omega, t)_a &= g(\omega, t)_T - g(\omega, t)\hat{\eta} \\
\hat{\eta} &= g(\omega, t)^+ g(\omega, t)_T \\
g(\omega, t) &= \begin{bmatrix} 1 & g(\omega, t_1)_e \\ \vdots & \vdots \\ 1 & g(\omega, t_T)_e \end{bmatrix}
\end{aligned}
\tag{33.17}
$$

$+$ denotes the pseudoinverse. Eqn. **33.17** is implicitly estimating baseline power and the contribution from evoked power and removing them from the total power. In other words, $\hat{\eta}$ is a 2-vector estimate of $g(\omega)_b$ and $(1+\eta)$. After these components have been removed the only components left must be structural in nature:

$$
g(\omega, t)_a \approx (|\Gamma(\omega, t)|^2 - |\Gamma(\omega, t_0)|^2) g(\omega)_\beta \tag{33.18}
$$

Plate 48 shows that the effect of trial-to-trial variability on induced responses disappears when using adjusted power. This means one can unambiguously attribute adjusted responses to structural mechanisms. The ERP-adjusted response removes evoked response components, including those mediated by structural changes. However, structurally mediated induced components will not be affected unless they have the same temporal expression. The usefulness of adjusted power in an empirical setting will be addressed in future work.

DISCUSSION

We have divided neuronal mechanisms into dynamic and structural, which may correspond to driving and modularity neurotransmitter systems respectively. These two sorts of effects are related to evoked and induced responses in M/EEG. By definition, evoked responses exhibit phase-locking to a stimulus whereas induced responses do not. Consequently, averaging over trials discounts both ongoing and induced components and evoked responses are defined by the response averaged over trials. Evoked responses may be mediated primarily by driving inputs. In M/EEG, driving inputs affect the state of measured neuronal assemblies, i.e. the dendritic currents in thousands of pyramidal cells. In contradistinction, structural effects, mediated by modulatory inputs, engage neural mechanisms which affect neuronal states, irrespective of whether they are phase-locked to the stimulus or not. These inputs are expressed formally as time-varying parameters of the state equations modelling the systems. Although the ensuing changes in the parameters may be slow and enduring, their effects on ongoing or evoked dynamics may be expressed as fast or high-frequency dynamics. We have considered a further cause of induced oscillations, namely trial-to-trial variability of driving inputs. As suggested in Truccolo *et al.* (2002), these can modelled by varying latency and gain. We have shown that gain variations have no effect on the ERR but increase induced responses in proportion to evoked responses. Secondly, we show that jitter in latency effectively smoothes the evoked responses and transfers energy from evoked to induced power, preferentially at higher frequencies.

The conclusions of this work, summarized in Figure 33.19, provide constraints on the interpretation of evoked and induced responses in relation to their mediation by dynamic and structural mechanisms. This is illustrated by the work of Tallon-Baudry and colleagues, who have looked at non-phase-locked episodes of synchronization in the gamma-band (30–60 Hz). They have emphasized the role of induced responses in feature-binding and top-down mechanisms of perceptual synthesis. The top-down aspect is addressed by their early studies of illusory perception (Tallon-Baudry *et al.*, 1996), where the authors, 'tested the stimulus specificity of high-frequency oscillations in humans using three types of visual stimuli: two coherent stimuli (a Kanizsa and a real triangle) and a non-coherent stimulus'. They found an early phase-locked 40 Hz component, which did not vary with stimulation type and a second 40 Hz component, appearing around 28 ms, which was not phase-locked to stimulus onset. This shows a nice dissociation between early evoked and late induced responses. The induced component was stronger in response to a coherent triangle, whether real or illusory and: 'could reflect, therefore, a mechanism of feature binding based on high-frequency synchronization'. Because it was late, the induced response can only be caused by structural mechanisms (see Figure 33.19). This is consistent with the role of top-down influences and the modulatory mechanisms employed by backward connections in visual synthesis (Maunsell and Van, 1983; Bullier et al., 2001; Albright and Stoner, 2002).

Classical ERP/ERF research has focused on dynamic perturbations (Coles and Rugg, 1995). On the other hand, studies of event-related synchronization (ERS) or desynchronization (ERD) are more concerned with structural effects that may be mediated by modulatory systems (Pfurtscheller and Lopes da Silva, 1999). Practically speaking, we have shown that it is not always possible to distinguish between dynamic and structural effects when inferring the causes of evoked and induced oscillations. However, certain features of induced oscillations might provide some hints: (i) induced oscillations in high frequencies concomitant with evoked responses in low frequencies may indicate a jittering of inputs; (ii) induced oscillations that are temporally dissociated from evoked responses are likely to be due to modulatory or structural effects. Finally, we have introduced the notion of adjusted power that can be unambiguously associated with structural effects.

CONCLUSION

Neural-mass models afford a straightforward approach to modelling the activity of populations of neurons. Their main assumption is that the state of the population can be approximated using very few state variables (generally limited to mean membrane currents, potentials and firing rates). Given a macroscopic architecture, describing the overall connectivity between populations of a given cortical area and between different cortical areas, it is possible to simulate the steady-state dynamics of the system or even the transient response to a perturbation of extrinsic input or connectivity. Consequently, neural-mass models are useful to describe and predict the macroscopic electrical activity of the brain. Since the early 1970s, they have been used to address several important issues, e.g. alpha rhythms (Lopes da Silva *et al.*, 1997), olfactory responses (Freeman, 1987), and focal attention (Suffczynski *et al.*, 2001). They are now being introduced into neuroimaging to understand the underlying neuronal mechanisms of fMRI and PET data (Horwitz and Tagamets, 1999; Almeida and Stetter, 2002; Aubert and Costalat, 2002).

Despite their relative simplicity, neural-mass models can exhibit complex dynamical behaviour reminiscent of the real brain. In David and Friston (2003), we have shown that physiologically plausible synaptic kinetics lead to the emergence of oscillatory M/EEG-like signals covering the range of theta to gamma bands. To emulate more complex oscillatory M/EEG dynamics, we have proposed a generalization of the Jansen model that incorporates several distinct neural populations that resonate at different frequencies. Changing the composition of these populations induces a modulation of the spectrum of simulated M/EEG signals.

We have investigated the consequence of coupling two remote cortical areas. It appears that the rhythms generated depend critically upon both the strength of the coupling and the propagation delay. As the coupling directly modulates the contribution of one area to another, the spectrum of the driven area, in the case of a unidirectional coupling, is obviously a mixture of the source and target spectra. More interestingly, a reciprocal coupling engenders more marked modifications of the M/EEG spectrum, which can include strong oscillatory activity. Bi-directional coupling is important because of the high proportion of reciprocal connections in the brain. The most robust consequence of coupling is phase synchronization of remote oscillations.

Obviously neural-mass models do not describe exactly how neural signals interact. These models represent a summary of underlying neurophysiological processes that cannot be modelled in complete detail because of their complexity. In particular, the model we used does not accommodate subcortical structures such as the reticular nuclei of the thalamus, which is thought to be involved in the genesis of delta and alpha oscillations of the EEG (Steriade, 2001). Despite these limitations, neural-mass models are useful in helping to understand some macroscopic properties of M/EEG signals, such as non-linearities (Stam *et al.*, 1999) and coupling characteristics (Wendling *et al.*, 2000). They can also be used to reconstruct *a posteriori* the scenario of inhibition/excitation balance during epileptic seizures (Wendling *et al.*, 2002). Moreover, fitting simple models to actual M/EEG data, as described above, allows one to determine empirically likely ranges for some important physiological parameters (Valdes *et al.*, 1999).

In David *et al.* (2004), we investigated the sensitivity of measures of regional interdependencies in M/EEG data, illustrating an important practical use of neural-mass models. It is known that some interactions among cortical areas are reflected in M/EEG signals. In the literature, numerous analytic tools are used to reveal these statistical dependencies. These methods include cross-correlation, coherence (Clifford Carter, 1987), mutual information (Roulston, 1999), non-linear correlation (Pijn *et al.*, 1992), non-linear interdependencies or generalized synchronization (Arnhold *et al.*, 1999), neural complexity (Tononi *et al.*, 1994), synchronization likelihood (Stam and van Dijk, 2002), phase synchronization (Tass *et al.*, 1998; Lachaux *et al.*, 1999), etc. These interdependencies are established in a way that allows one to make inferences about the nature of the coupling. However, it is not clear which aspects of neuronal interactions are critical for causing the frequency-specific linear and non-linear dependencies observed. Using the model described in this chapter, we have estimated how synaptic activity and neuronal interactions are expressed in M/EEG data and establish the construct validity of various indices of non-linear coupling. This particular topic has not been addressed in this chapter because we have focused more on emergent behaviour and mechanisms.

Another important application of neural-mass models is the study of event-related dynamics, which has been the focus of this chapter. We have shown that it is possible to construct hierarchical models for M/EEG signals. To that end, we have assumed an architecture for cortical regions and their connections. In particular, we have used the Jansen model (Jansen and Rit, 1995) for each source, and a simplified version of the connection rules of (Felleman and Van Essen, 1991) to couple these sources. We have shown that neural-mass models (Nunez, 1974; Jansen and Rit, 1995; Lopes da Silva *et al.*, 1997; Stam *et al.*, 1999; Valdes *et al.*, 1999; Robinson *et al.*, 2001; Suffczynski *et al.*, 2001; Rennie *et al.*, 2002; Wendling *et al.*, 2002; David and Friston, 2003) can reproduce a large variety of M/EEG signal characteristics: ongoing (oscillatory) activity, event-related activity (ERP/ERF, ERS/ERD). We have tried to highlight the relationships that exist between ERP/ERF and M/EEG oscillations on the one hand, evoked and induced responses on the other hand. One utility of neural-mass models is their ability to pinpoint specific neuronal mechanisms underlying normal or pathological activity. Effort is needed to incorporate them, more systematically, in M/EEG analyses to enable enquiry into mechanistic questions about macroscopic neuronal processes.

In the next chapter, we look at the inversion of dynamic models, using relatively simple models for fMRI. Later we will apply the same inversion to the models described in this chapter. The resulting approach to M/EEG data (David *et al.*, 2006a; Kiebel *et al.*, 2006) means we can frame our questions in a mechanistic and biologically grounded way, using neural-mass models.

REFERENCES

Abeles M (1991) *Corticonics: neural circuits of the cerebral cortex.* Cambridge University Press, Cambridge

Albright TD, Stoner GR (2002) Contextual influences on visual processing. *Annu Rev Neurosci* **25**: 339–79

Allman J, Miezin F, McGuinness E (1985) Stimulus specific responses from beyond the classical receptive field: neurophysiological mechanisms for local-global comparisons in visual neurons. *Annu Rev Neurosci* **8**: 407–30

Almeida R, Stetter M (2002) Modelling the link between functional imaging and neuronal activity: synaptic metabolic demand and spike rates. *NeuroImage* **17**: 1065–79

Angelucci A, Levitt JB, Lund JS (2002) Anatomical origins of the classical receptive field and modulatory surround field of single neurons in macaque visual cortical area V1. *Prog Brain Res* **136**: 373–88

Arieli A, Sterkin A, Grinvald A *et al.* (1996) Dynamics of ongoing activity: explanation of the large variability in evoked cortical responses. *Science* **273**: 1868–71

Arnhold J, Grassberger P, Lehnertz K *et al.* (1999) A robust method for detecting interdependences: application to intracranially recorded EEG. *Physica D* **134**: 419–30

Aubert A, Costalat R (2002) A model of the coupling between brain electrical activity, metabolism, and hemodynamics: application to the interpretation of functional neuroimaging. *NeuroImage* **17**: 1162–81

Azouz R, Gray CM (1999) Cellular mechanisms contributing to response variability of cortical neurons in vivo. *J Neurosci* **19**: 2209–23

Baillet S, Mosher JC, Leahy RM (2001) Electromagnetic brain mapping. *IEEE Signal Process Mag* 14–30

Basar E (1980) *EEG-brain dynamics: relation between EEG and brain evoked potentials.* Elsevier, New York

Breakspear M (2002) Nonlinear phase desynchronization in human electroencephalographic data. *Hum Brain Mapp* **15**: 175–98

Breakspear M, Terry JR (2002a) Detection and description of non-linear interdependence in normal multichannel human EEG data. *Clin Neurophysiol* **113**: 735–53

Breakspear M, Terry JR (2002b) Nonlinear interdependence in neural systems: motivation, theory, and relevance. *Int J Neurosci* **112**: 1263–84

Breakspear M, Terry JR, Friston KJ (2003) Modulation of excitatory synaptic coupling facilitates synchronization and complex dynamics in a biophysical model of neuronal dynamics. *Network* **14**: 703–32

Bullier J, Hupe J, James AC *et al.* (2001) The role of feedback connections in shaping the responses of visual cortical neurons. *Prog Brain Res* **134**: 193–204

Chance FS, Abbott LF, Reyes AD (2002) Gain modulation from background synaptic input. *Neuron* **35**: 773–82

Chawla D, Friston KJ, Lumer ED (2001) Zero-lag synchronous dynamics in triplets of interconnected cortical areas. *Neural Netw* **14**: 727–35

Clifford Carter G (1987) Coherence and time delay estimation. *Proc IEEE* **75**: 236–55

Coles MGH, Rugg MD (1995) Event-related brain potentials: an introduction. In *Electrophysiology of mind*, Rugg MD, Coles MGH (eds). Oxford University Press, Oxford, pp 1–26

Crick F, Koch C (1998) Constraints on cortical and thalamic projections: the no-strong-loops hypothesis. *Nature* **391**: 245–50

David O, Cosmelli D, Friston KJ (2004) Evaluation of different measures of functional connectivity using a neural-mass model. *NeuroImage* **21**: 659–73

David O, Friston KJ (2003) A neural-mass model for M/EEG: coupling and neuronal dynamics. *NeuroImage* **20**: 1743–55

David O, Garnero L, Cosmelli D *et al.* (2002) Estimation of neural dynamics from M/EEG cortical current density maps: application to the reconstruction of large-scale cortical synchrony. *IEEE Trans Biomed Eng* **49**: 975–87

David O, Harrison L, Friston KJ (2005) Modelling event-related responses in the brain. *NeuroImage* **25**: 756–70

David O, Kiebel SJ, Harrison LM *et al.* (2006a) Dynamic causal modelling of evoked responses in EEG and MEG. *NeuroImage* **30**: 1255–72

David O, Kilner JM, Friston KJ (2006b) Mechanisms of evoked and induced responses in M/EEG. *NeuroImage* **15**: 1580–91

DeFelipe J, Alonso-Nanclares L, Arellano JI (2002) Microstructure of the neocortex: comparative aspects. *J Neurocytol* **31**: 299–316

Edelman GM (1993) Neural Darwinism: selection and re-entrant signaling in higher brain function. *Neuron* **10**: 115–25

Engel AK, Fries P, Singer W (2001) Dynamic predictions: oscillations and synchrony in top-down processing. *Nat Rev Neurosci* **2**: 704–16

Felleman DJ, Van Essen DC (1991) Distributed hierarchical processing in the primate cerebral cortex. *Cereb Cortex* **1**: 1–47

Freeman WJ (1978) Models of the dynamics of neural populations. *Electroencephalogr Clin Neurophysiol* Suppl 34: 9–18

Freeman WJ (1979) Nonlinear gain mediating cortical stimulus-response relations. *Biol Cybernet* **33**: 237–47

Freeman WJ (1987) Simulation of chaotic EEG patterns with a dynamic model of the olfactory system. *Biol Cybernet* **56**: 139–50

Friston K (2002) Functional integration and inference in the brain. *Prog Neurobiol* **68**: 113–43

Friston KJ (1997a) Another neural code? *NeuroImage* **5**: 213–20

Friston KJ (1997b) Transients, metastability, and neuronal dynamics. *NeuroImage* **5**: 164–71

Friston KJ (2000a) The labile brain. I. Neuronal transients and nonlinear coupling. *Philos Trans R Soc Lond B Biol Sci* **355**: 215–36

Friston KJ (2000b) The labile brain. III. Transients and spatio-temporal receptive fields. *Philos Trans R Soc Lond B Biol Sci* **355**: 253–65

Friston KJ (2001) Brain function, nonlinear coupling, and neuronal transients. *Neuroscientist* **7**: 406–18

Friston KJ, Harrison L, Penny W (2003) Dynamic causal modelling. *NeuroImage* **19**: 1273–302

Friston KJ, Penny W, Phillips C *et al.* (2002) Classical and Bayesian inference in neuroimaging: theory. *NeuroImage* **16**: 465–83

Fuentemilla L, Marco-Pallares J, Grau C (2005) Modulation of spectral power and of phase resetting of EEG contributes differentially to the generation of auditory event-related potentials. *NeuroImage* **30**: 909–16

Haskell E, Nykamp DQ, Tranchina D (2001) Population density methods for large-scale modelling of neuronal networks with realistic synaptic kinetics: cutting the dimension down to size. *Network* **12**: 141–74

Horwitz B, Tagamets MA (1999) Predicting human functional maps with neural net modeling. *Hum Brain Mapp* **8**: 137–42

Jansen BH, Agarwal G, Hegde A *et al.* (2003) Phase synchronization of the ongoing EEG and auditory EP generation. *Clin Neurophysiol* **114**: 79–85

Jansen BH, Rit VG (1995) Electroencephalogram and visual evoked potential generation in a mathematical model of coupled cortical columns. *Biol Cybernet* **73**: 357–66

Jirsa VK, Haken H (1997) A derivation of a macroscopic field theory of the brain from the quasi-microscopic neural dynamics. *Physica D* **99**: 503–26

Jirsa VK, Kelso JA (2000) Spatiotemporal pattern formation in neural systems with heterogeneous connection topologies. *Phys Rev E Stat Phys Plasmas Fluids Relat Interdiscip Topics* **62**: 8462–65

Jones EG (2000) Microcolumns in the cerebral cortex. *Proc Natl Acad Sci USA* **97**: 5019–21

Kaneko K, Tsuda I (2003) Chaotic itinerancy. *Chaos* **13**: 926–36

Kiebel SJ, David O, Friston KJ (2006) Dynamic causal modelling of evoked responses in M/EEG with lead field parameterization. *NeuroImage* **30**: 1273–84

Kiebel SJ, Tallon-Baudry C, Friston KJ (2005) Parametric analysis of oscillatory activity as measured with EEG/MEG. *Hum Brain Mapp* **26**: 170–77

Klimesch W, Schack B, Schabus M *et al.* (2004) Phase-locked alpha and theta oscillations generate the P1-N1 complex and are related to memory performance. *Brain Res Cogn Brain Res* **19**: 302–16

Kloeden PE, Platen E (1999) *Numerical solution of stochastic differential equations*. Springer Verlag, Berlin

Kolev V, Yordanova J (1997) Analysis of phase-locking is informative for studying event-related EEG activity. *Biol Cybernet* **76**: 229–35

Lachaux J-P, Rodriguez E, Martinerie J *et al.* (1999) Measuring phase synchrony in brain signals. *Hum Brain Mapp* **8**: 194–208

Le Van Quyen M, Foucher J, Lachaux J *et al.* (2001) Comparison of Hilbert transform and wavelet methods for the analysis of neuronal synchrony. *J Neurosci Methods* **111**: 83–98

Linsker R (1990) Perceptual neural organization: some approaches based on network models and information theory. *Annu Rev Neurosci* **13**: 257–81

Lopes da Silva FH, Hoeks A, Smits H *et al.* (1974) Model of brain rhythmic activity. The alpha-rhythm of the thalamus. *Kybernetik* **15**: 27–37

Lopes da Silva FH, Pijn JP, Velis D *et al.* (1997) Alpha rhythms: noise, dynamics and models. *Int J Psychophysiol* **26**: 237–49

Lumer ED, Edelman GM, Tononi G (1997) Neural dynamics in a model of the thalamocortical system. I. Layers, loops and the emergence of fast synchronous rhythms. *Cereb Cortex* **7**: 207–27

Makeig S, Westerfield M, Jung TP *et al.* (2002) Dynamic brain sources of visual evoked responses. *Science* **295**: 690–94

Maunsell JH, Van E (1983) The connections of the middle temporal visual area (MT) and their relationship to a cortical hierarchy in the macaque monkey. *J Neurosci* **3**: 2563–86

McAdams CJ, Maunsell JH (1999) Effects of attention on orientation-tuning functions of single neurons in macaque cortical area V4. *J Neurosci* **19**: 431–41

Miller KD (2003) Understanding layer 4 of the cortical circuit: a model based on cat V1. *Cereb Cortex* **13**: 73–82

Murphy PC, Duckett SG, Sillito AM (1999) Feedback connections to the lateral geniculate nucleus and cortical response properties. *Science* **286**: 1552–54

Nunez PL (1974) The brain wave equation: a model for the EEG. *Math Biosci* **21**: 279–97

Nunez PL, Srinivasan R (2005) *Electric fields of the brain*, 2nd edn. Oxford University Press, New York

Penny WD, Kiebel SJ, Kilner JM *et al.* (2002) Event-related brain dynamics. *Trends Neurosci* **25**: 387–89

Pfurtscheller G, Lopes da Silva FH (1999) Event-related M/EEG synchronization and desynchronization: basic principles. *Clin Neurophysiol* **110**: 1842–57

Pijn JP, Velis DN, Lopes da Silva FH (1992) Measurement of inter-hemispheric time differences in generalised spike-and-wave. *Electroencephalogr Clin Neurophysiol* **83**: 169–71

Rennie CJ, Robinson PA, Wright JJ (2002) Unified neurophysical model of EEG spectra and evoked potentials. *Biol Cybernet* **86**: 457–71

Robinson PA, Rennie CJ, Rowe DL (2002) Dynamics of large-scale brain activity in normal arousal states and epileptic seizures. *Phys Rev E Stat Nonlin Soft Matter Phys* **65**: 041924

Robinson PA, Rennie CJ, Wright JJ *et al.* (2001) Prediction of electroencephalographic spectra from neurophysiology. *Phys Rev E* **63**: 021903

Rodriguez E, George N, Lachaux JP *et al.* (1999) Perception's shadow: long-distance synchronization of human brain activity. *Nature* **397**: 430–33

Roulston MS (1999) Estimating the errors on measured entropy and mutual information. *Physica D* **125**: 285–94

Salinas E, Thier P (2000) Gain modulation: a major computational principle of the central nervous system. *Neuron* **27**: 15–21

Shah AS, Bressler SL, Knuth KH *et al.* (2004) Neural dynamics and the fundamental mechanisms of event-related brain potentials. *Cereb Cortex* **14**: 476–83

Sherman SM, Guillery RW (1998) On the actions that one nerve cell can have on another: distinguishing 'drivers' from 'modulators'. *Proc Natl Acad Sci USA* **95**: 7121–26

Singer W, Gray CM (1995) Visual feature integration and the temporal correlation hypothesis. *Annu Rev Neurosci* **18**: 555–86

Smith MR, Nelson AB, Du LS (2002) Regulation of firing response gain by calcium-dependent mechanisms in vestibular nucleus neurons. *J Neurophysiol* **87**: 2031–42

Stam CJ, Pijn JP, Suffczynski P *et al.* (1999) Dynamics of the human alpha rhythm: evidence for non-linearity? *Clin Neurophysiol* **110**: 1801–13

Stam CJ, van Dijk BW (2002) Synchronization likelihood: an unbiased measure of generalized synchronization in multivariate data sets. *Physica D* **163**: 236–51

Steriade M (2001) Impact of network activities on neuronal properties in corticothalamic systems. *J Neurophysiol* **86**: 1–39

Suffczynski P, Kalitzin S, Pfurtscheller G *et al.* (2001) Computational model of thalamo-cortical networks: dynamical control of alpha rhythms in relation to focal attention. *Int J Psychophysiol* **43**: 25–40

Tallon-Baudry C, Bertrand O (1999) Oscillatory gamma activity in humans and its role in object representation. *Trends Cogn Sci* **3**: 151–62

Tallon-Baudry C, Bertrand O, Delpuech C *et al.* (1996) Stimulus specificity of phase-locked and non-phase-locked 40 Hz visual responses in human. *J Neurosci* **16**: 4240–49

Tass P, Wienbruch C, Weule J *et al.* (1998) Detection of *n:m* phase locking from noisy data: application to magnetoencephalography. *Phys Rev Lett* **81**: 3291–94

Tass PA (2003) Stochastic phase resetting of stimulus-locked responses of two coupled oscillators: transient response clustering, synchronization, and desynchronization. *Chaos* **13**: 364–76

Thomson AM, Deuchars J (1997) Synaptic interactions in neocortical local circuits: dual intracellular recordings in vitro. *Cereb Cortex* **7**: 510–22

Tononi G, Sporns O, Edelman GM (1994) A measure for brain complexity: relating functional segregation and integration in the nervous system. *Proc Natl Acad Sci USA* **91**: 5033–37

Treue S, Martinez Trujillo JC (1999) Feature-based attention influences motion processing gain in macaque visual cortex. *Nature* **399**: 575–79

Truccolo WA, Ding M, Knuth KH *et al.* (2002) Trial-to-trial variability of cortical evoked responses: implications for the analysis of functional connectivity. *Clin Neurophysiol* **113**: 206–26

Tsuda I (2001) Toward an interpretation of dynamic neural activity in terms of chaotic dynamical systems. *Behav Brain Sci* **24**: 793–810

Turrigiano GG, Nelson SB (2004) Homeostatic plasticity in the developing nervous system. *Nat Rev Neurosci* **5**: 97–107

Valdes PA, Jimenez JC, Riera J *et al.* (1999) Nonlinear EEG analysis based on a neural-mass model. *Biol Cybernet* **81**: 415–24

Van Rotterdam A, Lopes da Silva FH, van den EJ *et al.* (1982) A model of the spatial-temporal characteristics of the alpha rhythm. *Bull Math Biol* **44**: 283–305

Varela F (1995) Resonant cell assemblies: a new approach to cognitive functions and neuronal synchrony. *Biol Res* **28**: 81–95

Varela F, Lachaux J-P, Rodriguez E *et al.* (2001) The brainweb: phase synchronization and large-scale integration. *Nat Rev Neurosci* **2**: 229–39

Wendling F, Bartolomei F, Bellanger JJ *et al.* (2002) Epileptic fast activity can be explained by a model of impaired GABAergic dendritic inhibition. *Eur J Neurosci* **15**: 1499–508

Wendling F, Bellanger JJ, Bartolomei F *et al.* (2000) Relevance of nonlinear lumped-parameter models in the analysis of depth-EEG epileptic signals. *Biol Cybernet* **83**: 367–78

Whittington MA, Traub RD, Kopell N *et al.* (2000) Inhibition-based rhythms: experimental and mathematical observations on network dynamics. *Int J Psychophysiol* **38**: 315–36

Wilson HR, Cowan JD (1972) Excitatory and inhibitory interactions in localized populations of model neurons. *Biophys J* **12**: 1–24

CHAPTER

34

Bayesian inversion of dynamic models

K. Friston and W. Penny

INTRODUCTION

In this chapter, we look at the inversion of dynamic models. We use, as an example, the haemodynamic model presented in Chapter 27. The inversion scheme is an extension of the Bayesian treatments reviewed in Part 4. Inverting haemodynamic models of neuronal responses is clearly central to functional magnetic resonance imaging (fMRI) and forms the basis for dynamic causal models for fMRI based on neuronal networks (see Chapter 41). However, the principles of the inversion described in this chapter can be applied to any analytic, deterministic dynamic system and we will use exactly the same scheme for dynamic models of magnetoencephalography/electroencephalography (M/EEG) later. In this chapter, we focus on the inversion of a single model to find the conditional density of the model's parameters that can then be used for inference, on parameter space. In the next chapter, we will focus on inference about models themselves (i.e. inference on model space), in terms of Bayesian model comparison, selection and averaging.

This chapter is about estimating the conditional or posterior distribution of the parameters of deterministic dynamical systems. The scheme conforms to an EM search for the maximum of the conditional or posterior density. The inclusion of priors in the estimation procedure ensures robust and rapid convergence and the resulting conditional densities enable Bayesian inference about the model parameters. The approach is demonstrated using an input-state-output model of the haemodynamic coupling between experimentally designed causes or factors in fMRI studies and the ensuing blood oxygenation-level-dependent (BOLD) response (see Chapter 27). This example represents a generalization of current fMRI analysis models that accommodates non-linearities and in which the parameters have an explicit physical interpretation.

We focus on the identification of deterministic non-linear dynamical models. Deterministic here refers to models where the dynamics are completely determined by the state of the system. Random fluctuations or stochastic effects enter only at the point that the system's outputs or responses are observed.[1] By considering a voxel as an *input-state-output* system one can model the effects of an input (i.e. stimulus function) on some state variables (e.g. flow, volume, deoxyhaemoglobin content etc.) and the output (i.e. BOLD response) engendered by the changing state of the voxel. The aim is to identify the posterior or conditional distribution of the parameters, given the data. Knowing the posterior distribution allows one to characterize an observed system in terms of the parameters that maximize their posterior probability (i.e. those parameters that are most likely given the data) or, indeed, make inferences about whether the parameters are bigger or smaller than some specified value.

By demonstrating the inversion of haemodynamic models, we establish the key role of biophysical models of evoked brain responses in making Bayesian inferences about experimentally induced effects. Including parameters that couple experimentally changing stimulus or task conditions to the system's states enables this inference. The posterior or conditional distribution of these parameters can then be used to make inferences about the efficacy of experimental inputs in eliciting measured responses. Because the parameters we want to make an inference about have an explicit physical interpretation, in the context of the haemodynamic model used, the face validity of the ensuing inference is grounded in physiology. Furthermore, because the experimental effects are parameterized in terms of processes that have natural

[1] There is another important class of models where stochastic processes enter at the level of the state variables themselves (i.e. deterministic noise). These are referred to as stochastic dynamical models.

Statistical Parametric Mapping, by Karl Friston et al.
ISBN–13: 978-0-12-372560-8 ISBN–10: 0-12-372560-7

biological constraints, these constraints can be used as priors in a Bayesian scheme.

Part 4 focused on *empirical* Bayesian approaches in which the priors were derived from the data being analysed. In this section, we use a *fully* Bayesian approach, where the priors are assumed to be known and apply it to the haemodynamic model described in Friston *et al.* (2000). In Friston *et al.* (2000), we presented a haemodynamic model that embedded the Balloon-Windkessel model (Buxton *et al.*, 1998; Mandeville *et al.*, 1999) of flow to BOLD coupling to give a complete dynamical model of how neuronally mediated signals cause a BOLD response. This previous work used a single input-single output (SISO) system by considering only one input. Here we generalize the approach to multiple input-single output (MISO) systems. This allows for a response to be caused by multiple experimental effects and the estimation of causal efficacy for any number of explanatory variables (i.e. stimulus functions). Later (Chapter 41), we will generalize to multiple input-multiple output systems (MIMO) such that interactions among brain regions at a neuronal level can be addressed.

An important aspect of the model is that it can be reduced, exactly, to the model used in classical statistical parametric mapping (SPM)-like analyses, where one uses stimulus functions, convolved with a canonical haemodynamic response function, as explanatory variables in a general linear model. This classical analysis is a special case that obtains when the model parameters of interest (the efficacy of a stimulus) are treated as fixed effects with flat priors and the remaining biophysical parameters enter as known canonical values with infinitely small prior variance (i.e. high precision). In this sense, the current approach can be viewed as a Bayesian generalization of conventional convolution models of fMRI. The advantages of this generalization rest upon the use of a non-linear observation model and its Bayesian inversion. The fundamental advantage, of a non-linear MISO model over linear models, is that only the parameters linking the various inputs to haemodynamics are input-specific. The remaining parameters, pertaining to the haemodynamics *per se*, are the same for each voxel. In conventional analyses the haemodynamic response function, for each input, is estimated in a linearly separable fashion (usually in terms of a small set of temporal basis functions), despite the fact that the form of the impulse response function in relation to each input is the same. In other words, a non-linear model properly accommodates the fact that many of the parameters shaping input-specific haemodynamic responses are shared by all inputs. For example, the components of a compound trial (e.g. cue and target stimuli) might not interact at a neuronal level but may show subadditive effects in the measured response, due to non-linear haemodynamic saturation. In contradistinction to conventional linear analyses, the analysis proposed in this section could, in principle, disambiguate between interactions at the neuronal and haemodynamic levels. The second advantage is that Bayesian inferences about input-specific parameters can be framed in terms of whether the efficacy for a particular cause exceeded some specified threshold or, indeed the probability that it was less than some threshold (i.e. infer that a voxel did not respond). The latter is precluded in classical inference. These advantages should be weighed against the difficulties of establishing a valid model and the computational expense of identification.

Overview

This chapter is divided into four sections. In the first, we reprise briefly the haemodynamic model and motivate the four differential equations that it comprises. We will touch on the Volterra formulation of non-linear systems to show the output can always be represented as a non-linear function of the input and the model parameters (see also Chapter 39). This non-linear function is used as the basis of the observation model that is subject to Bayesian identification. This identification requires priors which, here, come from the distribution, over voxels, of parameters estimated in Friston *et al.* (2000). The second section describes these priors and how they were determined. Having specified the form of the non-linear observation model and the prior densities on the model's parameters, the third section describes the estimation of their posterior densities. The ensuing scheme can be regarded as a Gauss-Newton search for the maximum posterior probability (as opposed to the maximum likelihood as in conventional applications) that embeds the EM scheme in Appendix 3. This description concludes with a note on integration, required to evaluate the local gradients of the objective function. This effectively generalizes the EM algorithm for linear systems so that it can be applied to non-linear models.

Finally, we demonstrate the approach using empirical data. First, we revisit the same data used to construct the priors using a single input. We then apply the technique to the fMRI study of visual attention used in other chapters, to make inferences about the relative efficacy of multiple experimental effects in eliciting a BOLD response.

A HAEMODYNAMIC MODEL

The haemodynamic model considered here and in Chapter 27 was presented in detail in Friston *et al.* (2000). Although relatively simple, it is predicated on a substantial amount of careful theoretical work and empirical validation (e.g. Buxton *et al.*, 1998; Mayhew *et al.*, 1998; Hoge *et al.*, 1999; Mandeville *et al.*, 1999). The model is a SISO system with a stimulus function as input (that is supposed to elicit a neuronally mediated flow-inducing signal) and BOLD response as output. The model has six parameters and four state variables each with its corresponding differential equation. The differential or state equations express how each state variable changes over time as a function of the others. These state equations and the output non-linearly (a static non-linear function of the state variables) specify the form of the forward or generative model. The parameters determine any specific realization of the model. In what follows we review the state equations, the output non-linearity, extension to a MISO system and the Volterra representation.

The state equations

Assuming that the dynamical system linking synaptic activity and rCBF is linear (Miller *et al.*, 2000) we start with:

$$\dot{f} = s \qquad \textbf{34.1}$$

where $f(t)$ is inflow and $s(t)$ is some flow inducing signal. The signal is assumed to subsume many neurogenic and diffusive signal subcomponents and is generated by neuronal responses to the input (the stimulus function) $u(t)$:

$$\dot{s} = \varepsilon u(t) - \kappa_s s - \kappa_f(f_{in} - 1) \qquad \textbf{34.2}$$

ε, κ_s and κ_f are parameters that represent the efficacy with which input causes an increase in signal, the rate-constant for signal decay or elimination and the rate-constant for autoregulatory feedback from blood flow. The existence of this feedback term can be inferred from post-stimulus undershoots in rCBF and the well-characterized vasomotor signal (V-signal) in optical imaging (Mayhew *et al.*, 1998). Inflow determines the rate of change of volume through:

$$\begin{aligned} \tau\dot{v} &= f - f_0(v) \\ f_0(v) &= v^{1/\alpha} \end{aligned} \qquad \textbf{34.3}$$

This says that normalized venous volume changes reflect the difference between inflow $f(t)$ and outflow $f_0(t)$ from the venous compartment with a time constant (transit-time) τ. Outflow is a function of volume that models the balloon-like capacity of the venous compartment to expel blood at a greater rate when distended (Buxton *et al.*, 1998). It can be modelled with a single parameter (Grubb *et al.*, 1974) α based on the Windkessel model (Mandeville *et al.*, 1999). The change in normalized total deoxyhaemoglobin voxel content $q(t)$ reflects the delivery of deoxyhaemoglobin into the venous compartment minus that expelled (outflow times concentration):

$$\begin{aligned} \tau\dot{q} &= f\frac{E(f)}{E_0} - \frac{f_0 q}{v} \\ E(f) &= 1 - (1 - E_0)^{1/f} \end{aligned} \qquad \textbf{34.4}$$

where $E(f)$ is the fraction of oxygen extracted from inflowing blood. This is assumed to depend on oxygen delivery and is consequently flow-dependent. This concludes the state equations, where there are six unknown parameters, namely efficacy ε, signal decay κ_s, autoregulation κ_f, transit time τ, Grubb's exponent α and resting net oxygen extraction by the capillary bed E_0.

The output non-linearity

The BOLD signal $y(t) = g(v, q)$ is taken to be a static non-linear function of volume and deoxyhaemoglobin content:

$$\begin{aligned} y(t) &= g(v, q) = V_0\,(k_1(1 - q) + k_2(1 - q/v) + k_3(1 - v)) \\ k_1 &= 7E_0 \\ k_2 &= 2 \\ k_3 &= 2E_0 - 0.2 \end{aligned} \qquad \textbf{34.5}$$

where V_0 is resting blood volume fraction. This signal comprises a volume-weighted sum of extra- and intravascular signals that are functions of volume and deoxyhaemoglobin content. A critical term in the output equation is the concentration term $k_2(1 - q/v)$, which accounts for most of the non-linear behaviour of the haemodynamic model. The architecture of this model is summarized in Figure 34.1.

Extension to a MISO

The extension to a multiple input system is trivial and involves extending Eqn. **34.2** to cover n inputs:

$$\dot{s} = \varepsilon_1 u(t)_1 + \cdots + \varepsilon_n u(t)_n - \kappa_s s - \kappa_f(f - 1) \qquad \textbf{34.6}$$

The model now has $5 + n$ parameters: five biophysical parameters κ_s, κ_f, τ, α and E_0 and n efficacies $\varepsilon_1, \ldots \varepsilon_n$. Although all these parameters have to be estimated, we are only interested in making inferences about the efficacies. Note that the biophysical parameters are the same for all inputs.

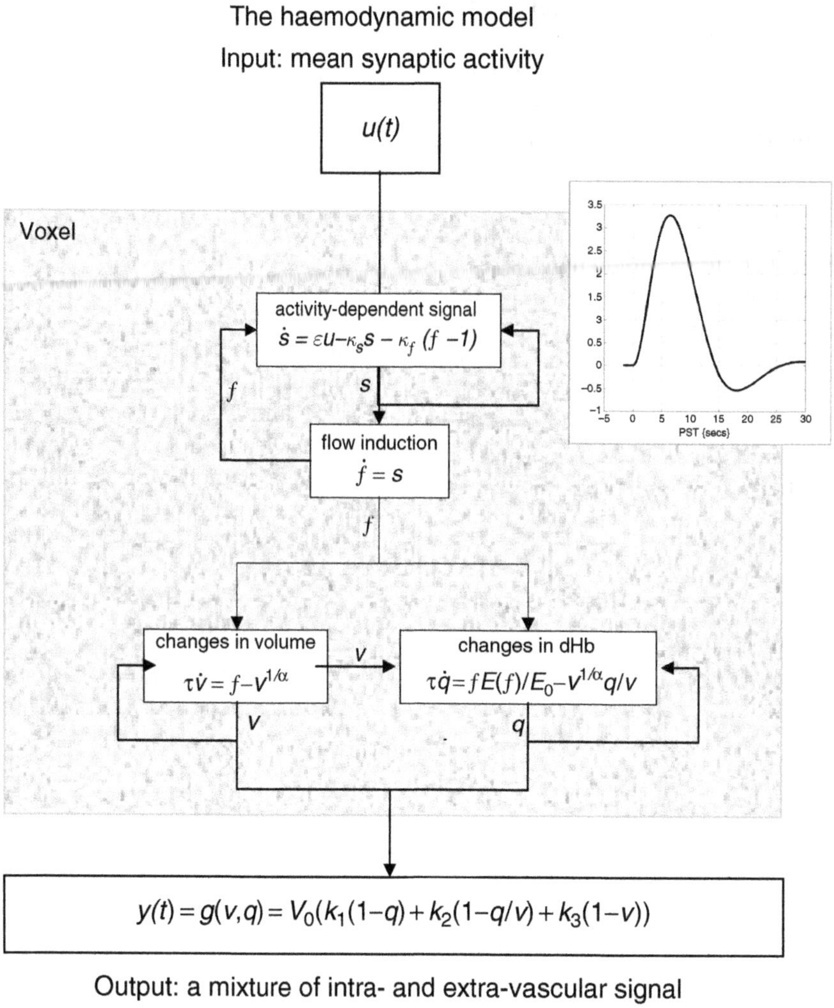

FIGURE 34.1 Schematic illustrating the architecture of the haemodynamic model. This is a fully non-linear single-input $u(t)$, single-output $y(t)$ state model with four state variables s, f, v and q. The form and motivation for the changes in each state variable, as functions of the others, are described in the main text.

The Volterra formulation

In our haemodynamic model, the state variables are $x = x_1, \cdots, x_4 = s, f, v, q$ and the parameters are $\theta = \kappa_s, \kappa_f, \tau, \alpha, E_0, \varepsilon_1, \ldots, \varepsilon_n$. The state equations and output non-linearity specify a multiple input-single output (MISO) model:

$$
\begin{aligned}
\dot{x} &= f(x, u) \\
y &= g(x) \\
\dot{x}_1 &= f_1(x, u) = \varepsilon_1 u(t)_1 + \cdots + \varepsilon_n u(t)_n - \kappa_s x_1 - \kappa_f (x_2 - 1) \\
\dot{x}_2 &= f_2(x, u) = x_1 \\
\dot{x}_3 &= f_3(x, u) = \frac{1}{\tau}(x_2 - x_3^{1/\alpha}) \\
\dot{x}_4 &= f_4(x, u) = \frac{x_2 1 - (1 - E_0)^{1/x_2}}{\tau E_0} - \frac{x_3^{1/\alpha} x_4}{\tau x_3} \\
y(t) &= g(x) = V_0(k_1(1 - x_4) + k_2(1 - x_4/x_3) + k_3(1 - x_3))
\end{aligned}
\tag{34.7}
$$

This is the state-space representation. The alternative Volterra formulation represents the output as a non-linear convolution of the input, critically without reference to the state variables (see Bendat, 1990). This series can be considered a non-linear convolution that obtains from a functional Taylor expansion of the response or outputs. The reason this is a functional expansion is that the inputs are a function of time. This means the coefficients of the expansion are also functions of time. These are the system's generalized kernels; for a single input this expansion can be expressed as:

$$y(t) = h(\theta, u)$$
$$= \kappa_0 + \sum_{i=1}^{\infty} \int_0^{\infty} \ldots \int_0^{\infty} \kappa_i(\sigma_1, \ldots \sigma_i) u(t-\sigma_1) \ldots u(t-\sigma_i) d\sigma_1 \ldots d\sigma_i \quad \textbf{34.8}$$

$$\kappa_i(\sigma_1, \ldots \sigma_i) = \frac{\partial^i y(t)}{\partial u(t-\sigma_1) \ldots u(t-\sigma_i)}$$

where $\kappa_i(\sigma_1, \ldots \sigma_i)$ is the i-th generalized convolution kernel (Fliess *et al.*, 1983). Eqn. **34.8** expresses the output as a function of the input and the parameters whose posterior distribution we require. In other words, the kernels represent a re-parameterization of the system. The kernels are a time-invariant characterization of the input-output behaviour of the system and can be thought of as generalized high-order convolution kernels that are applied to a stimulus function to produce the observed BOLD response. Integrating Eqn. **34.7** and applying the output non-linearity to the state variables is the same as convolving the inputs with the kernels as in Eqn. **34.8**. Both give the system's response in terms of the output. In what follows, the response is evaluated by integrating Eqn. **34.7**. This means the kernels are not required. However, the Volterra formulation is introduced for several reasons. First, it demonstrates that the output is a non-linear function of the inputs $y(t) = h(\theta, u)$. This is critical for the generality of the estimation scheme described here. Secondly, it provides an important connection with conventional analyses using general linear convolution models (see below). Finally, we can use the kernels to characterize evoked responses.

PRIORS

Bayesian inversion requires informative priors on the parameters. Under Gaussian assumptions, these prior densities can be specified in terms of their expectation and covariance. These moments are taken here to be the sample mean and covariance, over voxels, of the parameter estimates reported in Friston *et al.* (2000). Normally, priors play a critical role in inference; indeed the traditional criticism levelled at Bayesian inference reduces to reservations about the validity of the priors employed. However, in the application considered here, this criticism can be discounted. This is because the priors, on those parameters about which inferences are made, are relatively flat. Only the five biophysical parameters have informative priors.

In Friston *et al.* (2000), the parameters were identified as those that minimized the sum of squared differences between the Volterra kernels implied by the parameters and those derived directly from the data. This derivation used ordinary least square estimators, exploiting the fact that Volterra formulation is linear in the unknowns, namely the kernel coefficients. The kernels can be thought of as a re-parameterization of the model that does not refer to the underlying state representation. In other words, for every set of parameters, there is a corresponding set of kernels (see Friston *et al.*, 2000 and below for the derivation of the kernels as a function of the parameters). The data and Volterra kernel estimation are described in detail in Friston *et al.* (1998). In brief, we obtained long fMRI time-series from a single subject at 2 tesla and a short TR of 1.7 s. After discarding initial scans (to allow for magnetic saturation effects) each time-series comprised 1200 volume images with 3 mm isotropic voxels. The subject listened to monosyllabic or bi-syllabic concrete nouns (i.e. 'dog', 'radio', 'mountain', 'gate') presented at five different rates (10, 15, 30, 60 and 90 words per minute) for epochs of 34 s, intercalated with periods of rest. The presentation rates were repeated according to a Latin Square design.

The distribution of the five biophysical parameters, over 128 voxels, was computed to give our prior expectation η_θ and covariance C_θ. Signal decay κ_s had a mean of about 0.65 per second giving a half-life $t_{1/2} = \ln 2/\kappa_s \approx 1$ s. Mean feedback rate κ_f was about 0.4 per second. Mean transit time τ was 0.98 s. Under steady state conditions Grubb's parameter α is about 0.38. The mean over voxels was 0.326. Mean resting oxygen extraction E_0 was about 34 per cent and the range observed conformed exactly to known values for resting oxygen extraction fraction (between 20 and 55 per cent). Figure 34.2 shows the covariances among the biophysical parameters along with the correlation matrix (left-hand insert). The correlations suggest a high correlation between transit time and the rate constants for signal elimination and autoregulation.

The priors for the efficacies are taken to be relatively flat with an expectation of zero and a variance of 16 per second. Here, the efficacies are assumed to be independent of the biophysical parameters with zero covariance. A variance of 16, or standard deviation of 4, corresponds to time constants in the range of 250 ms. In other words, inputs can elicit flow-inducing signal over a wide range of time constants from infinitely slowly to very fast (250 ms) with about the same probability. A 'strong' activation usually has an efficacy of about 0.5 per second. Notice that, from a dynamical perspective, a large response depends upon the speed of the response not the percentage change. Equipped with these priors we can

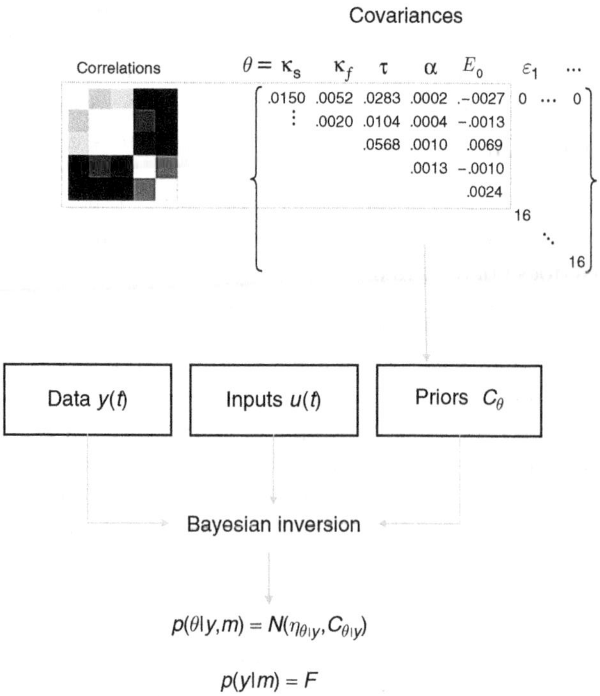

FIGURE 34.2 Prior covariances for the five biophysical parameters of the haemodynamic model in Figure 34.1. Left panel: correlation matrix showing the prior correlations among the biophysical parameters in image format (white = 1). Right panel: corresponding covariance matrix. These priors represent the sample covariances of the parameters estimated by minimizing the difference between the Volterra kernels implied by the parameters and those estimated empirically, using ordinary least squares as described in Friston *et al.* (2000).

now pursue a fully Bayesian approach to estimating the parameters given any data and experimental inputs.

SYSTEM IDENTIFICATION

This section describes Bayesian inference procedures for non-linear observation models, with additive noise, of the form:

$$y = h(\theta, u) + \varepsilon \qquad \textbf{34.9}$$

under Gaussian assumptions about the errors ε. These models can be adopted for any analytic dynamical system due to the existence of the equivalent Volterra series expansion above. Assuming the posterior density of the parameters is approximately Gaussian, the problem reduces to finding its first two moments, the conditional mean $\eta_{\theta|y}$ and covariance $C_{\theta|y}$.

The observation model can be made linear by expanding Eqn. **34.9** about a working estimate $\eta_{\theta|y}$ of the conditional mean:

$$\begin{aligned} h(\theta, u) &\approx h(\eta_{\theta|y}) + J(\theta - \eta_{\theta|y}) \\ J &= \frac{\partial h(\eta_{\theta|y})}{\partial \theta} \end{aligned} \qquad \textbf{34.10}$$

such that $y - h(\eta_{\theta|y}) \approx J(\theta - \eta_{\theta|y}) + \varepsilon$, where $\varepsilon \sim N(0, C_\varepsilon)$ is a Gaussian error term. The covariance of the errors is hyperparameterized in terms of a mixture of covariance components $C_\varepsilon = \sum \lambda_i Q_i$ where, usually, there is only one component that encodes serial correlations among the errors. This linear model can now be placed in the EM scheme described in Chapter 22 (Figure 22.4) and Appendix 3 to give an **E**-step that updates the conditional density of the parameters, $p(\theta|y, \lambda) = N(\eta_{\theta|y}, C_{\theta|y})$ and an **M**-step that updates the ML (maximum likelihood) estimate of the hyperparameters, λ.

Until convergence {

E-step

$$J = \frac{\partial h(\eta_{\theta|y})}{\partial \theta}$$

$$\bar{y} = \begin{bmatrix} y - h(\eta_{\theta|y}) \\ \eta_\theta - \eta_{\theta|y} \end{bmatrix}, \quad \bar{J} = \begin{bmatrix} J \\ I \end{bmatrix}, \quad \bar{C}_\varepsilon = \begin{bmatrix} \sum \lambda_i Q_i & 0 \\ 0 & C_\theta \end{bmatrix}$$

$$C_{\theta|y} = (\bar{J}^T \bar{C}_\varepsilon^{-1} \bar{J})^{-1} \qquad \textbf{34.11}$$

$$\eta_{\theta|y} \leftarrow \eta_{\theta|y} + C_{\theta|y} \bar{J}^T \bar{C}_\varepsilon^{-1} \bar{y}$$

M-step

$$P = \bar{C}_\varepsilon^{-1} - \bar{C}_\varepsilon^{-1} \bar{J} C_{\theta|y} \bar{J}^T \bar{C}_\varepsilon^{-1}$$

$$\frac{\partial F}{\partial \lambda_i} = -\frac{1}{2} tr\{PQ_i\} + \frac{1}{2} \bar{y}^T P^T Q_i P \bar{y}$$

$$\left\langle \frac{\partial^2 F}{\partial \lambda_{ij}^2} \right\rangle = -\frac{1}{2} tr\{PQ_i PQ_j\}$$

$$\lambda \leftarrow \lambda - \left\langle \frac{\partial^2 F}{\partial \lambda^2} \right\rangle^{-1} \frac{\partial F}{\partial \lambda}$$

}

This EM scheme is effectively a *Gauss-Newton* search for the posterior mode or maximum *a poseriori* (MAP) estimate of the parameters. The relationship between the **E**-step and a conventional Gauss-Newton ascent can be seen easily in terms of the derivatives of their respective objective functions. For conventional Gauss-Newton this function is the *log-likelihood*:

$$\begin{aligned} \ell &= \ln p(y|\theta) \\ &= -\frac{1}{2}(y - h(\theta))^T C_\varepsilon^{-1} (y - h(\theta)) + \ldots \end{aligned}$$

$$\frac{\partial \ell}{\partial \theta} = J^T C_\varepsilon^{-1}(y - h(\eta_{ML}))$$

$$-\frac{\partial^2 \ell}{\partial \theta^2} = J^T C_\varepsilon^{-1} J$$

$$\eta_{ML} \leftarrow \eta_{ML} + (J^T C_\varepsilon^{-1} J)^{-1} J^T C_\varepsilon^{-1}(y - h(\eta_{ML})) \qquad \textbf{34.12}$$

This is a conventional Gauss-Newton scheme. By simply augmenting the log-likelihood with the log prior we get:

$$L = \ln p(y, \theta) = \ln p(y|\theta) + \ln p(\theta)$$

$$= -\frac{1}{2}(y - h(\theta))^T C_\varepsilon^{-1}(y - h(\theta))$$

$$-\frac{1}{2}(\eta_\theta - \theta)^T C_\theta^{-1}(\eta_\theta - \theta) + \ldots$$

$$\frac{\partial L}{\partial \theta} = J^T C_\varepsilon^{-1}(y - h(\eta_{\theta|y})) + C_\theta^{-1}(\eta_\theta - \eta_{\theta|y}) \qquad \textbf{34.13}$$

$$-\frac{\partial^2 L}{\partial \theta^2} = J^T C_\varepsilon^{-1} J + C_\theta^{-1}$$

$$\eta_{\theta|y} \leftarrow \eta_{\theta|y} + (J^T C_\varepsilon^{-1} J + C_\theta^{-1})^{-1}(J^T C_\varepsilon^{-1}(y - h(\eta_{\theta|y})) + C_\theta^{-1}(\eta_\theta - \eta_{\theta|y}))$$

This is identical to the update for the conditional expectation in the **E**-step.

In short, the only difference between the **E**-step and a conventional Gauss-Newton ML search is that priors are included in the objective function converting it from log-likelihood into $L = \ln p(y, \theta) = p(\theta|y) + p(y)$, which is proportional to the log-posterior. In the special context of EM, $L = \ln p(\theta, y)$ is the same as the variational energy optimized in variational Bayes (see Appendix 4).

The use of EM rests upon the need to find not only the conditional mean but also the hyperparameters of unknown variance components. The **E**-step finds the current MAP estimate that provides the next expansion point for the Gauss-Newton search and the conditional covariance required by the **M**-step. The **M**-step then updates the ReML (restricted maximum likelihood) estimates of the covariance hyperparameters that are required to compute the conditional moments in the **E**-step. Technically, Eqn. **34.11** is a *generalized* EM (GEM) because the **M**-step increases the log-likelihood of the hyperparameter estimates, as opposed to maximizing it.

Relationship to other procedures

The procedure presented above represents a fairly obvious extension to conventional Gauss-Newton searches for the parameters of non-linear observation models. The extension has two components: first, maximization of the *posterior* density that embodies priors, as opposed to the likelihood. This allows for the incorporation of prior information into the solution and ensures uniqueness and convergence. Second, it covers the estimation of unknown covariance components. This is important because it accommodates unknown and non-spherical errors. The overall approach furnishes a relatively simple way of obtaining Bayes estimators for non-linear systems with unknown additive observation error. Technically, the algorithm represents a *posterior mode analysis* for non-linear observation models using EM. It can be regarded as approximating the posterior density of the parameters by replacing the conditional mean with the mode and the conditional precision with the curvature of the energy (at the current expansion point). This is known as the Laplace approximation. Covariance hyperparameters are then estimated, which maximize the expected log-likelihood of the data under this posterior density. This quantity is known as the variational free energy in statistical physics and is the quantity optimized in variational Bayes. This is important from two perspectives. First, the variational free energy is a lower bound approximation to the marginal log-likelihood or log-evidence for a model (e.g. model m; see Figure 34.2). The log-evidence plays a central role in comparing different models, as we will see in the next chapter. Second, the EM scheme above can be considered a special case of variational Bayes, in which one assumes the conditional density of the hyperparameters is a point mass. This perspective is developed more fully in Appendix 4.

Posterior mode estimation is an alternative to full posterior density analysis, which avoids numerical integration (Fahrmeir and Tutz, 1994: 58) and has been discussed extensively in the context of *generalized linear models* (e.g. Santner and Duffy, 1989). The departure from Gaussian assumptions in generalized linear models comes from non-Gaussian likelihoods, as opposed to non-linearities in the observation model considered here, but the issues are similar. Posterior mode estimation usually assumes the error covariances and priors are known. If the priors are unknown constants then empirical Bayes can be employed to estimate the required hyperparameters.

It is important not to confuse this application of EM with Kalman filtering. Although Kalman filtering can be formulated in terms of EM and, indeed, posterior mode estimation, Kalman filtering is used with completely different observation models – *state-space models*. State-space models comprise a *transition* equation and an *observation* equation (cf. the state equation and output non-linearity above) and cover systems in which the underlying state is hidden and is treated as a stochastic variable. This is not the sort of model considered here, in which the inputs (experimental design) and the ensuing states are known. This means that the conditional densities can be computed for the entire time series simultaneously (Kalman filtering updates the conditional density recursively, by stepping through the time series). If we treated the inputs as unknown and random, then the state equation could be re-written as a stochastic differential equation (SDE) and a transition equation derived from it, using local linearity assumptions (see Appendix 2 for details). This would form the basis of a state-space model. This approach may be useful for accommodating deterministic noise in the haemodynamic model but, in this treatment, we consider the inputs to be fixed. This means that the only random effects enter at the level of the observation or output non-linearity. In other words, we are assuming that the measurement error in fMRI is the principal source of random fluctuations in our measurements and that the haemodynamic response *per se* is determined by known inputs. This is the same assumption used in conventional analyses of fMRI data.

A note on integration

To iterate the EM, the local gradients $J = \partial h/\partial\theta$ have to be evaluated. This involves evaluating $h(\theta, u)$ around the current expansion point with the generalized convolution of the inputs for the current conditional parameter estimates according to Eqn. **34.8** or, equivalently, the integration of Eqn. **34.7**. The latter can be accomplished efficiently by capitalizing on the fact that stimulus functions are usually sparse. In other words, inputs arrive as infrequent events (e.g. event-related paradigms) or changes in input occur sporadically (e.g. boxcar designs). We can use this to evaluate $y(t) = h(\eta_{\theta|y}, u)$ at the times the data were sampled using a bilinear approximation to Eqn. **34.7**. The Taylor expansion of $\dot{x}(t)$ about $x(0) = x_0 = [0, 1, 1, 1]^T$:

$$\begin{aligned}\dot{x} &\approx f(x_0, 0) + \frac{\partial f}{\partial X}(x - x_0)\\ &+ \sum_i u_i\left(\frac{\partial^2 f}{\partial X \partial u_i}(x - x_0) + \frac{\partial f}{\partial u_i}\right)\end{aligned} \qquad 34.14$$

has a bilinear form, following a change of variables (equivalent to adding an extra state variable, which is a contact term):

$$\begin{aligned}\dot{X}(t) &\approx AX + \sum_i u(t)_i B_i X\\ X &= \begin{bmatrix} 1 \\ x \end{bmatrix}\\ A &= \begin{bmatrix} 0 & 0 \\ \left(f(x_0, 0) - \frac{\partial f}{\partial x}x_0\right) & \frac{\partial f}{\partial x}\end{bmatrix}\\ B_i &= \begin{bmatrix} 0 & 0 \\ \left(\frac{\partial f}{\partial u_i} - \frac{\partial^2 f}{\partial x \partial u_i}X_0\right) & \frac{\partial^2 f}{\partial x \partial u_i}\end{bmatrix}\end{aligned} \qquad 34.15$$

This bilinear approximation is important because the Volterra kernels of bilinear systems have closed-form expressions (see Appendix 2; Eqn. **A2.8**). This means that the kernels can be derived analytically, and quickly, to provide a characterization of the impulse response properties of the system. The integration of Eqn. **34.15** is predicated on its solution over periods $\Delta t = t_{k+1} - t_k$ within which the inputs are constant:

$$\begin{aligned}X(t_{k+1}) &= \exp(J\Delta t)X(t_k)\\ y(t_{k+1}) &= g(x(t_{k+1}))\\ J &= A + \sum_i u(t_k)_i B_i\end{aligned} \qquad 34.16$$

This quasi-analytical integration scheme can be an order of magnitude quicker than straightforward numerical integration, depending on the sparsity of inputs.

Relation to conventional fMRI analyses

Note that, if we treated the five biophysical parameters as known canonical values and discounted all but the first order terms in the Volterra expansion, the following linear model would result:

$$\begin{aligned}h(u, \theta) &= \kappa_0 + \sum_{i=1}^{n} \int_0^t \kappa_1(\sigma)u(t - \sigma)_i d\sigma = \sum_{i=1}^{n} \kappa_1 * u(t)_i\\ &\approx \kappa_0 + \sum_{i=1}^{n} \varepsilon_i \frac{\partial \kappa_1}{\partial \varepsilon_i} * u(t)_i\end{aligned} \qquad 34.17$$

where * denotes convolution and the second expression is a first-order Taylor expansion around the expected values of the parameters. This is exactly the same as the general linear model adopted in conventional analysis of fMRI time series, if we elect to use just one (canonical) haemodynamic response function (HRF) to convolve our stimulus functions with. In this context the HRF plays the role of $\partial\kappa_1/\partial\varepsilon_i$ in Eqn. **34.17**. This partial derivative is shown in Figure 34.3 (upper panel) using the prior expectations of the parameters and conforms closely to the sort of HRF used in practice. Now, by treating the efficacies as fixed effects (i.e. with flat priors), the MAP and ML estimators reduce to the same thing and the conditional expectation reduces to the Gauss-Markov estimator:

$$\eta_{ML} = (J^T C_\varepsilon^{-1} J)^{-1} J^T C_\varepsilon^{-1} y \qquad 34.18$$

where J is the design matrix. This is precisely the estimator used in conventional analyses when whitening strategies are employed.

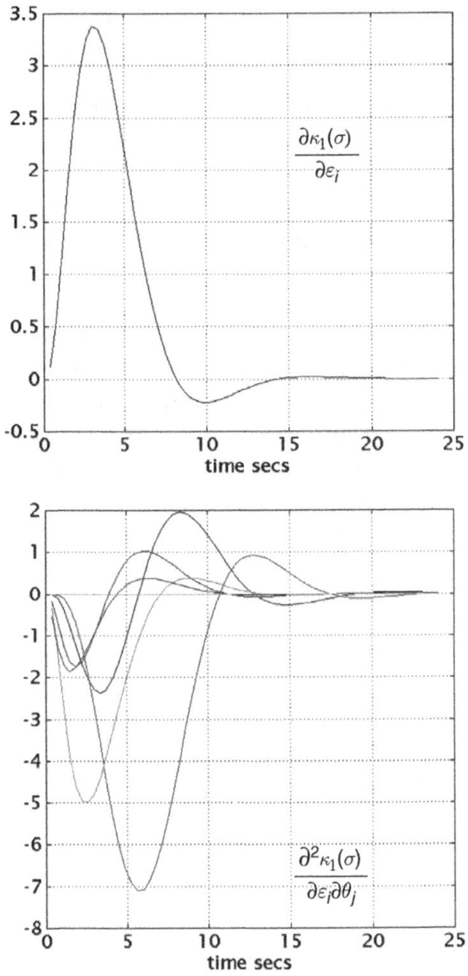

FIGURE 34.3 Partial derivatives of the kernels with respect to parameters of the model evaluated at their prior expectation. Upper panel: first-order partial derivative with respect to efficacy. Lower panels: second-order partial derivatives with respect to efficacy and the biophysical parameters. When expanding around the prior expectations of the efficacies, the remaining first- and second-order partial derivatives are zero.

Consider now the second-order Taylor approximation that obtains when we do not know the exact values of the biophysical parameters and they are treated as unknown:

$$h(\theta, u) \approx \kappa_0 + \sum_{i=1}^{n}\left(\varepsilon_i \frac{\partial \kappa_1}{\partial \varepsilon_i} * u(t)_i + \frac{1}{2}\sum_{j=1}^{5} \varepsilon_i \theta_j \frac{\partial^2 \kappa_1}{\partial \varepsilon_i \partial \theta_j} * u(t)_i\right) \quad 34.19$$

This expression is precisely the general linear model proposed in Friston *et al* (1998) and implemented in our software. In this instance, the explanatory variables comprise the stimulus functions, each convolved with a small temporal basis set corresponding to the canonical $\partial\kappa_1/\partial\varepsilon_i$ and its partial derivatives with respect to the biophysical parameters. Examples of these second-order partial derivatives are provided in the lower panel of Figure 34.3. The unknowns in this general linear model are the efficacies ε_i and the interaction between the efficacies and the biophysical parameters $\varepsilon_i\theta_j$. Of course, the problem with this linear approximation is that generalized least squares estimates of the unknown coefficients $\beta = \varepsilon_1, \ldots, \varepsilon_1\theta_1, \ldots, \varepsilon_1\theta_2, \ldots$ are not constrained to factorize into stimulus-specific efficacies ε_i and biophysical parameters θ_j that are the same for all inputs. Only a non-linear estimation procedure can do this.

In conventional linear models that use a temporal basis set to model the response functions (e.g. a canonical form and various derivatives), one obtains an ML or generalized least squares estimate of [functions of] the parameters in some subspace defined by the basis set. Operationally, this is like specifying priors but of a very particular form. This form can be thought of as uniform priors on the space spanned by the basis set and zero elsewhere. In this sense, basis functions implement hard constraints that may not be very realistic but provide for efficient estimation. The soft constraints implied by the Gaussian priors in the EM approach are more plausible but are computationally more expensive to implement.

Summary

This section has described a non-linear EM that can be viewed as a Gauss-Newton search for the conditional mode of the parameters of deterministic dynamical system, with additive Gaussian noise. We have seen that classical approaches to fMRI data analysis are special cases that ensue when considering only first-order kernels and adopting flat or uninformative priors. Put another way, the scheme can be regarded as a generalization of existing procedures that is extended in two important ways. First, it uses a biologically informed generalized or non-linear convolution model and second, it moves the estimation from a classical into a Bayesian frame.

EMPIRICAL ILLUSTRATIONS

A single input model

In this, the first of the two examples, we revisit the original data set on which the priors were based. This constitutes a single-input study where the input corresponds to the aural presentation of single words, at different rates, over epochs. The data were subject to a conventional event-related analysis, where the stimulus function comprised trains of spikes indexing the presentation of each word. The stimulus function was convolved with a canonical HRF and its temporal derivative. The data were highpass filtered by removing low-frequency components modelled by a discrete cosine set. The resulting SPM{t}, testing for activations due to words, is shown in Figure 34.4 (left-hand panel) thresholded at $p = 0.05$ (corrected).

A single region in the left superior temporal gyrus was selected for analysis. The input comprised the same stimulus function used in the conventional analysis and the output was the first eigenvariate of highpass filtered time series, of all voxels, within a 4 mm sphere, centred on the most significant voxel in the SPM{t} (marked by an arrow in Figure 34.4). The error covariance components Q comprised an identity matrix modelling white or an independent and identically distributed (IID) component and a second with exponentially decaying off-diagonal elements modelling an AR(1) component (see Friston *et al.* (2002) and Chapter 10). This models serial correlations among the errors. The results of the estimation procedure are shown in the right-hand panel in terms of the conditional distribution of the parameters and the conditional expectation of the first- and second-order kernels. The kernels are a function of the parameters and their derivation using a bilinear approximation is described in Friston *et al.* (2000) and Appendix 2. The upper-right panel shows the first-order kernels for the state variables (signal, inflow, deoxyhaemoglobin content and volume). These can be regarded as impulse response functions detailing the response to a transient input. The first- and second-order output kernels for the BOLD response are shown in the lower-right panels. They concur with those derived empirically in Friston *et al.* (2000). Note the characteristic undershoot in the first-order kernel and the

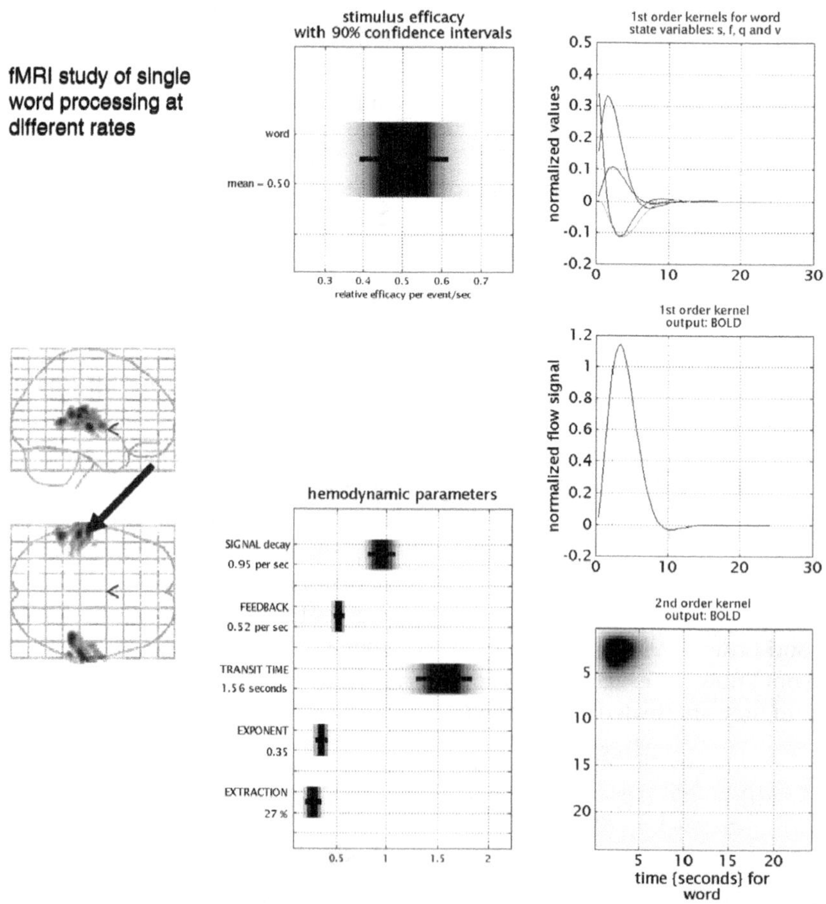

FIGURE 34.4 An SISO example: Left panel: conventional SPM{t} testing for an activating effect of word presentation. The arrow shows the centre of the region (a sphere of 4 mm radius) whose response entered into the Bayesian inversion. The results for this region are shown in the right-hand panel in terms of the conditional distribution of the parameters and the conditional expectation of the first- and second-order kernels. The upper-right panel shows the first-order kernels for the state variables (signal, inflow, deoxyhaemoglobin content and volume). The first- and second-order output kernels for the BOLD response are shown in the lower-right panels. The left-hand panels show the conditional or posterior distributions. The conditional density for efficacy is presented in the upper panel and those for the five biophysical parameters in the lower panel. The shading corresponds to the probability density and the bars to 90 per cent confidence intervals.

pronounced negativity in the upper left of the second-order kernel, flanked by two off-diagonal positive regions at around 8 s. These lend the haemodynamics a degree of refractoriness when presenting paired stimuli less than a few seconds apart and a superadditive response with about 8 s separation. The left-hand panels show the conditional or posterior distributions. The density for the efficacy is presented in the upper panel and those for the five biophysical parameters are shown in the lower panel. The shading corresponds to the probability density and the bars to 90 per cent confidence intervals. The values of the biophysical parameters are all within a very acceptable range. In this example, the signal elimination and decay appears to be slower than normally encountered, with the rate constants being significantly larger than their prior expectations. Grubb's exponent here is closer to the steady state value of 0.38 than the prior expectation of 0.32. Of greater interest is the efficacy. It can be seen that the efficacy lies between 0.4 and 0.6 and is clearly greater than zero. This would be expected given we chose the most significant voxel from the conventional analysis. Notice there is no null hypothesis here and we do not even need a p-value to make the inference that words evoke a response in this region. An important facility, with inferences based on the conditional distribution and precluded in classical analyses, is that one can infer a cause did not elicit a response. This is demonstrated next.

A multiple input model

In this example, we turn to a data set used in previous sections, in which there are three experimental causes or inputs. This was a study of attention to visual motion. Subjects were studied with fMRI under identical stimulus conditions (visual motion subtended by radially moving dots) while manipulating the attentional component of the task (detection of velocity changes). The data were acquired from normal subjects at 2 tesla using a VISION (Siemens, Erlangen) whole body MRI system, equipped with a head volume coil. Here we analyse data from the first subject. Contiguous multislice fMRI images were obtained with a gradient echo-planar sequence (TE = 40 ms, TR = 3.22 s, matrix size = $64 \times 64 \times 32$, voxel size $3 \times 3 \times 3$ mm). Each subject had four consecutive 100-scan sessions comprising a series of 10-scan blocks under five different conditions D F A F N F A F N S. The first condition (D) was a dummy condition to allow for magnetic saturation effects. F (Fixation) corresponds to a low-level baseline where the subjects viewed a fixation point at the centre of a screen. In condition A (Attention), subjects viewed 250 dots moving radially from the centre at 4.7 degrees per second and were asked to detect changes in radial velocity. In condition N (No-attention), the subjects were asked simply to view the moving dots. In condition S (Stationary), subjects viewed stationary dots. The order of A and N was swapped for the last two sessions. In all conditions, subjects fixated the centre of the screen. In a pre-scanning session the subjects were given five trials with five speed changes (reducing to 1 per cent). During scanning there were no speed changes. No overt response was required in any condition.

This design can be reformulated in terms of three potential causes, photic stimulation, visual motion and directed attention. The F-epochs have no associated cause and represent a baseline. The S-epochs have just photic stimulation. The N-epochs have both photic stimulation and motion, whereas the A-epochs encompass all three causes. We performed a conventional analysis using boxcar stimulus functions encoding the presence or absence of each of the three causes, during each epoch. These functions were convolved with a canonical HRF and its temporal derivative to give two repressors for each cause. The corresponding design matrix is shown in the left panel of Figure 34.5. We selected a region that showed a significant attentional effect in the lingual gyrus. The stimulus functions modelling the three inputs were the box functions used in the conventional analysis. The output corresponded to the first eigenvariate of highpass filtered time-series from all voxels in a 4 mm sphere centred on 0, −66, −3 mm (Talairach and Tournoux, 1988). The error covariance basis was the identity matrix (i.e. ignoring serial correlations because of the relatively long TR). The results are shown in the right-hand panel of Figure 34.5 using the same format as Figure 34.4. The critical thing here is that there are three conditional densities, one for each input efficacy. Attention has a clear activating effect with more than a 90 per cent probability of being greater than 0.25 per second. However, in this region neither photic stimulation nor motion in the visual field evokes any real response. The efficacies of both are less than 0.1 and are centred on zero. This means that the time constants of the visually evoked response could range from about 10 s to never. Consequently, these causes can be discounted from a dynamical perspective. In short, this visually unresponsive area responds substantially to attentional manipulation *showing a true functional selectivity*. This is a crucial statement because classical inference does not allow one to infer any region does not respond and therefore precludes inference about the specificity of regional responses. The only reason one can say 'this region responds *selectively* to attention' is because Bayesian inference allows one to say 'it does *not* response to photic stimulation with random dots or motion'.

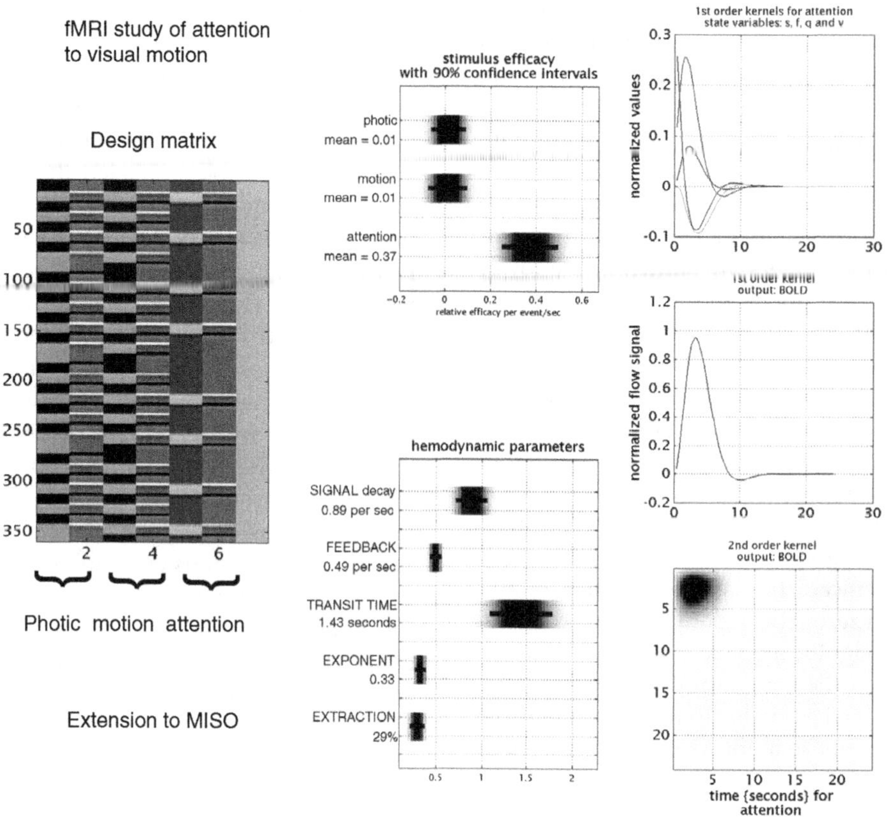

FIGURE 34.5 A MISO example using visual attention to motion. The left panel shows the design matrix used in the conventional analysis and the right panel shows the results of the Bayesian analysis of a lingual extrastriate region. This panel has the same format as Figure 34.4.

CONCLUSION

In this chapter, we have looked at an EM implementation of the Gauss-Newton method for estimating the conditional or posterior distribution of the parameters of a deterministic dynamical system. The inclusion of priors in the estimation procedure ensures robust and rapid convergence and the resulting conditional densities enable Bayesian inference about the model's parameters. We have examined the coupling between experimentally designed causes or factors in fMRI studies and the ensuing BOLD response. This application represents a generalization of existing linear models to accommodate dynamics and non-linearities in the transduction of experimental causes to measured output in fMRI. Because the model is predicated on biophysical processes the parameters have a physical interpretation. Furthermore, the approach extends classical inference about the likelihood of the data to more plausible inferences about the parameters of the model given the data. This inference provides confidence intervals based on the conditional density.

Perhaps the most important extension of the scheme described in this chapter is to MIMO systems where we deal with multiple regions or voxels at the same time. The importance of this extension is that one can incorporate interactions among brain regions at the neuronal level. This furnishes a framework for the dynamic causal modelling of functional integration in the brain (see Chapters 41 and 42). This chapter focused on inference on the space of parameters. In the next chapter, we look at inference on the space of models.

REFERENCES

Bendat JS (1990) *Nonlinear System Analysis and Identification from Random Data*. John Wiley and Sons, New York

Buxton RB, Wong EC, Frank LR (1998) Dynamics of blood flow and oxygenation changes during brain activation: The Balloon model. *Mag Res Med* **39**: 855–64

Fahrmeir L, Tutz G (1994) *Multivariate statistical modelling based on generalised linear models*. Springer Verlag Inc., New York, pp 355–56

Fliess M, Lamnabhi M, Lamnabhi-Lagarrigue F (1983) An algebraic approach to nonlinear functional expansions. *IEEE Trans Circuits Syst* **30**: 554–70

Friston KJ, Glaser DE, Henson RNA *et al.* (2002) Classical and Bayesian inference in neuroimaging: applications. *NeuroImage* **16**: 484–512

Friston KJ, Josephs O, Rees G *et al.* (1998) Non-linear event-related responses in fMRI. *Mag Res Med* **39**: 41–52

Friston KJ, Mechelli A, Turner R *et al.* (2000) Nonlinear responses in fMRI: the Balloon model, Volterra kernels and other hemodynamics. *NeuroImage* **12**: 466–77

Grubb RL, Rachael ME, Euchring JO *et al.* (1974) The effects of changes in PCO_2 on cerebral blood volume, blood flow and vascular mean transit time. *Stroke* **5**: 630–39

Hoge RD, Atkinson J, Gill B *et al.* (1999) Linear coupling between cerebral blood flow and oxygen consumption in activated human cortex. *Proc Natl Acad Sci* **96**: 9403–08

Mandeville JB, Marota JJ, Ayata C *et al.* (1999) Evidence of a cerebrovascular postarteriole windkessel with delayed compliance. *J Cereb Blood Flow Metab* **19**: 679–89

Mayhew J, Hu D, Zheng Y *et al.* (1998) An evaluation of linear models analysis techniques for processing images of microcirculation activity. *NeuroImage* **7**: 49–71

Miller KL, Luh WM, Liu TT *et al.* (2000) Characterizing the dynamic perfusion response to stimuli of short duration. *Proc ISRM* **8**: 580

Santner TJ, Duffy DE (1989) *The statistical analysis of discrete data.* Springer, New York

Talairach J, Tournoux P (1988) *A Co-planar stereotaxic atlas of a human brain.* Thieme, Stuttgart

CHAPTER

35

Bayesian model selection and averaging

W.D. Penny, J. Mattout and N. Trujillo-Barreto

INTRODUCTION

In Chapter 11, we described how Bayesian inference can be applied to hierarchical models. In this chapter, we show how the members of a model class, indexed by m, can also be considered as part of a hierarchy. Model classes might be general linear models (GLMs) where m indexes different choices for the design matrix, dynamic causal models (DCMs) where m indexes connectivity or input patterns, or source reconstruction models where m indexes functional or anatomical constraints. Explicitly including model structure in this way will allow us to make inferences about that structure.

Figure 35.1 shows the generative model we have in mind. First, a member of the model class is chosen. Then model parameters θ and finally the data y are generated. Bayesian inference for hierarchical models can be implemented using the belief propagation algorithm. Figure 35.2 shows how this can be applied for model selection and averaging. It comprises three stages that we will refer to as: (i) conditional parameter inference; (ii) model inference; and (iii) model averaging. These stages can be implemented using the equations shown in Figure 35.2.

Conditional parameter inference is based on Bayes' rule whereby, after observing data y, prior beliefs about model parameters are updated to posterior beliefs. This update requires the likelihood $p(y|\theta, m)$. It allows one to compute the density $p(\theta|y, m)$. The term conditional is used to highlight the fact that these inferences are based on model m. Of course, being a posterior density, it is also conditional on the data y.

Model inference is based on Bayes' rule whereby, after observing data y, prior beliefs about model structure are updated to posterior beliefs. This update requires the evidence $p(y|m)$. Model selection is then implemented by picking the model that maximizes the posterior probability $p(m|y)$. If the model priors $p(m)$ are uniform then this is equivalent to picking the model with the highest evidence. Pairwise model comparisons are based on Bayes factors, which are ratios of evidences.

Model averaging, as depicted in Figure 35.2, also allows for inferences to be made about parameters. But these inferences are based on the distribution $p(\theta|y)$, rather than $p(\theta|y, m)$, and so are free from assumptions about model structure.

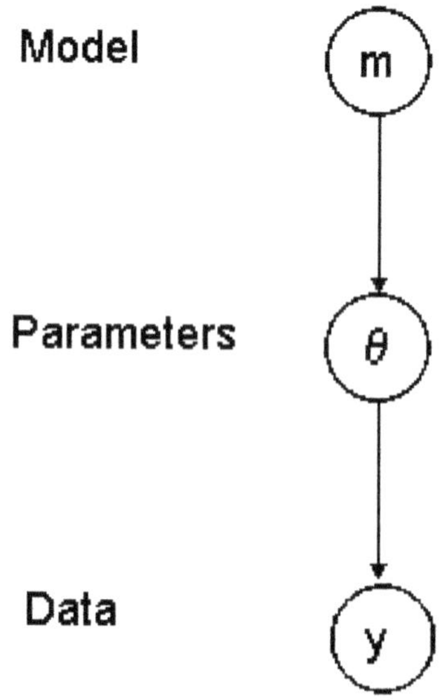

FIGURE 35.1 Hierarchical generative model in which members of a model class, indexed by m, are considered as part of the hierarchy. Typically, m indexes the structure of the model. This might be the connectivity pattern in a dynamic causal model or set of anatomical or functional constraints in a source reconstruction model. Once a model has been chosen from the distribution $p(m)$, its parameters are generated from the parameter prior $p(\theta|m)$ and finally data are generated from the likelihood $p(y|\theta, m)$.

Statistical Parametric Mapping, by Karl Friston et al.
ISBN–13: 978-0-12-372560-8 ISBN–10: 0-12-372560-7

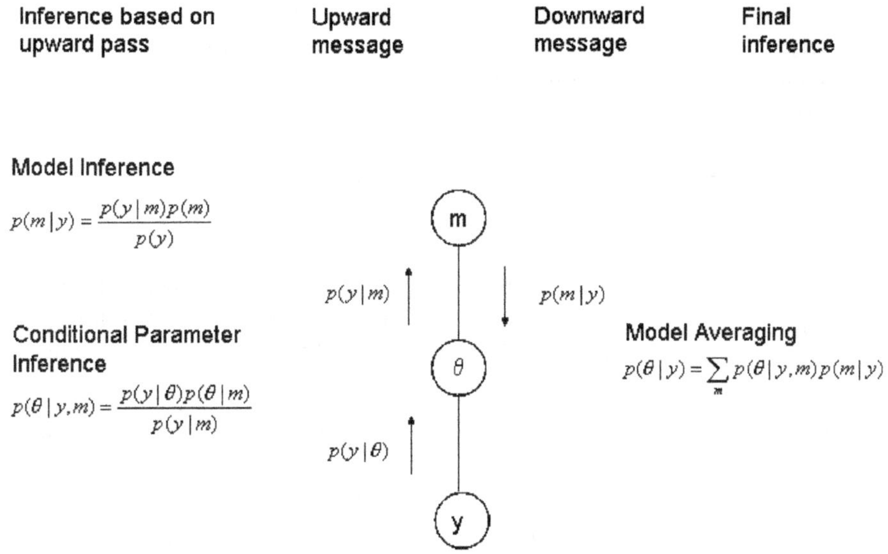

FIGURE 35.2 Figure 11.5 in Chapter 11 describes the belief propagation algorithm for implementing Bayesian inference in hierarchical models. This figure shows a special case of belief propagation for Bayesian model selection (BMS) and Bayesian model averaging (BMA). In BMS, the posterior model probability $p(m|y)$, is used to select a single 'best' model. In BMA, inferences are based on all models and $p(m|y)$ is used as a weighting factor. Only in BMA, are parameter inferences based on the correct marginal density $p(\theta|y)$.

This chapter comprises theoretical and empirical sections. In the theory sections, we describe (i) conditional parameter inference for linear and non-linear models, (ii) model inference, including a review of three different ways to approximate model evidence and pairwise model comparisons based on Bayes factors and (iii) model averaging, with a focus on how to search the model space using 'Occam's window'. The empirical sections show how these principles can be applied to DCMs and source reconstruction models. We finish with a discussion.

Notation

We use upper-case letters to denote matrices and lower-case to denote vectors. $\mathsf{N}(m, \Sigma)$ denotes a uni/multivariate Gaussian with mean m and variance/covariance Σ. I_K denotes the $K \times K$ identity matrix, 1_K is a $1 \times K$ vector of ones, 0_K is a $1 \times K$ vector of zeroes. If X is a matrix, X_{ij} denotes the i, jth element, X^T denotes the matrix transpose and $\mathrm{vec}(X)$ returns a column vector comprising its columns, $\mathrm{diag}(x)$ returns a diagonal matrix with leading diagonal elements given by the vector x, $\otimes$ denotes the Kronecker product and $\log x$ denotes the natural logarithm.

CONDITIONAL PARAMETER INFERENCE

Readers requiring a more basic introduction to Bayesian modelling are referred to Gelman *et al.* (1995) and Chapter 11.

Linear models

For linear models:

$$y = X\theta + e \quad \textbf{35.1}$$

with data y, parameters θ, Gaussian errors e and design matrix X, the likelihood can be written:

$$p(y|\theta, m) = \mathsf{N}(X\theta, C_e) \quad \textbf{35.2}$$

where C_e is the error covariance matrix. If our prior beliefs can be specified using the Gaussian distribution:

$$p(\theta|m) = \mathsf{N}(\mu_p, C_p) \quad \textbf{35.3}$$

where μ_p is the prior mean and C_p is the prior covariance, then the posterior distribution is (Lee, 1997):

$$p(\theta|y, m) = \mathsf{N}(\mu, C) \quad \textbf{35.4}$$

where

$$C^{-1} = X^T C_e^{-1} X + C_p^{-1} \quad \textbf{35.5}$$
$$\mu = C(X^T C_e^{-1} y + C_p^{-1} \mu_p)$$

As in Chapter 11, it is often useful to refer to precision matrices, C^{-1}, rather than covariance matrices, C. This is because the posterior precision, C^{-1}, is equal to the sum of the prior precision, C_p^{-1}, plus the data precision, $X^T C_e^{-1} X$. The posterior mean, μ, is given by the sum of the prior mean plus the data mean, but where each is weighted according to their relative precision. This linear Gaussian framework is used for the source reconstruction methods described later in the chapter. Here X is the lead-field matrix which transforms measurements from source space to sensor space (Baillet *et al.*, 2001).

Our model assumptions, m, are typically embodied in different choices for the design or prior covariance matrices. These allow for the specification of GLMs with different regressors or different covariance components.

Variance components

Bayesian estimation, as described in the previous section, assumed that we knew the prior covariance, C_p, and error covariance, C_e. This information is, however, rarely available. In Friston *et al.* (2002) these covariances are expressed as:

$$C_p = \sum_i \lambda_i Q_i \qquad 35.6$$
$$C_e = \sum_j \lambda_j Q_j$$

where Q_i and Q_j are known as 'covariance components' and λ_i, λ_j are hyperparameters. Chapter 24 and Friston *et al.* (2002) show how these hyperparameters can be estimated using parametric empirical Bayes (PEB). It is also possible to represent precision matrices, rather than covariance matrices, using a linear expansion as shown in Appendix 4.

Non-linear models

For non-linear models, we have:

$$y = h(\theta) + e \qquad 35.7$$

where $h(\theta)$ is a non-linear function of parameter vector θ. We assume Gaussian prior and likelihood distributions:

$$p(\theta|m) = \mathsf{N}(\mu_p, C_p) \qquad 35.8$$
$$p(y|\theta, m) = \mathsf{N}(h(\theta), C_e)$$

where m indexes model structure, θ_p is the prior mean, C_p the prior covariance and C_e is the error covariance.

The linear framework described in the previous section can be applied by locally linearizing the non-linearity, about a 'current' estimate μ_i, using a first order Taylor series expansion:

$$h(\theta) = h(\mu_i) + \frac{\partial h(\mu_i)}{\partial \theta}(\theta - \mu_i) \qquad 35.9$$

Substituting this into Eqn. **35.7** and defining $r \equiv y - h(\mu_i)$, $J \equiv \frac{\partial h(\mu_i)}{\partial \theta}$ and $\Delta\theta \equiv \theta - \mu_i$ gives

$$r = J\Delta\theta + e \qquad 35.10$$

which now conforms to a GLM (cf. Eqn. **35.1**). The 'prior' (based on starting estimate μ_i), likelihood and posterior are now given by:

$$p(\Delta\theta|m) = \mathsf{N}(\mu_p - \mu_i, C_p) \qquad 35.11$$
$$p(r|\Delta\theta, m) = \mathsf{N}(J\Delta\theta, C_e)$$
$$p(\Delta\theta|r, m) = \mathsf{N}(\Delta\mu, C_{i+1})$$

The quantities $\Delta\mu$ and C_{i+1} can be found using the result for the linear case (substitute r for y and J for X in Eqn. **35.5**). If we define our 'new' parameter estimate as $\mu_{i+1} = \mu_i + \Delta\mu$ then:

$$C_{i+1}^{-1} = J^T C_e^{-1} J + C_p^{-1} \qquad 35.12$$
$$\mu_{i+1} = \mu_i + C_{i+1}(J^T C_e^{-1} r + C_p^{-1}(\mu_p - \mu_i))$$

This update is applied iteratively, in that the estimate μ_{i+1} becomes the starting point for a new Taylor series expansion. It can also be combined with hyperparameter estimates, to characterize C_p and C_e, as described in Friston (2002). This then corresponds to the PEB algorithm described in Chapter 22. This algorithm is used, for example, to estimate parameters of dynamic causal models. For DCM, the non-linearity $h(\theta)$ corresponds to the integration of a dynamic system.

As described, in Chapter 24, this PEB algorithm is a special case of variational Bayes with a fixed-form full-covariance Gaussian ensemble. When the algorithm has converged, it provides an estimate of the posterior density:

$$p(\theta|y, m) = \mathsf{N}(\mu_{PEB}, C_{PEB}) \qquad 35.13$$

which can then be used for parameter inference and model selection.

The above algorithm can also be viewed as the E-step of an expectation maximization (EM) algorithm, described in section 3.1 of Friston (2002) and Appendix 3. The M-step of this algorithm, which we have not described, updates the hyperparameters. This E-step can also be viewed as a Gauss-Newton optimization whereby parameter estimates are updated in the direction of the gradient of the log-posterior by an amount proportional to its curvature (see e.g. Press *et al.*, 1992).

MODEL INFERENCE

Given a particular model class, we let the variable m index members of that class. Model classes might be GLMs where m indexes design matrices, DCMs where m indexes connectivity or input patterns, or source reconstruction models where m indexes functional or anatomical constraints.

Explicitly including model structure in this way will allow us to make inferences about model structure.

We may, for example, have prior beliefs $p(m)$. In the abscence of any genuine prior information here, a uniform distribution will suffice. We can then use Bayes' rule which, in light of observed data y, will update these model priors into model posteriors:

$$p(m|y) = \frac{p(y|m)p(m)}{p(y)} \qquad \mathbf{35.14}$$

Model inference can then proceed based on this distribution. This will allow for Bayesian model comparisons (BMCs). In Bayesian model *selection* (BMS), a model is selected which maximizes this probability:

$$m_{MP} = \underset{m}{\text{argmax}}[p(m|y)]$$

If the prior is uniform, $p(m) = 1/M$ then this is equivalent to picking the model with the highest evidence:

$$m_{ME} = \underset{m}{\text{argmax}}[p(y|m)]$$

If we have uniform priors then BMC can be implemented with Bayes factors. Before covering this in more detail, we emphasize that all of these model inferences require computation of the model evidence. This is given by:

$$p(y|m) = \int p(y|\theta, m)p(\theta|m)d\theta \qquad \mathbf{35.15}$$

The model evidence is simply the normalization term from parameter inference, as shown in Figure 35.2. This is the 'message' that is passed up the hierachy during belief propagation, as shown in Figure 35.2. For linear Gaussian models, the evidence can be expressed analytically. For non-linear models there are various approximations which are discussed in later subsections.

Bayes factors

Given models $m = i$ and $m = j$, the Bayes factor comparing model i to model j is defined as (Kass and Raftery, 1993, 1995):

$$B_{ij} = \frac{p(y|m=i)}{p(y|m=j)}$$

where $p(y|m=j)$ is the evidence for model j. When $B_{ij} > 1$, the data favour model i over model j, and when $B_{ij} < 1$, the data favour model j. If there are more than two models to compare then we choose one of them as a reference model and calculate Bayes factors relative to that reference. When model i is an alternative model and model j a null model, B_{ij} is the likelihood ratio upon which classical statistics are based (see Appendix 1).

A classic example here is the analysis of variance for factorially designed experiments, described in Chapter 13. To see if there is a main effect of a factor, one compares two models in which the levels of the factor are described by (i) a single variable or (ii) separate variables. Evidence in favour of model (ii) allows one to infer that there is a main effect.

In this chapter, we will use Bayes factors to compare dynamic causal models. In these applications, often the most important inference is on model space. For example, whether or not experimental effects are mediated by changes in feedforward or feedback pathways. This particular topic is dealt with in greater detail in Chapter 43.

The Bayes factor is a summary of the evidence provided by the data in favour of one scientific theory, represented by a statistical model, as opposed to another. Raftery (1995) presents an interpretation of Bayes factors, shown in Table 35-1. Jefferys (1935) presents a similar grading for the comparison of scientific theories. These partitionings are somewhat arbitrary but do provide descriptive statements.

Table 35-1 also shows the equivalent posterior probability of hypothesis i:

$$p(m=i|y) = \frac{p(y|m=i)p(m=i)}{p(y|m=i)p(m=i) + p(y|m=j)p(m=j)} \qquad \mathbf{35.16}$$

assuming equal model priors $p(m=i) = p(m=j) = 0.5$.

If we define the 'prior odds ratio' as $p(m=i)/p(m=j)$ and the 'posterior odds ratio' as:

$$O_{ij} = \frac{p(m=i|y)}{p(m=j|y)} \qquad \mathbf{35.17}$$

then the posterior odds are given by the prior odds multiplied by the Bayes factor. For prior odds of unity the posterior odds are therefore equal to the Bayes factor. Here, a Bayes factor of $B_{ij} = 100$, for example, corresponds to odds of 100-to-1. In betting shop parlance this is 100-to-1 'on'. A value of $B_{ij} = 0.01$ is 100-to-1 'against'.

TABLE 35-1 Interpretation of Bayes factors. Bayes factors can be interpreted as follows. Given candidate hypotheses *i* and *j*, a Bayes factor of 20 corresponds to a belief of 95 per cent in the statement 'hypothesis *i* is true'. This corresponds to strong evidence in favour of *i*.

B_{ij}	$p(m=i|y)(\%)$	Evidence in favour of model i
1 to 3	50–75	Weak
3 to 20	75–95	Positive
20 to 150	95–99	Strong
≥ 150	≥ 99	Very strong

Bayes factors in Bayesian statistics play a similar role to p-values in classical statistics. In Raftery (1995), however, Raftery argues that p-values can give misleading results, especially in large samples. The background to this assertion is that Fisher originally suggested the use of significance levels (the p-values beyond which a result is deemed significant) $\alpha = 0.05$ or 0.01 based on his experience with small agricultural experiments having between 30 and 200 data points. Subsequent advice, notably from Neyman and Pearson, was that power and significance should be balanced when choosing α. This essentially corresponds to reducing α for large samples (but they did not say *how* α should be reduced). Bayes factors provide a principled way to do this.

The relation between p-values and Bayes factors is well illustrated by the following example (Raftery, 1995). For linear regression models, one can use Bayes factors or p-values to decide whether to include an extra regressor. For a sample size of $N_s = 50$, positive evidence in favour of inclusion (say, $B_{12} = 3$) corresponds to a p-value of 0.019. For $N_s = 100$ and 1000 the corresponding p-values reduce to 0.01 and 0.003. If one wishes to decide whether to include multiple extra regressors the corresponding p-values drop more quickly.

Importantly, unlike p-values, Bayes factors can be used to compare models that cannot be nested.[1] This provides an optimal inference framework that can, for example, be applied to determine which haemodynamic basis functions are appropriate for functional magnetic resonance imaging (fMRI) (Penny *et al.*, 2006). They also allow one to quantify evidence in favour of a null hypothesis.

Computing the model evidence

This section shows how the model evidence can be computed for non-linear models. The evidence for linear models is then given as a special case. The prior and likelihood of the non-linear model can be expanded as:

$$p(\theta|m) = (2\pi)^{-p/2}|C_p|^{-1/2}\exp(-\frac{1}{2}e(\theta)^T C_p^{-1} e(\theta)) \quad 35.18$$

$$p(y|\theta, m) = (2\pi)^{-N_s/2}|C_e|^{-1/2}\exp(-\frac{1}{2}r(\theta)^T C_e^{-1} r(\theta))$$

where

$$e(\theta) = \theta - \theta_p \quad 35.19$$

$$r(\theta) = y - h(\theta)$$

are the 'parameter errors' and 'prediction errors'.

Substituting these expressions into Eqn. 35 and rearranging allows the evidence to be expressed as:

$$p(y|m) = (2\pi)^{-p/2}|C_p|^{-1/2}(2\pi)^{-N_s/2}|C_e|^{-1/2}I(\theta) \quad 35.20$$

where

$$I(\theta) = \int \exp(-\frac{1}{2}r(\theta)^T C_e^{-1} r(\theta) - \frac{1}{2}e(\theta)^T C_p^{-1} e(\theta))d\theta \quad 35.21$$

For linear models this integral can be expressed analytically. For non-linear models, it can be estimated using a Laplace approximation.

Laplace approximation

The Laplace approximation was introduced in Chapter 24. It makes use of the first order Taylor series approximation referred to in Eqn. **35.9**, but this time placed around the solution, θ_L, found by an optimization algorithm.

Usually, the term 'Laplace approximation' refers to an expansion around the maximum *a posteriori* (MAP) solution:

$$\theta_{MAP} = \underset{\theta}{\operatorname{argmax}}\,[p(y|\theta, m)p(\theta|m)] \quad 35.22$$

Thus $\theta_L = \theta_{MAP}$.

But, more generally, one can make an expansion around any solution, for example the one provided by PEB. In this case $\theta_L = \mu_{PEB}$. As we have described in Chapter 24, PEB is a special case of VB with a fixed-form Gaussian ensemble, and so does not deliver the MAP solution. Rather, PEB maximizes the negative free energy and so implicitly minimizes the Kullback-Liebler (KL)-divergence between the true posterior and a full-covariance Gaussian approximation to it. This difference is discussed in Chapter 24.

Whatever the expansion point, the model non-linearity is approximated using:

$$h(\theta) = h(\theta_L) + J(\theta - \theta_L) \quad 35.23$$

where $J = \frac{\partial h(\theta_L)}{\partial \theta}$. We also make use of the knowledge that the posterior covariance is given by:

$$C_L^{-1} = J^T C_e^{-1} J + C_p^{-1} \quad 35.24$$

For $C_L = C_{PEB}$ this follows directly from Eqn. **35.12**.

By using the substitutions $e(\theta) = (\theta - \theta_L) + (\theta_L - \theta_p)$ and $r(\theta) = (y - h(\theta_L)) + (h(\theta_L) - h(\theta))$, making use of the above two expressions, and removing terms not dependent on θ, we can write:

$$I(\theta) = \left[\int \exp(-\frac{1}{2}(\theta - \theta_L)^T C_L^{-1}(\theta - \theta_L))d\theta\right] \quad 35.25$$

$$\times \left[\exp(-\frac{1}{2}r(\theta_L)^T C_e^{-1} r(\theta_L) - \frac{1}{2}e(\theta_L)^T C_p^{-1} e(\theta_L))\right] \quad 35.26$$

[1] Model selection using classical inference requires nested models. Inference is made using step-down procedures and the 'extra sum of squares' principle, as described in Chapter 8.

where the first factor is the normalizing term of the multivariate Gaussian density. The algebraic steps involved in the above substitutions are detailed in Stephan *et al.* (2005). Hence:

$$I(\theta) = (2\pi)^{p/2}|C_L|^{1/2}\exp\left(-\frac{1}{2}r(\theta_L)^T C_e^{-1} r(\theta_L) - \frac{1}{2}e(\theta_L)^T C_p^{-1} e(\theta_L)\right) \quad 35.27$$

Substituting this expression into Eqn. **35.20** and taking logs gives the Laplace approximation to the log-evidence:

$$\begin{aligned}\log p(y|m)_L = &-\frac{N_s}{2}\log 2\pi - \frac{1}{2}\log|C_e| - \frac{1}{2}\log|C_p| \\ &+\frac{1}{2}\log|C_L| - \frac{1}{2}r(\theta_L)^T C_e^{-1} r(\theta_L) \\ &-\frac{1}{2}e(\theta_L)^T C_p^{-1} e(\theta_L)\end{aligned} \quad 35.28$$

When comparing the evidence for different models, we can ignore the first term as it will be the same for all models. Dropping the first term and rearranging gives:

$$\log p(y|m)_L = Accuracy(m) - Complexity(m) \quad 35.29$$

where

$$\begin{aligned}Accuracy(m) &= -\frac{1}{2}\log|C_e| - \frac{1}{2}r(\theta_L)^T C_e^{-1} r(\theta_L) \\ Complexity(m) &= \frac{1}{2}\log|C_p| - \frac{1}{2}\log|C_L| + \frac{1}{2}e(\theta_L)^T C_p^{-1} e(\theta_L)\end{aligned} \quad 35.30$$

Use of base-e or base-2 logarithms leads to the log-evidence being measured in 'nats' or 'bits' respectively. Models with high evidence optimally trade-off two conflicting requirements of a good model, that it fit the data and be as simple as possible.

The complexity term depends on the prior covariance, C_p, which determines the 'cost' of parameters. This dependence is worrisome if the prior covariances are fixed *a priori*, as the parameter cost will also be fixed *a priori*. This will lead to biases in the resulting model comparisons. For example, if the prior (co)variances are set to large values, model comparison will consistently favour models that are less complex than the true model.

In DCM for fMRI (Friston *et al.*, 2003), prior variances are set to fixed values so as to enforce dynamic stability, with high probability. Use of the Laplace approximation in this context could therefore lead to biases in model comparison. A second issue in this context is that, to enforce dynamic stability, models with different numbers of connections will employ different prior variances. Therefore the priors change from model to model. This means that model comparison entails a comparison of the priors.

To overcome these potential problems with DCM for fMRI, alternative approximations to the model evidence are used instead. These are the Bayesian information criterion (BIC) and Akaike information criterion (AIC) introduced below. They also use fixed parameter costs, but they are fixed between models and are different for BIC than AIC. It is suggested in Penny *et al.* (2004) that, if the two measures provide consistent evidence, a model selection can be made.

Finally, we note that, if prior covariances are estimated from data, then the parameter cost will also have been estimated from data, and this source of bias in model comparison is removed. In this case, the model evidence also includes terms which account for uncertainty in the variance component estimation, as described in Chapter 10 of Bishop (1995).

Bayesian information criterion

An alternative approximation to the model evidence is given by the Bayesian information criterion (Schwarz, 1978). This is a special case of the Laplace approximation which drops all terms that do not scale with the number of data points, and can be derived as follows.

Substituting Eqn. **35.27** into Eqn. **35.20** gives:

$$p(y|m)_L = p(y|\theta_L, m)p(\theta_L|m)(2\pi)^{p/2}|C_L|^{1/2} \quad 35.31$$

Taking logs gives:

$$\begin{aligned}\log p(y|m)_L = &\log p(y|\theta_L, m) + \log p(\theta_L|m) \\ &+\frac{p}{2}\log 2\pi + \frac{1}{2}\log|C_L|\end{aligned} \quad 35.32$$

The dependence of the first three terms on the number of data points is $O(N_s)$, $O(1)$ and $O(1)$. For the 4th term, entries in the posterior covariance scale linearly with N_s^{-1}:

$$\begin{aligned}\lim_{N_s\to\infty}\frac{1}{2}\log|C_L| &= \frac{1}{2}\log|\frac{C_L(0)}{N_s}| \\ &= -\frac{p}{2}\log N_s + \frac{1}{2}\log|C_L(0)|\end{aligned} \quad 35.33$$

where $C_L(0)$ is the posterior covariance based on $N_s = 0$ data points (i.e. the prior covariance). This last term therefore scales as $O(1)$. Schwarz (1978) notes that in the limit of large N_s, Eqn. **35.32** therefore reduces to:

$$\begin{aligned}BIC &= \lim_{N_s\to\infty}\log p(y|m)_L \\ &= \log p(y|\theta_L, m) - \frac{p}{2}\log N_s\end{aligned} \quad 35.34$$

This can be re-written as:

$$BIC = Accuracy(m) - \frac{p}{2}\log N_s \quad 35.35$$

where p is the number of parameters in the model. In BIC, the cost of a parameter, $-0.5 \log N_s$ bits, therefore reduces with an increasing number of data points.

Akaike's information criterion

The second criterion we use is Akaike's information criterion (AIC)[2] (Akaike, 1973). AIC is maximized when the approximating likelihood of a novel data point is closest to the true likelihood, as measured by the Kullback-Liebler divergence (this is shown in Ripley (1995)). The AIC is given by:

$$AIC = Accuracy(m) - p \tag{35.36}$$

Though not originally motivated from a Bayesian perspective, model comparisons based on AIC are asymptotically equivalent (i.e. as $N_s \to \infty$) to those based on Bayes factors (Akaike, 1983), i.e. AIC approximates the model evidence.

Empirically, BIC is biased towards simple models and AIC to complex models (Kass and Raftery, 1993). Indeed, inspection of Eqns **35.35** and **35.36** shows that for values appropriate for, e.g. DCM for fMRI, where $p \approx 10$ and $N_s \approx 200$, BIC pays a heavier parameter penalty than AIC.

MODEL AVERAGING

The parameter inferences referred to in previous sections are based on the distribution $p(\theta|y, m)$. That m appears as a dependent variable, makes it explicit that these inferences are contingent on assumptions about model structure. More generally, however, if inferences about model parameters are paramount one would use a Bayesian model averaging (BMA) approach. Here, inferences are based on the distribution:

$$p(\theta|y) = \sum_m p(\theta|y, m)p(m|y) \tag{35.37}$$

where $p(m|y)$ is the posterior probability of model m.

$$p(m|y) = \frac{p(y|m)p(m)}{p(y)} \tag{35.38}$$

As shown in Figure 35.2, only when these 'messages', $p(m|y)$, have been passed back down the hierarchy is belief propagation complete. Only then do we have the true marginal density $p(\theta|y)$. Thus, BMA allows for correct Bayesian inferences, whereas what we have previously described as 'parameter inferences' are conditional on model structure. Of course, if our model space comprises just one model there is no distribution.

BMA accounts for uncertainty in the model selection process, something which classical statistical analysis neglects. By averaging over competing models, BMA incorporates model uncertainty into conclusions about parameters. BMA has been successfully applied to many statistical model classes including linear regression, generalized linear models, and discrete graphical models, in all cases improving predictive performance (see Hoeting *et al.* 1999 for a review).[3] In this chapter, we describe the application of BMA to electroencephalography (EEG) source reconstruction.

There are, however, several practical difficulties with Eqn. **35.37** when the number of models and number of variables in each model are large. In neuroimaging, models can have tens of thousands of parameters. This issue has been widely treated in the literature (Draper, 1995), and the general consensus has been to construct search strategies to find a set of models that are 'worth taking into account'. One of these strategies is to generate a Markov chain to explore the model space and then approximate Eqn. **35.37** using samples from the posterior $p(m|y)$ (Madigan and York, 1992). But this is computationally very expensive.

In this chapter, we will instead use the Occam's window procedure for nested models described in Madigan and Raftery (1994). First, a model that is N_0 times less likely *a posteriori* than the maximum posterior model is removed (in this chapter we use $N_0 = 20$). Second, complex models with posterior probabilities smaller than their simpler counterparts are also excluded. The remaining models fall in Occam's window. This leads to the following approximation to the posterior density:

$$p(\theta|y) = \sum_{m \in C} p(\theta|y, m)p(m|y) \tag{35.39}$$

where the set C identifies 'Occam's window'. Models falling in this window can be identified using the search strategy defined in Madigan and Raftery (1994).

DYNAMIC CAUSAL MODELS

The term 'causal' in DCM arises because the brain is treated as a deterministic dynamical system (see e.g. section 1.1 in Friston *et al.* (2003) in which external inputs cause changes in neuronal activity which, in turn, cause

[2] Strictly, AIC should be referred to as an information criterion.

[3] Software is also available from http://www.research.att.com/volinsky/bma.html.

changes in the resulting fMRI, MEG or EEG signal. DCMs for fMRI comprise a bilinear model for the neurodynamics and an extended Balloon model (Buxton, 1998; Friston, 2002) for the haemodynamics. These are described in detail in Chapter 41.

The effective connectivity in DCM is characterized by a set of 'intrinsic connections' that specify which regions are connected and whether these connections are unidirectional or bidirectional. We also define a set of input connections that specify which inputs are connected to which regions, and a set of modulatory connections that specify which intrinsic connections can be changed by which inputs. The overall specification of input, intrinsic and modulatory connectivity comprise our assumptions about model structure. This, in turn, represents a scientific hypothesis about the structure of the large-scale neuronal network mediating the underlying cognitive function. Examples of DCMs are shown in Figure 35.5.

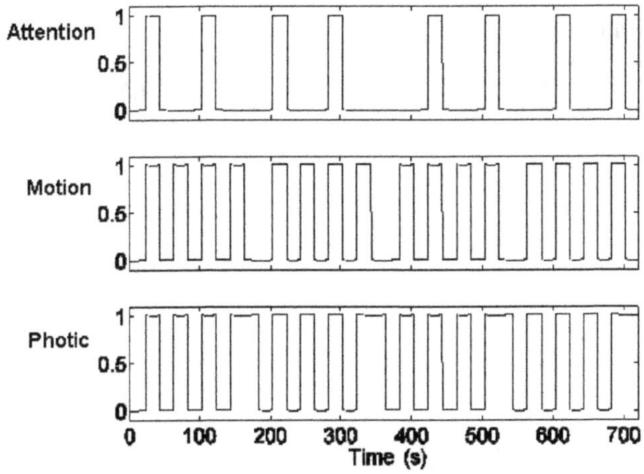

FIGURE 35.3 The 'photic', 'motion' and 'attention' variables used in the DCM analysis of the attention to visual motion data (see Figures 35.4 and 35.5).

Attention to visual motion

In previous work we have established that attention modulates connectivity in a distributed system of cortical regions that subtend visual motion processing (Buchel and Friston, 1997; Friston and Buchel, 2000). These findings were based on data acquired using the following experimental paradigm. Subjects viewed a computer screen which displayed either a fixation point, stationary dots or dots moving radially outward at a fixed velocity. For the purpose of our analysis we can consider three experimental variables. The 'photic stimulation' variable indicates when dots were on the screen, the 'motion' variable indicates that the dots were moving and the 'attention' variable indicates that the subject was attending to possible velocity changes. These are the three input variables that we use in our DCM analyses and are shown in Figure 35.3.

In this chapter, we model the activity in three regions V1, V5 and superior parietal cortex (SPC). The original 360-scan time series were extracted from the data set of a single subject using a local eigendecomposition and are shown in Figure 35.4.

We initially set up three DCMs, each embodying different assumptions about how attention modulates connections to V5. Model 1 assumes that attention modulates the forward connection from V1 to V5, model 2 assumes that attention modulates the backward connection from SPC to V5 and model 3 assumes attention modulates both connections. These models are shown in Figure 35.5. Each model assumes that the effect of motion is to modulate the connection from V1 to V5 and uses the same reciprocal hierarchical intrinsic connectivity.

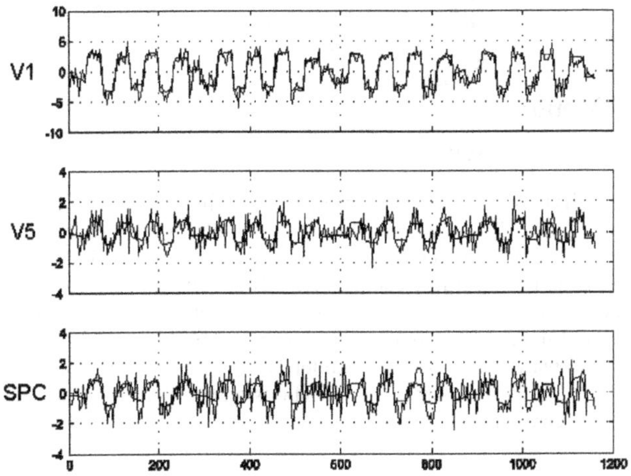

FIGURE 35.4 Attention data. fMRI time series (rough solid lines) from regions V1, V5 and SPC and the corresponding estimates from DCM model 1 (smooth solid lines).

We fitted the models and computed Bayes factors, shown in Table 35-2. We did not use the Laplace approximation to the model evidence, as DCM for fMRI uses fixed prior variances which compound model comparison. Instead, we computed both AIC and BIC and made an inference only if the two resulting Bayes factors were consistent (Penny *et al.*, 2004).

Table 35-2 shows that the data provide consistent evidence in favour of the hypothesis embodied in model 1, i.e. that attention modulates solely the forward connection from V1 to V5.

We now look more closely at the comparison of model 1 with model 2. The estimated connection strengths of the attentional modulation were 0.23 for the forward

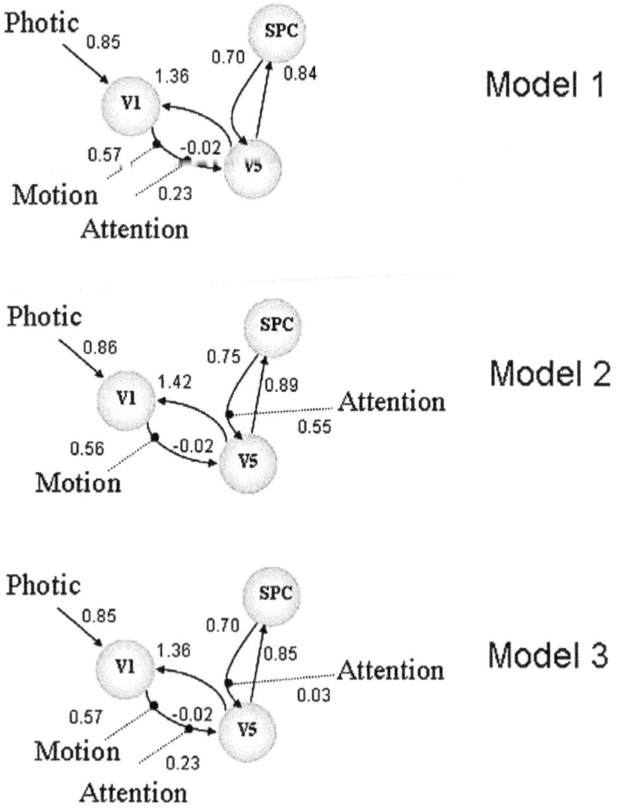

FIGURE 35.5 Attention models. In all models, photic stimulation enters V1 and the motion variable modulates the connection from V1 to V5. Models 1, 2 and 3 have reciprocal and hierarchically organized intrinsic connectitivty. They differ in how attention modulates the connectivity to V5, with model 1 assuming modulation of the forward connection, model 2 assuming modulation of the backward connection and model 3 assuming both. Solid arrows indicate input and intrinsic connections and dotted lines indicate modulatory connections.

TABLE 35-2 Attention data – comparing modulatory connectivities. Bayes factors provide consistent evidence in favour of the hypothesis embodied in model 1, that attention modulates (solely) the bottom-up connection from V1 to V5. Model 1 is preferred to models 2 and 3. Models 1 and 2 have the same number of connections so AIC and BIC give identical values

	B_{12}	B_{13}	B_{32}
AIC	3.56	2.81	1.27
BIC	3.56	19.62	0.18

connection in model 1 and 0.55 for the backward connection in model 2. This shows that attentional modulation of the backward connection is stronger than the forward connection. However, a breakdown of the Bayes factor B_{12} in Table 35-3 shows that the reason model 1 is favoured over model 2 is because it is more accurate. In particular, it predicts superior parietal cortex (SPC)

TABLE 35-3 Attention data. Breakdown of contributions to the Bayes factor for model 1 versus model 2. The largest single contribution to the Bayes factor is the increased model accuracy in region SPC, where 8.38 fewer bits are required to code the prediction errors. The overall Bayes factor B_{12} of 3.56 provides consistent evidence in favour of model 1

Source	Model 1 versus model 2 relative cost (bits)	Bayes factor B_{12}
V1 accuracy	7.32	0.01
V5 accuracy	−0.77	1.70
SPC accuracy	−8.38	333.36
Complexity (AIC)	0.00	1.00
Complexity (BIC)	0.00	1.00
Overall (AIC)	−1.83	3.56
Overall (BIC)	−1.83	3.56

activity much more accurately. Thus, although model 2 does show a significant modulation of the SPC-V5 connection, the required change in its prediction of SPC activity is sufficient to compromise the overall fit of the model. If we assume models 1 and 2 are equally likely *a priori* then our posterior belief in model 1 is 0.78 (from 3.56/(3.56+1)). Thus, model 1 is the favoured model even though the effect of attentional modulation is weaker.

This example makes an important point. Two models can only be compared by computing the evidence for each model. It is not sufficient to compare values of single connections. This is because changing a single connection changes overall network dynamics and each hypothesis is assessed (in part) by how well it predicts the data, and the relevant data are the activities in a distributed network.

We now focus on model 3 that has *both* modulation of forward and backward connections. First, we make a statistical inference to see if, within model 3, modulation of the forward connection is larger than modulation of the backward connection. For these data, the posterior distribution of estimated parameters tells us that this is the case with probability 0.75. This is a different sort of inference to that made above. Instead of inferring which is more likely, modulation of a forward or of a backward connection, we are making an inference about which effect is stronger when both are assumed present.

However, this inference is contingent on the assumption that model 3 is a good model. It is based on the density $p(\theta|y, m=3)$. The Bayes factors in Table 35-2, however, show that the data provide consistent evidence in favour of the hypothesis embodied in model 1, that attention modulates *only* the forward connection. Table 35-4 shows a breakdown of B_{13}. Here the largest contribution to the Bayes factor (somewhere between 2.72 and 18.97) is the increased parameter cost for model 3.

TABLE 35-4 Attention data. Breakdown of contributions to the Bayes factor for model 1 versus model 3. The largest single contribution to the Bayes factor is the cost of coding the parameters. The table indicates that both models are similarly accurate but model 1 is more parsimonious. The overall Bayes factor B_{13} provides consistent evidence in favour of the (solely) bottom-up model

Source	Model 1 versus model 3 relative cost (bits)	Bayes factor B_{13}
V1 accuracy	−0.01	1.01
V5 accuracy	0.02	0.99
SPC accuracy	−0.05	1.04
Complexity (AIC)	−1.44	2.72
Complexity (BIC)	−4.25	18.97
Overall (AIC)	−1.49	2.81
Overall (BIC)	−4.29	19.62

The combined use of Bayes factors and DCM provides us with a formal method for evaluating competing scientific theories about the forms of large-scale neural networks and the changes in them that mediate perception and cognition. These issues are pursued in Chapter 43 in which DCMs are compared so as to make inferences about inter-hemispheric integration from fMRI data.

SOURCE RECONSTRUCTION

A comprehensive introduction to source reconstruction is provided in Baillet *et al.* (2001). For more recent developments see Michel *et al.* (2004) and Chapters 28 to 30. The aim of source reconstruction is to estimate sources, θ, from sensors, y, where:

$$y = X\theta + e \qquad 35.40$$

e is an error vector and X defines a lead-field matrix. Distributed source solutions usually assume a Gaussian prior for:

$$p(\theta) = \mathrm{N}(\mu_p, C_p) \qquad 35.41$$

Parameter inference for source reconstruction can then be implemented as described in the section above on linear models. Model inference can be implemented using the expression in Eqn. **35.29**. For the numerical results in this chapter, we augmented this expression to account for uncertainty in the estimation of the hyperparameters. The full expression for the log-evidence of hyperparameterized models under the Laplace approximation is described in Trujillo-Barreto *et al.* (2004) and Appendix 4.

MULTIPLE CONSTRAINTS

This section considers source reconstruction with multiple constraints. This topic is covered in greater detail and from a different perspective in Chapters 29 and 30. The constraints are implemented using a decomposition of the prior covariance into distinct components:

$$C_p = \sum_i \lambda_i Q_i \qquad 35.42$$

The first type of constraint is a smoothness constraint, Q_{sc}, based on the usual L^2-norm. The second is an intrinsic functional constraint, Q_{int}, based on multivariate source prelocalization (MSP) (Mattout *et al.*, 2005). This provides an estimate, based on a multivariate characterization of the M/EEG data themselves. Thirdly, we used extrinsic functional constraints which were considered as 'valid', Q^v_{ext}, or 'invalid', Q^i_{ext}. These extrinsic constraints are derived from other imaging modalities such as fMRI. We used invalid constraints to test the robustness of the source reconstructions.

To test the approach, we generated simulated sources from the locations shown in Plate 49(a) (see colour plate section). Temporal activity followed a half-period sine function with a period of 30 ms. This activity was projected onto 130 virtual MEG sensors and Gaussian noise was then added. Further details on the simulations are given in Mattout *et al.* (2006).

We then reconstructed the sources using all combinations of the various constraints. Plate 50 shows a sample of source reconstructions. Table 35-5 shows the evidence for each model which we computed using the Laplace

TABLE 35-5 Log-evidence of models with different combinations of smoothness constraints, Q_{sc}, intrinsic constraints, Q_{int}, valid, Q^v_{ext} and invalid, Q^i_{ext}, extrinsic constraints

	Log-evidence	
1 constraint	Q_{sc}	205.2
	Q_{int}	208.4
	Q^v_{ext}	**215.6**
	Q^i_{ext}	131.5
2 constraints	Q_{sc}, Q_{int}	207.4
	Q_{sc}, Q^v_{ext}	214.1
	Q_{sc}, Q^i_{ext}	204.9
	Q_{int}, Q^v_{ext}	214.9
	Q_{int}, Q^i_{ext}	207.4
	Q^v_{ext}, Q^i_{ext}	213.2
3 constraints	$Q_{sc}, Q_{int}, Q^v_{ext}$	211.5
	$Q_{sc}, Q_{int}, Q^i_{ext}$	207.2
	$Q_{sc}, Q^v_{ext}, Q^i_{ext}$	214.7
	$Q_{int}, Q^v_{ext}, Q^i_{ext}$	212.7
4 constraints	$Q_{sc}, Q_{int}, Q^v_{ext}, Q^i_{ext}$	211.3

TABLE 35-6 Bayes factors for models with and without valid location priors, B_{21}, and with and without invalid location priors, B_{31}. Valid location priors make the models significantly better, whereas invalid location priors do not make them significantly worse

Bayes factor				
Model 1	Model 2	Model 3	B_{21}	B_{31}
Q_{sc}	Q_{sc}, Q^v_{ext}	Q_{sc}, Q^i_{ext}	7047	0.8
Q_{int}	Q_{int}, Q^v_{ext}	Q_{int}, Q^i_{ext}	655	0.4
Q_{sc}, Q_{int}	$Q_{sc}, Q_{int}, Q^v_{ext}$	$Q_{sc}, Q_{int}, Q^i_{ext}$	60	0.8

approximation (which is exact for these linear Gaussian models). As expected, the model with the single valid location prior had the highest evidence.

Further, any model which contains the valid location prior has high evidence. The table also shows that any model which contains both valid and invalid location priors does not show a dramatic decrease in evidence, compared to the same model without the invalid location prior. These trends can be assessed more formally by computing the relevant Bayes factors, as shown in Table 35-6. This shows significantly enhanced evidence in favour of models including valid location priors. It also suggests that the smoothness and intrinsic location priors can ameliorate the misleading effect of invalid priors.

MODEL AVERAGING

In this section, we consider source localizations with anatomical constraints. A class of source reconstruction models is defined where, for each model, activity is assumed to derive from a particular anatomical 'compartment' or combination of compartments. Anatomical compartments are defined by taking 71 brain regions, obtained from a 3D segmentation of the probabilistic MRI atlas (PMA) (Evans *et al.*, 1993) shown in Plate 51. These compartments preserve the hemispheric symmetry of the brain, and include deep areas like thalamus, basal ganglia and brainstem. Simple activations may be localized to single compartments and more complex activations to combinations of compartments. These combinations define a nested family of source reconstruction models which can be searched using the Occam's window approach described above.

The source space consists of a 3D-grid of points that represent the possible generators of the EEG/MEG inside the brain, while the measurement space is defined by the array of sensors where the EEG/MEG is recorded. We used a 4.25 mm grid spacing and different arrays of electrodes/coils are placed in registration with the PMA. The 3D-grid is further clipped by the grey matter, which consists of all brain regions segmented and shown in Plate 51.

Three arrays of sensors were used and are depicted in Plate 52. For EEG simulations a first set of 19 electrodes (EEG-19) from the 10/20 system is chosen. A second configuration of 120 electrodes (EEG-120) is also used in order to investigate the dependence of the results on the number of sensors. Here, electrode positions were determined by extending and refining the 10/20 system. For MEG simulations, a dense array of 151 sensors were used (MEG-151). The physical models constructed in this way, allow us to compute the electric/magnetic lead field matrices that relate the primary current density (PCD) inside the head, to the voltage/magnetic field measured at the sensors.

We now present the results of two simulation studies. In the first study, two distributed sources were simulated. One source was located in the right occipital pole, and the other in the thalamus. This simulation is referred to as 'OPR+TH'. The spatial distribution of PCD (i.e. the true θ vector) was generated using two narrow Gaussian functions of the same amplitude shown in Figure 35.6(a).

The temporal dynamics were specified using a linear combination of sine functions with frequency components evenly spaced in the alpha band (8–12 Hz). The amplitude of the oscillation as a function of frequencies is a narrow Gaussian peaked at 10 Hz. That is, activity is given by:

$$j(t) = \sum_{i=1}^{N} \exp(-8(f_i - 10)^2) \sin(2\pi f_i t) \quad \textbf{35.43}$$

where $8 \leq f_i \leq 12$ Hz. Here, f_i is the frequency and t denotes time. These same settings are then used for the second simulation study, in which only the thalamic (TH) source was used (see Figure 35.6(b)). This second

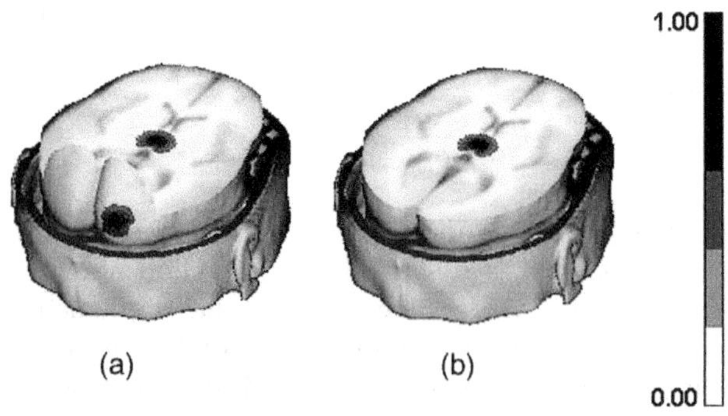

FIGURE 35.6 Spatial distributions of the simulated primary current densities. (a) Simultaneous activation of two sources at different depths: one in the right occipital pole and the other in the thalamus (OPR+TH). (b) Simulation of a single source in the thalamus (TH).

simulation is referred to as 'TH'. In both cases the measurements were generated with a signal-to-noise ratio (SNR) of 10.

The simulated data were then analysed using Bayesian model averaging (BMA) in order to reconstruct the sources. We searched through model space using the Occam's window approach described above. For comparison, we also applied the constrained low resolution tomography (cLORETA) algorithm. This method constrains the solution to grey matter and again uses the usual L^2-norm. The cLORETA model is included in the model class used for BMA, and corresponds to a model comprising all 71 anatomical compartments.

The absolute values of the BMA and cLORETA solutions for the OPR+TH example, and for the three arrays of sensors used, are depicted in Figure 35.7. In all cases, cLORETA is unable to recover the TH source and the OPR source estimate is overly dispersed. For BMA, the spatial localizations of both cortical and subcortical sources are recovered with reasonable accuracy in all cases. These results suggest that the EEG/MEG contains enough information for estimating deep sources, even in cases where such generators might be hidden by cortical activations.

The reconstructed sources shown in Figure 35.8 for the TH case show that cLORETA suffers from a 'depth biasing' problem. That is, deep sources are misattributed to superficial sources. This biasing is not due to masking effects, since no cortical source is present in this set of simulations. Again, BMA gives significantly better estimates of the PCD.

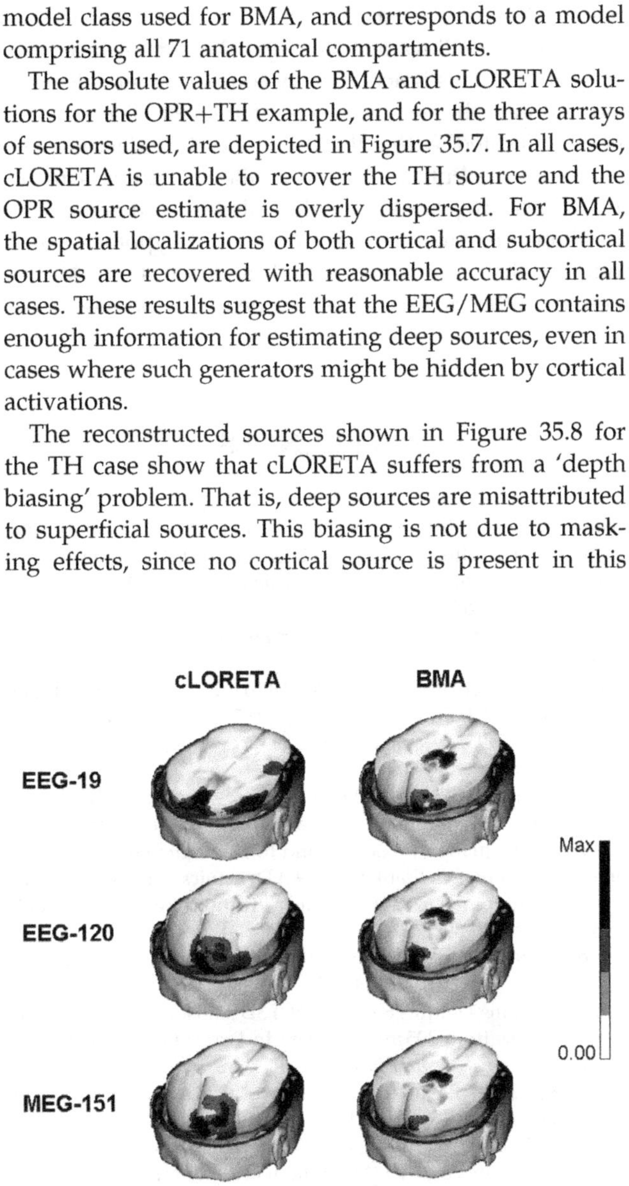

FIGURE 35.7 3D reconstructions of the absolute values of BMA and cLORETA solutions for the OPR+TH source case. The first column indicates the array of sensors used in each simulated data set. The maximum of the scale is different for each case. For cLORETA (from top to bottom): Max = 0.21, 0.15 and 0.05; for BMA (from top to bottom): Max = 0.41, 0.42 and 0.27.

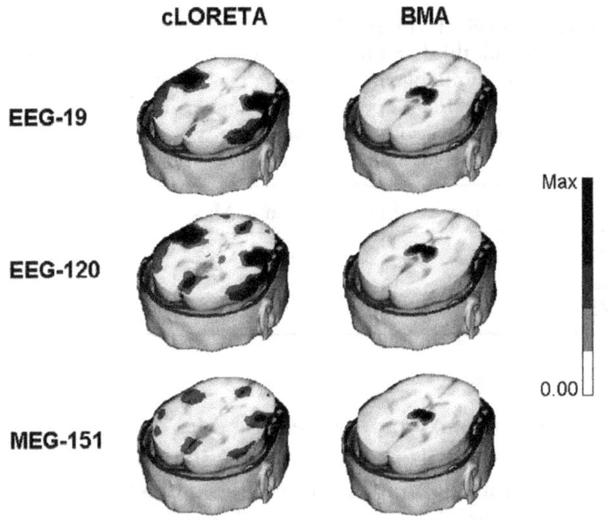

FIGURE 35.8 3D reconstructions of the absolute values of BMA and cLORETA solutions for the TH source case. The first column indicates the array of sensors used in each simulated data set. The maximum of the scale is different for each case. For cLORETA (from top to bottom): Max = 0.06, 0.01 and 0.003; for BMA (from top to bottom): Max = 0.36, 0.37 and 0.33.

Figures 35.7 and 35.8 also show that the reconstructed sources become more concentrated and clearer, as the number of sensors increases. Tables 35-7 and 35-8 show the number of models in Occam's window for each simulation study. The number of models reduces with increasing number of sensors. This is natural since more precise measurements imply more information available about the underlying phenomena, and then narrower and sharper model distributions are obtained. Consequently, as shown in the tables, the probability and, hence, the rank of the true model in the Occam's window increase for dense arrays of sensors.

Tables 35-7 and 35-8 also show that the model with the highest probability is not always the true one. This fact supports the use of BMA instead of using the maximum

TABLE 35-7 BMA results for the 'OPR+TH' simulation study. The second, third and fourth columns show the number of models, and minimum and maximum probabilities, in Occam's window. In the last column, the number in parentheses indicates the position of the true model when all models in Occam's window are ranked by probability

Sensors	Number of models	Min	Max	Prob true model
EEG-19	15	0.02	0.30	0.11(3)
EEG-120	2	0.49	0.51	0.49(2)
MEG-151	1	1	1	1

TABLE 35-8 BMA results for the 'TH' simulation study. The second, third and fourth columns show the number of models, and minimum and maximum probabilities, in Occam's window. In the last column, the number in parentheses indicates the position of the true model when all models in Occam's window are ranked by probability

Sensors	Number of models	Min	Max	Prob true model
EEG-19	3	0.30	0.37	0.30 (3)
EEG-120	1	1	1	1
MEG-151	1	1	1	1

posterior or maximum evidence model. In the present simulations, this is not critical, since the examples analysed are quite simple. But it becomes a determining factor when analysing more complex data, as is the case with some real experimental conditions (Trujillo-Barreto *et al.*, 2004).

An obvious question then arises. Why is cLORETA unable to exploit fully the information contained in the M/EEG? The answer given by Bayesian inference is simply that cLORETA, which assumes activity is distributed over all of grey matter, is not a good model. In the model averaging framework, the cLORETA model was always rejected due to its low posterior probability, placing it outside Occam's window.

DISCUSSION

Chapter 11 showed how Bayesian inference in hierarchical models can be implemented using the belief propagation algorithm. This involves passing messages up and down the hierarchy, the upward messages being likelihoods and evidences and the downward messages being posterior probabilities.

In this chapter, we have shown how belief propagation can be used to make inferences about members of a model class. Three stages were identified in this process: (i) conditional parameter inference; (ii) model inference; and (iii) model averaging. Only at the model averaging stage is belief propagation complete. Only then will parameter inferences be based on the correct marginal density.

We have described how this process can be implemented for linear and non-linear models and applied to domains such as dynamic causal modelling and M/EEG source reconstruction. In DCM, often the most important inference to be made is a model inference. This can be implemented using Bayes factors and allows one to make inferences about the structure of large scale neural networks that mediate cognitive and perceptual processing. This issue is taken further in Chapter 43, which considers inter-hemispheric integration.

The application of model averaging to M/EEG source reconstruction results in the solution of an outstanding problem in the field, i.e. how to detect deep sources. Simulations show that a standard method (cLORETA) is simply not a good model and that model averaging can combine the estimates of better models to make veridical source estimates.

The use of Bayes factors for model comparison is somewhat analagous to the use of F-tests in the general linear model. Whereas t-tests are used to assess individual effects, F-tests allow one to assess the significance of a set of effects. This is achieved by comparing models with and without the set of effects of interest. The smaller model is 'nested' within the larger one. Bayes factors play a similar role but, additionally, allow inferences to be constrained by prior knowledge. Moreover, it is possible simultaneously to entertain a number of hypotheses and compare them using the model evidence. Importantly, these hypotheses are not constrained to be nested.

REFERENCES

Akaike H (1973) Information theory and an extension of the maximum likelihood principle. In *Second international symposium on information theory*, Petrox BN, Caski F (eds). Akademiai Kiado, Budapest, p 267

Akaike H (1983) Information measures and model selection. *Bull Int Stat Inst* **50**: 277–90

Baillet S, Mosher JC, Leahy RM (2001) Electromagnetic brain mapping. *IEEE Signal Process Mag*, pp 14–30, November

Bishop CM (1995) *Neural networks for pattern recognition*. Oxford University Press, Oxford

Buchel C, Friston KJ (1997) Modulation of connectivity in visual pathways by attention: Cortical interactions evaluated with structural equation modelling and fMRI. *Cereb Cortex* **7**: 768–78

Buxton RB, Wong EC, Frank LR (1998) Dynamics of blood flow and oxygenation changes during brain activation: the Balloon model. *Magn Res Med* **39**: 855–64

Draper D (1995) Assessment and propagation of model uncertainty. *J Roy Stat Soc Series B* **57**: 45–97

Evans A, Collins D, Mills S *et al.* (1993) 3D statistical neuroanatomical models from 305 mri volumes. In *Proc IEEE Nuclear Science Symposium and Medical Imaging Conference*

Friston KJ (2002) Bayesian estimation of dynamical systems: an application to fMRI. *NeuroImage* **16**: 513–30

Friston KJ, Buchel C (2000) Attentional modulation of effective connectivity from V2 to V5/MT in humans. *Proc Natl Acad Sci USA* **97**: 7591–96

Friston KJ, Penny WD, Phillips C *et al.* (2002) Classical and Bayesian inference in neuroimaging: theory. *NeuroImage* **16**: 465–83

Friston KJ, Harrison L, Penny WD (2003) Dynamic causal modelling. *NeuroImage* **19**: 1273–302

Gelman A, Carlin JB, Stern HS *et al.* (1995) *Bayesian data analysis*. Chapman and Hall, Boca Raton

Hoeting JA, Madigan D, Raftery AE *et al.* (1999) Bayesian model averaging: a tutorial. *Stat Sci* **14**: 382–417

Jefferys H (1935) Some tests of significance, treated by the theory of probability. *Proc Camb Philos Soc* **31**: 203–22

Kass RE, Raftery AE (1993) Bayes factors and model uncertainty. Technical Report 254, University of Washington. http://www.stat.washington.edu/tech.reports/tr254.ps.

Kass RE, Raftery AE (1995) Bayes factors. *J Am Stat Assoc* **90**: 773–95

Lee PM (1997) *Bayesian statistics: an introduction*, 2nd edn. Arnold, London

Madigan D, Raftery A (1994) Model selection and accounting for uncertainty in graphical models using Occam's window. *J Am Stat Assoc* **89**: 1535–46

Madigan D, York J (1992) Bayesian graphical models for discrete data. Technical report, Department of Statistics, University of Washington, Report number 259

Mattout J, Pelegrini-Isaac M, Garnero L *et al.* (2005) Multivariate source prelocalisation: use of functionally informed basis functions for better conditioning the MEG inverse problem. *NeuroImage* **26**: 356–73

Mattout J, Phillips C, Penny WD *et al.* (2006) MEG source localisation under multiple constraints: an extended Bayesian framework. *NeuroImage* **30**: 753–67

Michel CM, Marraya MM, Lantza G *et al.* (2004) EEG source imaging. *Clin Neurophysiol* **115**: 2195–22

Penny WD, Stephan KE, Mechelli A *et al.* (2004) Comparing dynamic causal models. *NeuroImage* **22**: 1157–72

Penny WD, Flandin G, Trujillo-Barreto N (2006) Bayesian comparison of spatially regularised general linear models. *Hum Brain Mapp* (in press)

Press WH, Teukolsky SA, Vetterling WT *et al.* (1992) *Numerical Recipes in C*. Cambridge University Press, Cambridge

Raftery AE (1995) Bayesian model selection in social research. In *Sociological methodology*, Marsden PV (ed.). Cambridge, MA, pp 111–96

Ripley B (1995) *Pattern recognition and neural networks*. Cambridge University Press, Cambridge

Schwarz G (1978) Estimating the dimension of a model. *Ann Stat* **6**: 461–64

Stephan KE, Friston KJ, Penny WD (2005) Computing the objective function in DCM. Technical report, Wellcome Department of Imaging Neuroscience, ION, UCL, London

Trujillo-Barreto N, Aubert-Vazquez E, Valdes-Sosa P (2004) Bayesian model averaging in EEG/MEG imaging. *NeuroImage* **21**: 1300–19

Part 7

Connectivity

CHAPTER

36

Functional integration

K. Friston

INTRODUCTION

The next chapters are about functional integration in the brain. This chapter reviews the neurobiology of functional integration, in terms of neuronal information processing and frames the sorts of question that can be addressed with analyses of functional and effective connectivity. In fact, we use empirical Bayes (see Part 4) as the basis for understanding integration among the levels of hierarchically organized cortical areas. The next two chapters (Chapters 37 and 38) deal with functional and effective connectivity. Chapters 39 and 40 deal with complementary models of functional integration, namely the Volterra or generalized convolution formulation and state-space representations. The final chapters in this section cover dynamic causal modelling. For a more mathematical treatment of these models and their interrelationships, see Appendix 2.

In this chapter, we will review the empirical evidence for functional specialization and integration, with a special focus on extrinsic connections among cortical areas and how they define cortical hierarchies. We will then look at these hierarchical architectures, using ideas from theoretical neurobiology, which clarify the potential role of forward and backward connections. Finally, we see how neuroimaging can be used to test hypotheses that arise from this theoretical treatment. Specifically, we show that functional neuroimaging can be used to test for interactions between bottom-up and top-down influences on an area and to make quantitative inferences about changes in connectivity. This chapter is quite theoretical and is used to introduce constraints from neurobiology and computational neuroscience that provide a context for the models of functional integration presented in the subsequent chapters.

Interactions and context-sensitivity

In concert with the growing interest in contextual and extra-classical receptive field effects in electrophysiology (i.e. how the receptive fields of sensory neurons change according to the context a stimulus is presented in), a similar shift is apparent in imaging. Namely, the appreciation that functional specialization can exhibit similar extra-classical phenomena, in which a cortical area may be specialized for one thing in one context but something else in another (see McIntosh, 2000). These extra-classical phenomena have implications for theoretical ideas about how the brain might work. This chapter uses theoretical models of representational or perceptual inference as a vehicle to illustrate how imaging can be used to address important questions about functional brain architectures.

Many models of perceptual learning and inference require prior assumptions about the distribution of sensory causes. However, as seen in the previous chapters, empirical Bayes suggests that priors can be learned in a hierarchical context. The main point made in this chapter is that backward connections, mediating internal or generative models of how sensory data are caused, can construct empirical priors and are essential for perception. Moreover, non-linear generative models require these connections to be modulatory so that causes in higher cortical levels can interact to predict responses in lower levels. This is important in relation to functional asymmetries in forward and backward connections that have been demonstrated empirically.

Overview

We start by reviewing the two fundamental principles of brain organization, namely functional specialization

Statistical Parametric Mapping, by Karl Friston et al.
ISBN–13: 978-0-12-372560-8 ISBN–10: 0-12-372560-7

and functional integration and how they rest upon the anatomy and physiology of cortico-cortical connections in the brain. The second section deals with the nature and learning of representations from a theoretical or computational perspective. The key focus of this section is on the functional architectures implied by the theory. Generative models based on predictive coding rest on hierarchies of backward and lateral projections and, critically, confer a necessary role on backward connections. The theme of context-sensitive responses is used in the subsequent section to preview different ways of measuring connectivity with functional neuroimaging. The focus of this section is evidence for the interaction of bottom-up and top-down influences and establishing the presence of backward connections. The final section reviews some of the implications for lesion studies and neuropsychology. *Dynamic diaschisis* is introduced, in which aberrant neuronal responses can be observed following damage to distal brain areas that provide enabling or modulatory afferents.

FUNCTIONAL SPECIALIZATION AND INTEGRATION

The brain appears to adhere to two fundamental principles of functional organization, functional integration and functional specialization, where the integration within and among specialized areas is mediated by effective connectivity. The distinction relates to that between 'localizationism' and 'connectionism' that dominated thinking about cortical function in the nineteenth century. Since the early anatomic theories of Gall, the identification of a particular brain region with a specific function has become a central theme in neuroscience. However, functional localization *per se* was not easy to demonstrate: for example, a meeting that took place on 4 August 1881 addressed the difficulties of attributing function to a cortical area, given the dependence of cerebral activity on underlying connections (Phillips *et al.*, 1984). This meeting was entitled 'Localization of function in the cortex cerebri'. Goltz, although accepting the results of electrical stimulation in dog and monkey cortex, considered that the excitation method was inconclusive, in that the behaviours elicited might have originated in related pathways, or current could have spread to distant centres. In short, the excitation method could not be used to infer functional localization because localizationism discounted interactions, or functional integration among different brain areas. It was proposed that lesion studies could supplement excitation experiments. Ironically, it was observations on patients with brain lesions some years later (see Absher and Benson, 1993) that led to the concept of 'disconnection syndromes' and the refutation of localizationism as a complete or sufficient explanation of cortical organization. Functional localization implies that a function can be localized in a cortical area, whereas specialization suggests that a cortical area is specialized for some aspects of perceptual or motor processing, where this *specialization* can be anatomically *segregated* within the cortex. The cortical infrastructure supporting a single function may then involve many specialized areas whose union is mediated by the functional integration among them. Functional specialization and integration are not exclusive, they are complementary. Functional specialization is only meaningful in the context of functional integration and vice versa.

Functional specialization and segregation

The functional role, played by any component (e.g. cortical area, sub-area, neuronal population or neuron) of the brain, is defined largely by its connections. Certain patterns of cortical projections are so common that they amount to rules of cortical connectivity. 'These rules revolve around one, apparently, overriding strategy that the cerebral cortex uses – that of functional segregation' (Zeki, 1990). Functional segregation demands that cells with common functional properties be grouped together. This architectural constraint necessitates both convergence and divergence of cortical connections. Extrinsic connections, between cortical regions, are not continuous but occur in patches or clusters. This patchiness has a clear relationship to functional segregation. For example, the secondary visual area, V2, has a cytochrome oxidase architecture, consisting of thick, thin and interstripes. When recordings are made in V2, directionally selective (but not wavelength or colour selective) cells are found exclusively in the thick stripes. Retrograde (i.e. backward) labelling of cells in V5 is limited to these thick stripes. All the available physiological evidence suggests that V5 is a functionally homogeneous area that is specialized for visual motion. Evidence of this nature supports the notion that patchy connectivity is the anatomical infrastructure that underpins functional segregation and specialization. If it is the case that neurons in a given cortical area share a common responsiveness (by virtue of their extrinsic connectivity) to some sensorimotor or cognitive attribute, then this functional segregation is also an anatomical one. Challenging a subject with the appropriate sensorimotor attribute or cognitive process should lead to activity changes in, and only in, the areas of interest. This is the model upon which the search for regionally specific effects with functional neuroimaging is based.

The anatomy and physiology of cortico-cortical connections

If specialization rests upon connectivity then important organizational principles should be embodied in the neuroanatomy and physiology of extrinsic connections. Extrinsic connections couple different cortical areas, whereas intrinsic connections are confined to the cortical sheet. There are certain features of cortico-cortical connections that provide strong clues about their functional role. In brief, there appears to be a hierarchical organization that rests upon the distinction between *forward* and *backward* connections. The designation of a connection as forward or backward depends primarily on its cortical layers of origin and termination. Some characteristics of cortico-cortical connections are presented below and are summarized in Table 36-1. The list is not exhaustive but serves to introduce some important principles that have emerged from empirical studies of visual cortex:

- *Hierarchical organization* – the organization of the visual cortices can be considered as a hierarchy of cortical levels with reciprocal extrinsic cortico-cortical connections among the constituent cortical areas (Felleman and Van Essen, 1991). The notion of a hierarchy depends upon a distinction between reciprocal forward and backward extrinsic connections (see Figure 36.1).
- *Reciprocal connections* – although reciprocal, forward and backward connections show both a microstructural and functional asymmetry. The terminations of both show laminar specificity. Forward connections (from a low to a high level) have sparse axonal bifurcations and are topographically organized, originating in supra-granular layers and terminating largely in layer IV. Backward connections, on the other hand, show abundant axonal bifurcation and a more diffuse topography. Their origins are bilaminar-infragranular and they terminate predominantly in supra-granular layers (Rockland and Pandya, 1979; Salin and Bullier, 1995). Extrinsic connections show an orderly convergence and divergence of connections from one cortical level to the next. At a macroscopic level, one point in a given cortical area will connect to a region 5–8 mm in diameter in another. An important distinction between forward and backward connections is that backward connections are more divergent. For example, the divergence region of a point in V5 (i.e. the region receiving backward afferents from V5) may include thick and inter-stripes in V2, whereas the convergence region (i.e. the region providing forward afferents to V5) is limited to the thick stripes (Zeki and Shipp, 1988). Backward connections are more abundant than forward connections and transcend more levels. For example, the ratio of forward efferent connections to backward afferents in the lateral geniculate is about one to ten. Another important distinction is that backward connections will traverse a number of hierarchical levels, whereas forward connections are more restricted. For example, there are backward connections from TE and TEO to V1 but no monosynaptic connections from V1 to TE or TEO (Salin and Bullier, 1995).
- *Functionally asymmetric forward and backward connections* – functionally, reversible inactivation (e.g. Sandell and Schiller, 1982; Girard and Bullier, 1989) and neuroimaging (e.g. Büchel and Friston, 1997) studies suggest that forward connections are driving and always elicit a response, whereas backward connections can be modulatory. In this context, modulatory means backward connections modulate responsiveness to other inputs. At the single cell level, 'inputs from drivers can be differentiated from those of modulators. The driver can be identified as the transmitter of receptive field properties; the modulator can be identified as altering the probability of certain aspects of that transmission' (Sherman and Guillery, 1998).

The notion that forward connections are concerned with the segregation of sensory information is consistent

TABLE 36-1 Some key characteristics of extrinsic cortico-cortical connections

- The organization of the visual cortices can be considered as a hierarchy (Felleman and Van Essen, 1991)
- The notion of a hierarchy depends upon a distinction between forward and backward extrinsic connections
- This distinction rests upon laminar specificity (Rockland and Pandya, 1979; Salin and Bullier, 1995)
- Backward connections are more numerous and transcend more levels
- Backward connections are more divergent than forward connections (Zeki and Shipp, 1988)

Forward connections	Backward connections
Sparse axonal bifurcations	Abundant axonal bifurcation
Topographically organized	Diffuse topography
Originate in supra-granular layers	Originate in bi-laminar/infra-granular layers
Terminate largely in layer IV	Terminate predominantly in supra-granular layers
Postsynaptic effects through fast AMPA (1.3–2.4 ms decay) and $GABA_A$ (6 ms decay) receptors	Modulatory afferents activate slow (50 ms decay) voltage-sensitive NMDA receptors

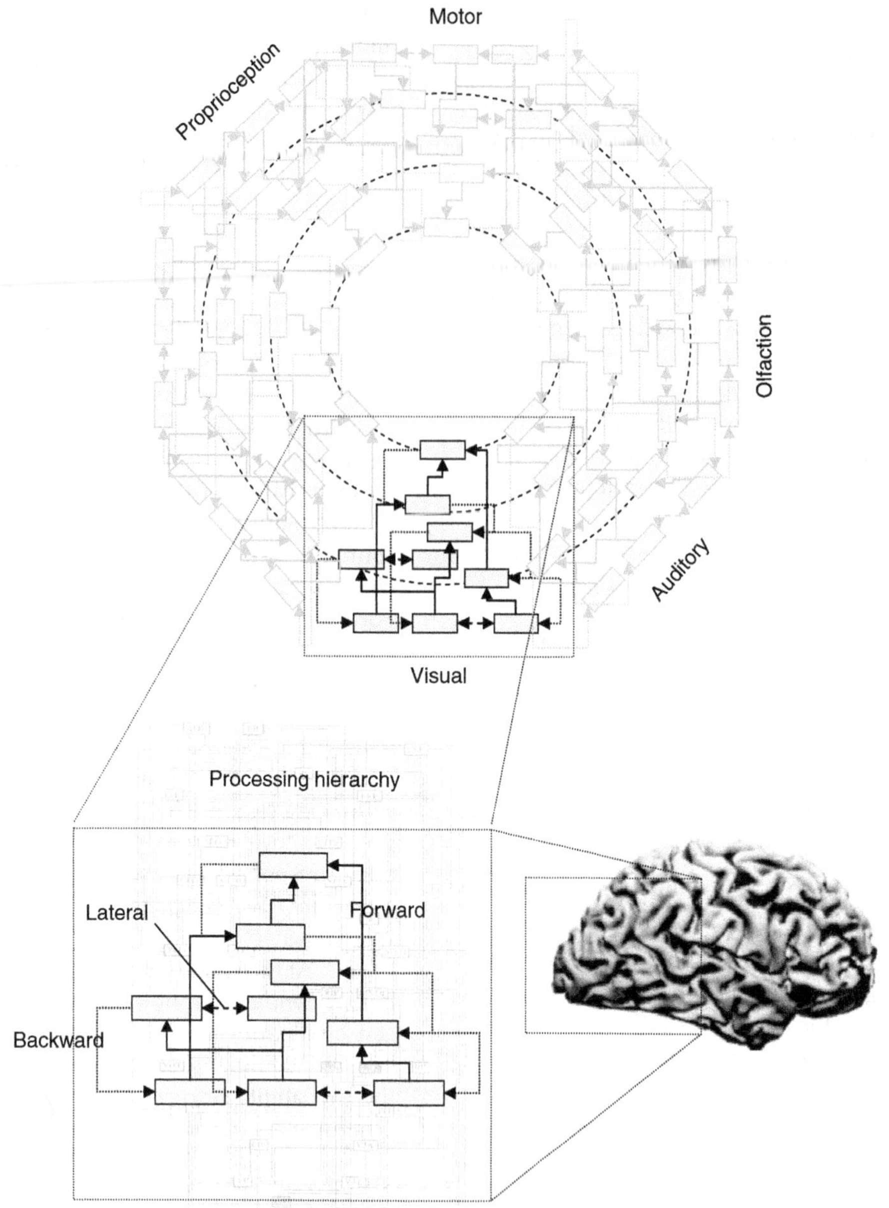

FIGURE 36.1 Schematic illustrating hierarchical structures in the brain and the distinction between forward, backward and lateral connections. This schematic is inspired by Mesulam's (1998) notion of sensory-fugal processing over 'a core synaptic hierarchy, which includes the primary sensory, upstream unimodal, downstream unimodal, hetero-modal, paralimbic and limbic zones of the cerebral cortex' (see Mesulam, 1998 for more details).

with their sparse axonal bifurcation, patchy axonal terminations and topographic projections. In contradistinction, backward connections are considered to have a role in mediating contextual effects and in the coordination of processing channels. This is consistent with their frequent bifurcation, diffuse axonal terminations and more divergent topography (Salin and Bullier, 1995; Crick and Koch, 1998). Forward connections mediate their post-synaptic effects through fast AMPA and GABA receptors. Modulatory effects can be mediated by NMDA receptors. NMDA receptors are voltage-sensitive, showing non-linear and slow dynamics (~50 ms decay). They are found predominantly in supra-granular layers, where backward connections terminate (Salin and Bullier, 1995). These slow time-constants again point to a role in mediating contextual effects that are more enduring than phasic sensory-evoked responses. The clearest evidence for the modulatory role of backward connections (mediated by slow glutamate receptors) comes from cortico-geniculate connections; in the cat lateral geniculate nucleus, cortical feedback is partly mediated by type 1 metabotropic glutamate receptors, which are located exclusively on distal segments of the relay-cell dendrites. Rivadulla *et al.* (2002) have shown that these backward afferents enhance the excitatory centre of the thalamic receptive field. 'Therefore, cortex, by closing this cortico-fugal loop, is able to increase the gain of its thalamic input within a focal spatial window, selecting key features of the incoming signal' (Rivadulla *et al.*, 2002) (see also Murphy and Sillito, 1987).

There are many mechanisms that are responsible for establishing connections in the brain. Connectivity results from interplay between genetic, epigenetic and activity- or experience-dependent mechanisms. *In utero*, epigenetic mechanisms predominate, such as the interaction between the topography of the developing cortical sheet, cell-migration, gene-expression and the mediating role of gene–gene interactions and gene products such as cell adhesion molecules. Following birth, connections are progressively refined and remodelled with a greater emphasis on activity- and use-dependent plasticity. These changes endure into adulthood with ongoing reorganization and experience-dependent plasticity that furnish behavioural adaptation and learning throughout life. In brief, there are two basic determinants of connectivity: *cellular plasticity*, reflecting cell-migration and neurogenesis in the developing brain; and *synaptic plasticity*, activity-dependent modelling of the pattern and strength of synaptic connections. This plasticity involves changes in the form, expression and function of synapses that endure throughout life. Plasticity is an important functional attribute of connections in the brain and is thought to underlie perceptual and procedural learning and memory. A key aspect of this plasticity is that it is generally associative.

- *Associative plasticity* – synaptic plasticity may be transient (e.g. short-term potentiation STP or depression STD) or enduring (e.g. long-term potentiation LTP or LTD) with many different time-constants. In contrast to short-term plasticity, long-term changes rely on protein synthesis, synaptic remodelling and infrastructural changes in cell processes (e.g. terminal arbours or dendritic spines) that are mediated by calcium-dependent mechanisms. An important aspect of NMDA receptors, in the induction of LTP, is that they confer an associative aspect on synaptic changes. This is because their voltage-sensitivity only allows calcium ions to enter the cell when there is conjoint presynaptic release of glutamate and sufficient postsynaptic depolarization (i.e. the temporal association of pre- and postsynaptic events). Calcium entry renders the postsynaptic specialization eligible for future potentiation by promoting the formation of synaptic 'tags' (e.g. Frey and Morris 1997) and other calcium-dependent intracellular mechanisms.

In summary, the anatomy and physiology of cortico-cortical connections suggest that forward connections are driving and commit cells to a prespecified response, given the appropriate pattern of inputs. Backward connections, on the other hand, are less topographic and are in a position to modulate the responses of lower areas to driving inputs from either higher or lower areas (see Table 36-1). For example, in the visual cortex Angelucci *et al.* (2002a; b) used a combination of anatomical and physiological recording methods to determine the spatial scale and retinotopic logic of intra-area V1 horizontal connections and inter-area feedback connections to V1. 'Contrary to common beliefs, these [monosynaptic horizontal] connections cannot fully account for the dimensions of the surround field [of macaque V1 neurons]. The spatial scale of feedback circuits from extrastriate cortex to V1 is, instead, commensurate with the full spatial range of centre-surround interactions. Thus, these connections could represent an anatomical substrate for contextual modulation and global-to-local integration of visual signals.'

Connections are not static but are changing at the synaptic level all the time. In many instances, this plasticity is associative. Backwards connections are abundant and are in a position to exert powerful effects on evoked responses, in lower levels, that define the specialization of any area or neuronal population. Modulatory effects imply the postsynaptic response evoked by presynaptic input is modulated by, or interacts with, another input. By definition this interaction must depend on non-linear synaptic or dendritic mechanisms.

Functional integration and effective connectivity

Electrophysiology and imaging neuroscience have firmly established functional specialization as a principle of brain organization in man. The functional integration of specialized areas has proven more difficult to assess. Functional integration refers to the interactions among specialized neuronal populations and how these interactions depend upon the sensorimotor or cognitive context. Functional integration is usually assessed by examining the correlations among activity in different brain areas, or trying to explain the activity in one area in relation to activities elsewhere. *Functional connectivity* is defined as correlations between remote neurophysiological events.[1]

[1] More generally any statistical dependency as measured by the mutual information

However, correlations can arise in a variety of ways. For example, in multiunit electrode recordings they can result from stimulus-*locked* transients, evoked by a common input or reflect stimulus-*induced* oscillations mediated by synaptic connections (Gerstein and Perkel, 1969). Integration within a distributed system is better understood in terms of *effective connectivity*. Effective connectivity refers explicitly to the influence that one neuronal system exerts over another, either at a synaptic (i.e. synaptic efficacy) or population level. It has been proposed that 'the [electrophysiological] notion of effective connectivity should be understood as the experiment- and time-dependent, simplest possible circuit diagram that would replicate the observed timing relationships between the recorded neurons' (Aertsen and Preil, 1991). This means effective connectivity is dynamic, i.e. activity-dependent, and depends upon a model of the interactions. Recent models of effective connectivity accommodate the modulatory or non-linear effects mentioned above. A more detailed discussion of these models is provided in subsequent chapters. In this chapter, the terms modulatory and non-linear are used almost synonymously. Modulatory effects imply the postsynaptic response evoked by one input is modulated, or interacts with, another. By definition this interaction must depend on non-linear synaptic mechanisms.

In summary, the brain can be considered as an ensemble of functionally specialized areas that are coupled in a non-linear fashion by effective connections. Empirically, it appears that connections from lower to higher areas are predominantly driving, whereas backward connections that mediate top-down influences are more diffuse and are capable of exerting modulatory influences. In the next section, we describe a theoretical perspective that highlights the functional importance of backward connections and non-linear interactions.

LEARNING AND INFERENCE IN THE BRAIN

This section describes the heuristics behind self-supervised learning based on *empirical Bayes*. This approach is considered within the framework of *generative models* and follows Dayan and Abbott (2001: 359–397) to which the reader is referred for background reading. A more detailed discussion of these issues can be found in Friston (2005).

First, we will reprise empirical Bayes in the context of brain function *per se*. Having established the requisite architecture for learning and inference, neuronal implementation is considered in sufficient depth to make predictions about the structural and functional anatomy that would be needed to implement empirical Bayes in the brain. We conclude by relating theoretical predictions with the four neurobiological principles listed in the previous section.

Causes, perception and sensation

Causes are simply the states of processes generating sensory data. It is not easy to ascribe meaning to these states without appealing to the way that we categorize things. Causes may be categorical in nature, such as the identity of a face or the semantic category of an object. Others may be parametric, such as the position of an object. Even though causes may be difficult to describe, they are easy to define operationally. Causes are quantities or states that are necessary to specify the products of a process generating sensory information. To keep things simple, let us frame the problem of representing causes in terms of a deterministic non-linear function:

$$u = g(v, \theta) \tag{36.1}$$

where v is a vector of causes in the environment (e.g. the velocity of a particular object, direction of radiant light etc.), and u represents sensory input. $g(v, \theta)$ is a function that generates data from the causes. θ are the parameters of the generative model. Unlike the causes, they are fixed quantities that have to be learned. We shall see later that the parameters correspond to connection strengths in the brain's model of how data are caused. Non-linearities in Eqn. **36.1** represent interactions among the causes. These can often be viewed as contextual effects, where the expression of a particular cause depends on the context established by another. A ubiquitous example, from early visual processing, is the occlusion of one object by another; in a linear world the visual sensation, caused by two objects, would be a transparent overlay or superposition. Occlusion is a non-linear phenomenon because the sensory input from one object (occluded) interacts, or depends on, the other (occluder). This interaction is an example of non-linear mixing of causes to produce sensory data. At a higher level, the cause associated with the word 'hammer' depends on the semantic context (that determines whether the word is a verb or a noun).

The problem the brain has to contend with is to find a function of the data that *recognizes* the underlying causes. To do this, the brain must undo the interactions to disclose contextually invariant causes. In other words, the brain must perform a non-linear un-mixing of causes and context. The key point here is that the non-linear mixing may not be invertible and recognition may be a fundamentally ill-posed problem. For example, no amount of un-mixing can recover the parts of an object that are occluded by another. The corresponding indeterminacy, in probabilistic learning, rests on the combinatorial explosion of ways in which stochastic generative models can generate input patterns (Dayan *et al.*, 1995). In what follows, we consider the implications of this problem. Put simply, recognition of causes from sensory data is the inverse of generating data from causes. If the generative model is not invertible, then recognition can only proceed if there is an explicit generative model in the brain.

The specific model considered here rests on empirical Bayes. This model can be regarded as a mathematical formulation of the long-standing notion (Locke, 1690) that: 'our minds should often change the idea of its sensation into that of its judgement, and make one serve only to excite the other'. In a similar vein, Helmholtz (1860) distinguishes between perception and sensation: 'It may often be rather hard to say how much from perceptions as derived from the sense of sight is due directly to sensation, and how much of them, on the other hand, is due to experience and training' (see Pollen, 1999). In short, there is a distinction between a percept, which is the product of recognizing the causes of sensory input, and sensation *per se*. Recognition, i.e. inferring causes from sensation, is the inverse of generating sensory data from their causes. It follows that recognition rests on models, learned through experience, of how sensations are caused.

Conceptually, empirical Bayes and generative models are related to 'analysis-by-synthesis' (Neisser, 1967). This approach to perception, from cognitive psychology, involves adapting an internal model of the world to match sensory input and was suggested by Mumford (1992) as a way of understanding hierarchical neuronal processing. The idea is reminiscent of Mackay's epistemological automata (MacKay, 1956) which perceive by comparing expected and actual sensory input, (Rao, 1999). These models emphasize the role of backward connections in mediating predictions of lower level input, based on the activity of higher cortical levels.

Generative models and perception

This section introduces a basic framework for understanding learning and inference. This framework rests upon generative and recognition models, which are functions that map causes to sensory input and vice versa. Generative models afford a generic formulation of representational learning and inference in a supervised or self-supervised context. There are many forms of generative models that range from conventional statistical models (e.g. factor and cluster analysis) to those motivated by Bayesian inference and learning (e.g. Dayan *et al.*, 1995; Hinton *et al.*, 1995). The goal of generative models is: 'to learn representations that are economical to describe but allow the input to be reconstructed accurately' (Hinton *et al.*, 1995). The distinction between reconstructing data and learning efficient representations relates directly to the distinction between inference and learning.

Inference vs. learning

Generative models relate unknown causes v and unknown parameters θ to observed sensory data u. The objective is to make *inferences* about the causes and *learn* the parameters. Inference may be simply estimating the most likely cause and is based on estimates of the parameters from learning. A generative model is specified in terms of a *prior* distribution over the causes $p(v; \theta)$ and the *generative* distribution or likelihood of the data given the causes $p(u|v; \theta)$. Together, these define the marginal distribution of data implied by a generative model:

$$p(u; \theta) = \int p(u|v; \theta)p(v; \theta)dv \qquad \textbf{36.2}$$

The conditional density of the causes, given the data, are given by the recognition model, which is defined in terms of the *recognition* or conditional distribution:

$$p(v|u; \theta) = \frac{p(u|v; \theta)p(v; \theta)}{p(u; \theta)} \qquad \textbf{36.3}$$

However, as considered above, the generative model may not be inverted easily and it may not be possible to parameterize this recognition distribution. This is crucial because the endpoint of learning is the acquisition of a useful recognition model that can be applied to sensory data. One solution is to posit an approximate recognition or conditional density $q(v)$ that is consistent with the generative model and that can be parameterized. Estimating the moments (e.g. expectation) of this density corresponds to *inference*. Estimating the parameters of the underlying generative model corresponds to *learning*. This distinction maps directly onto the two steps of expectation–maximization (EM) (Dempster *et al.*, 1977).

Expectation maximization

To keep things simple, assume that we are only interested in the first moment or expectation of $q(v)$ which we will denote by μ. This is the conditional mean or expected cause. EM is a coordinate ascent scheme that comprises an **E**-step and an **M**-step. In the present context, the **E**-step finds the conditional expectation of the causes (i.e. inference), while the **M**-step identifies the maximum likelihood value of the parameters (i.e. learning). Critically, both adjust the conditional causes and parameters to maximize the same thing.

The free energy formulation

EM provides a useful procedure for density estimation that has direct connections with statistical mechanics. Both steps of the EM algorithm involve maximizing a function of the densities above. This function corresponds to the negative free energy in physics (see Chapter 24 and the Appendices for more details):

$$F = \ln p(u;\theta) - KL\{q(v), p(v|u;\theta)\} \qquad \textbf{36.4}$$

This objective function has two terms. The first is the marginal likelihood of the data under the generative model. The second term is the Kullback-Leibler divergence[2] between the approximate and true recognition densities. Critically, the second term is always positive, rendering F a lower-bound on the expected log-likelihood of the data. This means maximizing the objective function (i.e. minimizing the free energy) is simply minimizing our surprise about the data. The **E**-step increases F with respect to the expected cause, ensuring a good approximation to the recognition distribution implied by the parameters θ. This is inference. The **M**-step changes θ, enabling the generative model to match the likelihood of the data and corresponds to learning:

$$\begin{aligned} &\text{Inference: } \mathbf{E} \quad && \mu = \max_{\mu} F \\ &\text{Learning: } \mathbf{M} \quad && \theta = \max_{\theta} F \end{aligned} \qquad \textbf{36.5}$$

The remarkable thing is that both inference and learning are driven in exactly the same way, namely to minimize the free energy. This is effectively the same as minimizing surprise about sensory data encountered. The implication, as we will see below, is that the same principle can explain phenomena as wide ranging as the mis-match negativity (MMN) in evoked electrical brain responses to Hebbian plasticity during perceptual learning.

Predictive coding

We have now established an objective function that is maximized to enable inference and learning in **E**- and **M**-steps respectively. Here, we consider how that maximization might be implemented. In particular, we will look at predictive coding, which is based on minimizing prediction error (Rao and Ballard, 1998). Prediction error is the difference between the data observed and that predicted by the inferred causes. We will see that minimizing the free energy is equivalent to minimizing prediction error. Consider any static non-linear generative model under Gaussian assumptions:

$$\begin{aligned} u &= g(v,\theta) + \varepsilon^{(1)} \\ v &= \eta + \varepsilon^{(2)} \end{aligned} \qquad \textbf{36.6}$$

where $Cov(\varepsilon^{(1)}) = \Sigma^{(1)}$ is the covariance of random fluctuations in the sensory data. Priors on the causes are specified in terms of their expectation η and covariance $Cov(\varepsilon^{(2)}) = \Sigma^{(2)}$. This form will be useful in the next section when we generalize to hierarchical models. For simplicity, we will approximate the recognition density with a point mass. From Eqn. **36.4**:

$$\begin{aligned} F &= -\frac{1}{2}\xi^{(1)T}\xi^{(1)} - \frac{1}{2}\xi^{(2)T}\xi^{(2)} - \frac{1}{2}\ln|\Sigma^{(1)}| - \frac{1}{2}\ln|\Sigma^{(2)}| \\ \xi^{(1)} &= \Sigma^{(1)^{-1/2}}(u - g(\mu,\theta)) \\ \xi^{(2)} &= \Sigma^{(2)^{-1/2}}(\mu - \eta) \end{aligned} \qquad \textbf{36.7}$$

The first term in Eqn. **37.7** is the prediction error that is minimized in predictive coding. The second corresponds to a prior term that constrains or regularizes conditional estimates of the causes. The need for this term stems from the ill-posed nature of recognition discussed above and is a ubiquitous component of inverse solutions.

Predictive coding schemes can be seen in the context of forward and inverse models adopted in machine vision (Ballard *et al.*, 1983; Kawato *et al.*, 1993). Forward models generate data from causes (cf. generative models), whereas inverse models approximate the reverse transformation of data to causes (cf. recognition models). This distinction embraces the ill-posed nature of inverse problems. As with all underdetermined inverse problems, the role of constraints is central. In the inverse literature, *a priori* constraints usually enter in terms of regularized solutions. For example: 'Descriptions of physical properties of visible surfaces, such as their distance and the presence of edges, must be recovered from the primary image inputs. Computational vision aims to understand how such descriptions can be obtained from inherently ambiguous and noisy inputs. A recent development in this field sees early vision as a set of ill-posed problems, which can be solved by the use of regularization methods' (Poggio *et al.*, 1985). The architectures that emerge from these schemes suggest that: 'Feedforward connections from the lower visual cortical area to the higher visual cortical area provide an approximated inverse model of the imaging process (optics)'. Conversely: 'while the back-projection connection from the higher area to the lower area provides a forward model of the optics' (Kawato *et al.*, 1993). This perspective highlights the importance of backward connections and the role of priors in enabling predictive coding schemes.

[2] A measure of the distance or difference between two probability densities.

Predictive coding and Bayes

Predictive coding is a strategy that has some compelling [Bayesian] underpinnings. To finesse the inverse problem posed by non-invertible generative models, constraints or priors are required. These resolve the ill-posed problems that confound recognition based on purely forward architectures. It has long been assumed that sensory units adapt to the statistical properties of the signals to which they are exposed (see Simoncelli and Olshausen, 2001 for review). In fact, the Bayesian framework for perceptual inference has its origins in Helmholtz's notion of perception as unconscious inference. Helmholtz realized that retinal images are ambiguous and that prior knowledge was required to account for perception (Kersten *et al.*, 2004). Kersten *et al.* (2004) provide an excellent review of object perception as Bayesian inference and ask a fundamental question: 'Where do the priors come from. Without direct input, how does image-independent knowledge of the world get put into the visual system?' In the next section, we answer this question and show how empirical Bayes allows priors to be learned and induced online, during inference.

Cortical hierarchies and empirical Bayes

The problem with fully Bayesian inference is that the brain cannot construct the prior expectation and variability, η and $\Sigma^{(2)}$ *de novo*. They have to be learned and, furthermore, adapted to the current experiential context. This calls for empirical Bayes, in which priors are estimated from data. Empirical Bayes harnesses the hierarchical structure of a generative model, treating the estimates at one level as priors on the subordinate level (Efron and Morris, 1973). Empirical Bayes provides a natural framework within which to treat cortical hierarchies in the brain, each level providing constraints on the level below. This approach models the world as a hierarchy of systems where supraordinate causes induce, and moderate, changes in subordinate causes. Empirical priors offer contextual guidance towards the most likely cause of the data. Note that predictions at higher levels are subject to the same constraints; only the highest level, if there is one in the brain, is unconstrained.

Next, we extend the generative model to cover empirical priors. This means that constraints, required by predictive coding, are absorbed into the learning scheme. This hierarchical extension induces extra parameters that encode the variability or precision of the causes. These are referred to as hyperparameters in the classical covariance component literature. Hyperparameters are updated in the **M**-step and are treated in exactly the same way as the parameters.

Hierarchical models

Consider any level i in a hierarchy whose causes $v^{(i)}$ are elicited by causes in the level above $v^{(i+1)}$. The hierarchical form of the generative model is:

$$\begin{aligned} u &= \\ v^{(1)} &= g(v^{(2)}, \theta^{(1)}) + \varepsilon^{(1)} \\ v^{(2)} &= g(v^{(3)}, \theta^{(2)}) + \varepsilon^{(2)} \\ v^{(3)} &= \ldots \end{aligned} \qquad 36.8$$

Technically, these models fall into the class of conditionally independent hierarchical models, when the stochastic terms are independent (Kass and Steffey, 1989). These models are also called *parametric empirical Bayes* (PEB) models because the obvious interpretation of the higher-level densities as priors led to the development of PEB methodology (Efron and Morris, 1973). Often, in statistics, these hierarchical models comprise just two levels, which is a useful way to specify simple shrinkage priors on the parameters of single-level models. We will assume the stochastic terms are Gaussian with covariance $\Sigma^{(i)}$. Therefore, the means and covariances determine the likelihood at each level:

$$p(v^{(i)}|v^{(i+1)}; \theta^{(i)}) = N(g^{(i)}, \Sigma^{(i)}) \qquad 36.9$$

This likelihood *also plays the role of an empirical prior on* $v^{(i)}$ at the level below, where it is jointly maximized with the likelihood $p(v^{(i-1)}|v^{(i)}; \theta^{(i-1)})$. This is the key to understanding the utility of hierarchical models; by inferring the generative distribution of level i one is implicitly estimating the prior for level $i-1$. This enables the learning of prior densities. The hierarchical nature of these models lends an important context-sensitivity to recognition densities not found in single-level models. Because high-level causes determine the prior expectation of causes in the subordinate level, they change the distributions upon which inference is based, in a data and context-dependent way.

Implementation

The biological plausibility of empirical Bayes in the brain can be established fairly simply. The objective function is now:

$$\begin{aligned} F &= -\frac{1}{2}\xi^{(1)T}\xi^{(1)} - \frac{1}{2}\xi^{(2)T}\xi^{(2)} - \ldots \\ &\quad -\frac{1}{2}\ln|\Sigma^{(1)}| - \frac{1}{2}\ln|\Sigma^{(2)}| - \ldots \\ \xi^{(i)} &= \mu^{(i)} - g(\mu^{(i+1)}, \theta^{(i)}) - \lambda^{(i)}\xi^{(i)} \\ &= (I + \lambda^{(i)})^{-1}(\mu^{(i)} - g(\mu^{(i+1)}, \theta^{(i)})) \\ \Sigma^{(i)} &= (I + \lambda^{(i)})^2 \end{aligned} \qquad 36.10$$

In neuronal models, the prediction error is encoded by the activities of units denoted by $\zeta^{(i)}$. These error units receive a prediction from units in the level above[3] via *backward* connections and *lateral* influences from the representational units $\mu^{(i)}$ being predicted. Horizontal interactions among the error units serve to de-correlate them (cf. Foldiak 1990), where the symmetric lateral connection strengths $\lambda^{(i)}$ hyperparameterize the covariances $\Sigma^{(i)}$, which are the prior covariances for level $i-1$.

The conditional causes and parameters perform a gradient ascent on the objective function:[4]

$$
\begin{aligned}
\mathbf{E} \quad & \dot{\mu}^{(i)} = \frac{\partial F}{\partial \mu^{(i)}} = -\frac{\partial \xi^{(i-1)T}}{\partial \mu^{(i)}} \xi^{(i-1)} - \frac{\partial \xi^{(i)T}}{\partial \mu^{(i)}} \xi^{(i)} \\
& \xi^{(i)} = \mu^{(i)} - g(\mu^{(i+1)}, \theta^{(i)}) - \lambda^{(i)} \xi^{(i)} \\
\mathbf{M} \quad & \dot{\theta}^{(i)} = \frac{\partial F}{\partial \theta^{(i)}} = -\left\langle \frac{\partial \xi^{(i)T}}{\partial \theta^{(i)}} \xi^{(i)} \right\rangle_u \\
& \dot{\lambda}^{(i)} = \frac{\partial F}{\partial \lambda^{(i)}} = -\left\langle \frac{\partial \xi^{(i)T}}{\partial \lambda^{(i)}} \xi^{(i)} \right\rangle_u - (1+\lambda^{(i)})^{-1}
\end{aligned}
\qquad \mathbf{36.11}
$$

Inferences, mediated by the **E**-step, rest on changes in units encoding expected causes $\mu^{(i)}$ that are mediated by forward connections from error units in the level below and lateral interactions with error units in the same level. Similarly, prediction error is constructed by comparing the activity of these units with the activity predicted by backward connections.

This is the simplest version of a very general learning algorithm. It is general in the sense that it does not require the parameters of either the generative or prior distributions. It can learn non-invertible, non-linear generative models and encompasses complicated hierarchical processes. Furthermore, each of the learning components has a relatively simple neuronal interpretation (see below).

IMPLICATIONS FOR CORTICAL INFRASTRUCTURE AND PLASTICITY

The scheme implied by Eqn. **36.11** has four clear implications for the functional architecture required to implement it. We review these in relation to cortical organization in the brain. A schematic summarizing these points is provided in Figure 36.2. In short, we arrive at exactly the same four points presented at the end of the first section.

- *Hierarchical organization* – hierarchical models enable empirical Bayesian estimation of prior densities and provide a plausible model for sensory data. Models that do not show conditional independence (e.g. those used by connectionist and infomax schemes) depend on prior constraints for inference and do not invoke a hierarchical cortical organization. The nice thing about the architecture in Figure 36.2 is that the responses of units at the i-th level $\mu^{(i)}$ depend only on the error at the current level and the immediately preceding level. Similarly the error units $\zeta^{(i)}$ are only connected to units in the current level and the level above. This hierarchical organization follows from conditional independence and is important because it permits a biologically plausible implementation, where the connections driving inference run only between neighbouring levels.
- *Reciprocal connections* – in the hierarchical scheme, the dynamics of units $\mu^{(i+1)}$ are subject to two, locally available, influences: a likelihood or recognition term mediated by forward afferents from the error units in the level below and an empirical prior conveyed by error units in the same level. Critically, the influences of the error units in both levels are mediated by linear connections with strengths that are exactly the same as the [negative] *reciprocal* connections from $\mu^{(i+1)}$ to $\zeta^{(i)}$ and $\zeta^{(i+1)}$. From Eqn. **36.11**:

$$
\begin{aligned}
\frac{\partial \dot{\mu}^{(i+1)}}{\partial \xi^{(i)}} &= -\frac{\partial \xi^{(i)}}{\partial \mu^{(i+1)}}^T \\
\frac{\partial \dot{\mu}^{(i+1)}}{\partial \xi^{(i+1)}} &= -\frac{\partial \xi^{(i+1)}}{\partial \mu^{(i+1)}}^T
\end{aligned}
\qquad \mathbf{36.12}
$$

Functionally, forward and lateral connections are reciprocated, where backward connections generate predictions of lower-level responses. Forward connections allow prediction error to drive units in supraordinate levels. Lateral connections, within each level, mediate the influence of error units on the predicting units and intrinsic connections $\lambda^{(i)}$ among the error units de-correlate them, allowing competition among prior expectations with different precisions (precision is the inverse of variance). In short, lateral, forward and backward connections are all reciprocal, consistent with anatomical observations.
- *Functionally asymmetric forward and backward connections* – although the connections are reciprocal, the functional attributes of forward and backward influences are different. The top-down influences of units $\mu^{(i+1)}$ on error units in the lower level $\xi^{(i)} = \mu^{(i)} - g(\mu^{(i+1)}, \theta^{(i)}) - \lambda^{(i)}\xi^{(i)}$ instantiate the forward model. These can be non-linear, where each unit in the higher

[3] Clearly, in the brain, backward connections are not inhibitory but, after mediation by inhibitory interneurons, their effective influence could be rendered so.

[4] For simplicity, we have ignored conditional uncertainty about the causes that would otherwise induce further terms in the **M**-step.

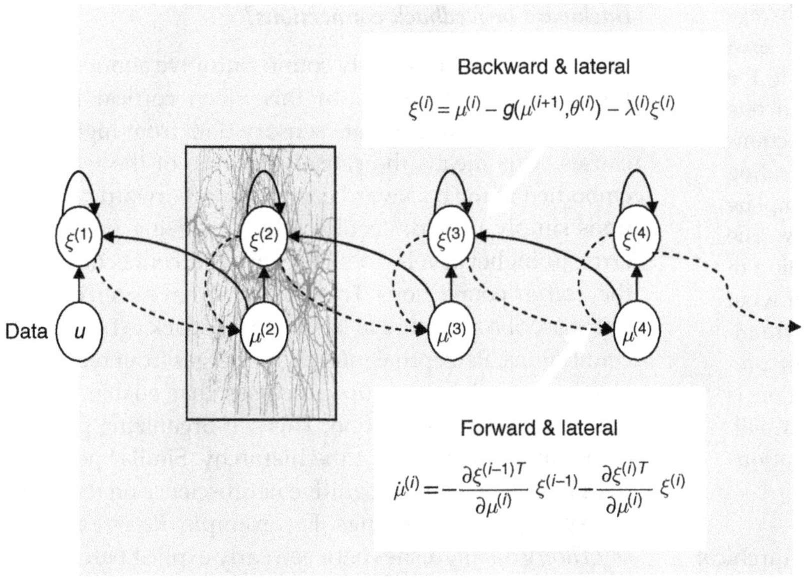

FIGURE 36.2 Schematic depicting a hierarchical predictive coding architecture. Here, hierarchical arrangements within the model serve to provide predictions or priors to representations in the level below. The upper circles represent error units and the lower circles functional subpopulations encoding the conditional expectation of causes. These expectations change to minimize both the discrepancy between their predicted value and the mismatch incurred by their prediction of the level below. These two constraints correspond to prior and likelihood terms respectively (see main text).

level *may modulate or interact with the influence of others*, according to the non-linearities in $g^{(i)}$. In contrast, the bottom-up influences of units in lower levels do not interact when producing changes at the higher level because, according to Eqn. **36.11**, their effects are linearly separable. This is a key observation because the empirical evidence, reviewed in the first section, suggests that backward connections are in a position to interact (e.g. through NMDA receptors expressed predominantly in supra-granular layers that are in receipt of backward connections). Forward connections are not. In summary, non-linearities, in the way sensory data are produced, necessitate non-linear interactions in the generative model. These are mediated by backward connections but do not require forward connections to be modulatory.

- *Associative plasticity* – changes in the parameters correspond to plasticity in the sense that the parameters control the strength of backward and lateral connections. The backward connections parameterize the prior expectations and the lateral connections hyperparameterize the prior covariances. Together they parameterize the Gaussian densities that constitute the empirical priors. The plasticity implied can be seen more clearly with an explicit model. For example, let $g(v^{(i+1)}, \theta^{(i)}) = \theta^{(i)}v^{(i+1)}$. In this instance:

$$\dot{\theta}^{(i)} = (1+\lambda^{(i)})^{-1}\langle \xi^{(i)}\mu^{(i+1)T}\rangle_u$$
$$\dot{\lambda}^{(i)} = (1+\lambda^{(i)})^{-1}(\langle \xi^{(i)}\xi^{(i)T}\rangle_u - I) \quad 36.13$$

This is just Hebbian or associative plasticity where the connection strengths change in proportion to the product of pre- and postsynaptic activity. An intuition about Eqn. **36.13** obtains by considering the conditions under which the expected change in parameters is zero (i.e. after learning). For the backward connections, this implies there is no component of prediction error that can be explained by estimates at the higher level $\langle \xi^{(i)}\phi^{(i+1)T}\rangle_u = 0$. The lateral connections stop changing when the prediction error is spherical or *IID* $\langle \xi^{(i)}\xi^{(i)T}\rangle_u = I$.

It is evident that the predictions of the theoretical analysis coincide almost exactly with the empirical aspects of functional architectures in visual cortices: hierarchical organization; reciprocity; functional asymmetry; and associative plasticity. Although somewhat contrived, it is pleasing that purely theoretical considerations and neurobiological empiricism converge so precisely.

Summary

In summary, perceptual inference and learning lends itself naturally to a hierarchical treatment, which considers the brain as an empirical Bayesian device. The dynamics of the units or populations are driven to minimize error at all levels of the cortical hierarchy and implicitly render themselves posterior modes (i.e. most likely values) of the causes given the data. In contradistinction to supervised learning, hierarchical prediction does not require any desired output. Unlike information theoretic approaches, they do not assume independent causes. In contrast to regularized inverse solutions, they do not depend on *a priori* constraints. These emerge spontaneously as empirical priors from higher levels.

The overall scheme sits comfortably with the hypothesis (Mumford, 1992):

> on the role of the reciprocal, topographic pathways between two cortical areas, one often a 'higher' area dealing with more abstract information about the world, the other 'lower', dealing with more concrete data. The higher area attempts to fit its abstractions to the data it receives from lower areas by sending back to them from its deep pyramidal cells a template reconstruction best fitting the lower level view. The lower area attempts to reconcile the reconstruction of its view that it receives from higher areas with what it knows, sending back from its superficial pyramidal cells the features in its data which are not predicted by the higher area. The whole calculation is done with all areas working simultaneously, but with order imposed by synchronous activity in the various top-down, bottom-up loops.

We have tried to show that this sort of hierarchical prediction can be implemented in brain-like architectures using mechanisms that are biologically plausible (Figure 36.3).

Backward or feedback connections?

There is something slightly counterintuitive about empirical Bayes in the brain. In this view, cortical hierarchies are trying to generate sensory data from high-level causes. This means the causal structure of the world is embodied in the backward connections. Forward connections simply provide feedback by conveying prediction error to higher levels. In short, forward connections are the *feedback* connections. This is why we have been careful not to ascribe a functional label like feedback to backward connections. Perceptual inference emerges from recurrent top-down and bottom-up processes that enable sensation to constrain perception. This self-organizing process is distributed throughout the hierarchy. Similar perspectives have emerged in cognitive neuroscience on the basis of psychophysical findings. For example, *Reverse Hierarchy Theory* distinguishes between early explicit perception and implicit low level vision, where: 'our initial conscious percept – vision at a glance – matches a high-level, generalized, categorical scene interpretation, identifying

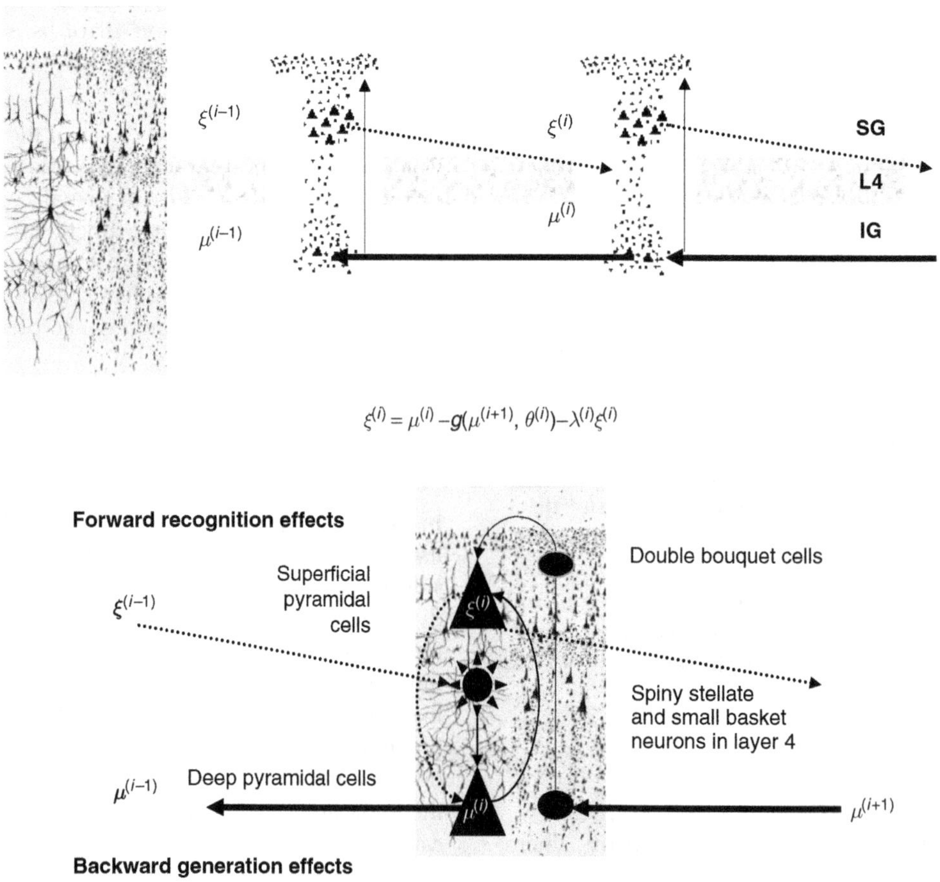

FIGURE 36.3 A more detailed picture of the influences among units. Here, we have associated various cells with different roles in the scheme described in the main text. The key constraint on this sort of association is that superficial pyramidal cells are the source of forward connections, which, according to the theory, should encode prediction error.

'"forest before trees"' (Hochstein and Ahissar, 2002). On the other hand, backward connections are responsible for predicting; lower level responses embed the generative or *forward* model. The effect of these predictions on lower-level responses is the focus of the next section.

ASSESSING FUNCTIONAL ARCHITECTURES WITH BRAIN IMAGING

In the previous section, we have seen one approach to understanding the nature of functional integration in the brain. We will use this framework to preview the different ways in which integration can be measured empirically, with a special focus on the interaction between forward and backward connections. The examples used will appear again in later chapters that take us from simple measure of statistical correlations among different brain areas though to dynamic causal models of cortical hierarchies.

Clearly, it would be nice to demonstrate the existence of top-down influences with neuroimaging. This is a slightly deeper problem than might be envisaged. This is because making causal inferences about effective connectivity is not straightforward (see Pearl, 2000). It is not sufficient to show regional activity is partially predicted by activity in a higher level to confirm the existence of backward connections because statistical dependency does, in itself, not permit causal inference. Statistical dependencies could easily arise in a purely forward architecture because the higher-level activity is caused by activity in the lower level. Although there are causal modelling techniques (i.e. dynamic causal modelling (DCM)) that can address this problem, we will start with a simpler approach and note that interactions between bottom-up and top-down influences cannot be explained by purely feed-forward architectures. An interaction, in this context, can be construed as an effect of backward connections on the driving efficacy of forward connections. In other words, the response evoked by the same driving bottom-up influence depends upon the context established by top-down influence. This interaction is used below simply as evidence for the existence of backward influences. There are instances of predictive coding that emphasize this phenomenon. For example, the 'Kalman filter model demonstrates how certain forms of attention can be viewed as an emergent property of the interaction between top-down expectations and bottom-up signals' (Rao, 1999).

This section focuses on the evidence for these interactions. From the point of view of functionally specialized responses, these interactions manifest as context-sensitive or contextual specialization, where modality-, category- or exemplar-specific responses, driven by bottom-up input are modulated by top-down influences induced by perceptual set. The first half of this section adopts this perceptive. The second part of this section uses measurements of effective connectivity to establish interactions between bottom-up and top-down influences. All the examples presented below rely on attempts to establish interactions by trying to change sensory-evoked neuronal responses through putative manipulations of top-down influences. These include inducing independent changes in perceptual set, cognitive [attentional] set, perceptual learning and, in the last section, through the study of patients with brain lesions

Context-sensitive specialization

If functional specialization is context-dependent then one should be able to find evidence for functionally specific responses, using neuroimaging, that are expressed in one context and not in another. The first empirical example provides such evidence. If the contextual nature of specialization is mediated by backwards connections then it should be possible to find cortical regions in which functionally specific responses, elicited by the same stimuli, are modulated by activity in higher areas. The second example shows that this is, indeed, possible. Both of these examples depend on factorial experimental designs.

Multifactorial designs

Factorial designs combine two or more factors within a task or tasks. Factorial designs can be construed as performing subtraction experiments in two or more different contexts. The differences in activations, attributable to the effects of context, are simply the interaction. Consider an implicit object recognition experiment, for example naming (of the object's name or the non-object's colour) and simply saying 'yes' during passive viewing of objects and non-objects. The factors in this example are implicit object recognition with two levels (objects versus non-objects) and phonological retrieval (naming versus saying 'yes'). The idea here is to look at the interaction between these factors, or the effect that one factor has on the responses elicited by changes in the other. In our experiment, object-specific responses are elicited (by asking subjects to view objects relative to meaningless shapes), with and without phonological retrieval. This 'two-by-two' design allows one to look at the interaction between phonological retrieval and object recognition. This analysis identifies not regionally specific activations but regionally specific *interactions*. When we performed this experiment, these interactions were evident in the left posterior, inferior temporal region and can be associated with the integration of phonology and object recognition (see Figure 36.4 and Friston *et al.*, 1996 for details). Alternatively, this region can be thought of as expressing recognition-dependent responses that are realized in, and only in, the context of having to name the object. These results can be construed as evidence of contextual specialization for object-recognition that depends upon modulatory afferents (possibly from temporal and parietal regions) that are implicated in naming a visually perceived object. There is no empirical evidence in these results to suggest that the temporal or parietal regions are the source of this top-down influence but, in the next example, the source of modulation is addressed explicitly using psychophysiological interactions.

Psychophysiological interactions

Psychophysiological interactions speak directly to the interactions between bottom-up and top-down influences, where one is modelled as an experimental factor and the other constitutes a measured brain response. In an analysis of psychophysiological interactions, one is trying to explain a regionally specific response in terms of an interaction between the presence of a sensorimotor or cognitive process and activity in another part of the brain (Friston *et al.*, 1997). The supposition here is that the remote region is the source of backward modulatory afferents that confer functional specificity on the target region. For example, by combining information about activity in the posterior parietal cortex, mediating attentional or perceptual set pertaining to a particular stimulus attribute, can we identify regions that respond to that stimulus when, and only when, activity in the parietal source is high? If such an interaction exists, then one might infer that the parietal area is modulating responses to the stimulus attribute for which the area is selective. This has clear ramifications in terms of the top-down modulation of specialized cortical areas by higher brain regions.

The statistical model employed in testing for psychophysiological interactions is a simple regression model of effective connectivity that embodies non-linear (second-order or modulatory effects). As such, this class of model speaks directly to functional specialization of a non-linear and contextual sort. Figure 36.5 illustrates a specific example (see Dolan *et al.*, 1997 for details). Subjects were asked to view degraded face and non-face control stimuli. The interaction between activity in the parietal region and the presence of faces was expressed most significantly in the right infero-temporal region not far from the homologous left infero-temporal region implicated in the object naming experiment above. Changes in parietal activity were induced experimentally by pre-exposure to un-degraded stimuli before some scans but not others. The data in the right panel of Figure 36.5 suggest that the infero-temporal region shows face-specific responses, relative to non-face objects, when, and only when, parietal activity is high. These results can be interpreted as a priming-dependent face-specific response, in infero-temporal regions that are mediated by interactions with medial parietal cortex. This is a clear example of contextual specialization that depends on top-down effects.

Effective connectivity

The previous examples, demonstrating contextual specialization, are consistent with functional architectures

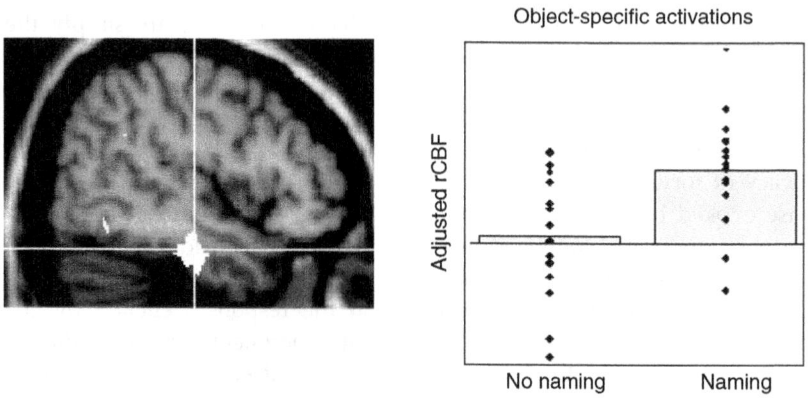

FIGURE 36.4 This example of regionally specific interactions comes from an experiment where subjects were asked to view coloured non-object shapes or coloured objects and say 'yes', or to name either the coloured object or the colour of the shape. Left: a regionally specific interaction in the left infero-temporal cortex. The SPM (statistical parametric map) threshold is $p < 0.05$ (uncorrected). Right: the corresponding activities in the maxima of this region are portrayed in terms of object recognition-dependent responses with and without naming. It is seen that this region shows object recognition responses when, and only when, there is phonological retrieval. The 'extra' activation with naming corresponds to the interaction. These data were acquired from six subjects scanned 12 times using PET.

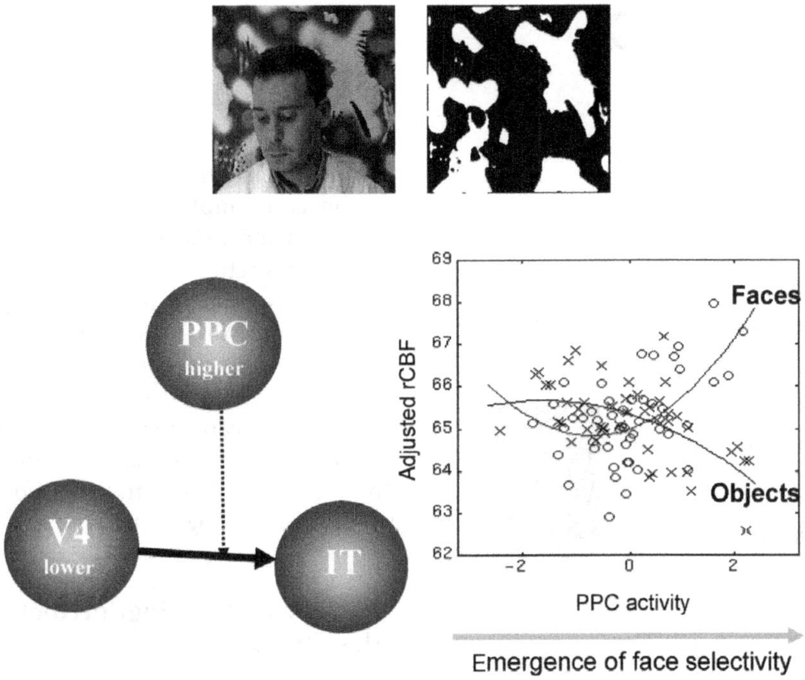

FIGURE 36.5 Top: examples of the stimuli presented to subjects. During the measurement of brain responses only degraded stimuli where shown (e.g. the right-hand picture). In half the scans, the subject was given the underlying cause of these stimuli, by presenting the original picture (e.g. left-hand picture) before scanning. This priming induced a profound difference in perceptual set for the primed, relative to non-primed, stimuli. Lower right: activity observed in a right infero-temporal region, as a function of mean-corrected posterior parietal cortex (PPC) activity. This region showed the most significant interaction between the presence of faces and activity in a reference location in the posterior medial parietal cortex. This analysis can be thought of as finding those areas that are subject to top-down modulation of face-specific responses by medial parietal activity. The crosses correspond to activity while viewing non-face stimuli and the circles to faces. The essence of this effect can be seen by noting that this region differentiates between faces and non-faces when, and only when, medial parietal activity is high. The lines correspond to the best second-order polynomial fit. These data were acquired from six subjects using PET. Lower left: schematic depicting the underlying conceptual model in which driving afferents from ventral form areas (here designated as V4) excite infero-temporal (IT) responses, subject to permissive modulation by PPC projections.

implied by empirical Bayes. However, they do not provide definitive evidence for an interaction between top-down and bottom-up influences. In this subsection, we look for direct evidence of these interactions using models of effective connectivity. This requires a plausible model of coupling among brain regions that can accommodate non-linear effects. We will illustrate the use of models based on the Volterra expansion and conclude with an example using DCM for event-related potentials (ERP). These examples change context using attention and perceptual learning respectively.

Non-linear coupling among brain areas

Linear models of effective connectivity assume that the multiple inputs to a brain region are linearly separable. This assumption precludes activity-dependent connections that are expressed in one context and not in another. The resolution of this problem lies in adopting non-linear models like the Volterra formulation. Non-linearities can be construed as a context-or activity-dependent modulation of the influence that one region exerts over another (Büchel and Friston, 1997). In the Volterra model, second-order kernels model modulatory effects. Within these models, the influence of one region on another has two components: the direct or *driving* influence of input from the first (e.g. hierarchically lower) region, irrespective of the activities elsewhere, and a *modulatory* component that represents an interaction with input from the remaining (e.g. hierarchically higher) regions. These are mediated by first- and second-order kernels respectively. The example provided in Figure 36.6 addresses the modulation of visual cortical responses by attentional mechanisms (e.g. Treue and Maunsell, 1996) and the mediating role of activity-dependent changes in effective connectivity.

The lower panel in Figure 36.6 shows a characterization of this modulatory effect in terms of the increase in V5 responses, to a simulated V2 input, when posterior parietal activity is zero (broken line) and when it is high (solid line). In this study, subjects were studied with fMRI under identical stimulus conditions (visual motion subtended by radially moving dots) while manipulating the attentional component of the task (detection of velocity changes). The brain regions and connections comprising

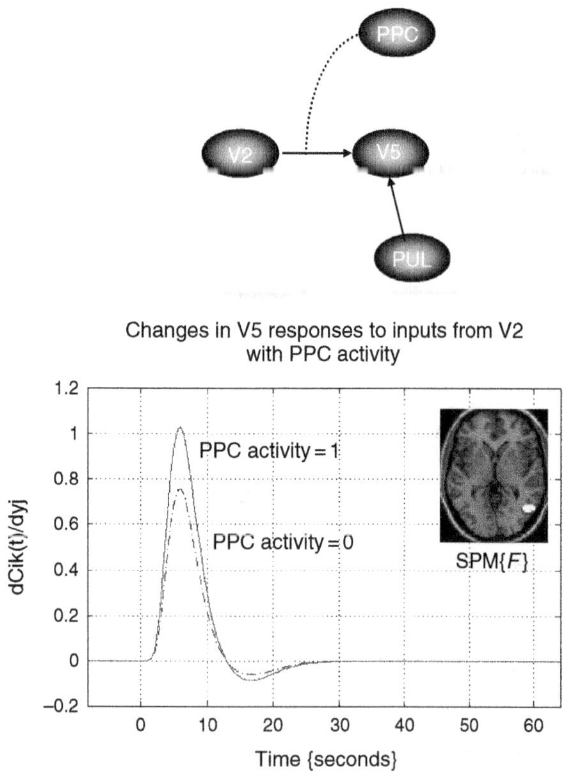

FIGURE 36.6 Upper panel: brain regions and connections comprising a model of distributed responses. Lower panel: characterization of the effects of V2 on V5 and their modulation by posterior parietal cortex (PPC). The broken line represents estimates of V5 responses when PPC activity is zero, according to a second order Volterra model of effective connectivity with input to V5 from V2, PPC and the pulvinar (PUL). The solid curve represents the same response when PPC activity is one standard deviation of its variation over conditions. It is evident that V2 has an activating effect on V5 and that PPC increases the responsiveness of V5 to these data. The insert shows all the voxels in V5 that evidenced a modulatory effect ($p < 0.05$ uncorrected). These voxels were identified by thresholding an SPM (statistical parametric map) of the *F*-statistic, testing for the contribution of second-order kernels involving V2 and PPC (treating all other terms as nuisance variables). The data were obtained with fMRI under identical stimulus conditions (visual motion subtended by radially moving dots) while manipulating the attentional component of the task (detection of velocity changes).

the model are shown in the upper panel. The lower panel shows a characterization of the effects of V2 data on V5 and their modulation by posterior parietal cortex (PPC) using simulated data at different levels of PPC activity. It is evident that V2 has an activating effect on V5 and that PPC increases the responsiveness of V5 to these inputs. The insert shows all the voxels in V5 that evidenced a modulatory effect ($p < 0.05$ uncorrected). These voxels were identified by thresholding statistical parametric maps of the *F*-statistic testing for second-order kernels involving V2 and PPC, while treating all other components as nuisance variables. The estimation of the Volterra kernels and statistical inference procedure are described in Friston and Büchel (2000).

These results suggest that top-down parietal influences may be a sufficient explanation for the attentional modulation of visually evoked extrastriate responses. More importantly, they are consistent with the functional architecture implied by predictive coding, because they establish the existence of functionally expressed backward connections. In our final example, we use perceptual learning to induce changes in connections and DCM to measure those changes. Unlike Volterra formulations of effective connectivity, dynamic causal models parameterize the coupling among brain areas explicitly. This means that one can make inference about directed influences that are causal and quantitative in nature. We will illustrate this using an example from electroencephalography.

Perceptual learning, prediction error and the MMN

The mismatch negativity (MMN) is a negative component of the ERP elicited by a change in some repetitive aspect of auditory stimulation. The MMN can be seen in the absence of attention and is generally thought to reflect pre-attentive processing in the temporal and frontal system (Näätänen, 2003). The MMN is elicited by stimulus change at about 100–200 ms after the stimulus, and is presumed to reflect an automatic comparison of stimuli to sensory memory representations encoding the repetitive aspects of auditory inputs. This prevailing theory assumes that there are distinct change-specific neurons in auditory cortex that generate the MMN. The alternative view is that preceding stimuli adapt feature-specific neurons. In this *adaptation hypothesis*, the response is delayed and suppressed on exposure to repeated stimuli, giving rise to the MMN (Jääskeläinen *et al.*, 2004).

The empirical Bayes scheme would suggest that a component of the event-related potential (ERP) corresponding to prediction error, is suppressed more efficiently after learning-related plasticity in backward and lateral connections (and implicitly forward connections by Eqn. **36.12**). This suppression would be specific for the repeated aspects of the stimuli and would be a selective suppression of prediction error. Recall that error suppression (i.e. minimization of free energy) is the motivation for plasticity in the **M**-step. The ensuing repetition suppression hypothesis suggests the MMN is simply the attenuation of evoked prediction error. As noted above, prediction error may be encoded by superficial pyramidal cells (see Figure 36.3), which are a major contributor to the ERP.

In summary, both the **E**-step and **M**-step try to minimize free energy; the **E**-step does this during perceptual inference, on a time-scale of milliseconds, and the **M**-step, during perceptual learning, over seconds or longer. If the ERP is an index of prediction error (i.e. free energy), the ERP evoked by the first, in a train of repeated stimuli, will decrease with each subsequent presentation. This decrease discloses the MMN evoked by a new (oddball) stimulus. In this view, the MMN is subtended by a *positivity* that increases with the number of standards.

DCM *and perceptual learning*

We elicited event-related potentials that exhibited a strong modulation of late components, on comparing responses to frequent and rare stimuli, using an auditory oddball paradigm. Auditory stimuli, 1000 or 2000 Hz tones with 5 ms rise and fall times and 80 ms duration, were presented binaurally. The tones were presented for 15 minutes, every 2 s in a pseudo-random sequence with 2000 Hz tones on 20 per cent of occasions and 1000 Hz tones for 80 per cent of the time (standards). The subject was instructed to keep a mental record of the number of 2000 Hz tones (oddballs). Data were acquired using 128 electrodes with 1000 Hz sample-frequency. Before averaging, data were referenced to mean earlobe activity and bandpass filtered between 1 and 30 Hz. Trials showing ocular artefacts and bad channels were removed from further analysis.

Six sources were identified using conventional procedures and used to construct four dynamic causal models (see Figure 36.7 and Chapter 42). To establish evidence for

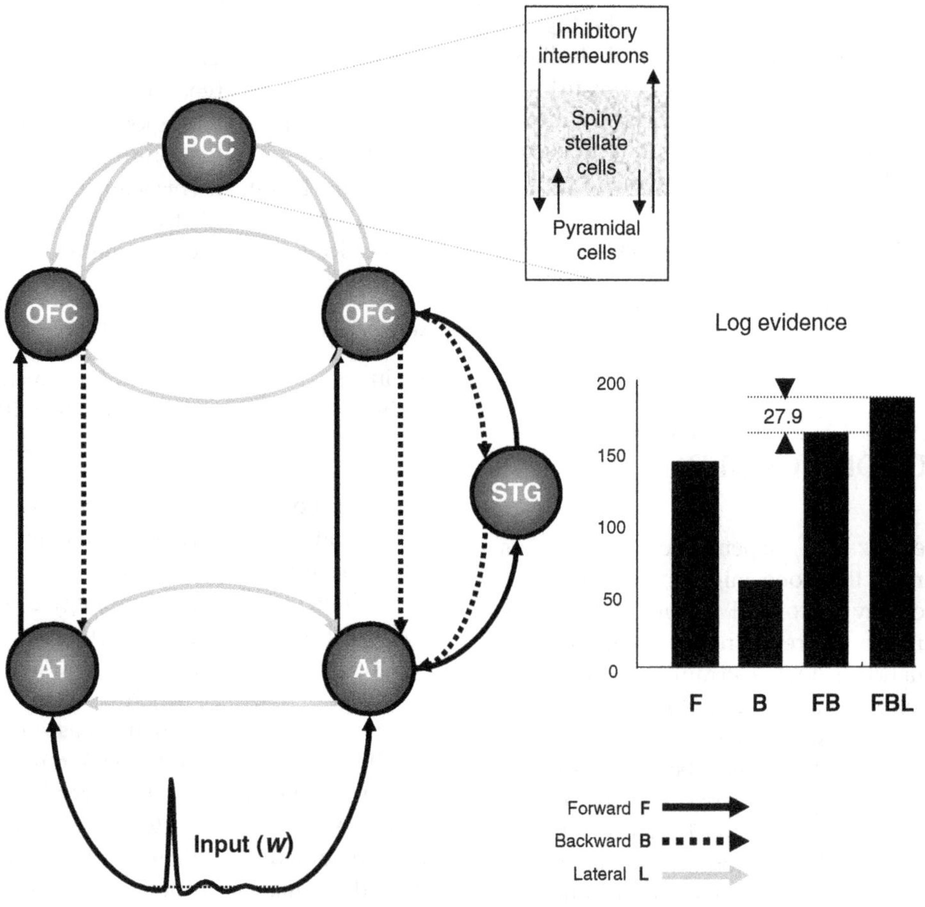

FIGURE 36.7 Left: schematic showing the extrinsic connections of a DCM (dynamic causal model) used to explain ERP data; a bilateral extrinsic input acts on primary auditory cortices which project reciprocally to orbitofrontal regions. In the right hemisphere, an indirect pathway was specified via a relay in the superior temporal gyrus. At the highest level, orbitofrontal and left posterior cingulate cortices were assumed to be laterally and reciprocally connected. Sources were coupled with extrinsic cortico-cortical connections following the rules of Felleman and van Essen (1991) – upper insert. A1: primary auditory cortex, OFC: orbitofrontal cortex, PCC: posterior cingulate cortex, STG: superior temporal gyrus (right is on the right and left on the left). The free parameters of this model included extrinsic connection strengths that were adjusted to explain best the observed ERPs. Critically, these parameters allowed for differences in connections between standard and oddball trials. Right: the results of a Bayesian model selection are shown in terms of the log-evidence for models allowing changes in forward (**F**), backward (**B**), forward and backward (**FB**) and forward, backward and lateral (**FBL**) connections. In this example, there is strong evidence that forward, backward and lateral connections change with perceptual learning.

changes in backward and lateral connections, above and beyond changes in forward connections, we employed Bayesian model selection (see Chapter 35). This entailed specifying four models that allowed for changes in forward, backward, forward and backward and in all connections. These, and only these, changes in extrinsic connectivity could explain the differences in the ERP, elicited by standard relative to oddball stimuli. The models were compared using the negative free energy as an approximation to the log-evidence for each model: if, after inversion, we assume the approximating conditional density is the true conditional density, the free energy reduces to the log-evidence (see Eqn. **36.4**). In Bayesian model selection, a difference in log-evidence of three or more can be considered as strong evidence for the model with the greater evidence, relative to the one with less. The log evidences for the four models are shown in Figure 36.7. The model with the highest evidence (by a margin of 27.9) is the DCM that allows for learning-related changes in forward, backward and lateral connections. These results provide clear evidence that changes in backward and lateral connections are needed to explain the observed differences in cortical responses.

In the final section, the implications of hierarchically organized connections are considered from the point of view of the lesion-deficit model and neuropsychology.

FUNCTIONAL INTEGRATION AND NEUROPSYCHOLOGY

If functional specialization depends on interactions among cortical areas, then one might predict changes in functional specificity in cortical regions that receive enabling or modulatory afferents from a damaged area. A simple consequence is that aberrant responses will be elicited in regions hierarchically below the lesion if, and only if, these responses depend upon input from the lesion site. However, there may be other contexts in which the region's responses are perfectly normal (relying on other, intact, afferents). This leads to the notion of a context-dependent region-specific abnormality, caused by, but remote from, a lesion (i.e. an abnormal response that is elicited by some tasks but not others). We have referred to this phenomenon as 'dynamic diaschisis' (Price *et al.*, 2001).

Dynamic diaschisis

Classical diaschisis, demonstrated by early anatomical studies and more recently by neuroimaging studies of resting brain activity, refers to regionally specific reductions in metabolic activity at sites that are remote from, but connected to, damaged regions. The clearest example is 'crossed cerebellar diaschisis' (Lenzi *et al.*, 1982), in which abnormalities of cerebellar metabolism are seen following cerebral lesions involving the motor cortex. Dynamic diaschisis describes the task-specific effects that a lesion can have on the *evoked responses* of a distant cortical region. The basic idea is that an otherwise viable cortical region expresses aberrant neuronal responses when, and only when, those responses depend upon interactions with a damaged region. This can arise because normal responses in any given region depend upon reciprocal interactions with other regions. The regions involved will depend on the cognitive and sensorimotor operations engaged at any particular time. If these regions include one that is damaged, then abnormal responses may ensue. However, there may be situations when the same region responds normally, for instance when its dynamics depend only upon integration with undamaged regions. If the region can respond normally in some situations then forward driving components must be intact. This suggests that dynamic diaschisis will only present itself when the lesion involves a hierarchically equivalent or higher area.

An example from neuropsychology

We investigated this possibility in a functional imaging study of four aphasic patients, all with damage to the left posterior inferior frontal cortex, classically known as Broca's area (Figure 36.8; upper panels). These patients had speech output deficits but relatively preserved comprehension. Generally, functional imaging studies can only make inferences about abnormal neuronal responses when changes in cognitive strategy can be excluded. We ensured this by engaging the patients in an explicit task that they were able to perform normally. This involved a key-press response when a visually presented letter string contained a letter with an ascending visual feature (e.g. h, k, l, or t). While the task remained constant, the stimuli presented were either words or consonant letter strings. Activations detected for words, relative to letters, were attributed to implicit word processing. Each patient showed normal activation of the left posterior middle temporal cortex that has been associated with semantic processing (Price, 1998). However, none of the patients activated the left posterior inferior frontal cortex (damaged by the stroke), or the left posterior inferior temporal region (undamaged by the stroke) (see Figure 36.4). These two regions are crucial for word production (Price, 1998). Examination of individual responses in this area revealed that all the normal subjects showed increased activity for words relative to consonant letter strings,

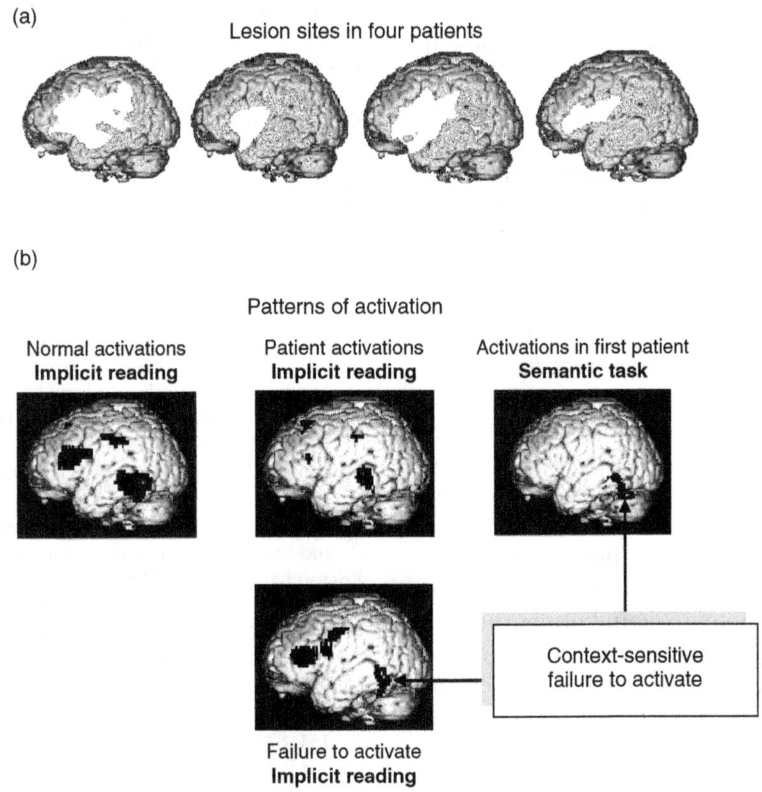

FIGURE 36.8 (a) Top: these renderings illustrate the extent of cerebral infarcts in four patients, as identified by voxel-based morphometry. Regions of reduced grey matter (relative to neurologically normal controls) are shown in white on the left hemisphere. The SPMs (statistical parametric maps) were thresholded at $p < 0.001$ uncorrected. All patients had damage to Broca's area. The first (upper left) patient's left middle cerebral artery infarct was most extensive encompassing temporal and parietal regions as well as frontal and motor cortex. (b) Bottom: SPMs illustrating the functional imaging results with regions of significant activation shown in black on the left hemisphere. Results are shown for normal subjects reading words (left); activations common to normal subjects; and patients reading words using a conjunction analysis (middle-top); areas where normal subjects activate significantly more than patients reading words, using the group times condition interaction (middle lower); and the first patient activating normally for a semantic task. Context-sensitive failures to activate are implied by the abnormal activations in the first patient, for the implicit reading task, despite a normal activation during a semantic task.

while all four patients showed the reverse effect. The abnormal responses in the left posterior inferior temporal lobe occurred even though this undamaged region lies adjacent and posterior to a region of the left middle temporal cortex that activated normally (see middle column of Figure 36.8(b)). Critically, this area is thought to mediate an earlier stage of word-processing than the damaged left inferior frontal cortex (i.e. is hierarchically lower than the lesion). From these results we can conclude that, during the reading task, responses in the left basal temporal language area rely on afferents from the left posterior inferior frontal cortex. When the first patient was scanned again, during an explicit semantic task, the left posterior inferior temporal lobe responded normally. The abnormal implicit reading related responses were therefore task-specific.

These results serve to illustrate the concept of dynamic diaschisis, namely, the anatomically remote and context-specific effects of focal brain lesions. Dynamic diaschisis represents a form of functional disconnection where regional dysfunction can be attributed to the loss of enabling inputs from hierarchically equivalent or higher brain regions. Unlike classical or anatomical disconnection syndromes, its pathophysiological expression depends upon the functional state at the time responses are evoked. Dynamic diaschisis may be characteristic of many regionally specific brain insults and may have implications for neuropsychological inference.

CONCLUSION

In conclusion, the function of any neuron, neuronal population or cortical area is context-sensitive. Functional integration, or interactions among brain systems, that employ forward (bottom-up) and backward (top-down) connections, mediate this adaptive specialization. A critical consequence is that hierarchically organized neuronal responses, in any given cortical area, can represent different things at different times. Although most models of perceptual learning and inference require priors on the causes of sensation, empirical Bayes suggests that these assumptions can be relaxed and that priors can be learned in a hierarchical context. We have tried to show that this hierarchical prediction can be implemented in brain-like architectures and in a biologically plausible fashion. The arguments in this chapter were developed under empirical or hierarchal Bayes models of brain function, where higher levels provide a prediction of the inputs to lower levels. Conflict between the two is resolved by changes in the higher-level representations, which are driven by the ensuing error in lower regions, until the mismatch is explained away. From this perspective, the specialization of any region is determined both by bottom-up inputs and by top-down predictions. Specialization is therefore not an intrinsic property of any region, but depends on both forward and backward connections with other areas. Because the latter have access to the context in which the data are generated they are in a position to modulate the selectivity of lower areas.

The theoretical neurobiology in this chapter has been used to motivate the importance of measuring effective connectivity, especially modulatory or non-linear coupling in the brain. These non-linear aspects will be a recurrent theme in subsequent chapters that discuss functional and effective connectivity from a conceptual and operational point of view.

REFERENCES

Absher JR, Benson DF (1993) Disconnection syndromes: an overview of Geschwind's contributions. *Neurology* **43**: 862–67

Aertsen A, Preil H (1991) Dynamics of activity and connectivity in physiological neuronal networks. In *Non linear dynamics and neuronal networks*, Schuster HG (ed.). VCH Publishers Inc., New York, pp 281–302

Angelucci A, Levitt JB, Walton EJ *et al.* (2002a) Circuits for local and global signal integration in primary visual cortex. *J Neurosci* **22**: 8633–46

Angelucci A, Levitt JB, Lund JS (2002b) Anatomical origins of the classical receptive field and modulatory surround field of single neurons in macaque visual cortical area V1. *Prog Brain Res* **136**: 373–88

Ballard DH, Hinton GE, Sejnowski TJ (1983) Parallel visual computation. *Nature* **306**: 21–26

Büchel C, Friston KJ (1997) Modulation of connectivity in visual pathways by attention: cortical interactions evaluated with structural equation modelling and fMRI. *Cerebr Cortex* **7**: 768–78

Crick F, Koch C (1998) Constraints on cortical and thalamic projections: the no-strong-loops hypothesis. *Nature* **391**: 245–50

Dayan P, Abbott LF (2001) *Theoretical neuroscience. Computational and mathematical modelling of neural systems.* MIT Press, [**36.3]

Dayan P, Hinton GE, Neal RM (1995) The Helmholtz machine. *Neural Comput* **7**: 889–904

Dempster AP, Laird NM, Rubin DB (1977) Maximum likelihood from incomplete data via the EM algorithm. *J Roy Stat Soc Series B* **39**: 1–38

Dolan RJ, Fink GR, Rolls E *et al.* (1997) How the brain learns to see objects and faces in an impoverished context. *Nature* **389**: 596–98

Efron B, Morris C (1973) Stein's estimation rule and its competitors – an empirical Bayes approach. *J Am Stat Assoc* **68**: 117–30

Felleman DJ, Van Essen DC (1991) Distributed hierarchical processing in the primate cerebral cortex. *Cereb Cortex* **1**: 1–47

Foldiak P (1990) Forming sparse representations by local anti-Hebbian learning. *Biol Cybern.* **64**: 165–70

Frey U, Morris RGM (1997) Synaptic tagging and long-term potentiation. *Nature* **385**: 533–36

Friston KJ, Price CJ, Fletcher P *et al.* (1996) The trouble with cognitive subtraction. *NeuroImage* **4**: 97–104

Friston KJ, Büchel C, Fink *et al.* (1997) Psychophysiological and modulatory interactions in neuroimaging. *NeuroImage* **6**: 218–29

Friston KJ, Büchel C (2000) Attentional modulation of V5 in human. *Proc Natl Acad Sci USA* **97**: 7591–96

Friston KJ (2005) A theory of cortical responses. *Philos Trans R Soc Lond B Biol Sci* **360**: 815–36

Gerstein GL, Perkel DH (1969) Simultaneously recorded trains of action potentials: analysis and functional interpretation. *Science* **164**: 828–30

Girard P, Bullier J (1989) Visual activity in area V2 during reversible inactivation of area 17 in the macaque monkey. *J Neurophysiol* **62**: 1287–301

Helmholtz H (1860/1962) *Handbuch der physiologischen optik* (English translation, Southall JPC, ed.), Vol. 3. Dover, New York

Hinton GE, Dayan P, Frey BJ *et al.* (1995) The 'Wake-Sleep' algorithm for unsupervised neural networks. *Science* **268**: 1158–61

Hochstein S, Ahissar M (2002) View from the top: hierarchies and reverse hierarchies in the visual system. *Neuron* **36**: 791–804

Jääskeläinen IP, Ahveninen J, Bonmassar G *et al.* (2004) Human posterior auditory cortex gates novel sounds to consciousness. *Proc Natl Acad Sci* **101**: 6809–14

Kass RE, Steffey D (1989) Approximate Bayesian inference in conditionally independent hierarchical models (parametric empirical Bayes models). *J Am Stat Assoc* **407**: 717–26

Kawato M, Hayakawa H, Inui T (1993) A forward-inverse optics model of reciprocal connections between visual cortical areas. *Network* **4**: 415–22

Kersten D, Mamassian P, Yuille A (2004) Object perception as Bayesian inference. *Annu Rev Psychol* **55**: 271–304

Lenzi GL, Frackowiak RSJ, Jones T (1982) Cerebral oxygen metabolism blood flow in human cerebral ischaemic infarction. *J Cereb Blood Flow Metab* **2**: 321–35

Locke J (1690/1976) *An essay concerning human understanding.* Dent, London

MacKay DM (1956) The epistemological problem for automata. In *Automata studies*, Shannon CE, McCarthy J (eds). Princeton University Press, Princeton, pp 235–51

McIntosh AR (2000) Towards a network theory of cognition. *Neural Networks* **13**: 861–70

Mesulam MM (1998) From sensation to cognition. *Brain* **121**: 1013–52

Mumford D (1992) On the computational architecture of the neocortex. II. The role of cortico-cortical loops. *Biol Cybernet* **66**: 241–51

Murphy PC, Sillito AM (1987) Corticofugal feedback influences the generation of length tuning in the visual pathway. *Nature* **329**: 727–29

Näätänen R (2003) Mismatch negativity: clinical research and possible applications. *Int J Psychophysiol* **48**: 179–88

Neisser U (1967) *Cognitive psychology*. Appleton-Century-Crofts, New York

Pearl J (2000) *Causality, models, reasoning and inference*. Cambridge University Press, Cambridge

Phillips CG, Zeki S, Barlow HB (1984) Localisation of function in the cerebral cortex: past present and future. *Brain* **107**: 327–61

Poggio T, Torre V, Koch C (1985) Computational vision and regularisation theory. *Nature* **317**: 314–19

Pollen DA (1999) On the neural correlates of visual perception. *Cereb Cortex* **9**: 4–19

Price CJ (1998) The functional anatomy of word comprehension and production. *Trends Cogn Sci* **2**: 281–88

Price CJ, Warburton EA, Moore CJ *et al.* (2001) Dynamic diaschisis: anatomically remote and context-sensitive human brain lesions. *J Cogn Neurosci* **13**: 419–29

Rao RP (1999) An optimal estimation approach to visual perception and learning. *Vision Res* **39**: 1963–89

Rao RP, Ballard DH (1998) Predictive coding in the visual cortex: a functional interpretation of some extra-classical receptive field effects. *Nature Neurosci* **2**: 79–87

Rivadulla C, Martinez LM, Varela C *et al.* (2002) Completing the corticofugal loop: a visual role for the corticogeniculate type 1 metabotropic glutamate receptor. *J Neurosci* **22**: 2956–62

Rockland KS, Pandya DN (1979) Laminar origins and terminations of cortical connections of the occipital lobe in the rhesus monkey. *Brain Res* **179**: 3–20

Salin P-A, Bullier J (1995) Corticocortical connections in the visual system: structure and function. *Psychol Bull* **75**: 107–54

Sandell JH, Schiller PH (1982) Effect of cooling area 18 on striate cortex cells in the squirrel monkey.*J Neurophysiol* **48**: 38–48

Sherman SM, Guillery RW (1998) On the actions that one nerve cell can have on another: distinguishing 'drivers' from 'modulators'. *Proc Natl Acad Sci USA* **95**: 7121–26

Simoncelli EP, Olshausen BA (2001) Natural image statistics and neural representation. *Annu Rev Neurosci* **24**: 1193–216

Treue S, Maunsell HR (1996) Attentional modulation of visual motion processing in cortical areas MT and MST. *Nature* **382**: 539–41

Zeki S (1990) The motion pathways of the visual cortex. In *Vision: coding and efficiency*, Blakemore C (ed.). Cambridge University Press, Cambridge, pp 321–45

Zeki S, Shipp S (1988) The functional logic of cortical connections. *Nature* **335**: 311–17

C H A P T E R

37

Functional connectivity: eigenimages and multivariate analyses

K. Friston and C. Büchel

INTRODUCTION

This chapter deals with imaging data from a multivariate perspective. This means that the observations at each voxel are considered jointly with explicit reference to the interactions among brain regions. The concept of functional connectivity is reviewed and is used as the basis for interpreting eigenimages. Having considered the nature of eigenimages and variations on their applications, we then turn to a related approach that, unlike eigenimage analysis, is predicated on a statistical model. This approach is called multivariate analysis of variance (MANCOVA) and uses canonical variate analysis to create canonical images. The integrated and distributed nature of neurophysiological responses to sensorimotor or cognitive challenge makes a multivariate perspective particularly appropriate, indeed necessary, for functional integration.

Functional integration and connectivity

A landmark meeting that took place on the morning of 4 August 1881 highlighted the difficulties of attributing function to a cortical area, given the dependence of cerebral activity on underlying connections (Phillips *et al.*, 1984). Goltz, although accepting the results of electrical stimulation in dog and monkey cortex, considered the excitation method inconclusive, in that the movements elicited might have originated in related pathways, or current could have spread to distant centres. Despite advances over the past century, the question remains: are the physiological changes elicited by sensorimotor or cognitive challenges explained by functional segregation, or by integrated and distributed changes mediated by neuronal connections? The question itself calls for a framework within which to address these issues. *Functional and effective connectivity* are concepts that are critical to this framework.

Origins and definitions

In the analysis of neuroimaging time-series, functional connectivity is defined as the *statistical dependencies among spatially remote neurophysiologic events*. This definition provides a simple characterization of functional interactions. The alternative is effective connectivity (i.e. *the influence one neuronal system exerts over another*). These concepts originated in the analysis of separable spike trains obtained from multiunit electrode recordings (Gerstein and Perkel, 1969). Functional connectivity is simply a statement about the observed dependencies or correlations; it does not comment on how these correlations are mediated. For example, at the level of multiunit microelectrode recordings, correlations can result from *stimulus-locked transients*, evoked by a common afferent input, or reflect *stimulus-induced oscillations*, phasic coupling of neuronal assemblies, mediated by synaptic connections. Effective connectivity is closer to the notion of a connection and can be defined as *the influence one neural system exerts over another*, either at a synaptic (cf. synaptic efficacy) or cortical level. Although functional and effective connectivity can be invoked at a conceptual level in both neuroimaging and electrophysiology, they differ fundamentally at a practical level. This is because the time-scales and nature of neurophysiological measurements are very different (seconds versus milliseconds and haemodynamic versus spike trains). In electrophysiology, it is often necessary to remove the confounding effects of stimulus-locked transients (that introduce correlations *not* causally mediated by direct neuronal interactions) to reveal an underlying connectivity. The confounding

Statistical Parametric Mapping, by Karl Friston et al.
ISBN–13: 978-0-12-372560-8 ISBN–10: 0-12-372560-7

effect of stimulus-evoked transients is less problematic in neuroimaging because propagation of dynamics from primary sensory areas onwards is mediated by neuronal connections (usually reciprocal and interconnecting). However, it should be remembered that functional connectivity is not necessarily due to effective connectivity (e.g. common neuromodulatory input from ascending aminergic neurotransmitter systems or thalamo-cortical afferents) and, where it is, effective influences may be indirect (e.g. polysynaptic relays through multiple areas).

EIGENIMAGES, MULTIDIMENSIONAL SCALING AND OTHER DEVICES

In what follows, we introduce a number of techniques (eigenimage analysis, multidimensional scaling, partial least squares and generalized eigenimage analysis) using functional connectivity as a reference. Emphasis is placed on the relationship among these techniques. For example, eigenimage analysis is equivalent to principal component analysis and the variant of multidimensional scaling considered here is equivalent to principal coordinates analysis. Principal components and coordinates analyses are predicated on exactly the same eigenvector solution and, from a mathematical perspective, are the same thing.

Measuring a pattern of correlated activity

Here we introduce a simple way of measuring the amount a pattern of activity (representing a connected brain system) contributes to the functional connectivity or variance-covariances observed in imaging data. Functional connectivity is defined in terms of statistical dependencies among neurophysiological measurement. If we assume these measurements conform to Gaussian assumptions, then we need only characterize their correlations or covariance (correlations are normalized covariances).[1] The point-to-point functional connectivity between one voxel and another is not usually of great interest. The important aspect of a covariance structure is the pattern of correlated activity subtended by (an enormous number of) pair-wise covariances. In measuring such patterns, it is useful to introduce the concept of a *norm*. Vector and matrix norms serve the same purpose as absolute values for scalar quantities. In other words, they furnish a measure of distance. One frequently used norm is the 2-norm, which is the length of a vector. The vector 2-norm can be used to measure the degree to which a particular pattern of brain activity contributes to a covariance structure. If a pattern is described by a column vector p, with an element for each voxel, then the contribution of that pattern to the covariance structure can be measured by the 2-norm of $Mp = \|Mp\|$. M is a (mean-corrected) matrix of data with one row for each successive scan and one column for each voxel:

$$\|Mp\|^2 = p^T M^T Mp \tag{37.1}$$

^{T} denotes transposition. Put simply, the 2-norm is a number that reflects the amount of variance-covariance or functional connectivity that can be accounted for by a particular distributed pattern. It should be noted that the 2-norm only measures the pattern of interest. There may be many other important patterns of functional connectivity. This fact begs the question: 'what are the most prevalent patterns of coherent activity?' To answer this question one turns to eigenimages or spatial modes.

Eigenimages and spatial modes

In this section, the concept of eigenimages or spatial modes is introduced in terms of patterns of activity defined above. We show that spatial modes are simply those patterns that account for the most variance-covariance (i.e. have the largest 2-norm).

Eigenimages or spatial modes are most commonly obtained using singular value decomposition (*SVD*). *SVD* is an operation that decomposes an original time-series, M, into two sets of orthogonal vectors (patterns in space and patterns in time) V and U where:

$$\begin{aligned} [U, S, V] &= SVD(M) \\ M &= USV^T \end{aligned} \tag{37.2}$$

U and V are unitary orthogonal matrices $U^TU = I$, $V^TV = I$ and $V^TU = 0$ (the sum of squares of each column is unity and all the columns are uncorrelated) and S is a diagonal matrix (only the leading diagonal has non-zero values) of decreasing singular values. The singular value of each eigenimage is simply its 2-norm. Because *SVD* maximizes the first singular value, the first eigenimage is the pattern that accounts for the greatest amount of the variance-covariance structure. In summary, *SVD* and equivalent devices are powerful ways of decomposing an imaging time-series into a series of orthogonal patterns that embody, in a step-down fashion, the greatest amounts of functional connectivity. Each eigenvector (column of V) defines a distributed brain system

[1] Clearly neuronal processes are not necessarily Gaussian. However, we can still characterize the second-order dependencies with the correlations. Higher-order dependencies would involve computing cumulants as described in Appendix 2.

that can be displayed as an image. The distributed systems that ensue are called *eigenimages* or *spatial modes* and have been used to characterize the spatiotemporal dynamics of neurophysiological time-series from several modalities including, multiunit electrode recordings (Mayer-Kress *et al.*, 1991), electroencephalography (EEG) (Friedrich *et al.*, 1991), magnetoencephalography (MEG) (Fuchs *et al.*, 1992), positron emission tomography (PET) (Friston *et al.*, 1993a) and functional magnetic resonance imaging (fMRI) (Friston *et al.*, 1993b). Interestingly, in fMRI, the application of eigenimages that has attracted the most interest is in characterizing functional connections while the brain is at 'rest' (see Biswal *et al.*, 1995).

Many readers will notice that the eigenimages associated with the functional connectivity or covariance matrix are simply principal components of the time-series. In the EEG literature, one sometimes comes across the Karhunen-Loeve expansion, which is employed to identify spatial modes. If this expansion is in terms of eigenvectors of covariances (and it usually is), then the analysis is formally identical to the one presented above.

One might ask what the column vectors of U in Eqn. **37.2** correspond to. These vectors are the time-dependent profiles associated with each eigenimage known as *eigenvariates*. They reflect the extent to which an eigenimage is expressed in each experimental condition or over time. Figure 37.1 shows a simple schematic illustrating the decomposition of a time-series into orthogonal modes. This is sometimes called spectral decomposition. Eigenvariates play an important role in the functional attribution of distributed systems defined by eigenimages. This point and others will be illustrated in the next section.

Mapping function into anatomical space – eigenimage analysis

To illustrate the approach, we will use the PET word generation study used in previous chapters. The data were obtained from five subjects scanned twelve times while performing one of two verbal tasks in alternation. One task involved repeating a letter presented aurally at one per 2 s (*word shadowing*). The other was a paced verbal fluency task, where subjects responded with a word that began with the heard letter (*word generation*). To facilitate inter-subject pooling, the data were realigned and spatially normalized and smoothed with an isotropic Gaussian kernel (full width at half maximum (FWHM) of 16 mm). The data were then subject to an analysis of covariance (ANCOVA) (with twelve condition-specific effects, subject-effects and global activity as a confounding effect). Voxels were selected using a conventional SPM{F} to identify those significant

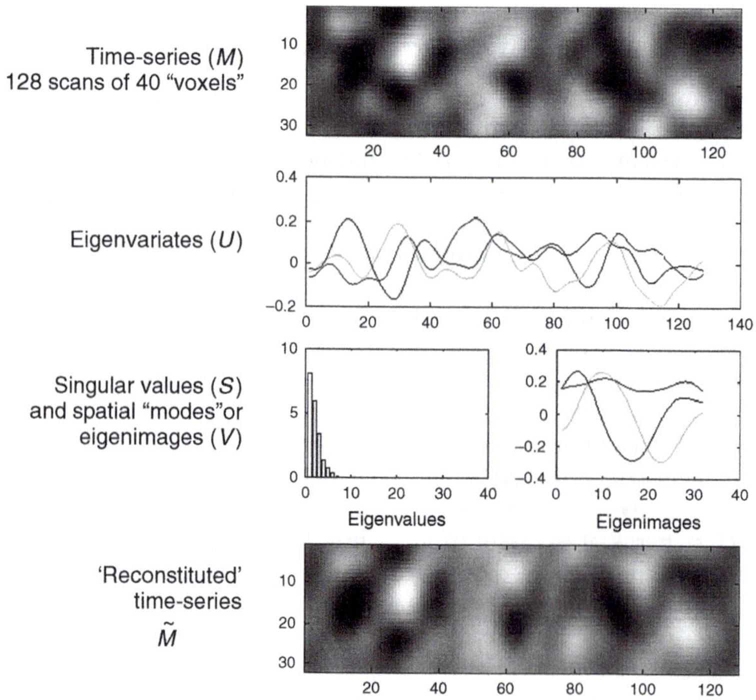

$$M = USV^T = s_1 U_1 V_1^T + s_2 U_2 V_2^T + \ldots$$

$$\tilde{M} = s_1 U_1 V_1^T + s_2 U_2 V_2^T + s_3 U_3 V_3^T$$

FIGURE 37.1 Schematic illustrating a simple spectral decomposition or singular-decomposition of a multivariate time-series. The original time-series is shown in the upper panel with time running along the x axis. The first three eigenvariates and eigenvectors are shown in the middle panels together with the spectrum [hence spectral decomposition] of singular values. The eigenvalues are the square of the singular values $\lambda = SS^T$. The lower panel shows the data reconstructed using only three principal components, because they capture most of the variance the reconstructed sequence is very similar to the original time-series.

at $p < 0.05$ (uncorrected). The time-series of condition-specific effects, from each of these voxels, were entered into a mean corrected data matrix M with twelve rows (one for each condition) and one column for each voxel.

M was subject to SVD as described above. The distribution of eigenvalues (Figure 37.2, lower left) suggests only two eigenimages are required to account for most of the observed variance-covariance structure. The first mode accounted for 64 per cent and the second for 16 per cent of the variance. The first eigenimage V_1 is shown in Figure 37.2 (top) along with the corresponding eigenvariate U_1 (lower right). The first eigenimage has positive loadings in the anterior cingulate, the left dorso-lateral prefrontal cortex (DLPFC), Broca's area, the thalamic nuclei and in the cerebellum. Negative loadings were seen bi-temporally and in the posterior cingulate. According to U_1, this eigenimage is prevalent in the verbal fluency tasks with negative scores in word shadowing. The second spatial mode (not shown) had its highest positive loadings in the anterior cingulate and bi-temporal regions (notably Wernicke's area on the left). This mode appears to correspond to a highly non-linear, monotonic time effect with greatest prominence in earlier conditions.

The *post hoc* functional attribution of these eigenimages is usually based on their eigenvariates, U. The first mode

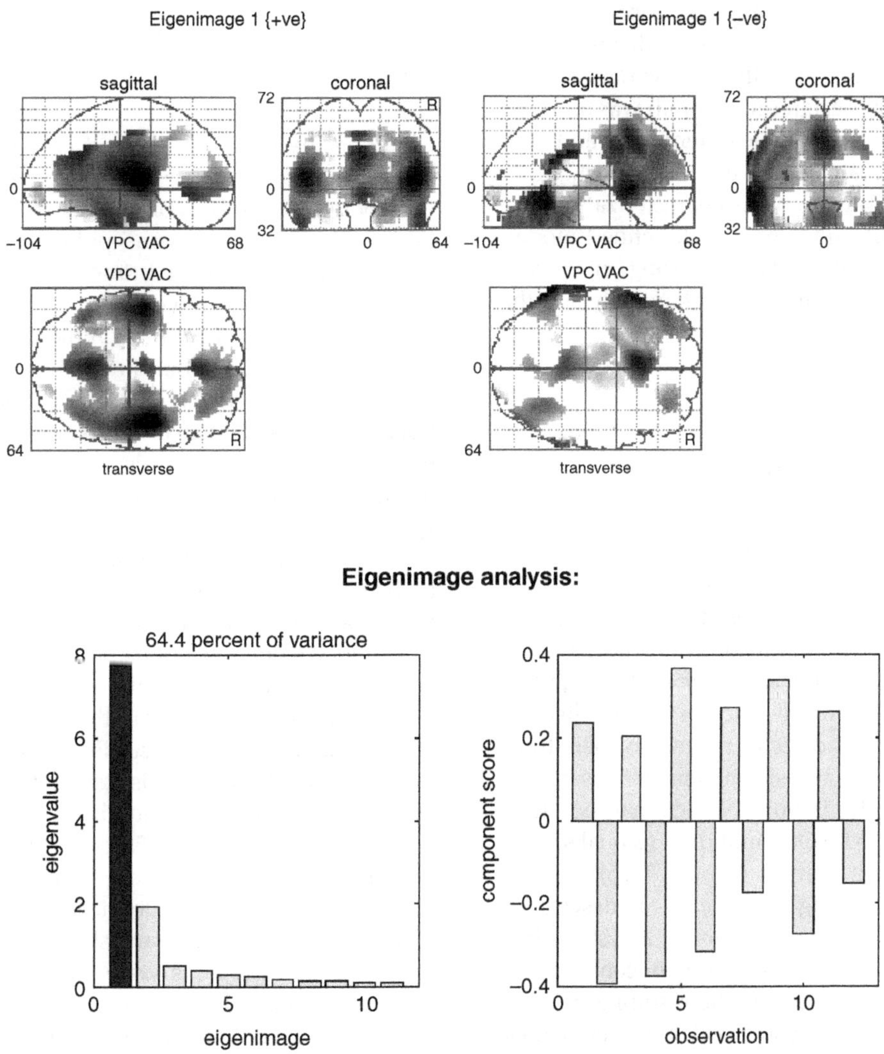

FIGURE 37.2 Eigenimage analysis of the PET activation study of word generation. Top: positive and negative components of the first eigenimage (i.e. first column of V). The maximum intensity projection display format is standard and provides three views of the brain (from the back, from the right and from the top). Lower left: eigenvalues (singular values squared) of the functional connectivity matrix reflecting the relative amounts of variance accounted for by the eleven eigenimages associated with these data. Only two eigenvalues are greater than unity and, to all intents and purposes, the changes characterizing this time-series can be considered two-dimensional. Lower right: the temporal eigenvariate reflecting the expression of this eigenimage over conditions (i.e. the first column of U).

may represent an *intentional* system critical for the intrinsic generation of words in the sense that the key cognitive difference between verbal fluency and word shadowing is the intrinsic generation as opposed to extrinsic specification of word representations and implicit mnemonic processing. The second system, which includes the anterior cingulate, seems to be involved in habituation, possibly of attentional or perceptual set.

There is nothing biologically important about the particular spatial modes obtained in this fashion, in the sense that one could rotate the eigenvectors such that they were still orthogonal and yet gave different eigenimages. The uniqueness of the particular solution given by *SVD* is that the first eigenimage accounts for the largest amount of variance-covariance and the second for the greatest amount that remains and so on. The reason that the eigenimages in the example above lend themselves to such a simple interpretation is that the variance introduced by experimental design (intentional) was substantially greater than that due to time (attentional) and both these sources were greater than any other effect. Other factors that ensure a parsimonious characterization of a time-series, with small numbers of well-defined modes, include the smoothness in the data and using only voxels that showed a non-trivial amount of change during the scanning session.

Mapping anatomy into functional space – multidimensional scaling

In the previous section, the functional connectivity matrix was used to define eigenimages or spatial modes. In this section, functional connectivity is used in a different way, namely, to constrain the proximity of two cortical areas in some functional space (Friston *et al.*, 1996a). The objective here is to transform anatomical space so that the distance between cortical areas is directly related to their functional connectivity. This transformation defines a new space whose topography is purely functional in nature. This space is constructed using multidimensional scaling or principal coordinates analysis (Gower, 1966).

Multidimensional scaling (MDS) is a descriptive method for representing the structure of a system. It is based on pair-wise measures of similarity or confusability (Torgerson, 1958; Shepard, 1980). The resulting multidimensional deployment of a system's elements embodies, in the proximity relationships, comparative similarities. The technique was developed primarily for the analysis of perceptual spaces. The proposal that stimuli be modelled by points in space, so that perceived similarity is represented by spatial distances, goes back to the days of Isaac Newton (1794).

Imagine k measures from n voxels plotted as n points in a k-dimensional space (k-space). If they have been normalized to zero mean and unit sum of squares, these points will fall on a $k-1$ dimensional sphere. The closer any two points are to each other, the greater their correlation or functional connectivity (in fact, the correlation is the cosine of the angle subtended at the origin). The distribution of these points is the functional topography. A view of this distribution that reveals the greatest structure is obtained by rotating the points to maximize their apparent dispersion (variance). In other words, one looks at the subspace with the largest 'volume' spanned by the principal axes of the n points in k-space. These principal axes are given by the eigenvectors of MM^T; i.e. the column vectors of U_1. From Eqn. **37.2**:

$$MM^T = U\lambda U^T \qquad \lambda = SS^T \tag{37.3}$$

Let Q be the matrix of desired coordinates derived by simply projecting the original data onto axes defined by U: where $Q = M^T U$. Voxels that have a correlation of unity will occupy the same point in MDS space. Voxels that have uncorrelated time-series will be $\pi/2$ apart. Voxels that are negatively, but completely, correlated will be maximally separated on the opposite sides of the MDS hyperspace. Profound negative correlations denote a functional association that is modelled in MDS functional space as diametrically opposed locations on the hypersphere. In other words, two regions with profound negative correlations will form two 'poles' in functional space.

Following normalization to unit sum of squares, over each column M (the adjusted data matrix from the word generation study above), the data matrix was subjected to singular value decomposition according to Eqn. **37.2** and the coordinates Q of the voxels in MDS functional space were computed. Recall that only two [normalized] eigenvalues exceed unity (see Figure 37.2; right), suggesting a functional space that is essentially two-dimensional. The locations of voxels in this two-dimensional subspace are shown in Plate 53(c) and (d) (see colour plate section) by rendering voxels from different regions in different colours. The anatomical regions corresponding to the different colours are shown in Plate 53(a) and (b). Anatomical regions were selected to include those parts of the brain that showed the greatest variance during the twelve conditions. Anterior regions (Plate 53(b)) included the medio-dorsal thalamus (blue), the DLPFC, Broca's area (red) and the anterior cingulate (green). Posterior regions (Plate 53(a)) included the superior temporal regions (red), the posterior superior temporal regions (blue) and the posterior cingulate (green). The corresponding functional spaces (Plate 53(c) and (d)) reveal a number of things about the functional topography elicited by this set of activation tasks. First, each anatomical region maps into a relatively localized portion of functional space. This preservation of local contiguity reflects the high correlations within anatomical regions, due in part to smoothness of the original data and to high degrees of intra-regional functional connectivity. Second, the anterior regions are almost in juxtaposition, as are posterior regions. However, the confluence of anterior and posterior regions forms two diametrically opposing poles (or one axis). This configuration suggests an anterior-posterior axis with prefronto-temporal and cingulo-cingulate components. One might have predicted this configuration by noting that the anterior regions had high positive loadings on the first eigenimage (see Figure 37.2), while the posterior regions had high negative loadings. Thirdly, within the anterior and posterior sets of regions certain generic features are evident. The most striking is the particular ordering of functional interactions. For example, the functional connectivity between posterior cingulate (green) and superior temporal regions (red) is high and similarly for the superior temporal (red) and posterior temporal regions (blue). Yet the posterior cingulate and posterior temporal regions show very little functional connectivity (they are $\pi/2$ apart or, equivalently, subtend 90 degrees at the origin).

These results are consistent with known anatomical connections. For example, DLPFC – anterior cingulate connections, DLPFC – temporal connections, bitemporal commissural connections and medio-dorsal thalamic – DLPFC projections have all been demonstrated in non-human primates (Goldman-Rakic, 1988). The medio-dorsal thalamic region and DLPFC are so correlated that one is embedded within the other (purple area). This is pleasing given the known thalamo-cortical projections to DLPFC.

Functional connectivity between systems – partial least squares

Hitherto, we have been dealing with functional connectivity between two voxels. The same notion can be extended to functional connectivity between two systems by noting that there is no fundamental difference between the dynamics of one voxel and the dynamics of a distributed system or mixture of voxels. The functional connectivity between two systems is simply the correlation or covariance between their time-dependent activities. The time-dependent activity of a system or pattern p_i is given by:

$$
\begin{aligned}
v_i &= Mp_i \\
C_{ij} &= v_i^T v_j = p_i^T M^T M p_j
\end{aligned}
\qquad \textbf{37.4}
$$

where C_{ij} is the functional connectivity between the systems described by vectors p_i and p_j. Consider functional connectivity between two systems in separate parts of the brain, for example the right and left hemispheres. Here the data matrices M_i and M_j derive from different sets of voxels and Eqn. **37.4** becomes:

$$
C_{ij} = v_i^T v_j = p_i^T M_i^T M_j p_j \qquad \textbf{37.5}
$$

If one wanted to identify the intra-hemispheric systems that showed the greatest inter-hemispheric functional connectivity (i.e. covariance), one would need to identify the pair of vectors p_i and p_j that maximize C_{ij} in Eqn. **37.5**. *SVD* finds another powerful application in doing just this:

$$
\begin{aligned}
[U, S, V] &= SVD(M_i^T M_j) \\
M_i^T M_j &= USV^T \\
U^T M_i^T M_j V &= S
\end{aligned}
\qquad \textbf{37.6}
$$

The first columns of U and V represent the singular images that correspond to the two systems with the greatest amount of functional connectivity (the singular values in the diagonal matrix S). In other words, *SVD* of the (generally asymmetric) cross-covariance matrix, based on time-series from two anatomically separate parts of the brain, yields a series of paired vectors (paired columns of U and V) that, in a step-down fashion, define pairs of brain systems that show the greatest functional connectivity. This particular application of *SVD* is also know as *partial least squares* and has been proposed for analysis of designed activation experiments where the two data matrices comprise an image time-series and a set of behavioural or task parameters, i.e. the design matrix (McIntosh *et al.*, 1996). In this application, the paired singular vectors correspond to a singular image and a set of weights that give the linear combination of task parameters that show the maximal covariance with the corresponding singular image. This is conceptually related to canonical image analysis (see next section) based on the generalized eigenvalue solution.

Differences in functional connectivity – generalized eigenimages

Here, we introduce an extension of eigenimage analysis using the solution to the generalized eigenvalue problem. This problem involves finding the eigenvector solution that involves two covariance matrices and can be used to find the eigenimage that is maximally expressed in one time-series relative to another. In other words, it can find a pattern of distributed activity that is most prevalent in one data set and least expressed in another. The example used to illustrate this idea is fronto-temporal functional disconnection in schizophrenia (see Friston *et al.*, 1996b).

The notion that schizophrenia represents a disintegration or fractionation of the psyche is as old as its name, introduced by Bleuler (1913) to convey a 'splitting' of mental faculties. Many of Bleuler's primary processes, such as 'loosening of associations' emphasize a fragmentation and loss of coherent integration. In what follows, we assume that this mentalist 'splitting' has a physiological basis and, furthermore, that both the mentalist and physiological disintegration have precise and specific characteristics that can be understood in terms of functional connectivity.

The idea is that, although localized pathophysiology in cortical areas may be a sufficient explanation for some signs of schizophrenia, it does not suffice as an explanation for the symptoms of schizophrenia. The conjecture is that symptoms, such as hallucinations and delusions, are better understood in terms of abnormal interactions or impaired integration between different cortical areas. This dysfunctional integration, expressed at a physiological level as abnormal functional connectivity, is measurable with neuroimaging and observable at a cognitive level as a failure to integrate perception and action that manifests as clinical symptoms. The distinction between a regionally specific pathology and a pathology of interaction can be seen in terms of a first-order effect (e.g. hypofrontality) and a second-order effect that only exists in the relationship between activity in the prefrontal cortex and some other (e.g. temporal) region. In a similar way, psychological abnormalities can be regarded as first order (e.g. a poverty of intrinsically cued behaviour in psychomotor poverty) or second order (e.g. a failure to integrate intrinsically cued behaviour and perception in reality distortion).

The generalized eigenvalue solution

Suppose that we want to find a pattern embodying the greatest amount of functional connectivity in control subjects, relative to schizophrenic subjects (e.g. fronto-temporal covariance). To achieve this, we identify an eigenimage that reflects the most functional connectivity in control subjects relative to a schizophrenic group, d. This eigenimage is obtained by using a generalized eigenvector solution:

$$\begin{aligned} C_i^{-1} C_j d &= d\lambda \\ C_j d &= C_i d\lambda \end{aligned} \qquad 37.7$$

where C_i and C_j are the two functional connectivity matrices. The generalized eigenimage d is essentially a pattern that maximizes the ratio of the 2-norm measures (see Eqn. **37.1**) when applied to C_i and C_j. Generally speaking, these matrices could represent data from two [groups of] subjects or from the same subject(s) scanned under different conditions. In the present example, we use connectivity matrices from control subjects and people with schizophrenia showing pronounced psychomotor poverty.

The data were acquired from two groups of six subjects. Each subject was scanned six times during the performance of three word generation tasks (A B C C B A). Task A was a verbal fluency task, requiring subjects to respond with a word that began with a heard letter. Task B was a semantic categorization task in which subjects responded 'man-made' or 'natural', depending on a heard noun. Task C was a word-shadowing task in which subjects simply repeated what was heard. In the present context, the detailed nature of the tasks is not important. They were used to introduce variance and covariance in activity that could support an analysis of functional connectivity.

The groups comprised six control subjects and six schizophrenic patients. The schizophrenic subjects produced fewer than 24 words on a standard (one minute) FAS verbal fluency task (generating words beginning with the letters 'F', 'A' and 'S'). The results of a generalized eigenimage analysis are presented in Figure 37.3. As expected, the pattern that best captures group differences involves prefrontal and temporal cortices and encodes negative correlations between left DLPFC and bilateral superior temporal regions (Figure 37.3; upper panels). The amount to which this pattern was expressed in each subject is shown in the lower panel using the appropriate 2-norm $\|d^T C_i d\|$. It is seen that this eigenimage, while prevalent in control subjects, is uniformly reduced in schizophrenic subjects.

Summary

In the preceding section, we have seen how eigenimages can be framed in terms of functional connectivity and the relationships among eigenimage analysis, multidimensional scaling, partial least squares and generalized eigenimage analysis. In the next section, we use the generative models perspective, described in the previous chapter, to take component analysis into the non-linear domain. In the subsequent section, we return to generalized eigenvalue solutions and their role in classification and canonical image analysis.

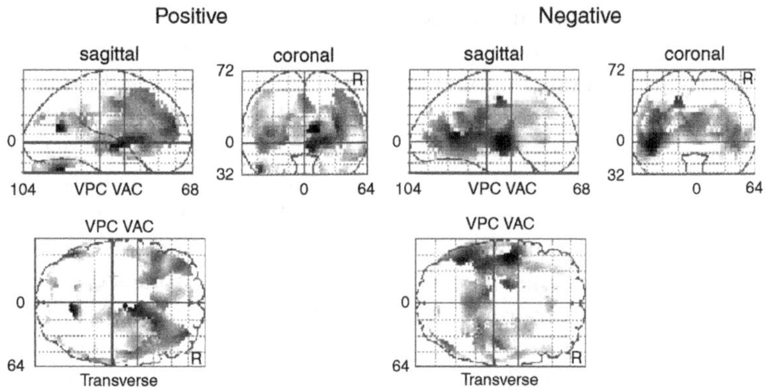

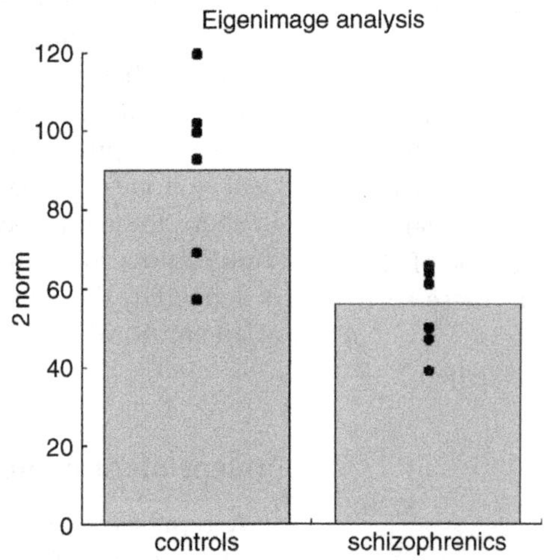

FIGURE 37.3 Generalized eigenimage analysis of schizophrenic and control subjects. Top left and right: positive and negative loadings of the first eigenimage that is maximally expressed in the control group and minimally expressed in the schizophrenic group. This analysis used PET activation studies of word generation with six scans per subject and six subjects per group. The activation study involved three word generation conditions (word shadowing, semantic categorization and verbal fluency), each of which was presented twice. The grey scale is arbitrary and each image has been normalized to the image maximum. The display format is standard and represents a maximum intensity projection. This eigenimage is relatively less expressed in the schizophrenic data. This point is made by expressing the amount of functional connectivity attributable to the eigenimage in (each subject in) both groups, using the appropriate 2-norm (lower panel).

NON-LINEAR PRINCIPAL AND INDEPENDENT COMPONENT ANALYSIS (PCA AND ICA)

Generative models

Recall from the previous chapter how generative models of data could be framed in terms of a *prior* distribution over causes $p(v, \theta)$ and a *generative* distribution or likelihood of the inputs given the causes $p(u|v; \theta)$. For example, factor analysis corresponds to the generative model:

$$\begin{aligned} p(v; \theta) &= N(0, 1) \\ p(u\,|v\,; \theta) &= N(\theta v, \Sigma) \end{aligned} \qquad 37.8$$

Namely, the underlying causes of inputs are independent normal variates that are mixed linearly and added to Gaussian noise to form inputs. In the limiting case of $\Sigma \to 0$, the model becomes deterministic and conforms to PCA. By simply assuming non-Gaussian priors one can specify generative models for sparse coding:

$$\begin{aligned} p(v; \theta) &= \prod p(v_i; \theta) \\ p(u\,|v; \theta) &= N(\theta v, \Sigma) \end{aligned} \qquad 37.9$$

where $p(v_i, \theta)$ are chosen to be suitably sparse (i.e. heavy-tailed), with a cumulative density function that corresponds to the squashing function below. The deterministic equivalent of sparse coding is ICA that obtains when $\Sigma \to 0$. These formulations allow us to consider simple extensions of PCA by looking at non-linear versions of the underlying generative model.

Non-linear PCA

Despite its exploratory power, eigenimage analysis is fundamentally limited because the particular modes obtained are uniquely determined by constraints that are biologically implausible. This represents an inherent limitation on the interpretability and usefulness of eigenimage analysis. The two main limitations of conventional eigenimage analysis are that the decomposition of any observed time-series is in terms of linearly separable components. Secondly, the spatial modes are somewhat arbitrarily constrained to be orthogonal and account, successively, for the largest amount of variance. From a biological perspective, the linearity constraint is a severe one because it precludes interactions among brain systems. This is an unnatural restriction on brain activity, where one expects to see substantial interactions that render the expression of one mode sensitive to the expression of others. Non-linear PCA attempts to circumvent these sorts of limitations.

The generative model implied by Eqn. **37.8**, when $\Sigma \to 0$, is linear and deterministic:

$$\begin{aligned} p(v;\theta) &= N(0,1) \\ u &= \theta v \end{aligned} \qquad \mathbf{37.10}$$

Here the causes v correspond to the eigenvariates and the model parameters to scaled eigenvectors $\theta = VS$. u is the observed data or image that comprised each row of M above. This linear generative model $g(v,\theta) = \theta v$ can now be generalized to any static non-linear model by taking a second-order approximation:

$$\begin{aligned} p(v;\theta) &= N(0,1) \\ u &= g(v,\theta) \\ &= \sum_i V_i v_i + \frac{1}{2}\sum_{ij} V_{ij} v_i v_j + \ldots \\ V_i &= \frac{\partial g}{\partial v} \\ V_{ij} &= \frac{\partial^2 g}{\partial v_i \partial v_j} \end{aligned} \qquad \mathbf{37.11}$$

This non-linear model has two sorts of modes; first-order modes V_i that mediate the effect of any orthogonal cause on the response (i.e. map the causes onto voxels directly) and second-order modes V_{ij}, which map interactions among causes onto the measured response. These second-order modes could represent the distributed systems implicated in the interaction between various experimentally manipulated causes. See the example below.

The identification of the first- and second-order modes proceeds using expectation maximization (EM) as described in the previous chapter. In this instance, the algorithm can be implemented as a simple neural net with forward connections from the data to the causes and backward connections from the causes to the predicted data. The **E**-step corresponds to *recognition* of the causes by the forward connections using the current estimate of the first-order modes and the **M**-step adjusts these connections to minimize the prediction error of the generative model in Eqn. **37.11**, using the recognized causes. These schemes (e.g. Kramer, 1991; Karhunen and Joutsensalo, 1994; Friston *et al.*, 2000) typically employ a 'bottleneck' architecture that forces the inputs through a small number of nodes (see the insert in Plate 54). The output from these nodes then diverges to produce the predicted inputs. After learning, the activity of the bottleneck nodes can be treated as estimates of the causes. In short, these representations obtain by projection of the input onto a low-dimensional curvilinear manifold that is defined by the activity of the bottleneck. Before looking at an empirical example we will briefly discuss ICA.

Independent component analysis

ICA represents another way of generalizing the linear model used by PCA. This is achieved, not through non-linearities, but by assuming non-Gaussian priors. The non-Gaussian form can be specified by a non-linear transformation of the causes $\tilde{v} = \sigma(v)$ that renders them normally distributed, such that when $\Sigma \to 0$, in Eqn. **37.9** we get:

$$\begin{aligned} p(\tilde{v};\theta) &= N(0,1) \\ v &= \sigma^{-1}(\tilde{v}) \\ u &= \theta v \end{aligned} \qquad \mathbf{37.12}$$

This is not the conventional way to present ICA, but is used here to connect the models for PCA and ICA. The form of the non-linear squashing function $\tilde{v} = \sigma(v)$ embodies our prior assumptions about the marginal distribution of the causes. These are usually supra-Gaussian. There exist simple algorithms that implicitly minimize the objective function F (see previous chapter) using the covariances of the data. In neuroimaging, this enforces an ICA of independent spatial modes, because there are more voxels than scans (McKeown *et al.*, 1998). In EEG, there are more time bins than channels and the independent components are temporal in nature. The distinction between *spatial* and *temporal* ICA depends on whether one regards Eqn. **37.12** as generating data over space or time (see Friston, 1998 for a discussion of their relative merits). The important thing about ICA, relative to PCA, is that the prior densities model independent causes not just uncorrelated causes. This difference is expressed in terms of statistical dependencies beyond second-order (see Stone, 2002 for an introduction to these issues).

An example

This example comes from Friston *et al.* (2000)[2] and is based on an fMRI study of visual processing that was designed to address the interaction between colour and motion systems.

We had expected to demonstrate that a 'colour' mode and 'motion' mode would interact to produce a second-order mode reflecting: (i) reciprocal interactions between extrastriate areas functionally specialized for colour and motion; (ii) interactions in lower visual areas mediated by convergent backwards connections; or (iii) interactions in the pulvinar mediated by cortico-thalamic loops.

Data acquisition and experimental design

A subject was scanned under four different conditions, in six scan epochs, intercalated with a low-level (visual fixation) baseline condition. The four conditions were repeated eight times in a pseudo-random order giving 384 scans in total or 32 stimulation/baseline epoch pairs. The four experimental conditions comprised the presentation of moving and stationary dots, using luminance and chromatic contrast, in a two by two-factorial design. Luminance contrast was established using isochromatic stimuli (red dots on a red background or green dots on a green background). Hue contrast was obtained by using red (or green) dots on a green (or red) background and establishing iso-luminance with flicker photometry. In the two movement conditions, the dots moved radially from the centre of the screen, at eight degrees per second to the periphery, where they vanished. This creates the impression of optical flow. By using these stimuli we hoped to excite activity in a visual motion system and one specialized for colour processing. Any interaction between these systems would be expressed in terms of motion-sensitive responses that depended on the hue or luminance contrast subtending that motion.

[2] Although an example of non-linear PCA, the generative model actually used augmented Eqn. **37.11** with a non-linear function of the second- order terms: $u = G(v) = \sum_i V_i v_i + \frac{1}{2} \sum_{ij} V_{ij}\theta(v_i v_j)$. This endows the causes with unique scaling.

Non-linear PCA

The data were reduced to an eight-dimensional subspace using SVD and entered into a non-linear PCA using two causes. The functional attribution of the resulting sources was established by looking at the expression of the corresponding first-order modes over the four conditions (right lower panels in Plate 54). This expression is simply the score on the first principal component over all 32 epoch-related responses for each cause. The first mode is clearly a motion-sensitive mode but one that embodies some colour preference, in the sense that the motion-dependent responses of this system are accentuated in the presence of colour cues. This was not quite what we had anticipated; the first-order effect contains what would functionally be called an interaction between motion and colour processing. The second first-order mode appears to be concerned exclusively with colour processing. The corresponding anatomical profiles are shown in Plate 54 (left panels). The first-order mode, which shows both motion and colour-related responses, shows high loadings in bilateral motion sensitive complex V5 (Brodmann areas 19 and 37 at the occipito-temporal junction) and areas traditionally associated with colour processing (V4 – the lingual gyrus). The second first-order mode is most prominent in the hippocampus, parahippocampal and related lingual cortices on both sides. In summary, the two first-order modes comprise: an extrastriate cortical system including V5 and V4 that responds to motion, and preferentially so when motion is supported by colour cues; and a [para]hippocampus-lingual system that is concerned exclusively with colour processing, above and beyond that accounted for by the first system. The critical question is where do these modes interact?

The interaction between the extrastriate and [para]hippocampus-lingual systems conforms to the second-order mode in the lower panels. This mode highlights the pulvinar of the thalamus and V5 bilaterally. This is a pleasing result in that it clearly implicates the thalamus in the integration of extrastriate and [para]hippocampal systems. This integration is mediated by recurrent cortico-thalamic connections. It is also a result that would not have been obtained from a conventional SPM analysis. Indeed, we looked for an interaction between motion and colour processing and did not see any such effect in the pulvinar.

Summary

We have reviewed eigenimage analysis and generalizations based on non-linear and non-Gaussian generative models. All the techniques above are essentially descriptive, in that they do not allow one to make any statistical inferences about the characterizations that obtain.

In the second half of this chapter, we turn to multivariate techniques that enable statistical inference and hypothesis testing. We will introduce *canonical images* that can be thought of as statistically informed eigenimages, pertaining to effects introduced by experimental design. We have seen that patterns can be identified using the generalized eigenvalue solution that are maximally expressed in one covariance structure, relative to another. Consider now using this approach where the first covariance matrix reflected the effects we were interested in, and the second embodied covariances due to error. This corresponds to canonical image analysis, and is considered in the following section.

MANCOVA AND CANONICAL IMAGE ANALYSES

In this section, we review multivariate approaches to the analysis of functional imaging studies. The analyses described use standard multivariate techniques to make statistical inferences about activation effects and to describe their important features. Specifically, we introduce multivariate analysis of covariance (MANCOVA) and canonical variates analysis (CVA) to characterize activation effects. This approach characterizes the brain's response in terms of functionally connected and distributed systems in a similar fashion to eigenimage analysis. Eigenimages figure in the current analysis in the following way: a problematic issue in multivariate analysis of functional imaging data is that the number of samples (i.e. scans) is usually very small in relation to the number of components (i.e. voxels) of the observations. This issue is resolved by analysing the data, not in terms of voxels, but in terms of eigenimages, because the number of eigenimages is much smaller than the number of voxels. The importance of the multivariate analysis that ensues can be summarized as follows:

- Unlike eigenimage analysis, it provides for statistical inferences (based on classical p-values) about the significance of the brain's response in terms of some hypothesis.
- The approach implicitly takes account of spatial correlations in the data without making any assumptions.
- The canonical variate analysis produces generalized eigenimages (canonical images) that capture the activation effects, while suppressing the effects of noise or error.
- The theoretical basis is well established and can be found in most introductory texts on multivariate analysis (see also Friston *et al.*, 1996c).

Although useful, in a descriptive sense, eigenimage analysis and related approaches are not generally considered as 'statistical' methods that can be used to make statistical inferences; they are mathematical devices that simply identify prominent patterns of correlations or functional connectivity. In what follows, we observe that multivariate analysis of covariance (MANCOVA) with canonical variate analysis combines some features of statistical parametric mapping and eigenimage analysis. Unlike statistical parametric mapping, MANCOVA is multivariate. In other words, it considers all voxels in a single scan as one observation. The importance of this multivariate approach is that effects, due to activations, confounding effects and error effects, are assessed both in terms of effects at each voxel *and interactions among voxels*. This means one does not have to assume anything about spatial correlations (cf. smoothness with random field models) to assess the significance of an activation effect. Unlike statistical parametric mapping, these correlations are explicitly included in the analysis. The price one pays for adopting a multivariate approach is that inferences cannot be made about regionally specific changes (cf. statistical parametric mapping). This is because the inference pertains to all the components (voxels) of a multivariate variable (not a particular voxel or set of voxels). Furthermore, because the spatial non-sphericity has to be estimated, without knowing the observations came from continuous spatially extended processes, the estimates are less efficient and inferences are less powerful.

Usually, multivariate analyses are implemented in two steps. First, the significance of hypothesized effects is assessed in terms of a p-value and secondly, if justified, the quantitative nature of the effect is determined. The analysis here conforms to this two-stage procedure. When the brain's response is assessed to be significant using MANCOVA, the nature of this response remains to be characterized. Canonical variate analysis is an appropriate way to do this. The canonical images obtained with CVA are similar to eigenimages, but are based on both the activation and error. CVA is closely related to de-noising techniques in EEG and MEG time-series analyses that use a generalized eigenvalue solution. Another way of looking at canonical images is to think of them as eigenimages that reflect functional connectivity due to activations, when spurious correlations due to error are explicitly discounted.

Dimension reduction and eigenimages

The first step in multivariate analysis is to ensure that the dimensionality (number of components or voxels) of the data is smaller than the number of observations. Clearly, this is not the case for images, because there are more voxels than scans; therefore the data have to be transformed. The dimension reduction proposed here is straightforward and uses the scan-dependent expression, Y, of eigenimages as a reduced set of components for each multivariate observation (scan). Where:

$$[U, S, V] = SVD(M)$$
$$Y = US \qquad \textbf{37.13}$$

As above, M is a large matrix of adjusted voxel-values with one column for each voxel and one row for each scan. Here, 'adjusted' implies mean correction and removal of any confounds using linear regression. The eigenimages constitute the columns of U, another unitary orthonormal matrix, and their expression over scans corresponds to the columns of the matrix Y. Y has one column for each eigenimage and one row for each scan. In our work, we use only the j columns of Y and U associated with eigenvalues greater than unity (after normalizing each eigenvalue by the average eigenvalue).

The general linear model revisited

Recall the general linear model from previous chapters:

$$Y = X\beta + \varepsilon \qquad \textbf{37.14}$$

where the errors are assumed to be independent and identically normally distributed. The design matrix X has one column for every effect (factor or covariate) in the model. The design matrix can contain both covariates and indicator variables reflecting an experimental design. β is the parameter matrix with one column vector of parameters for each mode. Each column of X has an associated unknown parameter. Some of these parameters will be of interest, the remaining parameters will not. We will partition the model accordingly:

$$Y = X_1\beta_1 + X_0\beta_0 + \varepsilon \qquad \textbf{37.15}$$

where X_1 represents a matrix of zeros or ones depending on the level or presence of some interesting condition or treatment effect (e.g. the presence of a particular cognitive component) or the columns of X_1 might contain covariates of interest that could explain the observed variance in Y (e.g. dose of apomorphine or 'time on target'). X_0 corresponds to a matrix of indicator variables denoting effects that are not of any interest (e.g. of being a particular subject or block effect) or covariates of no interest (i.e. 'nuisance variables', such as global activity or confounding time effects).

Statistical inference

Significance is assessed by testing the null hypothesis that the effects of interest do not significantly reduce the error variance when compared to the remaining effects alone (or alternatively the null hypothesis that β_1 is zero). The null hypothesis is tested in the following way. The sum of squares and products matrix (SSPM) due to error is obtained from the difference between actual and estimated values of the response:

$$S_R = (Y - X\hat{\beta})^T(Y - X\hat{\beta}) \qquad \textbf{37.16}$$

where the sums of squares and products due to effects of interest is given by:

$$S_T = (X_1\hat{\beta}_1)^T(X_1\hat{\beta}_1) \qquad \textbf{37.17}$$

The error sum of squares and products under the null hypothesis, i.e. after discounting the effects of interest are given by:

$$S_0 = (Y - X_0\hat{\beta}_0)^T(Y - X_0\hat{\beta}_0) \qquad \textbf{37.18}$$

The significance can now be tested with:

$$\Lambda = \frac{S_R}{S_0} \qquad \textbf{37.19}$$

This is Wilk's statistic (known as Wilk's Lambda). A special case of this test is Hotelling's T-square test and applies when one simply compares one condition with another, i.e. X_1 has only one column (Chatfield and Collins, 1980). Under the null hypothesis, after transformation, Λ has chi-squared distribution with degrees of freedom jh. The transformation is given by:

$$-(v - (j - h + 1)/2)\ln\Lambda \sim \chi^2_{jh} \qquad \textbf{37.20}$$

where v are the degrees of freedom associated with error terms, equal to the number of scans, n, minus the number of effects modelled, $rank(X)$. j is the number of eigenimages in the j-variate response variable and h is the degrees of freedom associated with the effects of interest, $rank(X_1)$. Eqn. **37.20** enables one to compute a p-value for significance testing in the usual way.

Characterizing the effect with CVA

Having established that the effects of interest are significant (e.g. differences among two or more activation conditions), the final step is to characterize these effects in terms of their spatial topography. This characterization employs canonical variates analysis or CVA. The objective is to find a linear combination (compound or contrast) of the components of Y, in this case the eigenimages, which best capture the activation effects, compared to error. More exactly, we want to find c_1 such that the variance ratio

$$\frac{c_1^T S_T c_1}{c_1^T S_R c_1} \quad 37.21$$

is maximized. Let $z_1 = Yc_1$ where z_1 is the first canonical variate and c_1 is a canonical image (defined in the space of the spatial modes) that maximizes this ratio. c_2 is the second canonical image that maximizes the ratio subject to the constraints $c_i^T c_j = 0$ (and so on). The matrix of canonical images $c = [c_1, \ldots c_h]$ is given by solution of the generalized eigenvalue problem:

$$S_T c = S_R c \lambda \quad 37.22$$

where λ is a diagonal matrix of eigenvalues. Voxel-space canonical images are obtained by rotating the canonical image in the columns of c back into voxel-space with the original eigenimages $C = Vc$. The columns of C now contain the voxel-values of the canonical images. The k-th column of C (the k-th canonical image) has an associated canonical value equal to the k-th leading diagonal element of λ. Note that the effect is a multivariate one, with j components or canonical images. Normally, only a few of these components have large canonical values and these are the ones reported. The dimensionality of the response, or the number of significant canonical images, is determined using the canonical values; under the null hypothesis the probability that the dimensionality is greater than D can be tested using:

$$(v - (j - h + 1)/2) \ln \prod_{j+D+1}^{j} (1 + \lambda_i) \sim \chi^2_{(j-D)(h-D)} \quad 37.23$$

It can be seen, by comparing Eqn. **37.23** to Eqn. **37.20** that there is a close relationship between Wilk's Lambda and the canonical values (see Appendix 1) and that the inference that the $D > 0$ is exactly the same as tests based on Wilk's Lambda, where $\Lambda^{-1} = \prod(1 + \lambda_i)$.

CVA, *linear discrimination and brain-reading*

Wilk's Lambda is actually quite important because it is a likelihood ratio test and, by the Neyman-Pearson lemma, the most powerful under parametric (Wishart) assumptions. As noted above, when the design matrix encodes a single effect this statistic reduces to Hotelling's t-square test. If the data are univariate, then Wilk's Lambda reduces to a simple F-test (see also Kiebel *et al.*, 2003). If both the data and design are univariate the F-test becomes the square of the t-test. In short, all parametric tests can be regarded as special cases of Wilk's Lambda. This is important because it is fairly simple to show (see Chawla *et al.*, 2000 and Appendix 1) that:

$$-\ln(\Lambda) = I(X, Y) = I(Y, X) \quad 37.24$$

where $I(X, Y)$ is the mutual information between the explanatory and response variables in X and Y respectively. This means that classical inference, using parametric tests, simply tests the null hypothesis $I(X, Y) = 0$, i.e. the two quantities X and Y are statistically independent. The importance of this lies in the symmetry of dependence. In other words, we can switch the explanatory and response variables around and nothing changes; MANCOVA does not care if we are trying to predict responses given the design matrix or whether we are trying to predict the design matrix given the responses. In either case, it is sufficient to infer the two are statistically dependent. This is one heuristic for the equivalence between CVA and linear discriminant analysis: Linear discrim inant analysis (LDA) and the related Fisher's linear discriminant are used in machine learning to find the linear combination of features that best separate two or more classes of objects or events. This linear combination is the canonical image. The resulting image or vector may be used as a linear classifier or in feature reduction prior to later classification. In functional imaging, classification has been called brain-reading (see Cox and Savoy, 2003) because one is trying to predict the experimental condition the subject was exposed to, using the imaging data. In short, MANCOVA, linear discriminant analysis, canonical variates analysis, canonical correlation analysis and kernel methods (e.g. support vector machines) that are linear in the observations are all based on the same model (see Appendix 1 for more details).

Relationship to eigenimage analysis

When applied to adjusted data, eigenimages correspond to the eigenvectors of S_T. These have an interesting relationship to the canonical images: On rearranging Eqn. **37.22**, we note that the canonical images are eigenvectors of $S_R^{-1} S_T$. In other words, an eigenimage analysis of an activation study returns the eigenvectors that express the most variance due to the effects of interest. A canonical image, on the other hand, expresses the greatest amount of variance due to the effects of interest *relative to error*. In this sense, a CVA can be considered an eigenimage analysis that is informed by the estimates of error and their correlations over voxels.

Serial correlations in multivariate models

CVA rests upon independent and identically distributed (IID) assumptions about the errors over observations.

Violation of these assumptions in fMRI has motivated the study of multivariate linear models (MLMs) for neuroimaging that allow for temporal non-sphericity (see Worsley *et al.*, 1997). Although this is an interesting issue, it should be noted that conventional CVA (with dimension reduction) can be applied after pre-whitening the time-series.

An example

We will consider an application of the above procedures to the word generation study in normal subjects, used above. We assessed the significance of condition-dependent effects by treating each of the twelve scans as a different condition. Note that we do not consider the word generation (or word shadowing) conditions as replications of the same condition. In other words, the first time one performs a word generation task is a different condition from the second time and so on. The (alternative) hypothesis adopted here states that there is a significant difference among the twelve conditions, but that this does not constrain the nature of this difference to a particular form. The most important differences will emerge from the CVA. Clearly, one might hope that these differences will be due to word generation, but they might not be. This hypothesis should be compared with a more constrained hypothesis that considers the conditions as six replications of word shadowing and word generation. This latter hypothesis is more directed and explicitly compares word shadowing with word generation. This comparison could be tested in a single subject. The point is that the generality afforded by the current framework allows one to test very constrained (i.e. specific) hypotheses or rather general hypotheses about some unspecified activation effect.[3] We choose the latter case here because it places more emphasis on canonical images as descriptions of what has actually occurred during the experiment. Had we chosen the former, we would have demonstrated significant mutual information between the data and the classification of each scan as either word shadowing or generation (cf. brain-reading for word generation).

The design matrix partition for effects of interest X_1 had twelve columns representing the conditions. We designated subject effects, time and global activity as uninteresting confounds X_0. The adjusted data were reduced to 60 eigenvectors as described above. The first 14 eigenvectors had (normalized) eigenvalues greater than unity and were used in the subsequent analysis. The resulting matrix data Y, with 60 rows (one for each scan) and 14 columns (one for each eigenimage) was subject to MANCOVA. The significance of the condition effects was assessed with Wilk's Lambda. The threshold for condition or activation effects was set at $p = 0.02$. In other words, the probability of there being no differences among the 12 conditions was 2 per cent.

Canonical variates analysis

The first canonical image and its canonical variate are shown in Figure 37.4. The upper panels show this system to include anterior cingulate and Broca's area, with more moderate expression in the left posterior inferotemporal regions (right). The positive components of this canonical image (left) implicate ventro-medial prefrontal cortex and bi-temporal regions (right greater than left). One important aspect of these canonical images is their highly distributed yet structured nature, reflecting the distributed integration of many brain areas. The canonical variate expressed in terms of mean condition effects is seen in the lower panel of Figure 37.4. It is pleasing to note that the first canonical variate corresponds to the difference between word shadowing and verbal fluency.

Recall that the eigenimage in Figure 37.2 reflects the main pattern of correlations evoked by the mean condition effects and should be compared with the first canonical image in Figure 37.4. The differences between these characterizations of activation effects are informative: the eigenimage is totally insensitive to the reliability or error attributable to differential activation from subject to subject, whereas the canonical image reflects these variations. For example, the absence of the posterior cingulate in the canonical image and its relative prominence in the eigenimage suggests that this region is implicated in some subjects but not in others. The subjects that engaged the posterior cingulate must do so to some considerable degree because the average effects (represented by the eigenimage) are quite substantial. Conversely, the medial prefrontal cortical deactivations are a more pronounced feature of activation effects than would have been inferred on the basis of the eigenimage analysis. These observations beg the question: 'which is the best characterization of functional anatomy?' Obviously, there is no simple answer but the question speaks of an important point. A canonical image characterizes a response *relative to error*, by partitioning the observed variance into effects of interest and a residual variation about these effects. Experimental design, a hypothesis, and the inferences that are sought determine this partitioning. An eigenimage does not entail any concept of error and is not constrained by any hypothesis.

[3] This is in analogy to the use of the SPM{F}, relative to more constrained hypotheses tested with SPM{t}, in conventional mass-univariate approaches.

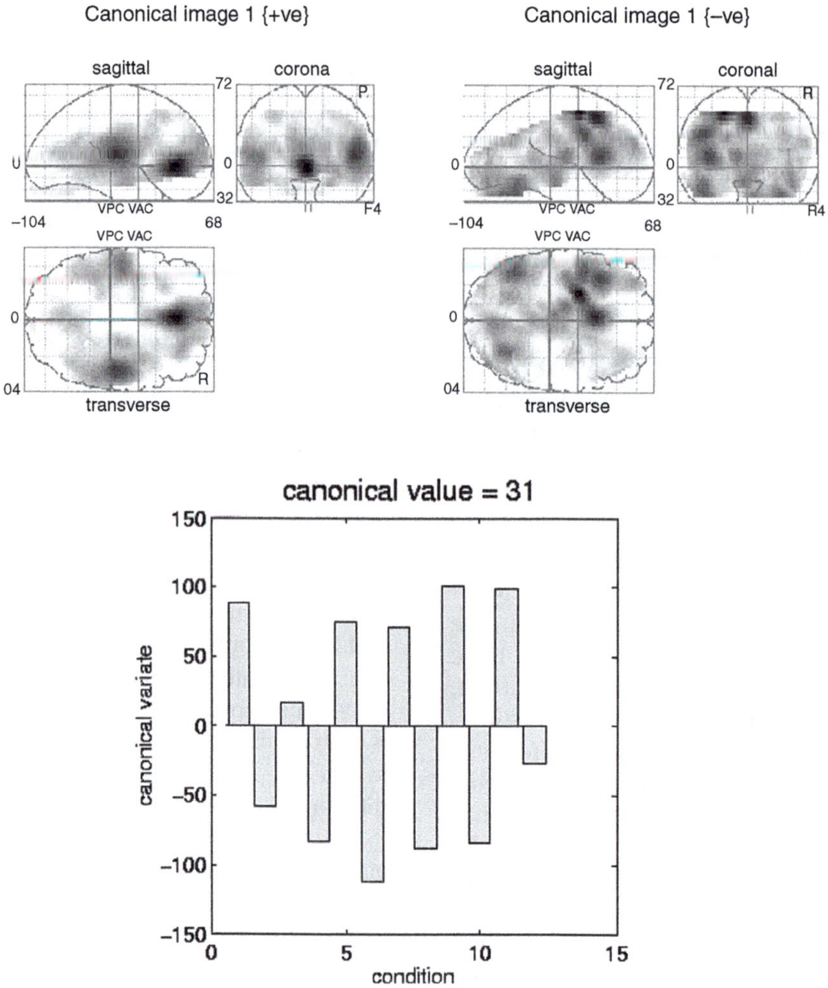

FIGURE 37.4 Top: the first canonical image displayed as maximum intensity projections of the positive and negative components. The display format is standard and provides three views of the brain from the front, the back and the right hand side. The grey scale is arbitrary and the space conforms to that described in the atlas of Talairach and Tournoux (1988). Bottom: the expression of the first canonical image (i.e. the canonical variate) averaged over conditions. The odd conditions correspond to word shadowing and the even conditions correspond to word generation. This canonical variate is clearly sensitive to the differences evoked by these two tasks.

Multivariate versus univariate models

Although multivariate linear models are important, this book focuses more on univariate models. There is a simple reason for this: any multivariate model can be reformulated as a univariate model by vectorizing the model. For example:

$$Y = X\beta + \varepsilon$$
$$[y_1, \ldots, y_j] = X[\beta_1, \ldots, \beta_j] + [\varepsilon_1, \ldots, \varepsilon_j] \qquad 37.25$$

can be rearranged to give a univariate model:

$$vec(Y) = (I \otimes X)vec(\beta) + vec(\varepsilon)$$
$$\begin{bmatrix} y_1 \\ \vdots \\ y_j \end{bmatrix} = \begin{bmatrix} X & & \\ & \ddots & \\ & & X \end{bmatrix} \begin{bmatrix} y_1 \\ \vdots \\ y_j \end{bmatrix} + \begin{bmatrix} \varepsilon_1 \\ \vdots \\ \varepsilon_j \end{bmatrix} \qquad 37.26$$

where $\otimes$ denotes the Kronecker tensor product. Here, $\mathrm{cov}(vec(\varepsilon)) = \Sigma \otimes V$, where Σ is the covariance among components and V encodes the serial correlations. In MLMs Σ is unconstrained and requires full estimation (in terms of S_R above). Therefore, any MLM and its univariate form are exactly equivalent, if we place constraints on the non-sphericity of the errors that ensure it has the form $\Sigma \otimes V$. This speaks to an important point: any multivariate analysis can proceed in a univariate setting with appropriate constraints on the non-sphericity. In fact, MLMs are special cases that assume the covariance factorizes into $\Sigma \otimes V$ and Σ is unconstrained. In neuroimaging there are obvious constraints on the form of Σ because this embodies the spatial covariances. Random field theory harnesses these constraints. MLMs do not and are therefore less sensitive.

Summary

This chapter has described multivariate approaches to the analysis of functional imaging studies. These use standard multivariate techniques to describe or make statistical inferences about distributed activation effects and characterize important features of functional connectivity. The multivariate approach differs fundamentally from statistical parametric mapping, because the concept of a separate voxel or region of interest ceases to have meaning. In this sense, inference is about the whole image volume, not any component. This feature precludes statistical inferences about regional effects made without reference to changes elsewhere in the brain. This fundamental difference ensures that mass-univariate and multivariate approaches are likely to be treated as distinct and complementary approaches to functional imaging data (see Kherif *et al.*, 2002).

In this chapter, we have used correlations among brain measurements to identify systems that respond in a coherent fashion. This identification proceeds without reference to the mechanisms that may mediate distributed and integrated responses. In the next chapter, we turn to models of effective connectivity that ground the nature of these interactions.

REFERENCES

Biswal B, Yetkin FZ, Haughton VM *et al.* (1995) Functional connectivity in the motor cortex of resting human brain using echoplanar MRI. *Mag Res Med* **34**: 537–41

Bleuler E (1913) Dementia Praecox or the group of schizophrenias. Translated into English in *The clinical roots of the schizophrenia concept*, Cutting J, Shepherd M, (eds) (1987). Cambridge University Press, Cambridge

Chatfield C, Collins AJ (1980) *Introduction to multivariate analysis*. Chapman and Hall, London, pp 189–210

Chawla D, Lumer ED, Friston KJ (2000) Relating macroscopic measures of brain activity to fast, dynamic neuronal interactions. *Neural Comput* **12**: 2805–21

Friedrich R, Fuchs A, Haken H (1991) Modelling of spatio-temporal EEG patterns. In *Mathematical approaches to brain functioning diagnostics*, Dvorak I, Holden AV (eds). Manchester University Press, New York

Fuchs A, Kelso JAS, Haken H (1992) Phase transitions in the human brain: spatial mode dynamics. *Int J Bifurcation Chaos* **2**: 917–39

Friston KJ, Frith CD, Liddle PF *et al.* (1993a) Functional connectivity: the principal component analysis of large (PET) data sets. *J Cereb Blood Flow Metab* **13**: 5–14

Friston KJ, Jezzard P, Frackowiak RSJ *et al.* (1993b) Characterising focal and distributed physiological changes with MRI and PET. In *Functional MRI of the brain*. Society of Magnetic Resonance in Medicine, Berkeley, pp 207–16

Friston KJ, Frith CD, Fletcher P *et al.* (1996a) Functional topography: multidimensional scaling and functional connectivity in the brain. *Cereb Cortex* **6**: 156–64

Friston KJ, Herold S, Fletcher P *et al.* (1996b) Abnormal frontotemporal interactions in schizophrenia. In *Biology of schizophrenia and affective disease*, Watson SJ (ed.). ARNMD Series **73**: 421–29

Friston KJ, Poline J-B, Holmes AP *et al.* (1996c) A multivariate analysis of PET activation studies. *Hum Brain Mapp* **4**: 140–51

Friston KJ (1998) Modes or models: a critique on independent component analysis for fMRI. *Trends Cogn Sci* **2**: 373–74

Friston KJ, Phillips J, Chawla D *et al.* (2000) Nonlinear PCA: characterising interactions between modes of brain activity. *Phil Trans R Soc Lond B* **355**: 135–46

Gerstein GL, Perkel DH (1969) Simultaneously recorded trains of action potentials: analysis and functional interpretation. *Science* **164**: 828–30

Goldman-Rakic PS (1988) Topography of cognition: parallel distributed networks in primate association cortex. *Annv Rev Neurosci* **11**: 137–56

Gower JC (1966) Some distance properties of latent root and vector methods used in multivariate analysis. *Biometrika* **53**: 325–28

Karhunen J, Joutsensalo J (1994) Representation and separation of signals using nonlinear PCA type learning. *Neural Networks* **7**: 113–27

Kherif F, Poline JB, Flandin G *et al.* (2002) Multivariate model specification for fMRI data. *NeuroImage* **16**: 1068–83

Kiebel SJ, Glaser DE, Friston KJ (2003) A heuristic for the degrees of freedom of statistics based on multiple variance parameters. *NeuroImage* **20**: 591–600

Kramer MA (1991) Nonlinear principal component analysis using auto-associative neural networks. *AIChE J* **37**: 233–43

Mayer-Kress G, Barczys C, Freeman W (1991) Attractor reconstruction from event-related multi-electrode EEG data. In *Mathematical approaches to brain functioning diagnostics*, Dvorak I, Holden AV (eds). Manchester University Press, New York

McIntosh AR, Bookstein FL, Haxby JV *et al.* (1996) Spatial pattern analysis of functional brain images using partial least squares. *NeuroImage* **3**: 143–57

McKeown MJ, Makeig S, Brown GG *et al.* (1998) Analysis of fMRI data by blind separation into independent spatial components. *Hum Brain Mapp* **6**: 160–88

Newton I (1794) *Opticks*. Book 1, part 2, prop. 6. Smith and Walford, London

Phillips CG, Zeki S, Barlow HB (1984) Localisation of function in the cerebral cortex. Past, present and future. *Brain* **107**: 327–61

Shepard RN (1980) Multidimensional scaling, tree-fitting and clustering. *Science* **210**: 390–98

Stone JV (2002) Independent component analysis: an introduction. *Trends Cogn Sci* **6**: 59–64

Talairach P, Tournoux J (1988) *A stereotactic coplanar atlas of the human brain*. Thieme, Stuttgart

Torgerson WS (1958) *Theory and methods of scaling*. Wiley, New York

Worsley KJ, Poline JB, Friston KJ *et al.* (1997) Characterizing the response of PET and fMRI data using multivariate linear models. *NeuroImage* **6**: 305–19

CHAPTER

38

Effective Connectivity

L. Harrison, K. Stephan and K. Friston

INTRODUCTION

In the previous chapter, we dealt with functional connectivity and different ways of summarizing patterns of correlations among brain systems. In this chapter, we turn to effective connectivity and mechanistic models that might mediate these correlations. This chapter can be regarded as an introduction to different models of effective connectivity covered in the remaining chapters of this Part.

Brain function depends on interactions among its components that range from individual cell compartments to neuronal populations. As described in Chapter 36, anatomical and physiological *in vivo* studies of connectivity suggest a hierarchy of specialized regions that process increasingly abstract features, e.g. from simple edges in V1, through colour and motion in V4 and V5 respectively, to faces in the fusiform gyrus. These specialized regions are connected, allowing distributed and recurrent neuronal information processing. This means their selective responses, or specialization, are a function of connectivity. An important theme in this chapter is that connections can change in a context-sensitive way. We refer to these changes as plasticity, to describe changes in the influence different brain systems have on each other. The formation of distributed networks, through dynamic interactions, is the basis of functional integration, which is time- and context-dependent. Changes in connectivity are important for development, learning, perception and adaptive response to injury.

Given the importance of changes in connectivity, we will distinguish between two classes of experimental factors or input to the brain. The first class evokes responses directly, whereas the second has a more subtle effect and induces input-dependent changes in connectivity that modulate responses to the first. We will refer to the second class as 'contextual'. For example, augmented neuronal responses associated with attending to a stimulus can be attributed to the changes induced in connectivity by attention. The distinction between inputs that evoke responses and those that modulate effective connectivity is the motivation for developing non-linear models that accommodate contextual changes in connection strength. We will focus on the simplest non-linear models, namely bilinear models.

This chapter is divided into three sections. In the first, we introduce a system identification approach, where brain responses are parameterized within the framework of a mathematical model. A bilinear model is derived to approximate generic non-linear dynamics. This is used to highlight the different ways in which experimental effects can be modelled and how the notion of effective connectivity emerges as a natural measure. We describe the different approaches to estimating functional integration, starting with static models in the second section and dynamic models in the third. We conclude with remarks on strategies and the features of models that have proved useful in modelling connectivity to date.

IDENTIFICATION OF DYNAMIC SYSTEMS

System identification is the use of observed data to estimate parameters of a model that represents a physical system. The model may be linear or non-linear, formulated in discrete or continuous time and parameterized in the time or frequency domain. The aim is to construct a mathematical description of a system's response to input. Models may be divided into two main categories: those that invoke hidden states and those that quantify relationships between inputs and outputs without hidden states, effectively treating the system as a black box (see Juang, 2001). Examples of the former include state-space models (SSM) and hidden Markov models (HMM),

Statistical Parametric Mapping, by Karl Friston et al.
ISBN–13: 978-0-12-372560-8 ISBN–10: 0-12-372560-7

whereas the latter include generalized convolution models (see Chapter 39) and autoregressive models (Chatfield, 1996 and Chapter 40).

There are two main requirements of biologically plausible models of functional integration: that they are dynamic and non-linear. They have to be dynamic, because the brain is a physical system whose state evolves continuously in time. This means that the current state of the brain affects its state in the future (we will see the issues involved in relaxing this requirement later). Models have to be non-linear, because biological systems depend on non-linear phenomena for much of their characteristic behaviour (Scott, 1999). Examples include the neuronal dynamics of action potentials (Dayan and Abbott, 2001), population dynamics in co-evolutionary systems (Glass and Kaplan, 2000) and limit cycles in physiological systems (Glass, 2001). The motivation for non-linear dynamic models is that their non-additive characteristics enable them to reproduce sufficiently complex behaviour, of the sort observed in biological systems. However, non-linear models are often mathematically intractable, calling for approximation techniques.

On the other hand, linear dynamic models can be analysed in closed form. Consequently, there exists a large body of theory for handling them: linear models adhere to the principle of superposition, which means that the system's response to input is additive. There are no interactions between different inputs or between inputs and the intrinsic states of the system. This means the response is a linear mixture of inputs. A system that violates this principle responds in a non-additive manner, i.e. with more or less than a linear mixture. Such a system is, by definition, non-linear. However, there is a price for the ease with which linear models can be analysed, because their behavioural repertoire is limited to exponential decay and growth, linear oscillation or a combination of these. Examples of subadditive responses are ubiquitous in physiology, e.g. saturation, where the effect of increasing input reaches a saturation point and further input does not generate an additional response (e.g. biochemical reactions or synaptic input).

A useful compromise is to make linear approximations to a generic non-linear model. These models have the advantage that they capture some essential non-linear features, while remaining mathematically tractable. This strategy has engendered bilinear models (Rao, 1992), where non-linear terms are limited to interactions that can be modelled as the product of two variables (inputs and states). Despite constraints on high-order non-linearities, bilinear models can easily model plasticity in effective connections. We will use a bilinear model to illustrate the concepts of linear and bilinear coupling and how they are used to model effective connectivity.

Approximating functions

We start with some basic concepts about approximating non-linear functions: consider a scalar, x, and a sufficiently smooth function $f(x)$. This function can be approximated in the neighbourhood of an expansion point, x_0, using the Taylor series expansion:

$$f(x) \approx f(x_0) + \frac{df}{dx}(x - x_0) + \frac{d^2 f}{dx^2}\frac{(x-x_0)^2}{2!} + \cdots + \frac{d^n f}{dx^n}\frac{(x-x_0)^n}{n!} \qquad 38.1$$

where the nth order derivatives are evaluated at x_0. These values are coefficients that scale the contribution of their respective terms. The derivatives are used as they map a local change in $(x-x_0)^n$ onto a change in $f(x)$. The degree of non-linearity of f determines the rate of convergence, weakly non-linear functions converging, rapidly. The series converges to the *exact* function as n goes to infinity. A simple example is shown in Figure 38.1.

Consider now bivariate non-linear functions, where $f(x, u)$ depends on two quantities x and u, the corresponding Taylor series including high-order terms involving products of x and u. The linear and bilinear approximations are given by f_L and f_B:

$$\begin{aligned} f_L(x, u) &= ax + cu \\ a &= \frac{\partial f}{\partial x} \\ c &= \frac{\partial f}{\partial u} \\ f_B(x, u) &= ax + bxu + cu \\ b &= \frac{\partial^2 f}{\partial x \partial u} \end{aligned} \qquad 38.2$$

For clarity, the expansion point $x_0 = 0$ and the series have been centred so that $f(x_0) = f(u_0) = 0$. The approximation, f_L depends on linear terms in x and u scaled by coefficients a and c, calculated from the first-order derivatives. The first-order terms of f_B are the same as f_L, however, a third term includes the product of x and u, which is scaled by b, a second-order derivative with respect to both variables. This term introduces the non-linearity. Note, we have not included quadratic terms in x or u. The resulting approximation has the appealing property of being both linear in x and u, but allowing for a modulation of x by u.

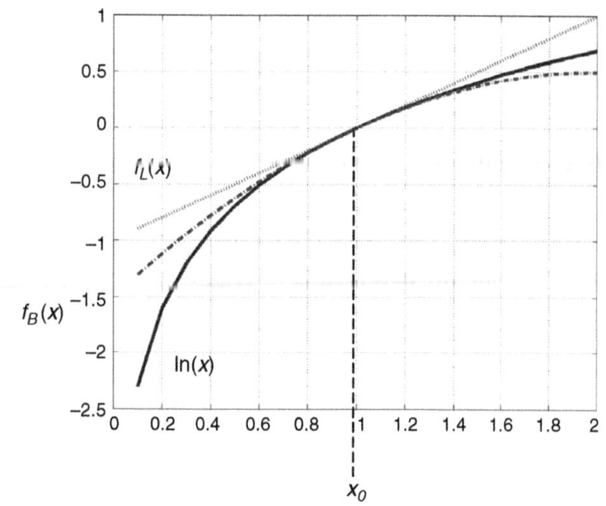

FIGURE 38.1 Approximations to $f(x) = \ln(x)$ about $x_0 = 1$ using a Taylor series expansion, where $f_L(x) = x - 1$ and $f_B(x) = (x-1) - 1/2(x-1)^2$ are the first- and-second-order approximations. The improvement for high-order approximations is apparent.

We now extend the above to vectors where $x = [x_1, \ldots, x_n]^T$ and $u = [u_1, \ldots, u_m]^T$. The linear and bilinear approximations can be written in matrix form, as:

$$f_L(x, u) = Ax + Cu$$
$$f_B(x, u) = Ax + Cu + \sum_j u_j B^j x \qquad 38.3$$
$$A = \frac{\partial f}{\partial x} \qquad B^j = \frac{\partial^2 f}{\partial x \partial u_j} \qquad C = \frac{\partial f}{\partial u}$$

The coefficients are now matrices as opposed to scalars. They look more complicated, but the same operations are applied to all the elements of the matrix coefficients, for example:

$$A = \frac{\partial f}{\partial x} = \begin{bmatrix} \frac{\partial f_1}{\partial x_1} & \cdots & \frac{\partial f_1}{\partial x_n} \\ \vdots & \ddots & \vdots \\ \frac{\partial f_n}{\partial x_1} & \cdots & \frac{\partial f_n}{\partial x_n} \end{bmatrix} \qquad 38.4$$

Linear dynamic models

The equations above can be used to model the dynamics of physical systems. Figure 38.2 shows a simple example. A system can be modelled by a number of states and inputs. The states are contained in $x(t)$, called the state vector, and inputs in $u(t)$, the input vector. The number of states and inputs in the model are given by n and m respectively. Each state defines a coordinate in state-space within which the behaviour of the system is represented as a trajectory. The temporal evolution of the states is modelled by a state equation, which is the first-order temporal derivative of the state-vector, written as $\dot{x}(t)$ and can therefore be approximated by a Taylor series as above:

$$\dot{x} = f_L(x, u) = Ax + Cu \qquad 38.5$$

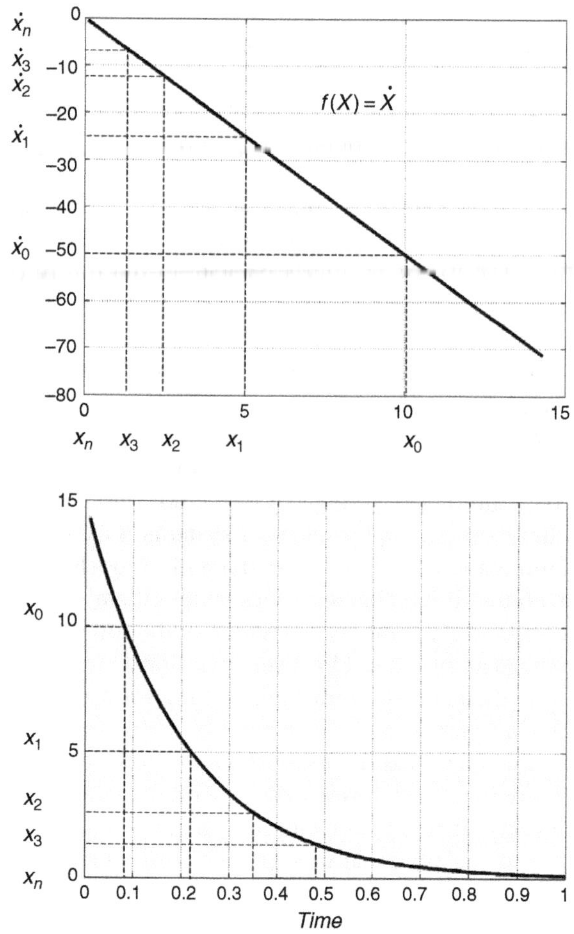

FIGURE 38.2 The function $f(x)$ models a simple one-state linear dynamical system, i.e. $\dot{x} = f(x)$. The state starts at the value x_0 where it *decreases* at the rate $\dot{x}_0$. After a period of time, the state has decreased to x_1 with a rate of $\dot{x}_1$. The state continues to decreases exponentially with time until $\dot{x}_n = 0$.

A linear dynamic system (LDS) is shown in Figure 38.3. The system consists of two states, $x_1(t)$ and $x_2(t)$, and inputs, $u_1(t)$ and $u_2(t)$, coupled through a state equation parameterized by matrices A and C.

A contains parameters that determine interactions among states (labelled inter-state in Figure 38.3) and the influence a state has on itself, while the elements of C couple inputs to states. The state equation provides a complete description of the dynamics of the system, otherwise known as the system's equation of motion. These models are sometimes called linear time invariant (LTI) systems, as A and C do not change with time.

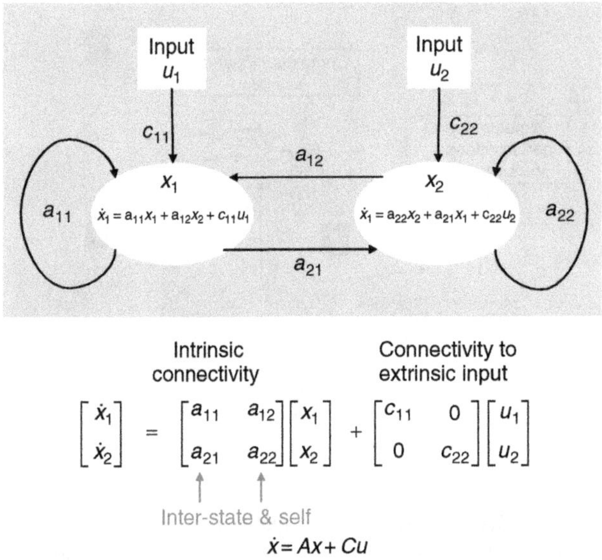

FIGURE 38.3 System with two states, x_1 and x_2, and inputs, u_1 and u_2, whose dynamics are determined by the time invariant matrices A and C in the state equation. The matrices describe the intrinsic connectivity and how states are connected to external inputs.

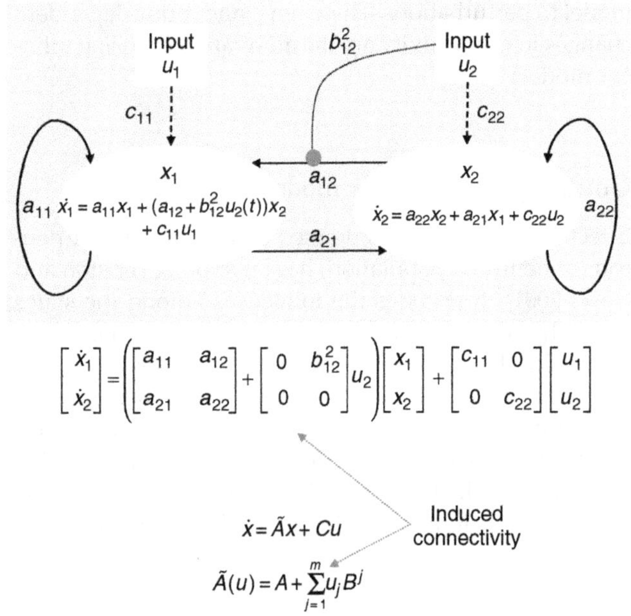

FIGURE 38.4 A bilinear model. Input u_2 interacts with x_2, rendering the matrix $\tilde{A}(u)$ state-dependent. This induces input-dependent modulation of the coupling parameters, and different responses to other inputs, i.e. u_1. All connections that are not shown correspond to zero in the coupling matrices.

Bilinear dynamic models

Linear models are useful because they are a good first-order approximation to many phenomena. However, they furnish rather restricted descriptions of non-linear systems. The model in Figure 38.4 has been augmented to illustrate how a bilinear model is formulated. The state equation adopts the same form as f_B, whose essential feature is the bilinear interaction between inputs and states:

$$\begin{aligned}\dot{x} &= f_B(x, u)\\ &= \left(A + \sum_{j=1}^{m} u_j B^j\right) x + Cu\\ &= \tilde{A}x + Cu\end{aligned} \qquad 38.6$$

The key difference is the addition of B^j. These scale the interaction among states and inputs when added to A and model input-dependent changes to the intrinsic connectivity of the network. This is illustrated in Figure 38.4 where the coupling coefficient a_{12} is modulated by the product $b^2_{12}u_2$. The modified matrix $\tilde{A}$ operates on the state vector and determines the response of the model. However, now $\tilde{A}$ changes with time because it is a function of time-varying input. This distinguishes it from the LTI model above.

These systems are simple to integrate over short periods of time, during which the input can be regarded as constant. In this context, the bilinear form reduces to a local linear form and Eqn. **38.6** can be re-written as:

$$\begin{aligned}\dot{z} &= Jz\\ J &= M + Nu\\ z &= \begin{bmatrix}1\\ x\end{bmatrix} \quad M = \begin{bmatrix}0 & 0\\ 0 & A\end{bmatrix} \quad N = \begin{bmatrix}0 & 0\\ C & B\end{bmatrix}\end{aligned} \qquad 38.7$$

Here $x(t)$ has been augmented with a constant, which allows the inputs and states to be treated collectively. This equation can be solved using standard linear techniques (Boas, 1983), such as the matrix exponential method (see Appendix 2).

It helps to consider a specific example. If u_2 is binary, the model in Figure 38.4 effectively comprises two LTI models. The model's behaviour can be characterized as a switching between two linear systems. The switch from one linear mode to the other will be determined by u_2. For instance, if the two linear models were linearly damped harmonic oscillators, each with different periods of oscillation, a switch from one state to another would be accompanied by changes in the oscillation of the states.

In bilinear models, inputs can be divided into two classes: perturbing and contextual. Perturbing inputs influence states directly (e.g. u_1 in Figure 38.4). The effects of these inputs are distributed according to the intrinsic connections of the model. Conversely, contextual inputs (e.g. u_2 in Figure 38.4) reconfigure the response of the model to perturbations. These time- and input-dependent changes in connectivity are the motivation for using bilinear models.

Coupling and bilinear models

Effective connectivity is defined as the influence a neuron (or neuronal population) has on another (Friston and Price, 2001). It encodes the influences among the states of a physical system, which is usually responding to external influences. At the neuronal level this is equivalent to the effect presynaptic activity has on postsynaptic responses, otherwise known as synaptic efficacy. Models of effective connectivity are designed to furnish a suitable measure of influence among interconnected components (or regions of interest) in the brain. In Figure 38.4 each state's equation is given by:

$$\begin{aligned}\dot{x}_1 &= a_{11}x_1 + (a_{12} + b_{12}^2 u_2)x_2 + c_{11}u_1 \\ \dot{x}_2 &= a_{22}x_2 + a_{21}x_1 + c_{22}u_2\end{aligned} \quad 38.8$$

Taking derivatives of $\dot{x}$ with respect to the states quantifies the coupling between the two states (i.e. regions):

$$\begin{aligned}\frac{\partial \dot{x}_1}{\partial x_2} &= \tilde{A}_{12}(u) = a_{12} + b_{12}^2 u_2 \\ \frac{\partial \dot{x}_2}{\partial x_1} &= \tilde{A}_{21}(u) = a_{21}\end{aligned} \quad 38.9$$

in terms of how one state causes changes in another. Generally, this coupling may be linear or non-linear. However, in bilinear models, it is described simply by the elements of $\tilde{A}(u) = A + \Sigma u_j B^j$. In our example, the coupling from u_1 to x_2 is linear and is a constant a_{21}. However, the interaction between u_2 and x_2 induces non-linearities in the network, rendering $\tilde{A}_{12}(u)$ a function of u_2. The influence x_2 has on x_1 therefore depends on u_2. This effect may be quantified by taking derivatives with respect to the contextual input:

$$\frac{\partial^2 \dot{x}_1}{\partial x_2 \partial u_2} = b_{12}^2 \quad 38.10$$

This second-order derivative may be a non-linear function of the states and input for arbitrary equations of motion, but it reduces to a constant in the bilinear model.

The first- and second-order derivatives quantify coupling among the states of a network and correspond to obligatory and modulatory influences in neuronal networks (Büchel and Friston, 2000). This distinction highlights the difference between perturbing and contextual inputs. As illustrated in Figure 38.4, input can be categorized by the way it affects the states of a system. Input

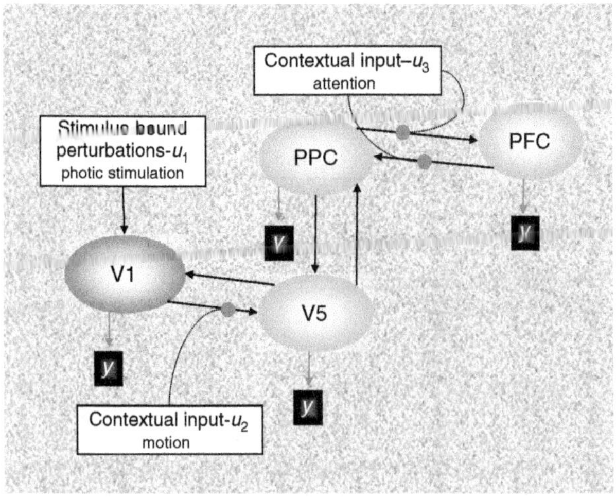

FIGURE 38.5 A schematic model of functional integration in the visual and attention systems. Sensory input has direct effect on the primary visual cortex, while contextual inputs, such as motion or attention, modulate pathways between nodes in the extended network. In this way, contextual input (e.g. induced by instructional set) may enable (or disable) pathways, which changes the response of the system to stimulus-bound inputs.

can perturb states directly or it can modulate intrinsic connectivity. These are perturbing and contextual effects respectively. Both evoke a response; however, contextual inputs do so vicariously by modulating the dynamics induced by perturbing inputs. Figure 38.5 is an illustrative model of responses evoked by visual motion that shows the sorts of architectures bilinear models can be applied to. This example uses photic stimulation as a perturbing input that evokes a response, which depends on the current connectivity. This connectivity changes with a contextual input, in this example the presence of motion in the visual field and attentional set (i.e. attending to motion). Changes in attention cause a reconfiguration of connectivity and can modify the response to photic stimulation. This is a dynamic casual model that has been inverted using real data (see Chapter 41).

Having described the basis for modelling the brain as a physical system and establishing a distinction between inputs that change states and those that change parameters (i.e. connections), we turn to some specific examples. We start with simple static models and generalize them to recover the dynamic models described in this section.

STATIC MODELS

The objective of an effective connectivity analysis is to estimate parameters that represent influences among regions that may change over time and with respect to experimental tasks. This rests on the inversion of a causal model. Identifying a complete and biologically plausible mathematical model requires a high level of sophistication. However, some progress can be made by modelling relationships in the data alone (among voxels or regions), without invoking hidden states and ignoring the consequent dynamics. This makes the maths much easier but discounts temporal information and is biologically unrealistic. We will call these models 'static' as they model instantaneous interactions among regions and ignore the influence previous states have on current responses.

Linear models

Linear models assume that each time sample is independent of the next. This is tenable for positron emission tomography (PET) data because the nature of the measurement is essentially steady state. Usually, PET data are acquired while holding brain states constant, using an appropriate task or stimulus and waiting until steady-state radiotracer kinetics can be assumed. Mathematically, this means the rate of change of the states is zero. However, in functional magnetic resonance imaging (fMRI), this assumption is generally violated (and certainly for electrophysiological measurements); this is the motivation for dynamic models of fMRI and electroencephalography (EEG) responses.

In static linear models, the activity of each voxel is modelled as a linear mixture of activity in all voxels plus some error. This can be expressed as a multivariate linear model:

$$Y = Y\beta + \varepsilon$$
$$Y = \begin{bmatrix} y_1 \\ \vdots \\ y_T \end{bmatrix} \qquad \textbf{38.11}$$

where y_t is a row vector containing all regional measurements at time t. The problem with this formulation is that the trivial solution $\beta = 1$ completely accounts for the data. We will see later how structural equation modelling deals with this by fixing some connections. Although unconstrained linear models cannot address connectivity *per se*, if we include non-linearities, they can be used to assess changes in connectivity.

Bilinear models

Bilinear models can be used to estimate *changes* in connectivity if the bilinear term includes a measure of neuronal activity. If the bilinear term represents an interaction between activity and an experimental factor, it can model the effects of contextual input. If the bilinear term is an interaction between neuronal activities, it can model modulatory interactions among populations of neurons. We will focus on the former: the model in Plate 55 (left-hand panel) (see colour plate section) is a simple bilinear model. It consists of an input u, which generates an output y. The non-linearity is due to an interaction between the input and output. This model comprises a linear and bilinear term parameterized by b_1 and b_2 respectively.

$$y = ub_1 + uyb_2 \qquad \textbf{38.12}$$

If the model contained the first term only, it would be linear and the relationship between input and output would be a straight line. The addition of the second term introduces non-linear behaviour. Plotting u and y for different values of b_2 demonstrates this. The input-output behaviour depends on b_2 and is revealed by the two curves in Plate 55 (right-hand panel). It can be seen that the sensitivity of changes in y to changes in u (i.e. the slope) depends on y. This sort of dependency was used in Friston *et al.* (1995) to demonstrate asymmetrical non-linear interactions between V1 and V2 during visual stimulation. In this example, the bilinear term comprised fMRI measures of neuronal activity in V1 and V2, corresponding to y and u respectively.

The bilinear term in Eqn. **38.12** contains the product of input and activity uy. This is referred to as a *psychophysiological interaction* (PPI). Although we will focus on psychophysiological interaction terms, all that follows can be applied to any bilinear effect. As noted above, these include the interactions between two physiological variates.

Psychophysiological interactions

Büchel *et al.* (1996) discuss a series of increasingly high-order interaction terms in general linear models. These are introduced as new explanatory variables enabling statistical parametric mapping (SPM) to estimate the magnitude and significance of non-linear effects directly. A special example of this is a psychophysiological interaction (Friston *et al.*, 1997) where the bilinear term represents an interaction between an input or psychological variable and a response or physiological variable y^i measured at the i-th brain region. Any linear model can be augmented to include a PPI:

$$Y = [X \quad u \times y^i]\beta + \varepsilon \qquad \textbf{38.13}$$

The design matrix partition $X = [u, y^i, \ldots]$ normally contains the main effect of experimental input and regional response. The PPI is the Hadamard product $u \times y^i$ and is obtained by multiplying the input and response vectors element by element. Both the main effects and interaction terms are included because the main effects have to be modelled to assess properly the additional explanatory power afforded by the bilinear or PPI term. Standard hypothesis testing can be used to estimate the significance of the PPI at any point in the brain with an SPM.

Plate 56 illustrates bilinear effects in real data. These data come from an fMRI study of the modulatory effects of attention on visual responses to radial motion (see Büchel and Friston, 1998 for experimental details). The aim was to quantify the top-down modulatory effect of attention on V1 to V5 connectivity. The model combines psychological data (attentional set) with physiological data (V1 activity) to model the interaction. The right-hand panel in Plate 56 show a regression analysis of the same data, divided according to attentional set, to demonstrate the difference in regression slopes subtending this PPI.

In this example, attention was modelled as a 'contextual' variable, while visual stimulation perturbed the system. The latter evoked a response within the context of the former, i.e. visual stimuli evoke different responses depending on attentional set, modelled as a change in connectivity. Here, attention reconfigures connection strengths among prefrontal and primary cortical areas (Mesulam, 1998). This bilinear effect may take any appropriate form in PPI models, including, for example, psychological, physiological or pharmacological indices. These models emphasize the use of factorial experimental designs (Friston *et al.*, 1997) and allow us to consider experimental inputs in a different light, distinguishing contextual input (one factor) from direct perturbation (another factor). PPI models have provided important evidence for the interactions among distributed brain systems and enabled inferences about task-dependent plasticity using a relatively simple procedure. The model we consider next was developed explicitly with effective connectivity or path analysis in mind, but adopts a different approach to the estimation of model parameters. This approach rests on specifying constraints on the connectivity.

Structural equation modelling

Structural equation modelling (SEM), or path analysis, is a multivariate method used to test hypotheses regarding the influences among interacting variables. Its roots go back to the 1920s, when path analysis was developed to quantify unidirectional causal flow in genetic data and developed further by social scientists in the 1960s (Maruyama, 1998). It was criticized for the limitations inherent in the least squares method of estimating model parameters, which motivated a general linear modelling approach from the 1970s onwards. It is now available in commercial software packages, including LISREL, EQS and AMOS. See Maruyama (1998) for an introduction to the basic ideas. Researchers in functional imaging started to use it in the early 1990s (McIntosh and Gonzalez-Lima, 1991, 1992a, b, 1994). It was applied first to animal autoradiographic data and later to human PET data where, among other experiments, it was used to identify task-dependent differential activation of the dorsal and ventral visual pathways (McIntosh *et al.*, 1994). Many investigators have used SEM since then. An example of its use to identify attentional modulation of effective connectivity between prefrontal and premotor cortices can be found in Rowe *et al.* (2002).

An SEM is a linear model with a number of modifications, which are illustrated in Figure 38.6: the coupling matrix, β, is 'pruned' to include only paths of interest. Critically, self-connections are precluded. The data matrix, Y, contains responses from regions of interest and possibly experimental or bilinear terms. The underlying model is a general linear model:

$$Y = Y\beta + \varepsilon \qquad \textbf{38.14}$$

where the free parameters, β, are constrained, according to the specified pruning or sparsity structure of connections. To simplify the model, the residuals ε are assumed to be independent. They are interpreted as driving each region stochastically from one measurement to another and are sometimes called innovations.

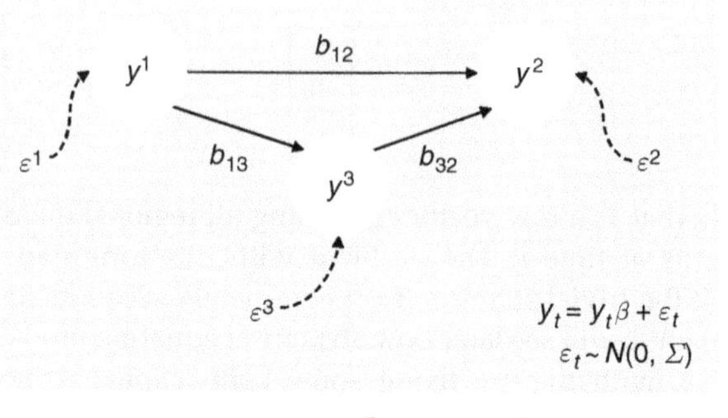

$$[y_t^1 \; y_t^2 \; y_t^3] = [y_t^1 \; y_t^2 \; y_t^3]\begin{bmatrix} 0 & b_{12} & b_{13} \\ 0 & 0 & 0 \\ 0 & b_{32} & 0 \end{bmatrix} + \varepsilon_t$$

FIGURE 38.6 An SEM is used to estimate path coefficients for a specific network of connections, after 'pruning' the connectivity matrix. The graphic illustrates a particular sparsity structure, which is usually based on prior anatomical knowledge. y_t may contain physiological or psychological data or bilinear terms (to estimate the influence of 'contextual' input). The innovations ε are assumed to be independent, and can be interpreted as driving inputs to each node.

Instead of minimizing the sum of squared errors, the free parameters are estimated using the sample covariance structure of the data. The rationale for this is that the covariance reflects the global behaviour of the data, i.e. capturing relationships among variables, in contrast to the former, which reflects the goodness of fit from the point of view of each region. Practically, an objective function is constructed from the sampled and implied covariance, which is optimized with respect to the parameters. The implied covariance, $\Sigma(\beta)$, is computed easily by rearranging Eqn. **38.14** and assuming some value for the covariance of the innovations, $\langle \varepsilon^T \varepsilon \rangle$:

$$\begin{aligned} Y(I-\beta) &= \varepsilon \\ Y &= \varepsilon(1-\beta)^{-1} \\ \Sigma &= \langle Y^T Y \rangle \\ &= (1-\beta)^{-T} \langle \varepsilon^T \varepsilon \rangle (1-\beta)^{-1} \end{aligned} \quad \textbf{38.15}$$

The sample covariance is:

$$S = \frac{1}{n-1} Y^T Y$$

where n is the number of observations and the maximum likelihood objective function is:

$$F_{ML} = \ln|\Sigma| - tr(S\Sigma^{-1}) - \ln|S| \quad \textbf{38.16}$$

This is simply the Kullback-Leibler divergence between the sample and the covariance implied by the free parameters. A gradient descent, such as a Newton-Raphson scheme, is generally used to estimate the parameters, which minimize this divergence. The starting values can be estimated using ordinary least square (OLS) (McIntosh and Gonzalez-Lima, 1994).

Inferences about changes in the parameters or path coefficients rest on the notion of nested, or stacked, models. A nested model consists of a free-model within which any number of constrained models is 'nested'. In a free model, all parameters are free to take values that optimize the objective function, whereas a constrained model has one, or a number of parameters omitted, constrained to be zero or equal across models (i.e. attention and non-attention). By comparing the goodness of fit of each model against the others, χ^2 statistics can be derived (Bollen, 1989). Hypotheses testing proceeds using this statistic. For example, given a constrained model, which is defined by the *omission* of a pathway, evidence for or against the pathway can be tested by 'nesting' it in the free model. If the difference in goodness of fit is unlikely to have occurred by chance, the connection can be declared significant. An example of a nested model that was tested by Büchel and Friston (1997) is shown in Figure 38.7.

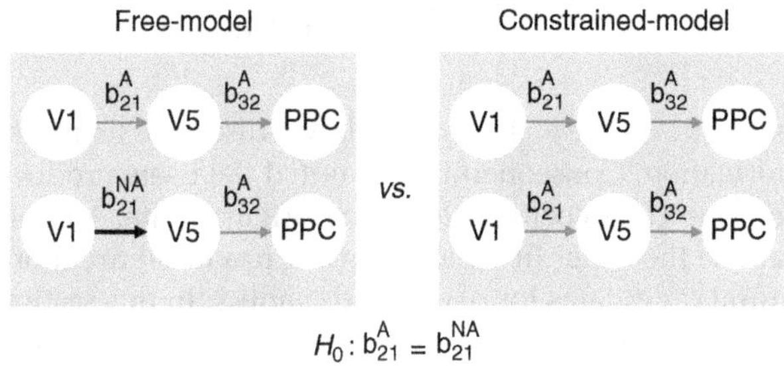

FIGURE 38.7 Inference about changes in connection strengths proceeds using nested models. Parameters from free and constrained models are compared with a χ^2 statistic. This example compares path coefficients during attention (A) and non-attention (NA), testing the null hypothesis that the V1 to V5 connections are the same under both levels of attention.

SEM can accommodate bilinear effects by including them as an extra node. A significant connection from a bilinear term represents a modulatory effect in exactly the same way as in a PPI. Büchel and Friston (1997) used bilinear terms in an SEM of the visual attention data set, to establish the modulation of connections by prefrontal cortex. In this example, the bilinear term comprised measurements of activity in two brain regions. An interesting extension of SEM has been to look at models of connectivity over multiple brains (i.e. subjects). The nice thing about this is that there are no connections between brains, which provide sparsity constraints on model inversion (see Mechelli *et al.*, 2002).

SEM shares the same limitations as the linear model approach described above, i.e. temporal information is discounted.[1] However, it has enjoyed relative success and become established over the past decade due, in part, to its commercial availability as well as its intuitive appeal. However, it usually requires a number of rather *ad hoc* procedures, such as partitioning the data to create nested models, or pruning the connectivity matrix to render the solution tractable. These problems are confounded with an inability to capture non-linear features and temporal dependencies. By moving to dynamic models, we acknowledge the effect of an input's history and embed *a priori* knowledge into models at a more plausible and mechanistic level. These issues will be addressed in the following section.

[1] There are extensions of SEM that model dynamic information by using temporal embedding. However, these are formally the same as the multivariate autoregressive models discussed in the next section.

DYNAMIC MODELS

The static models described above discount temporal information. Consequently, permuted data sets produce the same path coefficients as the original data. Models that use the order in which data are produced are more natural candidates for neuronal dynamics. In this section we will review Kalman filtering, autoregressive and generalized convolution models (see Chapters 39 and 40).

The Kalman filter

The Kalman filter is used in engineering to model dynamics (Juang, 2001). It is based on a state-space model that invokes an extra set of [hidden] variables to generate data. These models are useful because long-range temporal order, within observed data, is modelled through interactions among hidden states, instead of mapping input directly onto output (see below). The Kalman filter is an 'online' procedure consisting of two steps: prediction and correction (or update). The hidden states are estimated (prediction step) using the information up until the present, which is updated (correction step) on receipt of each new measurement. These two steps are repeated recursively as new information arrives. A simple example demonstrates intuitively how the filter works. This example is taken from Ghahramani (2002).

Consider a series of data, which are received sequentially. Say we wanted to calculate a running average with each new data point. The estimate of the mean (which we will call the 'state') after t values of x is $\hat{x}_t$, where:

$$\hat{x}_t = \frac{1}{t}\sum_t x_t \quad \text{and} \quad \hat{x}_{t-1} = \frac{1}{t-1}\sum_{t-1} x_{t-1} \Rightarrow$$
$$\hat{x}_t = \frac{t-1}{t}\hat{x}_{t-1} + \frac{1}{t}x_t \qquad 38.17$$
$$= \hat{x}_{t-1} + K(x_t - \hat{x}_{t-1})$$

where K is the Kalman Gain $= 1/t$. This example gives the basic idea behind Kalman filtering and illustrates its general form: a prediction and a weighted correction (for a less heuristic introduction, see Appendix 5). The filter balances two types of information from a prediction (based on previous data) and an observation, weighted by their respective precisions using Bayes' rule. If the measured data are not reliable, K goes to zero, weighting the prediction error less and relying more on the preceding prediction. Conversely, if sequential dependence is low, then K is large, emphasizing information provided by the data when constructing an estimate of the current state. The quantities required in the forward recursion are the Kalman Gain and the mean and covariance of the prediction. A backward recursive algorithm called the Kalman Smoother calculates the mean and covariance of the states using data from the future, which is a *post hoc* procedure to improve estimates.

Kalman filtering and smoothing are generally applied to state-space models in discrete time (see Appendix 5). To understand how the filter is applied we start with the familiar linear observation model:

$$y_t = x_t\beta_t + \varepsilon_t$$
$$\varepsilon_t \sim N(0, R) \qquad 38.18$$

Here y and x are univariate and are both known, e.g. blood oygenation-level-dependent (BOLD) activity from V1 and V5. However, β is now a *variable* parameter or state that is observed vicariously through BOLD responses and changes with time according to the update equation:

$$\beta_t = \beta_{t-1} + \eta_t$$
$$\eta_t \sim N(0, Q) \qquad 38.19$$

η_t is a further innovation. Given this model, the Kalman filter can be used to estimate the evolution of the hidden state or variable parameter, β. Note that if $\eta_t = 0$ then $\beta_t = \beta_{t-1}$. This is the static estimate from an ordinary regression analysis. In this form of states-space model, the states play the role of variable parameters. This is why this application of Kalman filtering is also known as variable parameter regression.

Büchel and Friston (1998) used Kalman filtering to measure effective connectivity between V1 and V5: by modelling the path coefficient as a hidden state, the filter disclosed fluctuations in the coupling, which changed with attention (even though attentional status did not enter the model). Figure 38.8 illustrates the model and plots the time-dependent estimate, β_t, for V1 to V5 connectivity (lower panel). This reveals fluctuations that match changes in attentional set; the light grey bars indicate periods of attention to visual motion and the dark grey bars, periods without attention. The connection is clearly much stronger during attention, suggesting that attention to motion has enabled the forward connections from V1 to V5. The results of this analysis show that there is task-dependent variation in inter-regional connectivity. Critically, this variation was estimated from the data. However, in general, these changes in connectivity are induced experimentally by known and deterministic causes.

In Chapter 41, we return to hidden-state models and re-analyse the attentional data set in a way that allows designed manipulations of attention to affect the hidden states of a causal model. In dynamic casual modelling, the states are dynamic variables (e.g. neuronal activity) and

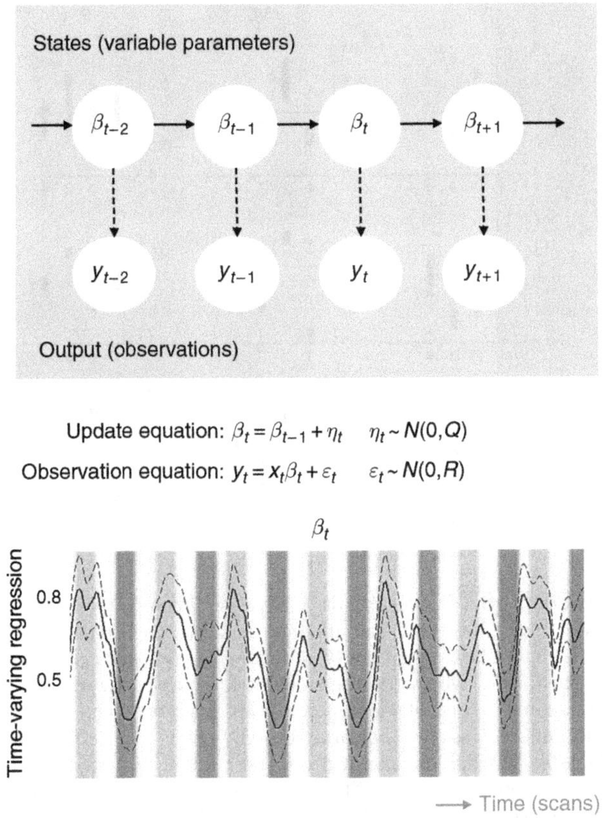

FIGURE 38.8 State-space model of the path coefficient between V1 and V5. Connection strengths are modelled as a hidden state (or variable parameter) that changes with time, according to the update equation (upper panel). Its evolution is estimated using the Kalman filter. This empirical example reveals fluctuations that match changes in attentional set (lower panel). The light grey bars indicate periods of attention to visual motion and the dark grey bars, periods without attention. The connection is clearly much stronger during attention, suggesting that attention to motion has enabled the forward connections from V1 to V5.

the effective connectivity corresponds to fixed parameters that can interact with time-varying states and inputs.[2] The remainder of this chapter focuses on approaches that do not refer to hidden states, such as autoregressive and generalized convolution models.

Multivariate autoregressive models

Sequential measurements often contain temporal information that can provide insight into the physical mechanisms generating them. A simple and intuitive model of temporal order is an autoregressive (AR) model, where the value of a variable at a particular time depends on preceding values.

The parameters of AR models comprise regression coefficients, at successive time lags, that encode sequential dependencies of the system in a simple and effective manner. This model can be extended to include several variables with dependencies among variables at different lags. These dependencies may be interpreted as the influence of one variable on another and can, with some qualification, be regarded as measures of effective connectivity. Models involving many variables are called multivariate autoregressive (MAR) models and have been used to measure dependencies among regional activities as measured with fMRI (Goebel *et al.*, 2003; Harrison *et al.*, 2003).

MAR models do not invoke hidden states. Instead, correlations among measurements at different time lags are used to quantify coupling. This incorporates history into the model in a similar way to the Volterra approach described below. MAR models are linear, but can be extended to include bilinear interaction terms (Penny *et al.*, 2005). To understand MAR we will build up a model from a univariate AR model and show that MAR models conform to general linear models (GLMs) with time-lagged explanatory variables.

Consider data at voxel i at time t modelled as a linear combination of previous values, plus an innovation:

$$y_t^i = [y_{t-1}^i, \ldots, y_{t-p}^i] \begin{bmatrix} w_1 \\ \vdots \\ w_p \end{bmatrix} + \varepsilon_t \qquad 38.20$$

w is a $p \times 1$ column vector containing the model parameters (AR coefficients). Here the explanatory variables are now preceding values over different time lags. We can extend the model to d-regions contained in the row vector:

$$y_t = \sum_{j=1}^{p} [y_{t-j}^1, \ldots, y_{t-j}^d] \begin{bmatrix} w_j^{11} & \cdots & w_j^{1d} \\ \vdots & \ddots & \\ w_j^{d1} & & w_j^{dd} \end{bmatrix} + [\varepsilon_1, \ldots, \varepsilon_d]$$

$$= \sum_{j=1}^{p} y_{t-j} W_j + \varepsilon \qquad 38.21$$

which has $d \times d$ parameters W_j at each time lag, describing interactions among all pairs of variables. This is simply a GLM whose parameters can be estimated in the usual way to give W, which is a $p \times d \times d$ array of AR coefficients (see Figure 38.9 for a schematic of the model). There are no inputs to the model, except for the errors, which play the role of innovations (cf. SEM). This means that experimentally designed effects have no explicit role (unless they enter through bilinear terms). However, the model attempts to identify relations between variables

[2] This should be contrasted with Kalman filtering, in which the connectivity itself was presumed to be a time-varying state.

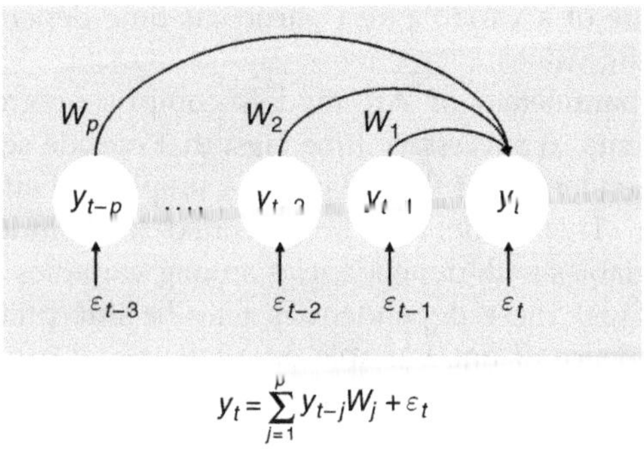

FIGURE 38.9 Temporal coupling can be modelled as a multivariate autoregressive process. The graphic shows time-lagged data where the arrows imply statistical dependence. The equation representing the model is show below. W_j comprise the autoregression coefficients and Y contains physiological or psychological data or interaction terms.

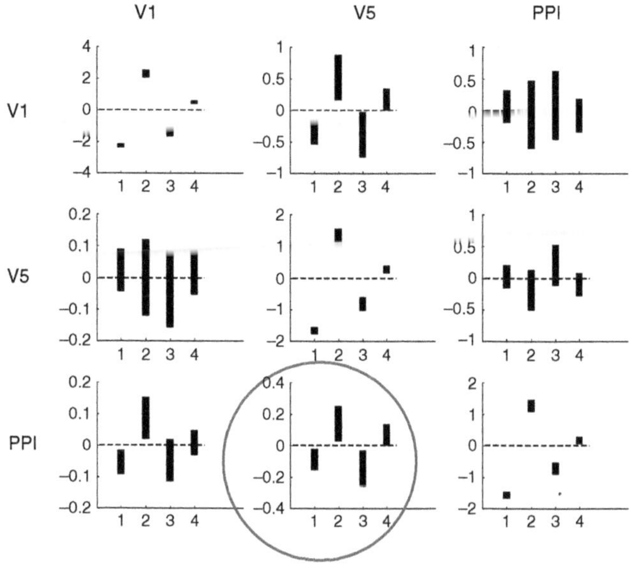

FIGURE 38.10 Results of a Bayesian inversion of a MAR model applied to the visual attention data set. Each panel shows posterior density estimates of W_j over time lags for each connection. The mean and two standard deviations for each posterior density are shown. Diagonal elements quantify autoregression and off diagonals crossregressions. The variates used were: V1, V5 and PPI; the PPI term was the Hadamard product $V1 \times PFC$ of activity in V1 and the prefrontal cortex (PFC). The circled estimates support coupling between V1 and V5 that depends on PFC activity in the past. This can be imputed from the fact that the regression coefficients coupling the $V1 \times PFC$ term to V5 were non-zero.

over time, which distinguishes it from static models of effective connectivity.

The value of p, or order of the model, becomes an issue when trying to avoid over-fitting. This is a common problem because a higher-order model will explain more variance in the data, without necessarily capturing the dynamics of the system any better than a more parsimonious model. A procedure for choosing an optimal value of is therefore necessary. This can be achieved using Bayesian inversion followed by model selection (Penny and Roberts, 2002 and Chapter 40).

We have used MAR to model the visual attention data with three regions. The aim was to test for a modulatory influence of PPC on V1 to V5 connectivity. To model this modulatory effect, we used a bilinear term, V1 × PPC as an extra variable in the MAR model and examined the regression coefficients coupling this term to V5. The results are shown in Figure 38.10 (see Chapter 40 for more details). The posterior densities of W_j are represented by the conditional mean and two standard deviations. The probability that an individual parameter is different from zero can be inferred from these conditional densities. Parameters coupling the PPI term to regional responses in V5 are circled and show one can be relatively certain they are not zero.

MAR models have not been used as extensively as other models of effective connectivity. However, they are an established method for quantifying temporal dependencies within time series (Chatfield, 1996). They are simple and intuitive models requiring no *a priori* knowledge of connectivity (cf. SEM).

Generalized convolution models

Up to now, we have considered models based on the general linear model and simple state-space models. The former may be criticized for not embracing temporal information within data, which the Kalman filter (an example of the latter) resolved by invoking hidden states. An alternative approach is to eschew hidden states and formulate a function that maps the history of input directly onto output. This can be achieved by characterizing the response (output) of a physical system over time to an idealized input (an impulse), called an impulse response function (IRF) or transfer function (TF). In the time domain, this function comprises a kernel that quantifies the idealized response. This is convenient as it bypasses any hidden states generating the data. However, it renders the system a 'black box', within which we have no model. This is both the method's strength and weakness.

Once the IRF has been characterized from experimental data, it can be used to model responses to arbitrary inputs. For linear systems, adherent to the principle of superposition, this reduces to convolving the input with the IRF. The modelled response depends on the input, without any reference to the interactions that may have produced it. An example, familiar to neuroimaging, is the haemodynamic response function (HRF) used to model the haemodynamic response of the brain to stimulus functions. However, we would like to cover non-linear models, which obtain by generalizing the notion of convolution models to include high-order interactions among inputs, an approach originally developed by Volterra in 1930 (Rieke *et al.*, 1997).

The generalized non-linear state and observation equations of any analytic system are, respectively:

$$\begin{aligned} \dot{x}(t) &= f(x(t), u(t)) \\ y(t) &= g(x(t), u(t)) \end{aligned} \qquad \mathbf{38.22}$$

These can be reformulated to relate output to input, $y(t) = h(u(t-\tau))$ without reference to the states $x(t)$. This mapping is a non-linear functional Taylor expansion:

$$\begin{aligned} y(t) = h_0 &+ \int_{-\infty}^{\infty} h_1(\tau_1)u(t-\tau_1)d\tau_1 \\ &+ \int_{-\infty}^{\infty}\int_{-\infty}^{\infty} h_2(\tau_1,\tau_2)u(t-\tau_1)u(t-\tau_2)d\tau_1 d\tau_2 + \cdots \\ &+ \int_{-\infty}^{\infty}\cdots\int_{-\infty}^{\infty} h_n(\tau_1,\cdots,\tau_n)u(t-\tau_1)\cdots u(t-\tau_n)d\tau_1\cdots d\tau_n \end{aligned} \qquad \mathbf{38.23}$$

where n is the order of the series and may take any positive integer to infinity. This is known as a Volterra series. Under certain conditions, h converges as n increases (Fliess *et al.*, 1983) and can provide a complete description of a system, given enough terms. To understand this we need to consider the Taylor series expansion as a means of approximating a general non-linear function (see Figure 38.1). Any sufficiently smooth non-linear function can be approximated, within the neighbourhood of an expansion point x_0, by scaling increasingly high-order terms, computed from derivatives of the function about x_0 (see Eqn. **38.1**). The Volterra series is a Taylor series expansion, where high-order terms are constructed from variables modelling interactions and scaled by time-varying coefficients. The Volterra series is a power-series expansion where the coefficients are now functions, known as kernels. The kernels are functions of time and, as the series involves functions of functions, they are known as functionals.

An increase in accuracy of the approximation is achieved by considering higher order terms, as demonstrated by deriving linear and bilinear models. The same is true of the linear and bilinear convolution models:

$$\begin{aligned} y(t) &\approx h_0 + \int_0^{\infty} h_1(\tau_1)u(t-\tau_1)\partial\tau_1 \\ y(t) &\approx h_0 + \int_0^{\infty} h_1(\tau_1)u(t-\tau_1)\partial\tau_1 \\ &+ \int_0^{\infty}\int_0^{\infty} h_2(\tau_1,\tau_2)u(t-\tau_1)u(t-\tau_2)\partial\tau_1\partial\tau_2 \end{aligned} \qquad \mathbf{38.24}$$

Note that the integrals in Eqn. **38.24** are from zero to infinity. This means the response depends only on past inputs. This renders the system casual (cf. the acausal system in Eqn. **38.23**). The HRF derives from a linear convolution model. The system's IRF and the occurrences of experimental trials are therefore h_1 and $u(t)$ respectively. The linear model complies with the principle of superposition: given two, the response is simply the sum of the two responses. By including the second-order kernel, non-additive responses can be modelled. Practically, this means that the timing of inputs is important in that different pairs of inputs may produce different responses. The Volterra formulation is a generalization of the convolution model, which convolves increasingly high-order interactions with multidimensional kernels to approximate the non-additive components of a system's response.

Kernels scale the effect each input, in the past, has had on the current response. As such, Volterra series have been described as 'power series with memory'. Sequential terms embody increasingly complex interactions among inputs up to arbitrary order. The series converges with increasing terms which, for weakly non-linear systems, is assumed to occur after the second-order term. A schematic, showing how two inputs can produce an output in a bilinear convolution model, is shown in Figure 38.11.

The first and second-order kernels quantify the linear and bilinear responses; this means they encode first- and second-order effective connectivity respectively (Friston, 2000). More specifically, the kernels map the input in the past to the current response:

$$\begin{aligned} h_1(\tau_1) &= \frac{\partial y(t)}{\partial u(t-\tau_1)} \\ h_2(\tau_1,\tau_2) &= \frac{\partial^2 y(t)}{\partial u(t-\tau_1)\partial u(t-\tau_2)} \\ &\cdots \end{aligned} \qquad \mathbf{38.25}$$

Having established the Volterra kernel as a measure of effective connectivity, we need to estimate them from

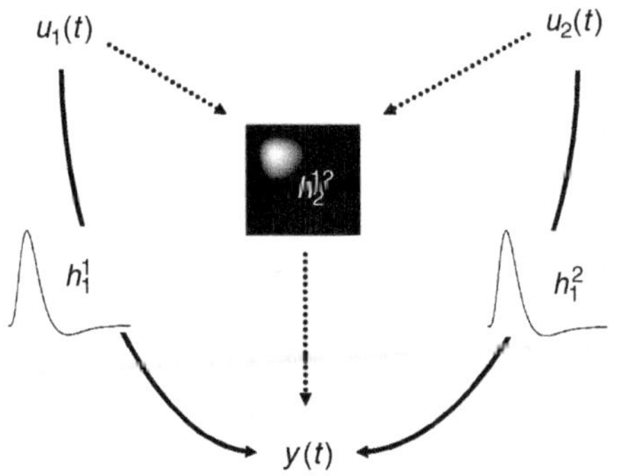

FIGURE 38.11 Schematic showing how two inputs can produce an output. Linear contributions from each input are formed by convolution with their respective first-order kernels h_1. In addition, non-additive responses, due to non-linear interactions between the two inputs and hidden states, are modelled by convolution with the second-order kernel, h_2. Note that the second-order kernel is two dimensional.

experimental data. By reformulating the model using an appropriate basis set, kernels can be reconstructed from estimated coefficients. The HRF is modelled well by gamma functions and this is the reason for choosing them to approximate Volterra kernels. Generally, the Volterra kernels for an arbitrarily non-linear dynamic system are difficult to compute unless the underlying generative process leading to the data is fairly well characterized.

A bilinear convolution model can be reformulated by convolving the inputs with the basis set b_i and using the set of convolved inputs in a GLM:

$$y(t) = \beta_0 + \sum_{i=1}^{n} \beta_i x_i(t) + \sum_{i=1}^{n}\sum_{j=1}^{n} \beta_{ij} x_i(t) x_j(t)$$
$$x_i(t) = \int b_i(\tau_1) u(t - \tau_1) \partial \tau_1 \qquad 38.26$$

The kernels are then recovered by:

$$h_0 = \beta_0$$
$$h_1(\tau_1) = \sum_{i=1}^{n} \beta_i b_i(\tau_1) \qquad 38.27$$
$$h_2(\tau_1, \tau_2) = \sum_{i=1}^{n}\sum_{j=1}^{n} \beta_{ij} b_i(\tau_1) b_j(\tau_2)$$

This method was applied to the attentional data set used previously. The model consisted of inputs from three regions: putamen, V1 complex and posterior parietal cortex (PPC), to V5, as shown in Figure 38.12. BOLD recordings from these regions were used as an index of neuronal activity, representing input to V5. The lower panel illustrates responses to simulated inputs, using the empirically determined kernels. It shows the response of V5 to

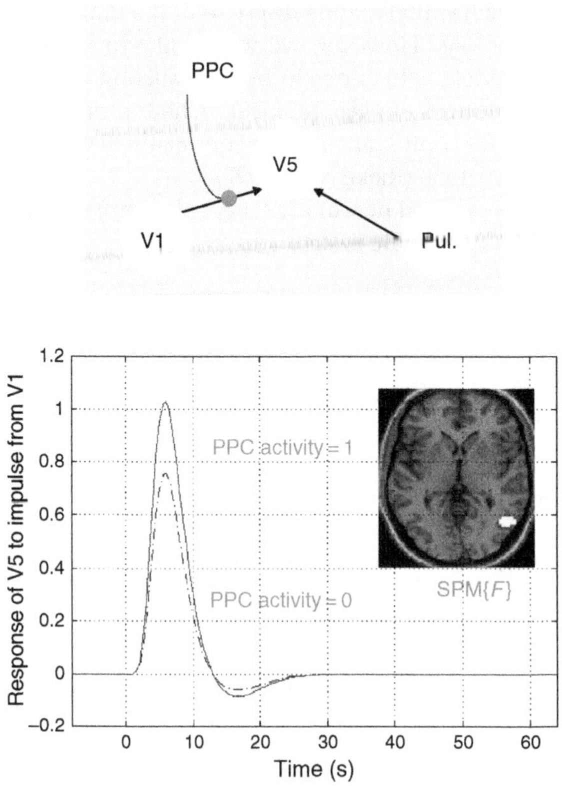

FIGURE 38.12 Top: brain regions and connections comprising a Volterra model of visually evoked responses. Bottom: characterization of the effects of V1 inputs on V5 and their modulation by posterior parietal cortex (PPC). Broken lines represent estimates of V5 responses when PPC activity is zero, according to a second-order Volterra model of effective connectivity with inputs to V5 from V1, PPC and the pulvinar (Pul.). The solid curve represents the same response when PPC activity is one standard deviation of its between-condition variation. It is evident that V1 has an activating effect on V5 and that PPC increases the responsiveness of V5 to these inputs. The insert shows all the voxels in V5 that evidenced a modulatory effect ($p < 0.05$ uncorrected). These voxels were identified by thresholding a statistical parametric map of the F-statistic, testing for the contribution of second-order kernels involving V1 and PPC (treating all other terms as nuisance variables). The fMRI data were obtained under identical stimulus conditions (visual motion subtended by radially moving dots) while manipulating the attentional component of the task (detection of velocity changes).

an impulse from V1 and provides a direct comparison of V5 responses to the same input from V1, with and without prefrontal cortex (PFC) activity. The influence of PFC is clear and reflects its enabling of V1 to V5 connectivity. This is an example of second-order effective connectivity.

The Volterra method has many useful qualities. It approximates non-linear behaviour within the familiar framework of a generalized convolution model. Kernels can be estimated using a GLM and inferences made under parametric assumptions. Furthermore, kernels contain the dynamic information we require to measure effective connectivity. However, kernels characterize an ideal response, generalized to accommodate nonlinear behaviour, which is effectively a summary of input-output behaviour. It should also be noted that Volterra series are only local approximations around an expansion point. Although they may be extremely good approximations for some systems, they may not be for others. For instance, Volterra series cannot capture the behaviour of periodic or chaotic dynamics (i.e. non-controllable systems). A major weakness of the method is that we have no notion of the internal mechanisms that generated the data, and this is one motivation for turning to dynamic causal models.

CONCLUSION

This chapter has described different methods of modelling inter-regional coupling using neuroimaging data. The development and application of these methods is motivated by the importance of changes in effective connectivity in development, cognition and pathology. We have portrayed the models incrementally, starting with linear regression models and ending with bilinear convolution models. In the next three chapters we will revisit bilinear models. Bilinear models cover plasticity induced by environmental and neurophysiological changes, while retaining mathematical tractability of linear models.

REFERENCES

Boas ML (1983) *Mathematical models in the physical sciences*. John Wiley & Sons, New York

Bollen KA (1989) *Structural equations with latent variables*. John Wiley & Sons, New York

Büchel C, Friston K (2000) Assessing interactions among neuronal systems using functional neuroimaging. *Neural Netw* **13**: 871–82

Büchel C, Friston KJ (1997) Modulation of connectivity in visual pathways by attention: cortical interactions evaluated with structural equation modelling and fMRI. *Cereb Cortex* **7**: 768–78

Büchel C, Friston KJ (1998) Dynamic changes in effective connectivity characterized by variable parameter regression and Kalman filtering. *Hum Brain Mapp* **6**: 403–08

Büchel C, Wise RJ, Mummery CJ *et al.* (1996) Nonlinear regression in parametric activation studies. *NeuroImage* **4**: 60–66

Chatfield C (1996) *The analysis of time series: an introduction*. CRC Press, Florida

Dayan P, Abbott L (2001) *Theoretical Neuroscience*. MIT, Cambridge, MA

Fliess M, Lamnabhi M, Lamnabhi-Lagarrigue F (1983) An algebraic approach to nonlinear functional expansions. *IEEE Trans Circ Syst* **30**: 554–70

Friston KJ (2000) The labile brain. I. Neuronal transients and nonlinear coupling. *Philos Trans R Soc Lond B Biol Sci* **355**: 215–36

Friston KJ, Buechel C, Fink GR *et al.* (1997) Psychophysiological and modulatory interactions in neuroimaging. *NeuroImage* **6**: 218–29

Friston KJ, Price CJ (2001) Dynamic representations and generative models of brain function. *Brain Res Bull* **54**: 275–85

Friston KJ, Ungerleider LG, Jezzard P *et al.* (1995) Characterizing modulatory interactions between areas V1 and V2 in human cortex: a new treatment of functional MRI data. *Hum Brain Mapp* **2**: 211–24

Ghahramani Z (2002) Unsupervised learning course (lecture 4). www.gatsby.ucl.ac.uk/~zoubin/index.html

Glass L (2001) Synchronization and rhythmic processes in physiology. *Nature* **410**: 277–84

Glass L, Kaplan D (2000) *Understanding nonlinear dynamics*. Springer Verlag, New York

Goebel R, Roebroeck A, Kim DS *et al.* (2003) Investigating directed cortical interactions in time-resolved fMRI data using vector autoregressive modeling and Granger causality mapping. *Mag Res Imag* **21**: 1251–61

Harrison L, Penny WD, Friston K (2003) Multivariate autoregressive modeling of fMRI time series. *NeuroImage* **19**: 1477–91

Juang J-N (2001) *Identification and control of mechanical systems*. Cambridge University Press, Cambridge

Maruyama GM (1998) *Basics of structural equation modelling*. Sage Publications, Inc., California

McIntosh AR, Gonzalez-Lima F (1991) Structural modelling of functional neural pathways mapped with 2-deoxyglucose: effects of acoustic startle habituation on the auditory system 7. *Brain Res* **547**: 295–302

McIntosh AR, Gonzalez-Lima F (1992a) The application of structural equation modelling to metabolic mapping of functional neural systems. In *Advances in metabolic mapping techniques for brain imaging of behavioural and learning functions*. Kluwer Academic Publishers, Dordrecht, pp 219–55

McIntosh AR, Gonzalez-Lima F (1992b) Structural modelling of functional visual pathways mapped with 2-deoxyglucose: effects of patterned light and footshock. *Brain Res* **578**: 75–86

McIntosh AR, Gonzalez-Lima F (1994) Structural equation modelling and its application to network analysis in functional brain imaging. *Hum Brain Mapp* **2**: 2–22

McIntosh AR, Grady CL, Ungerleider LG *et al.* (1994) Network analysis of cortical visual pathways mapped with PET. *J Neurosci* **14**: 655–66

Mechelli A, Penny WD, Price CJ *et al.* (2002) Effective connectivity and intersubject variability: using a multisubject network to test differences and commonalities. *NeuroImage* **17**: 1459–69

Mesulam MM (1998) From sensation to cognition. *Brain* **121**: 1013–52

Penny W, Ghahramani Z, Friston K (2005) Bilinear dynamical systems. *Philos Trans R Soc Lond B Biol Sci* **360**: 983–93

Penny W, Roberts SJ (2002) Bayesian multivariate autoregressive models with structured priors. *IEE Proc-Vis Image Signal Process* **149**: 33–41

Rao TS (1992) Identification of bilinear time series models 5. *Stat Sinica* **2**: 465–78

Rieke F, Warland D, de Ruyter van Steveninck R *et al.* (1997) Foundations. In *Spikes: exploring the neural code*. MIT, Cambridge, MA, pp 38–48

Rowe J, Friston K, Frackowiak R *et al.* (2002) Attention to action: specific modulation of corticocortical interactions in humans. *NeuroImage* **17**: 988–98

Scott A (1999) *Nonlinear science: emergence & dynamics of coherent structures*. Oxford University Press, Oxford

CHAPTER

39

Non-linear coupling and kernels

K. Friston

INTRODUCTION

This chapter revisits the Volterra formulation of effective connectivity from a conceptual and neurobiological point of view. Recall from the previous chapter that the generalized convolution representation, afforded by Volterra expansions, is just a way of describing the input-output behaviour of dynamic systems that have an equivalent state-space representation. In subsequent chapters, we will deal with state-space representations in more detail. Before proceeding to explicit input-state-output models, we will look at why the Volterra formulation is a useful summary of input-output relationships in this chapter and the role of multivariate autoregression models of output-output relationships in the next chapter.

The brain as a dynamic system

The brain can be regarded as an ensemble of connected dynamical systems and, as such, conforms to some simple principles relating the inputs and outputs of its constituent parts. The implications for the way we think about and measure neuronal interactions can be quite profound. These range from implications for which aspects of neuronal activity are important to measure and how to characterize coupling among neuronal populations, to implications pertaining to dynamic instability and complexity that is necessary for adaptive self-organization.

This chapter focuses on the first issue by looking at neuronal interactions, coupling and implicit neuronal codes from a dynamical perspective. In brief, by considering the brain in this light one can show that a sufficient description of neuronal activity must comprise activity at the current time *and its recent history*. This history constitutes a neuronal transient. Such transients represent an essential metric of neuronal interactions and, implicitly, a code employed in the functional integration of brain systems. The nature of transients, expressed coincidently in different neuronal populations, reflects their underlying coupling. A complete description of this coupling, or *effective connectivity*, can be expressed in terms of generalized convolution [Volterra] kernels that embody high-order or non-linear interactions. This coupling may be *synchronous*, and possibly oscillatory, or *asynchronous*. A critical distinction between synchronous and asynchronous coupling is that the former is essentially linear and the latter is non-linear. The non-linear nature of asynchronous coupling enables context-sensitive interactions that characterize real brain dynamics, suggesting that it plays an important role in functional integration.

Brain states are inherently labile, with a complexity and itinerancy that renders their invariant characteristics elusive. The basic idea pursued here is that the dynamics of neuronal systems can be viewed as a succession of transient spatiotemporal patterns of activity. These transients are shaped by the brain's infrastructure, principally connections, which have been selected to ensure the adaptive nature of the resulting dynamics. Although rather obvious, this formulation embodies a fundamental point, namely, that any description of brain state should have an explicit temporal dimension. In other words, measures of brain activity are only meaningful when specified over periods of time. This is particularly important in relation to fast dynamic interactions among neuronal populations that are characterized by synchrony. Synchronization has become a central theme in neuroscience (e.g. Gray and Singer, 1989; Eckhorn *et al.*, 1988; Engel *et al.*, 1991) and yet represents only one possible sort of interaction.

Statistical Parametric Mapping, by Karl Friston et al.
ISBN–13: 978-0-12-372560-8 ISBN–10: 0-12-372560-7

Ensemble dynamics, synchronization and self-organization

It is important to emphasize that this chapter is only about describing neuronal interactions; it is not about modelling them or trying to understand the underlying dynamical principles. Mechanistic or casual models of brain dynamics usually rest on mean-field assumptions and ensemble dynamics. The self-organization of coupled dynamic systems, through mean-field quantities, provides an established framework for looking at emergent behaviours and their neuronal correlates (see Harrison *et al.*, 2005 and Chapter 31). There is a growing body of work looking at the synchronization dynamics of coupled dynamical systems and their relevance to neuronal self-organization (e.g. Breakspear *et al.*, 2003). Others have used the discovery of chaotic itinerancy in high-dimensional dynamical systems (with and without a noise) to interpret neural activity in terms of high-dimensional transitory dynamics among 'exotic' attractors (see Tsuda, 2001). At a more macroscopic level, synergistics provides another framework to understand pattern-formation and self-organization. Here, the brain is conceived as a self-organizing system operating close to instabilities where its activities are governed by collective variables, the order parameters, which enslave the individual parts, i.e. the neurons. In this approach, the emphasis is on qualitative changes of behavioural and neuronal activities; using general properties of order parameters, at the phenomenological level, bi-stability, hysteresis and oscillations can be modelled (see Haken, 2006).

Unlike these more mechanistic approaches, this chapter is only concerned with how to describe and measure interactions. It is less concerned with how the interactions and dynamics are generated. However, there are some basic principles that place important constraints on the descriptions and measurements that could be used.

This chapter is divided into four sections. In the first, we review the conceptual basis of neuronal transients. This section uses the equivalence between two mathematical formulations of non-linear systems to show that descriptions of brain dynamics, in terms of neuronal transients and the coupling among interacting brain systems, are complete and sufficient. The second section uses this equivalence to motivate a taxonomy of neuronal codes and establish the relationship among neuronal transients, asynchronous coupling, *dynamic correlations* and non-linear interactions. In the third section, we illustrate non-linear coupling using magnetoencephalography (MEG) data. The final section discusses some neurobiological mechanisms that might mediate non-linear coupling.

NEURONAL TRANSIENTS

The assertion that meaningful measures of brain dynamics have a temporal domain is neither new nor contentious (e.g. von der Malsburg, 1985; Optican and Richmond, 1987; Engel *et al.*, 1991; Aertsen *et al.*, 1994; Freeman and Barrie, 1994; Abeles *et al.*, 1995; deCharms and Merzenich, 1996). A straightforward analysis demonstrates its veracity: suppose that one wanted to posit some quantities x that represented a complete and self-consistent description of brain activity. In short, everything needed to determine the evolution of the brain's state, at a particular place and time, was embodied in these measurements. Consider a component of the brain (e.g. a neuron or neuronal population). If such a set of variables existed for this component system, they would satisfy some immensely complicated non-linear state equation:

$$\dot{x} = f(x, u) \quad \textbf{39.1}$$

where x is a huge vector of state variables, which range from depolarization at every point in the dendritic tree to the phosphorylation status of every relevant enzyme, from the biochemical status of every glial cell compartment to every aspect of gene expression. $u(t)$ represents external forces or inputs conveyed by afferent from other regions. Eqn. **39.1** simply says that the evolution of state variables is a non-linear function of the variables themselves and some inputs. The vast majority of these variables are hidden and not measurable directly. However, there is a small number of derived measurements y that can be made (cf. phase-functions in statistical physics):

$$y = g(x, u) \quad \textbf{39.2}$$

such as activities of whole cells or populations. These activities could be measured in many ways, for example firing at the initial segment of an axon or local field potentials. The problem is that a complete and sufficient description appears unattainable, given that the underlying state variables cannot be observed directly. This is not the case. The resolution of this apparent impasse rests upon two things: first, a mathematical equivalence relating the inputs and outputs of a dynamical system and the fact that measurable outputs constitute the inputs to other cells or populations. In other words, there exists a set of quantities that serve a dual role as external forces or inputs to neuronal systems and a measure of their response (e.g. mean-field quantities in ensemble dynamics).

Input-state-output systems and Volterra series

Neuronal systems are inherently non-linear and lend themselves to modelling with non-linear dynamical systems. However, due to the complexity of biological systems, it is difficult to find analytic equations that describe them adequately. Even if these equations were known, the state variables are often not observable. An alternative approach to identification is to adopt a very general model (Wray and Green, 1994) and focus on the inputs and outputs. The Fliess fundamental formula (Fliess *et al.*, 1983) describes the causal relationship between the outputs and the recent history of the inputs. This relationship can be expressed as a Volterra series which expresses the output as a non-linear convolution of the inputs, critically without reference to the states. This series is simply a functional Taylor expansion of Eqn. **39.2**:

$$y(t) = h(u(t-\sigma)) = \sum_{i=0}^{\infty} \int_0^{\infty} \dots \int_0^{\infty} \kappa_i(\sigma_1, \dots \sigma_i) u(t-\sigma_1) \dots u(t-\sigma_i) d\sigma_1 \dots d\sigma_i$$

$$\kappa_i(\sigma_1, \dots, \sigma_i) = \frac{\partial^i y(t)}{\partial u(t-\sigma_1) \dots \partial u(t-\sigma_i)} \quad \textbf{39.3}$$

where $\kappa_i(\sigma_1, \dots, \sigma_i)$ is the i-th order kernel. Volterra series have been described as a 'power series with memory' and are generally thought of as a high-order or 'non-linear convolution' of the inputs to provide an output (see Bendat, 1990 for a fuller discussion).

Convolution and state-space representations

The Volterra expansion means that the output of any neuronal system is a function of the recent history of its inputs. The critical thing here is that we never need to know the hidden variables that describe the details of each cell's electrochemical and biochemical status. We only need to know the history of its inputs which, of course, are the outputs of other cells. Eqn. **39.3** is, in principle, a sufficient description of brain dynamics and requires the inputs $u(t-\sigma)$ at all times preceding the moment in question. These are simply neuronal transients. The degree of transience depends on how far back in time it is necessary to go fully to capture the brain's dynamics. For example, if we wanted to determine the behaviour of a cell in V1 (primary visual cortex) then we would need to know the activity of all connected cells in the immediate vicinity over the last millisecond or so to account for propagation delays down afferent axons. We would also need to know the activity in distant sources, like the lateral geniculate nucleus and higher cortical areas, some ten or more milliseconds ago. In short, we need the recent history of all inputs.

Transients can be expressed in terms of firing rates (e.g. chaotic oscillations, Freeman and Barrie, 1994) or individual spikes (e.g. syn-fire chains, Abeles *et al.*, 1995). Transients are not just a mathematical abstraction, they have real implications at a number of levels. For example, the emergence of fast oscillatory interactions among simulated neuronal populations depends upon the time-delays implicit in axonal transmission and the time-constants of postsynaptic responses. Another slightly more subtle aspect of this formulation is that changes in synaptic efficacy, such as short-term potentiation or depression, take some time to be mediated by intracellular mechanisms. This means that the interaction between inputs at different times that models these activity-dependent effects, again depends on the relevant history of activity.

Levels of description

The above arguments lead to a conceptual model of the brain as a collection of dynamical systems (e.g. cells or populations of cells), each represented as an input-state-output model, where the state remains hidden. However, the inputs and outputs are accessible and are causally related where the output of one system constitutes the input to another. A complete description therefore comprises the nature of these relationships (the Volterra kernels) and the neuronal transients that mediate them (the inputs). This constitutes a *mesoscopic* level of description that ignores the dynamics that are intrinsic to each system, but entails no loss of information about their interactions.

The equivalence, in terms of specifying the behaviour of a neuronal system, between microscopic and mesoscopic levels of description is critical. In short, the equivalence means that all the information inherent in unobservable microscopic variables that determine the response of a neuronal system is embedded in the history of its observable inputs and outputs. Although the microscopic level of description may be more mechanistically informative, neuronal transients are an equivalent description.[1]

[1] We have focused on the distinction between microscopic and mesoscopic levels of description. The *macroscopic* level is reserved for approaches, exemplified by synergistics (Haken, 1983), that characterize the spatiotemporal evolution of brain dynamics in terms of a small number of macroscopic order parameters (see Kelso, 1995 for an engaging exposition). Order parameters are created and determined by the cooperation of microscopic quantities and yet, at the same time, govern the behaviour of the whole system. See Jirsa *et al.* (1995) for a nice example.

Effective connectivity and Volterra kernels

The first conclusion so far is that neuronal transients are necessary to specify brain dynamics. The second conclusion is that a complete model of the influence one neuronal population exerts over another should take the form of a Volterra series.[2] This implies that a complete characterization of these influences (i.e. effective connectivity) comprises the Volterra kernels that are applied to the inputs to yield the outputs. Effective connectivity refers to: 'the influence that one neural system exerts over another, either at a synaptic [i.e. synaptic efficacy] or population level' (Friston, 1995a). It has been proposed (Aertsen and Preißl, 1991) that: 'the notion of effective connectivity should be understood as the experiment- and time-dependent, simplest possible circuit diagram that would replicate the observed timing relationships between the recorded neurons' (see the previous chapter).

Volterra kernels and effective connectivity

Volterra kernels are essential in characterizing the effective connectivity, because they represent the causal input-output characteristics of the system in question. In neurobiological terms they *are synonymous with effective connectivity*. From Eqn. **39.3**:

$$\kappa_1(\sigma_1) = \frac{\partial y(t)}{\partial u(t-\sigma_1)} \quad \kappa_2(\sigma_1, \sigma_2) = \frac{\partial^2 y(t)}{\partial u(t-\sigma_1)\partial u(t-\sigma_2)} \cdots \tag{39.4}$$

It is evident that the first-order kernel embodies the response evoked by a change in input at $t-\sigma_1$. In other words, it is a time-dependent measure of *driving* efficacy. Similarly, the second order kernel reflects the *modulatory* influence of the input at $t-\sigma_1$ on the response evoked by input at $t-\sigma_2$. And so on for higher orders.

If effective connectivity is the influence that one neural system exerts over another, it should be possible, given the effective connectivity and the input, to predict the response of a recipient population. This is precisely what Volterra kernels do. Any model of effective connectivity can be expressed as a Volterra series and any measure of effective connectivity can be reduced to a set of Volterra kernels. An important aspect of effective connectivity is its context-sensitivity. Effective connectivity is simply the 'effect' that input has on the output of a target system. This effect will be sensitive to other inputs, its own history and, of course, the microscopic state and causal architecture intrinsic to the target population. This intrinsic dynamical structure is embodied in the Volterra kernels. In short, Volterra kernels are synonymous with effective connectivity because they characterize the measurable effect that an input has on its target. An example of using Volterra kernels to characterize context-sensitive changes in effective connectivity was provided in the previous chapter (see Figure 38.12). This example used haemodynamic responses to changes in neuronal activity as measured with functional magnetic resonance imaging (fMRI).

NEURONAL CODES

Functional integration refers to the concerted interactions among neuronal populations that mediate perceptual binding, sensorimotor integration and cognition. It pertains to the mechanisms of, and constraints under which, the state of one population influences that of another. It has been suggested by many that functional integration among neuronal populations uses transient dynamics that represent a temporal code. A compelling proposal is that population responses, encoding a percept, become organized in time, through reciprocal interactions, to discharge in synchrony (von der Malsburg, 1985; Singer, 1994). The use of the term 'encoding' speaks directly of the notion of codes. Here, a neuronal code is taken to be a metric that reveals interactions among neuronal systems by enabling some prediction of the response in one population given the same sort of measure in another.[3] Clearly, from the previous section, neuronal transients represent the most generic form of code because, given the Volterra kernels, the output can, in principle, be predicted exactly. Neuronal transients have a number of attributes (e.g. inter-spike interval, duration, mean level of firing, predominant frequency etc.) and any of these could be contenders for a more parsimonious code. The problem of identifying possible codes can be reduced to

[2] An important qualification here is that each system is 'controllable'. Systems which are not 'controlled' have quasi-periodic or chaotic behaviours that are maintained by autonomous interactions among the states of the system. Although an important issue at the microscopic level, it is fairly easy to show that the mean field approximation to any ensemble of subsystems is controllable. This is because the Fokker-Planck equation governing ensemble dynamics has a point attractor (see Chapter 31).

[3] Although the term code is not being used to denote anything that 'codes' for something in the environment, it could be used to define some aspect of an evoked transient that has high mutual information with a stimulus parameter (e.g. Optican and Richmond, 1987; Tovee *et al.*, 1993).

identifying the form the Volterra kernels can take. If we know their form, then we can say which aspects of the input will cause a response. Conversely, it follows that the different forms of kernels should specify the various codes that might be encountered. This is quite an important point and leads to a clear formulation of what can and cannot constitute a code. We will review different codes in terms of the different sorts of kernels that could mediate them.

Instantaneous versus temporal codes

The first kernel characteristic that engenders a coding taxonomy is kernel depth. The limiting case is when the kernel's support shrinks to a point in time. This means that the only relevant history is the immediate activity of inputs (all earlier activities are 'ignored' by the kernel). In this case, the activity in any unit is simply a non-linear function of current activities elsewhere. An example of this is instantaneous rate coding.

Rate coding considers spike-trains as *stochastic processes* whose first order moments (i.e. mean activity) describe neuronal interactions. These moments may be in terms of spikes themselves or other compound events (e.g. the average rate of bursting, Bair *et al.*, 1994). Interactions based on rate coding are usually assessed in terms of cross-correlations. From the dynamical perspective, instantaneous rate codes are insufficient. This is because they predict nothing about a cell, or population, response unless one knows the microscopic state of that cell or population.

The distinction between rate and temporal coding (see Shadlen and Newsome, 1995; de Ruyter van Steveninck *et al.*, 1997) centres on whether the precise timing of individual spikes is sufficient to facilitate meaningful neuronal interactions. In temporal coding, the exact time at which an individual spike occurs is the important measure and the spike-train is considered as a *point process.* There are clear examples of temporal codes that have predictive validity, e.g. the primary cortical representation of sounds by the coordination of action potential timing (deCharms and Merzenich, 1996). These codes depend on the relative timing of action potentials and, implicitly, an extended temporal frame of reference. They therefore fall into the class of transient codes, where selective responses to particular inter-spike intervals are modelled by temporally extended second-order kernels. A nice example is provided by de Ruyter van Steveninck *et al.* (1997) who show that the temporal patterning of spike trains, elicited in fly motion-sensitive neurons by natural stimuli, can carry twice the amount of information than an equivalent [Poisson] rate code.

Transient codes: synchronous versus asynchronous

The second distinction, assuming the kernels have a non-trivial temporal support, is whether they comprise high-order terms or not. Expansions with just first-order terms are only capable of meditating linear or synchronous interactions. High-order kernels confer non-linearity on the influence of an input that leads to asynchronous interactions. Mathematically, if there are only first-order terms, then the Fourier transform of the Volterra kernel completely specifies the relationship (the transfer function) between the spectral density of input and output in a way that precludes interactions among frequencies, or indeed inputs. In other words, the expression of any frequency in a recipient system is predicted exactly by the expression of the same frequency in the source (after some scaling by the transfer function).

Synchronous codes

The proposal most pertinent to these forms of code is that population responses, participating in the encoding of a percept, become organized in time through reciprocal interactions so that they discharge in synchrony (von der Malsburg, 1985; Singer, 1994) with regular periodic bursting. It should be noted that synchronization does not necessarily imply oscillations. However, synchronized activity is usually inferred operationally by oscillations implied by the periodic modulation of cross-correlograms of separable spike trains (e.g. Eckhorn *et al.*, 1988; Gray and Singer, 1989) or measures of coherence in multichannel electrical and neuromagnetic time-series (e.g. Llinas *et al.*, 1994). The underlying mechanism of these frequency-specific interactions is usually attributed to phase-locking among neuronal populations (e.g. Sporns *et al.*, 1989; Aertsen and Preißl, 1991). The key aspect of these measures is that they refer to the extended temporal structure of synchronized firing patterns, either in terms of spiking (e.g. syn-fire chains, Abeles *et al.*, 1995; Lumer *et al.*, 1997) or oscillations in the ensuing population dynamics (e.g. Singer, 1994).

Many aspects of functional integration and feature-linking in the brain are thought to be mediated by synchronized dynamics among neuronal populations (Singer, 1994). Synchronization reflects the direct, reciprocal exchange of signals between two populations, whereby the activity in one population influences the second, such that the dynamics become entrained and mutually reinforcing. In this way, the binding of different features of an object may be accomplished, in the temporal domain, through the transient synchronization of oscillatory responses. This 'dynamical linking' defines their short-lived functional association. Physiological evidence is compatible with this theory (e.g. Engel *et al.*, 1991;

Fries *et al.*, 1997). Synchronization of oscillatory responses occurs within as well as among visual areas, e.g. between homologous areas of the left and right hemispheres and between areas at different levels of the visuomotor pathway (e.g. Engel *et al.*, 1991). Synchronization in the visual cortex appears to depend on stimulus properties, such as continuity, orientation and motion coherence.

The problem with synchronization is that there is nothing essentially dynamic about synchronous interactions *per se*. As argued by Erb and Aertsen (1992): 'the question might not be so much how the brain functions by virtue of oscillations, as most researchers working on cortical oscillations seem to assume, but rather how it manages to do so in spite of them'. In order to establish dynamic cell assemblies, it is necessary to create and destroy synchrony (see Breakspear *et al.*, 2003 for one mechanism). It is precisely these dynamic aspects that speak of changes in synchrony (e.g. Desmedt and Tomberg, 1994; Tass, 2005) and the asynchronous transitions between synchronous states as the more pertinent phenomenon. In other words, it is the successive reformulation of dynamic cell assemblies, through non-linear or asynchronous interactions, that is at the heart of dynamical linking (Singer, 1994).

Asynchronous codes

An alternative perspective on neuronal codes is provided by *dynamic correlations* (Aertsen *et al.*, 1994) as exemplified in Vaadia *et al.* (1995). A fundamental phenomenon, observed by Vaadia *et al.* (1995), is that, following behaviourally salient events, the degree of coherent firing between two neurons can change profoundly and systematically over the ensuing second or so (cf. induced responses in the EEG, Tallon-Baudry *et al.*, 1999; Lachaux *et al.*, 2000). One implication is that a complete model of neuronal interactions has to accommodate dynamic changes in correlations, modulated on time-scales of 100–1000 ms. Neuronal transients provide a simple explanation for temporally modulated coherence or dynamic correlation. Imagine that two neurons respond to an event with a similar transient. For example, if two neurons respond to an event with decreased firing for 400 ms, and this decrease was correlated over epochs, then positive correlations between the two firing rates would be seen for the first 400 of the epoch, and then fade away, exhibiting a dynamic modulation of coherence. In other words, the expression transient covariance can be formulated as covariance in the expression of transients. The generality of this equivalence can be established using singular value decomposition (SVD) of the joint-peristimulus time histogram (J-PSTH) as described in Friston (1995b). This is simply a mathematical device to show that dynamic changes in coherence are equivalent to the coherent expression of neural transients. In itself it is not important, in the sense that dynamic correlations are just as valid a characterization as neuronal transients and, indeed, may provide more intuitive insights into how this phenomenon is mediated (e.g. Riehle *et al.*, 1997). A more important observation is that J-PSTHs can be asymmetric about the leading diagonal. This means that coupled transients in two units can have a different temporal pattern of activity. This can only be explained by asynchronous or non-linear coupling.

Summary

In summary, the critical distinction between synchronous and asynchronous coupling is the difference between linear and non-linear interactions among units or populations.[4] This difference reduces to the existence of high-order Volterra kernels in mediating the input-output behaviour of coupled cortical regions. There is a close connection between asynchronous-non-linear coupling and the expression of distinct transients in two brain regions: both would be expressed as dynamic correlations or, in the EEG, as event-related changes in synchronization (e.g. induced oscillations (Friston *et al.*, 1997)). If the transient model is correct, then important transactions among cortical areas will be overlooked by techniques that are predicated on rate coding (e.g. correlations, covariance patterns, spatial modes etc.) or synchronization models (e.g. coherence analysis and cross-correlograms). Clearly, the critical issue is whether there is direct evidence for non-linear or asynchronous coupling that would render high-order Volterra kernels necessary.

EVIDENCE FOR NON-LINEAR COUPLING

Why is asynchronous coupling so important? The reason is that asynchronous interactions embody all the non-linear interactions implicit in functional integration and it is these that mediate the context-sensitive nature of

[4] The term 'generalized synchrony' has been introduced to include non-linear inter-dependencies (see Schiff *et al.*, 1996). Generalized synchrony subsumes synchronous and asynchronous coupling. An elegant method for making inferences about generalized synchrony is described in Schiff *et al.* (1996). This approach is particularly interesting from our point of view because it calls upon the recent history of the dynamics through the use of temporal embedding to reconstruct the attractors analysed.

neuronal interactions. Non-linear interactions among cortical areas render the effective connectivity among them inherently dynamic and contextual. Examples of context-sensitive interactions include the attentional modulation of evoked responses in functionally specialized sensory areas (e.g. Treue and Maunsell, 1996) and other contextually dependent dynamics (see Phillips and Singer, 1997). Whole classes of empirical phenomena, such as extra-classical receptive field effects, rely on non-linear or asynchronous interactions.

Non-linear coupling and asynchronous interactions

If the temporal structures of recurring transients in two parts of the brain are distinct, then the expression of certain frequencies in one cortical area should predict the expression of *different* frequencies in another. In contrast, synchronization posits the expression of the *same* frequencies. Correlations among different frequencies therefore provide a basis for discriminating between synchronous and asynchronous coupling.

Consider time-series from two neuronal populations or cortical areas. Synchrony requires that the expression of a particular frequency (e.g. 40 Hz) in one time-series will be coupled with the expression of the same frequency in the other. In other words, the modulation of this frequency in one area can be explained or predicted by its modulation in the second. Conversely, asynchronous coupling suggests that the power at a reference frequency, say 40 Hz, can be predicted by the spectral density in the second time-series at frequencies other than 40 Hz. These predictions can be tested empirically using standard time-frequency and regression analyses as described in Friston (2000). Figure 39.1 shows an example of this sort of analysis, revealing the dynamic changes in spectral density between 8 and 64 Hz over 16 s. The cross-correlation matrix of the time-dependent expression of different frequencies in the parietal and prefrontal regions is shown in the lower left panel. There is anecdotal evidence for both synchronous and asynchronous coupling. Synchronous coupling, based upon the co-modulation of the same frequencies, is manifest as hot-spots along, or near, the leading diagonal of the cross-correlation matrix (e.g. around 20 Hz). More interesting, are correlations between high frequencies in one time-series and low frequencies in another. In particular, note that the frequency modulation at about 34 Hz in the parietal region (second time-series) could be explained by several frequencies in the prefrontal region. The most profound correlations are with lower frequencies in the first time-series (26 Hz). Using a simple regression framework, statistical inferences can be made about the coupling within and between different frequencies (see Friston, 2000 for details). A regression analysis shows that coupling at 34 Hz has significant synchronous and asynchronous components, whereas the coupling at 48 Hz is purely asynchronous (middle and right peaks in the graphs), i.e. a coupling between beta dynamics in the pre-motor region and gamma dynamics in the parietal region.

THE NEURAL BASIS OF NON-LINEAR COUPLING

In Friston (1997), it was suggested that, from a neurobiological perspective, the distinction between non-linear [asynchronous] and linear [synchronous] interactions could be viewed in the following way. Synchronization emerges from the reciprocal exchange of signals between two populations, where each *drives* the other, such that the dynamics become entrained and mutually reinforcing. In asynchronous coding, the afferents from one population exert a *modulatory* influence, not on the activity of the second, but on the interactions within it (e.g. a modulation of effective connectivity or synaptic efficacies within the target population) leading to changes in the dynamics intrinsic to the second population. In this model, there is no necessary synchrony between the intrinsic dynamics that ensue and the temporal pattern of modulatory input. To test this hypothesis one would need to demonstrate that asynchronous coupling emerges when extrinsic connections are changed from driving connections to modulatory connections. Clearly, this cannot be done in the real brain. However, we can use computational techniques to create a biologically realistic model of interacting populations and test this hypothesis directly.

Interactions between simulated populations

Two populations were simulated using the neural-mass model described in Friston (2000). This model simulates entire neuronal populations in a deterministic fashion based on known neurophysiological mechanisms (see also Chapter 31). In particular, we modelled three sorts of synapse, fast inhibitory (GABA), fast excitatory (AMPA) and slower voltage-dependent synapses (NMDA). Connections intrinsic to each population used only GABA and AMPA-like synapses. Simulated glutaminergic extrinsic connections between populations used either driving AMPA-like synapses or modulatory NMDA-like synapses. Transmission delays

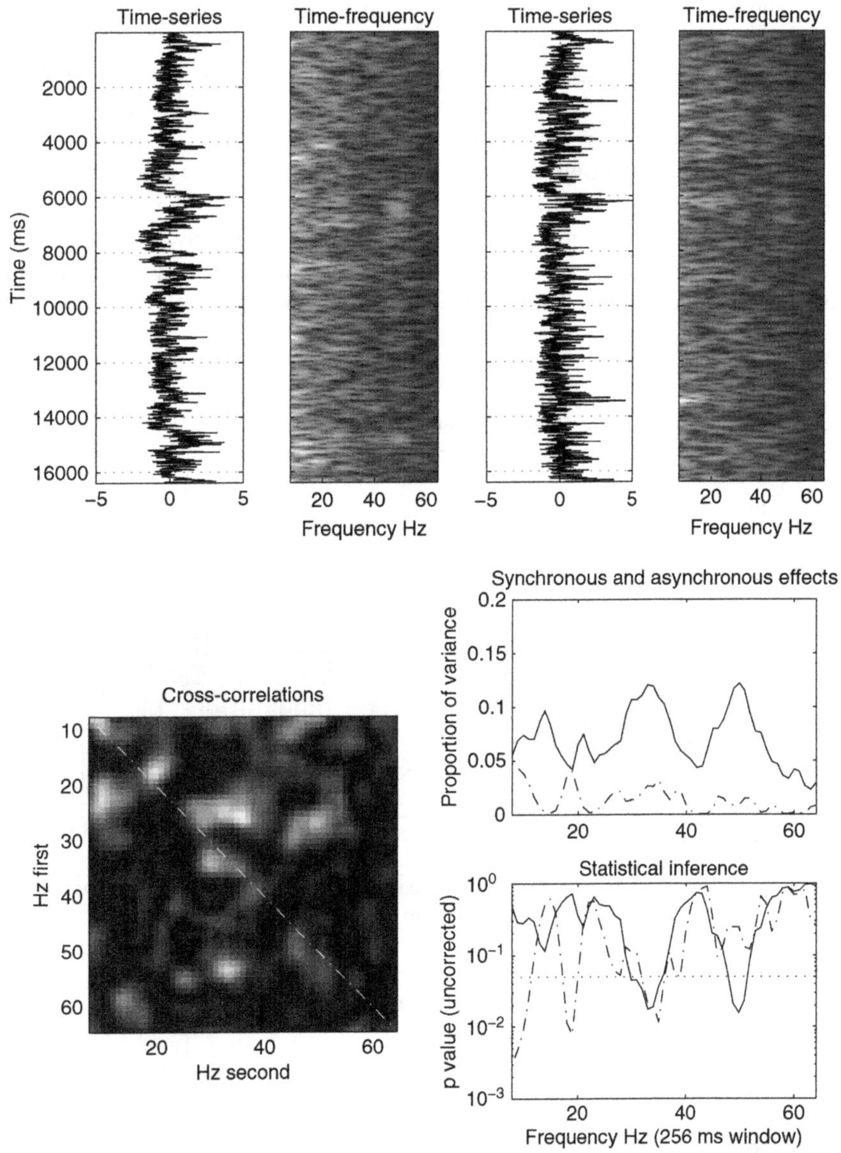

FIGURE 39.1 Time-frequency and regression analysis of MEG time-series designed to characterize the relative contribution of synchronous and asynchronous coupling. Neuromagnetic data were acquired from a normal subject using a KENIKRON 37-channel MEG system at one-millisecond intervals for periods of up to two minutes. During this time the subject made volitional joystick movements to the left, every two seconds or so. Paired epochs were taken from a left prefrontal and left parietal region. Top panels: the two times series (plots) and their corresponding time-frequency profiles (images). The first time-series comes from the left prefrontal region. The second comes from the left superior parietal region. Lower left panel: this is a simple characterization of the coupling among frequencies in the two regions and represents the (squared) cross-correlations of the time-varying expression of different frequencies from the upper panels. Lower right panels: these are the results of a linear regression analysis that partitions the variance in the second (parietal) time-series into components that can be attributed to synchronous (broken lines) and asynchronous (solid lines) contributions from the first (prefrontal) time series. The upper graph shows the relative contribution in terms of the proportion of variance explained and in terms of the significance using a semi-log plot of the corresponding p-values (lower graph). The dotted line in the latter corresponds to $p = 0.05$.

This example was chosen because it illustrates three sorts of coupling (synchronous, asynchronous and mixed). From inspection of the cross-correlation matrix, it is evident that power in the beta range (20 Hz) in the second time-series is correlated with a similar frequency modulation in the first, albeit at a slightly lower frequency. The resulting correlations appear just off the leading diagonal (broken line) on the upper left. The proportion of variance explained by synchronous and asynchronous coupling is roughly the same and, in terms of significance, synchrony supervenes (see upper graph). In contrast, the high correlations, between 48 Hz in the second time-series and 26 Hz in the first, are well away from the leading diagonal, with little evidence of correlations within either of these frequencies. The regression analysis confirms that, at this frequency, asynchronous coupling prevails. The variation at about 34 Hz in the parietal region could be explained by several frequencies in the prefrontal region. A formal analysis shows that both synchronous and asynchronous coupling coexist at this frequency (i.e. the middle peak in the graphs).

for extrinsic connections were fixed at 8 ms. By using realistic time constants the characteristic oscillatory dynamics of each population were expressed in the gamma range.

The results of coupling two populations with unidirectional AMPA-like connections are shown in the top of Figure 39.2 in terms of the simulated local field potentials (LFP). Occasional transients in the driving population were evoked by injecting a depolarizing current at random intervals (dotted lines). The tight synchronized coupling that ensues is evident. This example highlights the point that near-linear coupling can arise even in the context of loosely coupled, highly non-linear neuronal oscillators of the sort modelled here. Contrast these entrained dynamics under driving connections with those that emerge when the connection is modulatory or NMDA-like (lower panel in Figure 39.2). Here, there is no synchrony and, as predicted, fast transients of an oscillatory nature are facilitated by the low-frequency input from the first population (cf. the MEG analyses

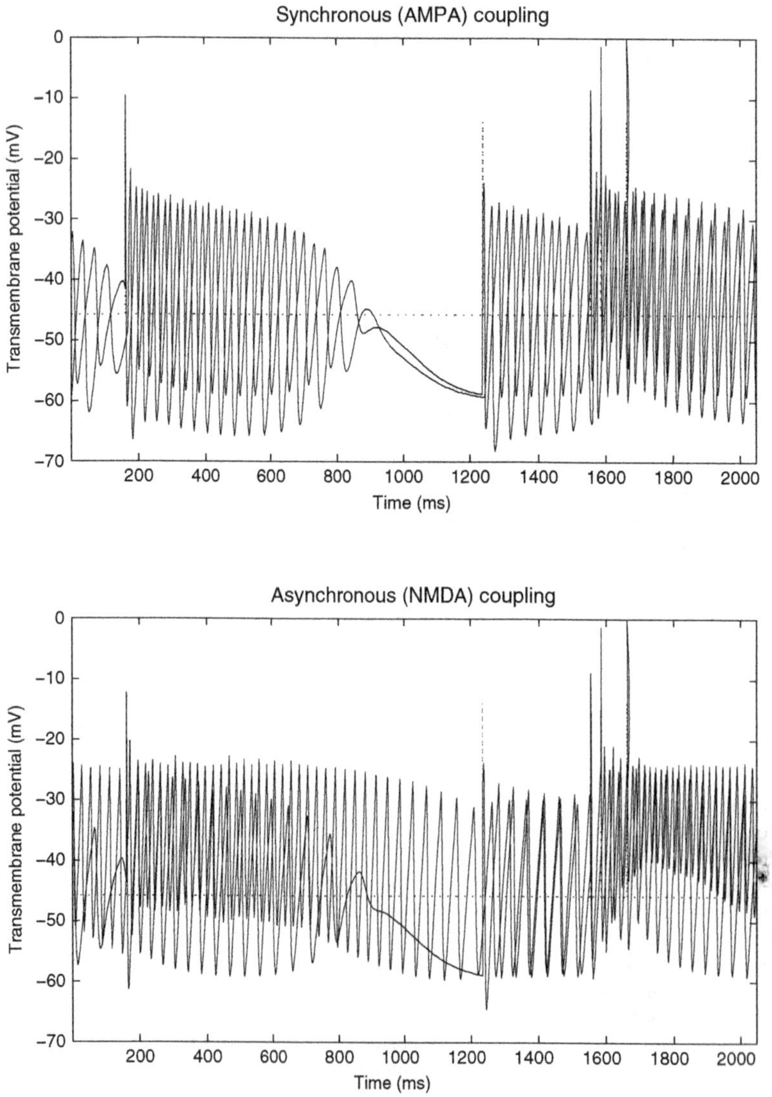

FIGURE 39.2 Simulated local field potentials (LFP) of two coupled populations using two different sorts of postsynaptic responses (AMPA and NMDA-like) to inputs from the first to the target population. The dotted line shows the depolarization effected by sporadic injections of current into the first population. The key thing to note is that, under AMPA-like or driving connections, the second population is synchronously entrained by the first. When the connections are modulatory or voltage-dependent (NMDA), the effects are much more subtle and resemble a frequency modulation. These data were simulated using a biologically plausible model of excitatory and inhibitory subpopulations. The model was deterministic with variables pertaining to the collective, probabilistic, behaviour of the subpopulations (cf. a mean-field treatment) (see Friston, 2000 for details).

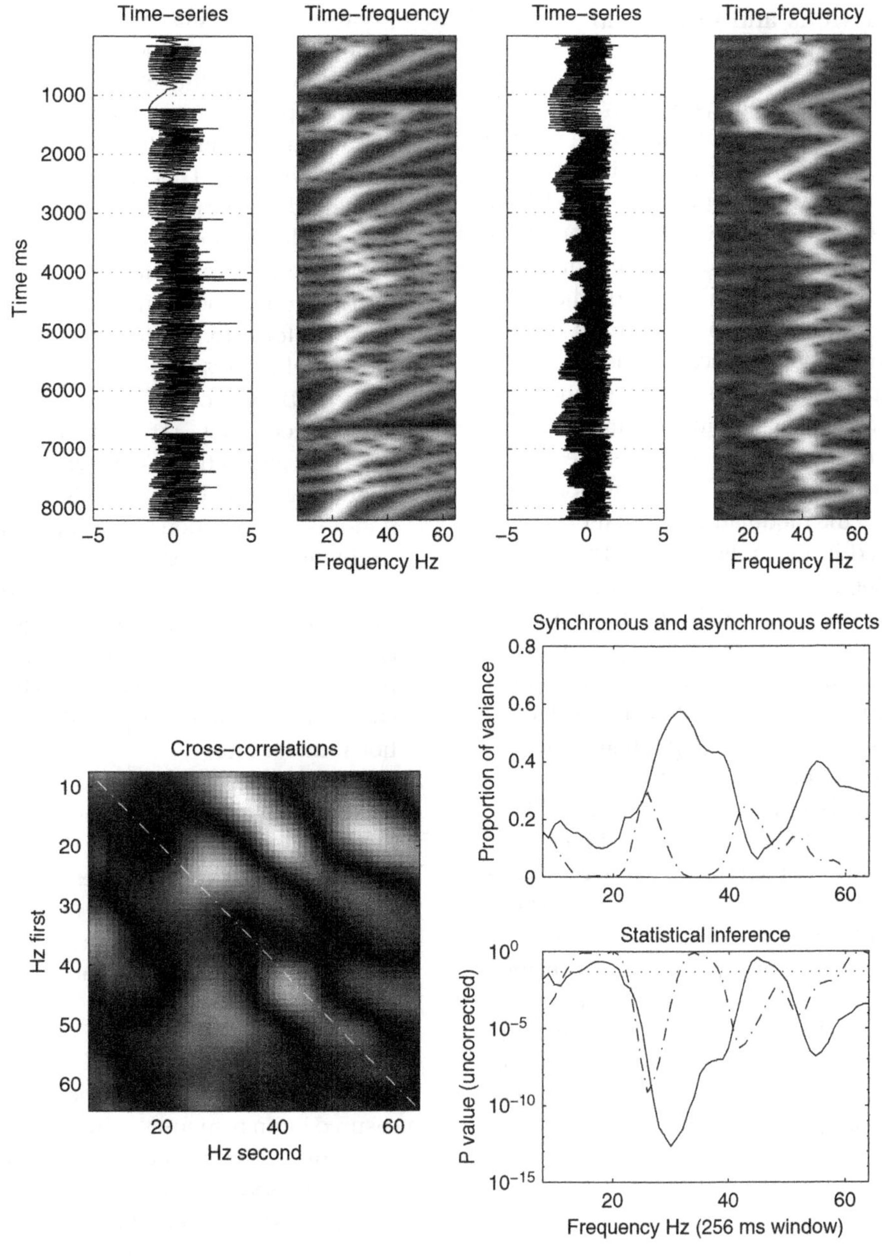

FIGURE 39.3 As for Figure 39.1, but here using the simulated data employing voltage-dependent NMDA-like connections. The coupling here includes some profoundly asynchronous [non-linear] components involving frequencies in the gamma range implicated in the analyses of real (MEG) data shown in Figure 39.1. In particular, note the asymmetrical cross-correlation matrix and the presence of asynchronous and mixed coupling implicit in the *p*-value plots on the lower right.

above). This is a nice example of asynchronous coupling that is underpinned by non-linear modulatory interactions between neuronal populations. The nature of the coupling can be characterized using the time-frequency analysis (identical in every detail) applied to the neuromagnetic data of the previous section. The results for the NMDA simulation are presented in Figure 39.3. The cross-correlation matrix resembles that obtained with the MEG data in Figure 39.1. Both in terms of the variance, and inference, asynchronous coupling supervenes at most frequencies but, as in the real data, mixed coupling is also evident. These results can be taken as a heuristic confirmation of the notion that modulatory, in this case voltage-dependent, interactions are sufficiently non-linear to account for the emergence of asynchronous coupling.

Modulatory interactions and non-linear coupling

In summary, asynchronous coupling is synonymous with non-linear coupling. Non-linear coupling can be framed in terms of the modulation of intrinsic interactions, within a cortical area or neuronal population, by extrinsic input offered by afferents from other parts of the brain. This mechanism predicts that the modulation of fast (e.g. gamma) activity in one cortical area can be predicted by much slower changes in other areas. This form of coupling is very different from coherence or other measures of synchronous coupling and concerns the relationship between the first-order dynamics in one area and the second-order dynamics (spectral density) expressed in another. In terms of the above NMDA simulation, transient depolarization in the modulating population causes a short-lived increased input to the second. These afferents impinge on voltage-sensitive NMDA-like synapses with time constants (in the model) of about 100 ms. These synapses open and slowly close again, remaining open long after an afferent volley. Because of their voltage-sensitive nature, this input will have no effect on the dynamics intrinsic to the second population unless there is already a substantial degree of depolarization. If there is then, through self-excitation and inhibition, the concomitant opening of fast excitatory and inhibitory channels, this will generally increase membrane conductance, decrease the effective membrane time constants and lead to fast oscillatory transients. This is what we observe in the lower panel of Figure 39.2. In relation to the MEG analyses, the implied modulatory mechanisms that may underpin this effect are entirely consistent with the anatomy, laminar specificity and functional role attributed to prefrontal efferents (Rockland and Pandya, 1979; Selemon and Goldman-Rakic, 1988).

CONCLUSION

In this chapter, we have dealt with some interesting and interrelated aspects of effective connectivity, neuronal codes, non-linear coupling, neuronal transients and dynamic correlations (e.g. induced oscillations). The key points can be summarized as follows:

- Starting with the premise that the brain can be represented as an ensemble of connected input-state-output systems (e.g. cellular compartments, cells or populations of cells), there is an equivalent input-output formulation in terms of a Volterra series. This is simply a functional expansion of each system's inputs that produces its outputs (where the outputs of one system are the inputs to another).
- The existence of this expansion suggests that the history of inputs, or neuronal transients, and the Volterra kernels are a complete and sufficient description of brain dynamics. This is the primary motivation for framing dynamics in terms of neuronal transients and using Volterra kernels to model effective connectivity.
- The Volterra formulation provides constraints on the form that neuronal interactions and implicit codes must conform to. There are two limiting cases: when the kernel decays very quickly; and when high-order kernels disappear. The first case corresponds to instantaneous codes (e.g. rate codes) and the second to synchronous interactions (e.g. synchrony codes).
- High-order kernels in the Volterra formulation speak to non-linear interactions and implicitly to asynchronous coupling. Asynchronous coupling implies coupling among the expression of different frequencies.
- Coupling among different frequencies is easy to demonstrate using neuromagnetic measurements of real brain dynamics. This implies that non-linear, asynchronous coupling is a prevalent component of functional integration.
- High-order kernels correspond to modulatory interactions that can be construed as a non-linear effect of inputs that interact with the intrinsic states of the recipient system. This implies that driving connections may be linear and engender synchronous interactions. Conversely, modulatory connections, being non-linear, may be revealed by asynchronous coupling and induce high-order kernels.

In the next chapter, we look at another way of summarizing and quantifying temporal dependencies among measured brain responses, using multivariate autoregression models. These models are formulated in discrete time, as opposed to the continuous time formulation used for generalized convolution models.

REFERENCES

Abeles M, Bergman H, Gat I *et al.* (1995) Cortical activity flips among quasi-stationary states. *Proc Natl Acad Sci USA* **92**: 8616–20

Aertsen A, Erb M, Palm G (1994) Dynamics of functional coupling in the cerebral cortex: an attempt at a model-based interpretation. *Physica D* **75**: 103–28

Aertsen A, Preißl H (1991) Dynamics of activity and connectivity in physiological neuronal networks. In *Non linear dynamics and neuronal networks*, Schuster HG (ed.). VCH publishers Inc., New York, pp 281–302

Bair W, Koch C, Newsome W *et al.* (1994) Relating temporal properties of spike trains from area MT neurons to the behaviour of the monkey. In *Temporal coding in the brain*, Buzsaki G, Llinas R, Singer W *et al.* (eds). Springer Verlag, Berlin, pp 221–50

Bendat JS (1990) *Nonlinear system analysis and identification from random data*. John Wiley and Sons, New York

Breakspear M, Terry JR, Friston KJ (2003). Modulation of excitatory synaptic coupling facilitates synchronization and complex dynamics in a biophysical model of neuronal dynamics. *Network* **14**: 703–32

deCharms RC, Merzenich MM (1996) Primary cortical representation of sounds by the co-ordination of action potential timing. *Nature* **381**: 610–13

de Ruyter van Steveninck RR, Lewen GD, Strong SP *et al.* (1997) Reproducibility and variability in neural spike trains. *Science* **275**: 1085–88

Desmedt JE, Tomberg C (1994) Transient phase-locking of 40 Hz electrical oscillations in prefrontal and parietal human cortex reflects the process of conscious somatic perception. *Neurosci Lett* **168**: 126–29

Eckhorn R, Bauer R, Jordan W *et al.* (1988) Coherent oscillations: a mechanism of feature linking in the visual cortex? Multiple electrode and correlation analysis in the cat. *Biol Cybernet* **60**: 121–30

Engel AK, Konig P, Singer W (1991) Direct physiological evidence for scene segmentation by temporal coding. *Proc Natl Acad Sci USA* **88**: 9136–40

Erb M, Aertsen A (1992) Dynamics of activity in biology-oriented neural network models: stability analysis at low firing rates. In *Information processing in the cortex. Experiments and theory*, Aertsen A, Braitenberg V (eds). Springer-Verlag, Berlin, pp 201–23

Fliess M, Lamnabhi M, Lamnabhi-Lagarrigue F (1983) An algebraic approach to nonlinear functional expansions. *IEEE Transactions on Circuits and Systems* **30**: 554–70

Freeman W, Barrie J (1994) Chaotic oscillations and the genesis of meaning in cerebral cortex. In *Temporal coding in the brain*, Buzsaki G, Llinas R, Singer W *et al.* (eds). Springer Verlag, Berlin, pp 13–38

Fries P, Roelfsema PR, Engel A *et al.* (1997) Synchronisation of oscillatory responses in visual cortex correlates with perception in inter-ocular rivalry. *Proc Natl Acad Sci USA* **94**: 12699–704

Friston KJ (1995a) Functional and effective connectivity in neuroimaging: a synthesis. *Hum Brain Mapp* **2**: 56–78

Friston KJ (1995b) Neuronal transients. *Proc Roy Soc Series B* **261**: 401–05

Friston KJ (1997) Transients metastability and neuronal dynamics. *NeuroImage* **5**: 164–71

Friston KJ (2000) The labile brain I: Neuronal transients and nonlinear coupling. *Philos Trans R Soc (Lond)* **355**: 215–36

Friston KJ, Stephan KM and Frackowiak RSJ (1997) Transient phase-locking and dynamic correlations: are they the same thing? Human Brain Mapping **5**: 48–57

Gray CM, Singer W (1989) Stimulus specific neuronal oscillations in orientation columns of cat visual cortex. *Proc Natl Acad Sci USA* **86**: 1698–1702

Haken H (1983) *Synergistics: an* introduction, 3rd edn. Springer, Berlin

Haken H (2006) Synergetics of brain function. *Int J Psychophysiol* Mar 7; (Epub ahead of print)

Harrison LM, David O, Friston KJ (2005) Stochastic models of neuronal dynamics. *Philos Trans R Soc Lond B Biol Sci* **360**: 1075–91

Jirsa VK, Friedrich R, Haken H (1995) Reconstruction of the spatiotemporal dynamics of a human magnetoencephalogram. *Physica D* **89**: 100–22

Kelso JAS (1995) *Dynamic patterns: the self-organisation of brain and behaviour*. MIT Press, Cambridge, MA

Lachaux JP, Rodriguez E, Martinerie J *et al.* (2000) A quantitative study of gamma-band activity in human intracranial recordings triggered by visual stimuli. *Eur J Neurosci* **12**: 2608–22

Llinas R, Ribary U, Joliot M *et al.* (1994) Content and context in temporal thalamocortical binding. In *Temporal coding in the brain*, Buzsaki G, Llinas R, Singer W *et al.* (eds). Springer Verlag, Berlin, pp 251–72

Lumer ED, Edelman GM, Tononi G (1997) Neural dynamics in a model of the thalamocortical System II. The role of neural synchrony tested through perturbations of spike timing. *Cereb Cortex* **7**: 228–36

Optican L, Richmond BJ (1987) Temporal encoding of two-dimensional patterns by single units in primate inferior cortex. II Information theoretic analysis. *J Neurophysiol* **57**: 132–46

Phillips WA, Singer W (1997) In search of common foundations for cortical computation. *Behav Brain Sci* **20**: 57–83

Riehle A, Grun S, Diesmann M *et al.* (1997) Spike synchronisation and rate modulation differentially involved in motor cortical function. *Science* **278**: 1950–53

Rockland KS, Pandya DN (1979) Laminar origins and terminations of cortical connections of the occipital lobe in the rhesus monkey. *Brain Res* **179**: 3–20

Schiff SJ, So P, Chang T *et al.* (1996) Detecting dynamical interdependence and generalised synchrony though mutual prediction in a neuronal ensemble. *Phys Rev E* **54**: 6708–24

Selemon LD, Goldman-Rakic PS (1988) Common cortical and subcortical targets of the dorsolateral prefrontal and posterior parietal cortices in the rhesus monkey: evidence for a distributed neural network subserving spatially guided behavior. *J Neurosci* **8**: 4049–68

Shadlen MN, Newsome WT (1995) Noise, neural codes and cortical organisation. *Curr Opin Neurobiol* **4**: 569–79

Singer W (1994) Time as coding space in neocortical processing: a hypothesis. In *Temporal coding in the brain*, Buzsaki G, Llinas R, Singer W *et al.* (eds). Springer Verlag, Berlin, pp 51–80

Sporns O, Gally, JA, Reeke GN *et al.* (1989) Reentrant signalling among simulated neuronal groups leads to coherence in their oscillatory activity *Proc Natl Acad Sci USA* **86**: 7265–69

Tallon-Baudry C, Kreiter A, Bertrand O (1999) Sustained and transient oscillatory responses in the gamma and beta bands in a visual short-term memory task in humans. *Vis Neurosci.* **16**: 449–59

Tass PA (2005) Estimation of the transmission time of stimulus-locked responses: modelling and stochastic phase resetting analysis. *Philos Trans R Soc Lond B Biol Sci* **360**: 995–99

Tovee MJ, Rolls ET, Treves A *et al.* (1993) Information encoding and the response of single neurons in the primate temporal visual cortex. *J Neurophysiology* **70**: 640–54

Treue S, Maunsell HR (1996) Attentional modulation of visual motion processing in cortical areas MT and MST. *Nature* **382**: 539–41

Tsuda I (2001) Toward an interpretation of dynamic neural activity in terms of chaotic dynamical systems. *Behav Brain Sci* **24**: 793–810

Vaadia E, Haalman I, Abeles M *et al.* (1995) Dynamics of neuronal interactions in monkey cortex in relation to behavioural events. *Nature* **373**: 515–18

von der Malsburg C (1985) Nervous structures with dynamical links. *Ber Bunsenges Phys Chem* **89**: 703–10

Wray J, Green GGR (1994) Calculation of the Volterra kernels of non-linear dynamic systems using an artificial neuronal network. *Biol Cybernet* **71**: 187–95

C H A P T E R

40

Multivariate autoregressive models

W. Penny and L. Harrison

INTRODUCTION

Functional neuroimaging has been used to corroborate functional specialization as a principle of organization in the human brain. However, disparate regions of the brain do not operate in isolation and, more recently, neuroimaging has been used to characterize the network properties of the brain under specific cognitive states (Buchel and Friston, 1997a; Buchel and Friston, 2000). These studies address a complementary principle of organization, functional integration.

Functional magnetic resonance imaging (fMRI) provides a unique opportunity to observe simultaneous recordings of activity throughout the brain evoked by cognitive and sensorimotor challenges. Each voxel within the brain is represented by a time-series of neurophysiological activity that underlies the measured blood oxygen-level-dependent (BOLD) response. Given these multivariate, voxel-based time-series, can we infer large-scale network behaviour among functionally specialized regions?

A number of methods have been proposed to answer this question including regression models (McIntosh *et al.*, 1994; Friston *et al.*, 1993, 1995, 1997), convolution models (Friston and Buchel, 2000; Friston, 2001) and state-space models (Buchel and Friston, 1998). Regression techniques, underlying for example the analysis of psychophysiological interactions (PPIs), are useful because they are easy to fit and can test for the modulatory interactions of interest (Friston *et al.*, 1997). However, this is at the expense of excluding temporal information, i.e. the history of an input or physiological variable. This is important as interactions within the brain, whether over short or long distances, take time and are not instantaneous. Structural equation modelling (SEM), as used by the neuroimaging community (McIntosh *et al.*, 1994; Buchel and Friston, 1997b) has similar problems.[1] Convolution models, such as the Volterra approach, model temporal effects in terms of an idealized response characterized by the kernels of the model (Friston, 2000). A criticism of the Volterra approach is that it treats the system as a black box, meaning that it has no model of the internal mechanisms that may generate data. State-space models account for correlations within the data by invoking state variables whose dynamics generate data. Recursive algorithms, such as the Kalman filter, may be used to estimate these states through time, given the data (Buchel and Friston, 1998).

This chapter describes an approach based on multivariate autoregressive (MAR) models. These are linear multivariate time-series models which have a long history of application in econometrics. The MAR model characterizes interregional dependencies within data, specifically in terms of the historical influence one variable has on another. This is distinct from regression techniques that quantify instantaneous correlations. We use MAR models to make inferences about functional integration from fMRI data.

The chapter is divided into three sections. First, we describe the theory of MAR models, parameter estimation, model order selection and statistical inference. We have used a Bayesian technique for model order selection and parameter estimation, which is introduced in Chapter 24 and is described fully in Penny and Roberts (2002). Secondly, we model neurophysiological data taken from an fMRI experiment addressing attentional modulation of cortical connectivity during a visual motion task (Buchel and Friston, 1997b). The modulatory effect of one region upon the responses to other regions is a second order interaction which is precluded

[1] There exist versions of SEM that do model dynamic information, see Cudeck (2002) for details of dynamic factor analysis.

Statistical Parametric Mapping, by Karl Friston et al.
ISBN–13: 978-0-12-372560-8 ISBN–10: 0-12-372560-7

in linear models. To circumvent this we have introduced bilinear terms (Friston *et al.*, 1997). Thirdly, we discuss the advantages and disadvantages of MAR models, their use in spectral estimation and possible future developments.

THEORY

Multivariate autoregressive models

Given a univariate time-series, its consecutive measurements contain information about the process that generated it. An attempt at describing this underlying order can be achieved by modelling the current value of the variable as a weighted linear sum of its previous values. This is an autoregressive (AR) process and is a very simple, yet effective, approach to time-series characterization (Chatfield, 1996). The order of the model is the number of preceding observations used, and the weights characterize the time-series.

Multivariate autoregressive models extend this approach to multiple time-series so that the vector of current values of all variables is modelled as a linear sum of previous activities. Consider d time-series generated from d variables within a system, such as a functional network in the brain, and where m is the order of the model. An MAR(m) model predicts the next value in a d-dimensional time-series, y_n as a linear combination of the m previous vector values:

$$y_n = \sum_{i=1}^{m} y_{n-i} A(i) + e_n \qquad \textbf{40.1}$$

where $y_n = [y_n(1), y_n(2), \ldots, y_n(d)]$ is the nth sample of a d-dimensional time-series, each $A(i)$ is a d-by-d matrix of coefficients (weights) and $e_n = [e_n(1), e_n(2), \ldots, e_n(d)]$ is additive Gaussian noise with zero mean and covariance R. We have assumed that the data mean has been subtracted from the time-series.

The model can be written in the standard form of a multivariate linear regression model as follows:

$$y_n = x_n W + e_n \qquad \textbf{40.2}$$

where $x_n = [y_{n-1}, y_{n-2}, \ldots, y_{n-m}]$ are the m previous multivariate time-series samples and W is a $(m \times d)$-by-d matrix of MAR coefficients (weights). There are therefore a total of $k = m \times d \times d$ MAR coefficients.

If the nth rows of Y, X and E are y_n, x_n and e_n respectively and there are $n = 1..N$ samples then we can write:

$$Y = XW + E \qquad \textbf{40.3}$$

where Y is an $(N - m)$-by-d matrix, X is an $(N - m)$-by-$(m \times d)$ matrix and E is an $(N - m)$-by-d matrix. The number of rows $N - m$ (rather than N) arises as samples at time points before m do not have sufficient preceding samples to allow prediction.

MAR models are fully connected in that each region is, by default, assumed connected to all others. However, by fitting the model to data and testing to see which connections are signficantly non-zero, one can infer a sub-network that mediates the observed dynamics. This can be implemented using Bayesian inference as described below.

These sub-networks are related to the concept of 'Granger causality' (Granger, 1969), which is defined operationally as follows. Activity in region X 'Granger' causes activity in region Y if any of the connections from X to Y, over all time lags, are non-zero. These causality relationships can be summarized by directed graphs as described in Eichler (2005). An example will be presented later on in the chapter.

Non-linear autoregressive models

Given a network model of the brain, we can think of two fundamentally different types of coupling: linear and non-linear. The model discussed so far is linear. Linear systems are described by the principle of superposition, which is that inputs have additive effects on the response that are independent of each other. Non-linear systems are characterized by inputs which interact to produce a response.

In Buchel and Friston (1997b), non-linear interactions were modelled using 'bilinear terms'. This is the approach adopted in this chapter. Specifically, to model a hypothesized interaction between variables $y_n(j)$ and $y_n(k)$ one can form the new variable:

$$I_n(j, k) = y_n(j) y_n(k) \qquad \textbf{40.4}$$

This is a 'bilinear variable'. This is orthogonalized with respect to the original time-series and placed in an augmented MAR model with connectivity matrices $\tilde{A}(i)$.

$$[y_n, I_n(j, k)] = \sum_{i=1}^{m} [y_{n-i}, I_{n-i}(j, k)] \tilde{A}(i) + e_n \qquad \textbf{40.5}$$

The relevant entries in $\tilde{A}(i)$ then reflect modulatory influences, e.g. a change of the connection strength between $y(j)$ and other time-series due to the influence of $y(k)$.

It should be noted that each bilinear variable introduces only one of many possible sources of non-linear behaviour into the model. The example above specifically models non-linear interactions between $y_n(j)$ and $y_n(k)$, however, other bilinear terms could involve, for instance, the time-series $y_n(j)$ and inputs $u(t)$. The inclusion of these terms is guided by the hypothesis of interest, e.g. does 'time' change the connectivity between earlier and later stages of processing in the dorsal visual pathway? Here $u(t)$ would model time.

Maximum likelihood estimation

Reformulating MAR models as standard multivariate linear regression models allows us to retain contact with the large body of statistical literature devoted to this subject (e.g. see Box and Tiao, 1992: 423). The maximum likelihood (ML) solution (e.g. see Weisberg, 1980) for the MAR coefficients is:

$$\hat{W} = (X^T X)^{-1} X^T Y \tag{40.6}$$

The maximum likelihood noise covariance, S_{ML}, can be estimated as:

$$S_{ML} = \frac{1}{N-k}(Y - X\hat{W})^T(Y - X\hat{W}) \tag{40.7}$$

where $k = m \times d \times d$. We define $\hat{w} = vec(\hat{W})$ where vec denotes the columns of $\hat{W}$ being stacked on top of each other (for more on the vec notation, see Muirhead, 1982). To recover the matrix $\hat{W}$ we simply 'un-stack' the columns from the vector $\hat{w}$.

The ML parameter covariance matrix for $\hat{w}$ is given by (Magnus and Neudecker, 1997: 321):

$$\hat{\Sigma} = S_{ML} \otimes (X^T X)^{-1} \tag{40.8}$$

where $\otimes$ denotes the Kronecker product (e.g. see Box and Tiao, 1992: 477) The optimal value of m can be chosen using a model order selection criterion such as the minimum description length (MDL) (e.g. see Neumaier and Schneider, 2000).

Bayesian estimation

It is also possible to estimate the MAR parameters and select the optimal model within a Bayesian framework (Penny and Roberts, 2002). This has been shown to give better model order selection and is the approach used in this chapter. The maximum-likelihood solution is used to initialize the Bayesian scheme.

In what follows $\mathsf{N}(m, Q^{-1})$ is a multivariate Gaussian with mean m and precision (inverse covariance) Q. Also, $\mathsf{Ga}(b, c)$ is the gamma distribution with parameters b and c defined in Chapter 24. The gamma density has mean bc and variance b^2c. Finally, $\mathsf{Wi}(s, B)$ denotes a Wishart density (Box and Tiao, 1992). The Bayesian model uses the following prior distributions:

$$\begin{aligned} p(W|m) &= \mathsf{N}(0, \alpha^{-1}I) \\ p(\alpha|m) &= \mathsf{Ga}(b, c) \\ p(\Lambda|m) &= |\Lambda|^{-(d+1)/2} \end{aligned} \tag{40.9}$$

where m is the order of the model, α is the precision of the Gaussian prior distribution from which weights are drawn and Λ is the noise precision matrix (inverse of R). In Penny and Roberts (2002), it is shown that the corresponding posterior distributions are given by:

$$\begin{aligned} p(W|Y, m) &= \mathsf{N}(\hat{W}_B, \hat{\Sigma}_B) \\ p(\alpha|Y, m) &= \mathsf{Ga}(\hat{b}, \hat{c}) \\ p(\Lambda|Y, m) &= \mathsf{Wi}(s, B) \end{aligned} \tag{40.10}$$

The parameters of the posteriors are updated in an iterative optimization scheme described in Appendix 40.1. Iteration stops when the 'Bayesian evidence' for model order m, $p(Y|m)$, is maximized. A formula for computing this is also provided in Appendix 40.1. Importantly, the evidence is also used as a model order selection criterion, i.e., to select the optimal value of m. This is discussed at length in Chapters 24 and 35.

Bayesian inference

The Bayesian estimation procedures outlined above result in a posterior distribution for the MAR coefficients $P(W|Y, m)$. Bayesian inference can then take place using confidence intervals based on this posterior (e.g. see Box and Tiao, 1992). The posterior allows us to make inferences about the strength of a connection between two regions. Because this connectivity can be expressed over a number of time lags, our inference is concerned with the vector of connection strengths, a, over all time lags. To make contact with classical (non-Bayesian) inference, we say that a connection is 'significantly non-zero' or simply 'significant' at level α if the zero vector lies outside the $1-\alpha$ confidence region for a. This is shown schematically in Figure 40.1.

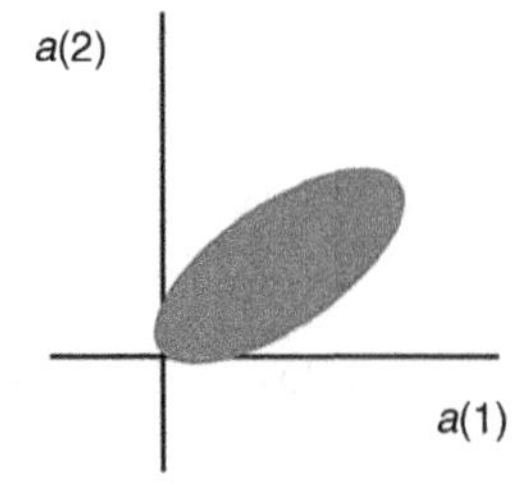

FIGURE 40.1 For a MAR(2) model the vector of connection strengths, a, between two regions consists of two values, $a(1)$ and $a(2)$. The probability distribution over a can be computed from the posterior distribution of MAR coefficients as shown in Appendix 40.1 and is given by $p(a) = \mathsf{N}(\mu, V)$. Connectivity between two regions is then deemed significant at level α if the zero-vector lies on the $1-\alpha$ confidence region. The figure shows an example $1-\alpha$ confidence region for a MAR(2) model.

APPLICATION

The responsiveness of dorsal visual brain regions in neuroimaging studies suggests attention is associated with changes in connectivity (Assad and Maunsell, 1995; O'Craven and Savoy, 1995). In this chapter, we use data from an fMRI study investigating attentional modulation of connectivity within the dorsal visual pathways (Buchel and Friston, 1997b). This provides a testbed for assessing how MAR models estimate changes in connectivity.

In brief, the experiment was performed on a 2T MRI scanner. The visual stimulus involved random dots moving radially outwards at a fixed rate. Subjects were trained beforehand to detect changes in velocity of radial motion. Attentional set was manipulated by asking the subject to attend to changes in velocity or to just observe the motion. Both of these states were separated by periods of 'fixation' where the screen was dark and only a fixation dot was visible. Each block ended with a 'stationary' condition in which a static image of the previously moving dots was shown. Unknown to the subjects, the radial velocity remained constant throughout the experiment such that the only experimental manipulation was attentional set.

Categorical comparisons using general linear model (GLM) analyses (see e.g. Chapter 8) were used to identify changes in brain activity dependent on attentional set. This revealed activations throughout right and left hemispheres in the primary visual cortex V1/2 complex, visual motion region V5 and regions involved in the attentional network including posterior parietal cortex (PPC) and in the right prefrontal cortex (PFC). Regions of interest (ROI) were defined with a diameter of 8 mm centred around the most significant voxel and a representative time-series was defined by the first eigenvariate of the region. For details of the experimental design and acquisition see Buchel and Friston (1997b). We analyse data from three subjects. Time-series from subject 1 are shown in Plate 57 (see colour plate section).

Inspecting the four time-series reveals a number of characteristics. The time-series from the V1/2 complex shows a dependence on the presentation of the moving image with a small difference between attention and non-attention. However, in the higher brain areas of PPC and PFC, attentional set is the dominant influence, with a marked increase in activity during periods of attention. The relative influence each region has on others is not obvious from visual inspection but, as we shall see, can be revealed from an MAR analysis.

Three models were tested using the regions and interaction terms shown below:

- Model 1: V1/V2, V5, PPC and PFC
- Model 2: V1/V2, V5 and $I_{v1,ppc}$
- Model 3: V5, PPC and $I_{v5,pfc}$

where $I_{v1,ppc}$ denotes an interaction between V1/V2 and PPC and $I_{v5,pfc}$ an interaction between V5 and PFC. These variables were created as described in the earlier section on non-linear autoregressive models. The interaction terms can be thought as 'virtual' nodes in a network.

For each type of model, we computed the model evidence as a function of model order m. The results in Figure 40.2 show that the optimal order for all three models was $m = 4$ (subject 1). A model order of $m = 4$ was then used in the results reported below.

Model 1 was applied to right hemisphere data only, to identify the functional network connecting key regions in the visual and attentional systems. Figure 40.3 shows connections in this model which, across all time lags, are significantly non-zero for subject 1. Over the three subjects, all V1/V2 to V5 connections ($\alpha < 0.0004$) and all PFC to PPC ($\alpha < 0.02$) connections were significant. We can therefore infer that activity in V1/V2 Granger causes activity in V5, and PFC Granger causes PPC. Also, the V5 to PPC connection was significant ($\alpha < 0.0009$) in two out of three subjects.

The second model was applied both to left and right hemisphere data. Figure 40.4 shows significantly non-zero connections for left hemisphere data from subject 1. This shows that activity in PPC changes the connectivity between V1/V2 and V5. The same was true for the other two subjects. For the right hemisphere data, however, only subject 1 showed an effect ($\alpha < 0.03$).

The third model was applied to right hemisphere data. Figure 40.5 shows significantly non-zero connections for data from subject 1. This shows that activity in PFC changes how PPC responds to V5. Subject 2 also showed this effect ($\alpha < 0.03$) but subject 3 did not.

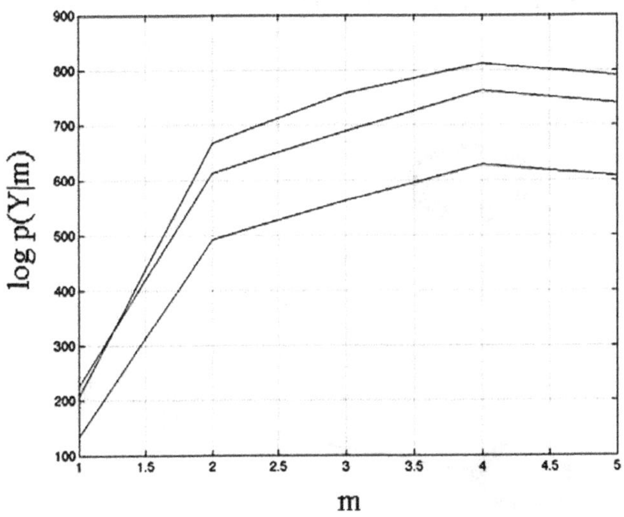

FIGURE 40.2 Plots of log-evidence, computed using the expression in Appendix 40.1, for each of the three MAR models for subject 1. The optimal order is $m = 4$ in each case.

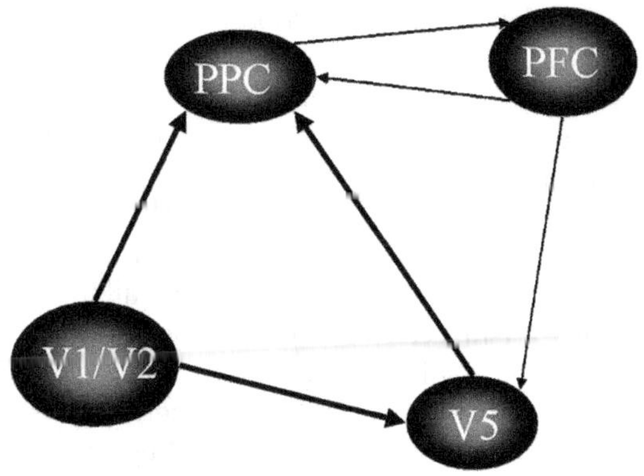

FIGURE 40.3 Inferred connectivity for model 1. Arrows indicate Granger causal relations. Thin arrows indicate $0.001 \le \alpha \le 0.05$ and thick, $\alpha \le 0.001$.

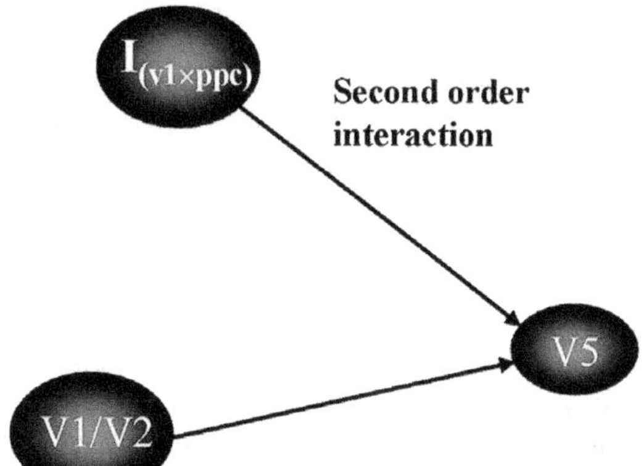

FIGURE 40.4 Inferred connectivity for model 2. Arrows show Granger causal relations ($0.001 \le \alpha \le 0.05$). The model supports the notion that PPC modulates the V1/V2 to V5 connection.

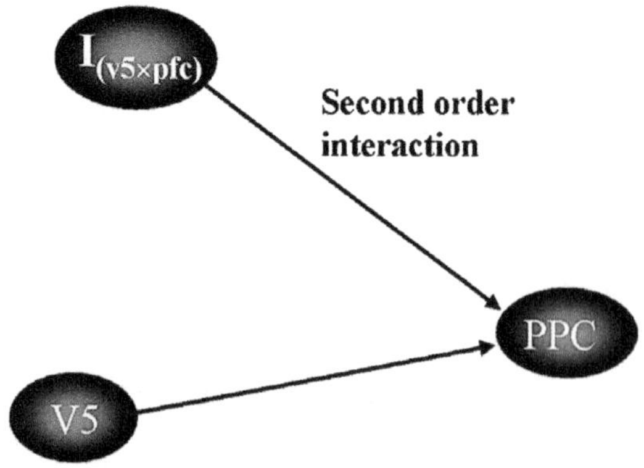

FIGURE 40.5 Inferred connectivity for model 3. Arrows show Granger causal relations ($0.001 \le \alpha \le 0.05$). The model supports the notion that PFC modulates the V5 to PPC connection.

DISCUSSION

We have proposed the use of MAR models for making inferences about functional integration using fMRI time-series. One motivation for this is that the previously dominant model used for making such inferences in the existing fMRI/PET (positron emission tomography) literature, namely structural equation modelling, as used in McIntosh *et al.* (1994) and Buchel and Friston (1997b), is not a time-series model. Indeed, inferences are based solely on the instantaneous correlations between regions – if the series were randomly permuted over time, SEM would give the same results. Thus SEM throws away temporal information.

Further, MAR models may contain loops and self-connections, yet parameter estimation can proceed in a purely linear framework, i.e. there is an analytic solution that can be found via linear algebra. In contradistinction, SEM models with loops require non-linear optimization. The reason for this is that MAR models do not contain instantaneous connections. The between-region connectivity arises from connections between regions at different time lags. Due to temporal persistence in the activity of each region, this captures much the same effect, but in a computationally simpler manner.

In this chapter, we applied Bayesian MAR models to fMRI data. Bayesian inferences about connections were then made on the basis of the estimated posterior distribution. This allows for the identification of a sub-network of connections that mediate the observed dynamics. These connections describe causal relations, in the sense of Granger (Granger, 1969).

This is in the spirit of how general linear models are used for characterizing functional specialization; all conceivable factors are placed in one large model and then different hypotheses are tested using *t*- or *F*-contrasts (Frackowiak *et al.*, 1997). We note that this approach is fundamentally different from the philosophy underlying SEM. In SEM, only a few connections are modelled. These are chosen on the basis of prior anatomical or functional knowledge and are interpreted as embodying causal relations. Thus, with SEM, causality is ascribed *a priori* (Pearl, 1998), but with MAR, causality can be inferred from data.

MAR models can also be used for spectral estimation. In particular, they enable parsimonious estimation of coherences (correlation at particular frequencies), partial coherences (the coherence between two time-series after the effects of others have been taken into account), phase relationships (Marple, 1987; Cassidy and Brown, 2000) and directed transfer functions (Kaminski *et al.*, 1997). MAR models have been used in this way to investigate functional integration from electroencephalography (EEG) and Electrocorticogram

(ECOG) recordings (Bressler *et al.*, 1999). This provides a link with a recent analysis of fMRI data (Muller *et al.*, 2001) which looks for sets of voxels that are highly coherent. MAR models provide a parametric way of estimating this coherence, although in this chapter we have reported the results in the time domain.

A further aspect of MAR models is that they capture only linear relations between regions. Following Buchel and Friston (1997b), we have extended their capabilities by introducing bilinear terms. This allows one to infer that activity in one region modulates connectivity between two other regions. Such inferences are beyond the current capabilities of dynamic causal models for fMRI (see Chapter 41).

It is also possible to include further higher-order terms, for instance, second-order interactions across different lags. Frequency domain characterization of the resulting models would then allow us to report bi-spectra (Priestley, 1988). These describe the correlations between different frequencies which may be important for the study of functional integration (Friston, 2000, see also Chapter 39).

A key aspect of our approach has been the use of a mature Bayesian estimation framework (Penny and Roberts, 2002). This has allowed us to select the optimal MAR model order. One promising direction for extending the model is to replace Gaussian priors with sparse priors. This would effectively remove most connections, allowing the model to be applied to a very large number of regions. This approach has been investigated in a non-Bayesian framework using penalized regression and pruning based on false discovery rates (Valdes-Sosa *et al.*, 2005).

APPENDIX 40.1

Bayesian estimation

Following the algorithm developed in Penny and Roberts, (2002), the parameters of the posterior distributions are updated iteratively as follows:

$$\begin{aligned}
\Lambda_D &= \hat{\Lambda} \otimes (X^T X) \qquad \textbf{40.11}\\
\hat{\Sigma}_B &= (\Lambda_D + \hat{\alpha} I)^{-1}\\
\hat{W}_B &= \hat{\Sigma}_B \Lambda_D \hat{W}\\
\frac{1}{\hat{b}} &= \frac{1}{2}\hat{W}_B^T \hat{W}_B + \frac{1}{2}\mathrm{Tr}(\hat{\Sigma}_B) + \frac{1}{b}\\
\hat{c} &= \frac{k}{2} + c\\
\hat{\alpha} &= \hat{b}\hat{c}\\
s &= N\\
B &= \frac{1}{2}(Y - X\hat{W}_B)^T (Y - X\hat{W}_B)\\
&\quad + \sum_n (I \otimes x_n)\hat{\Sigma}_B (I \otimes x_n)^T\\
\hat{\Lambda} &= sB^{-1}
\end{aligned}$$

The updates are initialized using the maximum-likelihood solution. Iteration terminates when the Bayesian log-evidence increases by less than 0.01 per cent. The log-evidence is computed as follows:

$$\begin{aligned}
\log p(Y|m) = \frac{N}{2}\log|B| &- KL_N(p(W|m), p(W|Y, m)) \qquad \textbf{40.12}\\
&- KL_{Ga}(p(\alpha|m), p(\alpha|Y, m)) + \log \Gamma_d(N/2)
\end{aligned}$$

where KL_N and KL_{Ga} denote the Kullback-Liebler (KL) divergences for normal and gamma densities defined in Chapter 24. Essentially, the first term in the above equation is an accuracy term and the KL terms act as a penalty for model complexity (see Chapters 24 and 35 for more on model comparison).

Testing the significance of connections

The connectivity between two regions can be expressed over a number of time lags. Therefore, to see if the connectivity is significantly non-zero, we make an inference about the vector of coefficients a, where each element of that vector is the value of a MAR coefficient at a different time lag. First, we specify $(k \times k)$ $(k = m \times d \times d)$ sparse matrix C such that

$$a = C^T w \qquad \textbf{40.13}$$

returns the estimated weights for connections between the two regions of interest. For an MAR(m) model, this vector has m entries, one for each time lag. The probability distribution is given by $p(a) = \mathsf{N}(\mu, V)$ and is shown schematically in Figure 40.1. The mean and covariance are given by:

$$\begin{aligned}
\mu &= C^T \hat{w} \qquad \textbf{40.14}\\
V &= C^T \hat{\Sigma}_B C
\end{aligned}$$

where $\hat{w} = vec(\hat{W}_B)$ and $\hat{\Sigma}_B$ are the Bayesian estimates of the parameters of the posterior distribution of regression coefficients from the previous section. In fact, $p(a)$ is just that part of $p(w)$ that we are interested in.

The probability α that the zero vector lies on the $1 - \alpha$ confidence region for this distribution is then computed as follows. We first note that this probability is the same as the probability that the vector m lies on the edge of the $1 - \alpha$ region for the distribution $\mathsf{N}(0, V)$. This latter probabilitiy can be computed by forming the test statistic:

$$d = \mu^T V^{-1} \mu \qquad 40.15$$

which will be the sum of $r = rank(V)$ independent, squared Gaussian variables. As such it has a χ^2 distribution:

$$p(d) = \chi^2(r) \qquad 40.16$$

This results in the same test for multivariate effects in general linear models described in Chapter 25. In the present context, if a are the autoregressive coefficients from region X to region Y, and the above test finds them to be significantly non-zero, then we can conclude that X Granger causes Y (Eichler, 2005).

REFERENCES

Assad JA, Maunsell RM (1995) Neuronal correlates of inferred motion in primate posterior parietal cortex. *Nature* **373**: 518–21

Box GEP, Tiao GC (1992) *Bayesian inference in statistical analysis*. John Wiley, New York

Bressler SL, Ding M, Yang W (1999) Investigation of cooperative cortical dynamics by multivariate autoregressive modeling of event-related local field potentials. *Neurocomputing* **26–27**: 625–31

Buchel C, Friston KJ (1997a) Characterizing functional integration. In *Human Brain Function*, Frackowiak RSJ, Friston KJ, Frith CD, *et al.* (eds). Academic Press, London, pp 127–40

Buchel C, Friston KJ (1997b) Modulation of connectivity in visual pathways by attention: cortical interactions evaluated with structural equation modelling and fMRI. *Cereb Cortex* **7**: 768–78

Buchel C, Friston KJ (1998) Dynamic changes in effective connectivity characterized by variable parameter regression and Kalman filtering. *Hum Brain Mapp* **6**: 403–08

Buchel C, Friston KJ (2000) Assessing interactions among neuronal systems using functional neuroimaging. *Neural Netw* **13**: 871–82

Cassidy MJ, Brown P (2002) Stationary and non-stationary autoregressive models for electrophysiological signal analysis and functional coupling studies. *IEEE Trans Biomed Eng* **49**: 1142–52

Chatfield C (1996) *The analysis of time series*. Chapman and Hall, London

Cudeck R (2002) *Structural Equation Modeling: Present and Future*. Scientific Software International, Lincolnwood

Eichler M (2005) Evaluating effective connectivity. *Phil Trans R Soc B* **360**: 953–67

Frackowiak RSJ, Friston KJ, Frith CD *et al.* (1997) *Human Brain Function*. Academic Press

Friston KJ (2000) The labile brain. I. Neuronal transients and nonlinear coupling. *Phil Trans R Soc London B* **355**: 215–36

Friston KJ (2001) Brain function, nonlinear coupling, and neuronal transients. *Neuroscientist* **7**: 406–18

Friston KJ, Buchel C (2000) Attentional modulation of effective connectivity from V2 to V5/MT in humans. *Proc Natl Acad Sci USA* **97**: 7591–96

Friston KJ, Buchel C, Fink GR *et al.* (1997) Psychophysiological and modulatory interactions in neuroimaging. *NeuroImage* **6**: 218–29

Friston KJ, Frith CD, Frackowiak RSJ (1993) Time-dependent changes in effective connectivity measured with PET. *Hum Brain Mapp* **1**: 69–79

Friston KJ, Ungerleider LG, Jezzard P (1995) Charcterizing modulatory interactions between V1 and V2 in human cortex: a new treatment of functional MRI data. *Hum Brain Mapp* **2**: 211–24

Granger C (1969) Investigating causal relations by econometric models and cross-spectral methods. *Econometrika* **37**: 424–38

Kaminski M, Blinowska K, Szelenberger W (1997) Topographic analysis of coherence and propagation of EEG activity during sleep and wakefulness. *Electroencephalogr Clin Neurophysiol* **102**: 216–27

Magnus JR, Neudecker H (1997) *Matrix differential calculus with applications in statistics and econometrics*. John Wiley

Marple SL (1987) *Digital spectral analysis with applications*. Prentice-Hall

McIntosh AR, Grady CL, Underleider LG *et al.* (1994) Network analysis of cortical visual pathways mapped with PET. *J Neurosci* **14**: 655–66

Muirhead RJ (1982) *Aspects of multivariate statistical theory*. John Wiley

Muller K, Lohmann G, Blosch V *et al.* (2001) On multivariate spectral analysis of fMRI time series. *NeuroImage*, **14**: 347–56

Neumaier A, Schneider T (2000) Estimation of parameters and eigenmodes of multivariate autoregressive models. *ACM Trans Math Softw* **27**: 27–57

O'Craven KM, Savoy RL (1995) Voluntary attention can modulate fmri activity in human MT/MST. *Neuron* **18**: 591–8

Pearl J (1998) Graphs, causality, and structural equation models *Sociol Meth Res* **27**: 226–84

Penny WD, Roberts SJ (2002) Bayesian multivariate autoregressive models with structured priors. *IEE Proc Vis Image Signal Process* **149**: 33–41

Priestley MB (1988) *Nonlinear and non-stationary time series analysis*. Harcourt Brace Jovanovich

Valdes-Sosa PA, Sanchez-Bornot J, Lage-Castellanos A *et al.* (2005) Estimating brain functional connectivity with sparse multivariate autoregression. *Phil Trans R Soc B* **360**: 969–81

Weisberg S (1980) *Applied linear regression*. John Wiley

CHAPTER

41

Dynamic Causal Models for fMRI

K. Friston

INTRODUCTION

In this chapter, we apply the system identification techniques described in Chapter 34 to the dynamic models of effective connectivity introduced in Chapter 38. In this chapter, we consider simple bilinear models for haemodynamic time-series (i.e. functional magnetic resonance imaging, fMRI). In the next chapter, we describe more elaborate models for electrical and magnetic responses as measured with electroencephalography and magnetoencephalography (EEG-MEG). By using a bilinear approximation, to any system's equations of motion, the parameters of the implicit causal model reduce to three sets. These comprise parameters that: mediate the influence of extrinsic inputs on the states; mediate regional coupling among the states; and [bilinear] parameters that allow the inputs to modulate that coupling.

We use this bilinear model for the analysis of effective connectivity using experimentally designed inputs and fMRI responses. In this context, the coupling parameters determine effective connectivity and the bilinear parameters reflect the changes in connectivity induced by inputs. The ensuing framework allows one to characterize fMRI experiments, conceptually, as an experimental manipulation of integration among brain regions (by contextual or trial-free inputs, like time or attentional set) that is disclosed using evoked responses (to perturbations or trial-bound inputs like stimuli). As with previous analyses of effective connectivity, the focus is on experimentally induced changes in coupling (cf. psychophysiological interactions). However, unlike previous approaches in neuroimaging, the causal model ascribes responses to designed deterministic inputs, as opposed to treating inputs as unknown and stochastic. To date, dynamic causal modelling (DCM) has been applied to a wide range of issues; ranging from category-effects (Mechelli *et al.*, 2003) through to affective prosody (Ethofer *et al.*, 2006) and rhyming (Bitan *et al.*, 2005).

Background

This chapter is about modelling interactions among neuronal populations, at a cortical level, using neuroimaging (haemodynamic) time-series. It presents the motivation and procedures for dynamic causal modelling of evoked brain responses. The aim of this modelling is to estimate, and make inferences about, the coupling among brain areas and how that coupling is influenced by changes in experimental context (e.g. time or cognitive set). Dynamic causal modelling represents a fundamental departure from existing approaches to effective connectivity because it employs a more plausible generative model of measured brain responses that embraces their non-linear and dynamic nature.

The basic idea is to construct a reasonably realistic neuronal model of interacting cortical regions or nodes. This model is then supplemented with a forward model of how neuronal or synaptic activity is transformed into a measured response. This enables the parameters of the neuronal model (i.e. effective connectivity) to be estimated from observed data. These supplementary models may be forward models of electromagnetic measurements or haemodynamic models of fMRI measurements. In this chapter, we will focus on fMRI. Responses are evoked by known deterministic inputs that embody designed changes in stimulation or context. This is accomplished by using a dynamic input-state-output model with multiple inputs and outputs. The inputs correspond to conventional stimulus functions that encode experimental manipulations. The state variables cover both the neuronal activities and other neurophysiological or biophysical variables needed to form the outputs. The outputs are measured electromagnetic or haemodynamic responses over the brain regions considered.

Intuitively, this scheme regards an experiment as a designed perturbation of neuronal dynamics that are distributed throughout a system of coupled anatomical nodes to change region-specific neuronal activity. These

Statistical Parametric Mapping, by Karl Friston et al.
ISBN–13: 978-0-12-372560-8 ISBN–10: 0-12-372560-7

changes engender, through a measurement-specific forward model, responses that are used to identify the architecture and time-constants of the system at the neuronal level. This represents a departure from conventional approaches (e.g. structural equation modelling and autoregression models, McIntosh and Gonzalez-Lima, 1994; Büchel and Friston 1997), in which one assumes the observed responses are driven by endogenous or intrinsic noise (i.e. innovations). In contrast, dynamic causal models assume the responses are driven by designed changes in inputs. An important aspect of dynamic causal models, for neuroimaging, pertains to how the experimental inputs enter the model and cause neuronal responses. We have established in previous chapters that experimental variables can elicit responses in one of two ways. First, they can elicit responses through direct influences on specific anatomical nodes. This would be appropriate, for example, in modelling sensory evoked responses in early visual cortices. The second class of input exerts its effect vicariously, through a modulation of the coupling among nodes. These sorts of experimental variables would normally be more enduring, for example, attention to a particular attribute or the maintenance of some perceptual set. These distinctions are seen most clearly in relation to existing analyses and experimental designs.

DCM and existing approaches

The central ideal DCM is to treat the brain as a deterministic non-linear dynamic system that is subject to inputs and produces outputs. Effective connectivity is parameterized in terms of coupling among unobserved brain states (e.g. neuronal activity in different regions). The objective is to estimate these parameters by perturbing the system and measuring the response. This is in contradistinction to established methods, for estimating effective connectivity from neurophysiological time-series, which include structural equation modelling and models based on multivariate autoregressive processes. In these models, there is no designed perturbation and the inputs are treated as unknown and stochastic. Multivariate autoregression models and their spectral equivalents like coherence analysis, not only assume the system is driven by stochastic innovations, but are usually restricted to linear interactions. Structural equation modelling assumes the interactions are linear and, furthermore, instantaneous in the sense that structural equation models are not time-series models. In short, dynamic causal modelling is distinguished from alternative approaches not just by accommodating the non-linear and dynamic aspects of neuronal interactions, but by framing the estimation problem in terms of perturbations that accommodate experimentally designed inputs. This is a critical departure from conventional approaches to causal modelling in neuroimaging and, importantly, brings the analysis of effective connectivity much closer to the analysis of region-specific effects: dynamic causal modelling calls upon the same experimental design principles to elicit region-specific interactions that we use in conventional experiments to elicit region-specific activations. In fact, as shown later, the convolution model, used in the standard analysis of fMRI time-series, is a special and simple case of DCM that arises when the coupling among regions is discounted. In DCM, the causal or explanatory variables that comprise the conventional design matrix become the inputs and the parameters become measures of effective connectivity. Although DCM can be framed as a generalization of the linear models used in conventional analyses to cover bilinear models (see below), it also represents an attempt to embed more plausible forward models of how neuronal dynamics respond to inputs and produce measured responses. This reflects the growing appreciation of the role that neuronal models may have to play in understanding measured brain responses (see Horwitz *et al.*, 2001 for a discussion).

This chapter can be regarded as an extension of previous work on the Bayesian identification of haemodynamic models (Friston 2002) to cover multiple regions. In Chapter 34, we focused on the biophysical parameters of a haemodynamic response in a single region. The most important parameter was the efficacy with which experimental inputs could elicit an activity-dependent vasodilatory signal. In this chapter, neuronal activity is modelled explicitly, allowing for interactions among the activities of multiple regions in generating the observed haemodynamic response. The estimation procedure employed for DCM is formally identical to that described in Chapter 34 (see also Appendix 4).

DCM and experimental design

DCM is used to test the specific hypothesis that motivated the experimental design. It is not an exploratory technique, as with all analyses of effective connectivity the results are specific to the tasks and stimuli employed during the experiment. In DCM, designed inputs can produce responses in one of two ways. Inputs can elicit changes in the state variables (i.e. neuronal activity) directly. For example, sensory input could be modelled as causing direct responses in primary visual or auditory areas. The second way in which inputs affect the system is through changing the effective connectivity or interactions. Useful examples of this sort of effect would be the attentional modulation of connections between parietal and extrastriate areas. Another ubiquitous example of contextual effects would be time. Time-dependent changes in connectivity correspond to plasticity. It is useful to regard experimental factors as inputs that belong to the class that produce evoked responses or to the class of contextual factors that induce changes in coupling (although, in principle, all inputs could do both). The first class comprises trial- or stimulus-bound perturbations, whereas the second establishes a context in which effects of the first sort evoke responses. This second class is typically trial-free and induced by task instructions or other contextual changes. Measured responses in high-order cortical areas are mediated by interactions among brain areas elicited by trial-bound perturbations. These interactions can be modulated by other set-related or contextual factors that modulate the latent or regional coupling among areas. Figure 41.1 illustrates this schematically. The important implication here, for experimental design in DCM, is that it should be multifactorial, with at least one factor controlling sensory perturbation and another factor manipulating the context in which the responses are evoked (cf. psychophysiological interactions).

In this chapter, we use bilinear approximations to any DCM. The bilinear approximation reduces the parameters to three sets that control three distinct things: first, the direct or extrinsic influence of inputs on brain states in any particular area; second, the regional or latent coupling of one area to another; and finally, changes in this coupling that are induced by input. Although, in some instances, the relative strengths of coupling may be of interest, most analyses of DCMs focus on the changes in coupling embodied in the bilinear parameters. The first class of parameters is generally of little interest in the context of DCM, but is the primary focus in classical analyses of regionally specific effects. In classical analyses, the only way experimental effects can be expressed is though a direct or extrinsic influence on each voxel because mass-univariate models (e.g. statistical parametric mapping, SPM) preclude coupling among voxels or its modulation.

Questions about the modulation of effective connectivity are addressed through inference about the bilinear parameters described above. They are bilinear in the sense that an input-dependent change in connectivity can be construed as a second-order interaction between the input and activity in the source, when causing a response in a target region. The key role of bilinear terms reflects the fact that the more interesting applications of effective connectivity address changes in connectivity induced by set or time. In short, DCM with a bilinear approximation allows one to claim that an experimental manipulation has 'activated a pathway' as opposed to a cortical region. Bilinear terms correspond to psychophysiological

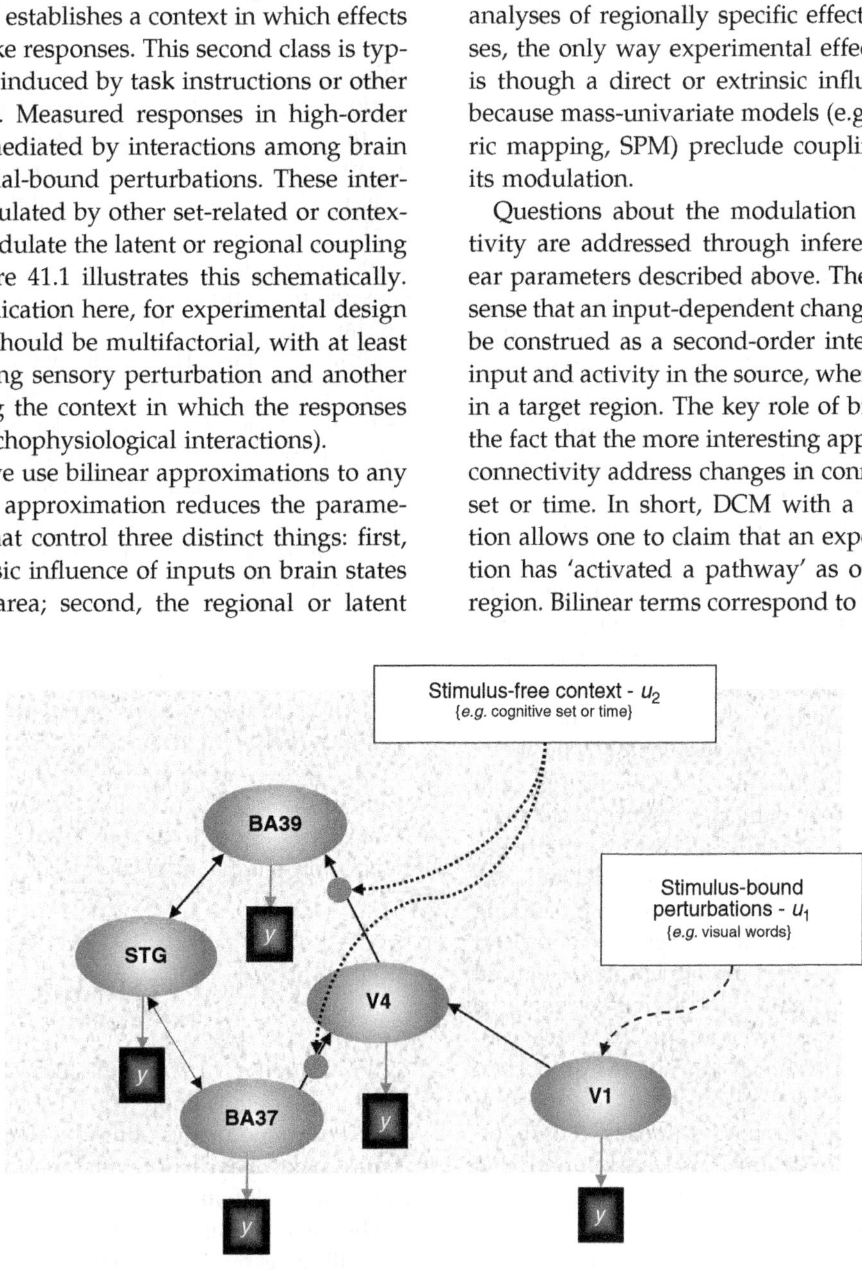

FIGURE 41.1 This is a schematic illustrating the concepts underlying dynamic causal modelling. In particular, it highlights the two distinct ways in which inputs or perturbations can elicit responses in the regions or nodes that comprise the model. In this example, there are five nodes, including visual areas **V1** and **V4** in the fusiform gyrus, areas 39 and 37 and the superior temporal gyrus (**STG**). Stimulus-bound perturbations designated u_1 act as extrinsic inputs to the primary visual area **V1**. Stimulus-free or contextual inputs u_2 mediate their effects by modulating the coupling between **V4** and **BA39** and between **BA37** and **V4**. For example, the responses in the angular gyrus (**BA39**) are caused by inputs to **V1** that are transformed by **V4**, where the influences exerted by **V4** are sensitive to the second input. The dark square boxes represent the components of the DCM that transform the state variables z_i in each region (neuronal activity) into a measured [haemodynamic] response y_i.

interaction terms in conventional regression analyses of effective connectivity (Friston *et al.*, 1997) and those formed by moderator variables (Kenny and Judd, 1984) in structural equation modelling (Büchel and Friston, 1997). Their bilinear aspect speaks again of the importance of multifactorial designs that allow these interactions to be measured and the central role of the context in which region-specific responses are formed (see McIntosh, 2000).

DCM and inference

Because DCMs are not restricted to linear or instantaneous systems, they are necessarily complicated and, potentially, need a large number of free parameters. This is why they have greater biological plausibility in relation to alternative approaches. However, this makes model inversion more dependent upon constraints. A natural way to embody the requisite constraints is within a Bayesian framework. Consequently, dynamic causal models are inverted using Bayesian schemes to furnish conditional estimators and inferences about the connections. In other words, the estimation procedure provides the probability distribution of a coupling parameter in terms of its mean and standard deviation. Having established this posterior density, the probability that the connection exceeds some specified threshold is easily computed. Bayesian inferences like this are more straightforward than corresponding classical inferences and eschew the multiple comparisons problem. The posterior density is computed using the likelihood and prior densities. The likelihood of the data, given some parameters, is specified by the forward model or DCM (in one sense all models are simply ways of specifying the likelihood of an observation). The prior densities on the coupling parameters offer suitable constraints to ensure robust and efficient estimation. These priors harness some natural constraints about the dynamics of coupled systems (see below), but also allow the user to specify which connections are present and which are not. An important use of prior constraints of this sort is the restriction of where inputs can elicit extrinsic responses. It is interesting to reflect that conventional analyses suppose that all inputs have unconstrained access to all brain regions. This is because classical models assume activations are caused directly by experimental factors, as opposed to being mediated by afferents from other brain areas.

Additional constraints on the intrinsic connections and their modulation by contextual inputs can also be specified, but they are not necessary. These additional constraints can be used to make a model more parsimonious, allowing one to focus on a particular connection. We will provide examples of this below. Unlike structural equation modelling, there are no limits on the number of connections that can be modelled because the assumptions and estimation scheme used by dynamic causal modelling are completely different, relying upon known inputs.

Overview

This chapter comprises a theoretical section and three sections demonstrating the use and validity of DCM. In the theoretical section, we present the concepts used in the remaining sections. The later sections address the face, predictive and construct validity of DCM respectively. Face validity entails the estimation, and inference procedure identifies what it is supposed to. The subsequent section on predictive validity uses empirical data from an fMRI study of single-word processing at different rates. These data were obtained consecutively in a series of contiguous sessions. This allowed us to repeat the DCM using independent realizations of the same paradigm. Predictive validity over the multiple sessions was assessed in terms of the consistency of the coupling estimates and their posterior densities. The final section on construct validity revisits changes in connection strengths among parietal and extrastriate areas induced by attention to optic flow stimuli. We have established previously attentionally mediated increases in effective connectivity using both structural equation modelling and a Volterra formulation of effective connectivity (Büchel and Friston, 1997; Friston and Büchel, 2000). Here we show that dynamic causal modelling leads to the same conclusions. This chapter ends with a brief discussion of dynamic causal modelling, its limitations and potential applications.

THEORY

In this section, we present the theoretical motivation and operational details upon which DCM rests. In brief, DCM is a fairly standard non-linear system identification procedure using Bayesian inversion of deterministic input-state-output dynamic models. In this chapter, the system can be construed as a number of interacting brain regions. We will focus on a particular form for the dynamics that corresponds to a bilinear approximation to any analytic system. However, the idea behind DCM is not restricted to bilinear forms, as we will see in the next chapter.

This section is divided into three parts. First, we describe the DCM itself, then consider the nature of priors on the parameters of the DCM and finally summarize the inference procedure. The estimation conforms to the posterior density analysis under Gaussian assumptions described in Chapter 34. In Chapter 34, we were primarily concerned with estimating the efficacy with which input elicits a vasodilatory signal, presumably mediated by neuronal responses to the input. The causal models here can be regarded as a collection of haemodynamic models, one for each area, in which the experimental inputs are supplemented with neural activity from other areas. The parameters of interest now cover not only the direct efficacy of experimental inputs, but also the efficacy of neuronal input from distal regions, i.e. effective connectivity (see Figure 41.1).

The Bayesian inversion finds the maximum or mode of the posterior density of the parameters (i.e. the most likely coupling parameters given the data) by performing a gradient assent on the log-posterior. The log-posterior requires both likelihood and prior terms. The likelihood obtains from Gaussian assumptions about the errors in the observation model supplied by the DCM. This likelihood or forward model is described in the next subsection. By combining the likelihood with priors on the coupling and haemodynamic parameters, described in the second subsection, one can form an expression for the posterior density that is used in the estimation.

Dynamic causal models

The dynamic causal model is a multiple-input multiple-output (MIMO) system that comprises m inputs and l outputs with one output per region. The m inputs correspond to designed causes (e.g. boxcar or stick stimulus-functions). The inputs are exactly the same as those used to form design matrices in conventional analyses of fMRI and can be expanded in the usual way when necessary (e.g. using polynomials or temporal basis functions). In principle, each input could have direct access to every region. However, in practice, the extrinsic effects of inputs are usually restricted to a single input region. Each of the l regions produces a measured output that corresponds to the observed blood oxygenation-level-dependent (BOLD) signal. These l time-series would normally be taken as the average or first eigenvariate of key regions, selected on the basis of a conventional analysis. Each region has five state variables. Four of these are of secondary importance and correspond to the state variables of the haemodynamic model first presented in Friston *et al.* (2000) and described in previous chapters. These haemodynamic states comprise a vasodilatory signal, normalized flow, normalized venous volume, and normalized deoxyhaemoglobin content. These biophysical states are required to compute the observed BOLD response and are not influenced by the states of other regions.

Central to the estimation of coupling parameters is the neuronal state of each region. This corresponds to average neuronal or synaptic activity and is a function of the neuronal states of other brain regions. We will deal first with the equations for the neuronal states and then briefly reprise the differential equations that constitute the haemodynamic model for each region.

Neuronal state equations

Restricting ourselves to the neuronal states $z = [z_1, \ldots, z_l]^T$ one can posit any arbitrary form or model for effective connectivity:

$$\dot{z} = F(z, u, \theta) \qquad \textbf{41.1}$$

where F is some non-linear function describing the neurophysiological influences exerted by inputs $u(t)$ and the activity in all brain regions on the evolution of the neuronal states. θ are the parameters of the model whose posterior density we require for inference. It is not necessary to specify the form of this equation because its bilinear approximation provides a natural and useful re-parameterization in terms of coupling parameters.

$$\begin{aligned} \dot{z} &\approx Az + \sum u_j B^j z + Cu \\ &= \left(A + \sum u_j B^j\right) z + Cu \end{aligned}$$

$$A = \left.\frac{\partial F}{\partial z}\right|_{u=0} \quad B^j = \frac{\partial^2 F}{\partial z \partial u_j} = \frac{\partial A}{\partial u_j} \quad C = \left.\frac{\partial F}{\partial u}\right|_{z=0} \qquad \textbf{41.2}$$

The Jacobian or connectivity matrix A represents the first-order coupling among the regions in the absence of input. This can be thought of as the regional coupling in the absence of experimental perturbations. Notice that the state, which is perturbed, depends on the experimental design (e.g. baseline or control state) and therefore the regional coupling is specific to each experiment. The matrices B^j are effectively the change in regional coupling induced by the j-th input. Because B^j are second-order derivatives these terms are referred to as bilinear. Finally, the matrix C encodes the extrinsic influences of inputs on neuronal activity. The parameters $\theta^c = \{A, B^j, C\}$ are the coupling matrices we wish to identify and define the functional architecture and interactions among brain regions at a neuronal level. Figure 41.2 shows an example of a specific architecture to demonstrate the relationship between the matrix form of the bilinear model and the underlying state equations for each region. Notice that the units of coupling are per unit time and therefore correspond to rates. Because we are in a dynamical setting, a strong connection means an influence that is expressed quickly or with a small time constant. It is useful to appreciate this when interpreting estimates and thresholds quantitatively. This will be illustrated below.

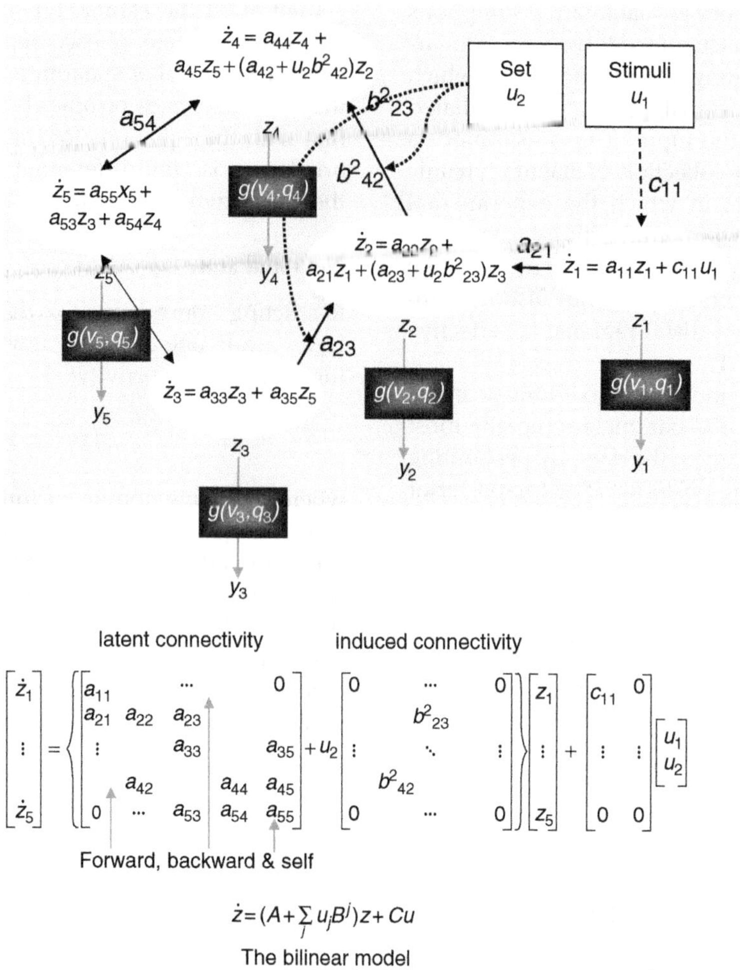

FIGURE 41.2 This schematic (upper panel) recapitulates the architecture in Figure 41.1 in terms of the differential equations implied by a bilinear approximation. The equations in each of the white areas describe the changes in neuronal activity z_i in terms of linearly separable components that reflect the influence of other regional state variables. Note particularly, how the second contextual inputs enter these equations. They effectively increase the intrinsic coupling parameters, a_{ij}, in proportion to the bilinear coupling parameters, b^k_{ij}. In this diagram, the haemodynamic component of the DCM illustrates how the neuronal states enter a region-specific haemodynamic model to produce the outputs y_i that are a function of the region's biophysical states reflecting deoxyhaemoglobin content and venous volume (q_i and v_i). The lower panel reformulates the differential equations in the upper panel into a matrix format. These equations can be summarized more compactly in terms of coupling parameter matrices A, B^j and C. This form is used in the main text and shows how it relates to the underlying differential equations that describe the state dynamics.

The evolution of neuronal activity in each region causes changes in volume and deoxyhaemoglobin to engender the observed BOLD response y as described next.

Haemodynamic state equations

The remaining state variables of each region are biophysical states (s, f, v, q), which form the BOLD signal and mediate the translation of neuronal activity into haemodynamic responses. Haemodynamic states are a function of, and only of, the neuronal state of each region. The state equations have been described in Chapters 27 and 34. These constitute a haemodynamic model that embeds the Balloon-Windkessel model (Buxton *et al.*, 1998; Mandeville *et al.*, 1999). In brief, an activity-dependent vasodilatory signal s increases flow f. Flow increases volume and dilutes deoxyhaemoglobin (v and q). The last two haemodynamic states enter an output non-linearity to give the observed BOLD response. A list of the biophysical parameters $\theta^h = \kappa, \gamma, \tau, \alpha, \rho$ is provided in Table 41-1 and a schematic of the haemodynamic model is shown in Figure 41.3.

The likelihood model

Combining the neuronal and haemodynamic states $x = z, s, f, v, q$ gives us a full forward model specified by

TABLE 41-1 Priors on biophysical parameters

Parameter	Description	Prior mean η_θ	Prior variance C_θ
κ	Rate of signal decay	0.65 per second	0.015
γ	Rate of flow-dependent elimination	0.41 per second	0.002
τ	Haemodynamic transit time	0.98 second	0.0568
α	Grubb's exponent	0.32	0.0015
ρ	Resting oxygen extraction fraction	0.34	0.0024

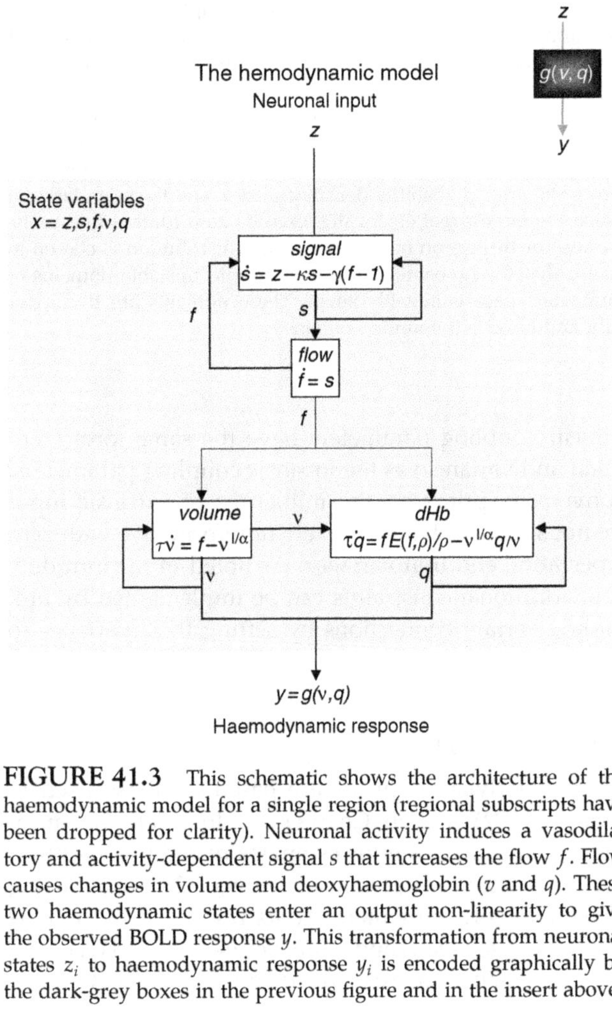

FIGURE 41.3 This schematic shows the architecture of the haemodynamic model for a single region (regional subscripts have been dropped for clarity). Neuronal activity induces a vasodilatory and activity-dependent signal s that increases the flow f. Flow causes changes in volume and deoxyhaemoglobin (v and q). These two haemodynamic states enter an output non-linearity to give the observed BOLD response y. This transformation from neuronal states z_i to haemodynamic response y_i is encoded graphically by the dark-grey boxes in the previous figure and in the insert above.

the neuronal bilinear state equation (Eqn. **41.2**) and the haemodynamic equations in Figure 41.3:

$$\dot{x} = f(x, u, \theta)$$
$$y = g(x, \theta) \qquad \mathbf{41.3}$$

with parameters $\theta = \{\theta^c, \theta^h\}$. For any set of parameters and inputs, the state equation can be integrated and passed through the output non-linearity to give the predicted response $h(u, \theta)$. This integration can be made quite expedient by capitalizing on the sparsity of stimulus functions commonly employed in fMRI designs (see Chapter 34; Eqn. **34.16**). Integrating Eqn. **41.3** is equivalent to a generalized convolution of the inputs with the system's Volterra kernels. These kernels are easily derived from the Volterra expansion of Eqn. **41.3** (Bendat, 1990 and Appendix 2):

$$h_i(u, \theta) = \sum_k \int_0^t \ldots \int_0^t \kappa_i^k(\sigma_1, \ldots, \sigma_k) u(t - \sigma_1), \ldots, u(t - \sigma_k) d\sigma_1, \ldots, d\sigma_k$$

$$\kappa_i^k(\sigma_1, \ldots, \sigma_k) = \frac{\partial^k y_i(t)}{\partial u(t - \sigma_1), \ldots, \partial u(t - \sigma_k)} \qquad \mathbf{41.4}$$

either by numerical differentiation or analytically through bilinear approximations (see Friston, 2002). κ_i^k is the k-th order kernel for region i. For simplicity, Eqn. **41.4** is for a single input. The kernels are simply a re-parameterization of the model. We will use these kernels to characterize regional impulse responses at neuronal and haemodynamic levels later.

The dynamic model can be made into a likelihood model by adding errors and confounding or nuisance effects $X(t)$ to give $y = h(u, \theta) + X\beta + \varepsilon$. Here β are the unknown coefficients for confounds. In the examples below, $X(t)$ comprises a low-order discrete cosine set, modelling low-frequency drifts and a constant term. The likelihood model specifies $p(y|\theta, \beta; u) = N(h(u, \theta) + X\beta, \Sigma(\lambda))$, where λ are some hyperparameters controlling the covariance of the errors $\Sigma(\lambda)$. To complete the specification of the generative model $p(y, \theta, \beta; u) = p(y|\theta, \beta; u)p(\theta)p(\beta)$ we need the priors $p(\theta)$ and $p(\beta)$. We will treat confounds as fixed effects, which means these have flat priors. The priors on the coupling and haemodynamic parameters are described next.

Priors

In this application, we use a fully Bayesian approach because there are clear and necessary constraints on neuronal dynamics that can be used to motivate priors on the coupling parameters and empirically determined priors on the biophysical haemodynamic parameters are relatively easy to specify. We will deal first with the coupling parameters.

Priors on the coupling parameters

It is self-evident that neuronal activity cannot diverge exponentially to infinite values. Therefore, we know that, in the absence of input, the dynamics must to return to a stable mode. We use this constraint to motivate a simple shrinkage prior on the coupling parameters that make large values and self-excitation unlikely. These priors impose a probabilistic upper bound on the intrinsic coupling, imposed by Gaussian priors that ensure its largest Lyapunov exponent is unlikely to exceed zero. The specification of priors on the connections is finessed by a re-parameterization of the coupling matrices A and B^j.

$$A \to e^{\sigma} A = e^{\sigma} \begin{bmatrix} -1 & a_{12} & \cdots \\ a_{21} & -1 & \\ \vdots & & \ddots \end{bmatrix} \quad B^j \to e^{\sigma} B^j = e^{\sigma} \begin{bmatrix} b^j_{11} & b^j_{12} & \cdots \\ b^j_{21} & \ddots & \\ \vdots & & \end{bmatrix} \tag{41.5}$$

This factorization into a non-negative scalar and normalized coupling matrix renders the normalized couplings adimensional, such that strengths of connections among regions are relative to their self-connections. From now on, we will deal with normalized parameters. Each connection has a prior Gaussian density with zero expectation and variance (see Friston *et al.*, 2002a, b):

$$C_a = \frac{l/(l-1)}{\phi(1-p)} \tag{41.6}$$

where ϕ is the inverse of the $\chi^2_{l(l-1)}$ cumulative distribution and p is a small probability that the system will become unstable if all the connections are the same. As the number of regions, l increases, the prior variance decreases. A Gaussian prior $p(\sigma) = N(0, \frac{1}{4})$ ensures that the temporal scaling $\exp(\sigma)$ is positive and covers dynamics that have a characteristic time constant of about a second.[1]

To provide an intuition about how these priors keep the system from diverging exponentially, a quantitative example is shown in Figure 41.4. Figure 41.4 shows the prior density of two connections that renders the probability of instability less than one in a hundred. It can be seen that this density lies in a domain of parameter space encircled by regions in which the maximum Lyapunov exponent exceeds zero (bounded by dotted white lines) (see the Figure legend for more details). Priors on the

[1] We will use the same device to place non-negative, log-normal priors on parameters in the next chapter dealing with neuronal kinetics. This parameterization is particularly useful in dynamic models because most of the parameters are rate constants, which cannot be negative.

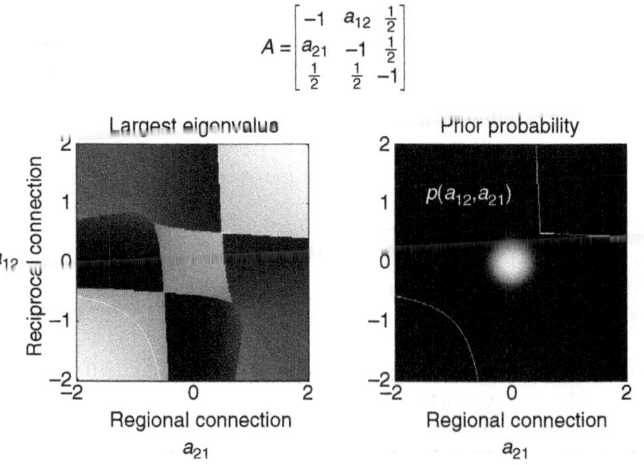

FIGURE 41.4 Prior probability density on the intrinsic coupling parameters for a specific intrinsic coupling matrix A. The left-hand panel shows the real value of the largest eigenvalue of A (the principal Lyapunov exponent) as a function of the connection from the first to the second region and the reciprocal connection from the second to the first. The remaining connections were held constant at 0.5. This density can be thought of as a slice through a multidimensional distribution over all connections. The right panel shows the prior probability density function and the boundaries at which the largest real eigenvalue exceeds zero (dotted lines). The variance or dispersion of this probability distribution is chosen to ensure that the probability of excursion into unstable domains of parameter space is suitably small. These domains are the upper right and lower left bounded regions.

bilinear coupling parameters have the same form (zero mean and variance) as the intrinsic coupling parameters. Conversely, priors on the influences of extrinsic input are not scaled and are relatively uninformative with zero expectation and unit variance. As noted in the introduction, additional constraints can be implemented by precluding certain connections by setting their variance to zero.

Haemodynamic priors

The haemodynamic priors are based on those used in Friston (2002) and in Chapter 34. In brief, the mean and variance of posterior estimates of the five biophysical parameters were computed over 128 voxels using the single-word presentation data presented in the next section. These means and variances (see Table 41-1) were used to specify Gaussian priors on the haemodynamic parameters.

Combining the prior densities on the coupling and haemodynamic parameters allows us to express the prior probability of the parameters in terms of their prior expectation and covariance $p(\theta) = N(\eta_\theta, C_\theta)$. Having specified the priors, we can proceed to estimation and inference.

Estimation and inference

Following the approach described in Chapter 34 we note:

$$
\begin{aligned}
y - h(u, \eta_{\theta|y}) &\approx J\Delta\theta + X\beta + \varepsilon \\
&= [J, X]\begin{bmatrix}\Delta\theta \\ \beta\end{bmatrix} + \varepsilon \qquad \textbf{41.7} \\
\Delta\theta &= \theta - \eta_{\theta|y}
\end{aligned}
$$

This local linear approximation then enters an expectation–maximization (**EM**) scheme as described previously:

Until convergence{

E-step

$$
\begin{aligned}
J &= \frac{\partial h(\eta_{\theta|y})}{\partial\theta} \\
\bar{y} &= \begin{bmatrix} y - h(\eta_{\theta|y}) \\ \eta_\theta - \eta_{\theta|y}\end{bmatrix}, \quad \bar{J} = \begin{bmatrix} J & X \\ 1 & 0 \end{bmatrix}, \quad \bar{C}_\varepsilon = \begin{bmatrix} \sum \lambda_i Q_i & 0 \\ 0 & C_\theta \end{bmatrix} \\
C_{\theta|y} &= (\bar{J}^T \bar{C}_\varepsilon^{-1} \bar{J})^{-1} \\
\begin{bmatrix}\Delta\eta_{\theta|y} \\ \eta_{\beta|y}\end{bmatrix} &= C_{\theta|y}\,(\bar{J}^T\bar{C}_\varepsilon^{-1}\bar{y}) \\
\eta_{\theta|y} &\leftarrow \eta_{\theta|y} + \Delta\eta_{\theta|y}
\end{aligned}
$$

M-step **41.8**

$$
\begin{aligned}
P &= \bar{C}_\varepsilon^{-1} - \bar{C}_\varepsilon^{-1}\bar{J}C_{\theta|y}\bar{J}^T\bar{C}_\varepsilon^{-1} \\
\frac{\partial F}{\partial\lambda_i} &= -\tfrac{1}{2}tr\{PQ_i\} + \tfrac{1}{2}\bar{y}^T P^T Q_i P\bar{y} \\
\left\langle \frac{\partial^2 F}{\partial\lambda_{ij}^2}\right\rangle &= -\tfrac{1}{2}tr\{PQ_iPQ_j\} \\
\lambda &\leftarrow \lambda - \left\langle\frac{\partial^2 F}{\partial\lambda^2}\right\rangle^{-1}\frac{\partial F}{\partial\lambda}
\end{aligned}
$$

}

These expressions are formally the same as Eqn. **34.11** in Chapter 34 but for the addition of confounding effects in X. These confounds are treated as fixed effects with infinite prior variance, which does not need to appear explicitly in the EM scheme.

Note that the prediction and observations encompass the entire experiment. They are therefore large $l \times n$ vectors whose elements run over l regions and n time bins. Although the response variable could be viewed as a multivariate times-series, it is treated as a single observation vector, whose error covariance embodies both temporal and interregional correlations $C_\varepsilon = V \otimes \Sigma(\lambda) = \sum \lambda_i Q_i$. This covariance is parameterized by some covariance hyperparameters λ. In the examples below, these correspond to region-specific error variances assuming the same temporal correlations $Q_i = V \otimes \Sigma_i$ in which Σ_i is an $l \times l$ sparse matrix with the i-th leading diagonal element equal to one.

Eqn. **41.8** enables us the estimate the conditional moments of the coupling parameters (and the haemodynamic parameters) plus the hyperparameters controlling error and represents a posterior density analysis under Gaussian assumptions. In short, the estimation scheme provides the approximating Gaussian posterior density of the parameters $q(\theta) = N(\eta_{\theta|y}, C_{\theta|y})$ in terms of its expectation and covariance. The expectation is also known as the posterior mode or maximum *a posteriori* (MAP) estimator. The marginal posterior probabilities are then used for inference that any particular parameter or contrast of parameters $c^T\eta_{\theta|y}$ (e.g. average) exceeded a specified threshold γ.

$$
p = erf\left(\frac{c^T\eta_{\theta|y} - \gamma}{\sqrt{c^T C_{\theta|y} c}}\right) \qquad \textbf{41.9}
$$

As above, *erf* is the cumulative normal distribution. In this chapter, we are primarily concerned with the coupling parameters θ^c and, among these, the bilinear terms. The units of these parameters are Hz or per second (or adimensional if normalized) and the thresholds are specified as such. In dynamical modelling, coupling parameters play the same role as rate-constants in kinetic models.

Relationship to conventional analyses

It is interesting to note that conventional analyses of fMRI data, using linear convolution models, are a special case of dynamic causal models using a bilinear approximation. This is important because it provides a direct connection between DCM and classical models. If we allow inputs to be connected to all regions and discount interactions among regions by setting the prior variances on A and B to zero, we produce a set of disconnected brain regions or voxels that respond to, and only to, extrinsic input. The free parameters of interest reduce to the values of C, which reflect the ability of input to elicit responses in each voxel. By further setting the prior variances on the self-connections (i.e. scaling parameter) and those on the haemodynamic parameters to zero, we end up with a single-input-single-output model at each and every brain region that can be reformulated as a convolution model as described in Friston (2002). For voxel i and input j the parameter c_{ij} can be estimated by simply convolving the input with $\partial\kappa_i^1/\partial c_{ij}$ where κ_i^1 is the first-order kernel mediating the influence of input j on output i. The convolved inputs are then used to form a general linear model that can be estimated in the usual way. This is precisely the approach adopted in classical analyses, in which $\partial\kappa_i^1/\partial C_{ij}$ is the haemodynamic response function. The key point here is that the general linear models used in typical data analyses are special cases of bilinear models but embody *more* assumptions. These assumptions enter through the use of highly precise priors that discount interactions among regions and prevent any variation in biophysical responses. Having described the theoretical aspects of DCM, we now turn to applications and its validity.

FACE VALIDITY – SIMULATIONS

In this section, we use simulated data to establish the utility of the bilinear approximation and the robustness of the estimation scheme described in the previous section. We deliberately chose an architecture that would be impossible to characterize using existing methods based on regression models (e.g. structural equation modelling). This architecture embodies loops and reciprocal connections and poses the problem of vicarious input, the ambiguity between the direct influences of one area and influences that are mediated through others.

A toy system

The simulated architecture is depicted in Figure 41.5 and has been labelled so that it is consistent with the DCM characterized empirically in the next section. The model comprises three regions: a primary (**A1**) and secondary (**A2**) auditory area and a higher-level region (**A3**). There are two inputs. The first is a sensory input u_1 encoding the presentation of epochs of words at different frequencies. The second input u_2 is contextual in nature and is simply an exponential function of the time elapsed since the start of each epoch (with a time constant of 8 s). These inputs were based on a real experiment and are the same as those used in the empirical analyses of the next section. The scaling of the inputs is important for the quantitative evaluation of the bilinear and extrinsic coupling parameters. The convention adopted here is that inputs encoding events approximate delta functions such that their integral over time corresponds to the number of events that have occurred. For event-free inputs, like the maintenance of a particular instructional set, the input is scaled to a maximum of unity, so that the integral reflects the number of seconds over which the input was prevalent. The inputs were specified in time bins that were a sixteenth of the interval between scans (repetition time; TR = 1.7 s).

The auditory input is connected to the primary area; the second input has no direct effect on activity but modulates the forward connections from **A1** to **A2** so that its influence shows *adaptation* during the epoch. The second auditory area receives input from the first and sends signals to the higher area (**A3**). In addition to reciprocal backward connections, in this simple auditory hierarchy a connection from the lowest to the highest area has

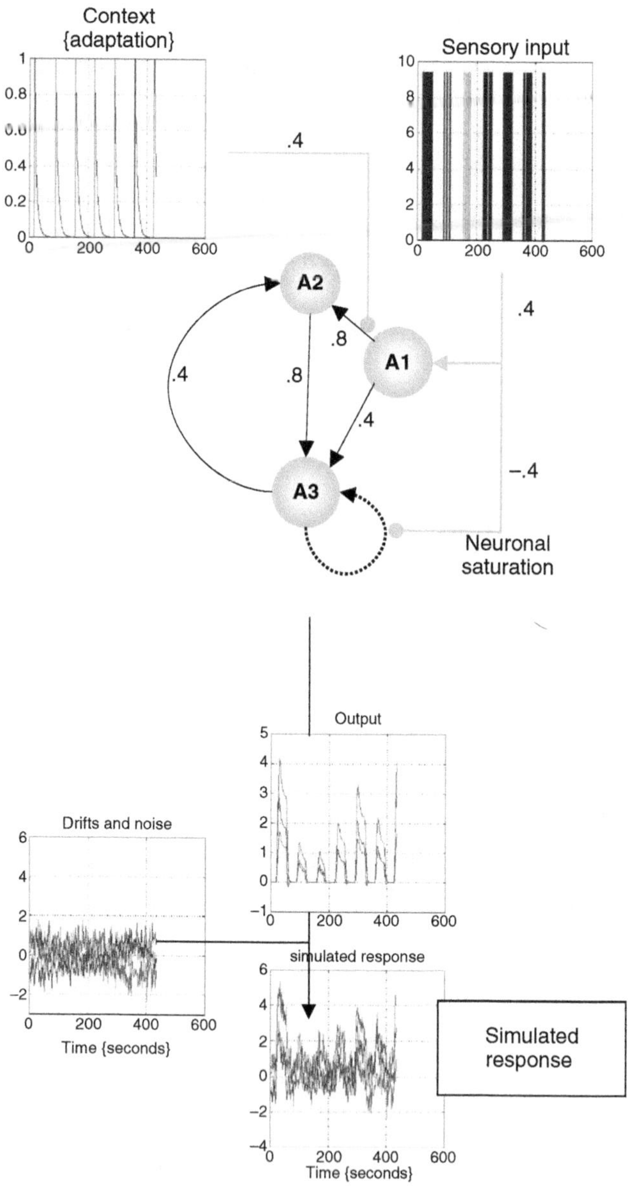

FIGURE 41.5 This is a schematic of the architecture used to generate simulated data. Non-zero regional connections are shown as directed black arrows with the strength or true parameter alongside. Here, the perturbing input is the presentation of words (sensory inputs) and acts as an intrinsic influence on **A1**. In addition, this input modulates the self-connection of **A3** to emulate saturation-like effects (see main text and Figure 41.6). The contextual input is a decaying exponential of within-epoch time and positively modulates the forward connection from **A1** to **A2**. The lower panel shows how responses were simulated by mixing the output of the system described above with drifts and noise as described in the main text.

been included. Finally, the first input (word presentation) modulates the self-connections of the third region. This influence has been included to show how bilinear effects can emulate non-linear responses. A bilinear modulation of the self-connection can augment or attenuate decay of synaptic activity, rendering the average response to streams of stimuli rate-dependent. This is because the bilinear effect will only be expressed if sufficient synaptic activity persists after the previous stimulus. This, in turn, depends on a sufficiently fast presentation rate. The resulting response emulates a saturation at high presentation rates or small stimulus onset asynchronies that has been observed empirically. Critically, we are in a position to disambiguate between neuronal saturation, modelled by this bilinear term, and haemodynamic saturation, modelled by non-linearities in the haemodynamic component of this DCM. A significant bilinear self-connection implies neuronal saturation above and beyond that attributable to haemodynamics. Figure 41.6 illustrates this neuronal *saturation* by plotting the simulated response of **A3** in the absence of saturation $B^1 = 0$ against the simulated response with $b^1_{3,3} = -0.4$. It is evident that there is a non-linear sub-additive effect at high response levels. It should be noted that true neuronal saturation of this sort is mediated by second-order interactions among the states (i.e. neuronal activity). However, as shown in Figure 41.6, we can emulate these effects by using the first extrinsic input as a surrogate for neuronal inputs from other areas in the bilinear component of the model.

Using this model we simulated responses using the values for A, B^1, B^2 and C given in Figure 41.7 and the

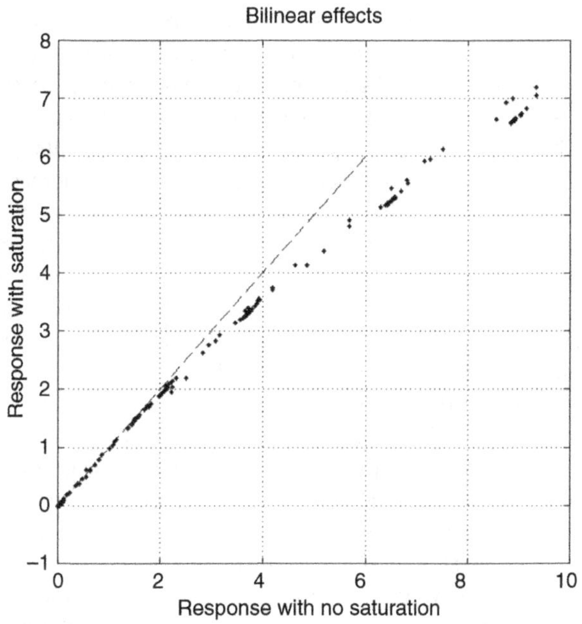

FIGURE 41.6 This is a plot of the simulated response with saturation against the equivalent response with no saturation. These simulated responses were obtained by setting the bilinear coupling parameter b^1_{33} labelled 'neuronal saturation' in Figure 41.5 to −0.4 and zero respectively. The key thing to observe is a saturation of responses at high levels. The broken line depicts the response expected in the absence of saturation. This illustrates how bilinear effects can introduce non-linearities into the response.

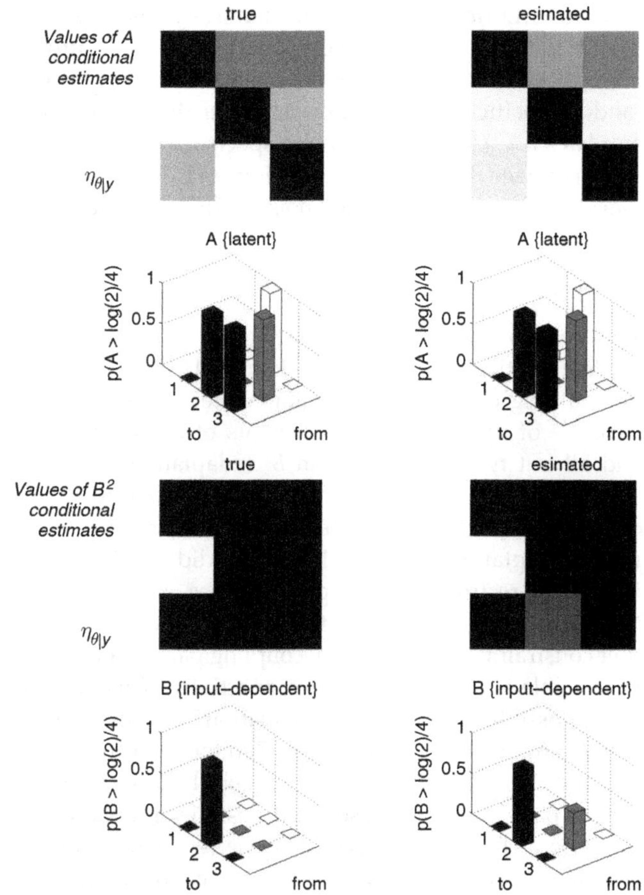

FIGURE 41.7 Results summarizing the conditional estimation based upon the simulated data of Figure 41.5. The upper panels show the conditional estimates and posterior probabilities pertaining to the regional coupling parameters. The lower panels show the equivalent results for bilinear coupling parameters mediating the effect of within-epoch time. Conditional or MAP estimates of the parameters are shown in image format with arbitrary scaling. The posterior probabilities that these parameters exceeded a threshold of ln(2)/4 per second are shown as three-dimensional bar charts. True values and probabilities are shown on the left, whereas the estimated values and posterior probabilities are shown on the right. This illustrates that the conditional estimates are a reasonable approximation to the true values and, in particular, the posterior probabilities conform to the true probabilities, if we consider values of 90 per cent or more.

prior expectations for the biophysical parameters given in Table 41-1. The values of the coupling parameters were chosen to match those seen typically in practice. This ensured the simulated responses were realistic in relation to simulated noise. After down-sampling these deterministic responses every 1.7 s (the TR of the empirical data used in the next section), we added known noise to produce simulated data. These data comprised time-series of 256 observations with independent or serially correlated Gaussian noise based on an AR(1) process. Unless otherwise stated, the noise had standard deviation of one half and was IID (independent and identically distributed). The drift terms were formed from the first six components of a discrete cosine set mixed linearly with normal random coefficients, scaled by one over the order. This emulates a $1/f^2$ plus white noise spectrum for the noise and drifts (see the lower panel of Figure 41.7 for an exemplar data simulation with IID noise of unit variance).

Exemplar analysis

The analysis described in the previous section was applied to the data shown in Figure 41.5. The priors on coupling parameters were augmented by setting the variance of the off-diagonal elements of B^1 (saturation) and all but two connections in B^2 (adaptation) to zero. These two connections were the first and second forward connections of this cortical hierarchy. The first had simulated adaptation, whereas the second did not. Extrinsic input was restricted to the primary area **A1** by setting the variances of all but c_{11} to zero. We placed no further constraints on the regional coupling parameters. This is equivalent to allowing full connectivity. This would be impossible with structural equation modelling. The results are presented in Figure 41.7 in terms of the MAP or conditional expectations of the coupling parameters (upper panels) and the associated posterior probabilities (lower panels). It can be seen that the regional coupling parameters are estimated reasonably accurately with a slight overestimation of the backward connection from **A3** to **A2**. The bilinear coupling parameters modelling adaptation are shown in the lower panels and the estimators have correctly identified the first forward connection as the locus of greatest adaptation. The posterior probabilities suggest inferences about the coupling parameters would lead us to the veridical architecture if we considered only connections whose half-life exceeded 4 s with 90 per cent confidence or more.

The MAP estimates allow us to compute the MAP kernels associated with each region, in terms of neuronal output and haemodynamics response. The neuronal and haemodynamic kernels for the three regions are shown in Plate 58 (upper panels) (see colour plate section). It is interesting to note that the regional variation in the form of the neuronal kernels is sufficient to induce differential onset and peak latencies, in the order of a second or so, in the haemodynamic kernels, despite the fact that neuronal onset latencies are the same. This difference in form is due to the network dynamics as activity is promulgated up the system and then re-enters lower levels. Notice also that the neuronal kernels have quite protracted dynamics compared to the characteristic neuronal time constants of each area (about a second). This enduring activity, particularly in the higher two areas is a product of the network dynamics. The MAP estimates also enable us to compute the predicted response (lower left panel) in each region and compare it to the true response without observation noise (lower right panel). This comparison shows that the actual and predicted responses are very similar.

In Friston *et al.* (2002b), we repeated this estimation procedure to explore the face validity of the estimation scheme over a range of hyperparameters like noise levels, slice timing artefacts, extreme values of the biophysical parameters etc. In general, the scheme proved to be robust to most violations assessed. Here we will just look at the effects of error variance on estimation because this speaks of some important features of Bayesian estimation and the noise levels that can be tolerated.

The effects of noise

In this sub-section, we investigate the sensitivity and specificity of posterior density estimates to the level of observation noise. Data were simulated as described above and mixed with various levels of white noise. For each noise level the posterior densities of the coupling parameters were estimated and plotted against the noise hyperparameter (expressed as its standard deviation) in terms of the posterior mean and 90 per cent confidence intervals. Figure 41.8 shows some key coupling parameters that include both zero and non-zero connection strengths. The solid lines represent the posterior expectation or MAP estimator and the broken lines indicate the true value. The grey areas encompass the 90 per cent confidence regions. Characteristic behaviours of the estimation are apparent from these results. As one might intuit, increasing the level of noise increases the uncertainty in the posterior estimates as reflected by an increase in the conditional variance and a widening of the confidence intervals. This widening is, however, bounded by the prior variances to which the conditional variances asymptote, at very high levels of noise. Concomitant with this effect is 'shrinkage' of some posterior means to their prior expectation of zero. Put simply, when the data become very noisy the estimation relies more heavily upon priors and the prior expectation is given more weight. This is why priors of the sort used here are referred to as 'shrinkage priors'. These simulations suggest that, for this level of evoked response, noise levels between zero and two permit the connection strengths to be identified with a fair degree of precision and accuracy. Noise levels in typical fMRI experiments are about one. The units of signal and noise are adimensional and correspond to percentage whole brain mean. Pleasingly, noise did not lead to false inferences in the sense that the posterior densities always encompassed the true values, even at high levels of noise (Figure 41.8).

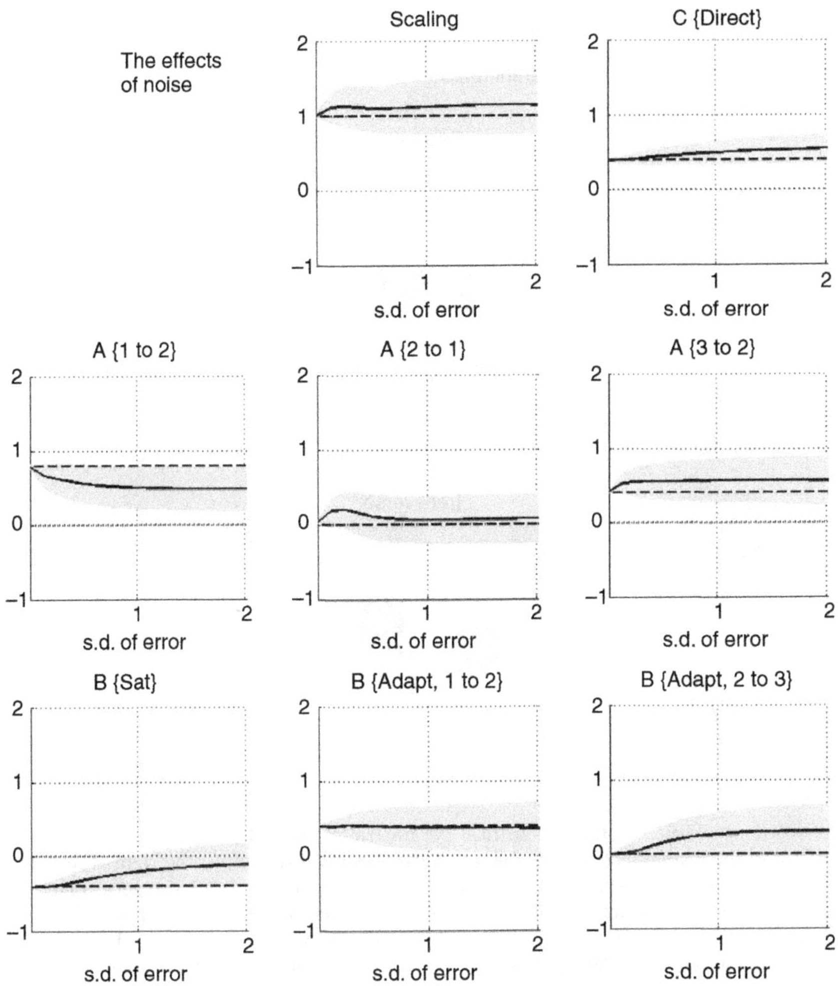

FIGURE 41.8 Posterior densities as a function of noise levels: the analysis, summarized in the previous two figures, was repeated for simulated data sequences at different levels of noise ranging from 0 to 2 units of standard deviation. Each graph shows the conditional expectation or MAP estimate of a coupling parameter (solid line) and the 90 per cent confidence region (grey region). The true value for each parameter is also shown (broken line). The top row shows the temporal scaling parameter and the extrinsic connection between the first input and the first area. The middle row shows some regional coupling parameters and the bottom row bilinear parameters. As anticipated, the conditional variance of these estimators increases with noise, as reflected by a divergence of the confidence region with increasing error.

PREDICTIVE VALIDITY – AN ANALYSIS OF SINGLE WORD PROCESSING

In this section, we illustrate the predictive validity of DCM by showing that reproducible results can be obtained from independent data. The data set we used was especially designed for these sorts of analyses, comprising over 1200 scans with a relatively short TR of 1.7 s. This necessitated a limited field of coverage, but provided relatively high temporal acuity. The paradigm was a passive listening task, using epochs of single words presented at different rates. These data have been used previously to characterize non-linear aspects of haemodynamics (e.g. Friston *et al.*, 1998, 2000, 2002a). Details of the experimental paradigm and acquisition parameters are provided in the legend to Figure 41.9. These data were acquired in consecutive sessions of 120 scans enabling us to analyse the entire time-series or each session independently. We first present the results obtained by concatenating all the sessions into a single data sequence. We then revisit the data, analysing each session independently to provide ten independent conditional estimates of the coupling parameters, to assess reproducibility and mutual predictability.

Analysis of the complete time-series

Three regions were selected using maxima of the SPM{F} following a conventional SPM analysis (see Figure 41.9). The three maxima were those that were closest to the primary and secondary auditory areas and Wernicke's area, using the anatomic designations provided in the atlas of Talairach and Tournoux (1988). Region-specific time-series comprised the first eigenvariate of all voxels within a 4 mm-radius sphere centred on each location. The anatomical locations are shown in Figure 41.9. As in the simulations, there were two inputs corresponding to a delta-function for the occurrence of an aurally presented word and a parametric input modelling within-epoch adaptation. The outputs of the system were the three principal eigenvariates from each region. As in the previous section, we allowed for a fully connected system.

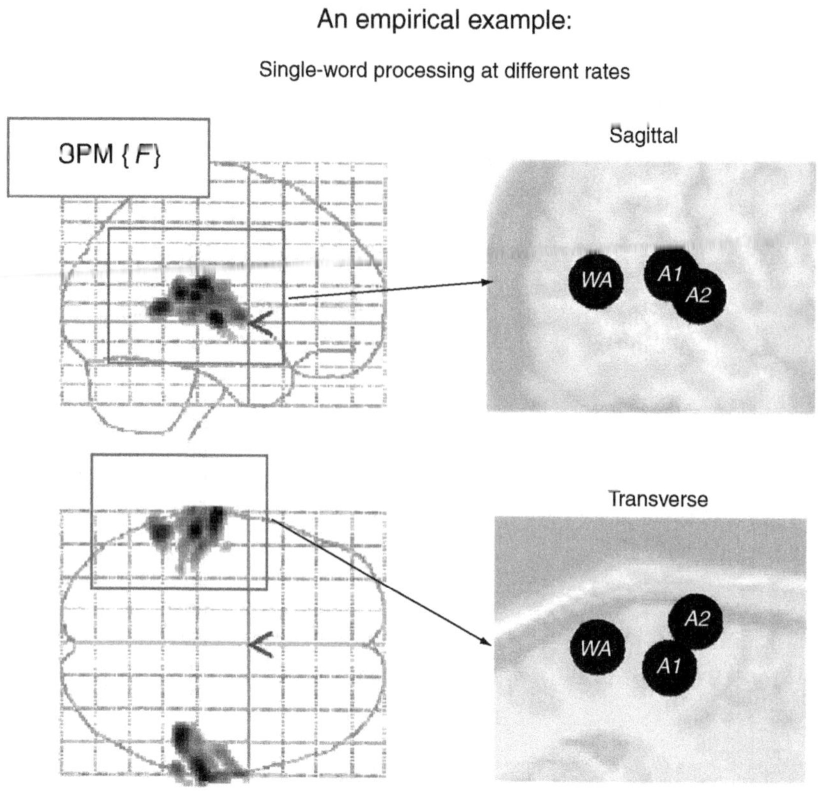

FIGURE 41.9 Region selection for the empirical word processing example: statistical parametric maps of the F-ratio, based upon a conventional SPM analysis, are shown in the left panels and the spatial locations of the selected regions are shown on the right. These are superimposed on a T1-weighted reference image. The regional activities shown in the next figure correspond to the first eigenvariates of a 4 mm-radius sphere centred on the following coordinates in the standard anatomical space of Talairach and Tournoux (1988). Primary auditory area **A1**; −50, −26, 8 mm. Secondary auditory area **A2**; −64, −18, 2 mm and Wernicke's area **WA**; −56, −48, 6 mm. In brief, we obtained fMRI time-series from a single subject at 2 tesla using a Magnetom VISION (Siemens, Erlangen) whole-body MRI system, equipped with a head volume coil. Contiguous multislice T2*-weighted fMRI images were obtained with a gradient echo-planar sequence using an axial slice orientation (TE = 40 ms, TR = 1.7 s, 64 × 64 × 16 voxels). After discarding initial scans (to allow for magnetic saturation effects) each time-series comprised 1200 volume images with 3 mm isotropic voxels. The subject listened to monosyllabic or bisyllabic concrete nouns (i.e. 'dog', 'radio', 'mountain', 'ate') presented at five different rates (10, 15, 30, 60 and 90 words per minute) for epochs of 34 s, intercalated with periods of rest. The five presentation rates were successively repeated according to a Latin Square design. The data were smoothed with a 5 mm isotropic Gaussian kernel. The SPM{F} above was based on a standard regression model using word presentation rate as the stimulus function and convolving it with a canonical haemodynamic response and its temporal derivative to form regressors.

In other words, each region was potentially connected to every other region. Generally, one would impose constraints on highly unlikely or implausible connections by setting their prior variance to zero. However, we wanted to demonstrate that dynamic causal modelling can be applied to connectivity graphs that would be impossible to analyse with structural equation modelling. The auditory input was connected to **A1**. In addition, auditory input entered bilinearly to emulate saturation, as in the simulations. The contextual input, modelling putative adaptation, was allowed to exert influences over all regional connections. From a neurobiological perspective an interesting question is whether plasticity can be demonstrated in forward connections or backward connections. Plasticity, in this instance, entails a time-dependent increase or decrease in effective connectivity and would be inferred by significant bilinear coupling parameters associated with the second input.

The inputs, outputs and priors on the DCM parameters entered the Bayesian inversion as described above. Drifts were modelled with the first 40 components of a discrete cosine set, corresponding to X in Eqn. **41.8**. The results of this analysis, in terms of the posterior densities and Bayesian inference, are presented in Figures 41.10 and 41.11. Bayesian inferences were based upon the probability that the coupling parameters exceeded 0.0866. This corresponds to a half-life of 8 s. Intuitively, this means that we only consider the influences, of one region on another, to be meaningful if this influence is expressed within a time frame of 8 s or less. The results show that the most probable architecture, given the inputs and data, conforms to a simple hierarchy of forward

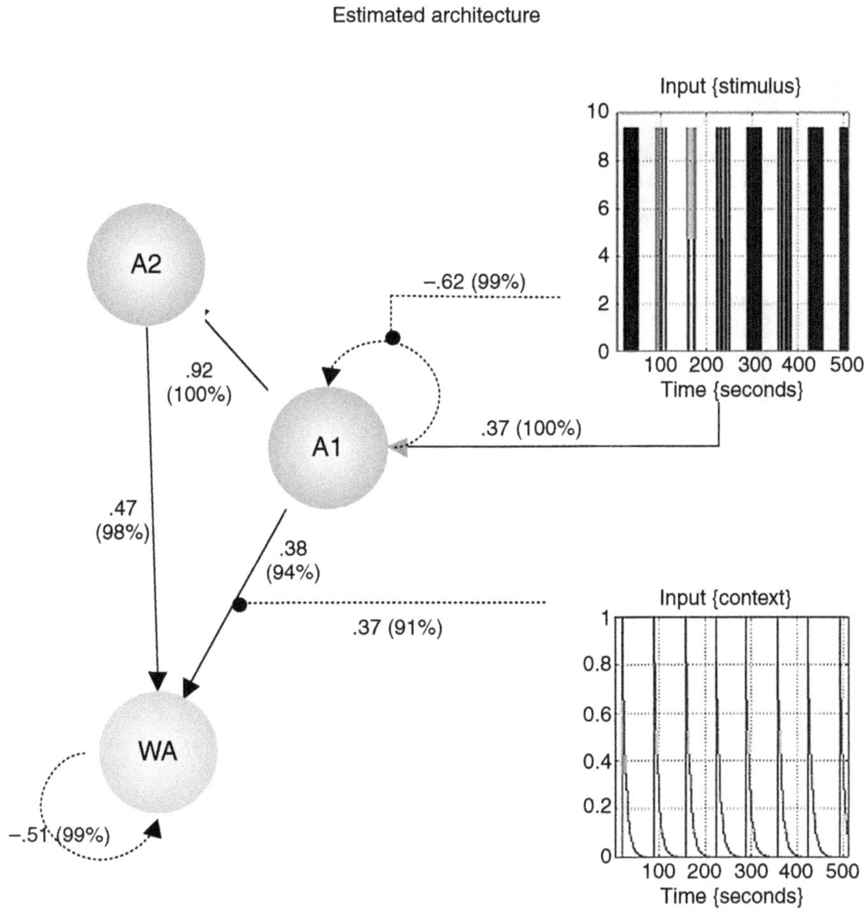

FIGURE 41.10 Results of a DCM analysis applied to the data described in the previous figure. The display format follows that of Figure 41.5. The coupling parameters are shown alongside the corresponding connections. The values in brackets are the percentage confidence that these values exceed a threshold of ln(2)/8 per second.

connections where **A1** influences **A2** and **WA**, whereas **A2** sends connections just to **WA** (Figure 41.10). Although backward connections between **WA** and **A2** were estimated to be greater than our threshold with 82 per cent confidence they are not shown in Figure 41.11 (which is restricted to posterior probabilities of 90 per cent or more). Saturation could be inferred in **A1** and **WA** with a high degree of confidence with b^1_{11} and b^1_{33} being greater than 0.5. Significant plasticity or time-dependent changes were expressed predominantly in the forward connections, particularly that between **A1** and **A3**, i.e. $b^2_{13} = 0.37$. The conditional estimates are shown in more detail in Figure 41.11, along with the conditional fitted responses and associated kernels. A full posterior density analysis for a particular contrast of effects is shown in Figure 41.11(a) (lower panel). This contrast tested for the average plasticity over all forward and backward connections and demonstrates that we can be virtually certain plasticity was greater than zero.

This analysis illustrates three things. First, the DCM has defined a hierarchical architecture that is the most likely given the data. This hierarchical structure was not part of the prior constraints because we allowed for a fully connected system. Second, the significant bilinear effects of auditory stimulation suggest there is measurable neuronal saturation above and beyond that attributable to haemodynamic non-linearities. This is quite significant because such disambiguation is usually impossible given just haemodynamic responses. Finally, we were able to show time-dependent decreases in effective connectivity in forward connections from **A1**. Although this experiment was not designed to test for plasticity, the usefulness of DCM, in studies of learning and priming, should be self-evident.

Reproducibility

The analysis above was repeated identically for each 120-scan session to provide ten sets of Bayesian estimators. Drifts were modelled with the first four components of a discrete cosine set. The estimators are presented graphically in Figure 41.12 and demonstrate extremely consistent results. In the upper panels, the intrinsic connections are shown to be very similar; again reflecting a hierarchical architecture. The conditional means and 90 per cent confidence regions for two connections are shown in Figure 41.12(a). These connections included the forward

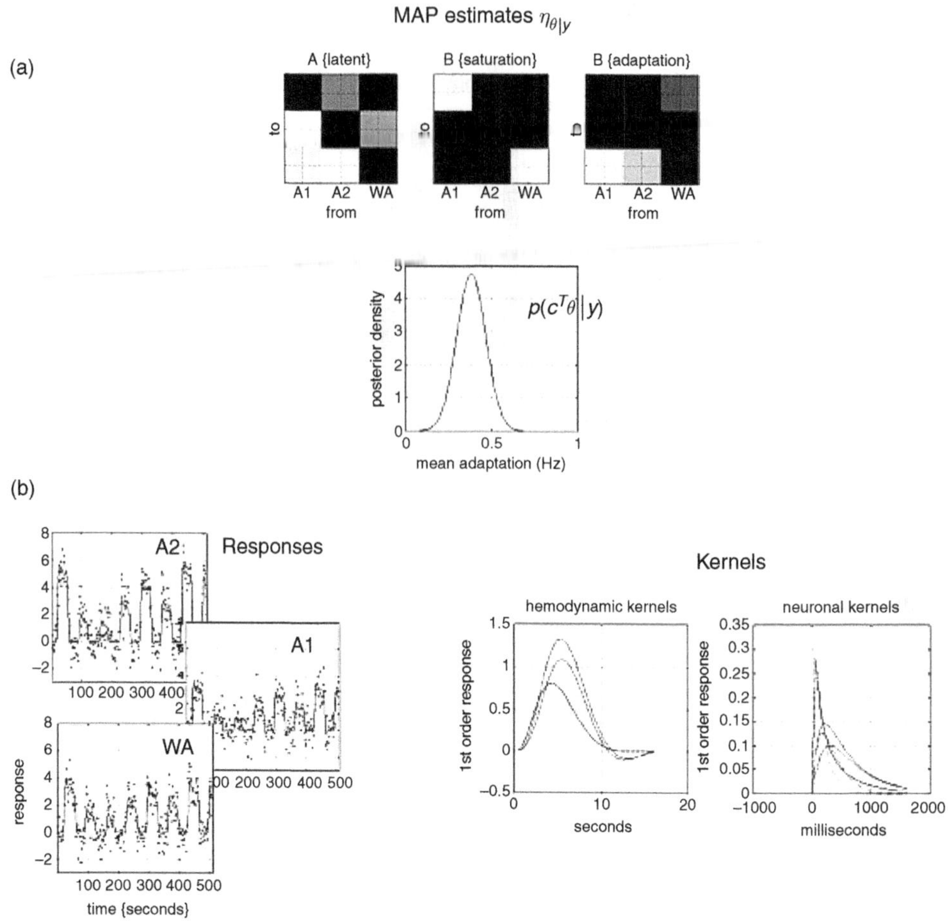

FIGURE 41.11 This figure provides a more detailed characterization of the conditional estimates. The images in the top row are the MAP estimates for the regional and bilinear coupling parameters, pertaining to saturation and adaptation. The middle panel shows the posterior density of a contrast of all bilinear terms mediating adaptation, namely the modulation of regional connections by the second time-dependent experimental effect. The predicted responses based upon the conditional estimators are shown for each of the three regions on the lower left (solid lines) with the original data (dots) after removal of confounds. A re-parameterization of the conditional estimates, in terms of the first-order kernels, is shown on the lower right. The haemodynamic (left) and neuronal (right) kernels should be compared with the equivalent kernels for the simulated data in Plate 58.

connection from **A1** to **A2** that is consistently strong. The backward connection from **WA** to **A2** was weaker, but was certainly greater than zero in every analysis. Equivalent results were obtained for the modulatory effects or bilinear terms, although the profile was less consistent (Figure 41.12(b)). However, the posterior density of the contrast testing for average time-dependent adaptation or plasticity is relatively consistent and again almost certainly greater than zero, in each analysis.

To illustrate the stability of hyperparameter estimates, the standard deviations of observation error are presented for each session over the three areas in Figure 41.13. As typical of studies at this field strength the standard deviation of noise is about 0.8–1 per cent whole brain mean. It is pleasing to note that the session-to-session variability in hyperparameter estimates was relatively small, in relation to region-to-region differences.

In summary, independent analyses of data acquired under identical stimulus conditions, on the same subject, in the same scanning session, yield remarkably similar results. These results are biologically plausible and speak of time-dependent changes, following the onset of a stream of words, in forward connections among auditory areas.

CONSTRUCT VALIDITY – AN ANALYSIS OF ATTENTIONAL EFFECTS ON CONNECTIONS

In this final section, we address the construct validity of DCM. In previous chapters, we have seen that attention positively modulates the backward connections in a distributed system of cortical regions mediating attention

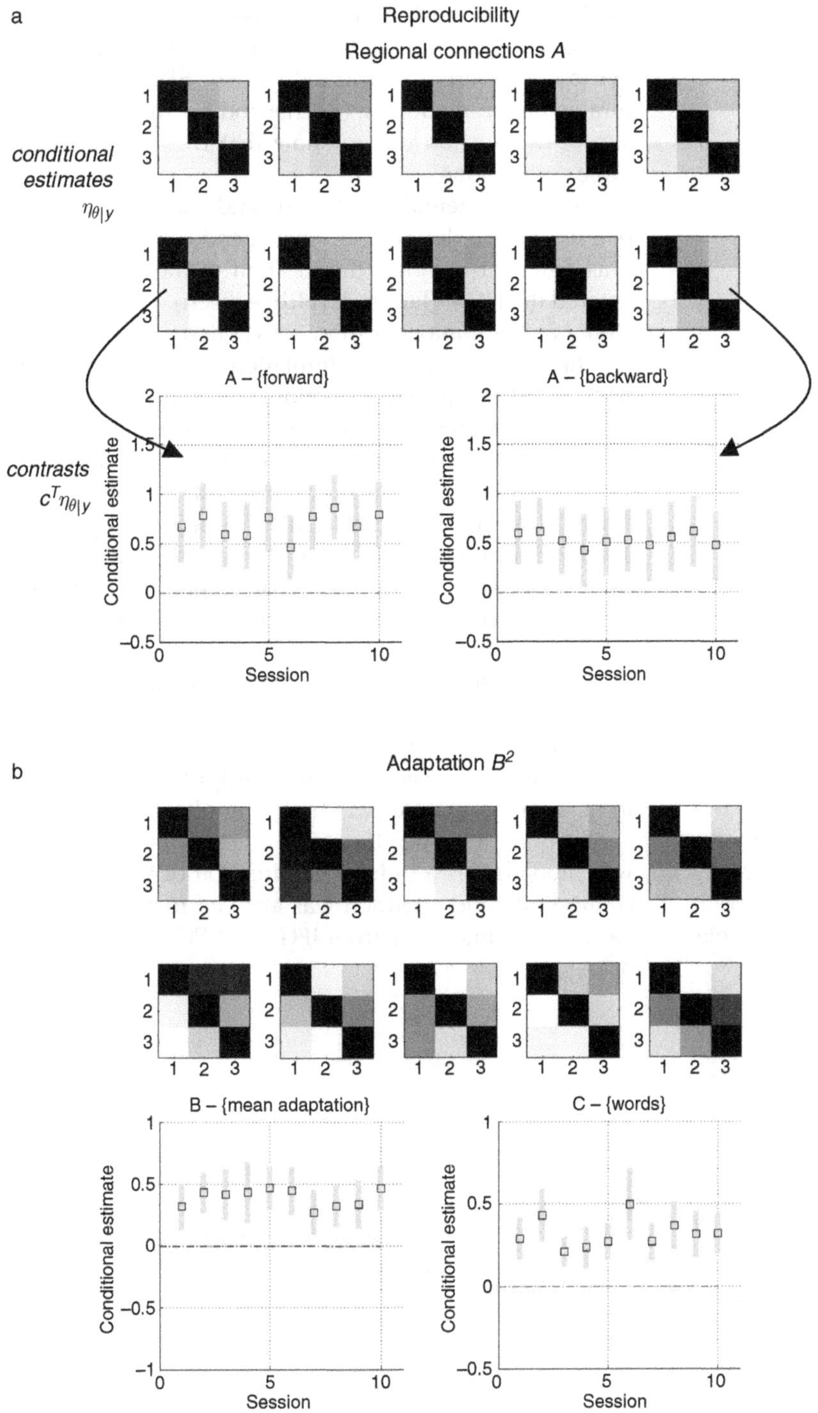

FIGURE 41.12 Results of the reproducibility analyses: (a) results for the regional parameters. The profile of conditional estimates for the 10 independent analyses described in the main text are shown in image format, all scaled to the maximum. The posterior densities, upon which these estimates are based, are shown for two selected connections in the lower two graphs. These densities are displayed in terms of their expectation and 90 per cent confidence intervals (grey bars) for the forward connection from **A1** to **A2**. The equivalent densities are shown for the backward connection from **WA** to **A2**. Although the posterior probability that the latter connections exceeded the specified threshold was less than 90 per cent, it can be seen that this connection is almost certainly greater than zero. (b) Equivalent results for the bilinear coupling matrices mediating adaptation. The lower panels here refer to the posterior densities of a contrast testing for the mean of all bilinear parameters (left) and the extrinsic connection to **A1** (right).

to radial motion. We use the same data in this section. In brief, subjects viewed optic flow stimuli comprising radially moving dots at a fixed velocity. In some epochs, subjects were asked to detect changes in velocity (that did not actually occur). This attentional manipulation was validated *post hoc* using psychophysics and the motion after-effect. Analyses using structural equation modelling (Büchel and Friston, 1997) and a Volterra formulation of effective connectivity (Friston and Büchel, 2000) have established a hierarchical backwards modulation of effective connectivity, where a higher area increases the effective connectivity among two subordinate areas. These analyses have been extended using variable parameter regression and Kalman filtering (Büchel and Friston, 1998) to look at the effect of attention directly on interactions between V5 and the posterior

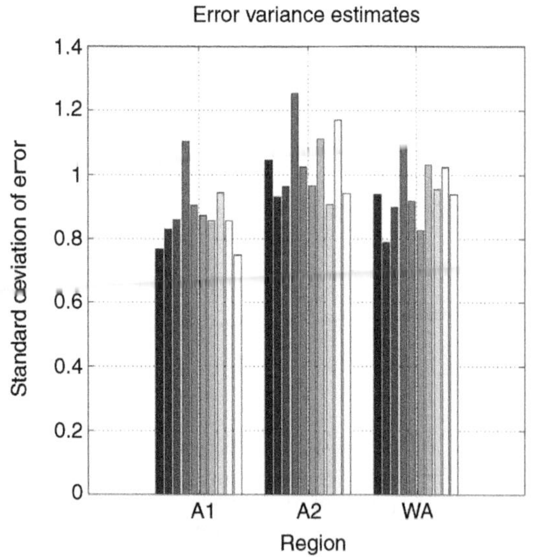

FIGURE 41.13 Hyperparameter variance estimates for each region and analysis: these estimates provide an anecdotal characterization of the within- and between-area variability, in hyperparameter estimates, and show that they generally lie between 0.8 and 1 (adimensional units corresponding to per cent whole-brain mean).

parietal complex. Even simple analyses, such as those employing psychophysiological interactions, point to the same conclusion that attention generally increases the effective connectivity among extrastriate and parietal areas. In short, we have established that the superior posterior parietal cortex (SPC) exerts a modulatory role on V5 responses using Volterra-based regression models (Friston and Büchel, 2000) and that the inferior frontal gyrus (IFG) exerts a similar influence on SPC using structural equation modelling (Büchel and Friston, 1997). The aim of this section was to show that DCM leads to the same conclusions.

The experimental paradigm and data acquisition parameters are described in the legend to Figure 41.14b. This figure also shows the location of the regions that entered the DCM (Figure 41.14(b) – insert). These regions were based on maxima from conventional SPMs testing for the effects of photic stimulation, motion and attention. As in the previous section, regional time courses were taken as the first eigenvariate of spherical volumes of interest centred on the maxima shown in the figure. The inputs, in this example, comprise one sensory perturbation and two contextual inputs. The sensory input was simply the presence of photic stimulation and the first contextual input was presence of motion in the visual field. The second contextual input, encoding attentional set, was unity during attention to speed changes and zero otherwise. The outputs corresponded to the four regional eigenvariates in Figure 41.14(b). The intrinsic connections were constrained to conform to a hierarchical pattern, in which each area was reciprocally connected to its supraordinate area. Photic stimulation entered at, and only at, V1. The effect of motion in the visual field was modelled as a bilinear modulation of the V1 to V5 connectivity and attention was allowed to modulate the backward connections from IFG and SPC.

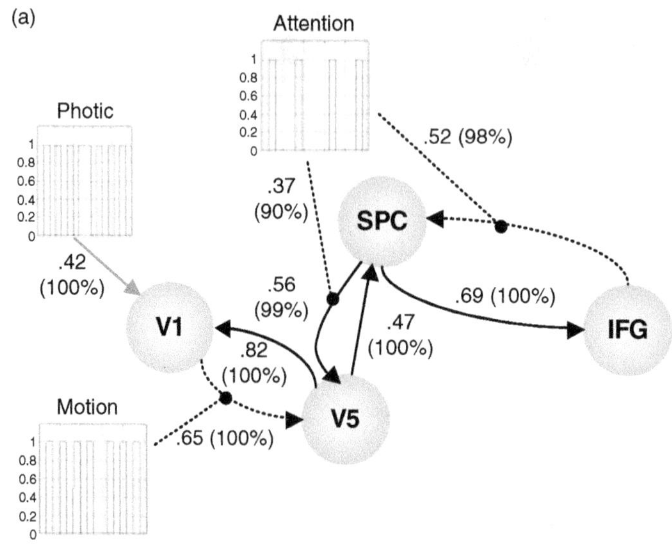

FIGURE 41.14 Results of the empirical analysis of the attention study. (a) Functional architecture based upon the conditional estimates displayed using the same format as Figure 41.10. The most interesting aspects of this architecture involved the role of motion and attention in exerting bilinear effects. Critically, the influence of motion is to enable connections from **V1** to the motion sensitive area **V5**. The influence of attention is to enable backward connections from the inferior frontal gyrus (**IFG**) to the superior parietal cortex (**SPC**). Furthermore, attention increases the latent influence of **SPC** on **V5**. Dotted arrows connecting regions represent significant bilinear affects in the absence of a significant regional coupling.

(b)

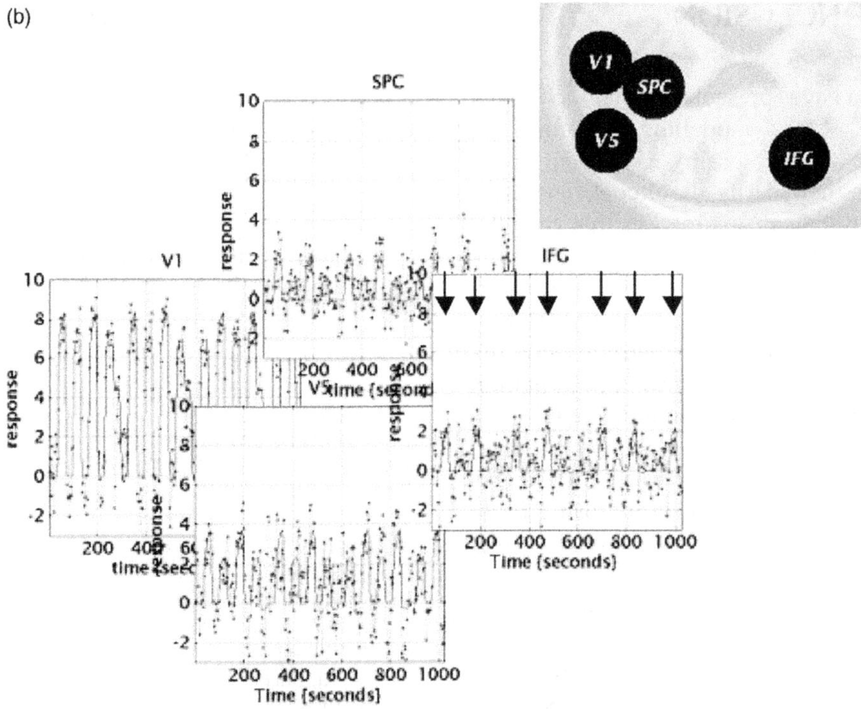

FIGURE 41.14 (*Continued*) (b) Fitted responses based upon the conditional estimates and the adjusted data are shown using the same format as in Figure 41.11. The insert shows the location of the regions, again adopting the same format in previous figures. The location of these regions centred on the primary visual cortex **V1**; 6, −84, −6 mm: motion-sensitive area **V5**; 45, −81, 5 mm. Superior parietal cortex, **SPC**; 18, −57, 66 mm. Inferior frontal gyrus, **IFG**, 54, 18, 30 mm. The volumes from which the first eigenvariates were calculated corresponded to 8 mm-radius spheres centred on these locations.

Subjects were studied with fMRI under identical stimulus conditions (visual motion subtended by radially moving dots) while manipulating the attentional component of the task (detection of velocity changes). The data were acquired from normal subjects at 2 tesla using a Magnetom VISION (Siemens, Erlangen) whole-body MRI system, equipped with a head volume coil. Here we analyse data from the first subject. Contiguous multislice T2*-weighted fMRI images were obtained with a gradient echo-planar sequence (TE = 40 ms, TR = 3.22 s, matrix size = 64 × 64 × 32, voxel size 3 × 3 × 3 mm). Each subject had four consecutive 100-scan sessions comprising a series of 10-scan blocks under five different conditions D F A F N F A F N S. The first condition (D) was a dummy condition to allow for magnetic saturation effects. F (Fixation) corresponds to a low-level baseline where the subjects viewed a fixation point at the centre of a screen. In condition A (Attention), subjects viewed 250 dots moving radially from the centre at 4.7 degrees per second and were asked to detect changes in radial velocity. In condition N (No attention), the subjects were asked simply to view the moving dots. In condition S (Stationary), subjects viewed stationary dots. The order of A and N was swapped for the last two sessions. In all conditions, subjects fixated the centre of the screen. In a pre-scanning session the subjects were given five trials with five speed changes (reducing to 1 per cent). During scanning there were no speed changes. No overt response was required in any condition.

The results of the DCM are shown in Figure 41.14(a). Of primary interest here is the modulatory effect of attention that is expressed in terms of the bilinear coupling parameters for this third input. As hoped, we can be highly confident that attention modulates the backward connections from IFG to SPC and from SPC to V5. Indeed, the influences of IFG on SPC are negligible in the absence of attention (dotted connection in Figure 41.14(a)). It is important to note that the only way that attentional manipulation can affect brain responses was through this bilinear effect. Attention-related responses are seen throughout the system (attention epochs are marked with arrows in the plot of IFG responses in Figure 41.14(b)). This attentional modulation is accounted for by changing just two connections. This change is, presumably, instantiated by instructional set at the beginning of each epoch. The second thing this analysis illustrates is how functional segregation is modelled in DCM. Here one can regard V1 as a 'segregating' motion from other visual information and distributing it to the motion-sensitive area V5. This segregation is modelled as a bilinear 'enabling' of V1 to V5 connections when, and only when, motion is present. Note that, in the absence of motion, the V1 to V5 connection was trivially small (in fact the MAP estimate was −0.04). The key advantage of entering motion through a bilinear effect, as opposed to a direct effect on V5, is that we can finesse the inference that V5 shows motion-selective responses with the assertion that these responses are mediated by afferents from V1. The two bilinear effects above represent two important aspects of functional integration that DCM was designed to characterize.

CONCLUSION

In this chapter, we have presented dynamic causal modelling. DCM is a causal modelling procedure for dynamical systems in which causality is inherent in the differential equations that specify the model. The basic idea is to treat the system of interest, in this case the brain, as an input-state-output system. By perturbing the system with known inputs, measured responses are used to estimate various parameters that govern the evolution of brain states. Although there are no restrictions on the parameterization of the model, a bilinear approximation affords a simple re-parameterization in terms of effective connectivity. This effective connectivity can be latent or, through bilinear terms, model input-dependent changes in coupling. Parameter estimation proceeds using fairly standard approaches to system identification that rest upon Bayesian inference.

Dynamic causal modelling represents a fundamental departure from conventional approaches to modelling effective connectivity in neuroscience. The critical distinction between DCM and other approaches, such as structural equation modelling or multivariate autoregressive techniques, is that the input is treated as known, as opposed to stochastic. In this sense, DCM is much closer to conventional analyses of neuroimaging time-series because the causal or explanatory variables enter as known fixed quantities. The use of designed and known inputs in characterizing neuroimaging data with the general linear model or DCM is a more natural way to analyse data from designed experiments. Given that the vast majority of imaging neuroscience relies upon designed experiments, we consider DCM a potentially useful complement to existing techniques. We develop this point and the relationship of DCM to other approaches in Appendix 2.

In the next chapter, we consider DCMs for EEG. Here the electromagnetic model mapping neuronal states to measurements is simpler than the haemodynamic models used in fMRI. Conversely, the neuronal component of the model is much more complicated and realistic. This is because there is more temporal information in EEG.

REFERENCES

Bendat JS (1990) *Nonlinear system analysis and identification from random data*. John Wiley and Sons, New York

Bitan T, Booth JR, Choy J *et al.* (2005) Shifts of effective connectivity within a language network during rhyming and spelling. *J Neurosci* **25**: 5397–403

Büchel C, Friston KJ (1997) Modulation of connectivity in visual pathways by attention: cortical interactions evaluated with structural equation modelling and fMRI. *Cereb Cortex* **7**: 768–78

Büchel C, Friston KJ (1998) Dynamic changes in effective connectivity characterised by variable parameter regression and Kalman filtering. *Hum Brain Mapp* **6**: 403–08

Buxton RB, Wong EC, Frank LR (1998) Dynamics of blood flow and oxygenation changes during brain activation: the Balloon model. *Mag Res Med* **39**: 855–64

Ethofer T, Anders S, Erb M *et al.* (2006) Cerebral pathways in processing of affective prosody: a dynamic causal modeling study. *NeuroImage* **30**: 580–87

Friston KJ (2002) Bayesian estimation of dynamical systems: an application to fMRI. *NeuroImage* **16**: 513–30

Friston KJ, Büchel C (2000) Attentional modulation of effective connectivity from V2 to V5/MT in humans. *Proc Natl Acad Sci USA* **97**: 7591–96

Friston KJ, Büchel C, Fink GR *et al.* (1997) Psychophysiological and modulatory interactions in neuroimaging. *NeuroImage* **6**: 218–29

Friston KJ, Josephs O, Rees G *et al.* (1998) Nonlinear event-related responses in fMRI. *Mag Res Med* **39**: 41–52

Friston KJ, Mechelli A, Turner R *et al.* (2000) Nonlinear responses in fMRI: the Balloon model, Volterra kernels and other hemodynamics. *NeuroImage* **12**: 466–77

Friston KJ, Penny W, Phillips C *et al.* (2002a) Classical and Bayesian inference in neuroimaging: theory. *NeuroImage* **16**: 465–83

Friston KJ, Harrison L, Penny W (2002b) Dynamic causal modelling. *NeuroImage* **19**:1273–302

Horwitz B, Friston KJ, Taylor JG (2001) Neural modeling and functional brain imaging: an overview. *Neural Netw* **13**: 829–46

Kenny DA, Judd CM (1984) Estimating nonlinear and interactive effects of latent variables. *Psychol Bull* **96**: 201–10

Mandeville JB, Marota JJ, Ayata C *et al.* (1999) Evidence of a cerebrovascular postarteriole Windkessel with delayed compliance. *J Cereb Blood Flow Metab* **19**: 679–89

McIntosh AR (2000) Towards a network theory of cognition. *Neural Netw* **13**: 861–70

McIntosh AR, Gonzalez-Lima F (1994) Structural equation modelling and its application to network analysis in functional brain imaging. *Hum Brain Mapp* **2**: 2–22

Mechelli A, Price CJ, Noppeney U *et al.* (2003) A dynamic causal modeling study on category effects: bottom-up or top-down mediation? *J Cogn Neurosci* **15**: 925–34

Talairach J, Tournoux P (1988) *A co-planar stereotaxic atlas of a human brain*. Thieme, Stuttgart

CHAPTER

42

Dynamic causal models for EEG

K. Friston, S. Kiebel, M. Garrido and O. David

INTRODUCTION

Neuronally plausible, generative or forward models are essential for understanding how event-related fields (ERFs) and potentials (ERPs) are generated. In this chapter, we describe the dynamic causal modelling (DCM) of event-related responses measured with electroencephalography (EEG) or magnetoencephalography (MEG). This approach uses a biologically informed causal model to make inferences about the underlying neuronal networks generating responses. The approach can be regarded as a neurobiologically constrained source reconstruction scheme, in which the parameters of the reconstruction have an explicit neuronal interpretation. Specifically, these parameters encode, among other things, the coupling among sources and how that coupling depends upon stimulus attributes or experimental context. The basic idea is to supplement conventional electromagnetic forward models, of how sources are expressed in measurement space, with a model of how source activity is generated by neuronal dynamics. A single inversion of this extended forward model enables inference about both the spatial deployment of sources and the underlying neuronal architecture generating them. Critically, this inference covers long-range connections among well-defined neuronal subpopulations.

In Chapter 33, we simulated ERPs using a hierarchical neural-mass model that embodied bottom-up, top-down and lateral connections among remote regions. In this chapter, we describe a Bayesian procedure to estimate the parameters of this model using empirical data. We demonstrate this procedure by characterizing the role of changes in cortico-cortical coupling, in the genesis of ERPs using two examples. In brief, in the first example, ERPs recorded during the perception of faces and houses are modelled as distinct cortical sources in the ventral visual pathway. We will see that category-selectivity, as indexed by the face-selective N170, can be explained by category-specific differences in forward connections from sensory to higher areas in the ventral stream. These changes allow one to identify where, in the processing stream, category-selectivity emerges. The second example uses an auditory oddball paradigm to show that mismatch negativity can be explained by changes in connectivity. Specifically, using Bayesian model selection, we will assess changes in backward connections, above and beyond changes in forward connections. In accord with theoretical predictions, we will see strong evidence for learning-related changes in both forward and backward coupling. These examples illustrate DCM for ERPs to address category- or context-specific coupling among cortical regions.

Background

ERFs and ERPs have been used for decades as magneto- and electrophysiological correlates of perceptual and cognitive operations. However, the exact neurobiological mechanisms underlying their generation are largely unknown. Previous studies have shown that ERP-like responses can be reproduced by perturbations of model cortical networks (Jansen and Rit, 1995; Rennie *et al.*, 2002; Jirsa, 2004; David *et al.*, 2005). Here we show that changes in connectivity, among distinct cortical sources, are sufficient to explain stimulus- or set-specific ERP differences.

Functional versus effective connectivity

The aim of dynamic causal modelling (Friston *et al.*, 2003) is to make inferences about the coupling among brain regions or sources and how that coupling is influenced by experimental factors. DCM uses the notion of *effective connectivity*, defined as the influence one neuronal system exerts over another. DCM represents a departure from existing approaches to connectivity because it employs

Statistical Parametric Mapping, by Karl Friston et al.
ISBN–13: 978-0-12-372560-8 ISBN–10: 0-12-372560-7

an explicit generative model of measured brain responses that embraces their non-linear causal architecture. The alternative to causal modelling is simply to establish statistical dependencies between activity in one brain region and another. This is referred to as *functional connectivity*. Functional connectivity is useful because it rests on an operational definition and eschews any arguments about how dependencies are caused. Most approaches in the EEG and MEG literature address functional connectivity, with a focus on dependencies that are expressed at a particular frequency of oscillations (i.e. coherence) (see Schnitzler and Gross, 2005 for a nice review). Recent advances have looked at non-linear or generalized synchronization in the context of chaotic oscillators (e.g. Rosenblum *et al.*, 2002) and stimulus-locked responses of coupled oscillators (see Tass, 2004). These characterizations often refer to phase-synchronization as a useful measure of non-linear dependency. Another exciting development is the reformulation of coherence in terms of autoregressive models. A compelling example is reported in Brovelli *et al.* (2004), who were able show that: 'synchronized beta oscillations bind multiple sensorimotor areas into a large-scale network during motor maintenance behaviour and carry Granger causal influences from primary somatosensory and inferior posterior parietal cortices to motor cortex'. Similar developments have been seen in functional neuroimaging with fMRI (e.g. Harrison *et al.*, 2003; Roebroeck *et al.*, 2005).

These approaches generally entail a two-stage procedure. First, an electromagnetic forward model is inverted to estimate the activity of sources in the brain. Then, a *post-hoc* analysis is used to establish statistical dependencies (i.e. functional connectivity) using coherence, phase-synchronization, Granger influences or related analyses such as (linear) directed transfer functions and (non-linear) generalized synchrony. DCM takes a very different approach and uses a forward model that explicitly includes long-range connections among neuronal subpopulations underlying measured sources. A single Bayesian inversion allows one to infer on the coupling parameters of the model (i.e. effective connectivity) that mediate functional connectivity. This is like performing a biological informed source reconstruction with the added constraint that the activity in one source has to be caused by activity in others, in a biologically plausible fashion. This approach is much closer in sprit to the work of Robinson *et al.* (2004) who show that, 'model-based electroencephalographic (EEG) methods can quantify neurophysiologic parameters that underlie EEG generation in ways that are complementary to and consistent with standard physiologic techniques'. DCM also speaks of the interest in neuronal modelling of ERPs in specific systems. See, for example, Melcher and Kiang (1996), who evaluate a detailed cellular model of brainstem auditory evoked potentials (BAEP) and conclude: 'it should now be possible to relate activity in specific cell populations to psychophysical performance since the BAEP can be recorded in behaving humans and animals' (see also Dau, 2003). Although the models presented in this chapter are more generic than those invoked to explain the BAEP, they share the same ambition of understanding the mechanisms of response generation and move away from phenomenological or descriptive quantitative EEG measures.

Dynamic causal modelling

The central idea behind DCM is to treat the brain as a deterministic non-linear dynamical system that is subject to inputs, and produces outputs. Effective connectivity is parameterized in terms of coupling among unobserved brain states, i.e. neuronal activity in different regions. Coupling is estimated by perturbing the system and measuring the response. This is in contradistinction to established methods for estimating effective connectivity from neurophysiological time-series, which include structural equation modelling and models based on multivariate autoregressive processes (Mcintosh and Gonzalez-Lima, 1994; Büchel and Friston, 1997; Harrison *et al.*, 2003). In these models, there is no designed perturbation and the inputs are treated as unknown and stochastic. Although the principal aim of DCM is to explain responses in terms of context-dependent coupling, it can also be viewed as a biologically informed inverse solution to the source reconstruction problem. This is because estimating the parameters of a DCM rests on estimating the hidden states of the modelled system. In ERP studies, these states correspond to the activity of the sources that comprise the model. In addition to biophysical and coupling parameters, the DCM's parameters cover the spatial expression of sources at the sensor level. This means that inverting the DCM entails a simultaneous reconstruction of the source configuration and their dynamics.

Implicit in the use of neural-mass models is the assumption that the data can be explained by random fluctuations around population dynamics that are approximated with a point mass (i.e. the mean or expected state of a population). This is usually interpreted in relation to the dynamics of an ensemble of neurons that constitute sources of signal. However, in the context of modelling ERPs and ERFs, there is also an ensemble of trials that are averaged to form the data. The mean-field-like assumptions that motivate neural-mass models can be extended to cover ensembles of trials. This sidesteps questions about the trial-to-trial genesis of ERPs. However, we have previously addressed these questions using the same neural-mass model used in this chapter (David *et al.*, 2005), by dissociating 'the components of event-related potentials (ERPs) or event-related fields (ERFs) that can be explained by a linear superposition of trial-specific responses and those engendered non-linearly (e.g. by phase-resetting)' (see David *et al.*, 2005 and Chapter 33 for further details).

DCM has been validated previously with functional magnetic resonance imaging (fMRI) time-series (Friston *et al.*, 2003; Riera *et al.*, 2004). fMRI responses depend on haemodynamic processes that effectively lowpass filter neuronal dynamics. However, with ERPs this is not the case and there is sufficient information, in the temporal structure of evoked responses, to enable precise conditional identification of quite complicated DCMs. We will use a model (David *et al.*, 2005 and Chapter 33) that embeds cortical sources, with several source-specific neuronal subpopulations, into hierarchical cortico-cortical networks.

This chapter is structured as follows. In the theory section, we review the neural-mass model used to generate M/EEG-like evoked responses. This section summarizes Chapter 33, in which more details about the generative model and associated dynamics can be found. The next section provides a brief review of Bayesian estimation, conditional inference and model comparison that are illustrated in the subsequent section. An empirical section then demonstrates the use of DCM for ERPs by looking at changes in connectivity that were induced, either by category-selective activation of different pathways in the visual system, or by sensory learning in an auditory oddball paradigm.

THEORY

Intuitively, the DCM scheme regards an experiment as a designed perturbation of neuronal dynamics that are distributed throughout a system of coupled anatomical nodes or sources to produce region-specific responses. This system is modelled using a dynamic input-state-output system with multiple inputs and outputs. Responses are evoked by deterministic inputs that correspond to experimental manipulations (i.e. presentation of stimuli). Experimental factors (i.e. stimulus attributes or context) can also change the parameters or causal architecture of the system producing these responses. The state variables cover both the neuronal activities and other neurophysiological or biophysical variables needed to form the outputs. In our case, outputs are those components of neuronal responses that can be detected by MEG-EEG sensors.

In neuroimaging, DCM starts with a reasonably realistic neuronal model of interacting cortical regions. This model is then supplemented with a forward model of how neuronal activity is transformed into measured responses, here, MEG-EEG scalp-averaged responses. This enables the parameters of the neuronal model (i.e. effective connectivity) to be estimated from observed data. For MEG-EEG data, the supplementary model is a forward model of electromagnetic measurements that accounts for volume conduction effects (e.g. Mosher *et al.*, 1999 and Chapter 28). We first review the neuronal component of the forward model and then turn to the modality-specific measurement model.

A neural-mass model

The majority of neural-mass models of MEG-EEG dynamics have been designed to generate spontaneous rhythms (Lopes da Silva *et al.*, 1974; Jansen and Rit, 1995; Stam *et al.*, 1999; Robinson *et al.*, 2001; David and Friston, 2003) and epileptic activity (Wendling *et al.*, 2002). These models use a small number of state variables to represent the expected state of large neuronal populations, i.e. the neural mass. To date, event-related responses of neural-mass models have received less attention (Jansen and Rit, 1995; Rennie *et al.*, 2002; David *et al.*, 2005). Only recent models have embedded basic anatomical principles that underlie extrinsic connections among neuronal populations.

The cortex has a hierarchical organization (Crick and Koch, 1998; Felleman and Van Essen, 1991), comprising forward, backward and lateral processes that can be understood from an anatomical and cognitive perspective (Engel *et al.*, 2001). The direction of an anatomical projection is usually inferred from the laminar pattern of its origin and termination (see Chapter 36 for more details). The hierarchical cortical described in Chapter 33 is used here as a DCM. In brief, each source is modelled with three neuronal subpopulations. These subpopulations are interconnected with intrinsic connections within each source. The sources are interconnected by extrinsic connections among specific subpopulations. The specific source and target subpopulations define the connection as forward, backward or lateral. The model is now reviewed in terms of the differential equations that embody its causal architecture.

Neuronal state equations

The model (David *et al.*, 2005) embodies directed extrinsic connections among a number of sources, each based on the Jansen model (Jansen and Rit, 1995), using the connectivity rules described in Felleman and Van Essen (1991). These rules, which rest on a tri-partitioning of the cortical sheet into supra- and infra-granular layers and granular layer 4, have been derived from experimental studies of monkey visual cortex. We assume these rules

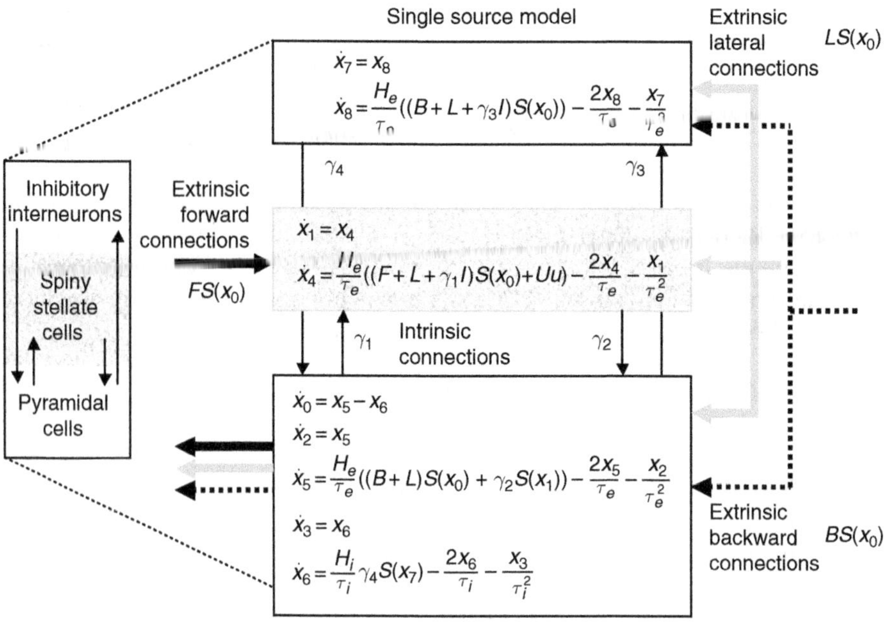

FIGURE 42.1 Schematic of the DCM used to model a single source. This schematic includes the differential equations describing the dynamics of the source's states. Each source is modelled with three subpopulations (pyramidal, spiny-stellate and inhibitory interneurons). These have been assigned to granular and agranular cortical layers, which receive forward and backward connections respectively.

generalize to other cortical regions (but see Smith and Populin, 2001 for a comparison of primary visual and auditory cortices). Under these simplifying assumptions, directed connections can be classified as: bottom-up or forward connections that originate in agranular layers and terminate in layer 4; top-down or backward connections that connect agranular layers; lateral connections that originate in agranular layers and target all layers. These long-range or extrinsic cortico-cortical connections are excitatory and comprise the axonal processes of pyramidal cells.

The Jansen model (Jansen and Rit, 1995) emulates the MEG-EEG activity of a cortical source using three neuronal subpopulations. A population of excitatory pyramidal (output) cells receives inputs from inhibitory and excitatory populations of interneurons, via intrinsic connections (intrinsic connections are confined to the cortical sheet). Within this model, excitatory interneurons can be regarded as spiny stellate cells found predominantly in layer 4. These cells receive forward connections. Excitatory pyramidal cells and inhibitory interneurons occupy agranular layers and receive backward and lateral inputs. Using these connection rules, it is straightforward to construct any hierarchical cortico-cortical network model of cortical sources (Figure 42.1).

The ensuing DCM is specified in terms of its state equations and an observer or output equation:

$$\begin{aligned}\dot{x} &= f(x, u, \theta)\\ h &= g(x, \theta)\end{aligned} \qquad 42.1$$

where x are the neuronal states of cortical areas, u are exogenous inputs and h is the output of the system. θ are quantities that parameterize the state and observer equations. The state equations $f(x, u, \theta)$ (Jansen and Rit, 1995; David and Friston, 2003; David *et al.*, 2005) for the neuronal states are:[1]

$$\begin{aligned}\dot{x}_7 &= x_8\\ \dot{x}_8 &= \frac{H_e}{\tau_e}((B+L+\gamma_3 I)S(x_0)) - \frac{2x_8}{\tau_e} - \frac{x_7}{\tau_e^2}\\ \dot{x}_1 &= x_4\\ \dot{x}_4 &= \frac{H_e}{\tau_e}((F+L+\gamma_1 I)S(x_0)+Uu) - \frac{2x_4}{\tau_e} - \frac{x_1}{\tau_e^2}\\ \dot{x}_0 &= x_5 - x_6\\ \dot{x}_2 &= x_5\\ \dot{x}_5 &= \frac{H_e}{\tau_e}((B+L)S(x_0)+\gamma_2 S(x_1)) - \frac{2x_5}{\tau_e} - \frac{x_2}{\tau_e^2}\\ \dot{x}_3 &= x_6\\ \dot{x}_6 &= \frac{H_i}{\tau_i}\gamma_4 S(x_7) - \frac{2x_6}{\tau_i} - \frac{x_3}{\tau_i^2}\end{aligned} \qquad 42.2$$

The states x_i are column vectors of the i-th state over all sources. Each represents a mean transmembrane potential or current of one of the three subpopulations. The state equations specify the rate of change of voltage as a function of current and specify how currents change as a function of voltage and current. The depolarization

[1] Propagation delays on the connections have been omitted for clarity, here and in Figure 42.1. See Appendix 42.1 for details of how delays are incorporated.

of pyramidal cells $x_0 = x_2 - x_3$ represents a mixture of potentials induced by excitatory and inhibitory (depolarizing and hyperpolarizing) currents respectively. This pyramidal potential is the presumed source of observed MEG-EEG signals.

Figure 42.1 depicts the states by assigning each subpopulation to a cortical layer. For schematic reasons we have lumped superficial and deep pyramidal units together, in the infra-granular layer. The matrices F, B, L encode forward, backward and lateral extrinsic connections respectively. From Eqn. **42.2** and Figure 42.1 it can be seen that the state equations embody the connection rules above. For example, extrinsic connections mediating changes in mean excitatory (depolarizing) current x_8, in the supra-granular layer, are restricted to backward and lateral connections. Interactions, among the subpopulations, depend on the constants $\gamma_1, \ldots, \gamma_4$, which control the strength of intrinsic connections and reflect the total number of synapses expressed by each subpopulation.

The remaining constants in the state equation pertain to two operators, on which the dynamics rest. The first transforms the average density of presynaptic inputs into the average postsynaptic membrane potential. This transformation is equivalent to a convolution with an impulse response or kernel:

$$\kappa(t)_e = \begin{cases} \frac{H_e}{\tau_e} t \exp(-t/\tau_e) & t \geq 0 \\ 0 & t < 0 \end{cases} \quad \mathbf{42.3}$$

where subscript 'e' stands for excitatory and 'i' is used for inhibitory synapses. H controls the maximum post-synaptic potential and τ represents a lumped rate-constant. The second operator S transforms the potential of each subpopulation into firing rate, which is the input to other subpopulations. This operator is assumed to be an instantaneous sigmoid non-linearity:

$$S(x) = \frac{1}{1+\exp(-rx)} - \frac{1}{2} \quad \mathbf{42.4}$$

where $r = 0.56$ determines its form. Figure 42.2 shows examples of these synaptic kernels and sigmoid functions. Note that the output of the firing rate function can be negative. This ensures that the neuronal system has a stable fixed-point, when all the states are equal to zero. Because the states approximate the underlying population or density dynamics, the fixed-point corresponds to the system's equilibrium or steady state. This means all the state variables can be interpreted as the deviation from steady-state levels. A DCM, at the neuronal level, obtains by coupling sources with extrinsic connections as described above. A typical three-source DCM is shown in Figure 42.3 and is the example used in Chapter 3.

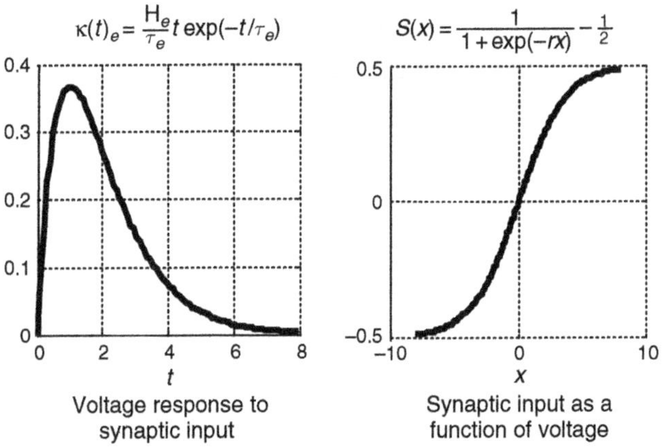

FIGURE 42.2 Left: form of the synaptic impulse response function, converting synaptic input (discharge rate) into mean transmembrane potential. Right: the non-linear static transformation of transmembrane potential into synaptic input. In this figure, the constants are set to unity, with the exception of $r = 0.56$. See main text for details.

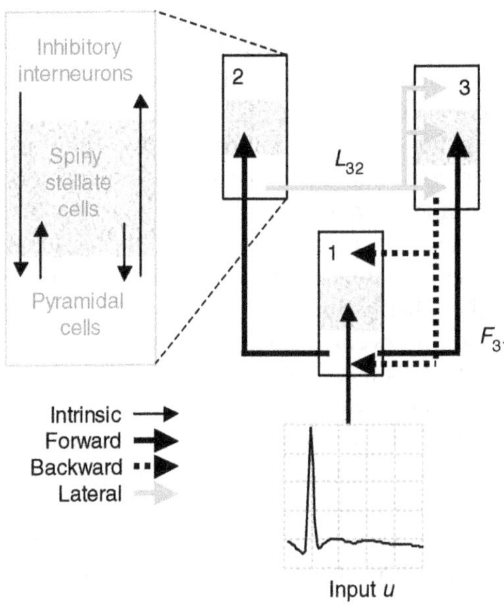

FIGURE 42.3 Typical hierarchical network composed of three cortical areas. Extrinsic inputs evoke transient perturbations around the resting state by acting on a subset of sources, usually the lowest in the hierarchy. Interactions among different regions are mediated through excitatory connections encoded by coupling matrices.

Event-related responses

To model event-related responses, the network receives inputs via input connections U. These connections are exactly the same as forward connections and deliver inputs u to the spiny stellate cells in layer 4. In the present context, inputs u model afferent activity relayed by subcortical structures and is modelled with two components:

$$u(t) = b(t, \eta_1, \eta_2) + \sum \theta_i^c \cos(2\pi(i-1)t) \quad \mathbf{42.5}$$

TABLE 42-1 Prior densities of parameters for connections to the *i*-th source from the *j*-th, in the *k*-th ERP

	$F_{ijk} = F_{ij}G_{ijk}$	$F_{ij} = 32\exp(\theta^F_{ij})$	$\theta^F_{ij} \sim N(0, \frac{1}{2})$
	$B_{ijk} = B_{ij}G_{ijk}$	$B_{ij} = 16\exp(\theta^B_{ij})$	$\theta^B_{ij} \sim N(0, \frac{1}{2})$
Extrinsic coupling parameters	$L_{ijk} = L_{ij}G_{ijk}$	$L_{ij} = 4\exp(\theta^L_{ij})$	$\theta^L_{ij} \sim N(0, \frac{1}{2})$
		$G_{ijk} = \exp(\theta^G_{ijk})$	$\theta^G_{ij} \sim N(0, \frac{1}{2})$
		$U_i = \exp(\theta^U_i)$	$\theta^U_i \sim N(0, \frac{1}{2})$
Intrinsic coupling parameters	$\gamma_1 = 128$	$\gamma_2 = \frac{4}{5}\gamma_1 \quad \gamma_3 = \frac{1}{4}\gamma_1$	$\gamma_4 = \frac{1}{4}\gamma_1$
Conduction delays (ms)	$\Delta_{ii} = 2$	$\Delta_{ij} = 16\exp(\theta^\Delta_{ij})$	$\theta^\Delta_{ij} \sim N(0, \frac{1}{16})$
Synaptic parameters (ms)	$T_i = 16$	$T^{(i)}_e = 8\exp(\theta^T_i)$	$\theta^T_i \sim N(0, \frac{1}{16})$
	$H_i = 32$	$H^{(i)}_e = 4\exp(\theta^H_i)$	$\theta^H_i \sim N(0, \frac{1}{16})$
	$u(t) = b(t, \eta_1, \eta_2) + \sum \theta^c_i \cos(2\pi(i-1)t)$		$\theta^c_i \sim N(0, 1)$
Input parameters (s)	$\eta_1 = 96\exp(\theta^\eta_1)$		$\theta^\eta_1 \sim N(0, \frac{1}{16})$
	$\eta_2 = 1024$	$\exp(\theta^\eta_2)$	$\theta^\eta_2 \sim N(0, \frac{1}{16})$

The first is a gamma density function $b(t, \eta_1, \eta_2) = \eta_2^{\eta_1} t^{\eta_1 - 1} \exp(-\eta_2 t)/\Gamma(\eta_1)$ with shape and scale constants η_1 and η_2 (see Table 42-1). This models an event-related burst of input that is delayed by η_1/η_2 s, with respect to stimulus onset and dispersed by subcortical synapses and axonal conduction. Being a density function, this component integrates to unity over peristimulus time. The second component is a discrete cosine set modelling systematic fluctuations in input, as a function of peristimulus time. In our implementation, peristimulus time is treated as a state variable, allowing the input to be computed explicitly during integration.

Critically, the event-related input is exactly the same for all ERPs. This means the effects of experimental factors are mediated by ERP-specific changes in connection strengths. This models experimental effects in terms of differences in forward, backward or lateral connections that confer a selective sensitivity on each source, in terms of its response to others. The experimental or ERP-specific effects are modelled by coupling gains:

$$
\begin{aligned}
F_{ijk} &= F_{ij}G_{ijk} \\
B_{ijk} &= B_{ij}G_{ijk} \\
L_{ijk} &= L_{ij}G_{ijk}
\end{aligned}
\qquad \mathbf{42.6}
$$

Here, G_{ijk} encodes the k-th ERP-specific gain in coupling to the i-th source from the j-th. By convention, we set the gain of the first ERP to unity, so that subsequent ERP-specific effects are relative to the first.[2] The reason we model experimental effects in terms of gain, as opposed to additive effects, is that by construction, connections are always positive. This is assured, provided the gain is also positive.

The important point here is that we are explaining experimental effects, not in terms of differences in neuronal responses, but in terms of the neuronal architecture or coupling that generates those responses. This is a fundamental departure from classical approaches, which characterize experimental effects descriptively, at the level of the states (e.g. a face-selective difference in ERP amplitude around 170 ms). DCM estimates these response differentials, but only as an intermediate step in the estimation of their underlying cause, namely changes in coupling.

Eqn. **42.2** defines the neuronal component of the DCM. These ordinary differential equations can be integrated (see Appendix 42.1) to generate pyramidal depolarizations, which enter the observer function to generate the predicted MEG-EEG signal.

Observation equations

The dendritic signal of the pyramidal subpopulation of each source is detected remotely on the scalp surface in MEG-EEG. Critically, the mapping between pyramidal activity and scalp data is linear:

$$
h = g(x, \theta) = LKx_0 \qquad \mathbf{42.7}
$$

where L is a lead-field matrix (i.e. forward electromagnetic model), which accounts for passive conduction of the electromagnetic field (Mosher *et al.*, 1999). If the spatial properties (orientation and position) of the source are known, then the lead-field matrix L is also known. In this case, $K = diag(\theta^K)$ is a leading diagonal matrix, which controls the contribution θ^K_i of pyramidal depolarization

[2] In fact, in our implementation, the coupling gain is a function of any set of explanatory variables encoded in a design matrix, which can contain indicator variables or parametric variables. For simplicity, we limit this chapter to categorical (ERP-specific) effects.

to the i-th source density. If the orientation is not known then λ becomes a function of free parameters encoding the location and orientation of the source (see Kiebel *et al.*, 2006 for details). For simplicity we will assume a fixed location and orientation for each source but allow the orientation to be parallel or anti-parallel (i.e. θ^K can be positive or negative). The rationale for this is that the direction of current flow induced by pyramidal cell depolarization depends on the relative density of synapses proximate and distal to the cell body.

Dimension reduction

For computational reasons, it is sometimes expedient to reduce the dimensionality of the sensor data, while retaining the maximum amount of information. This is assured by projecting the data onto a subspace defined by its principal eigenvectors or spatial modes Ω. The projection is applied to the data and lead-field:

$$\begin{aligned} y &\leftarrow \Omega y \\ L &\leftarrow \Omega L \\ \varepsilon &\leftarrow \Omega \varepsilon \end{aligned} \qquad 42.8$$

In the examples below, the data were projected onto the first three spatial modes following a singular value decomposition of the scalp data, between 0 and 500 ms. Reduction using principal eigenvariates preserves information in the data; in our examples about 70 per cent. Generally, one uses a small number of modes, noting that the dimension of the subspace containing predicted responses cannot exceed the number of sources.

The likelihood model

In summary, our DCM comprises a state equation that is based on neurobiological heuristics and an observer based on an electromagnetic forward model. By integrating the state equation and passing the ensuing states through the observer we generate a predicted measurement. This corresponds to a generalized convolution of the inputs to generate an output $h(\theta)$. This generalized convolution furnishes an observation model for the vectorized data[3] y and the associated likelihood:

$$\begin{aligned} y &= vec(h(\theta) + X\theta^X) + \varepsilon \\ p(y\,|\,\theta, \lambda) &= N(vec(h(\theta) + X\theta^X), diag(\lambda) \otimes V) \end{aligned} \qquad 42.9$$

Measurement noise ϵ is assumed to be zero mean and independent over channels, i.e. $Cov(\varepsilon) = diag(\lambda) \otimes V$, where λ is an unknown vector of channel-specific variances. V represents the error's temporal autocorrelation matrix, which we assume is the identity matrix. This is tenable because we down-sample the data to about 8 ms. Low-frequency noise or drift components are modelled by X, which is a block diagonal matrix with a low-order discrete cosine set for each ERP and channel. The order of this set can be determined by Bayesian model selection (see below). In this chapter, we used three components for the first study and four for the second. The first component of a discrete cosine set is simply a constant.

This model is fitted to data using Bayesian inversion. This involves maximizing the variational free energy with respect to the conditional moments of the free parameters θ. The parameters are constrained by a prior specification of the range they are likely to lie in (Friston, 2003). These constraints, which take the form of a prior density $p(\theta)$, are combined with the likelihood $p(y|\theta, \lambda)$, to form a posterior density $p(\theta|y, \lambda) \propto p(y|\theta, \lambda)p(\theta)$ according to Bayes' rule. It is this posterior or conditional density we want to approximate. Gaussian assumptions about the errors in Eqn. **42.9** enable us to compute the likelihood from the prediction error. The only outstanding quantities we require are the priors.

[3] Concatenated column vectors of data from each channel.

Priors

Under Gaussian assumptions, the prior distribution $p(\theta_i)$ of the i-th parameter is defined by its mean and variance. The mean corresponds to the prior expectation. The variance reflects the amount of prior information about the parameter. A tight distribution (small variance) corresponds to precise prior knowledge. Critically, nearly all the constants in our DCM are positive. To ensure positivity, we place Gaussian priors on the log of these constants. This is equivalent to a log-normal prior on the constants *per se*. For example, the forward connections are parameterized as $F_{ij} = \exp(\theta^F_{ij})$, where $p(\theta^F_{ij}) = N(\mu, \sigma^2)$. We will use this notion for the other parameters as well. A relatively tight or informative log-normal prior obtains when $\sigma^2 \approx \frac{1}{16}$. This allows for a scaling around the prior expectation of up to a factor of two. Relatively flat priors, allowing for an order of magnitude scaling, correspond to $\sigma^2 \approx \frac{1}{2}$. The ensuing log-normal densities are shown in Figure 42.4 for a prior expectation of unity (i.e. $\mu = 0$).

The parameters of the state equation can be divided into five subsets: (i) *extrinsic connection* parameters, which specify the coupling strengths among areas; (ii) *intrinsic* connection parameters, which reflect our knowledge about canonical microcircuitry within an area; (iii) *conduction* delays; (iv) *synaptic* parameters controlling the dynamics within an area; and (v) *input* parameters, which control the subcortical delay and dispersion of event-related responses. Table 42-1 shows how the constants

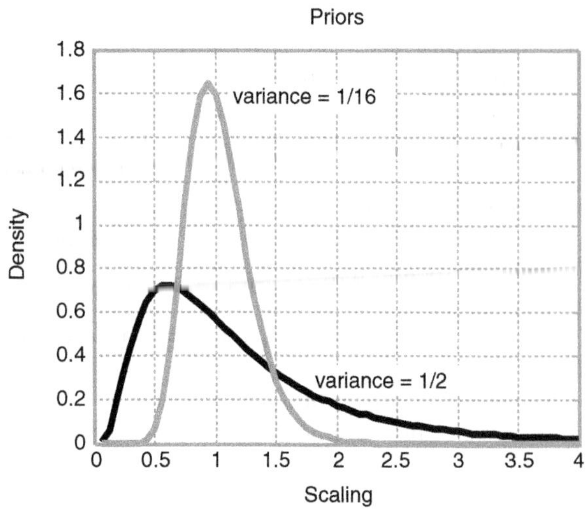

FIGURE 42.4 Log-normal densities on $\exp(\theta)$ entailed by Gaussian priors on θ with a prior expectation of zero and variances of 1/2 and 1/16. These correspond to fairly uninformative and informative priors respectively.

of the state equation are parameterized in terms of θ. It can be seen that we have adopted relatively uninformative priors on the extrinsic coupling $\sigma^2 \approx \frac{1}{2}$ and tight priors for the remaining constants $\sigma^2 \approx \frac{1}{16}$. Some parameters (intrinsic connections and inhibitory synaptic parameters) have infinitely tight priors and are fixed at their prior expectation. This is because changes in these parameters and the excitatory synaptic parameters are almost redundant, in terms of system responses. The priors in Table 42-1 conform to the principle that the parameters we want to make inferences about, namely extrinsic connectivity, should have relatively flat priors. This ensures that the evidence in the data constrains the posterior or conditional density in an informative and useful way (see David *et al.*, 2006 for more details).

Summary

In summary, a DCM is specified through its priors. These are used to specify how regions are interconnected, which regions receive subcortical inputs, and which cortico-cortical connections change with the levels of experimental factors. Usually, the most interesting questions pertain to changes in cortico-cortical coupling that explain differences in ERPs. These rest on inferences about the coupling gains $G_{ijk} = \exp(\theta^G_{ijk})$. This section has covered the likelihood and prior densities necessary for conditional estimation. For each model, we require the conditional densities of two synaptic parameters per source $\{\theta^{\tau}_i, \theta^H_i\}$, ten input parameters $\{\theta^{\eta}_1, \theta^{\eta}_2, \theta^c_1, \ldots, \theta^c_8\}$ and the extrinsic coupling parameters, gains and delays $\{\theta^F, \theta^B, \theta^L, \theta^G, \theta^U, \theta^{\Delta}\}$. The next section reviews conditional estimation of these parameters, inference and model selection.

BAYESIAN INFERENCE AND MODEL COMPARISON

In this section, we cover model inversion and selection. Model selection was introduced in Chapter 35 and will be dealt with in greater depth in the next chapter, using DCMs of fMRI data. For a given DCM, say model m, parameter estimation corresponds to approximating the moments of the posterior distribution given by Bayes' rule:

$$p(\theta|y, m) = \frac{p(y|\theta, m)p(\theta, m)}{p(y|m)} \qquad \mathbf{42.10}$$

The estimation procedure employed in DCM is described in Chapter 34 and Friston (2002). The posterior moments (conditional mean η and covariance Σ) are updated iteratively using variational Bayes under a fixed-form Laplace (i.e. Gaussian) approximation to the conditional density $q(\theta) = N(\eta, \Sigma)$. This is equivalent to an expectation maximization (EM) algorithm that employs a local linear approximation of Eqn. **42.9** about the current conditional expectation. The **E**-step conforms to a Fisher-scoring scheme (Press *et al.*, 1992) that optimizes the variational free energy $F(q, \lambda, m)$ with respect to the conditional moments. In the **M**-step, the error variances λ are updated in exactly the same way. The estimation scheme can be summarized as follows:

Repeat until convergence

$$\begin{aligned} &\textbf{E}\text{-step} \quad q \leftarrow \max_q F(q, \lambda, m) \\ &\textbf{M}\text{-step} \quad \lambda \leftarrow \max_\lambda F(q, \lambda, m) \\ &F(q, \lambda, m) = \langle \ln p(y|\theta, \lambda, m) + \ln p(\theta|m) - \ln q(\theta) \rangle_q \\ &\qquad = \ln p(y|\lambda, m) - D(q||p(\theta|y, \lambda, m)) \end{aligned} \qquad \mathbf{42.11}$$

Note that the free energy is simply a function of the log-likelihood and the log-prior for a particular DCM and $q(\theta)$. $q(\theta)$ is the approximation to the posterior density $p(\theta|y, \lambda, m)$ we require. The **E**-step updates the moments of $q(\theta)$ (these are the variational parameters η and Σ) by maximizing the variational free energy. The free energy is the log-likelihood minus the divergence between the real and approximate conditional density. This means that the conditional moments or variational parameters maximize the log-likelihood while minimizing the discrepancy between the true and approximate conditional density. Because the divergence does not depend on the covariance parameters, minimizing the free energy in the **M**-step is equivalent to finding the maximum likelihood estimates of the covariance parameters. This scheme is identical to that employed by DCM for fMRI, the details of which can be found in the previous chapter (see also Friston, 2002; Friston *et al.*, 2003 and Appendix 2).

Conditional inference

Inference on the parameters of a particular model proceeds using the approximate conditional or posterior density $q(\theta)$. Usually, this involves specifying a parameter or compound of parameters as a contrast $c^T\eta$. Inferences about this contrast are made using its conditional covariance $c^T\Sigma c$. For example, one can compute the probability that any contrast is greater than zero or some meaningful threshold, given the data. This inference is conditioned on the particular model specified. In other words, given the data and model, inference is based on the probability that a particular contrast is bigger than a specified threshold. In some situations, one may want to compare different models. This entails Bayesian model comparison.

Model comparison and selection

Different models are compared using their evidence (Penny *et al.*, 2004). The model evidence is:

$$p(y|m) = \int p(y|\theta, m)p(\theta|m)d\theta \qquad \textbf{42.12}$$

The evidence can be decomposed into two components: an accuracy term, which quantifies the data fit; and a complexity term, which penalizes models with a large number of parameters. Therefore, the evidence embodies the two conflicting requirements of a good model, that it explains the data and is as simple as possible. In the following, we approximate the model evidence for model m, with the free energy, after convergence. This rests on the assumption that λ has a point mass at its maximum likelihood estimate (equivalent to its conditional estimate under flat priors); i.e. $\ln p(y|m) = \ln \langle p(y|\lambda, m)\rangle_\lambda$. After convergence the divergence is minimized and:

$$\ln p(y|m) \approx F(q, \lambda, m) \qquad \textbf{42.13}$$

(see Eqn. **42.11**). The most likely model is the one with the largest log-evidence. This enables Bayesian model selection. Model comparison rests on the likelihood ratio of the evidence for two models. This ratio is the Bayes factor B_{ij}. For models i and j:

$$\ln B_{ij} = \ln p(y|m = i) - \ln p(y|m = j) \qquad \textbf{42.14}$$

Conventionally, strong evidence in favour of one model requires the difference in log-evidence to be about three or more. We have now covered the specification, estimation and comparison of DCMs. In the next section, we will illustrate their application to real data using two examples of how changes in coupling can explain ERP differences.

EMPIRICAL STUDIES

In this section, we illustrate the use of DCM by looking at changes in connectivity induced in two different ways. In the first experiment, we recorded ERPs during the perception of faces and houses. It is well known that the N170 is a specific ERP correlate of face perception (Allison *et al.*, 1999). The N170 generators are thought to be located close to the lateral fusiform gyrus, or fusiform face area (FFA). Furthermore, the perception of houses has been shown to activate the parahippocampal place area (PPA) using fMRI (Aguirre *et al.*, 1998; Epstein and Kanwisher, 1998; Haxby *et al.*, 2001; Vuilleumier *et al.*, 2001). In this example, differences in coupling define the category-selectivity of pathways that are accessed by different categories of stimuli. A category-selective increase in coupling implies that the region receiving the connection is selectively more sensitive to input elicited by the stimulus category in question. This can be attributed to a functional specialization of the region receiving the connection. In the second example, we use an auditory oddball paradigm, which produces mismatch negativity (MMN) or P300 components in response to rare stimuli, relative to frequent (Linden *et al.*, 1999; Debener *et al.*, 2002). In this paradigm, we attribute changes in coupling to plasticity underlying the perceptual learning of frequent or standard stimuli.

In the category-selectivity paradigm, there are no necessary changes in connection strength; pre-existing differences in responsiveness are simply disclosed by presenting different stimuli. This can be modelled by differences in forward connections. However, in the oddball paradigm, the effect only emerges once standard stimuli have been learned. This implies some form of perceptual or sensory learning. We have presented a quite detailed analysis of perceptual learning in the context of empirical Bayes in Chapter 36 (see also Friston, 2003). We concluded that the late components of oddball responses could be construed as a failure to suppress prediction error, after learning the standard stimuli. Critically, this theory predicts that learning-related plasticity should occur in backward connections generating the prediction, which are then mirrored in forward connections. In short, we predicted changes in forward and backward connections when comparing ERPs for standard and oddball stimuli.

In the first example, we are interested in where category-selective differences in responsiveness arise in a forward processing stream. We use inferences based on the conditional density of forward coupling-gain, when comparing face and house ERPs, to address this question. Backward connections are probably important in mediating this selectivity but exhibit no learning-related changes *per se*. In the second example, our question is more categorical in nature, namely, are changes in backward and lateral connections necessary to explain ERP differences between standards and oddballs, relative to changes in forward connections alone? We illustrate the use of Bayesian model comparison to answer this question. (See the figure legends for a description of the data acquisition, lead-field specification and pre-processing.)

Category-selectivity: effective connectivity in the ventral visual pathway

ERPs elicited by brief presentation of faces and houses were obtained by averaging trials over three successive 20-minute sessions. Each session comprised thirty blocks of faces or houses only. Each block contained twelve stimuli presented for 400 ms every 2.6 s. The stimuli comprised 18 neutral faces and 18 houses, presented in greyscale. To maintain attentional set, the subject was asked to perform a one-back task, i.e. indicate, using a button press, whether or not the current stimulus was identical to the previous.

As reported classically, we observed a stronger N170 component during face perception in the posterior temporal electrodes. However, we also found other components, associated with house perception, which were difficult to interpret on the basis of scalp data. It is generally thought that face perception is mediated by a hierarchical system of bilateral regions (Haxby *et al.*, 2002): a core system of occipito-temporal regions in extrastriate visual cortex (inferior occipital gyrus, IOG; lateral fusiform gyrus or face area, FFA; superior temporal sulcus, STS) that mediates the visual analysis of faces, and an extended system for cognitive functions. This system (intra-parietal sulcus, auditory cortex, amygdala, insula, limbic system) acts in concert with the core system to extract meaning from faces. House perception has been shown to activate the parahippocampal place area (PPA) (Aguirre *et al.*, 1998; Epstein and Kanwisher, 1998; Haxby *et al.*, 2001; Vuilleumier *et al.*, 2001). In addition, the retrosplenial cortex (RS) and the lateral occipital gyrus are more activated by houses, compared to faces (Vuilleumier *et al.*, 2001). Most of these regions belong to the ventral visual pathway. It has been argued that the functional architecture of the ventral visual pathway is not a mosaic of category-specifics modules, but rather embodies a continuous representation of information about object attributes (Ishai *et al.*, 1999).

DCM specification

We tested whether differential propagation of neuronal activity through the ventral pathway is sufficient to explain the differences in measured ERPs. On the basis of a conventional source localization and previous studies (Allison *et al.*, 1999; Ishai *et al.*, 1999; Haxby *et al.*, 2001, 2002; Vuilleumier *et al.*, 2001), we specified the following DCM (Plate 59, see colour plate section): bilateral occipital regions close to the calcarine sulcus (V1) received subcortical visual inputs. From V1 onwards, the pathway for house perception was bilateral and connected to RS and PPA using forward and backward connections. The pathway engaged by face perception was restricted to the right hemisphere and comprised connections from V1 to IOG, which projects to STS and FFA. In addition, bilateral connections were included, between STS and FFA, as suggested in Haxby *et al.* (2002). These connections constituted our DCM mediating ERPs to houses and faces. Face- or house-specific ERP components were hypothesized to arise from category-selective, stimulus-bound, activation of forward pathways. To identify these category-selective streams, we allowed the forward connections, in the right hemisphere, to change with category. Our hope was that these changes would render PPA more responsive to houses, while the FFA and STS would express face-selective responses.

Conditional inference

The results are shown in Figure 42.5 in terms of predicted cortical responses and coupling parameters. Using this DCM, we were able to replicate the functional anatomy, disclosed by the above fMRI studies: the response in PPA was more marked when processing houses versus faces. This was explained, in the model, by an increase in forward connectivity in the medial ventral pathway from RS to PPA. This difference corresponded to a coupling-gain of over fivefold. Conversely, the model exhibited a much stronger response in FFA and STS during face perception, as suggested by the Haxby model (Haxby *et al.*, 2002). This selectivity was due to an increase in coupling from IOG to FFA and from IOG to STS. The face-selectivity of STS responses was smaller than in the FFA, the latter mediated by an enormous gain of about ninefold ($1/0.11 = 9.09$) in sensitivity to inputs from IOG. The probability, conditional on the data and model, that changes in forward connections to the PPA, STS and FFA were greater than zero, was essentially 100 per cent in all cases. The connections from V1 to IOG showed no selectivity. This suggests that category-selectivity

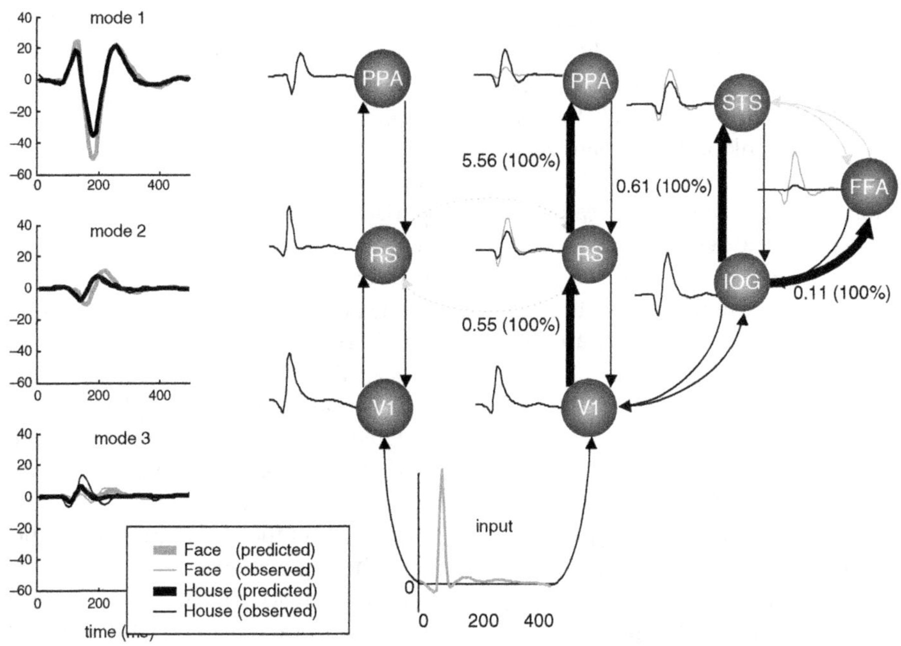

FIGURE 42.5 DCM results for the category-selectivity paradigm: Left: predicted (thick) and observed (thin) responses in measurement space. These are a projection of the scalp or channel data onto the first three spatial modes or eigenvectors of the channel data (faces: grey; houses: black). The predicted responses are based on the conditional expectations of the DCM parameters. The agreement is evident. Right: reconstructed responses for each source and changes in coupling for the DCM modelling category-specific engagement of forward connections, in the ventral visual system. As indicated by the predicted responses in PPA and FFA, these changes are sufficient to explain an increase response in PPA when perceiving houses and, conversely, an increase in FFA responses during face perception. The coupling differences mediating this category-selectivity are shown alongside connections, which showed category-specific differences (highlighted by solid lines). Differences are the relative strength of forward connections during house presentation, relative to faces. The per cent conditional confidence that this difference is greater than zero is shown in brackets. Only changes with 90 per cent confidence or more are reported and these connections are highlighted in bold.

The data reported in this and subsequent figures were acquired from the same subject, in the same session, using 128 EEG electrodes and 2048 Hz sampling. Before averaging, data were referenced to mean activity and bandpass filtered between 1 and 20 Hz. Trials showing ocular artefacts (~30 per cent) and 11 bad channels were removed from further analysis. To compute the lead field for each source we used a distributed source reconstruction procedure based on the subject's anatomical MRI and described in David *et al.* (2006). Following dimension reduction to the three principal eigenvariates, the data were down-sampled to 8 ms time bins. These reduced channel data were then used to invert the DCM.

emerges downstream from IOG, at a fairly high level. Somewhat contrary to expectations (Vuilleumier *et al.*, 2001), the coupling from V1 to RS showed a mild face-selective bias, with an increase of about 80 per cent ($1/0.55 = 1.82$).

Note how the ERPs of each source are successively transformed and delayed from area to area. This reflects the intrinsic transformations within each source, the reciprocal exchange of signals between areas and the conduction delays. These transformations are mediated by intrinsic and extrinsic connections and are the dynamic expression of category selectivity in this DCM. The conditional estimate of the subcortical input is also shown in Figure 42.5. The event-related response input was expressed about 96 ms after stimulus onset. The accuracy of the model is evident in the left panel of Figure 42.5, which shows the measured and predicted responses in sensor space, after projection onto their three principal eigenvectors.

Auditory oddball: effective connectivity and sensory learning

Auditory stimuli, 1000 or 2000 Hz tones with 5 ms rise and fall times and 80 ms duration, were presented binaurally for 15 minutes, every 2 s in a pseudo-random sequence; 2000 Hz tones (oddballs) occurred 20 per cent of the time (120 trials) and 1000 Hz tones (standards) 80 per cent of the time (480 trials). The subject was instructed to keep a mental record of the number of 2000 Hz tones.

Late components, characteristic of rare events, were seen in most frontal electrodes, centred on 250 ms to 350 ms post-stimulus. As reported classically, early components (i.e. the N100) were almost identical for rare and frequent stimuli. Using a conventional reconstruction algorithm (see figure legend), cortical sources were localized symmetrically along the medial part of the upper bank of the Sylvian fissure, in the right middle temporal gyrus, left medial and posterior cingulate, and bilateral orbitofrontal cortex (see insert in Plate 60). These locations are in good agreement with the literature: sources along the upper bank of the Sylvian fissure can be regarded as early auditory cortex, although they are generally located in the lower bank of the Sylvian fissure (Heschls gyrus). Primary auditory cortex has major inter-hemispheric connections through the corpus callosum. In addition, these areas project to temporal and frontal lobes following different streams (Romanski *et al.*, 1999; Kaas and Hackett, 2000). Finally, cingulate activations are often found in relation to oddball tasks, either auditory or visual (Linden *et al.*, 1999).

DCM specification

Using these sources and prior knowledge about the functional anatomy of the auditory system, we constructed the following DCM (see Plate 60): an extrinsic (thalamic) input entered bilateral primary auditory cortex (A1) which was connected to ipsilateral orbitofrontal cortex (OF). In the right hemisphere, an indirect forward pathway was specified from A1 to OF through the superior temporal gyrus (STG). All these connections were reciprocal. At the highest level in the hierarchy, OF and left posterior cingulate cortex (PC) were laterally and reciprocally connected.

Model comparison

Given these nodes and their connections, we created four DCMs that differed in terms of which connections could show putative learning-related changes. The baseline model precluded any differences between standard and oddball trials. The remaining four models allowed changes in forward F, backward B, forward and backward FB and all connections FBL, with the primary auditory sources. The results of a Bayesian model comparison are shown in Plate 60, in terms of the respective log-evidences (referred to the baseline model with no coupling changes). There is very strong evidence for conjoint changes in backward and lateral connections, above and beyond changes in forward or backward connections alone. The FB model supervenes over the FBL model that was augmented with plasticity in lateral connections between A1. This is interesting because the FBL model had more parameters, enabling a more accurate modelling of the data. However, the improvement in accuracy did not meet the cost of increasing the model complexity and the log-evidence fell by 4.224. This means there is strong evidence for the FB model, in relation to the FBL model. Put more directly, the data are $e^{4.224} = 68.3$ times more likely to have been generated by the FB model than the FBL model. The results of this Bayesian model comparison suggest the theoretical predictions were correct.

Conditional inference

The conditional estimates and posterior confidences for the FB MODEL are shown in Figure 42.6 and reveal a profound increase, for rare events, in all connections. We can be over 95 per cent confident these connections increased. As above, these confidences are based on the conditional

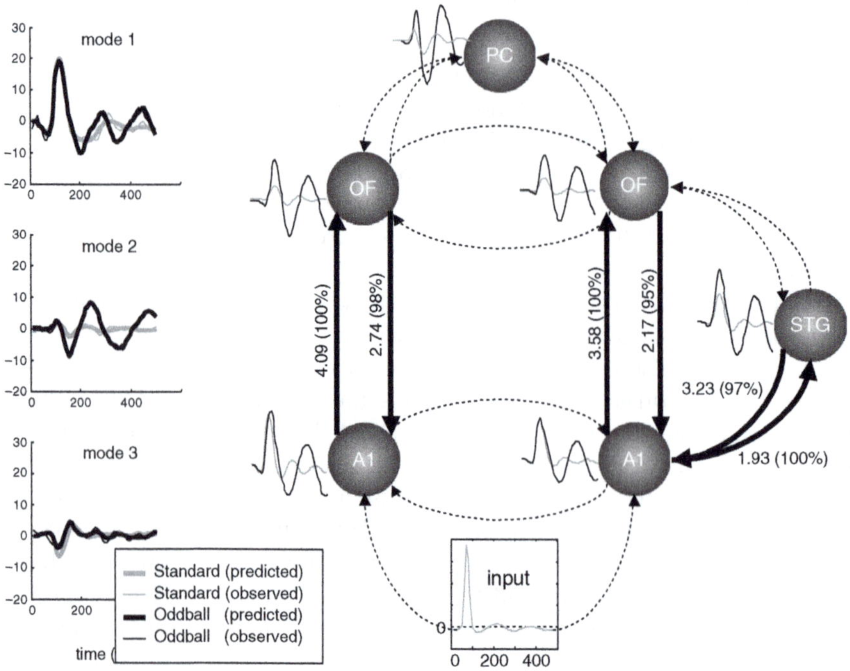

FIGURE 42.6 DCM results for the auditory oddball (FB model). This figure adopts the same format as Figure 42.5. Here the oddball-related responses show many components and are expressed most noticeably in mode 2. The mismatch response is expressed in nearly every source (black: oddballs, grey: standards), and there are widespread learning-related changes in connections (solid lines: changes with more than 90 per cent conditional confidence). In all connections, the coupling was stronger during oddball processing, relative to standards.

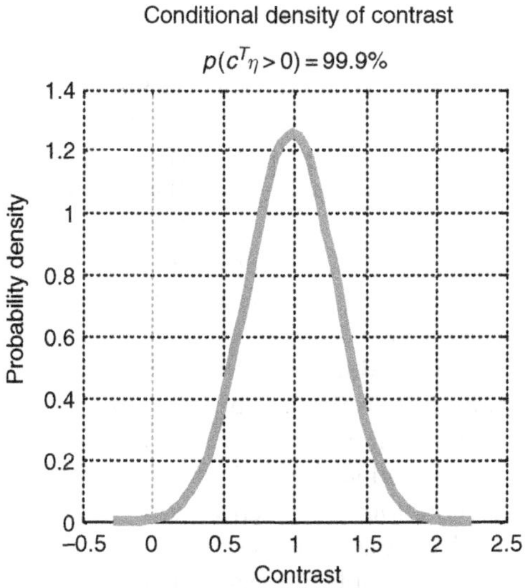

FIGURE 42.7 Conditional density of a contrast averaging over all learning-related changes in backward connections. It is evident that change in backward connections is unlikely to be zero or less given our data and DCM.

density of the coupling-gains. The conditional density of a contrast, averaging over all gains in backward connections, is shown in Figure 42.7. We can be 99.9 per cent confident this contrast is greater than zero. The average is about one, reflecting a gain of about $e^1 \approx 2.7$, i.e. more than a doubling.

These changes produce a rather symmetrical series of late components, expressed to a greater degree, but with greater latency, at hierarchically higher levels. In comparison with the visual paradigm above, the subcortical input appeared to arrive earlier, around 64 ms after stimulus onset. The remarkable agreement between predicted and observed channel responses is seen in the left panel, again shown as three principal eigenvariates.

In summary, this analysis suggests that a sufficient explanation for mismatch responses is an increase in forward and backward connections with primary auditory cortex. This results in the appearance of exuberant responses after the N100 in A1 to unexpected stimuli. This could represent a failure to suppress prediction error, relative to predictable or learned stimuli, which can be predicted more efficiently.

CONCLUSION

We have described a Bayesian inference procedure in the context of DCM for ERPs. DCMs are used in the analysis of effective connectivity to provide posterior or conditional distributions. These densities can then be used to assess changes in effective connectivity caused by experimental manipulations. These inferences, however, are contingent on assumptions about the architecture of the model, i.e. which regions are connected and which connections are modulated by experimental factors. Bayesian model comparison can be used to adjudicate among competing models, or hypotheses, as demonstrated above. The approach can be regarded as a neurobiologically constrained source reconstruction scheme, in which the parameters of the reconstruction have an explicit neuronal interpretation, or as a characterization of the causal architecture of the neuronal system generating responses. We have seen that it is possible to test mechanistic hypotheses in a more constrained way than classical approaches because the prior assumptions are physiologically informed.

The DCM in this chapter used a neural-mass model that embodies long-range cortico-cortical connections by considering forward, backward and lateral connections among remote areas (David *et al.*, 2005). This allows us to embed neural mechanisms generating MEG-EEG signals that are located in well-defined regions. This may make the comparison with fMRI activations easier than alternative models based on continuous cortical fields (Robinson *et al.*, 2001; Liley *et al.*, 2002). However, it would be interesting to apply DCM to cortical field models.

Frequently asked questions

In presenting this work to our colleagues, we have encountered a number of recurrent questions. We use these questions to frame our discussion of DCM for ERPs.

- *How do the results change with small changes in the priors?* Conditional inferences are relatively insensitive to changes in the priors. This is because we use relatively uninformative priors on the parameters about which inferences are made. Therefore, confident inferences about coupling imply a high conditional precision. This means that most of the conditional precision is based on the data (because the prior precision is very small). Changing the prior precision will have a limited effect on the conditional density and the ensuing inference.
- *What are the effects of wrong network specification (e.g. including an irrelevant source or not including a relevant source or the wrong specification of connections)?* This is difficult to answer because the effects will depend on the particular data set and model employed. However, there is a principled way in which questions of this sort can be answered. This uses Bayesian model comparison: if the contribution of a particular source or connection is in question, one can compute the log-evidence for two models that do and do not contain the source or connection. If it was important, the differences in log-evidence will be significant. Operationally,

the effects of changing the architecture are reformulated in terms of changing the model. Because the data do not change, these effects can be evaluated quantitatively in terms of the log-evidence (i.e. likelihood of the data given the models in question).

- *How sensitive is the model to small changes in the parameters?*
This is quantified by the curvature of the free energy with respect to parameters. This sensitivity is, in fact, the conditional precision or certainty. If the free energy changes quickly as one leaves the maximum (i.e. conditional mode or expectation), then the conditional precision is high. Conversely, if the maximum is relatively flat, changes in the parameter will have a smaller effect and conditional uncertainty is higher. Conditional uncertainly is a measure of the information about the parameter in the data.
- *What is the role of source localization in DCM?*
It has no role. Source localization refers to inversion of an electromagnetic forward model. Because this is only a part of the DCM, Bayesian inversion of the DCM implicitly performs the source localization. Having said this, in practice, priors on the location or orientation (i.e. spatial parameters) can be derived from classical source reconstruction techniques. In this chapter, we used a distributed source reconstruction to furnish spatial priors on the DCM. However, these priors do not necessarily have to come from a classical inverse solution. Our evaluations of DCM, using somatosensory evoked potentials (whose spatial characteristics are well known) suggest that the conditional precision of the orientation is much greater than the location. This means that one could prescribe tight priors on the location (from source reconstruction, from fMRI analyses, or from the literature) and let DCM estimate the conditional density of the orientation (Kiebel *et al.*, 2006).
- *How do you select the sources for the DCM?*
DCM is an inference framework that allows one to answer questions about a well-specified model of functional anatomy. The sources specify that model. Conditional inferences are then conditional on that model. Questions about which is the best model use Bayesian model selection as described above. In principle, it is possible to compare an ensemble of models with all permutations of sources and simply select the model that has the greatest log-evidence. We will illustrate this in a forthcoming multisubject study of the MMN in normal subjects.
- *How do you assess the generalizability of a DCM?*
In relation to a particular data set, the conditional density of the parameters implicitly maximizes generalizability. This is because the free energy can be reformulated in terms of an accuracy term that is maximized and a complexity term that is minimized (Penny *et al.*, 2004). Minimizing complexity ensures generalization. This aspect of variational learning means that we do not have to use *ad-hoc* measures of generalization (e.g. splitting the data into training and test sets). Generalization is an implicit part of the estimation. In relation to generalization over different data sets, one has to consider the random effects entailed by different subjects or sessions. In this context, generalization and reproducibility are a more empirical issue.
- *How can you be sure that a change in connectivity is not due to a wrong model?*
There is no such thing as a wrong model. Models can only be better or worse than other models. We quantify this in terms of the likelihood of each model (i.e. the log-evidence) and select the best model. We then usually make conditional inferences about the parameters, conditional on the best model. One could of course argue that all possible models have not been tested, but at least one has a framework that can accommodate any alternative model.
- *What is the basis for the claim that the neural-mass models and DCMs are biologically grounded?*
This is based largely on the use of the Jansen and Rit model (1995) as an elemental model for each source. We deliberately chose an established model from the EEG literature for which a degree of predictive and face validity had already been established. This model has been evaluated in a range of different contexts and its ability to emulate and predict biological phenomena has been comprehensively assessed (Jansen and Rit, 1995; David *et al.*, 2005 and references therein). The biological plausibility of the extrinsic connections has been motivated at length in David and Friston (2003), where we show that a network of Jansen and Rit sources can reproduce a variety of EEG phenomena.
- *Why did we exclude thalamus from our models?*
Because it was not necessary to answer the question we wanted to ask. In the models reported in this chapter, the effects of subcortical transformations are embodied in the parameters of the input function. If one thought that cortico-subcortical interactions were important, it would be a simple matter to include a thalamic source that was invisible to measurement space (i.e. set the lead field's priors to zero). One could then use Bayesian model comparison to assess whether modelled cortico-thalamic interactions were supported by the data.
- *Does DCM deal with neuronal noise?*
No. In principle, DCM could deal with noise at the level of neuronal states by replacing the ordinary differential equations with stochastic differential equations. However, this would call for a very different estimation scheme in which there was conditional uncertainty about the [hidden] neuronal states. Conventionally, these sorts of systems are estimated using a recurrent Bayesian update scheme such as Kalman or Particle

filtering. We are working on an alternative (dynamic expectation maximization), but it will be some time before it will be applied to DCM.

- *Why are the DCMs for EEG and fMRI not the same?* This is an important question, especially if one wants to use a DCM to explain both fMRI and EEG responses in the context of multimodal fusion. The DCM for EEG is considerably more complicated than the models used previously for fMRI. In fact, the bilinear form for the dynamics in fMRI is formally the same as the bilinear approximation to the state equations used in this chapter. The reason that DCM for EEG rests on more complicated models is that there is more conditional information in electromagnetic data about the parameters. This means that more parameters can be estimated efficiently (i.e. with greater conditional certainty). It would be perfectly possible to replace the bilinear approximation in DCMs for fMRI with the current neuronal model. However, Bayesian model comparison would show that the bilinear approximation was much better because it is not over-parameterized for fMRI data. Conversely, model comparison using both fMRI and EEG data should select the detailed model used here.
- *Why try to explain evoked responses solely by a change in effective connectivity?* In DCM, most of the biophysical parameters are rate or connectivity parameters that fall into three groups: (i) extrinsic connections among areas; (ii) intrinsic connections within an area (i.e. among the three sub-populations); and (iii) within subpopulation (i.e. the rate or time constants governing self-inhibition or adaptation). We have chosen to explain experimental differences in terms of coupling changes between areas. This is motivated by theoretical considerations that suggest sensory and perceptual learning involves experience-dependent changes in extrinsic forward and backward connections. However, the DCM machinery could easily be adapted (by a different choice of priors on the parameters) to explain differences in terms of changes in intrinsic connections, or even time-constants within a subpopulation. Furthermore, using Bayesian model comparison we can compare models to ask, for example, whether changes in intrinsic or extrinsic connections are the most likely explanation for observed responses.

SUMMARY

In this chapter, we have focused on the changes in connectivity, between levels of an experimental factor, to explain differences in the form of ERFs-ERPs. We have illustrated this through the analysis of real ERPs recorded in two classical paradigms: ERPs recorded during the perception of faces versus houses and the auditory oddball paradigm. We were able to differentiate two streams within the ventral visual pathway corresponding to face and house processing, leading to preferential responses in the fusiform face area and parahippocampal place area respectively. These results concur with fMRI studies (Haxby *et al.*, 2001;Vuilleumier *et al.*, 2001). We have shown how different hypotheses about the genesis of the MMN could be tested, such as learning-related changes in forward or backward connections. Our results suggest that bottom-up processes have a key role, even in late components such as the P300. This finding is particularly interesting as top-down processes are usually invoked to account for late responses.

APPENDIX

Here we describe the approximate integration of delay differential equations of the form:

$$\dot{x}_i(t) = f_i(x_1(t-\tau_{i1}), \ldots, x_n(t-\tau_{in})) \qquad \textbf{42.A1}$$

for n states $x = [x_1(t), \ldots, x_n(t)]^T$, where state j causes changes in state i with delay τ_{ij}. By taking a Taylor expansion about $\tau = 0$ we get, to first order:

$$\begin{aligned}\dot{x}_i &= f_i(x) - \sum_j \tau_{ij}\partial f_i/\partial \tau_{ij} \\ &= f_i(x) - \sum_j \tau_{ij} J_{ij} \dot{x}_j\end{aligned} \qquad \textbf{42.A2}$$

where $J = \partial f/\partial x$ is the system's Jacobian. 42.A2 can be expressed in matrix form as:

$$\dot{x} = f(x) - (\tau \times J)\dot{x} \qquad \textbf{42.A3}$$

where $\times$ denotes the Hadamard or element-by-element product. On rearranging 42.A3, we obtain an ordinary differential equation that can be integrated in the usual way:

$$\begin{aligned}\dot{x}_i &= D^{-1} f(x) \\ D &= I + \tau \times J\end{aligned} \qquad \textbf{42.A4}$$

REFERENCES

Aguirre GK, Zarahn E, D'Esposito M (1998) An area within human ventral cortex sensitive to 'building' stimuli: evidence and implications. *Neuron* **21**: 373–83

Allison T, Puce A, Spencer DD *et al.* (1999) Electrophysiological studies of human face perception. I: Potentials generated in occipitotemporal cortex by face and non-face stimuli. *Cereb Cortex* **9**: 415–30

Brovelli A, Ding M, Ledberg A *et al.* (2004) Beta oscillations in a large-scale sensorimotor cortical network: directional influences revealed by Granger causality. *Proc Natl Acad Sci USA* **101**: 9849–54

Büchel C, Friston KJ (1997) Modulation of connectivity in visual pathways by attention: cortical interactions evaluated with structural equation modelling and fMRI. *Cereb Cortex* **7**: 768–78

Crick F, Koch C (1998) Constraints on cortical and thalamic projections: the no-strong-loops hypothesis. *Nature* **391**: 245–50

Dau T (2003) The importance of cochlear processing for the formation of auditory brainstem and frequency following responses. *J Acoust Soc Am* **113**: 936–50

David O, Friston KJ (2003) A neural mass model for MEG-EEG: coupling and neuronal dynamics. *NeuroImage* **20**: 1743–55

David O, Garnero L, Cosmelli D *et al.* (2002) Estimation of neural dynamics from MEG-EEG cortical current density maps: application to the reconstruction of large-scale cortical synchrony. *IEEE Trans Biomed Eng* **49**: 975–87

David O, Harrison L, Friston KJ (2005) Modelling event-related responses in the brain. *NeuroImage* **25**: 756–70

David O, Kiebel SJ, Harrison LM *et al.* (2006) Dynamic causal modeling of evoked responses in EEG and MEG. *NeuroImage*, Feb 8; (Epub ahead of print)

Debener S, Kranczioch C, Herrmann CS *et al.* (2002) Auditory novelty oddball allows reliable distinction of top-down and bottom-up processes of attention. *Int J Psychophysiol* **46**: 77–84

Engel AK, Fries P, Singer W (2001) Dynamic predictions: oscillations and synchrony in top-down processing. *Nat Rev Neurosci* **2**: 704–16

Epstein R, Kanwisher N (1998) A cortical representation of the local visual environment. *Nature* **392**: 598–601

Felleman DJ, Van Essen DC (1991) Distributed hierarchical processing in the primate cerebral cortex. *Cereb Cortex* **1**: 1–47.

Friston K (2003) Learning and inference in the brain. *Neural Netw* **16**: 1325–52

Friston KJ (2002) Bayesian estimation of dynamical systems: an application to fMRI. *NeuroImage* **16**: 513–30

Friston KJ, Harrison L, Penny W (2003) Dynamic causal modelling. *NeuroImage* **19**: 1273–302

Harrison L, Penny WD, Friston K (2003) Multivariate autoregressive modeling of fMRI time series. *NeuroImage* **19**: 1477–91

Haxby JV, Gobbini MI, Furey ML *et al.* (2001) Distributed and overlapping representations of faces and objects in ventral temporal cortex. *Science* **293**: 2425–30

Haxby JV, Hoffman EA, Gobbini MI (2002) Human neural systems for face recognition and social communication. *Biol Psychiatr* **51**: 59–67

Ishai A, Ungerleider LG, Martin A *et al.* (1999) Distributed representation of objects in the human ventral visual pathway. *Proc Natl Acad Sci USA* **96**: 9379–84

Jansen BH, Rit VG (1995) Electroencephalogram and visual evoked potential generation in a mathematical model of coupled cortical columns. *Biol Cybernet* **73**: 357–66

Jirsa V (2004) Information processing in brain and behavior displayed in large-scale scalp topographies such as EEG and MEG. *Int J Bifurcation Chaos* **14**: 679–92

Kaas JH, Hackett TA (2000) Subdivisions of auditory cortex and processing streams in primates. *Proc Natl Acad Sci USA* **97**: 11793–99

Kiebel SJ, David O, Friston KJ (2006) Dynamic causal modeling of evoked responses in EEG/MEG with lead-field parameterization. *NeuroImage* **30**: 1273–84

Liley DT, Cadusch PJ, Dafilis MP (2002) A spatially continuous mean field theory of electrocortical activity. *Network* **13**: 67–113

Linden DE, Prvulovic D, Formisano E *et al.* (1999) The functional neuroanatomy of target detection: an fMRI study of visual and auditory oddball tasks. *Cereb Cortex* **9**: 815–23

Lopes da Silva FH, Hoeks A, Smits H *et al.* (1974) Model of brain rhythmic activity. The alpha-rhythm of the thalamus. *Kybernetik* **15**: 27–37

McIntosh AR, Gonzalez-Lima F (1994) Structural equation modelling and its application to network analysis in functional brain imaging. *Hum Brain Mapp* **2**: 2–22

Melcher JR, Kiang NY (1996) Generators of the brainstem auditory evoked potential in cat. III: Identified cell populations. *Hear Res* **93**: 52–71

Mosher JC, Leahy RM, Lewis PS (1999) EEG and MEG: forward solutions for inverse methods. *IEEE Trans Biomed Eng* **46**: 245–59

Penny WD, Stephan KE, Mechelli A *et al.* (2004) Comparing dynamic causal models. *NeuroImage* **22**: 1157–72

Press WH, Teukolsky SA, Vetterling WT *et al.* (1992) *Numerical recipes in C*. Cambridge University Press, Cambridge, MA

Rennie CJ, Robinson PA, Wright JJ (2002) Unified neurophysical model of EEG spectra and evoked potentials. *Biol Cybernet* **86**: 457–71

Riera JJ, Watanabe J, Kazuki I *et al.* (2004) A state-space model of the hemodynamic approach: nonlinear filtering of BOLD signals. *NeuroImage* **21**: 547–67

Robinson PA, Rennie CJ, Rowe DL *et al.* (2004) Estimation of multiscale neurophysiologic parameters by electroencephalographic means. *Hum Brain Mapp* **23**: 53–72

Robinson PA, Rennie CJ, Wright JJ *et al.* (2001) Prediction of electroencephalographic spectra from neurophysiology. *Phys Rev E Stat Nonlin Soft Matter Phys* **63**: 021903

Roebroeck A, Formisano E, Goebel R (2005) Mapping directed influence over the brain using Granger causality and fMRI. *NeuroImage* **25**: 230–42

Romanski LM, Tian B, Fritz J *et al.* (1999) Dual streams of auditory afferents target multiple domains in the primate prefrontal cortex. *Nat Neurosci* **2**: 1131–36

Rosenblum MG, Pikovsky AS, Kurths J *et al.* (2002) Locking-based frequency measurement and synchronization of chaotic oscillators with complex dynamics. *Phys Rev Lett* **89**: 264102

Schnitzler A, Gross J (2005) Normal and pathological oscillatory communication in the brain. *Nat Rev Neurosci* **6**: 285–96

Smith PH, Populin LC (2001) Fundamental differences between the thalamocortical recipient layers of the cat auditory and visual cortices. *J Comp Neurol* **436**: 508–19

Stam CJ, Pijn JP, Suffczynski P *et al.* (1999) Dynamics of the human alpha rhythm: evidence for non-linearity? *Clin Neurophysiol* **110**: 1801–13

Tass PA (2004) Transmission of stimulus-locked responses in two oscillators with bistable coupling. *Biol Cybernet* **91**: 203–11

Vuilleumier P, Armony JL, Driver J *et al.* (2001) Effects of attention and emotion on face processing in the human brain: an event-related fMRI study. *Neuron* **30**: 829–41

Wendling F, Bartolomei F, Bellanger JJ *et al.* (2002) Epileptic fast activity can be explained by a model of impaired GABAergic dendritic inhibition. *Eur J Neurosci* **15**: 1499–508

CHAPTER

43

Dynamic Causal Models and Bayesian selection

K. E. Stephan and W. D. Penny

INTRODUCTION

Ever since the description of localizable lesion and excitation effects in the nineteenth century, neuroscience has revolved around the dual themes of functional specialization and functional integration. Functional specialization refers to the notion that local neural populations are specialized in certain aspects of information processing, whereas functional integration refers to the interactions among distinct populations. A variety of techniques exist for understanding functional specialization of brain regions, which rest on either measuring neuronal responses (e.g. neuroimaging, invasive recordings) or observing the consequences of selectively disabling particular parts of the brain (e.g. anatomical or pharmacological lesions; transcranial magnetic stimulation). Understanding the principles of functional integration requires two things: first, simultaneous measurements of activity in *multiple* interacting brain regions and second, a formal model designed to test assumptions about the causal mechanisms which underlie these interactions (see Chapters 36 and 38 and Friston, 2002).

To specify the structure of a neural system model, one needs to specify three things: (i) which areas are elements of the system; (ii) which anatomical connections exist; and (iii) which experimentally designed inputs affect system dynamics, either by directly evoking activity in specific areas (e.g. visual stimuli evoking responses in primary visual cortex) or by altering the strength of specific connections (e.g. changes of synaptic strengths during learning), and where these inputs enter the system. A key issue is how to test competing hypotheses about the organization of a system. Given experimental measurements of the system of interest, an ideal solution is to formalize each hypothesis in terms of a specific model and use a Bayesian model selection (BMS) procedure that takes into account both model fit and model complexity (Pitt and Myung, 2002; Penny *et al.*, 2004; see also Chapter 35). In principle, model selection can concern any aspect of a system so long as different models are compared using the same data.[1] In this chapter, we focus on how model selection can be used to decide which of several experimentally controlled inputs change particular connection strengths. To illustrate this, we deal with inter-hemispheric integration in the ventral visual pathway during a letter decision task. After summarizing BMS as implemented for dynamic causal modelling (DCM) for functional magnetic resonance imaging (fMRI), we present the results from a combined DCM and BMS analysis of a single subject data set (Stephan *et al.*, 2005). Inter-hemispheric integration is a particularly instructive example for the usefulness of BMS; this is because a variety of competing theories about its functional principles and mechanisms exist, all of which make fairly precise predictions about what experimental factors should be the primary cause of changes in inter-hemispheric connection strengths and what pattern of effective connectivity should be observed.

Bayesian model selection in DCM for fMRI

A generic problem encountered by any kind of modelling is the question of model selection: given some observed data, which of several alternative models is

[1] This means that in DCM for fMRI, where the data vector results from a concatenation of the time series of all areas in the model, only models can be compared that contain the same areas. In this case, model selection cannot be used to address whether or not to include a particular area in the model. In contrast, in DCM for ERPs, the data measured at the sensor level are independent of how many neuronal sources are assumed in a given model. Here, model selection could also be used to decide which sources should be included.

Statistical Parametric Mapping, by Karl Friston et al.
ISBN–13: 978-0-12-372560-8 ISBN–10: 0-12-372560-7

the optimal one? This problem is not trivial because the decision cannot be made solely by comparing the relative fit of the competing models. One also needs to take into account the relative complexity of the models as expressed, for example, by the number of free parameters in each model. Model complexity is important to consider because there is a trade-off between model fit and generalizability (i.e. how well the model explains different data sets that were all generated from the same underlying process). As the number of free parameters is increased, model fit increases monotonically, whereas beyond a certain point, model generalizability decreases. The reason for this is 'over-fitting'; an increasingly complex model will, at some point, start to fit noise that is specific to one dataset and thus become less generalizable across multiple realizations of the same underlying generative process.[2]

Therefore, the question: 'What is the optimal model?' can be reformulated more precisely as: 'What is the model that represents the best balance between fit and complexity?'. In a Bayesian context, the latter question can be addressed by comparing the evidence, $p(y|m)$, of different models. According to Bayes' theorem:

$$p(\theta|y,m) = \frac{p(y|\theta,m)p(\theta|m)}{p(y|m)} \tag{43.1}$$

the model evidence can be considered as a normalization constant for the product of the likelihood of the data and the prior probability of the parameters, therefore:

$$p(y|m) = \int p(y|\theta,m)p(\theta|m)d\theta \tag{43.2}$$

Here, the numbers of free parameters (as well as the functional form) are subsumed by the integration. Unfortunately, this integral cannot usually be solved analytically, therefore, an approximation to the model evidence is needed.

In the context of DCM, one potential solution could be to make use of the Laplace approximation, i.e. to approximate the model evidence by a Gaussian that is centred on its mode. As shown by Penny *et al.* (2004), this yields the following expression for the natural logarithm of the model evidence: $\eta_{\theta|y}$ denotes the maximum *a posteriori* (MAP) estimate, $C_{\theta|y}$ is the posterior covariance of the parameters, C_ε is the error covariance, η_θ is the prior mean of the parameters, C_θ is the prior covariance and $h(u, \eta_{\theta|y})$ is the prediction by the model given the known system inputs u and MAP estimate $\eta_{\theta|y}$:

$$\begin{aligned} \ln p(y|m) &= accuracy(m) - complexity(m) \\ &= \left[-\frac{1}{2}\ln|C_\varepsilon| - \frac{1}{2}\varepsilon_y^T C_\varepsilon^{-1}\varepsilon_y\right] \\ &\quad - \left[\frac{1}{2}\ln|C_\theta| - \frac{1}{2}\ln|C_{\theta|y}| + \frac{1}{2}\varepsilon_\theta^T C_\theta^{-1}\varepsilon_\theta\right] \\ \varepsilon_y &= y - h(u,\eta_{\theta|y}) \\ \varepsilon_\theta &= \eta_{\theta|y} - \eta_\theta \end{aligned} \tag{43.3}$$

This expression reflects the requirement, as discussed above, that the optimal model should represent the best compromise between model fit (accuracy) and model complexity. The complexity term depends on the prior density, for example, the prior covariance of the intrinsic connections. This is problematic in the context of DCM for fMRI because this prior covariance is defined in a model-specific fashion to ensure that the probability of obtaining an unstable system is small (specifically, this is achieved by choosing the prior covariance of the intrinsic coupling matrix A such that the probability of obtaining a positive Lyapunov exponent of A is $p < 0.001$; see Friston *et al.*, 2003 for details). Consequently, the comparison of models with different numbers of connections is conflated with a comparison of different priors. Alternative approximations to the model evidence, which depend less on the priors, are useful for DCMs of this sort.

Suitable approximations, which do not depend on the prior density, are afforded by the Bayesian information criterion (BIC) and Akaike information criterion (AIC), respectively. As shown by Penny *et al.* (2004), for DCM these approximations are given by:

$$\begin{aligned} BIC &= accuracy(m) - \frac{p_\theta}{2}\ln n \\ AIC &= accuracy(m) - p_\theta \end{aligned} \tag{43.4}$$

where p_θ is the number of parameters and n is the number of data points (scans). If one compares the complexity terms of BIC and AIC, it is obvious that BIC pays a heavier penalty than AIC as soon as one deals with eight or more scans (which is virtually always the case for fMRI data):

$$\frac{d_\theta}{2}\ln n > p_\theta \Rightarrow n > e^2 \approx 7.39 \tag{43.5}$$

Therefore, BIC will be biased towards simpler models, whereas AIC will be biased towards more complex models. This can lead to disagreement between the two approximations about which model should be favoured. We adopt the convention that, for any pairs of models

[2] Generally, in addition to the number of free parameters, the complexity of a model also depends on its functional form (Pitt and Myung, 2002). This is not an issue for DCM, however, because here competing models usually have the same functional form.

m_i and m_j, a model is selected if, and only if AIC and BIC concur (see below); the decision is then based on the approximation which gives the smaller *Bayes factor* (BF):

$$BF_{ij} = \frac{p(y|m_i)}{p(y|m_j)} \quad \textbf{43.6}$$

Just as conventions have developed for using p-values in frequentist statistics, there are conventions for Bayes factors. For example, Raftery (1995) suggests interpretation of Bayes factors as providing weak (BF < 3), positive (3 ≤ BF < 20), strong (20 ≤ BF < 150) or very strong (BF ≥ 150) evidence for preferring one model over another.

We conclude this section with a short comment on group studies with DCM and BMS. When trying to find the optimal model for a group of individuals, it is likely that the optimal model will vary, at least to some degree, across subjects. An overall decision for N subjects can be made as follows. First, because model comparisons from different individuals are statistically independent, we can compute a *group Bayes factor* (GBF) by multiplying all N individual Bayes factors (where k is an index across subjects):

$$GBF_{ij} = \prod_k BF_{ij}^k \quad \textbf{43.7}$$

In principle, this is sufficient to select the optimal model across a group of subjects. However, the magnitude of the group Bayes factor may not always be easily interpretable since its value depends on N. It can therefore be useful to compute an *average Bayes factor* (ABF) which is the geometric mean of the group Bayes factor:

$$ABF_{ij} = \sqrt[N]{GBF_{ij}} \quad \textbf{43.8}$$

A problem that both GBF and ABF may suffer from is the presence of outliers.[3] For example, imagine you compare models m_i and m_j in a group of 11 subjects where, for 10 subjects, one finds $BF_{ij} = 10$ for a given comparison of interest and for a single subject one finds $BF_{ij} = 10^{-12}$. In this case, we would obtain $GBF_{ij} = 10^{-2}$ and $ABF_{ij} \approx 0.66$. Since these results are driven by a single outlier, it is doubtful whether one should conclude on the basis of these values that Y is the better model. A heuristic that precludes this sort of thing is the *positive evidence ratio* (PER), i.e. the number of subjects where there is positive (or stronger) evidence for model m_i divided by the number of subjects with positive (or stronger) evidence for model m_j:

$$PER_{ij} = \frac{|k : BF_{ij}^k > 3|}{|k : BF_{ji}^k > 3|} \quad \textbf{43.9}$$

where $k = 1, \ldots, N$ and $|\cdot|$ denotes set size.

Overall, BMS is a powerful procedure to decide between competing hypotheses represented by different DCMs. These hypotheses can concern any part of a model, e.g. the pattern of intrinsic connections or which inputs affect the system and where they enter. In the next section, we present an example that demonstrates how the combination of DCM and BMS can be applied to questions of inter-hemispheric integration.

INTER-HEMISPHERIC INTEGRATION IN THE VENTRAL STREAM

Theories of inter-hemispheric integration

Currently, three major theories of inter-hemispheric integration are entertained (see Stephan *et al.*, 2006a for a review). The oldest is that of *information transfer* between the hemispheres (e.g. Poffenberger, 1912). In the context of lateralized tasks with hemisphere-specific inputs (e.g. peripheral visual presentation), this theory predicts that transfer of sensory information should be asymmetrically enhanced from the non-dominant to the dominant hemisphere to ensure efficient processing in the specialized hemisphere (e.g. Nowicka *et al.*, 1996; Endrass *et al.*, 2002). In terms of effective connectivity, it predicts a task-dependent increase in influence of the non-dominant on the dominant hemisphere, but only when stimulus information is initially restricted to the non-dominant hemisphere.

A more recent and similarly influential concept has been the notion of *inter-hemispheric inhibition* (Kinsbourne, 1970). It has been argued that the regulatory mechanisms that 'coordinate, select, and integrate the processes subserved by each hemisphere' will also require a range of inter-hemispheric inhibitory mechanisms 'to achieve unified performance from a bilateral system capable of producing simultaneous and potentially conflicting outputs' (Chiarello and Maxfield, 1996). With regard to connectivity, inter-hemispheric inhibition predicts a task-dependent and symmetric pattern of negative connection strengths between hemispheres. It is important to note, however, that this does not necessarily mean that two areas, which affect each other by task-dependent inter-hemispheric inhibition, show decreased activity during

[3] Of course, the problem of outliers is not specific to Bayesian inference with DCMs but is inherent to all group analyses. In random-effects analyses, for example, subjects are assumed to derive from a Gaussian distribution about a typical subject (see Chapter 11, Hierarchical Models). This assumption is clearly violated in the presence of outliers.

that task. As can be easily shown in simulations, it is perfectly possible that task-dependent regional activations coexist with task-dependent inter-hemispheric inhibition if other, e.g. intra-hemispheric, influences onto the areas in question are positive.

The third major concept of inter-hemispheric integration concerns *hemispheric recruitment* or *processing mode setting*, i.e. whether information processing is restricted to a single hemisphere or distributed across both hemispheres. Several behavioural studies have shown that, if the neural resources in the hemisphere receiving a stimulus are insufficient for optimal processing, the benefits of distributing the processing load across both hemispheres are likely to outweigh the costs of transcallosal information transfer (see Banich, 1998 for review). Given a sufficiently demanding task, this recruitment of an additional hemisphere even occurs during lateralized tasks when the dominant hemisphere receives the stimulus (Belger and Banich, 1998). This additional recruitment of the non-dominant hemisphere requires tight cooperation, i.e. functional coupling, of both hemispheres, regardless of which hemisphere initially received the stimulus. Two components are likely to be expressed in terms of task-dependent changes in effective connectivity: an increase of the influence from the dominant to the non-dominant hemisphere that reflects the 'recruitment' of the non-dominant hemisphere, and an increase of connection strengths in the opposite direction, induced by the non-dominant hemisphere 'returning' the results of the computations delegated to it by the dominant hemisphere. Altogether, this cooperation is expected to be expressed either in terms of a symmetric task-dependent increase of connection strength between homotopic areas or, if 'recruitment' and 'return' processes are spatially segregated, an asymmetric task-dependent increase of connection strength between different areas.

Next, we review an experiment in which inter-hemispheric integration was necessary and could have been orchestrated by any of the three mechanisms described above. Using data from a single subject, we provide an example of how BMS can be used to disambiguate different theories of inter-hemispheric integration with DCM.

An fMRI study of inter-hemispheric integration

In a previous fMRI study on the mechanisms underlying hemispheric specialization, we investigated whether lateralization of brain activity depends on the nature of the sensory stimuli or on the nature of the cognitive task performed (Stephan *et al.*, 2003). For example, microstructural differences between homotopic areas in the left and right hemisphere have been reported, including visual (Jenner *et al.*, 1999) and language-related (Amunts *et al.*, 1999) areas. Within a given hemisphere, these differences could favour the processing of certain stimulus characteristics and disadvantage others and might thus support stimulus-dependent lateralization in a bottom-up fashion (Sergent, 1982). On the other hand, processing demands, mediated through cognitive control processes, might determine, in a top-down fashion, which hemisphere obtains precedence in a particular task (Levy and Trevarthen, 1976; Fink *et al.*, 1996). To decide between these two possibilities, we used a paradigm in which the stimuli were kept constant throughout the experiment, and subjects were alternately instructed to attend to certain stimulus features and ignore others (Stephan *et al.*, 2003). The stimuli were concrete German nouns (each four letters in length) in which either the second or third letter was printed in red (the other letters were black). In a letter decision (LD) task, the subjects had to ignore the position of the red letter and indicate whether or not the word contained the target letter 'A'. In a spatial decision (SD) task they were required to ignore the language-related properties of the word and to judge whether the red letter was located left or right of the word centre; 50 per cent of the stimuli were presented in the non-foveal part of the right visual field (RVF) and the other 50 per cent in the non-foveal part of the left visual field (LVF).

The results of a conventional fMRI data analysis were clearly in favour of the top-down hypothesis: despite the use of identical word stimuli in all conditions, comparing spatial to letter decisions showed strongly right-lateralized responses in the parietal cortex, whereas comparing letter to visuo-spatial decisions showed strongly left-lateralized responses, involving classical language areas in the left inferior frontal gyrus and visual areas in the left ventral visual stream, e.g. in the fusiform gyrus (FG), middle occipital gyrus (MOG) and lingual gyrus (LG) (Plate 61, see colour plate section). Notably, the LG areas also showed a main effect of visual field, whereas task-by-visual field interactions were not significant at the group level due to spatial variability of their locations across subjects (see Discussion).

Constructing a basic DCM

Through the conjoint use of lateralized and demanding tasks with peripheral visual representation, we ensured that inter-hemispheric integration was a necessary component of the cognitive processes in this paradigm. We now want to demonstrate how to use DCM to investigate which theory of inter-hemispheric integration is most likely to account for our data. We focus on the ventral stream of the visual system which, as shown

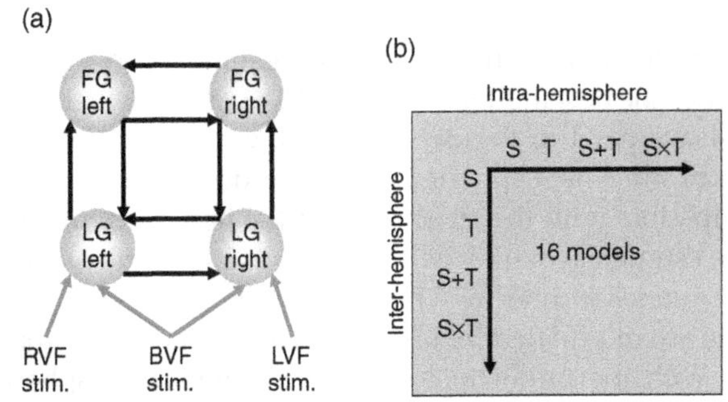

FIGURE 43.1 (a) Basic model that comprises the left and right lingual gyrus (LG) and left and right fusiform gyrus (FG). The areas are reciprocally connected (black arrows). Driving inputs are shown as grey arrows. RVF stimuli directly affect left LG activity and LVF stimuli directly affect right LG activity, regardless of task. Individual stimuli lasted for 150 ms, therefore, these inputs are represented as trains of events (delta-functions). During the instruction periods, bilateral visual field input was provided for 6 s; this was modelled as a boxcar input affecting LG in both hemispheres. (b) Schema showing how the 16 models tested were constructed by combining four different types of modulatory inputs for inter- and intra-hemispheric connections respectively.

in Plate 61, is preferentially involved in letter decisions in this experiment. For simplicity, we use a four-area model comprising LG and FG in both hemispheres. First, we need to define a model comprising these four areas (Figure 43.1(a)). Starting with the direct (driving) inputs to the system, we model the lateral stimulus presentation and the crossed course of the visual pathways by allowing all RVF stimuli to affect directly left LG activity and all LVF stimuli to directly affect right LG activity, regardless of task. Each stimulus lasted for 150 ms only; therefore these inputs are represented as trains of short events (delta functions). The induced activity then spreads through the system according to the connections of the model. For visual areas, a general rule is that both intra- and inter-hemispheric connections are reciprocal and that homotopic regions in both hemispheres are linked by inter-hemispheric connections (Segraves and Rosenquist, 1982; Cavada and Goldman-Rakic, 1989; Kötter and Stephan, 2003).[4]

Deriving a set of alternative models

Note that up to this point there are few, if any, plausible alternatives for how a DCM of inter-hemispheric integration between LG and FG, respectively, should be constructed. However, the important issue is how experimental factors affect transcallosal information in this system during the LD task. This is directly related to the predictions from the three theories described above. Instead of testing exclusively for these three predictions, we compare a larger, systematically derived set of models which will include those predicted by the three theories.

Given that some areas also showed a main effect of visual field, one could assume that inter-hemispheric interactions between visual areas are primarily determined by the visual field of stimulus presentation, independent of task demands: for example, whenever a stimulus is presented in the LVF and stimulus information is received initially by the right visual cortex, this information is transmitted transcallosally to the left visual cortex. Vice versa, whenever a stimulus is presented in the RVF, stimulus information is transmitted transcallosally from left to right visual cortex. In this scenario, the task performed is assumed to have no influence on callosal couplings, and task effects could, as shown in a previous analysis of connectivity, be restricted to modulate functional couplings *within* hemispheres (Stephan *et al.*, 2003). This model is subsequently referred to as the S-model, for stimulus-dependent.

Alternatively, one might expect that callosal connection strengths depend more on which task is performed than on which visual field the stimulus is presented in. That is, right → left and/or left → right callosal connections could be altered during the LD task. This type of modulation is predicted by two theories described above, the inter-hemispheric inhibition and hemispheric recruitment theories; however, they predict opposite directions of the change in coupling. This model is referred to as the T-model, for task dependent.

As a third hypothesis, it is conceivable that both visual field and task exert an influence on callosal connection strengths, but *independently* of each other. This is the S+T-model. As a fourth and final option, one might postulate that task demands modulate callosal connections, but *conditional* on the visual field, i.e. right → left connections are only modulated by LD during LVF stimulus presentation (LD|LVF) whereas left → right connections are only modulated by LD during RVF stimulus presentation (LD|RVF). This is the S × T-model.

Although less interesting in the present context, the same questions about the nature of modulatory inputs arise with respect to the intra-hemispheric connections. Therefore, to perform a thorough model comparison, we systematically compared all 16 combinations of how inter- and intra-hemispheric connections are changed in the four ways (S, T, S + T, S × T) described above (note that in the models presented here, we only allowed for modulation of the intra-hemispheric forward connections, i.e. from LG → FG; see Figures 43.2, 43.3).

[4] To be precise, we should point out that in primates left and right area V1 only have extremely sparse callosal connections with each other; this is in contrast to other visual areas like V2, which are strongly linked by homotopic callosal connections (Kennedy *et al.*, 1986; Abel *et al.*, 2000).

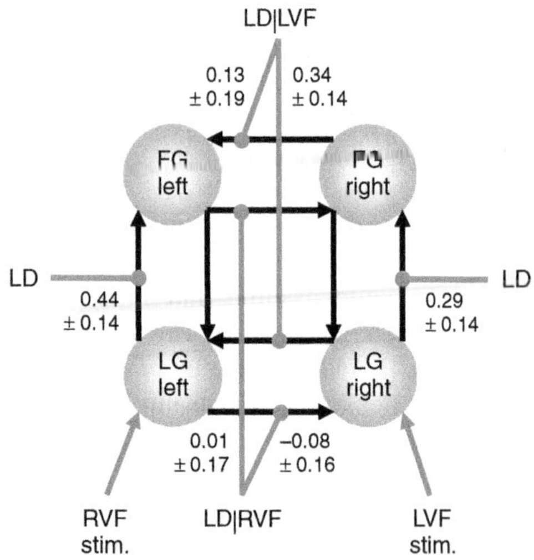

FIGURE 43.2 This figure shows the maximum *a posteriori* (MAP) estimates of the parameters (± square root of the posterior variances; units: Hz) for the S × T-T model which, for the particular subject studied, proved to be the best of all models tested. For clarity, only the parameters of interest, i.e. the modulatory parameters of inter- and intra-hemispheric connections, are shown and the bilateral visual field input has been omitted.

Figure 43.1(b) summarizes the combinatorial logic that resulted in 16 different models which were fitted to the same data. In the following, we refer to these 16 models by first listing the modulation of inter- and then intra-hemispheric connections. For example, the T-S-model is one where the callosal connections are modulated by the letter decision task and the intra-hemispheric connections are modulated by the visual field of stimulation.

Once the best model is identified, we can do two things. First, we need to decide whether any of the three theories of inter-hemispheric integration described above is compatible with the structure and the pattern of parameter estimates of the best model. For example, if the best model is one in which inter-hemispheric connections are modulated by task only, this would be compatible with the predictions by both the inter-hemispheric inhibition and hemispheric recruitment theories; however, inter-hemispheric inhibition predicts decreases and hemispheric recruitment predicts increases in callosal coupling (see above). Second, we can use the posterior density of the parameters of the optimal model to make statistical inferences about the strength of callosal coupling and its modulation. This would allow us to quantify how certain we are that contextual modulation of callosal connections is present at all levels within the ventral stream hierarchy, or whether it is regionally specific. With the exception of some EEG (electroencephalography) studies (Nowicka *et al.*, 1996; Schack *et al.*, 2003), which have a rather low spatial resolution, this is a largely unexplored issue.

Results

Here we report the results from fitting the 16 DCMs described above to the fMRI data of a single subject from the study by Stephan *et al.* (2003). The time-series were

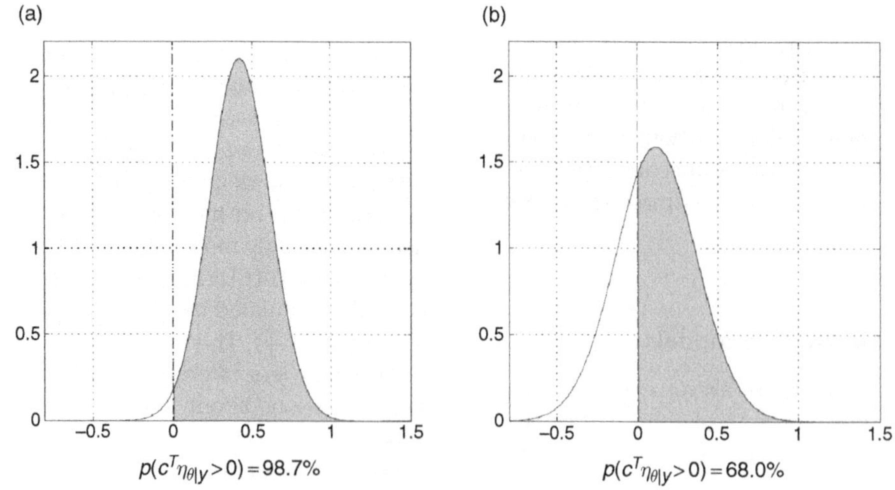

FIGURE 43.3 Asymmetry of callosal connections with regard to contextual modulation in the model shown in Figure 43.2. The plots show the probability that the modulation of the right → left connection by task conditional on left visual field stimulation is stronger than the modulation of the left → right connection by task conditional on right visual field stimulation. (a) For the callosal connections between left and right LG, we can be very confident that this asymmetry exists: the probability of the contrast being larger than zero is $p(c^T\eta_{\theta|y} > 0) = 98.7\%$. (b) For the callosal connections between left and right FG, we are considerably less certain about the presence of this asymmetry: the probability of the contrast being larger than zero is $p(c^T\eta_{\theta|y} > 0) = 68.0\%$.

selected based on local maxima of the 'main effect of task' contrast in the SPM (statistical parametric map). However, the regional time-series showed a main effect of task (LD > SD), a main effect of VF, a simple main effect (LD > SD separately for LVF and RVF presentation) and task-by-VF interactions (see Discussion).

The BMS procedure indicated that the optimal model was the S × T-T-model, i.e. modulation of inter-hemispheric connections by task, conditional on the visual field of stimulus presentation, and modulation of intra-hemispheric connection strengths by the task only (Figure 43.2). Table 43-1 shows the Bayes factors for the comparison of the S × T-T-model with the other models. The AIC and BIC approximations agreed for all comparisons. The second-best model was the T-S × T-model (i.e. the 'flipped' version of the S × T-T-model). The Bayes factor of comparing the S × T-T-model with the T-S × T-model was 2.33 which, according to the criteria summarized by Penny *et al.* 2004 (see their Table 1), could be interpreted as weak evidence in favour of the S × T-T-model. All other comparisons gave Bayes factors larger than 3, representing positive, strong or very strong evidence in favour of the S × T-T-model (see Table 43-1).

The optimal S × T-T-model has a structure (in terms of modulatory inputs) predicted by the information transfer hypothesis (see above). It remains to be tested, however, whether the pattern of modulatory parameters also matches the predictions from that hypothesis, i.e. whether for the left-lateralized LD task the modulation of right → left connections by LD|LVF is significantly higher than the modulation of the left → right connections by LD|RVF. Figure 43.2 shows the maximum *a posteriori* (MAP) estimates $\eta_{\theta|y}$ of the modulatory parameters (± standard deviation, i.e. the square root of the posterior variances) for the S × T-T-model. The modulatory parameter estimates indeed indicated a strong hemispheric asymmetry: both at the levels of LG and FG, the MAP estimates of the modulation of the right → left connections are much larger than those of the left → right connections. But how certain can we be about the presence of this asymmetry? This issue can be addressed by means of contrasts $c^T\eta_{\theta|y}$ of the appropriate parameter estimates (c^T is a transposed vector of contrast weights). The contrasts comparing modulation of the right → left connection by LD|LVF versus modulation of the left → right connection by LD|RVF are shown in Figure 43.3 (separately for connections at the level of LG and FG, respectively). These plots show $p(c^T\eta_{\theta|y} > 0)$, i.e. our certainty about asymmetrical modulation of callosal interactions in terms of the probability that these contrasts exceed a value of zero. For the particular subject shown here, we can be very certain (98.7 per cent) that modulation of the right LG → left LG connection by LD|LVF is larger than the modulation of the left LG → right LG connection by LD|RVF (cf. Figure 43.2). In contrast, we are much less confident (68.0 per cent) that this asymmetry also exists for callosal connections between right and left FG.

TABLE 43-1 Bayes factors (middle column) for the comparison of the best model with each of the other 15 models (left column)

S × T-T versus	BF	Evidence in favour of S × T-T
S+T-S+T	477.31	very strong
S+T-S × T	60.83	strong
S+T-T	110.84	strong
S+T-S	479.03	very strong
S × T-S+T	3.92	positive
S × T − S × T	4.48	positive
S × T-S	46267.47	very strong
T-S+T	19.96	positive
T − S × T	2.33	weak
T − T	3.43	positive
T − S	29.74	strong
S-S+T	16.85	positive
S − S × T	4.81	positive
S-T	5.59	positive
S-VF	1.35E+13	very strong

The right column lists the interpretation of the evidence in favour of the S × T-T model according to the criteria in Raftery (1995); see the summary by Penny *et al.* (2004).

DISCUSSION

In this chapter, we have shown how BMS can help to decide between different cognitive-neurobiological hypotheses, each represented by a specific model. Together with BMS, DCM is a powerful tool to assess which experimental manipulations (e.g. stimulus type, induction of cognitive set, learning processes etc.) have a significant impact on the dynamics of the network under investigation. By representing experimental factors as external inputs in the model, modelled effects can be interpreted fairly directly in neurobiological terms: any given DCM specifies precisely *where* inputs enter and whether they are *driving* (i.e. exert their effects through direct synaptic responses in the target area) or *modulatory* (i.e. exert their effects through changing synaptic responses in the target area to inputs from another area). This distinction, made at the level of neural populations, has a nice correspondence to empirical observations that single neurons can either have driving or modulatory effects on other neurons (Sherman and Guillery, 1998).

In our empirical example, we demonstrated that, for the particular task and subject studied, inter-hemispheric integration in the ventral visual stream conforms to the principle of inter-hemispheric information transfer. This conclusion rests on two findings: (i) the measured data are best explained by a model in which inter-hemispheric interactions depend on task demands, but conditional on the visual field of stimulus presentation; and (ii) there is a hemispheric asymmetry in context-dependent transcallosal interactions, with modulation of connections from the non-dominant (right) to the dominant (left) hemisphere being considerably stronger than modulation of the connections in the opposite direction. Importantly, this asymmetry was not equally pronounced for all visual areas studied. It was particularly strong for the callosal connections between left and right LG: performance of the letter decision task specifically enhanced the strength of the influence of the right on the left LG, but only if the stimulus was presented in the left visual field and thus the information was initially only available in the right hemisphere. The reversed conditional effect, i.e. modulation of left LG → right LG by LD|RVF, was much weaker (and actually slightly negative, see Figure 43.2). This result means that enhancement of callosal connections was only necessary if stimulus information was initially represented in the 'suboptimal', i.e. right, hemisphere.

Two short technical comments may be useful for a full understanding of the particular model structure chosen here. First, one may be concerned about the presence of correlations between inputs (in the extreme case, the presence of identical inputs entering the model at multiple sites). Concerning model selection, such correlations are not an issue when using Eqn. **43.3** (which takes into account the posterior covariance of the parameters), but may be problematic when using approximations such as AIC or BIC (see Eqn. **43.4**). This can be easily investigated for any given model using synthetic data generated from known parameters and adding observation noise. For the particular model here, we investigated this issue and found that even at high levels of noise: (i) the model selection reliably chose the correct model; and (ii) the estimates for the contrasts of interests (i.e. comparisons of modulations of callosal connections) were not biased, i.e. did not deviate significantly from the true values (Stephan *et al.*, unpublished data).

A second question concerns the interpretations of the DCM results in relation to the SPM results. One may argue that the type of statistical contrast used to decide which regions to sample from pre-determines the outcome of the model selection procedure. This is only true, however, for simple models. As an example, let us imagine a two-area T-T-model that only consists of left and right LG. In this model, the driving of right LG by LVF and the modulation of the right LG → left LG connection by the LD task corresponds to modelling a task-by-VF interaction in left LG. However, it would be misleading to conclude from this that the full four-area T-T-model is only appropriate to explain time-series dominated by task-by-VF interactions. In the full T-T-model with its complex connectivity loops, both main effects and interactions can be modelled, depending on the relative strengths of the modulatory inputs and the intrinsic connections. For example, as can be demonstrated with simulations, choosing a strongly positive modulation of the intra-hemispheric forward connections by LD, setting the strength of callosal modulations by LD to low positive values and choosing strong intra-hemispheric backward projections primarily models a main effect of task in FG and also in LG (because of the back-projections). Overall, the model parameters need to be fitted to explain optimally the overall constellation of main effects and interactions in *all* areas of the model. This precludes straightforward predictions about the outcome of model selection in situations where the model structure is complex and where regional time-series are influenced by various experimental effects, as in the present data set.

We conclude by emphasizing the importance of model selection. Model selection is essentially the same as hypothesis testing, in the sense that every hypothesis can be framed in terms of the difference between two models. This makes model selection central to the scientific process. Furthermore, model selection can also play an important role in clinical neuroscience, e.g. in psychiatry where diseases like schizophrenia often comprise heterogeneous phenotypes. BMS could be used to find subgroups of patients that differ in terms of DCMs that optimally explains the measured neurophysiological data. If the neural models are sophisticated enough to distinguish between the effects of different transmitter receptors, it might also be possible to obtain predictions for the optimal pharmacological treatment of individual patients (see Stephan 2004; Stephan *et al.*, 2006b).

REFERENCES

Abel PL, O'Brien BJ, Olavarria JF (2000) Organization of callosal linkages in visual area V2 of Macaque monkey. *J Comp Neurol* **428**: 278–93

Amunts K, Schleicher A, Bürgel U *et al.* (1999) Broca's region revisited: cytoarchitecture and intersubject variability. *J Comp Neurol* **412**: 319–41

Banich MT (1998) The missing link: the role of interhemispheric interaction in attentional processing. *Brain Cogn* **36**: 128–57

Belger A, Banich MT (1998) Costs and benefits of integrating information between the cerebral hemispheres: a computational perspective. *Neuropsychology* **12**: 380–98

Cavada C, Goldman-Rakic PS (1989) Posterior parietal cortex in rhesus monkey: I. Parcellation of areas based on distinctive limbic and sensory corticocortical connections. *J Comp Neurol* **287**: 393–421

Chiarello C, Maxfield L (1996) Varieties of interhemispheric inhibition, or how to keep a good hemisphere down. *Brain Cogn* **30**: 81–108

Endrass T, Mohr B, Rockstroh B (2002) Reduced interhemispheric transmission in schizophrenia patients: evidence from event-related potentials. *Neurosci Lett* **320**: 57–60

Fink GR, Halligan PW, Marshall JC *et al.* (1996) Where in the brain does visual attention select the forest and the trees? *Nature* **382**: 626–28

Friston KJ (2002) Beyond phrenology: what can neuroimaging tell us about distributed circuitry? *Annv Rev Neurosci* **25**: 221–50

Friston KJ, Harrison L, Penny W (2003) Dynamic causal modelling. *NeuroImage* **19**: 1273–302

Jenner AR, Rosen GD, Galaburda AM (1999) Neuronal asymmetries in primary visual cortex of dyslexic and nondyslexic brains. *Ann Neurol* **46**: 189–96

Kennedy H, Dehay C, Bullier J (1986) Organization of the callosal connections of visual areas V1 and V2 in the macaque monkey. *J Comp Neurol* **247**: 398–415

Kinsbourne M (1970) The cerebral basis of lateral asymmetries of attention. *Acta Psychol* **33**: 193–201

Kötter R, Stephan KE (2003) Network participation indices: characterizing component roles for information processing in neural networks. *Neural Netw* **16**: 1261–75

Levy J, Trevarthen C (1976) Metacontrol of hemispheric function in human split-brain patients. *J Exp Psychol Hum Percept Perform* **2**: 299–312

Nowicka A, Grabowska A, Fersten E (1996) Interhemispheric transmission of information and functional asymmetry of the human brain. *Neuropsychologia* **34**: 147–51

Penny WD, Stephan KE, Mechelli A *et al.* (2004) Comparing dynamic causal models. *NeuroImage* **22**: 1157–72

Pitt MA, Myung IJ (2002) When a good fit can be bad. *Trends Cogn Sci* **6**: 421–25

Poffenberger AT (1912) Reaction time to retinal stimulation with special reference to the time lost in conduction through nerve centres. *Arch Psychol* **23**: 1–73

Raftery AE (1995) Bayesian model selection in social research. In *Sociological methodology*, Marsden PV (ed.). Blackwell publishing Cambridge, MA, pp 111–96

Schack B, Weiss S, Rappelsberger P (2003) Cerebral information transfer during word processing: where and when does it occur and how fast is it? *Hum Brain Mapp* **19**: 18–36

Segraves MA, Rosenquist AC (1982) The afferent and efferent callosal connections of retinotopically defined areas in cat cortex. *J Neurosci* **2**: 1090–1107

Sergent J (1982) Role of the input in visual hemispheric asymmetries. *Psychol Bull* **93**: 481–512

Sherman SM, Guillery RW (1998) On the actions that one nerve cell can have on another: distinguishing 'drivers' from 'modulators'. *Proc Natl Acad Sci USA* **95**: 7121–26

Stephan KE, Marshall JC, Friston KJ *et al.* (2003) Lateralized cognitive processes and lateralized task control in the human brain. *Science* **301**: 384–86

Stephan KE (2004) On the role of general systems theory for functional neuroimaging. *J Anat* **205**: 443–70

Stephan KE, Penny WD, Marshall JC *et al.* (2005) Investigating the functional role of callosal connections with dynamic causal models. *Ann NY Acad Sci* **1064**: 16–36

Stephan KE, Fink GR, Marshall JC (2006a) Mechanisms of hemispheric specialization: insights from analyses of connectivity. *Neuropsychologia* (in press)

Stephan KE, Baldeweg T, Friston KJ (2006b) Synaptic plasticity and dysconnection in schizophrenia. *Biol Psychiatr* **59**: 929-39

Appendices

A P P E N D I X

1

Linear models and inference

K. Friston

INTRODUCTION

In this appendix, we gather together different perspectives on inference with multivariate linear models. In brief, we will see that all inference, whether it is based on likelihood ratios (i.e. classical statistics), canonical variates analysis, linear discriminant analysis, Bayesian analysis or information theory, can be regarded as tests of the same thing, namely, the statistical dependence between one set of variables and another. This is useful to appreciate because it means there is no difference between inference using generative models (i.e. functions of causes that generate data) and inference based on recognition models (i.e. functions of data that recognize causes). This equivalence rests on the invertability of linear models, which means that recognition or classification models are simply the inverse of generative or forward models. In other words, there is no difference between reverse-correlation methods (e.g. Hansen *et al.*, 2004; Hasson *et al.*, 2004) or brain-reading (Cox and Savoy, 2003) and conventional analyses (e.g. Friston *et al.*, 1996; Worsley *et al.*, 1997; Kherif *et al.*, 2002), provided the models are linear.

In what follows, we look at the problem of establishing statistical dependencies from an information theory point of view and then revisit the same issue from a classical, a Bayesian and finally a multivariate perspective.

INFORMATION THEORY AND DEPENDENCY

The aim of classification is to find some function of data y that can be used to classify or predict their causes x. Conversely, the aim of hypothesis testing is to show that hypothetical causes predict data. For simplicity, we will assume both x and y represent s independent samples drawn from multivariate distributions. The ability to predict one, given the other, rests on the statistical dependencies between x and y that are quantified by their mutual information. Irrespective of the form of these dependencies, the mutual information is given by the difference between both their entropies and the entropy of both:

$$\begin{aligned} I(x, y) &= H(y) + H(y) - H(x, y) \\ &= H(y) - H(y|x) \end{aligned} \quad \textbf{A1.1}$$

This is the difference between the entropy, or information, of one minus the information given the other. Under Gaussian assumptions, we have only to consider moments of the densities to second-order (i.e. covariances). Their Gaussian form allows us to express the densities in terms of $\Sigma_y \otimes I_s$, the covariance of $vec(y)$ and its conditional covariance $\Sigma_{y|x} \otimes I_s$.

$$\begin{aligned} H(y) &= \langle -\ln p(y) \rangle = \frac{1}{2} \ln |\Sigma_y \otimes I_s| \\ H(y|x) &= \langle -\ln p(y|x) \rangle = \frac{1}{2} \ln |\Sigma_{y|x} \otimes I_s| \\ I(x, y) &= \frac{s}{2} (\ln |\Sigma_y| - \ln |\Sigma_{y|x}|) \\ &= \frac{s}{2} \ln |\Sigma_{y|x}^{-1} \Sigma_y| \end{aligned} \quad \textbf{A1.2}$$

Here and throughout, constant terms have been ignored. The mutual information can be estimated using sample covariances. From Eqn. **A1.1**, and using standard results for the determinant of block matrices:

$$\begin{aligned} I(x, y) &= \frac{s}{2} (\ln |y^T y| - \ln |y^T y - y^T x (x^T x)^{-1} x^T y|) \\ H(x) &= \frac{s}{2} \ln |\Sigma_x| = \frac{s}{2} \ln |x^T x| \\ H(y) &= \frac{s}{2} \ln |\Sigma_y| = \frac{s}{2} \ln |y^T y| \\ H(x, y) &= \frac{s}{2} \ln \left| \begin{bmatrix} x^T x & y^T x \\ x^T y & y^T y \end{bmatrix} \right| \end{aligned} \quad \textbf{A1.3}$$

Statistical Parametric Mapping, by Karl Friston et al.
ISBN–13: 978-0-12-372560-8 ISBN–10: 0-12-372560-7

Comparison of Eqn. **A1.2** and Eqn. **A1.3** shows:

$$\begin{aligned}\Sigma_y &= y^T y \\ \Sigma_{y|x} &= y^T y - y^T x(x^T x)^{-1} x^T y\end{aligned} \quad \mathbf{A1.4}$$

In fact, we will see below that these are the maximum likelihood estimates of the covariances. It is sometimes useful to express the mutual information or predictability in terms of linearly separable components using the generalized eigenvector solution, with a leading diagonal matrix of eigenvalues v:

$$\begin{aligned}\Sigma_y c &= \Sigma_{y|x} c v \\ c^T c &= I \\ I(x, y) &= \frac{s}{2} \ln |\Sigma_{y|x}^{-1} \Sigma_y| \\ &= \frac{s}{2} \ln |v| \\ &= \frac{s}{2} \ln v_1 + \frac{s}{2} \ln v_2 + \cdots\end{aligned} \quad \mathbf{A1.5}$$

where c_i are the generalized eigenvectors, which define orthogonal mixtures of y that express the greatest mutual information with x. This information is simply $\frac{s}{2} \ln v_i$. We will see below that c_i are canonical vectors. Practically speaking, one could use these vectors to predict x, using one or more canonical variates $v_i = yc_i$.

OTHER PERSPECTIVES

Classical inference

In a classical setting, we test a null hypothesis. This calls for a model comparison, usually of a null model against an alternate model. We will assume a linear mapping between the causes and data:

$$y = x\theta + \varepsilon \quad \mathbf{A1.6}$$

where ε is some well-behaved error term. The null hypothesis is $\theta = 0$. Following the Neyman-Pearson Lemma, classical statistics uses the maximum likelihood ratio, which, in this context, is:

$$\begin{aligned}L &= \frac{p(y|x, \hat{\theta})}{p(y)} \Rightarrow \\ \ln L &= \ln p(y|x, \hat{\theta}) - \ln p(y) \\ \hat{\theta} &= \max_{\theta} p(y|x, \theta) = (x^T x)^{-1} x^T y\end{aligned} \quad \mathbf{A1.7}$$

The maximum likelihood value of the parameters is the usual least squares estimator (this is because we assumed the errors are IID (independent and identically distributed) and can be derived simply by solving $\partial \ln p(y|x, \theta)/\partial\theta = 0$. The maximum log-likelihoods, under the null and alternate hypotheses are:

$$\begin{aligned}\ln p(y) &= -\frac{s}{2} \ln |R_0| - \frac{1}{2} vec(y)^T (R_0^{-1} \otimes I_s) vec(y) \\ R_0 &= \max_{R_0} \ln p(y) = y^T y \\ \ln p\left(y|x, \hat{\theta}\right) &= -\frac{s}{2} \ln |R| - \frac{1}{2} vec(r)^T (R^{-1} \otimes I_s) vec(r) \\ r &= y - x\hat{\theta} \\ R &= \max_{R} \ln p(y|x, \hat{\theta}) = r^T r \\ &= y^T y - y^T x(x^T x)^{-1} x^T y\end{aligned} \quad \mathbf{A1.8}$$

$R_0 = \Sigma_y$ and $R = \Sigma_{y|x}$ are the sum of squares and products (SSQP) of the residuals under the null and alternate hypotheses respectively. The maximum likelihood expression for R_0, like the parameters, is obtained easily by solving $\partial \ln p(y)/\partial R_0^{-1} = 0$. Similarly for R, substituting Eqn. **A1.8** into Eqn. **A1.7** shows that the log-likelihood ratio statistic is simply the mutual information:

$$\ln L = \frac{s}{2}(\ln |R_0| - \ln |R|) = \frac{s}{2}(\ln |\Sigma_y| - \ln |\Sigma_{y|x}|) = I(x, y) \quad \mathbf{A1.9}$$

In this context, the likelihood ratio is known as Wilk's Lambda $\Lambda = L^{-1}$. When the dimensionality of y is one, this statistic is the basis of the F-ratio. When the dimensionality of x is also one, the square root of the F-ratio is the t-statistic. Classical inference uses the null distribution of the log-likelihood ratio to reject the null hypothesis that $\theta = 0$ to infer that $I(x, y) > 0$.

A Bayesian perspective

A Bayesian perspective on the predictability issue would call for a comparison of two models, with and without x as a predictor. This would proceed using the differences in log-evidence or marginal likelihoods between the alternative and null models:

$$\ln p(y|x) - \ln p(y) = -\frac{s}{2} \ln |R| + \frac{s}{2} \ln |R_0| = I(x, y) \quad \mathbf{A1.10}$$

In the context of linear models, this is simply the mutual information.

A multivariate perspective

In linear multivariate models, such as canonical variates analysis, canonical correlation analysis, and linear discriminant function analysis, one is trying to find a mixture of y that affords the best discrimination, in relation to x. This proceeds by maximizing the length of a vector projected onto the subspace of y, which can be explained by x, relative to its length in the subspace that cannot. More formally, subject to the constraint $c^T c = I$, we want:

$$c = \max_c \frac{c^T T c}{c^T R c} \qquad \text{A1.11}$$
$$T = \theta^T x^T x \theta$$

where T is referred to as the SSQP due to treatments. This is the null space of the residual SSQP. This means the total SSQP of y decomposes into the orthogonal covariance components:

$$R_0 = T + R \qquad \text{A1.12}$$

The canonical vectors c are the principal generalized eigenvectors:

$$Tc = Rc\lambda$$
$$c^T c = I$$

However, from Eqn. **A1.12**:

$$\begin{aligned} (R_0 - R)c &= Rc\lambda \\ R_0 c - Rc &= Rc\lambda \\ R_0 c &= Rc(\lambda + I) = \\ \Sigma_y c &= \Sigma_{y|x} c(\lambda + I) \end{aligned} \qquad \text{A1.13}$$

which has exactly the same form as Eqn. **A1.5**. In other words, the canonical vectors are simply the mixtures that express the greatest mutual information with x; the amount of information is $\frac{s}{2}\ln(\lambda_i + 1)$, where $\lambda_i = v_i - 1$ is the i-th canonical value. From Eqn. **A1.5**, Eqn. **A1.9** and Eqn. **A1.13** we get:

$$\begin{aligned} -\ln \Lambda &= I(x, y) \\ &= \frac{s}{2}\ln v_1 + \frac{s}{2}\ln v_2 + \cdots \\ &= \frac{s}{2}\ln(\lambda_1 + 1) + \frac{s}{2}\ln(\lambda_2 + 1) + \cdots \end{aligned} \qquad \text{A1.14}$$

Tests for the dimensionality of the subspace are based on the canonical values $\ln(\lambda_i + 1)$ (see Chapter 37).

SUMMARY

In the context of linear mappings under Gaussian assumptions, the heart of inference lies in the generalized eigenvalue solution. This solution finds pairs of generalized eigenvectors that show the greatest statistical dependence between two sets of multivariate data. The generalized eigenvalues encode the mutual information between the i-th variate and its corresponding vector. The total information is the log-likelihood ratio, or log-Bayes factor, comparing models with and without a linear mapping. Special cases of this quantity are Wilk's Lambda, Hotelling's T-square, the F-ratio and the t-statistic, upon which classical inference is based.

REFERENCES

Cox DD, Savoy RL (2003) Functional magnetic resonance imaging (fMRI) 'brain reading': detecting and classifying distributed patterns of fMRI activity in human visual cortex. *NeuroImage* **192**: 61–70

Friston KJ, Stephan KM, Heather JD *et al.* (1996) A multivariate analysis of evoked responses in EEG and MEG data. *NeuroImage* **3**: 167–74

Hansen KA, David SV, Gallant JL (2004) Parametric reverse correlation reveals spatial linearity of retinotopic human V1 BOLD response. *NeuroImage* **23**: 233–41

Hasson U, Nir Y, Levy I *et al.* (2004) Intersubject synchronization of cortical activity during natural vision. *Science* **303**: 1634–40

Kherif F, Poline JB, Flandin G *et al.* (2002) Multivariate model specification for fMRI data. *NeuroImage* **16**: 1068–83

Worsley KJ, Poline JB, Friston KJ *et al.* (1997) Characterizing the response of PET and fMRI data using multivariate linear models. *NeuroImage* **6**: 305–19

APPENDIX

2

Dynamical systems

K. Friston

INTRODUCTION

This appendix reviews models of dynamical systems. The review is framed in terms of analyses of functional and effective connectivity, in which we focus on the nature and form of the models and less on estimation or inference issues. The aim is to relate various models and to make their underlying assumptions transparent.

As we have seen in the preceding chapters, there are a number of models for estimating effective connectivity using neuroimaging time-series. By definition, effective connectivity depends on a causal model, through which it is defined operationally (Friston, 1995). This appendix reviews the principal models that could be adopted and how they relate to each other. We consider dynamic causal models (DCM), generalized convolution models (GCM), coherence analyses, structural equation models (SEM), state-space models (SSM) and multivariate autoregression-moving average (ARMA) models. In brief, we will show that they are all special cases of each other and try to emphasize their points of contact. However, some fundamental distinctions arise that guide the selection of the appropriate models in different situations. We now review these distinctions.

Coupling among inputs, outputs or hidden states?

The first distinction rests upon whether the model is used to explain the coupling between the inputs and outputs, among different outputs or among the system's states (e.g. neuronal activity in different ensembles). In terms of models, this distinction is between input-output models, e.g. multiple-input-single-output models (MISO) or multiple-input-multiple-output models (MIMO) and explicit input-*state*-output models. Usually, the input-output approach is concerned with the non-linear transformation of inputs, enacted by a system, to produce its outputs. This is like trying to establish a statistical dependence of the outputs on the inputs, without any comment on the mechanisms mediating this dependency. In some instances (e.g. ARMA and coherence analyses), dependences among different outputs are characterized (cf. functional connectivity).

Conversely, the input-state-output approach is generally concerned with characterizing the coupling among hidden variables that represent the states of the system. These states are observed vicariously through the outputs (Figure A2.1). Inferring the coupling among states induces the need for a causal model of how states affect each other and form outputs (cf. effective connectivity). Examples of input-output models include the Volterra formulation and generalized coherence analyses in the spectral domain. An example of a model that tries to estimate coupling among hidden states is DCM. In short, input-output models of coupling can proceed without reference to the hidden states. Conversely, interactions among hidden states require indirect access to the states through some model of the causal architecture of the system. In the next section, we start by reviewing input-output models and then turn to input-state-output models.

Deterministic or stochastic inputs?

The second key distinction is when the input is known (e.g. DCM) and when it is not (e.g. ARMA and SEM). This distinction depends on whether the inputs enter as known and/or deterministic quantities (e.g. experimentally designed causes of evoked responses) or whether we know (or can assume) something about the statistics

Statistical Parametric Mapping, by Karl Friston et al.
ISBN–13: 978-0-12-372560-8 ISBN–10: 0-12-372560-7

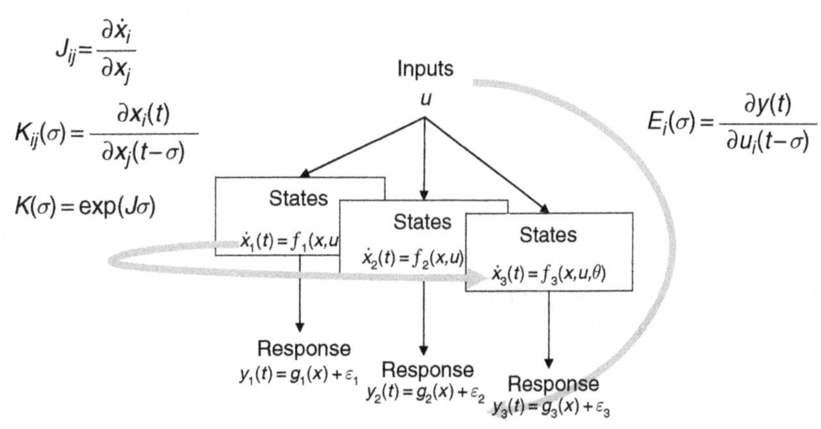

FIGURE A2.1 Schematic depicting the difference between analyses that address input-output behaviours and those that refer explicitly to interactions among coupled states.

of the input (i.e. its statistics up to second or higher orders). Most models of the stochastic variety assume the inputs are Gaussian, IID (independent and identically distributed) and stationary. Some stochastic models (e.g. coherence) use local stationarity assumptions to estimate high-order moments from observable but noisy inputs. For example, polyspectral analysis represents an intermediate case in which the inputs are observed but only their statistics are used.

Stationarity assumptions in stochastic models are critical because they preclude full analyses of evoked neuronal responses or transients that, by their nature, are non-stationary. On the other hand, there are situations where the input is not observable or under experimental control. In these cases, approaches like ARMA and SEM can be used if the inputs can be regarded as stationary. The distinction between deterministic and stochastic inputs is critical in the sense that it would be inappropriate to adopt one class of model in a context that calls for the other.

Connections or dependencies?

The final distinction is in terms of what is being estimated or inferred. Recall that functional connectivity is defined by the presence of statistical dependences among remote neurophysiological measurements. Conversely, effective connectivity is a parameter of a model that specifies the causal influences among states. It is useful to distinguish *inferences* about statistical dependencies and *estimation* of effective connectivity in terms of the distinction between functional and effective connectivity. Examples of approaches that try to establish statistical dependences include coherence analyses and ARMA. This is because these techniques do not presume a model of how hidden states interact to produce responses. They are interested only in establishing dependences among outputs over different frequencies or time lags. Although ARMA may employ some model to assess dependences, this is a model of dependences among outputs. There is no assertion that outputs *cause* outputs. Conversely, SEM and DCM try to estimate the model parameters and constitute analyses of effective connectivity proper.

EFFECTIVE CONNECTIVITY

Effective connectivity is the influence that one system exerts over another at a unit or ensemble level. This should be contrasted with functional connectivity, which implies a statistical dependence between two systems that could be mediated in any number of ways. Operationally, effective connectivity can be expressed as the response induced in an ensemble, unit or node by input from others, in terms of partial derivatives of the target activity x_i, with respect to the source activities. First- and second-order connections are then:

$$K_{ij}(\sigma_1) = \frac{\partial x_i(t)}{\partial x_j(t-\sigma_1)}$$

$$K_{ijk}(\sigma_1, \sigma_2) = \frac{\partial^2 x_i(t)}{\partial x_j(t-\sigma_1)\partial x_k(t-\sigma_2)}, \ldots \qquad \text{A2.1}$$

First-order connectivity embodies the response evoked by a change in input at $t - \sigma_1$. In other words, it is a time-dependent measure of *driving* efficacy. Second-order connectivity reflects the *modulatory* influence of the input at $t - \sigma_1$ on the response evoked at $t - \sigma_2$. And so on for higher orders. Note that, in this general formulation, effective connectivity is a function of inputs over the recent past.[1] Furthermore, implicit in Eqn. **A2.1** is the fact that effective connectivity is causal, unless σ_1 is allowed to be negative. It is useful to introduce the dynamic equivalent, in which the response of the target is expressed in terms of *changes in* activity:

$$J_{ij} = \frac{\partial \dot{x}_i}{\partial x_j} \qquad \frac{\partial J_{ij}}{\partial x_k} = \frac{\partial^2 \dot{x}_i}{\partial x_j \partial x_k} \ldots \qquad \textbf{A2.2}$$

where $\dot{x}_i = \partial x_i / \partial t$. In this dynamic form, all influences are causal and instantaneous. In this appendix, we will call the $K(\sigma)$ effective connections and J coupling. We will see later that they are related by $K(\sigma) = \exp(J\sigma)$ and that effective connectivity can be formulated in terms of Volterra kernels. Before considering specific models of effective connectivity, we will review briefly their basis (see also Chapter 38).

Dynamical systems

A plausible model of neuronal systems is a non-linear dynamical model that corresponds to an analytic multiple-input-multiple-output (MIMO) system. The state and output equations of a analytic dynamical system are:

$$\begin{aligned} \dot{x}(t) &= f(x, u, \theta) \\ y(t) &= g(x) + \varepsilon \end{aligned} \qquad \textbf{A2.3}$$

Typically the inputs $u(t)$ correspond to designed experimental effects (e.g. stimulus functions in functional magnetic resonance imaging, fMRI), or represent stochastic fluctuations or system perturbations. Stochastic observation error $\varepsilon \sim N(0, \Sigma)$ enters linearly in this model. For simplicity, the expressions below deal with single-input-single-output (SISO) systems, and will be generalized later. The measured response y is some non-linear function of the states of the system x. These state variables are usually unobserved or hidden (e.g. the configuration of all ion channels, the depolarization of every dendritic compartment etc.). The parameters of the state equation embody effective connectivity, either in terms of mediating the coupling between inputs and outputs (MISO models of a single region) or through the coupling among state variables (MIMO models of multiple regions). The objective is to estimate and make inferences (usually Bayesian) about these parameters, given the outputs and possibly the inputs.

[1] In contrast, functional connectivity is model-free and simply reflects the mutual information $I(x_i, x_j)$. In this appendix we are concerned only with models of effective connectivity.

Sometimes this requires one to specify the form of the state equation. A ubiquitous and useful form is the bilinear approximation; expanding around x_0:

$$\begin{aligned} &\dot{x}(t) \approx Ax + uBx + Cu \\ &y = Lx \\ &A = \frac{\partial f}{\partial x}, \quad B = \frac{\partial^2 f}{\partial x \partial u}, \quad C = \frac{\partial f}{\partial u}, \quad L = \frac{\partial g}{\partial x} \end{aligned} \qquad \textbf{A2.4}$$

For simplicity, we have assumed $x_0 = 0$ and $f(0) = g(0) = 0$. This bilinear model is sometimes expressed in a more compact form by augmenting the states with a constant:

$$\begin{aligned} &\dot{X} = (M + uN)X \\ &y = HX \\ &X = \begin{bmatrix} 1 \\ x \end{bmatrix} \quad M = \begin{bmatrix} 0 & 0 \\ f & A \end{bmatrix} \quad N = \begin{bmatrix} 0 & 0 \\ C & B \end{bmatrix} \quad H = [g \; L] \end{aligned} \qquad \textbf{A2.5}$$

(see Friston, 2002). Here the coupling parameters comprise the matrices $\theta = A, B, C, L$. We will use the bilinear parameterization when dealing with MIMO models and their derivatives below. We will first deal with MISO models, with and without deterministic inputs.

INPUT-OUTPUT MODELS

Models for deterministic inputs – The Volterra formulation

In this section, we review the Volterra formulation of dynamical systems. This formulation is important because it allows the input-output behaviour of a system to be characterized in terms of kernels that can be estimated without knowing the states of the system.

The Fliess fundamental formula (Fliess *et al.*, 1983) describes the causal relationship between the outputs and the history of the inputs. This relationship conforms to a Volterra series which expresses the output as a generalized convolution of the input, critically without reference to the states. This series is simply a functional Taylor expansion of the outputs with respect to the inputs (Bendat, 1990). The reason it is a *functional* expansion is that the inputs are a function of time:

$$\begin{aligned} &y(t) = h(u, \theta) + \varepsilon \\ &h(u, \theta) = \sum_i \int_0^t \ldots \int_0^t \kappa_i(\sigma_1, \ldots, \sigma_i) u(t - \sigma_1), \ldots, \\ &\qquad u(t - \sigma_i) d\sigma_1, \ldots, d\sigma_i \\ &\kappa_i(\sigma_1, \ldots, \sigma_i) = \frac{\partial^i y(t)}{\partial u(t - \sigma_1), \ldots, \partial u(t - \sigma_i)} \end{aligned} \qquad \textbf{A2.6}$$

where $\kappa_i(\sigma_1, \ldots, \sigma_i)$ is the i-th order kernel. In Eqn. **A2.6**, the integrals are over the past or history of the inputs. This renders the model causal. In some situations an acausal model may be appropriate (e.g. in which the kernels have non-zero values for future inputs; see Friston and Büchel, 2000). One important thing about the Volterra expansion is that it is linear in the unknowns, enabling relatively simple unbiased estimates of the kernels. In other words, Eqn. **A2.6** can be treated as a general linear observation model enabling all the usual estimation and inference procedures (see Chapter 38 for an example). Volterra series are generally thought of as a high-order or generalized non-linear convolution of the inputs to provide an output. To ensure the kernels can be estimated efficiently, they can be expanded in terms of some appropriate basis functions $q_j^i(\sigma_1, \ldots, \sigma_i)$ to give the general linear model:

$$\begin{aligned} y(t) &= \sum_{ij} \beta_j^i h_j^i(u) + \varepsilon \\ h_j^i(u) &= \int_0^t \ldots \int_0^t q_j^i(\sigma_1, \ldots, \sigma_i) u(t-\sigma_1), \ldots, \\ &\qquad u(t-\sigma_i) d\sigma_1, \ldots, d\sigma_i \\ \kappa_i(\sigma_1, \ldots, \sigma_i) &= \sum_j \beta_j^i q_j^i(\sigma_1, \ldots, \sigma_i) \end{aligned} \qquad \mathbf{A2.7}$$

The Volterra formulation is useful as a way of characterizing the influence of inputs on the responses of a region. The kernels can be regarded as a re-parameterization of the bilinear form in Eqn. **A2.4** that encodes the impulse response to input. The kernels for the states are:

$$\begin{aligned} \kappa_0 &= X(0) \\ \kappa_1(\sigma_1) &= e^{\sigma_1 M} N e^{-\sigma_1 M} X(0) \\ \kappa_2(\sigma_1, \sigma_2) &= e^{\sigma_2 M} N e^{(\sigma_1-\sigma_2)M} N e^{-\sigma_1 M} X(0) \\ \kappa_2(\sigma_1, \sigma_2, \sigma_3) &= \ldots \end{aligned} \qquad \mathbf{A2.8}$$

The kernels associated with the output follow from the chain rule:

$$\begin{aligned} h_0 &= H\kappa_0 \\ h_1(\sigma_1) &= H\kappa_1(\sigma_1) \\ h_2(\sigma_1, \sigma_2) &= H\kappa_2(\sigma_1, \sigma_2) + \kappa_1(\sigma_1)^T \partial H / \partial X \kappa_1(\sigma_2) \\ h_2(\sigma_1, \sigma_2, \sigma_3) &= \ldots \end{aligned} \qquad \mathbf{A2.9}$$

(see Friston, 2002 for details). If the system is fully non-linear, then the kernels can be considered local approximations. If the system is bilinear they are globally exact. It is important to remember that the estimation of the kernels does not assume any form for the state equation and completely eschews the states. This is the power and weakness of Volterra-based analyses.

The Volterra formulation can be used directly in the assessment of effective connectivity if we assume the measured response of one region constitutes the input to another, i.e. $u_i(x) = y_j(t)$. In this case, the Volterra kernels have a special interpretation; they are synonymous with effective connectivity. From Eqn. **A2.6**, the first-order kernels are:

$$\kappa_1(\sigma_1)_{ij} = \frac{\partial y_i(t)}{\partial y_j(t-\sigma_1)} = K_{ij}(\sigma_1) \qquad \mathbf{A2.10}$$

Extensions to multiple inputs (MISO) models are trivial and allow for high-order interactions among inputs to a single region to be characterized. This approach was used in Friston and Büchel (2000) to examine parietal modulation of V2 inputs to V5, by making inferences about the appropriate second-order kernel. The advantage of the Volterra approach is that non-linearities can be modelled and estimated in the context of highly non-linear transformations within a region and yet model inversion proceeds in a standard linear setting. However, one has to assume that the inputs conform to measured responses elsewhere in the brain. This may be tenable for some electrophysiological data, but the haemodynamic responses measured by fMRI make this a more questionable approach. Furthermore, there is no causal model of the interactions among areas that would otherwise offer useful constraints on the inversion. The direct application of Volterra estimation, in this fashion, simply examines each node, one at a time, assuming the activities of other nodes are veridical measurements of the inputs to the node in question. In summary, although the Volterra kernels are useful characterizations of the input-output behaviour of single nodes, they are not constrained by any model of interactions among regions. Before turning to DCMs that embody these interactions, we will deal with the SISO situation in which the input is treated as stochastic.

Models for stochastic inputs – coherence and polyspectral analysis

In this section, we deal with systems in which the input is stochastic. The aim is to estimate the kernels (or their spectral equivalents) given only statistics about the joint distribution of the inputs and outputs. When the inputs are unknown, one generally makes an assumption about their distributional properties and assumes [local] stationariness. Alternatively, the inputs may be measurable but too noisy to serve as inputs in a Volterra expansion.

In this case, they can be used to estimate the input and output densities in terms of high-order cumulants or polyspectral density. The n-th order cumulant of the input is:

$$c_u\{\sigma_1, \ldots, \sigma_{n-1}\} = \langle u(t)u(t-\sigma_1), \ldots, u(t-\sigma_{n-1})\rangle \quad \textbf{A2.11}$$

where we have assumed here, and throughout, that the expectation $E(u(t)) = 0$. It can be seen that cumulants are a generalization of autocovariance functions. The second-order cumulant is simply the autocovariance function of lag and summarizes the stationary second-order behaviour of the input. Cumulants allow one to formulate the Volterra expansion in terms of the second-order statistics of input and outputs. For example:

$$\begin{aligned} c_{yu}\{\sigma_a\} &= \langle y(t)u(t-\sigma_a)\rangle \\ &= \sum_i \int_0^t \ldots \int_0^t \kappa_i(\sigma_1, \ldots, \sigma_i)\, \langle u(t-\sigma_a)u(t-\sigma_1)\ldots \\ &\quad u(t-\sigma_i)\rangle\, d\sigma_1 \ldots d\sigma_i \\ &= \sum_i \int_0^t \ldots \int_0^t \kappa_i(\sigma_1, \ldots, \sigma_i)\times \\ &\quad c_u\{\sigma_a - \sigma_1, \ldots, \sigma - \sigma_i\} d\sigma_1 \ldots d\sigma_i \end{aligned} \quad \textbf{A2.12}$$

This equation says that the cross-covariance between the output and the input can be decomposed into components that are formed by convolving the i-th order kernel with the input's $i+1$-th cumulant. The important thing about this is that all cumulants, greater than second order, of Gaussian processes are zero. This means that if we can assume the input is Gaussian then:

$$c_{yu}\{\sigma_a\} = \int_0^t \kappa_i(\sigma_1)c_u\{\sigma_a - \sigma_1\}d\sigma_1 \quad \textbf{A2.13}$$

In other words, the cross-covariance between the input and output is simply the autocovariance function of the inputs convolved with the first-order kernel. Although it is possible to formulate the covariance between inputs and outputs in terms of cumulants, the more conventional formulation is in frequency space using polyspectra. The n-th polyspectrum is the Fourier transform of the corresponding cumulant:

$$g_u(\omega_1, \ldots, \omega_{n-1}) = \left(\frac{1}{2\pi}\right)^{n-1} \int \ldots \int c_u\{\sigma_1, \ldots, \sigma_{n-1}\} e^{-j(\omega\sigma_1, \ldots, \omega\sigma_{n-1})} d\sigma_1, \ldots, d\sigma_{n-1} \quad \textbf{A2.14}$$

Again, polyspectra are simply a generalization of spectral densities. For example, the second polyspectrum is spectral density and the third polyspectrum is bi-spectral density. It can be seen that these relationships are generalizations of the Wiener-Khinchine theorem, relating the autocovariance function and spectral density through the Fourier transform. Introducing the spectral density representation:

$$u(t) = \int s_u(\omega)e^{-j\omega} d\omega \quad \textbf{A2.15}$$

we can now rewrite the Volterra expansion as:

$$\begin{aligned} h(u, \theta) &= \sum_i \int_{-\pi}^{\pi} \ldots \int_{-\pi}^{\pi} e^{j(\omega_1 +, \ldots, +\omega_i)t} \\ &\times \Gamma_1(\omega_1, \ldots, \omega_i)s_u(\omega_1), \ldots, s_u(\omega_i)d\omega_1, \ldots, d\omega_i \end{aligned} \quad \textbf{A2.16}$$

where the functions

$$\begin{aligned} \Gamma_1(\omega_1) &= \int_0^\infty e^{-j\omega_1\sigma_1}\kappa_1(\sigma_1)d\sigma_1 \\ \Gamma_2(\omega_1, \omega_2) &= \int_0^\infty \int_0^\infty e^{-j(\omega_1\sigma_1 + \omega_2\sigma_2)}\kappa_2(\sigma_1, \sigma_2)d\sigma_1 d\sigma_2 \\ &\ldots \end{aligned}$$

are the Fourier transforms of the kernels. These functions are called *generalized transfer functions* and mediate the expression of frequencies in the output given those in the input. Critically, the influence of higher order kernels, or equivalently generalized transfer functions means that a given frequency in the input can induce a *different* frequency in the output. A simple example of this would be squaring a sine wave input to produce an output of twice the frequency. In the Volterra approach, the kernels were identified in the time domain using the inputs and outputs directly. In this section, system identification means estimating their Fourier transforms (i.e. the transfer functions) using second and higher order statistics of the inputs and outputs. Generalized transfer functions are usually estimated through estimates of polyspectra. For example, the spectral form for Eqn. **A2.13**, and its high-order counterparts are:

$$\begin{aligned} g_{uy}(-\omega_1) &= \Gamma_1(\omega_1)g_u(\omega_1) \\ g_{uuy}(-\omega_1, -\omega_2) &= 2\Gamma_2(\omega_1, \omega_2)g_u(\omega_1)g_u(\omega_2) \\ &\vdots \\ g_{u\ldots y}(-\omega_1, \ldots, -\omega_n) &= n!\Gamma_n(\omega_1, \ldots, \omega_n)g_u(\omega_1)\ldots g_u(\omega_n) \end{aligned} \quad \textbf{A2.17}$$

Given estimates of the requisite [cross]-polyspectra, these equalities can be used to provide estimates of the transfer

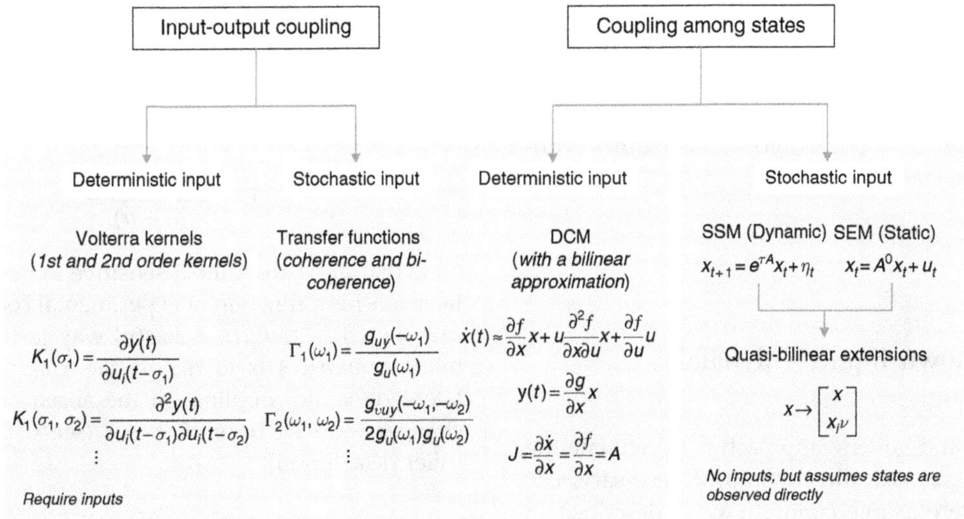

FIGURE A2.2 Overview of the models considered in this chapter. They have been organized to reflect whether they require known inputs or not and whether the model is a time-series model or not.

functions (see Figure A2.2). These equalities hold when the Volterra expansion contains just the n-th order term and are a generalization of the classical results for the transfer function of a linear system (i.e. the first equality in Eqn. **A2.17**). The importance of these results, in terms of effective connectivity, is the implicit meaning conferred on *coherence* and *bi-coherence* analyses. Coherence is simply the second-order cross spectrum $g_{uy}(\omega)$ between the input and output and is related to first-order effects (i.e. the first-order kernel or transfer function) through Eqn. **A2.17**. Coherence is therefore a surrogate for first-order or linear connectivity. Bi-coherence or the cross-bi-spectrum $g_{uuy}(\omega_1, \omega_2)$ is the third-order cross-polyspectrum and implies a non-zero second-order kernel or transfer function. Bi-spectral analysis was used (in a simplified form) to demonstrate non-linear coupling between parietal and frontal regions using magnetoencephalography (MEG) in Chapter 39. In this example, cross-bi-spectra were estimated, in a simple fashion, using time-frequency analyses.

Summary

In summary, Volterra kernels (generalized transfer functions) characterize the input-output behaviour of a system. The n-th order kernel is equivalent to n-th order effective connectivity when the inputs and outputs conform to processes that mediate interactions among neuronal systems. If the inputs and outputs are known, or can be measured precisely, the estimation of the kernels is straightforward. In situations where inputs and outputs are observed less precisely, kernels can be estimated indirectly through their generalized transfer functions using cross-polyspectra. The robustness of kernel estimation, conferred by expansion in terms of temporal basis functions, is recapitulated in the frequency domain by smoothness constraints during estimation of the polyspectra. The spectral approach is limited because it assumes the system contains only the kernel of the order estimated and stationariness. The intuition behind the first limitation relates to the distinction between parameter estimation and variance partitioning in standard regression analyses. Although it is perfectly possible to estimate the parameters of a regression model given a set of non-orthogonal explanatory variables, it is not possible uniquely to partition variance in the output caused by these explanatory variables.

INPUT-STATE-OUTPUT MODELS

In this section, we address models for multiple interconnected nodes (e.g. brain regions) where one can measure their responses to input that may or may not be known. Although it is possible to extend the techniques of the previous sections to cover MIMO systems, the ensuing inferences about the influence of input to one node on the response of another are not sufficiently specified to constitute an analysis of effective connectivity. This is because these influences may be mediated in many ways and are not parameterized in terms of the effective connectivity among the nodes themselves. A parameterization that encodes this inter-node coupling is therefore required. All the models discussed below assume some form or model for the interactions among the state variables in one or more nodes and attempt to estimate the parameters of this model, sometimes without observing the states themselves.

Models for known inputs – dynamic causal modelling

The most direct and generic approach is to estimate the parameters of Eqn. **A2.3** directly and, if necessary, use them to compute effective connectivity as described in Eqn. **A2.1** and Eqn. **A2.2**. Although there are many forms one could adopt for Eqn. **A2.3**, we will focus on the bilinear approximation, which is possibly the most parsimonious but useful non-linear approximation available. Furthermore, as shown below, the bilinear approximation re-parameterizes the state equations of the model directly in terms of effective connectivity. Dynamic causal modelling does not necessarily entail the use of a bilinear model. Indeed, DCMs can be specified to any degree of biological complexity and realism supported by the data. There are examples in this book where the parameters of the state equation are already effective connectivity or coupling parameters, for example, the extrinsic and intrinsic connections in neural-mass models of event-related potentials (ERPs) (see Chapter 42). However, we focus on bilinear approximations here because they represent the simplest form to which all DCMs can be reduced. This reduction allows analytic derivation of kernels and other computations, like integrating the state equation, to proceed in an efficient fashion.

Each region may comprise several state variables whose causal interdependencies are summarized by the bilinear form in Eqn. **A2.4**. Here the coupling parameters of the state equation are the matrices M and N. For a given set of inputs or experimental context, the bilinear approximation to any state equation is:

$$\begin{aligned}\dot{X}(t) &= JX(t)\\ X(t+\sigma) &= e^{J\sigma}X(t)\\ J &= M+\sum_i N_i u_i\end{aligned} \qquad \textbf{A2.18}$$

Notice that there are now as many N matrices as there are inputs. The bilinear form reduces the model to first-order connections that can be modulated by the inputs. In MIMO models, the coupling is among the states such that first-order effective connections are simply:

$$\begin{aligned}J &= \frac{\partial \dot{X}}{\partial X}\\ K &= \frac{\partial X(t)}{\partial X(t-\sigma)} = e^{J\sigma}\end{aligned} \qquad \textbf{A2.19}$$

Note that these are context-sensitive in the sense that the Jacobian J is a function of experimental context or inputs $u(t) = u_1(t), \ldots, u_m(t)$. A useful way to think about the bilinear matrices is to regard them as the intrinsic or latent dynamic coupling, in the absence of input, and changes induced by each input (see Chapter 41 for a fuller description):

$$\begin{aligned}J(0) &= M = \begin{bmatrix} 0 & 0 \\ f(0) & A \end{bmatrix}\\ \frac{\partial J}{\partial u_i} &= N_i = \begin{bmatrix} 0 & 0 \\ C_i & B_i \end{bmatrix}\end{aligned} \qquad \textbf{A2.20}$$

The latent coupling among the states is A. Often, one is more interested in B_i as embodying changes in this coupling induced by different cognitive set, time or drugs. Note that C_i enters as the input-dependent component of coupling to the constant term. Clearly, it would be possible to introduce other high-order terms to model interactions among the states, but we will restrict ourselves to bilinear models for simplicity.

Dynamic causal modelling has two parts: first, specification of the state and output equations of an ensemble of region-specific state variables. If necessary, a bilinear approximation to any state equation reduces the model to first-order coupling and bilinear terms that represent the modulation of that coupling by inputs. Second, inversion of the DCM allows one to estimate and make inferences about inter-regional connections and the effect of experimental manipulations on those connections.

As mentioned above, the state equations do not have to conform to the bilinear form. This is important because the priors may be specified more naturally in terms of the original biophysical parameters of the DCM, as opposed to the bilinear form. A nice example of this is the use of log-normal priors to enforce positivity constraints on the rate constants of ERP models in Chapter 42. Furthermore, the choice of the state variables clearly has to reflect their role in mediating the effect of inputs on responses and the interactions among areas. In the simplest case, the state variables could be reduced to mean synaptic activity per region, plus any biophysical state variables needed to determine the output (e.g. the states of haemodynamic models for fMRI). Implicit in choosing such state variables is the assumption that they model all the dynamics to the level of detail required. Mean field models and neural-mass models may be useful here in motivating the number of state variables and the associated state equations (see Chapter 31). Operationally, issues of parameterization and number of state variables can be resolved with Bayesian model selection and is directed principally by the nature of the data.

Summary

In summary, DCM is the most general and direct approach to identifying the effective connectivity among the states of MIMO systems. The identification of DCMs usually proceeds using Bayesian inversion to estimate the posterior mode or most likely parameters of the model given the data. The state equations can be arbitrarily complicated and non-linear, however, there will be an optimal level of model complexity that is supported by the data (and identified using Bayesian model selection). The simplest model is probably a bilinear approximation to causal influences among state variables. This serves to minimize the complexity of the model by parameterizing the model in terms of first-order coupling and its changes with input (the bilinear terms). In the next section, we deal with the situations in which the input is unknown. This precludes DCM with deterministic systems, because the likelihood of the responses cannot be computed unless we know what caused them.

Models for stochastic inputs – SEM and regression models

When the inputs are unknown, and the statistics of the outputs are considered to second order, one is effectively restricted to linear or first-order models of effective connectivity. Although it is possible to deal with discrete-time bilinear models, with white noise inputs, they have the same covariance structure as ARMA (autoregressive moving average) models of the same order (Priestley, 1988: 66). This means that to distinguish between linear and non-linear models, one would need to study moments higher than second order (cf. the third-order cumulants in bi-coherence analyses). Consequently, we will focus on linear models of effective connectivity, under white stationary inputs. There are two important classes of model here: structural equation models and ARMA models. Both are finite parameter linear models that are distinguished by their dependency on dynamics. In SEM, the interactions are assumed to be instantaneous, whereas in ARMA the dynamics are retained.

An SEM can be derived from any DCM by assuming the inputs vary slowly in relation to neuronal and haemodynamics. This is appropriate for positron emission tomography (PET) experiments and possibly some epoch-related fMRI designs, but not for event-related designs in ERP or fMRI. Note that this assumption pertains to the inputs or experimental design, not to the time constants of the outputs. In principle, it would be possible to apply DCM to a PET study.

Consider a linear approximation to any DCM where we can observe the states precisely and there was only one state variable per region:

$$\begin{aligned} \dot{x} &= f(x,u) \\ &= Ax + u = (A^0 - 1)x + u \\ y &= g(x) = x \end{aligned} \qquad \textbf{A2.21}$$

Here, we have discounted observation error but allow stochastic inputs $u \sim N(0,Q)$. To make the connection to the SEM more explicit, we have expanded the connectivity matrix into off-diagonal connections and a leading diagonal matrix, modelling unit decay $A = A^0 - 1$. For simplicity, we have absorbed C into the covariance structure of the inputs Q. As the inputs are changing slowly relative to the dynamics, the change in states will be zero at the point of observation and we obtain the regression model used by SEM:

$$\begin{aligned} \dot{x} &= 0 \Rightarrow \\ (1 - A^0)x &= u \\ x &= (1 - A^0)^{-1}u \end{aligned} \qquad \textbf{A2.22}$$

(see Chapter 38). The more conventional motivation for Eqn. **A2.22** is to start with an instantaneous regression equation $x = A^0x + u$ that is formally identical to the second line above. Although this regression model obscures the connection with dynamic formulations, it is important to consider because it is the basis of commonly employed methods for estimating effective connectivity in neuroimaging to data. These are simple regression models and SEM.

Simple regression models

$x = A^0x + u$ can be treated as a general linear model by focusing on one region at a time, for example the first, to give (cf. Eqn. **38.11** in Chapter 38):

$$x_1 = [x_2, \ldots, x_n] \begin{bmatrix} A_{12} \\ \vdots \\ A_{1n} \end{bmatrix} + u_1 \qquad \textbf{A2.23}$$

The elements of A can then be solved in a least squares sense by minimizing the norm of the unknown stochastic inputs u for that region (i.e. minimizing the unexplained variance of the target region given the states of the remainder). This approach was proposed in Friston (1995) and has the advantage of providing precise estimates of connectivity with high degrees of freedom. However, these maximum likelihood estimators assume, rather implausibly, that the inputs are orthogonal to the states and, more importantly, do not ensure the inputs to different regions conform to the known covariance Q. Furthermore, there is no particular reason that the input variance should be minimized just because it is unknown. Structural equation modelling overcomes these limitations at the cost of degrees of freedom for efficient estimation

Structural equation modelling

In SEM, estimates of A^0 minimize the difference (KL divergence) between the observed covariance among the [observable] states and that implied by the model and assumptions about the inputs.

$$\begin{aligned}\langle xx^T\rangle &= \langle (1-A^0)^{-1}uu^T(1-A^0)^{-1T}\rangle \\ &= (1-A^0)^{-1}Q(1-A^0)^{-1T}\end{aligned} \quad \text{A2.24}$$

This is critical because the connectivity estimates implicitly minimize the discrepancy between the observed and implied covariances among the states induced by stochastic inputs. This is in contradistinction to the instantaneous regression approach (above) or ARMA analyses (below) in which the estimates simply minimize unexplained variance on a region-by-region basis. It should be noted that SEM can be extended to embrace dynamics by temporal embedding. However, these models then become formally the same as autoregressive-moving average models, which are considered below. Estimation of the effective connectivity in SEM, in the context of designed experiments (i.e. in neuroimaging) is rather poorly motivated. This is because one throws away all the information about the temporal pattern of designed inputs and uses only $Q = \langle uu^T\rangle$. In many applications of SEM, the inputs are discarded and Q is assumed to be a leading diagonal or identity matrix.

Quasi-bilinear models – psychophysiological interaction and moderator variables

There is a useful extension to the regression model implicit in Eqn. **A2.22** that includes bilinear terms formed from known inputs that are distinct from stochastic inputs inducing [co]variance in the states. Let these known inputs be denoted by v. These usually represent some manipulated experimental context, such as cognitive set (e.g. attention) or time. These deterministic inputs are also known as moderator variables in SEM. The underlying quasi-bilinear DCM, for one such input, is:

$$\dot{x} = (A^0 - 1)x + Bvx + u \quad \text{A2.25}$$

Again, assuming the system has settled at the point of observation:

$$\begin{aligned}\dot{x} &= 0 \\ (1 - A^0 - Bv)x &= u \\ x &= A^0x + Bvx + u\end{aligned} \quad \text{A2.26}$$

This regression equation can be used to form least squares estimates as in Eqn. **A2.23**, in which case the additional bilinear regressors vx are known as *psychophysiological interaction* (PPI) terms (for obvious reasons). The corresponding SEM or path analysis usually proceeds by creating extra 'virtual' regions whose dynamics correspond to the bilinear terms. This is motivated by rewriting the last expression in Eqn. **A2.26** as:

$$\begin{bmatrix} x \\ vx \end{bmatrix} = \begin{bmatrix} A^0 & B \\ 0 & 1 \end{bmatrix}\begin{bmatrix} x \\ vx \end{bmatrix} + \begin{bmatrix} u \\ 0 \end{bmatrix} \quad \text{A2.27}$$

It is important to note that psychophysiological interactions and moderator variables in SEM are exactly the same thing and both speak of the importance of bilinear terms in causal models. Their relative success in the neuroimaging literature is probably due to the fact that they model changes in effective connectivity that are generally much more interesting than the connection strengths *per se*. Examples are changes induced by attentional modulation, changes during procedural learning and changes mediated pharmacologically. In other words, bilinear components afford ways of characterizing *plasticity* and, as such, play a key role in methods for functional integration. It is for this reason we focused on bilinear approximations as a minimal DCM in the previous section.

Summary

In summary, SEM is a simple and pragmatic approach to effective connectivity when dynamical aspects can be discounted, a linear model is sufficient, the state variables can be measured precisely and the input is unknown but stochastic and stationary. These assumptions are imposed by ignorance about the inputs. Some of these represent rather severe restrictions that limit the utility of SEM in relation to DCM or state-space models considered next. The most profound criticism of linear regression and SEM in imaging neuroscience is that they are models for interacting brain systems in the context of unknown input. The whole point of designed experiments is that the inputs are known and under experimental control. This renders the utility of SEM for designed experiments somewhat questionable.

MULTIVARIATE ARMA MODELS

ARMA (autoregressive-moving average) models can be represented as *state-space* (or Markov) models that provide a compact description of any finite parameter linear model. From this state-space representation, multivariate autoregressive (MAR) models can be derived and estimated using a variety of well-established techniques (see Chapter 40). We will focus on how the state-space representation of linear models can be derived from the dynamic formulation and the assumptions required in this derivation. Many treatments of dynamic systems consider the dynamic formulation in terms of a state-equation, a continuous state-space model (SSM). We preserve the distinction because there is an important asymmetry in the sense that one can always derive a discrete SSM from a DCM. However, there is no necessary mapping from an SSM to a DCM. This is critical for causal inference because only DCMs are causal in the control theory sense (see below).

Assume a linear DCM in which inputs comprise stationary white Weiner processes $u \sim N(0, Q)$ that are offered to each region in equal strength (i.e. $C = 1$). This renders Eqn. **A2.3** a linear stochastic differential equation (SDE):

$$\begin{aligned} \dot{x} &= Ax + u \\ y &= Lx \end{aligned} \qquad \textbf{A2.28}$$

The value of x at some future lag comprises a deterministic and a stochastic component η that obtains by regarding the effects of the input as an accumulation of local linear perturbations:

$$\begin{aligned} x(t+\tau) &= e^{\tau A}x(t) + \eta \\ \eta &= \int_0^\tau e^{\sigma A}u(t+\sigma)d\sigma \end{aligned} \qquad \textbf{A2.29}$$

Notice that the stochastic part can be regarded as convolving the random state-fluctuations with the system's first-order kernel. Using the assumption that the input is uncorrelated, the covariance of the stochastic part is:

$$W = \langle \eta\eta^T \rangle = \int_0^\tau e^{\sigma A} Q e^{\sigma A^T} d\sigma \qquad \textbf{A2.30}$$

It can be seen that when the lag is small in relation to the Lyapunov exponents, $eig(A)$ we get $e^{\sigma A} \approx 1$ and $W \approx Q\tau$. By incorporating the output transformation and observation error, we can augment this model to furnish a state-space model with system matrix $F = e^{\tau A}$, input matrix $G = \sqrt{W}$ and observation matrix L:

$$\begin{aligned} x_t &= Fx_{t-1} + Gz_t \\ y_t &= Lx_t + \varepsilon_t \end{aligned} \qquad \textbf{A2.31}$$

where z is an innovation representing dynamically transformed stochastic input. If we knew L and were interested in inferring on the hidden states, we would normally turn to Bayesian filtering (e.g. Kalman filtering) as described in Appendix 5. However, we will assume that we are more interested in inference about the coupling implied by the system matrix. In this case, we can reformulate the state-space model and treat it as an ARMA model.

Critically, every state-space model has an ARMA representation and vice versa. For example, if $L = 1$, we can eliminate the hidden states to give:

$$y_t - Fy_{t-1} = Gz_t + \varepsilon_t - F\varepsilon_{t-1} \qquad \textbf{A2.32}$$

This is simply an ARMA(1,2) model that can be inverted using the usual procedures (see Chapter 40). The autoregressive part is on the left and the moving average of the innovations is on the right. Critically, Eqn. **A2.32** formulates the dynamics in terms of, and only of, the response variable and random terms. Although it is always possible to derive an ARMA representation from a DCM (through the state-space representation), the reverse mapping is not necessarily defined. Having said this, ARMA models can be useful in establishing the presence of coupling even if the exact form of the coupling is not specified (cf. Volterra characterizations).

In summary, discrete-time linear models of effective connectivity can be reduced to multivariate AR (or, more generally ARMA) models, whose coefficients can be estimated given only the states (or outputs) by assuming the inputs are white and Gaussian. They therefore operate under similar assumptions as SEM but are time-series models.

A note on causality

There are many schemes for inverting state-space models of the sort in Eqn. **A2.31**. Inference on the system matrix could be considered in the light of functional connectivity, however, the regression coefficients are not really measures of effective connectivity. This is because there is no necessary mapping to the parameters of a DCM. In other words, although one can always map from the parameters of a causal model to its state-space representation $F \leftarrow e^{\tau A}$, the inverse mapping does not necessarily exist (a simple intuition here is that the log of a negative number is not real).

An interesting aspect of inference on the system matrix (i.e. regression coefficients) is the use of model comparison to compare the regression of one channel on another. Because these coefficients encode statistical dependence at different temporal lags, this model comparison is often framed in causal terms, by appeal to temporal precedence (e.g. Granger causality). However, for many, this rhetoric represents a category error because the regression coefficients cannot have the attribute 'causal'. This is because causal is an attribute of the state equation that implies $A = \partial f/\partial x$ is real. This Jacobian is not defined in the state-space or ARMA representations because the mapping $\frac{1}{\tau}\ln(F) \rightarrow \partial f/\partial x$ does not necessarily exist. It is quite possible to infer Granger causality that is acausal when a cause is observed after its effect. fMRI presents a good example of an acausal system, because of the delay imposed on the expression of neuronal dynamics (which are causal) at the level of haemodynamics (which are not). For example, one region, with a long haemodynamic latency, could cause a neuronal response in another that was expressed, haemodynamically, before the source. This example demonstrates that one cannot estimate effective connectivity or coupling using just the outputs of a system (e.g. observed fMRI responses).

CONCLUSION

We have reviewed a series of models, all of which can be formulated as special cases of DCMs. Two fundamental distinctions organize these models. The first is whether they model the coupling of inputs to outputs or whether they model interactions among hidden states. The second distinction (see Figure A2.2) is that between models that require the inputs to be known, as in designed experiments and those where the input is not under experimental control but can be assumed to be well behaved. With only information about the density of the inputs (or the joint density of the inputs and outputs) the models of connectivity that can be used are limited; unless one uses moments greater than second-order, only linear models can be estimated.

Many methods for non-linear system identification and causal modelling have been developed in situations where the system input is not under experimental control and, in the case of SEM, for static data. Volterra kernels and DCMs may be especially useful in neuroimaging because we deal explicitly with time-series data generated by designed experiments.

REFERENCES

Bendat JS (1990) *Nonlinear system analysis and identification from random data*. John Wiley and Sons, New York

Fliess M, Lamnabhi M, Lamnabhi-Lagarrigue F (1983) An algebraic approach to nonlinear functional expansions. *IEEE Trans Circ Sys* **30**: 554–70

Friston KJ (1995) Functional and effective connectivity in neuroimaging: a synthesis. *Hum Brain Mapp* **2**: 56–78

Friston KJ (2002) Bayesian estimation of dynamical systems: an application to fMRI. *NeuroImage* **16**: 465–83

Friston KJ, Büchel C (2000) Attentional modulation of V5 in humans. *Proc Natl Acad Sci USA* **97**: 7591–96

Priestley MB (1988) *Non-linear and non-stationary time series analysis*. Academic Press, London

APPENDIX

3

Expectation maximization

K. Friston

INTRODUCTION

This appendix describes expectation maximization (EM) for linear models using statistical mechanics (Neal and Hinton, 1998). We connect this formulation with classical methods and show the variational free energy is the same as the objective function maximized in restricted maximum likelihood (ReML). In Appendix 4, we show that EM itself is a special case of variational Bayes (Chapter 24).

The EM algorithm is ubiquitous in the sense that many estimation procedures can be formulated as such, from mixture models through to factor analysis. Its objective is to maximize the likelihood of observed data $p(y|\lambda)$, conditional on some hyperparameters, in the presence of unobserved variables or parameters θ. This is equivalent to maximizing the log-likelihood:

$$\ln p(y|\lambda) = \ln \int p(\theta,|\lambda)d\theta \geq$$
$$F(q, \lambda) = \int q(\theta)\ln p(\theta, y|\lambda)d\theta - \int q(\theta)\ln q(\theta)d\theta \qquad \text{A3.1}$$

where $q(\theta)$ is *any* density on the model parameters (Neal and Hinton, 1998). Eqn. **A3.1** rests on Jensen's inequality that follows from the concavity of the log function, which renders the log of an integral greater than the integral of the log. F corresponds to the [negative] free energy in statistical thermodynamics and comprises two terms: the energy and entropy. The EM algorithm alternates between maximizing F and, implicitly, the likelihood of the data, with respect to the distribution $q(\theta)$ and the hyperparameters λ, holding the other fixed:

$$\textbf{E}\text{-step: } q(\theta) \leftarrow \max_q F(q(\theta), \lambda)$$

$$\textbf{M}\text{-step: } \lambda \leftarrow \max_\lambda F(q(\theta), \lambda)$$

This iterative alternation performs a coordinate ascent on F. It is easy to show that the maximum in the **E**-step obtains when $q(\theta) = p(\theta|y, \lambda)$, at which point Eqn. **A3.1** becomes an equality. The **M**-step finds the ML estimate of the hyperparameters, i.e. the values of λ that maximize $p(y|\lambda)$ by integrating $\ln p(\theta, y|\lambda) = \ln p(y|\theta, \lambda) + \ln p(\theta|\lambda)$ over the parameters, using the current estimate of their conditional distribution. In short, the **E**-step computes sufficient statistics (in our case the conditional mean and covariance) of the unobserved parameters to enable the **M**-step to optimize the hyperparameters, in a maximum likelihood sense. These new hyperparameters re-enter into the estimation of the conditional density and so on until convergence.

The E-step

For linear models, under Gaussian (i.e. parametric) assumptions, the **E**-step is trivial and corresponds to evaluating the conditional mean and covariance as described in Chapter 22:

$$y = X\theta + \varepsilon$$
$$\overline{y} = \begin{bmatrix} y - X\theta \\ \eta_\theta \end{bmatrix} \overline{X} = \begin{bmatrix} X \\ I \end{bmatrix} \overline{C}_\varepsilon = \begin{bmatrix} \sum \lambda_i Q_i & 0 \\ 0 & C_\theta \end{bmatrix}$$
$$\eta_{\theta|y} = C_{\theta|y}\overline{X}^T\overline{C}_\varepsilon^{-1}\overline{y}$$
$$C_{\theta|y} = (\overline{X}^T\overline{C}_\varepsilon^{-1}\overline{X})^{-1} \qquad \text{A3.2}$$

where the prior and conditional densities are $p(\theta) = N(\eta_\theta, C_\theta)$ and $q(\theta) = N(\eta_{\theta|y}, C_{\theta|y})$. This compact form is a result of absorbing the priors into the errors by augmenting the linear system. As described in Chapter 22, the same augmentation is used to reduce hierarchal models with empirical priors to their non-hierarchical form. Under local linearity assumptions, non-linear models can be reduced to a linear form as described in Chapter 34. The resulting conditional density is used to estimate the hyperparameters of the covariance components in the **M**-step.

Statistical Parametric Mapping, by Karl Friston et al.
ISBN–13: 978-0-12-372560-8 ISBN–10: 0-12-372560-7

The M-step

Given that we can reduce the problem to estimating the error covariances of the augmented system in Eqn. **A3.2**, we only need to estimate the hyperparameters of the error covariances (which contain the prior covariances). Specifically, we require the hyperparameters that maximize the first term of the free energy (i.e. the energy) because the entropy does not depend on the hyperparameters. For linear systems, the free energy is given by (ignoring constants):

$$\begin{aligned}
\log p(\theta, y\,|\,\lambda) &= -\frac{1}{2}\ln|C_\varepsilon| - \frac{1}{2}(\overline{y}-\overline{X}\theta)^T C_\varepsilon^{-1}(\overline{y}-\overline{X}\theta). \\
\int q(\theta)\ln p(\theta, y\,|\,\lambda)\,d\theta &= -\frac{1}{2}\ln|C_\varepsilon| - \frac{1}{2}r^T C_\varepsilon^{-1} r \\
&\quad - \frac{1}{2}tr\{C_{\theta|y}\overline{X}^T C_\varepsilon^{-1}\overline{X}\} \\
\int q(\theta)\log q(\theta) &= -\frac{1}{2}\ln|C_{\theta|y}| \\
F &= \frac{1}{2}\ln|C_\varepsilon^{-1}| - \frac{1}{2}r^T C_\varepsilon^{-1} r \\
&\quad - \frac{1}{2}tr\{C_{\theta|y}\overline{X}^T C_\varepsilon^{-1}\overline{X}\} + \frac{1}{2}\ln|C_{\theta|y}|
\end{aligned} \tag{A3.3}$$

where the residuals $r = \overline{y} - \overline{X}\eta_{\theta|y}$. By taking derivatives with respect to the error covariance we get:

$$\frac{\partial F}{\partial C_\varepsilon^{-1}} = \frac{1}{2}C_\varepsilon - \frac{1}{2}rr^T - \frac{1}{2}\overline{X}C_{\theta|y}\overline{X}^T \tag{A3.4}$$

When the hyperparameters maximize the free energy this gradient is zero and:

$$C(\lambda)_\varepsilon = rr^T + \overline{X}C_{\theta|y}\overline{X}^T \tag{A3.5}$$

(cf. Dempster *et al.*, 1981: 350). This means that the ReML error covariance estimate has two components: that due to differences between the data and its conditional prediction; and another due to the variation of the parameters about their conditional mean, i.e. their conditional uncertainty. This is not a closed form expression for the unknown covariance because the conditional covariance is a function of the hyperparameters. To find the ReML hyperparameters, one usually adopts a Fisher scoring scheme, using the first and expected second partial derivatives of the free energy:

$$\begin{aligned}
\Delta\lambda &= -E\left(\frac{\partial^2 F}{\partial\lambda_{ij}^2}\right)^{-1}\frac{\partial F}{\partial\lambda_i} \\
\frac{\partial F}{\partial\lambda_i} &= tr\left(\frac{\partial F}{\partial C_\varepsilon^{-1}}C_\varepsilon^{-1}Q_i C_\varepsilon^{-1}\right) \\
&= -\frac{1}{2}tr\{PQ_i\} + \frac{1}{2}\overline{y}^T P^T Q_i P\overline{y} \\
\frac{\partial^2 F}{\partial\lambda_{ij}^2} &= \frac{1}{2}tr\{PQ_iPQ_j\} - \overline{y}^T PQ_iPQ_jP\overline{y} \\
E\left(\frac{\partial^2 F}{\partial\lambda_{ij}^2}\right) &= -\frac{1}{2}tr\{PQ_iPQ_j\} \\
P &= C_\varepsilon^{-1} - C_\varepsilon^{-1}\overline{X}C_{\theta|y}\overline{X}^T C_\varepsilon^{-1}
\end{aligned} \tag{A3.6}$$

Fisher scoring corresponds to augmenting a Gauss-Newton scheme by replacing the second derivative or curvature with its expectation. The curvature or Hessian is referred to as Fisher's information matrix[1] and encodes the conditional prediction of the hyperparameters. In this sense, the information matrix has a close connection to the degrees of freedom in classical statistics. The gradient can be computed efficiently by capitalizing on any sparsity structure in the constraints and by bracketing the multiplications appropriately. This scheme is general in that it accommodates almost any form for the covariance through a Taylor expansion of $C\{\lambda\}_\varepsilon$.

Once the hyperparameters have been updated they enter the **E**-step as a new error covariance estimate to give new conditional moments which, in turn, enter the **M**-step and so on until convergence. A pseudo-code illustration of the complete algorithm is presented in Figure 22.4 of Chapter 22. Note that in this implementation one is effectively performing a single Fisher scoring iteration for each **M**-step. One could postpone each **E**-step until this search converged, but a single step is sufficient to perform a coordinate ascent on F. Technically, this renders the scheme a generalized EM or GEM algorithm.

It should be noted that the search for the maximum of F does not have to employ Fisher scoring or indeed the parameterization of C_ϵ used above. Other search procedures, such as quasi-Newton searches, are commonly employed (Fahrmeir and Tutz, 1994). Harville (1977) originally considered Newton-Raphson and scoring algorithms, and Laird and Ware (1982) recommend several versions of EM. One limitation of the linear

[1] The derivation of the expression for the information matrix uses standard results from linear algebra and is most easily seen by differentiating the gradient, noting:

$$\frac{\partial P}{\partial\lambda_j} = -PQ_jP$$

and taking the expectation, using

$$E(tr(PQ_iP\overline{y}\overline{y}^T PQ_j)) = tr\{PQ_iPC_\varepsilon PQ_j\} = tr\{PQ_iPQ_j\}$$

hyperparameterization described above is that it does not guarantee that C_ϵ is positive definite. This is because the hyperparameters can take negative values with extreme degrees of non-sphericity. The EM algorithm employed by *multistat* (Worsley *et al.*, 2002) for variance component estimation in multisubject fMRI studies, uses a slower but more stable algorithm that ensures positive definite covariance estimates.

In Appendix 4, we will revisit this issue and look at linear hyperparameterizations of the precision. The common aspect of all these algorithms is that they (explicitly or implicitly) optimize free energy. As shown next, this is equivalent to restricted maximum likelihood.

RELATIONSHIP TO ReML

ReML or *restricted maximum likelihood* was introduced by Patterson and Thompson in 1971, for estimating variance components in a way that accounts for the loss in degrees of freedom that result from estimating fixed effects (Harville, 1977), i.e. that accounts for conditional uncertainty about the parameters. It is commonly employed in standard statistical packages (e.g. SPSS). Under the present model assumptions, ReML is formally identical to EM. One can regard ReML as embedding the **E**-step into the **M**-step to provide a single log-likelihood objective function: substituting $C_{\theta|y} = \left(\overline{X}^T C_\varepsilon^{-1} \overline{X}\right)^{-1}$ into the expression for the free energy gives:

$$F = -\frac{1}{2}\ln|C_\varepsilon| - \frac{1}{2}r^T C_\varepsilon^{-1} r - \frac{1}{2}\ln|\overline{X}^T C_\varepsilon^{-1}\overline{X}| \qquad \textbf{A3.7}$$

This is the ReML objective function (see Harville, 1977: 325). Critically, its derivatives with respect to the hyperparameters are exactly the same as those in the **M**-step.[2] Operationally, the **M**-step can be re-formulated to give a ReML scheme by removing any explicit reference to the conditional covariance using:

$$P = C_\varepsilon^{-1} - C_\varepsilon^{-1}\overline{X}(\overline{X}^T C_\varepsilon^{-1}\overline{X})^{-1}\overline{X}^T C_\varepsilon^{-1} \qquad \textbf{A3.8}$$

The resulting scheme is formally identical to that described in Section 5 of Harville (1977). Because one can eliminate the conditional density, one could think of ReML as estimating the hyperparameters in a subspace that is *restricted* in the sense that the estimates are conditionally independent of the parameters. See Harville (1977) for a discussion of expressions, comparable to the terms in Eqn. **A3.7** that are easier to compute, for particular hyperparameterizations of the variance components.

Having established ReML is a special case of **EM**, in Appendix 4, we take an even broader perspective and look at EM as a special case of variational Bayes.

REFERENCES

Dempster AP, Rubin DB, Tsutakawa RK (1981) Estimation in covariance component models. *J Am Stat Assoc* **76**: 341–53

Fahrmeir L, Tutz G (1994) *Multivariate statistical modelling based on generalised linear models*. Springer-Verlag Inc., New York, pp 355–56

Harville DA (1977) Maximum likelihood approaches to variance component estimation and to related problems. *J Am Stat Assoc* **72**: 320–38

Laird NM, Ware JH (1982) Random effects models for longitudinal data. *Biometrics* **38**: 963–74

Neal RM, Hinton GE (1998) A view of the EM algorithm that justifies incremental, sparse and other variants. In *Learning in graphical models*, Jordan MI (ed.). Kluwer Academic Press, Dordrecht, pp 355–68

Worsley KJ, Liao CH, Aston J *et al.* (2002) A general statistical analysis for fMRI data. *NeuroImage* **15**: 1–15

[2] Note

$$\frac{\partial \ln|\overline{X}^T C_\varepsilon^{-1}\overline{X}|}{\partial \lambda_i} = tr\left((\overline{X}^T C_\varepsilon^{-1}\overline{X})^{-1}\frac{\partial \overline{X}^T C_\varepsilon^{-1}\overline{X}}{\partial \lambda_i}\right)$$

$$= -tr\{C_{\theta|y} X^T C_\varepsilon^{-1} Q_i C_\varepsilon^{-1}\overline{X}\}$$

APPENDIX

4

Variational Bayes under the Laplace approximation

K. Friston, J. Mattout, N. Trujillo-Barreto, J. Ashburner and W. Penny

INTRODUCTION

This is a rather technical appendix, but usefully connects most of the estimation and inference schemes used in previous chapters, by showing they are all special cases of a single variational approach. This synthesis is underpinned by reference to the various applications we have considered in detail, so that their motivation and interrelationships are more evident.

We will derive the variational free energy under the Laplace approximation, with a focus on accounting for additional model complexity induced by increasing the number of model parameters. This is relevant when using the free energy as an approximation to the log-evidence in Bayesian model averaging and selection. By setting restricted maximum likelihood (ReML) in the larger context of variational learning and expectation maximization, we show how the ReML objective function can be adjusted to provide an approximation to the log-evidence for a particular model. This means ReML can be used for model selection, specifically to select or compare models with difference covariance components. This is useful in the context of hierarchical models, because it enables a principled selection of priors. Deriving the ReML objective function, from basic variational principles, discloses the simple relationships among variational Bayes, EM and ReML. Furthermore, we show that EM is formally identical to a full variational treatment when the precisions are linear in the hyperparameters.

Background

This chapter starts with a very general formulation of inference using approaches developed in statistical physics. It ends with a treatment of a specific objective function used in restricted maximum likelihood (ReML) that renders it equivalent to the free energy in variational learning. This is important because the variational free energy provides a bound on the log-evidence for any model, which is exact for linear models. The log-evidence plays a central role in model selection, comparison and averaging (see Penny *et al.*, 2004 and Trujillo-Barreto *et al.*, 2004, for examples in neuroimaging).

Although this appendix focuses on the various forms for the free energy, we use it to link variational Bayes (VB), EM and ReML using the Laplace approximation. This approximation assumes a fixed Gaussian form for the conditional density of the parameters of a model and is used implicitly in ReML and many applications of EM. Bayesian inversion using VB is ubiquitous in neuroimaging (e.g. Penny *et al.*, 2005 and Chapter 24). Its use ranges from spatial segmentation and normalization of images during pre-processing (e.g. Ashburner and Friston, 2005) to the inversion of complicated dynamical causal models of functional integration in the brain (Friston *et al.*, 2003 and Chapter 34). Many of the intervening steps in classical and Bayesian analysis of neuroimaging data call on ReML or EM under the Laplace approximation. This appendix provides an overview of how these schemes are related and illustrates their applications with reference to specific algorithms and routines we have referred to in this book. One interesting issue that emerges from this treatment is that VB reduces exactly to EM, under the Laplace approximation, when the precision of stochastic terms is linear in the hyperparameters. This reveals a close relationship between EM and full variational approaches.

Conditional uncertainty

In previous chapters, we have described the use of ReML in the Bayesian inversion of electromagnetic models to

Statistical Parametric Mapping, by Karl Friston et al.
ISBN–13: 978-0-12-372560-8 ISBN–10: 0-12-372560-7

localize distributed sources in electroencephalography (EEG) and magnetoencephalography (MEG) (e.g. Phillips *et al.*, 2002; Chapters 29 and 30). ReML provides a principled way of quantifying the relative importance of priors that replaces alternative heuristics like L-curve analysis. Furthermore, ReML accommodates multiple priors and provides more accurate and efficient source reconstruction than its precedents (Phillips *et al.*, 2002). We have also used ReML to identify the most likely combination of priors using model selection, where each model comprises a different set of priors (Mattout *et al.*, 2006). This was based on the fact that the ReML objective function is the free energy used in expectation maximization and is equivalent to the log-evidence $F^\lambda = \ln p(y|\lambda, m)$, conditioned on λ, the unknown covariance component parameters (i.e. hyperparameters) and the model m. The covariance components encoded by λ include the prior covariances of each model of the data y. However, this free energy is not a function of the conditional uncertainty about λ and is therefore insensitive to additional model complexity induced by adding covariance components (i.e. priors). In what follows we show how F^λ can be adjusted to provide the variational free energy, which, in the context of linear models, is exactly the log-evidence $\ln p(y|m)$. This rests on deriving the variational free energy for a general variational scheme and treating expectation maximization (EM) as a special case, in which one set of parameters assumes a point mass. We then treat ReML as the special case of EM, applied to linear models.

Overview

This appendix is divided into six sections. In the first, we summarize the basic theory of variational Bayes and apply it in the context of the Laplace approximation (see also Chapter 24). The Laplace approximation imposes a fixed Gaussian form on the conditional density, which simplifies the ensuing variational steps. In this section, we look at the easy problem of approximating the conditional covariance of model parameters and the more difficult problem of approximating their conditional expectation or mode using gradient ascent. We consider a dynamic formulation of gradient ascent, which generalizes nicely to cover dynamic models and provides the basis for a temporal regularization of the ascent. In the second section, we apply the theory to non-linear models with additive noise. We use the VB scheme that emerges as the reference for subsequent sections looking at special cases. The third section considers EM, which can be seen as a special case of VB in which uncertainty about one set of parameters is ignored. In the fourth section, we look at the special case of linear models where EM reduces to ReML. The fifth section considers ReML and hierarchical models. Hierarchical models are important because they underpin parametric empirical Bayes (PEB) and other special cases, like relevance vector machines. Furthermore, they provide a link with classical covariance component estimation. In the final section, we present some toy examples to show how the ReML and EM objective functions can be used to evaluate the log-evidence and facilitate model selection.

VARIATIONAL BAYES

Empirical enquiry in science usually rests upon estimating the parameters of some model of how observed data were generated and making inferences about the parameters (or model). Estimation and inference are based on the posterior density of the parameters (or model), conditional on the observations. Variational Bayes is used to evaluate these posterior densities.

The variational approach

Variational Bayes is a generic approach to posterior density (as opposed to posterior mode) analysis that approximates the conditional density $p(\vartheta|y, m)$ of some model parameters ϑ, given a model m and data y. Furthermore, it provides the evidence (or marginal likelihood) of the model $p(y|m)$ which, under prior assumptions about the model, furnishes the posterior density $p(m|y)$ of the model itself.

Variational approaches rest on minimizing the Feynman variational bound (Feynman, 1972). In variational Bayes, the free energy represents a bound on the log-evidence. Variational methods are well established in the approximation of densities in statistical physics (e.g. Weissbach *et al.*, 2002) and were introduced by Feynman within the path integral formulation (Titantah *et al.*, 2001). The variational framework was introduced into statistics through ensemble learning, where the ensemble or variational density $q(\theta)$ (i.e. approximating posterior density) is optimized to minimize the free energy. Initially (Hinton and von Camp, 1993; MacKay, 1995), the free energy was described in terms of description lengths and coding. Later, established methods like EM were considered in the light of variational free energy (Neal and Hinton, 1998; see also Bishop, 1999). Variational learning can be regarded as subsuming most other learning schemes as special cases. This is the theme pursued here, with special references to fixed-form approximations and classical methods like ReML (Harville, 1977).

The derivations in this appendix involve a fair amount of differentiation. To simplify things we will use the notation $f_x = \partial f / \partial x$ to denote the partial derivative of the function f, with respect to the variable x. For time derivatives we will also use $\dot{x} = x_t$.

The log-evidence can be expressed in terms of the free energy and a divergence term:

$$\begin{aligned} \ln p(y|m) &= F + D(q(\vartheta) \| p(\vartheta|y, m)) \\ F &= \langle L(\vartheta) \rangle_q - \langle \ln q(\vartheta) \rangle_q \\ L &= \ln p(y, \vartheta) \end{aligned} \quad \textbf{A4.1}$$

Here $-\langle \ln q(\vartheta) \rangle_q$ is the entropy and $\langle L(\vartheta) \rangle_q$ the expected energy. Both quantities are expectations under the variational density. Eqn. **A4.1** indicates that F is a lower-bound approximation to the log-evidence because the divergence $D(q(\vartheta) \| p(\vartheta|y, m))$ is always positive. In this, note all the energies are the negative of energies considered in statistical physics. The objective is to compute $q(\vartheta)$ for each model by maximizing F, and then compute F itself, for Bayesian inference and model comparison respectively. Maximizing the free energy minimizes the divergence, rendering the variational density $q(\vartheta) \approx p(\vartheta|y, m)$ an approximate posterior, which is exact for linear systems. To make the maximization easier, one usually assumes $q(\vartheta)$ factorizes over sets of parameters ϑ^i:

$$q(\vartheta) = \prod_i q^i \quad \textbf{A4.2}$$

In statistical physics this is called a mean-field approximation. Under this approximation, the Fundamental Lemma of variational calculus means that F is maximized with respect to $q^i = q(\vartheta^i)$ when, and only when:

$$\begin{aligned} \delta F^i = 0 \Leftrightarrow \frac{\partial f^i}{\partial q^i} &= f^i_{q^i} = 0 \\ f^i &= F_{\vartheta^i} \end{aligned} \quad \textbf{A4.3}$$

δF^i is the variation of the free energy with respect to q^i. From Eqn. **A4.1**:

$$\begin{aligned} f^i &= \int q^i q^{\backslash i} \ln L(\vartheta) d\vartheta^{\backslash i} - \int q^i q^{\backslash i} \ln q(\vartheta) d\vartheta^{\backslash i} \\ f^i_{q^i} &= I(\vartheta^i) - \ln q^i - \ln Z^i \\ I(\vartheta^i) &= \langle L(\vartheta) \rangle_{q^{\backslash i}} \end{aligned} \quad \textbf{A4.4}$$

where $\vartheta^{\backslash i}$ denotes the parameters not in the i-th set. We have lumped terms that do not depend on ϑ^i into $\ln Z^i$, where Z^i is a normalization constant (i.e. partition function). We will call $I(\vartheta^i)$ the variational energy, noting its expectation under q^i is the expected energy. The extremal condition in Eqn. **A4.2** is met when:

$$\begin{aligned} \ln q^i &= I(\vartheta^i) - \ln Z^i \Leftrightarrow \\ q(\vartheta^i) &= \frac{1}{Z^i} \exp(I(\vartheta^i)) \end{aligned} \quad \textbf{A4.5}$$

If this analytic form were tractable (e.g. through the use of conjugate priors), it could be used directly. See Beal and Ghahramani (2003) for an excellent treatment of conjugate-exponential models. However, we will assume a Gaussian fixed-form for the variational density to provide a generic scheme that can be applied to a wide range of models.

The Laplace approximation

Under the Laplace approximation, the variational density assumes a Gaussian form $q^i = N(\mu^i, \Sigma^i)$ with variational parameters μ^i and Σ^i, corresponding to the conditional mode and covariance of the i-th set of parameters. The advantage of this is that the conditional covariance can be evaluated very simply. Under the Laplace assumption:

$$\begin{aligned} F &= L(\mu) + \frac{1}{2} \sum_i (U^i + \ln |\Sigma^i| + p^i \ln 2\pi e) \\ I(\vartheta^i) &= L(\vartheta^i, \mu^{\backslash i}) + \frac{1}{2} \sum_{j \neq i} U^j \\ U^i &= tr(\Sigma^i L_{\vartheta^i \vartheta^i}) \end{aligned} \quad \textbf{A4.6}$$

$p^i = \dim(\vartheta^i)$ is the number of parameters in the i-th set. The approximate conditional covariances obtain as an analytic function of the modes by differentiating Eqn. **A4.6** and solving for zero:

$$\begin{aligned} F_{\Sigma^i} &= \frac{1}{2} L_{\vartheta^i \vartheta^i} + \frac{1}{2} \Sigma^{i^{-1}} = 0 \Rightarrow \\ \Sigma^i &= -L(\mu)^{-1}_{\vartheta^i \vartheta^i} \end{aligned} \quad \textbf{A4.7}$$

Note that this solution for the conditional covariances does not depend on the mean-field approximation, but only on the Laplace approximation. Substitution into Eqn. **A4.6** means $U^i = p^i$ and:

$$F = L(\mu) + \sum_i \frac{1}{2} (\ln |\Sigma^i| + p^i \ln 2\pi) \quad \textbf{A4.8}$$

The only remaining quantities required are the variational modes which, from Eqn. **A4.5** maximize $I(\vartheta^i)$. The leads to the following compact variational scheme, under the Laplace approximation:

until convergence

for all i

$$\mu^i = \max_{\vartheta^i} I(\vartheta^i)$$

$$\Sigma^i = -L(\mu)^{-1}_{\vartheta^i \vartheta^i}$$

end

end **A4.9**

The variational modes

The modes can be found using a gradient ascent based on:

$$\dot{\mu}^i = \frac{\partial I(\mu^i)}{\partial \vartheta^i} = I(\mu^i)_{\vartheta^i} \qquad \textbf{A4.10}$$

It may seem odd to formulate an ascent in terms of the motion of the mode in time. However, this is useful when generalizing to dynamic models (see below). The updates for the mode obtain by integrating Eqn. **A4.10** to give:

$$\begin{aligned} \Delta\mu^i &= (\exp(tJ) - I)J^{-1}\dot{\mu}^i \\ J &= \frac{\partial \dot{\mu}^i}{\partial \vartheta^i} = I(\mu^i)_{\vartheta^i \vartheta^i} \end{aligned} \qquad \textbf{A4.11}$$

When t gets large, the matrix exponential disappears, because the curvature is negative definite and we get a conventional Gauss-Newton scheme:

$$\Delta\mu^i = -I(\mu^i)^{-1}_{\vartheta^i \vartheta^i} I(\mu^i)_{\vartheta^i} \qquad \textbf{A4.12}$$

Together with the expression for the conditional covariance in Eqn. **A4.7**, this update furnishes a variational scheme under the Laplace approximation:

until convergence

for all i

until convergence

$$I(\mu^i)_{\vartheta^i_k} = L(\mu)_{\vartheta^i_k} + \frac{1}{2}\sum_{j \neq i} tr(\Sigma^j L_{\vartheta^j \vartheta^j \vartheta^i_k})$$

$$I(\mu^i)_{\vartheta^i_k \vartheta^i_l} = L(\mu)_{\vartheta^i_k \vartheta^i_l} + \frac{1}{2}\sum_{j \neq i} tr(\Sigma^j L_{\vartheta^j \vartheta^j \vartheta^i_k \vartheta^i_l}) \qquad \textbf{A4.13}$$

$$\Delta\mu^i = -I(\mu^i)^{-1}_{\vartheta^i \vartheta^i} I(\mu^i)_{\vartheta^i}$$

end

$$\Sigma^i = -L(\mu)^{-1}_{\vartheta^i \vartheta^i}$$

end

end

Note that this scheme rests on, and only on, the specification of the energy function $L(\vartheta)$ implied by a generative model.

Regularizing variational updates

In some instances deviations from the quadratic form assumed for the variational energy $I(\vartheta^i)$ under the Laplace approximation can confound a simple Gauss-Newton ascent. This can happen when the curvature of the objective function is badly behaved (e.g. when the objective function becomes convex, the curvatures can become positive and the ascent turns into a descent). In these situations, some form of regularization is required to ensure a robust ascent. This can be implemented by augmenting Eqn. **A4.10** with a decay term:

$$\dot{\mu}^i = I(\mu^i)_{\vartheta^i} - \nu\Delta\mu^i \qquad \textbf{A4.14}$$

This effectively pulls the search back towards the expansion point provided by the previous iteration and enforces a local exploration. Integration to the fixed point gives a classical Levenburg-Marquardt scheme (cf. Eqn. **A4.11**):

$$\begin{aligned} \Delta\mu^i &= -J^{-1}\dot{\mu}^i \\ &= (\nu I - I(\mu^i)_{\vartheta^i \vartheta^i})^{-1} I(\mu^i)_{\vartheta^i} \\ J &= I(\mu^i)_{\vartheta^i \vartheta^i} - \nu I \end{aligned} \qquad \textbf{A4.15}$$

where ν is the Levenburg-Marquardt regularization. However, the dynamic formulation affords a simpler alternative, namely temporal regularization. Here, instead of constraining the search with a decay term, one can abbreviate it by terminating the ascent after some suitable period $t = \nu$; from Eqn. **A4.11**:

$$\begin{aligned} \Delta\mu^i &= (\exp(\nu J) - I)J^{-1}\dot{\mu}^i \\ &= (\exp(\nu I(\mu^i)_{\vartheta^i \vartheta^i}) - I) I(\mu^i)^{-1}_{\vartheta^i \vartheta^i} I(\mu^i)_{\vartheta^i} \\ J &= I(\mu^i)_{\vartheta^i \vartheta^i} \end{aligned} \qquad \textbf{A4.16}$$

This has the advantage of using the local gradients and curvatures while precluding large excursions from the expansion point. In our implementations $\nu = 1/\eta$ is based on the 2-norm of the curvature η for both regularization schemes. The 2-norm is the largest singular value and, in the present context, represents an upper bound on rate of convergence of the ascent (cf. a Lyapunov exponent).[1] Terminating the ascent prematurely is reminiscent

[1] Note that the largest singular value is the largest negative eigenvalue of the curvature and represents the largest rate of change of the gradient locally.

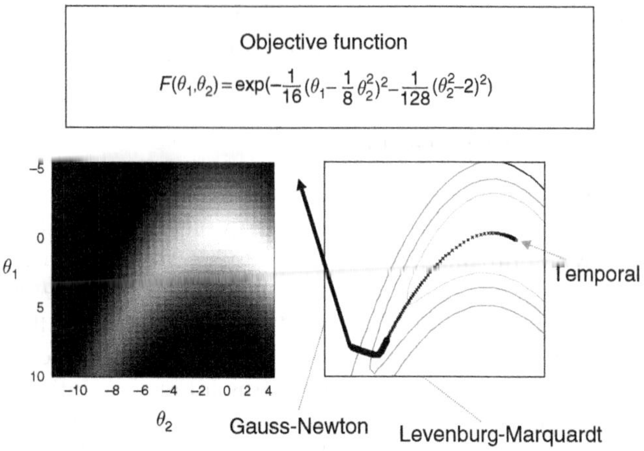

FIGURE A4.1 Examples of Levenburg-Marquardt and temporal regularization. The left panel shows an image of the landscape defined by the objective function $F(\theta_1, \theta_2)$ of two parameters (upper panel). This landscape was chosen because it is difficult for conventional schemes exhibiting curvilinear valleys and convex regions. The right panel shows the ascent trajectories, over 256 iterations (starting at 8, −10), superimposed on a contour plot of the landscape. In these examples, the regularization parameter was the 2-norm of the curvature evaluated at each update. Note how the ascent goes off in the wrong direction with no regularization (Gauss-Newton). The regularization adopted by Levenburg-Marquardt makes its progress slow, in relation to the temporal regularization, so that it fails to attain the maximum after 256 iterations.

of 'early stopping' in the training of neural networks in which the number of weights far exceeds the sample size (e.g. Nelson and Illingworth, 1991). It is interesting to note that 'early stopping' is closely related to ridge regression, which is another perspective on Levenburg-Marquardt regularization.

A comparative example using Levenburg-Marquardt and temporal regularization is provided in Figure A4.1 and suggests temporal regularization is better, in this example. Either approach can be implemented in the VB scheme by simply regularizing the Gauss-Newton update if the variational energy $I(\vartheta^i)$ fails to increase after each iteration. We prefer temporal regularization because it is based on a simpler heuristic and, more importantly, is straightforward to implement in dynamic schemes using high-order temporal derivatives.

A note on dynamic models

The second reason we have formulated the ascent as a time-dependent process is that it can be used to invert dynamic models. In this instance, the integration time in Eqn. **A4.16** is determined by the interval between observations. This is the approach taken in our variational treatment of dynamic systems, namely, dynamic expectation maximization or DEM (introduced briefly in Friston *et al.*, 2005 and implemented in spm DEM.m). DEM produces conditional densities that are a continuous function of time and avoids many of the limitations of discrete schemes based on incremental Bayes (e.g. extended Kalman filtering). In dynamic models the energy is a function of the parameters and their high-order motion, i.e. $I(\vartheta^i) \to I(\vartheta^i, \dot{\vartheta}^i, \ldots, t)$. This entails the extension of the variational density to cover this motion, using generalized coordinates $q(\vartheta^i) \to q(\vartheta^i, \dot{\vartheta}^i, \ldots, t)$. Dynamic schemes are important for the identification of stochastic dynamic casual models. However, the applications considered in this book are restricted to deterministic systems, without random fluctuations in the hidden states, and so we will focus on static models.

Having established the operational equations for VB under the Laplace approximation, we now look at their application to some specific models.

VARIATIONAL BAYES FOR NON-LINEAR MODELS

Consider the non-linear generative model with additive error $y = G(\theta) + \varepsilon$. Gaussian assumptions about the errors or innovations $p(\varepsilon) = N(0, \Sigma(\lambda))$ furnish a likelihood $p(y|\theta, \lambda) = N(G(\theta), \Sigma(\lambda))$. In this example, we can consider the parameters as falling into two sets $\vartheta = \{\theta, \lambda\}$ such that $q(\vartheta) = q(\theta)q(\lambda)$, where $q(\theta) = N(\mu^\theta, \Sigma^\theta)$ and $q(\lambda) = N(\mu^\lambda, \Sigma^\lambda)$. We will also assume Gaussian priors $p(\theta) = N(\eta^\theta, \Pi^{\theta-1})$ and $p(\lambda) = N(\eta^\lambda, \Pi^{\lambda-1})$. We will refer to the two sets as the parameters and hyperparameters. These likelihood and priors define the energy $L(\vartheta) = \ln p(y|\theta, \lambda) + \ln p(\theta) + \ln p(\lambda)$. Note that Gaussian priors are not too restrictive because both $G(\theta)$ and $\Sigma(\lambda)$ are non-linear functions that can embody a probability integral transform (i.e. can implement a re-parameterization in terms of non-Gaussian priors).

Given n samples, p parameters and h hyperparameters:

$$
\begin{aligned}
&L(\theta) = \\
&\quad -\frac{1}{2}\varepsilon^T\Sigma^{-1}\varepsilon + \frac{1}{2}\ln|\Sigma^{-1}| - \frac{n}{2}\ln 2\pi \\
&\quad -\frac{1}{2}\varepsilon^{\theta T}\Pi^\theta\varepsilon^\theta + \frac{1}{2}\ln|\Pi^\theta| - \frac{p}{2}\ln 2\pi \\
&\quad -\frac{1}{2}\varepsilon^{\lambda T}\Pi^\lambda\varepsilon^\lambda + \frac{1}{2}\ln|\Pi^\lambda| - \frac{h}{2}\ln 2\pi \\
&\varepsilon = G(\mu^\theta) - y \\
&\varepsilon^\theta = \mu^\theta - \eta^\theta \\
&\varepsilon^\lambda = \mu^\lambda - \eta^\lambda
\end{aligned}
\qquad \textbf{A4.17}
$$

and

$$
\begin{aligned}
L_\theta &= -G_\theta^T \Sigma^{-1} \varepsilon - \Pi^\theta \varepsilon^\theta \\
L_{\theta\theta} &= -G_\theta^T \Sigma^{-1} G_\theta - \Pi^\theta \\
L_{\lambda i} &= -\frac{1}{2} tr(P_i(\varepsilon\varepsilon^T - \Sigma)) - \Pi_{i\bullet}^\lambda \varepsilon^\lambda \\
L_{\lambda\lambda ij} &= -\frac{1}{2} tr(P_{ij}(\varepsilon\varepsilon^T - \Sigma)) - \frac{1}{2} tr(P_i \Sigma P_j \Sigma) - \Pi_{ij}^\lambda \\
P_i &= \frac{\partial \Sigma^{-1}}{\partial \lambda_i} \qquad P_{ij} = \frac{\partial^2 \Sigma^{-1}}{\partial \lambda_i \partial \lambda_j}
\end{aligned} \tag{A4.18}
$$

Note that we have ignored second-order terms that depend on $G_{\theta\theta}$, under the assumption that the generative model is only weakly non-linear. The requisite gradients and curvatures are:

$$
\begin{aligned}
&I(\theta)_{\theta k} = L(\theta, \mu^\lambda)_{\theta k} + \frac{1}{2} tr(\Sigma^\lambda A^k) && I(\lambda)_{\lambda i} = L(\mu^\theta, \lambda)_{\lambda i} + \frac{1}{2} tr(\Sigma^\theta C^i) \\
&I(\theta)_{\theta\theta kl} = L(\theta, \mu^\lambda)_{\theta\theta kl} + \frac{1}{2} tr(\Sigma^\lambda B^{kl}) && I(\lambda)_{\lambda\lambda ij} = L(\mu^\theta, \lambda)_{\lambda\lambda ij} + \frac{1}{2} tr(\Sigma^\theta D^{ij}) \\
&A_{ij}^k = -G_{\theta\bullet k}^T P_{ij} \varepsilon && C^i = -G_\theta^T P_i G_\theta \\
&B_{ij}^{kl} = -G_{\theta\bullet k}^T P_{ij} G_{\theta\bullet l} && D^{ij} = -G_\theta^T P_{ij} G_\theta
\end{aligned} \tag{A4.19}
$$

where $G_{\theta\bullet k}$ denotes the k-th column of G_θ. These enter the VB scheme in Eqn. **A4.13**, giving the two-step scheme:

until convergence

  until convergence

$$
\begin{aligned}
\Sigma^{\theta^{-1}} &= G_\theta^T \Sigma^{-1} G_\theta + \Pi^\theta \\
L(\mu)_\theta &= -G_\theta^T \Sigma^{-1} \varepsilon - \Pi^\theta \varepsilon^\theta \\
I(\mu)_{\theta k} &= L(\mu)_{\theta_k} + \frac{1}{2} tr(\Sigma^\lambda A^k) \\
I(\mu)_{\theta\theta kl} &= -\Sigma_{kl}^{\theta^{-1}} + \frac{1}{2} tr(\Sigma^\lambda B^{kl}) \\
\Delta\mu^\theta &= -I(\mu)_{\theta\theta}^{-1} I(\mu)_\theta
\end{aligned}
$$

  end

  until convergence

$$
\begin{aligned}
\Sigma_{ij}^{\lambda^{-1}} &= \frac{1}{2} tr(P_{ij}(\varepsilon\varepsilon^T - \Sigma) + P_i \Sigma P_j \Sigma) + \Pi_{ij}^\lambda \\
I(\mu)_{\lambda i} &= -\frac{1}{2} tr(P_i(\varepsilon\varepsilon^T - \Sigma + G_\theta \Sigma^\theta G_\theta^T)) - \Pi_{i\bullet}^\lambda \varepsilon^\lambda \\
I(\mu)_{\lambda\lambda ij} &= -\Sigma_{ij}^{\lambda^{-1}} - \frac{1}{2} tr(\Sigma^\theta G_\theta^T P_{ij} G_\theta) \\
\Delta\mu^\lambda &= -I(\mu)_{\lambda\lambda}^{-1} I(\mu)_\lambda
\end{aligned}
$$

  end

end **A4.20**

The negative free energy for these models is:

$$
\begin{aligned}
F = &-\frac{1}{2}\varepsilon^T \Sigma^{-1} \varepsilon + \frac{1}{2} \ln|\Sigma^{-1}| - \frac{n}{2} \ln 2\pi \\
&-\frac{1}{2}\varepsilon^{\theta T} \Pi^\theta \varepsilon^\theta + \frac{1}{2} \ln|\Pi^\theta| + \frac{1}{2} \ln|\Sigma^\theta| \\
&-\frac{1}{2}\varepsilon^{\lambda T} \Pi^\lambda \varepsilon^\lambda + \frac{1}{2} \ln|\Pi^\lambda| + \frac{1}{2} \ln|\Sigma^\lambda|
\end{aligned} \tag{A4.21}
$$

In principle, these equations cover a large range of models and will work provided the true posterior is unimodal (and roughly Gaussian). The latter requirement can usually be met by a suitable transformation of parameters. In the next section, we consider a further simplification of our assumptions about the variational density and how this leads to expectation maximization.

EXPECTATION MAXIMIZATION FOR NON-LINEAR MODELS

There is a key distinction between θ and λ in the generative model above: the parameters λ are hyperparameters in the sense, like the variational parameters, they parameterize a density. In many instances, their conditional density *per se* is uninteresting. In variational expectation maximization, we ignore uncertainty about the hyperparameters and assume $q(\lambda)$ is a point mass (i.e. $\Sigma^\lambda = 0$). In this case, the free energy is effectively conditioned on λ and reduces to:

$$
\begin{aligned}
F^\lambda &= \ln p(y|\lambda) - D(q(\theta)||p(\theta|y, \lambda)) \\
&= \\
&-\frac{1}{2}\varepsilon^T \Sigma^{-1} \varepsilon + \frac{1}{2} \ln|\Sigma^{-1}| - \frac{n}{2} \ln 2\pi \\
&-\frac{1}{2}\varepsilon^{\theta T} \Pi^\theta \varepsilon^\theta + \frac{1}{2} \ln|\Pi^\theta| + \frac{1}{2} \ln|\Sigma^\theta|
\end{aligned} \tag{A4.22}
$$

Here, $F^\lambda \le \ln p(y|\lambda)$ becomes a lower bound on the log likelihood of the hyperparameters. This means the variational step updating the hyperparameters maximizes the likelihood of the hyperparameters $\ln p(y|\lambda)$ and becomes an **M**-step. In this context, Eqn. **A4.20** simplifies because we can ignore the terms that involve Σ^λ and Π^λ to give:

until convergence

  until convergence: **E-step**

$$
\begin{aligned}
\Sigma^{\theta^{-1}} &= G_\theta^T \Sigma^{-1} G_\theta + \Pi^\theta \\
\Delta\mu^\theta &= -\Sigma^{\theta^{-1}} (G_\theta^T \Sigma^{-1} \varepsilon + \Pi^\theta \varepsilon^\theta)
\end{aligned}
$$

  end

until convergence: **M-step**

$$I(\mu)_{\lambda i} = -\frac{1}{2}tr(P_i(\varepsilon\varepsilon^T - \Sigma + G_\theta\Sigma^\theta G_\theta^T))$$

$$I(\mu)_{\lambda\lambda ij} = -\frac{1}{2}tr(P_{ij}(\varepsilon\varepsilon^T - \Sigma + G_\theta\Sigma^\theta G_\theta^T) + P_i\Sigma P_j\Sigma)$$

$$\Delta\mu^\lambda = -I(\mu)_{\lambda\lambda}^{-1}I(\mu)_\lambda$$

end

end **A4.23**

Expectation maximization is an iterative parameter re-estimation procedure devised to estimate the parameters and hyperparameters of a model. It was introduced as an iterative method to obtain maximum likelihood estimators with incomplete data (Hartley, 1958) and was generalized by Dempster *et al.* (1977) (see Appendix 3 for more details). Strictly speaking, EM refers to schemes in which the conditional density of the **E**-step is known exactly, obviating the need for fixed-form assumptions. This is why we used the term variational EM above.

In terms of the VB scheme, the **M**-step for $\mu^\lambda = \max I(\lambda)$ is unchanged because $I(\lambda)$ does not depend on Σ^λ. The remaining variational steps (i.e. **E**-steps) are simplified because one does not have to average over the conditional density $q(\lambda)$. This ensuing scheme is that described in Friston (2002) for non-linear system identification (see Chapter 34) and is implemented in spm_nlsi.m. Although this scheme is applied to time-series, it actually treats the underlying model as static, generating finite-length data-sequences. This routine is used to identify haemodynamic models in terms of biophysical parameters for regional responses and dynamic causal models (DCMs) of distributed responses in a variety of applications, e.g. functional magnetic resonance imaging (fMRI) (Friston *et al.*, 2003 and Chapter 41), EEG (David *et al.*, 2005 and Chapter 42), MEG (Kiebel *et al.*, 2006), and mean-field models of neuronal activity (Harrison *et al.*, 2005 and Chapter 31).

A formal equivalence

A key point here is that VB and EM are exactly the same when $P_{ij} = 0$. In this instance the matrices A, B and D in Eqn. **A4.19** disappear. This means the VB-step for the parameters does not depend on Σ^λ and becomes formally identical to the **E**-step. Because the VB-step for the hyperparameters is already the same as the **M**-step (apart from the loss of hyperpriors) the two schemes converge. One can ensure $P_{ij} = 0$ by adopting a hyperparameterization, which *renders the precision linear in the hyperparameters*, e.g. a linear mixture of precision components Q_i (see below). This resulting variational scheme is used by the SPM5 version of spm_nlsi.m for non-linear system identification.

Hyperparameterizing precisions

One can ensure $P_{ij} = 0$ by adopting a hyperparameterization, where the precision is linear in the hyperparameters, e.g. a linear mixture of precision components Q_i. Consider the more general parameterization of precisions:

$$\Sigma^{-1} = \sum_i f(\lambda_i)Q_i$$
$$P_i = f'(\lambda_i)Q_i \qquad \textbf{A4.24}$$
$$P_{ij} = \begin{cases} 0 & i \neq j \\ f''(\lambda_i)Q_i & i = j \end{cases}$$

where $f(\lambda_i)$ is any analytic function. The simplest is $f(\lambda_i) = \lambda_i \Rightarrow f' = 1 \Rightarrow f'' = 0$. In this case VB and EM are formally identical. However, this allows negative contributions to the precisions, which can lead to improper covariances. Using $f(\lambda_i) = \exp(\lambda_i) \Rightarrow f'' = f' = f$ precludes improper covariances. This hyperparameterization effectively implements a log-normal hyperprior, which imposes scale-invariant positivity constraints on the precisions. This is formally related to the use of conjugate [gamma] priors for scale parameters like $f(\lambda_i)$ (cf. Berger, 1985), when they are non-informative. Both imply a flat prior on the log-precision, which means its derivatives with respect to $\ln f(\lambda_i) = \lambda_i$ vanish (because it has no maximum). In short, one can either place a gamma prior on $f(\lambda_i)$ or a normal prior on $\ln f(\lambda_i) = \lambda_i$. These hyperpriors are the same when uninformative.

However, there are many models where it is necessary to hyperparameterize in terms of linear mixtures of covariance components:

$$\Sigma = \sum_i f(\lambda_i)Q_i$$
$$P_i = -f'(\lambda_i)\Sigma^{-1}Q_i\Sigma^{-1} \qquad \textbf{A4.25}$$
$$P_{ij} = \begin{cases} 2P_i\Sigma P_j & i \neq j \\ 2P_i\Sigma P_i + \frac{f''(\lambda_i)}{f'(\lambda_i)}P_i & i = j \end{cases}$$

This is necessary when hierarchical generative models induce multiple covariance components. These are important models because they are central to empirical Bayes (see Chapter 22). See Harville (1977) for comments on the usefulness of making the covariances linear in the hyperparameters, i.e. $f(\lambda_i) = \lambda_i \Rightarrow f' = 1 \Rightarrow f'' = 0$.

An important difference between these two hyperparameterizations is that the linear mixture of precisions is conditionally convex (Mackay and Takeuchi, 1996), whereas the mixture of covariances is not. This means there may be multiple optima for the latter. See Mackay and Takeuchi (1996) for further covariance hyperparameterizations and an analysis of their convexity. Interested readers may find the material in Leonard and Hsu (1992) useful further reading.

The second key point that follows from the variational treatment is that one can adjust the EM free energy to approximate the log-evidence, as described next.

Accounting for uncertainty about the hyperparameters

The EM free energy in Eqn. **A4.22** discounts uncertainty about the hyperparameters because it is conditioned upon them. This is a well-recognized problem, sometimes referred to as the overconfidence problem, for which a number of approximate solutions have been suggested (e.g. Kass and Steffey, 1989). Here, we describe a solution that appeals to the variational framework within which EM can be treated.

If we treat EM as an approximate variational scheme, we can adjust the EM free energy to give the variational free energy required for model comparison and averaging. By comparing Eqn. **A4.21** and Eqn. **A4.22**, we can express the variational free energy in terms of F^λ and an extra term from Eqn. **A4.18**:

$$\begin{aligned} F &= F^\lambda + \frac{1}{2}\ln|\Sigma^\lambda| \\ \Sigma^\lambda_{ij} &= -L(\mu)^{-1}_{\lambda\lambda} \end{aligned} \qquad \textbf{A4.26}$$

Intuitively, the extra term encodes the conditional information (i.e. entropy) about the model's covariance components. The log-evidence will only increase if an extra component adds information. Adding redundant components will have no effect on F. This term can be regarded as additional Occam factor (Mackay and Takeuchi, 1996). Adjusting the EM free energy to approximate the log-evidence is important because of the well-know connections between EM for linear models and restricted maximum likelihood. This connection suggests that the ReML objective function could also be used to evaluate the log-evidence and therefore be used for model selection. We now consider ReML as a special case of EM.

RESTRICTED MAXIMUM LIKELIHOOD FOR LINEAR MODELS

In the case of general linear models $G(\theta) = G\theta$ with additive Gaussian noise and no priors on the parameters (i.e. $\Pi^\theta = 0$) the free energy reduces to:

$$\begin{aligned} F^\theta &= \ln p(y|\lambda) - D(q(\theta)||p(\theta|y,\lambda)) \\ &= -\frac{1}{2}\varepsilon^T\Sigma^{-1}\varepsilon + \frac{1}{2}\ln|\Sigma^{-1}| - \frac{n}{2}\ln 2\pi + \frac{1}{2}\ln|\Sigma^\theta| \end{aligned} \qquad \textbf{A4.27}$$

Critically, the dependence on $q(\theta)$ can be eliminated using the closed form solutions for the conditional moments:

$$\begin{aligned} \mu^\theta &= \Sigma^\theta G^T\Sigma^{-1}y \\ \Sigma^\theta &= (G^T\Sigma^{-1}G)^{-1} \end{aligned}$$

to eliminate the divergence term and give:

$$\begin{aligned} F^\theta &= \ln p(y|\lambda) \\ &= -\frac{1}{2}tr(\Sigma^{-1}Ryy^TR^T) + \frac{1}{2}\ln|\Sigma^{-1}| - \frac{n}{2}\ln 2\pi \\ &\quad -\frac{1}{2}\ln|G^T\Sigma^{-1}G| \\ \varepsilon &= Ry \\ R &= I - G(G^T\Sigma^{-1}G)^{-1}G^T\Sigma^{-1} \end{aligned} \qquad \textbf{A4.28}$$

This free energy is also known as the ReML objective function (Harville, 1977). ReML or *restricted maximum likelihood* was introduced by Patterson and Thompson, in 1971, as a technique for estimating variance components, which accounts for the loss in degrees of freedom that result from estimating fixed effects (Harville, 1977). The elimination makes the free energy a simple function of the hyperparameters and, effectively, the EM scheme reduces to a single **M**-step or ReML-step:

until convergence: ReML-step

$$\begin{aligned} L(\mu)_{\lambda i} &= -\frac{1}{2}tr(P_iR(yy^T - \Sigma)R^T) \\ \langle L(\mu)_{\lambda\lambda ij}\rangle &= -\frac{1}{2}tr(P_iR\Sigma P_jR\Sigma) \\ \Delta\mu^\lambda &= -\langle L(\mu)_{\lambda\lambda}\rangle^{-1}L(\mu)_\lambda \end{aligned} \qquad \textbf{A4.29}$$

end

Notice that the energy has replaced the variational energy because they are the same: from Eqn. **A4.6** $I(\vartheta) = L(\lambda)$. This is a result of eliminating $q(\theta)$ from the variational density. Furthermore, the curvature has been replaced by its expectation to render the Gauss-Newton descent a Fisher-Scoring scheme using:

$$\langle Ryy^TR^T\rangle = R\Sigma R^T = R\Sigma \qquad \textbf{A4.30}$$

To approximate the log-evidence, we can adjust the ReML free energy after convergence as with the EM free energy:

$$\begin{aligned} F &= F^\theta + \frac{1}{2}\ln|\Sigma^\lambda| \\ \Sigma^\lambda_{ij} &= -\langle L(\mu)_{\lambda\lambda}\rangle^{-1} \end{aligned} \qquad \textbf{A4.31}$$

The conditional covariance of the hyperparameters uses the same curvature as the ascent in Eqn. **A4.29**. Being able to compute the log-evidence from ReML is useful because ReML is used widely in an important class of models, namely hierarchical models reviewed next.

RESTRICTED MAXIMUM LIKELIHOOD FOR HIERARCHICAL LINEAR MODELS

Parametric empirical Bayes

The application of ReML to the linear models of the previous section did not accommodate priors on the parameters. However, one can absorb these priors into the error covariance components using a hierarchical formulation. This enables the use of ReML to identify models with full or empirical priors. Hierarchical linear models (see Chapters 11 and 22) are equivalent to parametric empirical Bayes models (Efron and Morris, 1973) in which empirical priors emerge from conditional independence of the errors $\varepsilon^{(i)} \sim N(0, \Sigma^{(i)})$:

$$\begin{aligned} y^{(1)} &= & y^{(1)} &= \varepsilon^{(1)} \\ \theta^{(1)} &= G^{(1)}\theta^{(2)} + \varepsilon^{(1)} & &+ G^{(1)}\varepsilon^{(2)} \\ \theta^{(2)} &= G^{(2)}\theta^{(3)} + \varepsilon^{(2)} \equiv & &+ G^{(1)}G^{(2)}\varepsilon^{(3)} \\ &\vdots & &\vdots \\ \theta^{(n)} &= \varepsilon^{(n)} & &+ G^{(1)}\cdots G^{(n-1)}\theta^{(n)} \end{aligned} \qquad \mathbf{A4.32}$$

In hierarchical models, the random terms model uncertainty about the parameters at each level and $\Sigma(\lambda)^{(i)}$ are treated as prior covariance constraints on $\theta^{(i)}$. Hierarchical models of this sort are very common and underlie all classical mixed effects analyses of variance.[2] ReML identification of simple two-level models like:

$$\begin{aligned} y^{(1)} &= G^{(1)}\theta^{(2)} + \varepsilon^{(1)} \\ \theta^{(2)} &= \varepsilon^{(2)} \end{aligned} \qquad \mathbf{A4.33}$$

is a useful way to impose shrinkage priors on the parameters and covers early approaches (e.g. Stein shrinkage estimators) to recent developments, such as relevance vector machines (e.g. Tipping, 2001). Relevance vector machines represent a Bayesian treatment of support vector machines, in which the second-level covariance $\Sigma(\lambda)^{(2)}$ has a component for each parameter. Most of the ReML estimates of these components shrink to zero. This means the columns of $G^{(1)}$ whose parameters have zero mean and variance can be eliminated, providing a new model with sparse support.

Estimating these models through their covariances $\Sigma^{(i)}$ with ReML corresponds to empirical Bayes. This estimation can proceed in one of two ways: first, we can augment the model and treat the random terms as parameters to give:

$$y = J\theta + \varepsilon$$

$$y = \begin{bmatrix} y^{(1)} \\ 0 \\ \vdots \\ 0 \end{bmatrix} \quad J = \begin{bmatrix} K^{(2)} & \cdots & K^{(n)} \\ -I & & \\ & \ddots & \\ & & -I \end{bmatrix} \quad \varepsilon = \begin{bmatrix} \varepsilon^{(1)} \\ \varepsilon^{(2)} \\ \vdots \\ \varepsilon^{(n)} \end{bmatrix} \quad \theta = \begin{bmatrix} \varepsilon^{(2)} \\ \vdots \\ \theta^{(n)} \end{bmatrix}$$

$$K^{(i)} = \prod_{j=1}^{i} G^{(j-1)}$$

$$\Sigma = \begin{bmatrix} \Sigma^{(1)} & & \\ & \ddots & \\ & & \Sigma^{(n)} \end{bmatrix} \qquad \mathbf{A4.34}$$

with $G^{(0)} = I$. This reformulation is a non-hierarchical model with no explicit priors on the parameters. However, the ReML estimates of $\Sigma(\lambda)^{(i)}$ are still the empirical prior covariances of the parameters $\theta^{(i)}$ at each level. If $\Sigma^{(i)}$ is known *a priori*, it simply enters the scheme as a known covariance component. This corresponds to a full Bayesian analysis with known or full priors for the level in question.

spm_peb.m uses this reformulation and Eqn. **A4.29** for estimation. The conditional expectations of the parameters are recovered by recursive substitution of the conditional expectations of the errors into Eqn. **A4.33** (cf. Friston, 2002). spm_peb.m uses a computationally efficient substitution:

$$\frac{1}{2}tr(P_iR(yy^T - \Sigma)R^T) = \frac{1}{2}y^TR^TP_iRy - \frac{1}{2}tr(P_iR\Sigma R^T) \qquad \mathbf{A4.35}$$

to avoid computing the potentially large matrix yy^T. We have used this scheme extensively in the construction of posterior probability maps or PPMs (Friston and Penny, 2003 and Chapter 23) and mixed-effect analysis of multi-subject studies in neuroimaging (Friston *et al.*, 2005). Both these examples rest on hierarchical models, using hierarchical structure over voxels and subjects respectively.

Classical covariance component estimation

An equivalent identification of hierarchical models rests on an alternative and simpler reformulation of Eqn. **A4.30**

[2] For an introduction to EM algorithms in generalized linear models see Fahrmeir and Tutz (1994). This text provides an exposition of EM and PEB in linear models, usefully relating EM to classical methods (e.g. ReML p. 225).

in which all the hierarchically induced covariance components $K^{(i)T}\Sigma^{(i)}K^{(i)T}$ are treated as components of a compound error:

$$\begin{aligned} y &= \varepsilon \\ y &= y^{(1)} \\ \varepsilon &= \sum_{i=1}^{n} K^{(i)}\varepsilon^{(i)} \\ \Sigma &= \sum_{i=1}^{n} K^{(i)T}\Sigma^{(i)}K^{(i)T} \end{aligned} \qquad \textbf{A4.36}$$

The ensuing ReML estimates of $\Sigma(\lambda)^{(i)}$ can be used to compute the conditional density of the parameters in the usual way. For example, the conditional expectation and covariance of the i-th level parameters $\theta^{(i)}$ are:

$$\begin{aligned} \mu^{\theta(i)} &= \Sigma^{\theta(i)}K^{(i)^T}\tilde{\Sigma}^{-1}y \\ \Sigma^{\theta(i)} &= (K^{(i)^T}\tilde{\Sigma}^{-1}K^{(i)} + \Sigma^{(i)^{-1}})^{-1} \\ \tilde{\Sigma} &= \sum_{j\neq i} K^{(j)T}\Sigma^{(j)}K^{(j)T} \end{aligned} \qquad \textbf{A4.37}$$

where $\tilde{\Sigma}$ represents the ReML estimate of error covariance, excluding the component of interest. This component $\Sigma^{(i)} = \Sigma(\lambda)^{(i)}$ is treated as an empirical prior on $\theta^{(i)}$. spm_reml.m uses Eqn. **A4.29** to estimate the requisite hyperparameters. Critically, it takes as an argument the matrix yy^T. This may seem computationally inefficient. However, there is a special but very common case where dealing with yy^T is more appropriate than dealing with y (cf. the implementation using Eqn. **A4.35** in spm_peb.m).

This is when there are r multiple observations that can be arranged as a matrix $Y = [y_1, \ldots, y_r]$. If these observations are independent, then we can express the covariance components of the vectorized response in terms of Kronecker tensor products:

$$\begin{aligned} y &= vec\{Y\} = \varepsilon \\ \varepsilon &= \sum_{i=1}^{n} I \otimes K^{(i)}\varepsilon^{(i)} \\ \text{cov}\{\varepsilon^{(i)}\} &= I \otimes \Sigma^{(i)} \end{aligned} \qquad \textbf{A4.38}$$

This leads to a computationally efficient scheme employed by spm_reml.m, which uses the compact forms:[3]

$$\begin{aligned} L(\mu)_{\lambda i} &= -\frac{1}{2}tr((I\otimes P_iR)(yy^T - I\otimes\Sigma)(I\otimes R^T)) \\ &= -\frac{r}{2}tr(P_iR(\frac{1}{r}YY^T - \Sigma)R^T) \\ \langle L(\mu)_{\lambda\lambda ij}\rangle &= -\frac{1}{2}tr(I\otimes P_iR\Sigma P_jR\Sigma) \\ &= -\frac{r}{2}tr(P_iR\Sigma P_jR\Sigma) \end{aligned} \qquad \textbf{A4.39}$$

Critically, the update scheme is a function of the sample covariance of the data $\frac{1}{r}YY^T$ and can be regarded as a covariance component estimation scheme. This can be useful in two situations: first, if the augmented form in Eqn. **A4.33** produces prohibitively long vectors. This can happen when the number of parameters is much greater than the number of responses. This is a common situation in underdetermined problems. An important example is source reconstruction in electroencephalography, where the number of sources is much greater than the number of measurement channels (see Chapters 29 and 30 and Phillips *et al.*, 2005 for an application that uses spm_reml.m in this context). In these cases one can form conditional estimates of the parameters using the matrix inversion lemma and again avoid inverting large $(p \times p)$ matrices:

$$\begin{aligned} \mu^{\theta(i)} &= \Sigma^{(i)}K^{(i)^T}\tilde{\Sigma}^{-1}Y \\ \Sigma^{\theta(i)} &= \Sigma^{(i)} - \Sigma^{(i)}K^{(i)^T}\tilde{\Sigma}^{-1}K^{(i)}\Sigma^{(i)} \\ \tilde{\Sigma} &= \sum_{i=1}^{n} K^{(i)T}\Sigma^{(i)}K^{(i)T} \end{aligned} \qquad \textbf{A4.40}$$

The second situation is where there are a large number of realizations. In these cases, it is much easier to handle the second-order matrices of the data YY^T than the data Y itself. An important application here is the estimation of non-sphericity over voxels in the analysis of fMRI time-series (see Chapter 22 and Friston *et al.*, 2002 for this use of spm_reml.m). Here, there are many more voxels than scans and it would not be possible to vectorize the data. However, it is easy to collect the sample covariance over voxels and partition it into non-spherical covariance components using ReML.

In the case of sequential correlations among the errors $\text{cov}\{\varepsilon^{(i)}\} = V \otimes \Sigma^{(i)}$, one simply replaces YY^T with $YV^{-1}Y^T$. Heuristically, this corresponds to sequentially whitening the observations before computing their second-order statistics. We have used this device in the Bayesian inversion of models of evoked and induced responses in EEG/MEG (Chapter 30 and Friston *et al.*, 2006).

In summary, hierarchical models can be identified through ReML estimates of covariance components. If the response vector is relatively small, it is generally more expedient to reduce the hierarchical form by augmentation, as in Eqn. **A4.34**, and use Eqn. **A4.35** to compute the

[3] Note that we have retained the residual forming matrix R, despite the fact that there are no parameters. This is because, in practice, one usually models confounds as fixed effects at the first level. The residual forming matrix projects the data onto the null space of these confounds.

gradients. When the augmented form becomes too large, because there are too many parameters, reformulation in terms of covariance components is computationally more efficient because the gradients can be computed from the sample covariance of the data. The latter formulation is also useful when there are multiple realizations of the data because the sample covariance, over realizations, does not change in size. This leads to very fast Bayesian inversion. Both approaches rest on estimating covariance components that are induced by the observation hierarchy. This enforces a hyperparameterization of the covariances, as opposed to precisions.

MODEL SELECTION WITH REML

We conclude with a brief demonstration of model selection using ReML and its adjusted free energy. In these examples we use the covariance component formulation (spm_reml.m), noting exactly the same results would be obtained with augmentation (spm_peb.m). We use a simple hierarchical two-level linear model, implementing shrinkage priors, because this sort of model is common in neuroimaging data analysis and represents the simplest form of empirical Bayes. The model is described in Figure A4.2.

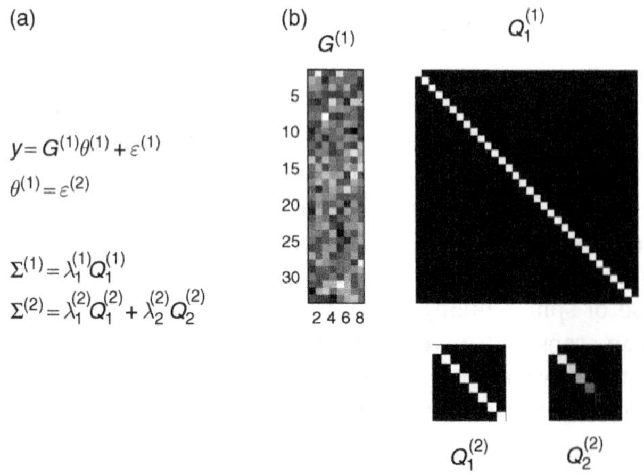

FIGURE A4.2 A hierarchical linear model. (a) The form of the model with two levels. The first level has a single error covariance component, while the second has two. The second level places constraints on the parameters of the first, through the second-level covariance components. Conditional estimation of the hyperparameters, controlling these components, corresponds to an empirical estimate of their prior covariance (i.e. empirical Bayes). Because there is no second-level design matrix the priors shrink the conditional estimates towards zero. These are known as shrinkage priors. (b) The design matrix and covariance components used to generate 128 realizations of the response variable y, using hyperparameters of one for all components. The design matrix comprised random Gaussian variables.

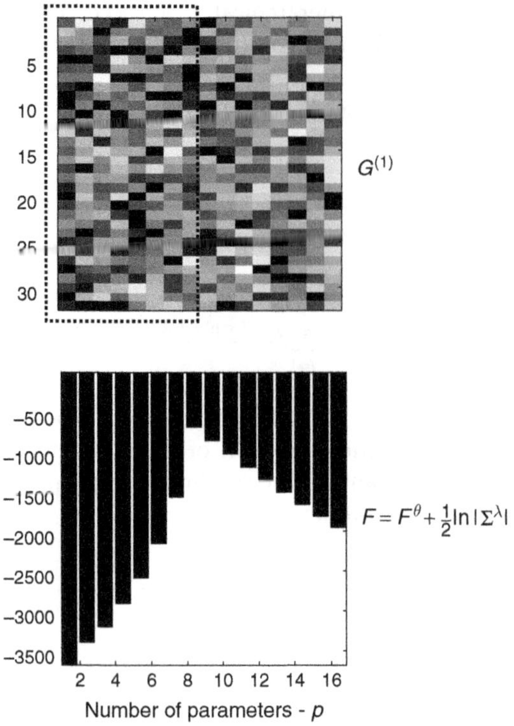

FIGURE A4.3 Model selection in terms of parameters using ReML. The data generated by the eight-parameter model in Figure A4.2 were analysed with ReML using a series of models with an increasing number of parameters. These models were based on the first p columns of the design matrix above. The profile of free energy clearly favours the model with eight parameters, corresponding to the design matrix (dotted line in upper panel) used to generate the data.

The free energy can, of course, be used for model selection when models differ in the number and deployment of parameters. This is because both F and F^{θ} are functions of the number of parameters and their conditional uncertainty. This can be shown by evaluating the free energy as a function of the number of model parameters, for the same data. The results of this sort of evaluation are seen in Figure A4.3 and demonstrate that model selection correctly identifies a model with eight parameters. This was the model used to generate the data (Figure A4.2).

The critical issue is whether model selection works when the models differ in their hyperparameterization. To illustrate this, we analysed the same data, produced by two covariance components at the second level, with models that comprised an increasing number of second-level covariance components (Figure A4.4). These components can be regarded as specifying the form of empirical priors over solution space (e.g. spatial constraints in a source reconstruction problem). The results of these simulations show that the adjusted free energy F correctly identified the model with two components. Conversely, the unadjusted free energy F^{θ} rose

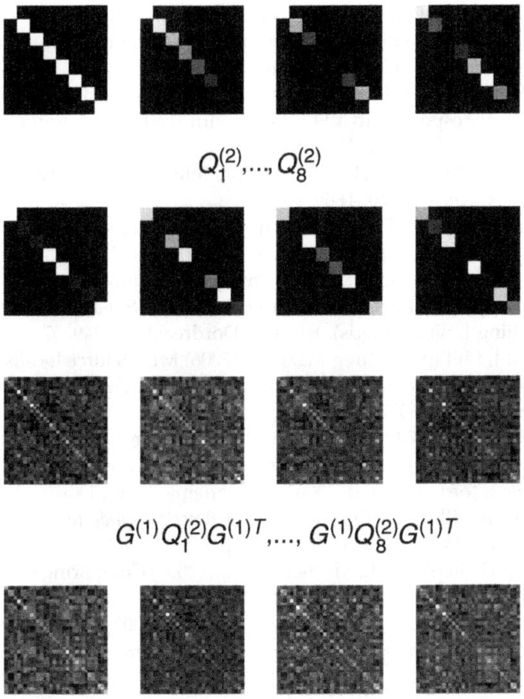

FIGURE A4.4 Covariance components used to analyse the data generated by the model in Figure A4.2. The covariance components are shown at the second level (upper panels) and after projection onto response space (lower panels) with the eight-parameter model. Introducing more covariance components creates a series model with an increasing number of hyperparameters, which we examined using model selection in Figure A4.5. These covariance components were leading diagonal matrices, whose elements comprised a mean-adjusted discrete cosine set.

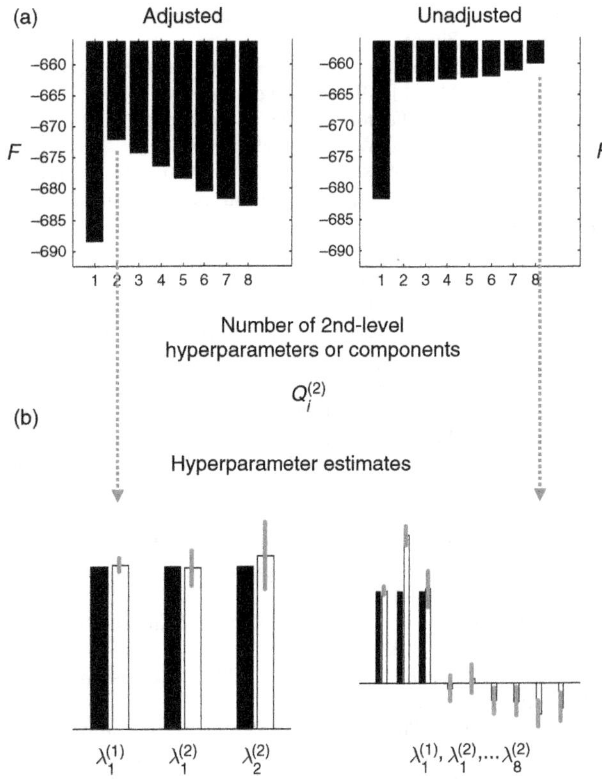

FIGURE A4.5 Model selection in terms of hyperparameters using ReML. (a) The free energy was computed using the data generated by the model in Figure A4.2 and a series of models with an increasing number of hyperparameters. The ensuing free energy profiles (adjusted – left; unadjusted – right) are shown as a function of the number of second-level covariance components used (from Figure A4.4). The adjusted profile clearly identified the correct model with two second-level components. (b) Conditional estimates (white) and true (black) hyperparameter values with 90 per cent confidence intervals for the correct (3-component, left) and redundant (9-component, right) models.

progressively as the number of components and accuracy increased (Figure A4.5).

The lower panel in Figure A4.5 shows the hyperparameter estimates for two models. With the correctly selected model, the true values fall within the 90 per cent confidence interval. However, when the model is over-parameterized, with eight second-level components, this is not the case. Although the general profile of hyperparameters has been captured, this suboptimum model has clearly overestimated some hyperparameters and under-estimated others.

Conclusion

We have seen that restricted maximum likelihood is a special case of expectation maximization and that expectation maximization is a special case of variational Bayes. In fact, nearly every routine used in neuroimaging analysis is a special case of variational Bayes, from ordinary least squares estimation to dynamic causal modelling. We have focused on adjusting the objective functions used by EM and ReML to approximate the variational free energy under the Laplace approximation. This free energy is a lower bound approximation (exact for linear models) to the log-evidence, which plays a central role in model selection and averaging. This means one can use computationally efficient schemes like ReML for both model selection and Bayesian inversion.

REFERENCES

Ashburner J, Friston KJ (2005) Unified segmentation. *NeuroImage* **26**: 839–51

Beal MJ, Ghahramani Z (2003) The variational Bayesian EM algorithm for incomplete data: with application to scoring graphical model structures. In *Bayesian statistics*, Bernardo JM, Bayarri MJ, Berger JO *et al.* (eds). OUP, Milton Keynes, ch 7

Berger JO (1985) *Statistical decision theory and Bayesian analysis*, 2nd edn. Springer, Berlin

Bishop C (1999) Latent variable models. In *Learning in graphical models*, Jordan M (ed.). MIT Press, London

David O, Harrison L, Friston KJ (2005) Modelling event-related responses in the brain. *NeuroImage* **25**: 756–70

Dempster AP, Laird NM, Rubin DB (1977) Maximum likelihood from incomplete data via the EM algorithm. *J Roy Stat Soc Series B* **39**: 1–38

Efron B, Morris C (1973) Stein's estimation rule and its competitors – an empirical Bayes approach. *J Am Stat Assoc* **68**: 117–30

Fahrmeir L, Tutz G (1994) *Multivariate statistical modelling based on generalised linear models*. Springer-Verlag Inc., New York, pp 355–56

Feynman RP (1972) *Statistical mechanics*. Benjamin, Reading, MA

Friston KJ (2002) Bayesian estimation of dynamical systems: an application to fMRI. *NeuroImage* **16**: 513–30

Friston K (2005) A theory of cortical responses. *Philos Trans R Soc Lond B Biol Sci* **360**: 815–36

Friston KJ, Penny W, Phillips C *et al.* (2002) Classical and Bayesian inference in neuroimaging: theory. *NeuroImage* **16**: 465–83

Friston KJ, Penny W. (2003) Posterior probability maps and SPMs. *NeuroImage* **19**: 1240–49

Friston KJ, Harrison L, Penny W (2003) Dynamic causal modelling. *NeuroImage* **19**: 1273–302

Friston KJ, Stephan KE, Lund TE *et al.* (2005) Mixed-effects and fMRI studies. *NeuroImage* **24**: 244–52

Friston KJ, Henson R, Phillips C *et al.* (2006) Bayesian estimation of evoked and induced responses. *Hum Brain Mapp* in press

Harrison LM, David O, Friston KJ (2005) Stochastic models of neuronal dynamics. *Philos Trans R Soc Lond B Biol Sci* **360**: 1075–91

Hartley H (1958) Maximum likelihood estimation from incomplete data. *Biometrics* **14**: 174–94

Harville DA (1977) Maximum likelihood approaches to variance component estimation and to related problems. *J Am Stat Assoc* **72**: 320–38

Hinton GE, von Camp D (1993) Keeping neural networks simple by minimising the description length of weights. In *Proc COLT-93* pp 5–13

Kass RE, Steffey D (1989) Approximate Bayesian inference in conditionally independent hierarchical models (parametric empirical Bayes models). *J Am Stat Assoc* **407**: 717–26

Kiebel SJ, David O, Friston KJ (2006) Dynamic causal modelling of evoked responses in EEG/MEG with lead field parameterization. *NeuroImage* **30**: 1273–84

Leonard T, Hsu JSL (1992) Bayesian inference for a covariance matrix. *Ann Stat* **20**: 1669–96

Mackay DJC (1995) Free energy minimisation algorithm for decoding and cryptoanalysis. *Electron Lett* **31**: 445–47

Mackay DJC, Takeuchi R (1996) Interpolation models with multiple hyperparameters. In *Maximum entropy & Bayesian methods*, Skilling J, Sibisi S (eds). Kluwer, Dordrecht, pp 249–57

Mattout J, Phillips C, Rugg MD *et al.* (2006) MEG source localisation under multiple constraints: an extended Bayesian framework. *NeuroImage* **30**: 753–67

Neal RM, Hinton GE (1998) A view of the EM algorithm that justifies incremental sparse and other variants. In *Learning in graphical models*, Jordan MI (ed.). Kulver Academic Press, Dordrecht

Nelson MC, Illingworth WT (1991) *A practical guide to neural nets*. Addison-Wesley, Reading, MA, pp 165

Penny WD, Stephan KE, Mechelli A *et al.* (2004) Comparing dynamic causal models. *NeuroImage* **22**: 1157–72

Penny WD, Trujillo-Barreto NJ, Friston KJ (2005) Bayesian fMRI time series analysis with spatial priors. *NeuroImage* **24**: 350–62

Phillips C, Rugg M, Friston KJ (2002) Systematic regularisation of linear inverse solutions of the EEG source localisation problem. *NeuroImage* **17**: 287–301

Phillips C, Mattout J, Rugg MD *et al.* (2005) An empirical Bayesian solution to the source reconstruction problem in EEG. *NeuroImage* **24**: 997–1011

Tipping ME (2001) Sparse Bayesian learning and the relevance vector machine. *J Machine Learn Res* **1**: 211–44

Titantah JT, Pierlioni C, Ciuchi S (2001) Free energy of the Fröhlich Polaron in two and three dimensions. *Phys Rev Lett* **87**: 206406

Trujillo-Barreto N, Aubert-Vazquez E, Valdes-Sosa P (2004) Bayesian model averaging. *NeuroImage* **21**: 1300–19

Weissbach F, Pelster A, Hamprecht B (2002) High-order variational perturbation theory for the free energy. *Phys Rev Lett* **66**: 036129

APPENDIX

5

Kalman filtering

K. Friston and W. Penny

INTRODUCTION

Bayesian inversion of state-space models is related to Bayesian belief update procedures (i.e. recursive Bayesian filters). The conventional approach to online Bayesian tracking of states in non-linear or non-Gaussian systems employs extended Kalman filtering or sequential Monte Carlo methods, such as particle filtering. These implementations of Bayesian filters approximate the conditional densities of hidden states in a recursive and computationally expedient fashion, assuming that the parameters and hyperparameters of the system are known. We start with systems (dynamic models) that are formulated in continuous time:

$$\begin{aligned} y &= g(x) + z \\ \dot{x} &= f(x, v) \end{aligned} \qquad \textbf{A5.1}$$

where the innovations $z(t)$ and causes $v(t)$ are treated as random fluctuations. As we will see below this is converted into a state-space model in discrete time before application of the filter. Kalman filters proceed recursively in two steps: prediction and update. The prediction uses the Chapman-Kolmogorov equation to compute the density of the hidden states $x(t)$ conditioned on the response up to, but not including, the current observation $y_{\rightarrow t-1}$:

$$p(x_t|y_{\rightarrow t-1}) = \int p(x_t|x_{t-1})p(x_{t-1}|y_{\rightarrow t-1})dx_{t-1} \qquad \textbf{A5.2}$$

This conditional density is then treated as a prior for the next observation and Bayes' rule is used to compute the conditional density of the states, conditioned upon all observations $y_{\rightarrow t}$. This gives the Bayesian update:

$$q(x_t) = p(x_t|y_{\rightarrow t}) \propto p(y_t|x_t)p(x_t|y_{\rightarrow t-1}) \qquad \textbf{A5.3}$$

Critically, the conditional density covers only the hidden states. This is important because it precludes inference on causes and the ability to de-convolve inputs from outputs. This is a key limitation of Bayesian filtering. However, Kalman filtering provides the optimal solution when the assumptions of the underlying model hold and one is not interested in causes or inputs. We now consider in more detail the operation equations for the extended Kalman filter. The extended Kalman filter is a generalization of the Kalman filter, in which the linear operators of the state equations are replaced by the partial derivatives of $f(x, v)$ with respect to the states.

THE EXTENDED KALMAN FILTER

This section provides a pseudo-code specification of the extended Kalman filter based on van der Merwe *et al.* (2000). To clarify the notation, we will use $f_x = \partial f / \partial x$. Eqn. **A5.1** can be re-written, using local linearization, as a discrete-time state-space model. This is the formulation treated in Bayesian filtering procedures:

$$\begin{aligned} y_t &= \mathbf{g}_x x_t + z_t \\ x_t &= \mathbf{f}_x x_{t-1} + w_{t-1} \\ \mathbf{g}_x &= g(x_t)_x \\ \mathbf{f}_x &= \exp(f(x_t)_x) \\ z_t &= z(t) \\ w_{t-1} &= \int \exp(f_x \tau) f_v v(t-\tau) d\tau \end{aligned} \qquad \textbf{A5.4}$$

For simplicity, we assume $\Delta t = 1$. The key thing to note here is that process noise w_{t-1} is simply a convolution of

Statistical Parametric Mapping, by Karl Friston et al.
ISBN–13: 978-0-12-372560-8 ISBN–10: 0-12-372560-7

the causes $v(t)$. This is relevant for Kalman filtering and related non-linear Bayesian tracking schemes that assume w_{t-1} is a well-behaved noise sequence. The covariance of process noise is:

$$\begin{aligned}\langle w_t w_t^T \rangle &= \int \exp(f_x \tau)\Omega \exp(f_x \tau)^T d\tau \\ &\approx \Omega \\ &= f_v R f_v^T\end{aligned} \quad \textbf{A5.5}$$

where R is the covariance of $v(t)$. We have assumed $v(t)$ has no temporal correlations and that the Lyapunov exponents of f_x are large relative to the time-step. The prediction and update steps are:

for all t

Prediction step

$$x_t = \mathbf{f}_\mathbf{x} x_{t-1}$$
$$\Sigma_t^x = \Omega + \mathbf{f}_x \Sigma_{t-1}^x \mathbf{f}_x^T$$

Update or correction step

$$K = \Sigma_t^x g_x^T (\Sigma + g_x \Sigma_t^x g_x^T)^{-1}$$
$$x_t \leftarrow x_t + K(y - g(x_t))$$
$$\Sigma_t^x \leftarrow (I - K g_x)\Sigma_t^x$$

end **A5.6**

Where Σ is the covariance of observation noise. The Kalman gain matrix K is used to update the prediction of future states and their conditional covariance, given each new observation. We actually use $x_t = x_{t-1} + (f_x - I)$ $f_x^{-1} f(x_{t-1})$. This is a slightly more sophisticated update that uses the current state as the expansion point for the local linearization. As mentioned in Chapter 37, Kalman filtering is also known as variable parameter regression, when the hidden state plays the role of a parameter (see Büchel and Friston, 1998).

REFERENCES

Büchel C, Friston KJ (1998) Dynamic changes in effective connectivity characterised by variable parameter regression and Kalman filtering. *Hum Brain Mapp* **6**: 403–08

van der Merwe R, Doucet A, de Freitas N *et al.* (2000) The unscented particle filter. Technical Report CUED/F-INFENG/TR 380

APPENDIX

6

Random field theory

K. Worsley and K. Friston

INTRODUCTION

This appendix details the technical background behind topological inference using random field theory. We treat the general case of several statistical parametric maps (SPMs) in terms of conjunctions. A conjunction is defined here as the occurrence of the same event at the same location in n independent SPMs. The standard results for a single SPM are the special case of $n = 1$. The SPMs or images are treated, under the null hypothesis, as smooth isotropic 3D random fields of test statistics, and the event occurs when the image exceeds a fixed high threshold. We give a simple approximation to the probability of a conjunction occurring anywhere in a fixed region, so that we can test for a local increase in the images at the same unknown location in all images; this can be regarded as a generalization of the split-t test. This is the corollary to a more general result on the expected intrinsic volumes (i.e. Minkowski functionals) of the set of points where a conjunction occurs.

THEORY

Let $X_i(t)$ be the value of image i at location t, and let x be a fixed threshold. The set of points where a conjunction occurs is:

$$C = \{t \in S : X_i(t) > x\} \qquad \textbf{A6.1}$$

for $1 \leq i \leq n$. We are interested in the probability that C is not empty $P(C \neq \emptyset)$, i.e. the probability that all images exceed the threshold at some point inside the volume S, or that the maximum of $\min X_i(t)$ exceeds x. If the images are independent stationary random fields, then the expected Lebesgue measure or volume of C is:

$$\begin{aligned} \langle |C| \rangle &= p^n |S| \\ p &= P(X_i(t) > x) \end{aligned} \qquad \textbf{A6.2}$$

Our main result is that Eqn. **A6.2** holds if the Lebesgue measures are replaced by a vector of intrinsic volumes (i.e. Minkowski functionals), and p is replaced by a matrix of Euler characteristic intensity functions for the random field. This gives Eqn. **A6.2** as a special case, and other interesting quantities, such as the expected surface area of C, which comes from the $D-1$ dimensional intrinsic volume. But the component of most interest to us is the zero-dimensional intrinsic volume, or Euler characteristic (EC). For high thresholds, the expected EC of C is a very accurate approximation to the probability we seek, namely, that C is not empty (Adler, 2000).

INTEGRAL GEOMETRY

In this section, we state some results from integral geometry and stereology that will be used to prove our main result. Let $\mu_i(A)$ be the i-th intrinsic volume of a set $A \subset \Re^D$, scaled so that it is invariant under embedding of A into any higher dimensional Euclidean space, where:

$$\begin{aligned} \mu_i(S) &= \frac{1}{s_{D-i}} \int_{\partial A} \det{}_{D-1-i}(Q)dt \\ s_i &= \frac{2\pi^{i/2}}{\Gamma(i/2)} \end{aligned} \qquad \textbf{A6.3}$$

where s_i is the surface area of a unit sphere in $\Re^i$, $\det_j(Q)$ is the sum of the determinant of all $j \times j$ principal minors of Q, which is the $D-1 \times D-1$ curvature matrix of ∂A, the

Statistical Parametric Mapping, by Karl Friston et al.
ISBN–13: 978-0-12-372560-8 ISBN–10: 0-12-372560-7

boundary of A. Note that $\mu_0(A)$ is the EC by the Gauss-Bonnet theorem, and $\mu_{D-1}(A)$ is half the surface area of A. $\mu_D(A) = |A|$ is its volume or Lebesgue measure. For example, the intrinsic volumes of a ball of radius r are:

$$\mu_0(S) = 1 \quad \mu_1(S) = 4r \quad \mu_2(S) = 2\pi r^2 \quad \mu_3(S) = (4/3)\pi r^3 \qquad \textbf{A6.4}$$

Other examples are provided in Plate 62 (see colour plate sections) for some common search volumes. We shall also use the result that any functional $\psi(A)$ that obeys the additivity rule:

$$\psi(A \cup B) = \psi(A) + \psi(B) - \psi(A \cap B) \qquad \textbf{A6.5}$$

is a linear combination of intrinsic volumes. Let $A, B \subset \Re^D$, then the Kinematic Fundamental Formula of integral geometry relates the integrated EC of the intersection of A and B to their intrinsic volumes.

$$\int \mu_0(A \cap B) = s_2 \ldots s_D \sum_{i=0}^{D} \frac{\mu_i(A)\mu_{D-i}(B)}{c_i^D}$$
$$c_i^D = \frac{\Gamma(\frac{1}{2})\Gamma(\frac{D+1}{2})}{\Gamma(\frac{i+1}{2})\Gamma(\frac{D-i+1}{2})} \qquad \textbf{A6.6}$$

where the integral is over all rotations and translations of A, keeping B fixed.

Gaussian field

$$\rho_0(z) = \int_z^\infty \frac{1}{(2\pi)^{\frac{1}{2}}} e^{-u^2/2} du$$
$$\rho_1(z) = \frac{\lambda^{\frac{1}{2}}}{2\pi} e^{-z^2/2}$$
$$\rho_2(z) = \frac{\lambda}{(2\pi)^{\frac{3}{2}}} e^{-z^2/2} z$$
$$\rho_3(z) = \frac{\lambda^{\frac{3}{2}}}{(2\pi)^2} e^{-z^2/2}(z^2 - 1)$$

t field with ν degrees of freedom, $\nu \geq d$

$$\rho_0(z) = \int_z^\infty \frac{\Gamma\left(\frac{\nu+1}{2}\right)}{(\nu\pi)^{\frac{1}{2}}\Gamma\left(\frac{\nu}{2}\right)} \left(1 + \frac{u^2}{\nu}\right)^{-\frac{1}{2}(\nu+1)} du$$
$$\rho_1(z) = \frac{\lambda^{\frac{1}{2}}}{2\pi} \left(1 + \frac{z^2}{\nu}\right)^{-\frac{1}{2}(\nu-1)}$$
$$\rho_2(z) = \frac{\lambda}{(2\pi)^{\frac{3}{2}}} \frac{\Gamma\left(\frac{\nu+1}{2}\right)}{\left(\frac{\nu}{2}\right)^{\frac{1}{2}}\Gamma\left(\frac{\nu}{2}\right)} \left(1 + \frac{z^2}{\nu}\right)^{-\frac{1}{2}(\nu-1)} z$$
$$\rho_3(z) = \frac{\lambda^{\frac{3}{2}}}{(2\pi)^2} \left(1 + \frac{z^2}{\nu}\right)^{-\frac{1}{2}(\nu-1)} \left(\frac{\nu-1}{\nu} z^2 - 1\right)$$

F field with k and ν degrees of freedom, $k + \nu > d$

$$\rho_0(z) = \int_z^\infty \frac{\Gamma\left(\frac{\nu+k}{2}\right)}{\Gamma\left(\frac{\nu}{2}\right)\Gamma\left(\frac{k}{2}\right)} \frac{k}{\nu} \left(\frac{ku}{\nu}\right)^{\frac{1}{2}(k-2)} \left(1 + \frac{ku}{\nu}\right)^{-\frac{1}{2}(\nu+k)} du$$
$$\rho_1(z) = \frac{\lambda^{\frac{1}{2}}}{(2\pi)^{\frac{1}{2}}} \frac{\Gamma\left(\frac{\nu+k-1}{2}\right) 2^{\frac{1}{2}}}{\Gamma\left(\frac{\nu}{2}\right)\Gamma\left(\frac{k}{2}\right)} \left(\frac{kz}{\nu}\right)^{\frac{1}{2}(k-1)} \left(1 + \frac{kz}{\nu}\right)^{-\frac{1}{2}(\nu+k-2)}$$
$$\rho_2(z) = \frac{\lambda}{2\pi} \frac{\Gamma\left(\frac{\nu+k-2}{2}\right)}{\Gamma\left(\frac{\nu}{2}\right)\Gamma\left(\frac{k}{2}\right)} \left(\frac{kz}{\nu}\right)^{\frac{1}{2}(k-2)} \left(1 + \frac{kz}{\nu}\right)^{-\frac{1}{2}(\nu+k-2)} \times \left[(\nu-1)\frac{kz}{\nu} - (k-1)\right]$$
$$\rho_3(z) = \frac{\lambda^{\frac{3}{2}}}{(2\pi)^{\frac{3}{2}}} \frac{\Gamma\left(\frac{\nu+k-3}{2}\right) 2^{-\frac{1}{2}}}{\Gamma\left(\frac{\nu}{2}\right)\Gamma\left(\frac{k}{2}\right)} \left(\frac{kz}{\nu}\right)^{\frac{1}{2}(k-3)} \left(1 + \frac{kz}{\nu}\right)^{-\frac{1}{2}(\nu+k-2)} \times \left[(\nu-1)(\nu-2)\left(\frac{kz}{\nu}\right)^2 - (2\nu k - \nu - k - 1)\left(\frac{kz}{\nu}\right) + (k-1)(k-2)\right]$$

FIGURE A6.1 Euler characteristic intensities for some common statistical fields. These are functions of the field's roughness λ. When roughness is one, these are the Euler characteristic densities referred to in previous chapters.

RANDOM FIELDS

If $X_i(t)$ is an isotropic random field with excursion set $A = \{t : X(t) \geq x\}$ then:

$$\langle \mu_0(A \cap S) \rangle = \sum_{i=0}^{D} \rho_i \mu_i(S) \qquad \textbf{A6.7}$$

for some constants ρ_i. This follows from the fact that the functional $\langle \mu_0(A \cap S) \rangle$ obeys the additivity rule Eqn. **A6.5**, since μ_0 does, so it must be a linear combination of the intrinsic volumes. The coefficients ρ_i are called Euler characteristic (EC) intensities in $\Re^i$, and can be evaluated for a variety of random fields (Worsley, 1994; Cao and Worsley, 1999; Adler, 2000; Worsley *et al.*, 2004). Figure A6.1 provides the expressions for some common statistics used in SPMs. The EC intensity is a function of the image roughness λ defined here as $\mathrm{cov}(\partial X/\partial t) = \lambda I$.

Eqn. **A6.7** is fundamental to inference on SPMs because the expected EC, $\psi_0 = \langle \mu_0(A \cap S) \rangle$ is an accurate approximation to the probability of getting an excursion set by chance (i.e. the adjusted p-value). Its form helps understand how this probability arises: the expected EC receives contributions, $\rho_i \mu_i(S)$ from each dimension of the search volume. Each contribution is the product of an EC intensity and an intrinsic volume. If we were dealing with statistically flat images with $\lambda = 1$, these quantities would correspond to EC densities and resolution elements (resel) counts respectively (see Chapter 19). The EC density represents the expected EC per resel and the resel count represents the resel number at each dimension. We will now generalize this for multiple images.

We can extend Eqn. **A6.7** to higher intrinsic volumes by the lemma in Worsley and Friston (2000):

$$\langle \mu_i(A \cap S) \rangle = \sum_{j=i}^{D} c_i^j \rho_{j-i} \mu_j(S) \qquad \textbf{A6.8}$$

Theorem: *Let R_k be the upper triangular matrix whose (i,j) elements are $\rho_{j-ik}c_i^j$ if $j \geq i$ and 0 otherwise; i.e.*

$$R_k = \begin{bmatrix} \rho_{0k}c_1^1 \cdots \rho_{0k}c_1^D \\ \vdots \quad \ddots \quad \vdots \\ 0 \quad \cdots \rho_{0k}c_D^D \end{bmatrix} \quad \text{and} \quad \mu(S) = \begin{bmatrix} \mu_0(S) \\ \vdots \\ \mu_D(S) \end{bmatrix}$$

where ρ_{ik} is the EC intensity of $X_k(t)$ in $\Re^i$ then:

$$\langle \mu(C) \rangle = [\psi_0, \ldots, \psi_D]^T = \left(\prod_{i-1}^{n} R_i \right) \mu(S) \qquad \textbf{A6.9}$$

Proof: the proof follows by induction on n. From Eqn. **A6.8**, we see that it is clearly true for $n = 1$. Let A_k be the excursion set for $X_k(t)$ so that $C = A_1 \cap \ldots A_n \cap S$. If the result is true for $n = k$ then by first conditioning on A_{k+1} and replacing S by $A_{k+1} \cap S$, we get:

$$\langle \mu(A_1 \cap \ldots A_k \cap (A_{k+1} \cap S)) \rangle = \left(\prod_{i=1}^{k} R_i \right) \langle \mu(A_{k+1} \cap S) \rangle$$

$$= \left(\prod_{i=1}^{k} R_i \right) R_{k+1} \mu(S) \qquad \textbf{A6.10}$$

by the result for $n = 1$. This completes the proof.

Comparing Eqn. **A6.9** with Eqn. **A6.2**, we see that they have the same form, where the volumes are replaced by vectors of intrinsic volumes, and the probability is replaced by the matrix of weighted EC intensities. The last element $\psi_D = \langle |C| \rangle$ is the same as the Lebesgue measure and the first element ψ_0 is the expected EC of the set of conjunctions. This is the approximation to the probability of a conjunction anywhere in S, for high thresholds x, we require.

EXAMPLE

We shall apply the result in Eqn. **A6.9** to some $D = 3$ dimensional functional magnetic resonance imaging (fMRI) data. The purpose of the experiment was to determine those regions of the brain that were consistently stimulated by all subjects, while viewing a pattern of radially moving dots. To do this, subjects were presented with a pattern of moving dots, followed by a pattern of stationary dots, and this was repeated 10 times, during which a total of 120 3D fMRI images were obtained at the rate of one every 3.22 s. For each subject i and at every point $t \in \Re^3$, a test statistic $X_i(t)$ was calculated

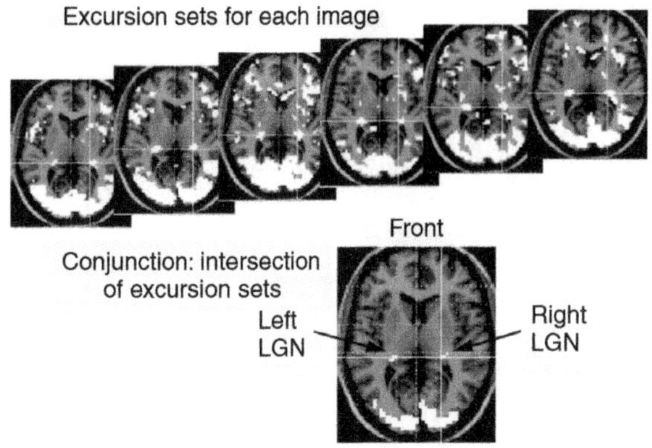

FIGURE A6.2 Conjunction of six SPM{t} from a visual task (only one slice of the 3D data is shown). The excursion sets of each are shown in white on a background of brain anatomy (top). The set of conjunctions C is the intersection of these sets (bottom). The visual cortex at the back of the brain appears in C, but the most interesting feature is the appearance of the lateral geniculate nuclei (LGN) (arrows).

for comparing the fMRI response between the moving dots and the stationary dots. Under the null hypothesis of no difference, $X_i(t)$ was modelled as an isotropic Gaussian random field with zero mean, unit variance and roughness $\lambda = 4.68\,\text{cm}^{-1}$. A threshold of $x = 1.64$, corresponding to an uncorrected level 5 per cent test, was chosen, and the excursion sets for each subject are shown in Figure A6.2 together with their intersection, which forms the set of conjunctions C. The search volume was the whole brain area that was scanned, which was an approximate spherical region with a volume of $|S| = 1226\,cm^3$. Finally, the approximate probability of a conjunction, calculated from Eqn. **A6.9**, was 0.0126. We can thus conclude, at the 1.26 per cent level, that conjunctions have occurred in the visual cortex, and more interesting, the lateral geniculate nuclei (see Figure A6.2).

REFERENCES

Adler RJ (2000) On excursion sets, tube formulae, and maxima of random fields. *Ann Appl Probab* **10**: 1–74

Cao J, Worsley K (1999) The geometry of correlation fields with an application to functional connectivity of the brain. *Ann Appl Probab* **9**: 1021–57

Worsley KJ (1994) Local maxima and the expected Euler characteristic of excursion sets of chi-squared, F and t fields. *Adv Appl Probab* **26**: 13–42

Worsley KJ, Friston KJ (2000) A test for a conjunction. *Stat Probab Lett* **47**: 135–40

Worsley KJ, Taylor JE, Tomaiuolo F *et al.* (2004) Unified univariate and multivariate random field theory. *NeuroImage* **23**: S189–95

Index

Edwards Brothers Malloy
Ann Arbor MI. USA
December 19, 2013